The Guide to **Simulations/Games** for education and training

ASSOCIATE EDITORS
4TH EDITION OF THE GUIDE

PAST EDITORS

The Guide to **Simulations/Games** for education and training

4th Edition

Robert E. Horn
and
Anne Cleaves
Editors

SAGE PUBLICATIONS Beverly Hills London

For information address:

SAGE PUBLICATIONS, INC.
275 South Beverly Drive
Beverly Hills, California 90212

SAGE PUBLICATIONS LTD
28 Banner Street
London EC1Y 8QE, England

Printed in the United States of America

Library of Congress Cataloging in Publication Data

Main entry under title:

The Guide to simulations/games for education and training.

Edition for 1973 by D. W. Zuckerman.
Includes bibliographical references and index.
1. Educational games. 2. Education–Simulation methods. I. Horn, Robert E. II. Cleaves, Anne. III. Zuckerman, David W. The guide to simulations/games for education and training.
LB1029.G3H67 1980 371.3'078 79-19823
ISBN 0-8039-1375-3

FOURTH EDITION
FIRST PRINTING

The fourth edition was supported by a grant from the Exxon Education Foundation.

CONTENTS

Introduction 7

Part I. Evaluative Essays

Communication Games and Simulations: An Evaluation
Brent D. Ruben 11

Community Land Use Game: An Evaluation
Joseph L. Davis 26

Computer Simulation Games: Exemplars
Ronald E. Anderson 37

Domestic Politics Games and Simulations: An Evaluation
Dorothy R. Dodge 47

Ecology/Land Use/Population Games and Simulations: An Evaluation
Larry Schaefer, Harry O. Haakonsen, and Orah Elron 58

Economics Games and Simulations: An Evaluation
Victor Pascale and Theodore Ehrman 69

End of the Line: An Evaluation
Andrew W. Washburn and Maurice E. Bisheff 80

Energy and Environmental Quality Games and Simulations: An Evaluation
Charles A. Bottinelli 86

Frame Games: An Evaluation
Sivasailam Thiagarajan and Harold D. Stolovitch 98

Futures Games and Simulations: An Evaluation
Charles M. Plummer 108

Health and Health Care Games and Simulations: An Evaluation
Amy E. Zelmer and A.C. Lynn Zelmer 132

History Games and Simulations: An Evaluation
Bruce E. Bigelow 141

International Relations Games and Simulations: An Evaluation
Leonard Suransky 162

Marketing Games: An Evaluation
Anthony J. Faria 177

Military History Games: An Evaluation
Martin C. Campion 187

Policy Negotiations: An Evaluation
Barbara Steinwachs 204

Religion Games and Simulations: An Evaluation
Lucien E. Coleman 214

Self-Development Games and Simulations: An Evaluation
Harry Farra 223

Sex Roles Games and Simulations: An Evaluation
Cathy S. Greenblat and Joan Gaskill Baily 239

SIMSOC: An Evaluation
Richard A. Dukes 248

Simulations of Developing a Society: An Evaluation
Frank P. Diulus 262

Social Studies Games and Simulations: An Evaluation
Michael J. Rockler 267

Total Enterprise Business Games: An Evaluation
J. Bernard Keys 277

Urban Gaming Simulations: An Evaluation
Mary Joyce Hasell 286

Part II. Academic Listings

Addictions 307
Communication 308
Community Issues 315
Computer Simulations 326
Domestic Politics 348
Ecology 365
Economics 372
Education 385

Frame Games 396
Futures 401
Geography 409
Health 412
History 418
Human Services 441
International Relations 446
Language Skills and Art 460
Legal System 464
Mathematics 469
Military History 477
Practical Economics 525
Religion 531
Science 540
Self-Development 546
Social Studies 556
Urban 577

Part III. Business Listings

Total Enterprise—Computerized 595
Total Enterprise—Manual 600
Production, Logistics, Operations 604
Personnel Development 609
Marketing 632
Finance 640
Specific Industry 642
Business for School 648

Part IV. Resources

Simulation/Gaming Periodicals 653
Simulation/Gaming Centers 654
Author Index 659
Game Index 672
Producer Index 680
About the Authors and Editors 689

INTRODUCTION

ABOUT THIS EDITION

This is the fourth edition of the *Guide*. Since we began monitoring the field of simulation and gaming for education and training about ten years ago, in 1969, we have seen the field flourish and mature. The number of listings in the first three editions increased threefold. While about one-fourth of the listings are new to this edition, we found that at least an equivalent number of simulations listed in earlier editions are no longer in print, or their designers and producers are no longer in business. This new edition, then, documents a relative degree of stabilization in the field—for the time being, at least.

As the field of simulation has reached a new stage, so has the *Guide*. The book has always been edited for the potential game user rather than for the theorist or designer, but this edition is designed much more as a consumer report than any that have come before. The format of the listings now describes the simulation structure and process, in contrast to the original format, which was essentially a tabulation of game elements. Most significant, however, is the collection of twenty-four essays evaluating and comparing simulations in various subject areas.

COVERAGE

Simulations, Games, Exercises

In our first edition, we defined the categories the *Guide* covers as follows: "For our purposes, a game is simply an activity undertaken by a player or players whose actions are constrained by a set of explicit rules particular to that game and by a predetermined end point. . . . The simulations which interest us differ from ordinary games only in that their elements comprise a more or less accurate representation or model of some external reality with which the players interact in much the same way they would interact with the actual reality."

In general, the *Guide* contains only simulations and games that have specific educational purposes. We have omitted pure simulations used for research and experimentation by a limited, specialized audience (though some do find their way into these pages, as training materials are developed from them and students use them). Many computer simulations used for policy making in business, industry, and government could serve broader educational functions, but if they are not accessible to the broader group of potential users, we do not list them.

To some degree, we have included structured exercises that are not games or simulations when they deal with communication and group dynamics—essential aspects of work with people. These appear in the sections on communication, education, self-development, and business simulations. Likewise, we have included in-basket exercises, in which the participant takes over a job and a full in-basket and makes decisions on the basis of material in the various memos, letters, and reports.

We have not, for the most part, listed ordinary, nonsimulation games. Any that do appear have a specific educational purpose. We have not included sports games or games whose sole purpose is amusement. Nor have we listed simple role-play situations except as they take on gamelike characteristics, or if they are part of a package that includes a simulation or game.

Criteria for Listings

Accessibility of a simulation to its potential users is the primary criterion we applied to our listings. In no case is any simulation or game listed unless we received from its producer or designer in response to our update inquiries definite confirmation that it was available for sale or distribution as of January 1979.

Listings in earlier editions of the *Guide* began with kindergarten and went up. We have raised the age level to junior high school and up, though the range for an occasional simulation may dip into the upper elementary grades.

We have, with this edition, generally restricted our listings to simulations that are offered through publishers or other institutions, though there are exceptions that prove the rule. Though this policy somewhat limits the number of presently existing simulations we cover, it will guarantee potential users a high rate of success in getting hold of simulations for some time beyond the immediate present. Our experience with updating listings from earlier editions proved the sense in this criterion: Individual game designers seem to be even more nomadic than the population at large. Organizations are somewhat more stable.

Finally, cost is an element we consider in our listings. Large, complex simulations are priced accordingly and involve considerable trainer time. Likewise, computer simulations involve the costs of human and computer time beyond the price of their programs and documentation. Anyone investigating available simulations of these types understands this. In general, however, the prices of the simulations and games listed in the *Guide* are low to relatively moderate. All prices are given as they stood in January 1979, and are probably best used as an indicator of the approximate price at any time thereafter.

Subject Areas

The field of simulation and gaming is an open invitation to, and from, educators and trainers whose imagination is caught by the multitude of possibilities that occur within and among people, groups, and structures in the world as it is, was, seems, or might be. Simulations are no easier to classify than the world they model. Most sections have a list of cross-references that should help readers use our categories rather than be restricted by them, but even the cross-references serve ultimately as a welcome to browse and consider many things.

Different instructors on different occasions might use a single given simulation in a study of economics, history, social structure, power. Some readers of the *Guide* will find an interest cluster in the sections on health, human services, addictions, and communication simulations. Others may follow links through the human services, community issues, and urban sections. Others will pursue a theme through the sections on community issues, domestic politics, economics, and urban simulations. And so on.

The urban section focuses on large, areawide simulations with multifunctional government activities. Most are designed for use at the college, university, or professional level. Many employ a computer. However, less complex simulations of communities appear in other sections according to whether their primary aspects are ecological, economic, political, or oriented to community action or service.

The section on computer-based simulations covers subject matters across a whole range. Except for the urban and business sections, which include them, listings of computer-based simulations appear in this separate section. A special note: Computer gaming software is growing thick and fast in conjunction with developments in hardware. Because of the commitment of time and/or money it can take to get a program running, we have taken care to see that we list only simulations that have proved workable and transferable. The essay on computer simulations will provide interested readers with all the leads they need to investigate today's phenomena, among which tomorrow's classics will emerge.

THE EVALUATION ESSAYS

Unique to this edition of the *Guide* is the collection of twenty-four essays evaluating currently available simulations and games, which was made possible by a grant from the Exxon Education Foundation. The task of the authors of the comparative essays was to select a number of simulations in a given field, to evaluate them on the basis of their analysis of the materials in a format that would help prospective users identify and compare significant aspects of the simulation. In addition, you will find in-depth essays on four particularly interesting, major simulations. Since each such essay was written by someone who has run the particular simulation many times and is thoroughly familiar with it, these essays are a rich source of descriptive material and practical advice.

CREDIT

The domestic politics and economics sections contain a number of simulation descriptions from the *Robert A. Taft Institute of Government Study on Games and Simulations in Government, Politics, and Economics,* for which we wish to express our appreciation to the authors and to Marilyn Chelmstrom, Executive Director of the Institute.

The urban and community issues sections contain descriptions from *The State-of-the-Art in Urban Gaming Models,* a report prepared by Environmetrics, Inc. for the Office of the Secretary for the Environment and Urban Issues, Department of Transportation, in 1971.

ACKNOWLEDGEMENTS

Finally, our thanks to the people who worked with us to bring this, the most complex edition of the *Guide,* into being. Tracy Marks worked with Robert Horn to set up and coordinate the evaluation project. She initiated and for some time conducted the process of collecting and reviewing simulations. Don Davis replaced her, not only reviewing new simulations as they came in but re-reviewing those listed in earlier editions according to the format and standards that had evolved in the meantime. Dena Davis (no relation) did the updating, a task of enormous scope and detail. Particular appreciation goes to Margot Holtzman, who helped coordinate the project and whose wonderful efficiency and spirit were essential contributions to the process and product.

We especially want to express our warm respect for our contributing authors. Their work shows a commitment to education and to simulation gaming, as an educational technique and as an approach to understanding lived experience, that welcomes one to explore the field.

—Anne Cleaves
—Robert E. Horn

Part I
EVALUATIVE ESSAYS

COMMUNICATION GAMES AND SIMULATIONS
An Evaluation

by Brent D. Ruben

INTRODUCTION

Within the field of communication, as in so many others, the use of experiential methods of instruction has increased remarkably during the past decade. Many of the reasons are clear. For one thing, simulation games offer a means for the instructor or facilitator to immerse learners in a range of communication situations within the context of the classroom; time and space can be telescoped and the risks and consequences of "winning" and "losing" controlled. And, beyond providing a common experience that learners can use as a basis for study, experience-based methods provide an opportunity to bridge the troublesome gap between theory and practice in fields where both are crucial.

An Historical Perspective

To understand fully the current state of gaming in communication, it is helpful to have some understanding of the recent history of instructional development in the field. The present popularity of experiential learning comes primarily from two influences, one theoretical and the other historical. Especially during the late 1950s and through the 1960s, a number of people in communication—as in a number of other fields—began to question the validity (or what was often termed "relevance") of communication education at all levels. Motivated both by an academic concern about what they sensed to be a widening gap between theory and practice and by a growing outcry for improved institutions of learning, a number of prominent communication scholars began a national dialogue on communication, mass communication, and speech communication that flourished through the mid-1970s. During this period of inquiry, John Dewey's *Education and Experience* and many of his other significant contributions were rediscovered. The work of others such as Jerome Bruner and Carl Rogers, too, had a significant impact on many concerned with communication instruction.

For those who became convinced that students—in and outside formal institutions—could be taught communication more effectively, and might further be helped to apply this learning to their own lives, experiential instructional techniques of all sorts seemed worth exploring.

A second influence came from various subdivisions of the discipline itself. Within speech, for example, even the very earliest efforts to teach public speaking and debate involved writing and delivering speeches and preparing for, and participation in, debates. Similarly, journalism students learned about photography, writing, and editing by performing the activities and having their work evaluated by an instructor or classmates. While the resemblance is seldom articulated, it is clear that many aspects of the philosophy underlying such methods were similar to that now undergirding the broadened contemporary interest in communication simulation and gaming. In a very real sense, writing or editing for a student newspaper, participating in debate competition, or delivering a speech to an audience composed of one's instructor and peers may be viewed as a simulation, the quality of which can be profitably evaluated with the same criteria that might apply to assessing other simulations.

Together, these two influences have led an increasing number of communication educators—particularly in speech and in interpersonal, group, organizational, and intercultural communication—to experiment with and refine methods of experiential instruction in communication.

Simulation gaming is used at virtually all educational levels of communication and in nearly all imaginable contexts, including the classroom, the management seminar, the community action group, the factory, and the overseas training program. Available games and exercises run the gamut from simple exercises that introduce learners to a single topic in a ten- to fifteen-minute activity to complex activities in which participants may spend up to a semester or a year in simulations of societal communication dynamics.

Our goal is to review a number of these simulation games and exercises and to provide a framework with which readers may better evaluate, compare, contrast, and select appropriate activities according to their instructional needs.

Selection Criteria

The nineteen activities we review fall into four categories that represent four major subdivisions of the communication field in which simulation and gaming are widely in use. These are:

(1) intrapersonal communication
(2) interpersonal and group communication
(3) organizational and mass communication
(4) cross-cultural communication.

Within each area, we selected simulation games that in our view, are: (1) *significant* (focus on critical aspects of communication); (2) *valid* (are consistent with communication theory and research); (3) *reliable* (have a high probability of resulting in similar outcomes from one iteration to the next); (4) *flexible* (can be used with persons of diverse background, education, and experience); (5) *popular* (are generally well-recognized as important and useful); (6) *accessible* (are easy to construct or acquire); (7) *inexpensive* (can be produced with minimal purchase or duplication costs); and (8) do not duplicate activities reviewed in other articles here.

The Approach

Each of the four sections of this article focuses on a major subdivision within the field of communication. Each section is organized in two parts. The first, introductory, part briefly surveys that area of communication. The second part presents evaluative reviews of the selected games, simulations, and exercises. This part consists of a brief summary of the structure, dynamics, and requirements of each activity, with comments on particular advantages and limitations where these apply. The first table in each section rates the activities according to selection criteria. The second presents a summary comparison of the activities in terms of their primary topic, the number and level of participants, their relative complexity, the time needed to run and debrief them, the presence or absence of debriefing guidelines, and the level of facilitator background and preparation they require.

Communication: A Definition

What is communication? When is a simulation, game, or exercise about communication and when is it not? As with so many other fields in the social-behavioral sciences, there are no hard and fast boundaries. Communication as a discipline concerns human interaction, so in some sense nearly every simulation could be considered relevant for communication study

However, for the purposes of this article, communication is considered to be that life process by which people symbolically relate themselves to one another and their environment. Thus, communication is fundamentally concerned with how people know and comprehend their world, and how and with what consequences they try to share their understanding with other people. The games we review in this article were designed to replicate some aspects of this basic life process in the classroom or workshop.

INTRAPERSONAL COMMUNICATION

The most basic facet of communication is that process through which an individual selects and interprets messages from his or her environment. This message-selection-interpretation is called *intrapersonal communication.* Intrapersonal communication concerns the way people sense, make sense of, and act back upon their environment and the people in it. Accordingly, the focus of intrapersonal communication, and hence intrapersonal communication games, is not so much how people *communicate to* others as how people are *communicated-with* in their environment.

Four simulation exercises that concern aspects of intrapersonal communication are: *The Learning Game, The Memory Game, Zif,* and *Listening Triads.* All are quite simple to administer, and costs are nil or minimal; they can be used without purchasing commercially packaged kits.

The Learning Game

The Learning Game initially examines the process by which people perceive and learn to pattern their experiences. Subsequently, the activity explores the problem of changing one's intrapersonal framework once established. Game materials consist of a mimeographed booklet composed of four pages, each with the numbers 1 to 60 scattered about. On the first three pages, which are identical, the numbers are arranged so all odd numbers are at the top half of the page and even numbers at the bottom. Within these areas they are otherwise randomly distributed. On the fourth page, odd numbers are on the left and even on the right.

In forty-five-second intervals (timed by the instructor), participants, working with pencils in their individual booklets, connect the numbers, beginning with 1, drawing a line to 2, then to 3, and so forth, in a "dot-to-dot" manner. At the end of the fixed time the instructor records on a flip chart (or equivalent) the last number reached by each participant. Participants have another forty-five-second interval to connect the numbers on the second page of the booklet. Again, the last numbers reached are recorded. The instructor holds a brief discussion to reveal (to those who have not yet discovered it) that odd numbers are on the top and even on the bottom. Participants have another forty-five-second interval for connecting numbers, with completion scores recorded. Without further discussion, participants are told to turn to page four and repeat the process. Again, completion scores are recorded. The instructor can plot a curve on the flip chart for the group's average last number on each of the four trials.

Drawing on the data from the curve, debriefing focuses on the learning and patterning. During each of the first three trials, completion rates improved as the trial-and-error learning gave way to pattern perception and skill development. Communication patterns or habits, even when established with only several experiences, can easily become rigid and resistant to change. This is nicely demonstrated by the drop-off in completion rate when "change" is introduced in round four (though the change is only a slight, and systematic, variation).

The Learning Game deals in a simple way with a complex communication topic of far-reaching consequence in both theoretical and practical terms. Its structure is fairly simple, and outcomes are highly predictable. It can be used with virtually any number of participants of junior high through adult levels.

One advantage of the activity is that the instructor needs relatively little theoretical or practical experience to conduct

TABLE 1 Overview of Simulations Games

Game		Summary
Intrapersonal Communication	*Learning Game*	A paper-pencil activity that demonstrates the dynamics of learning and the difficulty of changing patterns once learned.
	Memory Game	An innovative game designed to demonstrate the role of past experience in perception, attention, memory, and recall.
	Zif	A simple demonstration-game useful as a basis for exploring the nature of category formation and meaning processes.
	Listening Triads	A popular game that demonstrates the difficulty and importance of listening and provides a context for improving listening skills.
Interpersonal and Group Communication	*Telephone*	Telephone, or Rumor Clinic, simulates the dynamics that occur as messages are passed from one person to another, and another, and another.
	One-Way Two-Way Feedback	A game designed to present the concept of feedback in communication and demonstrate the differences in outcomes between two-way interactions and interactions where the listeners are not active participants.
	Cooperation Squares	A game that stresses the nature and importance of cooperation and cooperative behavior in group problem-solving situations.
	Lost on the Moon (Consensus)	A simulation that creates the dynamics of group decision making, problem solving, and consensus seeking as individuals strive to integrate their individual views with opinions of other group members.
	Prisoner's Dilemma	A game that explores the dynamics and outcomes of interaction between two groups who must choose between patterns of mutual trust and cooperation on the one hand, or distrust and competition on the other.
	Power	An intriguing game in which participants enact their personal goals and feelings for other players on game board with play following few rules.
Organizational and Mass Communication	*Hollow Squares*	A simulation game in which one group devises a plan for completion of a puzzle by a second group not involved in the planning, with a third group observing the entire dynamic.
	Lock-A-Block	An organizational communication simulation game in which several competing groups endeavor to replicate a toy model following guidelines and winning criteria established by participants serving on a panel of judges.
	Trio	A game stressing interviewing, writing, and evaluation skills.
	Interact II	A minimally-structured simulation of interpersonal, organization, and mass communication processes in which groups produce communication products for distribution to, and evaluation by, other participants.
Intercultural Communication	*Survival*	A game stressing the nature of culture, cultural development and evolution, and cross-cultural interaction, as participants deal with the problem of being survivors from a plane crash on an isolated island.
	Agitania, Meditania, Solidania	A role-play simulation in which participants from different hypothetical cultures with different communication styles confront one another interpersonally.
	Lobu-Abu	A simulation game stressing the role of language in cross-cultural adaptation as "Lobu-speaking" participants must interact with "Abu-speaking" participants.
	Market Day	A simulation activity designed to create the dynamics of the market as found in most Third World countries by providing participant groups with varying levels of purchasing power.
	Hypothetica	A simulation useful for exploring the nature and dynamics of development and underdevelopment, and the communication processes that occur as a group of participants from a hypothetical region of a hypothetical country try to exploit a newly discovered natural resource.

the activity. The major points are self-evident to participants, and post-play discussion will proceed in a lively manner with minimal facilitation. Exploration of the exercise's implications can be enhanced, however, if the instructor has some knowledge in areas such as pattern learning, perception, habit formation, and personal change.

With sophisticated audiences, the greatest strengths of *The Learning Game*—its simplicity and directness—may also be a weakness. The instructor can compensate for this by using the activity to launch a generalized discussion on learning and change, rather than as an end in itself.

TABLE 2 Intrapersonal Communication Simulation Games Overall Rating

	Selection Criteria						
	Significance	*Validity*	*Reliability*	*Flexibility*	*Popularity*	*Accessibility*	*Cost*
Learning Game	4	5	5	4	3	5	2r
Memory Game	5	4	4	4	1	1	2p
Zif	4	4	4	4	3	5	1n
Listening Triads	5	5	5	5	5	5	1n

Key: High = 5; Moderately High = 4; Moderate = 3; Moderately Low = 2; Low = 1; n = no expenditure required; p = purchase required; r = reproduction required

The Memory Game

The Memory Game, like *The Learning Game,* concerns perception and habit formation, but in a more complex and holistic fashion. Game materials are miscellaneous household items that can be collected, borrowed, or purchased. The instructor assembles objects that symbolize a range of activities common to some or all participants. Some might have religious associations (like a rosary or dreidel), others vocational associations (like a book or tape measure), others might relate to hobbies (like guitar strings, a piece of leather, an art brush), and others to interest areas or habits (like a sugar cube, cup and saucer, *Playboy* or *Playgirl* magazine, matches or cigarettes, plant mister). The objects are placed on a covered table in front of the room. People come up to the table in groups of five to seven. They are told that the table (which is still covered) has a number of items on it, and that when the table is uncovered they are look it over and try to notice and remember as many items as possible, so they can list them back at their seats. Each group (enough groups so all participants view the items) in turn has two minutes to look over the items before the table is recovered and participants are told to return to their seats and list as many items as they can.

After all lists are done, the instructor may initiate debriefing by asking participants to look their own lists over to determine whether there are any patterns. Because most persons perceive the task as simply as a test of memory, their attention is focused simply on trying to remember as many items as possible. Purposely, though, there are many more items that anyone can remember. Thus, in one way or another, each person *selects* some subset from the total possible range. Though participants may initially say that their selections were "unconscious," further discussion will suggest the importance of each person's life experience in determining what items were noticed and recalled. The rosary, for example, is most typically listed by someone who is either Catholic or well acquainted with the religion. That same person may not notice or recall the dreidel, a symbol that will most typically be remembered by a person of the Jewish faith. Each person's listing of items–and the groups' collective listing–provides the basis for an involved and personalized discussion of the role of

TABLE 3 Intrapersonal Communication Simulation Games Use Characteristics

	Primary Topic	Level (*Kindergarten, Post Grad*)	Number of Participants (*No, Variable*)	Basic Structure (*Simple, Average, Complex*)	Run Time (*Minutes, Hours*)	Debriefing Guide (*Included or Not*)	Debriefing Time (*Minutes, Hours*)	Theoretical Preparation (*Critical, Useful, Not necessary*)	Behavioral Simulation Facilitation Experience (*Critical, Useful, Not necessary*)
The Learning Game	learning, perception, change	8-PG	V	S	15m	I	30m	N	N
The Memory Game	memory, learning, perceptual patterning, experience	11-PG	5-25	A	30m	I	30m-1h	U	U
Zif	meaning, perception, category formation	10-PG	V	S	30m	I	30m	U	N
Listening Triads	listening, feedback, speaking	2-PG	V	S	30m	I	30m	U	U

experience and values in perceptual patterning and, more generally, of intrapersonal communication habit formation.

The game focuses on a subtle yet critical aspect of intrapersonal communication in a reasonably straightforward fashion. In a very general sense, outcomes of the game are highly predictable; the lists participants generate reflect their own life experiences and values. In another sense, outcomes vary considerably from one person or play of the game to the next as a function of the experiences and values of the persons involved.

Because of the relatively more complex nature of this game compared with *The Learning Game,* a bit more sophistication in debriefing behavioral simulations is required for most effective use, and some theoretical preparation in learning, socialization, or perception is probably desirable. Discussion also requires more time than *The Learning Game.*

Zif

Zif, like *The Learning Game,* can be used easily with any number of participants. The game focuses on meaning and the processes by which people evolve meaning for aspects of the environment. The exercise involves participants in trying to discover the meaning of "Zif," a label for a category of geometric forms. The only materials needed are a flip chart (or equivalent), paper, and pencils. On the flip chart the instructor draws between nine and twelve geometric forms, each of which has some structural characteristics in common with several or all of the others (for example: vertical lines, right angles, four corners, or three corners). The instructor has already selected one particular configuration (like a right angle or a vertical line with a left-branching appendate) that appears in some of the forms but not in others, and arbitrarily calls that "Zif." The task of participants is to discover the referent for the term. They may gather information from the instructor only by asking questions that can be answered with "yes" or "no."

Postplay discussion focuses on the "trial-and-error" process by which participants discovered the meaning of "Zif," relating this to learning and meaning-formation processes in "real-life" situations. The activity may also be used to focus on "unlearning" and change, in a manner like that used in *The Learning Game,* by repeating the process as the instructor selects a second configuration also defined as "Zif."

The major point of *Zif* comes through to participants with little facilitated discussion. Perhaps even more than with *The Learning Game,* extrapolation of the outcomes from the game situation to the "real world" is more readily accomplished if the instructor is familiar with theories of meaning and meaning formation, though this is probably not essential.

Outcomes of the game are quite predictable from one iteration to the next, participant involvement is quite high, and it can be used profitably with high school through adult participants. With modification, it would probably be suitable for younger audiences.

Listening Triads

Listening Triads deals with what is perhaps one of the most critical aspects of communication, and it does so in a simple, straightforward format requiring no materials or dollar expenditure. Participants are grouped by threes. One person in each group is designated the speaker, the second the listener, and the third the judge. The speaker talks without interruption for a specified number of minutes (two to five) on a topic selected by the instructor (for instance, why he or she chose this job, major, or workshop; his or her major; his or her communication strengths, goals, or weaknesses, and so forth), while the listener and judge listen attentively. After the set time has elapsed, the listener summarizes in as much detail as possible what the speaker said. The judge may interrupt at any point if he or she believes the listener has made an error or has left out something important. After the summary and judge's evaluation, roles are exchanged. The listener becomes the judge, the judge the speaker, and the speaker the listener. After completion of this cycle, roles are exchanged a third time so each participant serves once in each role.

Discussion following the activity takes place at two levels. First, the members of the triad may be asked to recall and analyze their feelings and their performance among themselves. Second, the group as a whole may be asked to summarize the issues raised by the activity and its implication for other communication situations.

As with the other intrapersonal communication games, extensive theoretical knowledge or preparation is not required, though it can be useful. Probably of equal or greater importance for maximum impact of this exercise is some experience in conducting and debriefing behavioral games, because the success of this activity is, in part, contingent on creating an atmosphere in which participants can comfortably examine the strengths and weaknesses of their own listening and speaking behavior. For example, it is not uncommon for participants serving as the listener to become quite embarrassed when they discover how little they can recall of what a speaker said. In such instances, the instructor needs to encourage participants to confront and explore this reality seriously, but in a fashion that does not lead to depression, mistrust, or defensiveness, all of which may inhibit self-reflection and learning.

Outcomes of the activity are highly predictable, and it can be used profitably with second grade students through adults.

INTERPERSONAL AND GROUP COMMUNICATION

While intrapersonal communication is concerned with how people relate to their environment in general, interpersonal communication centers on how people relate to selected persons within the environment. Specifically, interpersonal communication deals with the dynamics by which relationships are initiated, develope, evolve and grow, or terminate. A good deal of the time people spend with others is in groups. Among the primary topics with which group communication deals are how groups operate; the nature and patterns of involvement and noninvolvement, participation, leadership, and group climate; the establishment of trust and cooperation; and the nature of power.

The six simulation games centering on interpersonal and group communication that we discuss and compare here are

TABLE 4 Interpersonal and Group Communication Simulation Games Overall Assessment

	Criteria						
	Significance	*Validity*	*Reliability*	*Flexibility*	*Popularity*	*Accessibility*	*Cost*
Telephone (Rumor Clinic)	4	4	5	4	4	5	1n
One-Way Two-Way Feedback	4	4	4	3	4	5	1n
Cooperation Squares	3	3	5	3	5	4	2r
Lost on the Moon (Consensus)	5	5	4	5	4	4	1r
Prisoner's Dilemma	5	4	4	3	4	4	1n
Power	5	5	4	2	1	3	3p

Key: High = 5; Moderately High = 4; Moderate = 3; Moderately Low = 2; Low = 1; p = purchase required; r = reproduction required; n = no expenditure required

Telephone (or *Rumor*), *One-Way Two-Way Feedback, Cooperation Squares, Lost on the Moon* (or *Consensus*), *Prisoner's Dilemma,* and *Power.* In general, these activities are more complex and have a wider range of potential outcomes than those reviewed in the previous section. As a result, they require more facilitator skill and preparation.

Telephone

Telephone (or *Rumor,* as it is often called) is simply an academic version of the age-old kids' game in which one person initiates a story and tells another, who tells another, who tells another, and so on. When running it as an instructional simulation, instead of having the first person make up a story, the instructor uses a controlled stimulus such as a film, picture, or paragraph. The stimulus is presented to the first individual for an appropriate period of time. Next, one at a time, four to six persons who were not exposed to the original stimulus are brought into the room to hear the description from the person who went before. In turn, each passes an account along to the next. Other participants observe the exchanges and make notes on what happens as the message is passed along. Videotaping the chain is also useful.

Postgame discussion focuses on the nature of rumor and, more generally, on the inevitable dynamic that occurs when information is transferred from person to person. Discussion guides suggest emphasizing how some facts get added, others subtracted, and others distorted. Finally, discussion centers on the implications of this process for everyday communication, stressing the merit of first-hand information, care in observing and describing an event, and so on.

This exercise has a number of advantages. It is simple to run, treats an important aspect of interpersonal communication, can be used with any number of participants, requires little instructor preparation, costs little or nothing, and has highly predictable outcomes. And, by varying stimulus material, the activity may be used with any level of participant.

One-Way Two-Way Feedback

One-Way Two-Way Feedback is also a popular game which, from a theoretical point of view, builds nicely on *Telephone.* It emphasizes the importance of feedback—checking out with a speaker one's perception of what one thought was said.

The exercise has two parts. During each, a participant (generally a volunteer) is given a drawing of several connected geometric shapes (a right triangle on top of a circle on top of a right-sloped parallelogram, for example). The volunteer's task is to describe the drawing so the other participants (who cannot see the drawing) can reproduce it on their paper exactly as it is on the original sheet.

During part one, only "one-way" communication is allowed. The volunteer gives directions, but no questions, comments, or reactions—verbal or nonverbal—are permitted from the audience. During part two, a volunteer describes another geometric figure of comparable difficulty. This time, "two-way" communication is permitted. For each phase, the instructor records the amount of time required, the describer's estimate of the percentage of persons who correctly reproduced the figure, and the actual percentage who did so correctly. Discussion guides indicate that postplay discussion ought to focus on the differences—advantages and disadvantages—of "one-way" versus "two-way" communication in the game and in communication situations in general.

Predictably, the one-way phase takes less time, leads to grossly exaggerated estimates of correct responses by the describer, and is typically reported to be a highly satisfying experience by the describer. The two-way phase takes considerably longer, leads to overly conservative estimates of correct reproductions, many more correct reproductions, and is frequently a frustrating experience for the describer. Thus, the one-way phase is less time-consuming or more efficient, more satisfying to the describers, but less effective and leads to less accurate information transfer. The applicability of these propositions to a variety of different interpersonal situations becomes the final focal point for discussion in what is always an interesting and involving activity.

The game does not make great demands on the instructor in terms of ability to facilitate behavioral learning, but a minimal theoretical base is useful for extrapolating the game experience to broader spheres.

Cooperation Squares

Cooperation Squares is probably one of the most popular simulation games in group communication. It is designed to demonstrate the value of cooperation in group problem-solving, and it makes the point dramatically and unambiguously.

The game is played in groups of five persons seated around a table. Each receives an envelope containing cardboard pieces, and the group is told that the task is to construct five equal squares without talking. Because no one person has all the pieces needed to complete one square, the solution requires sharing and cooperation among group members.

Post-play discussion begins with an analysis of what happened and explores which strategies led to quickest completion of the five squares. Initially, the instructor explores with the group the fairly predictable tendency of participants to define the nature of the task as *individual* rather than *group.* Individuals often spend much time trying in vain to assemble a square with their own pieces and only after a considerable time explore the possibility that they could and should exchange pieces with others in the group.

Some skill at behavioral facilitation is useful in conducting *Cooperation Squares,* because conflicts that may require some debriefing sometimes develop during play. The instructor needs little theoretical background since the theme of cooperation and its importance in various activities is a familiar topic.

The materials for the game can be easily made in an hour or so, depending on the number of participants; with enough puzzle sets, up to forty persons (in groups of five) can participate. As with several games discussed so far, the major advantage may also be a liability in some circumstances. In this fairly straightforward lesson on the values of cooperation in group problem solving, there is some risk of oversimplification. Tasks are usually not finite, they may require skills and knowledge that are not equally distributed among group members, and cooperation is sometimes dysfunctional (as among members of the Watergate "team"). These issues however, can be discussed in debriefing.

Lost on the Moon

Lost on the Moon, which also focuses on group communication and decision-making, gives participants an opportunity to learn about the process individuals go through to reach group consensus on matters in which there may be substantial differences of opinion among individuals.

The game has two parts. Initially, participants receive a list of items to rank according to their assessment of the items' relative importance for survival on the Moon. (Other sorts of lists, such as value statements on education, male-female relations, leadership, may also be used.) Next, participants come together and are given the task of arriving at a group consensus on the ranking of the items.

Discussion guides indicate that post-play discussion focuses on patterns of decision-making, comparing (either by "eye ball" or using mathematics) each initial individual's ranking with the final group ranking. A small discrepancy suggests either that the individual's preferences were coincidentally well matched with the group's or that the person was highly influential in decision-making. Discussion may also focus on roles played, group climate, leadership patterns, norms, levels of participation and involvement, at the instructor's discretion and as a function of the instructor's goals and the group dynamics that emerge.

Though the basic structure and operating dynamics are quite simple, the outcomes are both rich and varied. Data generated from even the most unimaginative iteration of the simulation are extremely useful for exploring many, if not most, of the critical components of group communication.

From both a behavioral and theoretical perspective, the game requires considerably more instructor skill than most of the games discussed so far. Essentially, the instructor must develop a debriefing strategy based on his or her own observations of the activity. Also, debriefing often requires theoretical and interpersonal skill in facilitating learning from conflicts that develop during the game.

The activity is exceptionally versatile. Items to be ranked may be easily varied according to the participant group's

TABLE 5 Interpersonal and Group Communication Simulation-Games Use Characteristics

	Primary Topic	Level (*K*indergarten, *P*ost *G*rad)	Number of Participants (*N*o, *V*ariables)	Basic Structure *S*imple, *A*verage,	Run Time (*M*inutes, *H*ours)	Debriefing Guide (*I*ncluded, *N*ot)	Debriefing Time (*M*inutes, *H*ours)	Theoretical Preparation (*C*ritical, *U*seful, *N*ot necessary)	Behavioral Simulation Facilitation Experience (*C*ritical, *U*seful, *N*ot *N*ecessary)
Telephone (Rumor)	Rumor, information transfer	K-PG	V	S	30m	I	15m	U	N
One-Way Two-Way Feedback	Feedback	2-PG	V	S	30m	I	15m	U	N
Cooperation Squares	Cooperation	8-PG	V groups of 5	A	30m	I	30m	N	U
Lost on the Moon (Consensus)	Group Decision-making	8-PG	V groups of 5-9	A	45m	I	45m	U	C
Prisoner's Dilemma	Trust	12-PG	V matched groups of 5-10	A	45m	I	30m	C	C
Power	Interpersonal power and motivation	12-PG	V groups of 5-10	A	1-6h	I	1-2h	C	C

interests, level, and previous experience. Beyond minimal reproduction, there are virtually no costs involved, preparation time is minimal, and time paramenters may be varied subject to circumstances.

Prisoner's Dilemma

Prisoner's Dilemma deals with trust, cooperation and competition, and interpersonal and intergroup relations. The structure of the game is somewhat more complex than others discussed so far, and the outcomes may vary considerably from one iteration to the next, though certain themes surface with nearly every play.

The game requires that participants be divided into two groups, each of which is provided a separate room. With the goal of "winning as much as you can," game play consists of ten rounds in which each team decides between two alternative letters (A or B for one group, X or Y for the other) in an effort to win as many points as possible. Depending on the combination they select (see the payoff matrix), both teams may win points, one may win and the other lose, or both may lose. In each round, then, a team has a choice between selecting the letter that may lead to cooperation or one that further heightens the likelihood of competition.

TABLE 6 Payoff Matrix

Choices		Outcomes	
Team 1	*Team 2*	*Team 1*	*Team 2*
A	X	+3	+3
A	Y	−6	+6
B	X	+6	−6
B	Y	−3	−3

At two points in the game, representatives from each team may meet—at the discretion of the teams—to discuss mutual concerns. Scores are totaled for all ten rounds.

Discussion guides suggest focusing on the group's assessment of who won, the criteria being used to arrive at the judgment, what each team thought the other team's goal was, and so forth.

Typically, groups define the task as "beat the other team" rather than an alternative interpretation of "win as many points as you can." This precludes their seeing the goal as a cooperative one—both teams winning as many points as possible. This, of course, could be accomplished only if both groups adopted an "AX" strategy. The reasons groups did not select such a definition of the goal (where this is the case) leads predictably to a discussion of trust, cooperation, and competition—all crucial components in interpersonal, group, and intergroup relations.

A reasonable theoretical background and some preparation are useful for helping participants see the analogy between the game and such things as courtship behavior, communication between countries, the international arms race, and so on. This is particularly important since, as noted earlier, interaction patterns, participant responses, and overall outcomes may vary from one play to the next.

Predictably, different factions develop within at least one of the groups. One subgroup often takes a highly competitive posture and another wants to initiate cooperation with the other group. Where these splits develop—and particularly where one faction asserts itself to the exclusion of the others—instructor skills in behavioral facilitation are useful, if not crucial. Helping participants see how their individual behaviors contributed to cooperative and competitive outcomes can be as important a part of debriefing as demonstrating in an intellectual fashion how trust, cooperation, and competition operate "in theory." For this reason, debriefing time for the game is extensive.

Getting the most out of the game not only requires a skilled instructor but reasonably bright and sophisticated participants, as well, in that many of the most important linkages to the "real world" are subtle and abstract. In all, *Prisoner's Dilemma* is potentially one of the richest simulations discussed so far in terms of both theoretical and personal learning; but outcomes are less predictable than some others and necessary facilitation skills are greater.

Power

Power is one of the most interesting simulation games in interpersonal and group communication. It is virtually unstructured, and for the most part the rules, roles, and outcomes emerge totally as a consequence of the motives, needs, actions, and reactions of participants.

The game was designed to explore the place of power and motives in interpersonal behavior. Basically, the activity is quite simple. Materials consist of a game board composed of colored squares dealt out and arranged like a checker board, with playing pieces that each participant fashions with cardboard, paper, pencil and scissors. Players going in turn may add as many pieces to the game board as they wish, anywhere they wish. During a turn a player may also move, remove, or destroy any or all of his or her own or the other players' pieces. Once a player has completed a turn, he or she may not alter the game board in any way until the next turn. The game ends only when all players have withdrawn from the game.

With these minimal ground rules, amazingly elaborate roles, rules, interaction patterns, norms, and confrontations emerge as players fashion and display an identity and negotiate with other players for psychological and physical space on the game board.

There is virtually no predictability in outcomes from one iteration of the game to the next. Often, participants become very involved emotionally with the game and the behavior of other participants—in a manner reminiscent of aspects of an encounter group. As a result, instructor skills in facilitating personal learning and interpersonal conflict are essential for satisfactory debriefing and meaningful use of the game. *Power* is not an appropriate game for indiscriminate use. Beneficial play requires reasonably serious-minded and intellectually sophisticated participants, not to mention an instructor with a high tolerance for ambiguity. As Table 5 notes, the time requirements for *Power* vary widely, as participants determine the length of the game. Generally, however, the game and debriefing require at least two or three hours.

ORGANIZATIONAL AND MASS COMMUNICATION

Organizational communication considers intrapersonal, interpersonal, and group communication phenomena as they operate in more complex environments. Information flow, work and social roles, and authority, status, and power relations are among the topics of central concern. Two simulation games—*Hollow Squares* and *Lock-a-Block*—are discussed and compared in this section.

Mass communication similarly involves aspects of intrapersonal, interpersonal, and group, as well as organizational, communication. In its simplest forms, mass communication centers on the process by which an individual (or group or organization) designs, packages, and transmits messages for intended consumption by a large, heterogeneous, and anonymous audience. We review two games that focus on mass communication topics, *Interact II* and *Trio.*

Hollow Squares

Hollow Squares is an inexpensive, relatively flexible, and popular game that focuses on dimensions of intergroup supervision and coordination. The activity involves two task groups. The planning group has access to the information needed to complete a task, which is the assembly of a cardboard puzzle. A second group does not have this information and must rely in its efforts to construct the puzzle on instructions it receives from the planning group. A third group serves as observers, recording the events that transpire and later initiating discussion of the entire process.

Debriefing begins with a recounting of events, plans, and strategies by the observers, with comments and questions from members of the other two groups. Typically, members of each team make a number of unnecessary and limiting assumptions about the task, and these become the initial focus of discussion. Later discussion focuses on the relations that developed within and between groups, the planning process, authority relations and outcomes, theories and strategies for supervision, and so on.

TABLE 7 Organizational and Mass Communication Simulation-Games Overall Assessment

	Criteria						
	Significance	*Validity*	*Reliability*	*Flexibility*	*Popularity*	*Accessibility*	*Cost*
Hollow Squares	3	4	4	4	4	4	2r
Lock-a-Block	4	4	4	5	4	3	3p
Trio	3	4	4	5	2	3	1n
Interact II	4	4	4	3	2	3	3p

Key: High = 5; Moderately High = 4; Moderate = 3; Moderately Low = 2; Low = 1; p = purchase required; 4 = reproduction required; n = no expenditure required

Hollow Squares can be used with up to twenty-five persons from grade twelve through postgraduate level. Run time is one to two hours, with thirty minutes to one hour needed for debriefing. Little or no theoretical preparation is needed to use the game, and skill at behavioral facilitation is not necessary. Necessary materials can be assembled in an hour or so, depending on the number of participants.

Lock-a-Block

Lock-a-Block is a game with a reasonably simple structure that generates a great deal of enthusiasm among players while surfacing numerous critical communication issues.

Game materials are several sets of styrofoam blocks (or Lincoln Logs, Tinker Toys or Giant Tinker Toys, or other building materials). The instructor constructs a model or shape using about one-fifth of the blocks. Participants are divided into three to five groups of four to seven persons, and each group is assigned a different section of the room in which to carry out its activities. Participants are told that each group is to build a replica of the instructor's model with the unassembled blocks. Each group selects one person to serve as a judge. The judges (three to five, according to the number of groups) are told that they have three tasks: (1) make all rules by which group competition will proceed (for example, How many blocks can a group take at a time? How many persons may come up to look at the model at once?) (2) enforce all rules; and (3) select a winning group.

After the construction and judging phases are completed, debriefing begins. Discussion guides suggest that discussion may focus on the judges' leadership style and its effect on the various groups, the extent to which judges delegated authority to groups, the degree of division of labor within groups, relations between groups, participation within groups, leadership styles within groups, and so forth.

Though the outcomes of the game vary somewhat from one play to the next, depending on the posture of the judges and the sorts of rules, enforcement procedures, and winning criteria they select, the patterns of interaction are quite predictable. Characteristically, judges take little account of the needs of the groups who selected them as they develop rules. The power often goes to their heads, and when this occurs some resentment by group members generally follows. This often leads to very fruitful discussions of communication in power and authority relationships.

Lock-A-Block can be most effectively used with up to thirty persons from high school through postgraduate level. The activity is of average complexity, requires about fifteen minutes to set up, forty-five minutes to one and one-half hours to play, and thirty minutes to one hour to debrief.

Theoretical preparation in the areas of intergroup relations, organizational dynamics, or group and organizational communication is useful, as is some degree of skill in behavioral facilitation. The latter is particularly important for helping participants explore the nature and effects of their own behavior within the group activity.

Trio

Trio is a mass communication exercise that involves interviewing, writing, and being an audience. It is very similar in structure to *Listening Triads* but uses a written rather than spoken mode of interaction. Participants are grouped by threes. Each triad member in turn interviews each of the other two. The interview focuses on gathering biographical data, which forms the basis for short biographical sketches. After the interviews and biographies are completed, they are distributed to the persons about whom they were written. Each of the three persons selects from the two biographies the one that is most accurate and appealing and gives both authors written or oral feedback. Discussion after the exercise concerns interviewing theory and practice, including the relation between the sorts of questions one asks of the types of answers that result, alternate question-asking techniques, and so forth. The exercise also provides a basis for exploring dimensions of audience reaction, specifically, what factors led to the selection of some biographies and the rejection of others, and what is the relative importance of "content" compared with style.

The activity can be used most effectively with high school or undergraduate university students. It requires no preparation time, takes about two hours to play and thirty minutes to one hour to debrief. Theoretical preparation in the topics and skills the exercise addresses is useful, but no behavioral facilitation skills are necessary.

Interact II

Interact II was developed by the author. The assessments herein are based on comments by Fredric Powell, Department of Speech Communication, State University of New York, Brockport. Interact is a complex simulation game that focuses on mass communication along with interpersonal, group, and organizational dynamics. Participants are organized in five to fifteen groups of five to ten persons each. During the game—which may last from one to two semesters—each group prepares, produces, and distributes a series of communication broadcasts or publications on predetermined topics.

On specified dates, the products are "aired" or distributed to all other participants, who serve as members of the audience and evaluate the products by assigning points and providing narrative feedback. Evaluation points are totaled and given in bulk to the producing group, which may divide them among its members as they deem appropriate.

Interact II makes a provision for participants to "quit" their mass communication company and work for another, start their own company, or work with the Executive Council (which governs the entire simulation).

Much of the debriefing for *Interact II* occurs in tandem with play as participants work through the various problems of organizing a group of individuals to produce a product that will appeal to its audience. Because of the duration and intensity of the simulation, behavioral facilitation skills are crucial to its effective use. Helping participants analyze and see personal relevance in the various experiences and problems they and their groups encounter is both critical and often difficult to accomplish because of the complexity and multiplicity of the dynamics that occur. Theoretical background in various aspects of communication is useful to the facilitation process, though perhaps less critical than for other complex simulation games, in that the parallels between the *Interact II* dynamics and those of the "real world" are quite apparent to participants. Debriefing effectiveness hinges on the instructor's ability to help participants "understand" and deal with the dynamics in a personally illuminating, productive, and transferable manner.

Interact II is best suited to university-level participants. It

TABLE 8 Organizational and Mass Communication Simulations-Games Use Characteristics

	Primary Topic	Level (*K*indergarten *P*ost *G*rad)	Number of Participants (*No.*, *V*ariable)	Basic Structure (*S*imple, *A*verage, *C*omplex)	Run Time (*M*inutes, *H*ours)	Debriefing Time (*M*inutes, *H*ours)	Debriefing Guide (*I*ncluded, *N*ot)	Theoretical Preparation (*C*ritical, *U*seful, *N*ot necessary)	Behavioral Simulation Facilitation Experience (*C*ritical, *U*seful, *N*ot necessary)
Hollow Squares	intra and inter-group interaction	12-PG	V to 25	A-C	1-2h	30m-1h	I	N	U
Lock-a-Block	intergroup coordination	12-PG	V to 30	A	45m-1/2h	30m-1h	I	U	U
Trio	interviewing	10-G	V groups of 3	S	1h	1h	I	U	N
Interact II	intergroup, intraorganizational and mass communication	12-G	V	C	4h to several semesters	Variable	N	C	C

can be used as a course unto itself or in conjunction with other courses for some portion of a semester, with as many as 150 participants.

INTERCULTURAL COMMUNICATION

Intercultural communication is an increasingly popular area. Work in this area focuses on the communication dynamics that occur during interaction between two or more persons with differing cultural (or subcultural) backgrounds. Five intercultural communication simulation games that deal with facets of intercultural communication are reviewed in this section. They are *Survival; Agitania, Meditania, Solidania; Lobu-Abu; Market Day;* and *Hypothetica*.

TABLE 9 Intercultural Communications Simulations-Games Overall Assessment

	Criteria						
	Significance	*Validity*	*Reliability*	*Flexibility*	*Popularity*	*Accessibility*	*Cost*
Survival	5	4	4	4	2	2	1n
Agitania, Meditania, Solidania	5	4	4	5	3	2	1n
Lobu-Abu	5	4	4	5	2	3	2r
Market Day	3	4	5	5	1	2	3p
Hypothetica	5	5	4	5	2	3	2r

Key: High = 5; Moderately High = 4; Moderate = 3; Moderately Low = 2; Low = 1; p = purchase required; r = reproduction required; n = no expenditure required

Survival

Survival is a simulation game designed to increase participant awareness of the nature of culture and the process by which culture develops and evolves. It casts participants in the roles of survivors of an airliner crash on an isolated and uninhabited island, where they learn they are destined to remain for many years. Their initial task is to devise a plan of action—what needs to be done and who will do it. At various times, the instructor enters the room in which the "survivors" are located and asks questions that require the group to examine the probable evolution of the newly formed society six months, one year, and five years hence. In this way the group can be guided toward considering such topics as orientation to work, time, sexuality, religion, ethics, law, gender roles, and so on, as it defines what the ideal culture might be.

Some inevitable extrapolation occurs as participants respond to the experience—which is initially rather ambiguous—and speculate about how the culture might evolve. Discussion at the completion of the game is directed toward exploring specific cultural dimensions of the sort listed previously.

Survival generates particular interest and enthusiasm when used with groups (of five to fifteen members) of serious high school through postgraduate-level participants who have some degree of tolerance for ambiguity. The game structure is relatively simple and depends for its outcomes on the interest and involvement of participants. The activity requires from one and one-half to four hours to play, with thirty minutes to one hour for debriefing. Neither theoretical preparation nor behavioral facilitation skills are needed.

Agitania, Meditania, Solidania

Agitania, Meditania, Solidania is an enjoyable and informative role-play game in which participants assume roles as members of one of three cultural groups, each with its own distinctive cultural communication characteristics. Through interaction with members of the "other cultures," participants gain some experience in the problems and challenges encountered cross-culturally regarding such differences as orientation to time, physical contact, money, work, or greetings. At the same time, the game sensitizes participants to some crucial dimensions of cross-cultural communication, and through discussion raises questions about stereotyping and its effects.

The role-play can be used most effectively with fifteen to forty-five participants of high school through university level. Time required for play is one to two hours, with thirty minutes to one hour for debriefing. Theoretical preparation in the area of intercultural relations is critical for most effective use of this game, but facilitation skills are not needed.

Lobu-Abu

Lobu-Abu also splits participants into two cultural groups, but unlike *Agitania, Meditania, Solidania* it emphasizes language and language differences across cultures. One culture speaks Lobu, the other Abu.

A list of English words and their Abu equivalents is given to the Abu speakers. The Lobu speakers get a list of the same words with their Lobu equivalents. Each member of the Abu-speaking group is paired with a member of the Lobu-speaking group. Everyone becomes highly involved in attempting in turn, to get their counterparts to perform a simple task (from a list provided by the instructor) with instructions given only in the speaker's "native" language. Getting one's meaning across in Lobu to an Abu-speaking person can be reasonably complex. In a simple way, the game provides participants some first-hand experience in the problems of interacting with someone who does not share one's language. The game and subsequent discussion also sensitizes participants to the importance of nonverbal communication in intercultural information exchange and, more generally, highlights aspects of the language-learning process.

The activity is highly motivating for participants and can be used with junior high through university-level participants. It requires about one to one and one-half hours to play and thirty minutes to one hour to debrief.

Market Day

Market Day is a simulation game intended to recreate the economic and social conditions of markets as they exist in many areas of the world. The game establishes an economic order and involves participants in bargaining and negotiating for food and services.

The design of the simulation is straightforward. Various tropical fruits, vegetables, drink, and other foods are procured, and booths are set up and operated by instructors or their confederates. Other confederates serve as beggars, pickpockets, magicians, and other roles characteristic of a marketplace population. Participants are given colored chips that represent varying levels of income. Some are given enough chips to be, in effect, a wealthy class; they can buy all the food they want. Others are given the number of chips appropriate to a "middle class." A third group is forced into a "poverty" role, since they receive few chips and can buy little at the marketplace.

As participants move from booth to booth, they are confronted by the "reality" of their own economic positions and the discrepancies between their own condition and the apparent affluence or poverty of others. A wide range of potential behavior and feelings may result, including helplessness, resentment, superiority, hostility, begging, charity, donations of "conscience money," bargaining, negotiating, robbery, and so forth.

Postplay discussion focuses on these and other attitudes and behavior on the market itself, and on markets as they operate around the world.

The game can be used, with highly predictable outcomes, with persons of all ages and will easily accommodate up to seventy-five or eighty participants. It requires no behavioral facilitation skill, though some theoretical base in economic development and its cultural implications is certainly an asset.

Hypothetica

Hypothetica focuses on economic and cultural development from a communication perspective. The simulation places participants in one of several regional groups in an imaginary Third World nation. As the game opens, participants learn that valuable resources have just been discovered in one of the regions and that the national assembly (composed of representatives from each region) must decide whether and how the country should exploit the find.

For the region in which the resources are located, the decision is usually quite clear, as it is for the group that inhabits the coastal region from which shipping will take place. Both groups see immediate benefits. Persons in the other regions often see potential problems that may accompany economic development, and these issues become the core of the valuable dialogue between regional representatives.

An advantage of the game is its flexibility. Game materials can be adapted to highly sophisticated participants or used with participants totally unfamiliar with communication and development issues. The game consistently generates enthusiasm and involvement and often leads to challenging discussions of the pluses and minuses of "progress," "growth," "development," and "advancement." For the instructor, particular knowledge of development is crucial for getting the most from postplay discussion, as is some ability to facilitate learning in situations where heated conflicts may surface.

Hypothetica can be used with up to thirty or forty university through postgraduate participants. Time requirements vary as a function of participant actions, though two to three hours is generally a minimum.

Sources for Additional Information on Simulation Games in Communication

In the previous pages we have discussed a number of games in the categories of intrapersonal, interpersonal and group, organizational and mass, and intercultural communication.

TABLE 10 Intercultural Communication Simulations-Games Use Characteristics

	Primary Topic	Level (*K*indergarten-*P*ost *G*rad)	Number of Participants (No., *V*ariable)	Basic Structure (*S*imple, *A*verage, *C*omplex)	Run Time (*M*inutes, *H*ours)	Debriefing Time (*M*inutes, *H*ours)	Debriefing Guide (*I*ncluded, *N*ot)	Theoretical Preparation (*C*ritical, *U*seful, *N*ot necessary)	Behavioral Simulation Facilitation Experience (*C*ritical, *U*seful, *N*ot necessary)
Survival	cultural development, evolution and interaction	12-PG	5-15	A	1 1/2-2h	30m-1h	I	N	U
Agitania, Meditania, Solidania	cross-cultural differences and stereotyping	12-G	15-45	A	1-2h	30m-1h	I	C	N
Logu-Abu Market Day	cross-cultural adaptation and language learning	10-G	V groups of 2	A	1-1 1/2h	30m	I	V	N
Market Day	economic and cultural differences	K-PG	V	C	2-4h	15m-1h	I	V	U
Hypothetica	development and social action	12-PG	10-60	C	1h-	1-2h	I	V	U

Obviously, these are only some of the many simulations, games, and structured exercises available within these areas. Because of length considerations, we could give only thumbnail sketches of the actual structure, rules, procedures, and outcomes of these games. Table 11, however, lists sources of further information about each game. Where the game is available from more than one source, each is listed. These sources provide detailed descriptions of game materials, guides for postplay discussion, and, in many cases useful suggestions for variations of the activities.

TABLE 11 Intrapersonal Communication Simulation/Game Sources

Game	Sources
Learning Game	1. *Human Communication Handbook: Simulations and Games*, page 41 B. D. Ruben & R. W. Budd Hayden Book Company 50 Essex Street Rochelle Park, NJ 07669 $6.95
	2. *1978 Annual Handbook for Group Facilitators*, page 9 J. E. Jones & J. W. Pfeiffer University Associates Press 7596 Eads Avenue La Jolla, CA 92037 paperbound, $12.50, looseleaf notebook $29.50
Memory Game	1. *Human Communication Handbook: Simulations and Games, Volume 2,* page 50 B. D. Ruben Hayden Book Company 50 Essex Street Rochelle Park, NJ 07669 $8.95
Zif	1. *Human Communication Handbook: Simulations and Games, Volume 2,* page 56 B. D. Ruben Hayden Book Company 50 Essex Street Rochelle Park, NJ 07662 $8.95
Listening Triads	1. *The Dynamics of Human Communication: A Laboratory Approach*, page 199 G. E. Meyers and M. T. Meyers McGraw-Hill Book Company Princeton Road Hightstown, NJ 08520 $9.50
	2. *A Handbook of Structured Experiences for Human Relations Training, Volume I,* page 12 J. E. Jones and J. W. Pfeiffer University Associates Press 7596 Eads Avenue La Jolla, CA 92037 $6.00
One-Way Two-Way Feedback	1. *Managerial Psychology* H. J. Leavitt University of Chicago Press 5801 Ellis Avenue Chicago, IL 60637 $9.95
	2. *A Handbook of Structured Experiences for Human Relations Training, Volume 1*, page 13 J. E. Jones and J. W. Pfeiffer University Associates Press 7596 Eads Avenue La Jolla, CA 92037 $6.00
Cooperation Squares	1. *Ten Interaction Exercises for the Classroom* D. J. Mial and S. Jacobson National Training Labs Washington DC Original Source NA
	2. *Communication Games*, page 109 K. R. Krupar Free Press % The Macmillan Co. 866 Third Avenue New York, NY 10022 $3.95
	3. *Human Communication Handbook: Simulations and Games*, page 74 B. D. Ruben and R. W. Budd Hayden Book Company 50 Essex Street Rochelle Park, NJ $6.95
	4. *A Handbook of Structured Experiences for Human Relations Training, Volume 1*, page 25 J. W. Pfeiffer and J. E. Jones University Associates Press 7596 Eads Avenue La Jolla, CA 92037 $6.00
	5. *The Dynamics of Human Communication: A Laboratory Approach*, page 361 G. E. Myers and M. T. Myers McGraw-Hill Book Co. Princeton Road Hightstown, NJ 08520 $9.50
Prisoner's Dilemma	1. *A Handbook of Structured Experiences for Human Relations Training, Volume 3*, page 52 J. W. Pfeiffer and J. E. Jones University Associates Press 7596 Eads Avenue La Jolla, CA 92037 $6.00
	2. *Human Communication Handbook: Simulations and Games*, page 76 B. D. Ruben and R. W. Budd Hayden Book Company 50 Essex Street Rochelle Park, NJ 07669 $6.95
	3. "Trust" J. Boulogne 13965-64 Avenue Surrey, BC., Canada V3W IY7 $8.75

TABLE 11 Intrapersonal Communication Simulation/Game Sources (Cont)

	4. *A Handbook of Structured Experiences for Human Relations Training, Volume 2*, page 62 J. W. Pfeiffer and J. E. Jones University Associates Press 7596 Eads Avenue La Jolla, CA 92037 $6.00
Prisoner's Dilemma (Cont)	5. *The Dynamics of Human Communication: A Laboratory Approach*, page 351 G. E. Meyers and M. T. Myers McGraw-Hill Book Company Princeton Road Hightstown, NJ 08520 $9.50
Power	1. *Great Game and Symbol Company* A. Amberstone and W. Amberstone Westminster Road Brooklyn, NY NA
	2. *Human Communication Handbook: Simulations and Games*, page 113 B. D. Ruben and R. W. Budd Hayden Book Company 50 Essex Street Rochelle Park, NJ $6.95
Telephone	1. *Nothing Never Happens: Exercises to Trigger Group Discussions*, page 67 K. G. Johnson, J. J. Senatore, M. C. Liebig, and G. Minor Glencoe Press 17337 Ventura Blvd. Encino, CA 91316 $10.95
	2. *Human Communication Handbook: Simulations and Games*, page 62 B. D. Ruben Hayden Book Company 50 Essex Street Rochelle Park, NJ $6.95
	3. *A Handbook of Structured Experiences for Human Relations Training, Volume 2*, page 14 J. W. Pfeiffer and J. E. Jones University Associates Press 7596 Eads Avenue La Jolla, CA 92037 $6.00
Lost on the Moon-Consensus	1. *The Dynamics of Human Communication: A Laboratory Approach*, page 334 G. E. Myers and M. T. Myers McGraw-Hill Book Company Princeton Road Hightstown, NJ 08520 $9.50
	2. *Nothing Never Happens: Exercises to Trigger Group Discussions*, page 57 K. G. Johnson, J. J. Senatore, M. C. Liebig, and G. Minor Glencoe Press 17337 Ventura Blvd. Encino, CA 91316 $10.95
	3. *Human Communication Handbook: Simulations and Games*, page 82 B. D. Ruben and R. W. Budd Hayden Book Company 50 Essex Street Rochelle Park, NJ $6.95
Lost on the Moon-Consensus (Cont)	4. *Human Communication Handbook: Simulations and Games, Volume 2*, page 87 B. D. Ruben Hayden Book Company 50 Essex Street Rochelle Park, NJ $8.95
	5. *A Handbook of Structured Experiences for Human Relations Training, Volume 4*, page 51 J. W. Pfeiffer and J. E. Jones University Associates Press 7596 Eads Avenue La Jolla, CA 92037 $6.00
Hollow Square	1. *A Handbook of Structured Experiences for Human Relations Training, Volume 2*, page 32 J. W. Pfeiffer and J. E. Jones University Associates Press 7596 Eads Avenue La Jolla, CA 92037 $6.00
	2. *Human Communication Handbook: Simulations and Games*, page 97 B. D. Ruben and R. W. Budd Hayden Book Company 50 Essex Street Rochelle Park, NJ $6.95
Lock-A-Block	1. *Human Communication Handbook: Simulations and Games*, page 95 B. D. Ruben and R. W. Budd Hayden Book Company 50 Essex Street Rochelle Park, NJ $6.95
Trio	1. *Symposium on Simulations for Communication Education* A. D. Talbott and M. S. MacLean Jr. School of Journalism University of Iowa Iowa City, IA NA
	2. *Human Communication Handbook: Simulations and Games, Volume 2* Brent D. Ruben Hayden Book Company 50 Essex Street Rochelle Park, NJ $8.95

TABLE 11 Intrapersonal Communication Simulation/Game Sources (Cont)

Interact II	1. *Interact II* B. D. Ruben Avery Publishing Group 89 Baldwin Terrace Wayne, NJ $5.95
Survival	1. Pri Notowidigdo Briefing Center Canadian Internation Development Agency 200 Rue Principal Hull, Quebec, Canada NA
	2. *Human Communication Handbook: Simulations and Games, Volume 2*, page 103 B. D. Ruben Hayden Book Company 50 Essex Street Rochelle Park, NJ 07662 $8.95
	3. *Fig Leaf* D. Bensen Parthenon Publishing Company 2628 Old Lebanon Road Nashville, TN 37214 $5.95
Agitania, Meditania, Solidania	1. *Human Communication Handbook: Simulations and Games*, page 117 B. D. Ruben and R. W. Budd Hayden Book Company 50 Essex Street Rochelle Park, NJ 07662 $6.95
Lobu-Abu	1. *Human Communication Handbook: Simulations and Games, Volume 2*, page 141 B. D. Ruben and R. W. Budd Hayden Book Company 50 Essex Street Rochelle Park, NJ 07662 $8.95
Market Day	1. *Human Communication Handbook: Simulations and Games, Volume 2*, page 130 B. D. Ruben Hayden Book Company 50 Essex Street Rochelle Park, NJ 07662 $8.95
Hypothetica	1. *Human Communication Handbook: Simulations and Games*, page 121 B. D. Ruben and R. W. Budd Hayden Book Company 50 Essex Street Rochelle Park, NJ 07662 $6.95

COMMUNITY LAND USE GAME
An Evaluation

by Joseph L. Davis

INTRODUCTION

Confessions of a *CLUG*er

A great disappointment of my personal and professional life is that I have had the opportunity to play *CLUG* only once. The *Community Land Use Game,* created by Allan G. Feldt, was a mainstay of The Center for Simulation Studies as far back as 1971 when I first became involved with it. It was also the first simulation I ever played. The interest I developed in simulations after playing *CLUG* led me to intern with the Center, where one of my first tasks was to learn to run it. In the last seven years I have run *CLUG* in one form or other, estimating conservatively, three hundred times. Unfortunately, I have never had an opportunity to play it again. But in those roughly two thousand hours of "*CLUG*-ing" I have developed some definite opinions and insights on this very elegant simulation.

My background is in political science. And while I have run the model for political science classes more often than any other groups, my work with the Center has allowed me to present *CLUG* to a vast array of groups ranging from junior high school age through senior citizens and, in subject matter interest, from graduate classes in social work to county planning authorities to I.B.M. middle management executives.

Purpose

It is my purpose to: (1) present my insights into *CLUG*; (2) explain its adaptability as well as its limitations; (3) help the reader to determine the suitability of this model for particular simulation or teaching needs; (4) give some suggestions for running the game (as well as a few tricks of the trade); and (5) try to give a bit of the "flavor" of *CLUG*.

What is *CLUG*?

CLUG stands for the *Community Land Use Game* (originally the *Cornell Land Use Game*). It is a noncomputer based, nonzero-sum simulation designed by Allan G. Feldt of the Cornell Graduate School of Economics as a model for showing the economic development of American cities. A complete kit is available for $75.00 from the Institute of Higher Education Research and Services (Box 6293, University, Alabama 35486). Player's manuals are $6.95 each from The Free Press (Department F, Riverside, New Jersey 08075).

The simulation is played by ten to thirty participants divided evenly into five teams. Each team has a certain amount of resources in the form of money or credit and has the option of spending the money in a variety of ways. *CLUG* is played on a board that is a 14" x 14" grid representing a particular area of land. Teams have the opportunity to invest in various types of land use that, when placed at specific locations on the map in the form of building blocks, Legos, or another medium of construction, form a physical development that eventually grows into a community. The types of land uses available for investment include several varieties of industrial, commercial, and residential buildings. These buildings have certain fixed construction prices and income values that depend on how well they are used.

CLUG is played for an extended period, usually six to seven hours, in a series of rounds, each round representing from one to five years. Hence, the more rounds that are played the more years are played through, and the more the community can develop. In each round participants go through a set, specified, and ordered series of steps that break down the types of decisions the community needs to make into a logical sequence and also simplify the rather complex economic proceedings of the game.

CLUG Steps of Play

The rules for *CLUG* are available in the player's manual. There are no hidden rules or secret dealings between the game operator and any players. Everything is available and knowable. The only limit on information for participants is time and the complexity of the simulation. For this reason, the game is played in a series of rounds. The first round is played very slowly so people can catch on, but because the same steps and processes are played in every round, players can pick up the process quickly. By the third or fourth round, the game proceeds quite rapidly.

The steps of play for one round of the basic *CLUG* model follow:

Editor's Note: *CLUG* is listed in the urban section.

(1) Land is bought and sold between team members or from the "bank," which holds all unowned land. The *CLUG* board is numbered at the top and side so any square of land on the grid can be located by coordinating the two axes. Land bids can be recorded and marked on the board so that the same piece is not bid on twice. Competitive sealed bids are submitted, with the highest bidder getting the piece of land.

(2) After land bids there is a sort of community meeting that involves all participants. Two important decisions to make are locating utility lines and setting the tax rate. Utility lines, which are necessary before any building can be put on any land a team owns, start from a point on the board marked as the power plant, run down the boundary lines of grid squares (which also denote the road system), and, of course, all must connect.

The cost of utilities is assessed from all teams out of the property taxes. Teams are taxed on the price of land they buy plus the value of any buildings they construct (depreciated by five percent each round). After all teams have voted on utilities (which must be passed by three teams out of the five), the community sets a tax rate. The tax rate pays for utility construction and maintenance and for such community services as fire and police protection figured on a cost-per-resident charge.

(3) After the council meeting there is a construction phase in which teams have an opportunity to build whatever they wish from the types of land uses available at the costs indicated in the player's manual. Teams can construct as much as they wish if they own the land, have utilities for it, and the money for construction.

The game is designed, and the economics of the model are so balanced, that it is almost impossible for one team to build all it needs to be self-sufficient. Consequently, teams need to interact to buy labor or store goods or to find jobs for their residents.

(4) In the next step of play we designate the places of employment for all residents and all jobs to see who is working, who is unemployed, and what businesses are not open in the community. The manual tells how many employees each type of business needs.

(5) All commercial enterprises that have been built in this round and now have workers (that is, are open) may set their prices and try to pick up customers among the residences built on the board.

(6) After these agreements have been made, the game operator pays all the industries for their industrial income.

(7) Then all the stores collect their charges from the customers, and all labor collects income from their employers.

(8) All this time, a second game operator, called the community accountant or community tax assessor, has been keeping a duplicate record of all the transactions on the playing board—the buildings constructed and the land sold—and has been figuring the tax rate on the assessed evaluations of the individual teams and of the whole community. This information is reported, usually on a blackboard, and the game operator collects taxes from each team.

(9) The game operator then collects transportation charges incurred by the various teams in that round. Because the game is played on a map, and movement across it is movement of physical distance, players incur transporation charges as they go to work, to the stores, and to the marketplace.

(10) The collection of taxes and transportation costs marks the end of the round, and we immediately begin with buying and selling land in the next round.

Renovation

This event occurs only in rounds divisible by five. Since we are playing through time, and buildings are being depreciated at five percent per round to make them cheaper to operate, renovation gives teams the opportunity to fix up their existing structures. The incentive to spend money in renovation is based on a probability table stating that the older a building gets without renovation, the greater the chance it will be destroyed by natural or man-made disaster. Every fifth round all teams must roll the dice on all of their buildings, and if they roll a losing number they lose the building.

Debriefing

The game proceeds through numerous rounds. I try to get people through two renovations, which means ten rounds, so they have a lot of development to look back on. The game ends not at any set point but unexpectedly, because the "real world" does not really end.

The last section of the simulation, the debriefing, is the period in which we sit back, look at what has been happening throughout the simulation, reflect on it, and try to learn from this experience.

An Elegant Simulation

It is clear from its title that *CLUG* deals with land use in a community setting. But one of its primary advantages is that the issues and knowledge that emerge from it go far beyond determining the use of land for economic ends. In discussing *CLUG*'s usefulness I find myself in a situation like that of a Shakespearian scholar. After extensive use of this model, in which I have found many gems of insight, I am not certain whether all of them were actually designed into it (no offense to Dr. Feldt) or whether the basic design was so good and so flexible that the model can address the additional knowledge I wish to bring out of it even though they may not have been part of the original intent.

CLUG—THE PROCESS

CLUG is a person-to-person simulation. Its interaction take place on three levels—among members of each team, between the teams, and between the teams and the outside world as portrayed by the game operator, the rules, and the elements of chance.

Interaction Within Teams

CLUG provides no structure or instructions for the types of behavior that occur within each team. However, this level of interaction should not be ignored. How a team organizes its

FIGURE 1 *CLUG* **Interaction Diagram**

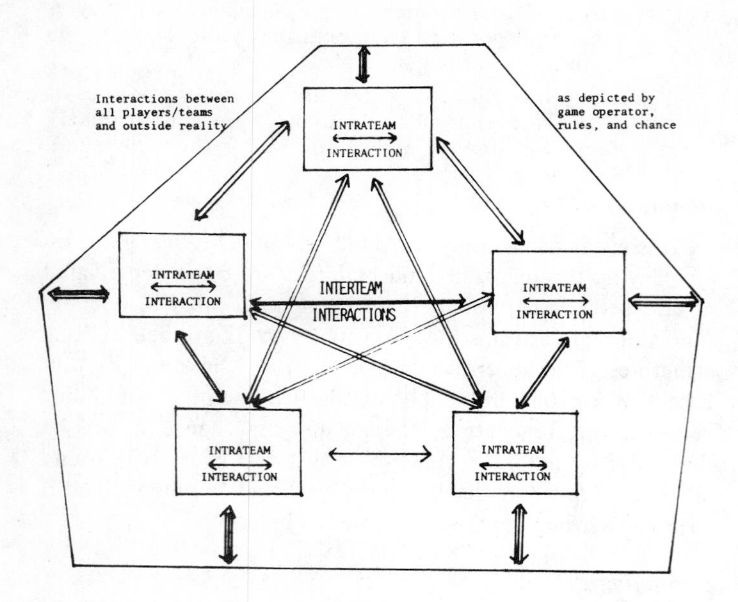

resources and its activities can be a very important part of learning.

I have played *CLUG* with many groups that were looking primarily not at urban processes but at problems of interpersonal communication and cooperation. *CLUG* is an excellent model for exploring these dimensions within a real world situation. Its complexity and economic realism allow one to look at interpersonal problems not in a vacuum, where cooperation is easy because there is not much cost, but in a life-like context in which cooperation and communication take place over some *thing* and have consequences that exist beyond the interaction of the moment.

CLUG confronts players with a complex situation and strict time pressure. They must sort out for themselves how they will accomplish tasks, in what order, and which will take priority, while others may not get done at all. It is always good to spend part of the debriefing discussing how teams came to decision-making and how that decision-making affected their play.

There doesn't seem to be one best style of intrateam interaction; some teams are democratic, some are autocratic. What seems to work is whatever the particular participants on that team are comfortable with. However, whatever the style of interaction, I believe one rule is hard and fast. Not only learning but also the enjoyment in playing *CLUG* are a function of participant involvement.

Interaction Between Teams

A large part of the interaction, and that which is most clearly spelled out in the *CLUG* rules, is the activity among teams. *CLUG* is a nonzero-sum simulation, and the activities among teams resemble the very common prisoner's dilemma model, though the reward and payoff of the prisoner's dilemma are modified by "legitimate" reasons for not cooperating. The competition that arises among teams in the games can be self-defeating (as is clear whenever an imbalance of labor and jobs or an excess of one particular type of land use, such as stores, arises). It is, however, this very competition that sparks a great deal of the interest and activity in the game. Thus we are faced with a real world dilemma where conflicting values come into play so that the most perfect theoretical play of the game is not necessarily the most successful, desirable, or even the most possible.

CLUG follows the prisoner's dilemma model in that the "winnings" of one particular team are partially out of their control and are controlled by other teams over which the first has little influence. The interdependence of the five teams necessitates some level of cooperation. However, I have found that the best cities and the best plays result neither from absolute cooperation—which would mean the abolition of teams and uniting everybody and all the money into one glorious group (which also usually limits the number of active participants)—nor in complete cut-throat competition in all areas. It is necessary to find a balance between competition and cooperation that continues the interests of all teams and participants by allowing them to pursue their own goals and rewards but does not become self-destructive for the teams by playing strictly to defeat other teams.

The opportunities for cooperation are many. In the economic dimension alone, all types of land use have some dependence on other types. For instance, industry and commercial enterprises need labor, which requires residences. If a team cannot build its own, it needs to hire workers from some other team. Commercial enterprises need customers; two that require customers and labor residences are the local store and the central store. These stores (plus industries) also need to buy office space. Thus anyone building a local store, a central store, or an office needs to have cooperation in the form of customers from the other people on the board or profit cannot be made. All of this very basic cooperation forms the community.

There are also many subtle opportunities to compete—both healthy and counterproductive. What I would call healthy competition is the attempt by each team to maximize its resources and to use the success of other teams to measure its own success. But there are also opportunities for many self-defeating types of competition. These make fine discussion topics in the debriefing because they almost always occur. They include such tactics as pulling workers out of another's business without enough warning so they must either close down for a round or shift workers from another place, thus causing a ripple effect through all the teams. Another is competition for utility lines that ends in producing too many lines simply because you won't let anybody else have any unless you get some, too; then everybody has to pay. Likewise, the construction of too many residences or local stores, for instance, causes undue competition between teams.

These are destructive forms of competition because in the end everyone loses. It may seem that pulling labor out of somebody else's business to work on your own is good—it not only gives you the workers you need but keeps someone else from making a profit in that round. This is a false notion of winning; the economic hardship it places on the other team will eventually get back to you.

All the activity in the game, no matter whose team it is in, no matter what table it is on, affects everybody else because there is just so much money in the system. If you do something to keep someone else from maximizing profit, the total value of the community cannot grow by whatever percentage is stifled. All players in the game see themselves as members of a community, in which what is good for someone as a team member may not be good for him or her as a member of the community.

Interaction Between Teams and Rules

The third level of interaction in *CLUG,* the relation between all the players and the outside reality as depicted by the game operator and the rules, is one of the reasons *CLUG* is such a good simulation.

In a continuum of simulations (see Figure 2), at one extreme would be models completely determined and where all activity is fixed before the game is played. Here we would have games such as parlor games or children's games where you roll dice, move so many spaces, pick up a card, and do what it tells you. At the other extreme of that continuum are simulations; players determine all activity and no rules limit what occurs—a form of unregimented play. Most simulations on the market fall between these two extremes.

CLUG clearly belongs toward the left of this scale, because much of the activity in the game is determined by participant decisions rather than by specific game rules directing behavior.

There seems to be a trade-off between the amount of freedom the model allows participants to create their situation and the manageability of the model by the game operator. Obviously, the more determined the model, the easier it is to comprehend and control the situation. Thus, a parlor game can

FIGURE 2 Continum of Simulations and Game Types

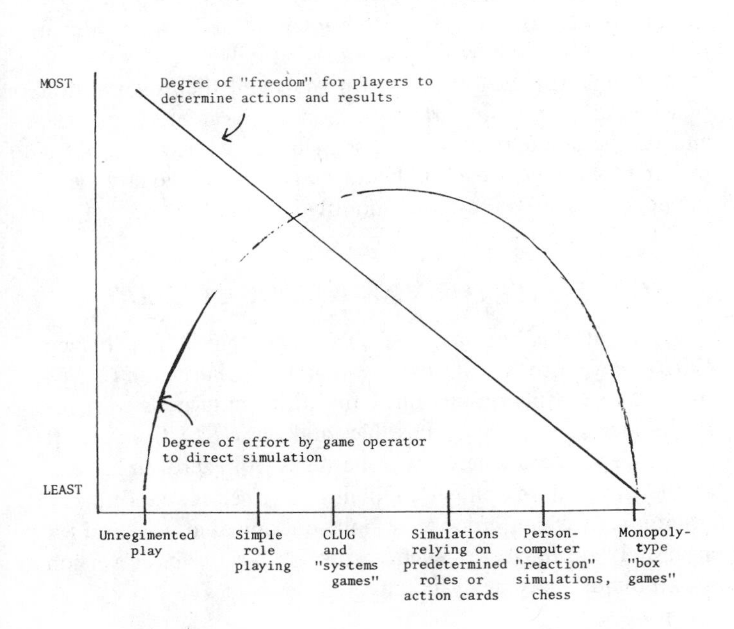

provide a simple page of instructions and any participant or group can pick it up and run it successfully the first time. A more open-ended game, on the other hand, requires more judgment and monitoring from the game operator to fit participants' activities into the constraints of the model. An open-ended simulation such as *CLUG* is not free-form play and does have specific subject matter, so the role of the game operator becomes one of translator and judge to fit whatever participants want to happen into the *CLUG* language.

I prefer simulations that give participants a large measure of freedom to design their own situation. One of the greatest lessons that can come out of such a simulation is an understanding that participants do create their situation and then have to live with the consequences. This learning is possible in *CLUG* because we are compressing time and can look at cause and effect relation among activities. In a situation in which all participant activity is determined and the results of that activity are preprogrammed, participants may have strong feelings of being manipulated. *CLUG* has a very low level of manipulation because so much of the activity in it is created by the participants themselves. The constraints, the limitations on what players can do, are determined by their previous decisions and previous rounds rather than by the dictates of the instruction manual or the game operator.

Rules of Play—The *CLUG* Language

The rules of *CLUG* sketch an economic reality by translating realistic cost, profit, and investment rates into a model. No attempt is made to create costs that are unrealistic in relation to the "outside world." So in one sense the limitations the rules place on the model are nothing more than the boundaries of the universe and the laws of supply and demand that become the foundation on which the activity of *CLUG* takes place. Except for learning the language of *CLUG* finances, the rules are really few and quite flexible.

If one can accept the economic base of *CLUG* as realistic, which is not difficult because the game is well researched and well tested, then the participants are free to test and probe the limits of this universe. Any other limitations on participants are of their own creation through decisions in previous rounds.

For instance, suppose a team has spent all of its money on unprofitable land uses and finally learns in the fourth round how to make some money. However, it has only $12,000 left to invest in, say, an industry that costs $48,000. It would be unrealistic to allow the team to build that industry for the $12,000 it has left. This would violate the translation of relative costs in the *CLUG* language. It was the fault not of the game but the team that ran out of money.

This low level of manipulation in *CLUG* frees the game operator from one of the biggest problems that she or he faces in any simulation—the sense of manipulating the group. It also gives the participants a greater feeling of owning their simulation activity, which always helps drive home the lessons. The results of *CLUG,* whatever they may be, are mostly the responsibility of the players, and not that of the designer of the model or the game operator.

There are people who do question the rules of *CLUG* to begin with—that it proposes a capitalistic system, a certain level of supply and demand, and a growth economy. This is the "universe" the designer decided to simulate. In my experience any attempts to alter this base cannot be easily accommodated in the model. Once one questions the supply of money or the existence of different groups that have to make decisions within a common framework, one has gone beyond the range of what *CLUG* can do. In my experience those groups that have tried to change this base have done so at the expense of player participation and jeopardized the logic of the model.

Modifications of the Interpersonal Process

As I will show later in discussing possible variations on *CLUG*, the content can be changed or added to quite radically, without any problems. The nature of the group processes, the interaction among the various levels, is less open to modification.

The first level of interaction (within teams) is a valuable aspect of the simulation, and, while it is completely unstructured, should take place. I would discourage an attempt to play *CLUG* with one player per team. I would also discourage playing with teams so large that interaction within them cannot occur effectively. I consider six players per team to be the maximum for effective interaction within the team; three to four is best.

Likewise, I believe that the second level (interteam interaction) cannot be dissolved without serious effects on the nature of the game. While I have run models in which groups have voted to dissolve teams and create, as it were, one classless society within which to build their city, this is usually done at the expense of participation by many players. Essentially, what happens is that the same sort of intergroup behavior takes place but it now occurs informally.

The role of the outside world and the third level of interaction is probably the most flexible part of *CLUG*, and many of the variations that have been designed have increased the role of these outside forces to accentuate some particular aspect of learning in the model.

Though we will cover some rule modifications in discussing content variations, I believe that the nature of the rules need to be kept economical, logical, and realistic lest *CLUG* fall into the range of manipulative games and lose some of its effect.

Computer *CLUG* and Group Interaction

The earliest games of *CLUG* I ran used a computer. Participants would make their decisions, we would put these decisions into the computer, and the computer would give readouts of the economic level of investments, returns on investments, cash on hand, and so forth, for each team. This was an extremely accurate and esoteric version of *CLUG*, because groups were simply looking at columns of figures, and it actually slowed down the game. It also lowered the level of intergroup activity and made for a much more studied and cool game. This was good for some groups, but for others it did not touch the level of feeling that many people have about cost and money, because participants were simply dealing with columns of figures. When we took away the computer and used paper money, it was amazing to see the difference in the level of activity, the level of trust, and the level of involvement among participants.

If you are interested in exploring the process that groups go through in making their decisions on economic matters, and you wish to explore some of the assumptions (right and wrong), groups have about economic activity, the noncomputerized version of *CLUG* is preferred. The computer becomes helpful and, indeed, essential in some of the more complex modifications. But for the basic models, the figuring is so simple (and essential if participants are to understand the system) that using a computer can detract from understanding the economics of *CLUG* by placing it all in a "black box."

Process-Oriented Groups

I have played *CLUG* a number of times with groups that were really not interested in urban planning, urban problems, land use, or the economic aspects of city development. They were, rather, interested in types of human behavior that take place within a human system in which the limits of reality modify the ideal behavior that can occur among groups. Rather than discussing notions of trust or honesty in a vacuum, *CLUG* allows participants to address these and other conceptions within the framework of a real world.

I highly recommend the use of *CLUG* as a stimulus for process-oriented group discussion. While it is quite time consuming, the payoffs can be tremendous; only in a realistic systems model like *CLUG* can the complexity of human processes be simulated with anything close to accuracy. Because *CLUG* is a nonzero-sum simulation, no interaction àt any level specifically follows a win or lose pattern.

Participants playing *CLUG* possess a wide range of goals, expectations, and values. While it may fit your particular use of the model to address those values before play, I much prefer to allow the teams, within themselves and with the other teams, to develop their own goals and their own notion of winning. A very important portion of the debriefing at the end of the simulation is the discussion of who won, what it meant to win, what each individual's and team's goals were at the beginning, how they were modified by the process.

CONTENT—THE VARIATIONS OF *CLUG*

I said at the beginning of this essay that I have played *CLUG* only once. I will now say that I have *never* run *CLUG*. In my use of this simulation, I find it advantageous to make the model as realistic as possible, and basic *CLUG* starts off with a clean board where everyone starts from zero. In looking at the urban development problems of America, we find few situations in which cities were built on land that was used for absolutely nothing else—such as the clean-slate, basic version which builds a city in a vacuum.

Clug-Alum

If you are working with a design group that is specifically interested in the economic interface with some other aspect of city development and you are trying to find the best path of development given some economic realities, the basic *CLUG* might be your cup of tea. We at the Center, however, have found that even for the introductory run of *CLUG* to participants who have never seen the model before, the *CLUG-ALUM* version (*Agricultural Land Use Modification*), which begins with some preurban economic interest, gives a much more economically realistic vision of where cities began in this country.

Other Variations

Like the *ALUM* version, most other variations on *CLUG*, have attempted to increase its verisimilitude to the real world. Many of these variations have been excellent. Allan Feldt's *CLUG* playing manual explains several that are fine examples of such adaptations.

One of the easiest can be used to emphasize certain aspects of urban development that are missing from the basic economic model. This is done by introducing a political sector, as the Center for Simulation Studies has done in its POLIS version. This includes public facilities as well as an external economy modification to show variation in industrial income due to the attachment of the community to the larger national economy. This version, along with *CLUG-ALUM,* is in Feldt's book.

Many other modifications can be made to the *CLUG* game by putting geographical limitations on the playing board. A great deal can be done to model the geography of a particular area simply by locating hills, rivers, bays, or other geographical features that place land-use limitations on the board.

Table 1 explains some of the modifications possible with *CLUG*. By modifications we generally mean additions to, rather than changes within, the economic model. Never try to use the more complex modifications in the initial game with a group. It is far better to have participants first play basic *CLUG* or one of the simpler modifications. After they have mastered that, they can address a specific subject with one of the more complex versions.

Most of these complex versions appear in Feldt's manual. Those I have listed in the table I know personally to have been well thought out and well worked out. All game operators are inventors by nature and should feel free to experiment with *CLUG*; but I would like to warn those who think it would be easy simply to change one or two variables to make something that more closely fits their needs. As I have already pointed out, the model has been extensively worked out to make a realistic economic balance. Any modifications that are attempted must be scaled to yield realistic effects. For instance, in the *ALUM* modification, which introduces farms, the cost of the farm house, the income from the farm, the amount of land to maintain the farm, and the cost of transportation of farm goods to the market, all had to be worked out in dollar values that would fit the *CLUG* model.

TABLE 1 Variations on CLUG used by The Center for Simulation Studies

Variation	Subjects Added to the Model Listed Above
Basic CLUG	• Economic interactions in the private sector of labor, industry and commerce • Economic interrelation of land use, location, community growth (economic, physical, and financial), and taxes
Topographical Modifications	• Geographical limiting factors and their influence on land use, physical growth, and urban economics
CLUG-ALUM (Agricultural Land Use Modification)	• Effect of farming (rural land use) on land valuation and urban use
External Economy Modification	• Effect of a variable national economy on the physical and fiscal life of a community • Introduction of "Headlines" as elements of chance that impose unpredictable events the community must deal with
POLIS	• Introduction of politically differentiated power among teams • Existence of public land uses (schools, etc.) • Election of a "mayor" from among the teams • Problem of pollution as a byproduct of industrialization • Suburban development
Richland	• Welfare and unemployment as a community expense • Population increases by a probabilistic table (out of team control) • Economic differentiation of teams (not all start with equal resources)
Inter-regional Relations	• Interactions among several such communities

Repeat Performances

CLUG has the advantage of being a simulation that can be played over and over without becoming stale and without being easily psyched-out by participants. *CLUG* has no secret rule or twist that make repetitions less meaningful than the original play. (The classic example here is that excellent simulation *Starpower* by my friend Garry Shirts, which can be played repeatedly; but it suffers in subsequent plays if only some of the participants have played before. All plays of this simulation except the first provide participants with an entirely different experience: In the first, the shift of power from the game operator to one group changes the whole nature of *Starpower.*) Even if some participants have played *CLUG* before and others have not, it is very difficult for those who have played before to skew the game to their advantage and to the disadvantage of the others. This is due both to the system's complexity and to its openness, where all rules are available to participants whether or not they have played the model before. Because of the nature of the interaction and the interdependence of all participants and teams, those who have never played before can even find themselves frustratingly dependent on people who have never played. They may find it almost impossible to convince others that they, the experienced players, know what is best in a particular situation.

RUNNING *CLUG*

CLUG is not the easiest game to run, and the variations invariably increase the responsibility and the activity of the game operator to the point that some of the more complex versions require as much from the game operator as they do from all other participants. My suggestion for anyone who wishes to run *CLUG* successfully is to play it and play it until you understand it. There is great internal logic to the rules, so once you can comprehend it all as a system you should have no difficulty in presenting it to other people. This ability to comprehend the system and the interactions of the parts is useful not only in advancing the flow of the simulation but in recognizing significant lessons that can be fed back to the group in debriefing.

As with many simulations, *CLUG* tends to take on the flavor of how the game operator is running it. I would like to emphasize that an interventionist, didactic style is not necessary. The game is self-explanatory and very logical. All of the rules are available to all of the participants. There are no tricks. The most successful style I have encountered for running *CLUG* is to be a walking instruction manual.

The game operator's role in *CLUG* is that of bank, construction company, transportation company, tax collector, state government, and mother nature. But with all those tasks it is still possible to maintain a low profile. In fact, it is important to do so lest you seem to be manipulating the group even though you are only providing a vehicle to maintain the flow of activity.

Introducing *CLUG*

The number of rules and the lengthy introduction can be rather intimidating to participants who suffer from information overload before starting to play; some tend to turn off. The only participants I have ever had not enjoy *CLUG* were those who could not "buy into" the system and, as it were, suspend reality to take on the simulation. This is not difficult for most people to do because the "gaminess" of the model gets them through the process of assimilating the rules, but a game operator must be particularly sensitive to participants who do seem to be turning off. Those who cannot buy into the system truly miss an opportunity.

I suggest dealing with *CLUG*'s intimidating complexity in the following ways:

(1) Present all of the information available in the manual as an overview during the introduction.
(2) Reinforce the notion that all this information is readily available in the manual.
(3) Assure participants that the simulation will proceed step by step and there will be ample time for questions at each step in each round.
(4) Reassure participants that the game is repetitive; if they miss something in one round they will get it in a future round.
(5) Stress the logic of the system and emphasize that if all else fails participants should simply ask themselves what this means in the real world.

Most participants do buy in, and they are almost unanimously sold on at least the play of the game. The suspension of reality and the creation of a simulated reality in which things have importance are very powerful in *CLUG*. The end of each game is often punctuated with groans from those who want to play on. I measure the success of a simulation partially on its fun, and I measure fun on the ability of participants to really get into the activity and care about what is happening. To me, any simulation that can take a group of intelligent adults and cause them to wheel and deal, care and cry over a handful of paper money and a pile of wooden blocks for seven hours, has something going for it.

Time

Timekeeping is a very important and often very controversial task for the game operator. *CLUG* is a long simulation. Though many runs of the game have been broken up into smaller time periods, I have still found that the best lessons come from playing it continuously for a period of five to seven hours.

The more rounds you play, the more time you are playing through. The city grows up and older as you play. The more rounds you play, the more city you have, the more decisions you can look at, the more history you have lived through. Thus at the end of a long continuous play of *CLUG*, while all of the decisions that have been made are still fresh in everyone's mind, participants can look back over their history, chart decisions through the growth of the city, see how early decisions affected the simulation later on, and have physical proof of how their own decisions limited or provided opportunities for future activity.

The biggest problem for people who are running and playing *CLUG* is the time it takes. But given the complexity of the system that is being simulated and the completeness with which the problem is addressed, such a time commitment is the price you must pay. I have successfully chopped the game up into shorter periods, though something is always lost in the break in continuity. It is important, whether the game be played in two three-hour periods, in a succession of fifty-minute classes, or in whatever time available, that you play through the full number of hours (if not more to make up for time lost in restarting for each period). A good guide for basic *CLUG*, as well as for some of the more basic adaptations, is to try to get through ten rounds. This allows you to accomplish two renovations (which come up every five rounds) and establishes enough development so there are truly some patterns of decisions to discuss. By all means get through at least the first renovation (round five) and a round for "recovery."

Pace

In discussing the time frame of *CLUG*, one must talk not only of the length of time for running the whole simulation but also the pace maintained throughout. It is up to the game operator to maintain the pace, and that pace has a direct effect on the simulation's outcome as well as the players' understanding of the simulation. By varying the speed at which *CLUG* is played, you can affect the actual product of the game simply by changing the mood in which participants feel they can attack this system and deal with problems.

At one extreme, if *CLUG* is run slowly and reflectively, participants can experiment with it, try some types of urban planning, and play out alternatives they have discussed and would like to try with this simulated system. At the other extreme is what I call the free market *CLUG.* The game operator constantly presses participants to make quick decisions. Thus, the additional pressure of time forces participants to make immediate decisions with limited information. While it is possible to rush participants too much, some time pressure is necessary to explore how decisions were made as cities began in this country. The simplicity of the decisions on the *CLUG* board in relation to the real-world demand that a much shorter time span be allowed for those decisions than in real world.

If you wish to simulate the passage of time and create a dynamic system, you must remember that time never stops, and if you are playing *CLUG* in a "realistic" time frame there cannot be any time-outs for planning or organization. Those activities are good, but they should be carried within the realistic time frame you are simulating in the game. In other words, if some of your team has gone to do some physical planning for the game board and they come back two rounds later and find that people have built things that had not been planned for in the new master plan—well, that is realistic.

Leadership Style

Aside from this, all that the game operator can do to facilitate the smooth running of *CLUG* is to know the model extremely well. This includes not only running it a number of times but also playing or at least studying and playing it by yourself.

I will not attempt to minimize the difficulty in trying to introduce and run *CLUG.* The process is so complex and the possibilities of interactions are so many that it is difficult to keep everyone's attention while you try to introduce any particular part of it. This is not meant to scare anyone out of running the simulation but only to advise you to stay cool if you find yourself explaining the same thing twenty-five times. A number of the participants are learning a whole new language, and what you say may seem to be slightly different in a different context even though you are essentially explaining the same thing. I usually reach a point at which I get out of the expert business and simply refer participants to the proper page or table in their instruction manual. Then they may take some active role in seeking information. Then I verify it with them to make sure they got the right information.

This may sound negative, but I cannot repeat too often that the rewards in using *CLUG* are so great that they far outweigh the difficulty you might find in introducing it, or the difficulty participants might find in first getting into it. They will (trust me) really get into it and eventually figure most of it out.

CLUG—A GAME?

CLUG and Monopoly

I have gone through several stages in my maturity with *CLUG.* One of the hardest things to get over was getting upset when somebody would, before or after playing *CLUG,* shrug it off and say, "It's just a big game of *Monopoly.*" I went through a long period of defending *CLUG* against such slander. *Monopoly* is not a bad game, but it is different from *CLUG.*

First, if we go back to our spectrum of games from the closed to the open model, *Monopoly* would definitely fall near the closed end; most of its activity is conducted through chance (the roll of the dice or picking cards). The only free and open activity that can take place is bargaining to trade land and to build monopolies. *CLUG,* on the other hand, is a much more open game, and the economic activities simply become the setting within which all sorts of free interaction take place among participants.

Goals

The basic difference between *Monopoly* and *CLUG* is the more mathematical distinction that *Monopoly* is a zero-sum game in which participants play over the rules to beat each other. *CLUG* is a nonzero-sum game in which participants play over each other to beat the rules. In a zero-sum game the object is to win and defeat other participants. In a nonzero-sum game the object is to beat the "system," the limits of the game.

The clear test for this is to ask, "What is the goal of *Monopoly*?" The answer, which is printed right on the inside cover of the box, is to "*win*"; more specifically, to beat all opponents by ending up with all the money and property. On the other hand, if you were to ask, "What is the goal of *CLUG*?" the answer is not so easy. But it is clearly not the same. The goal may very well be to see the process, to look at the types of things that happen. The most likely goal is to build a city, and building a city is not the same as vanquishing all the other players and ending up with a monopoly. Even if the teams admit that their goals are to make as much money as possible, it is not the same as gaining a monopoly, for it can be proven within the model that monopolistic practices may cause one to make less money than other modes of behavior.

In *CLUG,* as in all nonzero-sum simulations, you are trying to maximize your resources (money, power, property, or whatever else the game is being played about). Because the source of resources in *CLUG* is not the other teams, and your income is derived primarily from industrial and farm income (paid to you not by other teams but by the bank, the outside world), you will not maximize your resources simply by vanquishing other teams. In fact, if you do vanquish them, their money simply disappears from the community.

What happens in a nonzero-sum game is that while you pursue your goal of maximizing your resources (and other teams are doing the same), you may get in each other's way, but at other times you may find each other helpful. You are playing not against each other but against the limits of the game, whatever factors determine how you can go about maximizing your resources—how much money you can get for this or must pay for that. You are playing against the model. You are playing against the economic reality that says you can only get $22,000 per round for partial industry. So if you want to make more you have to have more partial industries, and they cost $48,000 each. The *Monopoly* mentality gets into *CLUG* when people start playing *CLUG* as a zero-sum game to vanquish other teams. They will find eventually, if not during the game then certainly during the debriefing, that with this strategy everyone loses.

But for all these theoretical distinctions there is something about *CLUG* that reminds people of *Monopoly*. I think it is the money. When I first started running *CLUG* as a computerized version, we did not have the paper money. But when we discontinued use of the computer and introduced paper money we found that as it greatly enhanced the interaction among the players, it also gave *CLUG* the flavor of *Monopoly*.

The *CLUG-ALUM* version contains an arbitrary ruling that has no logical, economic foundation within the model. Each team is allowed to bid on only three squares of land each round. Though I can find no logic for that within the economic foundation of *CLUG*, it is a very good rule and I suggest using it in all versions of *CLUG* because the *Monopoly* tendency in players is so great. I have played basic versions without this ruling in which a team, given $100,000, spends $85,000-$90,000 on land because its members take a look at the board, they take a look at the money, and they say to themselves: "Aha, *Monopoly*!" and try to buy up the whole board. But there are 196 squares on the grid, and it is very difficult for any team to "own" the board, particularly early in the game. In fact, when teams do spend all their money on land we simulate how cities did *not* grow and develop in the South of the United States at the turn of the century because everybody was so land poor they could not invest in anything else. Obviously, if all teams invest ninety to ninety-five percent of their money in land, little construction can take place. The arbitrary rule of a three-square limit, then, gets participants into a game without the Monopoly mentality of blowing all of their resources in the first few minutes.

I think my greatest reason for being upset when people say *CLUG* is just like *Monopoly* is that it short-changes *CLUG*, which has so much more to offer than economic aggrandizement. Granted there are many similiarities: it is an economic game, it is a game involving property and the use of land. Different teams seem to be competing, and all have paper money to spend. But *CLUG* proceeds very differently. There is no automatic rejuvenation as there is in Monopoly, no passing "GO" and collecting $200, no chance or community chest cards to give you more money. In fact, the money you start with, which varies depending on the version of *CLUG*, is all you get for the whole game. The only way you get any more is by investing it in a way that returns a profit.

The only advice I can give others who run into participants who say, "Oh, it's just like *Monopoly*," is that it might be worthwhile to explore what they mean. They may be comparing *CLUG* with *Monopoly* simply because they are unable to compare it with anything else they have played. If a participant is playing as if *CLUG* were *Monopoly* or claims to have used that strategy throughout the game, you can always ask whether or not he or she "won" and if the *Monopoly*-like strategy worked. Due to the complex interactions among teams, if everyone played with a *Monopoly* mentality you would have a classic no-win prisoner's dilemma.

The Element of Chance

Chance has some role in *CLUG* to account for the unforeseen in this model of reality. Dice are used as an indicator of chance. For things that operate by chance, the *CLUG* manual provides a probability table that gives the corresponding percentages of the chance that certain numbers will come up on the dice to the probability that an event will happen.

In basic *CLUG*, chance occurs in only a few specific places. One that causes considerable concern is renovation. As we play through time, buildings grow older. This is represented primarily through the depreciation of building values. Every five rounds players get a chance to pay to "renovate" buildings. They then have to roll dice and determine, according to a probability table, whether that building was destroyed by some disaster. The game assumes that as buildings grow older their chance of being destroyed increases.

Because buildings depreciate at five percent per round, the chance of losing a building increases by five percent per round. Various numerical combinations that come up on the dice can be placed in a table so that for a building of any age there are always certain losing numbers that correspond to the percentage of probability.

Though this activity can degenerate into a crap shoot with little thought or meaning, I think that renovation is a critical and necessary part of *CLUG* and can be justified on several grounds:

(1) Remember that *CLUG* rounds equal several years, so the depreciation of a building over twenty rounds is not twenty years but more like sixty to one hundred.

(2) Because renovation is the only time since construction that any money is spent on buildings, there has been no maintenance expenditure in those years and their deterioration would be significant.

(3) Understand that chance, seemingly so important in renovation, is almost completely at the team's control. Only 5.7 percent (the remaining probability of loss on a fully renovated building) chance is not controllable by the teams.

The devastation of renovation is excessive for several reasons. First, it all comes at once, every five rounds. That is probably unrealistic but most practical for running the game (and, significantly, for teams who know exactly when fate will take a hand). Second, there is no seeming protection from the loss (insurance), but I always point out that just because the game does not provide insurance doesn't mean a team could not do it for other teams. However, if I were an insurance company I surely would charge high premiums for those who would not maintain their buildings. Last, I'm afraid we are not in a generation of crap shooters. The most common tactic I hear is, "Let's bring it down to round two so we just can't roll *one* number. Why spend more money to change that one number?" Even though *CLUG* renovation tables clearly list the percentage of probability of loss for each age, players continue to look only at the number *not* to roll. It does not register with players that that "one number," if it is a seven, is three times easier to roll than the "one number" if it is a three.

The best way to handle renovation is to make sure it is not a surprise. Players are too overloaded with information at the beginning to think about renovation, but by round four they should be pretty comfortable with their tasks. You can bring it up then and give them a full round to plan and save for it.

I don't think you can ever completely overcome the crap shoot mentality that takes place around renovation. People are going to take chances until they get burned. This is why I always like to play *CLUG* long enough to get through two renovations, because by the second those who were burned by the first will have learned their lesson and will take a more realistic view. Those who got away without renovations the first time and let their buildings get older find that by the second renovation their property is so old that the probability of loss is very high unless they relent and renovate.

Renovation should not be overlooked during debriefing because it is a good place to point out group decision-making and its affect on the future activities of the team and the community's well-being.

As modifications are added to *CLUG* to accentuate a point or to make the game more realistic, the element of chance usually increases, if you expand the role of the outside world, which is beyond the players' control. For instance, in *POLIS* we add an "External Economy" (chosen by the roll of dice), which determines industry income; air pollution alerts, whose probability increases with the number of industries (and is determined by dice); and "headlines" or government/social actions from the outside world that affect the game. These and other chance factors can contribute to the reality of the game but do increase the (perhaps realistic) feeling of manipulation.

One area I have been toying with for years that *should* be determined by probability, at least for *CLUG* models that purport to simulate modern America, is population. Most *CLUG* models are "factory towns" where there is complete control of the population according to how many residences are built. I believe it would be more realistic to have some Malthusian type of table that determines the population growth each round according to the number of team residents in previous rounds.

TROUBLESHOOTING *CLUG*

For those of you who have not played *CLUG* but have become interested in it through this essay, I suggest that the best first move is to find a place to play *CLUG* under the leadership of someone who has run it before. I am confident that you can pick it up from a book and eventually run it yourself, but because the model is so time consuming to begin with (not to mention complex) I would not advise your spending days struggling with the model when it is much easier to grasp by seeing it operated properly.

For those of you who have run it before and have been dissatisfied or who have used it infrequently because of its complexity, let me suggest a few hints to perhaps make it run better:

(1) Have everything set up before the participants arrive so that as the game begins you can concentrate fully on explaining it.

(2) Make sure everyone is present before you go through the introduction.

(3) I have found in most circumstances that it is *not* beneficial to hand out the explanatory material ahead of time because it tends to confuse and intimidate people. Introduce all of the factors involved in the model quickly and without questions before play, and end that discussion with a run-through of all the steps in a round. At that point answer questions, but leave answers to basic "how to" questions for the appropriate point in the first playing round.

(4) Remind participants that they are playing in teams and that the most efficient and effective way to operate will be to have some division of labor.

(5) Suggest that teams make notes and keep records for their own benefit.

(6) Allow participants the freedom to play the game as they will. If they are making some obviously stupid mistake, question them about whether or not they really want to do that, but if they do, let them go ahead. This is a learning experience, and they will learn more by making this mistake and having to live with it, than they will if you save them at the last moment.

(7) Stay steady through the whole game. If you assume one mode of leadership, keep that mode throughout the whole game so people will not expect the other shoe to drop at any moment.

(8) Take every step in each round in order, and always stay in that order. Within each step, allow teams to proceed in an order, and stick to that order. For instance, start construction in the first round with the blue team, then move to green, orange, red, and yellow. In the second round, start with the green team, then orange, red, yellow, and blue. Rotate them but keep the same sequence so everyone knows in what order things will be done.

(9) Decide which economic decisions are most important for your run of the game and feel free to simplify other procedures as long as you do not change economic values. For instance, in the basic game, property value is determined by the price paid for a square of land, plus the price paid for the four adjacent squares of land divided by five. I usually just value the price of land at the original price paid for that land because the change in the assessed evaluation yielded by the very time-consuming process of averaging adds or subtracts maybe four or five hundred dollars in a five hundred thousand dollar city evaluation. Over ten rounds it is not worth that extra effort. Likewise, unless you are investigating competition in commercial enterprises and the value of consumer goods, it is convenient and perfectly acceptable to have stores set one price for everybody and hold to that price for five rounds, if at the same time you advise shoppers that if they agree to shop at a store they must also stay with it until the fifth round.

(10) And, finally, just stay as cool as possible, keeping your suggestions for what the city should do it a minimum so that at the end of the game it is their city and not yours. For this reason I suggest you play CLUG a few times to get it out of your system. It can be fun to run, too, but if you are running it you should not also be playing it.

Best Uses—How to Get the Most Out of Your *CLUG*

As I stated earlier, I have run *CLUG* with every imaginable type of group. While I can't think of a single instance in which the game was not appropriate, it was clearly more successful with some groups than with others. There is only one type of group I would suggest *not* playing *CLUG* with, and that is social group that is getting together for fun. While *CLUG* is fun, it is also a lot of work, and unless there is some commitment that keeps people's attention, the simulation will fall apart.

Assuming that you have people who are there to learn something, *CLUG* can be appropriate in at least the following areas:

(1) groups of all ages and backgrounds who are interested in basic economics on an urban or national level and in the relation of supply and demand;

(2) any political science (or other) group interested in urban affairs and the urban political process;

(3) planning, architecture, and design groups interested in the special setup of urban areas;

(4) any groups interested in the history and development of urban American;

(5) any group of practioners in the political system who are citizens interested in learning the problems of urban America on an area-wide basis;

(6) any group interested in systems and systems dynamics;

(7) any group interested in decision making, problem solving, or negotiation skills;

(8) any group interested in group dynamics, goal setting, or value orientations and looking for a reality check.

Mixing groups or types will not destroy the game but of course will mean that teams or individuals will be playing with different goals in mind. Sometimes this can be useful, but sometimes it can be counter-productive.

People need no special expertise or background to play *CLUG.* As a rule of thumb I ask that people be at least eleven or twelve years old, not only because the game takes a long time and requires a long attention span but also because a player needs to feel comfortable with numbers. People do not have to be good at mathematics (for the most part, *CLUG* involves only simple addition and subtraction), but an ability to internalize numbers and to think in numbers is essential in playing *CLUG.* There is no maximum age nor any minimum educational requirement for adults to play this simulation.

I do require that everyone who comes to the simulation must play. Not only is it nerve-wracking for a game operator to have people watching, but it is also boring to sit through the game for six hours and not really understand what is going on. As I pointed out earlier, what people can get out of the model is strictly what they put into it. To get the most out of *CLUG* you definitely need a commitment from the players to go through the whole process and to take it seriously. This doesn't mean that it has to be work and not fun, but people must make a commitment not to leave before the end. It is always good at the beginning of the simulation to get this verbal commitment again from participants so they realize the length of time and the other requirements that are going to be asked of them.

Beyond that, the only other suggestion I can give game operators for getting the most from their model is to be observant of what is going on; and, if necessary, to take notes on things you notice happening between teams and within the game. Finally, spend an extended time debriefing the model so you can discuss the many facets of the game and the many areas it touches. The emphasis of the debriefing will, of course, vary with the type of group, their goals, and their reason for coming and playing the game.

Conclusion

If it is not clear yet, I believe *CLUG* to be a most fascinating, flexible, and useful simulation. It has aged well, while others have shown markings of their era of development. Despite its length and complexity, is a sterling example of one the best learning tools in the field of simulation.

COMPUTER SIMULATION GAMES
Exemplars

by Ronald E. Anderson

The idea of using a "board" for parlor games stimulated the creation of hundreds of new games for serious learning as well as fun; likewise, the idea of using the computer for games has fostered the creation of hundreds of games, and we have only seen the beginning. New technology, be it a board, dice, a computer printer, or a TV screen, not only provides a more lively structure for modifying old games but stimulates the evolution of entirely new forms of playing games. I certainly agree with Greenblat and Uretsky (1977) that "there is probably no single development that has had as significant an impact on simulation-gaming in the past century as the advent of the computer and the advances of this technology in recent years."

Pedagogical Contributions of Computerized Simulation

It is well known that computerized video games have conquered the arcades and filled every toy store. It is less well known that computerized simulation games have begun to offer remarkably new pedagogical innovations. The capacity to carry on "dialogues" provides the most dramatic new impact. Through person-computer conversations it is possible not only to simulate interpersonal interactions for educational purposes but also to replicate a tutor-student interaction. The computer program becomes a patient tutor which individually responds to each student depending upon his or her sequence of decisions. The two key features of electronic computers that give them conversational potential are feedback and the ability to display and process language. These skills, combined with large data storage, rapid computation, and control of input/output devices such as graphical display screens, open up vast new opportunities for simulation and gaming techniques.

Some of the inherent pedogogical advantages of computerized simulation games have been outlined by Roberts (1976) and others. In brief, they are:

1. Artificiality can be minimized by the introduction of many variables and complex systems.
2. Clerical work is reduced, allowing participants to concentrate on decisions more central to the basic goals.
3. Competition among students is reduced because they are typically playing against "nature" rather than against each other.
4. Computer games may be less threatening because the student is not performing for peers or for an instructor and hence feels freer to experiment and explore.

In summary, Roberts argues that computer simulation games eliminate some of the problems with "role-playing, board games" while increasing students' sense of control over their future:

> Games seem able to create this sense of efficacy by bringing real-world problems into the classroom and allowing the students the opportunities to practice making real-world decisions. The empirical research indicates a correlation between this sense of efficacy and school success.

Oregon

A good example of the compelling quality of computer simulation games is a program called *Oregon.* This game simulates a trip over the Oregon Trail from Missouri during the mid 1850s. The instructions are given in frame 1.1 and a sample run is listed in frame 1.2. The main challenge is to survive the full 2,000 miles by spending money wisely, by typing "hunt" words, and to a certain extent by chance. The simulation is designed for elementary and secondary history or social studies courses and is intended to give students a better idea for what the westward journey was like for the rugged individuals who attempted it.

Oregon appeals to students, who often play the game over and over. The program is accessed about 400 times a day on the MECC (Minnesota Educational Computing Consortium) statewide computer system, which makes it about 10 times more popular than any other computer simulation game or learning package on the system. The secret of its success in the marketplace is that students learn something about early American history while they are having fun.

Author's Note: The author wishes to express appreciation to Pete Trotter, Janet Frederick, Harold J. Peters, Dan Klassen, Alfred Bork, Catherine Dunnagan, and David Cook for comments on an earlier draft of this essay.

FRAME 1.1 Instructions for Playing OREGON

```
:LIB,OREGON
:RNH
DO YOU NEED INSTRUCTIONS  (YES/NO)? YES


THIS PROGRAM SIMULATES A TRIP OVER THE OREGON TRAIL FROM
INDEPENDENCE, MISSOURI TO OREGON CITY, OREGON IN 1847.
YOUR FAMILY OF FIVE WILL COVER THE 2040 MILE OREGON TRAIL
IN 5-6 MONTHS --- IF YOU MAKE IT ALIVE.

YOU HAD SAVED $900 TO SPEND FOR THE TRIP, AND YOU'VE JUST
   PAID $200 FOR A WAGON.
YOU WILL NEED TO SPEND THE REST OF YOUR MONEY ON THE
   FOLLOWING ITEMS

      OXEN - YOU CAN SPEND $200-$300 ON YOUR TEAM
             THE MORE YOU SPEND, THE FASTER YOU'LL GO
                BECAUSE YOU'LL HAVE BETTER ANIMALS

      FOOD - THE MORE YOU HAVE, THE LESS CHANCE THERE
                IS OF GETTING SICK

      AMMUNITION - $1 BUYS A BELT OF 50 BULLETS
             YOU WILL NEED BULLETS FOR ATTACKS BY ANIMALS
                AND BANDITS, AND FOR HUNTING FOOD

      CLOTHING - THIS IS ESPECIALLY IMPORTANT FOR THE COLD
                WEATHER YOU WILL ENCOUNTER WHEN CROSSING
                THE MOUNTAINS

      MISCELLANEOUS SUPPLIES - THIS INCLUDES MEDICINE AND
                OTHER THINGS YOU WILL NEED FOR SICKNESS
                AND EMERGENCY REPAIRS

YOU CAN SPEND ALL YOUR MONEY BEFORE YOU START YOUR TRIP -
OR YOU CAN SAVE SOME OF YOUR CASH TO SPEND AT FORTS ALONG
THE WAY WHEN YOU RUN LOW.  HOWEVER, ITEMS COST MORE AT
THE FORTS.  YOU CAN ALSO GO HUNTING ALONG THE WAY TO GET
MORE FOOD.
WHENEVER YOU HAVE TO USE YOUR TRUSTY RIFLE ALONG THE WAY,
YOU WILL BE TOLD TO TYPE IN A WORD (ONE THAT SOUNDS LIKE A
GUN SHOT).  THE FASTER YOU TYPE IN THAT WORD AND HIT THE
'RETURN' KEY, THE BETTER LUCK YOU'LL HAVE WITH YOUR GUN.

AT EACH TURN, ALL ITEMS ARE SHOWN IN DOLLAR AMOUNTS
EXCEPT BULLETS
WHEN ASKED TO ENTER MONEY AMOUNTS, DON'T USE A '$'.

GOOD LUCK!!!


HOW GOOD A SHOT ARE YOU WITH YOUR RIFLE?
  (1) ACE MARKSMAN,  (2) GOOD SHOT,  (3) FAIR TO MIDDLIN'
         (4) NEED MORE PRACTICE,  (5) SHAKY KNEES
ENTER ONE OF THE ABOVE -- THE BETTER YOU CLAIM YOU ARE, THE
FASTER YOU'LL HAVE TO BE WITH YOUR GUN TO BE SUCCESSFUL.
? 5                                      -3-
```

FRAME 1.2 Sample "Conversation" Playing OREGON

```
HOW MUCH DO YOU WANT TO SPEND ON YOUR OXEN TEAM ? 22
NOT ENOUGH
HOW MUCH DO YOU WANT TO SPEND ON YOUR OXEN TEAM ? 2222
TOO MUCH
HOW MUCH DO YOU WANT TO SPEND ON YOUR OXEN TEAM ? 222
HOW MUCH DO YOU WANT TO SPEND ON FOOD ? 333
HOW MUCH DO YOU WANT TO SPEND ON AMMUNITION ? 22
HOW MUCH DO YOU WANT TO SPEND ON CLOTHING ? 44
HOW MUCH DO YOU WANT TO SPEND ON MISCELLANEOUS SUPPLIES ? 55
AFTER ALL YOUR PURCHASES, YOU NOW HAVE  24  DOLLARS LEFT

MONDAY MARCH 29 1847

TOTAL MILEAGE IS 0
FOOD            BULLETS          CLOTHING          MISC. SUPP.      CASH
 333             1100              44               55               24
DO YOU WANT TO (1) HUNT, OR (2) CONTINUE
? 1
TYPE WHAM
? WHAM

RIGHT BETWEEN THE EYES---YOU GOT A BIG ONE!!!!
FULL BELLIES TONIGHT!
DO YOU WANT TO EAT (1) POORLY  (2) MODERATELY
OR (3) WELL ? 2
RIDERS AHEAD.  THEY LOOK HOSTILE
TACTICS
(1) RUN  (2) ATTACK  (3) CONTINUE  (4) CIRCLE WAGONS
? 3
RIDERS WERE HOSTILE--CHECK FOR LOSSES
WILD ANIMALS ATTACK!
TYPE BANG
? BANG

NICE SHOOTIN' PARDNER---THEY DIDN'T GET MUCH

MONDAY APRIL 12 1847

TOTAL MILEAGE IS 161
FOOD            BULLETS          CLOTHING          MISC. SUPP.      CASH
 371              939              44               40               24
DO YOU WANT TO (1) STOP AT THE NEXT FORT, (2) HUNT, OR (3) CONTINUE
? 1
ENTER WHAT YOU WISH TO SPEND ON THE FOLLOWING
FOOD? 0
AMMUNITION? 0
CLOTHING? 0
MISCELLANEOUS SUPPLIES? 3
DO YOU WANT TO EAT (1) POORLY  (2) MODERATELY
OR (3) WELL ? 2
LOSE YOUR WAY IN HEAVY FOG---TIME IS LOST

MONDAY APRIL 26 1847

TOTAL MILEAGE IS 302        -4-
```

Definitions and Criteria

A word on definitions is in order before moving on to evaluation. Simulations and games in the computer context are evolving at an extremely rapid rate, and it is still too early to predict what forms will develop and become institutionalized. In the early days of computer simulation and gaming—that is, 5 to 10 years ago—simulation games were clearly identified as such. Now they may not be so labeled because they may be modules or components within a larger package of "courseware," the term for instructional packages. The Control Data Corporation PLATO system contains many such courseware units and illustrates the heterogeneity of simulations and games. Frame 2.1 shows a simulation game called *Dart* from the PCP mathematics material. The game is designed to assist in learning fractions. The student–player enters a fraction and then the computer program sends a dart flying across the screen at the place on the vertical scale. If it is the right fraction (or reasonably close) the dart will hit a balloon and burst it. *Dart* is an excellent example of computer-based animation and its possibilities for instructional design.

In Control Data's PLATO terminology, *Dart* is neither a simulation nor a game—it is simply a courseware unit. Although it simulates dart throwing, we would not call it a simulation because its intent is not to understand dart throwing; and while it can be played, some would not call it a game because if the player wins, nobody loses. Needless to say, much disparity exists over definitional boundaries for the terms *simulation, game,* and *simulation game.* The terms are evolving and it is impossible to predict how they will come to be defined.

Rather than take a precise, dogmatic position, I prefer a generous approach to definition. But since this discussion is an evaluative review, I am limiting myself to instructional simulation games. I have been open to and considered any simulation or game that has a computer component and is definitely instructional in character; these I call "computer simulation games."

Excluded from this review are noninstructional (or minimally instructional) games such as *Star Trek* (see, for example, Creative Computing [May-June 1975] pp. 40ff.), which resides on most computer systems. Also excluded from this review are problem-solving units that do not explicitly feature either a game or a simulation of a real-world process. Thus, I have not included such excellent instructional computer units as *Bertie* (available from CONDUIT), which teaches formal

FRAME 2.1 Display of DART Game on PLATO Screen

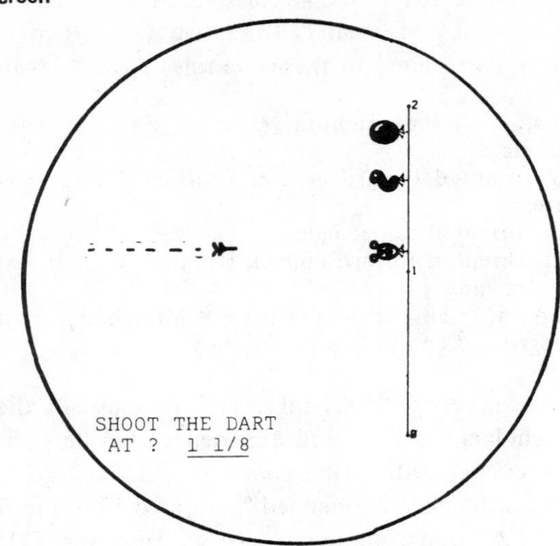

Note: the dart is flying toward the lower balloon and will burst it because the student typed in the correct answer of "1 1/8".

logic by checking student-entered deductions, because they are not clearly games or they do not contain an operating model of a naturally occurring process.

Still another exclusion is any program that solves problems by standard mathematical or statistical procedures alone. A simulation game may incorporate mathematical solutions, but most importantly it constitutes a process model that can be experimented with and is dynamic.

Finally, this *Guide* imposes further restrictions of scope. First, no simulation games are considered that are not advanced enough to be useful for high school students. The target audience is secondary and higher education. Second, I have excluded all business and management games because that area is reviewed separately.

In summary, I have restricted myself to evaluation of instructional computer simulation games, for practical purposes, which means I am interested in (1) any computer simulation of naturally occurring processes or (2) any game that directly depends upon the computer and, in either case, the simulation or game *must* be instructional and interactive in that students can experiment with the programs.

SOURCES OF COMPUTER SIMULATION GAMES

Using the criteria just discussed, the preparation for this review took me to numerous sources in search of any available computer simulation games. As mentioned earlier, I was not considering business games or those targeted below the secondary school level. Although I contacted many knowledgeable experts, searched computerized bibliographies (ERIC, for example), and consulted many published materials, the primary sources were:

Journals

- *Simulation and Games* (Sage Publications, 275 South Beverly Drive, Beverly Hills, CA 90212)
- *Creative Computing* (P.O. Box 789-M, Morristown, NJ 07960)
- *Pipeline* (CONDUIT, Box 388, Iowa City, Iowa 52240)
- *Journal of Experiential Simulation and Gaming* [formerly *Simulation/Gaming*] (Elsevier North-Holland, Inc., 52 Vanderbilt Ave., New York, NY 10017)

Proceedings of Annual Conferences

- "Conference on Computers in the Undergraduate Curricula" (CCUC/1, CCUC/2, ... , CCUC/9)
- North American Simulation and Gaming Association (NASAGA)

Computer Center Software Libraries

- MECC (Minnesota Educational Computing Consortium) 2520 Broadway Dr., Lauderdale, MN 55113. (MECC has been an acknowledged leader in instructional computing.)

Computer Company Software Inventories

- DEC (Digital Equipment Corporation, Maynard MA 01754)
- HP (Hewlett-Packard, 11000 Wolfe Rd., Cupertino, CA 95014)
- CDC, Project PLATO (Control Data Corp., 8100 34th Ave. So., Bloomington, MN 55440)

These sources appear in order of current usefulness. That is, I found the journals most useful and the computer companies least useful.

For those interested in updating this article by reading the current literature, I recommend the following journals: *Pipeline, Simulation and Games,* and *Creative Computing. Pipeline* concentrates upon higher education, while *Creative Computing* is oriented more toward elementary and secondary education. *Simulation and Games* is mixed in emphasis and has a special section devoted to simulation reviews. While it is impossible to predict the publishing industry, these periodicals currently cover most of the developments in the field of simulations and games for computer-based instruction.

EVALUATION

When one attempts to select the "best," the "top 10," or even "exemplary" instructional simulation games, the overriding concern must necessarily be *quality.* Furthermore, the numerous dimensions of relevant quality such as validity, pedagogy, involvement, and so forth, must all be weighted and combined to arrive at an overall evaluation of quality. Thus my first question is what aspects or dimensions of quality are the most important.

Theorists and reviewers of simulation and games (cf. Boocock, 1972; Inbar and Stoll, 1972; Elder, 1973) emphasize such factors as validity, flexibility, impact, accuracy, plausibility, and so forth. These considerations generally concern either the substantive worth of the underlying model on the one hand or the user impact on the other. Quantitative assessment, however, with one exception (Anderson, 1976), has focused upon user or student ratings (cf. Dukes and Waller, 1976; Liggett, 1977). While user ratings are extremely important for measuring impact, many other considerations may be equally important] (Orbach, 1977). Substantive premises, clarity of documentation, and transferability of materials must also be taken into account.

CONDUIT has developed an evaluation scheme that attempts to take most relevant factors into consideration in reviewing computer simulation games and other instructional materials. They established a peer review system like that of professional journals and publishing houses. CONDUIT's goal is to assemble high-quality computer-based instructional units and deliver them to the educational community. Any computer simulation game of any merit within higher education is considered, but only the best are packaged and disseminated. A series editor in each of 11 disciplines coordinates the review procedure while a central staff performs any required technical modifications and manages the production of the packages. Anderson (1976) statistically analyzed the CONDUIT ratings of over 150 reviews of 80 packages, finding that five factors (dimensions) summarized the evaluation considerations. These five dimensions of quality are:

1. substantive pedagogy
2. presentation of concepts and theories
3. presentation of methods and techniques
4. student orientation
5. transfer expectancy.

Based upon these findings and upon less quantitative assessments, a slightly improved rating form was designed (see frame 3.1).

FRAME 3.1 CONDUIT Evaluation Rating Form

PART II:
EVALUATION

Column 1:
Rate this package on each of the selected characteristics listed below by circling the appropriate number. Please complete this entire column before working on column 2.

Column 2:
Indicate the importance of each feature for this instructional package. Circle the appropriate number.

6 Exceptional
5 Very good
4 Good
3 Fair
2 Poor
1 Very Poor
A Item not applicable
B Insufficient information, can't evaluate
C Not qualified to evaluate

Column 2: 4 Critical, 3 Important, 2 Optional, 1 Inappropriate

Column 1				Column 2
			A. SUBSTANTIVE CONTENT	
6 5 4 3 2 1	A B C	1.	Definition of key concepts	4 3 2 1
6 5 4 3 2 1	A B C	2.	Discussion of underlying assumptions	4 3 2 1
6 5 4 3 2 1	A B C	3.	Validity of principles, theories	4 3 2 1
6 5 4 3 2 1	A B C	4.	Discussion of relevant literature	4 3 2 1
6 5 4 3 2 1	A B C	5.	Overall substantive content quality	4 3 2 1
			B. DOCUMENTATION/TEXTUAL MATERIALS	
6 5 4 3 2 1	A B C	1.	Clarity of information in textual materials	4 3 2 1
6 5 4 3 2 1	A B C	2.	Completeness of instructor guides	4 3 2 1
6 5 4 3 2 1	A B C	3.	Adequacy of instructions for operating programs	4 3 2 1
6 5 4 3 2 1	A B C	4.	Overall quality of documentation	4 3 2 1
			C. SUPPORT OF THE TEACHING PROCESS	
6 5 4 3 2 1	A B C	1.	Ease of integration with course procedures	4 3 2 1
6 5 4 3 2 1	A B C	2.	Potential for improving instructor's ability to communicate principles and theories	4 3 2 1
6 5 4 3 2 1	A B C	3.	Potential for improving instructor's ability to communicate methods and techniques	4 3 2 1
6 5 4 3 2 1	A B C	4.	Potential for teaching how to interpret and apply results	4 3 2 1
6 5 4 3 2 1	A B C	5.	Overall instructional quality	4 3 2 1
			D. STIMULATION OF STUDENT INTEREST	
6 5 4 3 2 1	A B C	1.	Potential for capturing student interest	4 3 2 1
6 5 4 3 2 1	A B C	2.	Challenge to student creativity	4 3 2 1
6 5 4 3 2 1	A B C	3.	Student choice in patterns of use	4 3 2 1
6 5 4 3 2 1	A B C	4.	Appropriateness for student-initiated work	4 3 2 1
6 5 4 3 2 1	A B C	5.	Overall contribution to student motivation	4 3 2 1
			E. COMPUTER TECHNIQUES/MATERIALS	
6 5 4 3 2 1	A B C	1.	Soundness of computer programming methods	4 3 2 1
6 5 4 3 2 1	A B C	2.	Completeness of technical documentation	4 3 2 1
6 5 4 3 2 1	A B C	3.	Portability (machine-independence of computer program)	4 3 2 1
6 5 4 3 2 1	A B C	4.	Ease of program use	4 3 2 1
6 5 4 3 2 1	A B C	5	Overall quality of computer techniques	4 3 2 1
6 5 4 3 2 1	A B C		OVERALL EVALUATION OF PACKAGE	4 3 2 1

In the design and implementation of computer simulation games there are five major social roles, each with its own dominant orientations of quality. (The five sections of the CONDUIT form correspond to these five roles.) The roles are:

1. Scientist, scholar–oriented toward communication through materials
2. Author–oriented toward communication through written materials
3. Teacher–oriented to pedagogy
4. Student–oriented toward appeal, stimulation, and assistance in learning
5. Computer specialist–oriented toward interchangeability and integrity of computing techniques.

CONDUIT reviewers generally fill several roles; usually they are scientist/scholars, authors, and experienced teachers with some past experience with using computers in the teaching process. Such academically inclined persons will typically place a low value upon computer concerns. However, CONDUIT engages computer specialists to evaluate every candidate package and give their stamp of approval before the academics review it. This preliminary technical evaluation is simply to make sure that the transportability packaging will not be outrageously costly.

The final CONDUIT product must be quite high on all five standards: (1) substance, (2) communication, (3) pedagogy, (4) appeal, and (5) transferability. Like CONDUIT, I used the same five criteria to select a set of exemplary computer simulation games to highlight in this article. All the simulation games highlighted in the next section are reasonably high or satisfactory on all five standards.

EXEMPLARY COMPUTER SIMULATION GAMES

Figure 1 lists the computer simulation games selected as exemplary. This means that these game or simulation packages are outstanding as prototypes of what can be done instructionally with computer techniques. The selection does *not* mean that these are the most reputable or the very best 17 computer simulation games across all disciplines. In fact, one can note that, with one exception, no more than one package has been selected from any given discipline.

Some will be disappointed that no simulations were selected from such innovative projects as PLATO or PCDP (Physics Computer Development Project at the Irvine campus of the University of California). The latter project, under the direction of Alfred Bork, has pioneered the development of dialogue units that include simulations of processes that are graphically and dynamically displayed for each student. The problem with such packages is not their substantive or pedagogical quality but rather their poor transportability. Interactive graphics is not yet standardized within the computing field, and some of the display technology is still rather expensive. Because of the transferability problem, simulation games that use advanced graphical facilities are not listed among the exemplary packages, although they are included among the reviews in the section on computer games.

Some packages use graphical displays that can be made with a simple printing unit such as a teletype, which minimizes the transferability problem; *Limits,* which will be discussed later, and *Demographics* are cases in point. *Demographics* is not listed among the exemplary simulation games because it is first and foremost a data base query/analysis unit; the simulation component is quite minor.

SUMMARY DESCRIPTIONS

This section expands the brief phrases of Figure 1 into full paragraphs discussing each exemplary computer simulation game in turn.

Change Agent

The process of social diffusion of either information or innovation has long been of interest to social scientists interested in simulation. *Change Agent* embodies the generally accepted principles of social diffusion and turns the situation into a game in which the student tries to diffuse the innovation as quickly as possible. The scenario for the game is a rural village of 100 farms, and the student plays the role of a professional change agent who attempts to speed up the adoption of an important new farming method. *Change Agent* is appropriate for sociology and community courses as well as mass communication.

FIGURE 1 Summary and Subject of Exemplary Computer Simulation Games

Name of Package	Summary and Subject
Change Agent	social diffusion game experiments with adoption processes in a rural village (mass communication)
Chebo	social interaction simulation whereby actors move as a function of attitudes (social psychology)
Cognitive Psychology	series of laboratory exercises for cognitive or experimental psychology (psychology)
Critical Incidents	tutorial scenarios for training teachers in handling classroom incidents (education)
Energy (U.S. Policy)	policy analysis simulation of social, economic, and environmental impacts (policy)
EXPERSIM	generates data from any process model, e.g., model of a psychology experiment (psychology and other subjects)
Ghetto	revision of a board game to increase empathy with ghetto residents (urban culture)
IDGAME	game of minimizing costs in determining inorganic unknowns (chemistry)
INS2	group decision-making, role-playing game on international relations (political science)
INTERP	exercises for learning about the wave theory of light (physics)
Limits	exercises for speculating upon limits to growth due to system dynamics (economics)
Linkover	exercises with simulation of genetic mapping (biology)
Oregon	game of chance and typing skill for teaching American history at pre-college level (history)
Quantitative Experimental Analysis	simulations of different theoretical distributions demonstrating difficult concepts (statistics)
SIMSEARCH	tutorial exercises for decisions in designing social research (sociology)
Spatial Marketing	simulation and tutorials on locational analysis (geography)
Wheels	exercise in consumer decisions regarding automobile purchase and upkeep (home economics)

Chebo

Chebo, which stands for "checkerboard," is a model developed and programmed by Sakoda (1971) but described in Lehman (1978). *Chebo* operates a simple model of social interaction based upon some elementary principles of field theory and attitude/attraction theory. In brief, actors move closer or apart depending upon their current interpersonal distances, their attitudes toward members of their own group, and their attitudes toward members of another group. Such interaction is depicted in frame 4.1, in which one group, the social workers, is represented by open circles and the other group, lost souls, is represented by solid circles. At cycle 0, they are randomly dispersed around the checkerboard space, but by cycle 15, the final structure, the social workers are huddled in the center and the lost souls are spread around the outer boundary. This occurred largely because the initial starting conditions defined the social workers as having positive attitudes toward themselves as well as toward the outgroup, while the lost souls (solid) held negative attitudes both toward themselves and toward the outgroup. The model is included

FRAME 4.1 CHEBO Output Showing Interaction Between the Social Workers and the Lost Souls

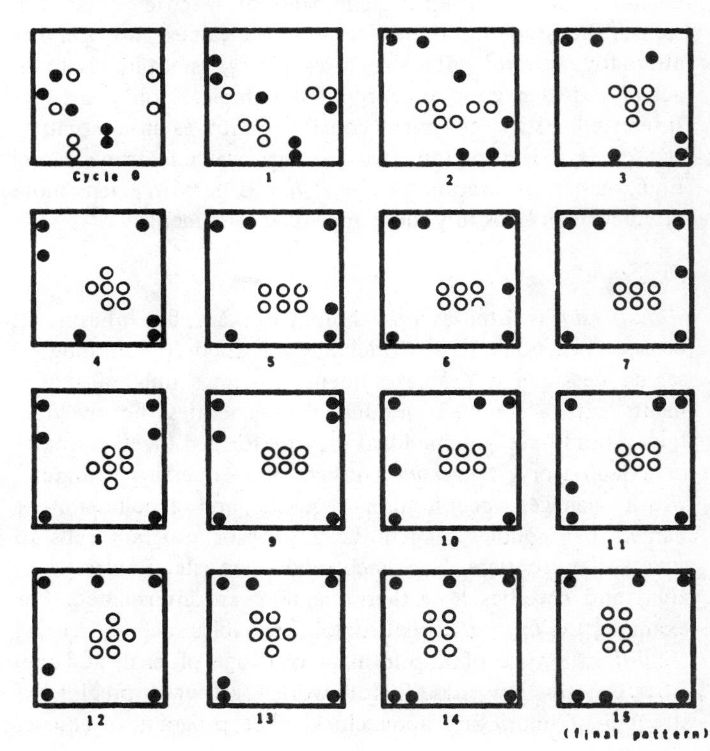

because it demonstrates a variety of interesting social structures evolving from a very simple process.

Cognitive Psychology

Cognitive Psychology was written by Bewley for students in cognitive psychology or experimental psychology. It consists of a series of separate programs to simulate live experiments in each of these areas: pattern recognition, short-term memory, long-term memory, discrimination-net learning, concept learning, problem solving, and decision making in mixed-motive games. The package is modular, and an instructor can select a single simulation program for a specific laboratory exercise.

Critical Incidents

Critical Incidents (my name for "Simulated Incidents in Teaching") is for training teachers in handling human relations problems in the classroom. The underlying concepts and the terminology are based upon Transactional Analysis (TA), so some minimal reading or training in TA is required. The package provides 22 scenarios of classroom incidents and then asks the student (usually inservice teachers) how he or she would handle the incident. The program allows the teacher or teacher-to-be to try out various "theories" or solutions hypothetically, thus avoiding the emotional stress of a real-life situation. *Critical Incidents* is best described as a role-playing game rather than a simulation model. It is tutorial in format and the simulated events are determined by the implicit theories of TA.

Energy

Energy, which is short for *United States Energy, Environment and Economics Problem: A Public Policy,* simulates the economic and environmental impacts of specific policy decisions. The student inputs decisions on values and specifies alternative governmental structures. *Energy* is not highly interactive but is a good prototype of complex policy models. Interested instructors might consider *Policy* as an alternative. (*Policy* is a Huntington Two package available from Digital Equipment Corporation.) While *Policy* is narrower, it is more interactive and possibly more appealing to students.

Expersim

Expersim is listed as a psychology package, but it has much broader potential. It was originally designed for teaching research design and enables students to run simulated experiments with a computer module that generates the raw data they would have obtained had they performed the experiment in a laboratory. *Expersim* consists of a supervisor program, which manages specific user requests, and a collection of models or modules, which the supervisor program calls to generate appropriate data. Each model consists of several variables and specifies how these variables are interrelated. For example, the *Imprint* model contains variables such as "rearing conditions," type of imprinting target, age of bird, and specifies exactly how these factors work together to predict the strength of imprinting upon chicks. The program, of course, does not require the student to buy a supply of chicks. In addition to the imprinting model, *Expersim* has incorporated models for (1) the etiology of schizophrenia, (2) motivation in performing routine tasks, (3) effects of drugs on learning among rats, and (4) social facilitation effects upon performance. Other models from fields other than psychology have been developed, but their quality and availability are not yet well known.

Ghetto

Ghetto is a translation of the well-known board game into a computer game that one person can play. Some of the social and entertaining features of the board game are lost, so this is a controversial use of the computer. None the less, it is a prototype of what can be done with computerized versions of board games, and students have found it fun to take the roles of people living in a large urban ghetto. *Ghetto* is *not* designed for ghetto residents; rather it is designed to sensitize middle-class students to the social world of the ghetto. Someone has yet to design a game called *Middle Class* that would sensitize the children of the ghetto to the complex world of the middle class.

Idgame

In the field of chemistry, *Idgame* is a well-known and highly recommended computer simulation game for learning about organic qualitative analysis. Students run the program to generate simulated laboratory tests for the purpose of identifying unknowns. Students can run a number of simulated tests, but each test costs the student points. Thus the students compete with one another to minimize costs while obtaining enough data to be sure of the right answer. Other recommended simulation packages in chemistry include *Rkinet* and *Titration,* and both are available through CONDUIT. The PLATO system contains a novel simulation of titration processes (see frame 5.1) in which the addition of various chemicals can be graphically simulated; for instance, the density of the liquid in the bottle is visually represented by the degree of shading on the screen. Within chemistry there are many activities that lend themselves to instructional simulation games on the computer, and a number of new materials are likely to be available soon.

INS2

Inter-Nation Simulation, originally developed by Guetzkow, is a well-known role-playing game for a large number of participants. *INS2* is a new computerized version of this game; it includes (1) programs to aid the instructor in managing the game and (2) programs for the students to test the implications of their decisions before proceeding. Although the game requires a long time (four to eight one-hour sessions), it has established an important place for itself in the pedagogy of teaching international relations and policy decision making.

Interp

Interp has been chosen to represent physics, not for its elegance but for the appropriate simplicity of the model and

FRAME 5.1

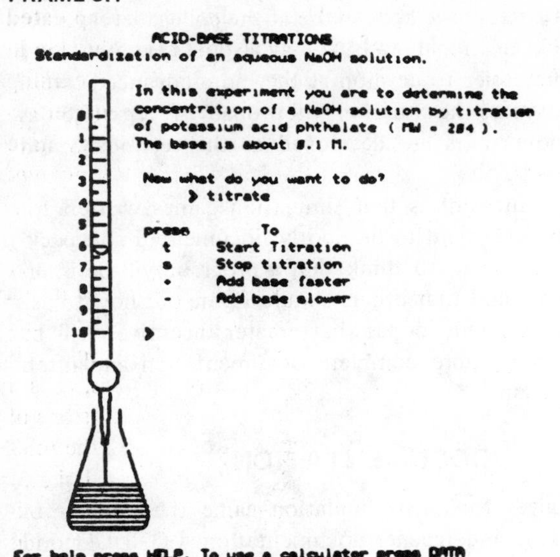

FRAME 6.1 Sample Printout & Graph from LIMITS

```
STANDARD RUN (1=YES,0=NO) ?0
WHAT IS THE END YEAR ?2100
VARIABLE        CHANGE:PERCENT,YEAR
--------        ------------------
R                 ?0,0
D                 ?0,0
F                 ?0,0
R                 ?-60,1980
O                 ?0,0
X                 ?0,0

DO YOU WANT (1)A GRAPH, (2)A TABLE, OR (3)BOTH ?3

THE WORLD IN THE YEAR 1970 :
THE POPULATION IS 3.6 BILLION.
FOOD SUPPLY IS 2000 CALORIES/PERSON/DAY.
AVERAGE INCOME: 800 DOLLARS/PERSON/YEAR.
THE DEATH RATE IS: 15 / 1000.

          SCALES IN HUNDREDS OF PERCENT OF 1970 VALUES

F   0          .5          1          1.5          2
R   0          .25         .5         .75          1
O   0          2.5         5          7.5          10
X   0          5           10         15           20
P   0          .75         1.5        2.25         3
```

In this run, we wanted to change the resource usage rate. So we typed 0 for a non-standard run.

Resource usage rate has been changed by a negative 60% in 1980 to simulate recycling of our natural resources. Recycling slows the rate at which we use up resource reserves.

Select GRAPH and TABLE outputs.

X has gone off the scale. Pollution has risen to more than 20 times its 1970 value.

YEAR	FOOD CAL.	NAT RES PERCENT	IND OUT TRILLION	POLLUT PERCENT	POP BILLION
1975	2048	97.3	2.044	128	4.037
2000	2929	85.6	6.632	449.8	5.679
2025	4000	62.1	11.199	1255.8	6.661
2050	1856	36.5	4.556	4023.7	2.712
2075	4000	25.7	4.322	282.3	2.744
2100	4000	15.2	3.852	248.2	3.426

```
THE WORLD IN THE YEAR 2100 :
THE POPULATION IS 3.426 BILLION.
FOOD SUPPLY IS 4000 CALORIES/PERSON/DAY.
AVERAGE INCOME: 2023.66 DOLLARS/PERSON/YEAR.
THE DEATH RATE IS: 10 / 1000.
```

We wanted a complete run of the program, so we selected the last end year possible, 2100.

thoroughness of the documentation. The unit was developed under the auspices of the Chelsea College science project in London, which has produced a number of other simulations as well. *Interp* is an aid to learning about the wave theory of light, specifically interference and diffraction phenomena. The model allows students to investigate the effects of the direction and distance of secondary source factors upon the intensity distribution. As already mentioned, there are the many less transportable physics packages available from the Irvine project.

Limits

Limits is a Huntington Two simulation set up for students to get extrapolations of trends in population, industrial growth, resource consumption, pollution, and agricultural production. The simulation is a simplification of the Forrester-Meadows Limits-to-Growth model. Various input parameters can be adjusted and graphs of the implications printed out (see frame 6.1 for an example). The model, of course, is speculatory and controversial, but the simulation without a doubt helps to teach how variables interact in a complex system.

Linkover

Linkover is a biology package that simulates genetic mapping. Another interesting package from Chelsea College, it is designed to allow students to plan and execute a series of experiments to generate the data necessary to construct an accurate genetic map. Biology has a number of other good simulations including *Coexist, Ecological Modeling, Evolut, Compete,* and *Enzkin,* all of which are available from CONDUIT.

Oregon

Oregon has already been described. I would simply point out that because elementary and secondary students love to play *Oregon* so much, its pedagogical value is sometimes overlooked. A close reading of the documentation and the references cited will suggest a rich number of ways that the game can be integrated into the history classroom.

Quantitative Experimental Analysis

Quantitative Experimental Analysis provides a method of clarifying the difficult statistical concepts of estimation, distribution parameters, convergence, and power and significance. Most important, the unit is designed to demonstrate the connections between theoretical distributions and typical, empirical data. The programs simulate different theoretical distributions, some of which are known to the students and others of which are not known.

Simsearch

Simsearch, like *Critical Incidents,* is not based upon a formal simulation but provides hypothetical scenarios for sequential decision making. *Simsearch* poses problems of a sociological nature and then cycles the student through a number of decisions which allow him or her to construct a full-scale research design. The package is particularly useful in demonstrating the hierarchical nature of design decisions, that is, the way decisions sequentially reduce the range of the remaining choices. It stresses both the importance of careful research planning and the close link between theory and method. The six design exercises deal with research problems in the areas of deviant behavior, family, sex-role socialization, sociology of sport, minority group relations, and the sociology of youth. *Simsearch* simulates research in each of these areas by making the student feel like someone who is managing a research project and constructing design plans.

Spatial Marketing

Spatial Marketing is a highly flexible simulation model that gives the student the ability to replicate the location of markets, the transportation system, the population distribution, and various economic and cultural parameters of a region. The strength of the package comes from the sophistication of the model, the thorough tutorial materials within the package, and the elegant displays of geographic phenomena. The package was developed by Ellinger (1977).

Wheels

Wheels, a simulation of automobile purchasing by a teenager, is the least academic of all the exemplary units. It is truly practical, however, in that it teaches about consumer decision making, including insurance, financing, and automative upkeep. Neither the manual nor the program requires a high level of sophistication or literacy. As a prototypical simulation game it suggests many ways that the computer could be used for mass, continuing education.

INTENDED AUDIENCE, COST, AND SOURCE

As already discussed and as shown in figure 2, most of the computer simulation games are targeted to the college or secondary/college audience. This probably results from the fact that computers have been in use at the college level longer than at lower educational levels. It may also be true that many simulations are easier to develop at the more advanced levels. In the next 10 years the situation will probably reverse itself as more and more units are developed at the secondary and elementary levels.

One disappointment is that simulation games written for the secondary level tend to be poorly documented and packaged. Developers seem to think that a program will be internally compelling and that others should figure out how to use it for themselves. One hopes that greater incentives will be found to support more complete documentation and interchange of programs.

DOCUMENTATION

Figure 3 gives for each simulation game the number of pages of student and teacher documentation. CONDUIT and others generally supply programmer notes with most packages to aid in the installation of programs. Teacher guides also sometimes contain programmer information. As one can see from scanning the right-hand column of Figure 3, most simulation game packages are self-contained in that they do not typically require specific articles or books.

Most of the packages have both a student and a teacher booklet or guide, and this is a very valuable practice. Sometimes the program can contain and print out documentation student users need, but usually students need more. Of course, there is the inevitable trade-off between fuller documentation and higher cost of materials for teacher and student alike.

IMPLEMENTATION CONSIDERATIONS

Even though the exemplary packages were selected to be relatively transferable from one computer to another, varia-

FIGURE 2 Intended Audience, Cost, and Source

Name	Intended Audience	Cost	Source
Change Agent	h.s.-college	$30.00	CONDUIT[1]
CHEBO	college-graduate	$15.00	James Sakoda[2]
Cognitive Psychology	college	$50.00	CONDUIT
Critical Incidents	teacher training	$60.00	CONDUIT
Energy (U.S. Policy)	college	$10.00	CONDUIT
EXPERSIM	college	$85.00	CONDUIT
Ghetto	h.s.-college	$50.00	CONDUIT
IDGAME	college	$65.00	CONDUIT
INS2	h.s.-college	$95.00	CONDUIT
INTERP	college	$25.00	CONDUIT
Limits	h.s.	$3.00	Digital Equipment Corp.[3]
Linkover	college	$20.00	CONDUIT
Oregon	h.s.	$15.00	Minnesota Educational Computing Consortium[4]
Quantitative Experimental Analysis	college	$90.00	CONDUIT
SIMSEARCH	college	–	CONDUIT
Spatial Marketing	college	$95.00	CONDUIT
Wheels	h.s.	$46.00	Paul S. Amidon and Associates[5]

1. CONDUIT, Box 388, Iowa City, Iowa 52240
2. James Sakoda, Department of Sociology, Brown University, Providence, RI 02912
3. Digital Equipment Corp., 146 Main Street, Maynard, MA 91754
4. Minnesota Educational Computing Consortium, 2510 Broadway Drive, Lauderdale, MN 55113
5. Paul S. Amidon and Associates, 1966 Benson Ave., St. Paul, MN 55116

FIGURE 3 Computer Simulation Games: Documentation

Name	Student Guide (pages)	Teacher Guide (pages)	Other
Change Agent	11	12	–
CHEBO	None	–	described in Lehman (1978)
Cognitive Psychology	102	96	–
Critical Incidents	10	30	ref. on Transactional Analysis
Energy (U.S. Policy)	52	–	–
EXPERSIM	under revision	under revision	
Ghetto	20	–	–
IDGAME	52	–	–
INS2	57	100	programmer's guide
INTERP	11	15	–
Limits	41	19	55 p. resource handbook
Linkover	11	15	–
Oregon	–	25	–
Quantitative Experimental Analysis	–		–
SIMSEARCH	–	?	–
Spatial Marketing	–	?	–
Wheels	23	–	–

FIGURE 4 Computer Simulation Games: Implementation Considerations

Name	Language	Lines of Program	Special Issues
Change Agent	BASIC	376	–
CHEBO	FORTRAN	about 800	listed in App. A of Lehman, 1977
Cognitive Psychology	BASIC	about 1,200	6 programs
Critical Incidents	BASIC	about 15,000	26 programs
Energy (U.S. Policy)	FORTRAN	605	batch program
EXPERSIM	FORTRAN	?	extremely large
Ghetto	BASIC	about 2,000	needs random access data files
IDGAME	FORTRAN	1,500	3 data files
INS2	BASIC	about 3,600	12 data files
INTERP	BASIC	284	–
Limits	BASIC	240	–
Linkover	BASIC	380	–
Oregon	BASIC	670	–
Quantitative Experimental Analysis	?	?	–
SIMSEARCH	BASIC	?	–
Spatial Marketing	BASIC	?	–
Wheels	BASIC	?	–

tions in computer systems make complete standardization impossible. Figure 4 gives some technical characteristics that are of interest to installers. Only four of the simulation games are written in the Fortran language; the remainder are in the Basic language. The number of lines of the program indicates size, not complexity. For instance, *Critical Incidents* has about 15,000 lines of program, but because the structure of the program consists almost entirely of branching and of printing tutorial text, the package is relatively easy to install, even on small computers, as long as programs can be chained and conveniently stored. (*Critical Incidents* consists of 26 separate programs, and these may have to be stored in separate files.) *Ghetto* is probably the least transferable package listed, due to its complex internal structure and its requirement for random access files.

These data on internal program characteristics are provided to give a sense of the variety of program issues. None of the issues in the exemplary package set is serious enough to be cause for rejection. If there is a need for any one of these units, chances are the effort to implement will be extremely minor. We would not make such a claim, however, for the typical package outside this highly select subset.

CONCLUSION

The many simulation and gaming possibilities opened up by the widespread availability of computers has generated a diverse collection of simulations and games. But with the exception of *INS2,* all of them contain at least one instructional unit that can be assigned as a single laboratory or homework exercise. Almost all of the computer simulation games can be played by a single student. So it would seem that computer technology is definitely having an individualizing impact upon instruction. For the most part this is desirable because more and better learning is possible. I anticipate, therefore, the development of many more such self-contained simulation game packages for both high school and college students.

I also predict another type of institutionalization of instructional simulation and gaming. Simulations and games will be built into large courses of instruction called *courseware.* In such cases students will probably be less aware of the underlying technical and pedagogical strategies. Thus one may become less aware of how he or she using simulations and games even though they are being used more. With this declining visibility of simulation and game units, the assessment of quality may become more difficult. The tendency will be to evaluate larger units or even entire computer-based courses. In such cases the individual simulation or game modules may not receive close scrutiny.

On the other hand, with the increasing institutionalization of computer simulation games in instructional design, the methodologies will improve and expertise will expand. There should be many high quality, creative new developments with computer simulation games.

REFERENCES

Adams, Dennis M. 1973. *Simulation Games: An Approach to Learning.* Worthington, Ohio: C.A. Jones Publishing Company.

Anderson, Ronald E. 1976. Technical standards and performance standards in the pursuit of quality. *Behavioral Research Methods and Instrumentation* 8,2: 211-17.

Baker, Justine C. 1975. *The Computer in the School.* Bloomington, Ind.: Phi Delta Kappa Educational Foundation.

Boocock, Sarane S. 1972. Validity-testing of an inter-generational relations game. *Simulation and Games* 3,1: 29-40.

Boocock, Sarane S., and Schild, E.O. 1968. *Simulation Games in Learning.* Beverly Hills, Calif.: Sage.

Dukes, Richard L., and Waller, Suzan J. 1976. Validity-testing of an inter-generational relations game. *Simulation and Games* 7, 1: 75-96.

Elder, C.D. 1973. Problems in the structure and use of educational simulation. *Sociology of Education* 46: 335-54.

Ellinger, Robert S. 1978. Spatial Marketing Simulation: A new package. *Pipeline* 3, 3: 26-31.

Greenblat, Cathy Stein, and Uretsky, Myron. 1977. "Simulation in social science." *American Behavioral Scientist* 20, 3: 411-26.

Inbar, Michael, and Stoll, Clarice, eds. 1972. *Simulation and Gaming in Social Science.* New York: Free Press.

Lehman, Richard S. 1977. *Computer Simulation and Modeling: An Introduction.* Hillsdale, N.J.: Lawrence Erlbaum Associates.

Liggett, Helen. 1977. An evaluation instrument for use with urban simulation games. *Simulation and Games* 8, 2: 155-88.

Maidment, Robert, and Bronstein, Russell H. 1973. *Simulation Games: Design and Implementation.* Columbus, Ohio: Merrill.

Orback, Eliezer. 1977. Some theoretical considerations in the evaluation of instructional simulation games. *Simulation and Games* 8, 3: 341-60.

Rawitsch, Dan. 1978. Oregon Trail. *Creative Computing* 4, 3: 132-39.

Roberts, Nancy. 1976. Simulation Gaming: A Critical Review. Eric Document No. ED 137 165.

Sakoda, James M. 1971. The Checkerboard Model of Social Interaction. *Journal of Mathematical Sociology* 1: 119-32.

Tansey, P.J. and Unwin, Derick. 1969. *Simulation and Gaming in Education.* London: Methuen.

DOMESTIC POLITICS GAMES AND SIMULATIONS
An Evaluation

by Dorothy R. Dodge

THE USES OF SIMULATION IN THE SOCIAL SCIENCES

The social science disciplines tend to highlight patterns of human interaction. Sociology focuses on social interrelationships. Economics raises issues of distribution and management of economic activity and production. Geography explores questions of spatial and resource allocations and use. Political science analyzes process and procedures of decision-making behavior.

Simulations and games offer a method for illustrating such complex varieties of human interaction. Simulation models have proven effective in all of the social science disciplines, both as an instructional technique and for research. When an instructor wants students to study some aspect of the complex network of human interactions–whether in family relationships, the lobbying tactics of pressure groups, corporate management, or legislative coalition building–simulation models are one technique available. Simulations permit active student participation in exploring a variety of human behaviors and provide an opportunity for exposure to a wide range of human experiences not necessarily familiar to the student.

Role-Playing Models

One instructional simulation style is the *role-playing* model. Students benefit from the opportunity to place themselves in someone else's shoes and pursue whatever goal the model prescribes as one the role position desires. The model forces students to operate within the capabilities and limitations assigned to the role. Role playing tends to increase student awareness of the dilemmas others often face in attempting to achieve their goals. It can also make more apparent to students the virtual impossibility of attaining goals for those who find themselves in underprivileged role positions in society.

Problem-Solving Models

A second style of instructional model is a *problem-solving* simulation. Almost any problem in the social sciences is amenable to modeling. Problem-solving models afford an instructor the opportunity to introduce students to the background or history of a given social issue or problem. Students can explore the conflicts and divisions among various societal groups that favor or oppose a series of proposed resolutions. Problem-solving models may be included as an exercise in any instructional unit that involves analysis or discussion of issues under debate by society. Students might research an assigned set of materials providing background for the problem before they play the problem-solving model. Or the instructor may prefer to use the simulation to stimulate additional research after it has been played.

A frequent result of participation in a problem-solving simulation is that students undertake additional research on their own to find data they lacked during play or seek to answer questions they may have found perplexing or especially interesting. Use of a problem-solving model not only tends to encourage reading and research, but it is one method for illustrating the conflicts and issues surrounding the resolution of a given problem.

Bargaining Model

A third style of instructional model involves *bargaining-behavior* simulation. Current attention to the "games people play" reflects growing popular awareness of the role of bargaining behavior in societal interrelationships as well as continuing interest in the concept of power. Contemporary literature often describes power as a thing or "object" someone has or can obtain. Eric Hoffer makes the point that "power can't be put in cans." Political science, on the other hand, describes power as a relationship among individuals attempting to achieve whatever their goals or desires dictate. Power in this sense is seen as the ability to get another individual to do as you wish. This may mean getting someone to start an action, cease an action, or just continue an action that fits your interests or goals.

Influence and power sometimes are used synonymously. In other cases, power is often described as a threat-promise relationship, while influence is a process that involves persuasion or instruction. Influence style avoids the threat of sanction by suggesting "what's best for another person."

Editor's Note: All of the simulations discussed in this essay appear in the domestic politics section.

Power and influence are associated with gaining control over some "object of value" desired by interacting individuals. Such "objects" may be tangible—such as money, resources, or military armaments—or they may be intangible—involving approval, love, affection, acceptance, or prestige. Interaction among individuals occurs as they negotiate to obtain, divide, increase their share of, or deny each other the object involved. The possible relationships among individuals struggling over some object of value are frequent subjects for simulation bargaining models.

Objects of value provide players with a potential base for power or influence, but their ability to cause someone to do what they wish depends also on their skill in interaction and in communicating. Without struggle over an object of value, interaction would not occur. With such an object, the result of the power interaction will depend on a large number of additional factors that a bargaining model permits students to explore.

The contemporary domestic and international political scenes provide numerous examples of power interactions occurring over tangible objects of value such as oil, foodstuffs, mineral deposits, unemployment compensation, tax relief, social security benefits, or civic awards and recognition. Politics has been described as the art of "who gets what," or "how you cut the pie." The bargaining model focuses on that "art" and the factors involved in getting the "piece" you desire.

Models may involve a simple two-person bargaining situation involving purchase of a car or a complex multinational conference over nuclear disarmament.

Game theory problems, too, are the subject of bargaining models. The basic purpose of game theory is to study a competitive situation among some number of players when the outcome or payoff from the bargaining for any player depends on the mutual choices made by all players. Game mathematics hypothesizes that rational players will follow a strategy of minimizing their potential losses in relation to what other players may decide to do, rather than attempting to maximize their gain at the risk of total loss. Rational play is seen as minimizing the possible losses, not winning as much as possible. Game models permit analysis of whether actual players will move to a strategy of losing as little as possible or pursue a risk-taking decision in an attempt to win "big." Such bargaining models also explore the function of threats or promises on choices and the importance of communication among players.

Process Models

A final style of instructional model involves analysis of the *structures, processes,* or *procedures* of decision making. Models of legislatures, political parties, congressional committees, or the United Nations are examples. Such models attempt to replicate the formal structures and procedures selected institutions in their decision-making behavior. These models are effective for exposing students to formal decision-making processes, legislative behavior, or parliamentary law. Rather than studying a flow chart of structure and outlines of formal process, students participate in a model that replicates the structure of the real-world institution and from the experience learn the formal channels for decision making and how things get done.

The four styles of instructional model offer the social science teacher effective methods for presenting and exploring a wide variety of social interrelationships and behaviors. Although many other instructional techniques might be employed, simulation or game models directly involve students in a participatory situation rather than a passive "looking" or "listening" experience. When they are included in instructional units, simulations and games give students variety in the style of learning experience. Students should be cautioned, however, that simulations or games are hypothetical models; real-world interactions and behavior may differ significantly from the simulation depending on the reliability of the model.

SIMULATIONS SELECTED FOR ANALYSIS

The simulations discussed here were selected as examples of the four instructional styles of models. They also represent a variety of subject content within a general area of political process.

These simulations and games have as their subject matter (content) a wide variety of local, state, or national interrelationships. They also illustrate the four instructional styles of models discussed above.

INSTRUCTOR'S ROLE

Instructors understandably avoid instructional techniques that require elaborate preparation or complex computation. Simulation models that are not intended as research designs generally attempt to avoid both demands on the instructor.

Preparation

In any simulation exercise the instructor should allow adequate time to introduce students to the purpose of the model or to the problem or situation involved, and to define all the terms and concepts the model employs. After students have read the manual, cards, or foldouts provided, they should have an opportunity to discuss and resolve questions about rules and procedures for play. If the instructor requires additional research, these assignments should be explained and discussed.

Supervision

While the model is being played, the instructor's chief responsibility is to supervise the series of steps modeled and to observe time constraints. For example, the model may prescribe for step one ten minutes to fill out issue position sheets; for step two, a ten-minute negotiation period between players; and for step three, a city council session to vote on a given issue. The instructor as supervisor must be careful that time limits and steps are followed so student players may observe the pressures and procedures the model illustrates.

TABLE 1 Selected Simulations

Simulation	Summary Description
American Government Simulation Series	A series of five exercises about the structure and process of U.S. government, involving: 1. constitutional convention 2. congressmen at work 3. presidential election campaigning 4. budgetary politics 5. presidential decision making that may be used as a unit or separately
Amnesty	A simulation dealing with Vietnam war issues, the problems of amnesty, and the decisions of the President's Clemency Board
Budget	A simulation illustrating procedures in the formation of the national budget and the conflicting interests involved in budget decisions
City Hall	A simulation focusing on local governmental structures, decision-making procedures, and voter behavior in urban elections
Election	A game providing step-by-step exposure to primary and general election process and structures and modeling individual public office holding and career advancement
Hat in the Ring	A game centering on presidential nominating procedures, the steps involved in becoming a candidate for President of the U.S. (campaign planning, presidential primaries, and national convention), and the influence of campaign financing on candidate success or failure
Metropolitics	A simulation exploring four structural styles of urban government: 1. single unified government 2. two-level county approach 3. special district 4. neighborhood governments
Parksburg	A simulation of crisis decision making in a local urban area about garbage service and city financial limitations
Taxes	A simulation of the influence or role of taxation decisions and policies on election results for city council members favoring one or another set of tax policies
Women's Liberation	A role-playing model focusing on the concerns of the women's liberation movement and goals of groups within it

TABLE 2 Types of Models

Simulation	Role Playing	Bargaining	Structural or Process	Problem Solving
American Government	X		X	X
Amnesty				X
Budget	X		X	X
City Hall	X		X	
Election			X	
Hat in the Ring	X	X		
Metropolitics	X	X	X	
Parksburg	X	X	X	
Taxes	X	X		X
Women's Liberation	X			

Debriefing

A very important instructional experience occurs during debriefing. All of the models reviewed in this essay discuss or refer to debriefing activity for students. Discussion questions exploring the issues the simulations raise are listed in the instructor's manual, or at the end of the players' instruction sheets if there is no separate instructor's manual. The *American Government* series provides a detailed discussion and research exercise for debriefing accompanied by an extensive bibliography. *Metropolitics* and *Parksburg* include a series of discussion topics concerning urban politics. *Amnesty* suggests an extensive research project as well as discussion topics. The other models provide one or two discussion items for debriefing.

The debriefing period, in addition to model discussion topics, might profitably explore the following questions:

(1) What strategy did each player adopt in playing the model?
(2) Was this strategy effective, or would another strategy have been more effective?
(3) Did other players follow effective strategies and play the simulation well?
(4) Did any aspects of the model process or interaction behaviors suggest any examples from the local community "real world"?
(5) Were there any aspects of the model that appeared to differ from local community process or behavior?
(6) What did the players feel they learned from the experience of playing the model?

These questions permit the instructor to explore any misunderstandings concerning the model and inaccurate perceptions of the "real world" that may have resulted from the playing experience. Debriefing should involve careful exploration of model accuracy or reliability to avoid false impressions that the model replicates the real world. The instructor may point out that models are intended as reflections of some aspects of the "real world," but that no model claims to replicate it.

PURPOSE OF THE SIMULATION

The simulations discussed here model a wide variety of problems.

American Government

This series of exercises is designed to give students laboratory experiences in the field of American politics. The five issues or exercises are:

TABLE 3 Instructor Role

Simulation	Supervise Steps and Explain Roles	Evaluate Student Work	Check Scoring and Rules	Compute Results	Additional Uses and Revision Possibilities
American Government	X	X	X		X
Amnesty	X	X		X	X
Budget	X	X		X	
City Hall	X				X
Election	X				
Hat in the Ring	X	X			
Metropolitics	X				
Parksburg	X		X		
Taxes	X			X	
Women's Liberation	X		X		X

(1) *Founding Fathers,* explaining the Constitutional Convention;
(2) *Congress,* illustrating congressional process and the issues of representation of constituent interests;
(3) *Presidential Election,* analyzing the strategy candidates may employ to achieve office;
(4) *Budget,* detailing the steps involved in federal budget making; and
(5) *Congressional Committee,* illustrating behavioral patterns of congressional committees.

Amnesty

This simulation is designed to explore the issues that surrounded the granting of amnesty after the Vietnam war. The model emphasizes research into the history of U.S. involvement in Vietnam, rather than the structure or process of amnesty proceedings.

Budget

This simulation models the steps in the formation of the national budget and the conflicts among various factions and interests over budgeting policy. Final budget decisions are shown to be limited both by the resources of the national economy and by past policy and budgetary decisions.

City Hall

This simulation models the structure of local government and the process and procedures involved in local decision-making behavior and policy decisions.

Election

This simulation leads students through a step-by-step analysis of the structure of primaries and the election process in becoming an office holder at local, state, and national levels in the United States.

Hat in the Ring

This simulation details the steps and strategies involved in becoming nominated for the office of President of the United States and, finally, running as a candidate for that office. The primary emphasis is on financial requirements and strategy, rather than issues or coalition building.

Metropolitics

This simulation explores the issues involved in urban governmental reorganization and the styles of urban governmental structure that are currently functioning in the United States. The model also explores the process of population reorganization at the local level.

Parksburg

This crisis model illustrates the negotiation and bargaining that occur when garbage collection service is curtailed as a result of financial limitations.

Taxes

This simulation exposes students to the effect of taxation policies on individuals with a variety of income, property ownership, and career levels in a community. It explores why individual interests support one style of taxation rather than another.

Women's Liberation

This role-playing model involves students in exposure to and analysis of the goals and desires represented by six of the major groups functioning in the women's liberation movement.

STYLES OF INTERACTION

Related to a simulation's intended or expressed purpose as an instructional technique are the styles of interaction programmed in the model. The interaction most frequently programmed as the first step is the formation of individual or special-interest positions after careful reading of rules and model role statements or background materials. As the second step, the model prescribes some formal device for articulating these individual or special-interest positions. This permits all players to become aware of the conflicting or competing interest positions in the simulation, and identifies the students playing the various role positions that represent the individual or special interests programmed in the model.

Negotiation and bargaining interactions follow this exposure to competing or conflicting interests as players attempt to achieve the goals specified for their roles and to "win" or gain "high scores." Aggregation of interests may occur as the next step if coalition building, log-rolling, or trade-offs are programmed as interaction styles required to reach individual goals and score well. Many models do not include additional styles of interaction; they intend only to illustrate the interest articulation and aggregation behaviors of society. However, if the model is designed to illustrate system process and structure as well, additional steps may require players, after they form coalitions or aggregates, to learn how to use the modeled structures and the channels for interest input. To win, players must attempt to gain system attention for their position. A final step would be decision making, which ends with the formal structure announcing a decision on the issue modeled. The simulations discussed here vary in the styles of interactions programmed.

PREPARATION BY STUDENTS TO PLAY

Simulations vary in the amount of time and research they require of students to master their role assignments. Formal classroom discussion and explanation is always necessary to assure that everyone in the class understands the rules and operation of the simulation.

Some simulations also recommend or require that students do additional research before they begin to play the model.

BASIC FACTORS

Table 6 summarizes the basic factors of age level, number of players, class time for play, and cost of the simulations.

Age Level

The age level for which a simulation is suitable for use in the classroom is important, because a model at too advanced a level will be ineffective for younger students and an elementary level tends to be boring for the older and adult players.

Number of Players

The flexibility of a model in accommodating class size is an important consideration. Classes may be divided into teams, each playing separately and comparing results at the end. It is unwise to try to add roles to accommodate all student numbers unless an active and meaningful role can result. Students who have little to do learn little during play and are frequently bored. Playing with too few students may force the omission of roles that may be essential for playing the model.

Class Time

The amount of time required in the class schedule to prepare, play, and debrief a model is important in instructional planning. Instructors may not wish to devote a significant amount of class time to a model, or they may wish to make a model a major activity.

Cost

The cost of a model may be an important factor in deciding whether to use it. Library or instructional assistance funds may cover the cost, or students may be asked to pay for the player manuals to cover costs. Costs of virtually all the models discussed here range from $25.00 to $30.00.

PACKAGED MATERIALS

The packaging of a simulation usually influences the ease of its operation as well as its cost. Models generally provide an instructor's sheet or manual that explains the model, requirements for duplicating materials, room arrangements, and any research requirements instructors should prepare before play. Players are provided with a participant's manual, sheet, or background foldout. Roles are detailed on individual role sheets (to keep role data secret) or on role descriptions that are provided for all players. Any scoring sheets, cards, or forms are either provided in quantity or as masters for duplication. Some models also may provide filmstrips, maps, or overheads to assist play.

THE SETTING

These simulations do not require elaborate special facilities, but instructors should be aware that various activities may

TABLE 4 Styles of Interaction

Simulation	Individual or Special-Interest Articulation	Negotiation and Bargaining	Coalition Building, Log-rolling, and Trade-Offs	Process or Structure Exposure and Interest Input	Decision Making	Evaulation of How Well Programmed
American Government	X	X	X	X	X	Well
Amnesty	X				X	Moderate
Budget	X	X	X			Moderate
City Hall	X	X	X			Moderate
Election	X				X	Well
Hat in the Ring	X	X	X			Moderate
Metropolitics	X	X	X	X	X	Well
Parksburg	X	X	X	X		Well
Taxes	X	X	X			Moderate
Women's Liberation	X	X	X			Moderate

TABLE 5 Student Preparation

Simulation	Preparation Time and Research Requirements	Comments
American Government	Time: 1 hour No additional research	The five exercises are intended as illustrations of a variety of political behaviors and processes of American politics. The model is written simply and requires only that students read the manuals for each exercise and make themselves familiar with their roles and the rules of procedure. Forms are included for the various steps that make up the exercises.
Amnesty	Time: 3 to 4 hours Additional research required. Time allotted depends on instructor.	Students take a value orientation questionnaire and class themselves according to simulation identities. Library research is required to make the students familiar with amnesty viewpoints related to the simulation identities. Before the amnesty board hearings, each student must write a paper discussing the consequences of "my action." The time needed for research and development of viewpoints depends on student ability and experience with writing identity and viewpoint papers.
Budget	Time: 3 hours on model Additional research required before play. Time allotted depends on need to train students in library research.	Students have a guide of about 38 pages explaining the model problem, rules of procedure, and roles. Players receive individual identity cards, goal cards, individual score cards, voting power records, and job descriptions. Students may need several hours to become familiar with role goals, constraints, and requirements. Appropriation forms also require class explanation and discussion to ensure understanding. The job description forms require student research of budget materials. Students may need time for training in how to use library materials and research indices.
City Hall	Time: ½ hour No additional research	Student manual provides background detail for 38 role positions and the rules of procedure. Limited class discussion is required to clarify roles.
Election	Time: ½ hour or more No additional research	Players draw opportunity cards and use strategy in attempting to gain a high score in career activity, nomination for President or Vice President, and, finally, success in election. No elaborate explanation is required. The game is suitable for low-ability students.
Hat in the Ring	Time: ½ hour No additional research	Students get foldout sheet for roles and positions. Rules are simple, involving random chance using a deck of cards. No computation or strategy selection is necessary. Limited class time is required.
Metropolitics	Time: ½ hour No additional research	Students get role envelopes detailing their marital status, income, social status, residence, and views toward political issues. Chips represent each player's influence. Rules are simple and require no computation or additional class explanation time.
Parksburg	Time: ½ hour No additional research	A manual explains the simulation problem and provides detailed role statements. Little class time is required to discuss role or rules. Problems for further research and an additional bibliography are provided. The instructor may decide whether to require research after the simulation.
Taxes	Time: ½ hour No additional research	Students have an identity chart, and they draw numbers for roles and tax positions. Those with similar positions meet before play to form factions for political action. Rules are simple and require little explanation.
Women's Liberation	Time: ½ hour No additional research	Students get an explanation sheet, a copy of women's liberation proposals, and special instructions explaining rules and procedures. After discussion of these materials, the class is divided into six groups to discuss their priority solutions and strategy for the voting session. Rules are simple and require no computation.

TABLE 6 Basic Factors

Simulation	Age Level	Number of Players	Time	Cost
American Government	Grade 9 to college freshmen	32 to 49 roles provided	If all 5 exercises are used and homework is assigned, unit may last 2 to 3 weeks. Each exercise requires several hours.	$1.95, $1.55 with school discount, per student handbook for each exercise; $1.35, $1.08 with school discount, instructor's guide for each exercise.
Amnesty	High School	15 to 40	Four to five hours, plus extensive library research and instruction.	$10.00
Budget	Grades 7-12	25 to 40 roles provided	Fifteen periods plus research on budgeting issues and background	$14.00
City Hall	Grade 9 to college freshmen	14 to 48	Four to seven 45-minute periods	$10.85
Election	Grades 6-12	4; may have teams	One-half to one hour	$9.95
Hat in the Ring	Grades 5-12	3 to 27 in teams of 3 to 9	Two to four hours in 45-minute periods	$9.95
Metropolitics	Grades 8-12 Adult	18 to 35 roles provided	One to two hours	$25.00
Parksburg	Grades 9-12	38; may be varied	Four to five hours in 45-minute steps	**
Taxes	High School	varies; instructor must adapt	Six hours in a series of opinion-formation and voting steps	$10.00
Women's Liberation	High School	varies with class, but class must provide roles	Three hours, plus time for research and library instruction	$3.00

**Must be purchased as part of Old Cities, New Politics package ($82.00 list, $75.00 school).

TABLE 7 Packaged Materials

Simulation	Instructor Material	Participant Material	Role Descriptions	Scoring Records	Film Strips	Overheads	Additional Bibliography
American Government	Manual	Manual	Sheets	Forms			X
Amnesty	Manual	Foldout	Foldout	Forms			X
Budget	Manual	Manual	In Manual	Forms			X
City Hall	Manual	Manual	In Manual	Forms			
Election	Sheet	Cards	Cards	Cards			
Hat in the Ring	Foldout	Foldout	Foldout				
Metropolitics	X	X	Sheets	Forms			X
Parksburg	Manual	Manual	In Manual	Forms	X	X	X
Taxes	Manual	Foldout	Foldout	Forms			X
Women's Liberation		Manual	In Manual	Form			X

require special classroom arrangements. One style of simulation is played around a table. Generally, this style uses a small number of players (four to six is common) and involves dice or card decks for drawing activities or instructional data.

Another style of model programs a town meeting, United National General Assembly, or a legislative session and requires only one large room with table space for each player. The room may need to be reorganized to give the presiding group a table for this general conference style.

A third, and the most common, style combines the general conference with a series of small-group caucuses, negotiation or bargaining interactions, or group issue sessions. These submeetings require an arrangement that will enable groups to meet privately and with secrecy if the model requires it. The

TABLE 8 The Setting

Simulation	Table	One Large Room	One Large Room with Small Conference Rooms
American Government			X
Amnesty		X	
Budget		X	
City Hall			X
Election	X		
Hat in the Ring	X		
Metropolitics			X
Parksburg			X
Taxes		X	
Women's Liberation		X	

most effective setting is a large room and several small rooms. If additional rooms are not available, the classroom may be arranged to place subgroups in corners or small circles apart from the general conference area, although this arrangement may be less effective if secrecy is required. The noise level in one large room may make it more difficult to negotiate effectively, as well.

RULES

Recent simulations for instruction use simplified rules and procedures so students can concentrate on play and interaction rather than on learning complex rules or computation techniques. All the models discussed here have reduced rules to a minimum and use scoring only as a device to encourage students to play roles effectively; the desire to win or do well will influence efforts to play well.

ROLES

Role playing is an important learning experience in a simulation or game. It presents students with problems and capabilities or constraints that are not necessarily part of their own backgrounds or lifestyles. Playing an unemployed, minority urban dweller who is denied city service may be as alien an experience for a middle-class suburban student as playing the chairman of a city council or the President of the United States.

The models discussed here employ three styles of role playing. The first style specifies no role background or identity. Students are assigned issue preferences or are asked to define their own attitudes toward an issue and play the simulation from these perspectives. Because the roles are loosely defined, the potential exists for students to play themselves rather than a "role."

The second style prescribes roles as functions such as voters, council members, party leaders, candidates for office, or citizens. The model provides students with the goals for these groups and their functions. Interpretations of these functions and the ways to play such roles may vary from player to player because the simulation programs flexibility in playing the roles.

The third style involves individual role identities. Individual participants receive names and backgrounds, including residence, career, income, race, religion, family status, educational level, and issue preference. Each role has specific goals and varying capabilities for achieving them. Scoring relates to success in goal achievement.

Individual role statements are the least flexible style and assure "role playing" by students; they require more elaborate

TABLE 9 Rules and Scoring

Simulation	Rules and Criteria of Success	Scoring Style
American Government	Simple process and voting rules. Score by success on issues and votes.	Tables and instructor's evaluation of play.
Amnesty	Simple process rules. No score except success in influencing board decisions.	Whether issues are included.
Budget	Simple process rules. Score by success on issues.	Prescribed votes and totals on score forms.
City Hall	Simple process rules. Scoring based on success of candidates and issues.	Score sheet for candidate success.
Election	Simple process rules. Scoring derives from opportunity cards and random drawing.	Career score form and nomination.
Hat in the Ring	Simple process rules. Score by random drawings from deck of cards.	Score form.
Metropolitics	Simple process rules. Scoring based on success of issues and candidates.	Role card, alliance cards, influence chips, and final score total.
Parksburg	Simple process and voting rules. Score by success on issues.	Role and issue strategy forms.
Taxes	Simple process rules. Success based on issues and candidates selected in final vote.	Score sheet for taxes and positions.
Women's Liberation	Simple process rules. Success with issues included in group and final conference.	Whether issues are included.

TABLE 10 Roles

Simulation	Attitude or Issues Role	Functional Roles	Individual Identity Roles	Role Flexibility	Role Activity	Role Reliability	Role Unity
American Government			X	Controlled	Good	Good	X
Amnesty	X			Loose	Medium to low	Medium to low	
Budget			X	Controlled	Medium	Good	X
City Hall		X		Medium	Medium	Medium	X
Election	X			Loose	Medium to low	Low	
Hat in the Ring	X			Loose	Medium to good	Medium	
Metropolitics			X	Controlled	Good	Good	X
Parksburg			X	Controlled	Good	Good	X
Taxes			X	Controlled	Medium	Medium	X
Women's Liberation	X			Loose	Medium	Medium	

modeling by the author and careful analysis of the roles. This style is the most effective, but it is omitted in less complex simulation models. The variety of roles in a model relates to these styles of role programming. More complex models provide greater variety.

An important consideration for the instructor is whether the model roles provide activity for every student. Models that leave blocks of time with little activity for some students may be ineffective as class learning experiences. Students with little to do become bored or frustrated and learn less from the experience. Most of the models provide reasonable activity for all players, but the instructor should check those listed as medium or low on activity for class suitability.

MODEL RELIABILITY

Simulation models that are used for research or laboratory experiments seek to model replications of the "real world." Such models are complex and often require computer assistance and programming. Even with elaborate modeling, however, replication is difficult at best. Student-played models generally warn that they should be presented as *illustrations* but not *replications* of the "real world" system they model. The debriefing period is valuable as a time to analyze and discuss the simulation and its variance from the "real world," as well as the results of the simulation play and their similarity or dissimilarity to what has occured in the "real world."

Caution is always required in discussing model reliability. No model claims to be a replication of the real-world situation modeled. Real-world structures, behaviors, processes, and competing interests are complex and often poorly understood or researched, and social science literature reveals wide disagreements among social scientists over any of these factors. Gaps in research and knowledge exist over process, structure, bargaining behavior, and coalition-building tactics. Modelers cannot be expected to replicate the real world when the literature in the field is unsure of what the real world may involve. Two plus two is not difficult to model, but social science affords little opportunity for such precision as a total of four. The problem of reliability should be discussed thoroughly in class to avoid and correct misunderstandings or misleading acceptance of the simulation as a replication of the "real world."

CONCLUSION

Any of these models may be employed usefully when they are appropriate to course content or instructional unit goals. Each involves a different set of roles, processes, and problems.

All of the models provide students with opportunities to play roles and experience the constraints and frustrations of a hypothetical person placed in the model's decision-making setting or problem. All of the models also afford some experience with strategy development and coalition-building tactics as individual players attempt to achieve their goals. *American Government, Budget, Metropolitics, Parksburg,* and *Taxes* give more background information in their exposure to bargaining behavior and the myriad conflicting interests and viewpoints represented by the hypothetical individual roles and by the variety of interest or pressure group positions included in the models. The other models provide players with less detailed information concerning role descriptions and pressure or interest group concerns.

A lack of background material leaves more to students' imaginations or ingenuity in relation to possible coalition-building conflicts and stumbling blocks. Because students are frequently unaware of the competing interests that operate in their local communities and have had limited experience with the political decision-making process, models that specify individual role backgrounds and group interests may be more helpful in introducing them to possible pressures and counter-pressures involved in determining policy or solving problems. Detailed role specifications of the economic, social, educational, and political interests involved in a model require more extensive research and model development by the model designer. Such detail also tends to involve longer manuals, role sheets, and instructions, thus increasing price. However, the additional cost and the additional time students must put into preparing their roles before effective play can begin is more than justified by the controlled experience they receive from

TABLE 11 Summary Analysis of Models

Simulation	Model
American Government	Illustration and partial representation of American political systems. Does not use contemporary issues. Modelers warn that political science has only partially validated hypotheses, and "reality" is difficult to determine.
Amnesty	Analysis of attitude toward amnesty and Vietnam war policy. No coalition building or political skills programmed. Attitude and role loosely controlled and may tend toward students playing "themselves."
Budget	Model of coalition building and voting power. Tends to ignore communication channels or skills of the players. Budget decision-making model not applicable to other areas of decision making.
City Hall	Emphasis on voting issues and coalition building, but not on political philosophy or issue content.
Election	Simple game based on gambling and random chance. Illustrates structure, but not coalition building, problem solving, or issue content.
Hat in the Ring	Illustrates steps in nomination for office. Based on gambling and random chance, not reality.
Metropolitics	Structure and process of urban political styles of government. Illustrates complex urban issues and coalition building. Complex structural problems and not easy for all students to grasp.
Parksburg	Illustrates political bargaining and power bases of various groups. Weak on political concepts. Emphasis on political skill rather than problem resolution.
Taxes	Illustrates the influence and effect of taxation on individuals and why certain tax policies are preferred by some individuals and not by others. Little exploration of coalition building or system process.
Women's Liberation	Illustrates issues involved in women's liberation movement and six groups active in the movement. Illustrates women's concerns. Ignores power, legal positions, or other issues.

playing the detailed model. Greater role detail assures the instructor that students will envision the roles they are playing in relation to certain prescribed factors rather than relying on their imaginations, which may run rampant or land them in confusion with the result that their role play is ineffective or restrained.

Some of the simulations model structure and process. *American Government* focuses upon congressional and elective structures. *Budget* details steps in budget formulation. *City Hall* looks at local decision-making processes. *Metropolitics* centers upon urban structural styles. Finally, *Election* provides primary and general election processes. The others either do not explore structure and formal process or do so only incidentally.

This brief discussion of content, role playing, coalition building or bargaining behavior, and structure and process in the models re-emphasizes the necessity that the instructor select a model designed to give students the experience desired. All of the models have strengths and limitations as instructional units. No single model necessarily provides the subject matter (content) and all the structural bargaining and role-playing experiences one might seek. The instructor may need to decide whether subject matter is more important than exposure to structure, whether exposure to structure is more important than experience with detailed role playing, and so forth.

All of the models require careful debriefing and discussion of the limits of the model and the divergence from the "real world." It might be argued that simulation models are not useful, given both their limitations in modeling replications of the real world and the lack of precision that exists in the data of the social sciences. However, one might as well reach the opposite conclusion. Given the complexity of social and political structures, processes, and behaviors, simulation models permit students to explore some factors of a decision-making experience in a laboratory setting and to analyze the model for player behavior, strategy, and success or failure in achieving goals. Students may compare and contrast this laboratory experience with the local community. Any differing or complementary factors they find between the model and the real world may not only suggest the model's level of reliability, but also may foster greater understanding of the complexities that occur in the real world.

The use of simulations for instruction should be encouraged for this opportunity to provide greater research and analysis of the "real world." Students comment frequently that a simulation experience has been valuable and has promoted their interest, additional research, and an appreciation of real world factors and complexities. If this is a frequent instructional result of the use of simulations as an educational technique, their limitations are more than compensated.

Sources

American Government
William Coplin and Leonard Stitelman
1969

Science Research Associates, Inc.
155 North Wacker Drive
Chicago, Illinois 60606

Amnesty
Charles L. Kennedy
1974

Interact
Box 262
Lakeside, California 92040

Budget
Charles L. Kennedy
1973

Interact
Box 262
Lakeside, California 92040

City Hall
Judith Gillespie
1972

Ginn & Co., Xerox Education Center,
P.O. Box 2649
Columbus, Ohio 43216

Election
Joseph and Marlene Young
1974

Educational Games Co.
363 Peekskill, New York 10566

Hat in the Ring
Paul A. Theis & Donald M. Zahn
1971

Division EMC Corp.
180 East Sixth Street
St. Paul, Minnesota 55101

Metropolitics
R. Garry Shirts
1970

Simile II
P.O. Box 910
Del Mar, California 92014

Parksburg
Faith Dunne
1971

Olcott-Forward, Inc.
Pleasantville, New York 10570

Taxes
David Rosser
1974

Interact
Box 262
Lakeside, California 92040

Women's Liberation
David J. Boin
1974

Edu-Game
P.O. Box 1144
Sun Valley, California 91352

ECOLOGY/LAND USE/POPULATION GAMES AND SIMULATIONS

An Evaluation

by Larry Schaefer, Harry O. Haakonsen, and Orah Elron

"Environmental and/or population education programs embody several features which make games and role-playing attractive as educational techniques. These include: the strong attitudinal component of program objectives; the complexity of the systems with which these programs deal; and the resistance to change which pervade the social and economic fabric of many societies."[1] Furthermore, ecology, land use, and population issues are naturals for simulation. Ecology is the study of the totality or pattern of *relations* between organizing populations and their environments. Land use issues focus on the cybernetic *relationships* of ecology and the relationship of social, economic, cultural, and normative factors to the environment. In a parallel pattern, simulations as a learning method foster the experiencing of *interrelatedness.*

The Purpose of this Essay

In this essay, we will compare, contrast, and evaluate eight simulation games according to content, process, and suitability for classroom use as well as for their instructional value. The games focus on land use, population, and ecology. The essay elaborates several citeria for game selection ranging from clarity of rules to flexibility and realism. We conclude with a summary of the games' strengths and limitations in a wide range of categories.

Definition

Several types of teaching techniques have evolved to bridge the gap between the conceptual level, at which traditional teaching occurs, and real-life situations. They can be arranged on a continuum as follows:

conceptual description (i.e.) lecture discussion)	case study	open-ended case situation	role play	simu-lation	real life

Editor's Note: Listings for all simulations and games discussed in this essay are in the ecology section, except for *Baldicer* in social studies, and *Predator* in science.

As one moves along this continuum from conceptual level toward real life, increasing amounts of realism enter the learning situation. Experiences most often classified as educational "games" fall into the categories of role play and simulation.

The *role play* emphasizes interpersonal exchange, with students assuming assigned roles that may differ considerably from their real-life roles. The role involves a situation that is only briefly outlined, and the role player has considerable freedom to elaborate on the role identity in interaction with other participants. Role play can be very emotionally involved, and it is particularly useful for learning about processes in human interaction and for considering the player's emotional responses to various situations.

A *simulation* is a more structured situation that duplicates certain real-world conditions but usually "telescopes" the time dimension. Participants play roles, but the teaching focus is shifted somewhat away from interpersonal interaction toward the issues and processes involved in the simulation and toward consideration of factors such as power, communication, persuasion, and planning strategy trade-offs. A simulation that achieves closure, usually through scoring or win criteria, comes closest to the popular concept of a "game."

Games enable us to compress the time and space components of real-life situations. This is especially important in ecological and land use issues in which the real-life consequences of decisions are often felt only years later. A "game" gives a player a chance to learn the consequences of actions and situations without experiencing them. Not only can a considerable amount of cognitive information be transmitted but also attitudes and values can be clarified and modified. Through the simulation mode, students primarily strengthen process skills such as problem solving, planning, and decision making. These features contribute to make games, role playing, and simulations an attractive teaching technique in environmental education.

William Stapp, a leader in environmental education, has identified problem solving, values clarification, and community problem solving as three crucial processes in environmental education. These three processes involve identifying and evaluating alternatives and decision making. It is clear that these same skills and processes are important components of educational "games." In a symbiotic manner, educational "games" are a powerful method for implementing environmental education.

Criteria for Selection

The eight games we have selected deal with ecology, population, and land use issues. Each of the games has been observed, played, and evaluated in several classroom situations. As of January 1979, all are available commercially or from government agencies. All are priced at or below $25. Most important, the selection of games provides a cross-section of exposure in the areas of ecology, population, and land use. These games not only develop content and process skills but they also are fun to play.

TABLE 1 The Simulation Game

Simulation Games	Purpose	Game Effectiveness in Achieving Purpose	Summary
Algonquin Park	To develop a land use plan that will provide maximum benefit to the people of a region.	Very Effective	Players role play a multiple-use conflict about a parcel of wilderness land. Cottage owners, miners, lumbermen, and resource planners consider the economic and ecological factors required to produce a land-use plan that the Ministry of Natural Resources will accept.
Balance	To create awareness, transmit knowledge, and stimulate research on environmental problems from the point of view of the family.	Very Effective	Representing members of four contemporary families, players have to choose among economic, social, and family interests versus ecological values. They experience how age and family role influence their decisions.
Baldicer	To experience the interdependence of the world economy.	Effective	Participants play the role of food coordinator for 150 million people. In each round, food coordinators must earn enough food for their people. After a period of training, bargaining, and purchasing, coordinators calculate status and determine whether they are "alive" and may continue to play. Those who "die" take on a new role as the World Conscience, seeking to dramatize the plight of hunger to remaining players.
Ecopolis	To generate a better understanding of environmental problems, in particular pollution and over-population. To motivate real-life projects that attack environmental problems.	Effective	Representing citizen groups, players solve ecological problems of land use in their community. They are also encouraged to engage in real-life ecological projects that earn them bonus "gasps" in the game.
Land Use	To expose players to a wide range of potential uses for land and the types of information that should be considered in selecting a single, best land-use plan.	Effective	Players as members of a community develop a land-use plan for one square mile of land on the urban fringe. Players select alternative use patterns, develop plans, make proposals to the community, and vote on a plan.
Planet Management	To develop an understanding of the issues involved in managing a simplified ecosystem.	Satisfactory	Players as managers of the imaginary planet Aariun try to improve living conditions by investments in projects represented by perforated computer cards. Thus, they observe realistic results of their choices.
Predator	To understand food chains and food webs.	Satisfactory	Based upon the natural food, relationships in a temperate zone forest, predator is a card deck with each card representing a plant or animal. The several possible games are based on old favorites such as War, Concentration, and Rummy.
Redwood Controversy	To involve players as representatives of interest groups in designing a land-use plan and arguing the case before a legislative body.	Very Effective	Players assume roles of senators and expert witnesses in debating and determining the size and use patterns for the Redwood National Park.

PRELIMINARY CONSIDERATIONS

Purpose and Summary Description

Table 1 states the purpose of each game reviewed in this essay, indicates its effectiveness in achieving that purpose, and gives a short summary description. Using Table 1, one can make a preliminary determination of which game or games accord with one's purposes and educational objectives.

Age Level, Group Size, Playing Time

One of the first considerations in selecting a game for classroom use is its suitability for the age, ability, and size of the class or group. A second criterion is how the time a game requires matches the time available. The playing time of a game can be influenced by many factors: the teacher's experience in game management, student exposure to simulation experiences, modifications of the game to meet special class needs, and the diverse factors that influence student ability to adjust to new educational strategies. Table 2 summarizes these considerations and should aid in choosing games appropriate for the situation.

Complexity

In Table 3, we rank the simulations according to their complexity.

The simplest games require little reading and a minimum of prior classroom preparation. The more complex games require better mathematical, reading, organizational and intellectual skills. We rank *Balance* as complex because it involves two different simulations, complex multiple roles, organizational skills, and outside projects and research. *Predator* has a fairly simple format based on familiar card games such as war, rummy, and concentration. *Baldicer* is rated moderately complex because it requires mathematical skills.

TABLE 3 Complexity

Simple	Easy	Moderate	Complex
Predator	*Land Use* *Ecopolis*	*Redwood Controversy* *Baldicer* *Planet Management* *Algonquin Park*	*Balance*

ISSUES

One of the most important concerns involved in choosing an educational game is the nature of the issues it presents and resolves. The range of issues in the eight games falls into four major categories: ecological, economic, social, and environmental quality. Table 4 indicates what issues are dealt with in each game and whether a given issue is of primary, secondary, or optional significance. Few ecology, population, or land use games are single-issue games.

SCOPE AND DEPTH OF CONTENT

Another concern in the choice of an educational game is the scope and depth of the content. This may include already existing material in game manuals and additional information. One may wish to modify the instructional value of the game. Does one want a game that primarily teaches content, or would one prefer a game that develops process skills such as problem solving? Does one prefer games that include all the required content and information for play, or would one rather have a skeletal structure that can be strengthened with

TABLE 2 Some Basic Considerations

Simulation Game and Publisher	Grade Level	Playing Time	No. of Players	No. of Groups
Algonquin Park David Dagg Pembrook, Ontario	grade 7 – adult	7 – 8 50-minute periods	5 – 35	5
Balance Interact	grades 7 – 12	15–20 days in periods of 45 min. (there is also a 3 hour version)	18 – 35 (class)	4
Baldicer John Knox Press	grade 7 – adult	6 – 8 25-minute rounds or 1½ – 3 hours	10 – 20, more with modifications	1
Ecopolis Interact	grades 5 – 8	12–15 days in 45-minute periods	18 – 35 (class)	2
Land Use U.S. Forest Service	grade 9 – adult	2 – 3 hours	15 – 20	2 – 6
Planet Management Houghton Mifflin	grade 7 – adult	2 or 3 45-minute periods	2 – 10	5 maximum
Predator Ampersand Press	grades 4 – 8	30 minutes per round	3 – 49	–
Redwood Controversy Houghton Mifflin	grade 7 – adult	1½ – 2½ hours	optimum 21 range	–

TABLE 4 Issues

Issues	Algonquin Park	Balance	Baldicer	Ecopolis	Land Use	Planet Management	Predator	Redwood Controversy
ECOLOGICAL								
ecology	P	P	O	P	S	P	S	P
land use	P	P	O	P	P	S	—	P
population	P	P	P	P	P	P	S	P
extinction	O	S	—	S	O	—	S	S
food chain and web	S	—	—	—	O	—	P	S
ECONOMIC								
agriculture/forestry	S	S	P	O	O	P	—	P
industrial	S	O	—	O	O	P	—	—
research	—	—	—	—	O	P	—	—
income	S	O	S	—	—	P	—	—
benefit/costs	P	O	—	—	S	—	—	P
development costs	P	O	S	—	S	—	—	S
utilities	S	P	—	O	S	—	—	—
SOCIAL								
education	O	—	—	—	O	P	—	—
family interests	—	P	—	S	—	—	—	—
generation gap	—	P	—	—	—	—	—	—
public health	—	—	—	—	O	P	—	—
recreation	S	S	—	S	O	S	—	P
food and hunger	—	O	P	—	—	P	—	—
ENVIRONMENTAL QUALITY								
benefit/cost analysis	P	O	—	—	P	S	—	P
water quality	O	—	—	O	O	—	—	O
air quality	O	P	—	O	O	—	—	O
mass transit	—	O	—	S	O	S	—	—
land use	P	P	—	P	P	P	—	P
parks	P	S	—	P	O	—	—	P
housing	S	O	—	O	O	—	—	P

KEY:
P — primary issue
S — secondary issue
O — optional issue at discretion of game director.

inclusions that are germaine to the instructional situation? Table 5 should help one determine what games involve scope and depth appropriate for the situation.

Baldicer is rated very low in scope and depth of content, but we give the game an overall rating of very good. This apparent discrepancy can be explained in light of *Baldicer's* strength in simulating broad concepts such as inequitable distribution of world resources and the interdependence of the world economy. These concepts are difficult for most students to grasp, but *Baldicer* illustrates them clearly and forcefully.

Two games that demonstrate broad scope and good depth of content are *Algonquin Park* and *Redwood Controversy*. Both are extraordinarily useful to classes or groups exploring the issues of park formation or land-use decision making. Each game contains excellent background material on the issue of preserving open space, endangered plants, or areas and on strategies for designing and evaluating alternative land use plans. *Land Use*, produced by the U.S. Forest Service, is much shallower in content, though well designed. As a result, it can be played in less time and may function mainly as an introduction to the land use decision-making process.

TABLE 5 Scope and Depth of Content

Very Low (no material)	Low	Moderate	High
Baldicer	*Predator*	*Balance*	*Algonquin Park*
	Land Use	*Redwood Controversy*	
	Planet Management	*Ecopolis*	

Content Learning, Awareness, Social Skills

What are a teacher's aims in choosing a particular game? Is the teacher particularly interested in content learning, in fostering social skills and interactions, or in creating ecological awareness as learning motivation? Table 6 should help teachers in this aspect of game selection.

ROLE PLAYING

With the exception of the card game *Predator*, all the ecological games we review here involve some kind of role playing. Several factors concerning role play might influence

TABLE 6 Educational Objectives

Simulation Games	Content Learning	Awareness Creating	Developing Social Skills
Algonquin Park	S	P	P
Balance	S	P	P
Baldicer	—	P	P
Ecopolis	P	P	P
Land Use	S	P	P
Planet Management	—	P	—
Predator	P	—	—
Redwood Controversy	S	P	P

Key:
P = primary
S = secondary

TABLE 7 Role Types

Highly structured roles (lengthy and detailed)	Unstructured roles (brief-only individual task given)	The same task is given to all players
Ecopolis	*Baldicer*	*Planet Management*
Balance	*Land Use*	
Algonquin Park		
Redwood Controversy		

TABLE 8 Roles in Simulations

Simulation Games	Roles
Algonquin Park	Committee on Land Use, Forestry Management Company, Land Developers and Planners, Lawyers representing cottage owners, Mining Company
Balance	Four families, each composed of father, mother, young adult and adolescent
Baldicer	Food coordinators representing areas of the world having different quantities of food; two game directors and members of the world conscience
Ecopolis	Interested citizens: business persons, construction worker, outdoor person, elementary, high school and college students, teacher, homemaker, county commissioners, politician, secretary, writer
Land Use	Decision makers, county Board of Commissioners, member of a committee, members of interest groups and citizens
Planet Management	Representatives of management committee
Predator	No roles
Redwood Controversy	Senators, conservation group members, lumbering interest group members

the teacher's decision about whether a game is appropriate to a particular class or group of players.

Structured Versus Nonstructured Roles

Roles assigned in education games may be highly detailed or skeletal in structure. A highly detailed role (generally consisting of two or more descriptive paragraphs) requires the player to identify with a clearly outlined personality. Such roles influence players' attitudes during the entire game—the opinions they express, their votes, and their final decisions. The main purpose of this kind of role playing is to foster better understanding of other people's attitudes in a given situation based upon their backgrounds and needs. For instance, adolescents might understand their own parents better after having played the father, the mother, and the young adult in the game of *Balance.* A player may learn to understand the great difficulty politicians have in coping with the task of decision making by playing *Algonquin Park* or *Redwood Controversy.*

In contrast to these very detailed roles are the short, unstructured roles that outline in only a few sentences the tasks with which the players have to cope. In these cases, players are asked to fulfill given roles according to their own interpretation. Their opinions and decisions in the simulation are strongly influenced by their own ideas and attitudes. Here, the main purpose of the role play is to foster a high level of responsibility that in real life would rest on the shoulders of a few representatives of society. For example, the players in *Baldicer* are in charge of the food supply for a whole country, while in *Planet Management* they are responsible for establishing the living standards of an entire planet. Making a decision in one field often leads to most unexpected results in another. This kind of role play often develops an understanding of the complex interaction of ecological processes.

Students playing the four games in the first group learn to understand the factors that influence the decisions of those people they represent in their roles, whereas the players in the two other groups of games have to cope with their given task according to their own interpretation of the role. To understand these differences better, we will look at the roles these games include (see Table 8).

Interest Factor of Roles

Many people, and surely all children, like to role play. But is the role given in a certain game unduly restraining a player's self-expression? Are the roles in a game interesting to the players? Is it possible to adapt the roles in a game to the problems of one's own community so players can cope with a

TABLE 9 Role Interest, Involvement, Flexibility

Simulation Games	Interest Factor	Personal Involvement	Flexibility to Adaptation
Algonquin	very interesting	high	high
Balance	good	adequate	moderate
Baldicer	very interesting	moderate	moderate
Ecopolis	good	adequate	moderate
Land Use	very interesting	moderate	moderate
Planet Management	very interesting	moderate	low
Predator	no roles	—	—
Redwood Controversy	most enjoyable	high	high

situation much closer to real life? How involved do players become in the roles they play? Table 9 will help one deal with these factors.

In all the games, each player receives an individual role description and task, but in some of the games there is also a group role to be played. Individual role descriptions only are given in *Balance, Ecopolis,* and *Planet Management.* Players have individual and group roles in *Algonquin Park, Baldicer, Land Use,* and *Redwood Controversy.*

Changing Roles

In some games the player's role stays the same throughout the game. In three, however, players change roles at different game sessions. This enables them to identify with different aspects of the same question and thus helps them be more tolerant and understanding of the roles people fulfill in real life.

RULES

The clarity, organization, and completeness of rules are very important factors in the choice of a game. When rules are hard to understand or when too much time is spent in explanations, participants' interest and motivation may be lost. If it is only the instructor who has to read a lot of pages to prepare for the game, this does not necessarily interfere with the participants' readiness to play. A careful analysis of rules is a prime responsibility of the game director; their quality may not be apparent at first glance.

TABLE 10 Role Changes

Same Role for Entire Game	Roles Change During Game
Planet Management	*Ecopolis*
Baldicer	*Balance*
Redwood Controversy	*Land Use*
Algonquin Park	

TABLE 11 Clarity and Organization of Rules

Rules for the Instructor	Number of Pages Devoted to Rules	Rating of Clarity, Organization, and Completeness: Inadequate	Adequate	Good	Outstanding
Algonquin Park	4			X	
Balance	40		X		
Baldicer	15				X
Ecopolis	34			X	
Land Use	4				X
Planet Management	2				X
Predator	3		X		
Redwood Controversy	3				X
Rules for the Players	**# of Pages**	**Inadequate**	**Adequate**	**Good**	**Outstanding**
Algonquin Park	1			X	
Balance	10			X	
Baldicer	3		X		
Ecopolis	12			X	
Land Use	1			X	
Planet Management	1	X			
Predator	—				
Redwood Controversy	2				X

The instructor rules for *Baldicer, Land Use, Planet Management,* and *Redwood Controvery* are outstanding for clarity and explicitness. Other games might take some experience to play fluently.

It is advisable to leave a certain amount of flexibility to each group of players, encouraging them to adapt the existing rules of a game to their specific needs. The amount of instruction a game gives to the *players* is sufficient if it allows them to play and evaluate their steps without requiring the instructor to reread detailed instructions at each step.

Table 11 shows the number of pages devoted to game rules and our evaluation of the clarity and complexity of the rules for the instructor and for the players.

DEBRIEFING

One of the most important aspects of any well-designed educational game is the debriefing session. During this period, game participants step out of their simulated roles and settings into their normal group roles. They are then in a position to discuss roles they have played, decisions they have made, techniques that proved to be successful or ineffective, conflicts between personal values and assigned roles, anxieties created by game participation, the validity and realism of the simulation, and possible extensions of the game to solving group problems. These and a host of other matters may become focal points for maximizing the instructional value of participating in the gaming session.

Structuring the debriefing session is an important task of the game director, and a good set of guidelines can be very useful in starting and maintaining this important activity. Some games provide very thorough debriefing guidelines. Others give a limited number of suggestions. Still others give no guidelines at all. Table 12 summarizes our rating of the debriefing guidelines provided with each game in terms of quality and quantity.

TABLE 12 Debriefing Guidelines

Inadequate (no debriefing questions)	Adequate	Thorough
Predator	*Algonquin Park*	*Baldicer*
	Balance	*Planet Management*
	Ecopolis	
	Land Use	
	Redwood Controversy	

STRATEGIES FOR IMPROVING CLASSROOM IMPLEMENTATION

In reviewing and selecting games, one will undoubtedly be thinking of ways to integrate the game into the curriculum. As a part of the process it would be wise to consider strategies for improving classroom implementation.

Baldicer is a solid, usable game. In playing it, however, we have found that it helps to place posters describing game rules and activities around the room as convenient reference points for students. We have also found that the enthusiasm of students often makes it difficult to get them settled quickly so play can move to the next round, and so one poster reads "NOISE POLLUTION FINE–5 BALDICERS." A quick reference to the noise pollution poster stops unwanted discussion immediately.

Game designers and producers frequently include large quantities of detail in the rules section. As a result, students must often spend long periods of time analyzing the rules before play can proceed. In many cases, however, the rules can be summarized in a few sentences, and if these abbreviated rules are put on a card in the game box, students can start playing much more rapidly. As their need to know about the intricacies of the rules develops, they can make coherent use of the detailed rules provided in the game.

Slides, films, and other media are often effective devices for introducing or following up a simulation activity. *Redwood Controversy* and *Algonquin Park* can be enhanced greatly by exposing students to the recreation, industrial, commercial, and environmental components of the areas in which the games are set.

As one reviews games, keep thinking of ways to improve implementation of the gaming session. Creativity can improve all games, and some that are basically mediocre may in fact be used in ways that make them worthwhile instructional activities.

FLEXIBILITY

Of the multitude of factors that affect the part a game plays in the instructional process, one that can be extremly significant is game flexibility.

A rigidly structured game must be played with a set number of players and specific rules and resources. A flexible game permits the director to modify one or more of the following components: roles, rules, issues, locations, number of players, age of players, and game resources. Many of these factors have been described in previous tables in this essay.

TABLE 13 Game Flexibility

Inflexible	Slight Alteration Possible	Flexible
Planet Management	*Ecopolis*	*Algonquin Park*
	Predator	*Balance*
	Redwood Controversy	*Baldicer*
		Land Use

As a case in point, such a simple factor as the number of players that can use a board game can dramatically affect the game's usefulness. If a game requires exactly four players, it functions only with multiples of four, no matter how many sets of the game are available. In a group of 15 students, 3 would be excluded from playing. However, if a game can incorporate 4 or 5 players, then a greater number can play. As shown in Table 2, the game we review in this essay can accommodate a wide range of player numbers.

Table 13 classifies the eight games into three groups according to their flexibility.

A game director can change games such as *Baldicer, Balance,* and *Land Use* a great deal to make them more effective for use by a specific group. Real geographical areas and their food-famine profiles can be substituted for the nonspecific regions assigned in the *Baldicer.* In addition, special outside assignments can help players (that is, countries) work hard and increase their available food. *Land Use* can be restructured to address local land use planning problems so it becomes extremely relevant to local community groups in and out of the classroom.

A game that is flexible provides a creative director with a wide range of instructional options. Modifications are time consuming, but the result of creative change can frequently benefit the players, the director, and the group as a whole.

SKILLS, ACTIVITIES, AND INTERACTIONS

In choosing a game, one will need to determine whether it offers practice in skills and activities that suit one's educational purposes. As Table 14 shows, the eight games vary considerably in the number and intensity of skills their participants practice.

Sequence of Activities

Most simulation games consist of several rounds of activities that evolve during play. Table 15 presents an overview of the sequence of activities and interactions in each of the eight ecological games we are reviewing.

RESOURCES AND SCORING

Game resources are often an important consideration in determining which games are appropriate for a group. The handling of money, tokens, and other player resources may prove to be a worthwhile and interesting experience for some students and an impediment to progress for others. One should select a game carefully to see that the resources it includes maximize player involvement.

TABLE 14 Skills and Activities

Skills/Activities	*Algonquin Park*	*Balance*	*Baldicer*	*Ecopolis*	*Land Use*	*Planet Management*	*Predator*	*Redwood Controversy*
Information Processing								
analysis	S	S		S	P	P		
gathering information	S	P	S	S				S
planning goals, strategies	P	S		S	P	P		P
extensive reading and research	S	O		S	O			O
individual problem solving	O		P		O			O
Proposals and Lawmaking								
designing/revising proposals	P	S	S	S	P			P
promoting proposals	P	P	S	P	P	P		P
opposing proposals	P	P		P	P	O		P
writing proposals	P	O	O		P			P
voting on proposals, laws	P	P		P	S			P
Group Activities								
group problem solving	P	S	P	S	P			P
debate persuasive argument	P	P		P	P			P
speaking group presentation	P	S		P	P			P
small group discussion	P	P	S		P			P
class large-group discussion	O	S		P	O			O
challenging before a group	S	S						S
Human Relations								
competition	P		P		P			P
persuading/influencing	P	P	P	P	P			P
bargaining/negotiating	P		P		O			O
helping/supporting	S	S	P	S				O
interviewing	O	O		O				
electing	P				S			
forming coalitions	O	S	S	S	O			O
Role Playing								
specific individual	S	P		P				P
work function			P	S		P		P
group member	P	P			P			P
switching roles		P		P				
Resource Management								
survival		S	P	S		S		
maximizing resources	P	S	P	S	P			P
managing resources	P	S	P	S	P	P		P
Evaluating (Exclusive of Debriefing)								
self-evaluation		S		S		P		
evalution of peers		P		P	S	S		
Miscellaneous Activities								
map reading	S	S		S	S			S
land-use planning	P	P		P	P	P		P
graphing/charting	O				O	P		O
decision-making skills	P	P		P	P	P		P
mathematics skills	P	S	P	S				

Key:
P — Primary
S — Secondary
O — Optional

TABLE 15 Sequence of Activities

Simulation Game	Sequence of Activities
Algonquin Park	(1) Read roles, (2) intergroup discussion, (3) prepare basic strategies, (4) prepare presentation of strategies, (5) prepare defense of plant, (6) interest groups report, (7) government committee completes land-use policy, (8) secret election of committee, (9) assess each group's fate, (10) discuss activities and goals
Balance	(1) The valley 150 years ago, small animals, simulations, (2) players receive roles and student guide, (3) introduce real-life ecology projects for earning "gasps," (4) family groups meet, (5) air pollution: large-group representations and votes, class family discussion, real self discussion, work on research and planning, (6) water and power: same sequence as step 5, (7) land use: same sequence as step 5, (8) debriefing, evaluation, discussion
Baldicer	Six to eight rounds of six phases each: (1) work period, (2) social forces factor cards (chance), (3) tally period, (4) planning, (5) purchasing and trading, (6) assessment; debriefing after last round
Ecopolis	Ecosystem of Ecopolis and survival simulation; start of real-life projects for earning "gasps"; Crisis 1. Our Land in Crisis—group one discusses, group two evaluates; Crisis 2. Over-Population—group two discusses, group one evaluates; forum preparations to choose main ecological problem; forum day; scoring and essay evaluation
Land Use	(1) Individuals list possible uses for hypothetical parcel of land, (2) decision-making groups determine specific use in stated categories, (3) groups evaluate impact of their proposed uses, (4) groups develop proposal and presentation for their concepts, (5) Board of Commissioners holds public hearing for proposal review, (6) evaluate proposals and approve one, (7) group discusses plans, processes, and ramifications of proposals
Planet Management	Open class discussion of projects that would make Earth a better place to live; ten rounds repeating following steps: (1) read project guide to choose up to five projects (invest up to $10 million), (2) record projects and money spent, (3) chance factor (coin flipping) leads to card computer, (4) evaluate computerized results, (5) record results on game ledger and/or graph; final group evaluation of "winner" and class discussion on future of life on planet Clarion
Predator	Several formats available (similar to rummy, concentration, solitaire); timed version: (1) deal cards, (2) first player chooses another and asks for showdown, (3) at signal, both lay down one card, (4) if one card "eats" another, that player takes eaten card, stand-off if neither "eats other," (5) steps 2-4 repeat with next player, (6) also possible to challenge player and take specific card, (7) repeat process for 30 minutes
Redwood Controversy	(1) Introduction and background, (2) role assignment, (3) witnesses testify before Senate, (4) Senators make first statement of intention to vote, (5) recess for discussion, (6) voting (2/3 to pass), (7) discussion and re-voting until one proposal passes, (8) debriefing and discussion

Scoring is a major concern in game selection. Some games are designed to foster intense competition among players; each move advances or retards the individual's or group's progress in accumulating the points, money, or tokens that place them in a specific position relative to other players or groups. Winning or losing becomes the focus of action for every individual. Competition and scoring can be a strong motivating device for most students, but we advise caution. One should determine the impact of winning or losing on the prospective players.

Many games make use of feedback schemes, group discussions, and proposal development to stimulate and maintain player interest. The way these games are constructed makes explicit individual and group scoring techniques unnecessary as motivational devices.

TABLE 16 Resources and Scoring

Simulation Games	Resources	Scoring
Algonquin Park	No resources	No scoring
Balance	GASPS (Goal and Satisfaction Points)	Score sheet; individual and family scoring personal evaluation. Highest wins.
Baldicer	BALDICERS (Balanced Diet Certificates). Units of food unequally distributed. Food Machines and Super Food Machines	Winners are survivors.
Ecopolis	GASPS (Goal and Satisfaction Points)	Score sheet; individual scoring highest GASPS win.
Land Use	No resources	No scoring
Planet Management	BOX — monetary resources, project cards	At the end of fifty years (ten rounds) teams assess their own progress and compare their planetary status with other teams. Class decides criteria for winning.
Predator	No resources	Player with most cards wins.
Redwood Controversy	No resources	No scoring

Table 16 summarizes the resources and scoring systems the eight environmental games use.

MODEL VALIDITY

Because simulation games are, by definition, models of real-life situations, it is important to consider the validity of a game and its model before selecting it as an instructional resource. One can determine model validity by analyzing the game's structure to see how closely it fits reality. Are the roles of significant interest groups and decision makers realistically presented (as discussed in Table 9)? Are decision-making activities structured so the game's approaches to problem analysis, proposed design, and evaluation fit the schemes used in real environmental decision-making activities? Is chance inserted into the game in a way that represents the nonpredictable influences that affect decisions, or is it used inappropriately to accomplish some other purpose such as speeding up the game? Do the game's techniques enhance player activities as realistic, informed, active participation, or do the techniques distort the process that is being simulated?

Table 17 describes our review panel's interpretation of the overall validity of the games.

The games we rated high for model validity are simulations that involve students in problem analysis, proposal design, small and large group discussions, persuasive argumentation, and evaluation of alternative problem solutions. The issues they address are true to life, and the simulations are carefully designed and effectively implemented.

TABLE 17 Model Validity

Adequate	High
Baldicer	*Algonquin Park*
Land Use	*Balance*
	Ecopolis
	*Predator**
	Redwood Controversy
	Planet Management

*Predator is a game with high model validity, but it is not a role-playing simulation game.

Land Use is rated as adequate because of the limited amount of information it presents to describe the situations it simulates. By adding background material and role information, one can improve the validity and realism of the game. *Baldicer* is also rated adequate because the assignment of resources (Baldicers) identifies only three discrete groups, and this does not reflect the food distribution patterns of the real world. In addition, all roles are identically and simply defined. As the game develops, population units faced with starvation enter into negotiation and active, sometimes hostile, schemes to obtain resources. Finally, the game play leads to outcomes that fail to reflect the wide spectrum of real-world responses to insufficient food resources. This weakness in model validity is balanced by the high interest it generates and by the high level of participant interaction.

PACKAGING

Games come in assorted colors, sizes, boxes, folders, envelopes, manuals, and kits (and unfortunately, packaging often influences sales more than game content). In our opinion, games should be packaged in strong containers. Playing boards, role cards, and other resources should be durable and of good quality. Costs should be kept to a minimum so that supplying games for an entire group does not become a monumental financial problem. *Planet Management,* for example, has an 89-page Data Book which should be available in multiple copies to evaluate each round, but only one copy comes with the game set.

When buying a game, one generally expects to find all of the required playing materials in the purchased package. If any material must be duplicated, the package should include high-quality masters for reproduction. If additional materials are needed, the instructor's manual should include a prominent and concise statement of what they are and where one can get them.

Table 18 provides information on the packaging, cost, completeness, and durability of the games we have reviewed.

TABLE 18 Packaging

Game	Kind of Packaging	Cost Per Kit or Book	Completeness	Durability & Reusability
Algonquin Park	Large envelope containing all game materials – kit.	$15.00, $10.00 if prepaid	complete	average
Balance	Manual	$14.00	requires duplication	average
Baldicer	Kit – Box containing manuals, work yield inventories, score sheets and posters	$25.00	complete	average
Ecopolis	Manual	$14.00	requires duplication	average
Land Use	Manual	$.95	requires duplication	average
Planet Management	Kit – Box containing manual, punch cards, computer data.	$22.56	complete	high
Predator	Deck of cards	$5.00	complete	average
Redwood Controversy	Kit – Box includes game manual, role cards and maps	$14.25	complete	average

CONCLUSION

When all is said and done, game selection is a highly individualized process. A game that proves eminently successful with one group can be a crashing bore with another. The differences frequently center around the entering levels of the players, the relevance of the game to situations the players encounter, and the players' needs, interests, and abilities.

Spend an hour or two analyzing a game with the criteria we have outlined in this essay before deciding to use it. Then compare the needs and abilities of the group with the instructional possibilities the game provides. Areas that need modification should be identified and changed. Then devise a classroom implementation scheme and work out debriefing guidelines in detail. After one has taken these steps, the characteristics of the game will be familiar, and productive play may begin.

Ultimately, objective evaluation and subjective evaluation meet. There are aspects of games that can be analyzed in detail, while others simply elicit like or dislike. Table 19 summarizes the strengths and limitations we have found in these eight simulation games on population, ecology, and land use.

Note

1. Katherine Finseth and Larry Schaefer, "Simulation Gaming and Personal Decision-Making," *Current Issues in Environmental Education–II,* ed. R. Marlette, (Columbus, Ohio: ERIC Center for Science, Math and Environmental Education, 1976), p. 35.

TABLE 19 Strengths and Limitations

Simulation Game	Strengths	Limitations
Algonquin Park	wide range of well-defined roles can be played with a wide range of players well-defined content and information for play develops environmental awareness strong emphasis on the individual role in decision making lengthy and detailed role description	requires seven to eight periods for play
Balance	model validity high role play interesting flexibility high changing roles enable players to see different points of view real-life projects are part of the game	rules and introduction lengthly evaluation complicated takes several weeks to play simulation within simulation might not appeal to students
Baldicer	rules and time limits well defined very effective in creating awareness effective in illustrating interdependence of world economy flexible high player involvement game director's rules outstanding debriefing guidelines thorough	does not work in small groups depth of content for players roles limited requires four to five class periods or two hours to play
Ecopolis	model validity high adaptable to actual problems interesting challenges real-life projects are part of the game	takes several weeks to play evaluation complicated simulation within simulation might not appeal to students
Land Use	realistic involvement in developing land use plan simple to introduce and play very effective in creating awareness effective development of social skills rules and time limits well defined	requires minimum of fifteen players limited exposure to the content of land use management limited information on situations game purports to simulate
Planet Management	rules very clear and simple model validity high game graphs helpful in post-game discussion	flexibility low use of card computer slows game considerably role play limited
Predator	single-issue simulation inexpensive	single-issue simulation limited player involvement
Redwood Controversy	strong emphasis on ecological issues effective in developing environmental awareness teaches social skills through active player participation well-defined, interesting roles rules are well written and easily followed excellent model validity	requires a large group of players—optimum twenty-one many players only marginally involved in several stages of simulation

ECONOMICS GAMES AND SIMULATIONS
An Evaluation

by Victor Pascale and Theodore Ehrman

INTRODUCTION

If you want to provide economic information to a group, whether it is in a classroom, business, or governmental setting, consider doing so through the highly motivational instructional method of a simulation game experience.

Goals

Simulation gaming provides content information in the form of facts, principles, concepts, and generalizations. It also generates process information in the form of decision-making, analytical and social skills, as well as attitudes. You must be able to choose the particular simulation or game that will best achieve the specific goals and objectives you desire. My aim is to provide you with the kinds of information you need about ten economic simulation games that will help you make the best possible selection for your purposes. I shall review and evaluate these exercises, providing data about the following factors: producers, cost, game description, level of game play, number of players, playing time, physical setting, prior preparation time, subject matter, goals and objectives, social skills, game administration, limitations, and special recommended adaptation. To conclude, I will provide a general evaluation of each simulation game.

What are Economic Simulation Games?

An economic simulation game is an activity in which participants interact within an artificially produced environment that re-creates some aspect of economic reality. The participants, or players, assume the roles of individuals or groups that exist in a particular economic system. Each of the players or groups is striving to achieve a goal. The rules of the simulation game limit the range or possible responses of each player, as well as the number of players. Essentially, the ultimate objective of the players is to reach their goals.

Uses of Economic Simulation Games

These simulation games are extremely useful for business, government, and social organizations, as well as at the levels of secondary, college, and adult education. The simulation games may be used for one or more of the following purposes:

(a) *as a body of knowledge,* to teach specific facts and principles;
(b) *as a training technique,* to prepare individuals in the dynamic process of decision making, analyzing, creating feedback, and planning;

TABLE 1 The Ten Economic Simulations

Simulation Game	Summary Description
Stocks and Bonds (board game)	The purpose of this game is to create the economic atmosphere of Wall Street. Participants have the opportunity to invest in ten securities and attempt to accumulate as much wealth as possible. The participant who amasses the greatest fortune is considered the winner.
Stock Market (board game)	Simulates the stock market exchange, including all dynamic economic forces. Margin buying, conversion of securities, selling short, and warrants are essential parts of this game.
Economic System (simulation game)	Simulates an economy that involves levels of production, consumption, investments, and natural resources. Players represent workers, manufacturers, and farmers. Participants attempt to satisfy their needs for food and manufactured goods.
Tightrope (simulation game)	Simulates the economic stability and growth of a nation. The participants form a group named the Economic Advisory Council, which makes decisions dealing with the fiscal and monetary policies of that country.
Executive Decision (board game)	Simulates the economic conditions that involve corporate management and big business. Players enact the roles of top-level executives who make decisions on what to produce. Prices are determined by the law of supply and demand.
Railroad (simulation game)	Simulates the railroad competition of the 1870s as four independent lines compete for the transportation business of carrying ore from the mines to the mills.
The National Economy (simulation game)	Simulates a nation's economic atmosphere and conditions. Each industry attempts to maximize profits and minimize losses.
Collective Bargaining (simulation game)	Simulates the atmosphere and conditions that prevail during the negotiation of a new labor contract. In the bargaining

TABLE 1 The Ten Economic Simulations

Simulation Game	Summary Description
	sessions, union representatives and management deal with such economic issues as (1) wages; (2) additional holidays and vacations; (3) greater pension, health insurance, and other fringe benefits.
The Firm (simulation game)	A simulation of the economic conditions and problems encountered in running a business. Each store owner's major objective is to accumulate as high a profit as possible to out-distance the competition.
Scarcity and Allocation (simulation game)	Simulates conditions on a desert island on which participants try to increase their standards of living. The island contains the basic necessities, such as vegetation and wildlife.

TABLE 2 Suitability for Group Levels

Simulation Game	Business/ Government Organizations	College	High School
Stocks and Bonds	not suitable	least suitable	most suitable
Stock Market	not suitable	suitable	most suitable
Economic System	not suitable	least suitable	most suitable
Tightrope	most suitable	most suitable	suitable
Executive Decision	not suitable	not suitable	most suitable
Railroad	least suitable	least suitable	most suitable
The National Economy	least suitable	most suitable	most suitable
Collective Bargaining	most suitable	most suitable	most suitable
The Firm	least suitable	suitable	most suitable
Scarcity and Allocation	not suitable	not suitable	most suitable

(c) *as a role experience* that allows an individual to undertake an event or position to gain greater awareness and knowledge; and
(d) *as a means of further developing social skills* in terms of communication, cooperation, competition, and understanding social rules.

Table 1 presents a summary description of each simulation game.

Criteria For Selection

Simulation games should provide the kind of knowledge that participants can apply to new situations and problems; game operators expect participants to acquire concepts and generalized techniques for dealing with new problems and new materials. Instruction in economics should be directed toward increasing the store of dependable and reliable knowledge, encouraging the formation of concepts, developing generalizations, and enhancing participants' social skills so that when they encounter a new problem or situation, they will be better able to select an appropriate technique for analyzing it and will bring to bear the necessary information (facts, concepts, and generalizations) for a solution. The simulation games selected had to incorporate the above factors, but include economic behavior and goals as well. The standards for inclusion permitted a wide enough range, but also enough focus to serve specific training and educational needs.

A final selection criterion was price. All the simulation games are available commercially for less than $50.00 as of January 1980.

BASIC CONSIDERATIONS

Age Level, Group Size, Playing Time

You should carefully consider the age, ability, size, and available time of your group before selecting a simulation game. Tables 2 and 3 provide the basic information you will need to make a wise selection.

Instructional Purpose and Preparation

Table 4 indicates the stated instruction purpose of each simulation game and how successfully it is carried out. Table 5 indicates the amount of time and the knowledge of subject matter essential in preparing the various simulation games. This table is divided into two components—the preparation involved for the instructor and for the participants.

Complexity

Table 6 ranks the simulations according to their complexity as follows:

TABLE 3 Simulation Games: Some Basic Considerations

	Playing Time (Hours)			No. of Players			
Simulation/Game	Minimum	Best	Maximum	Minimum	Best	Maximum	No. of Groups
Stocks and Bonds	1-1/2	3	4-1/2	2	24	30	6-8
Stock Market	1	8	12	1	6	12	6
Economic System	7	20	25	9	12	26	13
Tightrope	3	5	8	1	15	no limit	—
Executive Decision	1	3	5	2	6	12	no limit
Railroad	1	1-1/2	2	5	15	24	4
The National Economy	2-1/2	3-1/2	5	any number	any number	any number	3
Collective Bargaining	1	2	3	6	24	36	6
The Firm	3	5	7	4	20	no limit	no limit
Scarcity and Allocation	1	2	3	any number	any number	any number	no limit

TABLE 4 Stated Purpose and How Well Carried Out

Simulation/ Game	Purpose	Carried Out
Stocks and Bonds	To understand the terminology and differences between types of stocks	highly successful
Stock Market	To understand the different securities and how prices are determined	highly successful
Economic System	To understand the basic economic questions of what to purchase, how and for whom to produce	highly successful
Tightrope	To understand how the government uses fiscal and monetary policy	moderately successful
Executive Decision	To understand the laws of supply and demand	highly successful
Railroad	To understand certain competitive and oligopolistic monopolies	highly successful
The National Economy	To understand the components of national income and how they are related	highly successful
Collective Bargaining	To understand the process of collective bargaining	highly successful
The Firm	To understand the explicit and implicit cost of a business enterprise	somewhat successful
Scarcity and Allocation	To understand how people handle their alternative and opportunity costs	moderately successful

Easy: one-dimensional with little or no verbal skills; simple and repetitive tasks, with single goals and objectives and limited subject matter.

Moderate: two-dimensional approach, involving some verbal skills; greater task difficulty, with more goals and objectives and increased mastery of subject matter.

Complex: multidimensional approach involving abstract strategies, detailed planning, high verbal skills, and comprehensive subject matter, with many goals and objectives.

RULES

When rules are direct and concise, the game operator can use time effectively and efficiently so participants' motivational and involvement levels are not only maintained but, in many cases, increased. Tables 7 and 8 rate the ten economic simulation games for clarity and organization.

ECONOMIC ISSUES

In considering a particular simulationgame, you should carefully examine it to see what issues are involved in order to fully determine its suitability to the needs and goals you have established. Table 9 indicates the major economic issues in each simulation game.

TABLE 5 Instructor and Participant Preparation

Simulation/ Game	Instructor	Participants
Stocks and Bonds	(1 hour) knowledge of stocks and securities	no prior knowledge required
Stock Market	(1 hour) knowledge of the stock and securities exchange	(1 hour) knowledge of the stock and securities exchange
Economic System	(1 hour) knowledge of factors of production, law of supply and demand	no prior knowledge required
Tightrope	(5 hours) knowledge of fiscal and monetary policy	(10 hours) need instruction in fiscal and monetary policy
Executive Decision	(1 hour) knowledge of factors of production, law of supply and demand	no prior knowledge required
Railroad	(1 hour) knowledge of formation of large-scale corporations	no prior knowledge required
The National Economy	(1 hour) knowledge of national income determination	no prior knowledge required
Collective Bargaining	(1 hour) knowledge of labor relations	(1 hour) knowledge of labor relations
The Firm	(1 hour) knowledge of resources, capital, and management	(1 hour) knowledge of resources, capital, and management
Scarcity and Allocation	(1 hour) knowledge of factors of production, economic system	no prior knowledge required

TABLE 6 Complexity

Easy	Moderate	Complex
The Firm	*Collective Bargaining*	*Tightrope*
Scarcity and Allocation	*Stock Market*	*The National Economy*
Executive Decision	*Economic System*	
Railroad		
Stocks and Bonds		

TABLE 7 Clarity of Rules

Simulation/ Game	Less Than Adequate	Adequate	Good	Outstanding
Stocks and Bonds			X	
Stock Market				X
Economic System				X
Tightrope	X			
Executive Decision			X	
Railroad				X
The National Economy		X		
Collective Bargaining			X	
The Firm			X	
Scarcity and Allocation			X	

TABLE 8 Organization of Rules

Simulation/Game	Less Than Adequate	Adequate	Good	Outstanding
Stocks and Bonds				X
Stock Market				X
Economic System				X
Tightrope	X			
Executive Decision				X
Railroad				X
The National Economy			X	
Collective Bargaining			X	
The Firm			X	
Scarcity and Allocation			X	

TABLE 9 Economic Issues

Simulation/Game	Economic Issues
Stocks and Bonds	Law of supply and demand Stocks and securities
Stock Market	Law of supply and demand Stock Market
Economic System	*Law of supply and demand Production and consumption Investment and improvement Natural resources
Tightrope	Fiscal policy Monetary policy
Executive Decision	Factors of production Law of supply and demand
Railroad	Monopoly Free enterprise system
The National Economy	Investment Production Consumption Inflation Employment
Collective Bargaining	Role of labor Role of management
The Firm	Management and profits Production and distribution
Scarcity and Allocation	*Factors of Production

*Indicates, when more than one game treats an issue, which game does so most effectively.

CONTENT

A simulation game may be content or process oriented. You must decide whether your primary purpose is to create a content-centered learning experience involving facts, concepts, and so forth, or process-centered, involving attitudes, decision making, and analytical and verbal skills.

Table 10 classifies each simulation game according to its orientation toward content, process, or a combination.

Instructors establish goals or specific information that they want to impart. Table 11 indicates the information content of each simulation game to help you choose among these exercises.

TABLE 10 Content/Process Orientation

Content-Centered	Combination	Process-Centered
Tightrope	*Economic System* *Collective Bargaining* *Stocks and Bonds*	*Stock Market* *Executive Decision* *Railroad* *The National Economy* *The Firm* *Scarcity and Allocation*

TABLE 11 Content Information

Simulation/Game	Topics Incorporated	
Stocks and Bonds	Security Bond Interest Stock Dividend Split	Round lot Bear market Bull market Market price
Stock Market	Common stock Preferred stock Bonds Warrants Broker Buying on margin	Selling short Selling long Stock splits Bull market Market price
Economic System	Farmer Worker Manufacturer Factory Commissioner Mine owners Production	Consumption Resources Investment Marginal utility Marketing Tax revenues
Tightrope	Federal Reserve Act Income tax laws Corporation tax laws Great Depression World War II 1950s economic conditions 1960s economic conditions Monetary policy Fiscal policy Gross national product Money supply Whole price index	
Executive Decision	Raw materials Finished product Corporation Executive Labor Management	Law of supply Law of demand Equilibrium price Price fluctuation Decision making
Railroad	First transcontinental railroad completed Hepburn Act Transportation Act Interstate Commerce Act of 1887 Vanderbilt Competition Profit Price determination Market Fixed costs Rebates Variable costs	
The National Economy	Capitalism Production Profit	Consumer goods Producer goods Optimum level

TABLE 11 Content Information

Simulation/Game	Topics Incorporated	
	Loss Unemployment Inflation Investment Luxury goods	Deficit spending—Keynesian Economics Total Annual Income Law of diminishing goods
Collective Bargaining	Strike Fringe benefits Wage package Pension Insurance	Collective bargaining Work stoppage Business stewards Negotiations One-year contracts
The Firm	Firm Profit Loss Salary Inventory Entrepreneur Buyer Short run Long run	Marginal cost Production Public accountant Manager Law of supply Law of demand Fixed cost Variable cost Financing loans
Scarcity and Allocation	Savings Opportunity cost Capital formation	Scarcity Allocation

ECONOMIC AWARENESS

Each of these simulation games imparts, in some form, a certain economic awareness. A game operator who has identified this aspect of the learning experience is better equipped to handle it in a way most comfortable for and beneficial to participants.

Table 12 lists the major economic awarenesses that are set forth in each simulation game.

ROLES

Role playing in these particular simulation games ranges from a single- to a multidimensional approach. The single-dimensional roles involve brief data in the form of name, age, employment, and position; that is, they are roles in name only, and are not very flexible. Multidimensional roles incorporate duties, responsibilities, tasks, and so forth. These roles have depth and are extremely flexible. Participants' reactions to the various roles in the ten simulation games ranged from high to minimal involvement. The instructor's role may vary from supervisor to supervisor/participant to full participant.

Table 13 indicates the various roles for each game, the flexibility level of the roles, and the role of the instructor.

Interest Factor of Roles

Besides identifying the roles, one has to ask how well they motivate participants, whether they are challenging, whether they sustain interest, and whether participants get involved in them. Table 14 rates the motivational level of each of the economic simulation games.

TABLE 12 Economic Awareness

Simulation/Game	Economic Awareness
Stocks and Bonds	– Individuals play an important part in our stock market. – Information concerning a stock or bond affects its price.
Stock Market	– Stock Market is largely based upon the principles of supply and demand. – Stock Market is a risk-taking endeavor where wealth is both accumulated and lost.
Economic System	– An aggressive approach may be rewarding in that it facilitates the achievement of certain goals. – The development of certain strategies may lead to more effective play.
Tightrope	– The influence of governmental spending policy (Keynesian Approach) in bolstering the economy. – The influence of a "laissez-faire" approach and how it affects the economy.
Executive Decision	– The effect of decision making upon the success and failure of a business enterprise. – The role the "profit motive" plays in generating economic activity. – Prices can be influenced by the law of supply and demand.
Railroad	– Competition may or may not be beneficial to a society.
The National Economy	– The need for governmental regulation to protect the economy.
Collective Bargaining	– Each person at the bargaining table brings personal values and views that influence the course of negotiations. – Cooperation and compromise are essential of negotiation.
The Firm	– The free enterprise system can operate effectively when the factors of production are employed wisely. – Management skills are needed for a successful business venture. – A business firm is an extremely complex enterprise.
Scarcity and Allocation	– People will use "saved time" for many different economic purposes.

SKILLS, ACTIVITIES, AND INTERACTIONS

Table 15 presents the various skills and activities employed in each simulation game. These elements of simulation gaming play a major part in shaping the level of participant involvement.

Interaction

A simulation game may generate an interaction based on the individual with the system, one-to-one competition, partnership, or group relationship. You must determine the kind

TABLE 13 Roles

Simulation/Game	Participant Roles	Flexibility	Instructor's Role
Stocks and Bonds	Investors, bankers, stock clerk	Moderate	Supervisor
Stock Market	Investors, bankers, stock clerk	High	Supervisor
Economic System	Farmer, manufacturer, worker, mine operator, road and police commissioners	High	Supervisor
Tightrope	Representatives of economic interests	Moderate	Supervisor/Participant
Executive Decision	Manufacturers, bankers, market rater	Low	Supervisor
Railroad	Railroad executives, mine operators	Moderate	Supervisor/Participant
The National Economy	Representatives of consumers, luxury goods, capital and/or producer goods	High	Supervisor
Collective Bargaining	Representatives of management, shop stewards	High	Supervisor
The Firm	President, vice-president, accountant	Low	Supervisor
Scarcity and Allocation	Roles are Open-ended determined by players	Low	Supervisor

TABLE 14 Motivational Levels of Roles

Low	Average	Above Average	High
Tightrope	*The National Economy*	*Railroad*	*Economic System*
Scarcity and Allocation	*The Firm*	*Collective Bargaining*	*Executive Decision*
			Stocks and Bonds
			Stock Market

of interaction that best suits your needs. Table 16 indicates the various kinds of interactions that each simulation game may create. Some employ several types of interactions.

Competition, Compromise, Cooperation

Competition, compromise, and cooperation are three types of interaction that appear frequently during play of the economic simulation games. Some use more of one type than another. Table 17 will help you determine the levels of competition, compromise, and cooperation in each game.

SCORING

The end results of a simulation game may be determined in a quantitative or qualitative manner. Quantitative scoring determines a winner according to which participant or participants have the highest numerical scores in points, money, and so forth.

Qualitative scoring determines a winner by descriptive terms; that is, through achieving a certain goal or goals. You must determine which scoring system is more suitable to your needs.

REALITY AND SIMULATION GAMING

The economic simulation games attempt to capture different aspects of our economic system. Actually, each is a re-creation, involving an activity in which participants interact within an artificially produced environment. When that environment closely resembles the actual situation, we may say that it closely depicts reality. In selecting a simulation game, you must determine your own needs concerning "reality testing." If your priority demands a very close resemblance between a simulation game and reality, you should choose one with a high rating; whereas if your priorities are otherwise, a low- or medium-rating game may be suitable. Table 19 indicates the ratings of the economic simulation games and their similarity to economic reality.

DEBRIEFING OR POSTGAME DISCUSSION

The debriefing of any simulation game is extremely important. It is generally during this period that participants have the opportunity to analyze, reflect on facts and concepts, and make generalizations about their experience. A good guideline for this process can be extremely useful for the game operator. Table 20 indicates the ratings of the debriefing guidelines of each simulation game.

FLEXIBILITY

If a particular economic simulation game does not meet your specific needs, certain ones lend themselves more to adaptation and modification than others. The process of adaptating and modifying a game involves designing changes to add a new dimension or reshape an old one to make a simulation game a more effective instructional device. Table 21 indicates the flexibility of each simulation game, as well as some suggestions for possible adaptations or modifications. You may also wish to develop your own changes.

MATERIALS AND COST

The economic feasibility of a simulation game is extremely important. Will the dollars you spend produce the greatest return for you? Table 22 indicates not only the kind of packaging, cost per simulation game, completeness, and durability, but an evaluation rating of best buys.

TABLE 15 Skills and Activities

	Stocks and Bonds	*Stock Market*	*Economic System*	*Tightrope*	*Executive Decision*	*Railroad*	*The National Economy*	*Collective Bargaining*	*The Firm*	*Scarcity and Allocation*
Information Processing										
analysis	VS	VS	VS	VS	VS	VS	VS	VS	VS	VS
gathering information				S				VS		
planning goals, strategies	VS	VS	VS	VS	VS	VS	VS	VS	VS	VS
research				VS				VS		
extensive reading				VS				VS		
news reporting/writing										
individual problem solving	VS		VS		VS	VS	VS		VS	VS
Group Activities										
group problem solving			VS	VS		VS	VS	VS		
debate				VS				VS		
public speaking				VS				VS		
small group discussion			VS	VS		VS	VS	VS	VS	
class/large group discussion				VS				VS		
Human Relations										
competition	M		VS		VS	VS		VS		
persuading/influencing		S	VS	VS			VS	VS		
bargaining/negotiating			VS	VS		VS	VS	VS		
helping/supporting	M		VS	VS		VS	VS	VS		
interviewing										
listening	VS	VS	VS	VS	VS			VS		
hiring and firing/supervising			VS							
deceiving; exploiting			VS					VS		
forming coalitions			VS	VS		VS		VS		
Role Playing										
specific individual			VS					VS		
a group member			VS	VS		VS	VS	VS	VS	
switching roles			VS							
Resource Management										
record keeping		VS	VS				VS		VS	VS
survival			VS				VS			VS
chance	VS	VS								
maximizing resources	VS	VS	VS	VS	VS	VS	VS	VS	VS	VS
managing resources	VS	VS	VS	VS	VS	VS	VS	VS	VS	VS
Evaluation (exclusive of debriefing)										
self-evaluation	VS	VS	VS	VS	VS	VS	VS	VS	VS	VS
evaluation of peers	VS	VS	VS	VS	VS	VS	VS	VS	VS	VS

Key: VS—very significant
S—significant
M—minor

TABLE 16 Kinds of Interaction

Simulation/ Game	Individual	One-to-One	Partnership	Group
Stocks and Bonds	X			
Stock Market	X			
Economic System	X	X	X	X
Tightrope				
Executive Decision	X			
Railroad			X	X
The National Economy				X
Collective Bargaining		X	X	X
The Firm				X
Scarcity and Allocation	X			

TABLE 17 Competition, Compromise, Cooperation

Simulation/ Game	Competition	Compromise	Cooperation
Stocks and Bonds	Low	Low	Moderate
Stock Market	Low	Low	Moderate
Economic System	High	High	High
Tightrope	Low	High	High
Executive Decision	Moderate	Low	Low
Railroad	High	Moderate	Moderate
The National Economy	Moderate	High	High
Collective Bargaining	High	High	Moderate
The Firm	Moderate	Low	Low
Scarcity and Allocation	Low	Low	Low

TABLE 18 Resources and Scoring

Simulation/Game	Participants' Starting Resources	Scoring
Stocks and Bonds	Equal amounts of money to invest	All stocks and bonds are converted into money. The participant with the greatest amount wins.
Stock Market	Equal amounts of money to invest ($2000 and ten shares of blue chip and speculative stock)	At the end of the twelfth turn, the winner is the player who has the greatest net worth.
Economic System	Capital Money Food Consumer products for each player	The participant with the most points at the end of the game wins.
Tightrope	No economic resources	Success is measured qualitatively in terms of improved decision making, as well as the various fiscal and monetary movements made.
Executive Decision	Equal amounts of money to invest	Winner is determined by converting all assets into money, and the participant with the largest sum is declared winner.
Railroad	Unequal amounts of capital, land, and money	The group that accumulates the most money wins.
The National Economy	Equal amounts of capital for each role	A winner or loser may be determined quantitatively by profits.
Collective Bargaining	Unequal amounts of capital	A winner is determined by dollars, which are converted into points.
The Firm	Equal amounts of labor, capital, and land	A winner is determined by net worth.
Scarcity and Allocation	Equal amounts of capital, labor, and food	The participant who has accumulated the most hours saved or materials gained through the twelve rounds is declared the winner

TABLE 19 Similarity to Economic Reality

Simulation/Game	Low	Moderate	High
Stocks and Bonds		X	
Stock Market		X	
Economic System			X
Tightrope		X	
Executive Decision	X		
Railroad			X
The National Economy		X	
Collective Bargaining			X
The Firm	X		
Scarcity and Allocation	X		

TABLE 20 Debriefing Guidelines

Totally Inadequate (No debriefing questions)	Inadequate (Few questions)	Adequate (8-12 questions)	Thorough (Several pages of questions and guidelines)
Stocks and Bonds	*Scarcity and Allocation* *Executive Decision* *Collective Bargaining* *The Firm* *The National Economy*		*Tightrope* *Stock Market* *Economic System* *Railroad*

TABLE 21 Flexibility and Possible Modifications

Simulation/Game	Inflexible	Slight Alterations Possible	Flexible	Highly Flexible	Adaptation/Modification
Stocks and Bonds			X		*Prices*: Instead of being predetermined, prices could be determined by law of supply and demand.
Stock Market				X	*Prices* could reflect current stock market prices.
Economic System				X	*Structure of Classroom*: Open area with 7 large tables, since movement of participants is essential. *Winners*: Instead of having one winner, there can be three, one in each category (farmer, worker, manufacturer.) *Taxes*: Incorporate an income tax system, using the various forms, so partipants experience a new dimension of learning.
Tightrope		X			*Expansion*: Changing general economic advisors to specific advisors (such as labor, manufacturing, banking) would create distinct points of view, which will not only be represented but pursued, thus presenting many new areas and challenges to participants, requiring them to utilize more of their analytical abilities. *Point System*: The instructor may assign numerical point values to Complex decisions may warrant more points, while less complex decisions receive a lower point value. This would give participants a quantitive measure of the effect of their decisions.
Executive Decision				X	*Time Flexibility*: Changing time sequence may add dimensions to this particular simulation to face participants with new problems, crises, and conflicts, which generate new approaches in their decision making.
Railroad		X			*Expansion*: Increase number of railroad groups from 4 to 8. Stimulate further competition in game by using the concept of oligarchy.
The National Economy		X			
Collective Bargaining		X			*Price*: Financial figures could be revised to reflect current economic price and cost levels.
The Firm	X				
Scarcity and Allocation	X				

CONCLUSION

Now that we have reviewed the different aspects of each simulation game, you may wish to view the total evaluation of each game, as shown in Table 23.

Sources

Economic Decision Games—
Collective Bargaining,
The Firm'
The National Economy
Scarcity and Allocation
1968
Erwin Rausch

Didactic Systems, Inc.
Box 457
Cranford, New Jersey 07016
$13.30

Economic System
James S. Coleman and
T. Robert Harris
1969

Bobbs-Merrill, Education Division
4300 West 62nd Street
Indianapolis, Indiana 46268
$25.00

Executive Decision
Sidney Sackson
1971

TABLE 22 Packaging

Simulation/Game	Kind of Packaging	Best Buy ($ Value)	Cost per Kit or Book	Completeness	Durability and Reusability
Stocks and Bonds	Kit	Fair	$12.00	Complete	High
Stock Market	Kit	Best	$12.00	Complete	High
Economic System	Kit	Best	$25.00	Complete	High
Tightrope	Pamphlet	Good	$2.00	Must duplicate materials	Average
Executive Decision	Kit	Fair	$12.00	Complete	High
Railroad Game	Booklet	Best	$1.25	Must duplicate materials	Low
The National Economy	Pamphlets for 30	Good	$13.30	Must duplicate materials	Low
Collective Bargaining	Pamphlets for 30	Good	$13.30	Must duplicate materials	Average
The Firm	Pamphlets for 30	Low	$13.30	Must duplicate materials	Average
Scarcity and Allocation	Pamphlets for 30	Low	$13.30	Must duplicate materials	Average

TABLE 23 Total Evaluation of Simulation Games

	Stocks and Bonds	*Stock Market*	*Economic System*	*Tightrope*	*Executive Decision*	*Railroad*	*The National Economy*	*Collective Bargaining*	*The Firm*	*Scarcity and Allocation*
Content	2	2	2	0	2	2	2	1	1	1
Roles	1	(1)	2	1	1	1	1	2	1	1
Rules	2	2	2	1	2	2	1	2	2	2
Skills	1	1	2	0	1	1	2	2	0	1
Validity	2	2	2	1	1	2	2	2	0	1
Flexibility	1	2	2	0	1	1	1	1	0	0
Debriefing	0	2	2	1	0	2	0	0	0	0
Realistic	2	2	2	0	2	1	0	1	1	1
Durability	2	2	2	1	2	0	0	1	1	1
Expense	1	2	2	2	1	2	2	2	0	0
TOTAL	14	18	20	7	13	14	11	14	6	8
Overall Rating	Good	Outstanding	Outstanding	Adequate	Good	Good	Good	Good	Adequate	Adequate

Key: 0 = Inadequate
1 = Adequate
2 = Excellent
(1) = Information unavailable or irrelevant

Avalon Hill
4517 Harford Road
Baltimore, Maryland 21214
$12.00

Railroad
Fred W. Newmann and
Donald W. Oliver
1967

Xerox Education Publications
245 Long Hill Road
Middletown, Conneticut 06457
$1.25

Stock Market
1970

Avalon Hill
4517 Harfor Road
Baltimore, Maryland 21214
$12.00

Stocks and Bonds
Hooper, Brooks, and
Hoffman
1964

Avalon Hill
4517 Harford Road
Baltimore, Maryland 21214
$ 12.00

Tightrope
Larry Baskind, Ann Buddington,
Melvin Erickson, Ala Kay Hill,
and Geraldine Murphy
1969

El Paso Public Schools
Purchasing Agent
Box 1710
El Paso, Texas 79999
$2.00

END OF THE LINE
An Evaluation

Andrew W. Washburn and Maurice E. Bisheff

INTRODUCTION

If you are looking for a social simulation game designed to explore the experience of giving "help" to and attempting to get help as an elderly person, you must consider *The End of the Line* in your selection. It is a well-designed simulation game offering participants opportunities to directly experience the dilemmas of building helping relationships in complex social systems.

The End of the Line is about social service agencies and the elderly. Its stated intent is "to give participants a feel for what it's like to grow old and what it's like to help people who are growing old." The game is not designed for the elderly themselves but for those working with the elderly. Its goal is to sensitize service providers to the aging process and to explore more effective strategies for providing services to the elderly.

The game was designed by Frederick L. Goodman, Professor of Education at the University of Michigan, in collaboration with the university's Institute of Gerontology and Extension Gaming Service. An earlier simulation of Goodman's, *The Helping Hand Strikes Again,* assigns a task to a small group of players who then require help. The remaining players are assigned responsibility for helping the first group, and the dynamics of helping are reflected in transactions between the two groups.

In 1971, the Institute of Geronotology received a grant to provide a series of workshops on the issues and techniques of providing protective services for the elderly. Drawing on the *Helping Hand* game, Goodman collaborated with the institute to design *The End of the Line* to simulate various processes of aging as they affect social service agency attempts to aid the aging citizen. The exercise was subsequently field tested and fine tuned by the Extension Gaming Service. This done, the often thankless task of documenting the work was accomplished.

The simulation's instruction manual is now available ($5.50) from Publications, Institute of Gerontology, University of Michigan, 520 E. Liberty, Ann Arbor, Michigan 48109. A complete kit with instructions is available ($75.00) from the Institute of Higher Education Research and Services, Box 6293, University, Alabama 35486.

The overall purpose of *The End of the Line* is to give players an opportunity to "emphathize" with the role of an aging person and/or service provider personnel and with the personal and organizational dilemmas involved in such processes. The game provides participants with opportunities to experience and possibly alter their attitudes toward senior citizens and the complexities of providing assistance to them. *The End of the Line* provides a powerful, emotionally charged experience that leads to rich postgame discussions.

YOUR FIRST CONTACT WITH THE END OF THE LINE: THE MAJOR ELEMENTS

Imagine, if you will, showing up for your SOC 101 class one day, having missed the prior class and not knowing what is to take place. As you enter the room you immediately sense that something different is about to happen. The classroom has been rearranged and new objects are present (see Figure 1).

A number of chairs are scattered in the center of the room. Each one has materials on it. On closer inspection, you find the materials are a wooden tray with white cord(s) attached, paper clips, a scoresheet of some sort with a six-by-six matrix on it, and a pencil attached to it with a string. Tables labeled AGENCY X, AGENCY Y, and AGENCY Z surround the chairs. Fellow classmates are taking their seats. At the front of the room, the instructor finishes placing masking tape on the floor in the shape of a two-by-six matrix. Beside the instructor on the blackboard is a large six-by-six matrix (see Figure 2) with numbers in it. "What have I gotten myself into?" you ask, just as your eye catches the poster on the side wall that reads:

TO STAY ALIVE:

1. 1 RE-LINE (MARKED IN FOLDER)
2. 1 MONEY CLIP (RESTING IN PROPERTY TRAY ON CHAIR)
3. 1 WHITE ROPE (CONNECTING PROPERTY TRAY TO CITIZEN)

Editor's Note: *The End of the Line* is listed in the human services section.

FIGURE 1 Suggested Room Set Up

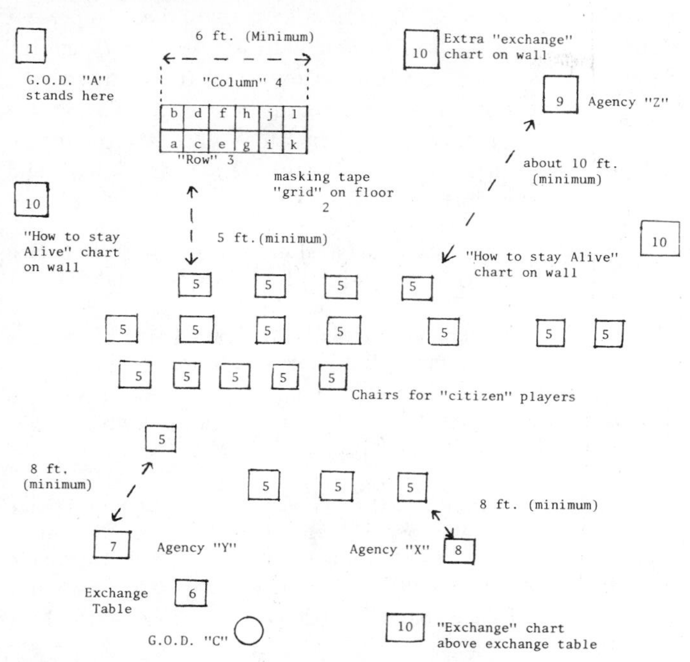

NOTE: Three-quarters of the "Citizens" (numbered 5) must be able to reach 1, 2, 3, and 4 with their white cords. At least one-half of the "Citizens" should be able to reach the exchange table 6 with their white cords.

FIGURE 2 Blackboard Matrix

	0	1	2	3	4	5
0	0,+1	+3, 0	-3,-4	+2,+3	-3,+3	-4,-3
1	+1, 0	+2,+2	-1, 0	+1,+3	-3,-3	0,+1
2	0,+3	-2,-2	0,+1	+2,-2	-1,-1	-1,+2
3	+3,+3	+1, 0	+3,-2	-1,-1	+3,+3	-1,+2
4	0, 0	-1,-1	+3,+2	+3, 0	+2,-2	-1,+2
5	0,+2	+2,+4	-2,-3	+3,+1	-4,-4	+2,-1

"Oh my god!" you whisper to yourself. As you prepare to dash through the exit, the instructor speaks: "I am the Game Overall Director. That's G.O.D. And this is *The End of the Line.*" Curiously unwilling to move at this point, you listen as the Game Introduction begins.

Game Introduction: All About Staying Alive

The instructor explains what a gaming-simulation is and the purpose of this particular exercise. You begin to get excited about becoming involved in this simulation and perhaps learning more about the elderly and how to help them. You are told that the people in the chairs are "citizens" of the community. Their objective is "to stay alive." The wall chart lists the requirements for staying alive. The wooden trays are "property trays." One money clip must remain in the tray at all times. Citizens may not move their chairs and must hold their own personal matrix forms. The form indicates the number of personal relationships a citizen possesses. Citizens are allowed to get up and move around as far as the white and blue cords attached to their wrists allow. One white rope is necessary to stay alive.

The Game of Life

To obtain more resources, players may choose to go with a partner to the Matrix Game (The Game of Life) at the front of the room. There they show G.O.D. from zero to five fingers at the appointed time. One partner "calls" the row, the other calls the column. Depending on the values in the cells of the large matrix, each partner gains or loses re-lines, which are half-inch orange slashes penned by G.O.D. in designated boxes on the personal matrix form.

The Grim Reaper Enters

The Grim Reaper (another G.O.D.) circulates around the community requiring citizens to pick a card. The cards may cause the player to give up one or more re-lines, one or more clips, or one or more white ropes. Through this process, simulated "aging and disability" occurs. Another card may cause the Grim Reaper to make slits in the personal matrix form. Enough slits or cuts and the string attached to the pencil is severed. This is important because the pencil can no longer be used by the citizen to keep track of changes in the large matrix at the front of the room. Without the pencil, the citizen begins to lose his/her ability to "remember" the reality of the matrix game, and "senility" sets in.

The Helping Agencies

The players at the Agency tables have different tasks. They are attempting to help the elderly citizens in any way they

FIGURE 3 The Game Artifacts: Their Meaning and Use*

WHITE CORDS
represents maximum physical mobility

PROPERTY TRAYS
represents minimum property resources of each player

PAPER CLIPS
represents fluid financial resources (money)

RE-LINES
represents the abstract ability to relate or to form relationships

BLUE CORDS
represent purchasable mobility

PENCIL
represents the ability to remember

ORIGINAL MATRIX SHEET
the original "state of the world" is depicted in the form of the Matrix

BOARDS
represents the ability to remember data; and a place to record re-lines

*Adapted from Frederick L. Goodman, The End of the Line, Resources in Aging, No. 4 (Ann Arbor, Mich.: Institute of Gerontology, University of Michigan and Wayne State University, 1975).

can. Direct service agencies X and Y receive funding (paper clips) from Agency Z based on Agency Z's set of criteria. Agency helpers cannot move further from their tables than their blue cords allow. The blue cords are temporary and are collected at the end of each round.

The Exchange Table

Blue cords, clips, re-lines, and paste-on re-lines are available from the exchange table operated by yet another G.O.D. A sign hangs on the wall above the exchange table:

EXCHANGE

1 MONEY CLIP ⟷ 1 RE-LINE
(if purchased with another person)
OTHERWISE: 2 CLIPS ⟶ 1 RE-LINE
2 RE-LINES ⟶ 1 CLIP
5 MONEY CLIPS ⟶ 1 PASTE-ON RE-LINE
2 MONEY CLIPS ⟶ 1 BLUE CORD

The Game Begins

The instructor demonstrates with two volunteers how to play the matrix game and then answers questions from participants.

The first round begins. Many things are happening at once:

- Agency X uses its blue cords to visit Agency Z and ask for more money clips.
- Several pairs of citizens immediately rush to the matrix game and begin calling the "row" and "column" to gain the maximum relines.
- Other citizens, unable to reach the matrix game (due to lack of sufficient cords), begin yelling for assistance from Agency X ("more blue cords!"). Still others just sit, oblivious to it all and possibly confused.
- Agency Y begins asking citizens to identify their needs but is unable to understand clearly the citizens' responses. They stop listening and begin designing a form labeled "Request for Assistance."
- Two citizens who had persevered enough to prompt Agency X to lend them some blue cords finally make it to the matrix only to learn that their selected matrix cell has already been called by someone else.
- The Grim Reaper makes his (or her) rounds and asks a citizen to pick a card. The card is drawn, turned over, and two re-lines are lost as indicated by the card.
- In an effort to teach the exchange table to buy a paste-on re-line for a friend, a citizen accidentally goes too far and pulls her property tray off the chair. The Grim Reaper immediately confiscates the money clip that was in the tray and makes the citizen replace it before continuing.

After twenty minutes, G.O.D. announces the end of Round One. Blue cords (loaners) are collected by G.O.D. and returned to the Exchange Table. Then G.O.D. funds Agency Z for the upcoming round and displays a new chart indicating *changes* in the old matrix. G.O.D. emphasizes that these represent changes in the original chart and are *not* the new payoffs. Some citizens are too far back in the room to see the specific changes. Others are unable to record the changes, having lost their pencils to the Grim Reaper. They will either count on friends to help them or take a chance themselves. Now Round Two begins. This time:

- A citizen is accused of stealing money clips from citizens' property trays while they were at the Matrix Game.
- "Bad" numbers are occasionally called at the Matrix Game, causing one or both partners to lose re-lines.
- In an attempt to reach a citizen who needs help, a member of Agency X accidentally causes two citizens' cords to go taut and to pull property trays off chairs onto the floor.
- Agency Z announces the "Great Community" Assistance Program. It will begin as soon as Agencies X and Y submit proposals describing the problem, their approach, the anticipated results, and a detailed budget.
- Several citizens band together in an attempt to share resources.
- Agency Y announces it has the updated matrix for anyone who needs it. Unfortunately, the agency makes mistakes in the last column, causing several citizens to return from the Matrix Game with fewer relines than they had when they went to the Game. A "credibility gap" develops and spreads between Agency Y and the community.
- Several citizens, who had abundant resources from the beginning, continue to be satisfied with good conversation and observing what is going on around them. One citizen is then approached by the Grim Reaper who, by chance, takes the next-to-the-last white cord. Panic, frustration, then resignation set in as the citizen forms a new view of what is important.
- Unnoticed, one rather quiet citizen is approached by the Grim Reaper and loses all remaining white cords.

Round Two ends and each agency makes brief announcements to the community. Community members voice a number of complaints. Agency Z members announce their "Second Leap Forward" program and begin handing out money clips without any specified process, attempting to "get the clips into the hands of those who need it most." Those citizens who have done well at the Matrix Game are not interested in the agency proclamations except when agencies suggest they should share their wealth with less fortunate community members. Yelling and name-calling ensues. No one has yet noticed the "dead" citizen.

Round Three begins and the situation deteriorates still further:

- One citizen commits "suicide" because "there was no worthwhile reason to go on living."
- The Grim Reaper takes the lives of several other citizens.
- Only a few select "elderly" still have an up-to-date record of the matrix.

The round ends. Agencies make their new announcements. The community works with the agencies to design a plan for improving the community. The agencies reorganize, and Agency Z announces a cut-back in available funds. The instructor breaks in at this point and ends the game. Everyone breathes a sigh of relief and excitement is in everyone's eyes.

Debriefing . . . "It Is Over"

The instructor initiates a free exchange of views for about an hour on what just transpired. Players are eager to describe their experiences in the simulation and begin relating their experiences to areawide agencies on aging. Current legislation relating specifically to the needs of the elderly is raised and discussed in terms of its probable physical, social, economic, and psychological effect on the elderly in America. Comparative attitudes toward the elderly in other societies are discussed.

People still have things to say as the class ends. Conversations continue in small groups outside the classroom. People have gained insights. Next week's scheduled class discussion of social service delivery to the aged will be alive with thoughts resulting from the experience of *The End of the Line.*

SO YOU WANT TO RUN *THE END OF THE LINE:* VIEW FROM THE G.O.D.

After committing yourself (at least mentally) to presenting *The End of the Line,* it would be normal if the proverbial shiver ran down the G.O.D.'s spine. Presenting a simulation is risky. "It may not go well." "I may forget part of the rules." "I'm not *in charge* during the game." "What if the faculty hears that I'm playing games in my class?"

The Game Kit and Manual

Fortunately, it usually turns out fine, and the G.O.D. has a distinct advantage in increasing that probability with an excellent game director's manual and game kit. The kit contains *everything* needed to present the game: an operator's manual, wooden trays, wall charts, tape, cords, markers, scissors, clips, name badges, agency signs, and a supply of forms for multiple plays of the game. If they are lost, all the materials can be replaced without too much trouble. The only consumable materials are the personal matrix forms, and they can be easily reproduced. (One 8.5-by-11 sheet per citizen is needed for each presentation of the exercise.)

The operator's manual is a delight. Unlike some simulations, disseminated by word of mouth with minimal materials and an "afterthought" manual, *The End of the Line* comes with a comprehensive instruction manual that includes:

- an introduction to simulating gaming
- an overview of the game
- a description of models used to develop *The End of the Line*
- a complete description of the room setup, participant starting positions, game resources, and staffing tasks
- complete directions for introducing, running, and debriefing the game
- game variations and suggestions
- a complete description of all materials used during the simulation
- a selected gaming bibliography
- a selected bibliography in aging.

The game and the game materials are rich in symbolism. They add to the potential impact of the experience and are fully described in the operator's manual. The manual is well organized and well thought out—a blessing when one is presenting a simulation replete with highly complex artifacts and rules. Numerous photographs and charts assist in transferring the game designer's intent to the potential user.

Setting Up the Game

Although the manual suggests that a run of the game can be set up in half an hour, COMEX (Center for Multidisciplinary Educational Exercises, University of Southern California) has found that it takes considerably more time. One must first untangle cords from the previous game. Arranging the gaming room and making adjustments for the facility, the number of players, the specific objectives, and the context within which the game will be played can easily consume an hour.

Number of Players

In our opinion, the game is clearly designed for adults with serious interests in the study of social services, the elderly, or in helping relationships generally. Players need not possess any special knowledge before playing *The End of the Line.*

The optimal number of participants for running the exercise is 30: 20 become community members, while the remaining 10 are divided among the three agencies. In our experience, the minimal number is 20 players, with 17 in the community and 3 in the agencies. The designer's recommendations appear in chart 3. We have run the exercise with as many as 40 participants, and it has worked well. The difficulty with such a large group is that the logistics of game directing become very difficult. Some players lose their identity in a larger group, so that more than normal numbers withdraw and drop out.

We find that a critical mass (16 persons) is necessary in the community to allow players in these roles a reasonable chance of developing links with other players, the agencies, and the Grim Reaper. Depending on one's objectives in running the exercise, the number of agencies played (one, two, or three) will vary. We typically place 2 or 3 persons in each agency to help insure active discussion of strategies and plans.

FIGURE 4 Number of Players

ROLE	MINIMUM	IDEAL	MAXIMUM
Citizen	10	20	40
Agency Z	2 2	4	6
Agency X	2	3	6
Agency Y	2	3	6
TOTALS	16	30	58

SOURCE: Goodman (1975).

Directing the Game

When the game starts, we usually talk about what simulation games do and explain the purpose of the exercise. We distinguish between natural (from G.O.D.) and man-made laws (which the players can establish themselves). We then proceed to explain G.O.D.'s laws and emphasize the requirements of staying alive in the community.

Most players feel initially overloaded or overwhelmed by the complexity of the artifacts, so we try to get players into a round as soon as possible. Too much detail simply increases player anxiety. The game introduction should not last longer than *15 minutes,* including a physical demonstration (walk-through) of the Matrix Game with one participant at the front of the room.

We feel it is important that the G.O.D. have some experience in directing or facilitating simulation exercises and an adequate understanding of group dynamics. Because of its complexity, *The End of the Line* is not an exercise for newcomers to simulation gaming. We suggest an informal walk-through with friends to acquaint G.O.D. with the logistics and mechanics of running the exercise.

The Game Director and the Grim Reaper (Assistant G.O.D.) must work as a team. The Game Director works with the Game of Life and must manage matrix payoffs efficiently. To some extent the Grim Reaper sets the tone and pace of the game by the speed at which he or she moves through the community. We usually have a volunteer manage the Exchange Table.

Both G.O.D.s will be the targets and the receptacles of player activities designed to seduce them into compromising their roles. For example, players are likely to bargain with or flee from the Grim Reaper. To insure the smooth running of the exercise, it is important that Game Directors stay in role and complete their tasks each round, not getting enmeshed in the anxiety and excitement around them.

The G.O.D. must exercise judgment in halting the exercise. With an optimal size group (30) one to one and a half hours of playing time is usually sufficient. A minimum of four game rounds is usually required for a successful simulation. One clue for when to terminate the exercise is simply the number "dead" in the community. You may, for example, place the dead in a special hospital area, assign them (as we do) to agencies, or give them consultant roles for the rest of the game.

Debriefing

The exercise requires at least 45 minutes for debriefing following the exercise. The debriefing period allows participants an opportunity to review their experiences and to withdraw or "get out of the game."

One method of debriefing is to ask prepared questions of the citizens and agencies that link the game's objectives with real-life experience. Our preference is to, first, enable members to share their experience in an open-ended way. Later, or in subsequent class sessions, we begin to relate theories or concepts to the experience. A first question in the debriefing might be, "Well, what did you experience in this exercise?" This offers people an opening to express feelings. Normally, we do not comment but simply let others share their experiences without criticism or probing questions.

As the debriefing sessions continue, we may begin asking more pointed questions, having citizens and agencies assess their roles and the strategies used in the game and examine how they relate to models of the real world. Examples of debriefing questions suggested by Goodman include:

- What was your reaction on losing your ropes to the Grim Reaper?
- Was it possible to win at the Matrix Game without a pencil?
- How helpful were the agencies?
- Did the agencies know what the citizens needed better than the citizens themselves?
- What does it mean to help or be helped?

Excellent debriefings result from presentations of *The End of the Line.* The complexity of the exercise, coupled with the richness of the materials, provides a common experiential reference point for group discussions. Program strategies, agency versus community perceptions of help, and citizen relations in the community are also interesting topics for discussion following the exercise.

Game Adaptations

The underlying model in this game assumes that a community begins with an aggregate or collection of individuals who have the opportunity to collaborate with others to assist in their individual survival. How the players define their goals and strategies to stay alive will determine whether a cohesive community, small "neighborhoods," or isolated pairs emerge. In addition, the actions of the direct helping agencies, X and Y, will assist in facilitating the nature of community self-reliance or dependence on agency resources. Thus the community's shape and texture are not predefined but are determined by the interactional processes that take place. Whether agency help is really helpful to the community in reaching its goals is always a major question this model poses.

Nonetheless, *The End of the Line* is almost morbid, and dramatically portrays the struggle to stay alive and create meaning in an almost hopeless situation. While the simulation is true to its stated objective, it elicits powerful overt and covert individual, group, and intergroup processes that replicate critical issues of dependency and interdependency, fantasy and reality, and collaboration and competition in providing and receiving human services.

As a result, one of the simulation's strengths is its ability to provide opportunities for learning at different levels of social experience. When used in graduate public administration classes, for example, the goal of the exercise is often to study the impact of intergovernmental relations on service delivery. Agency X and Y become direct service agencies. In one version both agencies are funded by Agency Z (the federal government) and must comply with Z's requirements for a needs assessment, program plan, service delivery, and evaluation schedule. Here the game demonstrates the skewing effect of agency politics on service delivery patterns to consumers. Players in the agency are forced to justify their use of resources, make decisions about resource priorities (who needs services the most), and develop effective service delivery strategies within political and bureaucratic constraints. When COMEX presented this version to a special Department of Health, Education and Welfare invitational conference on human services, the postgame discussion lasted two and a half hours, longer than the exercise itself. Participants also ranked the game highest among all components of the three-day conference.

Another application of the exercise is to the study of direct interface between a human services agency and the community. In this version, Agencies X and Y receive a certain number of resources from G.O.D. and have autonomy to use them at their discretion. One agency often emerges as a traditional bureaucratic organization with community members completing forms, justifying statements of need, and being served on the basis of categories established by the agency. The other agency, viewing its function as promoting self-reliance in the community, often only provided information and referral services for citizens.

A third application of *The End of the Line* is to the study of psychosocial processes of aging by emphasizing the dynamics of the community and personal struggles to survive while losing resources and facing the possibility of death. The game becomes a powerful experience for participants in what it is like to grow old and be in need of help.

It is important, then, for G.O.D. to clearly state the goal or task of the exercise at the outset to provide participants with a reference point for understanding their experiences. Due to the game's emotional impact, G.O.D. must specify, first, how the simulation integrates the particular class or training program goals and, second, that participation in the game is optional.

OVERALL ASSESSMENT

The game's major strength *and* limitation is its complexity. The rules are elaborate and difficult for players to remember, and the game itself is not easy to introduce. Yet complex life is a reality for those who are aging, particularly those with disabilities. The game's artifacts are brilliantly conceived metaphors (a pencil as one's memory, the Matrix Game as work, ropes as limits on physical and social mobility) and make intuitively obvious dynamics explicitly visible. All these elements add to the game's complexity and rich potential for in-depth learning.

Most simple simulations fail to provide substantive topics for debriefing, while others *more* complex than *The End of the Line* also fail due to their clutter, numerous rules, and artifacts requiring even lengthier preparation and instructions.

All too often, game designers spend their energies in culling reality, in deciding which roles, interactions, and issues are not important to simulate. What results too often overlooks the effect of critical forces at work in the individual, in the group, or in intergroup dynamics during the game. *The End of the Line* differs in this regard. It offers a well-thought-out laboratory to study how complex community processes influence the lives of community members.

We believe that both adult and high school students would benefit from experiencing this gaming-simulation. Challenges of survival are different for those of us in different life stages, and this simulation provides a vivid experience of one life phase in our adult development. Our experiences indicate that high school students cope with the challenges in the game similarly to adults, and their reactions and insights seem to be just as dramatic.

Participants report their shock in realizing what it is like to have limited mobility, to be so dependent on others—or even an agency—for survival, and to feel somewhat helpless in the face of probable decline. In addition, most participants express a new regard for the position of the elderly in society, and more empathy for the problems of day-to-day living. It is for these reasons that we find *The End of the Line* a valuable learning experience.

References

Coppard, L.C., Goodman, F.L., eds. "The End of the Line." In *Urban Gaming Simulation,* pp. 123-132. Ann Arbor, Michigan: School of Education, 1977.

Goodman, F.L., *The End of the Line.* Resources in Aging, no. 4. Ann Arbor, Michigan: Institute of Gerontology, University of Michigan and Wayne State University, 1975.

ENERGY AND ENVIRONMENTAL QUALITY GAMES AND SIMULATIONS
An Evaluation

by Charles A. Bottinelli

INTRODUCTION

Earth Day, 1970, has been recognized generally as the dawning of our society's awareness of man's impact on the fragile life-support systems of Spaceship Earth. Yet, aside from the avalanche of rhetoric since then, it appears that we have progressed but a short distance from awareness of the problems to a concerted action directed at their solutions. Environmentalists, responsible for sounding the first alarm, have been involved in countless high-speed, head-on collisions with the proponents of growth, generating much heat (low quality) yet little light and an enormous credibility gap for the public at large.

During past crises, society has turned toward educators for help in changing attitudes, in preparing students for new roles in the work force, and in adapting to new ethics and behaviors consistent with changing societal goals, needs, and policies. The latest summons for help is the energy/environmental dilemma, seen by many to be a precursor of a series of worsening crises, impacting all sectors and aspects of society for the next several decades. But school districts themselves have fallen victim to inexorably increasing energy costs, having paid out $2 billion more for their energy in 1977 than they did in 1973.

It has become increasingly evident in the past several years that energy and environment must become central themes in all disciplines and at all grade levels—from science to home economics and from kindergarten through adult-hood. So, you might ask, "Is there an effective way of teaching concepts, analyzing attitudes and viewpoints, and promoting individual values related to a positive energy/environment ethic?"

If you have had experiences with environmental simulations and games, you probably already know that this instructional strategy holds considerable promise and may be most appropriate for teaching the complex, interdependent, and cross-disciplinary facets of energy/environmental education. Yet the explosion in environmental energy literature and rhetoric is paralleled by an incredible increase in educational simulations dealing with this topic. With so many simulations and games from which to choose, how can you find the ones most suitable for your students' needs? Which ones are really outstanding, and which of those best meet your objectives? This leads us to the purpose of this essay.

Purpose

In this essay we will compare, contrast, evaluate, and rate eight of the best environmental quality/energy simulation games according to several criteria, including the most important: observation of actual play under conditions prescribed by the game developers. We will examine closely the realism of the simulation or game; that is, is the game situation an acceptable model of its real-life counterpart? We will look at cost factors, the issues considered, playing time, age level, and the rest of the evaluative gamut in an effort to aid you in your decisions.

Definition

Environmental simulation games, in general, are instructional strategies that purportedly model a system or systems related to population, resources (including food and energy), and environmental quality for learning or training purposes. Subsets of environmental simulation games, the environmental quality and energy games, emphasize the environmental impacts of collective human activities on the ecosphere and attempt to make players familiar with the many issues, attitudes, and facts involved in solving problems related to air and water pollution, energy production, and demand and resource depletion.

Table 1 alphabetically lists and briefly describes the eight environmental quality and energy simulation games selected for analysis in this essay.

Uses

If you are an educator, are affiliated with a community or governmental agency, or work with a grass-roots environmental/legislative organization, you will find these simulation games appropriate for the following applications:

(a) As a *teaching/learning strategy* in which participants learn about real-life environmental issues by reacting to a model, making related decisions, and checking the outcomes and impacts of those decisions;

(b) as a *communications/interaction sensitizing tool* that effectively brings participants with different backgrounds or perspectives on the issues together to share other points of view and to empathize with others, thus paving a path toward future cooperation.

Selection Criteria

The eight games we have chosen represent the gamut of general concepts inherent in the environmental quality and energy controversies. They are commercially available and have been so for a number of years; thus, they have all been played and evaluated (at least by the developers) a number of times. With one exception, they are priced at less than $30.00. Finally, they appear to facilitate the kinds of interactions necessary to a basic understanding of the central issues with audiences ranging from secondary classes to adults, with little modification. Although there are many fine environmental simulations designed for use with a computer, we felt that, because many groups have no easy computer access, we would omit these simulations from this discussion.

TABLE 1 The Eight Energy/Environment Quality Simulation Games

Simulation Game	Summary Description
Dead River	Players as private and public officials, selected as members of a regional water council, propose water quality standards for the restoration of an interstate waterway based on their own needs and economics.
Energy-Environment Game	Players as members of the "Governor's Commission on Energy and the Environment" represent conservationists, the utilities, and lay sectors of a region, and assess the needs for additional power facilities.
Energy-X	Players as "special advisers to the President" represent various regions of the country in an attempt to allocate, according to region needs, a finite quantity of the hypothetical energy source, Energy X, found in a large meteorite.
Gomston	Players as representatives of local and state agencies attempt to clean up a hypothetical city plagued by severe pollution problems.
No Dam Action	Players as representatives of county and state agencies deal with the problems and issues related to flood control of a local river.
Pollution	Players as factory owners and members of community councils attempt to maximize profits while minimizing air, water, and noise pollution in a farm-town region.
Pollution Game	Players as influential citizens make decisions and initiate actions related to minimizing air and water pollution while maximizing profits.
Pollution-Negotiating A Clean Environment	Players representing business, government, citizens, and conservation interests negotiate to obtain a quality environment while still satisfying personal and/or corporate goals.

PRELIMINARY CONSIDERATIONS

Basic Factors

In choosing a suitable simulation or game for your class or group, you should consider three basic factors: age/grade level, group size, and the amount of time you are willing to devote to the activity. Additionally, you need to know the basic format; that is, is it a "board" game, which restricts the number of participants who may gather around a single playing area and necessitates the use of multiple copies for a group, or is it a "class" game in which an entire group uses one copy?

Table 2 summarizes these factors for each of the eight simulations. Although each simulation or game is flexible within limits for each of the factors, use Table 2 as a basic guide. (Our analyses may vary slightly from individual instruction manuals or from synopses found elsewhere in the *Guide.*)

Complexity

We have ranked the simulations and games according to complexity in Table 3. Generally, reading level determines complexity; games with little or no reading are best suited for the junior high/middle school grade levels, while games concentrating heavily on reading comprehension are best for senior high, college, and adult groups. Your group's mathematical ability (computational, chart and graph interpretation) may, simultaneously, limit a game's suitability. Additionally, games that are loosely structured and open-ended or that

TABLE 2 Preliminary Considerations

Simulation Game	Format*	Grade Level	Playing Time (hrs)	Number of Players	Number of Subgroups
Dead River	C	10-12, college, adult	4-5	10-30	5-6
Energy-Environment Game	C	10-12, college, adult	5-7	20-40	4
Energy-X	C	9-12, college, adult	4-5	20-40	9
Gomston	C	8-12, college, adult	5-6	10-40	10-13
No Dam Action	C	10-12, college, adult	7-9	12-30+	12
Pollution	B	7-9	½-1½	12-16/Board	**
Pollution Game	B	7-9	3/4-1½	4 Board	**
Pollution-Negotiating A Clean Environment	C	7-12, college, adult	1-3	4-32	4

*B = Board Game, C = Class Game
**The number of subgroups will vary in the case of board games, depending upon the availability of extra copies of the game. For class games, these numbers of ranges indicate the number of subgoups into which the players are divided.

model complicated processes are most appropriate for more mature audiences. Table 3 takes these factors into account, showing ascending order of complexity within the given categories.

TABLE 3 Complexity*

Simple	Moderate	Complex
Pollution Game	*Gomston*	*Energy-Environment Game*
Pollution	*Energy-X*	*No Dam Action*
	Dead River	
	Pollution-Negotiating A Clean Environment	

*Titles listed in descending order of complexity within columns.

ISSUES

Perhaps there are particular issues you would like your group to confront, or perhaps you want a simulation game that will allow your students to react to a number of environmental issues simultaneously. Table 4 is a breakdown of all the issues many environmental educators consider essential for a general exposure to the environmental quality and energy controversies. Please note that there is considerable overlap and that issues may be treated more or less significantly by different simulation games.

You may use Table 4 in two ways, depending on your objectives:

(1) Do you wish to use a simulation game that offers

TABLE 4 Issues

Issues	*Dead River*	*Energy-Environment Game*	*Energy-X*	*Gomston*	*No Dam Action*	*Pollution*	*Pollution Game*	*Pollution—NCE*	Totals
Environmental Quality									EQ
aesthetics	X	X		X	X	X		X	6
air quality		X		X		X	X	X	5
flood control					X			X	2
land erosion				X				X	2
land use		X			X			X	3
noise						X			1
policy	X			X	X	X			4
population impact		X	X	X				X	4
preservation		X							1
timbering								X	1
transportation			X	X			X		3
solid waste	X			X				X	3
urban environment		X			X				5
eutrophication	X			X					2
chemical pollution	X	X		X		X	X	X	6
thermal pollution	X	X		X	X		X	X	6
management	X			X	X		X		4
recreation	X				X		X		3
Energy									E
alternative sources		X	X						2
fossil fuel plants		X		X					2
nuclear energy		X					X	X	3
policy		X	X						2
plant sitings		X							1
resources/depletion		X	X					X	3
radioactive waste									0
consumption patterns			X						1
supply/demand		X	X						2
growth projections			X						1
electrical consumption		X	X						2
Related Economics									ECON
cost/benefit	X	X	X	X	X	X			6
economies of scale		X	X		X			X	4
supply/demand			X		X				2
abatement taxation	X			X		X	X	X	5
trade-offs	X	X	X	X	X		X	X	7
employment				X					1
TOTAL NUMBER OF ISSUES CONSIDERED	11	19	13	16	13	7	12	16	35
PERCENTAGE	31	54	37	46	37	20	34	46	100

exposure to a large number of environmental quality and energy issues?

If so, read down the simulation game columns. Note, for instance, that the *Energy-Environment Game* includes coverage of over 50 percent of the issues listed; in contrast, *Pollution* incorporates only 20 percent of the issues.

(2) Do you wish to locate simulation games that address a specific issue of interest to you and your students?

If so, read across the issue rows. For example, if your target issue is "aesthetics," you will have little difficulty in finding appropriate simulations that address this issue, since six of the eight incorporated "aesthetics." However, should your target be "radioactive waste," you will note that none of the eight focuses on this topic. Omission of this issue in energy simulation games is inexcusable given its potential impact in the social and environmental realms, and it should be addressed by simulation developers.

In view of the numbers of environmental quality and energy issues covered, we consider the *Energy–Environment Game, Gomston,* and *Pollution–Negotiating a Clean Environment* to offer very good coverage of representative issues in environmental/energy education.

TABLE 5 Depth of Content*

Low	Moderate	High
Pollution-Negotiating A Clean Environment	*Dead River*	*No Dam Action*
Pollution Game	*Gomston*	*Energy-X*
Pollution	*Energy-Environment Game*	

*A simulation or game with low depth of content includes from one to three pages of content-related information in the instruction guides for teachers and/or students; moderate, four to six pages; high, more than six pages (including bibliographies and audiovisual materials). Titles are listed in descending order of depth within columns.

CONTENT

Cognitive Base

One of the primary concerns of every educator who uses simulation games is the depth of content the activity provides. You must decide whether you wish to stress information, process, the affective realm, or a reasonable balance. Are you interested in a simulation or game with a heavy emphasis on environmental "facts" or one that will provide a cursory introduction to the issues? Table 5 can give you some useful guidelines.

Affective Base

Research indicates that the affective realm may be the forte of simulation games. Perhaps you want your students to be exposed to different attitudes, viewpoints, and values related to energy and environmental quality. To help them understand the range of attitudes that surrounds any given environmental issue, you might want to choose a simulation whose model includes a diversity of conflicting points of view and values; for example, business/industry versus environmentalism versus citizen views, or wildlife manager versus the polluter.

Perhaps the most significant value, among several that occur repeatedly in most of the eight simulations, relates to the free enterprise system and how its success and future are dependent on compromises (tradeoffs) among environmentalists, industry representatives, business managers, and law makers. Most of the simulations present this value implicitly in the game process; a few (for example, *Energy-Environment Game, No Dam Action,* and *Gomston*) treat it explicitly by explaining its significance on several role profile cards and by urging players to adopt the value in their roles. Either way, students begin to recognize the value of the free enterprise system in our society and its interdependence with environmental quality and economics.

The *Energy-Environment Game,* in our opinion, does the best job of introducing values, viewpoints, and attitudes related to the energy/environment crisis, accomplishing this task, in part, through filmstrip/tape cassette interviews with six people who view the environmental quality/energy dilemma from quite different perspectives.

Other, more traditional, values that appear to a greater or lesser extent in this collection of simulation games include those inherent in the philosophies of "bigger is better," "growth is good," "maximum good for the greatest number," and "preserve the environment at all costs." Of course, these are real values held in society, and the conflicts arising from them emerge in natural, uncontrived ways in most of these simulations.

Once the values were dissected into their elements during the normal course of play, we found that students often were able to rebuild a more positive environmental ethic from them. This ethic usually emerged in the debriefing sessions, and several components of it were easy to discern:

(1) Except for a continuous supply of sunshine, "Spaceship Earth" must complete its journey using the "supplies" placed on board when it was "launched."
(2) Human beings cannot perform one action without some consequence(s) of that action occurring somewhere else in the ecosphere.
(3) Mankind is a *part of* many overlapping ecosystems, is subject to their governing principles, and should not be considered separately from them.
(4) Everything must go somewhere, and nothing is "free."
(5) Uncontrolled growth on a planet of finite resources cannot be long sustained; therefore, more controls can be expected to be imposed on our free enterprise system.

We consider the promotion of this kind of environmental ethic to be one of the fundamental criteria for success in environmental quality/energy simulations. Table 6 rates each game according to the estimated percentage of game time spent on clarification of these values and on the development of this kind of ethic.

TABLE 6 Percentage of Time Allocated to Affective Considerations

Low (0-10%)	Adequate (11-25%)	Moderate (26-75%)	High (over 75%)
Pollution	*Energy-X*	*Dead River*	*Energy-Environment Game*
Pollution Game		*Pollution-Negotiating A Clean Environment*	*No Dam Action*
		Gomston	

ROLE-PLAYING

Here are additional questions you need to resolve before choosing the right simulation game for your needs. If the simulation game you are considering includes provisions for participants to assume roles, what kinds of roles are included? Does each participant respresent another individual, a group of individuals, or is his or her participation purely functional (that is, is the participant given a separate assignment with special rules)? How structured are the roles? How open-ended are they?

Consider these examples:

(1) Individual Role: Regional Director, Universe Club (*Energy-Environment Game*)

"You are a sociology professor at the State University, but a major portion of your time is devoted to being Regional Director of the Universe Club. You feel your duty is to lead the club on "battles" against major contributors to environmental damage, particularly the public utility companies. The pollution they generate from fossil-fueled plants is clearly a legitimate target for your attack. Of course, you are quite violently opposed to nuclear plants as well."

(2) Group Role: Steel and Chemicals (*Gomston*)

"Congratulations! You are the representatives of the steel and chemical firms in the Gomston area. As you know your firms are causing much of the air and water pollution in and around Gomston. Even though you must resolve your own pollution problems by installing new equipment and techniques, it might be wise for you to draw attention away from yourselves by attacking the other groups for doing very little about their pollution."

(3) Functional Role: (*No Dam Action*)

"In *No Dam Action,* there is a variety of groups, from formal decision-making groups to largely policy-oriented groups. While each group has its own frame of reference and purpose, these junctions dominate:

- create study groups or task forces
- issue news releases
- conduct elections or replace resigned officials
- organize petitions
- keeping track of scoring data."

Table 7 indicates the kinds of roles in each of the simulation games.

TABLE 7 Kinds of Roles

Individual	Group	Group, Functional	Individual, Group, Functional
*Pollution Game***	*Pollution-Negotiating A Clean Environment*	*Gomston*	*No Dam Action*
Energy-Environment Game	*Energy-X*		
*Pollution***	*Dead River*		

**No Dam Action* includes individual, group, and functional roles with considerable overlap.

**Students play the roles of influential citizens, but the game includes no role descriptions. However, participants may write their own, if so desired.

The large diversity of roles the eight simulations represent is indicated in Table 8 by role title.

TABLE 8 Role Diversity

Simulation Game	Titles of Roles Included
Dead River	Taxpayers Association
	State Eco-Action Agency
	Federal Environmental Policy Agency
	Regional industry
	Valley Recreational Development Association
Energy-Environment Game	Governor's Commission on Energy & Environment
	Electric power company officials
	Lay citizens
	Conservation & Environmental Agency representatives
	Commerce, industry, and professional respresentatives
Energy-X	Project control officials
	Representatives of eight U.S. geographic areas
Gomston	Mayor
	City officials
	Coal, petroleum industry representatives
	Transportation agency
	Steel, chemical representatives
	Utilities' representatives
	Antipollution group
	Agriculture, lumber interests
	Forestry, wildlife officials
	Chamber of Commerce
	News media
	Local geographer
	State government officials
No Dam Action	Representatives from the following interests:
	utilities
	mining
	paper & chemicals
	Army Corps of Engineers
	health office
	law
	university
	engineering
	journalism
	religion

Table 8 Role Diversity (Cont)

Simulation Game	Titles of Roles Included
	labor sanitation banking real estate forestry local businesses accounting farming
Pollution	Community representatives Factory owners
Pollution Game	Influential citizens
Pollution-Negotiating A Clean Environment	Business representatives Governmental agencies Citizens Conservation groups

Role Structure/Flexibility

Depending on your purposes and your students' maturity levels, you may wish to choose a simulation with more or less role structure. For example, if you want your students to develop a close empathy with persons represented by the role profiles, you will need to choose simulation with highly structured role profiles. On the other hand, if you want a high degree of flexibility and open-endedness, you will want to select simulation with less structured, less-developed role profiles. You might also have your students design their own role profile cards for the less structured simulations or to augment the diversity of roles in those that are more structured.

Table 9 indicates the degree of structure and role flexibility inherent in the role descriptions of each simulation game.

Role Interest Levels

A criticism we found recurring quite frequently is related to those simulations listed in the first column of Table 9. Many students in the age groups recommended by the developers do not have sufficient background to play an unstructured role adequately. We found that unless the game facilitator supplied additional role information to the participants, play proceeded in a rather chaotic, disorganized manner. Our experience indicates that younger players, especially, appreciate a high degree of structure, because most have little or no preconceived notion of the responsibilities, viewpoints, attitudes, and values espoused by a prototypical role profile.

Additionally, since the effectiveness of most role-play simulations is greatly enhanced if players can assume roles that are divorced from their own identities, we found that allowing players to wear tags with names they made up provided an enthusiasm that helped carry the game process over rough spots. Most of the games do not provide name tags, although some do provide name plates that carry the title of the group or agency represented. Even for those simulations with a high degree of role structure, we would advise the facilitator to allow players to "refine" their roles by providing details (for instance, hypothetical names, marital status, number of children, aspirations, and so on) that do not appear on the profile cards.

TABLE 9 Role Structure/Flexibility*

Low Structure High Flexibility	Moderate Structure and Flexibility	High Structure Low Flexibility
Pollution	*Energy-X*	*Energy-Environment Game*
Pollution Game	*Gomston*	*No Dam Action*
Pollution-Negotiating A Clean Environment	*Dead River*	

*Titles listed in descending order within columns.

Unfortunately, even though a simulation may provide role structure adequate for your needs, there is no guarantee that students will find the roles interesting. And if the roles are boring and unexciting, players cannot be expected to portray them with enthusiasm. This will detract from your instructional objectives and ultimately doom a simulation game to failure.

Our observations of and interactions with players involved with the eight energy and environmental quality simulations indicate the following:

(1) Players new to the simulation technique appear to generate more interest and enthusiasm (and, by inference, more direct learning) with simulations exhibiting a moderate to high degree of role structure.

(2) Older, more experienced students seem to appreciate more latitude in the structure of the roles they assume.

Accordingly, Table 10 presents the interest factor inherent in the roles associated with each of the simulations. (Our test groups may not be representative of other student groups, so you should use our judgments only as rough guidelines.)

TABLE 10 Interest Levels of Roles*

Low (Least Enjoyable)	Average	Above Average	High (Most Enjoyable)
Pollution	*Pollution-Negotiation A Clean Environment*	*Dead River*	*Energy-Environment Game*
Pollution Game		*Gomston*	*No Dam Action*
		Energy-X	

*Titles listed in descending order within columns.

RULES

Clarity, Organization, and Comprehensiveness

One of the most important of your considerations deals with the game manual. How organized, how complete, and how clear is the presentation of the procedures you must follow? In most cases the rules are outlined in the facilitator's manual. In a few simulations, however, you may find procedures and instructions listed on the players' role profiles, on the backs of name plates, or in special "Players' Manual." Although it is not necessary (or possible) for manuals to cover every situation with a specific rule, instructors usually want comprehensive treatment so they can make spur of the mom-

TABLE 11 Manual Ratings, Procedures, and Preparation Time

Simulation Games	Overall rating of manual for clarity, organization, completeness. Total number of pages in manual appear in parentheses.			Number of pages devoted to procedures	Preparation Time (Hrs)
	Adequate	Good	Outstanding		
*Dead River**	X (20)			3	2-3
*Energy-Environment**			X (28)	18	3-4
Energy-X			X (8)	4	1½-2
Gomston			X (24)	5	2-2½
No Dam Action	X (18)			4	3-4
Pollution		X (3)		2	3/4-1
Pollution Game		X (2)		1	½-3/4
Pollution-NCE		X (18)		12	1-1½

*Players' or team manuals provided.

ent rulings within the game's context during play. You need to keep in mind, however, that the more complex simulations necessitate more procedural guidelines which, in turn, may increase preparation time significantly. (Preparation time is the total time it takes the instructor and players to become familiar with the simulation model, procedures, and arrangements.)

Most of the simulations we evaluate here have instruction manuals that are adequate to give the facilitator a mental impression of the model and procedures on a cursory first reading. However, we became frustrated and impatient with the manuals for *No Dam Action* and *Dead River* because their formats interspersed procedure with background content information, making each more difficult to understand quickly.

Table 11 shows our ratings of the instruction manuals, the number of pages the manual devotes to procedure, and the approximate preparation times involved from the time the activity is removed from the shelf to the first operational move by players.

ACTIVITIES, SKILLS, AND INTERACTIONS

Are a simulation's activities, skills, and interactions consistent with your instructional objectives? Perhaps you are interested in having players participate in certain activities, reinforce or hone specific skills as they interact individually or in groups.

Table 12 lists patterns of generalized activities that occur in the sequence indicated (some are repetitive for a varying number of rounds) for each of the eight simulation games.

Table 13 indicates the kinds and degrees of specific skills and activities that participants practice in each of the simulations. This chart is organized using Bloom's Taxonomy as the vertical scale and the eight simulations as the horizontal. You should find this chart very useful in that the levels of skills displayed are hierarchical in ascending order, and each higher level necessarily incorporates those below it. The simplest skill, knowledge, involves information collection and processing; evaluation is the most complex.

Now that you are aware of the kinds of skills and activities represented in our eight simulation games, you will be interested in knowing to what extent they are practiced. Table 14

TABLE 12 Activity Sequences

Simulation Game	Activity Sequence
Dead River	1. Consensus on objectives 2. Determining share of costs 3. Analyzing cost/benefits 4. Presenting alternatives 5. Preparing a team proposal 6. Voting
Energy-Environment Game	1. Problem analysis by special interest groups 2. Public hearing #1 3. Task force meetings 4. Public hearing #2 5. Special interest group meeting
Energy-X	1. Fact finding and planning 2. Proposals/presentations 3. Project control decision making 4. Appeals
Gomston	1. Analysis and study of pollution sources 2. Team meetings 3. Town meetings 4. Decision making
No Dam Action	1. "Quest" activities 2. Intrateam analysis of background statements 3. Interteam analysis of resource reports 4. Establishing priorities 5. Decision making
Pollution	1. Developing pollution abatement programs 2. Analyzing results 3. Repeat cycle
Pollution Game	1. Determining ownership of properties 2. Adjusting water/air pollution indices 3. Adjusting taxation fees for properties 4. Elections 5. Repeat cycle
Pollution-Negotiating A Clean Environment	1. Intergroup negotiations 2. Voting 3. Analyzing consequences 4. Repeat cycle

TABLE 13 Specific Skills/Activities Represented (Ascending Order)

Skills/Activities	*Dead River*	*Energy-Environment Game*	*Energy-X*	*Gomston*	*No Dam Action*	*Pollution*	*Pollution Game*	*Pollution-NCE*
self-performance			s	s				
peer-performance			m					
proposals	s	vs	s	vs	vs			vs
6. Evaluation (exclusive of debriefing)								
planning strategies		s	vs	s	vs			s
planning goals		vs		s	vs			
writing/revising	s			s	s			
formulating models				m	vs			
making decisions	s	vs	vs	vs	vs	s	s	vs
5. Synthesis								
individual problem solving							m	
group problem solving	s	vs	vs	vs	vs	m		s
reviewing proposals	s	vs	vs	vs	vs			vs
4. Analysis								
voting on proposals	s	vs	s	vs	vs			vs
proselytizing	s	vs	vs	vs	vs			vs
negotiating	s	vs	vs	vs	vs	s		vs
challenging	s	vs	vs	vs	s			vs
role playing	s	vs	s	s	vs			vs
hypothesizing			s	vs	vs			
predicting		s	s		s	s	m	
inferring	s	s	vs	s	s	s	m	
3. Application								
large group discussion	s	vs	vs	vs	vs			s
small group discussion	s	vs	vs	vs	vs	m	m	vs
writing				s				
public speaking	m	s	s	s	s			vs
explaining			vs		vs	m		
interpreting data	s	vs	vs	s	vs		m	
2. Comprehension								
interviewing		s	m	m	s			s
reading/researching	s	s	vs	s	vs			m
recording data	vs				s			
observing	s		s	s		m	s	
collecting information	s	vs	vs	vs	vs			s
1. Knowledge								

Key: m—minor
s—significant
vs—very significant

gives our rating according to the variety of skills and activities used and the extent to which participants practice them.

If one of your goals is to provide a social environment in which participants develop communication skills through group work, you may find Table 15 helpful. It lists the kinds of social interactions each of the simulations exhibits.

TABLE 14 Scope and Extent of Practice Skills/Activities

Extent Practiced	Few	Some	Many
	Number of skills/activities practiced		
High		*Pollution-Negotiating A Clean Environment*	*Energy-Environment* *Energy-X* *Gomston* *No Dam Action*
Moderate	*Pollution*		*Dead River*
Low		*Pollution Game*	

TABLE 15 Kinds of Interaction

Small Groups	Small & Large Groups
Pollution	*Energy-Environment Game*
Pollution Game	*Gomston*
	No Dam Action
	Dead River
	Pollution-NCE
	Energy-X

RESOURCES AND SCORING

Additional concerns you will have in choosing an appropriate simulation relate to resources and scoring. For example, must the players keep track of money, chips, tokens, points, or some index throughout the simulation? Or, as in many environmental quality/energy simulations, are there no identifiable resources? What determines scoring? Do some players gain resources at the expense of others? Is the scoring procedure easily followed, or is it very involved? Table 16 summarizes resources and scoring for each of the eight simulations.

TABLE 16 Resources and Scoring

Simulation Game	Resources	Scoring
Dead River	No identifiable resources except influence differences among members of the Regional Council	Points allotted to teams according to various criteria on a Team Performance Score Sheet
Energy-Environment Game	No identifiable resources except influence differences on role profile cards	No explicit scoring procedure; Governor's Commission votes on proposals
Energy-X	No identifiable resources except influence differences among Region Representatives and Project Control officials	No explicit scoring procedure; Project votes on proposals
Gomston	No identifiable resources except influence differences among mayor, governmental officials and business/agency representatives	No explicit scoring procedure; win if problems are solved
No Dam Action	Influence points on role profiles are assigned to each participant and are unequally distributed	Two scoring methods suggested: 1. Ratings of involvement of groups/individuals by peer groups 2. Accounting of "risked" influence points
Pollution	Money and pollution tokens, equally distributed at start	According to amount of money earned, pollution indicators, and pollution tokens
Pollution Game	Money, election cards, and properties, equally distributed at start	According to amount of money accumulated and pollution levels
Pollution-NCE	Unequal distribution of influence	According to election results and associated Quality of Life and Personal/Corporate Goal Satisfaction scores

Kinds of Scoring Procedures

Table 17 indicates the kinds of scoring procedures the eight simulations use. Please bear in mind that in many energy/environmental quality simulations, it is unnecessary to have a scoring procedure. Usually, a class, group, team, or individual "scores" by proposing solutions to energy/environmental problems, by cooperating or competing with other groups or individuals, and, finally, by winning a vote of confidence for the solution proposed. Intrinsic student/group feedback is still provided at each step, and the effectiveness of individual or group performance becomes evident when proposals are either accepted or rejected. In many cases the entire class may "win" or "lose" according to the "quality of life" resulting from their interactions.

TABLE 17 Scoring Procedures

No explicit scoring; players vote on proposals	Individual Scoring	Group Scoring	Class Scoring
Energy-Environment Game	*No Dam Action**	*Dead River*	*Gomston**
Energy-X	*Pollution*	*No Dam Action**	*Pollution-NCE**
*Gomston**	*Pollution Game*	*Pollution-NCE**	

*Two kinds of scoring procedures overlap as indicated.

Evaluation of Scoring Procedures

We found that players had considerable difficulty learning the scoring procedures in *Polution–Negotiating a Clean Environment,* as evidenced by an excess of questions during play. This is easily understood, because the procedures involve more computation than a simple accounting of money or tokens. We would suggest either allotting more time to explain scoring for these simulations streamlining the scoring procedures according to your students' abilities.

In general, we observed that complicated and drawn-out scoring procedures tended to detract from the smooth flow of simulation sequences and that simulations with no explicit scoring procedures appeared to function more effectively.

MODEL VALIDITY

For each simulation you consider, you should ask yourself: "How closely does the simulation model approximate reality; that is, how valid (realistic) is this model?" By definition, educational simulations are simplified representations of reality. How much simplification are you willing to accept in exchange for ease of facilitation? Is what has been left out crucial to your instructional objectives? To what extent does a simulation's reliance on the random or chance factor distort the simulated process?

Table 18 gives our ratings of model validity based on discussion of the experience with our test instructors.

We are critical of the validities of *Pollution* and *The Pollution Game,* because of their excessive dependence on chance, although the latter compensates somewhat with "council

TABLE 18 Model Validity (Realism)*

Low	Adequate	High
Pollution	*Dead River*	*Energy-Environment Game*
Pollution Game		*Energy-X*
		Gomston
		No Dam Action
		Pollution-NCE

*Titles listed in descending order within columns.

meetings" and "elections." In general, the most realistic and valid simulations are those that involve the interactions of diverse social and community groups in a problem-solving approach. In fact, the five listed in the "High" column of Table 18 were so realistic that we observed most players exhibiting high levels of frustration–frustration often observed in real-life situations.

For example, in the *Energy-Environment Game,* the Chairman of the Governor's Commission on Energy and the Environment reportedly walked out of the "chambers" because "there was no way we could satisfy everyone." In *Gomston,* where special interest groups are pitted sharply against environmental interests, the student playing the role of Gomston's mayor remarked, "We should wipe this mess off the map and start over somewhere else!" Many agreed.

DEBRIEFING

At the conclusion of the simulation sequence, it is important to provide an opportunity for players to discuss their feelings about the roles they played, the interactions within and among groups and between individuals, and the relevance, realism, and outcomes of the activity. This postsimulation discussion, often termed debriefing, should not be overlooked by the facilitator. It is an extremely important part of any simulation and is vital for closure.

Many simulation games provide a list of suggestions, guidelines, and/or questions that can help you adequately debrief the players. Some do not provide guidelines or questions, but may suggest the importance of debriefing, leaving it up to you to work out the details. A few ignore debriefing altogether. Obviously, in this case, you should take a few minutes of preparation time to draft a list of questions for debriefing.

Table 19 indicates our ratings of the eight simulations based on whether (and to what extent) they include provisions for debriefing.

TABLE 19 Provisions for Debriefing*

Totally Inadequate (No guidelines)	Inadequate (1-2 paragraphs)	Adequate (several paragraphs)	Thorough (1 or more pages)
Dead River	*Energy-X*	*No Dam Action*	*Energy-Environment Game*
Pollution Game	*Gomston***		
Pollution-NCE	*Pollution*		

*Titles in descending order within columns.
**Provides several instruments for evaluation.

FLEXIBILITY

A high degree of flexibility in a simulation can be an important added attribute, because it will allow you a degree of ease in modifying format, procedures, roles, issues, rules, resources, and group size according to your objectives, needs, and interests. For example, a board game such as *The Pollution Game* is inflexible with respect to group size and roles because only from four to six players may physically gather around it. On the other hand, *The Energy-Environment Game* is so flexible that one could relatively easily provide for up to 64 participants, because it includes an extra set of role profiles (two players using a pair of identical profiles). If an instructor preferred 64 separate roles, it would not be difficult to have students write another set.

In fact, those simulations that deal with group problem solving of local issues (for example, *Gomston* and *No Dam Action*) are the most flexible on nearly all counts. It is important to remember an earlier caveat, however: Less structure (more flexibility) requires greater maturity in participants.

Table 20 ranks each of the simulations for overall flexibility.

TABLE 20 Flexibility*

Inflexible	Some modifications possible	Flexible	Highly Flexible
Pollution Game	*Pollution*	*Dead River*	*Energy-Environment*
		Energy-X	*Gomston*
		Pollution-NCE	*No Dam Action*

*Titles in descending order within columns.

MATERIALS AND COST

Final considerations in choosing any simulation relate to packaging, cost, expendable supplies (reusability), and completeness. If you have little time to spend on pregame preparation, you will want a simulation packaged as a complete kit that includes ample supplies of all the necessary materials, especially support components such as literature, score sheets, and maps. Because in many schools different teachers must share the same resource materials, you are probably seeking simulations that are durably packaged, inexpensive or at least reasonably priced, and have expendable copy that can be inexpensively mimeographed or xeroxed once the originals have been mutilated.

Table 21 summarizes cost and packaging information for each of the eight simulations.

We consider *Energy-Environment Game, Energy-X,* and *Gomston* to be the most reasonably priced kits for the money. They contain all the necessary materials, even including a filmstrip and audio tape cassette, as well as ample supplies of printed support materials. On the other hand, we believe *No Dam Action* and *Pollution* to be grossly overpriced, given the quality of materials and the packaging. *Pollution* arrives in an envelope with game pieces, manual, and a folding plastic game "board" more suitable as a table cloth. We found that after repeated handling the plastic smudged very easily.

We found that the 88-page manual for *Pollution–Negotiating a Clean Environment,* transparencies and pages in detachable binding, did not hold up well. The loose pages are easily misplaced.

CONCLUSION

Now, having considered well over a dozen criteria applied to the selection of energy/environmental quality simulation games, you are faced with a decision: Which best meets your objectives, given preparation and playing time, group size requirements, and other limiting factors?

We offer one final table, Table 22, summarizing each simulation in terms of our ratings, which, in turn, were based on direct observation of each simulation game in operation. We used a five-point rating scale; an overall average rating appears in the last row of the table.

We consider the *Energy-Environment Game, Gomston, No Dam Action,* and *Energy-X* to be *excellent* simulations. *Dead River* is a *very good* simulation; *Pollution–Negotiating a Clean Environment* is *good*; and the remaining two are more or less *fair.* We award top honors in the energy category to the *Energy-Environment Game* and first prize for the best environmental quality simulation to *Gomston.* Both combine a knowledge of fundamental principles with a strong concern for the energy-environmental dilemmas confronting the United States, and both do so in ways that are highly motivational and fun–no small task. Above all, they teach in convincing fashion that there are no simple answers to complicated issues, that tradeoffs are inevitable, and that people, individually and collectively, still have a voice and a responsibility in determining future directions and goals for society.

Although we gave *Energy-X* and *No Dam Action* excellent ratings, we feel they are slightly less effective and less comprehensive than our first choices.

If you've been searching for your first "polluted river" simulation, we suggest that *Dead River* will serve your purposes well with its high validity, flexibility, and quality-to-cost ratio.

Pollution–Negotiating a Clean Environment, although exhibiting moderate to high degrees of model validity and flexibility, was deficient in the content and roles categories. Therefore, we rated it "good" on those items.

Pollution and *The Pollution Game* were relatively ineffective in their development and definition of roles, in simulating reality, in providing sufficient structural flexibility, and in maintaining high levels of participant interest. We rated them inadequate to fair.

Nevertheless, don't let our ratings dissuade you from a closer examination of those simulations that did not make high marks. The importance of various facets of each of the simulations we evaluated will differ with different purposes, objectives, and needs. You should select simulations according to your own requirements and the interests of your students. Remember, a small amount of your time spent in clarifying

TABLE 21 Packaging, Cost, Completeness and Reusability

Simulation Game	Packaging	Cost/Package	Completeness	Reusability
Dead River	kit	$13.00	complete	average
Energy-Environment	kit	1st edition was $20.00	requires some duplication (ditto)	high
Energy-X	kit	$19.95	requires some duplication (ditto)	high
Gomston	kit	$26.50	complete	high
No Dam Action	kit	$145.00	complete	high
Pollution	envelope	$26.00	complete	average
Pollution Game	kit	not set	complete	average
Pollution-NCE	manual with detachable binding	$27.00	complete	low

TABLE 22 Summary Ratings

Criterion	*Dead River*	*Energy-Environment Game*	*Energy-X*	*Gomston*	*No Dam Action*	*Pollution*	*Pollution Game*	*Pollution-NCE*
Content	3	4	5	3	4	2	1	2
Roles	3	5	3	4	5	1	1	3
Manuals	2	5	4	5	3	2	1	4
Skills	4	5	5	5	5	1	1	4
Validity	4	5	4	5	5	1	1	4
Flexibility	4	5	4	5	5	2	1	4
Debriefing	1	5	2	3	4	3	1	1
Completeness	5	4	4	5	5	5	5	5
Durability	3	4	4	5	5	3	3	1
Quality/Cost Ratio	5	5	4	5	1	1	3	4
Interest	4	5	5	5	3	2	1	3
Overall Rating (Average)	3.5	4.7	4.0	4.6	4.1	2.1	1.7	3.2

Key:
5–excellent
4–very good
3–good
2–fair
1–inadequate

your purposes and examining the pros and cons of each simulation may be well worth your effort in the long term, because, first, you are making an investment in a learning tool that will be used repeatedly by many students, and second, while a picture is worth a thousand words, a good simulation can be worth more than a thousand pictures!

Sources

Dead River
E. Nelson Swinerton
1973

Union Printing Company
17 West Washington Street
Athens, Ohio 45701
$13.00

Energy-Environment Game
Raymond A. Montgomery, Jr., and Toby H. Levine
1973

Edison Electric Institute
90 Park Avenue
New York, N.Y. 10016
about $25.00?

Energy X
Norman S. Warns, Jr.
1974

Ideal School Supply Company
11000 South Lavergne
Oak Lawn, Illinois 60453
$19.95

Gomston
Norman S. Warns, Jr.
1973

Ideal School Supply Company
11000 South Lavergne
Oak Lawn, Illinois 60453
$26.50

No Dam Action
R. G. Kleitsch

System's Factors, Inc.
1940 Woodland Avenue
Duluth, MN 55803
$145.00

Pollution
Ron Faber and Judith Platt
1973

Games Central, Abt Publications
55 Wheeler Street
Cambridge, Mass. 02138
$26.00

Pollution Game
Educational Research Council of America
1979

Carolina Biological Supply
2700 York Road
Burlington, NC 27215
price not extablished

Pollution: Negotiating a Clean Environment
Paul A. Twelker
1971

Simulation Systems
Box 46
BlackButte Ranch, OR 97759
$27.00

FRAME GAMES
An Evaluation

by Sivasailam Thiagarajan and Harold D. Stolovitch

This essay deals with a special type of games called "frame games." It is organized according to the following outline:

(1) the concept of frame game
(2) an overview of frame games evaluated in this chapter
(3) an analysis and comparison of the selected frame games on the basis of eight important characteristics
(4) overall evaluations of selected frame games
(5) advantages and disadvantages of frame games

The content and organization of this essay reflect its prime purpose: to make you a more informed selector and user of frame games. Specifically, these are the objectives of the essay in terms of what you should be able to do:

(1) explain the concept of frame games and discriminate a frame game from other similar related activities;
(2) describe a few currently available frame games;
(3) list and describe different attributes of a frame game; Apply this knowledge to analyze a frame game and to make a more discriminating choice of frame games to fulfill your needs;
(4) describe some of the advantages and disadvantages of using frame games so you can accentuate the former and reduce the latter.

THE CONCEPT OF FRAME GAME

The concept of frame game is fairly easy to explain. In fact, you probably have had some experience with this concept. Here are a few examples which illustrate frame games in action.

- A grade-school teacher proudly demonstrates innumerable variations of Bingo games designed to teach addition facts, matching of words with pictures, initial consonants, and the like.
- A social studies teacher uses a *Monopoly*-like game to teach the process through which a legislative bill becomes a law or about life in communist countries.
- A teacher of Spanish uses a vocabulary game that looks suspiciously like the dictionary game you played at John's party last week.
- You purchase a slickly packaged chemistry game only to realize that it is nothing more than a glossy modification of *Old Maid,* which you used to play as a child.

If you take any game, you can usually analyze it into two major divisions: content and structure.

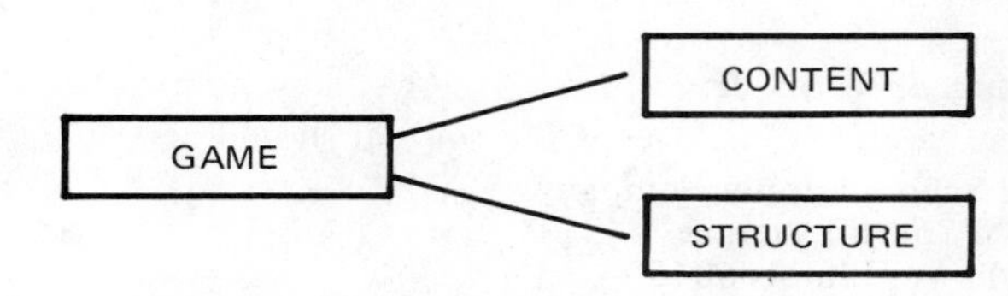

As an example, let us look at the popular game *Rummy*. The content of Rummy involves cards that vary in two different ways: suits and values. The structure involves a number of rules for initiating, continuing, and terminating the play of the game and includes the fundamental rule for collecting sets of cards that have the same suit and a sequence of values or the same value but different suits. By dislodging the original content you can identify the *frame* of the game. By loading new content on this frame, you create a new game that can help you achieve your own specific objectives:

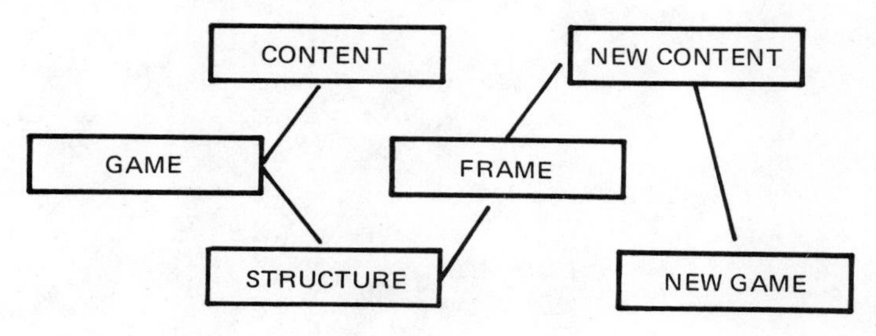

To give an example of how this can be done, let us assume that you are a high school teacher of English. Let us say you have figured out that the content-free frame of Rummy involves the creation of sets from elements that vary along two dimensions. You therefore decide to create new content by typing excerpts from English literature from six different

---

Editor's Note: Listings for all the games discussed in this essay are in the Frame Games Section, except for *Encapsulation,* which is in the Communication Section, and *Making a Change,* in the Community Issues Section.

periods on blank index cards. You also decide to use four different forms of writing for each period. So now you have created your own *Literature Rummy* in which players attempt to collect sets of cards from the *same* period with *different* forms or vice versa.

This is definitely a crude example, and we hope that we have not inspired you to create endless variations of Rummy to teach everything to everyone. But it does illustrate the process of frame gaming and the fundamental axiom that all games are frame games. However, for the purposes of this essay we use a more restrictive definition of frame games. We define frame games as *those games that are deliberately developed to provide a content-free instructional structure on which can be loaded locally relevant content.*

Here is an actual frame game (reprinted with the kind permission of the publisher, Educational Technology Publications, 1978) from a longer article by Thiagarajan that illustrates the critical attributes of our definition of frame games.

The Press Conference Game

Number of players: Ten to thirty

Approximate Time Requirement: Two to three hours

Materials: Index cards of four or five different colors, pencils

Step-by-Step Directions for Play:

(1) *Needs Analysis.* Before the play of the game, divide the content of your lecture into a convenient number of subtopics. Although the logic of the content should determine the exact nature of these subtopics, you will have to take these two factors into consideration.

- The game lasts longer if you have more subtopics. If you have only limited time, use fewer subtopics.
- The number of subtopics is also the number of teams. Choose the number so that your teams do not have too many or too few members to permit collaborative learning.

In general, three to five subtopics result in effective play.

When the players arrive, announce the topic and the subtopics. Give each player index cards of as many different colors as there are subtopics. Let's assume that you have four subtopics. Each player receives a blue, white, pink, and yellow index card. Explain that each color stands for one of the subtopics and specify which color goes with which subtopic. Ask the players to write one or more questions on each subtopic on the appropriate color-coded card. These are questions for which the player would like an answer before the end of the game.

(2) *Team Work.* After about five minutes, check to see if all players have finished writing their questions. Collect the question cards. Divide the players into as many teams as there are subtopics. Give each team all the cards of a specific color. Ask the teams to carefully review the questions, eliminate redundant ones, add more if necessary, and organize the questions in a logical order. Leave the players alone for about fifteen minutes.

(3) *Press Conference.* Station yourself behind a real or simulated podium and announce that you are an international authority on the topic and that you are now ready to conduct a press conference. Randomly select one team to be the inquisitive reporters. They have fifteen minutes to question you on their subtopic, using the edited list of questions in any way they want to. Tell the team to cut you off politely if your answers are too lengthy or rambling because of the time limit. Warn the other teams to listen carefully to the questions and answers and to take copious notes because their score in the game will depend on how effectively they store, retrieve, and process the information.

(4) *Information Processing.* At the end of the allotted time, stop the press conference. Ask the teams to prepare a succinct summary of the main points made by the expert. This (legibly) written summary is to be produced within fifteen minutes. The questioning team does not prepare such a summary. Instead, it designs a checklist for evaluating the summaries from the other teams on the basis of such criteria as comprehensiveness, brevity, inclusion of main points, and elimination of trivial points.

(5) *Reporting and Ranking.* Collect the summary reports from all teams. Read aloud each summary without identifying the team that produced it. After reading all the reports, ask the evaluating team to rank order the summaries and to divulge secretly their decisions to you. Give the score of three points to the top-ranked summary, two to the next one, and one point to the third one. Write down these scores on the summary sheets but do not announce them yet.

(6) *Recycling and Concluding.* Repeat the previous three steps as often as needed so that every team plays the role of the reporters probing you on the other subtopics. At the end of the last round, let each team retrieve its summary reports and add up the scores. Declare the team with the highest total score to be the winners, but do not make a big fuss about it. Conduct a debriefing session and respond to any leftover questions from individual players.

This sample frame game has been developed expressly to provide a content-free instructional structure on which a wide range of new content can be loaded.

OVERVIEW OF THE FRAME GAMES ANALYZED

This essay provides a comparative analysis of fourteen frame games. All of these fourteen have been selected on the basis of the following criteria.

- These frame games are currently available in written form from their designers or publishers.
- These frame games are fairly inexpensive: None of them costs more than $35, most of them cost less than $10. Some have been published in professional periodicals that are easily accessible to the potential user.
- These frame games are selected to represent a large number of game designers. It is true that four of the frame games are of our own creation, but there being two of us, we have each separately made our evaluative comments of the other's materials.
- The selected frame games represent a broad range of levels of complexity. They are suitable for a wide range of players

from upper elementary school through adult, with an emphasis on the adult end.

• We have selected only those games that we have ourselves run with a group of players or in which we have participated as players. Because of this requirement, we are unable to include an excellent set of frame games by Cathy Greenblat and Richard Duke; we were not able to get hold of copies of these games for player testing.

Given below are brief descriptions of the fourteen frame games we have selected for our comparative evaluation. These descriptions highlight the content-free nature of the games and provide the base for their analysis in the later sections of this chapter.

Confrontation

During each round, players receive a card that specifies a confrontation situation and assigns roles. Two players are in adversary roles and the third is a mediator. Adversary players choose one of five possible positions on the issue and compare their choices. Adversaries discuss/debate their positions and with the help of the mediator reach a common ground. They exchange poker chips depending on the shift from their initial positions.

Encapsulation

This frame game's materials come with six different content area adaptation kits: labor versus management, black versus white, affluent versus deprived, career versus homemaking, student versus teacher, and parent versus child. The two primary players assume adversary roles in a culture-clash simulation. They sit across from each other with their bias boards, gradually revealing how they perceive themselves and each other. They attempt to resolve their conflicts through a conference that is mediated by a facilitator. Each adversary is also under the influence of peer group members who keep sending notes imploring the adversary not to compromise. There are also participant and nonparticipant observers. In addition, there is a structured debriefing session at the end of the conference.

Facts in Five

Players receive a playcard with a five-by-five matrix that has different categories along the columns and letters of the alphabet along the rows. Players fill each cell of the matrix with a key word beginning with the specified letter that fits the appropriate category. Players' words are scored with extra points for originality.

GAMEgame IV

Each team creates a list of five important items related to a selected theme. A common list is created by the game leader. Teams secretly write down their top choice from the common list and are rewarded for reaching consensus. This procedure is repeated until the top five items are identified.

GAMEgame VI

The game leader prepares cards with individual opinions about a topic or an issue. Players write four personal opinions on blank cards. These cards are randomly distributed to all players. Individual players exchange opinion cards at a discard table and with each other. They form coalitions with others of similar opinions and reduce their total number of cards to five. Each group writes a summary statement of its philosophic stance and selects an appropriate name for itself.

Making a Change

During the first session, members of each team select a problem and within that team evaluate each other's problems, identify resisting and encouraging forces, and draw up a list of questions to be answered about the problem. During the second session, each team completes the listing of significant forces and shares it with another team. Using structured forms, teams provide feedback for each other. The same procedure is applied preparing a suitable plan of action. The revised plan of action becomes the starting point for making a change.

Planning Exercise

A problem is presented to all teams. One team is given the role of evaluators; all others have to create a solution to the problem. The evaluation team develops and shares criteria for judging solutions. Each team presents its solution to the evaluation team, which provides appropriate feedback. During the rebuttal phase, each team clarifies any misunderstandings and summarizes the strengths and weaknesses of the solution from other teams. All strong points are consolidated into a superplan.

Policy Negotiations

Three teachers and three school board members are engaged in negotiating a large number of possible issues. Other players represent social agencies and newspapers. Each core member has certain influence points that can be used in any of the following four ways during each round: (1) to vote on a specific issue on top of the agenda, (2) to move another agenda item toward the top, (3) to acquire more prestige in the player's own constituency, and (4) to store the influence with a social agency. Outcomes of each round affect the prestige of each player, which in turn affects the future rounds of the game.

Press Conference Game

See the description of this game in an earlier discussion.

Pro's and Con's

Players decide individually to agree, disagree, or remain neutral about eight issues listed on a card. The initial votes are recorded on a score sheet. Issues are arranged by the players in order of priority and each issue is discussed for ten to fifteen minutes in a specific role. (Each player is provided with a flipcard packet that specifies the positive, negative, or facilitative role.) At the conclusion of all discussions, there is another round of recording agreements, disagreements, and neutral stands toward the issues. This is followed by a structured debriefing session.

SciFi

Each team receives an envelope with a problem written on its face. The team writes a solution on an index card, puts it in the envelope, and passes the envelope to the next team. This procedure is repeated until the problem envelopes return to the original teams. Teams remove all solutions and rank them from the best to the worst. The team that contributed the largest number of "best" solutions wins the game.

System I

The two axes of the information display grid represent two dimensions in any given subject area. The tiles contain data that can be fitted into this grid. Players attempt to place their tiles in appropriate positions on the grid. Each player has a partial list of correct answers, and, therefore, players verify the correctness and give score points to each other.

Teams-Games-Tournament

T-G-T is a tournament structure that accommodates any instructional game. Each team has four players, one above average, two average, and one below average. Within each team, players help each other, tutoring the weaker ones. Once a week players compete with each other, not as one team against the other, but as players of approximately equal ability against each other. Winners are bumped up to meet tougher players the following week; losers are matched with less-skilled ones. Scores are converted into tournament points and fed back to the players.

They Shoot Marbles, Don't They?

Five core players participate in this game, which begins with very few rules and adds more, created by the players themselves. During each round, there is a bargaining session and a shooting session. In the bargaining session, players form coalitions to divide up the pot of marbles. During the shooting session, they shoot marbles, attempting to hit job marbles while avoiding the trouble marbles. Players who do not belong to the coalition also attempt to knock down a wooden tower and thus nullify the agreement. These core players are incorporated into a larger infrastructure which includes the police, the court, and government officials.

A COMPARISON OF FRAME GAMES

This section analyzes the fourteen frame games and compares them on the basis of nine critical attributes. The organization of this section uses the following sequence:

- a description of a critical characteristic of frame games
- illustrative examples from selected frame games
- a table comparing the selected characteristics of the fourteen frame games

Loadability

The most unique feature of frame games is the ease with which a variety of new content can be loaded onto the stable structure. Before loading the new content, the old one has to be unloaded. We use the term *loadability* to refer to the process of both unloading the old content and loading the new one.

Some frame games come with no content at all in their original version. Other things being equal, this *skeletal* frame game has the highest loadability because there is no need to unload anything. *Press Conference* and *System 1* are examples of this kind of structure with no content showing.

Encapsulation and *Pro's and Con's* are examples of another type of loadability in which the basic game structure comes prepackaged with alternative contents. The rules and the equipment for *alternative-contents* frame games can be used to accommodate different contents supplied by the developer. Following the model, the game user can create the user's own new content for loading.

Sometimes a frame game appears in *alternative versions,* each with its own content. The game user can compare these versions, identify the stable structure and varying content, and proceed to unload the old content and then reload with the new. For example, *GAMEgame IV* has appeared in many different versions: *Make Policy, Not Coffee* (Frick, 1974) deals with the status of women; *Afar* (Thiagarajan, 1976) deals with future forecasting; *Indicator Hunt* (Thiagarajan, 1974b) deals with the analysis of affective educational goals; *Energy Resources Game* (Thiagarajan, 1974a) deals with the conservation of energy; and *Policy Council Game* (Stolovitch and Thiagarajan, 1979) deals with the problems of hunger in developing countries.

Some frame games are presented in a single version with a *sample content.* However, the developer makes the basic structure of the game explicit and comments on its use as a frame. For example, *GAMEgame VI* is presented as a game for exploring people's opinions toward the use of simulations and games. The rules of the game are presented in general terms and illustrated through sample play related to the content. The author also comments on the use of the game structure to explore similar instructional topics.

Some complex frame games become confusingly abstract when the developer attempts to describe it either in a skeletal fashion or even with sample content. These games have to be played for one or two rounds before the game leader and the players become familiar with its mechanics. They can then cooperatively redesign the game by loading locally relevant content (while making necessary changes to the rules). Goodman is the leading exponent of this type of *priming game* approach to frame games. Both *They Shoot Marbles, Don't They?* and *Policy Negotiations* are deliberately designed to be redesigned by players.

TABLE 1 Comparison of the Loadability of Selected Frame Games

Frame Game	Skeletal	Alternative Contents	Alternative Versions	Sample Content	Priming Game	Recycle
Confrontation			X	X		
Encapsulation		X				
Facts in Five		X				
GAMEgame IV			X	X		
GAMEgame VI			X	X		
Making a Change	X	X				X
Planning Exercise	X	X				X
Policy Negotiations				X	X	
Press Conference	X					
Pro's and Con's	X	X				
SciFi	X			X		
System 1	X	X				
Teams-Games-Tournament	X					
They Shoot Marbles	X				X	

Finally, many frame games get automatically loaded with new content in the process of playing them for the first time. Twelker's games–*Planning Exercise* and *Making a Change* –illustrate this recycling approach to loadability. In the former, for example, players begin with a problem and evaluate solutions. The selected solution can then become the starting problem for the replay of the game.

Table 1 indicates which forms of loadability are exemplified by each of the fourteen selected frame games.

Loader

Related to the loadability of a frame game is the question of who does the loading. Although in the last analysis all frame games permit loading by the user of the game, there are three basic types of people (and various combinations) who may load the game.

• *Loading by the game developer.* The frame game *Encapsulation* is an example of a game which comes with "factory-loaded" content. With this particular game, there are actually six different loads (labor and management, black and white, affluent and deprived, career and homemaking, student and teacher, parent and child) that share some common attributes. Other examples of frame games with preloaded content include *Pro's and Con's* and *Confrontation. System 1* also comes with preloaded content, but this is more in the way of suggested examples to the teacher-user than complete specifications.

• *Loading by the game leader.* Many frame games depend on the game leader (teacher or trainer) to load the new content before play. *Teams-Games-Tournament,* for example, requires the leader to load specific games. The game leader initiates *Planning Exercise* by selecting an appropriate problem area.

• *Loading by players.* Some frame games are dependent on the players themselves for loading new content. *Making a Change* is this type of frame game in which players cooperatively select an appropriate problem area. An effective variation of the player-load theme involves different players (or teams) working on the content loaded by one another. In later phases of *Making a Change,* for example, each team evaluates and provides feedback on the forms completed by another team. An effective example of players operating on each other's content is *SciFi.* The problem identified by one team receives appropriate solutions from *all* other teams during the play of this game.

• *Combination loads.* It should be obvious that different people may take partial responsibility in loading a frame game. *Press Conference,* for example, requires loading by players to create the initial sets of questions and then loading by the game leader (or outside experts) to provide the "answers." *GAMEgame IV* uses a more comprehensive combination: It comes with suggested opinion cards from the developer; the game leader is encouraged to prepare a set of opinion cards before the game; the first activity of the game requires each player to contribute four additional opinion cards. Priming games (*Policy Negotiations* and *They Shoot Marbles*) are deliberately designed to be reloaded by the players after the initial rounds of playing the game with the developer's content.

Table 2 indicates who does the loading in each of the fourteen selected frame games.

TABLE 2 Comparison of the Loader of Selected Frame Games

Frame Game	Developer Load	Leader Load	Player Load (self)	Player Load (others)
Confrontation	X		X	
Encapsulation	X			
Facts in Five	X	X	X	
GAMEgame IV		X	X	
GAMEgame VI	X	X		X
Making a Change			X	X
Planning Exercise		X		X
Policy Negotiations	X		X	
Press Conference		X	X	X
Pro's and Con's	X			
SciFi				X
System 1	X	X		
Teams-Games-Tournament		X		
They Shoot Marbles	X		X	

Flexibility

"Loadability" refers to the ease with which a frame game may be adapted to handle new content. "Flexibility" refers to the ease with which a frame game may be adapted to suit the physical resources and constraints present in the use-context. The three major elements that contribute to the flexibility of a frame game are briefly discussed below.

• *Number of players.* A highly flexible frame game can be played by any number of players. A low level of flexibility is indicated by a frame game that requires an exact number—no more, no less—to successfully play the game. None of the frame games described here has rigid requirements in terms of numbers of players. Generally speaking, a game that has a smaller minimum number of players as a requirement is more flexible than one that requires a large minimum number of players. This is so because a group can always be split into subunits, each playing the game independently. The fewer the number of players required to form each subunit, the easier it is to include everyone. In this respect, *GAMEgame IV* (which requires a minimum of ten players), *Planning Exercise* (a minimum of sixteen players), and *Press Conference* (a minimum of ten players) are less flexible than *System 1, Facts in Five,* and *Confrontation,* all of which can be played by two or three players.

• *Time requirement.* Games with a *brief* minimum time requirement are more flexible than those with longer *minimum* time requirements. You can always replay the former games any number of times to fill up your available space. Thus *System 1* (minimum time requirement of twenty minutes) and *Facts in Five* (minimum time requirement of five minutes) are much more flexible than *Planning Exercise* and *They Shoot Marbles,* which require at least a couple of hours to play. In addition, some frame games (for instance, *GAMEgame IV* and *Confrontation*) have a standard format for playing each round so the number of rounds may be increased or decreased to fit the available time, whereas frame games like *Making a Change* have a nonrepeated progression.

• *Material requirements.* Frame games that require mere paper and pencil for play are much more flexible than those that require elaborate equipment and special game materials. Although most of the frame games reviewed here are extremely flexible in this respect, *They Shoot Marbles* is an exception. It requires an elaborate set of paraphernalia (game surface, marbles, Lego blocks, timers, and so forth) to set up the game. Some of the commercial games (*Encapsulation* and *Facts in Five*) tend to require specialized materials, whereas noncommercial ones are less expensive and require more easily

TABLE 3 Comparison of the Flexibility of Selected Frame Games

Frame Game	Number of Players	Time Requirement	Time Flexibility Rating	Materials
Confrontation	3 or multiples of 3	45 to 90 mins.	high	Confrontation cards poker chips
Encapsulation	3 to 12	45 to 90 mins.	medium	"Bias" board, role cards, forms
Facts in Five	2 or more (in teams, if necessary)	5 minutes per round	high	Deck of cards with classes and categories, playcards, master score cards, timer
GAMEgame IV	3 to 30 players in 3 to 6 teams	30 to 60 mins.	high	No special materials
GAMEgame VI	10 to 60	30 to 60 mins.	high	Opinion cards (blank and pre-printed)
Making a Change	12 or more players in teams of about 5	Two sessions of approximately two hours each, with time for indep. study in between	low	Participant's manual and different forms
Planning Exercise	16-36 players in 3-7 teams	3 to 5 hours	low	No special materials
Policy Negotiations	6 to 30	2 to 4 hours	low	Poker chips, list of issues, and score board
Press Conference	10 to 30	90 mins. to 3 hours	low	Index cards of 4 different colors
Pro's and Con's	3 to 15	2 to 3 hours	high	Issue cards, score sheets, flip card rings, observer's form
SciFi	3 to 30 players (individual players up to 7; after that 3 to 6 equal teams)	30 to 90 mins.	high	Envelopes for problems and index cards for solutions
System 1	2 to 9	20 to 40 mins.	high	Information display grids, plastic tiles and adhesive paper for preparing data units, and storage units
Teams-Games-Tournament	12 to 60 divided into teams of four players	Regular class period, once a week for a semester	medium	Materials required for the games used in the tournament
They Shoot Marbles	12 to 50	1 to 3 hours	low	Games surface, marbles, Lego blocks, wooden blocks, cylinder, etc.

available materials.

Table 3 compares the flexibility of the fourteen selected games. All the information provided is factual except for the time-flexibility rating. This is an expert judgment on our part which takes into consideration a number of factors including our experiences with the game.

Purposes

Frame games may be designed to serve a variety of purposes. One primary purpose of the games reviewed is to provide instruction. Within this broad goal, the game may help the players achieve different types of specific objectives. Other frame games may have a noninstructional intent: They are designed to provide an efficient organization for group activities. Brief descriptions and illustrations of the different purposes of frame games are provided below.

Instructional Purposes

• *Awareness objectives.* Some frame games produce results in the borderline area between the cognitive and affective domains where players are sensitized to the presence of various factors. In *GAMEgame IV,* for example, players become aware of the wide range of opinions toward the use of simulations and games, including those they had never imagined possible. *Confrontation* and *Encapsulation* are examples of frame games in which the players acquire insights into influences that govern their behaviors.

• *Lower-level cognitive objectives.* Some frame games help players learn basic facts and figures. *System 1* is a game that can sugarcoat a dull didactic drill. Even though *Facts in Five* and *Teams-Games-Tournament* could be used for other instructional purposes, they are examples of frame games that are suited for this kind of lower-level learning.

• *Higher-level cognitive objectives.* Frame games may also help players synthesize and apply various skills and concepts for solving problems. *Press Conference,* for instance, requires players to gather, structure, and effectively summarize information. *Facts in Five* and *System 1* are also examples of games that may be used to help players acquire such higher-level cognitive skills.

Noninstructional Purposes

• *Planning.* An obvious example of a frame game that structures the planning activities of a group is Paul Twelker's *Planning Exercise.* This frame game enables a team to identify a problem, create alternative solutions, and evaluate their relative merits. Twelker's other game, *Making a Change,* is also an example of a frame game designed for planning purposes.

• *Group decision making.* Planning is just one example of a wide range of group decision making activities that may be rendered more systematic through the use of a frame game. *Policy Negotiations* is an example of a frame game that organizes the decision-making process so that various players have a say on what decisions are made and how they are made. Goodman's other game, *They Shoot Marbles, Don't They?* structures the way in which a group of players makes decisions to govern the group's activities.

• *Evaluation.* Some frame games are used for enabling a group to jointly evaluate an idea or object. In *GAMEgame VI,* the group is required to brainstorm a list of ideas and then to arrange them in a collaborative effort in order of priority. *SciFi* requires different players to come up with alternative solutions to different problems and lets the players themselves decide on the relative efficiency of each solution.

Table 4 indicates the purposes of the fourteen selected frame games. Whereas it is true that any given frame game may be used to serve any purpose in the hands of a skilled game leader, we have indicated only the primary purposes of each frame game.

Levels of Simulation

Just like any other instructional game, a frame game may or may not be a simulation of some aspects of reality. The following is a brief discussion of different levels of simulation among the selected frame games.

• *Nonsimulation frame games.* These frame games make no attempt to re-create reality. They are usually games with a cognitive instructional intent at the lower levels. Often they help the players master facts and figures and arrange them in various categories. *Facts in Five* and *System 1* are examples of this type of nonsimulation frame game.

TABLE 4 Comparison of the Purposes of Selected Frame Games

	Instructional Purposes			Noninstructional Purposes		
	Awareness	Lower Cognitive	Higher Cognitive	Planning	Decision Making	Evaluation
Confrontation	X					
Encapsulation	X				X	
Facts in Five		X	X			
GAMEgame IV					X	X
GAMEgame VI	X				X	
Making a Change				X	X	X
Planning Exercise				X	X	X
Policy Negotiations	X				X	
Press Conference	X	X	X			
Pro's and Con's	X				X	
SciFi					X	X
System 1		X	X			
Teams-Games-Tournament		X	X			
They Shoot Marbles	X				X	

• *Pseudo-simulation frame games.* Very often a frame game of the nonsimulation variety is clothed in a simulation that has no relevance to the instructional intent. *Press Conference,* for example, has some elements that represent what happens during an actual press conference. However, this representation has nothing at all to do with the instructional intent of the game, which is to communicate various facts related to a subject area.

• *Simulation frame games. Confrontation* is an example of a simulation frame game in which critical elements of a confrontation situation are depicted in a game format to help players achieve insights and skills related to the interpersonal dynamics of such a situation. *Policy Negotiations* is another simulation frame game in which various critical elements of the way a group makes a decision are realistically portrayed in a compressed-time model. *They Shoot Marbles* is another simulation game related to the process by which people relate to each other in a society. In comparison with the other simulation frame games, *Marbles* is a highly abstract simulation.

• *Operational frame games.* Most of the frame games used for noninstructional purposes are procedures for problem solving in groups. *Planning Exercise,* for example, is not meant to *simulate* how groups solve problems in real life. It is to be used for actual problem solving. *SciFi* is another example of a frame game that may be used for putting a systematic procedure for group interaction into operation.

Table 5 compares the levels of simulation of the selected frame games.

Complexity of Rules

The number and complexity of rules vary considerably from one frame game to another. Frames such as those contained in *Facts in Five* and *SciFi* have fairly simple rules that can be explained in a matter of minutes. On the other hand, frames like those in *They Shoot Marbles* and *Making a Change* have sets of complex interrelated rules. Even with elaborate explanations, the players may have to get into the game with a only a partial understanding of the rules and then wait to obtain "on-the-game" clarification.

Table 6 compares the number and complexity of rules for the selected frame games. Incidentally, there is no value judgment implied in a frame game's having complex rules. It is our belief that all the selected frame games have the optimum complexity of rules and that, given adequate preparation and appropriate setting, all the selected games work well.

TABLE 6 Comparison of the Complexity of Rules of the Selected Frame Games

Frame Game	Simple	Medium	Complex
Confrontation		X	
Encapsulation		X	
Facts in Five	X		
GAMEgame IV		X	
GAMEgame VI		X	
Making a Change			X
Planning Exercise			X
Policy Negotiations			X
Press Conference		X	
Pro's and Con's	X		
SciFi	X		
System 1		X	
Teams-Games-Tournament	X	X	X
They Shoot Marbles			X

Scoring

One type of rules related to frame games deals with the scoring system and the determination of winners and losers. There are two specific aspects of scoring systems which vary among the selected frame games, and these are briefly described below.

• *Number of criteria for winning.* A frame game may emphasize just one criterion for winning or it may focus on more than one. *SciFi* and *Facts in Five* have single win criteria. In the former, the player or team that writes the most top-ranked solutions is the winner. In the latter, the player or team that has the most cells correctly filled is the winner. *Confrontation* and *GAMEgame IV* use multiple criteria. In the former, scores are compared to decide who has accumulated the most poker chips, which group has made more compromises, and which group has solved its problems most effi-

TABLE 5 Comparison of Levels of Simulation of Selected Frame Games

Frame Game	Nonsimulation	Pseudo-simulation	Simulation	Operational Game
Confrontation			X	
Encapsulation			X	
Facts in Five	X			
GAMEgame IV				X
GAMEgame VI	X			
Making a Change				X
Planning Exercise				X
Policy Negotiations			X	
Press Conference		X		
Pro's and Con's			X	
SciFi				X
System 1	X			
Teams-Games-Tournament	X	X		
They Shoot Marbles			X	

ciently. In the latter, teams may win either for being able to psych out the other teams' selections or for including most of the top items in their original list.

• *Zero-sum and non-zero-sum frame games.* In games like *GAMEgame IV* and *SciFi,* there is a single winner (for each category, if there is more than one). In games like *Press Conference* and *Confrontation,* it is possible for more than one person to win in each category. The former type, in which it is not possible for more than one person (or team) to win, is called a zero-sum game (in a simplified sense). The other type, which potentially permits everyone to win, is called a non-zero-sum game.

Table 7 compares the scoring systems of the fourteen selected frame games. As you may notice, some frame games (for instance, *Encapsulation* and *They Shoot Marbles*) de-emphasize the scoring system completely.

Advantages and Limitations of Frame Games

We do not claim that the fourteen games we reviewed represent the best that is currently available in the area of frame games. It is our opinion, however, that these fourteen are excellent examples of the versatility and diversity of frame games.

The advantages and limitations of instructional games have been listed and discussed so frequently that it would be superfluous to repeat them here. However, we would like to briefly point out some unique advantages and disadvantages of *frame* games.

On the positive side, frame games help the user in these ways.

• Designing a game from scratch is a time-consuming and unpredictable task. However, with the frame game approach, a teacher or trainer can "design" a game in a fraction of the time and with guaranteed results (because of the testing and revisions the original frame has undergone).

• Very often the teacher or trainer does not find a game that precisely meets the local needs and objectives. With the frame game approach, the game can be custom tailored to incorporate the exact contents desired.

• For the learner, frame games frequently provide an opportunity to participate in loading new content. Players probably learn more from such participation in the creation of the game than in actual play.

• Once the rules of a frame game have been mastered, it is easier to learn how to play the new loads. Thus the learner spends less time mastering the mechanics of a game and more time exploring the content when different versions of the same frame game are played.

Frame games are efficient and powerful tools, and therein lies the major danger in their use. The simplicity of frame games brings into action Kaplan's Law of Hammerability ("Give a kid a hammer and he will find hundreds of things that need hammering"). In earlier workshops we used to extol the virtues of the children's game Slapjack. We lived to regret our contagious enthusiasm when we received hundreds of loads on the frame for the next three years. Teachers and trainers were using the game to teach shape discrimination in kindergarten and sonar-blip discrimination for the crew of a nuclear submarine, bovine-respiratory-ailment discrimination for veterinary doctors, and discrimination of different styles of acting for theater majors. There is nothing more damaging to an innovation than looking for suitable problems to apply it to, and nothing more disillusioning than the rigid appreciation of a flexible tool. We hope that in this essay we were able to convince you of the merits of frame games in general and a few specific ones in particular without creating an obsessive need to use them.

Sources

Confrontation
Harold D. Stolovitch
1979

TABLE 7 Comparison of the Scoring Systems of Selected Frame Games

Frame Game	Win Criteria		Type of Game		Scoring De-emphasized
	Single	Multiple	Zero-sum	Non-zero-sum	
Confrontation	X		X	X	
Encapsulation					X
Facts in Five	X		X		
GAMEgame IV		X	X		
GAMEgame VI		X	X		
Making a Change					X
Planning Exercise					X
Policy Negotiations		X	X		
Press Conference		X		X	
Pro's and Con's					X
SciFi	X		X		
System 1	X		X		
Teams-Games-Tournament	X		X		
They Shoot Marbles					X

Instructional Alternatives
4423 East Trailridge Road
Bloomington, IN 47401
$2.00

Encapsulation
1972

Creative Learning Systems, Inc.
936 C Street
San Diego, CA 92101
master set $35.00

Facts in Five
R. A. Onanian
1966

Avalon Hill
4517 Harford Road
Baltimore, MD 21214
$12.00

GAMEgames IV and VI
Sivasailam Thiagarajan
1976, 1978

Instructional Alternatives
4423 East Trailridge Road
Bloomington, IN 47401
each $2.00

Making a Change
Paul A. Twelker and
Kent T. Layden
1974

Simulation Systems
Box 46
Black Butte Ranch, OR 97759
$4.00

Planning Exercise
Paul A. Twelker
1971

Simulation Systems
Box 46
Black Butte Ranch, OR 97759
$3.00

Policy Negotiations
Frederick L. Goodman
1974

Institute of Higher Education Research
and Services
Box 6293
University, AL 35486

Pro's and Con's
1976

Creative Learning Systems, Inc.
936 C Street
San Diego, CA 92101
$19.95

SciFi
Diane Dormant
1976

Instructional Alternatives
75¢

System 1
Instructional Simulations, Inc.
1969

Griggs Educational Service
1731 Barcelona Street
Livermore, CA 94550

Teams-Games-Tournament
David L. DeVries and
Keith J. Edwards
1973

Center for Social Organization of Schools
The Johns Hopkins University
3505 N. Charles Street
Baltimore, MD 21218
$3.00

They Shoot Marbles, Don't They?
Frederick L. Goodman
1973

Institute of Higher Education Research
and Services
Box 6293
University, AL 35486

References

FRICK, K. B. (1974) "Make policy, not coffee." Simulation/Gaming/News 1: 13.

STOLOVITCH, H. and S. THIAGARAJAN (1979) Frame Games. Englewood Cliffs, NJ: Educational Technology Publications.

THIAGARAJAN, S. (1976) "Alternative futures analysis and review (AFAR): an operational game for predicting desirable futures. Viewpoints 52: 2.

--- (1974a) "ERG: energy resources game." OPT: The Magazine on People and Things 1.

--- (1974b) "Indicator hunt: a goal analysis game." Educational Technology 14: 4.

FUTURES GAMES AND SIMULATIONS
An Evaluation

by Charles M. Plummer

Future-oriented simulations/games move us away from the-die-is-cast past and present viewpoints toward the fluid shape of things to come. Out of awareness of the possibility of change we develop a desire to create preferred events.

If we were to inventory the controls we might want over time and events, there might appear to be some justification in shifting our orientation to a more distant future. We cannot change past events, and present events are occurring at a rapidly accelerating rate; food, energy, ecological, and population crises may suggest that the power to control events is out of reach.

Arthur C. Clark, in *Profiles of the Future* (1963), has detailed powers we might like to have over time, compiling controls without regard to their feasibility. He suggests that our ideal list of powers should include ways to (a) see, reconstruct, change, and travel into the past; (b) speed up or slow down the present; and (c) see and travel into the future. Impressive powers, we might agree, but certainly impossible.

The methods of simulation/gaming give us a chance to capture some aspects of powers over time which Clark has suggested. Through realistic re-creation of events, simulation/gaming can often provide participants with dramatic experiences in seeing, reconstructing, or traveling into a different epoch. When students reenact historical events during simulation/gaming—where the events, resources, or goals assume values the students have selected—they may thereby experience changing the past, creating their own "history."

There are many examples of the unique advantages of simulating events that might happen in the future. Many of us have seen television programs showing how simulation training prepares astronauts to land a spaceship on the moon and handle space travel emergencies. Many have learned survival strategies from fire drills. Some nations possibly have been deterred from initiating a nuclear holocaust because of conclusions they reached from observing their simulations of the consequences of war.

While science fiction books and films give us opportunities to observe possible future events, the simulation/gaming approach moves a step beyond to create an environment within which the participant may actively engage and interact with the potential events. These alternative future events may not only be seen, but they may be traveled into through the time-transcending system of simulation/gaming. Concrete experiences with hypothetical events become reality. Although the events being simulated are hypothetical, the experiences of the simulation are real. We become engaged in a process of making believe for real.

Dynamic Modeling of Alternative Futures

Designers of simulations/games involving human interaction, when attempting to dynamically model alternative futures, proceed through a process of design, development, and testing that generally includes the following major phases:

(1) *Awareness is directed to probable, preferable, or possible future events.* Initially, a set of likely or hypothetical events receives attention. Growing from an awareness of these multiple alternative future possibilities and their potential consequences, we systematically focus upon a few future events that may pose the most important consequences. Attitudes toward these events are typically strongly positive or negative, depending upon the desirability of the conditions envisioned.

(2) *Components of a model of events are specified and organized.* A framework, structure, or model of reality is constructed in an attempt to portray accurately the most crucial features of the events of interest. At this point, we proceed through two simultaneous stages:

   (a) We create an abstract or symbolic model that imitates the real or possible events, and perhaps changes their scale, in an attempt to direct attention to specific elements.

   (b) We attempt to preserve the organization of the interreleationships among the elements in the model in an attempt to faithfully reflect the patterns of events to be simulated.

(3) *An operable simulation/game is designed which gives concrete form to possible events.* Goals, constraints, rules, roles, assumptions, values of resources, and other structural elements are specified as limits within which

Editor's Note: *Prospects* is listed in the self-development section, and *Edplan* and *Edventure II* in education. The rest of the simulations and games discussed in this essay are listed in the futures section.

simulation participants are free to operate. Although the simulation design process may be a deliberate attempt to represent a model as accurately as possible, it will also be the result of a number of compromises and arbitrary decisions. A simulation should always be understood to inherently reflect the knowledge, preferences, and perspectives of its creator. To acknowledge bias in the design process is not to attribute ulterior motives to the designer, but is merely to recognize that it is often not possible to incorporate as many features, or to portray them as accurately, as one may wish. Of course, distortions may be deliberately incorporated, and in some cases may even be the central purpose.

(4) *Operation of the simulation/game is initiated.* The initial structure of the simulation/game is superimposed as a created environment surrounding participants. Their actions subsequently feed back information that is actively or passively accommodated, or responded to, when it interacts with the framework of the simulation.

(5) *Feedback created by participants in the simulation/game becomes the stimulus for subsequent actions during the simulation operation.* The behavior of participants is a crucial input which interacts with the simulation structure to produce a successful simulation/game enactment.

(6) *Operation of the simulation/game is repeated, replicating or varying simulation elements.* Greater confidence may be placed in the validity of the simulation, or the interrelationships of its elements, when it is repeated under exactly the same conditions, with the effects noted. Carefully controlled manipulation of a few variables of interest may also offer evidence of validity if predicted consequences occur.

(7) *Events created by the simulation enactments are observed and/or evaluated while the model is operating.* Observable actions, or any products created by the simulation enactments, can provide a basis for future modifications, or for evaluating questions of interest.

(8) *Termination of simulation/gaming occurs at some stage.* The dynamic modeling of alternative futures through simulation/gaming is concluded.

(9) *Generalization from the simulation/game through the model to the future events of original interest may be attempted.* The validity of generalizing to future events must rest ultimately upon either (a) the degree to which predictions match real events or (b) the extent to which control over present events is enhanced through engaging in the process of future-oriented simulation.

THE SIMULATIONS

We will compare, categorize, and evaluate twelve futures research-based simulations in terms of their underlying models, educational purposes, procedures, content, and outcomes. First, we will analyze their educational uses, briefly describe each simulation/game, and note some preliminary factors to be considered in a selection decision, including age level, playing time, and the number of players and groups the simulation/game accommodates. Second, we will present the stated educational objectives from each simulation/game manual, as well as an analysis of the knowledge and skills participants learn or apply. Third, we will summarize the sequence of major activities, types of roles and role descriptions, and an evaluation of the roles and rules of each simulation. Fourth, we will review the resources most relevant to participants' success and describe the scoring systems. Fifth, we will review the packaging, cost, completeness, and durability of the simulations. Sixth, we will analyze the debriefing procedures and comment on the adaptability of the simulation/game to other purposes and/or audiences. Finally, we will present an overall rating of the simulations based on all these dimensions to help users select and evaluate those most appropriate for their purposes.

All simulations/games selected for review are future-oriented, have been played a number of times, are available within the United States, and cost less than $100 (in most cases, less than $20). All incorporate the findings and/or techniques of futures research, and each can be applied to achieve one or more stated educational objectives. All are creatively designed and are highly arousing and motivating, or fun to play, or both.

Educational Uses

These simulations/games may serve one or more of the following educational uses:

(1) *Teaching analytical skills through estimating probabilities:* Explore or estimate the probabilities that hypothetical events or trends are likely to happen through using opinions of experts in several rounds of consensus-building.

(2) *Exploring possibilities:* Create a climate of suspended judgment in which possibilities may be seriously explored. Futures simulation/gaming is particularly helpful when the possibilities to be considered may appear remote, irrelevant, or threatening to the target group, or when people's opinions about the issues or events are so positively or negatively charged that they find it difficult to consider alternative points of view or even new information.

(3) *Learning futures research skills:* Learn some of the skills involved in conducting futures research through experience in applying delphi methodology, cross-impact analysis, and other futures research techniques as a simulation participant.

(4) *Clarifying preferabilities:* Develop a clarification of values and preferences regarding highly desirable future events or goals, regardless of the likelihood that they will happen or be attained.

(5) *Developing future-focused role images:* Develop a more future-focused role image through opportunities to think clearly and specifically about one's own future and to begin planning immediately to take action to achieve desired outcomes. Broadening perspective in space and time encourages individuals to think more creatively and imaginatively, so they will not expect their personal futures simply to mirror the past.

(6) *Increasing optimism:* Increase feelings of control over the future through simulated opportunities to confront personally relevant issues before they happen in the real world. The chance to try out alternative courses of action, especially with situations often perceived as outside one's sphere of control or influence, may help

TABLE 1 Educational Uses

Educational Uses	*Futuribles*	*Dynamic Modeling of Alternative Futures*	*Simulating the Values of the Future*	*Space Patrol*	*Hybrid Delphi Game*	*Future Planning Games*	*Cope*	*Utopia*	*Prospects*	*Edplan*	*Edventure II*	*2000 A.D. Futura City*
1. Teaching analytical skills through estimating probabilities	H	H	H	H	H	M	H	M	H	M	H	H
2. Exploring possibilies	H	H	M	H	H	H	H	H	H	M	H	M
3. Learning future research skills	M	H	M	L	H	M	H	H	M	L	L	M
4. Clarifying preferabilities	H	H	H	L	H	H	M	H	H	H	H	H
5. Developing future-focused role images	H	M	M	M	M	H	M	H	H	L	H	L
6. Increasing optimism	H	M	M	H	M	H	L	H	H	L	H	M
7. Experiencing decision-making opportunities/ confronting moral or ethical dilemmas	H	H	M	H	H	H	H	H	H	M	M	H

Key: H = Highly useful
M = Moderately useful
L = Limited usefulness

develop optimism, strategy, and skills in approaching real situations.

(7) *Experiencing decision making opportunities; confronting moral or ethical dilemmas:* Decision-making strategies and techniques may be applied to moral and ethical problem situations. Group debate, discussion, persuasion, and consensus-building may help clarify facts, inferences, and implications, as well as clarifying underlying values and assumptions.

We have reviewed each of these simulations/games in terms of a variety of these potential educational uses, and the reader can make detailed comparisons by reviewing tables on their educational uses (Table 1), summary descriptions of their content and process (Table 2), their objectives (Table 9), their issues (Table 7), possible values promoted (Table 6), and model characteristics (Table 8).

Futuribles and *Prospects* receive, respectively, seven and six ratings of highly useful. *Edplan* is the most limited, receiving a high rating only for the purpose of clarifying preferabilities. Although *Simulating the Values of the Future* is cited as highly useful for only two purposes, it achieves them extremely well and is clearly one of the best simulations/games for those purposes.

TABLE 2 Summary Descriptions of Simulations/Games

Simulation/Game	Description
Futuribles	A game engaging the player in prediction, clarification of, or commitment to possible, preferable, and/or probable future projections in 19 categories, with 288 cards.
Dynamic Modeling of Alternative Futures	A series of five simulation/games for adults employing the methods of delphi, in-basket exercise, dynamic modeling, and a "game as reality" paradigm. Topics include predicting desirable futures, assessing needs, identifying creativity, simulating a lifetime, and modifying a behavior.
Simulating the Values of the Future	A delphi simulation designed by Olaf Helmer to disclose the interaction of technology and values. Participants as "experts" consider probability and difficulty weightings in 20 areas and 186 specific events, and achieve consensus. Evaluating the desirability of the events is then conducted from the perspectives of five different groups.
Space Patrol	A sophisticated science fiction war game provides a simulation design and rule structure for generating "Star Trek" or Star Wars types of scenarios. Precise directions help users create their own scenarios; characters (aliens and creatures); determine success probabilities; account for gravity and movement; and provide for contact, combat, and recovery.
Hybrid Delphi Game	Participants rank the desirability of future events, achieving consensus through a delphi process on most desirable events. Users could readily put in alternative future events for evaluation, discussion, and debate.
Future Planning Games	Brief classroom activities and attitude questionnaires have students discuss and debate issues within different games on future issues: constructing a political philosophy, planning tomorrow's society, focusing on the ecology crisis, constructing a life philosophy, planning tomorrow's prisons, determining America's role in the world, determining family and sexual roles, dealing with death, protecting minority rights, examining American values, dealing with developing nations, determining economic values, and preventing crime and violence. Supporting paperbacks on opposing viewpoints are available

TABLE 2 Summary Descriptions of Simulations/Games (Cont)

Simulation/Game	Description
Cope	Participants are "born" into the future city of Technopolis and experience five ten-year periods, earning "Creative Work Units" for solving problems of living, which include (1) thinking about future alternatives, (2) solving society's problems, (3) learning a FUTURESPEAK abbreviated language, (4) evaluating technology, and (5) coping with rapid change.
Utopia	As part of a communal experimental group in Sunrise Valley, students develop an "ideal" society and clarify their own ideologies. Through exposure to alternatives for organizing society's politics, technology, economics, and morality, students conduct research and present "ideal" solutions which the group evaluates.
Prospects	A self-administered simulation leading one to disclose future career goals, evaluate career and training needs, and maintain and enhance the relevance of one's skills, to avoid professional obsolescence.
Edplan	Focusing on the politics of educational decision-making, players may assume roles as various interests in a community converge on the issue of allocating a future school system budget. Through debating the emphasis on components of the new school program, participants meet, formulate presentations and lobby, participate in a school board meeting, have city and federal agencies review their requests, have elections for city council and school board, and determine outcomes for the school.
Edventure II	This career guidance game for students projects them into a role profile character with varied levels of strengths, needs, experience/education, savings, and surplus income. Through their selections from an "Education Menu Book," they plan their characters educational life from 1981 to 2000 and experience the consequences in terms of life satisfaction, surplus income, and savings for retirement.
2000 A.D. Futura City	This self-contained multimedia kit combines narrative readings, handouts, a cassette and filmstrips, displays, and simulation materials to form an opportunity for students to clarify values, analyze trend indicators, evaluate priorities, interpret scenarios, conduct trend extrapolation, and experience group interaction. Twenty to thirty participants role play the future planning of an urban city afflicted with many problems of decay and alienation. In the process of reading carefully selected and edited readings, they generate an urban plan, present it to the Constituent Assembly, reevaluate and bargain, debate, and vote. Simulation concludes with a structured debriefing and evaluation.

PRACTICAL CONSIDERATIONS

Age Level, Group Size, Playing Time, Flexibility in Outcomes/Issues

Table 3 presents the simulation/game age levels, playing time, number of players, and number of groups, along with the range of possible outcomes and issues. *Prospects,* intended primarily to aid users in avoiding obsolescence in their chosen careers is classified at the highest age level. The playing times for the simulations/games range from 15 minutes for *Futuribles* to 17 hours for *Cope,* a simulation that organizes classroom activities for four weeks of 50-minute periods. Although among the longest in duration, *Cope* and *Utopia* are extremely well organized in specifying a calendar of activities, teacher instructions, and student projects. Numbers of players and groups range from a single player in a self-administered mode, as in *Prospects,* to as many as 30 players in ten groups, as in the original enactment of *Simulating the Values of the Future.* (To apply that simulation with as many players would require extensive preparation of additional materials, however.)

You can readily determine the flexibility of the simulations/games by reviewing Table 3 on possible outcomes and issues. The most flexible is clearly *Futuribles.* Packaged as a deck of cards, it is highly portable, nominal in cost, and embraces the spectrum of possible approaches to the future in the outcomes sought and issues treated. It is the most adaptable simulation game for application with a variety of age levels, given differing time constraints. You can keep the issues simple or readily make them more complex. The least adaptable simulation/game reviewed is *Edplan,* in that its large-group format and its highly political planning arena restrict the range of issues, alternatives, and outcomes.

Complexity

The complexity of the simulations is reflected in Table 4 by the placement of the simulations along a continuum from simple to complex. This generally parallels their age range, with the more complex simulations most effectively applied at the age of 16 or older. The simulations involving the greatest complexity—especially in the manual of instructions and in the amount of computations necessary to implement or score—include *Space Patrol and Simulating the Values of the Future.* However, although they are complex by comparison with the other simulations, the calculations require simple addition, subtraction, multiplication, or division, which can be done with a pocket calculator. At the other end of the continuum, procedures for playing *Futuribles* and *Future Planning Games* are clear and easily understood. These games can be equally effective with 12-year-olds and adults, with only minor adaptations.

SEQUENCE OF MAJOR ACTIVITIES

For most of these simulations/games, the sequence of activities frequently involves presentation of rules and objectives, distribution of player materials (role descriptions, scorecards), information input, several rounds of play, evaluation, and

TABLE 3 Age Level, Duration, Number of Players/Groups, Range of Outcomes/Issues

		Duration		Players			Outcomes		Issues	
Simulation	Age Level	Degree of Flexibility	Playing Time	Degree of Flexibility	Number of Players	Number of Groups	Degree of Flexibility	Range of Possible Outcomes	Degree of Flexibility	Range of Possible Outcomes
Futuribles	12 yrs+	High: 15 minutes to 4 hours		Moderate: 2 to 8, with larger group possible (more decks required)		1-6	High: 1 to 14 rounds of play, players may each produce lists of different dimensions/concerns		High: Includes opportunities to confront future probabilities, feelings, cause and effect, values, before and after, scenarios, inventions, visions, priorities	
Dynamic Modeling of Alternative Futures	High School+	High: 15 minutes to 3 hours		Moderate: 10 to 30+		1-15	High: Lists of probabilities/desirabilities of Future Events; Simulated Life Career; Decisions on Placement of Children into Gifted Program; Case Study solutions to behavior modification problems.		High: Generating lists of possible and desirable future events; determining access to equal opportunity issues in behavior control and engineering	
Simulating the Values of the Future	High School+	Moderate: 5 hours specified, but can be adapted to 2 hours		Moderate: 2 to 30+		2-10	Moderate: List of decisions, estimate of social consequences and their evaluation		High: Probability of future events, planning and decisons regarding creating events; analysis and evaluation of social consequences in terms of 6 societal groups	
Space Patrol	High School+	Moderate: 2 to 6 hours		Low: 2 to 6		1-4	Moderate: Enjoyment of a science fiction war game; could use as a means of generating roles and scenarios in designing a game		High: Issues involve decisons creating characters, selecting ranges of success probabilities, degree of gravity, movement, equipment, contact, recovery, combat, and scenario construction.	
Hybrid Delphi Game	High School+	High: 1½ to 3 hours		Moderate: 6 to 36		1-12	Moderate: task of listing probabilities and desirabilities for up to 90 events, with consensus list of 15 most desirable, provides for inputing own events.		Moderate: Decision of game user on additions/deletions of proposed events; very controversial debates on desirability and probability of events can reflect many alternative values/beliefs.	
Future Planning Games	12 yrs+	Moderate: 30 minutes to 1½ hours		High: 2 to 36		2-8	High: Review of alternative scenarios and solutions to reveal controversial social problems; completed attitude questionnaires; individual/group reports; related projects possible based on supporting books.		High: Prison reform, political philosophy, America's international role and policies, ecology, crisis, type of society and national priorities, family and sexual roles, dealing with death, constructing a life philosophy.	
Cope	High School+	Low: 50-minute class sessions each day for 4 weeks (20 sessions)		Low: entire class (6 to 36 possible)		2-10	Moderate: Students can obtain experience in alternative roles of humanists, engineers, computer assistants, innovative technology specialists, life quality technicians, communication/language specialists, food production and control specialists, transportation and production specialists.		High: What is a *good* and *bad* future? Can we adapt quickly enough to keep up with the pace of change? Can we *cause* the kind of future we prefer?	

TABLE 3 Age Level, Duration, Number of Players/Groups, Range of Outcomes/Issues (Cont)

Simulation	Age Level	Duration: Degree of Flexibility	Duration: Playing Time	Players: Degree of Flexibility	Players: Number of Players	Number of Groups	Outcomes: Degree of Flexibility	Outcomes: Range of Possible Outcomes	Issues: Degree of Flexibility	Issues: Range of Possible Outcomes
Utopia	High School+	Low: 50 minute class sessions each day for 2½ weeks (13 sessions)		Low: entire class (8 to 35 possible)		2-8	Moderate: Group decisons on moral, economic, technological, and political systems; individual written reports; written proposals for group consensus decisions, oral reports		High: How does one *construct* a moral, economic, technological, and political system? How does one *enforce* it? How does one judge disputes, and what standards/criteria are applied? How does one control outcomes, and socialize people to accept the system? Are minority rights protected?	
Prospects	College+	Low: 4½ hours with follow-up session 1 month later		Moderate: 1 or more; 1 booklet per participant		0	Moderate: A future career profile, action plans, list of ideal and and expected jobs, review of related issues; broad range of alternative roles can be considered and evaluated.		Moderate: What should my future career be? What are my strengths/weaknesses? What are my ideal and expected jobs? Do relevant others share my perspectives? Are my action plans realistic?	
Edplan	High School+	Moderate 1½ to 3 hours		Low 29 to 36		6-11	Low: Presentations of viewpoints from role positions of a school system, administrators, school board, city council, PTA, federal aid representative, and taxpayers. Experience in debate, lobbying, and elections.		Low: The game is largely a political view of planning, demonstrating how various interest groups interact to determine an educational budget. Different political factions can be substituted.	
Edventure II	12 yrs+	Moderate: 1½ to 5 hours		High: 1 to 45		0	Low: Experience with with educational planning, and its consequences from 1981 to 2000; a record of "satisfaction," income, and savings resulting from 19 years' decisions; different educational experiences could be offered.		Moderate: The impact of earlier decisions upon later alternatives becomes clear; the possible relationships of education, work, and leisure to both income and satisfaction is explored in a context of constraints and opportunities.	
2000 A.D. Futura City	14 yrs+	Moderate: 3 to 10 50-minute class sessions (5 to 8 hours)		Low 20 to 30		8	Low: After research presentations and debate, an urban plan is presented to a simulated "Constituent Assembly" and voted upon, with rounds of re-evaluation and bargaining; factors to be considered in decision making can be modified.		Moderate: What kind of future do you want? What future events and possibilities do futurists envision? Can you invent your own future? How should a city plan to use two tracts of land?	

TABLE 4 Complexity

Complex				Simple
Space Patrol	*Dynamic Modeling of Alternative Futures*	*2000 A.D. Futura City*	*Edplan*	*Futuribles*
Simulating the Values of the Future		*Prospects*	*Edventure II*	*Future Planning Games*
		Cope	*Hybrid Delphi Game*	
		Utopia		

TABLE 5 Major Activities in Sequence

Simulation	Activities
Futuribles	(1) form groups; (2) deal cards; (3) get familiar with cards; (4) player chooses card to place face up on basis of probabilities (most likely), feelings, cause and effect, values, before and after, scenarios, inventions, leverage, priorities; (5) draw new card; (6) process continues in turn; (7) end when all cards drawn.
Dynamic Modeling of Alternative Futures	*Delphi*: (1) receive future forecasts; (2) assign probabilities; (3) receive feedback; (4) revise estimates; (5) consider new items; (6) recycle until all items done. *Alternative Future Analysis and Review*: (1) teams generate lists of 5 desirable future events; (2) compile master list of 10; (3) each team secretly selects most desirable event; (4) team receives as many points as number of teams choosing same event; (5) events given desirability ranks; (6) generate list of top five events. *Identification of Gifted Simulation*: (1) futuristic multimedia scenario; (2) review application blanks; (3) make individual selections; (4) make group decisions; (5) feedback in one of 2 modes: (a) information only or (b) detailed didactic debriefing; (6) administration of criterion measures/attitude scales. *Label Game*: (1) read rules and distribute game board materials; (2) players begin at start, toss die, move markers; (3) labeling rings awarded; (4) labeling stigma/consequences result in tracks; (5) achieve differential success; (6) debriefing discussion of implications. *Toward Walden Two*: (1) read rules; (2) select reward goals; (3) sign game contract; (4) information input by one of three self-selected modes; (5) form teams; (6) select case study; (7) designate "observers" and "earner" roles; (8) "earners" discuss solution while observers record their behavior; (9) earn points; (10) player's points compared to reward costs; (11) rewards purchased and enjoyed; (12) debriefing and discussion; (13) recycling.
Simulating the Values of the Future	(1) introductory address on "Interaction and Technology and Values"; (2) divide into 10 groups of three types (a) planners, (b) social predictors, (c) evaluation committee; (3) planners allocate resources to raise probabilities of events to raise GNP or promote better values/freedom; (4) consensus in goup; (5) final resource allocation; (6) determine probabilities associated with events; (7) make social predictions; (8) compute importance and likelihood of social consequences; (9) achieve consensus; (10) evaluation committees determine preferences and desirability of alternative futures; (11) Final plenary assessment session in which weighting is conducted for population sectors; (12) dual debriefing of outcomes and methods.
Space Patrol	(1) introduction; (2) prepare materials; (3) generate and select scenario; (4) select role; (5) create characters and aliens with 13 dimensions; (6) select equipment; (7) enact contact/recovery/combat scenario; (8) use Space Patrol Tables to compute outcomes.
Hybrid Delphi Game	(1) instructions; (2) individuals read and rank for desirability 90 statements about 20-year-distant future events; (3) share results; (4) form groups of 3 known least well; (5) groups negotiate consensus list of 15 most desirable futures; (6) total group consensus of 16 most desirable; (7) discussion.
Future Planning Games	(1) introduction; (2) compare alternative viewpoints on roles, policies, and decisions; (3) disclose personal attitude toward viewpoints through rating on 11-point continuum; (4) design or decision-making exercise; (5) group decision-making on controversial issues and specify recommendations; (6) formulate an integrated, philosophical position on issue.
Cope	(1) organize materials and room; (2) overview; (3) future role diagnosis; (4) students review guide and write questions; (5) related information search for "library of future"; (6) reading; (7) discuss coping with change; (8) move into city, get ID card, fill out Creative Work Unit (CWU) data card, read Creative Production Module (CPM 1) handouts; (9) CPM 1 Job Training; (10) CPM 1 Tasks; (11) presentation/evaluation; (12) allocate/record CWUs; (13) CPM II Job Training; (14) analyze bulletins; (15) CPM II Tasks; (16) allocation of CWUs; (17) CPM III Job Training; (18) begin accelerated computer clock; (19) analyze bulletin; (20) allocate/record CWUs; (21) decision point on future of micro-society; (22) debriefing of issues using Cope evaluation debriefing guidelines.
Utopia	(1) overview of Unit Time Chart; (2) students read Phase I guide; (3) class grouped into subunits; (4) initiate Sunrise (Commune) Log in response to Situation 1; (5) Complete situations 2, 3, 4, 5; (6) panel discussion; (7) students read Phase II guide; (8) form four systems/groups: morality, economic, technological, political; (9) groups divide research responsibilities for individual reports; (10) work day; (11) individual reports to group, groups report to class; (12) conclave; (13) Sunrise evaluation; (14) allocate subsystem duties; (15) develop subsystem proposal reports (16) groups receive reports; (17) spokespersons give recommendations for declaration of Utopian Commitment; (18) committee writes commitment; (19) discussion; (10) debriefing.
Prospects	(1) overview; (2) review of one's past and present; (3) shared review; (4) looking to the future; (5) some reality testing; (6) relevant others identified who can influence outcomes.
Edplan	(1) set up game; (2) read scenario; (3) assign roles; (4) designate group leadership; (5) group meetings; (6) free period; (7) school board meeting; (8) formulate budget; (9) lobbying; (10) City Council and Federal Aid Representative consider funding requests; (11) City Council and School Board elections; (12) campaign speeches; (13) elections.
Edventure II	(1) overview, rules, role descriptions, 20-year history form, and educational institutions; (2) course enrollment; (3) record initial situation on role profiles; (4) education and plan work; (5) purchase education or select work; (6) compute final outcomes in satisfaction points, surplus income, and savings.
2000 A.D. Futura City	(1) Filmstrip, background reading, layout and distribution of materials; (2) game introduction; (3) set objectives and priorities in group; (4) urban plan; (5) presentations to Constituent Assembly; (6) re-evaluation; (7) group bargaining; (8) assembly debate; (9) final adjustments; (10) vote; (11) debriefing; (12 evaluate group performance.

debriefing. Detailed information on the major activities in sequence appears in Table 5.

VALUES

Participants in simulations are often implicitly encouraged to value certain points of view, often not as a direct result of the simulation's stated objectives, but as a result of the content it includes (and excludes) and of the processes or viewpoints participants feel are encouraged and rewarded when they dynamically interact with the simulation/game. We have evaluated each of the simulations/games in terms of the degree (high, medium, or low) to which they may promote nine possible values in Table 6. *Futuribles, Future Planning Games,* and *Utopia* rate high on six of the nine values. To the extent that these values are desirable to the user, these simulations/games would be excellent choices. *Space Patrol,* on the other hand, promotes only two values to any great extent, and these are expected from a science fiction game—valuing technological advancements as solutions to problems and valuing speculation about the future for present entertainment and enjoyment.

ISSUES

These futures research-based simulations/games confront a variety of issues ranging from the obviously future-oriented issues of predicting events or determining priorities for which technological innovations should be developed, to more present-oriented issues of controversial lifestyle and political choices, as well as emergent issues in educational administration, organizational and urban planning, and proposal writing. *Futuribles, Edplan,* and *2000 A.D.*, most issue-laden, confront seven or more of the 13 issues listed in Table 7.

THE MODELS

An underlying model provides the structural basis for a simulation. The model will always indicate the most crucial features of the events that are of interest and will attempt to show precisely the relationship of parts to the whole. Although most simulations represent real-world events (flying an airplane, passing a bill through Congress), this cannot be the case with future-oriented simulations, which deal with hypothetical future events, or futuribles (future possibilities), as De Jouvenal has called them in *The Art of Conjecture* (1967). In Table 8 we present a brief description of the underlying

TABLE 6 Possible Values Promoted

Values	*Futuribles*	*Dynamic Modeling of Alternative Futures*	*Simulating Values of of the Future*	*Space Patrol*	*Hybrid Delphi Game*	*Future Planning Games*	*Cope*	*Utopia*	*Prospects*	*Edplan*	*Edventure II*	*2000 A.D. Futura City*
Clarification of own values regarding the future	H	H	H	L	H	H	H	H	H	H	H	H
Valuing personal planning for the future	H	L	L	L	M	H	M	H	H	L	H	L
Valuing utopian ideals as serious future alternatives	H	M	M	L	M	H	M	H	M	L	M	M
Valuing participatory group process and consensus building activities as decision-making strategies in planning for the future	L	H	H	L	H	H	M	H	L	H	L	H
Valuing knowledge of probabilities of future events (prediction)	L	M	H	L	M	L	H	L	L	L	L	M
Valuing technological advancements as solutions to problems	M	M	H	H	M	L	M	L	L	L	L	H
Valuing alternative social or interpersonal structures as solutions to problems	H	H	L	L	M	H	M	H	M	M	H	H
Valuing speculation about the future for its present entertainment and enjoyment	H	L	L	H	L	L	L	L	L	L	L	L
Valuing a comprehensive vision of tomorrow	H	M	M	L	M	H	M	H	L	L	M	L

Key: H = High
M = Medium
L = Low

TABLE 7 Issues

Issues	*Futuribles*	*Dynamic Modeling of Alternative Futures*	*Simulating Values of the Future*	*Space Patrol*	*Hybrid Delphi Game*	*Future Planning Games*	*Cope*	*Utopia*	*Prospects*	*Edplan*	*Edventure II*	*2000 A.D. Futura City*
Determining priorities for which technological innovations should be developed	S	S/F	S		S	O	S	S				
Forecasting the probability of future events	S	S	S		O	O	O	O				O
Determining which group should have the most power to determine the future	O	O	S	S	S	S	S	S		S		S
Systematically studying future possibilities through learning to apply futures research methods (Delphi, Trend Extrapolation, etc.)	S	S	S		S		S	S	S	S		S
Exploring science fiction fantasy	O		O	S	O	O	O	O				O
Clarifying personally desirable future events	S	S	S		S	S	O	S	S/F	S	S	O
Presently experiencing a scenario of a possible future event	O	S		S		S	S	S/F		S/F	S/F	S/F
Exploring educational administration	O	S/F			O	O			O	S	S	
Confronting controversial present issues through exploring alternative future social and interpersonal lifestyles and/or social policies	O	S	O		S	S	S	S	S	S	S	S
Planning a personally successful professional and educational career which avoids obsolescence and insures productivity and satisfaction						O			S		S	
Writing proposals for innovative ideas that may receive government funding						O	O	O		S		S
Urban planning and design which is highly responsive to human needs and values						O	O	F				S
Organizational planning and decision making	O	S/F	S		O				O	S		S

Key: S = Issue significant
O = Issue optional and may be included in variations of simulation game
F = Frame game allowing players to design and insert issues

models for the simulations/games. These models range from real-world controversial social issues that are likely to continue to be important issues in the future (*Future Planning Games*), to the science fiction scenarios generated in creative fantasy (*Space Patrol*).

STATED EDUCATIONAL OBJECTIVES AND ANALYSES OF KNOWLEDGE AND SKILLS

One of the most essential factors in selecting a simulation/game is the type of educational objective it is designed to achieve. We have presented this information in two ways. First, we have summarized the stated objectives as specified in

(text continued on page 120)

TABLE 8 Models of the Simulations/Games

Simulation	Model Characteristics	Evaluation of Models
Futuribles	Possible future trends/events/conditions selected from those generated in Delphi studies constitute the possible future situation modeled.	This cards game's rounds provide an excellent model of future studies techniques in simple form. The range of real-world situations is comprehensive in treating 19 areas.
Dynamic Modeling of Alternative Futures	Delphi. Experts predictions of 25 to 50 potential future developments serve as the model for the game.	Validity of the model depends entirely upon the particular group of experts chosen.
	Alternative Futures Analysis and Review. Participants generate lists of desirable future events based on their own experiences.	Validity of the model is determined by the real-world information base of the particular participants.
	Identification of the Gifted and Talented. A 2000 A.D. scenario of a hypothetical "Presidential Advisory Board for the Identification of Gifted and Talented Youth" presents participants with a decision-making task involving the selection and admission of gifted and talented students into an elite educational program.	A very realistic underlying model that closely parallels the adoption process or the selection of foster parents.
	The Label Game. A formal model of the dynamic concept of a hypothetical life career of special education students is stated, based explicitly and closely upon theory and research findings from Mercer and others. The consequences of attaching lables like "mentally retarded" to children early in their lives are displayed through a flow chart specifying decisions and alternative consequences and long-term effects.	The model of the life career of the exceptional child is made operational as a game board, and players interact with the hypothetical future consequences through playing the simulation game. An excellent simplification of a complex model.
	Toward Walden Two. Basic learning theory principles upon which B.F. Skinner based the utopian community of "Walden Two" is the formal model for this simulation game. A "game as reality" paradigm is illustrated specifying the relationship among theory, prediction, reality, the simulation game, and game behavior. A flow chart specifies the sequence of simulation procedures; participants' observations, and concept forms are specified; and analysis of outcomes is presented.	A brief summary statement of operant conditioning principles is stated. The application of these principles to a case study of a problem child is very realistic and highly useful in teacher training.
Simulating the Values of the Future	The process of social planning and decision making for the future is simulated in a role-playing task involving prioritizing resources, estimating societal consequences of decisions, and arriving at a moral evaluation of alternative futures when participants are confronted with 20 potential future developments derived from potential developments, including: (1) fertility control; (2) 100-year life span; (3) personality control drugs; (4) incapacitating rather than lethal weapons; (5) sophisticated teaching machine; (6) ocean farming; (7) controlled thermonuclear reactors; (8) continued automation in commerce and industry; (9) artificial life; (10) weather control; (11) general immunization; (12) genetic control; (13) man-machine symbiosis; (14) household robots; (15) preservation of privacy; (16) wide-band communications systems, (17) continued space exploration; (18) advanced techniques of opinion control, thought manipulation, and propaganda; (19) continued trend toward urbanization; and (20) ova/sperm banks.	Empirical anchors exist in the futurist literature for many, if not most, of the alternative future possibilities that constitute the underlying model of possible events. The bias of events envisioned involves advanced technology as major components of the future.
Space Patrol	Although this is a science fiction game, the elaborately detailed tables and charts specify alternative dimensions of characters, equipment, characteristics of their environment (gravity, buildings), artifacts, weapons (fangs, shape-changing), success probabilities, landing zones, encounter (type and degree of surprise), resources, and minerals. A special table is designed for scenario generation.	The model underlying the content is science fiction, and is analogous to the episodes of the "Star Trek" and "Space 1999" television series, and the movie *Star Wars*.

TABLE 8 Models of the Simulations/Games (Cont)

Simulation	Model Characteristics	Evaluation of Models
Hybrid Delphi Game	A simulation of the diversity of real-world viewpoints is conducted through selecting a list of 90 events that could occur in 1996 A.D., placing them on a questionnaire, and having participants assign a numberical desirability index (low 10%, medium 50%, or highly desirable 90%) to the events, and then negotiate a consensus list of 15 most desirable futures within small groups and then within the entire group.	The validity of the possible future events can only be determined by comparison with actual future events. The predictions offered for consideration represent a mixture of events that are possible or probable in some cases, and, in other instances, either highly desirable or undesirable events, both with a low probability of occurring.
Future Planning Games	Many diverse real-world models and controversial issues underlie these games, including alternative family structures (communal), polygynous, monogamous, same-sex, professional parents, single parenthood); writing a marriage contract; determining national priorities; selecting among capitalist, welfare, or socialist societies; alternative prison models; selecting lifestyle components (materialist, christian, humanist, athiest, guru, hippie); teenager's dilemma of abortion, keeping the baby, or adoption; removal of life support system from car accident victim in coma; prearranging grandpa's funeral; determining crimes punishable by death penalty; constructing an international philosophy based on alternative models (cooperation, competition, cold war, peaceful coexistence, United Nations, regionalism, world law, or protracted conflict models); choosing a social philosophy (capitalist, socialist, reformer, black reformer, radical, conservative); dealing with a civil war in India through neutrality, economic assistance, or military intervention; limiting population through methods of abortion, abstention, free birth control, severe taxation for children, creation of suicide assistance agency, liberalization of capital punishment, execution of nonproductive, free sterilization, free exportation for those giving up citizenship, no assistance to starving, or government education program.	The real-world situations that are the models for the games generally have two characteristics: (1) they represent in many cases controversial alternatives, and (2) the alternative considered could be found to have a real-world referent. Although the positions are simplified for the game, enough complexity and reality are retained to permit serious consideration of issues, and lively discussion.
Cope	The classroom becomes Technopolis, a city where students live through five time periods from 2000 to 2040 A.D. Period 1: The city is a leisurely intellectual community whose citizens research the future. Period 2: A computer called COMCON helps with information and material problems and asks citizens to provide human input for problems. Period 3: COMCON, having assimilated all international and intergalaxy computer systems, directs all human activity and requires citizens to learn computer forms and drastically increase their productivity. Period 4: Citizens are forced to learn a new language, FUTURESPEAK, and must compete with COMCON to create ever more sophisticated technology. Period 5: COMCON has grown impatient with human inefficiency, tells citizens that human beings are apparently obsolete, and asks citizens to choose between a life of uncaring bliss or one of constant struggle.	The science fiction model of a computer directing all human activity is an increasingly common theme. The simulation alters real-world events to accelerate change. The simulation of rapid change could be carried out in considerably less time, however. The scenario and procedures have compromised the model to integrate educational experiences, which is accomplished with a reasonable degree of success.
Utopia	Participants consider what they would do if they were subjects in a professor's experiment in an isolated, self-sufficient utopian community of SUNRISE. The group has received unlimited resources for technical, medical, or industrial teams, and individuals have up to one million dollars each. All members must agree to construct an ideal society with moral, economic, tech-	While having educational value in considering alternative futures, the provision of one million dollars to each participant in the utopian experiment removes a substantial degree of relevance to real-world utopian experiments, where the issue of economic viability is crucial.

TABLE 8 Models of the Simulations/Games (Cont)

Simulation	Model Characteristics	Evaluation of Models
	nological, and political components to assure the survival of the society. There are eight subcategories within each of these four areas. Participants conduct research on components of their major systems and draft ideal solutions to problems in each subsystem. The proposed solutions are considered by the whole group, which votes on the principles for their new society. The new society is tested by several crisis situations outlined on the teacher's guide.	
Prospects	This self-analysis program in self-planning attempts to aid individuals in evaluating career and training needs with the intent of producing a plan for implementation in the real world within one to two months. The simulation seeks to prevent obsolescence through having individuals (a) examine their career, (b) evaluate their training needs, and (c) develop specific action plans to maintain and enhance the relevance of their skills.	The model is very close to the reality of career development decisions individuals face, and the simulation provides a highly cost-effective self-help tool.
Edplan	A new budget must be designed by a school board for a school system in a small city in a rural area for presentation and approval by the city council. Participants, acting in the roles of principals, superintendent, teachers' union, student council president, taxpayers, PTA members, school board members, city council, and Federal Aid Representative, act to frame a proposal most responsive to their special interests. The determination of educational policy in the context of a political climate surrounding a financial issue is simulated through a highly interactive simulation involving bargaining, negotiating, lobbying, and voting.	Although simulating the political element in school planning, the model leaves out many significant nonpolitical factors including sufficient time for the development of both a rationale and procedures for implementing the proposed new educational programs.
Edventure II	The game simulates the education market place of the future (1981 to 2000). Learners are able to select from a wide range of options (college, work, self-improvement or hobby-related experiences), and compete with one another to a limited extent in gaining admission to desired courses. In large part, it is a "buyer's market," since potential learners can get just about what they want each year over a 20-year educational venture/life span. The educational model underlying the simulation is the assumption or prediction that increased leisure and higher standards of living will create greater demands for giving education in a great diversity of fields. The design of the game assumes the adoption by 1981 of federal higher education voucher credits which are worth two years' tuition of any full-time course of the part-time equivalent. Participants select courses or work options each year for 20 years and maintain a running tally of consequence outcomes, determining whether or not they have won or lost by the number of satisfaction points generated and/or surplus income and savings accumulated.	Although the simulation requires decision-making and incorporates excellent procedures for providing feedback, the model itself is highly limited, in that the real-world situation being simulated is essentially the decision situation of selecting courses at the beginning of a college semester or quarter. The simulation works well within that limited context.
2000 A.D. Futura City	Central City is an imaginary urban center beset with many problems afflicting American cities presently. The primary task of the game for participants, through acting in roles, is to construct a future city through effective planning and coordination of the development of two land tracts adjacent to the city. Receipt of a federal grant is contingent upon the participants' design of an	The model underlying the game is anchored very well in current urban planning procedures. Although a simplified model, the simulation/game involves high school students actively in many of the decisions and considerations faced by real urban planners.

TABLE 8 Models of the Simulations/Games (Cont)

Simulation	Model Characteristics	Evaluation of Models
	urban plan and a plan for the new property. The task of planning Futura City is to be achieved through participation of diverse groups having conflicting vested interests. Presentation of the plan to a constituent assembly, reevaluation, bargaining, debate, and a final vote conclude the simulation, resulting in consequences for major systems such as housing, transportation, and industry.	

the written materials accompanying the simulation/game to help clarify the designer's *intended* educational purposes, and we have rated the degree to which these objectives are achieved. Second, we have analyzed the degree to which a variety of skills are *actually* involved in participation. We have classified these skills into eight general categories, which summarize an analysis of the knowledge and skills developed through participation. Those simulations/games involving the highest levels of the component skills are listed after each category:

(1) Research: *Utopia, Cope, Dynamic Modeling of Alternative Futures, Prospects, Futuribles*
(2) Computation: *Simulating the Values of the Future*
(3) Predicting probabilities: *Cope, Simulating the Values of the Future, Futuribles, Dynamic Modeling of Alternative Futures, Prospects*
(4) Generating alternative possibilities: *Dynamic Modeling of Alternative Futures, Utopia, Futuribles, Cope, Prospects*
(5) Clarifying preferabilities: *Futuribles, Dynamic Modeling of Alternative Futures, Simulating the Values of the Future, Future Planning Games, Cope, Utopia*
(6) Modes of interaction: *Future Planning Games, Utopia, Edplan, 2000 A.D. Futura City, Hybrid Delphi Game, Simulating the Values of the Future*
(7) Interpersonal interaction: *Utopia, Hybrid Delphi Game, Future Planning Games, Edplan, Dynamic Modeling of Alternative Futures, Simulating the Values of the Future*
(8) Evaluation: *Dynamic Modeling of Alternative Futures, Cope, Future Planning Games, Simulating the Values of the Future, Utopia, Hybrid Delphi Game*

Table 9 covers the intended educational objectives and the success with which they are realized. Table 10 presents an analysis of the skills and activities the simulations/games involve.

TABLE 9 Stated Objectives of Simulations/Games

Simulation	*Objectives*	*Degree of Success in Achieving Stated Objectives*
Futuribles	To help individuals (1) get acquainted with future possibilities; (2) share feelings about the future; (3) clarify own values regarding the future; (4) create visions of "hoped for" futures; (5) write scenarios of how the future may happen; (6) select priorities for group long-range planning; (7) choose preferred futures for daily work, service, and for other areas; and (8) anticipate the future with "joy and commitment."	High
Dynamic Modeling of Alternative Futures	Delphi: Achieve consensus through anonymously and objectively estimating the likelihood of particular events occurring within a given field by some future date. *Alternative Futures Analysis and Review*: Achieve group consensus through public and personal means on those future events within a field that are most desirable or preferable to the participants in the simulation. *Identification Simulation*: Increase awareness of the educational needs for the gifted and talented through the use of a one-hour activity which demonstrates the inappropriateness of Intelligence Quotient scores used alone as an identification criterion. *Label Game*: Examines models of the life of the handicapped on which the game was patterned and describes the translation of those models into game format; provides an affective experience in the negative social psychological consequences in life as a function of various assigned "labels." *Toward Walden Two*: Basic components of an operant conditioning-based social system are enacted and observed in a classroom simulation. Participants learn basic principles of operant conditioning while trying to apply the principles to others, within the context of a token economy, i.e., micro-society based on operant conditioning.	High
Simulating the Values of the Future	Through the operations research techniques of simulation and the use of expert judgment, the participant (a) makes decisions affecting the character of the environment, (b) estimates the societal consequences of those decisions, and (c) evaluates the desirability of these consequences.	High

TABLE 9 Stated Objectives of Simulations/Games (Cont)

Simulation	Objectives	Degree of Success in Achieving Stated Objectives
Space Patrol	Participants create the dimensions of a science fiction adventure—its scenario, roles, alien capabilities, and rules—and then experience these fantastic possibilities through their presentation in a role-playing war-game framework. Each participant "faces tribulations that would titillate a Flash Gordon and baffle a Captain Kirk." Force, wit, and intelligence "are the weapons of redemption." The objective is entertainment.	Medium
Hybrid Delphi Game	Participants experience the very real conflicts in values, priorities, and directions that arise during negotiation of desirable futures; determination of who benefits from a particular future, and exploring how one gets from the present to the future.	Medium
Future Planning Games	While students confront controversial social issues existing presently, or potentially impinging upon their planning of their personal future roles, the primary objectives sought are: values disclosure; values clarification; constructing a life philosophy; problem solving; debate/discussion; research of data to support positions; an introductory exposure to such tasks as writing a marriage contract, determining societal growth/no growth priorities, and resource allocation; philosophy; formulating foreign policy on a specific issue; dealing with death; and planning prisons.	High
Cope	Through creative or productive task requirements participants may be led to (1) think more about the different alternatives we might have for life in the future; (2) solve some problems that arise personally, for the city, and the universe; (3) learn a completely new (but not entirely foreign) language to help them communicate with other game participants; (4) evaluate new forms of technology for their effects on human beings; and (5) attempt to cope with an exceedingly rapid pace of change.	Medium
Utopia	*Knowledge*: (1) understanding the four systems basic to any society and some of the elements of each system and (2) understanding the problems of human motivation and behavior that forming an ideal society poses. *Skills*: (1) using the library for research, (2) using parliamentary procedure moves to promote one's ideas, (3) using oral language in an organized fashion in discussions with, and presentations to, small groups. *Attitudes*: (1) appreciating the complexity and challenge of forming an ideal society; (2) desiring to work toward an ideal society, regardless of the difficulties; and (3) feeling that individual contributions are important in a society, even if they are not totally accepted.	High
Prospects	To help participants examine their career, evaluate their training needs, and develop specific action plans to maintain and enhance the relevance of their skills, decreasing the chances of potential obsolescence. The process provides an opportunity to learn how to give and receive helpful feedback.	High
Edplan	It provides insight into the political nature of educational planning and decision-making; provides a working example of politics in action; and provides an example of how justifiable are all possible points of view on some educational issues.	Medium
Edventure II	This career guidance game is designed to dynamically demonstrate four principles: (1) the range of educational options open to people now and as of 1981 is very large, going far beyond conventional academic programs and training for trades; (2) obtaining an education is a life-long process; it does not stop after high school or after college; (3) types of educational offerings must be suited in content and instructional style to the types of individual learner in order for learning to be successful; (4) failure as well as success is to be anticipated as a possible outcome of any educational situation.	High
2000 A.D. Futura City	To give students an opportunity to: (1) demonstrate their understanding of future planning; (2) become involved in futurist decision-making activities, including (a) clarification of values, (b) analysis of trend indicators, (c) evaluation of priorities, (d) scenario interpretation, (e) trend extrapolation, (f) group interaction; (3) understand the different roles of key urban groups and their influence in shaping the future of America's cities.	Medium

TABLE 10 Knowledge, Skills, and Activities

Knowledge, Skills, Activities	*Futuribles*	*Dynamic Modeling of Alternative Futures*	*Simulating Values of the Future*	*Space Patrol*	*Hybrid Delphi Game*	*Future Planning Games*	*Cope*	*Utopià*	*Prospects*	*Edplan*	*Edventure II*	*2000 A.D. Futura City*
Research												
Gathering information	—	—	—	—	—	L	H	H	H	—	—	—
Extensive reading	—	L	L	H	—	M	H	H	L	—	L	M
Alert observation	—	H	—	—	—	—	—	—	—	—	—	—
Skillful listening	L	H	L	—	—	—	—	—	—	—	—	—
Generating hypotheses	H	M	L	—	—	L	M	H	—	—	—	—
Measurement/ experimentation	—	H	—	—	—	L	—	L	—	—	—	—
Convergent thinking	H	H	H	M	M	H	H	H	H	M	H	H
Divergent thinking	H	H	H	H	H	H	H	H	H	—	—	M
Preparing reports/ writing	—	L	—	—	—	M	H	H	L	—	—	—
Memory	—	L	—	—	—	—	L	—	—	—	—	—
Computation												
Calculating mathematically	—	L	L	L	—	—	L	—	—	—	L	—
Predicting Probabilities												
Trend extrapolation	M	M	H	—	—	M	H	—	H	—	—	L
Cross impact analysis	M	M	H	L	—	L	H	H	H	—	—	L
Delphi consensus	H	H	H	—	—	L	H	—	—	—	—	—
Generating Alternative Possibilities												
Brainstorming	M	H	L	—	—	—	M	H	—	—	—	—
Wishful scenarios	H	H	L	H	—	L	M	H	H	H	—	L
Alternative pathways	M	H	L	H	—	L	H	M	H	L	H	M
Clarifying Preferabilities												
Planning goals/ strategies	H	H	H	H	—	H	H	H	H	H	H	H
Values clarification	H	H	H	H	H	H	H	H	H	H	M	L
Delphi event desirability consensus	H	H	H	—	H	H	H	H	—	—	—	M
Relevance tree analysis	L	L	L	—	—	—	—	—	—	—	—	—
Modes of Interaction												
Small group discussion	H	H	H	H	H	H	H	H	L	H	L	H
Debate	—	H	H	—	H	H	L	H	—	H	—	H
Large group discussion	—	—	H	—	H	H	L	H	—	H	—	H
Parliamentary procedure	—	—	—	—	—	H	L	H	—	H	—	H
Voting	L	H	H	—	H	H	L	H	—	H	—	H
Interpersonal Interaction												
Persuading	L	H	H	—	H	H	M	H	—	H	—	M
Bargaining/negotiating	—	H	H	—	H	H	M	H	—	H	—	H
Helping/supporting	L	M	L	—	M	M	M	H	L	L	L	L
Coalition formation	—	M	M	L	M	M	L	H	—	H	—	L
Evaluation												
Criterion-referenced test	—	H	—	—	—	—	M	—	—	—	H	H
Norm-referenced test	—	H	—	—	—	—	—	—	—	—	—	—
Evaluation of simulation	—	H	H	L	—	—	H	L	—	—	—	—
Self-evaluation	H	H	H	L	H	H	H	H	H	H	L	L
Evalutation of peers	—	H	—	L	H	H	H	H	—	L	L	M
Descriptive disclosure of present status	H	L	L	—	L	H	—	—	H	—	—	—

Key: H = High
M = Medium
L = Low

ROLES

Types of Roles

Descriptions of the types of roles appear in Table 11. Of all the simulations/games, the *Future Planning Games* will provide players with roles and role-playing opportunities closest to most individuals' real-life experiences or general information. Because these games deal with controversial social issues, they often produce high levels of arousal, debate, and discussion. At the other extreme, *Space Patrol* broadens the range of roles and character types to provide for weird extraterrestrial life forms one might imagine encountering. Simulation/gaming designers will find the rich variety of dimensions of people and situations in *Space Patrol* potentially useful in structuring their thinking about scenarios for their own simulations. *Prospects* does not involve role playing, since the individual players apply this career planning simulation to themselves.

TABLE 11 Types of Roles

Simulation	Roles
Futuribles	Explorers; Planners; Students; Self
Dynamic Modeling of Alternative Futures	Self; "Experts" on a given topic; member of Presidential Advisory Board for identification of Gifted and Talented Youth; Individuals "labeled" either normal or stigmatizing labels which emphasis disabilities and not ability—congenitally defective, emotionally disturbed, mentally retarded; operant conditioning behavioral engineers
Stimulating the Values of the Future	Planning Group; Social Predictors; Evaluation Committee Members; Teenagers; Housewifes; Middle class Employed; Persons over 65; Cultural Elite; Poor
Space Patrol	Game Master; Aliens and Creatures of the possible types: mollusk, plant arthropod, amphibian, reptile, mammal, avian exotic, mechanical polymorph, crystalline, gaseous energy, Characters may be specially constructed along the following dimensions, with the number of variations possible specified: Metabolism (9), Special Capabilities (11 including increased hearing, touch, sensitivity, telescopic vision), Psionics (7 including empathy, telepathy, telekinesis, clairvoyance, mind control, teleportation); General Shape (10 including monoped to octoped, wings, tentacles, etc.); Sex (5 types); Size (13 sizes); Cyborg replacements (13 including limbs, structure, or sensory abilities); Special Weapons and Resources; Attitudes (5 levels from friendly to hostile)
Hybrid Delphi Game	Self
Future Planning Games	Member of Presidential Commission on Moral issues; Member of family facing euthanasia decision; Social Philosopher; Neighborhood Protection Committee Member; ERA Advisory Commission Member; 2050 A.D. Member of National Commission on Ecology; Warden; Prison Designer; Member, Immigration Review Board; Designer of a mythical All-American Hero; Supporters of Populist Party or Reformist Party, debating Social Security issues; 2300 A.D. Economic System Planner; Foreign Policy Commission member, faced with specifying a developing nations policy decision or deciding a war issue; six alternative world orders, and two alternative philosophies of cooperation/competition; Cabinet Member making U.S. budget decisions; Person choosing capitalist, welfare, or socialist society; Member of a family of following type: communal, polygamous, monogamous, same-sex couples, professional parents, single parenthood; materialist, Christian, humanist, atheist, guru, hippie
Cope	Intellectuals; Engineers; Human Services Specialists; Mechanical Services Technicians; Innovative Technologist Specialist; Life Quality Technician; Computer Services Assistants; Director of Human Services; Director of Mechanical Services; Director of Innovative Technology; Director of Life Quality
Utopia	Subjects in the "Sunrise" Cabin Communal Experiment; "Expert" Planners for Morality, Economic, Technological, and Political Systems; Parliamentary Procedure Chairperson; Committee for the "Declaration of Utopian Commitment"
Prospects	Self
Edplan	Principal of Elementary School; High School Principal; Superintendent of Schools; Teacher's Bargaining Association Representative; Student Council President; Taxpayer; PTA Member; School Board Member; City Councillor; Federal Aid Representative; School Board Member
Edventure II	Female, 28, Divorcee: Male, 41, Part-Time Worker; Female, 54, Housewife with Grown Children; Male, 56, Supermarket Owner; Female, 55, Widow of Astronaut; Male, 32, Retailer; Female 18, High School Graduate; Male, 27, B.A. in Physics; Male, 49, Divorcee; Female 58, Widow; Male, 19, Factory Worker; Female, 52, Spinster Elementary Teacher; Male, 29 Assistant Personnel Manager; Female, 19, Single Parent; Female, 36, Stenographer
2000 A.D. Futura City	Chairman, Constituent Assembly; Professional City Planners; Business and Industry Group Members; Black Activists; Private Land Developers and Construction Firms; Civil Rights Organization; White Middle-class Ethnic Families; Environmental Protection Organization; Citizens Group for Urban Life Renewal

Role Descriptions

Role playing in a simulation is greatly influenced by the way goals, character, and personality are described. Limited detail provides for interpretation and elaboration of the role according to the player's whims, while lengthy detail can be supplied to ensure the performance of a role in a manner faithful to the character upon which the role was based. Role descriptions presented barriers to effective role playing in *2000 A.D. Futura City,* because the role definitions were inadequate for players who are unfamiliar with such roles as city planners, land developers, or environmental protection agency representatives. In the case of the other simulations/games, the level of detail provided was generally appropriate.

Role Chracteristics and Flexibility

Three role-playing factors help ensure the successful enactment of a simulation game. First, all players should be equally or evenly occupied with meaningful tasks. Second, players should strongly identify with their roles, acting them out with total involvement. Third, the roles should be interesting both to role players and to those with whom they interact. On these dimensions of evenness, involvement, and interest value, none of the simulations/games rate low. *Cope, Utopia, Edplan,* and *2000 A.D. Futura City* rate average in the areas of both evenness and involvement because they all have patterns of notably strong and weak roles, with some roles having high degrees of involvement, while others had very limited degrees of involvement. On the other hand, a game director can easily modify roles in all four to add details or distribute tasks to ensure more effective participation by all players. When role characteristics and role flexibility are jointly considered, there are four simulations that generally invole players evenly in interesting roles and in roles which can be easily modified, added, or deleted. These are *Futuribles, Space Patrol, Future Planning Games,* and *Edventure II.* Table 13 rates role characteristics and role flexibility for all simulations/games.

TABLE 12 Role Descriptions

Brief: Several Sentences	Detailed: 1-2 Paragraphs	Lengthy: 1 Page or More
Dynamic Modeling of Alternative Futures	*Futuribles*	*Space Patrol*
Hybrid Delphi	*Simulating the Values of the Future*	*Cope*
2000 A.D. Futura City	*Future Planning Games*	*Utopia*
	Edventure II	*Prospects*
		Edplan

INTERACTIONS

In consideration of students' opportunities to learn social interaction skills or to practice leadership roles, as well as of game directors' preferences for interactions of various types or intensity, we have reviewed the nature of the interactions in these simulations/games. Table 14 displays the range. *Prospects,* a self-administered exercise, involves very limited interaction when there is any at all, while, at the other end of the scale, *Simulating the Values of the Future, Edplan,* and *2000 A.D. Futura City* involve intense interactions in the forms of alternative roles, frequent attempts to persuade, and several different types of interaction, including within and between small groups, as well as within and between large groups.

RULES

The rules for most of the simulations are generally good to outstanding in their clarity, organization, and completeness, as Table 15 reflects, though three are related as only adequate on these counts. *Simulating the Values of the Future, Space Patrol,* and *2000 A.D. Futura City* are fairly complex simulations, but their rules lack the precision and organization to make them easy to follow. This does not render them unplay-

TABLE 13 Role Characteristics and Flexibility

Role Characteristics	*Futuribles*	*Dynamic Modeling of Alternative Futures*	*Simulating Values of the Future*	*Space Patrol*	*Hybrid Delphi Game*	*Future Planning Games*	*Cope*	*Utopia*	*Prospects*	*Edplan*	*Edventure II*	*2000 A.D. Futura City*
Evenness: Degree to which all players are equally occupied	High	High	Avg.	Avg.	High	High	Avg.	Avg	N/A	Avg.	High	Avg.
Involvement: Degree to which players become involved with roles	High	High	High	High	High	High	Avg.	Avg.	High	Avg.	High	Avg.
Interest value of roles	High	Avg.	High	High	High	High	High	High	High	Avg.	High	High
Role Flexibility												
Ease with which new roles can be *added*	Easy	Avg.	Avg.	Easy	Diff.	Easy	Easy	Easy	Diff.	Easy	Easy	Easy
Ease with which existing roles can be *deleted*	Easy	Diff.	Avg.	Easy	Diff.	Diff.	Easy	Easy	Diff.	Easy	Easy	Diff.
Ease with which existing roles can be *modified*	Avg.	Easy	Easy	Easy	Easy	Easy	Easy	Easy	Diff.	Easy	Diff.	Easy

TABLE 14 Interactions

Types of Interactions	Futuribles	Dynamic Modeling of Alternative Futures	Simulating Values of the Future	Space Patrol	Hybrid Delphi Game	Future Planning Games	Cope	Utopia	Prospects	Edplan	Edventure II	2000 A.D. Futura City
Individual-individual	X	X	X	X	X	X	X	X	X	X	X	X
Within small groups (3-7)	X	X	X	X	X	X	X	X		X		X
Between small groups		X	X		X	X	X	X		X		X
Within large groups (7+)		X	X		X	X	X	X		X		X
Between large groups			X							X		X
Intensity of Interaction												
Frequent interaction	X	X	X	X	X	X	X	X		X	X	X
Frequent trading				*								
Frequent persuasive attempts/ compromising/ bargaining	X	X	X		X	X		X		X		X
Several alternative roles	X	X	X	X		X	X	X		X	X	X
Same individual in several different roles	X	*	*	X		*	*	*		X		*
Frequent research-based oral presentations before small/large groups		*					X	X				X

Key: X = Characteristic present
* = Possible variation

able, but it does lengthen the preparation time before play can begin.

PLAYER/ROLE RESOURCES SCORING

The procedures for scoring are generally clear and adequate, as the summary in Table 16 shows. This table also notes the general background that might enhance the effectiveness of players. When you consider using one of these simulations/ games, give careful attention to the backgrounds of the potential participants. By reviewing the information in Table 16 on the types of player resources most relevant to success, a game director can try to compensate before play begins for any significant variations in player's knowledge, experience, or ability that may compromise the interest, enjoyment, and educational value of the simulation. You might either take special care in assigning individuals to roles or you might modify the complexity of role tasks and the nature of goals to prevent unfair advantages or disadvantages from previous experience from entering into play as disruptive factors.

SIMULATION/GAME PACKAGING

If any of the future-oriented simulations/games appear useful and appropriate for your purposes on the basis of previous

TABLE 15 Rules

Simulation/Game	Number of Pages Devoted to Rules	Clarity, Organization, and Completeness: Adequate	Good	Outstanding
Futuribles	14 (½-page size)			X
Dynamic Modeling of Alternative Futures	Averages about 6 pages each simulation		X	
Simulating the Values of the Future	11	X		
Space Patrol	25	X		
Hybrid Delphi Game	4		X	
Future Planning Games	2		X	
Cope	21		X	
Utopia	7		X	
Prospects	34			X
Edplan	10		X	
Edventure II	8		X	
2000 A.D. Futura City	2	X		

TABLE 16 Player Resources and Scoring

Simulation	Most Relevant Player Resources	Scoring
Futuribles	Experience knowledge of predicted future trends	No explicit criteria. Result is values/priorities/predictions clarification. Readily usable in decision making.
Dynamic Modeling of Alternative Futures	Knowledge and experience; mastery of content/process during play	Participants receive points for predictive accuracy, decision-making, skill in movement on game board, or skill in mastering content of simulation, with rewards contingent upon performance.
Simulating the Values of the Future	Power of some roles determined by differential weighting; knowledge of predicted technology; computational accuracy	Computation and comparison of initial with final probabilities, likelihood and consequences. Individual compares group results with his/her intent, reviews overall consequences.
Space Patrol	Imagination and fantasy; skills in bargaining strategy	Detailed computations and weights tables available to facilitate scenario generation; success probability tables; shield rating tables; ranged weapons table detailing accuracy and power; tables for detailing effects of landing zones, types of encounters, and attributes of aliens or objects encountered.
Hybrid Delphi Game	Knowledge and experience; knowledge of predicted future trends	Comparison of initial individual selections with final group selections; degree of success at persuasion within groups of 3.
Future Planning Games	No particular prior skills; power of some roles determined by differential weighting	Comparison of initial individual selections and decisions with final group selections; degree of success at persuasion/consensus building; degree of success of one's team in debates.
Cope	Library study; persuasive presentation of ideas	Points awarded for quality of ideas in describing future in human or technological terms; degree of success in job training and task performance; degree of success in information gathering, assimilation, organization, and utilization; degree of success in coping with accelerating change; degree of success in mastering simple computer language.
Utopia	Library study; persuasive presentation of ideas	Degree of success in achieving consensus on moral, economic, technological, and political systems; degree of success in library research; degree of unanimity in class on utopian commitment; degree to which systems are defensible in the face of crucial evaluation questions/criteria.
Prospects	Knowledge of one's professional career goals and present strengths/weaknesses	Degree to which the action plan for career development is implemented during two months following the simulation
Edplan	Skill in the political process and persuasive presentation of ideas/positions	Degree to which one retains powerful position, or has candidates who win, during successive rounds of voting. Computation of relative degree of power takes into account degree of power allocated to individual's role.
Edventure II	Experience selecting elected courses to achieve career goals	Learner determines success during the 20 year simulated educational venture by computing satisfaction points, surplus income, and savgins, and comparing those figures to other players. To avoid losing, must accumulate $20,000 by game's end.
2000 A.D. Futura City	Knowledge and experience with urban planning procedures; study and presentation of positions	The winning group will have outlined priorities and objectives for Futura City closest to those adopted by the Assembly in the final urban plan.

evaluation criteria, you will hope that the package will be economical, durable, reusable, and ready for you to use. Those simulations that come ready-to-use are *Futuribles, Future Planning Games, Prospects, Edplan, Edventure II,* and *2000 A.D. Futura City*. The last three are kits, the most expensive in this selection. *Cope* and *Utopia* require some simple duplication. The remaining simulations/games come in book or booklet form and require construction and duplication. Although this process involves time and some additional expense for duplication, these costs are not excessive, and these simulations remain on balance very cost effective. They have relatively reasonable purchase prices, especially considering that one is getting both a book and one or more simulations/games for the price of a book alone. Table 17 summarizes this information.

TABLE 17 Cost and Packaging

Simulation and Publisher	Kind of Packaging	Cost per Kit or Book	Completeness	Durability
Futuribles (World Future Society)	kit	$9.45	complete	excellent
Dynamic Modeling of Alternative Futures (School of Education, Indiana Univ.)	book	$2.00	construction and duplication	average
Simulating the Values of the Future (The Free Press)	book	$4.95	construction and duplication	average
Space Patrol (Gamescience)	book	$5.00	construction	average
Hybrid Delphi Game (R. Saroff)	booklet	$5.50	duplication	average
Future Planning Games (Greenhaven Press)	large fold outs	$.95 each	complete	average
Cope (Interact)	kit	$14.00	complete	average
Utopia (Interact)	kit	$10.00	complete	average
Prospects (Transnational Programs)	book	$6.50	complete	average
Edplan (Games Central)	kit	$30.00	complete	average
Edventure II (Games Central)	kit	$60.00	complete	average
2000 A.D. Futura City (Newsweek)	kit	$47.00	complete	good

DEBRIEFING

The degree to which students may achieve important simulation/game objectives may depend upon the opportunity they have to reflect and generalize from their experiences through a structured debriefing. The availability of printed guides for conducting the debriefing, evaluation forms or procedures, and guidelines for additional activities can be a great help to game directors in conducting an effective debriefing. Unfortunately, procedures for debriefing are lacking entirely for *Futuribles, Simulating the Values of the Future, Space Patrol,* and the *Hybrid Delphi Game*; they are very limited in *Edplan.* Evaluation forms or procedures are specified for only six of the twelve simulations/games.

Though *Futuribles, Space Patrol,* and the *Hybrid Delphi Game* are among the simulations most deficient in debriefing and evaluation procedures, all three offer some excellent guidelines for additional activities. The sections on "planner games" and "student uses" in *Futuribles* substantially contribute to achieving educational objectives. *Space Patrol*'s scenario generator gives the user outstanding freedom and flexibility in generating ideas for replays, and, since the game was designed to entertain, absence of debriefing procedures is not a very significant omission. The *Hybrid Delphi Game* suggests helpful modifications and extensions to play.

These simulation/gaming designers as a whole seem somewhat reluctant to look upon the past, even when that past is players' behavior during a future-oriented simulation/game. Thus, you should plan to give some additional preparation time to structure players' reflection upon their experience in your effort to help them achieve the intended purposes of these simulations.

SUMMARY

Evaluation

On the basis of the ratings previously presented, Table 19 presents a summary of the evaluations for 14 areas, with each listed dimension weighted equally to give the overall rating. A review of this chart will disclose that almost half of the future-oriented simulations/games have received an overall excellent rating: *Futuribles, Dynamic Modeling of Alternative Futures, Simulating the Values of the Future, Future Planning Games, Cope,* and *Utopia.* The four rated "good" overall were: *Hybrid Delphi Game, Prospects, Edplan,* and *2000 A.D. Futura City.* Two were rated average: *Space Patrol* and *Edventure II.*

In interpreting these ratings, you, as the potential user, should *always* begin with a clear idea of your own purposes. For example, although *Edventure II* was rated lower than others, it does an excellent job of achieving its stated objectives, and the packaging was rated excellent. By comparison, although rated among the top group overall, *Simulating the Values of the Future* has two important limitations–it is a complex simulation, and the user has to spend time assembling the materials before it can be played. This is to suggest that, before hastily selecting one simulation over another, you

TABLE 18 Debriefing

	Printed Guide for Directed Discussion	Evaluation Forms or Procedures	Guidelines for Additional Activities	Comments
Futuribles	no	no	yes	Although procedures are not specified for debriefing, by encouraging discussion of values during each round of play you can achieve objectives normally sought by a formal debriefing. Procedures offered for "planner games" and "student uses."
Dynamic Modeling of Alternative Futures	yes	yes	yes	Dependent variable measurement has been the most common form of debriefing employed in these simulations/games. Many have used a variety of measurement procedures for research and evaluation.
Simulating the Values of the Future	no	yes	yes	The evaluation proceeds in 3 phases: 1) evaluation of outcomes conducted from the perspectives of six groups (teenagers, housewives, middle-class employed persons over 65, cultural elite, and the poor); 2) evaluation of outcomes; and 3) critique of exercise.
Space Patrol	no	no	yes, scenario generator	The purpose is to entertain debriefing is not required.
Hybrid Delphi Game	no	no	yes	Although no debriefing procedures are specified, guiding the discussion during consensus-building may help achieve important objectives.
Future Planning Games	yes	yes	yes	Excellent discussion guides and evaluation forms are built into the participant's game materials, which require decisions or disclosure of attitudes/values.
Cope	yes	no	yes	A one-page printed set of debriefing questions is good; a bibliography of films, articles, and books helps the teacher enhance the simulation's educational impact.
Utopia	yes	yes	yes	Students receive a set of evaluation questions concerning the moral, economic, technological, and political systems they have established; teacher has selected bibliography for additional activities.
Prospects	yes	yes	yes	The exercise has debriefing/evaluation procedures built in at every step.
Edplan	yes, but very limited	no	no	Debriefing procedure is very inadequate, since it is a process having players read their role profiles aloud, with the group voting on whether the players achieved their objectives.
Edventure II	yes	no	no	A two-paragraph general guide is provided, offering one "opener" question; procedures generally inadequate.
2000 A.D. Futura City	yes	no	yes	Eleven debriefing questions stated; evaluation procedures for determining final scores are clear; excellent guides for additional activities.

would be advised to be very clear about your own intent, determine the importance to be placed upon the criteria and in what areas for making evaluative judgments, and then *review at least three* alternatives before arriving at a decision.

Comments

Our discussion of the simulations/games we have reviewed concludes with some informal comments centering on suitable conditions for using them, some possible limitations, and personal experiences with them, with the intent of helping you decide what you might like to try with your group or students.

TABLE 19 Evaluation Summary

	Futuribles	*Dynamic Modeling of Alternative Futures*	*Simulating Values of the Future*	*Space Patrol*	*Hybrid Delphi Game*	*Future Planning Games*	*Cope*	*Utopia*	*Prospects*	*Edplan*	*Edventure II*	*2000 A.D. Futura City*
Educational uses	H	H	M	M	H	H	M	H	H	L	M	M
Practical considerations: age level, group size, playing time, flexibility, in outcomes/issues	H	H	M	M	M	H	M	M	M	L	M	L
Complexity*	H	L	L	L	H	H	M	M	M	H	H	H
Values promoted	H	M	M	L	M	H	M	H	L	L	L	M
Issues	M	H	M	L	M	H	M	H	L	M	L	H
Accomplishment of stated objectives	H	H	H	M	M	H	M	H	H	M	H	M
Knowledge and skills												
Research	M	H	M	M	L	H	H	H	M	L	L	M
Computation		L	L	L			L				L	
Predicting probabilities	H	H	H	L	L	M	H	M	M	L	L	L
Generating alternative possiblities	H	H	M	M	L	L	H	H	M	M	M	M
Clarifying preferabilities	H	H	H	M	M	H	H	H	M	M	L	M
Modes of interaction	M	M	H	L	H	H	M	H	L	H	L	H
Interpersonal interaction	L	H	M	L	H	H	M	H	L	H	L	M
Evaluation	M	H	M	L	M	H	H	M	M	L	L	M
Role descriptions	M	L	M	H	L	M	H	H	H	H	M	L
Role characteristics	H	M	M	M	H	H	M	M	L	L	H	M
Types of interaction	L	M	H	L	M	M	M	M	L	H	L	H
Intensity of interaction	M	H	M	M	L	M	M	H	L	M	L	H
Rules clarity, organization, and completeness	H	M	L	L	M	M	M	M	H	M	M	L
Packaging	H	L	L	L	M	H	H	H	H	H	H	M
Debriefing	L	H	M	L	L	H	M	H	H	L	L	M
	Ex	Ex	Ex	Avg	Good	Ex	Ex	Ex	Good	Good	Avg	Good

Key: H = High
M = Medium
L = Low

*A high rating was assigned if the simulation/game was rated as simple.

Futuribles

This card game offers the user an excellent technique for getting participants to explore future possibilities in nonthreatening ways, revealing a great deal about their present values, attitudes, knowledge, and goals. It is an excellent basic futures introduction, as well as an enjoyable parlor game. For example, after not having seen an old friend for two years, an hour with *Futuribles* easily updated me on his personal past, present, and future at a breadth, intensity, depth, and detail that would have taken a day or two to emerge normally.

Dynamic Modeling of Alternative Futures

This 100-page publication presents detailed instructions for assembling five simulations/games, all of which the authors have empirically tested in a variety of teacher training settings. The simulations/games are generally well designed for playability and efficiency in achieving objectives that frequently concern the preparation or inservice training of teachers working with, or designing curriculum for, exceptional children. Although these simulations/games were primarily designed within the area of special education, at least four have had wider application with a variety of target audiences. All are especially effective in educating teachers on the subjects of the social psychology of mainstreaming handicapped children, using paraprofessionals, building staff consensus, training teachers in operant conditioning concepts, and identifying/assessing gifted/talented children and youth.

Simulating the Values of the Future

This simulation is a classic in the field, written by Olaf Helmer (a codesigner with Norman Dalkey of the delphi method for conducting futures research). This version is a written description of the simulation's methods and materials as it was conducted in Pittsburgh for five hours with 30 people in September 1966 as part of a conference on the effects of technological change. It appears in an excellent futures book which has a number of other outstanding articles by well-known futurists on the impact of technological change on past, present, and future American values. I first used it in abbreviated form in a graduate course I taught on Futures Simulation/Gaming. Although it involved somewhat laborious computation, it provided an excellent overview of issues, and the group remained highly motivated and task-oriented throughout, despite protests over calculations. I have heard the simulation praised by the Denison Simulation Center staff (Denison University, Granville, Ohio), who used it in undergraduate liberal arts courses; I have also observed its highly successful application as a university faculty/staff development exercise in that same setting.

Space Patrol

The serious war gamer, science fiction enthusiast or author, or simulation/gaming designer will find the *Space Patrol* manual a rich source of ideas for game scenarios and science fiction plots. Although sometimes difficult to follow and time-consuming to read, the manual contains a variety of attributes of beings, situations, and scenarios, that, when selected and/or varied (often through the use of multifaceted die or dice), provide the setting for many hours of science fiction and war game-based enjoyment. It is my firm conviction that movie, television, or paperback writers could readily use the manual as a device for modeling plots to stimulate and help organize their creative writing.

Hybrid Delphi Game

My introduction to this highly playable game was at a training session of the World Future Society dealing with futures simulations/games. Our participation was enthusiastic, debate was heated and intense, and consensus was achieved with a sense of satisfaction. We obtained an excellent practical experience in delphi methodology. The exercise is readily adaptable to different events the user may want to have the group consider.

Future Planning Games

Bender and McCuen, now directors at Greenhaven Press, were two high school social studies teachers who began designing games on controversial social issues to bring out assumptions and facts underlying opposing viewpoints. Their instructional development activities grew into a full-time profession as Greenhaven Press, publishing the Opposing Viewpoints Series of paperbacks on controversial issues, and the Future Planning Games. Their classroom experience, I'm sure, contributed substantially to the extremely useful, well-designed, and cost-effective format, with content focusing on important present and future social issues for junior and senior high school students.

Cope

One of the strategies most frequently recommended by simulation/gaming researchers and practitioners is to imbed a simulation/game in the context of a variety of traditional educational activities. The "learning through involvement" approach of *Cope* provides not simply for imbedding, but for using *Cope* as a four-week simulation/game, integrating a selected bibliography, time chart, detailed teacher role instructions, and handouts for students into an attempt to provide students with the experience of the simulated stresses and anxieties of trying to cope with rapid social change.

Utopia

One of the most difficult things to accomplish in exploring futures in educational settings is to get students to creatively and imaginatively think of their personal futures in terms of the future. It is often difficult for students to suspend the momentum of present and past experiences, which often eclipses their serious exploration of alternative future possibilities. This simulation of constructing an ideal society is similar in one or two respects to B. F. Skinner's Walden Two: simulation participants are subjects in a professor's experiment, and there is group participation in voting on principles that will form the basis for the new society. In objecting to a rule proposed by another player, one participant was overheard to remark, "I don't want to be a rat in your kind of Skinner Box!" But the simulation *Utopia* differs from *Walden Two* in many respects: participants are led to engage in a great deal of library research in exploring eight subcategories of occupations and 31 job possibilities. Thus, it is most likely that no single psychological theory will underlie the utopian society the students created. Another important difference is that *Utopia* is begun in a fertile valley donated by a wealthy industrialist, and participants have no economic worries, having each been given one million dollars. Walden Two has to make it on its own economically from the outset. This removal of economic reality in *Utopia* is at some times a strength, and at other times a weakness. While it may help students consider more far-fetched possibilities as alternative principles, it also removes a considerable amount of accountability and reality: The kind of principles selected as the basis for an ideal society are not likely to be the same between someone worried about providing the next meal for his or her family and a gentleman farmer who has considerable inherited wealth. The user of this simulation might want to consider an alternative scenario or introduce some pressing realities to "test" the adequacy of the principles drafted by the class, once they have survived the political selection process within *Utopia.*

Prospects

As a self-focused and self-administered career development tool, *Prospects* provides simulation users with opportunities to improve the personal planning of their professional career development at far less than the cost of a management consulting/career counseling firm. The printed materials readily provide a walk through the major decision and analysis steps. I believe that the simulation compares favorably to the first free "promotional" career counseling sessions provided to me by a management consulting firm when I was a Captain in the United States Air Force considering alternative professional careers. This firm offered me the opportunity to invest in myself by giving them $2000 or more to do an exhaustive psychometrically based evaluation, followed by professional career counseling, presumably leading to a rising career as an executive. (I respectfully declined to pursue a Ph.D.) I believe that the principles underlying the *Prospects* approach are very similar to those being offered in far more expensive career counseling packages, all of which are designed to invoke positive, self-fulfilling prophecies for the career advancement of the aspiring professional.

Edplan

The political dimensions of planning a school budget are dynamically simulated in this very actively involving role-playing simulation. The basic content is a worthwhile and, in some areas, even crucial topic with the current dilemmas of public school systems in adequately financing their educational programs. *Edplan* presents an excellent representation of political realities and pressures, and offers feedback to participants through votes on their positions, elections, and simulated review of proposals by a federal funding agency. One limitation inherent in the simulation's approach is its stress on the political factors in the process, at the expense of rational factors. One participant expressed some resentment during debriefing that "politics seems to be the only thing that's important in this game." Although the public forum of debate presents a lively simulated setting for consideration of issues, the underlying model is not always representative of all the factors that go into educational planning, and that is an inherent design limitation. If the underlying model of the simulation is acceptable to the user, the simulation will do a good job of creating that type of educational planning atmosphere.

Edventure II

A clear, easy-to-follow set of role descriptions and game administrator procedures are presented for simulating 20 years of future educational life. This simulation—built around a college course catalog and a registration/transcript record form, with planning for future education, work, and leisure, and with feedback on decisions—can involve middle or high school students in more seriously considering their long-term futures.

2000 A.D. Futura City

The self-contained multimedia kit reflects the highly polished professional "packaging" one might expect from the publisher of *Newsweek.* As a part of the kit on 2000 A.D., the Futura City simulation is relatively very short. The materials of the entire kit provide the structural components for a variety of future-oriented activities.

CONCLUSION

Whether we are designers or participants in simulations, our assumptions may easily forestall decisions or may sometimes launch untimely and misdirected actions. This may occur because we "know" our plans, our visions of tomorrow, are firmly rooted within the present and past, appearing—erroneously—right or wrong for the tasks at hand. With sophomoric arrogance we may become like the sorcerer's apprentice who tinkered with trends during his master's absence, flooding the laboratory.

But dynamic modeling of the future through simulation does offer the chance to transcend the constraints of real-world time and space; the social sciences do, in a sense, briefly become sorcery and harness the power of knowing the probable practical impacts of simulated alternative future events. With the enactment of a simulation, people participating within created scenario environments are led to make believe for real. The events of the simulation are real to them. At the same time, the simulation may stand for events having either hypothetical future existence or actual past realization. In either case, the underlying elements are often diffused and separated across distances of space and depths of time not easily traversed by human perception. We have at once both the symbol (simulation) and the thing it represents (hypothetical, or perhaps real, future events).

We must not confuse our gaming, modeling, and simulation with reality. Perhaps we may find these created realities much better than the booming, buzzing confusion of real and chaotic events, and we may prefer them. As soothsayers, perhaps we will be no better off as future story tellers—or "futstorians"—than the group of historians whose writings have been regarded by some as "a pack of lies told by a group of people who weren't even there." Nonetheless, it appears desirable to keep designing alternative futures simulations, because we need better operating manuals to run the social systems of Spaceship Earth. The fact that visions of tomorrow created by future-oriented simulations eventually will be found not to correspond with actual events does not eliminate the value of this activity. Future-oriented simulations/games may help us plan better for the future by encouraging us to experience parts of the future's structure before the real events occur.

Soren Kierkegaard observed, in *Purity of Heart is to Will One Thing,* that "the aim is a more reliable indication of the marksman's goal than the spot the shot strikes." Were we to hit the target's center every time, there would be no sport in taking aim.

HEALTH AND HEALTH CARE GAMES AND SIMULATIONS
An Evaluation

by Amy E. Zelmer and A.C. Lynn Zelmer

INTRODUCTION

Perhaps you are a community health worker who wants to teach a group of low-income mothers about meal planning for their families. Perhaps you are a trainer teaching firemen how to carry out cardiopulmonary resuscitation. Perhaps you are a nursing instructor trying to teach a number of students about planning nursing care for a group of patients—and you don't have unlimited clinical time. Perhaps you are a neurologist trying to teach medical students how to diagnose different types of disorders—and you don't always have patients with the required symptoms available.

In any of these very diverse situations, you may want to include a simulation or game as a part of your teaching strategy. There are materials available to meet the needs of teachers and learners varying from illiterates dealing with basic personal and community health matters to advanced technical/professional students.

At any point where patient safety and patient rights come into conflict with students' needs for practice, or where students must learn how to cope with unusual situations, simulations may help. Simulations often provide opportunities to apply principles in a problem-solving situation, and games provide opportunities for memory recall. Both learning situations are common in education about health and health care at any level.

Availability

So if simulations and games can provide such good learning experiences, where does one find them? There's the snag. There are undoubtedly hundreds of homemade games used by instructors in the health field, but few have made their way into the commercial world. In some ways that's unfortunate because it means that we have to reinvent materials for our own use, and that is time consuming, though locally prepared materials have the advantage that they can be tailored to particular learners.

Selection

The eleven simulations and games we review in this essay were selected from those commercially available to show you the widest possible range of types, audiences, and subject matter (see Table 1). That range makes it difficult to compare any two games, but it may give you some ideas for developing or modifying materials on your own. Each selection is representative of similar materials covering the same or other subjects. There is, for instance, a reasonably large selection of card games in nutrition and related areas. Some of the materials we review are not available as a ready-made package; we have included a few items that are available only in books, but we felt that you could always add your own playing pieces if the ideas were available.

We have *not* included in this essay materials that are chiefly concerned with sanitation or materials focusing on population (rather than family planning, which we have included). The essay on *The End of the Line* in this *Guide* also covers an exercise relevant to health workers.

BASIC CONSIDERATIONS: LEVEL, GROUP SIZE, PLAYING TIME

To an instructor, the real-life constraints of a group's level of experience, its size, and the time available will probably be important considerations. Please bear in mind that the information in Table 2 is only approximate. The way you introduce and follow up material can lengthen or shorten the playing time, and you can make some modifications of reading level to adapt materials for community groups.

Time

You can extend playing time by repeating an exercise (particularly games) for further practice. For example, *Nourish* can be used repeatedly, with fourteen different games drawing on the same deck. Because students can use these cards inde-

---

Editor's Note: Listings for all the simulations and games discussed in this essay are in the health section, except for *Brookside Manor* and *Everybody Counts!*, which are in the human services section, and *Community Target*, which is in addictions.

Author's Note: Comments regarding *Everybody Counts!* were based on examination of an earlier edition of the game, entitled *Smltd Lrng Xprnc*, and may not apply to the new edition.

TABLE 1 The Eleven Health and Health Care Simulations and Games

Title	Summary Description
Blood Money	Participants role play hemophiliacs and health care workers. The simulation uses chips to symbolize money, medical care, and blood—all of which are required to cope with the hemophiliac "attacks."
Brookside Manor	Participants role play residents and staff of a new home for the aged. They must resolve their differences regarding significance of personal belongings, which in turn affect or reflect opportunities for social relationships, privacy, and individual expression.
Clinical Simulations: Selected Problems in Patient Management	A book of twenty patient problems for medical students that uses an answer-disclosing technique; each problem can be used only once.
Community Target: Alcohol Abuse	Participants role play various community members trying to determine the extent of, and find solutions for, an "alcohol problem" in their community.
District Nutrition Game	A board game to help participants realize some of the important steps in planning a good community nutrition project.
Everybody Counts!	A series of exercises for health care workers to promote affective understanding of the handicapped.
Nourish	Deck of 144 cards, each showing a different food, in suits representing different nutrients. Can be used for 14 different card games to teach food values.
Nursing Crosswords	Word games to give students practice in recalling correct terms and definitions.
Planafam II	A game of chance and strategy for 3-10 players in which a lifetime reproductive cycle is lived out and the consequences of choice assessed.
Psychiatric Nurse-Patient Relationship Game	A structured role-playing situation for 2 players to practice communication skills and analysis of processes.
Resusci-Ann	An interactive model that enables individuals to practice cardiopulmonary resuscitation and to have feedback on their efforts.

pendently (outside of class time) such repeated use need not involve the whole class nor take up scarce class time.

In the simulations that call for players to assume roles (*Community Target, Blood Money, Brookside Manor*) you may wish to have players retain the same roles in repeated play. In this way they can try alternate strategies and build upon their previous learning. However, this is often difficult to do because it requires the cooperation of all players in limiting the variation in play. A more fruitful use of the time may be to have players reverse their roles (for instance, those who were "residents" become "staff" and vice versa in *Brookside Manor*) so they can experience and compare the constraints imposed by different roles within the same situation, particularly for how this affects their behavior.

In our experience, the more structured the exercise, the less value and the more player resistance there is to repetition. The *Psychiatric Nurse-Patient Relationship Game* generally cannot be repeated for this reason.

Number of Players

The number of players can vary quite considerably. We have generally found it useful to have players work together (even on *Nursing Crosswords*) so they can discuss answers and learn from each other. This is particularly useful for beginning or disadvantaged learners, but it may also be useful at advanced levels, as in the clinical simulations. It is also a good way to deal with unexpectedly large numbers of learners in some community situations and to make scarce materials go further.

In general, you should probably try to work through any simulation the first time with the recommended number of players and introduce variations only as you become familiar with the dynamics of the exercise.

At all costs, avoid having a number of passive "observers." It is useful in complex situations like *Blood Money* to have designated observers who can report back to the group as a whole during the debriefing period. However, they should be given specific guidelines for what to observe, otherwise they may miss key issues and report only on obvious actions of which the players themselves are already aware.

A useful rule of thumb is that if observers constitute more than twenty percent of the number of active participants, you should restructure the exercise by providing additional sets of materials, doubling roles (having two players share one role),

TABLE 2 Basic Considerations

Title	Level	Playing Time	Number of Players
Blood Money	adult; prof.	2½ hours or 3 1-hour periods	20-35
Brookside Manor	adult; health prof.; students	2 hours	15-60
Clinical Simulations	med. student	variable	1
Community Target	h.s.; adult	4 40-minute periods	20-30
District Nutrition Game	adult (low) reading level	20-60 minutes	2-4
Everybody Counts!	parents; teachers institutional staff	variable	5-30
Nourish	jr. high—adult	20 minutes	1-20
Nursing Crosswords	health prof. students	20-60 minutes per puzzle	1
Planafam II	adult; health prof. students	1-2 hours	3-10
Psychiatric Nurse-Patient Relationship Game	mental health prof. students	variable	2
Resusci-Ann	h.s.—adult	variable	1

or devising other activities for half the group. We cannot repeat too strongly that a large group of passive observers is inimical to the use of simulations and games as a learning method.

A related problem that sometimes arises, especially with community groups, is the reluctance of one or more individuals to participate—the "I'll just watch" syndrome. While we don't believe that anyone can, or indeed should, be forced to participate, we do believe that having some people opt out like this may put extra pressure on those who are willing to risk participation. We have generally handled this situation (it doesn't happen often) by asking the nonparticipants to become active observers with a specific task or by giving them another assignment altogether in some other location. Adults are often reluctant to appear "foolish" in front of their peers, and in groups that are not used to learning through simulations and games those who are willing to take the risk need to be supported. The problem, if any, generally disappears in a second experience for the group if handled as we suggest; otherwise you may have more people wanting to opt out in later activities.

ISSUES AND CONTENT

Most games and simulations are chosen because of the subject matter (content) inherent in the material or because of the issues they raise for further discussion. A simulation often involves more than one issue (for example, the *District Nutrition Game* brings together information about nutrition and community organization), and it is the interplay between the issues that provides the driving force for the simulation.

In general, the larger issues must be approached in the debriefing. The content of the simulation itself generally deals with only part of an issue (see Table 3) and while most players can probably make at least some transition to the larger issues without a debriefing, it will be helpful for many to have an issue-related discussion while the content is fresh in their minds. For example, the content of *Brookside Manor* centers around which personal belongings residents may take with them into a group home. The debriefing can develop a more general approach to the issues of safety, social needs, accommodation to group living, attitudes of cleaning staff or unions, the necessity for maintaining family ties, and so forth, so players can generalize from the specific content of *Brookside Manor* to other situations.

Similarly, in *Clinical Simulations* the student is required to work through diagnosis and treatment procedures for a specific situation. In the event that the student meets such a situation again, she or he will obviously have information from the simulation to apply; however, even more important will be whether she or he can apply the principles this simulation teaches to other similar but different situations. Again, discussion following the exercise to analyze the situations in terms of the principles involved may be an important part of the learning experience if the students do not (or cannot) do this on their own initiative.

It will be relatively simple for your to take some game formats and vary the subject matter (for example, *Nursing Crosswords*). Others have much less flexibility because the process of playing is part of the content (*Psychiatric Nurse-Patient Relationship Game*). Another section of this *Guide* deals with frame games, which are designed to let you insert your own material. If you have not been able to find a suitable health-related game, you might want to develop one of these for your own purposes.

Substitution in something like the *District Nutrition Game* would require considerable thought, because the relative values of the items substituted should be the same or the dynamics of the game will be altered. For example, if you were to substitute child safety content in the *District Nutrition Game* (which is basically a snakes and ladder format) you should be careful to locate major hazards on the big snakes and minor hazards on the small snakes.

In any card game, the content must lend itself to being sorted into sets (corresponding to suits), perhaps with numerical values. The more closely your deck of cards corresponds to a standard 52-card deck, the more possibilities there are for using your deck to play standard card games. This will eliminate much, though not all of the tedium of learning rules for many people, and it increases the possibility that the games can be used in informal situations such as clinic waiting rooms.

PROCESSES

Sometimes you will want to choose a simulation or game for the experience the process of playing the game gives the participants (*Everybody Counts!* is high on the scale in this respect). In other exercises the process itself is relatively unimportant except as a vehicle for the content. For example, with card decks it is often irrelevant whether participants play snap, rummy, or bridge; the important thing is to provide a vehicle to practice matching items, building sets of like items, or remembering values. The important consideration will be to find a vehicle that has an element of fun for participants and does not require a great deal of time for an explanation of mechanics.

RANGE AND DEPTH

How might these exercises fit into the teaching/learning process? Some are self-contained; that is, participants do not need any specialized knowledge before they start, they learn something during the exercise, and they do not necessarily need to discuss the experience afterward, although further discussion may consolidate or build upon this learning. The *District Nutrition Game* is perhaps the best example of this type.

Other simulations presuppose that a good deal of basic learning has already taken place and that the simulation will provide practice in applying that learning. *Nursing Crosswords* and the *Clinical Simulations* are examples of this type. Generally it is up to the instructor to choose the point at which to introduce such exercises into the students' learning.

Still other simulations are designed to introduce a topic to a relatively unknowledgeable group and to provide a basis for further discussion. It is with this type of simulation that

TABLE 3 Issues and Processes

	Blood Money	*Brookside Manor*	*Clinical Simulations*	*Community Target*	*District Nutrition Game*	*Everybody Counts!*	*Nourish*	*Nursing Crosswords*	*Planafam II*	*Psych. Nurse-Patient Relationships Game*	*Resusci-Ann*
Issues											
Alcoholism				P							
Community Organizing				S	P						
Organized Health Care	P								S		
Medical Care Costs	S										
Gerontology		S				S					
Institutions		P			P						
Family Planning									P		
Handicaps	S					P					
Nursing								S		P	
Nutrition							P				
Psychiatry			S							S	
Patients	S	S	S			S				S	
Specific Medical Diagnoses	P		P					P			
First Aid/ Emergency			S								P
Processes											
Recognition					P		P				
Recall			S				P	P		S	S
Role Playing—Empathy	P	P		S	S	P			P	P	
Negotiation	S	S									
Problem Solving	S		P	P							
Analysis									P	P	
Psycho-motor Skill practice						S					P

Key: P = primary importance
S = of secondary importance

further readings, references, and bibliographies will be most useful to the instructor and students. Of course, any instructor can, and probably should, add materials of particular relevance to his or her students, but Table 4 will give you some idea of what the materials already include.

All of the exercises that have no accompanying materials will need to be preceded or followed with considerable material. *Community Target* might be used to introduce a unit on the topic, although it could also be used after some initial presentation of the scope of the problem in the players' own community. *Resusci-Ann* is best used for practice after students have had some introduction to the theory and rationale for the resuscitation procedures. Although the recording models will give students some feedback on their performance, the instructor will generally have to show how to overcome deficiencies. The other three exercises in the left column of Table 4 should only be used after the instructor can assume that students have "learned" the content and need practice in its application. In these cases the instructor should be prepared

TABLE 4 Accompanying Materials

None	Moderate	Extensive
Community Target	*Brookside Manor*: very limited bibliography	*Blood Money*: part of a three-part kit of materials
Clinical Simulations		*District Nutrition Game*: part of a book dealing extensively with the issues
	Planafam II: some suggestions for discussion	*Everybody Counts!:* a detailed workshop manual
Nourish	*Psychiatric Nurse-Patient Relationship Game*: limited bibliography	
Nursing Crosswords		
Resusci-Ann		

to refer students to the appropriate items in their basic material if gaps in their knowledge become apparent.

The items listed in the "moderate" column require some additional user input. In general, their accompanying materials give a good starting point for discussion but need to be supplemented with more up-to-date materials because all deal with areas in which new materials are constantly appearing. The references given are generally appropriate for the suggested audience; if you intend to use the same material for a different audience (for instance, *Planafam II* with a professional group) you may want to choose readings more appropriate to that group.

VALUES

Almost all simulations in the health field have as part of their underlying value system a positive value on good health, perhaps to the exclusion of other values that may be important to some—different allocation of resources, for example. Many simulations and games in the health field do not make their value systems explicit (*Psychiatric Nurse-Patient Relationship*), while others are about value systems in conflict or potential conflict (*Community Target, Brookside Manor*).

An important part of the art of using simulations is helping participants identify their own value systems and those built into the simulation. The positive value of good health that is designed into most health simulations may be specific to North America and the twentieth century. However, it is a basic tenet of our health services. Many games also assume that health services are curative rather than preventative, and health educators, at least, would take exception to this. Your students may remember only the basic values of the exercise, forgetting the details. It is crucial therefore that you are aware of the overt and covert values of an exercise before you use it (for example, what are "good" foods? what is the "proper" behavior?).

As game designers ourselves, we have found that we are often not even aware of many of our built-in or covert values. Even when we are attempting to influence the participants' value system, the structure of the exercise may be counter to the desired behavior. A competetive game design, for example, does not function well as a vehicle for imparting cooperative team-building attitudes for medical staffs, and materials for an in-basket must be selected with care if delegation of responsibility/authority is a goal. While activities such as *Blood Money, District Nutriton Game,* and *Nourish* probably use competition to encourage participation, there is the risk that the content will get overlooked in the "game." Thus, we cannot overstress the value of a thorough briefing and debriefing.

ROLE PLAYING

Participants are required to play roles in some exercises (see Table 5). A role may be outlined in functional terms, that is, a job to be done to which the individual brings his or her own value system. Or a role may have more specific directions about the participant's attitude toward a topic, such as, "You have noted a rise in the consumption of alcohol among the students on your campus this year. You are extremely concerned over this situation and you feel there is an urgent need for an alcohol program in the schools." Individuals who are not experienced in role playing may have more difficulty with this type of role and may require more help from the instructor in thinking through how to act it.

In some instances, role-playing directions require the individual to display specific behaviors (for example, "Again, and throughout the reading lesson, incorrect answers are made fun of, are ridiculed or ignored"). This behavior may be difficult to carry through consistently, particularly if it conflicts with the individual's own value system. It may also provide an escape for that individual or other participants, who can say, "But that's not how it is in real life."

The moderator or instructor may also be required to do some role playing; in fact, the instructions quoted in the last paragraph are for the moderator of *Everybody Counts!* In deciding whether to use a particular simulation you may want to examine whether the roles assigned to the moderator and key participants can be carried off.

A role description needs to be complete enough to provide reasonable direction to the player for the purposes of the exercise. For complex "simulated patients," this might mean a complete patient file, X-rays, medical charts, tests, and so forth, as well as extensive briefing on the correct responses to specific stimuli. For most roles, however, the directions can be quite simple. Our observations suggest that a description of 50 to 150 words provide enough direction for most role-play activities when accompanied by the general game materials (community history, pregame readings). The role descriptions in *Community Target* are good in this regard. They allow players to use their own names, they provide age, sex, and personal information, and they give information relevant to the game complete with specific "facts" for use in the role.

Roles that are too detailed usually do not allow players to develop their own reactions to the situation. Roles that are farcical ("You are Dr. Dogood") or too short are, almost invariably, not functional. They promote excessive experimentation and may destroy the learning value of the exercise.

Sex differences sometimes affect the conduct of an exercise, as does the "role" age. School children and men seem to have considerable difficulty with sex reversal. Very young adults often react with extreme stereotypes if asked to portray elderly people. While the discussion of these difficulties may be fruitful, the role reversals may have damaged the learning possibilities in the role.

If you are planning to use a simulation requiring role playing with an already functioning group (as part of an in-service education program, for example) you may have particular difficulties. The group may not be able to shed real-life roles sufficiently to participate. This is especially true if there is much difference in status (in real life) among participants and if the agency has a rather rigid structure. This is not to say that you should not use role-playing simulations under such circumstances, but only that you should be aware of the hazards.

The role descriptions in the exercises we have discussed here are all generally adequate within the limitations noted above. Graduate nurses who have worked through the *Psychiatric Nurse-Patient Relationship Game* have generally felt that their responses were too limited by the roles, but this problem may not arise with beginning students who have less of their own experience to draw on. Because the role descriptions in both *Community Target* and *Everybody Counts!* require individuals to adopt particular viewpoints, more pregame help may be required from the instructor so individuals are comfortable with their roles and can portray them realistically.

MOTIVATION AND SCORING

Within a simulation or game, what acts as the driving force to keep players participating (and, one hopes, learning)? The motivation may come only from internal satisfaction of taking part in a process (*Community Target*), from persuading others to adopt one's point of view (*Nursing Crosswords*), or from winning points in an artificial scoring system.

External scoring systems are generally more crucial in games where the overall goal is to have students practice using new terms. Intrinsic motivation in the form of satisfaction either with one's performance or with learning (even if the actual experience is somewhat uncomfortable) becomes more important if the overall learning goals are related to a gain in empathy. Each exercise we consider in this section has a different mechanism; Table 6 summarizes them.

Chance factors have little place in learning exercises except to indicate events and possibilities not otherwise accounted for within the body of the simulation. Chance plays a relatively large part in two of the games; here the chance factor makes it more likely that participants with less skill and knowledge can "win." This may make it easier to keep those participants who would ordinarily be disadvantaged active in the game and thus still practicing.

MODEL VALIDITY

Does the simulation or game accurately represent the real world? If it doesn't, your students will be learning to cope

TABLE 5 Types of Roles in Role-Play Simulations

Roles	*Blood Money*	*Brookside Manor*	*Community Target*	*Everybody Counts!*	*Planafam II*	*Psychiatric Nurse-Patient Relationship Game*
Players						
• independent roles			X			X
• roles as group Members		X		X	X	
• functions	X	X				
Moderator				X		X

TABLE 6 Scoring and Chance Factors

Title	Scoring	Chance
Blood Money	chips representing money; disability and death indicators	Cards are used to determine events within game; probabilities are based on real-world experience
Brookside Manor	Participants rank order items and attempt to reach agreement	none
Clinical Simulations	No scoring as such; because one can see how many responses were requested, one can determine if student used most efficient line of inquiry and treatment	none
Community Target	Committee decides on issues based on previous discussion	none
District Nutrition Game	Board game; first to reach finish wins	Events totally controlled by chance
Everybody Counts!	None as such. Participants try to complete various tasks while handicapped	none
Nourish	Depends on game chosen	Depends on game chosen
Nursing Crosswords	Completed puzzles can be checked against answers	Minimal. Some words could be completed by guessing
Planafam II	None as such. Players may try to set their own goals for reproductive behavior within game	Playing cards and dominoes are used to determine events; probabilities based on real-world experience
Nurse-Patient Relationship Game	Progress sheets completed for each round indicate participants' achievement against total possible	none
Resusci-Ann	Recording tape show students' performance rated against requirements for effective resuscitation	none

with some nonexistent situation. As far as we and our colleagues have been able to judge, all of the information presented in these simulations is accurate—as far as it goes. That's an important qualification.

Most of the materials will go out of date quickly. *Planafam II*, for example, is already somewhat behind the times in that it is based on probabilities from the 1960s.

Most of the materials are culture bound. Though this is perhaps most obvious when we look at something from outside our own culture (like the *District Nutrition Game*), it applies equally to other materials. *Blood Money*, for example, will require extensive revision if it is to be used to best advantage in areas where medical insurance systems and blood banks operate on a different basis from the model used for this game.

All simulations and games are simplifications, and some options are omitted. Such limitations do not mean that these materials are "bad," only that, as with all teaching aids, they must be previewed before use and adapted to the instructor's and learners' particular needs.

MATERIALS, COST, AND DURABILITY

Some exercises come complete with all the pieces; you simply need to open the box and start. Others require that you duplicate materials and obtain playing pieces. The amount of preparation is generally balanced by the cost. Table 7 summarizes the cost and materials required, with comments on general durability.

You will find that all materials will last much longer if paper items are prepared on good-quality paper, and on standard size sheets. Heavily used materials should be printed or typed on card stock or laminated. You should keep all materials together, preferably in standard size boxes or envelopes that are suitable for storage on your shelf and small enough to carry from place to place for use. Paste a checklist of contents and auxiliary materials inside the lid of the box and refer to it before taking the exercise out for play. With exercises that require many copies of printed papers we often retain *only* a master copy and duplicate the required sets when needed (subject of course to copyright restrictions).

DEBRIEFING

Debriefing is that important part of the learning experience that follows completion of the exercise. While some of the games described in this essay can be used independently by students without any planned debriefing, their learning value may increase with some discussion of the issues. For example,

TABLE 7 Materials and Cost

Title	Cost	Materials Required	Comments
Blood Money	n/c	playing cards, poker chips, name tags, coding dots, pencils, pads, 80-90 pages duplicated	Soft cover manual opens flat for easy duplication, good print quality
Brookside Manor	$5.00	group ID cards, newsprint and felt pens; duplicate worksheets if more than 30 players or for repeat runs	Kit contains manual and players' worksheets
Clinical Simulations	$23.50	one book per person and special disclosing pen	Book is soft cover. Once answers are revealed case cannot be reused
Community Target	$6.00	30 role cards to duplicate, name tags	Manual is soft cover. Originals for roles rather faint
District Nutrition Game	$10.00 for book	dice or spinning top, game board, playing pieces	Soft cover book contains directions for construction and play
Everybody Counts!	$10-$15	Extensive list of required equipment including casts, wheelchairs, plastic letters, swim goggles, cassette, tapes, flashcards, overhead transparencies	New edition has not been examined
Nourish	$9.00	one deck of cards per group	Cards are good quality; main problem may be loss of some over time
Nursing Crosswords	$5.95	one copy of puzzle per participant (copyrighted materials cannot be photocopied without permission)	Book is soft cover, print is good quality
Planafam II	$.65 microfiche, $3.29 hard copy	large sheets of cardboard, 3 sets dominoes, 2 decks of cards, marking pens	"Hard copy" is xerox
Psychiatric Nurse-Patient Relationship Game	$15.95	Kit comes complete in burlap bag. "Progress sheet" required for each player	Painted oilcloth game board tends to peel. Printed materials rather flimsy, small print
Resusci-Ann	$1,025.25 (full price-recording); recording paper $3.10 each	Recording tape, alcohol swabs to clean mouthpiece required for each use	Seems to stand up well with repeated use

players who have used the *Nourish* cards may well benefit from discussing the principles of nutrition, local dietary patterns, and so on. As Table 4 indicates several exercises have no accompanying materials or suggestions for debriefing. The instructor must prepare his or her own discussion guide for these.

Several of the simulations *require* debriefing so participants are not left with unresolved issues or erroneous impressions. *Community Target, Brookside Manor, Planafam II, Psychiatric Nurse-Patient Relationship Game, Blood Money,* and *Everybody Counts!* all fall into this category, but only the latter two include adequate debriefing guidelines. All the other exercises appear to assume that an experienced instructor will be available to guide the debriefing. This assumption is not bad in itself, but simulations, like films or any other teaching method, need to be used as a part of an overall learning strategy if the most effective learning is to take place.

CONCLUSION

There are relatively few commercially available simulations in the health and health care field. Those that are available cover a wide range of topics and are aimed at very different levels of players. This makes comparison difficult, but Table 8 attempts to summarize the major strengths and limitations of each exercise we have reviewed.

Simulations can be a very useful teaching/learning strategy in the health field. We hope that much more development will take place in this area before the next edition of the *Guide* is prepared.

Sources

Blood Money: A gaming-Simulation of the Problems of Hemophilia and Health Care Delivery Systems
Cathy Stein Greenblat and John H. Gagnon

DHEW Publication No. (NIH) 76-1082
Superintendent of Documents
U.S. Government Printing Office
Washington, D.C. 20402

Brookside Manor: A Gerontological Simulation
Dorothy H. Coons and Justine Bykowski

Institute of Gerontology
The University of Michigan—Wayne State University
520 East Liberty
Ann Arbor, Michigan 48108

Clinical Simulations: Selected Problems in Patient Management
Christine McGuire, Lawrence M. Solomon, Phillip M. Forman

Prentice-Hall
Englewood Cliffs, New Jersey 07632

TABLE 8 Strengths and Limitations

Title	Strengths	Limitations
Blood Money	• provides rich experience for debriefing • can be used with wide range of lay adult to professional players simultaneously • good accompanying materials	• very complex • oriented to U.S. health care delivery system • requires considerable advance preparation
Brookside Manor	• uses deceptively simple framework to get at several vital issues	• best if very skilled facilitator available to lead debriefing
Clinical Simulations	• format relatively inexpensive	• can be used only once • very specific audience
Community Target	• deals with important social issue • suitable for use within classroom constraints	• requires knowledgeable leader • oriented to California situation
District Nutrition Game	• easily adapted to local conditions • minimal literacy required	• high chance factor
Everybody Counts!	• good suggestions for empathy exercises for a variety of handicapping conditions	• directions to leader may be too negative
Nourish	• format appeals to many • cards durable and attractive	• can be played without necessarily learning principles and facts
Nursing Crosswords	• format appeals to many • good range of topics	• not reusable • can be worked through rote without much learning
Planafam II	• deals with important social issue in realistic fashion	• requires fair amount of initial preparation • needs updating to incorporate recent advances in fertility control
Psychiatric Nurse-Patient Relationship Game	• can provide independent student practice	• thought by many players to be too mechanistic
Resusci-Ann	• only safe way of giving students appropriate practice and feedback	• expensive, single-purpose teaching tool

Community Target: Alcohol Abuse
David A. Sleet

Center for Health Games and Simulations
Department of Health Science and Safety
San Diego State University
San Diego, California 92182

"District Nutrition Game" in *Nutrition for Developing Countries*
Maurice King, et al.

Oxford University Press
200 Madison Avenue
New York, New York 10016

Everybody Counts! Explorations in Affective Understanding
Harry Dahl

The Council for Exceptional Children
1920 Association Drive
Reston, Virginia 22091

Nourish
Camille Freed Pfiefer and Mary Shaw Smith

Fun with Food
P.O. Box 954
Belmont, California 94002

Nursing Crosswords and Other Word Games
Sheryll Dempsey

Trainex Press
P.O. Box 116
Garden Grove, California 92642

Planafam II: A Game for Population Education
Katherine Finseth

ERIC Document ED 064 228
ERIC
855 Broadway
Boulder, Colorado 80302

The Psychiatric Nurse-Patient Relationship Game
Carolyn Chambers Clark

P.O. Box 132
Sloatsburg, New York 10974

Resusci-Ann
T. Delgarno

Safety Supply Company
6120 99th Street
Edmonton, Alberta T6E 3P2
Canada
or local supplier

HISTORY GAMES AND SIMULATIONS
An Evaluation

by Bruce E. Bigelow

INTRODUCTION

Simulations and the study of history belong together. In fact, simulation is the only way to study the past. At first glance, such a statement may strike you as absurd, and many professional historians would scoff at such an idea. But if we examine what history is, what we do when we study history, and the role of simulation in such study, we will find that the connection between history and simulation is much closer than it may initially appear.

What Is History?

The intricacies and nuances of this question notwithstanding, we can suggest at least some of the answer. First, history is the story of what happened in the past. How one tells that story, who one selects as the major characters, and how one analyzes the plot are factors that depend entirely upon who is telling the story. But that some aspect of the human story will emerge is one of the tenets of history. Further, many would argue that history demands answers to questions of why things happened as they did. That is, the study of the past is more than just a recitation of events but is rather the weaving together of a complex set of interrelationships from which we can learn how change occurred and, by extension, how it is still occurring or may continue to occur in the future. Still others contend that history goes beyond the study of events, persons, and institutions, that it rests fundamentally upon an understanding of the intangibles of human life, upon human values and their impact upon human behavior.

The Methodology of History

Definitions differ, and historians argue at length with one another about the "correct" understanding of their craft. But whatever the definition, the methodology remains the same. First, historians rely on primary materials–archival collections, public records, data sheets, and the like. And from those materials they try to reconstruct a set of events in which they did not take part. That is, they try to simulate the past. Historians can do no other than to simulate; the entire process of "doing history" rests upon a good ability to simulate.

Second, historians rely upon their own sense of human nature to extrapolate from the materials they have at hand to try to understand the motives or values of people in other times and places. Again, they are simulating. Again, they have no alternative. Most historians, of course, have not traditionally used games to communicate the simulations they create; words have done the job. But words are, after all, only symbols of abstract ideas. And whether one uses words or models to approach the past, simulation remains the essential tool.

The Purpose of This Essay

You might wonder why we should write a review of historical simulations, then, if all historians always simulate. To answer that, we must separate the kind of simulating professional historians have done in their research and writing from the simulations we will discuss in this essay and which are part of the process of group learning that takes place in the classroom.

The simulations we will compare and analyze are commercially available games, illustrative of different models and the range of uses to which simulation can be put as a means of communicating about history. We make no claim that these simulations are exhaustive or that they should replace the more traditional methods of simulating through words, whether written or spoken, that have marked historical thinking in the past. But we do argue that these simulations deserve every bit as much attention, scrutiny, and constructive criticism as those other methods. And we would urge that all teachers of history examine these tools, looking as we shall at their own goals and asking how well these simulations will help them achieve their ends.

The Problem with Historical Simulations

That said, we face the difficulty of deciding just what is an historical simulation. History is by nature an eclectic discipline. It encompasses the fields of economics, politics, sociology, psychology, science, and philosophy. While not every historian is an expert in each of these areas, the discipline as a whole recognizes that it cannot leave out any of them. Furthermore, history is not limited strictly to the distant past.

You cannot say with certainty that history stops in 1492, in 1789, or in 1979. In fact, the continuation of the human story dictates that history extends into the future. Accordingly, simulations dealing with each of these various disciplines and discussed in greater detail elsewhere in the guide could in some sense all serve the study of history as well. Likewise, simulations dealing with present events or with the future might with justification be subsumed under the historical rubric. We have chosen to limit our pool of simulations to only those which clearly deal with past events or developments, but we direct those of you interested in the fuller range of possibilities for historial simulations to consult other chapters in this volume.

TABLE 1 Historical Simulations

Simulation	Summary Description
Alpha Crisis	Players as ministers or rulers of fictional states modeled after the powers of Europe in 1914 attempt to respond to a crisis parelleling that of the assassination of Franz Ferdinand and the resulting Austrian ultimatum to Serbia.
American Constitutional Convention	Players as representatives to American Constitutional Convention discuss and attempt to resolve conflicts over future governmental structure of the United States.
The Ch'ing Game	Players are members of eighteenth-century Chinese society attempt to advance themselves within the ranks while adhering to the values of that society.
Congress of Vienna	Players as representative of five great powers at Congress of Vienna (1815) attempt through negotiations to resolve territorial disputes of the time.
Czar Power	Players as members of mid-nineteenth century Russian society attempt to resolve series of critical issues and to maximize points of various sorts for themselves.
Czech-mate	Players as delegates from eight European countries or territories in 1938 attempt to resolve in negotiation or war the question of the Sudetenland.
Destiny	Players as members of U.S. government or representatives of foreign states and interests attempt to resolve issues tied to the Spanish-American conflict in the Caribbean in 1898.
Gateway	Players as either immigrants or immigration officials experience the trials of passage to the new world and entrance at Ellis Island in the nineteenth century.
Grand Strategy	Players as members of ten national delegations respond to crisis of June 1914, and negotiate or go to war with one another.
The Haymarket Case	Players as witnesses, court officials, or defendants recreate the trial of anarchists charged with responsibility for the Haymarket Riot in Chicago in 1886.
Liberté	Players as members of various French social classes before and during the French Revolution re-enact critical segments from the events of 1788-1794.
Nuremberg	Players as witnesses, court officials, and defendants re-create the trial in Nuremberg in 1945 of Nazi officials charged with war crimes.
Origins of World War II	Players as spokesmen for major nations of Europe attempt through strategic placement of counters to respond to counterstrategies of opponents and augment their nations' power.
Panic	Players as members of various socioeconomic groups in the United States in 1929 respond to stock market crash and economic crises of the 1930s.
Scramble for Africa	Players as military and political officers representing various European nations seek to maximize national interests on the African continent after 1882.
Seneca Falls	Players represent delegates to Seneca Falls Women's Rights Conference of 1848 and attempt to resolve issues associated with early struggle for women's suffrage.
Trade-Off at Yalta	Players as delegates from Big Three powers in 1944 recreate setting of Yalta Conference and attempt to hammer out compromises on issues of postwar settlement.
Waging Neutrality	Players as representatives of trading and political interests in the United States and Europe attempt to deal with dilemmas posed for the United States by fighting in Europe in World War I.

Criteria for Selection

The games selected for special consideration in this essay represent a variety of games commercially available for use in the history classroom. They are all relatively inexpensive. They represent a good cross-section of the types of simulations available and illustrate not only games on the market but the range of possibilities within which historians may create their own simulations. Accordingly, we will analyze not only the strengths and limitations of these simulations themselves but also how we might adapt these games for varying use and learn from these simulations what prototypes are best used in the creation of our own models.

Types of Simulations Under Consideration

The two general types of simulations available to historians are board games and role-playing simulations. Of the former, most are military strategy games and are dealt with more particularly in the review of military history games. Most are of limited utility in the teaching of history because they concentrate so heavily on military tactics to the exclusion of nonmilitary factors in war. We have, however, selected a few examples of the board game variety for attention here.

Role-playing simulations make up the bulk of the games we will consider, primarily because they are more flexible than board games and therefore more useful for students of history. Further, role-playing games tend to stress the less quantifiable aspects of historical development and deal with values as well as the passage of information. In contrast to the board games, role-playing simulations derive their utility primarily from the process of play rather than the outcome. Accordingly, it is critical that a good role-playing simulation define roles clearly and provide the players with sufficient direction yet with enough flexibility that they can play their roles well and derive the lessons the simulation intends.

There are four basic types of role-playing simulation of interest to the historian. These are (1) generalized experience games whose focus spans an extended length of time and whose intent is to model the general experience of a wide number of people; (2) trial simulations, modeling the trials, either real or fictitious, of famous persons in history; (3) crisis simulations, which focus on revolutions or social and economic upheaval; (4) policy formation or diplomacy simulations in which the primary activity is negotiation among the players over land or issues. Some of the simulations we will consider fall under more than one of these categories and can be used, therefore, in several different contexts. Our discussion below of game activities and the characteristics of the roles called for in the games will help clarify just how these different functions mesh in any given simulation.

PRELIMINARY CONSIDERATIONS

The Uses of History Simulations

The simulations upon which we are focusing, as well as simulations in general, are designed to answer one or more of four basic questions. The first criterion, therefore, that you should consider in selecting a game for your own use is the goal you have in mind, what you would like the simulation to accomplish, what question you would like to have it help answer. These key questions are:

1. How can I spark a greater *interest* in my course, in the subject matter of the course, or in history in general?
2. How can I help the students to *identify* more closely with other people in other cultures in other times; how can I help them understand the values from which these other people reasoned?
3. How can I convey more clearly the *information* about a set of events, a set of relationships, or a long-range historical development?
4. How can I help the students *generalize* from the historical examples they have studied so they become able to apply the lessons of the past to new situations and to their own lives in the present and in the future?

As the information in Table 3 shows, most of our games promote interest in the subject. *Congress of Vienna, Liberté, Origins of WWII,* and *Scramble For Africa* rate the lowest, largely because of their simplistic or confusing format, which leads sometimes to boredom in players. The other games vary in their effectiveness in promoting identification, conveying information, and facilitating generalization, depending partly upon the character of the simulation and partly upon the complexity of the model used. *Alpha Crisis,* for instance, conveys straight data at a low level because it does not rely upon real-life models but uses instead artificial countries. However, this device enables it to teach complex relationships very well because the players are not distracted by a glut of factual detail. All in all, in selecting a game for possible use, you should first choose the goal you want and turn to those games that seem best able to fulfill your criteria.

Age Level, Group Size, Playing Time, Room Requirements

Taking a more practical approach, let us turn now to the basic physical requirements of these simulations so you can match them with the time you have available, the age of your students, the number of players you have to work with, and the room you will have at your disposal. Table 4 should help sort these factors out.

Objectives of the Simulations

A good simulation will tell you what its objectives are before you get started. That way you can evaluate easily whether it meets its own stated goals. As a general rule, all games produced by Interact get exceptional marks in this

TABLE 2 Types of Simulations

Board Games			Role-Playing Games			
Games of Military Strategy	Games of Diplomacy and Negotiation	Combined Board and Role-Playing Games	Generalized Experience Games	Trial Simulations	Crisis Simulations	Diplomacy or Conference Simulations
See essay and section on war games	*Origins of WWII*	*Grand Strategy*	*The Ch'ing Game*	*The Haymarket Case*	*Czar Power*[a]	*American Constitutional Convention*
		Scramble for Africa	*Czar Power*[a]	*Nuremburg*	*Destiny*[a]	
			Gateway	*Liberté*[a]	*Liberté*[a]	*Congress of Vienna*
			Waging Neutrality		*Panic*[a]	*Czech-mate*
			Panic[a]		*Alpha Crisis*	*Destiny*[a]
						Seneca Falls
						Trade-off at Yalta

a. May fill more than one category

TABLE 3 Uses of Simulation

Uses		Alpha Crisis	American Constitutional Convention	The Ch'ing Game	Congress of Vienna	Czar Power	Czech-mate	Destiny	Gateway	Grand Strategy	Haymarket Case	Liberté	Nuremberg	Origins of WWII	Panic	Scramble for Africa	Seneca Falls	Trade-Off at Yalta	Waging Neutrality
Stimulates interest	in course	E	E	E	S	E	E	E	E	M	M	M	E	S	E	S	E	E	E
	in subject matter	E	M	E	M	E	E	E	E	E	M	M	E	S	E	S	E	E	E
	in history	E	M	E	S	E	E	E	E	M	M	S	E	I	E	S	M	E	E
Promotes identification	with general situation	E	E	E	S	E	E	E	E	M	M	S	E	S	E	S	M	E	E
	with other values	S	M	E	I	M	M	M	E	S	M	S	E	I	M	I	E	M	E
	with other persons	M	E	E	S	E	M	E	E	S	M	M	E	I	E	S	M	E	E
	with another culture	M	S	E	I	M	M	M	M	I	NA	I	M	I	NA	I	NA	S	S
Conveys information	straight data—what happened	M	M	I	M	M	M	M	E	M	M	S	E	S	M	S	E	E	M
	cause-effect ideas—how it happened	M	M	M	S	M	M	M	M	S	M	S	E	S	M	S	M	E	E
	complex relationship—why it happened	E	E	E	I	E	E	E	E	I	S	S	E	I	M	I	M	M	M
Facilitates generalization	over a long time period	E	M	E	I	M	M	M	E	S	M	S	E	S	M	S	M	M	M
	about another culture	S	S	E	S	E	M	M	S	S	NA	I	M	I	NA	I	NA	M	M
	about a general type of experience	E	E	M	S	M	E	E	E	M	M	S	E	S	E	S	E	E	E

Key:
I = inadequate
S = satisfactory
M = moderately effective
E = very effective
NA = not applicable

respect. It is a company policy that its games articulate right at the beginning what they aim to accomplish in terms of information conveyed, understanding and awareness broadened, and skills developed. While they vary in their ability to carry through these objectives, Interact games set a high standard of goal setting for other simulations, either those commercially available or your own.

Some of the games under review here purposely limit their objectives; the makers of the *Haymarket Case,* for example, disclaim any desire to teach about the larger economic issues of the industrialization process in the United States. Accordingly, they cannot be held responsible for not including such considerations in their simulation. One might argue that the scope of a given game is insufficient to be of much use in the normal classroom, but one has to judge the simulations by their own standards as well as one's own. This is what Table 5 does. Please keep in mind that we are evaluating only how well each game measures up to its own goals; we shall examine the utility of those goals later.

Game Complexity

Although the complexity of these simulations is linked partially to the age group toward which they are directed, it might be useful to look in a general sense at their relative complexity. We have ranked games as easy if their format is relatively simple and straightforward and they do not require a great deal of preparation and outside research to play. The easier games are also more limited in scope than the complex ones. Those ranked as complex involve an intricate set of social relationships that is difficult to keep straight but yields significant results if studied well before playing the games.

TABLE 4 Basic Factors

Simulation	Age Level	Group Size	Playing Time	Room Requirements
Alpha Crisis	h.s., college, adult	20-55	4-10 hours	one large or six small rooms
American Constitutional Convention	h.s., college	13-98	4-15 hours	one large room or one medium-sized room and several smaller rooms
The Ch'ing Game	college, adult	17-60	8-10 hours	one large area and nine smaller rooms
Congress of Vienna	h.s., college	10-30	6-12 hours	medium-sized room
Czar Power	h.s., college, adult	13-50	3-20 hours	one large room and one small room or study area
Czech-mate	h.s., college, adult	8-50	10-15 hours	large room and several small meeting areas
Destiny	h.s., college		7-25 hours	
Gateway	h.s., college	6-50	16-20 hours	large room
Grand Strategy	jr.h., h.s.	10-32	3-10 hours	one large or several small rooms
Haymarket Case	h.s., college	22-45	5-6 hours	medium-sized room
Liberté	h.s., college	30-50	14-20 hours	large room
Nuremberg	h.s., college, adult	12-40	10-20 hours	medium-sized room
Origins of WWII	jr.h., h.s., college	2-15	2-5 hours	medium-sized room
Panic	h.s., college	18-40	20-25 hours	large room
Scramble for Africa	jr.h., h.s.	6-30	2-4 hours	medium-sized room
Seneca Falls	h.s., college	12-52	1-3 hours	medium-sized room
Trade-Off at Yalta	h.s., college, adult	12-30	2-5 hours	large room plus several small meeting rooms
Waging Neutrality	h.s., college	15-33	3-5 hours	large room

Necessary Preparation

Few history simulations can be played without some familiarity with the historical context the game is intended to model. Accordingly, it is critical for you to know roughly how much time is required to prepare adequately for the game you wish to play before you can actually begin. As you might expect, the more complex a game is, the more preparation is necessary to play it successfully. Similarly, you would expect that the preparation required of the instructor is in almost all cases more than that expected of the students. Usually this entails a more thorough knowledge of the subject matter and therefore derives from the instructor's training and prior awareness of the course material in general. However, even in the simulation itself, the instructor must keep tabs on all players and understand all the roles. He or she is often called upon to play God and exercise ultimate control of the game; in such cases it is essential that the game leader know precisely what decisions to make and why.

Preparation for these games is of these three general varieties: (1) material included in the game package designed as background information; (2) extra time spent in class discussion going over the events and relationships leading up to the time dealt with in the simulation; (3) outside reading from books either recommended by the game designers or chosen by the instructor.

Almost every simulation we discuss includes at least one page of background information, and for some that is enough. Others present a complex series of discussions of the personalities, the socioeconomic relations, the political institutions, the religious and philosophical beliefs of the simulated society; these require a detailed study of the game materials before players can profit from the game itself. In many cases also, both we and the simulation materials recommend that you supplement the materials in the packages, regardless of how good they may be, with outside reading and class lectures or discussions. In that way the simulation becomes much more integrated with the course you are teaching and thus much more effective as a teaching aid.

Let us add one final note. As we shall see more closely when we discuss the roles players take during the games, many of the simulations prescribe in detail the positions, perspectives, attitudes, and goals of the various characters who interact within the context of the games. For some of you this will be good, and we generally give those simulations that provide clear guidance on roles higher rating than those that are more ambiguous. However, for many others of you this ready-made role preparation will be frustrating in that the students (if they follow the format) will not have the opportunity to discover for themselves what a given country's aspirations were at a crucial time in history, what a given individual thought about the issues of the time, or what the critical social relationships

TABLE 5 Stated Purpose of Games and How Well Carried Out

Simulation	Purpose	Rating
Alpha Crisis	To introduce students to issues and dynamics of pre-WWI European diplomacy	E
American Constitutional Convention	To confront students with primary issues and perspectives in constitutional debate.	E
	To teach process of negotiation and compromise	M
The Ch'ing Game	To convey understanding of and empathy for the Chinese political structure and value system	E
Congress of Vienna	To teach process of negotiation and compromise to maximize points for one's country	M
Czar Power	To teach major issues confronting Russian society around 1870	E
Czech-mate	To teach different national responses to Sudeten crisis.	E
	To teach process of compromise and negotiation and avoid simple answers to problems of international relations.	E
Destiny	To teach the pro and con arguments of American involvement in Cuba in 1878.	E
	To teach the process of negotiation and persuasion in presidential decision-making.	E
Gateway	To teach reasons for immigration to America and experiences of immigrants.	E
	To teach reasons for discrimination against minority ethnic groups.	M
Grand Strategy	To teach national goals, resources, allignment.	M
	To teach process of diplomacy and compromise.	M
Haymarket Case	To acquaint students with labor issues of early industrialization.	M
	To confront issues of court procedure and free speech.	E
Liberté	To acquaint students with data, processes, and issues of French Revolution.	S
	To teach differences in decision-making between democracy and various forms of authoritarianism.	E
	To teach difficulties of making group decisions in time of crisis.	M
Nuremberg	To acquaint students with issues surrounding trial of Nuremberg defendants.	E
	To make students familiar with process of researching and presenting arguments.	E
Origins of WWII	To recreate the diplomatic conflict of 1930s that led to WWII.	S
Panic	To acquaint students with causes of Depression.	M
	To teach difficulties in dealing with Depression.	E
Scramble for Africa	To recreate circumstances that led to colonial empires in thirty years after 1882.	S
Seneca Falls	To understand the perspectives of men and women or roles of women's rights in the 1840s.	M
Trade-Off at Yalta	To acquaint students with major international problems.	E
	To teach ways of finding negotiated settlements.	E
Waging Neutrality	To increase comprehension of formation of policy and issues surrounding U.S. neutrality, with advantages and disadvantages of continuing such a policy.	E

TABLE 6 Game Complexity

Easy	Moderate	Complex
Grand Strategy	*Czech-mate*	*The Ch'ing Game*
The Haymarket Case	*Destiny*	*Czar Power*
Scramble for Africa	*Gateway*	*Waging Neutrality*
	Liberté	*Alpha Crisis*
	Nuremberg	
	Origins of WWII	
	Panic	
	Trade-Off at Yalta	
	American Constitutional Convention	
	Alpha Crisis	
	Seneca Falls	

or values of another society might have been. Some of our games lend themselves to modification in that you may wish to leave out some of the prepared introductory material and substitute your own, which may require a greater expenditure of time from students than would the original game outline. The simulations most easily adapted in this manner are noted in Table 7.

RULES

Laws are the rules of life; those who do not understand or who ignore the law risk disrupting the order of the society of which they are a part. Rules are equally necessary in simulations, and the clarity of rules is critical to the success of a simulation game. One of the cardinal tenets of simulation design is that rules should be complete—that is, that they should cover all important developments within the game—but that they should be kept as simple as possible. Rules that are superfluous or poorly stated serve only to confuse players and

TABLE 7 Preparation Necessary

	Minimal (1-2 pages of material and/or one day in class)	*Moderate* (3-10 pages of material and/or several days of lecture/discussion)	*Extensive* (Over 10 pages and/or over one week in class)	*Heavy* (large bibliography or entire course)
In-package preparation only	*Scramble for Africa* *Haymarket Case*[a] *The Alpha Crisis* *Waging Neutrality*	*Grand Strategy* *Liberté*[a] *Gateway* *Panic* *Origins of WWII* *Seneca Falls*	*American Constitutional Convention*	
Requires reading and in-class preparation	*Congress of Vienna*	*Czech-mate*[a] *Trade-Off at Yalta*	*Czar Power*[a] *Nuremberg*[a,b] *Destiny*[a,b]	*The Ch'ing Game*[a,b]

a. Bibliography suggessted
b. Allow for either in-class or reading preparation

cost valuable time either in explanation before play begins or in misunderstanding and erroneous play. We have judged the rules for these games, therefore, not simply on the basis of length, but on internal clarity, logic of organization, and succinctness. We stress that these evaluations reflect only the rules as they are stated in the game package and do not necessarily indicate the general complexity of the simulation as a whole. Thus, a simulation may rate well on its presentation of rules, as does *Congress of Vienna,* but be limited in scope and present its roles in elementary fashion.

ROLES

One of the most enjoyable and rewarding aspects of simulation in the study of history is role playing, actually assuming for a short time the character of another person, sometimes real, sometimes fictional, in the historical story. Each of us has a little of the actor within, and role play is a great way to involve students who are a bit shy about venturing opinions in an intellectual class discussion as well as to allow those more voluble students another outlet for their energies.

For role playing to achieve its potential, however, the game description must be clear and complete but concise and flexible. The roles must be appropriate to the situation and must allow all participants to take an active role in the game, unless, of course, frustration is an intended by-product of the playing of certain roles (as with some of the immigrant roles in *Gateway,* for example). In the following analysis, we will look not only at the variety of roles these games offer but at the ways the game descriptions present the roles, the characteristics of the roles as they are played, and the appropriateness of the roles to each simulation's historical context. We will look also at the role played by the instructor and the potential changes you might make in roles to make the game conform more to your own needs than it may in its present form.

TABLE 8 Rules

Simulation	Number of pages devoted to rules: Student	Instructor	Clarity	Logic	Succinctness	Overall
Alpha Crisis	11	7	G	G	E	G
American Constituional Convention		4	S	G	G	G
Ch'ing Game	26	15	E	G	E	E
The Congress of Vienna	5	2	E	E	E	E
Czar Power	16		E	E	G	E
Czech-mate	3	9	E	G	E	E
Destiny	8	8	E	E	E	E
Gateway	17	25	E	G	G	E
Grand Strategy	2	13	G	S	S	S
The Haymarket Case	16		G	S	G	G
Liberté	6	14	E	S	G	G
Nuremberg	7	11	E	E	E	E
Origins of WWII	8		G	S	G	G
Panic	5	17	E	E	G	E
Scramble for Africa	12	3	G	G	S	G
Seneca Falls	5	2	G	G	G	G
Trade-Off at Yalta	10	12	E	G	E	E
Waging Neutrality		14	E	E	E	E

Key: E = excellent
G = good
S = satisfactory

Kinds of Roles

Roles in these games may be of the following four general types: (1) generic roles based upon one's function in society—for example, a soldier, a head of state, a peasant, a merchant, a reporter; (2) generic roles based upon the group to which one belongs—for instance, a member of the American delegation to an international conference, a member of the revolutionary tribunal ruling a country after a coup d'état; (3) roles of fictional characters designed to typify individuals who might have filled such roles in the society or time simulated—for example, the judge at the trial of the Haymarket defendants, the king of country Alpha in *Alpha Crisis,* the governor of Kiangsu Province in the *Ch'ing Game,* or the ambassador from Spain in *Destiny*; (4) roles modeled after real historical figures whose personalities were critical to the turn of events simulated—for instance, Stalin in *Trade-Off at Yalta,* Alexander II in *Czar Power,* Elizabeth Cady Stanton in *Seneca Falls,* or Rudolf Hess in *Nuremberg.*

Our ultimate judgment of the effectiveness of the roles offered in these games will depend in large part upon the type of roles they use. Note that some games use more than one type of role.

Appropriateness of Roles

For the most part the types of roles chosen for these games are appropriate for the game model. There is one area, however, in which we would argue that the roles should have been of a different kind. In games designed to simulate a particular international conference or crisis situation, roles should be modeled as closely as possible upon the personalities of the individuals really present. It was important that Czar Alexander I and Metternich led their respective delegations to the Congress of Vienna, for instance, and the simulation that models that conference would be more effective if it had used them and their fellow leaders as the role models. That Neville Chamberlain and Adolf Hitler met to settle the Czech crisis of 1938 was critical, and *Czech-mate* would be a better game had it used these men more directly in its format. The national aspirations of Chancellor Bethmann-Hollweg of Germany, of Berchtold in Austria, or of Pasic in Serbia helped considerably to determine the direction of events in 1914, and *Grand Strategy* should have taken that more into account. In each instance, the use of real persons as the basis for roles would call for further study by the participants, but the simulations would accomplish their purpose better if such study were undertaken.

TABLE 9 Kinds of Roles

	Generic Roles Based Upon:			
Simulation	Function	Group	Fictional Characters	Real Figures
Alpha Crisis	X			X
American Const. Conv.				X
The Ch'ing Game			X	
Congress of Vienna	X	X		
Czar Power		X	X	
Czech-mate			X	
Destiny		X		
Gateway	X		X	
Grand Strategy	X	X		
The Haymarket Case	X		X	
Liberté	X			X
Nuremberg	X		X	X
Origins of WWII		X		
Panic	X	X		X
Scramble for Africa	X			
Seneca Falls		X		X
Trade-Off at Yalta			X	X
Waging Neutrality	X	X	X	

Variety of Roles

Table 10 lists the general roles these games call for. From this list you can see the range of roles and the ways participants interact within the context of the simulation.

TABLE 10 Variety of Roles

Simulation	Roles
Alpha Crisis	heads of state, ministers of foreign affairs, war, interior, and economy of six major countries, judges of International Court, reporters, managers, negotiators
American Constitutional Convention	delegates to Constitutional Convention or observers assigned to particular delegates
The Ch'ing Game	emperor, imperial aides, provincial governors, local magistrates, degree candidates for higher office, merchants, scribes and calligraphers, poets, messengers
Congress of Vienna	members of delegations from Austria, Prussia, Russia, France, and England
Czar Power	czar, state ministers, grand dukes, members of state council, provincial nobles, clergymen, Procurator of Holy Synod, peasants, workers, artisans, merchants
Czech-mate	heads of state, ambassadors-at-large, ministers of information and ministers of foreign affairs from England, France, USSR, Germany, Italy, Czechoslovakia, United States; representative from Sudetenland
Destiny	President McKinley, Spanish diplomat, members of the Cuban Junta, American businessmen, newspaper reporters, imperialists, anti-imperialists, members of Congress
Gateway	European immigrants, ships' officials, Ellis Island immigration officials, Congress and committee members, lobbyists, school board members, social studies teachers, citizens of white majority or ethnic minority groups
Grand Strategy	heads of state, ministers of war, and foreign ministers from Great Britain, France, Germany, Russia, Belgium, Italy, Serbia, Austria-Hungary, USA, and the Ottoman Empire

TABLE 10 Variety of Roles (Cont)

Simulation	Roles
The Haymarket Case	judge, anarchist defendants, attorneys and assistants, reporters, bailiff, police inspector, informer, witness to riot, mayor of Chicago, strikers, manager of McCormick Harvester Plant, jury members
Liberté	king, queen, nobles, clergy, bourgeoisie, peasants, members of National Assembly
Nuremberg	judges, defendants, attorneys, witnesses
Origins of WWII	representative decision-makers for France, Germany, Great Britain, the USA, and the USSR
Panic	bankers, laborers, farmers, women, social critics, businessmen, presidential candidates, senators
Scramble for Africa	diplomats, military commanders, leaders of missions in Africa from Britain, France, and Germany
Seneca Falls	Lucretia Mott, John Mott, Elizabeth Cody Stanton, Frederick Douglass, women delegates, men delegates
Trade-Off at Yalta	Roosevelt, Churchill, Stalin, Hopkins, Eden, Molotov, Stettinius, Kerr, Bohlen, Marshall, Byrnes, Harriman, Moran, Gromyko, Vishinsky, other advisers
Waging Neutrality	J. P. Morgan, industrial and farm brokers, purchasing agents for England and Germany, American shipping executives, German and British political and military representatives

Role Descriptions

Your interests may lead you to look for roles with deliberately brief descriptions so you can fill them in yourself or let the students fill in for themselves. On the other hand, you may be looking for detailed role descriptions that offer not only the general characteristics of the role but the attitudes the player ought to take toward certain issues or toward the other roles. Table 11 indicates the character of the descriptions offered in the simulation packages. Keep in mind, of course, that you are not obligated to use a role description just because it is included in the material you purchase.

TABLE 11 Detail in Role Descriptions

Very Brief (one sentence or less)	Moderate (several sentences to one paragraph)	Extensive (several paragraphs)
Congress of Vienna	*Czech-mate*	*The Haymarket Case*
Grand Strategy	*Liberté*	*The Ch'ing Game*
Panic	*Nuremberg*[a]	*Alpha Crisis*
Scramble for Africa	*Gateway*	*Waging Neutrality*
Origins of WWII	*Seneca Falls*[a]	*American Constitutional Convention*
		Destiny
		Czar Power
		Trade-Off at Yalta

Note: Role descriptions vary in length and quality.

General Role Characteristics

Some simulations suffer from an unevenness of the roles players take. That is, some players find that at points during the game they have little to do. When that happens, boredom usually sets in rapidly, interest in the simulation wanes, and the effectiveness of the exercise diminishes. It is critical, therefore, that simulations be designed to keep all the roles actively involved in the game at all stages of the simulation. The only exception involves cases in which you want some of the players to become frustrated with the originally assigned roles and break out of the mold into which they have been placed, thus simulating reaction like that of the individuals after which they are modeled in the real historical situation.

In some of the games under consideration, involvement and interest levels are only adequate or moderate in spite of the evenness of the roles because of limitations the game format imposes upon the players. If there is little to do, or if their actions are too regimented and predictable, participants tend to lose interest. The level of interest often depends on the ability of players to exercise their own freedom of action within the context of the role they play. When players' actions are too closely prescribed, as in *Haymarket,* or when freedom of action is relatively limited, as in *Congress of Vienna,* the interest level falls below that of some of the other games. However, you may wish to keep freedom of action under control and be willing to sacrifice a certain amount of involvement to do so. If so, you should select one of these games.

In other simulations, roles are uneven. That is, some players have more to do than others, the players with marginal roles often moving to the fringes of the game, both physically and emotionally. Peasants in the first part of *Liberté,* for example, have little to do beyond the rote exchange of goods, while the members of the bourgeoisie are kept much busier and, consequently, more involved and interested in the game.

Role Flexibility

Some of the difficulties we discussed in the preceding section can be ironed out by looking at the flexibility of the roles presented by the game designers or at the ability of the instructor to add, subtract, or otherwise modify roles to create greater evenness, eliminating roles that are marginally involving or adding roles that are not in the suggested game outline.

Instructor's Roles

In selecting a game for your own use, you should be aware not only of what is expected of participants during the game but what your own role will be. In these simulations, the instructor's duties vary considerably from minimal supervisory responsibilities to major interaction with the players during every part of play.

In almost all cases, the instructor serves as the court of last appeal, the final arbiter of disputes, and the unquestioned authority on rules and play. You must establish this authority right from the start and make sure that the players know your word is final. To make your credibility stick, however, you must know the rules. If you do not and a player catches you in

TABLE 12 General Role Characteristics

Simulation	Evenness of Involvement	Degree of Involvement	Interest Factor
Alpha Crisis	generally even	high	high
American Constitutional Convention	divided; observer roles less involving	high for delegates	high
The Ch'ing Game	even	high	high
Congress of Vienna	equivalent roles	adequate—limited by tasks	adequate
Czar Power	some unevenness due to difference in resources	high	high
Czech-mate	even	high	high
Destiny	even	high	high
Gateway	even	high	high
Grand Strategy	equivalent roles	moderate	moderate
The Haymarket Case	uneven—some roles more developed than others	some roles have little involvement—witnesses not involved except when on stand	high while testifying, but falls off when not actively involved
Liberté	uneven	some roles offer too little involvement at critical junctures	interest falls off for some roles as involvement diminishes, while it remains high for others
Nuremberg	uneven—some roles more developed than others	witnesses passively involved except when on stand	issues tend to keep interest high
Origins of WWII	even	moderate	moderate
Panic	even	high	high
Scramble for Africa	even	moderate	moderate
Seneca Falls	uneven	moderate	moderate
Trade-Off at Yalta	some unevenness	roles of delegation leaders more involving during negotiation sessions	high in general
Waging Neutrality	even	high	high

a major rule infraction, your authority will diminish quickly. We have usually found that establishing authority is a matter of self-assertion, complete familiarity with how the game is to be played, and humor. It is often an effective opener to take the role of the Game Overall Director (GOD).

In some of these simulations, such as the *Ch'ing Game,* the role of the instructor is so complex, the records to be kept so massive, and the potential appeals to the instructor's arbitration so numerous that more than one person should be assigned the task of overseeing the simulation. We have used students who have played the game before or who have a special expertise in simulation as assistants in such cases.

You may prefer to keep out of the simulation as much as you can, letting the students work out for themselves the difficulties and ambiguities of play. If so, you might gravitate to games in which the instructor's role is minimal. Table 14 indicates the general scope of instructor involvement in the simulations we are considering.

Keep in mind, however, that the critical factor in determining how active a role the instructor takes is you. You may vary your involvement in all these simulations, recognizing as you do that you must never relinquish the authority to step in at any time and redirect the game onto the right track if it becomes unduly bogged down on a tangential issue. Likewise, you must recognize the value of freedom for students to learn from their own mistakes. One of the most important aspects of simulation playing is the opportunity to make mistakes without having to face the consequences in real life, and students should have the chance to do so. Remember also that these games can be stimulating and learning experiences for the instructor, too. The attitude you display in explaining participation in the game will determine in large part the game's impact on the players.

ACTIVITIES AND INTERACTIONS

Major Game Activities

As we all learned as children, games are usually conducted in one of three ways. Either a game is the "cops and robbers" variety in which everyone acts at once, and play continues indefinitely until someone or everyone decides they have had enough; or it is the "Monopoly" type in which everyone takes

TABLE 13 Role Flexibility

Simulation	Freedom to Add New Roles	Freedom to Subtract Suggested Roles	Freedom to Change Suggested Roles
Alpha Crisis	moderate—process suggested	limited	minimal
American Constitutional Convention	none	moderate—may use only some of delegates	none
The Ch'ing Game	wide—process suggested	relatively wide—process suggested	fairly inflexible
Congress of Vienna	narrow—limited by game dynamics	relatively wide—process suggested	roles are minimally defined, opportunity to define more firmly
Czar Power	relatively wide	moderate, although some roles are essential	moderate
Czech-mate	narrow—limited by game	relatively wide—process suggested	opportunity to tighten loosely defined roles
Destiny	none	none	narrow
Gateway	relatively wide	relatively wide—process suggested	relatively wide since players help define roles
Grand Strategy	relatively wide	relatively wide	opportunity to tighten loosely defined roles
The Haymarket Case	moderate—process suggested	moderate—process suggested	narrow, although there is opportunity to change some of the prescriptive nature of role descriptions
Liberté	moderate	relatively narrow	fairly inflexible
Nuremberg	moderate	narrow	minimal descriptions of witness roles permits wide latitude
Origins of WWII	none	moderate—process suggested	none
Panic	moderate	moderate	fairly inflexible
Scramble for Africa	moderate—simply expand teams	moderate	none
Seneca Falls	none	none	opportunity to tighten loosely defined roles
Trade-Off at Yalta	narrow—limited by game format	moderate—process suggested	narrow
Waging Neutrality	narrow	narrow	inflexible

TABLE 14 Instructor Roles

Simulation	Roles
Alpha Crisis	to oversee and control game; to monitor press releases and agreements
American Constitutional Convention	to oversee game; as George Washington, to preside over convention debate; to select issues for debate
Ch'ing Game	to oversee game; to record all economic transactions; to regulate relations between roles to insure historical validity. Requires three supervisors to interject crisis situations.
Congress of Vienna	to preside at conferences; to oversee negotiations; to allocate points.
Czar Power	to oversee game and preside in conferences; to award and keep track of various types of points; to keep simulation moving on schedule
Czech-mate	to oversee simulation and regulate historical feasibility; to act for all powers not represented by other players; to interject natural disasters; to publicize activities of teams. Best with at least two instructors
Destiny	to oversee game; to allocate and keep track of points; to grade quizes and written exercises; to inject outside factors into simulation; to preside at congressional debate
Gateway	to oversee game; to calculate quiz grades and evaluate immigrant autobiographies; to control debate at congressional hearing
Grand Strategy	to oversee play; to facilitate conference if requested by players; to present participant moves; to control timing of negotiation sessions; to record military movements and conflicts and to announce victories, losses, and casualties
The Haymarket Case	to preside at court (as bailiff, not judge) and insure smooth running of game
Liberté	to oversee play; to administer quiz and evaluate responses
Nuremberg	to oversee play; to administer quiz and evaluate responses
Origins of WWII	to set up game and resolve conflicts or keep track of points
Panic	to oversee play; to conduct surveys and score sheets; to interject data cards
Scramble for Africa	to oversee game; to keep strict time restrictions; to monitor moves and scoring; to adjudicate conflicts
Seneca Falls	to oversee game and insure even flow
Trade-Off at Yalta	to oversee and supervise game; to insure that deliberations proceed according to accepted procedure; to pass along "secret" information
Waging Neutrality	to supervise game; to inject news items and chance into play; to coordinate open forums

turns, going round and round until a fixed time limit has elapsed or someone has "won"; or it is the "Diplomacy" type in which all players (or teams) make simultaneous decisions and play proceeds in rounds, each one of which reflects the decisions of the previous round. Table 15 divides the simulations we are considering into these three categories, and Table 16 details the activities in which the players are involved.

Interactions

The kinds of interactions called for in the games you use will influence greatly the kinds of learning which take place among the participants. Table 17 categorizes the simulations according to the ways in which players deal with one another.

The advantages of one-to-one or small-group interaction are that most players will be active most of the time and there is less likelihood of physical or emotional drop-out. However, games that stress small groups tend to be confusing to players because they have little opportunity to step back from their own activities and look at the game as a whole. Games that make extensive use of large groups or have the entire class debate an issue together make it easier for participants to monitor events, but some of the less vocal players may move into the background and participate only marginally.

TABLE 15 Types of Activities

Continuous Interactive Play, Often with Game Progressing by Stages	Sequential Turns	Rounds of Activity in Which All Interact in Specified Ways
Czech-mate	*The Haymarket Case*	*The Ch'ing Game*
Gateway	*Nuremburg*	*Congress of Vienna*
Liberté	*Seneca Falls*	*Grand Strategy*
Destiny		*Origins of WWII*
Panic		*Czar Power*
Waging Neutrality		*Trade-Off at Yalta*
		Alpha Crisis
		Scramble for Africa
		American Constitutional Convention

Issues and Player Initiative

Each of the simulations we are looking at focuses on a series of issues, most of which stem from the particulars of the historical circumstances that prompted the creation of the simulation. Some of these are limited in scope; some are rather wide in potential. Table 18 shows the major issues these simulations treat and notes which present the issues ready-made for players to consider and which require players to develop issues before they can begin work on possible solutions. In some cases the issues are presented in only outline form and the players must fill in the details before substantive solutions are possible. Whether your interests call for games in which issues are all ready for consideration or, alternatively, for games in which students give considerable attention to the process of problem development as well as problem solving, this table should help you select games that best fulfill your needs.

TABLE 16 Major Activities

Simulation	Sequence
Alpha Crisis	3 or more rounds of negotiations, reactions to previous activity, policy decisions.
American Constitutional Convention	1 to 5 rounds of 5 stages each: (1) state caucuses to decide approaches to issues; (2) convention meeting to make proposals to entire body; (3) informal discussion among delegates; (4) second stage of state caucuses to decide votes; (5) voting in full convention by state.
The Ch'ing Game	4 or 5 rounds of activities, determined by roles, which include preparing for and taking exams, trading goods and services, collecting and administering taxes, deciding questions of social propriety, and adjudicating disputes.
Czar Power	4 rounds of 5 stages each: (1) select 4 major problems for discussion; (2) players consider alternative solutions; (3) small groups discuss solution alternatives; (4) decision by Czar; (5) winning or loss of points. Other events as a purge or overthrow of Czar can also occur.
Czech-mate	Generally continuous play in response to Sudeten crisis may involve conferences or declaration of war. No set order of response.
Destiny	In introductory section players perform tasks related to information gathering; 3 rounds of negotiation, reaction to new events, arguments before President; Congressional debate and vote on President's decision.
Gateway	4 stages: (1) identification of immigrant roles and transition to the U.S.; (2) experience of passage through Ellis Island authority; (3) Congressional debate on immigration laws; (4) case study of integration of public school system.
Grand Strategy	3 or more rounds of Declaring sessions, in which action is announced, and Conferring sessions, in which decisions are made and negotiations undertaken.
The Haymarket Case	Prosecution and defense take turns questioning witnesses and defendants at trial followed by jury deliberation and decision.
Liberté	(1) economic phase: trade, pay, or collect taxes, try to earn points in a quiz; (2) legislative phase: discuss up to 12 issues; (3) trial phase: decide fate of Louis XVI; (4) terror phase: arbitrary arrests and "executions" by Robespierre.
Nuremberg	(1) assignment of roles and major research of evidence or arguments used by roles; (2) prosecution and defense attorneys take turns questioning witnesses and defendants, followed by judges' decision.
Origins of WWII	In each of 6 rounds each player in turn places political factors on various territories, attacks or defends territories, seeks to establish control of certain territories, and negotiates with others on future moves.

TABLE 16 Major Activities (Cont)

Simulation	Sequence
Panic	(1) assignment of roles, buying and selling stocks, market crash, and bank closing; (2) congressional hearings and voting on bills to solve economic crisis.
Scramble for Africa	6 to 10 rounds of three phases each: (1) negotiations; (2) write orders for movement of military and civilian units and sign treaties; (3) publish plans and resolve conflicts.
Seneca Falls	After initial speeches by principals at conference, participants take turns speaking and voting on four resolutions.
Trade-Off at Yalta	5 rounds of negotiation and conferring on five major issues of conference.
Waging Neutrality	2 rounds of 3 phases each: (1) commercial action: negotiate, trade, and conduct business; (2) operations meeting: separate groups decide on policy and strategy; (3) open forum: discusss issues in public.

RESOURCES AND SCORING

One of the critical factors to keep in mind when you select a simulation is the incentive-reward system built into the game. The major lessons of a simulation often come not from the particulars of content but from the way the game is played. If players recognize that a certain type of behavior and interaction produces positive results in a simulation, they will assume that a similar type of behavior would have produced similar results in the real-life prototype. If you want to teach that cheating the system was the only way of winning in a particular historical situation, you should select a game in which cheating can produce a win. If competition is a critical factor, then look for a simulation in which competition over resources is a fundamental part. If cooperation is the goal, then cooperation ought to be the method whereby players succeed. And, finally, if you wish to concentrate upon the process of play and to diminish the importance of "winning," then look for an incentive system that promotes that end. Players give their major attention to those aspects that promote their own interests within the context of a simulation, and you should be extremely conscious of this factor in your selection, use, or creation of simulations.

MODEL VALIDITY

In using historical simulations, one of your most important considerations will probably be whether the simulation does in fact do a good job of modeling the historical event or conditions you want to teach. There are three levels at which we might assess the historical validity of a game: (1) does it portray facts and major historical relations correctly, or does it mislead players by leaving out important figures or groups, by skewing the facts to produce an unrealistic outcome, or by giving erroneous information? (2) does it promote a valid analysis of the cause and effect relations of the historical situation, or does it mislead by oversimplifying conditions and suggesting monocausal solutions to complex questions? (3) does it lead to a valid understanding of the historical patterns of thinking and value systems, or does it allow players to transpose their modern values onto an historical stage where they are inappropriate?

We are critical of the validity of these games for different reasons. Let us outline some of the most important difficulties for you. First, we have categorized *Congress of Vienna, Grand Strategy,* and *Scramble for Africa* as limited because they concentrate too much on the territorial conflicts of their respective times. *Congress of Vienna* is the worst offender in this respect, the other two including some political issues in their format, but all of them leave out important considerations of ideology, personality, and economics.

Waging Neutrality is limited to some extent in its treatment of ideological issues, concentrating as it does upon economic considerations. While these were of critical importance to the American decision to remain neutral and later to enter World War I, they were not as overwhelming as the game would suggest.

Seneca Falls and *Haymarket* fall into the "limited" category because they are limited in scope. They fail to draw sufficient attention to the larger movements of which their foci are a part—the long-range struggle for women's rights in the former instance, and the labor and anarchist movements in the latter. In addition, *Haymarket* uses purposely distorted historical facts in the case. These alterations are minor—changing the names of streets and buildings, for instance, and not using the real defendants—but they seem unnecessary and draw attention away from the true information upon which the case hinged.

TABLE 17 Kinds of Interactions

Primarily One-to-One	Primarily Between Individuals and Small Groups	Primarly Between Small and Large Groups	Entirely in Large Groups
The Ch'ing Game	*Czar Power*	*Czech-mate*	*The Haymarket Case*
Origins of WWII	*Gateway (1)*	*Congress of Vienna*	*Gateway*
	Liberté (1)	*Grand Strategy*	*Nuremberg*
	Panic (1)	*Trade-Off at Yalta*	*Seneca Falls*
	Alpha Crisis	*American Constitutional Convention*	*Liberté (2)*
	Waging Neutrality (1)	*Destiny (1)*	*Waging Neutrality (2)*
	Scramble for Africa	*Panic (2)*	*Destiny (2)*

Key: (1)(2) Refer to the first or second phases of the simulations in question, since these differ considerably in the types of interactions called for.

TABLE 18 Issues and Player Initiative

Simulation	Major Issues	Amount of Player Initiative
Alpha Crisis	issues center on Austrian ultimatum to Serbia, conflict of international prestige and power	P
American Constitutional Convention	rights and powers of executive, legislative, and judiciary branches of government; federal state relations; procedure for amendment	S
The Ch'ing Game	social, economic, and political interactions among representative citizens of China	P
Congress of Vienna	territorial division of Europe after 1815	S
Czar Power	internal political dissidence; land distribution; foreign policy and internal policy and internal minorities; military needs; church-state relations; taxation; censorship; trade monopolies; government organization	S
Czech-mate	territorial and diplomatic response to crisis of 1938	P
Destiny	whether the United States should go to war with Spain in 1877-1878	P
Gateway	experience of nineteenth-century discrimination against immigrant ethnic groups	S
Grand Strategy	territorial and diplomatic response to crisis of 1914	P
The Haymarket Case	fate of the Haymarket defendants	S
Liberté	causes of and progress of French Revolution; legislative power in France; fate of Louis XVI and others	S
Nuremberg	fate of the Nuremberg defendants; responsibility of individuals to act on basis of their own values	S
Origins of WWII	diplomatic control of Europe in 1939	S
Panic	economic rise and fall; fluctuations of U.S. economy in 1920s and 1930s	I
Scramble for Africa	military and civilian control of and conflict over African cities after 1882	S
Seneca Falls	equal rights for women; equal job opportunities; elimination of double work standard; suffrage—all as applied in 1898	S
Trade-Off at Yalta	Polish political system; control of Germany; creation of United Nations	P
Waging Neutrality	U.S. neutrality; economic opportunities at time of war; role of economic factors in U.S. involvement in WWI	P

TABLE 19 Resources and Scoring

Simulation	Resources	Scoring
Alpha Crisis	no individual resources; power is military and economic, but only on paper	no specific scoring procedure; emphasis on play
American Constitutional Convention	no identifiable resources	no specific scoring procedure; emphasis on quality of play
The Ch'ing Game	money and land distributed unequally at start, prestige represented by social position	maximization of personal position—wealth is less important than influence
Congress of Vienna	no identifiable resources	teams can accumulate points depending upon treaties signed and distribution of territory
Czar Power	rubles, titles, land distributed unequally at start; positive and negative points for quality of life	players attempt to accumulate points measured against a common standard rather than against one another
Czech-mate	troops and ships	no specific scoring procedure; emphasis on process of negotiation; simple formula for determining winner if war breaks out
Destiny	no identifiable resources	players score Presidential Advice Points based on personal and group performances and on results of conferences and Presidential decisions
Gateway	no identifiable resources	no explicit scoring procedure beyond scores on quizes; emphasis on quality of play
Grand Strategy	troops and ships	simple formula for determining losses and wins, but emphasis on process of diplomacy and negotiation
The Haymarket Case	no identifiable resources	no explicit scoring procedure; emphasis on quality of role play
Liberté	money and land distributed unequally at start	points lost or gained by taxation, fate, scores on quizes; emphasis on relative improvement.

TABLE 19 Resources and Scoring (Cont)

Simulation	Resources	Scoring
Nuremberg	no identifiable resources	no explicit scoring procedure; emphasis on researching and playing roles
Origins of WWII	political factors distributed unequally at start	points for gaining understanding or control factors in territories designated by national objectives chart at beginning of game
Panic	wealth points distributed unequally at start	points lost or gained by speculation, investment, fate, performance on quizes, fate of congressional bills, taxation
Scramble for Africa	each country receives two military and two civilian units at start	points for control of certain tours by military and civilian units and for successful signing of treaties
Seneca Falls	no identifiable resources	no explicit scoring procedure; emphasis on quality of play
Trade-Off at Yalta	no identifiable resources	no specific scoring procedure; emphasis on quality of play
Waging Neutrality	money, cargo, destroyers distributed unequally	no specific scoring procedure; emphasis on quality of play

The difficulties with *Gateway* and *Trade-Off at Yalta* lie with the sequence in which the games are played. The first two sections of *Gateway*, in which the immigrants travel across the ocean and land at Ellis Island, are good historically, but the transition to the third and fourth parts is poor and tends to confuse players. *Yalta* falls down in its artificial division of the issues with which the delegates are confronted. The way the game is currently set up, players are unable to make deals across issues, whereas in the real situation such divisions were not present. This difficulty is easily remedied with a change in the rules to permit such negotiation and allow players to consider all the issues simultaneously, at least as they prepare their general strategy.

Liberté and *Origins of World War II* present more serious problems. *Origins of World War II* deals almost entirely with territorial issues, and while it makes use of markers symbolizing "political" and "understanding" factors, these are for all practical purposes still tied to territorial control as if they were military units. The game is intriguing as a game of pure strategy, but it teaches little about the true origins of war. It skips over any changes in the political climate from 1935 to 1940 and ignores internal factors such as the role of popular support for a particular policy, ideology, or internal political conflict, such as that in the USSR in the 1930s.

TABLE 20 Historical Validity

Low	Limited	High
Liberté	*Congress of Vienna*	*The Ch'ing Game*
Origins of WWII	*Gateway*	*Czech-mate*
	Grand Strategy	*Czar Power*
	The Haymarket Case	*Destiny*
	Scramble for Africa	*Nuremberg*
	Seneca Falls	*Panic*
	Trade-Off at Yalta	*Alpha Crisis*
	Waging Neutrality	*American Constitutional Convention*

Liberté poses several problems of historical validity. First, it leaves out the sans-culottes, the urban poor of eighteenth-century France, who played such a critical role in the Revolution, and lumps them in with the peasantry, thus perpetuating one of the most common misperceptions beginning students have about the French Revolution. Second, the game employs poor transition between sections, leaving it unclear how or why the Revolution developed in the way it did. And finally, it gives Robespierre almost arbitrary power to purge anyone he wants from the game, thus distorting both the political dynamics of the purge, in which Robespierre was only one of the many actors, and the reasons Robespierre himself had for leading the country in the direction he did. He was far from the totally arbitrary megalomaniac which the game presents. While the quizes help clear up some of these oversimplications, the play of the game leaves too much room for potential misunderstanding.

DEBRIEFING

The ultimate success or failure of a simulation exercise often depends upon the character of the debriefing session, the discussion after play of what happened, why it happened, the changes in feelings and attitudes players experienced during the game, the strengths and weaknesses of the simulation itself, and the simulation's historical validity. Some of these simulations provide a guide for the debriefing session, listing important questions to ask, issues to confront, and points to note. Others provide only sketchy suggestions. Still others leave out this component altogether, assuming that you will be able to fend for yourself.

While one cannot automatically equate quantative suggestions for evaluation and debriefing guidelines, the two are almost always congruent. Simulations are by nature less predictable in their outcomes than other types of classroom activity, and therefore they require careful listing of possible alternative outcomes and issues that might be confronted in a debriefing session. A simulation that does not take the time or devote the space in its published description to such examination does a poor job of handling the debriefing process.

Our simulations vary greatly in the amount of space they

devote to debriefing suggestions and, as a consequence, in the quality of their debriefing section. One cannot always predict from the general character of the simulation its approach to debriefing. Even games like the *Ch'ing Game* or *Panic,* which in most other respects are outstanding in their conceptualization and description, brush only lightly by debriefing with a few suggested issues and a general admonition to the instructor to hold a debriefing session. Others, such as *Haymarket,* even though less sophisticated than some of the other games in our listing, carefully and completely lay out the issues, questions, and problems an instructor might deal with in debriefing.

Whether or not debriefing comes as part of a packaged game, you should pay careful attention to the debriefing session, noting in the course of the game items or developments to which you would like to return later for analysis. Remember also that what does not happen in a game may be just as significant as what does. Likewise, a shortcoming that participants note after the simulation is over may result in their learning as much as they would have if the left-out factor had been incorporated into the exercise.

TABLE 21 Debriefing Outlines

Inadequate	*Fair*	*Adequate*	*Excellent*
(no guidelines)	(a few questions)	(several paragraphs of suggestions—over 10 questions)	(several pages of suggestions and questions)
Congress of Vienna *Scramble for Africa* *Origins of WWII* *Czar Power*	*The Ch'ing* *Panic* *American Constitutional Convention* *Seneca Falls*	*Grand Strategy* *Liberté* *Nuremberg* *Gateway* *Destiny*	*Czech-mate* *The Haymarket Case* *Alpha Crisis* *Waging Neutrality* *Trade-Off at Yalta*

FLEXIBILITY

Games that lend themselves to adaptation are often more useful in the long run than those that depend upon a rigid playing out of the game as presented in the package you buy. It is important, therefore, if you are going to invest in one of these games, to know how you might modify it to suit your own purposes or how much you can use the format it suggests to create your games. In Table 22 we will look not only at the built-in flexibility of each simulation concerning length, number of players, outcome, and issues, but we will also examine some of the ways changes might be made without destroying the game's dynamics and suggest which games are best suited as prototypes for developing simulations of your own.

PUBLICATION AND PACKAGING

In selecting a commercial game, you must pay attention not only to the content and character of the game itself but to the way it is put together. Simulations come in all sizes and shapes, some well packaged, durable, and reusable, some shoddy and easily worn out. Some games come only as des-

TABLE 22 Flexibility

Simulation	Flexibility: Length	No. of Players	Outcome	Issues	Possible Alterations	Value as Prototype
Alpha Crisis	L	H	M	L	might change role descriptions to correspond more to real persons; have students research real leaders and events	H
American Constitutional Convention	M	H	M	N	might allow students to develop issues more on their own rather than following historical pattern so directly	M
The Ch'ing Game	M	H	H	M	might add crises or change crises cited in the game	L
Congress of Vienna	M	M	M	N	might have student research roles, personalities of leaders, and national goals; allow players to select how their priorities rank and points may be won; inject issues other than territory	M
Czar Power	M	M	H	L	might have students select attitudes to be expressed by roles based on research; add more roles representing workers	H
Czech-mate	M	M	H	L	might have students research roles, personalities of leaders, and national goals; base roles on real leaders rather than prototypes	M

TABLE 22 Flexibility (Cont)

Simulation	Flexibility: Length	No. of Players	Outcome	Issues	Possible Alterations	Value as Prototype
Destiny	M	H	M	L	high flexibility in time students spend researching issues and roles	H
Gateway	L	H	H	M	might play first two sections only; have players use personal family background as basis for roles	M
Grand Strategy	M	H	H	L	might allow students to research national goals; pattern roles after real persons; inject other issues	M
The Haymarket Case	L	M	M	L	might allow players to determine more the character of play; have players research case as background for roles	M
Liberté	M	M	L	L	might play only third section on trial of Louis XVI or second section on legislation	L
Nuremberg	M	M	M	L	might try only some of defendants, concentrating on selected issues; specify more directly	H
Origins of WWII	L	L	M	L	might tie roles more to real leaders; add other countries; inject factors other than territorial control	M
Panic	L	H	H	M	might assign roles more specifically or have players pattern roles after real experiences	H
Scramble for Africa	L	M	H	N	might increase complexity by creating new roles or potential conflicts, by developing background information, or by modeling roles more realistically	L
Seneca Falls	L	H	M	L	might define roles more explicitly; stretch time factor; allow negotiation on issues	L
Trade-Off at Yalta	M	H	M	N	might use only some of issues; allow players to negotiate on all issues simultaneously; allow outside conferences among aides while major leaders negotiate in control session	M
Waging Neutrality	L	M	M	L	might inject more political or ideological issues; suggest impact of public opinion by adding roles	M

Key: H = High M = Moderate L = Low N = None

criptions, requiring you to supply all the props and paraphernalia necessary to run the game properly. Others contain everything from the rules to the pencils and name tags. And, last but not least, costs vary.

The games we selected for review here are within a general educational materials budget, but we recognize that not all of you will be in a position to afford the more expensive games. Consequently, price will influence your selection. In general, as in making any purchase, you should make sure that you are getting the most for your money in content and design. A poorly packaged and overpriced game is not worth getting even if it does meet other criteria outlined in the earlier part of this essay.

TABLE 23 Packaging

Simulation	Kind of Packaging	Approx. Cost	Completeness	Durability
Alpha Crisis	book	$2.00	d	high
American Constitutional Convention	student handbook	$1.55 each	d	average
The Ch'ing Game	book	$2.00	p, c, d	high
Congress of Vienna	pamphlet	£.50	d	average
Czar Power	kit	$68.50	complete	high
Czech-mate	manual	$14.00	d	average
Destiny	manual	$14.00	c, d	average
Gateway	manual	$14.00	d	average
Grand Strategy	kit	$39.00	complete	high
The Haymarket Case	manual	$15.00	d	average
Liberté	manual	$14.00	c, d	average
Nuremberg	manual	$14.00	d	average
Origins of WWII	kit	$10.00	complete	high
Panic	manual	$14.00	d	average
Scramble for Africa	pamphlet	$3.95	d	average
Seneca Falls	manual	$10.00	complete	average
Trade-Off at Yalta	kit	$35.00	complete	high
Waging Neutrality	manual	$14.00	c, d	average

Key: p = purchasing required c = construction required d = duplication required

OVERALL STRENGTHS AND WEAKNESSES

If none of these simulations is perfect, likewise, none is a total flop. In the closing section of this essay, therefore, it would be useful for us to review the games we are considering for their most critical strengths and limitations. In the long run, whether these characteristics are of critical importance depends primarily upon you. Major strengths to some will be only minor benefits to others; a glaring weakness to one will be a minor inconvenience to another. The strengths and limitations we have identified are shown in Table 24. We leave the final judgment up to you.

TABLE 24 Major Strengths and Limitations

Simulation	Strengths	Limitations
Alpha Crisis	confronts complex issues with minimum of confusion; roles very well developed.	no initial statement of purpose; rules not systematically laid out for instructor.
American Constituional Convention	presents excellent role analysis, giving perspective of delegates on variety of issues.	issues limited and simplified; format restricts freedom of initiative as students are called upon to follow historical pattern rather than confront issues openly.
The Ch'ing Game	good introductory material; roles carefully constructed and accurate; format carefully thought through and thoroughly researched; highly effective at achieving goal of simulating twelfth-century social realtions in China.	rules are highly complex and require high level of player understanding; requires purchase or construction of additional materials; heavy responsibilities for instructor.
Congress of Vienna	good mixture of formal and informal negotiation; rules clear and point allocation fair and unambiguous.	background to Congress is limited; issues entirely territorial, thus leaves out issues like Concert of Europe or Holy Alliance; tends to stress competition among countries rather than drive for great power balance; no discussion of debriefing process; roles not clearly defined and not based on real personalities.
Czar Power	good at addressing issues of Russian autocracy and critical problems of late nineteenth-century; roles well articulated; moves and point allocations complicated but easy to follow.	lone worker's role tends to underplay influence of urban workers; roles sometimes too prescriptive of actions.
Czech-mate	good introductory and background material; emphasis on process of negotiation, not winning war; clearly constructed and well outlined; good debriefing section.	roles not based on personalities of real leaders; tends to be heavy load on instructor; suggested schedule tends to produce overcrowded activity.
Destiny	excellent timing and format; background information good.	poorly defined attitudinal goals; shift of roles near end of game tends to diminish impact of presidential decision: students have hard time abandoning earlier positions.
Gateway	excellent goals; sections 1 and 2 well outlined and clearly developed.	poor transition among sections; sections 3 and 4 only superficially developed.

TABLE 24 Major Strengths and Limitations (Cont)

Simulation	Strengths	Limitations
Grand Strategy	stresses diplomacy and negotiation process, not strategy of war; nicely packaged; good debriefing section.	roles not tied to real-life prototypes; downplays influence of personalities; issues mainly territorial and political rather than ideological or internal: issues like Pan-Slavism, Young Turk Revolt, or Austria-Hungarian rivalry not discussed.
The Haymarket Case	case carefully outlined; roles strictly defined; gets well at issues of jurisprudence; good debriefing section.	fact sheet contains insufficient background information; role descriptions too detailed in direction of player actions; tends to be difficult to pick up attitudes of 1880s.
Liberté	excellently articulated goals; good intitial section on nature of resolutions; good trial and legislative sections.	tries to do too much in one game; poor role division: leaves out critical role of sans-culottes; economic activity in first section stilted; uneven transitions between stages; oversimplified explanation of Terror; information in quiz sections not effectively mirrored in play.
Nuremberg	excellent goals; good concentrations on issues; good role assignment (stresses not what to do but how to approach issues); rules well constructed.	minimal description of possible roles for witnesses leaves options too open-ended.
Origins of WWII	excellent as game of strategy.	territorial conquest too much the goal of game; thus bypasses nonquantifiable causes of war such as national pride, ideological conflict, internal affairs; rules complicated; no sense of change in political circumstances from 1935 to 1940; background essay historically misleading.
Panic	excellent goals; format and timing clear; gets well at both feelings and facts.	roles poorly defined and little background information.
Scramble for Africa	intriguing game of strategy.	poor general description; distorts motives of explorers by lumping riches, religion, and glory together in single category; concentrates too heavily on military conflict to detriment of political of ideological issues; rules initially confusing.
Seneca Falls	defines issues to be debated well.	roles not clearly defined; perspectives of leading characters vague; requires outside research or careful briefing by instructor; limited in scope of issues debated.
Trade-Off at Yalta	issues well articulated; good background information; format clear and rules easy to follow.	issues artificially divided; actions by players uneven.
Waging Neutrality	excellent focus on critical issues of commercial causes of American neutrality and later entrance in WWI; roles well defined; rules clear and well organized.	leaves out political issues and personality of Wilson or other leaders as factors.

CONCLUSION

Some of the criteria we have considered in this review are geared to match your requirements with simulations that best meet them. In such cases only you can rate which are better. In other cases, however, we are able to make an overall judgment, and we have attempted to do so in Table 25.

Of the eighteen simulations we have reviewed, we would rate nine as outstanding, with scores of 18 to 20. Four more we would rank as good, and five with scores of 10 or below as only adequate. Even the strongest have weak points, although in most cases they are not glaring, some of the exceptions being the debriefing segment of the *Ch'ing Game* or the role descriptions in *Panic.* However, these do not detract from the overall balance of the more outstanding games; that quality is what brings them their rating. The games to which we have assigned lower scores exhibit increasing imbalance or, in the cases of *Gateway, Haymarket,* or *Seneca Falls,* less well-developed characteristics across the board. These lower scores, however, should not keep you from considering the games we have designated as good or adequate. Time and budgetary considerations or more limited pedagogical goals may make these more appropriate for your needs than the more fully developed simulations.

When all is said and done, as we have stressed time and again in the course of this essay, your needs are what make a simulation truly worthwhile or not, and no matter how we might rate them, we cannot make an absolute judgment. As we noted in our opening remarks about the nature of history and

TABLE 25 Overall Evaluation

Simulation	Breadth of Content	Effectiveness in Carrying Out Stated Purpose	Rules	Role Description	Role Involvement	Internal Order and Conversion	Historical Validity	Debriefing	Flexibility	Packaging	Cost	General[a]	Total
Alpha Crisis	1	2	1	2	2	2	2	2	1	1	2	2	20
American Constitutional Convention	1	2	1	2	1	1	2	0	0	1	2	1	14
The Ch'ing Game	2	2	2	2	2	2	2	0	1	1	2	2	20
Congress of Vienna	1	1	2	0	1	2	1	0	0	0	2	0	10
Czar Power	2	2	1	2	2	2	2	0	1	2	0	2	18
Czech-mate	2	2	2	1	2	2	2	2	1	1	1	2	20
Destiny	1	2	2	2	2	2	2	1	1	1	1	2	19
Gateway	2	1	2	2	2	1	1	1	1	1	1	1	16
Grand Strategy	1	1	0	0	1	1	1	1	1	2	0	1	10
The Haymarket Case	1	1	1	2	1	2	1	2	0	1	1	1	14
Liberté	2	1	1	1	1	0	0	1	0	1	1	0	9
Nuremberg	2	2	2	1	1	2	2	1	1	1	1	2	18
Origins of WWII	0	1	1	0	1	1	0	0	2	2	2	0	10
Panic	2	2	2	0	2	2	2	1	2	1	1	2	19
Scramble for Africa	0	1	1	0	1	1	1	0	0	0	2	0	7
Seneca Falls	1	1	1	1	1	1	1	1	1	1	1	1	12
Trade-Off at Yalta	2	2	2	1	2	1	1	2	1	2	0	2	18
Waging Neutrality	2	2	2	2	2	2	1	2	0	1	1	2	19

a. Subjective and informed overall judgement

Key: 2 = very good
1 = good
0 = limited

the role simulation has played in the study of the discipline, we must all select those tools that best help us to investigate, learn, and communicate the history in which we are interested.

No simulation, no matter how well conceived, how lavishly packaged, or how good a bargain, can serve effectively unless it meets our goals. No simulation can be used effectively in a course if it is simply stuck on as a last-minute appendage to the syllabus. It must be integrated into the flow of the course; it must reinforce the ideas and issues upon which the course as a whole is built; it must complement the lectures, discussions, readings, or audiovisual activities in which the course participants are engaged. Its use therefore requires careful planning.

You may have to modify the games you buy; you may find even with the variety of games on the market, that none meets your needs and you have to create your own, either using some of these simulations as models or starting from scratch. In either case, the time you spend analyzing your needs, looking over the available simulations, and planning for the integration of simulation activities will be time well spent and will, if done properly, result in greater variety of context, greater interest in issues, and greater learning for some of the game participants. Simulation is, as we noted early on, the primary tool of the historian's trade; the use of these simulations only broadens the effectiveness of our apparatus and offers to make us in the long run more complete historians and more consummate teachers.

SOURCES

Alpha Crisis
William A. Nesbitt
1973

Center for International Programs
and Comparative Studies
The University of the State of New York
Albany, NY 12210
$2.00

American Constitutional Convention
Leonard Stitelman and
William Coplin
1969

Science Research Associates, Inc.
155 N. Wacker Dr.,
Chicago, Ill. 60606
Student Handbook, $1.95 each

The Ch'ing Game
Robert B. Oxnam
1972

Learning Resources in International Studies
60 E. 42nd St., Suite 123
New York, N.Y. 10017

Congress of Vienna
B. Barker
R. Boden
1973

Longman Group Ltd., Resources Unit
9-11 The Shambles
York, United Kingdom
£6.75

Czar Power
R. G. Klietsch
1971

Systems Factors, Inc.
1940 Woodland Ave.
Duluth, Minn. 55803
$68.50

Czech-mate
Daniel R. Place
1976

Interact
P.O. Box 262
Lakeside, Calif. 92040
$14.00

Destiny
Paul DeKock and
David Yount
1969

Interact
P.O. Box 262
Lakeside, Calif. 92040
$14.00

Gateway
Jay Mack
1974

Interact
P.O. Box 262
Lakeside, Calif. 92040
$14.00

Grand Strategy
Clark C. Abt and
Ray Glazier
1970, 1975

Games Central, Abt Publications
55 Wheeler St.
Cambridge, Mass. 02138
$39.00

The Haymarket Case
David DalPorto
1979, 1972

History Simulations
P.O. Box 1775
Santa Clara, Calif. 95051
$15.00

Liberté
Sister Marleen Brasefield
1970

Interact
P.O. Box 262
Lakeside, Calif. 92040
$14.00

Nuremberg
Arthur Pegas
1971

Interact
P.O. Box 262
Lakeside, Calif. 92040
$14.00

Origins of World War II
The Avalon Hill Company
1971

The Avalon Hill Co.
4517 Harford Rd.
Baltimore, Md. 21214
$12.00

Panic
Paul DeKock and
David Yount
1968

Interact
P.O. Box 262
Lakeside, Calif. 92040
$14.00

Scramble for Africa
B. Barker and
R. Boden
1973

Longman Group, Ltd., Resources Unit
19 West 44th St.
New York, N.Y. 10036
$3.95

Seneca Falls
Paul DeKock
1974

Interact
P.O. Box 262
Lakeside, Calif. 92040
$10.00

Trade-Off at Yalta
Daniel C. Smith
1972

Prentice-Hall Media
150 White Plains Rd.
Tarrytown, N.Y. 10591
$35.00

Waging Neutrality
Russ Durham and
Virginia Durham
1970

Simulation Systems
Box 46
Black Butte Ranch, Ore. 97759
$14.00

INTERNATIONAL RELATIONS GAMES AND SIMULATIONS
An Evaluation

by Leonard Suransky

INTRODUCTION

The Purpose of This Essay

In this essay we will compare, contrast, and evaluate a variety of twelve significant simulation games in the field of international relations according to their context and focus, process and suitability to your needs. We will then consider what the simulation games offer in terms of their areas of content, conceptual orientation, roles, rules, skills, activities, interactions, and debriefing possibilities.

Definition

International relations simulation games were designed when studies in the field indicated that if the workings of the elite decision-making circles of nation-states are understood, it is easier to understand both how historical conflicts were generated and how diplomats tussle with present-day and future dilemmas in our world. Because there is little access either to such corridors of power or to much of the secret information that guides and influences such decisions, simulations based on what is available proved an excellent, if imperfect, instrument to probe these matters.

Your players will most often find themselves representing the cabinet of a superpower, a rich or poor third-world state, a medium power in Europe, or a group aspiring to power, such as a liberation movement. Their roles will call on a diverse set of expertise, rudimentary and exploratory for beginners, sophisticated for professionals (who tend to agree that they learn more about, and gain new insight into, their field from a good simulation game). At times these decision-making roles are less specific and are played collectively by a team. Areas of activity can range from defense to economic management, from diplomatic negotiations to journalistic or espionage activities. Skills in decision-making (often under intense pressure and national jeopardy), in writing, oratory, negotiating, and planning are the order of the day.

Editor's Note: Listings for all the simulations discussed in this essay are in either the international relations section or in that subsection of the computer games section, except for *BaFá BaFá*, which is in the social studies section.

Some of the simulations, such as *Confrontation, SALT III,* and *MESG,* are explicitly content oriented, focusing on specific problems. Others, such as the *INS* games, present abstract, make-believe nation-states (Yora, Zena, Dorb), allowing players to generate their own insights about the differential behaviors of smaller nations versus superpowers, along with a variety of other concepts to neutralize stereotyped preconceptions they might otherwise bring with them.

Table 1 introduces you to the twelve games selected for this survey. These fall into two main categories: the short one- to three-hour game, which is flexible and convenient to insert into a variety of settings with ease, and the longer games, which are optimally used in conjunction with a course, conference, or retreat.

Uses

These games serve a variety of purposes for a range of different groups that include social or cultural meetings interested in exposure to the addictive thrill and fascination of international politics and intrigue; captive school students imposed upon from above (at last) to acquire greater literacy and proficiency about the wider world in "global education" courses; the more dedicated and research-oriented inquirers at the college level; and civil servants, politicians, and military personnel already involved in the field of international relations or diplomacy.

One can use a good simulation game in a number of ways:

a. As a *learning* tool, to motivate and stimulate participants through an exciting and memorable experience (often indelible in the case of the longer game), to teach basic facts and whet the appetite for further study, and to provide a shared event laden with insights and incentives for further discussion.

b. As a *training* tool, particularly in the disparate art of diplomacy, which demands skills in writing and speaking with clarity, a sense for fine nuances and persuasion, the ability to drive a hard bargain with a broad strategy in mind, and a sensitive appreciation of the misunderstandings, miscommunications, and misperceptions that bedevil the world of international relations.

c. As a *planning and projecting* tool, with the interaction and interplay among as many as ten teams generating

TABLE 1 The Twelve International Relations Simulations

Simulation	Summary Description
BaFá BaFá	Players develop two differing cultures; one easy-going, relaxed and patriarchal; the other, efficient, pluralistic, production-oriented. Cross-cultural visits explore the meaning of culture and the sources of hostility and chauvinism.
Confrontation: The Cuban Missile Crisis	Players occupy positions of U.S. and USSR Decision Makers facing dilemmas of the Cuban missile crisis. Shows dangers of a world system dependent on military competition and restraint of superpowers.
Crisis	Players representing six hypothetical nations vie for a newly discovered energy resource, "Dermatium," risking war as they make and break coalitions.
Grand Strategy	Players step back into history, reliving events leading up to the 1914 "Great War" in ten teams, confronting themes of imperialism, nationalism, and militarism.
Inter-Nation Simulation (INS)	Players take on roles of chief decision makers in five to seven nations of different sizes and forms of government. A statistical report on military, economic, population, and domestic status is furnished each round.
INS2	A computerized version of Inter-Nation Simulation with a variety of scenarios of hypothetical states, some based on real-world conflicts such as the Vietnam War.
MESG (Mideast Conflict Simulation Game)	Players participate in a game-within-a-course (one semester), focusing on authentic reconstructions of Arab, Israeli, Palestinian, or superpower cultures and personalities. Scenario updates Arab-Israeli conflict.
POLIS	Players on different campuses represent decision-makers in a multipolar world not limited to one region. Foreign policy conducted by computer/phone.
PRINCE	One or more players conduct U.S. foreign policy opposite computer-programmed France, USSR, India, Pakistan. Links international and domestic policy.
SALT III Negotiation (Arms Control and Disarmament)	Another game-within-a-course. The United States and the USSR meet over two months, getting familiarized with, planning, and negotiating intricate disarmament details.
SIMULEX IV	Enables teacher (and students) to design their own simulation for any conflict or crisis. Teaches substance and process of foreign policy.
Uses of the Sea	Players gain a general overview of the economic and political factors shaping attempts of nations to regulate the uses of the sea.

alternative strategies and alternative futures that are less likely to come to individual thinkers in a vacuum, and that may have relevance to the real world.

d. As a *testing* tool to verify and manipulate variables regarding current theory and research in international relations.

e. As a *communications* tool to bring conflicting nationals or subcultures into a form of dialogue with each other, although this is rare, given the highly emotional nationalistic tensions that abound in an international conflict. (This has been best achieved by introducing one or two foreign nationals onto their home team, along with U.S. players, but only if the former are restrained enough to realize the "game" nature of the exercise.)

f. As a *sensitizing* tool, to help players identify with another culture's way of thinking and become aware of the historical forces that sometimes possess them, as actors, in the realm of foreign policy confrontation or even in simple negotiation.

g. As a *sobering* tool, to undermine the often naive views of the citizenry at large about what is possible in the realm of foreign affairs. This enables players to appreciate the complexity and sometimes duplicity of international dealings, which are sovereign and therefore not subject to the supreme law-enforcing agencies present in the domestic/civic arena, and to realize that some conflicts and world problems are not necessarily possible to resolve at a particular time.

Table 1 will help you determine which simulations best suit the interests and level of understanding of your group; Tables 2 and 3 will help you choose the specific issues you are addressing; Table 7 will help you match an exercise with the skills and activities that are most important to you. More detailed information appears in the game descriptions themselves.

PRELIMINARY CONSIDERATIONS

Age, Level, Computer Assistance, Group Size, and Playing Time

You should give first consideration to the appropriateness of a game for the age group and sophistication of the players you intend to introduce to it. It is my contention that, other things being equal, the weekend-long games offer a qualitatively superior learning experience, even when spread over several weeks.

Complexity

Table 3 shows a ranking of the simulations according to their complexity. The "simple" games in table 3 require little reading and player participation in preparing role materials, team strategy, and so on, and little facilitator preparation.

The most intricate games, such as *POLIS, MESG, SALT III,* and *SIMULEX IV,* involve players in an ongoing research project to prepare their roles, strategies, and team contingency plans before actually playing the game. In this sense, players can be regarded as designing important segments around the existing frame of the game. (This could be avoided by an enterprising game facilitator willing to collect, edit, and update the roles, strategies, and contingency plans from a previous game for a group with limited time, running a relative short game of three to six hours with no prior time and study commitment. However, the learning experience would be less valuable and less intense.) An important thesis gleaned from widespread experience in the field of simulation gaming is that

TABLE 2 Simulation Games—Some Basic Considerations

Simulation	Age Level	Playing Time	Computer	No. of Players	No. of Groups
BaFá BaFá	h.s., college, adult	1-3 hr.	—	18-75	2
Confrontation	h.s., college, adult	2-9 hr.	—/x[a]	12-24 36 (or 1)	4
Crisis	h.s., college	1-3 hr.	—	18-36	6
Grand Strategy	h.s.	3-4 hr.	—	10-30	10
Inter-Nation Simulation	h.s., college adult	1½-16 hr.	—	24-50	5-6
INS 2	h.s., college, adult	5-10 hr.	x	24-50	5-9
MESG	h.s., college, adult	12-20 hr.	—/x[b]	50-60	9
POLIS	college, adult	16-30 hr.	x	50-100	4-12
PRINCE	college, adult	3-10 hr.	x	1-7	1
SALT III	college, adult	16-20 hr.	—	24	2
SIMULEX IV	h.s., college adult	3-6	—	25	5
Uses of The Sea	h.s., college, adult	3	—	16-24	8

a. A BASIC computer version is now available, with a FORTRAN version to follow. This makes a single-player game possible.
b. This game can also be run using CONFER, a computer conferencing system to relay messages electronically rather than by courier.

TABLE 3 Simplicity and Complexity of Simulations

Simple	Moderate	Complex	Intricate
Grand Strategy *Crisis*	*BaFá BaFá* *Confrontation* *Uses of the Sea*	*Inter-Nation Simulation* *INS2* *PRINCE*	*POLIS* *MESG* *SALT III* *SIMULEX IV*
15-30 minutes	15 minutes—1 hour	1—4 hours	4—12 hours (involves course preparation)

Preparation Time (dependent upon the facilitator's expertise)

the greatest benefits it can yield arise from involving players in the design of their own game.

These factors do not hold for the *INS* games because the hypothetical, non-real-world nature of their nation-states precludes the deep process of developing familiarity that is possible with the often abundant literature available on real-world cultures and political systems.

In the case of *Confrontation,* extensive background information, including reports from the press of the U.S.A., the U.S.S.R., and Cuba during the actual crisis, is available, and materials also include a variety of graded policy options and their responses from the antagonist.

Grand Strategy and *Crisis* present players with concise (if restricted) background information for each team.

One of the remarkable features of simulation games as learning tools is their essential "unfinishedness." Like folk songs or folk stories, they lend themselves to change, adaptation, and modification. Users will join this spirit by feeling free to insert their own issues, roles, and dilemmas into any of these frameworks, and indeed to merge and integrate two or more games, taking and leaving what suits their particular needs. Doing so by no means guarantees instant success, and anyone should plan to run several tests of any such modifications. However, modifying a game is second only to designing one from scratch for stimulating learning and creativity, and is to be heartily encouraged.

In this vein, *BaFá BaFá* has been simplified to *RaFá RaFá* for use with elementary school children, and it has also been modified to work as a game investigating racism (black-white cultures) and sexism (male-female cultures). *Inter-Nation Simulation* has been used for a wide variety of research purposes and for many varied teaching settings. *SALT III* has been simplified to fit in as a shorter three-week module of a semester course. The *MESG* framework has been adapted to accomodate a Southern Africa Conflict game, and the *POLIS* framework must remain pliant and responsive to the continuous flux of global politics.

CONTENT

A third concern you should have in choosing a simulation game is the scope and depth of the content material the simulation game manual furnishes. The games that fell into the categories of simple and moderately complex in Table 3 contain the necessary content material in one to five pages (with *Confrontation* having ten).

However, given the volatile nature of the international system in our time, the changing leadership and shifting alliances that affect the personnel and policies of even the most stable of states, our more complex games are structured differently. It appears to be the consensus among the designers of the more complex games that a very fruitful means of generating content for these games is through their relationship to a course, workshop, or weekend retreat. In this situation, it is both exciting and stimulating for students to build up their own dossiers of up-to-date information from available library resources. The absence of a good college or public library should not be a deterrent to pursuing these games, for as a last resort a collection of old and current news magazines will suffice.

Users of these more complex games report far more intensive and wider-ranging (often unsolicited) reading by their students. International quality newspapers (e.g., New York *Times, Washington Post, Le Monde,* Manchester *Guardian*) tend to become indispensable where they are available, sometimes converting students to more discerning taste in this regard long after the game has become a happy memory. Student evaluations often express appreciation at being offered this latitude to take over some responsibility for gathering, organizing, and analyzing their own data.

This experience of co-designing and co-defining the game,

before and during the play, tends to be an excellent way to introduce students to their own research and to a critical appraisal of the research of others in their textbooks, especially from the vantage point of their lived experiences in the game. Further, this is a good instance of the positive way the sometimes too rigid teacher/learner categories are blurred and even, at times, reversed to the educational benefit of all concerned.

VALUE ORIENTATION

All simulation games in some way carry certain value orientations, often subliminally. It is not unusual for a game designer to be unaware of some of the underlying assumptions built into a game. These assumptions are rooted in the designer's sociocultural context, a background it is sometimes difficult to be fully aware of and to transcend in designing a game. Consider a large group of inner-city social studies teachers who, after a mock run of *Crisis,* insisted that many of their students would refuse to *read* even the one-page scenario! (This might account for the growing use of cassette tapes to issue instructions, since they seem to command greater respect than the spoken or written word—a tendency we should not necessarily encourage.)

Similarly, concepts such as "democracy" that are implicit in many of these games need to be approached with sensitivity. I have played some of these games with black city folk who, given their lived experiences, have a more dubious and skeptical view of the meaning and function of a democratic system than their white counterparts in the suburbs.

International relations simulation games are, in themselves, a luxury of sorts for people who have never left their city or state, and for whom international relations is a concept far removed from their lived reality. This factor should not stand as an argument against introducing these games, but one should be attuned to the possible pitfalls on the way. The very notion of modern international relations presupposes the accepted "good" or necessity of learning to live with other nations. We live today in a world in which our growing interdependence is taken for granted, but it was not always so, as an historical game like *Grand Strategy* on World War I will illustrate, nor is the concept of interdependence necessarily a universally accepted truth.

Some of you may be turned away by the propensity for war in many of these games, though they are certainly not war games, but if anything games of diplomacy with generally unsophisticated and arbitrary means for arriving at outcomes to war.[1] However, the debriefing of a war, a terrorist/guerrilla

TABLE 4 Content

Issues	*BaFá BaFá*	*Confrontation*	*Crisis*	*Grand Strategy*	*Inter-Nation Simulation*	*INS2*	*MESG*	*POLIS*	*PRINCE*	*SALT III*	*SIMULEX IV*	*Uses of the Sea*
Domestic												
Democracy		M	M	S	S	S	S	S	M	S	F	
Authoritarianism	S	M	M	S	S	S	S	S		S	F	
Elections			O		S	S	F	F				
Coups			S	S	F	F	S	S			F	
Capitalism	M	M				O	M	M	M		F	O
Socialism		M				O	M	M			F	
Public opinion		M	O		S	S	S	M	S	M	F	F
Revolution			M	O	F	F	S	S			F	
Economy	M		S		S	S	M	O	M	S	M	S
Defense		S	S	S	S	S	S	S	S	S	S	S
Media		S	O			O	S	O	M		F	
Discrimination	S											
International												
Ideology	S	S	M	S	M	M	S	S		S	S	
Nationalism	S		S	S	S	S	S	S	S	S	S	S
Imperialism		S	S	S	O	O		O		S	F	S
Terrorism			O	M	O	O	S	F	O	F	O	
War and peace		S	S	S	S	S	S	S	S	S	S	
World war		S	S	S	S	S	S	S	S	S	S	
World economy		M	S		S	S	M	S	S	M	O	S
Arms limitation		S	O	O	M	M	M	S	O	S	F	
Deterrence		S			S	S	M	S		S	F	
Balance of power/terror		S	S	S	S	S	S	S	S	S	S	
Alliances			S	S	S	S	S	S	S	S	S	
International organization		M	S	S	S	S	S	S	M		S	S
Ecology			O									S

Key: S: issue significant
M: issue marginal
O: issue optional
F: frame game allows players or game director to design and insert issues

attack, and so forth, can far better come to grips with the cruelty, the wastefulness, and the sheer shock of great violence, than games that preclude the possibility of violent conflict. The tension engendered in *Confrontation: The Cuban Missile Crisis* as the world totters on the brink of a Third World War can be almost as salutary and educational to participants as it was to President Kennedy and his advisors at the time. Some students come away from such perilous situations, where they hold the responsibility for thousands or millions of lives, overawed, more respectful of decision-makers, and, in some cases, determined never to allow themselves to be in such a real-world situation.

I have been impressed with the great learning potential of these "simulated yet real" learning experiences, and most especially those experiences where plans and intentions have gone terribly wrong, have misfired, been misperceived, misunderstood, or misinformed as to certain facts. Learning through mistakes, from negative experiences, strikes me as being t e more profound learning I have witnessed in students, and simulation games are relatively safe environments in which such education can beneficially take place.

ROLE PLAYING

The majority of international relations games are peopled by rather stereotypical roles, which is not to say that these roles are any less interesting than those in other games. On the contrary, in these games the role players tend to represent those members of the cabinet interested or affected by foreign affairs, and the fact that these are a nation's top public decision-makers[2] tends to lend a certain aura of power and pomp to games in this particular field. A vicarious pride and sense of significance are attributed to being a foreign policy decision-maker. Indeed, this reflects the real world, in which many leaders will attest to their preference (over and above their domestic responsibilities), for the glamor of high diplomacy, the intrigue of highly secret bilateral or international conferences, state visits abroad, and so forth. An international relations simulation can even be said to be a cheap form of travel abroad when substantial effort goes into replicating the mood and atmosphere of a foreign culture!

Roles in four of our games are more specific. The roles in *Confrontation* are U.S. and U.S.S.R. advisors to their respective leaders, four people in each case ranging in their perceptions of the crisis from "doves" to "hawks."

Likewise, *PRINCE* involves one or more top foreign policy bureaucrats, and *SALT III* furnishes experts of another sort, twelve people representing the variety of military and political personnel involved in disarmament, including experts on naval weapons, strategic weapons, and Soviet and U.S. strategic nuclear doctrine.

The exception to the rule that roles are foreign policy decision-makers or negotiators of one sort or another is *BaFá BaFá.* This gem of a game is included in this selection for its excellent quality as an opener to a course, conference, or group investigating the pitfalls of approaching another culture. While the Alpha culture does have a ruler in the guise of the domineering patriarch, *BaFá BaFá* is overwhelmingly a game of the people, particularly of those lucky enough to be tourists visiting a foreign culture. (The game was originally designed for U.S. Navy personnel to better attune them to the differences they could expect when assigned to naval bases in Greece. Its goal was to help them avoid past misdemeanors and improprieties in such a setting.) The main charge of those playing *BaFá BaFá* is to very rapidly acquire a new culture and, after playing in their own culture for a while, to try (as much as they are inclined) interacting in the foreign culture for the briefest of visits. Not surprisingly, this leads to many misperceptions and stereotypings, the very stuff of what so often goes wrong in the relations between nations on the more elite levels. The observer in *BaFá BaFá* is sometimes called an ambassador, since he or she does not interact in the culture but rather visits it with a view to reporting to the people at home about the strange ways of the aliens.

The remaining eight games are peopled, depending on their numbers, with a head of state, foreign minister, defense minister, economics minister, opposition ministers (not less than two if this is to be a viable experience), and if there are extra people, journalists, public interest groups, and, in the more developed states, even representatives of multinational corporations wielding considerable nonpublic power. With three to five members on most nation-state teams, these roles are possible only for superpowers and tend to be less fulfilling because of their peripheral relationship to the game.

Among these eight games, we find that *CRISIS, Uses of the SEA*, and the *Inter-Nation Simulation* and *INSS 2* tend to use hypothetical, non-real-world nations; thus, these leaders function with their role title only. The other games, which represent real nations, represent the real-world leaders by name.

MESG, however, places a special emphasis on this role playing. Players are encouraged to spend the term building up an increasingly close identification with the person they represent. At times, people have been auditioned for their personal compatibility with the tone, personality, and temperament of the leader they are to portray. Players dress to enhance the imitation of their real-world counterparts, and if they feel able, they are encouraged to impersonate accents. In some cases, students have undertaken to study speeches their role person made to parliamentary or congressional assemblies in order to use similar word and language patterns. Why do some gamers discourage this approach? What are the advantages of it?

In the *INS* and *Crisis* games the countries names Algo, Erga, Yora, and Somme (*INS*) or Dolchaveet and Axiom (*Crisis*), attempt to preclude any possibility that subjective bias can creep into how players relate to their or another's country and their roles. The goal is to achieve a scientific objectivity through which participants will get an untrammeled view of a small or large state with certain resources and how it tends to behave, uninfluenced by preconceptions. (The designs of these fictitious states are, however, often based upon real-world states.) On the other hand, as *BaFá BaFá* points out, the very stuff of culture so often at the heart of international differ-

TABLE 5 Roles Available in Games

Ministers/ Decision-Makers	*BaFá BaFá*	*Confrontation*	*Crisis*	*Grand Strategy*	*Inter-Nation Simulation*	*INS2*	*MESG*	*POLIS*	*PRINCE*	*SALT III*	*SIMULEX IV*	*Uses of the Sea*
Control/ validation			O				x	x			x	
Head of state	x	x	x	x	x	x	x	x			x	x
Foreign/secretary of state			x	x	x	x	x	x			x	x
Defense/war			x	x	x	x	x	x			x	
Economics			D		x	x	D	O			x	
Interior (domestic affairs)			D		x	x	D	O			x	x
Intelligence (e.g. C.I.A.)					O	O	D	O		x	O	
Opposition					D/O	D/O	x					
Diplomat/ ambassador	x											
Foreign policy advisers		x	D					x	x			
State Dept./ foreign office desk heads								x	x	x		
Military advisors										x		
Journalists/ mass media					O	O	D/x	x			x	
Citizens	x						D	O				
Industry/ multinationals							D					
Guerrilla movement							x	D				
Spies							O					
World organization (e.g., UNO)			x		x	x	D	x				x

Key: D = Depends on number of participants/importance of state
O = Optional

ences, conflicts, and misunderstandings is sacrificed. *BaFá BaFá* in its simplicity enables us to experience how readily and almost inevitably we slip into chauvinism for our newly adopted make-believe culture (Alpha or Beta), and how readily discrimination against the out-group culture tends to build up.

The alternative to this approach is to acquaint players deeply with the culture they are to represent in a detail as fine as time allows. The goal is to generate a potentially more profound experience of the cross-purposes and misconnecting levels of intercourse that complicate international diplomacy, with a view to encouraging insights into the real-world feel of the conflict.

So, too, with roles. One can play head of state and learn about the particular responsibilities and other characteristics of such a leader of a small or large state, or one can intimately research a personality, and, in a quasi-theatrical manner, play that role. The latter method involves more preparation, and though it is more particularistic, to my mind it gives greater game-wide insight into a current real-world conflict. A game facilitator should be aware, especially when real-world figures are being portrayed, that an occasional player may need some help disentangling the "real" from the "game" self after the simulation.

The more abstract approach is to be preferred for conduct-

TABLE 6 Type of Role and Culture

Abstract Role Abstract Culture	Yourself in Role Abstract Culture	Yourself in Role[a] Real-World Culture	Real-World Role Real-World Culture
Inter-Nation Simulation	*BaFá BaFá*	*Confrontation*	*MESG*
INS2		*Grand Strategy*	*SIMULEX IV*
Crisis		*Prince*	
Uses of the Sea		*SALT III*	
		POLIS	

a. No attempt is made to impersonate a living or historic personality.

ing an examination of conflict in general, or whatever other research variables are under scrutiny.

RULES

Two types of rules appear in our selection of games. There are those that are short, concise, and sufficient unto themselves, in the case of the games of shorter duration. In longer games, those that lend themselves to use in some connection with a course or ongoing game run, the rules are less standardized and offer flexibility for input from the particular game facilitator.

BaFá BaFá, in both its first and second edition, offers very economical rules that run three or four pages for each culture with some additional pages to explain the reshuffling of the game for new use (which, note well, takes an hour to reorganize). The latest edition provides cassette tapes with instructions to players, though some game facilitators prefer to indoctrinate players into their new cultures themselves, sometimes handing each one a copy of the culture's rules to follow while the facilitator demonstrates the strange new gestures and mouths the new language.

Confrontation, with its glossy and professional-looking memo and response sheets and its editorial excerpts from the press of three nations, lends its materials an air of distinction and prestige. The five pages of rules are clear and explicit, though slightly confusing until implemented. Three filmstrips from three points of view can be used to introduce the Cuban missile crisis.

Crisis is neatly and compactly presented in separate players' and director's manuals. If possible, the rules should be tried out in a test run (or players appropriately warned and asked to be indulgent of their guinea-pig function) to allow for every game facilitator's need to adapt the game to his or her style. (This actually applies to any game, and beginner game directors/facilitators should never be discouraged by their first attempts!) For instance, if time permits beyond the fifty-minute basic period, you could try allowing more time for meetings, playing this by ear. I would also have messages, especially military moves, pass through the hands of those in charge of the "consequence" tally en route to their destinations. This will enable the facilitators to be aware of troop movements and other game developments ahead of time.

The revised (1975) edition of *Grand Strategy* makes for a more sophisticated game, with the new plasticized wall map of Europe in 1914 being an improvement over, and more practical than, the overhead transparency map. The twenty-three-page teacher's manual is clear and explicit. The scenario, rules, and national profiles for players are printed tersely and lucidly on large card folders, which include an unusual amount of necessary information and vital statistics in a pleasantly uncluttered manner.

The two versions of the *Inter-Nation Simulation* are among the most complex of the games presented here, and hence, so are their rules. Great care has been taken to build into the games variables such as consumer satisfaction and a sense of national security which, if ignored, may lead to the removal of the chief decision-maker either peacefully or by revolution. Players are made aware of the complex juggling of domestic and international political, economic, and police/military variables. The introduction of the computer in *INS2* lends itself to the constant computations that lie at the center of this game, and can take a considerable load from the game director's shoulders if she or he has a team of computer assistants.

The Middle East Conflict Simulation Game, *MESG,* resembles *SIMULEX IV* in that the "frame" rules are relatively simple and essentially involve assigning students to teams of their choice or design and enabling them to develop the materials they will need for play once the game begins in its separate country-team rooms. The goal of both games is to develop expertise in the particular area or areas of the simulated conflict, usually through a concurrent course. The *SIMULEX IV* edition this time includes a truly outstanding, wide-ranging, and self-sufficient introductory international relations course, replete with bibliographies. The game rules are clear and concise for both games.

POLIS is another complex game, especially since it attempts to enable ten or more teams at different geographical locations, sometimes as far away as Tokyo or Jerusalem, to play the same game, though their back-home situations and goals are very different. However, the designers of *POLIS* computer conferencing assure prospective players across the country that their system is so well established now that it should not daunt the prospective user. Because *POLIS* is usually played in conjuction with a course, its in-team rules resemble those of *MESG* and *SIMULEX IV,* and like them it assumes a good deal of student research. The game's great benefit is the possibility of a "multi-logue" across the country on a particular issue in foreign policy, an unusual sharing of perspectives and perceptions.

The rules for *PRINCE* are clearly presented, but more difficult to grasp than most, in that it seeks to put a microscope to the finer details of foreign policy. *PRINCE* not only carefully simulates the computerized input and influence of four other states on U.S. foreign policy but also distinguishes more finely between domestic, partisan, bureaucratic, and interest-group influence. The rules delineate how the game can be used in a variety of ways for more or less sophisticated purposes and students. These range from making students familiar with the world and pressures of the foreign policy establishment in the United States, to a research and policy analysis function.

SALT III probably requires more technical expertise and know-how in the labyrinths of disarmament politics, with its calculations about payloads, kilometric range, and the latest technologies, than any other game presented here. (This is, once again, most rewardingly done through the context of a course and student research.) However, in the latest simplified version the actual rules for two teams meeting formally and informally face-to-face to negotiate are straightforward and exceptionally clear.

Uses of the Sea is another such negotiation game played in a hypothetical world. Like *Confrontation,* it presents a limited set of actions to the players, and the consequences of their decisions are also available in the manual. Just as *SALT III* players will come away with a deep knowledge of the intricate world of the superpower arms race, so players of this game will become familiar with the mysteries of measuring territorial waters, continental shelves, and domestic and international marine policy. The maps of the fictitious fishing nations clearly depict their many demographic properties (oil and fish resources), shipping lanes, and the volume of trade, as well as the legal claims of these nations to maritime territory. Other information players need is captured in simple and explicit graphs.

SKILLS AND ACTIVITIES

Table 7 lists a broad range of skills and activities in our twelve games to help you select a game to meet your objectives by these criteria. Those games marked with "very significant" weight in areas such as analysis, information gathering, research, and reading are best used in the context of an ongoing course or conference. In feedback game directors have received from players they tend to be most highly evaluated for skills and content learned, insights gleaned, and overall impact.

SOME OBSERVATIONS PROS, CONS, IMPACT OF GAMES

BaFá BaFá is a gem of a short game. It is delightful, it can hardly go wrong, and it is an excellent introduction to what happens when two cultures come into contact and begin to size each other up and reassess their own culture in the light of the visit. The design challenge, as yet unmet, is for a game such as this that allows for and can accomodate a more permanent exchange and absorption of foreign immigrants, rather than the five to ten-minute "tourist trip" of this game.

Confrontation is a well-documented, tightly controlled design that offers an opportunity to live through the crisis dilemmas that brought us closer to World War III in our time than any other international event. It is a very good introductory experience to the area of international conflict politics but perhaps too tightly programmed and constraining for the sophisticated player. The games an excellent job, through its Soviet, Cuban, and United States team compartmentation, of stressing the role of perception and misperception in international affairs through the disparity in the three views of the same event it documents and presents. More interpersonal interaction between the teams would make this less predominantly a cognitive exercise.

Crisis is another popular game to introduce players to a wide range of "feel" for the world of international governmental intercourse. It strikes me as needing tight and firm stewardship from the game directors, especially given its propensity to lead to armed intervention as an extension of the negotiating route. It gives a lucid insight into the difficulty of balancing power and then rebalancing it once alliances break down, and it dangles the option of international cooperation before the players.

Grand Strategy is an early classic in the field and has weathered the years well. It is an excellent example of how history can be taught more excitingly, gripping student interest as it allows them to explore alternative routes to, or outcomes other than, World War I. It underscores a suggestion tossed forth by many of these games that crucial war or peace decisions can, at times, depend on one or more historical personalities on the world stage.

Inter-Nation Simulation is, in many ways, the grandfather of games in this field, having broken a lot of the first ground back in the late 1950s. More than most games, it constantly harasses national decision-makers with the broad arc of their responsibilities (financial, military, diplomatic, and domestic political), forcing them to mediate on many fronts. This is its strength, but, at times, a weakness in terms of its playability. If one desired and valuable aspect of simulation gaming is to be swept up in the fantasy of being not yourself but some far-off decision-maker tussling with grave problems, this suspension of disbelief is too often undermined here by the constant need to calculate, assess, and respond to the constant flow of paperwork. Such headaches are realistic and down to earth, but by my observation they preclude other important insights attained in games such as *MESG* and *SIMULEX IV* through a simpler, more flexible, and more thorough immersion in the world of international diplomacy. You cannot have both, and hard choices need to be made here between a more scientifically exact or theatrically impressionistic route to understanding.

INS2, a representation of the original *INS* (*Inter-Nation Simulation*) game, adjusted for the computer technology of the 1980s and with a growing range of new scenarios, has considerably affected the mood of the game. Hand calculators strike the observer as the order of the day, considerably speeding up the calculations needed each round. The modern laboratory for the play of *INS2* at its birthplace at the University of North Carolina–Asheville is drawing more than a dozen high school games a year, and there is a great clamor from neighboring states to get in on the action.

MESG, the Middle East Conflict Simulation Game, more than any other attempts to model itself most exactly on the latest developments in the Middle East and in the related world, even to the extent of replacing personalities, such as a Secretary of Defense Rumsfeld for a Schlesinger only days before the game. The game-within-a-course format lends itself to such a week-to-week monitoring of events. Players are urged to enter a role suited to their temperament, and (where possible) to "become" that person for the game's duration. The sometimes uncanny impersonations (achieved by only a minority of players and not required of them) help transport the game to the Middle East and generate an illusion of realness. Time periods and much of the activity of the game are unusually loose, so the game can take its shape from the world the student players create. The game's integration into a twelve-week course has generated the same echoes of participant praise and enthusiasm reported for *SALT III, POLIS, INS* and *SIMULEX IV.* The game is weak in simulating economic factors, though the opening of a war room has remedied a similar deficiency in the military sphere.

POLIS is a technological breakthrough, as yet rather expensive, but possibly indicative of things yet to come in the world of simulation games, given the technological prospects of communication satellites being cheaper than a cross-country telephone hook-up. Perhaps, just as CBers talk to each other today, tomorrow will see us more widely able to play games by the computer conferencing system over great distances, and no game could be more fitting than one about global politics. The potential for sharing expertise, viewpoints, and perceptions across the nation and beyond is remarkable. However, the geographical distances and the loss of human contact in bilateral conferences do remove a certain mystique that

TABLE 7 Skills and Activities

Skills/Activities	*BaFá BaFá*	*Confrontation*	*Crisis*	*Grand Strategy*	*Inter-Nation Simulation*	*INS2*	*MESG*	*POLIS*	*PRINCE*	*SALT III*	*SIMULEX IV*	*Uses of the Sea*
Information Processing												
Analysis		S			S	VS	VS	VS	VS	VS	VS	S
Gathering Information	M		VS	VS	M	M/VS	VS	VS	VS	VS	VS	
Planning goals, strategies		VS			VS	VS	VS	VS	VS	VS	VS	S
Research				O		–/VS	VS	VS	VS	VS	VS	
Book, journal reading				O		–/VS	VS	VS	VS	VS	VS	
Newspaper reading/ clipping						–/VS	VS	VS	S	VS	VS	
Policy formulation		VS	VS	VS	VS	VS	VS	VS	VS	VS	VS	VS
Diplomatic writing			S	VS	S	S	VS	S			VS	
Media journalism					VS	VS	S	O			S	
Budget computation			S		VS	VS	M	S	M			
Weapons analysis		M	S	VS	VS	VS	S	S		VS	S	
Historical analysis	M	S		VS		–/VS	VS	S		M	VS	
Group Activities												
Group problem solving	S	VS	S	S	S	S	S	S	O	VS	VS	S
Debate		VS	S	S	VS	VS	S	S	S	VS	S	VS
Negotiation	S		S	VS	VS	VS	VS	S	VS	VS	VS	VS
Public speaking	S	S	S	S	S	S	S			VS	S	VS
Chair meetings		S	S	O	S	S	S			S	S	S
Small group discussion	VS	VS	S	VS	VS	VS	VS	VS	VS	VS	VS	VS
Class/large group discussion	VS	S	S	S	S	S	S	S	M	VS	S	O
Challenging/criticizing others	M	VS	S	S	VS	VS	VS	S		VS	VS	VS
Bargaining	VS	S	S	VS	VS	VS	VS	S	S	VS	VS	VS
Delegating authority	M	M	S	S	S	S	VS	S		VS	S	
Human Relations												
Competition/ cooperation	VS	S	VS	VS	VS	VS	VS	VS		VS	VS	VS
Persuading/influencing	S	VS	S	VS	VS	VS	VS	S	VS	VS	VS	VS
Spying, espionage	S		S	S	S	S	S			O	S	
Supporting/ undermining	S		S	S	S	S	S	S		S	S	
Interviewing							M	M		O	O	
Electing	M		S	S		O	M	O			O	
Deceiving, exploiting	S	M	S	S	S	S	S	S		S	S	S
Empathizing	VS						S				S	S
Threatening		S	S	S	S	S	S	S		VS	S	S
Forming coalitions		S	VS	VS	VS	VS	VS	VS	S		VS	VS
Waging war/making peace		VS	VS	VS	VS	VS	VS	VS	S	VS	VS	M
Role Playing												
Specific individual				O			VS	S			VS	
Work function		VS	VS	VS	VS	VS	VS	VS	VS	VS	VS	VS
Foreigner	VS	S		VS		–/S	VS	VS		VS	VS	S
Group member	VS	S	VS	S	S	S	S	S	S	S	S	S
Imitating accents/ mannerisms	VS			O		O/VS						
Wearing costumes (nat'l dress)							VS					
Resource Management												
Survival		VS	VS	VS	S	S	S	S	S	VS	VS	VS
Maximizing resources	S		VS	VS	VS	VS	M	S	VS	VS	S	VS
Investment/foreign aid			VS		VS	VS	S	S	VS		S	
Balancing resources (guns/butter)			VS	S	VS	VS	M		VS	VS	M	VS
Miscellaneous Activities												
Game modification/ design							S	S	VS	S	S	
Videotaping/audio recording							S			O	O	
Computer operation skills						VS	O	VS	VS			

KEY:
vs = very significant
s = significant
m = minor
o = optional
– = not used

usually builds up for players in their separate, yet contiguous, country-rooms in the single-locality version of the international relations game. Ideally, a *POLIS* game should end with a weekend face-to-face finale, or at least a face-to-face debriefing.

PRINCE is an early attempt to model "bureaucratic politics" into a game and therefore fills an important gap in this field. The designers have used the game, based as it is on current theories of international relations, to test these ideas and to make changes in the game based on them. They claim that an entire course could be built around this kind of dialogue between the game and current theory. However, the game is flexible enough to be used for introductory purposes as well.

SALT III and similar arms limitation games are very much in vogue today, and likely to remain so for a few years to come, given the complexity and the importance of these negotiations. As a game, it tends to appear more restrained than the more active diplomacy games, yet it furnishes an opportunity for a microfocus on the intricately detailed technicality of hammering out such an agreement in its entirety. Rarely can this both evolve and be followed through in a weekend-long game of the *MESG* type, which takes hours of preparation to even get to a conference. Student assessments of the overall experience, for all its greater seriousness and decorum, are enthusiastic.

SIMULEX IV differs from *MESG* and other games like it in that not all teams are nation-states. This allows the game to encompass a major theme such as the Arab-Israeli conflict as well as several minor themes such as OPEC prices, North-South dialogue, East-West competition, and more. This is achieved by having one relatively homogeneous team such as Israel, an OECD (Western nations and Japan) team, a broad OPEC team, an Eastern bloc (with Cuba), and an UNCTAD Group of 77 team. In this way the game can represent most interests in the world and absorb a large number of students. Something is to be said, though, for limiting team numbers to around eight people, so one does not find oneself with a handful of students who are being left out of the central action. The role of a control team with the power to invalidate communications if they are found unrealistic appears to me a responsible and essential pedagogic posture. I am not impressed with the conviction that this role can be scrapped and that the game has its own mechanisms for righting wrongs. Of course, this is not the case in an abstract *INS* game in which there are no real-world criteria for reference.

Uses of the Sea is a tightly controlled game offering its players limited options. It does not pretend to be a game that encourages exploration of sea-related matters (managing large coastal oil slicks, for instance) beyond its set parameters. It limits its goal to giving students a general overview of the economic and political factors that shape the attempts of nations to regulate the uses of the sea. As such, it does a competent introductory job, touching on an as yet obscure issue that deserves, and will in time demand, more of our attention.

MODEL VALIDITY

Geographical Separation

The basic model of most international relations games, and of nine of the twelve we discuss here, differs from that of most other simulation games in the physical, geographical separation of teams, usually in separate rooms. This underscores the sovereign and separate nature of nation-states in the international political system many of these games simulate.

The simpler games, *Crisis* and *Grand Strategy,* are normally played within a single classroom, clustering the different nations in various parts of the room, but even these games could be spread out along a corridor. *BaFá BaFá,* though more an inter-nation cultural game than one of international politics and diplomacy, also borrows from the geographical separation model, placing its two cultures in separate rooms, and in more than two rooms when game facilitators have designed additional cultures. On the other end of the spectrum, *POLIS* is played by computer hook-up between separate cities and, indeed, countries!

The design model of separate rooms closed to other teams by the game rules introduces an element of mystery and tension. A wandering game director or message courier is often struck by the variety of divergent atmospheres and moods that emerge in different rooms. A given atmosphere reflects in part the nature of the social culture and the political regime of a nation-state, mixed with the coincidental idiosyncrasies of the role players themselves, especially when they wear costumes and otherwise decorate their "countries." It also reflects the team's spatial and organizational conception of the bureaucratic form its national decision-making unit should take. In some cases, players have gone to great lengths to probe the frontier of an adversary or ally, including the introduction of secret bugging devices, two-way walkie-talkies (sheepishly disavowed when discovered as nothing but a "hot line"), and somewhat perilous mountaineering feats along a narrow ledge between two windows. In one *MESG* run, a humorous control team donned "potty patrol" badges to alert players to the fact that their excessive negotiations in the bathroom (a favorite rendezvous) were under surveillance.

Confrontation and *PRINCE* simulate national foreign policy consultation and decision-making with little or no interaction among teams. In the case of *PRINCE,* the interaction is with the computer model of the other countries represented. *SALT III* is wholly a *conference* simulation with a very particular focus on negotiating an arms limitation treaty. Thus its preparation is pointed more narrowly and less superficially than the incidental conferences spawned by most of the other games.

Degrees of Abstraction

Another dimension of the model is whether a game is designed to reflect real-world nation-states and, in addition, the real-world personalities of their decision-makers or whether a game uses fictitious, abstract nation-states and nonpersonalized roles. The former will most often be chosen when a teacher aims to acquaint players with current world events (or an historical event). The abstract model (*INS, Crisis,* and *Uses of the Sea*) is preferred for a focus on conceptual and theoretical issues, and the model is assumed to be less biased in its abstraction. A user's teaching and research style preferences (an existential versus a behavioral approach, for instance) will often govern his or her choice between real-world versus abstract conflict and crisis games. Users are therefore encouraged to experiment with the model and games that best suit their and their students' needs.

Scope and Detail

Clearly, only a highly complex model could approximate the enormous intricacy of the international political system, and no such model exists. All of the models we have looked at are somewhat deficient; they proffer different strengths and limitation. Wedding the *SALT III* conference model to the *SIMULEX IV* international system and introducing *PRINCE*-type procedures to deal with domestic, bureaucratic politics would theoretically make for a richer game, but it would be virtually unplayable. These games were originally turned to, as we mentioned earlier, to acquaint students with the complexities of international diplomacy and decision making and to give students an experiential taste of a world that is otherwise inaccessible. For these games to succeed, they should not be overburdened with detail, but kept buoyant and light, and they will of their own accord generate many of the aspects and concepts of international politics, which can be brought out in perceptive and skillful debriefing. Where you want greater clarity, you could offer two or even three game experiences at different times, in one or more departments. (Very often, once "turned on," some students will seek out any other gaming opportunities on a given campus.)

Flexibility

Another issue in looking at a game model is the flexibility-inflexibility spectrum. Bob Beattie at Michigan University's Institute for Social Research has translated Bahram Farzanegan's *INS2* from BASIC to FORTRAN and incorporated it into Bob Parnes' CONFER computer-conferencing system, sometimes used for *MESG* runs at Michigan. In an interview, Beattie pointed out that as a computerized model based on Guetzkow's *Inter-Nation Simulation,* which was built at the height of the Cold War, *INS2* is best suited for big power problems. It does not readily allow for the introduction of nonstate actors such as guerrillas and their liberation struggles. It would also need to be remodeled to reflect the growing dependence of the industrialized powers on oil in our time, and to deal with other such resource allocation issues. More flexible games such as *MESG* and *SIMULEX IV* can be easily modified to reflect such developments, but at the loss of the precision and budgetary finesse possible to some degree in *INS2.* Users must make such choices according to their priorities.

TABLE 8 Eight Multipolar Inter-Nation Game Models

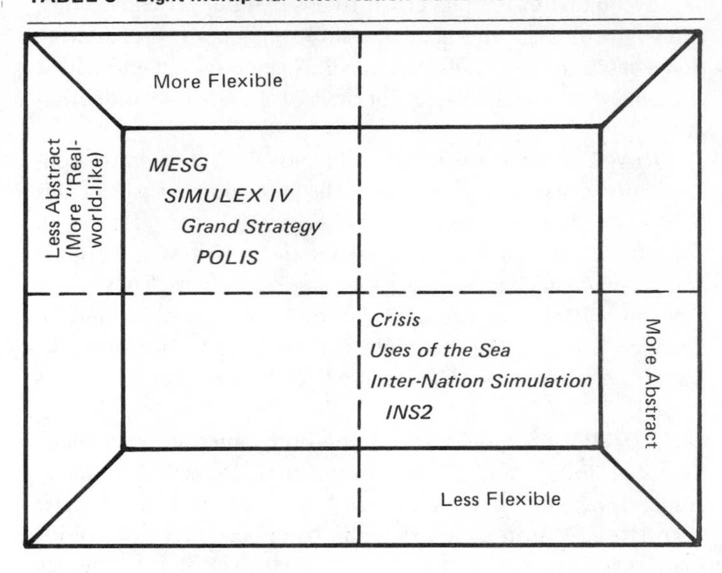

RESOURCES AND SCORING

Unlike many other simulation games, international relations games tend not to depend on artifacts to represent variables, goods, or currency. Likewise, these multination games do not usually lend themselves to scoring. Although players new to these games will tend to cast around for ways to score, to win or lose, the most valuable lesson they may come away with in their quest is a realization of the growing interdependence of nations in the international political system in an age of nuclear peril and dwindling food, energy, and other resources.

Most of the games handle their resources on paper or computer printout. BaFá BaFá is the exception, using packs of colored and numbered cards in the Beta culture, and colored chips and mysteriously designed cards in the Alpha culture. The Beta citizens also tally their scores with a vengeance unconsciously reflecting the materialistic and competitive inclinations of their culture, only to find out that these scores hold no reward or special significance at the end of the game. Therein may lie a useful debriefing lesson.

The two *INS* games are intricately programmed to "score" the fluctuating economic/resource, governmental, and military fortunes of the interacting nations.

Otherwise, scoring is not a major facet of these diplomatic games and is usually managed by political experts at central control.

Though these are not war games, but rather games about diplomacy and international politics, the potential for war is usually included to make these simulations more realistic. In cases of war or skirmish (which tend to be surprisingly rare), the control/umpire/war room team must make definite decisions and declare winners and losers, successes and defeats, and the human and material gains and losses.

The debriefing sessions can often be most salutary in emphasizing the longer term inappropriateness of such absolute concepts as winning and losing in the relations between states with their all-too-fragile citizenry and other valuable resources.

MATERIALS AND COST

The overwhelming majority of the twelve games selected for review here come in manual form, usually with director and participant manuals. In a few cases, cassettes, filmstrips, or a videotape supplement these. Often, the more expensive games can be constructed from scratch with the help of the manual, and this can often be a pleasant preliminary to playing the game, very much in the spirit of student design participation and teacher-student cooperation and collaboration. This dictum applies even to the more expensive computer games; even if one does not personally have the expertise, computer buffs on campus can be roped into an interdisciplinary project that has been known to make life-long converts to the field of simulation gaming, which so readily lends itself to computerization. (This collaboration is at times necessary when a computer program has to be translated into a different computer language.)

All the games we have described as best used in conjunction with a course (*INS2, MESG, POLIS, SALT III, SIMULEX IV*) involve students in research and the development of position papers, team strategies, and in-depth analyses of an aspect of a conflict before and/or after the game run. These activities cannot, by their very nature, be prepackaged, and the independent effort involved is at the heart of the high success rating students give these games-within-a-course.

In the case of *POLIS* long-distance telephone costs need to be budgeted into the game's cost for teams that are hundreds or thousands of miles away. Bob Noël claims that the top limit for East Coast players relaying their material after 11 P.M. or before 8 A.M. is $500.00 for a five-week game. An Israeli student team playing in the spring of 1979 budgeted $1200.00 for long-distance phone costs.

TABLE 9 Resources

Simulation	Resources
BaFá BaFá	Betas have different colored cards with different numbers on them. They strive to accumulate points by obtaining a royal flush in any one color. Only one of the Alpha cards has any social significance, the other two cards and the colored chips function only to provide an activity. Each player gets one visit to the aliens.
Confrontation	The noninteracting foreign policy advisers to their respective leaders work with preprogrammed memos, which give them an evolving scenario of the missile crisis depending on their decisions, chosen from prepared policy options. The consequences of these actions are found in the data book which contains graduated responses. The resources at their disposal range from diplomatic compromise to confrontation, with all the economic and military sanctions including the nuclear option, at the disposal of the two superpowers.
Crisis	Each player gets an index with an approximation of the relative overall strength of each of six fictitious nations. This includes data from a separate table of the strength of the armed forces of each state, as well as their natural resources and level of technology.
Grand Strategy	Profile of each of ten nations details its historical background, its differential present-day resources, military and other, and its strategic objectives. Teams with monarchs or emperors focus more power in their hands than that allowed democratic leaders. They have more power to make policy, declare war, issue ultimatums or negotiate. Most teams belong to one of two alliances which both enhances their stature, but also can restrain their ability to act unilaterally.
Inter-Nation Simulation INS2	All teams continually reassess and make decisions as to how to use the differential resources they start out with in the game, be they economic, military, political, etc. Depending on their management of these resources, they can make external and internal policy decisions, ranging from the peaceful and conciliatory through confrontation and aggression, to the annexation of another state.
MESG POLIS SIMULEX IV	Students in these "games-within-a-course" compile their economic, military, political press, cultural and other resources into dossiers gleaned from and reflecting their state's "real-world" data and position. In the game-play they use these resources, (sometimes more explicitly as when a war room in MESG is available), to further their diplomatic, economic and other international strategic ends. The spectrum of action ranges from friendly to hostile.
PRINCE	The relatively complex computer models in the game try to focus interactions between resources such as foreign aid, and national budgets and "affect" values, as well as the usual sources of power and influence. From the calculation come the responses to player moves.
SALT III	This is a negotiating game, with many of the resources taking the form of weapons technology and their alleged performance propensities. The negotiators for the two nuclear superpowers have some input into decisions on how to link this military, economic and technological, political and diplomatic power (a) to achieve a treaty, (development of new weapon systems or sabre-rattling which could lend pressure); (b) the obvious long-term influence in defining arms usage in years to come.
Uses of the Sea	Descriptions are given of the differential means of each of seven nations at the start of the game. These include fishing areas, oil reserves, volume of trade, number of vessels, scientific research capability, marine pollution, food production and level of industrialization. On the basis of these factors teams interact and attempt to regulate their marine policies.

DEBRIEFING

I consider the debriefing of a game to be as important as the playing process. Be sure to give it ample time and consideration. Having lived through an experiential event, your

TABLE 10 Simulation Game Packaging

Simulation and Producer	Kind of Packaging	Cost	Completeness	Durability & Reusability
BaFá BaFá (Simile II)	manual (make own)	$3.50	p,c,d	average
	kit (with cassettes)	$35.00	complete	average
Confrontation (Social Studies School Service)	manuals, filmstrips, and cassettes	$96.00	complete	high
	BASIC and FORTRAN Versions by co-designer	$20.00	–	
Crisis (Simile II)	25 person kit; set of	$25.00	complete	average
	consumable forms	$2.50	–	–
Grand Strategy (Games Central)	kit	$39.00	complete	high
Inter-Nation Simulation (Science Research Assoc)	kit	$82.00	complete	average
INS2 (Conduit)	kit (includes programmer's guide and computer software)	$95.00	complete	average
MESG (Simulectical Simulations)	manual	$5.00	p,c,d	average
	videotape	$75.00	p,c,d	high
MESG	CONFER Manual	$2.00	complete	average
POLIS	manual, operations guide, scenario	$100.00 per participating team plus phone costs	complete	average
PRINCE	participant's guide plus PRINCE model	$5.00 (bulk order)	p,c,d	average
PRINCE	book "Everyman's PRINCE," Duxbury Press, 1976	$6.95	c,d	high
SALT III	monograph	$3.00	d	average
SIMULEX IV (N.H. Council on World Affairs)	manual	$5.00	c,d	average
Uses of the Sea (Learning Resources in International Studies)	manual	$2.00	c,d	average

Key: p = requires purchasing
c = requires construction
d = requires duplication

players have much to reflect upon and share with each other.

Many game directors flinch at the prospect of debriefing a game. International relations games, as a rule, lend themselves to debriefing better than other games, because, in almost all plays, the teams have been separated in some way to give the illusion of geographic distance and nation-state sovereignty. (I strongly recommend where possible the separation of country teams at least along a corridor for the added tension it lends the game. Of necessity, this involves after-hours, evening, or weekend game runs.)

An international relations game that is not well or adequately debriefed by the directors will often be debriefed informally, since the players have so much to share and have been separated for so long. The only pity is that so much will be unheard by the community as a whole.

A good way to start a debriefing is to allow one member from each team (and another if there is disagreement), to give their perspectives on what they were trying to do (their goals), and what they perceived as happening in the game (their accomplishments and mishaps). This, in itself, usually allows much of what is most pressing to come out and generates counterclaims, perceptions, and unknown revelations in exchange.

A question about whether anything nefarious or weird went on will usually uncover some humorous attempt at skulduggery or espionage, along with gasps of amazement.

After the first round of letting off steam, which is essential to restoring psychological equilibrium to your players, especially after a long game, the line of questions should attempt to draw analogies to the real world. If possible, you should have an expert in the subject matter on hand to draw out these analogies and insights.

In a game-within-a-course format it is easily possible to spend four hours or more of class time (preferably in small groups) continuing the debriefing. Video or audio recordings of what a player said, of what went on in another country-team, of how a foreign minister slightly twisted his or her cabinet's brief or changed it for the ears of several different leaders on a tour of many nations, are fine triggers for further discussion.

Computer printouts generated by the *POLIS* system and the Ann Arbor CONFER system (which has been used for communication in the *MESG* game) present players with a virtual book of information that is sufficient for a whole semester's analysis in another course if desired, especially if one wants to apply the findings to existing theory. Professor Noël, the *POLIS* designer, mentioned when I interviewed him what he thought would be an ideal situation for these games-within-a-course: a three-trimester or full academic year program with one quarter or term to research roles and co-design the game; a second quarter to play the game (*POLIS* is played over four weeks rather than over an intensive weekend); and a third quarter to debrief and research the printout generated. I would add only that at this point students would be in excellent shape to round off the whole year-long experience with a far more sophisticated play of a weekend-long second game.

One of the most succinct and thorough summaries of the postgame discussion are the four debriefing tasks of prominent Scottish gamer Jacquetta Megarry.

> In the first place, there is often a need perceived at least by the participants to talk about the events of the game itself, though it is not always perceived by the director. Second, if the game was also a simulation, the director will want participants to relate the in-game events to the outside world, to question the fidelity of the model and to examine their position in relation to the value judgements embedded in the design. Third, it may be appropriate to open up a free-ranging discussion of emergent principles and wider issues raised by the game. Lastly, there may also be a stage in which the director collects evaluative reactions to the game, asks for comments, criticisms, suggestions for improvements and variations.[3]

These suggestions will supplement debriefing guidelines in the games, especially those that are less adequate.

Table 11 indicates our rating of the debriefing guidelines of each simulation game.

AFTERTHOUGHTS AND AFTERWORDS

In a debriefing frame of mind, it would seem fitting to end with a select cross-section of comments and learnings gathered from some of the experience of participants and teachers with these games.

TABLE 11 Debriefing Guidelines

Inadequte (few questions)	Adequate (8–12)	Thorough (several pages of questions and guidelines)
Inter-Nation Simulation	*BaFá BaFá*	*Confrontation*
INS2	*Crisis*	*MESG*
PRINCE	*Grand Strategy*	*POLIS*
Uses of the Sea	*SALT III*	*SIMULEX IV*

Teacher and Student

After initial skepticism and perhaps even a nervousness tinged with some fear about the propriety of games such as these for their students, many teachers have come to appreciate them. The change from a more conventional and at times dysfunctionally authoritarian teacher/student relationship to one that is more open, mutually beneficial, and educationally reciprocal, comes to be appreciated and even enjoyed. Frank Caldwell of Pepperdyne University, responsible for the monograph on *SALT III,* lauds the more active social role students have in their educational process and believes it accounts in part for the success of his course.

Bahram Farzanegan of the University of North Carolina–Asheville has had his students take *INS2* out of the university setting. They have taken it into scores of high schools, where, in conjunction with local teachers, these students have developed a remarkable learning network for the teachers and their students and, perhaps principally, for themselves. This culminates in an annual Inter-Scholastic Simulation Competition, a type of *INS2* super bowl, with glowing commendations.

Statements from students in answer to the question of whether the traditional teacher/student role has changed around a game are replicable in feedback from around the country. University of Michigan students have made comments like: "This question is ludicrous. Of course it has. I know that for myself I've never enjoyed such a relaxed relationship and still been able to respect and gain and give"; or, "The student takes the initiative and he learns himself instead of being taught;" or, "There was mutual respect, no airs of superiority or displays of condescension."

Lest we wax too euphoric, though, it would be wise to take note of Jacquetta Megarry's warning that games can and have been used in a manipulative and authoritarian manner.[4] They are all the more dangerous used so, because in a game a student's guard is down. Manipulation is the last thing one expects of simulation games, reputed to be so "democratic."

The Advent of Computers

Bob Beattie, a one-time student of the father of *INS,* Harold Guetzkow, subsequently his teaching assistant, and now the developer of a FORTRAN version of *INS2,* agrees with Bahram Farzanegan of BASIC *INS2* repute that the computer introduced into this game has done wonders to free the instructor from the role of overworked master calculator. Likewise, the students now have the facility to do interactive computing, which allows them to do such operations as simulating a variety of different budgets on the computer before making final decisions. Farzanegan talks too of shorter decision-making periods, a decrease from one or two hours to fifteen-minute periods.

Here too, though, caution and pedagogic concern and vigilance should be displayed. Beattie, for instance, believes that he sees students spending too much time on their budgets and not enough on the "big issues," compared with the manual *INS*.

William Coplin, the designer of *PRINCE*, has encountered a different criticism, related to the difficulty inherent in attempting to generalize and mathematize foreign policy-making issues in his game. This "rigidity" in *PRINCE* has prompted him to move to more *PRINCE*-like exercises to forecast decisions in the real world. These are discussed in his book, *Everyman's PRINCE*.

Experience at the University of Michigan with Bob Parnes' exciting CONFER system, which can be adapted as a communications medium in the *MESG* game, gives rise to some general cautions about the introduction of the computer into these games. Ideally, all players should have a computer terminal of their own in lieu of the typewriter or pen of earlier days. Clearly this is impossible, but with one terminal per team, the frustration of waiting to get onto the terminal, or waiting to receive rather than send messages, detracts from the major cooperative activity of cabinet decision-making, planning, and negotiating in the diplomatic realm. True, more messages may be generated overall in a game, but then few of these games are oriented to volume, and diplomacy, by its very nature, is classically a slow, careful, and painstaking process. Game directors should be sensitive to the subtle, but important, changes in mood and focus that come with this magnetic new technology, and weigh its pros and cons.[5]

The rapid CONFER-*MESG* and *POLIS* message relay systems tend to get bogged down in the validation function of control, which is, in fact, always a bottleneck, and there is an understandable technical inclination to want to bypass this function. Bob Noël, the designer of *POLIS*, says that attempts to do this, or to allow a split-second review of messages, have been tried with *POLIS*, but that it removes the important function of "using umpiring to do teaching" in the game.

This notion that teaching should continue during the play of the game is vital, especially when simulation games are so often academically rejected by colleagues for being "unrealistic" and inaccurate. In my experience, several small irregularities, and often a major one, have occurred in *MESG* when it is not rigorously validated. For instance, a notion once arose at an Arab summit conference and was pursued for one and one-half out of two days' play, that Syria should willingly sacrifice the Golan Heights for an Arab-Israeli peace settlement, with no prior concessions asked in return. This issue effectively sidetracked the entire game, when it should have been quickly invalidated, and it made any substantive peace progress impossible.

Finally, on the pro side, the computer printout at the end of a CONFER-*MESG* or *POLIS* run is invaluable and offers a mine of information for debriefing, teaching, and research, especially with its incredible possibilities for retrieval and indexing by issue, by person, or by team.

Attitude Change and Learning

Frank Caldwell tells the classic story that many of us have witnessed. In this case a very "hawkish," hard-line student, who tended to be skeptical and suspicious of the Soviets, played the Soviet military representative in *SALT III*. In a video interview he claimed that after he became a Soviet he felt "surrounded by the West" and developed a somewhat paranoid, completely different perspective of the Soviet dilemma.

Going one better, though, some time after the same game, a fellow U.S.S.R. negotiator graduated and was looking for a government job. In an interview with the CIA she was asked what she knew about Soviet weaponry. She answered, "*We*, at present, have the SS9, but are going to replace it with a more powerful. . . ." She was met by looks of horror and concern on the faces of her interviewers, and had to do some quick explaining about *SALT III*, what it was, and how hard it is to disengage oneself from one's role. She got the job. One wonders if perhaps that is why both the real U.S. and Soviet SALT delegations asked for copies of the final draft of the SALT treaty of that Stanford run of the game.

Simile II, the designers of *BaFá BaFá*, discuss the possibility in the debriefing of their game of a similar circumstance. They talk of moving students beyond stereotyping to description, and coming to appreciate the *similarities* between cultures, along with the differences, rather than emotively and chauvinistically labeling what is "other" bad, and what is "mine" as good.

It seems fitting to end with the words of a Michigan student playing an Israeli minister, since it is our students who have made so much of this essay feasible.

> I don't think I would have ever gained the personal insights into myself (and international politics) by conventional means such as books, lectures, and discussions. You can't get close enough to the situation unless you are in the situation. It is easy to read that the Israeli government has difficulty reaching a consensus. You can't fathom how difficult it is unless you feel the frustration.

Notes

1. The *MESG* game on the Middle East has a War Room, usually staffed by war-gamers who have, at times, given scintillating decisions on matters such as the guerrilla hijacking and wiring-up of an oil tanker later boarded by a rescue team of naval frogmen. This, however, could only have occurred with the sprinkling of ROTC students in some teams able to deliver intricate plans of attack and defense.

2. In the case of historical games or regimes that continue as effective monarchies, the prime minister, president, or chairman of the party are king, queen, or whatever.

3. Jacquetta Megarry, "Retrospect and Prospect," *Perspectives on Academic Gaming and Simulation*, ed. R. McAleese (London: Kogan Page, 1977), p. 198.

4. Ibid., pp. 195-8.

5. Prof. Olof Wärneryd (Lund University) is Sweden's "Mr. Simulation Game." Originally he used computers in his urban and regional planning games. Then he dumped the computer! Why? I asked him in a recent correspondence. His rationale is important, and acutely perceptive, applying to our international relations games as to others: A game participant (in *INS2* or *PRINCE*, for example) does not usually know the computer model underlying a given game. This lack of understanding leads to *alientation* in the player-machine relationship and, in turn, in relationships between players. In addition, he finds, this alienation tends to constrain player creativity.

MARKETING GAMES
An Evaluation

by Anthony J. Faria

INTRODUCTION

What do the terms *Compete, Marksim, Mia,* and *Markstrat* have in common? Give up? These are the names of four of the nearly fifty marketing simulation games currently being played by business students in their university courses, business managers in executive development programs, and management trainees being prepared for executive positions. Interested in further information? Of course you are. The fact that you are reading this chapter indicates that you are interested in business games in general and marketing games in particular. You are certainly not alone in this interest. The large number of marketing games currently being published certainly would not be available if there were not significant interest in marketing gaming. Furthermore, two recent surveys indicate that over 90% of all business schools use decision simulations somewhere in their programs. Finally, entire organizations devoted to business simulations have come into existence. An example of such an organization is the Association for Business Simulation and Experiential Learning (ABSEL), which holds an annual conference in April of each year.

The Purpose of This Essay

The purpose of this essay is to introduce you to some of the leading marketing simulation games. To this end, we shall compare and evaluate the eight most widely available marketing games with regard to their content, scope, complexity, and suitability for use in various settings. In order to narrow down the list of marketing games to a reasonable number for consideration in this article, we considered for analysis only what are called "functional" marketing games.

Categories of Business Games

Business simulation games can be divided into three categories. These are:

1. top management games
2. functional area games
3. concept games

Top management games put the participants in the role of the very top managers of the business firm. As such, participants are responsible for making decisions in each of the functional areas of the business firm. This would involve making production, marketing, financial and personnel decisions. In the functional area simulation, the participant is responsible for making decisions in only one of the functional areas of the firm. Thus, participants could be put either in the roles of a firm's marketing managers, the finance managers, or the production managers, but not in more than one of these areas. The concept simulation focuses on only one aspect of each functional area. Thus, concept simulations within the area of marketing would focus on only advertising decisions, or inventory decisions, or product planning decisions, for example.

The eight marketing games that we will be evaluating are all functional area games. Each of these games places the participants in the roles of the marketing managers of a business firm. The participants are responsible for all or most of the decisions made by the marketing managers of a business firm. This would include decisions in the areas of product, price, place, and promotion. We will refer to concept games briefly later in this article.

Selection Criteria

The initial selection criterion for inclusion in this essay, then, was that the marketing game be a functional area game, the most widely used type of marketing game.

Concept games are highly diversified, serving very specific purposes, so that it is impossible to make reasonable comparisons among them.

Further criteria for selecting these marketing games were their usage and their ease of availability. By usage we mean that there must be some evidence that reasonable numbers of universities and business firms are using the game. By ease of availability we mean that the game is being published and distributed by a well-known publisher. This provides the interested user with the assurance that all materials for the game can be provided on time and in the quantities needed. Thus, we did not consider marketing games distributed by their designer or the designer's school. As a result, the eight marketing games that you will be reading about in this essay are the eight leading functional marketing games currently commercially available.

How Does the Marketing Game Work?

Though most of you will be familiar with the mechanics of using marketing simulation games, please bear with me during this brief discussion for the benefit of those who are not.

Marketing simulation games put participants in the role of the marketing managers of a business firm. As such, the participants are responsible for making all or most of the decisions that the marketing managers of a business firm would make. The participants are generally grouped into teams of three or four members each. Each team represents a business firm selling certain products in competition with other business firms (made up of other game participants) selling similar products in the same markets. Most simulation games accommodate about five competing companies within each industry, so that about fifteen to twenty participants comprise an industry. If there are more participants, more than one industry can operate at one time.

Each marketing team (company) makes a round of decisions, which are submitted to the game administrator. The decisions involve such areas as product features, price, promotion, and product distribution. The decisions of all of the companies are evaluated (generally by computer model) and the results are fed back to each company. The results are in the form of sales volume, market share, revenue, profits/losses, and so forth. Based on its results, each team must adjust its strategy and make another round of decisions.

Each decision generally represents one quarter (three months) of business operation. Most marketing games are geared to eight or twelve rounds (two or three years) of decision-making competition. Decisions can be made either in class or outside of class.

Why Use a Marketing Game?

At this point, you may want to ask yourself why you should use a marketing game. Compared with the more traditional methods of teaching marketing principles and/or marketing decision-making and strategy development, the use of marketing simulation games is relatively new. The use of marketing games dates back only about twenty years compared with sixty years for the use of the case method and all the way back to 1902 for the lecture method in the first marketing course offered. In the short time since their introduction, however, the use of marketing games has increased rapidly due to the many advantages this approach has over the lecture and case methods.

What benefits can you expect to achieve with the use of a marketing game in your classes or training program? The benefits most generally attributed to the use of marketing games include the following:

1. a high level of participant interest and involvement
2. the opportunity to integrate concepts in a dynamic, responsive systems framework
3. the ability to incorporate the temporal dimension into classroom or training exercises
4. the experience of living with past decisions
5. the experience of group decision-making
6. the experience of objective setting, strategy formulation and execution
7. the exhibition of relationships among each of the functional areas of the business firm
8. the development of closer instructor-participant relationships

If any of these points represent objectives you are trying to achieve, please keep reading. The comparison and evaluation of the marketing games that follows will point up many other advantages and benefits to be gained from the use of marketing games.

PRELIMINARY CONSIDERATIONS

As we stated in the previous section, all of the marketing games described here are available from major publishers. You would order them as you would order a textbook for any class you are teaching. Each game has a teacher's manual that is available from the publisher. In addition, each game is computer scored, and the computer program (generally in *Fortran*

TABLE 1 General Game Information

Simulation Game	Publisher and Date of Publication	Cost per Student	Computer Language	Participants per Industry
Compete: A Dynamic Marketing Simulation	Business Publications, Inc. (1979)	$7.95	Fortran IV	9-20
Marketing Dynamics: Decision and Control	McGraw-Hill Book Company (1975)	$7.50	Fortran IV	12-16
Marketing in Action: A Decision Game	Richard D. Irwin, Inc. (1978)	$7.95	Fortran IV	9-24
Marketing Interaction: A Decision Game	PPC Books (1977)	$9.50	Fortran IV	9-20
Marketing Strategy: A Marketing Decision Game	Charles E. Merrill Publishing Company (1975)	$8.50	Fortran IV	9-20
Markism: A Marketing Decision Simulation	Harper & Row Publishers, Inc. (1964)	$7.50	Fortran	9-12
Markstrat	The Scientific Press (1977)	$8.35	BASIC	9-20
Operation Encounter	Goodyear Publishing Company, Inc. (1975)	$7.95	Fortran IV	9-20

IV or BASIC) is available on cards or tape from the publisher. You or your students, executives, or trainees have no knowledge of computers? It doesn't matter. None of these games requires any knowledge of computers or computer programming. So keep reading; more will be said about this later.

Table 1 summarizes some preliminary information about each of the games. The cost per student in Table 1 is the approximate price you could expect your university bookstore or other source to charge for the student manuals. There is no upper limit on the number of participants who could take part in each of these simulation competitions as more than one industry can be in operation at any time.

The Marketing Games

Table 2 provides a short description of the eight marketing games. Each game features different products and highlights different aspects of the firm's marketing mix. In all eight games, participants assume the roles of the marketing managers of a manufacturing concern selling consumer products. This is probably because most participants (particularly students) are more familiar with consumer than industrial products, and manufacturing concerns generally have a broader range of decision responsibilities than wholesalers or retailers.

Do you have any product or industry preferences for the game you would like to use? If your thing is electronic gadgets, you may want to consider *Compete.*[1] If you're a sports buff, you may be interested in *Marketing Dynamics.* Five of the games are very specific about the manufactured product or products, while three are not. The five that specifically name and describe the products are: *Compete* (TV videotape system, CB radio, and TV video game), *Marketing Dynamics* (cross-country, alpine, and water skis), *Marketing in Action* (cola, diet cola, and lemon-lime soft drinks), *Marketing Strategy* (three automobiles), and *Operation Encounter* (long-playing record albums). *Marketing Interaction* involves an unspecified clothing item, *Marksim* an unspecified consumer product, and *Markstrat* an unspecified electronic entertainment product.

There is, of course, some debate as to whether products should be specifically identified in marketing games. Some simulation users feel that specifically identifying the products makes the simulation more realistic and that it is more reasonable to expect participants to be able to establish sound strategies for known products. However, some simulation users feel that when known or identified products are involved participants too often will rely on their knowledge of the "real world" when making decisions in the simulation competition. According to this argument, it would be more appropriate, therefore, not to identify products and thus to encourage understanding of the simulated industry rather than any "real" industry. On the other hand, participants often feel uncomfortable when dealing with hazy product descriptions or unknown product areas. You will have to decide which of these alternatives you agree with as you decide on a marketing game to use.

TABLE 2 The Eight Marketing Games

Simulation Game	Summary Description
Compete: A Dynamic Marketing Simulation	Each team represents the top marketing managers of one of up to five companies manufacturing and selling to retailers any or all of three lines of audiovisual entertainment and communications products. Each marketing team is responsible for decision-making in the areas of product, price, place, and promotion.
Marketing Dynamics: Decision and Control	Each team represents the marketing managers of a manufacturer of skis (cross-country, alpine, and water) selling to retail outlets. Product selection decisions are highlighted. Each industry contains four companies.
Marketing in Action: A Decision Game	Each team represents the top management of one of three to six companies manufacturing one to three soft drinks sold on a regional basis to retail outlets. Product selection is an important decision area.
Marketing Interaction A Decision Game	Each team acts as the marketing managers of a manufacturer in the garment (product not otherwise specified) industry. Firms have choice of various channel systems to use in the distribution of their product. Each industry contains three to five companies.
Marketing Strategy: A Marketing Decision Game	The students comprise the marketing managers of one of five companies manufacturing three automobiles. The automobiles are sold through a dealership network for each of the three cars to three possible market segments.
Marksim: A Marketing Decision Simulation	Each firm represents the marketing managers for one of three companies in an industry manufacturing an unspecified consumer product. The product is sold through distribution centers and independent wholesalers to retail outlets.
Markstrat	Initially each company is one of five firms manufacturing two (more may be added after competition is underway) electronic entertainment products. The products are sold through three separate categories of retail outlets.
Operation Encounter	Each company in the competition is one of five firms in an industry producing and selling long-playing record albums to wholesale distributors. The selection of performing artists featuring different music styles is an important decision area.

Getting on the Computer

All of these games are computer scored. This simply means that a computer model has been devloped for each of these marketing games to evaluate the decisions of each participating team, determine their market performance, and print out each team's results. It should be noted that both *Marketing in Action* and *Marketing Dynamics* can be hand scored with calculators, but this is very time consuming and subject to error.

As we stated earlier, no computer knowledge or computer programming abilities are required to use these games. The computer is used simply as a fast and efficient method of scoring the participants' decisions. For each of the eight games, the computer program is supplied free, either on cards or tape, to the user.

In general, all the user needs to do with the computer is to get the program to the computer center or computer facilities and into the hands of someone who knows the computer system. This person will get the program fed into the system, where it will remain as long as you want to use it. Particularly good in this regard are *Compete* and *Markstrat.* The teacher's manuals for these two simulations provide step-by-step instructions on whom to see at the computer center, what to tell them, how to test the program to see that it is running properly, what to do if it is not, and whom to ask for help.

DECISION AREAS

In all likelihood, the most important factor influencing what marketing game you will use under particular circumstances and given your objectives is the type or range of decisions required in the game. Table 3 summarizes the decision areas involved in each game. An "X" in a cell corresponding to a particular decision area and game means that that particular game includes that decision. Let's see what each of the games involves.

Product

Remember, each team in each of the simulations is a manufacturer and, as such, has certain control over the product or product features they will be manufacturing. Some games, of course, allow more freedom of choice in the product area than others. For example, *Marketing Dynamics* and *Marketing in Action* would rate relatively high in terms of product choice decisions (indicated by an "H" in Table 3), while *Marketing Interaction* and *Marksim* would rate low in this regard. Table 4 explains briefly the product area decisions that are made in each of the games.

TABLE 3 Decisions Included in the Eight Games

	Simulation Games							
Decision Areas	*Compete*	*Marketing Dynamics*	*Marketing in Action*	*Marketing Interaction*	*Marketing Strategy*	*Marksim*	*Markstrat*	*Operation Encounter*
I. Product	M	H	H	L	M	L	M	M
A. Additions		X	X	X	X		X	X
B. Deletions	X	X	X	X	X		X	X
C. Improvements	X							
D. Changed features	X		X			X	X	
E. Sales forecasts	X	X	X		X	X	X	X
II. Price	H	M	M	H	M	L	M	M
A. Wholesale	X	X	X					X
B. Retail				X	X	X	X	
a. by period	X	X	X	X	X	X	X	X
b. by product	X	X	X	X	X		X	X
c. by region	X							
III. Place								
A. Channels				X		X		
B. Retail outlets					X		X	
C. Geographic area	X							
IV. Promotion	H	M	M	L	L	L	M	L
A. Advertising	X	X	X	X	X	X	X	X
a. Budget	X	X	X	X	X	X	X	X
b. Media	X			X				
c. By product	X		X		X		X	X
d. By region	X				X			
e. Advertising content	X							
B. Sales Force	X	X	X				X	
a. Size	X	X	X				X	
b. Geograhic allocation	X						X	
c. Method of payment	X							
d. Income level	X							
e. Sales presentation	X	X	X					
C. Sales Promotion	X					X		X
V. Brand Name Selection	X						X	X
VI. Marketing Research	H	H	H	L	M	L	H	H

H = high M = medium L = low X = decision area included

Price

The pricing area is generally a very important marketing mix element for most business firms. Thus, a marketing game that allows maximum freedom of decision-making in this area may be important to you. The two marketing games that allow participants maximum pricing flexibility are *Compete* and *Marketing Interaction. Compete,* for example, allows for nine separate pricing decisions, which may be altered in each round of competition. Pricing flexibility is most limited in *Marksim.* The pricing decisions made in each of the games are summarized in Table 5.

TABLE 4 Product Decisions in the Eight Games

Game	Summary of Product Area Decisions
Compete	Each company begins the game with three audiovisual products that are sold in three geographic regions. Any product can be deleted from the company offering entirely or from any of the regions. Product features may be changed to improve the products. Company management must develop a sales forecast for each product being sold in each region.
Marketing Dynamics	Each company begins the game with two products. The companies may choose to produce as many as eight separate types of skis from twenty-five possibilites available. Sales forecasts must be developed for each product produced.
Marketing in Action	Each company begins the game selling one cola soft drink. A diet cola and a lemon-lime soft drink can be added. Each product may be produced at any of five flavor levels, three sweetness levels, and two carbonation levels. Sales forecasts for each product sold are required.
Marketing Interaction	Each company begins the game with one unspecified consumer garment. After one year of operation, additional garments may be added to the product offering. No sales forecasts are developed.
Marketing Strategy	Each company may select any three from fifteen automobile models to produce. The models may be changed at the end of each year. Sales forecasts are required for each automobile produced.
Marksim	Each company manufactures one unspecified consumer product. The company may select what quality level the product level be. A sales forecast for the product is necessary.
Markstrat	Each company begins the game with two electronic entertainment products. Up to three additional products may be added to the original two. Sales forecasts are necessary for each product being sold.
Operation Encounter	Each company chooses to produce four types of long-playing records from twenty possibilities. Product offerings may be changed from time to time. Sales forecasts for each album produced are required.

TABLE 5 Pricing Decisions in the Eight Games

Game	Summary of Pricing Decisions
Compete	Nine wholesale prices (prices from manufacturer to retailers) are necessary. This represents a price for each of three products in each of three regions. Prices may be adjusted during each period of play.
Marketing Dynamics	A wholesale price must be established for each product sold. Prices may be adjusted during each period of competition.
Marketing in Action	A wholesale price (price to supermarkets) for each type of soft drink sold (by the case) is set each period. Prices may be adjusted on a month-by-month basis.
Marketing Interaction	A suggested retail price and channel markup must be set for each product. The suggested retail price is important in getting retailers to carry a firm's product. Price changes may be made but could result in lost sales when the manufacturer changes suggested retail price.
Marketing Strategy	A retail price (price to consumer) must be set for each car sold. Prices may be adjusted during each period of competition. In an advanced version, prices are also established by region.
Marksim	A retail price must be established each period for the firm's one product. The price may be changed at the beginning of each period.
Markstrat	Each company must establish recommended retail prices for each of its products. Prices may be adjusted each period.
Operation Encounter	Each company must set a wholesale price for each of its long-playing records. These prices may be changed during each period of competition.

Place

Place decisions pertain to the geographic area in which a company's products will be distributed, the channels of distribution that will be used, and/or the retail outlets through which the company's products will be sold. This undoubtedly represents the area in which the majority of the marketing games deviate most from reality by simplifying the distribution area. The reason for this, of course, is that the distribution activities of most firms are so complex that they are difficult to simulate without making the game too complex. In fact, three of the eight games, *Marketing Dynamics, Marketing in Action,* and *Operation Encounter,* do not allow for place decisions at all. Let's see what place or distribution decisions the others allow for.

Compete allows participants to select the geographic territories in which they will sell their products. A firm may choose to distribute in only selected geographic territories, for example, concentrating their resources for maximum effort in these territories. On the other hand, a firm may choose to distribute in all possible territories. This would spread out their resources but give the firm exposure to more potential customers. No decisions with regard to wholesalers or retailers that will be used are made, however. *Marketing Interaction* allows the companies to select from among twelve possible channels of distribution. Different channels represent the use of different wholesalers and agents and the reaching of different retail outlets. No geographic decisions are made, however.

Marketing Strategy allows the participants to determine how many dealerships will be used for each automobile produced. In an advanced version of this game, it is also possible to allocate dealerships by geographic territory. *Marksim* allows participants to determine what proportion of their products will be distributed through company-operated distribution centers versus independent wholesalers, but no geographic decisions are possible. *Markstrat* allows participants to determine how much sales effort their company will devote to each of three channels they are using but, again, does not allow for geographic decisions.

Promotion

The area of promotion includes all of the methods by which the business firm can communicate with its potential customers. Marketing literature generally identifies three forms of promotion. These are advertising, sales force activities, and sales promotional activities. Only one of the marketing games, *Compete,* allows participants to make decisions in each of these three areas. Two of the games, *Marketing Interaction* and *Marketing Strategy,* allow participants to make only advertising decisions while ignoring the sales force and sales promotional areas. Table 6 summarizes the promotional decisions required in each of the games.

Somewhat closely related to the area of promotion is the brand name or names that the business firm will give the products. After all, when a company advertises, it promotes the name of its products. Only three of the marketing games allow participants to select brand names for their products. These are *Compete, Markstrat,* and *Operation Encounter.* Of these three, only *Compete* provides a simple mechanism in the computer program by which you can evaluate the brand names, give them a weighting, and include this as part of the performance scoring. Why would you want to do this? For one thing, you may want to give your students, managers, or trainees an opportunity to select a brand name. Also, you might want to reward a well-conceived brand name selection. After all, wouldn't we all agree that Mustang is a much better brand name for an automobile than Edsel, and shouldn't that be rewarded? This adds a further touch of realism to the game and is an exercise that participants generally find interesting and rewarding.

TABLE 6 Promotional Decisions in the Eight Games

Game	Summary of Promotional Decisions
Compete	Participants establish the entire advertising budget and make allocations by product, region, and media. Participants also determine advertising content (i.e., the advertising message). Participants determine the sales force size, geographic allocation of salespeople, the method of paying sales force, the level of compensation, and how much time sales force devotes to each product. In addition, a budget must be established for sales promotional activities.
Marketing Dynamics	Participants establish only a total advertising budget. No allocation of dollars among products, media, or geographic territories is possible. Participants establish the total sales force size and the allocation of sales force time among the products. No sales promotional decisions are required.
Marketing in Action	Participants establish an advertising budget for each product that the company is selling. Participants also determine the total sales force size and percentage of time that the sales force will devote to each product. No sales promotional decisions are required.
Marketing Interaction	Participants determine expenditures for TV or three separate journals. No sales force or sales promotional decisions are made.
Marketing Strategy	Participants determine an advertising budget for each of their products. No sales force or sales promotional decisions are required. In an advanced version, the advertising budget must also be allocated by geographic region.
Marksim	Participants determine a total advertising budget and a total budget for sales promotional activities. No sales force decisions are made.
Markstrat	Participants must determine an advertising budget for each product sold. Participants determine the total sales force size and the allocation of salespeople by channel. No sales promotional decisions are required.
Operation Encounter	A total advertising budget and a total sales promotional budget must be established for each product sold. No sales force decisions are required.

Marketing Research

An important factor in game selection for many users is the amount of information available to participants. Why should you be concerned with this? It is important that participants do not guess or make decisions off the top of their heads. This would defeat the learning experience that the marketing game is supposed to provide. It is important, then, that the game participants have the opportunity to acquire proper information so they can make intelligent decisions. In fact, the decision area of what information to acquire may, itself, become an important decision.

Information for future decision-making is generally made available to participants in two forms. Some information is returned automatically every period to each competing company. We will say more about that later. The more important, and detailed, information for decision-making purposes takes the form of marketing research studies, which may be purchased out of company funds. The participants must determine what information is important to them for their decision-making purposes and whether it is worth what it will cost.

Table 7 summarizes the number of marketing research studies by study area available in the eight games. The numbers in the cells of the table show the number of studies available in the information area indicated. Because of the wide diversity of marketing research studies available in the eight games, it was not possible to be more specific in tabular form than this.

Among the eight marketing games, *Compete* has the highest number and greatest variety of marketing research studies available. Also ranking high in this regard are *Markstrat, Marketing in Action, Operation Encounter* and *Marketing Dynamics.* Lowest are *Marksim* and *Marketing Interaction.*

Compete offers nineteen separate research studies in such areas as economic forecasts, pricing studies, sales force studies, advertising studies, market share, and product information. The companies may also join an industry trade association that gives members periodic reports. The *Markstrat* simulation provides fifteen separate studies in many of the same areas but has no industry trade association. *Marketing in Action* makes available thirteen market research studies but no industry trade association. *Operation Encounter* provides twelve market research studies, including an industry trade association membership option. *Marketing Dynamics* also provides twelve studies and an industry newsletter.

What about the other games? *Marketing Strategy* provides eight studies in its regular version and eleven studies in its advanced version. This is adequate for most purposes. *Marksim* provides only five; this is not enough. *Marketing Interaction* provides only four different categories of market research studies, but there are several options within each category.

ADDITIONAL SELECTION VARIABLES

The decisions described in the previous section comprise some of the most important factors to consider in the selection of a marketing game. However, there may well be several other factors that you should consider before making your choice. First, what is the class or training program in which the game will be used? Table 8 indicates which of the eight games could be used in various classes or programs. A number of concept games that we will mention at the end of this essay would also be appropriate in many of these cases.

In addition to what we have already mentioned, there are still more factors you should consider in the selection of a marketing game. These include such things as participant output, instructor (or trainer) output, flexibility in usage, how realistic the game is, the readability and clearness of the game

TABLE 7 Number of Market Research Studies

Marketing Research Areas	*Compete*	*Marketing Dynamics*	*Marketing in Action*	*Marketing Interaction*	*Marketing Strategy*	*Marksim*	*Markstrat*	*Operation Encounter*
Sales forecasts	2	4	2		2	2	3	2
Product studies	2	6	4	1	2	1	6	2
Pricing studies	2	1	1	1				
Distribution studies	3		1	1	2		2	1
Advertising studies	6	1	3	1	2	2	2	5
Sales force studies	4		2				2	2
Total research studies	19	12	13	4	8	5	15	12

TABLE 8 Courses or Programs in Which the Eight Games May Be Used

	Simulation Games							
Courses/Programs	*Compete*	*Marketing Dynamics*	*Marketing in Action*	*Marketing Interaction*	*Marketing Strategy*	*Marksim*	*Markstrat*	*Operation Encounter*
Courses:								
Marketing principles	X	X	X	X	X	X		
Marketing cases/ problems	X	X	X	X	X	X	X	X
Marketing research	X	X					X	X
Sales management	X							
Advertising	X							
Product planning		X	X		X		X	
Consumer behavior								X
Distribution				X		X		
Marketing management	X	X	X	X	X		X	X
Programs:								
Management trainee	X	X	X	X	X		X	X
Sales training	X							
Executive development	X	X	X				X	

instructions, explanation of the model (and hints on improving game usage) in the teacher's manual, and general overall game quality.

Let's examine each of these issues which are summarized later, in Table 11, very briefly here.

Participant Output

In discussing participant output, we are concerned with the amount and clarity of the information automatically returned to each participating company at the end of each period. This does not include the market research studies, which we have already considered. Table 9 summarizes the types of participant output each game provides.

The best participant output is provided by *Markstrat* and *Compete.* The output they automatically return each period includes income statements (for *Compete,* regional income contributions as well), a balance sheet, market share for each product in each region on a quarterly and yearly basis, inventory updates, product feature or quality reports from the firm's research and development department, selected industry pricing information, and earnings per share information on a quarterly and yearly basis for all teams in the industry. The other games provide somewhat less information.

Instructor's (Trainer's) Output

Instructor's output refers to the amount of information provided to you by which you can evaluate the performance of the game participants. *Compete* and *Operation Encounter* rank the highest on this issue. These games provide you with financial, sales, and market share information for all competing firms (*Compete* provides this information on a regional as well as company-wide basis), a printout of all market research studies, as well as (for *Compete*) ratios and graphs evaluating current performance and illustrating trends. Table 10 summarizes the instructor (trainer) output each game provides.

Flexibility

Flexibility refers both to a game's ability to accommodate variable numbers of companies (and, therefore, participants) and to the game design, which allows you to modify the game to suit your tastes or requirements. In using games for university teaching purposes, it is often important to have this second form of flexibility so students cannot pass on information from one semester to the next. Certain games allow you to stop this practice by modifying or changing the environment of the game to create a new competition each semester.

Ranking high in flexibility are *Marketing in Action, Compete,* and *Marketing Interaction.* Each will accommodate variable numbers of participating companies. In addition, each of these games provides a separate parameter deck, apart from the main program, which allows you to modify the game in any way you desire by changing selected computer cards. The teacher's manual of each of these games provides an explanation of how to make any appropriate changes. *Marketing Dynamics* and *Marksim* provide the lowest level of flexibility.

Realism

Realism refers to how closely the game simulates "real world" business environment. None of the eight games has been rated "good" in Table 11 on this issue. The reason is, of course, that the real business environment is so complex. All of the games, therefore, make a number of simplifying assumptions so they will be understandable. Some, of course, are

TABLE 9 Survey of Participant Output

	Simulation Games							
Participation Output	*Compete*	*Marketing Dynamics*	*Marketing in Action*	*Marketing Interaction*	*Marketing Strategy*	*Marksim*	*Markstrat*	*Operation Encounter*
Balance sheet	X		X		X			X
Income statement	X	X	X	X	X	X	X	X
Regional (or product) contributions	X		X	X		X	X	X
Market share information	X	X			X	X	X	
Earnings per share information	X			X	X			X
Inventory updates	X	X				X	X	
R&D Feedback	X							X

TABLE 10 Summary of Instructor Output

	Simulation Games							
Instructor Output	*Compete*	*Marketing Dynamics*	*Marketing in Action*	*Marketing Interaction*	*Marketing Strategy*	*Marksim*	*Markstrat*	*Operation Encounter*
Sales summary	X	X	X	X	X	X	X	X
Market share summary	X	X		X	X	X	X	X
Financial summary	X		X		X	X	X	X
Decision area summary	X	X	X	X				
Charts/graphs	X							
Market research summary	X			X			X	X

more simplistic than others. If it were absolutely necessary to identify the most realistic among the eight games, *Compete* and *Markstrat* would be selected. The complex models for these two games create "realistically playing" competitive environments. *Marketing Dynamics, Marketing Interaction,* and *Marketing Strategy* create the least realistic (real world-like) environment. It should be noted, however, that the intention of these three games is to create a simple, easy-to-understand environment.

Readability

Readability refers to the clarity of the instructions to participants for play of the game and to the depth of the information. How easy is it to understand the game, and how well prepared are participants are for their first decision?

Four games have been given a rating of good on this issue. These are *Compete, Marketing in Action, Marketing Strategy,* and *Marksim.* Each of these games provides adequate, easy-to-read instructions. Two games, *Markstrat* and *Operation Encounter,* provide some difficulty in this regard. Their instructions are not entirely clear, and the participants are sometimes uneasy as they approach their first decision.

Teacher's Manual

A good teacher's manual can be very useful to you. A good teacher's manual should clearly explain the operation of the computer model, provide some advice on how to get the best results from the game, give some hints on how to improve the learning environment for your participants, explain how to evaluate the performance of the participants, and, in general, make the game as easy as possible for you to administer.

Which of the games does this best? The best teacher's manuals come with *Compete, Marketing in Action,* and *Markstrat.* The teacher's manual for *Marketing Interaction* seems to fall shortest from these ideals.

Overall Quality

The overall quality simply refers to how good a marketing game it is. This is, of course, a difficult assessment to make. Certainly, depending on your objectives, different games may rate best. However, based on all of the issues we have explored throughout this article, *Markstrat* and *Compete* would be rated as having the highest overall quality. This conclusion is based on the degree of realism of these two games, their playability, the aid they provide game administrators, the extent of decisions, and the output they provide.

Table 11 summarizes all of these additional points.

CONCEPT GAMES

You should keep one final point in mind before we go on to our concluding remarks. We have examined only eight from a multitude of existing marketing games. Many marketing games are concept games, that is, games that focus on one small aspect of the marketing process. For example, there are games like *ADMAG* for those interested in the area of advertising only, *Marketing a New Product* (for a focus on product planning only), and *The Sales Management Game* for that area. Of course, there are many other games beyond this. If you are interested in a marketing game of almost any kind, just look around. You can surely find one to suit your purposes.

CONCLUDING REMARKS

We have addressed many issues throughout this essay. The eight games we have described in detail are all good games. They have been designed to serve different purposes. Which should you use? That's up to you. You have to decide what you want from the use of a marketing game and select the game that best suits your purpose. This essay should have provided you with some useful guidelines for selecting a game. A few final comments—in alphabetical order—may be helpful.

If you want a game that provides the widest variety of marketing decisions and greatest availability of marketing research studies, select *Compete.* If you want a game that offers great product variety, has a simple model, and can be hand scored if necessary, select *Marketing Dynamics.* If you want a game that allows you to manipulate product characteristics in an attempt to identify the "right" product and involves a product everyone is familiar with (soft drinks), select *Marketing in Action.* You say that you want a game oriented to channel selection decisions that does not get you bogged down in sales forecasts? Try *Marketing Interaction.* Would you like to try a game that lets participants start out in a simple environment and, once they get the hang of it, progress to a

TABLE 11 Selected Issues in Game Choice

	Issues						
Simulation Game	*Participant Output*	*Instructor's Output*	*Flexibility*	*Realism*	*Readability*	*Teacher's Manual*	*Overall Quality*
Compete	G	G	G	A	G	G	G
Marketing Dynamics	A	B	B	B	A	A	A
Marketing in Action	A	A	G	A	G	G	A
Marketing Interaction	A	A	G	B	A	B	A
Marketing Strategy	A	A	A	B	G	A	A
Marksim	A	B	B	A	G	A	A
Markstrat	G	A	A	A	B	G	G
Operation Encounter	A	G	A	A	B	A	A

G = good A = average B = below average

more complicated environement? *Marketing Strategy* will let you do this. How about a game that serves as almost a mini-marketing textbook providing detailed instructions to the participant? *Marksim* will do this. Are you interested in teaching market segmentation and exploring product/market matchups? If so, use *Markstrat.* Would you like to get into show business? *Operation Encounter* will give you a taste of it.

In the selection of a marketing game, please keep this final point in mind. All of the decisions made in the various games as described throughout this essay are made for the same purpose. In each of the eight games, participants are competing for sales to the same markets. The companies are fighting for market share and profitable operations in a limited market. As you decide which marketing game to use, determine for yourself which game provides the most appropriate decision-making environment to insure that well-thought-out strategies, and not luck, will result in profitable operations and the determination of the best performing team.

Note

1. *Compete* was co-designed by the author. Most of the material for evaluating this game came from two published reviews: one by Dr. James B. Cloonan of Quantitative Decision Systems, Inc., Chicago, in the *Journal of Marketing*; the other by Dr. J. Daniel Cougar, University of Colorado, Colorado Springs, in *The Computing Newsletter.*

MILITARY HISTORY GAMES
An Evaluation

by Martin C. Campion

In this essay, I want to try to help you find one or more games in the area of military history that will be useful to you in your classroom. Because subject matter is less flexible in history courses than it is for games in other disciplines, I have cast my net as widely as possible by assuming that the course for which you are game shopping is a Western civilization survey course or a general Western military history course. With this assumption, you ought to find something appropriate here no matter what your actual course is. I might have posited some true world history courses, but it would have done no good, as there are no military history games that deal with non-Western history, that is, history beyond the scope of that traditional history of Western civilization, which moves in a straight line from Ancient Egypt and Greece to the modern European-shaped world.

Before we go into the specific games I will recommend, you may want to know something generally about military history simulations, their appearance, availability, and use.

WHAT'S IN A GAME

Military historians and theorists talk about tactics, grand tactics, strategy, and grand strategy as subdivisions of military operations. Tactics is the art of moving troops around on a battlefield and coordinating the use of their weapons. Grand tactics is the art of moving and fighting large formations (battalions, regiments, and divisions). Strategy is the art of conducting a campaign, which is a series of maneuvers and battles over an extended time but in a limited area, as in one European country. Grand strategy is the art of conducting a war. Strategy and grand strategy both deal with political as well as military values, and in grand strategy, at least, politics frequently becomes the most important ingredient. Of course, all of these categories have very indistinct boundaries.

We can find wargames in and between each of these categories. That is to say, they differ in scale. The typical wargame is played on a map about two feet by three feet in size (but some are two to five times this size). In one game, a map of this size shows a few city blocks. In another, it shows the Pacific Ocean.

In most cases, the map is printed with a hexagon grid, and usually the geography of the map has been slightly distorted to conform to the grid and aid its purpose, which is to measure movement and, in some cases, the ranges of weapons. During the game, all distances on the map are measured in hexagons, which are usually about three-fourths of an inch in size.

The maps show various kinds of terrain, simplified to aid the game. Tactical game maps show hills, streams, woods, swamps, villages, roads, canals, and bridges. As the scale grows smaller, streams, villages, and roads disappear to be replaced by rivers, cities, and railroads. Woods and hills tend to merge into a generalized "rough terrain." The smallest scale maps show only land masses, oceans and seas, mountains, and political boundaries.

The typical wargame comes with a sheet of die-cut cardboard counters. Each counter is one-half inch square and is usually printed with at least this information: kind of unit represented, combat power (expressed as single number), and movement ability (expressed as another number). Sometimes there is more information, like ranges of weapons in tactical games. In some games, all units have the same movement power and so only the combat power is printed on the counter. Sometimes the counter represents a historical unit, like a division or squad. Sometimes, in games of grand strategy, it represents an abstract point value of men, without showing what kind of units those men were organized in.

Finally, the typical wargame has a set of rules with accompanying charts and tables, which try to show how the counters are moved and fight on the hexagon grid. The rules are usually very explicit and legalistic. This is because the games are designed to be played by two people, without an umpire, and because wargamers, in the heat of battle, sometimes become outrageously literal in their interpretation of the rules. So rules are written like contracts and, to be understood, they have to be read with full attention. It is usually better not to expect any new gamers to read the rules. Games can be explained with the help of a map much more easily than they can be explained in writing alone. However, the person who thus introduces a new person to a wargame must know it thoroughly himself and must therefore have spent two to five hours learning the typical game.

AVAILABILITY AND USE

There are over two hundred military history simulations currently in print. None of them was designed for educational use, but many of them have appeared in classrooms, mine and those of other people who teach history in junior high schools, high schools, community colleges, and colleges. We use them for the same general reasons that other teachers use other kinds of games. Games give an immediacy to a subject that cannot be had in any other way. This immediacy is a particularly important quality for courses in history, a subject that often seems peculiarly remote, and therefore dull, to students. Games motivate students to do other, more traditional, tasks (like reading) on the same subjects as the games played. They help loosen up the classroom and help students and teachers get better acquainted with each other. Military history simulations are particularly useful for conveying to students an understanding of the limitations on the actions of people trying to do their best in a complex situation in which they do not have sufficient information to make purely logical decisions.

Military history simulations are thus used for the same reasons as other educational simulations, but because they have not been designed as classroom experiences, they present some unusual problems. Often they have to be expanded for the classroom from two-player games to multiplayer games. Teaching tactics, collateral reading, and debriefing exercises all have to be developed by the individual teacher. Furthermore, the games are much more complex than other educational games and have to be simplified or at least taught with more than ordinary patience.

There are many ways you might use military history games in your classes. You can use the published versions in assignments outside class if you are willing to finance and operate a lending library. They can be used in special "lab" periods outside regular class time, or they can be used in the regular class period by the whole class or by a smaller, perhaps voluntary, group. I use semivoluntary groups in my American history surveys, in "War in Western Civilization," and in "World War II." Students in these classes are expected to do eight to fifteen media reports on their experiences with films, filmstrips, tapes, or games. Because of the greater investment of time necessary to play games, reports on games are worth more than reports on other media.

Criteria for Selection

My first thought in selecting the twenty (nineteen, because one has gone out of print) "best" military history simulations was that I wanted as wide a spread of subjects as possible—I wanted to cover the course that I was selecting games for. A look at Table 1, which lists and briefly describes each game, will show you how well this desire was fulfilled. Also, I looked

TABLE 1 Chronological/Brief Description

Game	Dates	Summary Description
Raphia	217 B.C.	Battle game. Seleucid army of phalanxes, light infantry, cavalry, and elephants meet to Ptolemaic army with similar units. One leader on each side adds to combat effectiveness. Six-turn game. Victory points are given for destruction of enemy units.
The Conquerors: The Romans	200 – 197 B.C. or 192 – 189 B.C.	Grand strategy game with battles. The Romans at war with either Macedonians or Seleucids or with all three powers involved in shifting alliances. Both sides get varied income from taxation and looting. Money can be used to buy armies, fleets, and allies, and to supply armies. "Augury table" has events that help or hinder the Roman player. Either game lasts twenty-six game turns. Some battles are decided with separate battle game on separate battle board between varied armies. Victory is decided by victory points, amassed mainly by conquering cities.
Decline and Fall	375 – 450 A.D.	Four-player diplomatic and grand strategy game. The Roman Empires tries to hold onto its territory while being pressed by the Goths and the Vandals, who are in turn being pressed by the Huns. Players may make many kinds of deals with each other or go to war or both. Forces and victory conditions are quite different for each player. The Romans get points for retaining cities; the Goths and Vandals for looting cities and for settling territory; and the Huns for looting or destroying enemy forces. Play proceeds serially for fifteen complete turns.
Kingmaker	c. 1450 – c. 1490	Two- to seven-player diplomatic and grand strategy game. Each player represents a faction that is trying to have its candidate for king reign supreme in the English War of the Roses. Factions are formed and reinforced by the draw of cards. Drawing from a deck of event cards introduces many problems. The balance of power changes constantly as nobles are killed and their heirs turn up in the enemy's camp. Play is serial and ends when all potential kings are dead save one. The player who controls that one is the winner.
Conquistador	1495 – 1600	One- to five-player grand strategy game on exploring, exploiting, and fighting over the American continents. Players are, in order of importance, Spain, England, France, Portugal, and the German bankers. The game speeds up the involvement of England and France in the New World for greater competition. The aim of players is to make discoveries, control areas, and amass gold. The game is played serially for twenty-one complete turns.
Thirty Years War: Rocroi	1643	Battle game on the decisive battle between the French and the Spanish which symbolized the decline of Spain as a great power and the rise of France to preeminance.

TABLE 1 Chronological/Brief Description (Cont)

Game	Dates	Summary Description
		Leaders, infantry, artillery, and cavalry start the game deployed. The game lasts fourteen turns and is won by destroying enemy units, eliminating their leaders, and demoralizing their troops.
Frederick the Great	1756, 1757, 1758, or 1759	Strategy game on one to four of the most significant campaign years of the Seven Years' War in Central Europe. Prussian and Hanoverian forces in the middle are encircled by the more numerous but divided and poorly led forces of France, Austria, Sweden, and Russia. The powers of leaders deployed on the map influence greatly the ability of their armies to march, fight, and maintain morale. Armies spend much time besieging fortresses and worrying about supply lines and supply depots. Games last ten to eighteen turns and are won by eliminating enemy forces and capturing fortresses.
1776	1776 – 1781 in various combinations	Strategic or grand strategic game on the American Revolutionary War. Players may fight the whole war for two years or five years and be concerned with raising troops, garrisoning strategic towns, winter problems, and naval welfare. Or players may fight a simple military struggle in a limited area over five to nine turns (months). Militarily, the game uses infantry, cavalry, artillery, fortresses, entrenchments, supply units, magazines, decoys for semisecret movement, and bateaux for river movement. In the longer games, the British win by occupying strategic towns. Otherwise the Americans win by surviving.
Napoleon's Last Battles	1815	Two-player battle or strategy/battle game. Players take either Napoleon's French army or the opposing armies and fight either the last part of the Waterloo campaign or any of the four battles that occured during the campaign. In the battle games, infantry, brigades and regiments, cavalry divisions, and artillery units attempt for eight turns (hours) to destroy enemy units and cause their army to become demoralized or to disintegrate. In the large game, players become concerned about the qualities of leaders, reorganization of previously destroyed units, and supply, and the game lasts for thirty-six turns (three days).
The American Civil War	1861 – 1865	Two-player grand strategy game. The players manipulate leaders, armies, naval and riverine units, moving by land (marching or by railroad) or sea. Players must worry about finding good generals, raising troops, supplying armies, nonbattle attrition. The Union wins within sixteen turns (three months each) or the Confederacy wins by surviving.
Blue & Gray: Chickamauga	1863	Battle game on the nearly decisive defeat of the Union army in the West in 1863. Union and Confederate infantry, cavalry, and artillery maneuver over very rough terrain. The Union player attempts to pull his scattered army together and retreat in good order while the Confederate player tries to destroy him in detail or cut off his retreat. Both players try to eliminate enemy units. The game lasts for fifteen turns, representing two days and a night.
Diplomacy	1900 on	Seven-player diplomatic and grand strategy game of shifting alliances and eventual European domination. The game mirrors no particular historical situation but uses a map of early-twentieth century Europe, and its action often includes the formation of alliances and counter-alliances, naval arms races, conflicting spheres of interest, and other characteristics of that and other periods. Movement is simultaneous for an indeterminate length.
World War I	1914 – 1918	Two-player grand strategy game. The players use "combat resource points" to build armies and take losses. The seven major and seven minor powers have separate armies and combat resource points. Most units are infantry armies. The Germans have shock armies ("Stosstruppen") and the French one optional tank army. New nations enter the war by a timetable or by getting invaded. The game lasts for ten half-year turns.
Soldiers	1914 – 1915	One- or two-player tactical game. Each player commands a small allied or central power force in a portion of a major battle. Units are infantry companies, machine gun sections and companies, cavalry squadrons, and field gun and howitzer batteries or platoons. Counters in the mix represent British, French, Japanese, German, Belgian, Russian, and Austro-Hungarian units. The fourteen scenarios last from ten to twenty turns each.
World War II	1939 – 1945	Two- or three-player grand strategy game played on a map of Europe, North Africa, and the Middle East with counters representing the forces of six major and twelve minor powers. Counters represent only land units with abstract sea and air power. The game uses infantry, mechanized, garrison, parachute, and partisan units. Germans may vary their armed forces by varied spending of the points they get each turn. The

TABLE 1 Chronological/Brief Description (Cont)

Game	Dates	Summary Description
		extent of Russian reinforcements depends on the number of resource centers they hold. Other nations' reinforcements arrive on schedule. Political rules govern the entry of some neutrals and the creation of Vichy France. The game lasts twenty-three turns representing three months each.
The Fall of Tobruk	1942	Two-player battle game showing the tactical relationships that decided the Battle of Gazala outside Tobruk in 1942. Infantry, artillery, and tank units, the latter identified by the type of tank that predominated in the unit, maneuver on a board dominated by the man-made terrain of minefields. The game is often decided by the German ability to recover wrecked tanks faster than the Allies. The game lasts thirty-two turns (days).
Barbarossa	1941 – 1945	Two-player strategy game on the Russo-German War. German and Russian armored groups and armies fight either a single year's campaign or the whole war from the invasion to the beginning of the Battle of Germany. Reinforcements arrive by time table. As the war continues, the Russians benefit from rules changes that represent their increasing skill as well as power. Supply is very important. Each move represents one month.
Fast Carriers	1942	Two-player game on the great carrier battles of 1942 in the Pacific. Two small scale maps show the Coral Sea/Guadalcanal area and the Midway area. Counters represent individual ships and groups of six aircraft of a variety of types. Players maneuver fleets, search for enemy, send air strikes, and then fly in on a tactical board for the plane-to-plane and plane-to-ship combat. Scenarios last eighteen to twenty-four turns (three to four days).
Strikeforce One	1970s	A simple two-player game, supposedly about a Russian-American confrontation in a small part of Germany. Its only virtue is that it is fast and can be used to teach some basic wargaming ideas quickly, but it is very good for that task.

for games that dealt with a multiplicity of the factors that influenced military affairs without undue complexity. I wanted games that reflected a variety of the conditions of the simulated time: political, economic, strategic, tactical, emotional, and technological. Although the games would not be designed specifically for classroom use, they had to be adaptable to such use. They had to be accurate (within the appropriate scale), preferably more concrete than abstract in their representations of reality, and they had to be physically attractive if possible. Finally, they had to be games I knew well. For the most part, these games are among those I have used in my own classes. If not, they are games I have played extensively and are similar to games I have used in my classes.

In several cases, I have been tempted to abandon a game on this list in favor of newer games that looked as though they might be better. Generally, I have avoided the temptation because I was unable to arrange enough classroom time to be sure that the newer game was even as good as the older.

COMPARING THE GAMES: GENERAL CONSIDERATIONS

There are many things that have to be considered when picking out a game for a class. When you consulted Table 1 you received an idea of the number of games on my list that are appropriate for your course's subject. The next thing to consider is the difficulty of the games, their scales, and the number of people each can accommodate.

Complexity

Table 2 lists one game as "simple" and the rest as possessing various degrees of difficulty. *Strikeforce One* is included in my game list for one reason only. Although it has no historical content and is not even very interesting as a game, it is excellent as an introduction to most wargames in general and to most of the games on my list. So you might consider using it, no matter what the subject of your course, as a way of getting your class started on simulations and familiar with some of the basic concepts of wargames. The rest of the games are "difficult," but that need not unduly discourage you from using them—at least not out of fear that the students will be unable to play them.

It is necessary for you, as the teacher, to learn one of these games well if you want to use it, but it is not necessary for the students to learn the game equally well in order to play it under your direction. Indeed, it may be better, for instructional purposes, if the class does not know the game too well.

Normally, there comes a time when I am introducing a new class to simulations, right after I have explained the rules once over lightly, when the students start looking at me with a particular expression in their eyes. I know what they are thinking—sometimes they even say it. They are thinking, "You are crazy if you think we can play this thing." At that point I tell them, "It doesn't matter if you don't know what's going on. The historical characters didn't know exactly what was going on either. Your confusion will be analogous to theirs." Having said that, I push on and get them actually playing.

When the game starts, I may have to push them around a bit, explain several points about the rules several times, and explain possible courses of action to puzzled generals. I may even have to intervene to prevent the commission of silly and disastrous mistakes, like those that arise from ignoring the rules about supply. Finally, I may have to simply tell some of the students what to do. Whatever the need for intervention at the beginning of a game, that need declines and usually disappears very quickly.

One encouraging example may help. The first time I used *Conquistador,* probably the most difficult game on my list, I used it with a volunteer group from my American history survey. Because this is a lower-level course and I offered it for extra credit and it was near the end of the semester, I got a small selection of freshmen and sophomores, most of whom had not done very well in the other work of the course. As one might expect, they had to be guided carefully through the first few moves. By the end of the experience, however, they were all formulating their own strategies, still not very skillfully perhaps, but with a growing understanding of the game and of the reality it was supposed to reflect.

Scale

The other characteristic shown in Table 2 is the scale of the various games. The range is wide. The player can be involved in manipulating forces as diverse as the company units of *Soldiers* on the one hand, and the armies and army groups of *World War I* and *World War II* on the other.

The scale you prefer depends on what you think needs to be studied most. My preference is obvious from the table. Although most of the military history games available are grand tactical, campaign, and strategy games, most of the games I have picked are grand strategy and tactical games. I have picked probably a proportionate number of tactical games but a disproportionate number of other kinds. Most of my teaching of military history centers on the analysis of the conditions of warfare, including the interaction of weapons and tactics, in a certain period rather than the narrative of operations in that period. Grand strategy and strategy games give a better picture of the whole military situation of a period than do the other games, while weapons are dealt with best in tactical games.

NUMBER OF PLAYERS AND GAME EXPANSION

The table that shows the number of players (Table 3) requires some explanation. In their original form, even the games using the largest numbers of players would fit the requirements of only the smallest college and high school classes. Most of the games are designed for a very inadequate two players. Of course, you can simply divide the class into groups of two, and purchase enough copies of the game you want to use to go around, or set the different groups to playing different games. I have tried doing this and I don't recommend it for two main reasons. You tend to lose touch with what the students are doing and learning when many groups are operating, and you lose the opportunity to use the games to simulate more conditions of warfare than the original two-person games could fit in.

Later, I will suggest some ways of expanding games by adding a few simple rules. Right now I will only suggest two ways of adding players—by either the committee system or role expansion. The committee system allows two or three people (larger committees are too unwieldy) to function as one. Each person now has someone to talk to and help in formulating plans. The committee system is good for complex games and games that are already designed for a large number of players. Role expansion, the other option, means that a game position is divided into several functions and each player is given one function to perform. This system represents a significant improvement in realism over most original games. Instead of having one person act in many roles simultaneously (as commander-in-chief and several subordinates, for example), give several players these roles. This action leads to a very realistic decline in efficiency as communication problems arise and personalities clash.

Table 3 lists the number of players for which each game was designed, suggests which of the two systems is better for expanding the number of players, gives details on how the expansion could be implemented, and lists the expanded number of players.

The table does not indicate some of the possible difficulties of game expansion. Any time you attempt to use more than

TABLE 2 Games Classified by Simulation Scale and Difficulty

	Tactical	Grand Tactical	Campaign	Strategy	Grand Strategy
Simple		*Strikeforce One*			
Least difficult	*Soldiers*	*Raphia* *Thirty Years War: Rocroi* *Napoleon's Last Battles* *Blue & Gray: Chickamauga* *Fall of Tobruk*			*Decline and Fall* *Kingmaker* *Diplomacy* *World War I*
Moderately difficult	*The Congurors: The Romans*		*Napoleon's Last Battles*	*Barbarossa*	*The Conquerors: The Romans* *1776* *World War II*
Difficult	*The Fast Carriers*		*The Fast Carriers*	*Frederick the Great*	*The American Civil War*
Very difficult					*Conquistador*

TABLE 3 Game Player Expansion Suggestions

Game	No. of Players Designed For	Best Method of Expansion	Expanded Positions	Total Players in Expanded Game
Raphia	2	role expansion	Each army: king, commander of the center, commander of the right flank, commander of the left flank	8
The Conquerors: The Romans	2-3	role expansion	One player per army, fleet, legion, etc. Use leader counters with game.	10-20
Decline & Fall	4	role expansion-committee system	Divide Roman position into East and West; then use committee of three for each position. One should act as ambassador.	15
Kingmaker	2-7	committee system	Use committee of two or three for each position	6-21
Conquistador	2-5	committee system	Varied sizes of committees: Spain (four to five), England (three), France (three), Portugal (two), German Bankers (two)	14-15
Thirty Years War: Rocroi	2	role expansion	Use one student to identify with each leader counter in the game—four on each side.	8
Frederick the Great	2	role expansion	One student as commander of each major army, usually three Prussian, one Hanoverian, one French, two Austrian, one German, two Russian, or use one commander and one second-in-command.	10-20
Napoleon's Last Battles	2	role expansion	Many leader counters come with the game. At the least, each major army gets a player: Napoleon, Wellington, Blücher; then add the others defined in the game as "commanders"—Ney, Grouchy, and Orange; then add five French corps "officers," the British II Corps "officer" (Hill), nine British "officers" in charge of the large divisions, the Brunswick Corps, and the cavalry, and the four Prussian corps "officers." Expansion should be done proportionally if there are not enough students to fill any level.	3-27
1776	2	role expansion	As in *Frederick the Great*, but the number of armies is varied.	6-24
The American Civil War	2	role expansion-committee system	On each side, have one player in charge of the East, one of the west. The Union also should have a player in charge of amphibious campaigns. Then each side gets a player in charge of mediating (Lincoln or Davis). After this, each of the military players can use up to two assistants.	7-17
Blue & Gray: Chickamauga	2	role expansion	The Union operated with one commander, four infantry and one cavalry corps commanders, giving the Union six leaders. The Confederates had a commander, and five infantry and one cavalry corps commanders. On the second day the Confederates also used two wing commanders operating between the commander-in-chief and the corps commanders. This gives the Confederates a maximum of nine leaders.	15
Diplomacy	7	committee system	Each position could use up to three players. With all three negotiating with different foreign powers, some interesting cross purposes can develop.	7-21
World War I	2	role expansion	First separate the Western allies from Russia for a three-player game. Then add a separate Austro-Turkish, then an Italian, then separate Britain and France.	6
Soldiers	2	role expansion	Use one commander on each side and two or three subordinates depending on the scenario.	6-8
World War II	2	role expansion-committee system	First designate one player for each major power: Britain, France, Russia, Germany and Italy. The French player becomes the U.S. player later. After that add one or two assistants to each position.	5-15

TABLE 3 Game Player Expansion Suggestions (Cont)

Game	No. of Players Designed For	Best Method of Expansion	Expanded Positions	Total Players in Expanded Game
The Fall of Tobruk	2	role expansion	Use one commander-in-chief and three corps commanders on each side. (You can use the historical corps, except that you have to combine two Italian infantry corps.)	8
Barbarossa	2	role expansion	Use a Northern, Center, and Southern commander on each side, plus a commander-in-chief, and perhaps political leaders (Hitler and Stalin) for each side.	8-10
The Fast Carriers	2	role expansion	In each scenario, use one task force commander for each task force, one air officer for each carrier task force, one air officer for all ground-based air, and an overall commander.	14-18
Strikeforce One	2	committee system	Use two players on each side. After that just use more game sets.	4

eight or ten players in a single board game, you will have a traffic control or seating problem around the board. You may have to forbid the players to pull chairs up to the table, or you may have to time periods of access to the board. You undoubtedly will find that games will last longer with more people attempting to make decisions. If you attempt to implement role expansion, you may find some of the students trying to circumvent your system. So you will have to forbid political leaders to move armies, or corps commanders to move the units of other corps commanders.

In many games, you should have your players put themselves on the map. *Rocroi* uses eight leader counters, which can represent your eight players. For other games, you can make extra counters. (Blank counters are available in four color sheets from Simulations Publications, Inc.)

WHAT THE GAMES SIMULATE

Before you try to pick out a specific wargame to use, you might want to find out what areas of reality the game simulates. Tables 4 and 5 describe the twenty games on the list in terms of their representations of geography and human activity.

Time and Space

Table 4 details the chronological, spatial, and human scales of each game and lists the kinds of terrain that are shown on its board. I have used the game move as a kind of standard for the chronological scale, but you should know that the standard is an elastic one, because moves themselves consume varied amounts of real time. A move in a relatively simple game like *Rocroi* can be over in fifteen minutes, even with multiple commanders on each side. A move in *Conquistador* can easily consume an hour. *The Romans* and *Fast Carriers* both contain subgames in which lengthy battles are fought on tactical maps on a large scale while time in frozen in the main game.

The Hexagon Grid

The hexagon, which I am using as a spatial standard, is really a standard. Most wargame maps are printed with a hexagon grid, which is used to measure movement and range. On most of the grids, the hexagons measure about three-fourths of an inch from side to side. The hexagon grid is one of the things that makes wargaming possible. It smooths out the problems of moving through varied terrain. It obviates the use of any other kind of measuring device, which use would be difficult on a crowded mapboard. It makes it possible to represent various formations of men and various densities of armies without various sized counters.

The use of a hexagon grid also entails some disadvantages. Movement that is not in line with the pattern of the grid is distorted. The map itself must be distorted to take advantage of the pattern of the grid. Terrain features must completely fill a hexagon or be ineffective. Rivers and boundary lines are ordinarily forced to run along hexagon sides. Nevertheless, the advantages of the hexagon grid outweigh the disadvantages by a large amount.

However, there are two games on the list which do not use it, *Diplomacy* because movement and combat are very abstractly represented in the game and the precision of hexagons is not needed, and *Kingmaker* because the designer used an alternate system of making areas on the board different sizes to represent different degrees of difficulty in the terrain.

Terrain

The right part of Table 4 lists the kinds of terrain and man-made features shown on the various mapboards. Most representations of terrain involve some kind of compromise with reality. Rivers are ordinarily represented as all equally difficult. At the most, there are three widths of river, as on the *American Civil War* map. "Rough terrain" is a popular catchall as a terrain feature. On the *Barbarossa* map, "rough terrain" is shown for what are in reality mountains, marshes, hills, and forests. Other games treat important terrain features even more cavalierly. *World War II* contains no rivers. Whatever

TABLE 4 Time/Space/Unit Representation Scales/Terrain Representation

Game	Time per Move	Space per Hexagon	No. of People per Unit or Name of Most Common Unit	ocean/sea	rivers	streams/creeks	canal	lakes	forest woods	rough	hill/crest/slope	mountains	mountain passes	swamp/marsh	desert	escarpment	cities	towns	villages	ports/anchorages	roads	trails	railroads	ferries	bridges	fords	walls	fortress/fortified city	castles	minefields	chateaux	cathedrals	gold mines	treasure cities	resource centers
Raphia	uncertain	uncertain	500-3000							X																									
The Conquerors: The Romans	1 month	20 miles	700-900	X	X			X		X				X		X			X	X								X							
Decline and Fall	5 years	c. 130 miles	20,000-80,000	X	X							X		X	X		X										X	X							
Kingmaker	indeterminate	various sized areas	10-200	X					X								X	X		X	X							X	X			X			
Conquistador	5 years	c. 160 miles	uncertain	X	X			X	X	X		X																					X	X	
Thirty Years War: Rocroi	45 minutes	175 meters	75-100			X			X												X														
Frederick the Great	15 days	c. 16 miles	c. 2500		X			X				X	X															X							
1776	1 month	c. 19 miles	c. 500	X	X			X		X				X				X		X				X	X	X									
Napoleon's Last Battles	1 hour	480 meters	700-2000		X	X			X		X			X			X			X	X			X							X				
American Civil War	3 months	25 miles	uncertain	X	X					X				X			X			X			X	X	X			X							X
Blue and Gray: Chickamauga	c. 1-2 hours	400 meters	750-2200			X			X	X											X		X		X	X									
Diplomacy	6 months	various sized areas	uncertain	X																															X
World War I	6 months	70 kilometers	army	X						X							X						X					X							X
Soldiers	10 minutes	100 meters	company				X		X		X							X			X				X										
World War II	3 months	120 miles	army	X						X		X																							X
Fall of Tobruk	1 day	c. 2 miles	battalion/ company	X							X					X			X		X	X								X					
Barbarossa	1 month	c. 65 miles	army	X	X			X		X							X																		
Fast Carriers	4 hours	1000 yards/ 90 nautical miles	6 planes/ 1 ship	X																															
Strikeforce One	uncertain	uncertain	company						X									X																	

TABLE 5 Actions and Conditions Simulated/Degree of Abstraction and Concreteness

	Raphia	*The Conquerors: The Romans*	*Decline and Fall*	*Kingmaker*	*Conquistador*	*Thirty Years War: Rocroi*	*Frederick the Great*	*1776*	*Napoleon's Last Battles*	*American Civil War*	*Blue and Gray: Chickamauga*	*Diplomacy*	*World War I*	*Soldiers*	*World War II*	*Fall of Tobruk*	*Barbarossa*	*Fast Carriers*	*Strikeforce One*
I. Background of War																			
Population growth	0	0	2	0	2	0	0	0	0	0	0	0	0	0	0	0	0	0	0
Colonization	0	0	2	0	3	0	0	0	0	0	0	0	0	0	0	0	0	0	0
Resistance of natives to colonization	0	0	1	0	1	0	0	0	0	0	0	0	0	0	0	0	0	0	0
Diplomacy	0	1-3	3	3	3	0	0	0	0	0	0	5	1	0	2	0	0	0	0
Relations with equal allies	0	0/3	3	3	3	0	2	2	2	0	0	5	1	0	3	0	0	0	0
Relations with satellites	0	2	0	0	0	0	0	0	0	0	0	5	3	0	0	0	1	0	0
Need to prepare for peace	0	1	2	0	3	0	1	1	0	0	0	0	1	0	1	0	0	0	0
Unexpected outside events	0	1	1	3	2	0	0	0	0	1	0	0	0	0	0	0	0	0	0
Financial basis of war, taxation	0	2	2	0	3	0	1	1	0	1	0	2	3	0	1	0	0	0	0
Civilian population losses	0	1	1	0	3	0	0	0	0	0	0	0	0	0	0	0	0	0	0
Exploration of new lands	0	0	0	0	3	0	0	0	0	0	0	0	0	0	0	0	0	0	0
Missionary activities	0	0	0	0	1	0	0	0	0	0	0	0	0	0	0	0	0	0	0
Search for gold and treasure	0	0	0	0	3	0	0	0	0	0	0	0	0	0	0	0	0	0	0
Entry of neutrals into the war	0	2	0	0	0	0	0	0	0	1	0	4	4	0	3	0	0	0	0
Importance of political capitals	0	3	3	2	0	0	0	0	0	3	0	0	0	0	0	0	0	0	0
Political changes in major countries	0	1	2	3	1	0	0	0	0	2	0	0	2	0	2	0	0	0	0
II. Building the Tools of War																			
Raising armies	0	3	3	2	2	0	0	2	0	1	0	1	2	0	2	0	0	0	0
Raising fleets	0	3	3	2	2	0	0	2	0	1	0	1	2	0	0	0	0	0	0
Maintenance of armies and fleets	0	3	3	0	2	0	0	0	0	0	0	1	0	0	0	0	0	0	0
Building new fortresses	0	0	2	0	0	0	0	2	0	2	0	0	0	0	0	0	0	0	0
III. Campaigns																			
Land warfare	0	5	2	2	1	0	4	4	2	4	0	2	3	0	4	0	3	0	0
Sea warfare	0	2	1	0	1	0	0	2	0	2	0	2	0	0	0	0	0	4	0
Armored warfare	0	0	0	0	0	0	0	0	0	0	0	0	0	0	3	0	3	0	0
Air warfare	0	0	0	0	0	0	0	0	0	0	0	0	0	0	0	0	0	3	0
Air-ground warfare	0	0	0	0	0	0	0	0	0	0	0	0	0	0	0	0	2	2	0
Naval-air warfare	0	0	0	0	0	0	0	0	0	0	0	0	0	0	1	0	0	4	0
Need for maintenance supply	0	3	2	0	2	0	3	0	2	3	1	2	3	0	2	1	2	1	0
Need for attack supply	0	0	0	0	0	0	3	3	0	2	0	0	3	0	0	0	3	0	0
Need for defense supply	0	0	0	0	0	0	3	3	0	0	0	0	3	0	0	0	0	0	0
Supply by rail	0	0	0	0	0	0	0	0	0	3	0	0	0	0	0	0	2	0	0
Supply by river	0	0	0	0	0	0	0	3	0	3	0	0	0	0	0	0	0	0	0
Looting cities	0	3	3	0	0	0	0	0	0	0	0	0	0	0	0	0	0	0	0
Tribute, bribery, and ransom	0	2	2	1	0	0	0	0	0	0	0	0	0	0	0	0	0	0	0
Changing allegiances of mercenaries	0	0	2	2	0	0	0	0	0	0	0	0	0	0	0	0	0	0	0
Interruption of enemy supplies	0	2	0	0	0	0	3	3	2	3	1	0	2	0	2	1	2	0	0
Building supply depots	0	1	0	0	0	0	3	3	0	0	0	0	0	0	0	0	0	0	0
Prisoners of war and prisoner exchanges	0	0	0	1	0	0	2	0	0	0	0	0	0	0	0	0	0	0	0
Influence of terrain on movement	0	4	2	1	1	0	3	3	0	3	0	0	2	0	1	0	2	0	0
Influence of leaders on strategy	0	4	3	4	4	0	4	0	0	3	0	0	0	0	0	0	0	0	0
Rise of new leaders	0	4	2	4	0	0	0	0	0	3	0	0	0	0	0	0	0	0	0
Land movement of armies	0	4	2	4	3	0	4	4	4	4	0	0	3	0	2	0	2	0	0
Forced marching	0	3	0	2	0	0	3	3	0	0	0	0	0	0	0	0	0	0	0
Sea transportation of armies	0	4	3	4	3	0	0	4	0	3	0	2	2	0	2	0	0	1	0
River transportation of armies	0	0	0	0	0	0	0	4	0	2	0	0	0	0	0	0	0	0	0
Rail movement of armies	0	0	0	0	0	0	0	0	0	3	0	0	3	0	2	0	2	0	0
Necessity of occupying unfriendly areas	0	0	1	0	0	0	0	3	0	3	0	0	2	0	3	0	1	0	0
Partisan warfare	0	0	0	0	0	0	0	2	0	0	0	0	0	0	3	0	0	0	0
Garrisoning fortified places	0	4	4	4	0	0	4	3	0	4	0	0	3	0	0	0	0	0	0
Besieging fortified places	0	2	2	2	0	0	3	3	0	2	0	0	0	0	0	0	0	0	0
Using fortified camps and trenches	0	1	0	0	0	0	0	3	0	3	0	0	0	0	0	0	0	0	0
Need for naval bases	0	2	1	0	2	0	0	2	0	0	0	0	0	0	0	0	0	1	0

TABLE 5 Actions and Conditions Simulated/Degree of Abstraction and Concreteness

	Raphia	*The Conquerors: The Romans*	*Decline and Fall*	*Kingmaker*	*Conquistador*	*Thirty Years War: Rocroi*	*Frederick the Great*	*1776*	*Napoleon's Last Battles*	*American Civil War*	*Blue and Gray: Chickamauga*	*Diplomacy*	*World War I*	*Soldiers*	*World War II*	*Fall of Tobruk*	*Barbarossa*	*Fast Carriers*	*Strikeforce One*
Need for air bases	0	0	0	0	0	0	0	0	0	0	0	0	0	0	0	0	0	3	0
Limited strategic intelligence	0	1	0	1	0	0	2	3	0	0	0	0	0	0	0	0	0	5	0
Nonbattle attrition	0	2	1	2	3	0	3	3	0	3	0	1	0	0	0	0	0	0	0
Winter weather	0	2	0	0	0	0	2	3	0	3	0	0	2	0	2	0	2	0	0
Other weather	0	0	0	0	0	0	0	0	2	0	0	0	0	0	2	0	2	1	0
IV. Battles																			
Influence of leaders on tactics	2	4	0	0	0	2	1	0	5	1	0	0	0	0	0	1	0	0	0
Death of leaders	3	3	2	3	0	2	0	0	3	2	0	0	0	0	0	0	0	0	0
Influence of terrain on movement	1	1	0	0	0	3	0	0	4	0	3	0	0	4	0	3	0	0	2
Influence of terrain on combat	0	1	0	0	0	2	0	0	4	0	3	0	0	5	0	1	0	0	2
Arms characterstics																			
Infantry	3	3	0	0	0	2	0	0	3	0	2	0	0	4	0	2	0	0	0
Heavy infantry	3	4	0	0	0	0	0	0	0	0	0	0	0	0	0	0	0	0	0
Light infantry	3	2	0	0	0	0	0	0	0	0	0	0	0	0	0	0	0	0	0
Cavalry	3	3	0	0	0	3	0	0	3	0	2	0	0	4	0	0	0	0	0
Heavy cavalry	0	3	0	0	0	0	0	0	0	0	0	0	0	0	0	0	0	0	0
Light cavalry	0	2	0	0	0	0	0	0	0	0	0	0	0	0	0	0	0	0	0
Elephants	4	3	0	0	0	0	0	0	0	0	0	0	0	0	0	0	0	0	0
Artillery	0	0	0	0	0	3	0	0	3	0	3	0	0	4	0	2	0	0	0
Armor	0	0	0	0	0	0	0	0	0	0	0	0	1	0	1	4	1	0	0
Weapons interactions	1	2	0	0	0	1	0	0	2	0	0	0	0	4	0	3	0	3	0
Use of combined arms	2	2	0	0	0	2	0	0	2	0	0	0	0	4	0	4	0	0	0
Use of entrenchments and hasty fortifications	0	0	0	0	0	0	0	0	0	0	0	0	0	4	0	4	0	0	0
Loss of morale	4	2	0	0	0	3	0	0	4	0	2	0	0	0	0	0	0	0	0
Recovery of morale	4	0	0	0	0	0	0	0	4	0	2	0	0	0	0	0	0	0	0
Reorganization of destroyed units	0	0	0	0	0	0	0	0	3	0	0	0	0	0	1	3	0	0	0
Difference between regulars and militia	2	1	0	0	0	0	0	0	0	0	0	0	0	3	0	0	0	0	0
Influence of army organization on battle	1	2	0	0	0	1	0	0	4	0	0	0	0	0	0	0	2	0	0
Difficulty of coordinating allied armies	0	0	0	0	0	0	0	0	2	0	0	0	0	0	0	0	0	0	0
Disorganization of troops in combat	1	1	0	0	0	2	0	0	2	0	2	0	0	3	0	1	0	0	2
Use of specially trained troops	2	0	0	0	0	0	0	0	0	0	0	0	1	0	0	0	0	0	0
Overwhelming attacks	0	0	0	0	0	0	0	0	0	0	0	0	0	0	1	2	1	0	0
Suffering casualties in battle	1	2	2	3	1	1	2	2	2	2	1	0	2	2	2	2	1	4	2
Destruction of unit cohesion in battle	2	2	0	0	0	2	0	0	3	0	3	0	0	4	0	2	2	0	2
Capture and use of enemy weapons	0	0	0	0	0	3	0	0	0	0	0	0	0	0	0	0	0	0	0
Night activities	0	0	0	0	0	0	0	0	0	0	0	0	0	0	0	0	0	3	0

your reaction to particular design decisions by the designers of these games, you will be able, by consulting the chart, to find a game that is on the scale that you prefer to deal with and with a comfortable mix of terrain.

Degrees of Abstraction

Table 5 also describes the games, this time by listing some of the kinds of actions that are permitted in the game and some of the conditions which affect those actions, and indicating the degree of abstraction or concreteness with which each action or condition is represented. Turning reality into a game will always involve considerable abstraction, and therefore even what is described as "concrete" on the chart is only relatively so.

Some examples of the way the numbers are used will help you interpret the chart. When I give the area of diplomacy a "1" for abstract in *World War I* and a "5" for concrete in the game *Diplomacy,* I mean that in the first game allies have no chance of developing conflicting aims and that the ways they can relate to each other are rigidly controlled by the rules, while in the second game real differences of interest are built into the game and real diplomacy is necessary if any of the players are to cooperate with any others even for a time.

Quite often, abstract activities are those that take place off the board, as in *American Civil War.* "Entry of neutrals" is given a "1" because it is simply part of the victory conditions. If the game has progressed to the point at which Britain and France will enter the war on the Southern side, then the victory conditions simply say that the South has won and the entry of the former neutrals need not be actually shown.

The symbol for "not simulated" is "0." Things are not simulated mostly because they are anachronistic (as "air warfare" in ancient history games) or inappropriate to the scale of the game, as supply rules would be in *Rocroi,* a game that simulates a single day's battle. At other times, activities are simply assumed to be going on in the background or are taken care of by an abstract rule. For example, the use of railroads to supply armies has been an essential part of warfare since the middle of the nineteenth century, but the only game that makes this explicit is *American Civil War,* where the armies are tied to the rails in an appropriate fashion. In *World War I,* railroads are used explicitly only for the transportation of combat units, while supply is handled with an abstract rule. In *World War II,* railroads are not even on the map and supply is abstract. In *Barbarossa,* there are no rails printed on the map, but they are explicitly considered to be everywhere and fairly concrete supply units are moved by those omnipresent rails.

How do you evaluate a game after you learn how concrete it is in various areas? In one sense, the more concrete the better, but this generalization runs into practical difficulties. Concrete representation is more time consuming and difficult than abstract. For example, it is easier to assume that commanders are where they are supposed to be on the field and doing what they are supposed to be doing, as in *Chickamauga,* than it is to actually move commander counters around and use the very complex leadership provisions of *Napoleon's Last Battles.* Excessive concreteness would make games, and in fact has made some games, unmanageable and therefore uneducational.

The purpose of Table 5 is to allow you to identify where compromises with concreteness have been made and whether they have been made in the right places for you. If they haven't, you still might be able to use a game if you are willing to develop some new rules yourself. You can do this by making up your own rules or by adapting the rules of one game to another. For example, if you want to use *Chickamauga* but want to stress leadership, you might adapt the rules of *Napoleon's Last Battles* to *Chickamauga.*

INFORMATION AND RULES

Tables 6, 7, and 8 evaluate the hard information, the rules, and the other reading material furnished with each game.

Hard Information

Of the three, Table 6 is most likely to be misinterpreted. A game includes a lot of information about the historical situation on which it is based. The typical game includes a historical map for its game board, showing geographical features, including for some maps the political boundaries of the period. A grand strategy game will contain information on the economic strength of various powers as well as their military strength—a complete evaluation of their abilities to make war. Some games have very complete orders of battle, detailed unit rosters, including the names of units, their strengths according to a numerical system, the large units to which each small unit belongs, and where they were at the time the battle or war began. Some games include numerical assessments of the abilities of historical leaders by name, and some games detail the interactions of the weapons of the period.

All other things being equal, it is better for a game to simulate each of these aspects and simulate it accurately, but other things are never equal. Game designers have to subordinate all their material to the game itself. Furthermore, they do not always have very good information to work with, but they must invent it if they don't. I wrote above about the way physical and political geography is distorted to conform to the hexagon grid or to promote simplicity. The balance of military power is usually the most accurately simulated characteristics of a wargame, as one might guess, but details such as order of battle and army organization are frequently ignored, and weapons are dealt with extensively only in tactical games like *Soldiers,* or in games with a tactical subgame like *The Romans.* Economic and diplomatic conditions are sometimes inaccurate from a desire for simplification, but they are also sometimes inaccurate simply because they were outside the designer's area of interest and so he did not put much thought into formulating the rules about them.

As a potential user of these games, you should be aware of the weak points in those you are considering, because you will have to supplement the games with reading assignments and your own remarks. You should be particularly aware that your students are likely to take the information in the games more literally than it ought to be taken, so you must be ready to correct any misunderstanding that arises.

Rules

A few things about Table 7 ought to be called to your attention. The evaluation of the rules will give you some idea of how much time it will take you to prepare to introduce the game to your class. Generally, you should not count on your students reading the rules, or reading them with much understanding if they do. You yourself will find, when you prepare a game, that the rules do not make much sense until you actually start using them. To start a new game, therefore, you ought to plan on learning the game and playing it solitaire at least once. Then make the rules available to your students, but expect to give them a lot of help in the first few moves. The better students will read the rules with understanding, but probably only after the game has already started.

Non-Rule Reading Material

Non-rule reading material is often included in wargames, though not as often and not as extensively as I would like. Table 8 evaluates this material for your information. In only a few cases is this material lengthy enough and good enough to lighten your burden very much in preparing to use the game.

TABLE 6 Evaluation of Hard Information Presented in the Simulation

Game	Physical Geography	Political Boundaries	Balance of Military Power	Balance of Economic Power	Orders of Battle	Army Organization	Starting Locations of Forces	Roles of Leaders	Names of Leaders	Types of Weapons	Diplomatic Situation
Raphia	1	0	4	0	5	3	0	2	3	1	0
The Conquerors: The Romans	3	3	4	1	0	2	0	3	3	3	4
Decline and Fall	2	2	2	1	1	1	1	2	0	0	2
Kingmaker	2	4	1	0	1	2	0	2	4	0	2
Conquistador	3	0	3	2	0	0	0	3	5	1	1
Thirty Years War: Rocroi	1	0	5	0	4	3	3	2	5	2	0
Frederick the Great	3	3	4	0	0	0	2	3	0	0	4
1776	5	4	4	2	0	0	3	0	0	0	3
Napoleon's Last Battles	3	0	4	0	5	5	4	4	5	0	0
American Civil War	3	3	4	4	0	0	0	2	0	0	3
Blue and Gray: Chickamauga	2	0	4	0	3	3	2	0	0	0	0
Diplomacy	2	3	1	1	0	0	0	0	0	0	1
World War I	2	4	4	2	3	0	3	0	0	0	4
Soldiers	1	0	0	0	1	0	0	0	0	3	0
World War II	1	1	4	2	0	0	3	0	0	0	1
Fall of Tobruk	3	0	0	0	4	4	4	0	0	4	0
Barbarossa	2	2	3	0	1	0	0	0	0	0	2
Fast Carriers	2	0	0	0	4	0	4	0	0	3	0
Strikeforce One	0	0	0	0	0	0	0	0	0	0	0

TABLE 7 Evaluation of Rules

Game	No. of Words	No. of Illustrations of Play	No. of Charts/Tables	Readability of Rules	Clarity of Rules	Organization of Rules	Completeness of Rules	Accessibility of Charts/Tables
Raphia	4,500	2	3	B	B	C	B	C
The Conquerors: The Romans	29,000	5	19	C	B	B	B	A
Decline and Fall	5,400	8	5	A	A	B	A	C
Kingmaker	9,000	5	13	B	A	B	B	C
Conquistador	13,000	0	12	C	B	B	B	A
Thirty Years War: Rocroi	5,000	0	5	B	A	B	A	B
Frederick the Great	8,500	0	5	C	B	B	B	A
1776	10,000	0	8	B	B	B	B	A
Napoleon's Last Battles	13,000	3	3	B	A	C	B	A
The American Civil War	13,300	2	7	C	B	C	B	B
Blue and Gray: Chickamauga	5,600	1	2	B	A	B	A	A
Diplomacy	3,300	17	1	A	A	A	A	A
World War I	10,200	2	3	A	B	B	B	B
Soldiers	9,000	5	4	B	A	C	A	A
World War II	19,000	2	9	C	B	B	B	A
The Fall of Tobruk	4,000	0	6	A	B	B	C	B
Barbarossa	6,000	4	2	B	B	B	B	A
The Fast Carriers	23,000	3	8	C	B	B	A	B
Strikeforce One	3,300	22	1	A	A	A	A	A

You will be able to assemble historical background materials from history books available in your school library or other local libraries. Unfortunately, you cannot do as much for playing hints and designers' notes. Extensive material has been published on most of the games listed here but the material is not available in even the biggest libraries. Such material will be unavailable to you unless you can find a wargaming hobbyist in your neighborhood with a large file of wargaming publications.

GAME EXPANSION BY RULES ADDITIONS

Wargames have been designed to be played competitively, usually by two players, sometimes by more, but always by groups in which everyone is playing the game. This intention of the designer has mixed results from the point of view of the teacher who wants to use a game for teaching history. Because wargames are often played by people who are highly competitive, the rules are often written in a pseudo-legalistic prose that attempts to answer all the problems that might arise. In games as complex as wargames, rules can never answer all possible questions, but rules writers try to do this so game players will be able to proceed without endless arguments.

Your problems as a teacher wanting to use wargames are different from those of the competitive player. You want carefully written rules but you do not need to follow them slavishly. You are able to make interpretations or even new rules without consulting another, equal player. Furthermore, you have opportunities to put greater realism in the games you conduct that are not available in purely competitive play.

TABLE 8 Availability and Evaluation of Nonrule Reading Material Included with Game

Game	Historical Commentary		Playing Hints		Designers' Notes	
	No. Words	Value	No. Words	Value	No. Words	Value
Raphia	1,600	B	0	N/A	0	N/A
The Conquerors: The Romans	1,300	B	0	N/A	3,000	B
Decline and Fall	1,000	A	500	C	0	N/A
Kingmaker	3,200	A	0	N/A	1,700	A
Conquistador	0	N/A	0	N/A	0	N/A
Thirty Years War: Rocroi	275	C	0	N/A	440	A
Frederick the Great	0	N/A	0	N/A	0	N/A
1776	5,800	A	0	N/A	1,200	B
Napoleon's Last Battles	660	C	0	N/A	1,400	B
The American Civil War	0	N/A	2,300	A	2,300	B
Blue and Gray: Chickamauga	0	N/A	400	C	1,500	B
Diplomacy	0	N/A	0	N/A	0	N/A
World War I	0	N/A	0	N/A	0	N/A
Soldiers	1,300	B	0	N/A	0	N/A
World War II	0	N/A	0	N/A	300	C
Fall of Tobruk	0	N/A	0	N/A	0	N/A
Barbarossa	0	N/A	0	N/A	500	A
Fast Carriers	0	N/A	0	N/A	600	A
Strikeforce One	0	N/A	0	N/A	0	N/A

TABLE 9 Game Expansion: Chart of Suitable Additional Rules

Game	Inhibited Communications	Secret Movement	Political Inhibitions on Military Commanders	Economic Goals	Varied Victory Conditions for Allies	Creative Refereeing	Withholding Rules
Raphia	VS	N	N	N	N	S	S
The Conquerors: The Romans	VS	N	S	S	N	S	N
Decline and Fall	N	N	N	N	N	N	N
Kingmaker	N	N	N	N	N	N	N
Conquistador	N	N	N	S	N	S	N
Thirty Years War: Rocroi	VS	N	N	N	N	S	N
Frederick the Great	S	N	N	N	S	S	N
1776	VS	S	S	N	S	S	N
Napoleon's Last Battles	VS	S	S	N	S	S	N
American Civil War	N	N	N	N	N	S	S
Blue and Gray: Chickamauga	VS	VS	N	N	N	S	S
Diplomacy	N	N	N	N	N	N	N
World War I	N	N	VS	S	S	S	S
Soldiers	S	S	N	N	N	S	VS
World War II	N	N	VS	S	VS	S	S
Fall of Tobruk	N	S	N	N	N	S	S
Barbarossa	N	N	VS	VS	N	S	NS
Fast Carriers	S	VS	N	N	N	VS	S
Strikeforce One	N	N	N	N	N	N	N

Key: N = not suitable S = suitable VS= very suitable

In competitive play, all possibilities have to be known equally by both players. In most games, not only are all possibilities known perfectly but even the current locations of all forces are known perfectly by both players. The shock element that has been such a large part of human experience with warfare is excluded from most wargames, most of the time, in the name of fair competition. Your aim is not primarily competition (although you will use competition to help motivate your players) but education, and you can therefore change some of these conditions.

Table 9 refers to seven kinds of rules or referee activities that can be added to existing games to make them more realistic and educational in a classroom setting. Here I will explain each of them and suggest how each improves a published game.

Inhibited Communications

Inhibited communications rules assume that you have divided up the command on each side in a game so there are two or more players on each side. After that, you increase their problems significantly by not letting them talk to each other. This is particularly appropriate to battle games in the preradio era. Each player should have a personal marker on the map, and players should not be allowed to converse unless their markers are in the same space. Ordinarily they will communicate in writing with messages passed through the referee, who will estimate the time the message will take (and who might even "lose" the message completely). Students using this rule will directly experience the confusion that has made so much military history.

Secret Movement

Secret movement has obvious advantages in realism over completely open movement—in some games. It is not suitable in grand strategic games or very abstract games. It is not suitable in battle games where the entire battlefield could have been seen by one person. It is not very suitable, for playability reasons, in games in which the mapboard spaces are not numbered. *Soldiers* and *Fall of Tobruk* would be marked "VS" in the table were it not for this disability. To implement the rule, simply have the players move their units around by the numbers of the spaces. When appropriate, inform the players of what they can see. If they want to inform other people on their own side of what they see, they will have to write messages. In the games with unnumbered spaces, you can use three boards, one for you and one for each of the players.

Political Inhibitions

Some games come already equipped with some political inhibitions on military commanders, notably games that deal with World War II in Europe, where Hitler's clumsy interference with German arms is notorious, and probably exaggerated. At any rate, most generals, unless they are like Frederick or Napoleon, have to work under civilian direction, and this direction ought to be simulated more systematically in games. One way of doing this is to create a role of political leader and fill it from the class. If this is not feasible, you as referee can interfere from time to time with unexpected political directives. If you have extra players, you could consider enforcing your demands by firing your generals. Political rules are not suitable for very abstract games nor for games which simulate a very short period of time.

Economic Goals

Economic goals are also built into some games but not into all the places they would be appropriate. Economic rules are not suitable for battle games, but they are most suitable for grand strategy games, where they already exist to some extent. You can tie new economic rules directly into the military rules of the game by giving players direct military benefits for achieving economic goals. In *Barbarossa,* for example, the Germans can be tempted into the Caucasus by the promise of receiving a new Panzer army if they seize Batum. Later, they might be persuaded to defend certain areas by the threat of losing combat units or supply units if they lose the areas. These expedients are crude perhaps, but they carry the desired message.

Varied Victory Conditions for Allies

Another kind of opportunity arises when you start by separating what was originally united. In games where all allies (like Germany and Italy in World War II) are played by the same person, the victory conditions refer to that side, not to the individual countries. If you give different countries to different players, you also ought to make sure they have different goals. In the game *World War II,* Germany and Italy should be given different aims. The Italians should be required to try to conquer their own empire in the Mediterranean and, if things start to go badly, they ought to be required to try to change sides smoothly.

Creative Refereeing

Creative refereeing is more of an attitude than a new rule in itself. Your goal as referee ought to be to keep the game loose and faithful to the historical conditions (though not necessarily to the historical narrative). Your players should be told that you will not feel absolutely bound by the published rules. If the rules produce absurdities, then you will change them, even while the game is in progress. You should feel at liberty to bring up surprises, to inject new elements into the game, from time to time. Your students may be brought into the fun of changing rules too. Tell them that if any student proposes a rules change and supports his proposal with reasonable arguments, you will consider adopting the change. Naturally, students will suggest new rules that aid their own side in the struggle. You might not even tell the other side that the new rule has been adopted until the players who thought it up are actually in the process of using it. You might not even tell the people who suggested the new rule. If players suggest a new rule, you might tell them that you will consider it, but that they will have to actually try to use the new rule on the board to see if it works or not.

Withholding Rules

Withholding rules is an idea related to creative refereeing, but it refers to the designer's rules rather than your own. Withholding rules requires the players to learn vital parts of the game by experience rather than by reading. This is particularly suitable for games that deal with a whole war or the beginning of a war. It is especially appropriate for rules dealing with combat. A major condition of human experience with war has been that it is impossible to know what is going to work in combat until combat actually occurs, at least since the fourteenth century or so, when that process began in which each generation's technological and social changes vitiated the previous generation's experience with combat. The technique for withholding combat rules is simple. You teach the players how to cause combat but not how to resolve it. You keep the combat results table out of circulation and resolve all combat yourself, reporting to the players simply what happened to their units as a result of combat.

ALLIANCES AND SKILLS

Warfare, like other human activities, is a social activity requiring cooperation, planning, organization, and so forth. Wargames also demand the same kinds of skills. The extent to which a wargame demands certain skills depends in part on what kind of relationships are allowed in the game, so I have classified the games according to their kinds of alliances and their concomitant need for certain skills. Table 10 is written with the assumption that your classroom game will use the game expansion ideas outlined earlier in this article. The published versions of these games are more limited and need fewer skills.

A problem might arise especially in considering the free association games. You might ask, "Do we want to teach our students the 'skills' of intimidating and deceiving?" This somewhat misstates the problem. Games do not teach skills so much as use skills that already exist, and we all have the skills in question, generally kept under strict control in polite society. The game environment allows the players to act in that environment free of the inhibitions polite society imposes on "real" actions. The point of the game is not so much to exercise these skills as to *force* the participants to realize how the practice of these skills has played its part in history. Using a game is not the moral equivalent of actually using force and deception; it is a way of studying the use of force and deception.

GENERAL ASSESSMENTS

The last two tables are very subjective general assessments of the games recommended in this article. These assessments of accuracy and general merit cannot be defended, except to say that they arise from my eight years of personal experience in using wargames in teaching my courses, "War in Western Civilization" and "World War II." But while the assessments cannot be defended, they can be explained somewhat.

Strikeforce One places out of the running in both tables, but is still recommended—for one thing only. It is a good, quick introduction to other games; it is better than any more serious games as an introduction because of its simplicity and unusually full instructions and copious illustrations of play.

As far as the other assessments are concerned, you will notice immediately that low historical accuracy in Table 11 does not prevent a game from placing among "the best" in Table 12. Historical accuracy is a matter of detail, and both *Decline and Fall* and *Diplomacy* have little concern with historical details. *Decline and Fall* "generalizes" some fifteen or more large German tribal groups into two for the purposes of the game. But the game is good for classes, partly because it does ignore details, which allows players to concentrate on the interplay of force, trust, betrayal, and greed with which the Romans interact with the barbarians and the barbarians with

TABLE 10 Games Classified by Kinds of Alliances and Skills (using multiplayer versions as advocated above)

No Alliances (straightforward battle)	Historical Alliances Only	Free Association (number of different powers or sides)
Games		
Raphia	*Frederick the Great*	*The Conquerors: The Romans (3)*
Thirty Years War: Rocroi	*1776*	*Decline and Fall (4 or 5)*
The American Civil War	*Napoleon's Last Battles*	*Kingmaker (3 to 7)*
Blue and Gray: Chickamauga	*World War I*	*Conquistador (3 to 5)*
Soldiers	*World War II*	*Diplomacy (7)*
Fall of Tobruk		
Barbarossa		
Fast Carriers		
Strikeforce One		
Skills Implied		
Assessing complex information	*Assessing complex information*	*Assessing complex information*
Planning strategies and tactics	*Planning strategies and tactics*	*Planning strategies and tactics*
Implementing complex plans	*Implementing complex plans*	*Implementing complex plans*
Writing clear communications	*Writing clear communications (in some)*	*Persuading and negotiating*
Understanding written communications	*Understanding written communications (in some)*	*Intimidating*
	Cooperating	*Deceiving*
		Cooperating

TABLE 11 General Historical Accuracy of the Simulation

Nonexistent	Low	Medium	High
Strikeforce One	*Decline and Fall*	*Raphia*	*The Conquerors: The Romans*
	Diplomacy	*Kingmaker*	*Frederick the Great*
	World War I	*Conquistador*	*The American Civil War*
	World War II	*Thirty Years War; Rocroi*	*Soldiers*
		1776	*The Fast Carriers*
		Napoleon's Last Battles	
		Blue and Gray: Chickamauga	
		The Fall of Tobruk	
		Barbarossa	

TABLE 12 Overall Recommendation: The Best of the Best

The Best	Second Best	Third Best	Special Purpose
The Conquerors: The Romans	*Frederick the Great*	*Raphia*	*Strikeforce One*
Decline and Fall	*American Civil War*	*Conquistador*	
Kingmaker	*World War I*	*Thirty Years War: Rocroi*	
Diplomacy	*Fall of Tobruk*	*1776*	
Soldiers	*Barbarossa*	*Napoleon's Last Battles*	
	Fast Carriers	*Blue and Gray: Chickamauga*	
		World War II	

each other. *Diplomacy,* at first glance, looks like a historically accurate game because it has an accurate map, but the map is just background for the very unequal spaces into which the board is divided. More important, the economic-military balance of power that exists on the board as the game begins is like nothing at all in history. But the action of the game is historically accurate in the sense that it is full of analogies to historical actions.

There are no other large apparent discrepancies between the two tables. High historical accuracy assures a game of a place at least among the "second best." The "third best" are in that category for various reasons: *Conquistador* because of its complexity; *Raphia, Rocroi, Napoleon's Last Battles,* and *Chickamauga,* all because they are only battle games and so do not, I believe, teach as much about their eras as is possible in either strategy games or more tactical games; and the other three because I have had to spend more time than usual adapting them to use in classrooms.

Conclusion

Which game you use or can use will depend as much on your subject matter as on the quality of the available games. You cannot use *The Romans* or *Decline and Fall* in an American history survey. Nor can you use very many of these simulations in any one session of even the general military history course I posited at the beginning of this article. I find that I cannot fit in any more than two or three classroom simulations in a semester of fifteen weeks. So this list of games gives you plenty of choice for places to start and directions to go. If you still cannot find anything to your liking, by all means look among the rest of the two hundred games described in the wargames section. Many are much like the games described in this essay, but they may include games that come closer to the exact subject for which you are looking.

SOURCES

American Civil War
James F. Dunnigan
1974

Simulations Publications, Inc.
257 Park Avenue South
New York, N.Y. 10010
$10.00

Barbarossa
James F. Dunnigan
1971

Simulations Publications, Inc.
$9.00

Blue and Gray: Chickamauga
Irad B. Hardy, III, Christopher G. Allen, Thomas Walczyk, Edward Curran
1975

Simulations Publications, Inc.
$4.00

The Conquerors: The Romans
Richard H. Berg
Redmond A. Simonson
Frank Davis
1977

Simulations Publications, Inc.
$17.00

Conquistador
Richard Berg and
Greg Costykian
1976

Simulations Publications, Inc.
$9.00

Decline and Fall
Terrence P. Donnelly
1972

Philmar, Ltd.
47-53 Dace Rd.
London, E3 2NG, U.K.
£5.55

Diplomacy
Alan B. Calhamer
1971

Avalon Hill
4517 Harford Road
Baltimore, Md. 21214
$15.00

Fall of Tobruk
Frank Alan Chadwick
1975

Conflict Game Co.
P.O. Box 432
Normal, Ill. 61761
$11.98

Fast Carriers
James Dunnigan
1975

Simulations Publications, Inc.
$11.00

Frederick the Great
Frank Davis and
Edward Curran
1975

Simulations Publications, Inc.
$11.00

Kingmaker
Andrew McNeil
1974

Avalon Hill
$12.00

Napoleon's Last Battles
Kevin Zucker, Redmond A. Simonsen,
J. A. Nelson
1976

Simulations Publications, Inc.
$13.00

Raphia
Marc Miller and
Frank Chadwick
1977

Game Designer's Workshop
203 North St.
Normal, Ill. 61761
$5.00

1776
Randall C. Reed
1974

Avalon Hill
$12.00

Soldiers
David C. Isby
1972

Simulations Publications, Inc.
$9.00

Strikeforce One
Redmond Simonsen and
James F. Dunnigan
1975

Simulations Publications, Inc.
free

Thirty Years War: Rocroi
Brad E. Hessel, R. A. Simonsen,
T. Walczyk, L. D. Mosca, S. B. Patrick
1976

Simulations Publications, Inc.
$13.00

World War I
James F. Dunnigan
1975

Simulations Publications, Inc.
$4.00

World War II
James F. Dunnigan
1973

Simulations Publications, Inc.
$9.00

POLICY NEGOTIATIONS
An Evaluation

by Barbara Steinwachs

It was a cold Saturday morning in October when the director of Greenville's Community Center and the executive manager of its Chamber of Commerce joined the University of Michigan's Extension Gaming Service staff to set up a table in front of one of the main street downtown stores. We hung twelve posters on the store window behind us, one for each "alternative future idea" developed so far by the Citizen Planning Committee. We stacked dozens of piles of poker chips and spread out twelve little cards to correspond with the twelve posters. Along came the three citizens who had volunteered to help staff during the first hour. We were ready to go!

For this western Michigan community of 7,000, downtown is a hub of Saturday morning activity. Of the 300 persons who walked by during our morning operation, close to 180 played our minimodification of *Policy Negotiations.*

Had you been an experienced simulation gamer watching our street corner that morning, you probably would not have recognized *Policy Negotiations,* so far from the original sophisticated model was our version. Had you joined us "staffing" a game-in-the-streets for the first time ever, you could not but have picked up some of our sense of excited expectation, our eagerness to see if the game would play in what might as well have been Peoria.

Had you been a downtown shopper just walking by, you would have noticed that the sidewalk in front of you was blocked by someone holding out a handful of poker chips. "Here are your ten chips!" was all we said. To our surprise, almost everyone accepted them and by so doing unwittingly began to play. They considered, some at great length, the options on the window posters, and then "voted" their priorities for their town's future by placing their chips on the cards representing those ideas they most wanted to see brought to fruition. Many of them were still discussing the ideas with their family or friends as they moved on to continue their shopping.

We talked to almost all of the players personally, urging them to attend the town meeting coming up two weeks later. At the town meeting, we told them, they would get a chance to join a larger group of Greenville citizens and share their ideas on the priorities enumerated that morning.

The whole project, a joint venture of the University of Michigan Gaming Service and the Michigan Council for the Humanities, had begun months earlier when a small planning group of Greenville citizens joined our staff and a few academic humanists from the nearby area to identify those aspects of the community causing them concern. They had focused on two principal areas—the need for new educational alternatives for Greenville's schools and the need to provide a safer environment in the downtown area.

After gleaning input from the Saturday morning downtown minigame, as well as from conversations with other concerned citizens, the planning group "loaded up" the adaptable framework of *Policy Negotiations* with content they felt simulated the situation in their own community. This new version of the game was played at a "town meeting" during which the participants chose as they came in either to be themselves or to try on someone else's shoes for the evening. Here is the story of what happened as told by the local newspaper, *The Daily News and Belding Banner,* on November 8, 1976.

40 PARTICIPATE IN GREENVILLE GAMING

GREENVILLE—"If you vote for our proposal this time, we'll support your proposal in the next round."

"We're upset because we feel we're being ignored, so we're going to tear up downtown."

"You non-traditional young people have the most to gain from our plan to beautify the downtown area, because it will provide a nice place for you to go. But we need your votes to pass it."

These were some of the comments that were heard Saturday in an unusual town meeting at the Greenville Area Community Center. The meeting brought together a group of area residents and some academic humanists to discuss two basic issues facing the city-creation of more and better educational alternatives and providing a better and safer downtown.

About 40 participants used a game simulation called "Policy Negotiation" to examine ways they could get involved in governmental decision-making.

The citizens chose roles such as business people, school administrators, local police, traditional young

---

*Editor's Note: *Policy Negotiations* is listed in the frame games section.

people, senior citizens, juvenile court staff, parents of school-age children, school board, city council-mayor and non-traditional young people. The groups set goals for themselves, introduced options for solving the problems, and then tried to convince other groups to vote for their solutions. In the game, they had a $200,000 no-strings attached grant to finance their solutions, each of which had a price tag.

In the short time allotted for the simulation, the citizens decided to spend their money on initiating a process for the community to state and set priorities for five major objectives for their educational institutions and to seek outside professional evaluation of the police and fire departments. They also voted to develop a parent-teacher-student organization to stimulate communication about educational problems.

The gaming simulation was conducted by members of the University of Michigan's Extension Gaming Service staff in cooperation with the Community Center. The participants in the simulation included area citizens, the academic humanists and the U-M staffers.

"We had hoped there would be more people taking part, but we feel that those people who were here had a good experience," Bill Larkin, center manager, said.

The gaming format was used to facilitate discussion between townspeople and the experts, according to Barbara Steinwachs, U-M gaming specialist.

Playing a game tends to release people from their 'real life' roles temporarily and to give all participants a common reference point in discussing their perception of the issues, Steinwachs said.

The game also compresses the time element involved in deciding and trying various policies so that policies which might take six months to a year to initiate and put into action can be seen in a single evening, Art Bechhoefer of U-M pointed out.

Nearly all the participants felt they had made progress toward their goals in the simulation, but felt they had not had enough time to carry them out.

After each decision by the group, three "consequence determiners" decided what effect the decision would have had on the quality of life in Greenville, and what effect it had on the influence levels of the various groups.

The issues used in the game were chosen after a series of meetings between Gaming Service personnel, local citizens and community leaders. In October, an introductory street game in downtown Greenville drew 187 citizens, who used poker chips to vote on 12 hypothetical changes relating to Greenville's future growth.

The simulation was made possible by a grant from the National Endowment for the Humanities through the Michigan Council for the Humanities.

At a follow-up meeting one month later, citizens and humanists decided to try to meet in the future on their own to form a sort of real life counterpart to the game's team, "Citizens for New Alternatives in Education." Our part in the project ended at this time when responsibility was transferred to the citizens themselves.

I. WHY MIGHT THE GAME BE USED?

Because *Policy Negotiations* is a "frame" game—a game without content until the framework is "loaded" or filled up—it provides a highly adaptable structure for any content situation wherein diverse interest groups negotiate to set priorities among diverse policy options.

A. To Explore and Gain Insight

The game works well in academic settings where the goal is to understand something not already known. An urban planning class at the University of Michigan, for example, completed "building" a version begun by their instructors to consider the spectrum of zoning options facing a hypothetical U.S. city. After all agreed on the content, they assumed roles representing the interest groups they themselves had agreed were central to the situation. Within these new roles, they used their allotted influence (poker chips) to pass or defeat one or more policy options! They bargained, negotiated, and compromised—common endeavors whenever different interests converge around the same issues.

B. To Stimulate Dialogue and Communication

In more practical, less academic arenas, the game supports persons in need of a focus for their conversations and/or in need of a structure to bring them together. In an effort to improve rapport between school administrators and school board members, the Woodhaven School District near Detroit sought to provide a context within which they could gain a better understanding of each other's roles and explore certain issues of joint concern. A "retreat" day for administrators and board members was planned, and a simplified version of *Policy Negotiations* constructed as a practical and fun instrument for achieving their goals.

The planning committee determined some of the game's content in advance with the University of Michigan Extension Gaming Service staff. They decided to focus on issues of evaluation, accountability, and measurement for both students and staff within the district. The entire retreat session consisted of a simple introduction to simulation gaming in general and to *Policy Negotiations* in particular, further preparation by all participants of the final version of the game's content, a two-hour play of the game, and a debriefing of the game experience, followed by a prolonged discussion of the implications the game decisions held out to the real world situation. Participants took roles different from their own for the play and initial debriefing but put their real life hats back on for the final practical discussion, which related the ideas that emerged in the game to genuine areas of concern for the district. (For the game's starting content, see appendix I.)

C. To Facilitate Planning, Priority Setting, and Decision Making

Ordinarily, the game will not replace the more usual ways of making decisions, but it can be introduced as a strategic step to move along an ongoing process. The people involved play their real life roles in a game where the policy options are the real life options. Before play, there is some probing and clarification of the options (perhaps over many months), and everyone agrees to accept the game's outcomes (the passed and defeated policies) as a statement of the mind of the group, if not as the final decision.

Some time ago, a church body in Rochester, New York, formulated fifteen or so committees to meet over a year's time to study a broad spectrum of alternative concerns and develop future courses of action for the church to adopt. Near the end of the year's work, each committee's recommendations were collected to be "the content" of a day-long simplified *Policy Negotiations* game. Each real life committee was a team in the game, and their real life proposals were the game policy options. The bargaining and lobbying of game "play" resulted in several concrete proposals—some of them newly shaped—emerging with strong backing from coalition segments of the group. Other proposals became clearly identified as having found little favor by the group. Near the end of the day, a more traditional consensus-formulating exercise was used to prepare an actual set of recommendations for presentation to the church's formal decision-making body. While it is worth debating whether or not a year's deliberations should be brought to a focus in one day of intense negotiation, it is nevertheless clear that such a procedure provides at the very least an opportunity for movement and growth.

A grander, more sophisticated version of the game was used by the Center for Simulation Studies to help diverse interest groups in the Central West End of St. Louis reach agreement on the kind of housing to be constructed on a plot of land in their transitional neighborhood. Earl Mulley of the center has written in *Simulation/Gaming/News* (January 1974) that developers and other concerned community persons wanted to get a broadly based coalition to buy into the final plan to be presented to the board of aldermen, and to neutralize potential opposition from the extremes by exploring the implications of alternative plans.

Mulley describes how, in consultation with designer Fred Goodman, (professor of education, the University of Michigan), *Policy Negotiations* was chosen as a forum for bringing the diverse interest groups together to discuss options, during both the development and the use of the simulation. They devised a three-tiered model using concurrent play in three separate rooms. The first tier was built around the constituencies and issues of the local neighborhood, the second represented the larger area surrounding the neighborhood, and the third simulated the interests and activities of the board of aldermen, the mayor, and his staff. When, during game play, the Local Neighborhood group decided on the housing option it thought most desirable, the city council convened a hearing on the proposal, initiating a lively debate of the issues and alternatives.

Approximately 150 people from the neighborhood participated in the runs of the simulation. Business people, home owners, renters, two aldermen, persons from black organizations, liberals, and conservatives all were there. The payoff, as summarized by Mulley, was what in fact happened at the real hearing when the final redevelopment proposal, which reflected the basic consensus of the community people, was presented to the board of aldermen. Anticipated opposition to the type of plan submitted never materialized. Leaders of both factions in the community that had made rumblings about the redevelopment plan had been in the simulation and had discussed with other community people, face to face, the realities of the trade-offs involved in this particular piece of history. Though not ecstatically happy with the plan, they concluded that it certainly was one of the best options and should not be resisted. Because the issues and questions that arose at the real hearing had at some point been anticipated in the simulation, the plan submitted was passed with only one dissenting vote. And, to cap the experience, communication and cooperation on community affairs among those who participated has continued.

The use of *Policy Negotiations* frequently is not limited to any single one of the functions discussed in this section but serves them all. It can simultaneously help a group explore and gain insight about their concerns, stimulate in-depth dialogue and communication, and facilitate their efforts to plan, determine priorities, and make decisions.

II. LET'S ZERO IN ON THE BASIC GAME PROCESS

Policy Negotiations has come to be considered a "frame" game—a structure into which any appropriate content can be inserted, and a set of rules simulating informal negotiation and priority-setting process. The game can be adapted to reflect countless analogous situations where a variety of interest groups bring different concerns to the same set of issues.

Instead of a frame game, the game's designer Fred Goodman prefers to think of it as a "priming" game, priming the pump of game design, a "starter" from which the redesigners can begin to analyze both the content and the processes basic to the situation they are trying to simulate. It works as a jumping-off point, a common frame of reference, for analyzing a situation they are concerned about.

Before the "new" game can be used, there must be agreement on which content will be loaded into the skeletal structure and which set of rules will best put that content into operation. Usually both content and rules are modified before the job is done.

In any run of any version of the game, several teams—representing selected major interest groups—use their influence to pass or defeat one or more "policy options" by whatever negotiating techniques they can devise. Thousands of simulation games do this much, similarly requiring their players to become familiar with their roles, with a variety of possible policies (or issues or proposals) that might come under consideration during play, with the game artifacts, and with the order and rules for play. The original version of *Policy Negotiations* adds to this basic core a complex set of rules which models with more sophistication the negotiation process. In addition, it adds some fine nuances to the basic content framework.

A. Description of the Original Version

This original version is well described in some detail in Goodman's and Larry Coppard's *Urban Gaming/Simulation* (see section IV, A4). The description that follows is modified somewhat from theirs.

The scenario presents the behind-the-scenes bargaining, influence distribution, and negotiations between school board members and teachers' representatives on issues such as salary increases and class size. It is not a "collective negotiations" situation; the individuals are negotiating policies in a much more general way than would be the case in a formal bargaining arena. Three players (or teams of players) represent school board members, with conservative, inner city, and suburban constituencies, and three players or teams represent teachers' representatives, with backing from new teachers, old teachers, and special teachers. In addition, there are up to four other roles which exert influence on the negotiations from "outside." These roles are a labor union, a newspaper, a civil rights group, and a chamber of commerce.

Each team (role) starts play with a certain amount of "influence" (one's standing or power when compared with one's peers—that is, others with different constituencies but relatively equal standing vis à vis the same arena of issues) and "prestige" (one's standing, or popularity, with one's own constituency; the likelihood of being again chosen to represent that constituency).

Ordinarily, this version is played using Lego (a variety of very small, plastic, children's blocks) as its principal artifact. To keep track of what is happening, it is necessary to master the rather elaborate Lego system that visually displays the allocation and tallying of influence and prestige, as well as the current status (e.g., chances of passage) of the policy options.

The game is played in rounds. During each round the following occurs:

(1) Out of twenty or so specific policy options, one (e.g., a fringe benefit package for the teachers) comes to the top of the agenda for discussion and vote.

(2) Teams negotiate with one another and finally allocate their influence to one or more of the following ends:
  (a) to pass or defeat the policy option up for vote
  (b) to raise another policy option to the top of the agenda for the next round
  (c) to raise their prestige level with their constituency (increase chance of reelection)
  (d) to store their influence (with an "outside" social force) for use later.

(3) At the end of the round, the game facilitator tallies the influence and determines:
  (a) whether the policy option just considered passed or failed
  (b) which policy option will be at the top of the agenda next round
  (c) the effects of 3a above on each team's prestige and influence levels (e.g., if the fringe benefit package for teachers fails, the teachers might lose prestige)
  (d) the effects of 3a on the chances for passage of future issues (e.g., if the fringe benefit passes, the chances are lowered for a salary increase passage).

(4) An unexpected event in the "outside world" may affect the game. The game facilitator may introduce a "newspaper headline" which could affect the influence or prestige levels of players, or the chances of passage for an issue.

B. The Three Phases

When Fred Goodman, with help from his students and colleagues, first designed the version just described, he thought of it as sort of a total-context game about teachers negotiating with school boards. Not until some time later, when an early player pointed out its versatile framework, did the game's potential become apparent. Goodman was "exhilarated," he says in retrospect, by the growing understanding that people could learn far better by building games—thoroughly analyzing the many dimensions of a situation and then creating analogues—than by playing them. The concept of the "priming" game began.

What all this means is that there is no such thing as *The Policy Negotiations* game. Instead, there is a gaming process that requires at least two of the following phases.

(1) *Priming Game Phase.* The introduction, actual play, and debriefing of a "sample" version of the game. A number of alternative priming games have now been documented.

(2) *Redesign Phase.* The reloading of the priming game framework with a new body of content and, frequently, the reworking of some of the game's rules as well. The game-building process provides an unusually good forum for explicit analysis, debate, and exploration by a group of individuals. It might be done by a committee in an afternoon, drawing on the expertise they already possess, or by a class over a semester, drawing on assigned readings and personal interviews.

(3) *Replay Phase.* The playing of the "new" game, if this is useful. Although the redesign phase is valuable enough to be carried out as an end in itself, play can provide an opportunity to test ideas, stimulate more dialogue, supplement planning and strategy making, and facilitate the priority-setting or decision-making process.

Of course, a group other than the redesigners also could play the new game, but if they do, it should be remembered that the nature of the activity changes from a self-initiated one to a more presentational one. The redesign phase works so well because people ordinarily learn much more by building a game than by playing one. Playing one built by someone else is experiencing someone else's view of things—a valuable activity, but not the same as figuring out your own.

III. SUPPOSE YOU WANT TO USE THE GAME?

Because the *Policy Negotiations* game does not exist, deciding to use it means making a number of major decisions and doing a substantial amount of work.

A. Some Work To Do

(1) Plan the total context within which the game will be used and each of its phases.
(2) Hold a priming game session.
(3) Reload the game content and make any necessary rule changes.
(4) Perhaps hold a replay session.
(5) Facilitate the group sessions attendant on the above.

B. Some Decisions to Make

(1) What is the purpose of using the game?

(2) What priming game will be played? What will the debriefing focus on? Who will attend the priming game session, and how will the session be structured?

(3) Who will redesign the game (and thereby slant the analysis), and in what time span?

(4) Will there be a replay? If so, how many trial runs will be held in advance? Who will attend these trial runs, and how will the sessions be structured? Who will attend the replay session, and how will it be structured? (The question of who will attend the sessions is particularly sensitive if the game is intended to serve more than an academic purpose—if, for instance, it is to aid a group in setting priorities with consequences outside the scope of the game. When such applied goals exist, careful strategic planning must be employed, for questions such as the following become crucial. Shall all interest groups be represented each phase of the way, or will this give too much away too early? Are there certain real life roles who should not be invited for any stage? If so, should others try to play their roles during the game?)

(5) Should the participants play their real life roles, or "switch hats"? If they do not play their real life roles, how will it be determined which role each will play? How will everyone be helped to get into their game roles?

(6) Re: Game redesign

(a) What will the game focus on?
(b) What will be its arena, its sphere of concern?
(c) Will the new game model things as they currently are, or as they might be?
(d) Who are the principal decision makers in this arena (roles/teams in the game)?
(e) What is each team's starting relative influence? prestige level?
(f) What are the principal policy options to be considered?
(g) What are the consequences of certain policy options being passed? defeated?
(h) What major larger-world events (real or imaginary) might impact on the situation? How?
(i) Will the consequences be determined before or during play? By whom? How will these consequences be put into operation in the play of subsequent rounds?
(j) What rule changes must be made to use the game for the context at hand?

(7) Will some of the redesign work be completed before the replay session? How much? If some is to be done during the replay session by its participants, who will facilitate this?

(8) Who will facilitate all the group sessions described above?

(9) Who will make all these decisions?

(10) Who will decide who will make all these decisions?

C. Further Considerations

Because so much effort must go into using *Policy Negotiations,* let us draw back for a while and consider whether it should be used at all. The early part of this chapter elaborates the many uses to which the game can effectively be put, but these potentially positive outputs should be weighed against the time and energy required to move through the various preparation phases, as well as against the following five considerations.

(1) *Reliance on Role Playing.* Unless it is played in a real life situation in which the real holders of the roles play themselves, the game relies heavily on role playing. Providing detailed written role descriptions for participants to read does not do much to help them get into roles not their own. Of course, it is good to have people try on someone else's shoes for a while, but if the experience merely stimulates the acting out of stereotypes, more harm than good can result. This weakness can be overcome in part by players' researching their roles, or interviewing and "shadowing" their real life counterparts *before play.*

(2) *The Consequence-Determining (Feedback) System.* Inherent in the basic game model is another means for partially meeting the getting-into-roles problem: the consequence-determining, or feedback, system. The fact that the passage or defeat of policy options results in consequences for the players, and perhaps for the policies as well, helps keep people "true" to their roles. However, in the game's original version, and in the simpler variations I am familiar with, the feedback system is not strong enough either to define the roles or to hold people strictly to them.

Some elaborate attempts to meet this problem have been worked out, among them *Policyplan,* developed by Larry Coppard of the University of Michigan School of Education. Larry's feedback system works somewhat better, for it is more complex, relating player-generated goals with designer-generated impacts. (In most of his versions, a computer is required to manage the accounts.)

In the three-tiered version for the Central West End of St. Louis (discussed earlier), each of the three games had its own consequence-determining system. But in addition, as Earl Mulley describes it, all three were cross-normed to reflect the tugs and pulls of decisions up and down the system. A decision made by the city council, for example, could well affect the prestige and/or influence of groups in the other games. Thus the norming had to take into account eighteen teams and forty-eight issues, making for a substantial norming grid!

The feedback system is weaker for simpler versions of the game. It helps to have "consequence determiners" doing their work "live" as play evolves, for then they can make judgments based not only on their *opinions* of the kinds of impacts likely to occur, but also on what in fact *is happening* during play. Whether the consequence determining is done in advance or during play, however, one further problem remains. Consequence determiners are not players; they represent an objective view of things, a best guess as to what probably would happen *if.* But is it possible to find people to perform this task who will not, albeit inadvertently, lay their perspectives and biases on the game?

(3) *More Process Than Content?* Even when an effective feedback system is operating, and even when participants are

playing their "real world selves" (not role playing), the negotiation activity of the game can degenerate into process at the expense of content. A play of the game often yields more insight into influence-wielding and negotiation than into the subject matter at hand.

In *Urban Gaming/Simulation* (see section IV, A-4), Larry Coppard describes the phenomenon this way:

> In any application where we have had this bargaining and negotiating activity there's been a high level of interaction and a lot of public debate in the sessions. . . . [But] where we have emphasized the bargaining and negotiating, we have not been terribly satisfied with the quality of the reflection that the group has done . . . that is, they'll fight like the devil to get something into the platform and not pay a whole lot of attention to what impact that alternative may or may not have on the system [p. 427].

He suggests one way of meeting this problem:

> We have tended to de-emphasize the bargaining and negotiating in a number of the applications we've made to the point where we sometimes just strip out that whole process. We run the thing more as a discussion and say, "Let's just focus our attention on the impact exercise." We may look at four or five alternatives [policy options]. "O.K., what would happen if we did it this way, what would happen if we did it that way?"

Another approach is to carefully structure the order of play in the rounds to allow for the above as sort of a "time apart," but without eliminating the play value of the open negotiation periods. In addition, team caucus periods—which allow small groups to reflect and plan in relative calm—encourage close analysis and sharing on the core issues. Two documented games which do this in different ways are *Sitte* (see section IV, D3) and *Libra Public Schools* (IV, C5).

Probaly most importantly, involving the participants in the game design phase will shift the content focus to the time before play—where perhaps it better belongs. The heat of negotiation may not be the ideal place to scrutinize all sides of an issue but—in the real world at least—it frequently is the arena where, for better or worse, directions are set. Many a lucid and perfectly rational analysis has been set aside by decision makers under other sorts of causal pressures. *Policy Negotiations* is probably not the game to use if quiet, rational, "reflective" analysis is the only goal!

(4) *Frames as "Stackers" of the Design Deck.* Related to the above consideration is a point raised in *Urban Gaming/Simulation* (see section IV, A4). Game users, the authors say (p. 16), should be aware of the subtle way in which the frames themselves can convey bias. In providing a streamlined design process, the frame game presents choices of categories, types of rules, and forms of questions that subtly affect the development of the game eventually built into the frame.

This is particularly true for *Policy Negotiations,* a game relating to the process whereby decisions are made. Its title gives it away; the players "negotiate" policy—rather than, for example, being encouraged to reach consensus in a sharing, trust-filled forum. It is argued frequently that most real world decisions are reached this way—not by consensus methods, but rather by political negotiation. But is this the method to encourage? Should people be trained to continue to negotiate and make political trade-offs rather than to develop alternative methods?

(5) *Time and Resources Required.* Finally, there is the problem of time and money (or labor hours) as scarce resources! Several years of client requests for something which would help diverse interest groups engage in dialogue or set priorities in real world settings led the University of Michigan Gaming Service staff to develop simpler ways of using Fred Goodman's original version. When there was plenty of time to do the necessary work, and plenty of money to support it, and *if* a more elaborate simulation really facilitated the task at hand, then we worked from the basic model. But these "ifs" rarely applied. What our clients wanted was something that could be developed without very much time and effort and used in a relatively short session (frequently, only a half-day). Rather than a detailed simulation, they wanted to help a group air issues, get to understand each other's perspectives better, and/or reach some kind of group decisions.

We soon learned that a half-day is not long enough to accomplish these goals, particularly if a goodly portion of the group's time was spent learning what almost always was perceived by them as a complex Lego "language" and set of rules. So, gradually and with some regret, we began stripping down the original elegant game model. And, as we sensed that people were much more willing to engage in an activity with real life ramifications when they had had substantive input to that activity, we gradually moved as much of the redesign phase as time would allow right into the replay session.

In brief, the Gaming Service model eliminated everything except the basic teams (no "Outside Fources"), policy options (but no distinction between "internal" and "external"), and consequence determining or norming, (but usually only for influence levels, because ordinarily we drop historical propensity, prestige, and external events). In addition, we replaced the Lego artifacts with poker chips, 3 x 5 cards, newsprint, and markers. Frequently, we begin the game with only the purpose, scenario, teams, starting influence levels, and a few sample policy options having been decided in advance. Participants assume roles quickly, and each team then creates a couple of policy options which are written on newsprint and hung up around the room. The time saved by eliminating the need for lengthy instructions, and the ownership gained by having created the policies to be negotiated make for a session that usually meets clients' goals far better than a more complex game built in entirety before the session.

The Gaming Service's way of doing this has come to be called "Simplified *Policy Negotiations*" because it was written up under that name in a short paper a few years back.* It should be pointed out, however, that the Gaming Service, far from running any one specific version more than once, was always adapting, seeking new ways to make the exercise work more effectively in the practical arena.

IV. RESOURCES/DOCUMENTATION AVAILABLE ON POLICY NEGOTIATIONS

From all that has been said, it should be clear that there is no one piece of documentation covering everything for the user of *Policy Negotiations*. Although surely at least hundreds of different adaptations have been developed and used, many of these have been for one-time-only situations, and no documentation is available. Some, however, have been documented. Here is a list of the more noteworthy (or more available) that have crossed my desk in the last few years.

A. General

(1) One of the best sources remains the designer, Fred Goodman himself. In his inimitable way, he never really stops creating his games. He loves to juxtapose each new idea that crosses his path with something old, and in this way his energy and inventiveness go on breathing new life into old works, and they flower almost every season. *Policy Negotiations,* after it finally reached the Lego artifact version, was modified by Fred into a number of "pencil and paper variations," and even one making use of an overhead projector and assorted forms (*Policy Projections,* developed for the Ontario (Canada) Education Communications Authority).

He never modifies simply the artifacts, but the concepts of the game as well–one paper and pencil version explores just how much influence remains for someone who spends (or does not spend) his or her initial influence. He has been working for some time now with the game's norming–or feedback–system, trying to operationalize "probability norming"–some combination of what the consequence determiners think should happen with a more "objective" probabilistic system that directs the rolling of carefully calibrated dice.

No documentation on any of these variations is available at the time of this writing. For information, contact Fred Goodman directly (School of Education, the University of Michigan, Ann Arbor).

(2) Because of the intensive redesign phase required whenever the game is to be used, it is always better to talk with someone than to read something about the game if you are adapting it for the first time. Several persons worked closely with Fred in the early stages of the game's development and/or have used it in many variations over the decade since. Among them are the staffs of:

The Center for Simulation Studies
736 De Mun Avenue
Clayton, MO 63105

COMEX
Davidson Conference Center
University of Southern California
University Park
Los Angeles, CA 90007

(3) In addition, the staff of the University of Michigan's Extension Gaming Service in Ann Arbor adapted the game many times in the course of its existence until the Gaming Service closed in June 1979. Developed at the request of community, educational, and governmental clients, these variations were directed to nonacademic uses, ordinarily involving the "real world role holders" in play. Among those who might be contacted about these applications are Barbara Steinwachs (the author), Russell Stambaugh and Terry Anderson through the University of Michigan or Ken Smith, currently with the Legal Services Corporation in Washington, D.C.

(4) *"Policy Negotiations."* In *Urban Gaming/Simulation, A Handbook for Educators and Trainers,* edited by Larry C. Coppard and Frederick L. Goodman, pp. 407-420, Ann Arbor, Michigan: University of Michigan School of Education, 1979. A detailed description of the original version and its three phases for use is followed by extensive comment by the designer, Fred Goodman, and others who have used the game for a variety of purposes.

(5) "An Introduction to the Virtues of Gaming." Chapter 2 of *Educational Aspects of Gaming*; edited by P. J. Tansey, pp. 26-37. New York: McGraw-Hill, 1971. Fred Goodman authored this presentation of games as an alternative educational tool for those who value a discovery approach for which the educator is "guide." *Policy Negotiations* and Layman Allen's *Equations* are used as key examples.

(6) "Operational Gaming and Local Government," an unpublished paper (21 pages, 1974) by Ken Smith, formerly of the University of Michigan Extension Gaming Service and currently working in Washington, D.C. with the Legal Services Corporation. Available by contacting Barbara Steinwachs. After presenting some background material on operational gaming and its past uses in city government, a case study of an actual city policy issue and a hypothetical scenario is developed to indicate how *Policy Negotiations* might have been used to resolve the issue. Some observations stemming from the case study are discussed and a proposal made for future gaming applications in local government.

(7) "The Use of Games in the Regulatory Process," an unpublished paper (13 pages, 1976) by Arthur S. Bechhoefer, a doctoral candidate in Urban and Regional Planning at the University of Michigan. Available by contacting the author. The paper argues that the use of gaming simulation at critical points in the decision-making process by regulatory agencies offers the prospect of greater public participation in the formal public hearing process and a more rapid, satisfactory settlement of complex issues. A case study is presented, involving a possible application of *Policy Negotiations* in the licensing of nuclear power plants by the Nuclear Regulatory Commission.

(8) "Some Propositions about the Use of the *Policy Negotiations* Model in the Community Context": see Appendix II of this article.

B. The Original Version

(1) The original priming game, presenting negotiation among school board members and teachers' representatives, is included in a do-it-yourself kit along with four variations that originally appeared in *Urban Games: Four Case Studies in Urban Development* (now out of print), described in C2,

below. The kit is available for $14.00 from the Institute of Higher Education Research and Services, Box 6293, University, Alabama 35486.

C. Selected Variations

(1) See section A1 above for comments on variations by the designer.

(2) *Urban Games: Four Case Studies in Urban Development,* by Margaret Warne Monroe. Berkeley: University Extension, University of California, 1972. Meg Monroe begins the Operator's Manual with a fine expose of the use of operational games, and then describes, fluently and in careful detail, how to run the original version of *Policy Negotiations,* illuminating the various game components as she does so. In addition, she illustrates how the game can be used as an analytic tool, and then as a planning tool, for any specific social problem. She then presents a helpful guide to the redesign process. The manual concludes with sets of pregame positions, special rules, and the already–worked-out consequences ("feedback matrices") for the four priming games presented in the Player's Manual. The Player's Manual contains the scenario, role descriptions, and starting policy options (issues) for four individual variations on the original model: *Community Issues Game, Rapid Transit Game, Industrial Park Game,* and *Regional Shopping Center Game.* For availability, see section B1 above.

(3) Developed by the University of Michigan Extension Gaming Service Staff. (a) "Use of a Simplified POLICY NEGOTIATIONS structure by Groups to Build Their Own Games," an unpublished paper (10 pages, 1975) by Ken Smith and Ansell Horn. Available by contacting Barbara Steinwachs. This paper describes the simplified version discussed in section III (above) in a manner sufficiently detailed that persons familiar with *Policy Negotiations* should be able to get an idea of the overall game structure—and perhaps even feel brave enough to run it themselves.

(b) *Family Negotiations,* unpublished (8 pages, 1973), by Barbara Steinwachs. Available by contacting the author. This paper presents a body of content (but no game rules or process instructions) focusing on some issues facing a nuclear, middle-income family. The game can be used as is, to encourage exploration of role perspectives, or as a priming game. It functions well as a priming game because players—already familiar with families—can quickly enter into the roles and play the game to see how it works without getting bogged down in the content. Consequence determining (norming) must be done "live"; it has not been worked out in advance.

(c) Summary of Greenville Town Meeting *Policy Negotiations* session, unpublished (1976), by Barbara Steinwachs. Available by contacting the author. More detail on the community-planning project described at the beginning of this chapter.

(d) "The Power Elites: American National Policy-Making," unpublished (2-4 pages, 1976) by Robert D. Putnam, Professor of Political Science, University of Michigan (Ann Arbor). Available by contacting the author *or* Barbara Steinwachs. Notes present suggested starting actors and policy options, together with some interesting rule variations.

(e) "Gaming in City's Halls," an unpublished article (5 pages, 1974) by Ansell Horn and Ken Smith. Available by contacting Barbara Steinwachs. The story of the background and planning meetings for a *Policy Negotiations* session for Ann Arbor City Council members and City Department heads. The gaming session was developed to provide a congenial environment for local government personnel to meet, to listen, and to communicate with each other.

(4) "After *Policy Negotiations,* No Surprises at the Actual Hearing," an article in *Simulation/Gaming/News,* No. 10 (January 1974), pp. 9 & 10. Earl Mulley's description was abbreviated early in this article. The article concerns the use of a three-tiered *Policy Negotiations* by the Center for Simulation Studies to help diverse interest groups reach agreement on the kind of housing to be constructed in a transitional neighborhood in St. Louis' Central West End.

(5) *Libra Public Schools,* a simulation game by Robert L. Epps, John R. Kolkmeyer, Richard S. Rubin, and John A. Wilson, (Center for University Ministry, 1514 East Third, Bloomington, IN 47401), 1974. For documentation contact Robert Epps. This well-detailed modification is intended to provide skill in negotiation for teachers, school board members, and neutral mediators within a larger training workshop exploring collective bargaining. Among some very interesting changes from the original game is the addition of a collective bargaining component. In addition, this version formalizes and structures time for negotiations *within teams* and *within factions* (i.e., among three people who together are playing the one role of a particular individual member of a team).

(6) WESSPAC, a simulation gaming "package" composed of four gaming modules—"The Researcher Game," "The Architect Game," "The Developer Game," and "WESS-POWER" (a resident game)—all "nested" within a master game, WESPON (WESley Town POlicy Negotiations). Developed in 1975 by Thomas Douglas, Jack Nasar, Eugene Bazan and Luis Summers of Pennsylvania State University/University Park, PA 16902. For information contact the authors, and see "WESSPAC," a section of *Urban Gaming/Simulation* (see section IV, A4).

Simulating the planning, design and development of a neighborhood, this gaming package experiments with "interfacing games." The designers call it a "modular gaming strategy" in which several simple human–operated (noncomputerized) games are run concurrently or sequentially, linked together by participant interaction and a play-generated payoff network. *Policy Negotiations* serves as the keystone activity for the hierarchical series.

D. A Few Related Games

Related games are countless, ranging in complexity from the simple children's exercise described below to a computer-assisted process.

(1) Many exercises in building imaginary cities using boxes, blocks, and the like have been developed for children. One

was documented in the Los Angeles *Times* July 5, 1970. Its title: "Child Planners Build City of Dreams—More Space for Recreation" (by Dick Turpin). Sometimes children in "games" similar to this one take roles of urban decision makers; sometimes they do not. But in any case there is a good deal of (usually unstructured) informal negotiating.

(2) "Rules for Use of Decision Making Tool," unpublished notes (4 pages, 1972) by Mitch Rycus and Frank Ferguson. Available by contacting Mitch Rycus, the University of Michigan (Ann Arbor) Mental Health Research Institute. This is a simple *Policy Negotiations*-like procedure for setting priorities. Artifacts are pennies and allocation sheet ("decision board").

(3) *Sitte,* a simulation game by Hall T. Sprague and R. Garry Shirts, 1969. It is not presently available from its publisher, Simile II (P.O. Box 910, Del Mar, CA 92014), but a number of copies of the kits are in existence. Within a framework encouraging negotiation and coalition forming, conflicting interest groups use their influence to produce changes in an imaginary city. Proposals "pass" when substantial evidence of at least some coalition support exists. This works well enough under the checks-and-balances network built into the role interactions of the game, but not very well when the game is modified to other situations, unless great care is taken to preserve the checks and balances (see *Urban Gaming/Simulation,* pp. 480-488).

(4) WHIPP (Why Housing Is a Problem and a Priority) is a simulation game by Barbara Steinwachs, 1972. Available for $5 from Barbara Steinwachs, 1640 Argonne Place, N.W., Washington, D.C. 20009. Inspired by *Sitte,* this is a simple game in which participants assume roles representing diverse socioeconomic groups and endeavor to better the housing situation of an imaginary metropolis (see *Urban Gaming/Simulation,* pp. 577-581).

(5) *Policyplan,* a flexible set of exercises (usually gamed) developed by Larry C. Coppard of the University of Michigan Institute of Gerontology and David O. Moses of ERDA, Washington, D.C., 1972. In each of the many shapes the *Policyplan* process has taken as it has been applied to specific client-interest areas, it has provided a structure for exploring urban social issues and facilitating planning. Participants identify planning alternatives, explore their political feasibility, evaluate their impact, and often establish priorities among them. The feedback system ordinarily (but not necessarily) used is computer assisted. (See *Urban Gaming/Simulation,* pp. 421-428, for a fine critique by the designer.)

APPENDIX I: STARTING CONTENT FOR WOODHAVEN SCHOOL DISTRICT MODIFICATION

Principal Interest Groups	*Starting Influence*
Board members with city concerns	10
Board members with rural concerns	8
K-12 administrators	8
Building administrators	7
Representatives of teacher's union	6
Representatives of AFSCME	4
Citizens Committee	4
Formal parent groups	2

These primary actors had been enumerated in advance after some debate by the Planning Committee. The retreat participants accepted the breakdown intact, except for the last two groups, which they combined into one. They also decided to create a "students" team, but gave them no influence at all.

Starting Policy Options

(1) The middle school will pilot a pass-fail no report card student evaluation for one year.

COST	*SOURCE*
$3800	salary monies
$ 200	administrative monies

(2) The evaluation of the head custodians will include the efficient use of supplies, as determined by the building administrator (no cost).

(3) No student shall be retained for more than one school year during his/her K-8 program (no cost).

(4) Each staff member shall be formally observed without advance notice by his/her supervisor each school year (no cost).

Before play began, each team created two more policy options. All twenty options were written on newsprint and hung up around the room.

Tasks of the Consequence-Determiners

(1) Validate dollar amounts for each policy option created by participants.

(2) Determine if any legal entanglements might result from policy option decisions.

(3) Determine team influence allotment changes resulting from policy decisions.

(4) Determine team popularity changes resulting from policy option decisions—that is, would the team still be perceived by its constituency as representative?

Special Financial Rules

(1) On each new policy option written, note dollar amount necessary for one year's implementation—over and above current expenditures.

(2) Have this dollar amount validated by consequence determiners.

(3) Total amount actually spent may not exceed: $100,000 of salary monies and $8,000 of administrative monies.

APPENDIX II: SOME PROPOSITIONS ABOUT THE USE OF THE POLICY NEGOTIATIONS MODEL IN THE COMMUNITY CONTEXT

Earl S. Mulley
Center for Simulation Studies

The Center for Simulation Studies has used the *Policy Negotiations* model for more than four years in both academic and nonacademic contexts. Out of this experience we are willing to state certain propositions about the model. The validity of the propositions for us is rooted in user observations and reflections on the experience of the game play, immediately after the play itself and after some longer time period. It is on the basis of such positive and constructive feedback that we have continued to use the model, adapting and innovating when needed to fit the needs of client groups.

Propositions

(1) a tool to assist groups in analyzing and describing their situations both structurally and dynamically;

(2) a tool for sharpening the definition of the presuppositions,

character, and objectives of key interest groups in a system's decision-making process;

(3) an exercise in defining issue and policy alternatives, and exploring their various systemic implications;

(4) a means of surfacing role perceptions and feelings of members of a group or organization;

(5) a tool for anticipating strategies and tactics of conflicting interest groups in the process of policy formation;

(6) a device for making educated guesses about those forces outside the system that affect it;

(7) a means of discovering and even experimenting with negotiating and leadership skills and style;

(8) a means of creating empathy for those in another life situation by assuming their position and feeling the pressures exerted on that position by the system;

(9) a tool for testing the perceptions and data incorporated in the model against the strain or tension created in the model by the game play itself. (Where there is a fit, there is a learning; and where there is not a fit, there is a learning.)

Two Caveats

Because the people supply the context of the game out of their own experience, the simulation becomes the opportunity to sharpen and clarify, in the interaction and exchange process during and following the play, the varied perceptions of and feelings about the organization of which all the participants are a part. This process demands of the game director a broader range of skills than simple knowledge of the model. It is the conviction of the Center for Simulation Studies staff that the model is best used with a group as a part of a larger training or developmental program that has built into it those processes and skills to lead the group beyond the new insights and learning evoked by the *PN* model.

The *PN* model should be used within its limits, i.e., not with extremely complex organizations or with groups that want to use it as an idealistic "what if" excercise.

RELIGION GAMES AND SIMULATIONS
An Evaluation

by Lucien E. Coleman

INTRODUCTION

Interested in simulations and games in the area of religion? Well, as the old saw goes, "I have good news and bad news."

The bad news is that some of the old standbys listed in the religion section of the third edition of the *Guide (The Lovable Church Game, Agenda, Witherspoon Church, The Church Resources Game)* have gone out of print. The good news is that some interesting new products have come along to fill the gaps in our arsenals of religion-oriented materials.

The Purpose of This Essay

Religious educators make significant use of a wide spectrum of simulation gaming materials drawn from such fields as business, communication, ethics, sociology, and urban planning. Those materials are described in other sections of this *Guide.* The purpose of this essay is to describe, compare, evaluate, and offer suggestions for using ten representative games and simulations directly connected to religion. It should help you identify games that are potentially useful in your own work, become familiar with their strengths and limitations, and assess their value in terms of your own needs and resources. In addition, I include what I hope will be helpful suggestions for using these games, gleaned from numerous experiences with them in a variety of situations.

Definition

Religion is one of those ubiquitous terms that resist separation from other categories. Most religious educators would find it difficult to isolate religion from ethics, interpersonal relations, social issues, ecology, and a host of related areas. However, there are a number of simulations and games that are clearly related to the historical heritage or thought life of recognized religious traditions. These are the games examined in this essay. (A possible exception is *Ralph,* by Dennis Benson, which was originally used in a church setting but is not specifically tied to religious language.)

You will notice as you proceed through this essay that religion has been defined broadly. There are games related to Buddhism and Hinduism as well as to Christianity and Judaism.

Criteria for Selection

I would be very reluctant to claim that the ten games selected for inclusion in this chapter are the "best" in their category. For when you use evaluative terminology like that, you must also answer the question, "Best—for what purpose?" A game that serves very well as a vehicle for communicating concrete information might do less well as an instrument for attitudinal change. A game may be rich in ideational content but have poor potential for generating a variety of strategies. Some games are "best" for developing planning skills, whereas others are "best" for evoking appreciation of a religious tradition.

I can say, however, that every simulation and game included here has demonstrated its worth when used in appropriate settings for the right purposes. Each has been carefully examined and tested, and, to put it colloquially, there isn't a "clunker" in the lot.

Several considerations entered into their selection. First, all were commercially available through bookstores, distributors, or directly from producers, as of January 1979. Second, none costs more than $25. (I would have been willing to set the ceiling at $100, but it was not necessary to go that high.) Third, they represent a variety of purposes, content areas, and game types. This means that some games were purposely excluded, not because they lacked merit but because they closely resembled other games.

About half the games are marketed separately; but the rest are from very useful collections. For example, *Ralph* and *What Happened in the Garden* are from books that contain other games or simulation activities. *Formissia* is available in a boxed kit that includes five other items. Though a single game was chosen from each of these collections, other items of equal quality might just as easily have been selected.

Simulations, Games, and Other Things

Early on in this project, I fretted about the formal definitions of games, simulations, simulation-games, and the like, but finally decided that it was not terribly important. Such definitions are sometimes useful, but they should not be used to exclude good religious education tools which do not quite fit the labels. A stickler for definitions, for example, would not call *Nexus* a simulation game; but *Nexus* is unquestionably a useful vehicle for exploring spiritual gifts.

Keeping It Honest

The decision to include *Teaching Styles* and *Formissia* (with the concurrence of the general editors) presented a unique problem. It seemed a bit arbitrary to leave them out, simply because I had been involved in their development. On the other hand, no one can evaluate his own handiwork with complete objectivity. So we enlisted the services of two other persons to provide evaluative feedback on these games. Nancy Burgess, a specialist in single adult ministries, evaluated *Teaching Styles.* Billy Kruschwitz, field representative for the Foreign Mission Board, Southern Baptist Convention, provided input on *Formissia.* Each has made extensive use of the game designated.

Using the Tables

The tables in this essay will serve as quick-reference information sources for easy comparisons of the simulations and games described in the following pages. Table 1, for example, provides a capsule description of each game. Table 2 presents a summary of educational purposes that might be served by each. The text will explain the meaning of the data presented in subsequent tables.

GENERAL CHARACTERISTICS

When you begin the process of selecting a simulation or game for a given occasion, some very practical questions immediately come to mind: How many can play? How much skill and prior knowledge does the game require? For what age level is it suitable? How complicated is it? How much space is needed? How much time does it take?

Such questions are important, for a wrong answer to any one of them might preclude further consideration of a game. For example, a board game is hardly suitable for forty people unless you can provide multiple copies. And a game that requires three hours for playing and debriefing is not appropriate for a Sunday School hour, unless it can be spread over three consecutive sessions.

This section summarizes the general characteristics of our ten games and simulations under the headings of Player Characteristics, Time Requirements, Space Requirements, and Complexity.

Player Characteristics.

The *number* of players can be a critical factor in some games. For example, *Formissia* and *Persecution* tend to be dull with fewer than ten players, because it takes that many more to create enough interaction. *Leela* and *Rebirth,* on the other hand, would be boring with as many as ten players, because it would take a long time for everybody to complete a turn.

In some games, the maximum number of players is set by the artifacts in the game kit. *Gestapo,* for instance, includes 20 "value boards," one for each player; and *Teaching Styles* provides 120 cards, enough for a maximum of 30 players. There are a few games that will accommodate more players through "role splitting," an adaptation in which two players take the same role.

The *age level* of players is important not only because intellectual ability varies but also because characteristic patterns of social interaction change from one age level to another. Both *Gestapo* and *Persecution* are built on the theme of religious persecution; but *Persecution* is action-packed, whereas *Gestapo* is more cerebral and less active physically. Therefore, one might predict that junior high students would play *Persecution* with more enthusiasm.

But, do not observe the age level recommendations in Table 3 too rigidly. Some adults are uninhibited enough to have great fun playing a game designed primarily for youngsters; and some junior high students are bright enough to enjoy intellectually sophisticated games. And, by all means, feel free to experiment with intergenerational groupings. Simulation games can be great relationship builders and gap bridgers.

TABLE 1 Ten Simulations/Games for Religious Education

Simulation/ Game	Summary Description
Formissia	Players are missionaries in a foreign country, planning their work through negotiating proposals that compete for limited resources.
Gestapo	As German Jews during the Holocaust, players make value choices and struggle to survive under the oppression of the Nazis.
Leela	A board game of ancient Hindu origin in which players make uncertain pilgrimages from birth to a state of cosmic consciousness.
Nexus	Seated around a wheel-shaped playing board, players help one another identify and interpret their unique spiritual gifts.
Persecution	Players take the roles of Christians, pagans, Roman soldiers and magistrates, bishops, and slaves in a first-century scenario in which Christians must worship the emperor or face harsh consequences.
Ralph	Confined in a "survival shelter," players struggle with social, ethical, and theological problems as their "survival computer" generates crises.
Rebirth	Players strive to move from lower states of existence to Nirvana in this ancient Tibetan Buddhist board game.
Teaching Styles	Players examine their personal philosophies of teaching and learning as they evaluate, trade, and discard cards bearing a diversity of stimulator statements.
To Stay or Go	In the roles of Egyptian citizens, followers of Moses, and undecided Hebrews, players explore the pros and cons of leaving Egypt with Moses.
What Happened in the Garden?	Adam and Eve are on trial as players take the roles of prosecutors of defense attorneys.

TABLE 2 Educational Uses of the Ten Simulations/Games

This game . . .	May be used to . . .
Formissia	1) help learners develop vicarious identification with modern missionaries; 2) communicate information about the nature of modern missionary programs; 3) develop negotiation and planning skills.
Gestapo	1) introduce a unit of study on the Holocaust; 2) communicate specific information about the nature of the Nazi persecution of Jews from 1933 to 1945; 3) help learners appreciate the personal dimensions of that experience.
Leela	1) introduce a study of Hindu thought; 2) make learners familiar with Hindu terminology; 3) develop an intial conceptual framework for exploring Hindu thought.
Nexus	1) develop conceptual knowledge of the charismata listed in New Testament writings; 2) help learners identify spiritual gifts in themselves and others; 3) develop skills in interpersonal communication.
Persecution	1) help learners identify with first-century Christians; 2) kick off a study of early Christain history; 3) explore ethical problems (e.g., Is it okay to lie to save oneself and others?); 4) stimulate discussion of present-day challenges to faith.
Ralph	1) build group relationships; 2) explore ethical dilemmas; 3) analyze structural and dynamic factors in group life, including leadership; 4) stimulate discussion of relationship of science and religion.
Rebirth	1) introduce learners to Buddhist thought; 2) enrich a study of world religions.
Teaching Styles	1) kick off a unit on teaching in church training courses; 2) promote dialogue between church teachers and class members about the teaching-learning process; 3) help teachers identify the biases in their own styles of teaching.
To Stay or Go	1) help learners comprehend some of the human problems that might have been involved in the decision to leave Egypt; 2) enrich a biblical study of the Exodus; 3) help learners identify vicariously with the Hebrews of the Exodus.
What Happened in the Garden?	1) stimulate thinking about theological concepts such as the nature of sin, of man, of God; 2) promote dialogue on the nature of right and wrong and other ethical issues; 3) enrich or introduce a biblical study of Genesis 1-11.

The ten games reviewed in this essay require various levels, and various *kinds,* of skill and knowledge. For example, *What Happened in the Garden?* goes best with groups that have a reasonable degree of theological sophistication and creative imagination. But in *Nexus,* interpersonal communication skill is more important than these other characteristics.

Because the kinds of knowledge and skill required by these games are varied, it would be difficult to rate them on a single scale. So, in Table 3, you will find verbal descriptions, rather than quantitative ratings, in the "Knowledge and Skill" column.

Time Requirements

Time is often an important factor in the use of simulation and gaming for religious education, since so much scheduling is done in one- to two-hour periods. You should never schedule a game unless there is ample time for debriefing immediately after the game or in a subsequent session. Game leaders should also have a realistic perception of time needed to prepare materials.

Space Requirements

The playing areas that the ten games need fall into three general ranges. Board games require only enough space to seat participants around a table or in a small circle on the floor. Other games can be managed in regular classrooms or spaces that would accommodate equivalent groups in other kinds of learning activities. But a few require large rooms with enough space to permit lots of movement. For example, a playing area of 1,200 square feet would be none too large for a run of *Formissia,* and *Persecution* could well require even more room.

Complexity

Simulations and games may be arranged on a continuum from the simplest to the most complex. Simple games are quickly grasped, easily led, and require relatively stereotyped actions on the part of players. Complex games require more extensive explanation, experienced and skilled leaders, and usually call for a broad repertory of actions on the part of players. Characteristically, the rules for simple games are concise; rules for complex games are more extensive, sometimes requiring careful study and elaborate explanations by the leader.

In Table 6 our ten games and simulations are rated as simple, moderately complex, and very complex. These ratings refer to the relative difficulty of understanding the rules of play and the demands placed on the leader during play. Some games with fairly simple rules may become complicated in content (this is true of *Leela* and *Rebirth,* for example), but the categories in Table 6 refer only to rules and leadership, not to potential difficulty of content.

CONTENT AND PROCESS

Some games and simulations are designed to teach social skills, communication skills, or to evoke certain affective responses. These games are typically *process* oriented. They focus on the immediate experiences of the learners, their interactions as they play the game.

Other games and simulations are designed as vehicles for conveying information. These games are *content* oriented. They tend to emphasize cognitive learning, such as historical data, language meanings, or ethical principles.

TABLE 3 Player Characteristics

Game	No. of Players	Age-Level	Knowledge and Skill
Formissia	25-30 optimum	jr. high-adult	No special knowledge needed. Ability to negotiate and strategize helpful.
Gestapo	4-20 per kit	jr. high-adult	No special knowledge or skill.
Leela	4-6 optimum	jr. high-adult	Above average conceptual ability needed to grasp categories of Hindu thought.
Nexus	2-10 per kit	sr. high-adult	Works best if players know one another well and have a continuing relationship.
Persecution	25 optimum	jr. high-adult	Dramatic imagination and high level of tolerance for noise and action make it go better.
Ralph	Multiples of 6-8 (limited only by facilities)	jr. high-adult	No particular requirement, but interpersonal communication skills help.
Rebirth	4-6 optimum	sr. high-adult	Above average conceptual ability needed to understand and retain knowledge of Buddhist terminology. Game also requires patience.
Teaching Styles	16-30 optimum minimum of 4	college-adult	Some experience in Bible study groups is desirable, though not mandatory.
To Stay or Go	Indeterminate; but minimum of 10	jr. high-adult	Ability to formulate and present persuasive arguments; prior knowledge of Exodus 1-12
What Happened in the Garden?	12-36	college-adult	Prior familiarity with Genesis 1-3; ability to think theologically and to formulate logical arguments.

TABLE 4 Time Requirements

Game	Preparation	Play	Debriefing
Formissia	1 hour	1-1½ hours	20-40 minutes
Gestapo	10 minutes	45-60 minutes	1 hour
Leela	5 minutes	1-2 hours (variable)	30-60 minutes (variable)
Nexus	5-10 minutes	1-5 hours	1-1½ hours
Persecution	30-60 minutes	2-3 hours	30-60 minutes
Ralph	1-1½ hours	1½-2 hours	1 hour
Rebirth	5-10 minutes	2 hours or more (beginners take longer)	30-60 minutes (may be combined with game play)
Teaching Styles	5 minutes	30-60 minutes	30-60 minutes
To Stay or Go	10 minutes	1-1½ hours	30-60 minutes
What Happened in the Garden?	15-30	1-2 hours	30-60 minutes

TABLE 5 Space Requirements

Board Games	Games Requiring Regular Classrooms	Games Requiring Large Areas
Gestapo (4-8 players)	*Gestapo* (more than 8 players)	*Formissia*
		Persecution
Lella	*Teaching Styles*	*Ralph* (preferably several small rooms)
Nexus	*To Stay or Go*	
Rebirth	*What Happened in the Garden?*	

TABLE 6 Complexity

Simple	Moderately Complex	Very Complex
Gestapo	*Nexus*	*Formissia*
Leela	*Persecution*	
Ralph	*To Stay or Go*	
Rebirth	*What Happened in the Garden?*	
Teaching Styles		

Leela and *Rebirth* include extensive commentaries on the states of existence represented by the squares on the playing boards. The interaction between players is low key, informal, and, in fact, nonessential to the progress of the games. To a slightly lesser degree, *Gestapo* emphasizes presentation of historical information; interaction between players is not intensive. These are examples of content-oriented games.

Persecution, on the other hand, is a highly interactive game, involving persuasive communication, strategizing, and dramatization of roles. Though the game is related to a first-century historical setting, in a general sort of way, the emphasis is on the emotional responses and personal experiences of early

Christians under Roman persecution. There is little specific historical content. *Persecution* is primarily a process-oriented game.

Table 7 is a two-dimensional grid showing the relationship of each of our ten games to content and process. This graph is intended to portray only the relative emphases of these games as they compare with one another on these two scales. Because the placement of the games on the grid is contingent upon many factors which defy quantification, the writer has had to exercise a large measure of subjective judgment based on his experiences with the games.

TABLE 7 Content-Process Orientation

CONTENT (High → Low) / PROCESS (Low → High)	Low process		Medium process		High process
High	*Rebirth*				
		Leela		*Happened in Garden*	
				Nexus	
		Gestapo			*Teaching Styles*
			Stay or Go		
					Formissia
				Persecution	
Low				*Ralph*	

GENERATIVITY

One of the dullest board games I have ever played is a simple "race" game. Starting at "GO," the players take turns rolling dice and moving their respective tokens around a board until, eventually, one player's token reaches the finish line. That's it. Nothing more. No draw cards, no traps, no detours. Just roll the dice and count off the spaces. Ho hum. The main problem with that game is that it is "strategy poor." Players have no freedom to develop even the simplest of individual strategies. According to the rules of play, everyone does exactly the same thing in monotonous lockstep.

How vastly different is chess, or the Japanese favorite, Go. One can play those games for a lifetime and never tire of them. Why? Because they permit players to invent an almost endless variety of strategies. This ability of a game to beget alternative approaches to play might appropriately be described as "generativity." Chess is highly generative. The game described in the last paragraph is sadly lacking in generativity.

Compared to a 100-yard dash, football is a strategy-rich game. There are not many optional ways to run a 100-yard dash. But even though American-style football has been played for more than a century now, there probably are strategies yet to be invented.

Table 8 places the ten games reviewed in this chapter on a "generativity" continuum from low to high. Games on the low side of the continuum offer little latitude for making strategies. Games that are high in generativity are rich in opportunities for creative play.

(One word of cauton. Although there usually is a positive correlation between a game's ability to generate strategies and its interest quotient, it does not follow that all games that are

TABLE 8 Generativity

Low in Strategic Possibilities		GENERATIVITY		High in Strategic Possibilities
				Formissia
	Gestapo			
Leela				
		Nexus		
			Persecution	
		Ralph		
Rebirth				
	Teaching Styles			
			To Stay or Go	
				What Happened in the Garden?

low on this scale are necessarily dull. Sometimes, as in the case of *Rebirth,* the content of the game may capture the interest of players even though the rules of play are limited.)

DEPTH OF CONTENT

The depth and scope of content matter in these games varies widely. By content I mean information or subject matter, though it could be argued that affective experiences within a game are also content.

Gestapo, for instance, contains quite a bit of historical data relating to the Holocaust in Germany from 1933 to 1945. And the information within the game itself is augmented by an excellent bibliography of additional resources ranging from books to audiovisual materials. *Ralph,* on the other hand, seeks to establish a context for dialogue among players but does not present a body of specific information related to the issues it raises.

One indicator of the extent to which a game gets into subject matter is the amount of printed material devoted to such information. For example, *Rebirth* devotes 136 pages to explanations of the 104 squares on the gameboard, whereas *To Stay or Go* contains just two pages of persuasive arguments, pro and con, concerning the decision to join Moses in the Exodus. *Persecution* contains minimal historical material.

Let us make a distinction between the depth of content actually presented in game materials and the *potential* depth of study that might be generated by a game. *What Happened in the Garden?*, for instance, does not actually contain extensive resource information; but in the hands of a creative group of budding theologians with access to a good library it could lead to a quite sophisticated treatment of the universal issues it raises.

The ratings in Table 9 were based solely on information presented within the game materials. They have to do with "subject matter" content only, not with the complexity of the rules or the degree of sophistication inherent in the leaders' instructions.

TABLE 9 Depth of Content (Subject Matter)

Game	Rating	Comments
Formissia	moderate	30 player profiles based on actual personnel records of a present-day missionary agency.
Gestapo	moderate	80 information cards present factual data about the Holocaust. Bibliography contains enough sources for extended study.
Leela	extensive	108 pages of commentary on meanings of gameboard squares.
Nexus	moderate	18 pages of information on meaning and uses of spiritual gifts in player guides.
Persecution	minimal	25 fictitious role profiles to give players a "feel" for historical period.
Ralph	minimal	Recorded survival computer "voice print-out" merely guides group process.
Rebirth	extensive	136 pages of commentary on meaning of gameboard squares plus brief history of Tibetan Buddhism.
Teaching Styles	moderate	Statements on 120 stimulator cards designed to evoke personal responses on teaching; two pages on "inquiry" vs. "expository" styles.
To Stay or Go	moderate	Assumes prior knowledge of Exodus 1-12; presents two pages of "go" or "stay" arguments based on scriptures.
What Happened in the Garden?	moderate	Information primarily in explanation of roles, including lists of relevant scriptures.

PLAYER ROLES

Role playing is an important component of most simulation games. Some roles are highly abstract (Red Team, Blue Team); others are defined in terms of categories (bishops, soldiers, slaves); and some are defined specifically (Moses, Simon Peter).

Individual roles are assigned exclusively to individual players. In any given game only one person plays each individual role. In some cases, a detailed profile specifies the content of the role. But sometimes, especially when the role is that of a familiar historical character, the player must draw upon his or her own knowledge and imagination.

Group roles may also be based on historical groupings (followers of Moses, citizens of Egypt), but they are sometimes more abstract, as in the case of players in *Ralph,* who all become survivors of a world wide catastrophic event.

Functional roles are related to the operation of the game or simulation. In *Monopoly,* for instance, the banker plays such a role. In *Formissia,* there are designated group leaders (Glorious Leaders of Groups–GLOGs) who perform essential game functions.

TABLE 10 Roles and Role-Descriptions

Individual Roles	Group Roles	Functional Roles	Adequacy of Role Descriptions
Formissia		*Formissia*	Player profiles quite detailed.
	Gestapo		Each person plays individually, but roles are undifferentiated.
Persecution	*Persecution*		Some player profiles tend to be too brief.
	Ralph		All players are "survivors," a role defined only in scenario.
	To Stay or Go		Guidelines adequate.
What Happened in the Garden?	*What Happened in the Garden?*	*What Happened in the Garden?*	Group roles adequately defined in profiles; individual roles improvised by groups; "divine judges" have individual and functional roles

Table 10 presents information on the kinds of roles and the adequacy of role descriptions in six of the games. (*Leela, Nexus, Rebirth,* and *Teaching Styles* are omitted because they do not involve role playing.)

Games that include more than one kind of role are listed in each of the appropriate columns; therefore, some are listed more than once.

RULES

Rules are of paramount importance to a game. They establish the structure, set up patterns of interaction, determine allocation of resources, impose constraints, and guide the sequence of activities within the game.

The rules make the game. Without rules, there is no game. Without adequate rules, leader and players run into dead ends. Without clear rules, players are likely to find themselves in a welter of confusion.

Rules for simulation games fall into the two categories of rules for players and instructions for leaders. Rules are presented to players sometimes in printed handouts, sometimes orally by the game leader. Instructions for leaders should include directions for setting up the game, suggestions for introducing the game to players, and guidelines for debriefing, as well as procedures for operating the game.

Clarity and Adequacy of Rules

In all ten games, the rules were well written and well organized. One or two ambiguities emerged as we played *Gestapo,* but it was not difficult to improvise rules to cover these. *Persecution* also leaves a question or two unanswered. For example, the rules do not say whether or not the "secret" meeting place for Christians should be known only to selected players. Our outside evaluator pointed out that the instructions for running *Formissia* tend to be complicated and advises leaders to read through the rules several times before the game.

Mode of Presentation

Table 11 will tell you how rules for leaders and players are presented in the ten games, how extensive they are, and whether or not instructions for debriefing are included.

TABLE 11 Presentation of Rules

Game	Leaders' Instructions	Players' Rules	Guidelines For Debriefing
Formissia	17-page pamphlet	5-page pamphlet	detailed
Gestapo	4-page folder	given orally	8 questions
*Leela**	not needed	2 pages	(see footnote)
*Nexus***	9-page pamphlet	8-page pamphlet	(see footnote)
Persecution	4-page folder	given orally	adequate
Ralph	5 pages in book	recorded	sample debriefing
*Rebirth**	not needed	3 pages in book	(see footnote)
Teaching Styles	16-page booklet	given orally	3 pages of notes and questions
To Stay or Go	5 pages in book	printed handouts	5 suggested questions
What Happened in the Garden?	2 pages in book, gen. guidelines	printed handouts	5 suggested questions

*Leela and Rebirth are self-administering; no leader is needed; discussion may accompany play of game.
**Debriefing is possible but not essential, since the game itself involves extensive discussion among participants.

MATERIALS

The product is more important than the packaging, but the attractiveness and durability of game materials are worth considering, especially when you are putting out good money for them.

Attractiveness

All of the games reviewed here are attractively done. *Nexus* is of exceptional quality, and the printing in *Formissia* is creative. *Rebirth* not only is bound in a handsome paperback volume but also includes a beautiful four-color reproduction of a hand-painted Tibetan gameboard.

Durability

If you want to get the most from your investment, you will want to reuse game materials. Thus, durability is an important consideration. On the whole, the materials in the ten games are durable enough. But two or three potential problems are worth mentioning.

The individual player profiles in *Persecution* are printed on fairly light paper, and many of them came back creased, tattered, and wrinkled the first time the game was played. This is because players had to carry them around during very physical activity.

All of the games except *Leela, Ralph,* and *Rebirth* contain materials that must be handed to players. If you want to reuse these artifacts, you must take pains to explain that players should not "fold, spindle, or mutilate" them and to be sure they are collected at the conclusion of the game. This is especially important with one-of-a-kind materials, such as the individual profiles in *Formissia* and *Persecution.* If these are lost, you've had it. There's no way to replace them except to buy a new kit.

Packaging

Table 12 shows how each game is packaged.

TABLE 12 Packaging

Game	Book	Kit	Collection[1]	Nonprint Material
Formissia		X	X	
Gestapo		X		
Leela	X			
Nexus		X		
Persecution		X		Cassette recording
Ralph	X		X	Soundsheet
Rebirth	X			
Teaching Styles		X		
To Stay or Go	X		X	
What Happened in the Garden?	X		X	

[1]Packaged in books or kits containing more than one game.

SKILLS AND ACTIVITIES

When selecting games for particular instructional purposes, one needs to know what kinds of activities and skills will be required of the players. Table 13 has been prepared to give you an overview of the activities that are characteristic of each game.

CONCLUSION

Having examined our ten games point by point, it would be appropriate, perhaps, to weigh them against one another and come up with a ranking—good, better, and best. These games are so diverse, however, in content, structure, purpose, and style, that such a comparison would be virtually impossible. It would be like comparing oranges with apples—and grapes, and cucumbers, and pecans.

So, rather than pick a winner, let's try an overall rating of the games based on eight criteria: depth of content, interest level, generativity, clarity and adequacy of rules, pacing, flexibility, replayability, and packaging.

Depth of content was defined earlier as the extent to which the game goes into subject matter. The amount of information within the game materials is one index. The seriousness with which the material is treated is another.

Interest level has to do with the zest with which the players enter into the game. It is an answer to the question "Is it fun to play?" Let me confess to a large measure of subjectivity here. I can only base my judgments on my observations of people playing these games and, in some cases, on oral and written feedback in game debriefings.

Generativity, which was also defined earlier, has to do with a game's ability to generate a variety of strategies.

Clarity and adequacy of rules becomes an important criterion when you are trying to lead a new game. Clear rules are those that are easy to understand. Adequate rules are those that leave no doubts about how the game should be run.

Pacing refers to how well a game moves along. A poorly paced game tends to lag, it leaves too many players with nothing to do for long periods, or it becomes repetitious and monotonous.

Flexibility refers to a game's adaptability. Can it be adjusted, tinkered with, modified to fit a new situation or to accomplish a slightly different purpose? Can you improvise new content to fit the basic game model?

Replayability has to do with the possibility of playing a game more than once with the same players. If a game is not replayable, it is not necessarily a bad game. (*Starpower* is one of the most successful simulation games in the business; but most would agree that it is strictly a "one-shot" game.) On the other hand, replayable games are certainly more economical than others.

Packaging has been discussed already. Attractiveness, dura-

TABLE 13 Skills and Activities

Skills/activities	*Formissia*	*Gestapo*	*Lella*	*Nexus*	*Persecution*	*Ralph*	*Rebirth*	*Teaching Styles*	*To Stay or Go*	*What Happened in the Garden?*
Information Processing										
analysis	s		s	s		vs	s	s	s	vs
gathering information	m			m	m	m			s	s
planning goals & strategies	vs	m			m	s			s	vs
research									m	vs
extensive reading			s	m			vs			m
individual problem solving	m	m			vs	s		m	m	
Legislating Proposals										
making/revising proposals	s					s				
promoting proposals	vs					s				
opposing proposals	vs					m				
writing proposals	m									
voting on proposals	vs									
Group Activities										
group problem-solving	s	m			vs	vs		vs		vs
debate	m									vs
public speaking	s									vs
small group discussion	s			vs	s	vs		vs	s	s
listening				vs	s	s		m		
Human Relations										
competing	s								m	
persuading/influencing	vs	m			vs	vs		s	vs	vs
bargaining/negotiating	vs	m			s					
helping/supporting	m	m		vs	s	m				
interviewing				s					s	vs
deceiving/exploiting					m					
forming coalitions	vs	m			s			s		
Role Playing										
role playing an individual	vs				vs					s
role playing a work function	m				s					s
role playing a group role	m	s			vs	m			vs	vs
Resource Management										
survival		vs			vs	m				
maximizing resources	m	s			s					
managing resources	vs	s			s					
trading resources		m			s			m		
Evaluation (not in debriefing)										
self-evaluation		vs	vs	vs		m	vs	m		
evaluation of peers			s	vs		m	s			m
Miscellaneous Activities										
physical/mechanical activity	m				vs			m		m

Key: vs = very significant
s = significant
m = minor

TABLE 14 Overall Ratings of Games/Simulations

	Formissia	*Gestapo*	*Leela*	*Nexus*	*Persecution*	*Ralph*	*Rebirth*	*Teaching Styles*	*To Stay or Go*	*What Happened in the Garden?*
Content	2	3	5	4	2	2	5	3	3	3
Interest Level	5	3	3	3	4	5	3	5	4	4
Generativity	4	1	1	3	4	3	1	2	3	5
Rules	4	3	4	5	3	4	4	5	5	5
Pacing	3	3	2	3	4	5	2	5	4	4
Flexibility	5	3	1	4	4	3	1	5	3	4
Replayability	3	3	5	5	3	2	5	2	2	4
Packaging	4	3	4	5	3	5	5	4	4	5

Key: 1 = low; 2 = below average; 3 = average; 4 = above average; 5 = high

bility, and completeness of content are the criteria for judgment here.

Table 14 presents the results of these ratings. I hope this information will be of some help in your quest for good simulation gaming materials for religious education.

Good gaming!

SOURCES

Formissia (in Mission Games)
Lucien E. Coleman, Russell Bennett
1978

Brotherhood Commission
Southern Baptist Convention
158 Poplar Avenue
Memphis, TN 38104
$15.00

Gestapo
Raymond Zaverin, Audrey
Friedman Marcus, and
Leonard Kramish
1976

Alternatives in Religious Education, Inc.
3945 S. Oneida Street
Denver, CO 80237
$8.50

Leela
1975

Coward, McCann and Geoghegan, Inc.
200 Madison Avenue
New York, NY 10016
$4.95

Nexus
John Hendrix
1974

Broadman
127 Ninth Avenue
Nashville, TN 37234
$7.75

Persecution
James Buryska
1975

Contemporary Drama Service
Box 457
Downers Grove, IL 60515
$9.95

Ralph (in Gaming)
Dennis Benson
1971

Abingdon Press
Nashville, TN
$9.95

Rebirth
Mark Tatz and Jody Kent
1977

Anchor Press/Doubleday
Garden City, NY
$6.95

Teaching Styles
Lucien E. Coleman
1977

Broadman
127 Ninth Avenue
Nashville, TN 37234
$5.95

To Stay or Go (in Moses and the Exodus)
Jack Schaupp and
Donald L. Griggs
1974

Griggs Educational Service
1731 Barcelona Street
Livermore, CA 94550
$5.00

What Happened in the Garden? (in Using Biblical Simulations)
Donald E. Miller, Graydon F.
Snyder, and Robert W. Neff
1975

Judson Press
Valley Forge, PA 19481
$5.95

SELF-DEVELOPMENT GAMES AND SIMULATIONS
An Evaluation

by Harry Farra

INTRODUCTION

Why Self-Development?

The twentieth century opened awesome doors of personal and societal opportunities. But with all of this also came tension, frustration, incredible pressures, crises of all sorts, and a general threat to personal survival and purpose. We have long known that these are the best of times and the worst of times. We have been conditioned by a whole new vocabulary describing our existence: credibility gaps, fragmentation, alienation, valuelessness, purposelessness, boredom, lack of ethics, meaninglessness. Our shrunken spirits are starved for self-expression, nurture, satisfaction, freedom. From every avenue of society help has come in the form of books, courses, articles, and psychotherapeutic sessions, encounters, marathons, meditation strategies, and materials to instill better interpersonal awareness.

Many games and simulations have been developed along this line to help us remove the kinds of personal blockages that hinder growth and expression. They can provide us with skills and insights with which we may better understand ourselves and more effectively face reality.

Why Games and Simulations?

Games and simulations offer you an opportunity to play yourself free of personal hindrances and blockages. The play instinct is as old in human beings as the cave fire at which hunters enacted for family and friends their most recent success. Our Protestant ethic often divorced serious learning and playful fun, but we have lately rediscovered the subtle potency of learning through more playful educational models.

The rationale for using simulations and games to learn by rests on the theory of play. The principle is that learning is increased and enhanced when based on experiences that are fun, entertaining, playful, and gamelike. Psychologists, sociologists, and anthropologists have often emphasized the importance of the play instinct in culture, even in adult social contexts. In 1916, for example, Carl Seashore, in *Psychology in Daily Life,* asserted that "play is the principal instrument of growth. It is safe to conclude, without play, there would be no normal adult cognitive life; without play, no healthful development of the power of the will."

The Purpose of This Essay

This essay compares, evaluates, and rates on a number of scales ten self-development simulation games chosen for their effectiveness and variety. They are discussed and compared in areas that include content, learning objectives, roleplaying, the psychological schools or theories they represent, rules, skills, activities, debriefing, cost, and packaging. Finally, an overall rating graph sums up the strengths and weaknesses of all of the simulations.

Definition

Self-development simulation games are not easily categorized, since self-development and social development are often mixed. The line is rather thin, for example, between a communication game and a self-development game. A communication game can develop the self, and a self-development game can be geared toward improving communication. For the most part, however, a self-development simulation game is designed to bring about *personal* behavioral change.

The simulation games chosen here reflect a number of psychology-of-self approaches, such as balancing values, examining concepts of death and dying, learning how to make more effective decisions, and working through to clearer reality orientation. Some of the simulation games are obvious in their purpose, some very subtle; one is even "rigged" to produce a prejudiced response. Several of the games involve a very small group, whereas others can involve an entire class. Activities include role playing, movements on a game board, decision making, physical expression, testing, observing audiovisual presentations, and interpersonal bargaining.

Editor's Note: All simulations and games discussed in this section are listed in the self-development section, except for *Identity,* which is in social studies, and *Kidney Machine (1974 Annual Handbook for Group Facilitators),* which is in communication.

Criteria for Selection

A number of criteria were used to select the simulation games for review in this essay. All are presently available in the United States. They represent either a significant development in the self-help category or are so well known that they are considered to be a standard game or focus. All but one cost less than $100.00, and most cost less than $25.00. These simulation games were also chosen because they deal with a variety of topics, such as career guidance, facing death, working with values, learning to be O.K. The simulation games examined in this essay are: *Becoming a Person, Career Game, Cruel, Cruel World, Decisions, Hang Up, Identity, Kidney Machine, O.K., Play Yourself Free, Thanatos.*

Tables and Reference Materials

The reference tables begin with external features like packaging, playing time, age level, and number of players, and proceed to more substantive aspects like content analysis, sequential procedures, and model validity and effectiveness. Three major tables offer plot descriptions, a comparison of skills and activities, and the learning objectives of each of the simulation games. After this analysis, a final summing up of the comparative advantages of each simulation game appears in tabular form. The reference tables are, for the most part, self-explanatory. Where they need explanation, the text provides it.

Uses

Self-development simulation games are helpful to teachers, counselors, students, pastors, and enablers of all sorts. Some are complex and need good supervision and administration. Others can be handled in an informal setting of friends and peers. They can be used personally as learning tools, as training tools, as exercises to develop sensitivity, awareness, and values, and as worthy leisure time experiences. They are equally at home in church, classroom, or conference settings.

Their usefulness is determined by the honesty, openness, and intent with which they are played. In and of themselves they offer very little instantaneous help. Most provide simply a framework for viewing personal needs and capabilities. Thorough preparation, and previewing, is necessary to make any of them work. Keep in mind the necessity for reinforcement, debriefing, and follow-up strategies to make these simulation games most effective.

PRELIMINARY CONSIDERATIONS

Simulation Game Format

Of prime concern should be game aspects such as playing time, the number of players, the number of groups or teams, and the age level for which the game is designed. Most of the simulation games take about an hour to play. Some use only individual players; others use groups or individuals and groups.

TABLE 1 Self-Development Simulation/Games Plots

Simulation Game	Plot Description
Becoming a Person	Simulates how society discriminates against women. Teams acting as males and females choose alternatives in order to become fulfilled persons. But the game is "rigged"—"male" teams always score well; "females" do poorly.
Career Game	Used individually or with groups, this game involves the player in making decisions under a branching concept (process of elimination). The player discovers, at the end of a trail, a career choice he could not predict in advance.
Cruel, Cruel World	Goal is to achieve a "balanced" value profile. Players "balance" eight values in their own lives: affection, respect, skill, enlightenment, influence, wealth, well-being, and responsibility.
Decisions	Players make decisions for which they are awarded play money. The more consistent the decision is with predetermined goals, the more money.
Hang-Up	Players learn how racism affects the interaction of "stress situations," approached through various "hang-ups" and modified by "wild card" cop-outs.
Identity	Players learn seven identity areas: time, self-images, roles, work, sex, involvement, and values. Players then apply these identity areas to literature and life situations.
Kidney Machine	Players as members of a committee must decide which one person of five suggested will get to use a life-sustaining kidney machine and who must be allowed to die.
O.K.	Alternating between mini-lectures and exercises based on them, players learn the basic skills of transactional analysis such as: ego states, four phases of O.K.-ness, "strokes," games, discounts.
Play Yourself Free	Players learn to label and control emotions by overt intention rather than by accident or simple reaction through techniques of Rational Emotive Therapy.
Thanatos	Confronts players from a Christian perspective with the subject of death and dying, provides exercises to clarify personal views about death and descriptive situations that allow players to focus on death events and to look at alternatives.

Complexity

The complexity of simulation games can be difficult to categorize because games can often be played at different levels of sophistication. Nevertheless, we have tried to place our group of games on a continuum between simple and complex, according to their design and primary target level. Other factors we considered include the number of rounds, quantity of reading, playing time, ideational content, frustration factor, follow-up exercises, and how threatening the game

TABLE 2 Format

Simulation Game	Age Level	Playing Time (Hours)	No. of Players	No. of Groups
Becoming a Person	h.s., college, adult	1½-2	36	6
Career Game	8th-grade—college	5 class periods	class	individuals or class
Cruel, Cruel World	h.s.	1-2	3 or 4 players per game	individuals
Decisions	middle school	1	class	variable
Hang-Up	h.s., college, adult	1-2	individuals	
Identity	h.s.	enough time to cover a whole unit	class	various: individuals, partners, groups
Kidney Machine	h.s., college, adult	1	5-7 per group	any number
O. K.	h.s.	enough time to cover a whole unit	class	individuals and groups
Play Yourself Free	h.s., college, adult	variable	variable	1
Thanatos	h.s., college, adult	three phases or sessions	groups of ten	several small groups

might be to the participant. The following comments reflect how some of these factors affect certain games.

Identity and *O.K.* are the most complex, consisting of entire units of study in a classroom situation. Both are initiated by mini-lectures, involve take-home and classroom exercises by individuals and groups, and may consume up to several weeks. *Thanatos* demands that the players deal with a highly sensitive issue—death and dying—which could be treated rather flippantly or, on the other hand, make a player feel quite uneasy. *Becoming a Person* could be rather puzzling to players until that moment of recognition when they perceive that the game is rigged.

One of the interesting things about gaming is that participants can technically accomplish a game and sometimes be unaware of the underlying dynamic features. In other words, a game that is structured to be a rather complex experience may not be perceived that way. And players may respond to a seemingly simple game on a number of levels psychologically; for them it has become complex. In other words, complexity is not always an inherent feature of a game itself. I remember using a game called *Sensitivity* once in an acting class that precipitated more growth in my actors than any other assignment. It profoundly affected them by its implications, in spite of the fact that the game is rather simple.

PACKAGING

Certainly one main criterion in choosing a simulation game is its price and how sturdily it is packaged. Cost needs to be considered in relation to how well the game holds up under use, its initial cost compared with the long-range cost, the availability of replacement parts or materials, and whether the game includes all necessary materials or whether the user must provide supplementary materials like pencils, paper, paints, or scissors. If preparation time also includes duplicating materials, this can be costly in its own right. On the other hand, colorful packaging can catch the attention of players and whet their interest in becoming involved. For example, *A Balanced Life in a Cruel, Cruel World* has both a colorful box and a snappy title. *Hang-Up* catches the eye with its fourteen-inch-high cylindrical can.

Kidney Machine comes in the 1974 *Annual Handbook for Group Facilitators,* a book that contains other exercises and is well worth its cost. All the kits seem to be of reasonable quality, though a "kit" can have any number of component parts for the money. *Career Game* is probably overpriced, even considering that the kit does include a filmstrip and an audio tape.

Several of the simulation games require supporting materials like chalkboard, pencil, or paper. Some, like *Kidney Machine,* require duplicating of materials. This can run into considerable expense with large groups and continued use.

Two final packaging factors to note are storage and loss of pieces. Some of the kits, like *Play Yourself Free, Career Game, Hang-Up,* and *Cruel, Cruel World,* come in fairly large game boxes and obviously take a good deal more storage room than the book that contains *Kidney Machine.* A manuscript game like *Thanatos* can easily be filed (but also lost in a file). *Play Yourself Free* has several sets of cards that can eventually be lost, misplaced, or worn.

TABLE 3 Complexity

Easy	Moderate	Complex
Career Game	*Becoming a Person*	*Identity*
Decisions	*Cruel, Cruel World*	*O.K.*
Kidney Machine	*Hang-Up*	*Play Yourself Free*
	Thanatos	

OBJECTIVES

Beneath all the fun and playfulness of these games are their serious purposes and learning objectives. The major criterion in choosing a simulation game should be the compatibility of its objectives with your own. In this respect, let the consumer beware. Game makers have a tendency to claim more advantages and learning objectives for their games than what actually accrues. There are times, too, when a user can develop

TABLE 4 Packaging

Simulation Game and Publisher	Kind of Packaging	Cost per Kit or Book	Completeness	Durability and Reusability
Becoming a Person (Teleketics, Inc.)	kit	$7.95	s	average
Career Game (Educational Progress)	kit	$118.50	complete	high
Cruel, Cruel World (Pennant)	kit	$14.95	complete	high
Decisions (Innovative Education)	kit	$8.00	s	high
Hang-Up (Synectics)	kit	$18.75	complete	high
Identity (Interact)	manual	$14.00	d	high
Kidney Machine (University Associates)	book	$12.50	d	high
O. K. (Interact)	manual	$14.00	d	high
Play Yourself Free (Inst. for Rational Living)	kit	$16.75	complete	high
Thanatos (Simulation Sharing Service)	manual	$3.00	c, s	low

Key: p = purchasing other items
c = requires construction
d = requires duplication
s = requires support materials such as pencils, paper, chalkboard, timer

new objectives with a game. In table 5 we display what seem to be the relevant stated objectives of each game. It was sometimes necessary to condense these statements because of their cumulative length.

How Objectives Are Best Achieved

Not only is it important to examine and evaluate the various objectives, but it is also helpful to know how obvious these objectives are to the players. The success of objectives is often related to the way they are achieved. Learning objectives are more effective when realized with subtlety. Another aspect to consider is the chronology of objective achievement. An objective is sometimes not reached until the debriefing is concluded. Table 6 attempts to describe how the objectives for our self-development games are achieved.

Because *Becoming a Person* is rigged in favor of better scoring for one team, the learning objective is not achieved until the player realizes that there is a built-in factor of prejudice. This recognition may not come until the debriefing. *Thanatos* is played in three phases, and the learning may not be complete until all three phases have been accomplished. *Identity* and *O.K.* may not see all of these objectives realized until the entire unit is complete. *Hang-Up* is structured to bring the player back frequently to realizing the goals of the game; players can win only as others consciously guess their "hang-ups."

BACKGROUND AND ORIENTATION MATERIAL

Another important dimension in choosing a simulation game is the amount of information in the game manual. Some

TABLE 5 Game Objectives

Simulation Game	Objectives
Becoming a Person	1. understand discrimination against women. 2. experience feelings of discrimination. 3. discuss what it means to be a person in society.
Career Game	1. encourage students to think more systematically about career choices. 2. provide counselors with a career-guidance tool. 3. provide career information and motivation for further search. 4. increase student readiness to accept *realistic* guidance for career. 5. stimulate actual career decision and condition. 6. force students to challenge previous goals.
Cruel, Cruel World	1. understand how values are involved in real-life situations. 2. recognize values and how they are interrelated. 3. learn how values are enhanced or deprived by goals set. 4. learn that a balanced life can be achieved only through balancing eight particular values. 5. learn that this balance can be achieved by transacting values with himself or another. 6. learn to analyze own life.
Decisions	1. learn that goals and decisions about those goals are based on values. 2. learn that inconsistency leads to frustration.

TABLE 5 Game Objectives

Simulation Game	Objectives
	3. learn that consistency leads to living without tension.
Hang-Up	1. gain personal insight about racial attitudes and stereotypes. 2. grasp the emotional significance of a stressful situation in order to dramatize it. 3. project oneself perceptively into another player's dramatization.
Identity	1. gain knowledge of seven identity areas. 2. locate identity areas in others and oneself. 3. gain various group, individual, and dyadic skills connected with these identity areas and feelings in general. 4. develop a process for clarifying identity areas in which one is confused.
Kidney Machine	1. explore choices involving values. 2. study problem-solving procedures in groups. 3. examine the impact of individuals' values and attitudes on group decision making.
O.K.	1. learn how to understand inner feelings. 2. improve skills in relating with other people. 3. learn the major concepts of transactional analysis, how to identify these in self and others, and how to use these for greater self-understanding and interpersonal effectiveness.
Play Yourself Free	1. recognize that one has a choice of irrational or rational behavior. 2. recognize one's own self-talk and attitudes that cause one's feelings. 3. learn to combat and challenge attitudes that create painful emotions. 4. learn to feel neutral, calm, or happy despite bad or disappointing behavior by others toward one.
Thanatos	1. help one become aware of and constructively examine one's attitude toward death and dying. 2. clarify one's values about death. 3. learn to focus on death events and examine alternative responses from Christian perspective.

simulations are oriented toward skills, some to a single insight, and some are geared to involve the participant in a total growing experience. For the most part, simulation games in the self-development category tend to be rather slim in providing significant background information or resources. Less than half contain any bibliography.

Quantity of material cannot be a simple gauge of what a simulation's worth to you. It is relative to other factors like the purpose of the game, the quality of the information, and the model reality. But Table 7 will give you some indication of how much background or orientation information the games under consideration provide.

TABLE 7 Scope of Background and Orientation Material

Low (1-3 pages or implicit in game)	Adequate (10 or more pages)
Becoming a Person	*Career Game*
Cruel, Cruel World	*Decisions*
Kidney Machine	*Hang-Up*
	Identity
	O.K.
	Play Yourself Free
	Thanatos

VALUE ORIENTATION

The games under consideration vary greatly in the number of values they involve and their degrees of significance. The presentation of a single value may trigger all kinds of implications in the mind of an individual player. Because these simulations deal with self-development, most do have a value orientation of one sort or another, which often extends into interpersonal and group settings from a central internal, psychological concept. Some games describe in detail the values that underlie them, whereas others give no descriptive evidence of inherent values. Values may still be there, though they are not identified.

Cruel, Cruel World provides the most extensive discussion of values. Its very title suggests value in achieving a balanced life in a complex world, and the game is based on balancing eight categories of values. *Kidney Machine* is a precise model

TABLE 6 How Objectives are Best Achieved

Simulation Game	Covertly—Objectives Not Specifically Noticeable to Player	Overtly—Objectives Are Specifically Noticeable to Player	Debriefing Brings Out Objectives Significantly	Follow-Up or Preplay Exercises Strongly Reinforce Objectives
Becoming a Person	X		X	
Career Game	X			X
Cruel, Cruel World	X		X	
Decisions	X		X	X
Hang-Up		X	X	
Identity		X		X
Kidney Machine	X		X	
O.K.		X		X
Play Yourself Free	X	X		
Thanatos		X	X	

for teaching the relative worth of life in general and the social values attaching to certain stereotypes. *Hang-Up* deals with the volatile subject of racial prejudice and how deep-seated those feelings of tension are. *Decisions* rests on two linked perceptions of reality: when one's goals are inconsistent with one's everyday decisions, one may be frustrated in trying to cope, and when one's goals are consistent with one's everyday decisions, coping will probably be relatively easy. *Play Yourself Free*, attempting to provide skills of reality therapy, posits that once you really understand how your mind works in creating painful feelings, you can learn to control the mind and ultimately those hurtful feelings. The other simulations, such as *Thanatos*, *O.K.*, and *Identity*, also have significant value orientations.

Table 8 shows the variety of values the various games cover. There seems to be no specific pattern in terms of the *kinds* of values emphasized. The values range from being liked to beauty, from consistency to caring for other people. The table also indicates how the values are emphasized in the process of play. "Explicitly" means that the value is asserted, described, or discussed openly. "Implicitly" means that the value is handled in a subtle way, at a subconscious level, and is not specifically mentioned until, perhaps, the debriefing. Five of the games reveal some of their values explicitly and some implicitly. For instance, *Play Yourself Free* deals specifically with rational behaviors. But then there is a principle of entrapment in which the player, in defending choices, falls into irrational behaviors that the other players challenge.

A value as a value risks getting lost in the intricacies of the game. For example, I have found in using *Cruel, Cruel World* that because they are tied up with dice-rolling and peg-moving, the values sometimes lose their content and become only "items" to be manipulated.

TABLE 8 Values

Simulation Game	Value Areas Emphasized	How Emphasized: Explicitly or Implicitly
Becoming a Person	male-female social equality tolerance reality perception	implicitly
	life values: honesty, justice, perfection, fulfillment in life, uniqueness, goodness, pleasure, lawfulness, beauty	explicitly
Career Game	making realistic career goals seeking out more complete career information challenging previous career goals	explicitly
	building self-image through proper career selection	implicitly
Cruel, Cruel World	affection, respect, skill enlightenment, influence, wealth, well-being, responsibility; balance these	explicitly
Decisions	consistency	implicitly
	being liked, obeying parents, achieving or winning, getting even, caring about people, having money, having fun, being honest	explicitly
Hang-Up	empathy self-knowledge racial tolerance	explicitly
Identity	uses of time, self-image roles, work, sex, involvement, loyalty, ideological commitment, skills, knowledge, religion, love and family, power, health, freedom, objects, approval and acheivement, leadership and isolation, concern, confidence, resilience	explicitly
Kidney Machine	relative values people place on certain kinds of individuals in society	explicitly
O.K.	satisfaction with self and others	explicitly
Play Yourself Free	productive feelings; happiness, calm, motivation to change, acceptance, challenge; rational behavior	explicitly/ implicitly
Thanatos	feelings about death facing death realistically	explicitly

ROLE PLAYING

Kinds of Roles

Because they are dealing with individual behavioral change, self-development games tend to rely primarily on individual roles. Players are usually asked to envision themselves, as they are, but in different contexts or situations. For example, in *Thanatos*, you play yourself confronted with various situations that relate to death and dying. *Decisions* and *Hang-Up* do require specific role playing. *Hang-Up* also requires participants to role play stress situations, hang-ups, and comic stereotypes.

Table 9 provides a summary of role descriptions.

TABLE 9 Role Descriptions

Simulation Game	Roles
Becoming a Person	Team member seeking best human fulfillment alternatives.
Career Game	A person making career choices.
Cruel, Cruel World	A person seeking a balance of values.
Decisions	Banker, Dealer, Person making decisions.
Hang-Up	A person wishing to free self of prejudice. Frequent shifting from playing a black to playing a white.
Identity	Students in search of identity-solidifying skills.
Kidney Machine	Members of a decision-making hospital staff.
O.K.	Students seeking to understand themselves and how to relate to others better.
Play Yourself Free	A person wishing to develop better behavorial skills through reality therapy.
Thanatos	A person seeking to understand death and dying.

Role Description Detail

Because these simulations tend to let players be themselves, though in settings different from those they are used to, one could expect minimal role descriptions in the manuals. Table 10 shows that this holds true. In a number of games, like *Thanatos* and *Cruel, Cruel World,* the roles are only implied because it is assumed that players realize they play the game as themselves. Most of the simulations offer a modest description of roles. Three are somewhat lengthy in describing the context that the individual is being placed in. Although *Decisions* has two functional roles, dealer and banker, the real purpose of the game is tied to the nondescript individual and his or her decision-making process. *Hang-Up* roles are unusually structured. A player roles dice each turn. If the throw is even, the player is black; and if the throw is odd, white. This generates frequent alternation of racial status.

TABLE 10 Role Description Detail

Implied	Very Brief (1 Sentence)	Average (Several Sentences)	Lengthy Detailed (1-2 Paragraphs)
Cruel, Cruel World	*Becoming a Person*	*Career Game*	*Identity*
Decisions		*Kidney Machine*	*O.K.*
Thanatos		*Play Yourself Free*	
		Hang-Up	

SKILLS, ACTIVITIES, AND INTERACTIONS

Your choice of a simulation involves considering the kinds of skills and activities that might be compatible with your objectives. The self-development simulation games, while they vary greatly in the skills and activities they involve, often include such skills and activities as information processing, group strategies, role playing, team building, bargaining, self-evaluation, learning to cope, and working with values.

These kinds of skills and activities are mapped out in table 11.

Table 12 classifies the simulation games according to how many and how much various skills are practiced. Note that we can make only a general estimate of these two factors and that this table ought to be examined in relation to the main Skills and Activities table. The number of rounds played and the number of players or groups would certainly affect some of the outcomes in this category.

In terms of the degree to which skills are practiced, we rate some games high because of the extent to which they are played. *Identity* and *O.K.* are games that cover an entire unit and involve all kinds of follow-up and reinforcement activities.

The number of skills practiced varies considerably. *O.K.* and *Identity* require many skills because they are unit-long studies. *Cruel, Cruel World* is based on eight kinds of values and eleven concepts. *Play Yourself Free* necessitates grasping the various aspects of reality therapy theory. Two of the simulation games are based on learning only one or two skills but learning them well. *Kidney Machine* is geared primarily to group decision making and *Career Game* to choosing a career.

Your choice of a simulation game may often depend upon the focus of the skills in the game process. For example, if you are into a unit on team building, then a game with a primary focus on individual skills would not be suitable for you. The success of a number of simulations depends upon an individual's personal and internal understanding rather than some outer behavior. Into this category fall such games as *Thanatos, Cruel, Cruel World,* and *Hang-Up.* What happens within the individual is all-important. *O.K.,* on the other hand, focuses on better interpersonal relationships.

Interactions

Somewhat related to Table 13 on game process is Table 14, which displays the kinds of interactions that take place between players. Three simulations that allow a great deal of interactional flexibility are *Thanatos, O.K.,* and *Identity,* but many of the others are borderline. That is, they are adaptable to various kinds of interactional strategies, or they are vague to the extent that the kind of interactions in them depends on the natures of the participants.

Activities

The games in the self-development category exhibit considerable variety in "sequencing." Some, like *Career Game,* have logical sequences: one step prepares for the next. This

TABLE 11 Skills and Activities

Skills/Activities	*Becoming a Person*	*Career Game*	*Cruel, Cruel World*	*Decisions*	*Hang-Up*	*Identity*	*Kidney Machine*	*O.K.*	*Play Yourself Free*	*Thanatos*
Information Processing										
analysis		s		vs	vs	vs	s	vs	vs	vs
information gathering		vs				vs	s	s		
planning goals, strategies		vs	vs	vs				s	vs	
research		m				s				
reading (extensive)						m		m		
reporting-writing						m				
individual problem solving		vs		vs	vs	vs		vs	vs	
observation & identification						s		vs	s	
choosing	vs	vs		s		s	vs	s	vs	vs
Group Activities										
group problem solving	vs					s	vs	s	m	
public speaking						s				
small group discussion	s			s		s	vs	s		
class/large-group discussion						vs	vs	s		
challenging before group							m		vs	vs
Role Playing										
role playing specific individual		m			vs		s	m		vs
role playing group function	m									
role playing group member							vs	m		
switching roles					vs					
role playing emotional facets					vs					
Human Relations										
competition	m			s	s		s	m	s	
persuading/influencing										
bargaining/negotiating			m	s				s	vs	
helping/supporting			m		vs			s	vs	s
forming coalition										
team building	m						s			
Evaluation (Exclusive of Debriefing)										
self			vs	vs	vs	vs	s	vs	vs	vs
peer					vs				vs	m
Miscellaneous Activities										
competition with own goals	m	vs	s	vs	vs			vs	vs	
achieving balance	vs		vs	s	vs	s	s	vs	vs	vs
value clarification	vs	s	vs	vs	vs	vs	vs	vs	vs	vs
achieving insight	vs	s	s	vs	vs	vs	vs	vs	vs	vs
learning to cope		s	vs	vs	s	vs	m	vs	vs	vs
creativity				s		s	s	m	s	s

Key: vs = very significantly
s = significant
m = minor

TABLE 12 Practice of Skills

How Much Skills Are Practiced		How Many Different Skills Are Practiced		
Some	A Lot	Few	Some	Many
Becoming a Person	*Cruel, Cruel World*	*Career Game*	*Becoming a Person*	*Cruel, Cruel World*
Kidney Machine	*Decisions*	*Kidney Machine*	*Decisions*	*Identity*
Thanatos	*Identity*		*Hang-Up*	*O.K.*
	O.K.		*Thanatos*	*Play Yourself Free*
	Play Yourself Free			
	Career Game			
	Hang-Up			

TABLE 13 Game Process

Primarily Intrapersonal	Group	Mixed
Career Game	*Kidney Machine*	*Becoming a Person*
Cruel, Cruel World		*Decisions*
Hang-Up		*Identity*
Play Yourself Free		*O.K.*
Thanatos		

TABLE 14 Kinds of Interactions

Primarily Small Groups	Primarily Small & Large Groups	Flexible Groups*
Career Game	*Becoming a Person*	*Identity*
Cruel, Cruel World	*Kidney Machine*	*O.K.*
Decisons		*Thanatos*
Hang-Up		
Play Yourself Free		

*Groups form, disband, re-form, or individuals work on their own and then in groups.

also makes the "sequencing" tightly organized, and the game comes apart if you do not proceed step by step. Other games are loosely organized in terms of sequence. For example, *Play Yourself Free* (a four-in-one game) can be played dealing with either irrational or rational behavior (using either spinner system in the game). In *Kidney Machine*, the sequence of various items can be shifted. Some of the games incorporate a psychological structure as players move through the sequence of activities. That is, a player is put into a psychological state, set, or attitude before moving to the next item. To do *Kidney Machine* justice, for instance, one must feel a sense of crisis, then proceed to the next psychological state of evaluating human worth, and then to making life-and-death decisions.

Table 15 surveys the activities involved in each of the self-development simulation games.

RULES

In choosing a simulation game, you need to consider how much time it takes to introduce and implement. This depends in large part upon the clarity, simplicity, organization, and completeness of the rules. Here are some of the relevant questions you need to ask: What constraints do the rules place on the players? Are the rules frustrating? Are the rules easy to grasp and retain firmly without constant reference? Do the rules promote an argumentative spirit? How complicated are the rules? How long does it take to get the rules and directions across to participants?

Most of the games have rules that are clear, complete, and well organized. *Kidney Machine* is grasped quickly and easily. *Identity, O.K.,* and *Decisions* have rules that develop progressively throughout the game. *Decisions* needs an effective administrator to work through all dimensions of the game. *Identity* and *O.K.* are unit studies involving a number of activities that need a good deal of overseeing. A nice feature of *Hang-Up*

TABLE 15 Sequence of Activities

Becoming a Person	1. Teams divided into blue and pink representing males and females. 2. 4 rounds in which choices are made and scored. 3. Scoring rigged so blue generally wins. 4. Final discussion.
Career Game	1. See filmstrip and hear tape. 2. Players make career decisions one by one that lead them down certain tracks. 3. Success at planning scored. 4. Career cards offer further information to pursue. Chosen card tells player next card to choose. 5. Counselor—teacher gives help at end.
Cruel, Cruel World	1. Players set goals. 2. Pursue goals, working toward "balanced" life, achieved by balancing chosen values through transacting values with self or another. 3. Discussion. 4. Several rounds possible.
Decisions	1. Each player ranks eight values. 2. Each player establishes goals based on ranking. 3. Negotiate values. 4. Make decisions on basis of values. 5. Players learn consistency.
Hang-Up	1. Players are dealt "hang-up" cards. 2. Moves made on game board by dice roll. 3. Spaces on game board represent "stress situations" to be pantomimed. 4. Number on "stress situations" same as some numbered "hang-ups," which must also be pantomimed. 5. Alternatives to "hang-up" cards are "wild cards," which must also be pantomimed. 6. Players guess contents of pantomimes. 7. Debriefing.
Identity	1. Understanding seven identity areas through lecture, discussion, projects. 2. Applying seven identity areas to literature and life through lecture, discussion, projects. 3. Discussion along the way and at end.
Kidney Machine	1. Discussion of role of "values" in group problem solving. 2. Distribution of game information, biographical sheets, psychological sheets. 3. Small groups make value decisions. 4. Evaluation of choices and debriefing. 5. Can include role playing.
O.K.	1. Players learn ego states and about psychological stroking, psychological games and discounts, and how to stay O.K. 2. Debriefing.
Play Yourself Free	1. Choose character. 2. Discuss irrational behavior and guess which produce a certain negative emotion. 3. Learn to change irrational behavior to rational and move on to the road to reason.

TABLE 15 Sequence of Activities (Cont)

	4. Learn to challenge behaviors.
	5. Homework assignment.
Thanatos	1. Players sensitized to issue of death and dying in context of Christian beliefs.
	2. Clarify values about death.
	3. Focus on death events and alternative responses to them.
	4. Discussion follows each stage.

is a four-page brochure describing the game's logic and rationale.

Some of the games, like *Decisions, Hang-Up,* and *Play Yourself Free,* might need a rehearsal round to give participants their bearings.

One does notice that the rules for a game and the procedural directions for playing it sometimes blend in the instruction manual. *Decisions* has a section titled "rules," but it flows into the process of play and ends some twenty-five pages later. In *Thanatos* it is also difficult to pick out just what the ground rules are. *Play Yourself Free* has the same kind of problem. Without clear rules, it is difficult for a game administrator to identify the important constraints and keep them before a group.

Table 16 displays the number of pages given to explaining the rules and constraints for each game with an overall rating of the rules for clarity, completeness, and organization.

TABLE 16 Rules

Simulation Game	Number of Pages Devoted to Rules	Overall Rating for Clarity, Organization, Completeness: Adequate	Good	Outstanding
Becoming a Person	2		X	
*Career Game**	2			X
Cruel, Cruel World	2	X		
Decisions	25	X		
Hang-Up	6			X
*Identity**	6			X
Kidney Machine	1		X	
*O.K.**	6			X
Play Yourself Free	10	X		
Thanatos	2		X	

*Based on procedural directions rather than specific rules

RESOURCES AND SCORING

Another factor to look for in choosing the simulation game most suitable to your needs is scoring procedures and the kinds of resources available in the game for scoring, such as tokens, money, cards, buttons. How tangible are the resources? Is it possible that these resources blind the players to the true learning objectives of the game? Is the win/loss factor of the game made concrete for the players? What kind of feedback do players get to let them know how successfully they are performing? Do the players clearly understand what they have to do to score successfully?

Table 17 summarizes the resources and scoring procedures required for each simulation game.

TABLE 17 Resources and Scoring

Simulation Game	Resources	Scoring
Becoming a Person	no identifiable resources	Scoring based on points is biased and manipulated to demonstrate discrimination. Win/loss factor.
Career Game	no identifiable resources	Competition is not with others but the game itself. Points are scored. Can be replayed to better one's score.
Cruel, Cruel World	no identifiable resources	Scoring involves moves on a peg board. Win/loss factor on balancing values.
Decisions	play money and value cards	Score by making appropriate decisions.
Hang-Up	cards dealt represent hang-ups to get rid of	score by eliminating hang-up cards and moves on game board.
Identity	no identifiable resources	no explicit scoring
Kidney Machine	no identifiable resources	no explicit scoring
O. K.	no identifiable resources	no explicit scoring
Play Yourself Free	no identifiable rosources	Score points awarded by group consensus. Win/loss factor.
Thanatos	no identifiable resources	Score points. Win/loss factor.

Kinds of Scoring Procedures

Table 18 shows how the scoring of the simulations varies. There are some relevant questions to ask yourself at this point in choosing your simulation. Is scoring equally possible for all players? (*Becoming a Person* has as its object to violate this by allowing one team always to score better in order to simulate prejudice.) Is scoring by individuals or groups or both? Is scoring easy or difficult? Is the scoring earned, or does it come about by chance? To help you in this analysis, Table 18 divides the simulations into four groups.

TABLE 18 Kinds of Scoring

No Explicit Scoring	Individual Scoring	Group Scoring	Special
Identity	*Cruel, Cruel World*	*Becoming a Person*	*Career Game*
Kidney Machine	*Decisions*		
O.K.	*Hang-Up*		
	Play Yourself Free		
	Thanatos		

Scoring Procedures

Three of the games (*Kidney Machine, O.K.,* and *Identity*) have no explicit scoring system, but each one provides ongoing feedback throughout the game. Three more (*Play Yourself Free, Identity, O.K.*) have a number of follow-up activities that need some kind of evaluation. The debriefing sessions themselves provide feedback on how well the individual player has done.

In *Career Game,* individuals play against themselves, so the game can be played by a single student whose score is based on the appropriateness of his or her career decisions. In *Hang-Up,* one cannot score unless helped to score by others who guess what individual pantomimes represent.

Because these are self-development simulations, most of them have individuals scoring on an equal basis as they make an appropriate response. The competition tends to motivate the player to make personal improvement. This kind of motivation cannot be belittled.

We have already mentioned that the group scoring in *Becoming a Person* is rigged and that this unfair, prejudiced scoring system is the point of the game. Most participants have been very surprised to discover the built-in bias. As it is being played, the game seems trustworthy. Participants think they are after one learning value, when the game, seductively, teaches them another. The game always stimulates a good deal of discussion about prejudice and how it paralyzes personal growth.

MODEL VALIDITY

If you are interested in serious simulation gaming and in learning that counts, then you really need to see how well the particular simulation corresponds to reality. There are some simulation games that, although they have a lot of suspense and challenge, have little relation to reality. Without this kind of validity, players have difficulty making associations with their own lives. How much of reality is represented here? Is the model oversimplified or too complicated? What aspect of the reality is being represented in the game—content? process? attitudes? experience? Is the learning forced, or does it grow naturally out of the environment of the game? Table 19 shows our ratings of model validity and realism:

TABLE 19 Model Validity

Adequate	High
Kidney Machine	*Becoming a Person*
Thanatos	*Career Game*
	Cruel, Cruel World
	Decisions
	Hang-Up
	Identity
	O.K.
	Play Yourself Free

Kidney Machine is rated adequate because the kind of decision that has to be made, choosing who among a select group should live or die, is not likely to ever be in our province. *Thanatos* gears us to respond to death and dying, a subject we like to avoid. While some of the situations and attitudes may have no specific relation to where we are as individuals, the topic could certainly lead to wholesome discussion of death and the way we avoid it and rationalize it.

Hang-Up has some built-in features to constantly remind the players of the "realities" behind the game. And it has been my experience in playing this game that these protective features do sustain players in playing the game as a "slice of life."

Important Considerations

In connection with model validity, several important considerations arise:

(1) Does the game content represent an in-life experience that the player has had or will have? *Career Game,* for example, concerns career planning, a rather universal experience. *Kidney Machine,* on the other hand, presents a certain kind of crisis few of us will ever experience. Most of the games provide a field of experience well within possibility in our own lives.

(2) Does the game serve as a prototypical experience? Does it provide a kind of categorical experience? In this sense, *Kidney Machine* redeems itself. Although we may not ever face exactly the same crisis, the process of deciding between human beings is certainly a strong possibility. *Cruel, Cruel World* has this moral: you can survive in this world only as you learn to consistently balance your values. The game values here may not represent relevant ones for me, but the principle of balancing *some kinds* of values is relevant.

(3) Is the game structured in a way that makes participants lose sight of what game elements or dynamics symbolize? The producer of *Hang-Up* notes that "the potential of winning is central to any real game, but the drive to win can bring out so much competitive response that the growth of attitude awareness is subverted." I have found that participants can get so caught up in some games with collecting money, rolling dice, moving tokens on gameboards, turning spinners, challenging other players that the symbolized experience is no longer in conscious view. With *Decisions,* which involves a banking system of getting "bucks" and ridding oneself of IOU's to the bank, participants have sometimes lost sight of the point of the game. The player with the most "bucks" could conceivably learn the least about the concept of the game, although that would be difficult. In *Play Yourself Free,* players turn three kinds of spinners and two sets of cards, as well as defending their positions and challenging other players' positions. *Cruel, Cruel World,* as mentioned before, has wooden pegs, individual playing boards, two sets of cards, dice, and figures to move on a game board. One can easily lose sight of the fact that these game features are symbols of various aspects and values in life. In short, one can learn to play the game but miss the connection.

This is why effective debriefing is essential to good gaming. It places the game in perspective when the participants, in the heat and intensity of game playing, may have responded to the game only as a game and not as a representation of life and experience.

Most of the simulations we rate as high in their correspondence to reality, because, for the most part, they are able to overcome the potential problems related here.

DEBRIEFING

Debriefing is the heart of simulation gaming. One leader in the field of gaming once remarked that he would never consider running a simulation game unless it was followed by a debriefing session. Others have said that a simulation game is sure to fail in its learning objectives without some kind of debriefing, feedback, or follow-up. You will often find that if a set of debriefing questions is supplied with a particular simulation game, those debriefing questions not only are helpful but also provide an index or key to the real learning purposes of the simulation.

In choosing a simulation, then, you will want to know how extensive and perceptive the debriefing questions are that come with the game. You will find a whole range of debriefing responses. Some, you will discover to your regret, provide no debriefing suggestions at all. And you will often discover that you can think of better questions than those the game maker provides. Don't be afraid to pick, choose, and adapt the questions supplied with the simulation.

Try to have a structure to your debriefing session. For example, you might want to start with the content of the game, then discuss what happened in the game. You might follow this in an orderly way with an analysis of the players' feelings and what they learned, concluding with an evaluation of the game itself in terms of roles, reality, objectives, rules, interactions, and difficulty. The important thing is to be organized and have a debriefing plan.

TABLE 20 Debriefing Guidelines

Totally Inadequate (no debriefing questions)	Inadequate (few questions)	Adequate (8-12 questions)	Thorough (several pages of questions and guidelines)
Career Game	*Kidney Machine*	*Becoming a Person*	*Cruel, Cruel World*
Play Yourself Free	*O.K.*	*Decisions*	*Identity*
		Hang-Up	
		Thanatos	

Table 20 shows our ratings of the games according to the debriefing guidelines provided with the game.

As you can note from the table, almost half of the simulation games offer either no debriefing suggestions or very few. This means that you will have to build your own debriefing system for those games, and that will take time and thought. The game with the most comprehensive debriefing system is *Cruel, Cruel World.*

RESIDUAL EFFECTS AND LEARNING EFFECTIVENESS

A further important consideration to check in choosing a simulation game is the effectiveness of the learning experience. It accomplishes very little to learn something through a simulation game if a player is not as a result motivated to put that learning into practice or to make some kind of association with the real world. A further aspect of this concerns how permanently the principle of the particular game is learned. If a game teaches something that is forgotten a week later, we can question the learning effectiveness of that game. Obviously, part of the answer lies in the nature of the person who plays the game. But games can be built to reinforce learning and to provide such a selected gaming experience that the learning is made concrete.

It would help if the game producers would indicate the nature and extent of their field testing. Only one game in this survey does so—*Career Game.* Over 800 students played *Career* in the course of a year, 514 in classrooms and 286 independently under supervision. Less than 2 percent failed to take the game seriously, and the majority played the game out to the goal of three career choices.

Table 21 rates most of the simulations at least adequate or high in learning effectiveness. *O.K.* and *Identity* are the two most effective in this respect because they cover a number of weeks and have ongoing reinforcements, feedback, and take-home exercises and projects, as well as frequent debriefing.

The assignments in the table are not based on objectively measurable standards but rather on my own personal reactions to aspects like feedback later on from players, the effectiveness of the game in doing what it set out to do, how well the game went, the debriefing session, and sixteen years of college-teaching experience working with materials that either do or do not have "sticking" power. *Kidney Machine,* for example, almost always has good "sticking" power, residual effect. It immediately catches players' attention and fires discussion. I have talked to people years after playing *Kidney Machine,* and they still remember the experience and are able to relate its main features. *Thanatos* has some of the same effectiveness because it deals with such a personally sensitive topic.

TABLE 21 Residual Effects and Learning Effectiveness

Adequate	High
Becoming a Person	*Career Game*
Cruel, Cruel World	*Hang-Up*
Decisions	*Identity*
	Kidney Machine
	O.K.
	Play Yourself Free
	Thanatos

Learning Factors

Table 22 outlines more specifically some learning factors. Two obvious conclusions arise from an examination of this table. First, very few of the games lend themselves to being played solo. Second, the promotion of understanding of values, attitudes, and behavior is a strong feature of these games.

The range for skills is wide. Several of the games, like *Cruel, Cruel World, Identity, O.K.,* and *Play Yourself Free,* effectively develop both cognitive and affective skills. Others fall somewhere between these two points of the continuum for

TABLE 22 Learning Factors

Specific Learning Factors	*Becoming a Person*	*Career Game*	*Cruel, Cruel World*	*Decisions*	*Hang-Up*	*Identity*	*Kidney Machine*	*O.K.*	*Play Yourself Free*	*Thanatos*
Develop skills:										
Cognitive	M	E	E	S	S	E	S	E	E	M
Affective	E	NA	E	S	E	E	E	E	E	E
Provides for:										
Individual study	NA	E	NA	NA	NA	M	NA	M	NA	NA
Classroom use	E	E	E	E	E	E	E	E	E	E
Study, training, therapy groups	E	M	E	E	E	E	E	E	E	E
Fun time situations	M	NA	E	M	S	NA	E	NA	M	M
Promotes understanding of										
Values	E	E	E	E	E	E	E	E	E	E
Attitudes	E	M	M	M	E	E	E	E	E	E
Behaviors	E	M	M	M	S	E	NA	E	E	E
Conveys information about:										
Self	E	E	E	E	E	E	E	E	E	E
Others	M	S	E	E	E	E	E	E	E	M
Situations	S	M	NA	M	E	M	S	M	E	M
Facilitates:										
Principles acquisition	E	E	E	E	M	E	E	E	E	E
Behavioral change	E	M	M	M	E	E	S	E	E	E
Personal adjustment	E	M	E	M	E	E	S	E	E	E
Interpersonal sensitivity	E	NA	S	S	E	E	M	E	E	E

Key: S = satisfactory
M = moderately effective
E = very effective
NA = not applicable

kinds of skills developed and the effectiveness with which they are developed.

Because these are self-development games, they convey considerable information about self and others. Understanding specific situations does not appear to figure largely in the learning purposes of most of these games.

The games vary in their ability to facilitate such things as principles acquisition, behavioral change, personal adjustment, and interpersonal sensitivity. Several are rated very effective in all four categories. They are *Becoming a Person, Identity, O.K.*, and *Play Yourself Free.*

THEORETICAL SCHOOLS REPRESENTED

As a matter of interest you might want to know with what recognizable theoretical schools the various simulations tend to align themselves. You might have a particular preference for a certain psychological theory and would like to see how these simulations match up with your interests. Or if you see that a game falls into a certain category, you know that you will be able to find considerable background information for follow-up and debriefing work. It is good to see some of the theoretical schools wanting to make their theories practical and relevant through simulation gaming, an effective way to popularize some important and relevant psychological concepts.

Some of the games come directly out of a specific psychological school. For example, *Play Yourself Free* is based on reality therapy principles developed by the Institute for Rational Living. *O.K.* is based on the three ego states and the "I'm O.K., You're O.K." positions of transactional analysis. Several of the simulations are related to concepts of values clarification. Some loosely borrow concepts from various schools or seem to have not been discernibly influenced by any particular theoretical school. Table 23 points out some of the psychological influences behind the simulations.

FLEXIBILITY

The preceding pages have offered substantial information valuable in helping you choose a suitable simulation. A final consideration for you might be the degree to which you may adapt these simulations for different or more effective use. Most have some manipulable components or processes such as sequence of events, role playing, values, playing time, number of players. Some important questions to review are: Can it be adapted? How will the adaptation affect the outcome of the simulation? How long will it take to make the debriefing need to be adjusted to allow for the adaptation?

Three of the simulations, *Career Game, Hang-Up,* and *Becoming a Person,* have to be used generally as is. Since *Becoming a Person* is a rigged game, it would serve no purpose to modify parts of it. *Career Game* could function without the audiovisual aids, but not effectively. *Hang-Up* is extremely effective as is and I cannot see any modification in it improving the impact.

Five of the simulations allow for some variations in the playing. For example, *Hang-Up* provides blank cards for creating some original role-playing situations. *Play Yourself Free*

TABLE 23 Theoretical Schools

Simulation Game	Values Clarification	Sensitivity Training	Reality Therapy	Transactional Analysis	Cognitive Skills	Sex Discrimination	Kinesics	Psychodrama and Role Playing	Conflict Resolution	Not Discernable
Becoming a Person						X				
Career Game										X
Cruel, Cruel World	X									
Decisions	X				X					
Hang-Up		X						X		
Identity	X									
Kidney Machine	X									
O.K.				X						
Play Yourself Free			X							
Thanatos	X									

has some flexibility built in, but it also has three pages of follow-up exercises to be done before the next playing of the game. *Cruel, Cruel World* provides four suggestions on how to vary the play.

Kidney Machine is quite flexible and includes suggestions on how the game play can be varied. *Identity* and *O.K.* are highly flexible and contain a wealth of material. Both are intended to be unit-long studies and incorporate a good deal of content, many strategies, homework assignments, and observation exercises. These can be added to, deleted, or modified to suit your needs.

Table 24 indicates our evaluation of the simulations in terms of their flexibility. Keep in mind that a game's adaptability depends a great deal on the creative ingenuity of its administrator.

TABLE 24 Flexibility

Inflexible	Slight Alternations Possible	Flexible	Highly Flexible
Becoming a Person	*Cruel, Cruel World*	*Kidney Machine*	*Identity*
Career Game	*Decisions*		*O.K.*
	Play Yourself Free		
	Thanatos		
	Hang-Up		

CONCLUSION

We have covered many criteria in our evaluation. We hope we have helped you choose an appropriate simulation game or at least suggested to you the kinds of questions and factors to consider in choosing a game. To tie all of our evaluation together, we offer table 25, which makes an overall comparison of the simulation games.

TABLE 25 Overall Ratings of Simulation Games

	Becoming a Person	*Career Game*	*Cruel, Cruel World*	*Decisions*	*Hang-Up*	*Identity*	*Kidney Machine*	*O.K.*	*Play Yourself Free*	*Thanatos*
Roles	1	0	1	1	2	1	1	1	1	1
Rules	2	2	2	2	2	1	1	1	2	1
Skills	1	1	2	2	1	2	1	2	2	1
Validity	2	2	1	2	2	2	1	2	2	1
Flexibility	0	0	1	1	1	2	1	2	1	1
Debriefing	1	0	2	1	1	2	1	1	0	1
Completeness	1	2	2	2	2	1	1	1	2	1
Durability	1	2	2	2	2	2	2	2	2	0
Expense	F	E	F	F	F	F	F	F	F	I
Content	1	2	2	2	2	2	1	2	2	2
Learning	1	2	1	1	2	2	2	2	2	2
Other	1	1	2	2	2	2	1	2	2	1
Total	12	14	18	18	19	19	13	18	18	12
Overall Rating	Good	Good	Excellent	Excellent	Outstanding	Outstanding	Good	Excellent	Excellent	Good

Key: 0: not applicable or not relevant
1: adequate
2: excellent
I: inexpensive (under $5)
F: fairly priced ($10-$20)
E: expensive (over $100)

We have rated over half the simulations as excellent or outstanding. *Identity, Hang-Up,* and *O.K.* received the highest ratings. One of your favorite and most usable simulations may have been given a low rating. This is the assessment of one critic and should not be taken as the final word. These ratings are intended as a guideline to help you make a good decision about what simulation might best fit your needs.

A final table summarizes in a general way the strengths and limitations of the ten games.

Simulation games are to be taken seriously. The best way to fail in simulation gaming is to simply grab a game and play it without much thought and preparation. Simulating gaming often demands more preparation time than another mode of learning. In other words, gaming is not a shortcut. But once you gain some expertise in administering simulation games, a whole new world of learning possibilities will open for you and your lucky group.

TABLE 26 Strengths/Limitations Comparison

Simulation Game	Strengths	Limitations
Becoming a Person	1. Deals with a relevant issue; women's rights. 2. Ingeniously "rigs" the game. 3. Supplies ample debriefing questions.	1. Game can be played only once with the same group. 2. Game seems to involve a lot of smoke-screen activity to accomplish its "rigged" status.
Career Game	1. Offers careful, controlled instruction in career planning. 2. Can be used with an individual or groups. 3. Is reinforced with media such as filmstrip and cassette. 4. Provides learning features such as the "educational price of decisions" and how to read warnings along the way in choosing career. 5. Offers information on a number of careers. 6. Scoring helps motivate even when played solo. 7. Can be played across five class periods for reinforcement.	1. Expensive: around $120.00 2. While the game offers some of the complexities of career planning, it can hardly provide those dynamic, subconscious features that go into career decisions. It assumes that career planning is a conscious activity.
Cruel, Cruel World	1. Teaches the balancing of values. 2. Offers interesting game format. 3. Includes "negotiating" feature. 4. Provides long list of debriefing questions and suggests game variations.	1. Limited to four players at a time. 2. Increase in a value often depends on chance roll of dice. 3. Limited to eight values. 4. Values as values are lost in rolling dice, moving figures on a board, and moving pegs.
Decisions	1. Teaches consistency of goals and decisions. 2. Monetary system helps motivate decision making. 3. Case study situation provided for clarifying and illustrating. 4. Effective debriefing process.	1. Purpose of the game sometimes lost in monetary competitiveness. 2. Some aspects of game seem only loosely related to others.
Hang-Up	1. Game causes players to alternately play member of black or white race. 2. Game structures allow for getting rid of "hang-ups" only by help from others. 3. Stress situations and hang-ups to be pantomimed are very realistic. 4. Clear, organized rules.	1. Needs to include more specific help in pantomiming. 2. Game blends serious and comic situations—players need to be made to understand why comic situations are included.
Identity	1. Offers an entire unit of study. 2. Offers effective follow-up exercises. 3. Offers substantial debriefing. 4. Offers a multiplicity of reinforcement exercises and experiences. 5. Offers considerable theoretical information concerning identity areas.	1. Has very few game or simulation features; more an instructional model. Therefore, lacks international features of gaming. 2. To accomplish the full purpose, the entire unit has to be done.
Kidney Machine	1. Provides attention-getting format. 2. Creates dynamic value decisions. 3. Provides effective discussion and decision-making tool. 4. Very adaptable: biographical data can be changed.	1. Situation too hypothetical to be taken very seriously. 2. No single solution—depends on values group decides to set. 3. Too many variables in the biographical data and in the situation itself.

TABLE 26 Strengths/Limitations Comparison (Cont)

Simulation Game	Strengths	Limitations
O.K.	1. Offers good introduction to transactional analysis. 2. Provides an entire unit of study. 3. Excellent debriefing. 4. Offers a multiplicity of reinforcement exercises and experiences. 5. Substantial background and theoretical information surveyed.	1. Has very few game or simulation features—more an instructional unit; therefore, lacks the effective interactional features of gaming. 2. To accomplish the full purpose, the entire unit has to be done.
Play Yourself Free	1. Covers well the basics of reality therapy. 2. Offers four games in kit. 3. Has several levels of learning about rational and irrational behavior. 4. Creates a situation where participants fall into the same kinds of irrational behavior they're trying to isolate.	1. Takes some deep thought and keen perceptions. 2. Requires participants to think on several levels at same time; participants must be a select group. 3. Scoring is awarded by group consensus on best choices. Participants can become too involved in "persuading" group or feeling unjustly ruled against.
Thanatos	1. Handles an unusual topic—death. 2. Game factors are built-in phases. 3. Game in part builds upon another successful game Propaganda. 4. Includes value clarification and attitude application exercises. 5. Uses case study situation for clarity and relevance.	1. Because of content, game could make certain kinds of participants very uneasy. 2. Needs skilled administrator. 3. Designed for a church setting. Theology of Phase I may not be acceptable to some churches. 4. Scoring system seems to get in the way of the gaming; it's not that important to the game.

SEX ROLES GAMES AND SIMULATIONS
An Evaluation

by Cathy S. Greenblat and Joan Gaskill Baily

INTRODUCTION

Are you teaching a course or working in a program of Women's Studies? Do you wish to put a unit or module on changing sex roles into a social studies or social science course or curriculum? Or are you involved with a group of men or women in the community who wish to explore changing male-female relationships? For any of these audiences you may hope that a gaming-simulation will give you an exciting and provocative way to begin, prompting players to further explorations. Alternatively, you may hope that you can use a gaming-simulation to teach some of the principles of sexual stratification, differential opportunities and rewards, or alternative ways males and females can deal with each other in personal and social relationships.

If so, there are two strategies you can employ.

Sex Roles Issues in a Number of Games

First, there are a number of games dealing with other topics that can be discussed in terms of sex role dimensions. Any game that includes male and female roles contains implicit assumptions about sex roles. Thus, you can lead the postplay discussion of one of these games in terms of the differential decisions and responses of male and female players, irrespective of the sex role they played, or the differential decisions and responses of male *or* female players who played the *same* sex role.

Another Option: Games Specifically Dealing with Sex Role Issues

There is another and, we believe, a better option. In the past few years, several games dealing specifically and explicitly with sex roles, "women's issues," and relationships between men and women have been developed and are available to you. The purpose of this essay is to compare the best of these, giving some you guidance in selecting the one or ones most appropriate for your purposes and for the group that will be playing.

While this sounds optimistic, we must note at the outset that we have found more *poor* sex roles games than good ones. In our own teaching and again in preparation for this article, we have run and critically reviewed fourteen games, and though we had hoped to compare eight or nine here, we could not find more than four that met our minimum criteria for this review. We would have recommended two other items we have seen: *Make Policy, Not Coffee* and *Liberation: A Role-Playing Simulation*, but we have not included them here for several reasons. *Make policy, Not Coffee* appeared in the September 1975 issue of *Simulation/Gaming/News* and, as of this writing, is therefore very difficult to obtain. Also, both are relatively simple exercises, rather than games or simulations, and thus we have excluded them on the grounds that they are not really amenable to comparison on most criteria with the four included.

There is one further game that we consider valuable and that met our two criteria: *Herstory*. While we had initially planned to include it, we have decided not to because it really is several games and related exercises all in one and thus is not amenable to easy analysis of roles, rules, scoring procedures, and so forth. Instead, we have included *Seneca Falls*—a component part. We recommend that you look at the description of *Herstory* in the social studies section and consider using it if you are working with high school students and can devote a large amount of time to a game.

Criteria for Inclusion in This Essay

Although we have been concerned with such considerations as availability and reasonable cost, there are two critical dimensions to our recommending a gaming-simulation. First, it must contain a valid model of the dimension of reality it simulates; this is the "model" dimension. Second, it must be smoothly playable and operable; this is the "gaming" dimension. These two components do not necessarily go together; it is possible for a game to have an excellent underlying model but be unwieldy to operate, too long, and so forth. Alternatively, we have seen some games that are simple to operate and fun to play—but do not accurately model reality.

Editor's Note: *Access* is listed in the social studies section, *Job Crunch* in education, *The Marriage Game* in self-development, and *Seneca Falls* in history.

On one or both of these criteria, then, we have rejected the following games: *Encapsulation, ERA, Female Images, The Lib Game, Pro and Con, Woman and Man,* and *Women's Lib.* Our notes on these materials are perhaps instructive. They contain the following comments, related to the two dimensions of model and playability:

The probabilities of success are highly distorted.

The consequences are not true to life.

Information on the issues is too meager.

The model is a zero-sum game: as one sex gains the other automatically loses.

The images are so stereotyped that little empathy for either sex is promoted except about its shortcomings.

The instructions are confusing and hard to read.

The situation cards are unrelated to the list of identities.

Play is boring.

Events are too exaggerated to be taken seriously.

The game never ends!!!

Using This Essay to Select a Game

The model and the game elements, then, are the two criteria we used to select the games included in this overview. All four games—*Access, Job Crunch, The Marriage Game,* and *Seneca Falls*—have been judged to contain fairly accurate representations of some dimension of sex roles and to be games you can operate smoothly for a valuable and enjoyable experience for high school or college students or an adult group. None is very expensive.

While all qualify as "sex roles games," you will see in Table 1 that each deals with a different subtopic. If you wish to use a game to explore a particular dimension of sex roles, such as affirmative action or the distribution of power in a marriage relationship, our comparisons will be of limited value, for you will find only one game that is substantively appropriate. A reading of this essay will nonetheless give you a better idea of that game's characteristics and how it compares with others. If, on the other hand, your needs are more general, you will find four good games reviewed here, and you will find bases for selecting among them outlined.

TABLE 1 The Four Sex Roles Games

Simulation Game	Summary Description
Access	Players as members of two groups in a society have differential access to the society's rewards and resources.
Job Crunch	Players as college faculty, administrators, and students proceed through the hiring process and explore implications of affirmative action as they consider three candidates for one faculty position.
The Marriage Game	Players as single males and females make the decisions of courtship and early marriage, including whether to stay single, cohabit, or marry; jobs; arrangement of finances; allotment of leisure time; fertility; and sexual gratification. They may also consider whether to stay married or to divorce.
Seneca Falls	Players as delegates recreate the first Women's Rights Convention held at Seneca Falls in 1848.

A further sense of the differences between the four games will follow from a review of the issues each deals with, as Table 2 shows.

PRELIMINARY CONSIDERATIONS

Age level, Number of Players, and Playing Time

Three questions you are likely to ask as you try to select from available games are: What age level is it appropriate for? How many (or how few) players are required? How long does it take to play? These separate questions are partly linked, for if you are a high school teacher you will have different needs and constraints from those of a college teacher or someone working with a community group of adolescents or adults. Table 3 summarizes these dimensions.

Several of the games are playable by students of high school age. *Seneca Falls* is designed specifically for this age group, and both the number of players the game can accommodate and the amount of time to play have been devised to fit easily with the high school teacher's needs.

Access may also be played by high school students, and a typical size class can participate in one game run. The teacher who is sharply restricted to the fifty-minute period, however, may have difficulties running this game. While little time is

TABLE 2 Issues in the Sex Roles Games

Social Issues	*Seneca Falls*	*Access*	*Job Crunch*	*The Marriage Game*
Discrimination against women:				
Historical	vs	m	s	m
Contemporary	vs	vs	vs	vs
Domains of discrimination:				
Education	x	vs	vs	s
Family	vs	s	s	vs
Equal rights	vs	vs	vs	vs
affirmative action	m	m	vs	m
Relation between racism and sexism	s	m	vs	vs
Power, differential access to, by sex	vs	vs	vs	vs
Relations between social class and sex role options	m	m	m	vs
Interpersonal				
Marriage, costs and rewards	m	vs	s	vs
Parenthood, costs and rewards	m	m	m	vs
Consequences of pregnancy	m	m	m	vs
Sexual double standard	m	s	s	vs
Sexist attitudes	vs	vs	vs	vs
Career vs. family life	s	s	s	vs
Nature of Women; Male/Female Differences	vs	vs	vs	vs

Key: m = minor
s = significant
vs = very significant

TABLE 3 Sex Roles Simulation Games: Some Basic Considerations

Title	Age Level	Playing Time	No. of Players
Access	high school, college, adult	1 1/2-2 hours	10-35
Job Crunch	college, adult	2 1/2-3 hours	20-30
The Marriage Game	high school, college, adult	6-10 hours	2-60 (more with another operator
Seneca Falls	high school	3 class periods with study time between	15-35

required to explain the rules and get things moving, and the postplay discussion or debriefing can be deferred to the following day, our experience is that the game cannot easily be played in fifty minutes, nor can it be "broken" into two days of play without considerable loss of momentum and enthusiasm.

Although *The Marriage Game* has been marketed primarily for colleges, it is playable by high school students. The first round, which always takes about two hours as players make themselves familiar with the numerous decisions, charts, and scoring techniques, probably would need to be stretched to three fifty-minute periods for high school students. After that, they, like college students or adults, could average a round (representing a year) per period, thus playing a five-year period in seven to eight class periods, followed by several periods of discussion.

Access and *The Marriage Game* are appropriate for both college students and adults. The number of players for *Access* is quite flexible, making it usable with small (10 to 15) and medium-sized (25 to 30) classes. We have run it with groups as small as 10 and as large as 50. We find 10 possible, but the experience has not been nearly as powerful as with a somewhat larger group. With 50 players, the game was unwieldy, though it "worked." We do not advise using it with a group that size (neither do the designers). Play can easily be encompassed in the one hour and fifteen minute class period common on many campuses, with discussion at the next meeting; with a longer period, both play and discussion can take place the same day. Where this latter option is available, we recommend taking it. If the play and discussion must be on separate days, we have found it useful to give students a list of some of the questions for discussion before they leave the first day. This seems to help them focus their thinking about the experience while it is still fresh. (This strategy is advisable with any game in which the debriefing is delayed.)

The Marriage Game can be played in class periods of any length, though, as previously noted, the first round will require about 2 hours and subsequent ones about 50 minutes each. Most college instructors we know of have devoted 3 weeks of class to play and discussion of the game; many others have scheduled play over a weekend. The long period of time required will make the game unusable in some contexts. One instructor can run the game with anywhere from 2 to 50 or 60 participants; with larger classes, it is best to have an assistant operator or to have a few students play the first round in advance and then circulate to aid others as they play the first round in class. If you are very familiar with the game, you can, with the aid of a microphone, run the game for a very large group (one of us has run it for 250 participants in a gymnasium), but we do not recommend this unassisted form with a large group for any but the brave, committed, and desperate.

Job Crunch is also appropriate for college students and can accommodate a moderate-sized class. It is possible and easy to break it into modules if class periods are short, extending play over a two-day period, followed by a day of discussion. It probably should not be played with freshmen or those not somewhat aware of academic roles and academic politics. We found the game particularly appropriate for and interesting to graduate students, faculty, and administrators. Adults not familiar with university settings would probably not constitute a good player group, as "academic amateurs" may not know how to play the roles effectively. One such player in a group of people generally knowledgeable about the university world commented, "My lack of familiarity with the 'ins and outs' of hiring procedures and the criteria for determining the 'best qualified' candidate for the teaching position caused me anxiety, for the role demands such knowledge."

In short, whatever age group your players are, there are several choices of games available to you, but the choice may be somewhat narrowed by the time constraints and the number of participants in your group. None of these games, for example, is appropriate for a large class of college students in which you are willing to devote two periods to play and discussion—an unfortunate situation, since this is a common requirement in many colleges. With extra help and extra rooms, it is, of course, possible to divide a class and run several groups in a game simultaneously.

ROLE PLAYING

Kinds of Roles

The four games include several different types of roles. First, there are *historical* roles, found in *Seneca Falls.* Here some players are specific figures from that time (e.g., Elizabeth Cady Stanton), and others are unnamed persons who lived at that time and attended the convention. Second, there are *functional* roles, found in *Job Crunch*: President of the college, a black faculty member, and so forth. Third, there are *abstract group* roles as in *Access,* where players are members of the red territory or the green territory. Fourth, players may play *themselves.* In *The Marriage Game,* players usually play themselves, but slightly older (that is, college students play as if they have just graduated from college). Adults may play themselves at their present age, marital, and parental statuses or may "re-live" some of their young adulthood, playing as if they, too, had just graduated from college. In addition, players in *The Marriage Game* may play the opposite sex role—that is, males may play females and vice versa. Whatever these demographic characteristics, players play with their own values and preferences rather than having these assigned. Table 4 summarizes the roles in the four games.

TABLE 4 Roles in Sex Roles Simulations

Title	Roles
Access	Members of red territory; members of green territory
Job Crunch	Candidates for college teaching position; college faculty; college students; college administrators
The Marriage Game	Individual males and females
Seneca Falls	Nineteenth-century men and women attending the convention; historical figures: Lucretia Mott, John Mott, Elizabeth Cady Stanton, Frederick Douglass

Role Description Detail

Players in *Access* are given very brief descriptions of the roles they are to play; the game dynamic evolves through the structure of constraints placed upon different players. The rules are unequal for reds and greens, and the latter rapidly discover that they are permitted to do less and that they are unable to accumulate resources as easily or as extensively as the reds can.

The *Job Crunch* roles are generally one paragraph long and are well constructed. As an example, consider what is given to the player who is the President of the College:

> President of the College; Ph.D., Chicago, U.S. History; Former successful Director of Admissions. Ethical, indecisive, good-natured, uneasy in conflict situations and does not wish to offend anyone. Tends to respond to situations as they arise. Can be manipulated by strong personalities. Gives unpleasant tasks to the Provost, your second in command. Married. 4 children. Member of the Episcopal Church.

Job Crunch players are encouraged to take the initial few minutes of play to think about the person they are playing and define some additional components of that person's personality and preferences. This was fairly easily done by the graduate student and adult players we observed. Some college students and nonuniversity adults required assistance in getting started, because they knew something about what college faculty "were like" but not how they acted with their peers and what kinds of questions they asked candidates for a position.

The Marriage Game roles are fairly detailed, but they are defined by the players themselves, as they specify their likes and dislikes for leisure activities and consumption (in the "Catalogue") and specify their values in the first step of play.

Roles in *Seneca Falls* are fairly detailed and well defined. Players are expected to read something about the characteristics and attitudes of nineteenth-century persons between the time they are assigned roles and the beginning of play. They

TABLE 5 Role Description Detail

Very Brief: Approximately 1 Sentence	Lengthy, Detailed: 1-2 Paragraphs
Access	*Job Crunch*
	*The Marriage Game**
	Seneca Falls

*player defines detail

TABLE 6 Interest Factor of Roles

Medium	High
Seneca Falls (some roles)	*Seneca Falls* (some roles)
Access	*Job Crunch*
	The Marriage Game

are also provided with the "Declaration of Sentiments" and "Resolutions" from the original convention to help them formulate positions and arguments. Those who play historical figures go to the library for research; they are not given specific material for their roles.

Interest Factor of Roles

We found that players had above average interest in the roles in all of the games, and thus we have ranked them all high. *Seneca Falls* appears in two categories—"medium" because some students may not do much outside reading and thus not "get into" the roles as fully as the majority seem to, and "high" because some players do seem to get highly involved. The instruction sheet encourages them to come in costume and to elaborate upon what is given, using their imaginations and being creative. The teacher's guide makes the following suggestion:

> Encourage the groups to brainstorm for ideas. For example, one of the author's girl students made a tape recording of her baby brother crying. Whenever a particularly obnoxious boy student stood up to speak during the convention, she would turn on the casette recorder—concealed under the baby doll she was carrying—and compete with his speech.

Where students do such enterprising things, interest runs high.

Degree of Role Playing Required

Both *Access* and *The Marriage Game* require little role playing by participants. In *Access,* players are themselves but have constraints depending upon which territory they are assigned to; in *The Marriage Game,* as already indicated, players do not have to assume highly unfamiliar roles. Both *Seneca Falls* players and *Job Crunch* players, however, may have to assume roles quite different from their real-life ones, taking on different attitudes and performing unfamiliar tasks. We do not mean to indicate these are negative. Rather, you, as the game operator, need to assess how much the players' previous experience has prepared them to understand the role and perhaps give some advice on how to play it.

TABLE 7 Degree of Role Playing Required

Little	Considerable
Access	*Seneca Falls*
The Marriage Game	*Job Crunch*

Flexibility of Roles

A strong point we found in all the games was the opportunity to have players play sex roles opposite their own. In each game, males could play females, and vice versa. This was often a very potent learning experience for them, as they "saw" something of what it might feel like to be in the opposite position. It was our impression that this was especially valuable for males, who reported considerable growth in empathy, having experienced the limited options and rewards, and the stigmatization often given to women. (But note that all real-life males should not be made "Greens" in *Access*, as the meaning of the red/green territories is to be emergent, not explained or otherwise given away before play.)

ACTIVITIES, INTERACTIONS, AND SKILLS

Sequence of Activities

The games differ considerably in the activities in which players engage. *Access* and *The Marriage Game* include several "rounds" of play, whereas *Job Crunch* and *Seneca Falls* contain a continuous sequence of activities. These are summarized in Table 8.

TABLE 8 Sequence of Activities

Title	Sequence
Access	A. Divide players into red and green players. B. Send red and green players to different "territories." C. Each player receives a game "tip." D. Three rounds of: 1. Accumulate chips in each area; 2. Form partnerships if desired; 3. Enter "inner circle" game if eligible.
Job Crunch	A. Assume roles. B. "Cocktail party" where candidates are introduced and questioned by others. C. Caucus of faculty and caucus of administrators. D. Hearing. E. Vote on candidates.
The Marriage Game	A. Preplay preparation of materials from manual. B. Four to eight rounds of: 1. Assign value weights to six types of points; 2. Determine major statuses and their consequences: a. Marital status b. Parental status c. Job status 3. Make basic financial decisions: a. Total money available for the round b. Basic budget decisions c. Purchase goods and services d. Economic chance draw 4. Interact with others: a. Allocate free time b. Go on a vacation c. Obtain sexual gratification (optional) and check for conception and VD possibilities; 5. Obtain partner ratings; 6. Calculate final score for the round.
Seneca Falls	A. Assign roles. B. Students do library research on roles and general topic. C. Convention, including: 1. Speeches; 2. Debate on resolution; 3. Voting.

Types of Interactions

The games also vary in the types of interactions between players, as Table 9 shows.

TABLE 9 Types of Interactions

Title	Individual-Individual	Stable Dyads	Small Groups	Whole Group
Access	medium	—	low	—
Job Crunch	medium	—	medium	medium
The Marriage Game	medium	high	low	—
Seneca Falls	low	—	medium	high

In toto, then, the games differ in the degree of total interaction the player engages in. *Access* entails much activity, but we believe many people play without interacting much with others and hence have labeled it "low" in total interaction. Because so much of the action of *Seneca Falls* takes place during the "convention," when, people for the most part, speak one at a time, we have labeled it "medium." Both *Job Crunch* and *The Marriage Game* involve the players in considerable interaction with others, though, as noted above, players in the former engage in different types of interactions whereas players in the latter engage largely in dyadic interaction.

TABLE 10 Degree of Total Player Interaction

Low	Medium	High
Access	*Seneca Falls*	*Job Crunch*
		The Marriage Game

Skills and Activities

Because we do not believe that there are any skills that are automatically defined by gender nor any enterprise that one gender is automatically better at than the other, we do not see any of the skills used or enhanced by play in these games as being "sex roles" skills. Rather, as Table 11 reveals, these are general skills that may be developed or employed in games on any number of possible topics.

The activities, on the other hand, are related to the specific

TABLE 11 Skills and Activities

	Access	Job Crunch	The Marriage Game	Seneca Falls
Information Processing:				
Analysis	m	s	vs	s
Gathering/research	m	m	s	vs
Planning goals, strategy	s	vs	vs	vs
Proposals and Law Making:				
Problem solving	m	s	vs	vs
Lobbying	s	vs	vs	vs
Debating	m	vs	vs	vs
Running meetings/ parliamentary procedure	m	vs	m	vs
Voting	m	vs	s	vs
Public speaking	m	s	m	vs
Making proposals	m	vs	s	vs
Human Relations:				
Exploring feelings	s	s	vs	s
Cooperation/teamwork	vs	vs	vs	vs
Negotiation/mediation	s	vs	vs	vs
Leadership	s	vs	s	vs
Decision making	s	vs	vs	vs
Using power	vs	vs	vs	vs
Empathy	s	s	vs	s
Role Playing:	s	vs	s	vs
Resource Management/ Budgeting:	s	m	vs	m
Evaluation:				
Self-examination	vs	s	vs	s
Exploring personal goals	vs	s	vs	s
Evaluating arguments	m	vs	vs	s
Observations	m	vs	m	m
Evaluating others	m	s	vs	s
Miscellaneous:				
Reading	m	m	vs	vs
Mathematical skills	m	m	vs	m

Key: m = minor
s = significant
vs = very significant

topic of sex roles. Decision making is a skill needed in each game; the particular decision to be made depends on the specific sex role activity—choosing a marriage partner, for instance, or deciding how to support equal rights legislation.

RULES

Complexity of Rules

The games vary in both the complexity and the clarity of the rules given to players, and thus in how difficult it is to get the run of the game underway and keep it moving smoothly. The judgment of complexity is not a value judgment unless a game goes to an extreme of simplicity or complexity. By saying a game is "complex," we mean that it incorporates a large number of variables, interactions, and decisions and thus requires a number of explanations and a longer set of rules. *Seneca Falls,* for example, has extremely simple rules. *Access* and *Job Crunch* have rules of medium complexity. The rules for *The Marriage Game,* on the other hand, are complex.

TABLE 12 Complexity of Rules Given to Players

Low	Medium	High
Seneca Falls	*Access*	*The Marriage Game*
	Job Crunch	

Clarity of Rules Given to Players

What *is* critical, we believe, is the clarity with which the rules are presented to players. A game with simple rules that are presented unclearly will be difficult and frustrating to play. A game with complex rules presented in a well-organized and clear fashion will take some time to get fully underway but will not present great difficulties. We have rated *Job Crunch* medium in clarity, and *Access, The Marriage Game* and *Seneca Falls* high in clarity.

The potential user of *The Marriage Game* must be aware that despite the clarity of the rules, their complexity makes the first round difficult for most players. It is thus strongly urged that the operator "take" the players step by step through this first round, helping those who require assistance locating the proper chart, recording the consequences of decisions, and so forth. Following this initial round, players can proceed at their own pace with minimal direction. Thus, at any one time, some players will be at the end of the second round, others may be playing the third round, and an occasional couple may be beginning the fourth round. This is no problem, and the instructor is free to circulate, answering occasional questions and checking on players' progress.

TABLE 14 Clarity of Rules Given to Operator

Medium	High
Job Crunch	*Access*
The Marriage Game	*Seneca Falls*

Clarity of Rules Given to Operator

All of the games reviewed in this article ranked at least "medium" in the clarity of rules given to the operator. The *Job Crunch* manual would profit from more explicit guidelines to the operator about what she or he must do and might expect. The instructions to the operator in *The Marriage Game* are presented clearly; but without having seen the game in operation, the operator is likely to experience some initial confusion akin to that felt by players. A helpful solution is to find a friend and play one or two rounds; this will make you familiar enough with what players must do and will permit an easy operation. (We have seen undergraduate students who

TABLE 13 Clarity of Rules Given to Players

Medium	High
Job Crunch	*The Marriage Game*
Access	*Seneca Falls*

TABLE 15 Resources and Scoring

Title	Resources	Scoring
Access	Red chips, green chips, and gold stars earned through play but unequally available to players.	Players score according to the number of chips they accumulate.
Job Crunch	Noncandidates have votes to allot to candidates.	Candidates receive votes and highest one is offered the job.
The Marriage Game	Six kinds of points used as costs and rewards of decisions and interactions; financial assets, enjoyment from leisure activities and consumption, social esteem, sex gratification, ego support, and parental status.	Each player assigns value weights to the six kinds of points according to their importance to him/her. Scores for the round are thus weighted ones, reflecting success in maximizing rewards most valued. Scores are thus comparable, but there are no "winners" or "losers."
Seneca Falls	Convention delegates have votes to allot to proposals (to reflect the fact that there were more women than men at the convention, women's votes count double in the game).	Proposals are passed or defeated. Players do not get individual scores.

have played a few rounds introduce other students to play with no difficulty.)

RESOURCES AND SCORING

Given the different topics, it is no surprise that there are different resources given players in the four games. In each case, we have found the types of resources appropriate.

Two of the games present some problems in scoring. As we shall discuss in the section on the game models, we believe that there are some conceptual problems in the scoring procedures in *Access.* In *The Marriage Game* the problem of complex scoring is a shortcoming. The second (revised) edition of the game contains a much simpler mode of scoring than the original did; but given the number of decisions and the multiple consequences of most decisions, keeping score is still a time-consuming task for players, albeit a necessary one if the game message about the interrelationships of decisions is to be conveyed successfully.

DEBRIEFING GUIDELINES

However rich the experience of playing the game may be, you will find that the most important learning takes place during the postplay discussion and critique. During this phase, players need to describe what they did and why, analyze the game model, and compare what transpired in the game with what happens in the real-world system simulated.

In our minds, a good game designer provides the game operator with three things to make this important stage fruitful. First, the designer should provide an *Explicit* statement of what the model was, what assumptions about the real-world it includes, and why things are connected as they are. Second, the designer should provide a list of questions most important to discuss. Operators can add to these or omit any not relevant to the purpose for which the game is run, but at least they have a "starter set" to insure a good debriefing. And third, the designer should guide the operator and/or the participants to readings on the topic.

TABLE 16 Debriefing Aids

Title	Explicit Statement of the Model	Suggested Questions for Discussion	Guide to Further Sources
Access	no	yes	yes
Job Crunch	partial	partial	no
The Marriage Game	yes	yes	included in manual
Seneca Falls	no	partial	yes

Do these games meet these three tests? To varying degrees, as table 16 shows.

The biggest shortcoming of these games is in the lack of statements about the model by the designer. This is true not only of these games but of most. Robert Horn in the third edition of this *Guide* commented that *The Marriage Game* was one of the *only* games listed that provided such information. We consider this a sorry state, for operators may have to run a game several times and piece together assumptions from each run until they have a clear idea of what the designer was doing.

Access fails to include such a statement by the designers, but it includes a very well-written guide to discussion and a very helpful annotated list of sources.

Job Crunch also fails to include a model statement, but a paper by the designers available from the same source includes fruitful discussions of some of the decisions made in the design stage. We wish this paper were considered a piece of the game manual, rather than a separate, optional purchase. *Job Crunch* also contains an evaluation form for participants to fill out as soon as play is ended and use as the basis for discussion. We believe this contains some interesting questions but needs further work. You should plan on taking some preparatory time developing a guideline for discussion of the game with your group.

The Marriage Game manual contains a chapter describing the designers' views of marital decision making, why they included some elements and excluded others, and how each of these elements is simulated. A paper describing the design process is also included in the readings. The manual also contains a list of twenty questions for post-play discussion and a set of thirteen parallel readings, selected because they deal with major elements of the game.

Seneca Falls provides limited guides to debriefing, though a few readings are suggested.

These considerations have led us to the evaluation of the debriefing guidelines displayed in Table 17.

TABLE 17 Quality of Debriefing Guidelines

Adequate	Good	Excellent
Job Crunch	*Access*	*The Marriage Game*
Seneca Falls		

THE CONTENT AND VALIDITY OF THE MODEL

In the introduction we indicated that we consider the question of the model a critical one. No matter how well the game "works," no matter how much fun it is to play, if the model is inaccurate there are severe problems in use of the game. It is possible to use a bad simulation for instructional purposes, having players analyze what is wrong with the model, and we have often found this a most fruitful enterprise. This, however, requires considerable skill on the part of the operator, who must have done a critical analysis of the game in advance and must know a great deal about the subject matter to pinpoint the flaws accurately.

Because the accuracy of the model was one of the two selection criteria for inclusion in this review, all four of the games seem to us to be generally accurate simulations. Space limitations prevent our being able to elaborate upon the strengths of the models, but the criticisms that follow should be read in the context of our overall endorsement. We offer these criticisms so you may be on the alert for points that may need to be made in the debriefing.

The *Access* Model

In *Access,* the designers have attempted to model the differential access of men and women to status positions in the society. Reds, through their "employment," have access to high rewards and prestige. Greens have little access to the Red game, and while they have somewhat easier rules in the Green game, this advantage does not get them very far. Furthermore, Greens have little to offer Reds to get them to form partnerships, whereas Reds have much to offer Greens (difficult-to-gain red chips, which permit their survival in the game).

While a number of elements of this model seem to us accurate, there are two shortcomings. The first has to do with partnerships. In play, few Reds, unless specifically instructed to do so by a "tip," form partnerships with a Green. In real life, male-female partnerships with a traditional division of labor are a common occurrence. Although the game fairly accurately models the differential powers of males and females in the partnerships, it fails to model reality regarding the frequency of formation of such partnerships or the rewards to males (Reds) for doing so. In *Access,* Reds can do it all alone; in real life their male counterparts can, too, but often do not, for a variety of reasons. Men do not "play the Green game earning their green chips," but ally with women who do the cooking, washing, buying, decorating, entertaining, cleaning, and carpooling.

The second criticism is quite different: There should be more opportunity for some Greens to "play by the Red rules" or come close to it. Although Greens who choose to do so can play in the Red game, there is no way for a Green to "make it" in that game. Thus, one never finds a truly successful Green to compare with those who remain in their territory happily or unhappily allied to a Red or to those who remain struggling for red chips. Likewise, Reds do not learn anything about the advantages of allying with a Green who has different notions about the distribution of rights and responsibilities for earning both green chips and red ones (getting domestic chores done and earning income). In the present game, Greens receive high status only vicariously through their alliance with Reds; we believe the game should be modified to demonstrate the parallel to women who succeed in the labor force.

The *Job Crunch* Model

This model is quite accurate for conveying a general idea of the considerations involved in preferential hiring. The supplementary paper describes the elements that have been deliberately left out in the interest of playability, and it can be used to elicit other elements of the process. You should also note that in the limited time for play some of the backroom caucusing and politicking that takes place around hiring cannot flourish. The model is probably *not* generalizable to hiring practices in the nonacademic world—a fact the designers conscientiously note.

The Marriage Game Model

This game is based upon an exchange theory model that, roughly stated, holds that each decision an individual makes will accord that individual certain costs and certain rewards. Reviewers of the game who are experts in the sociology of the family have generally found the model to be extremely accurate, with the one exception being the problem noted by the designers: the absence of the strong emotions (both positive and negative) that often exist between marital partners, which affect their decision making. Except when the partners are in a real-life emotional relationship, this factor never operates strongly in the game, and postplay discussion must focus on the effects of its absence.

The *Seneca Falls* Model

Because most of this simulation depends either on historical documents that are included with game materials or on library research for the major historical roles, the only place the designer specifies how roles should be played is for the other men and women. It seems to us that he has done a disservice to nineteenth-century men by making all the male roles unanimously opposed to the convention motions. Several ministers are included among the males, and only they are permitted to speak as often as the "ladies." Were there no clergy sympathetic to women's rights? The guest speaker, Frederick Douglass, is to make the connection between abolition and women's rights. Were there no other men who also cared about the issue?

FLEXIBILITY

One of the marks of a useful and interesting game is that it be adaptable to changing circumstances and that it invite the modification of its component parts. The designers of *The Marriage Game* have paid the most attention to the issue of flexibility. The game book includes a whole section on several possible alterations of the game; charts for two variants are included (alteration of social class, equality for women).

None of the other designers specifically discuss possible variations of their games. *Job Crunch* would be the most difficult to change, because the roles are very specific. Although the general ideas might be used to build a game dealing with the business world, such an enterprise involves designing a new game, not modifying the present one.

Conceptually, *Access* invites alteration of its parts because of the abstract nature of the game. It would be possible for an enterprising operator to change the rules to see to what degree different outcomes occur. For example, if positive socialization tips were given to all Reds (males) and negative ones to all Greens (females), play would be different. Suggestions for modification, however, are not offered.

Because *Seneca Falls* is historically specific, the basic game cannot be readily altered, though, as we have already noted, different runs may be quite different, depending upon players' preparation and input.

PACKAGING

None of these games is expensive to run. *Access* consists of an operator's manual that contains player instructions. To complete a kit, materials that must be bought range from a jacks set to gold stars—standard inexpensive items. A kit costs approximately $20, and everything except the gold stars and some masking tape is reusable.

Job Crunch also consists of an operator's manual. A few name tags must be made and several sheets must be reproduced for each player.

Seneca Falls comes as a kit including separate four-page pamphlets for each player and a different pamphlet for the operator. Costumes and other materials (flags, banners, posters) to enliven the convention are player created. Except for the instruction sheets, materials therefore depend on the enthusiasm of the operator and the players.

The Marriage Game is packaged as a one hundred-plus-page book, one copy of which is required for each player (i.e., ten copies for ten players, fifty copies for fifty players). The book contains all the rules and all materials for play (except a few paper clips) for one player. In addition, thirteen readings are included for review in conjunction with play and to provide further material for use in discussing the issues raised by play. The book could be used for a second play, but for further use the materials will run out. Potential operators can obtain a free examination copy by writing to the publisher on official stationery.

TABLE 18 Flexibility

Slight Alterations Possible	Flexible	Highly Flexible
Job Crunch *Seneca Falls*	*Access*	*The Marriage Game* (includes section on several possible alterations of game and charts for two variants)

CONCLUSION

We find ourselves unable to compute an "overall score" for a final comparison of these games. Such an evaluation would indeed be misleading. Some games rated higher on one variable and lower on another. Because we do not believe all these variables are equally important, we do not feel adding up separate scores and comparing the results gives an accurate picture of overall quality. For example, the clarity of the rules or the clarity of the model is to our minds much more important than the game's flexibility.

Thus, the choice between these games depends upon the importance to you of the issues to be explored and the other needs to be filled by the game. All four are recommended and could be used successfully in a variety of contexts.

TABLE 19 Packaging

Title and Publisher	Kind of Packaging	Cost Per Kit or Book	Completeness	Amount of Replacement of Materials for Reuse
Access Simile II	operator's manual including player instructions	$5.00	p (about $20), c	little
Job Crunch Denison Simulation Center	operator's manual including instructions	$5.50	c, d	little
The Marriage Game Random House	book (1 per player required; includes operator's instructions	$6.95	complete	could be used second time
Seneca Falls Interact	pamphlets for players; pamphlet for operator	$10.00	complete	none; can be used many times

Key: p = requires purchase
c = requires construction
d = requires duplication

SIMSOC
An Evaluation

by Richard A. Dukes

A BRIEF DESCRIPTION OF SIMSOC

Three Major Features

SIMSOC has three major features:

(1) a set of seven Basic Groups and two types of Agencies,
(2) a set of four National Indicators, and
(3) a Bank

The major effects of these features upon one another are summarized in Figure 1.

FIGURE 1 Basic Interrelations Among Major Features of SIMSOC

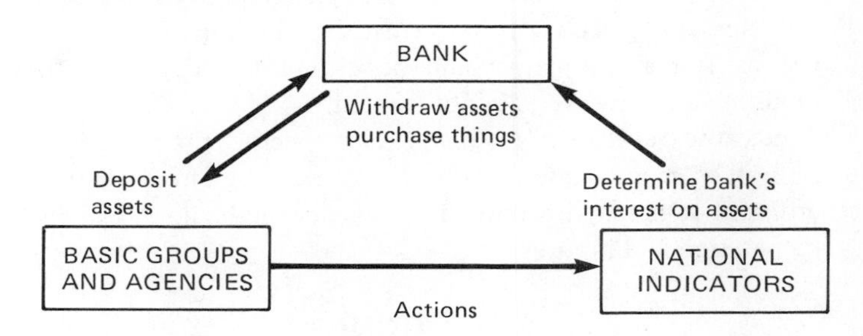

Basic Groups and Agencies

Table 1 presents Basic Groups and Agencies and the characteristics of each. Each group or agency is given basic resources to develop. The incomes of each group in later rounds are determined in part by the lowest National Indicator and in part by the group's performance. The incomes of the Agencies are not affected directly by the National Indicator.

The Four Regions

The simulated society in *SIMSOC* is divided into four regions—Blue, Green, Yellow, and Red. Each region contains 25 percent of the population of the society, but there are striking differences among the regions in the number of group heads, travel agencies, and subsistence agencies that each contains. Figure 2 presents the resources of each region.[1] As can be seen from the figure, the Red Region is a deprived one.

Gamson summarizes the problems of the Red Region:

> While the problem facing the better-off regions is that of the institutionalization of privilege, the deprived region (Red) faces the problem of survival, and under extremely difficult circumstances. The members not only lack subsistence but are isolated by the absence of any means of travel. They have no way of initiating communication with other members of society, and have little or no internal differentiation among themselves [Gamson, 1978a: 16].

TABLE 1 Basic Groups and Agencies of SIMSOC

Group	Location on Figure 3 (Row Letter/ Column Number)	Overall Group Objective	Resources	Income
BASIN (Basic Industry)	O-10	To expand its assests and Income as much as possible (Camson, 1978b: 11).	\$100-200 starting assets in Bank, Buying Passages and Identifying vowels in them.	Up to 125% of Investments in passages plus (10% to 120% interest rate** x 10% starting interest) on assets held in bank.
RETSIN (Retail Sales Industry)	S-10	Same as BASIN.	\$100-200 starting asset* in Bank. Buying and Solving angrams for profit.	Up to 150% Investment in angrams plus (10% to 120% interest rate** x 10% starting interest) on assets held in bank.

Editor's Note: *SIMSOC* is listed in the social studies section.

TABLE 1 Basic Groups and Agencies of SIMSOC (Cont)

Group	Location on Figure 3 (Row Letter/ Column Number)	Overall Group Objective	Resources	Income
MASMED (Mass Media)	H-19	To insure good communications across regions about what is happening in society (Gamson, 1978b: 16).	$30-60 Basic Income* One Permanent Transportation Certificate. Five travel tickets (first session only). Access to all information on National Indicators, Support Cards for Basic Groups totals on Individual Goal Declarations. Two free Broadcasts per session. Two pages free copy distributed per session.	(10% to 120% rate** x $2) for each MASMED subscription.
EMPIN (Employee Interests)	E-13	To see to it that members of SIMSOC who are not heads of Basic Groups have adequate subsistence and a fair share of the society's wealth.	$30-60 Basic Income*	(10% to 120% rate** x $2) for each EMPIN membership card turned in. Rate is determined by Lowert NATIONAL INDICATOR
JUDCO (Judicial Council)	G-3	To clarify and interpret the rules as honestly and conscientiously as possible (Gamson, 1978b: 18).	$30-60 Basic Income*	(10% to 120% rate** x Basic income)
POP (Party of the People)	C-9	To determine the major public policies followed by the society and to develop programs and mobilize supporters for this purpose; to be more influential than its rival (Gamson, 1978b: 14).	$40-80 Basic Income*	rate** x Number of Party Support Cards x Basic Income and Number of Participants x $2.50
SOP (Society Party)	B-4	Same as POP.	$40-80 Basic Income	Same as POP
Subsistence Agencies (Number varies with size of society)	0-30	Set by heads.	5 subsistence tickets per round	5 subsistence tickets per round
Travel Agencies (Number varies with size of society)	S-30	Set by heads.	5 travel tickets per round	5 travel tickets per round

*Amount of Starting Assets or Basic Income depends upon the size of the society. The larger the society, the larger the income.
**Rates are determined by the lowest National Indicator. The lower the lowest indicator, the lower the rate.

Summary of the Entire Simulated Society

Figure 3 displays the entire set of functional interdependencies for *SIMSOC.* References to features of this figure will be made in the coordinate notation of letters (rows) and numbers (columns) similar to that on a road map.

THE ESTABLISHMENT OF SOCIETY

The Beginning of SIMSOC

When *SIMSOC* begins, a society does not yet exist—participants themselves must establish one. In fact, the game structure provides many roadblocks to the creation of social order. These impediments, outlined succinctly by Gamson and Stambaugh (1978) are presented in the first column of Table 2. I have added an additional characteristic concerning force (#8) to this list. The second column of this figure presents the ways these impediments are expressed in the structure of the game. The third column presents a classic list of requirements a society must meet if it is to continue functioning (Inkeles, 1969: 73-129). *SIMSOC* has posed every one of these problems to its participants *plus* three more (#1, #9, #10) that involve stratification. These are not essential to the existence of a society, but they are absolutely essential to the way it functions. *SIMSOC,* then, appears to incorporate almost all of the most important points concerning the establishment of a society.

FIGURE 2 The Geography of SIMSOC: Simulated Society 3rd Ed 1978

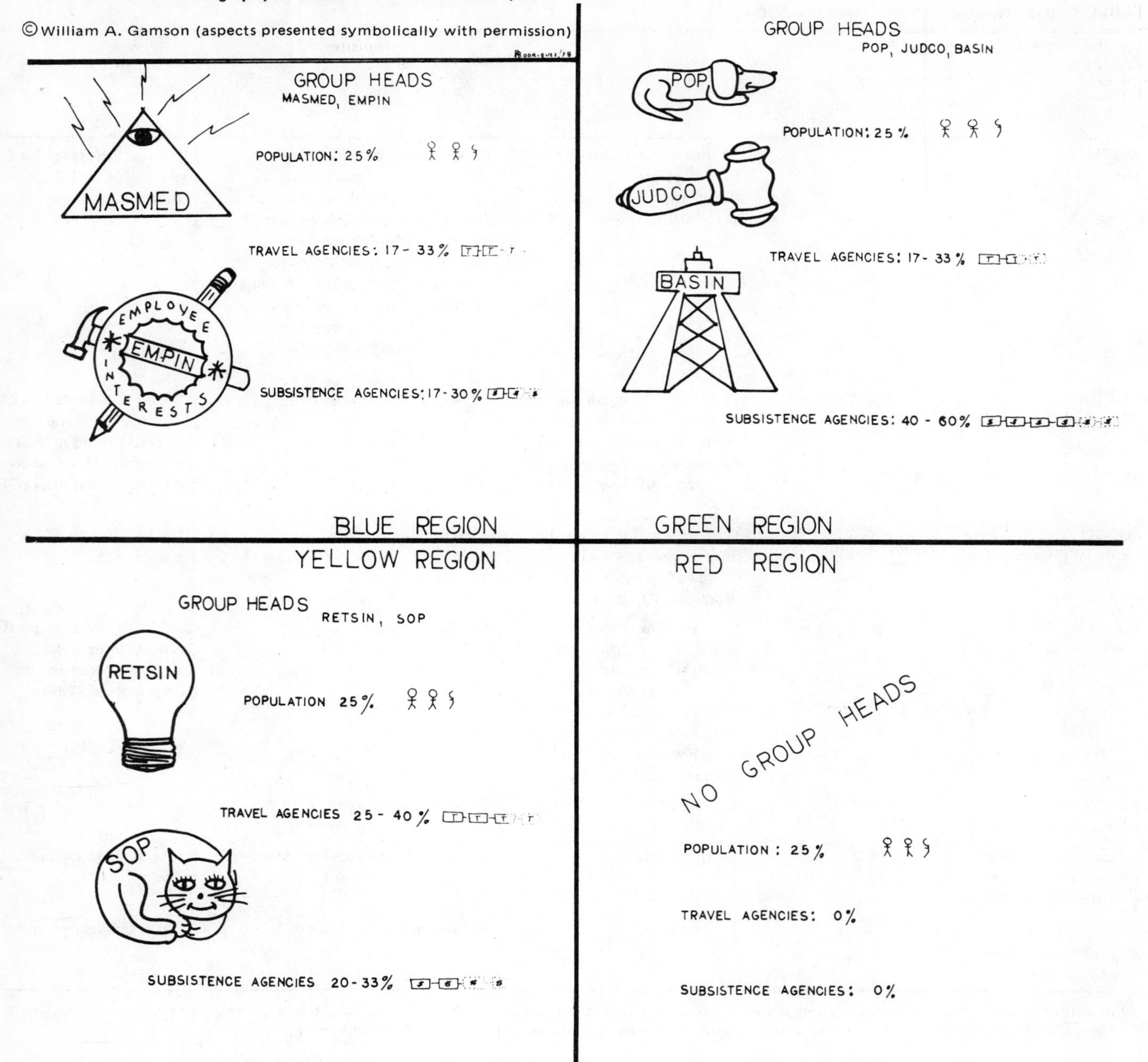

The Process of SIMSOC

"What I see as a central process in the game is essentially the creation of a society out of an aggregate of unconnected individuals" (interview with Gamson by Coppard and Goodman, 1977: 281).

Though play of *SIMSOC* has a highly emergent quality, it frequently progresses through a series of identifiable stages. The process of an entire play, of course, is a complicated affair, due not only to the complexity of the model but also to the fact that players determine to a large extent what will happen. Gamson and Stambaugh (1978) have presented these processes in schematic form (see Table 3). Their complete description of these processes appears as a precisely written 13-page passage and is beyond the scope of this evaluation. Interested readers should see Gamson and Stambaugh (1978: 135-148).

Following Gamson and Stambaugh's outline, students in a "typical" play at our institution invariably organize the basic groups on a regional basis (the organization phase in the diagram of typical play below). Those who are not group heads become employees. Members of the Red region are isolated from the mainstream and fight for subsistence. Red region comes close to collapse, but is "saved in the nick of time" by one or more of the privileged regions to obtain employment from Red's members and to avoid further decline of the National Indicators. Red accepts "help," but by this

TABLE 2 Impediments to Social Order in SIMSOC

Societal Conditions at the Beginning of the Game	Examples	Requisites for Continued Functioning of Society (Inkeles, 1969; 73-129)
1. Scarcity of subsistence	The society does not possess enough subsistence tickets for all participants to survive unless additional subsistence is bought from the game director.	Adequate physiological relationship of society to physical setting
2. Extreme inequality by region	Green region has relative abundance; blue and yellow regions are somewhat self-sufficient; red region has nothing.	Red: Adequate physiological relationship of region to physical setting regarding subsistence.
3. No group structure	Basic Groups of BASIN (0-(0), RETSIN (5-10), MASMED (H-19), JUDCO (G-3), SOP (B-4), POP (C-9), Travel (S-30) and Subsistence Agencies (0-30) have functions assigned to them (see figure 3) but they do not have members, roles, a division of labor, or a hierarchy of authority. They only have heads.	Adequate Role differentiation and role assignment
4. Communication barriers	The regions can communicate with each other *only* through MASMED or through players who purchase travel from one region to another.	Shared communication
5. No consensus on individual goals	Individuals are pursuing a wide range of personal goals, many of which are incompatible with the goals of other participants, Participants have listed their own personal goals without communicating with other players.	Shared goals
6. No consensus on means of individual goal attainment	Participants do not know how to go about fulfilling their personal goals. They simply have listed them for themselves most probably without thinking about how they will pursue those goals. Their individual pursuits are likely to be in conflict with the actions of others.	Adequate regulation of choice of means to goals.
7. Lack of government	The rules do not provide any authority for making collective decisions. This authority must be developed by the participants.	Effective control of desruptive behavior
8. No agreement concerning the proper use of force.	SIMFORCE may be used by anyone. SIMFORCE may arrest and kill other participants.	Adequate regulation of (negative) affective expression.
9. Extreme inequality by individuals	Heads of the Basic Groups of BASIN (0-10), RETSIN (5-10) MASMED (H-19), JUDCO (G-3), SOP (B-4), POP (C-9), and Travel (S-30) and Subsistence (O-30) Agency Heads have control over current resources and potential control over future resources. Other players possess only their own labor.	
10. Privilege is not legitimate	Participants receive their positions of privilege as heads of Basic Groups or Agencies through "luck of assignment" by game administrator not by earning them.	(As a real society evolves, this characteristic is covered under #6 above)

TABLE 3 Schematic Representation of SIMSOC Model

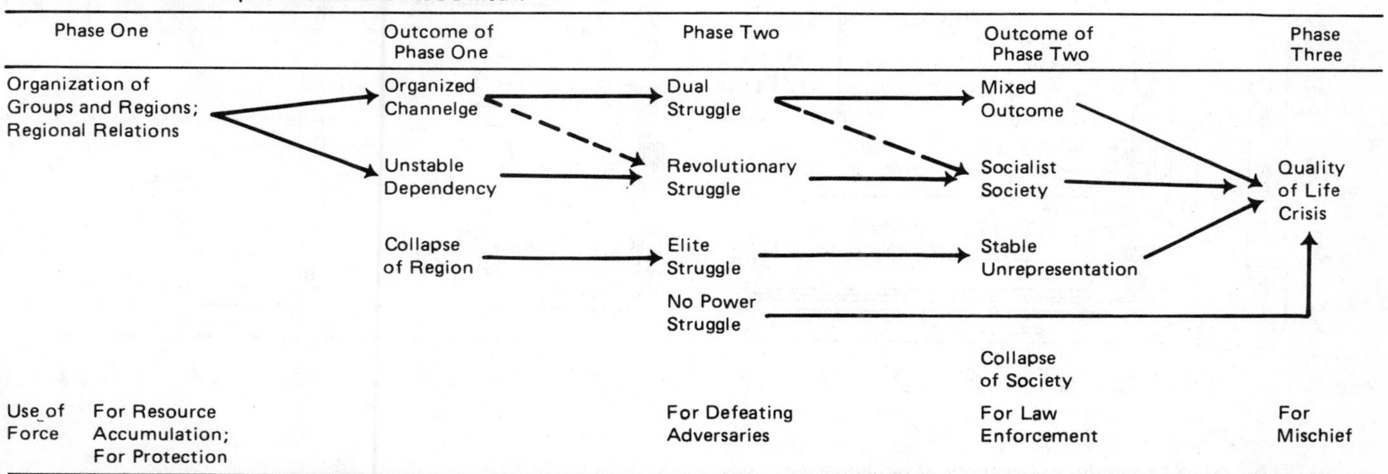

Key: continuous arrow = probable
broken arrow = sometimes
no arrow = rarely

FIGURE 3 Functional Interdependencies in SIMSOC

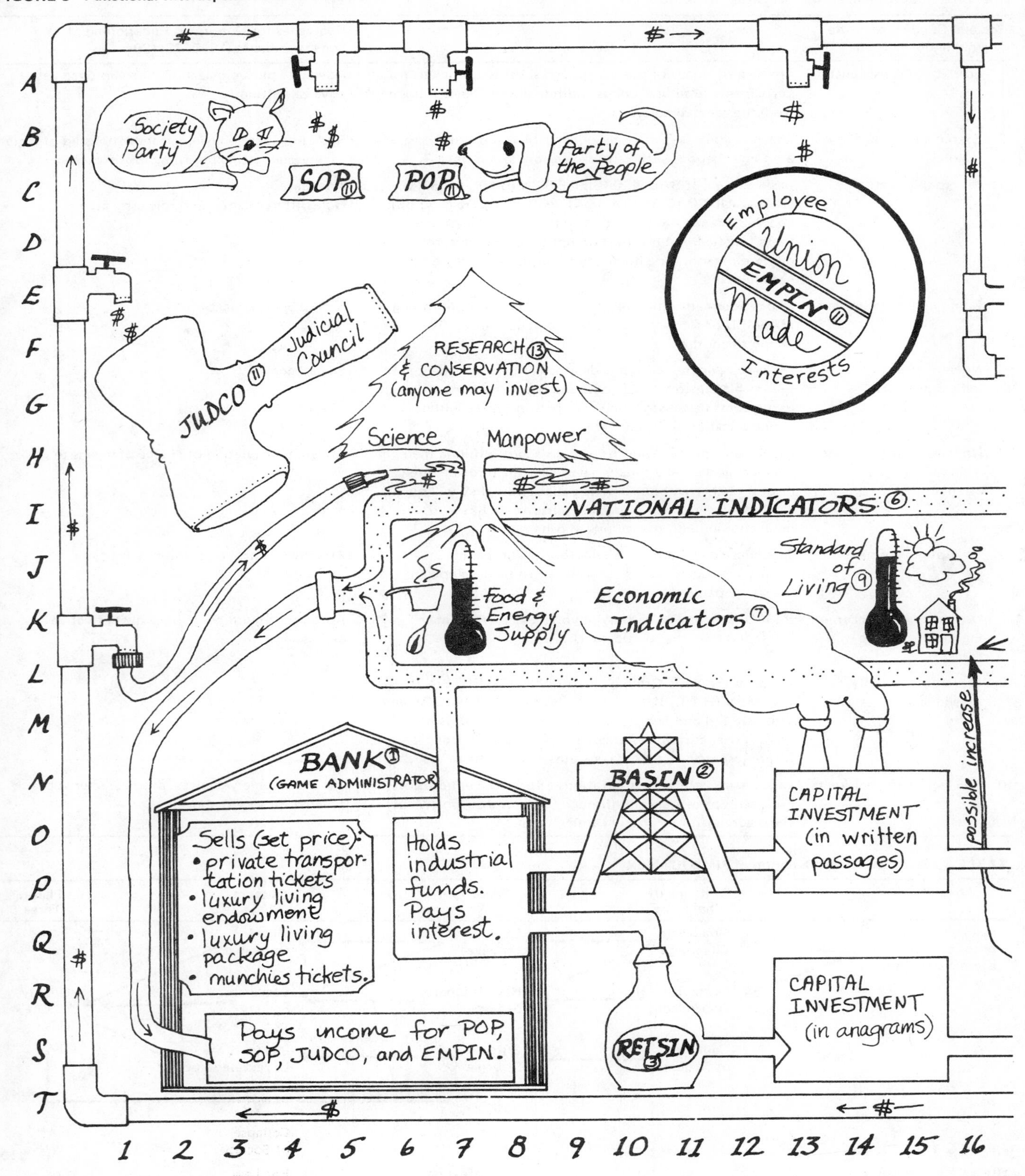

NOTE: See Notes section for explanations of various elements and relationships in this figure.

FIGURE 3 Functional Interdependencies in SIMSOC (Cont)

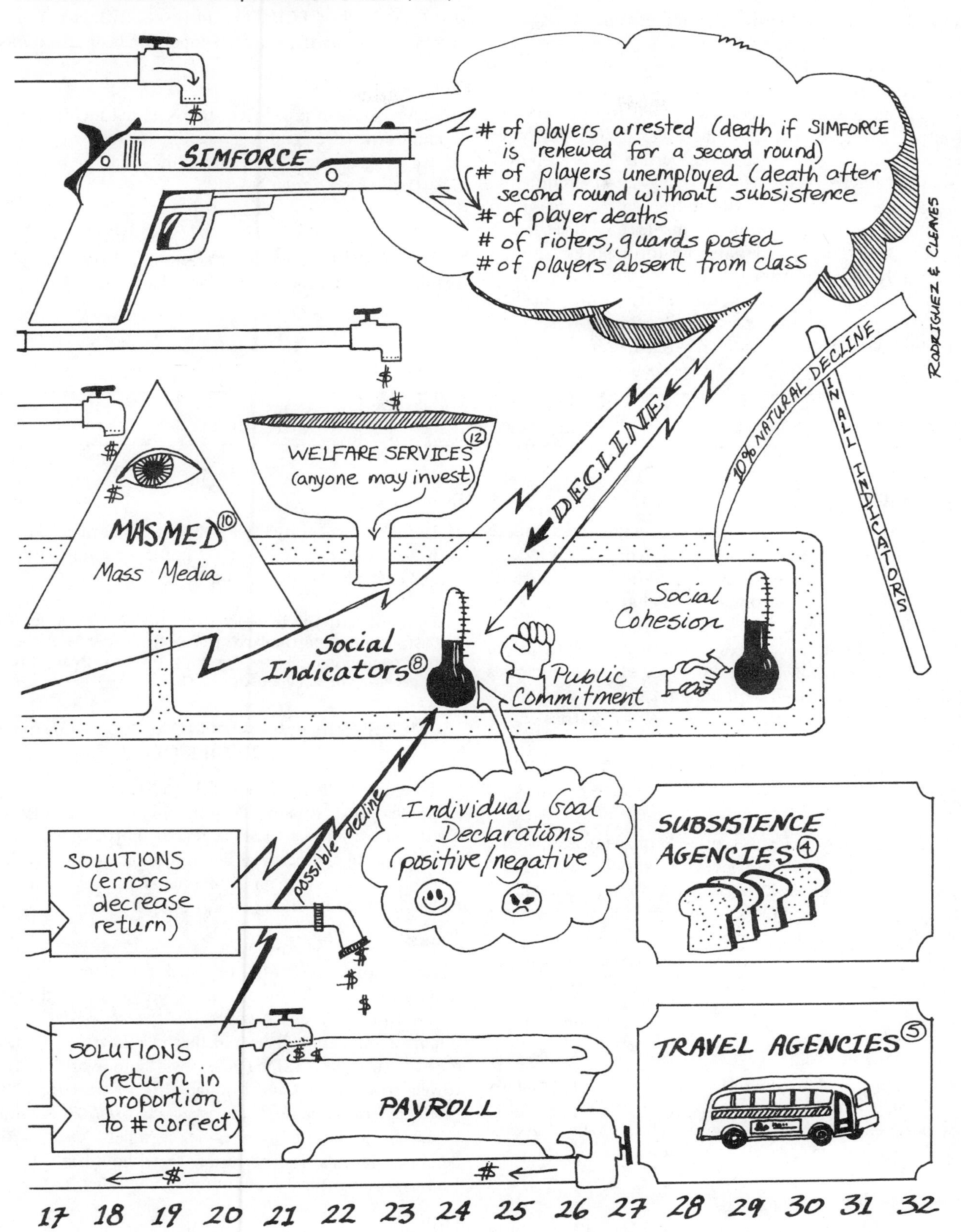

This figure appeared in an article written by Ron Stadsklev (1974) and is reproduced here with the author's permission.

time members are bitter about the failure of other regions to recognize their plight before the situation became so desperate (unstable dependency). Red's members feel used, but they continue to work mostly within the system of private ownership with some governmental control in which elite members of privileged regions are competing with each other for control of resources. If they can, Red members secretly play the other regions off against each other, often getting more resources than they otherwise would get, by lying about production, subsistence needs, and so forth to each privileged group separately and receiving duplicate payments (dual struggle and mixed outcome). Privileged groups use force on each other, but not on Red. As players begin to develop a surplus of resources, they begin to question the quality of life and their own goals (quality of life crisis). Red often uses force as mischief to show that they must be reckoned with.

A diagram of this "typical play" appears below.

Using SIMSOC with Different Groups

I have not used *SIMSOC* with groups other than college students. Gamson indicates that it has not worked as well with students who are not college bound, and efforts to reduce its complexity are not easily accomplished. Adult groups, he says, have produced lively games. Instructors who have used it in foreign countries note differences in the way students handle some political issues (for instance, the role of political parties is more or less pronounced depending on the country). Overall, though, play *seems* to be similar in various countries.

CHANGES IN THE THIRD EDITION OF SIMSOC

Subsistence

In the third edition instead of a Permanent Subsistence Ticket, participants may purchase a Luxury Living Endowment, a Luxury Living Package, and/or a number of Munchie Tickets (P-4.)[2] that may be exchanged for food and beverages at the refreshment table. The Standard of Living Indicator determines the total number of Munchie Tickets available to the society.

Group Income

Memberships in EMPIN (E-13), JUDCO (G-3), SOP (B-4), and POP (C-9), as well as subscription to MASMED (H-19), have increased importance as individual choices for players and as effects on the income of these groups and on the National Indicators (K-8, 14, 24, 28).

The Quality of Work

Productive labor now is more important, and differences in the quality of the work experience have been introduced. BASIN (0-10) now performs unskilled labor (O-13 to 20). RETSIN (S-10) performs skilled labor (S-13 to 20). In the third edition it is more important that members of EMPIN (E-13), SOP (B-4), POP (C-9), and MASMED (H-19) travel and persuade other participants to support their organizations.

Individual Goals

Individuals can now affect the National Indicator of Public Commitment (K-24) through meeting (or failing to meet) their individual goals and expressing this situation on Individual Goal Declaration Cards (N-25).

Simforce

Finally, the third edition has simplified the rules governing a Simforce (B-20) and has introduced a Simriot (C-25).

Summary

Though I have not yet played with the third edition, I agree with Gamson that there are "subtle and important differences" between the second and third editions:

> The most striking is likely to be the enlarged role of the basic groups, particularly the political parties. I believe this version is likely to sharpen processes already present in previous ones [Gamson, 1978a: 5].

MEASURING ANALYTICAL COMPONENTS OF SIMSOC

To describe *SIMSOC* more fully, I have applied to it Stadsklev's Measuring Analytical Components Model (1974) (see Figure 4). According to Stadsklev, the purpose of the model is to go beyond a superficial description and to assess the *character* of the four main components of a simulation game (social model, role taking, drill, game).

Game

In its entirety, *SIMSOC* tends to lean more toward being a game than a simulation because it provides for quite a bit of *competition between players for scarce resources.* It still retains a simulation flavor, in that resources are brought into society through BASIN (0-10) by "hard work" (extracting vowels from words) and through RETSIN (S-10) by "solving problems" (anagrams). These aspects of *SIMSOC* are similar to an ideal-type autonomous game in which "each participant demonstrates his best performance without any interference [that is, competition] from others" (Stadsklev, 1974: 88).

Drill

Within the structure of the game, any drill on things to be learned is covert. Drill is not systematically built into the game in any way that is manifest.

FIGURE 4 Stadsklev's Model Applied to SIMSOC

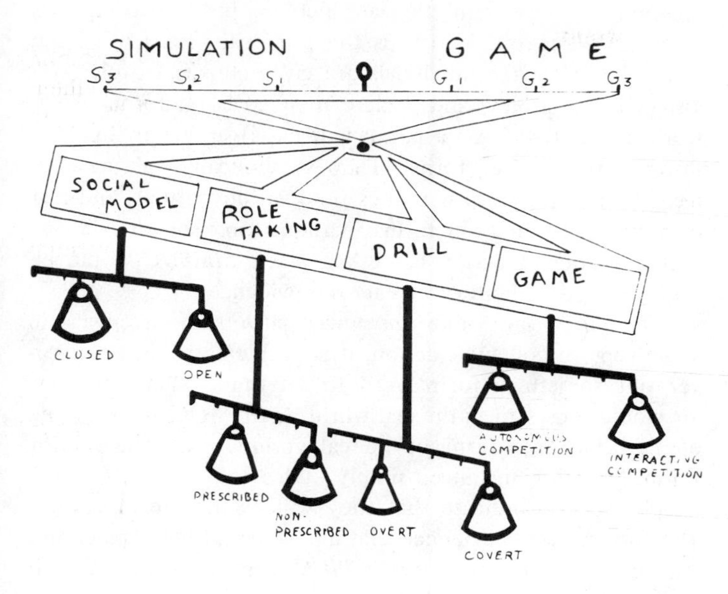

Role Taking

Roles are not taken in *SIMSOC* in any prescribed way. Players are given resources to develop, and groups are given goals, but players are not given any information on what their characteristics or attitudes should be like. In fact, players are encouraged to set their own individual goals.

Social Model

SIMSOC tends to be an open-model game. It is not a tightly organized structure with which players interact; rather the participants create their own structure within several broad parameters. The outcomes of *SIMSOC* are varied and somewhat unpredictable.

The Overall Model

Players interact with each other directly without any drill. They are not required to take roles in prescribed ways, and they create their own structure and outcomes. This mix of general characteristics allows *SIMSOC* to realize its objectives: to have players set their own goals, to temper the realization of these goals with loosely prescribed sets of group goals, and to evaluate outcomes of the exercise in terms of the total pattern of goal attainment. The *SIMSOC* model matches these objectives very well.

DISTINCTIVE FEATURES OF SIMSOC

SIMSOC has a number of features that make it unique and more complete than other exercises.

Macro Model of Society

Perhaps the most distinctive feature of *SIMSOC* is that it models an *entire functioning society*. As such, it presents students with a *macro* perspective that is very hard to teach in other ways. It seems to provide a framework within which fit the major topics of most social science courses.

SIMSOC's feature of National Indicators (K-8, 14, 24, 28), for instance, provides a framework within which almost any change in social organization can be seen to affect other parts of the society.

A player from a privileged region once commented, "So what if members of Red region remain unemployed, so long as they don't make trouble for us?" A cohort who understood the larger perspective of the society in regard to national indicators that *SIMSOC* was teaching bellowed, "Did I hear you say, 'So what'? Well, if they remain unemployed, Standard of Living [K-14], Social Cohesion [K-28], and Public Commitment [K-24] will all go down, and so will the incomes of all our groups. Then the only way to bring things back up will be to spend lots of money that we don't have on welfare [Welfare Services]. Why don't we just try to make this society work in the first place, instead of trying to patch it up later?"

I suspect that it would take far more time and effort to teach students about the larger societal (macro) relationships among employment, standard of living, public commitment, welfare, and social control through lectures and readings than that one brief instant in which a student brought together a number of abstract relationships and framed a social policy to deal with them. I'm not saying that *SIMSOC* always performs this learning task, but it does extend an invitation to students to bring things together in this way.

The Use of Force

Through the use of Simforce (B-20), Simriot (C-25), and Arrest, *SIMSOC* effectively structures the use of force in ways that do not detract from its emergent quality. *Starpower* (Shirts, 1969), for instance, also effectively models force, but, by contrast to *SIMSOC*, it allows force to emerge *without* structure from the game. Force in *Starpower* is generated by players themselves, and it is likely to be a result of righteous indignation on the part of nonprivileged groups. *SIMSOC*, unlike *Starpower*, communicates the idea that force can and *must* be bought. Because larger, more powerful Simforces cost more money, students begin to understand repression as well as riot. They begin to understand powers of policing and arrest as implements of the existing order rather than only as impartial upholders of an agreed-upon set of moral values. The typical first response by students to the idea of Simforce is one of those "Aha!" or "Oh, my God!" insight/reactions followed by a sinister chuckle when it occurs to them what they or their group can do with a Simforce.

SIMSOC Manuals

SIMSOC is packaged in book form. The manuals (Gamson, 1978a, 1978b) are the best of any I have seen. First, they are as complete as one could imagine. Through 5,000 runs and many revisions, most problem areas in play have been anticipated, and the manuals present solutions for them. When an *unanticipated* problem occurs, there is even a mechanism for solving it (see JUDCO below). Second, the coordinator's manual is clearly written. Though the overall exercise is complex and the rules are long, they are straightforward and complete. Third, the manual is well organized. Even so, its large size (180 pages) sometimes prevents quick access to needed information. I know of one person who uses color-coded index tabs to facilitate quick referencing. The manual contains a good index, so one can usually find things without much trouble.

The *SIMSOC* coordinator's manual is one of the few documents in the area of simulation and games that would allow an instructor without much previous experience in running gamed simulations to conduct the game in the classroom without previously seeing it run by someone else. However, prior experience with similar exercises by instructor and students is *desirable* for successful play in any gamed simulation, and this point cannot be exaggerated for *SIMSOC.* Run as the last in a course sequence that includes several shorter, less complicated exercises (all tied as intimately as possible to the theme of the course), *SIMSOC* may realize the greatest part of its vast potential.

Finally, each player has a set of all the necessary individual materials, and the pages are perforated so they can be torn out.

SIMSOC Readings

The manuals contain almost 100 pages of student readings that are grouped under six general headings (see Table 4).

TABLE 4 Readings Included in Simsoc Manuals

PART 1	ON THE USE OF SIMULATION GAMES
	Clark C. Abt The Reunion of Action and Thought
	John C. Raser What and Why Is a Simulation?
PART 2	SOCIAL PROTEST AND SOCIAL CHANGE
	Ron E. Roberts and Robert Marsh Kloss Social Movements: Between the Balcony and the Barricade
	Saul D. Alinsky The Process of Power
	Roland L. Warren Truth, Love, and Social Change
	William A. Gamson The Importance of Trust
PART 3	SOCIAL CONTROL
	Robert A. Dahl Legitimacy and Authority
	Gresham M. Sykes The Defects of Total Power
	William A. Gamson The Management of Discontent
PART 4	SOCIAL CONFLICT: ITS NATURE, FUNCTIONS, AND RESOLUTION
	Lewis Coser The Functions of Conflict
	Robert A. Dahl Conflict: A Paradigm
	Kenneth E. Boulding Conflict Resolution and Control
PART 5	INTERPERSONAL INFLUENCE AND LEADERSHIP
	Peter M. Blau Social Exchange, Power, and Leadership
	Sidney Verba Leadership: Affective and Instrumental
	Eugene A. Weinstein Toward a Theory of Interpersonal Tactics
PART 6	CREATING A BETTER SOCIETY
	Seymour B. Sarason The Creation of Settings
	Philip E. Slater The Pursuit of Loneliness

SIMSOC Forms

SIMSOC also includes an imposing array of forms that participants fill out at various times during the game. Table 5 presents a list of forms that are included in the participant's manual. Filling out a form is the principal means of taking action in *SIMSOC.* Individuals are given choices of roles, are assigned their roles, and declare their goals on forms. They work, move, travel (on a permanent basis), and enjoy luxuries through the use of forms. They symbolically subscribe to media, support political parties, join various groups, and riot or repress by filling out forms. Without some sort of evidence that a particular action has taken place, *SIMSOC* would be totally chaotic. The forms create this evidence.

While in a game like *Starpower* players often engage in some type of collective action, it is problematic for an observer—and sometimes for players—to determine what the behavior would represent in the real world. In *SIMSOC,* the meaning of the behavior in terms of the real world is clear. The Simriot Form (M), for instance, simply states that "the following people wish to indicate that they will participate in a riot." The form presents evidence that an event has taken place and indicates what it represents. *SIMSOC* seems to give up a small degree of spontaneity by having players fill out these forms, but it models such a wide range of events that forms are necessary.

Forms also allow for the bookkeeping that is necessary to compute the effects of events on the National Indicators. For example, the larger the *percentage* of rioters, the more Social Cohesion (K-28) declines. To compute the decline, the number of rioters must be gauged accurately so the percentage can be figured. The game administrator need only consult Form M to determine the number of rioters.

Players and game administrators often feel inundated by the number of forms, but both groups should be advised to "hang in there," and players should be counseled on the importance and function of the forms. (By the way, don't forms make the *real* world go around?)

TABLE 5 The Simsoc Forms

Form A:	Choice Sheet
Form B:	Assignment Sheet
Form C:	Moving Sheet
Form D:	Private Transportation Certificate
Form F:	Transfer of Certificates or Agencies
Form G:	Job Schedule (4 copies)
Form H:	Industry Manufacturing Form (3 copies)
Form I:	Withdrawal of Assets Form (8 copies)
Form J:	JUDCO Decision Form (3 copies)
Form K:	Minority-Group Member Action Sheet (2 copies)
Form L:	Simforce Action Form (3 copies)
Form M:	Simriot Form (2 copies)
Form N:	Guard Post Form (2 copies)
Form O:	Self-Test on SIMSOC Rules
Party Support Cards	
MASMED Subscriptions	
EMPIN Membership Cards	
Individual Goal Declaration Forms	

The Book Format

Because every participant purchases a copy of the participant's manual, the costs of the game are shared. This aspect of the game can be very attractive to schools or departments in which funds for instructional materials are limited.

The book format means that every student has a complete set of *new* materials. It also represents an investment by the student, so full use of the game and the readings by the instructor is at least implicitly encouraged (a worthwhile feature).

Books Are Used Up

Because the books are used up (as perforated pages are torn out), they cannot be used again, unless students hand-copy forms and do not tear out the originals. Consult your local bookstore about selling back "complete" manuals. There is no used book market for incomplete manuals, so if pages are torn out the student must bear the entire retail cost of the manual. Most students seem to feel that the game had better be worth it, and most instructors may feel increased pressure to have the game work.

Cost Ceiling

With the current marketing arrangement there never is a ceiling on costs. Most gamed simulations have very high one-time initial costs and must be repaired periodically. *SIMSOC* has no maintenance costs, but if the books are used up, the exercise costs $5.95 per player *each time it is played.*

Time Commitment

SIMSOC requires one of the largest time commitments of any simulation. This is balanced for most users by the fact that the subject matter (an entire, functioning society) seems to more than justify the commitment. Students do not perceive (subjectively, of course) the time commitment as much longer than a two-hour play of *High School* (Coleman and Seidner, 1972; see Table 6). Also, while the total length of time devoted to *SIMSOC* is long, the length of each session may vary, thus yielding some flexibility in scheduling the exercise. The coordinator's manual discusses various lengths and numbers of sessions (Gamson, 1978a: 33-35).

Physicial Movement and Communication

In *SIMSOC,* physical movement from one region to another is allowed only to those participants who have purchased travel privileges (S-30). Communication between regions is limited to travelers and to MASMED (H-19). The regular travelers in *SIMSOC* often have a lot to say about which policies are implemented, as they are the only ones who have a comparative knowledge of what is going on in society. Because of its nonprivileged status, Red region rarely sends out travelers, though it frequently receives them. Before travelers return from Red region, no one else in society knows how bad things are there. Red's condition regularly becomes critical before the other groups can organize to help. Due to lack of communication between groups, Red players do not easily forgive the other groups for not helping them sooner. Conditions are so bad (and boring) for Red in the first few rounds that any travelers sent out from Red (on borrowed funds) are not likely to return during the round!

It usually takes MASMED a few rounds to understand its own potential and to begin communicating effectively. For the first few rounds, players experience an *information shortage* as well as a shortage in subsistence.

Room Arrangements

Ideally, *SIMSOC* requires one of the most complicated room arrangements of almost any classroom exercise, but for most users one or two classrooms prove to be acceptable sites for the game. With separate room arrangements, Red region really becomes isolated. Though travel and communication must be restricted, I prefer that isolation not be total. The coordinator's manual devotes about two pages (pp. 31-33) to good discussion of various room arrangements.

Number of Players

SIMSOC can accommodate a wide range of numbers of players. From 15 to 60 may take part, making this one of the most flexible exercises on the market with regard to numbers of participants. One reason *SIMSOC* is so flexible is that it makes adjustments in resources (to ensure scarcity in the initial rounds) for three different size levels: size one had 15 to 32 players, size two has 33 to 47, and size three has 48 to 60.

An "undersized" *SIMSOC* contains fewer than 24 players. The chief rule change for the smaller society is the elimination of one of the regions. As the lower limit of players is approached and passed, things quiet down; there is not the same collective feeling one observes with more players. (I am currently considering using *SIMSOC* with a small class of 12, but there would be almost more roles than players. I am not sure how multiple roles for some players would affect play.)

Absentees

SIMSOC incorporates student absenteeism into the game, because absenteeism otherwise would undercut the scarcity assumption built into the game model. The Standard of Living Indicator (K-14) and Public Commitment Indicator (K-24) are lowered by two points for every participant who is absent on each day of play, a mechanism that is a very effective means of bringing classroom characteristics to bear on the game's equilibrium.

JUDCO

JUDCO (G-3) is the final arbiter on the meaning and interpretation of all rules. The establishment of JUDCO is a unique way of settling the ever-present problems of rule interpretation in simulations because it is accomplished *within* the game by the *players themselves.*

For instance, a Simforce once arrested someone and confiscated all his possessions. Later the arrestee was released, but his possessions were not returned to him. The rules specify that "the return of confiscated materials upon a person's release is an internal matter and is not specified by the rules" (Gamson, 1978b: 27). The arrestee and the head of the Simforce turned to JUDCO for a ruling. JUDCO ruled that all "private possessions" such as the arrestee's money could be kept by the Simforce but that tickets for his travel agency were part of the public domain and must be returned for the good of society.

Individual Goals

While the rules of *SIMSOC* provide group goals, participants are asked to choose and pursue their own individual goals. If players choose goals of being the most powerful player, the most popular player, or some other concrete goal, there are some "objective" criteria by which they can measure their goal attainment (how many resources one controls, or how many groups one belongs to). However, if players choose very personal and abstract goals such as to have fun or to be happy, suitable personal measures are not available, and players often don't have ready answers about the extent of their own goal attainment. Players can change their goals each round, so if they are uncomfortable with ambiguity in this matter they may choose more concrete goals. In the new version they can indicate satisfaction or dissatisfaction in meeting these goals. Such declarations (N-25) directly affect the National Indicator of Public Commitment (K-24).

Subsistence and Employment

One must acquire subsistence each round. Unemployment and death follow lack of subsistence. Resources in the game are allocated carefully according to the number of players so that a subsistence shortage confronts all SIMSOCs.

Travel and communication problems intensify the struggle to keep everyone alive, because meager resources cannot be distributed easily in an efficient manner to those who need them. In the early rounds scarcity is the order of the day, and the mock plea, "My kingdom for a horse," is *SIMSOC* folk humor that means players are really hungry. Generally, the privileged regions manage to keep almost all their own members fed and employed. Most often they seem altruistic about Red; it suffers more from benign neglect than from genocide.

Conspicuous Consumption

Participants can acquire consumable resources of food and drink at the Munchie Bar, an option that can be greatly affected by the time of day the class meets. Munchies are more important just before lunch than just after! Gamson cautions against making munchies so attractive that "even the most committed members of the society . . . use their Simbucks for personal consumption" (Gamson, 1978a: 51). However, munchies should be "attractive enough to offer an alternative means of using Simbucks for individuals who do not care much about societal or group goals" (Gamson, 1978a: 51). Instructors should be cautioned, however, to avoid making an uncomfortable mistake: Ignorant at the time of the difference between raw and roasted peanuts, I provided the wrong kind. In an effort to stimulate consumption, I conspicuously consumed enough to become quite sick, as did those players who were particularly susceptible to my "advertising."

Importing Currency

BASIN (O-10) and RETSIN (S-10) bring Simbucks into the society through "work." BASIN performs the unskilled labor of identifying the correct number of vowels in a passage (O-13 to 20), while RETSIN performs the skilled labor of solving anagrams (S-13 to 20). Both industries assume some risk in their capital investments in passages (BASIN) and anagrams (RETSIN). Their investments and their work affect the National Indicators (K-8 to 28). National Indicators, in turn, affect the interest on the assets of each industry that are held in the Bank (Q-7).

Summary

The sections above have contained a long list of *SIMSOC*'s distinctive features. These features make *SIMSOC* special in the already special world of gamed simulations. In summary, let us just say that *SIMSOC* is one of the few simulations to model an entire society. One student said, "Geez, it's all here, isn't it?" As such, *SIMSOC* seems to have *some* interest for almost every student and *much* interest for many. It tends to be most fun for players who really get involved, but provisions for apathy as well as absenteeism are tied to the National Indicators. Being built into the game and usable in game strategies, they are "acceptable" behaviors for students. The game does not stop because of them, and students are not made to feel guilty as if they had hurt play. One group in Red region decided to act indifferent as a strategy for getting more pay for work from industries in privileged regions. Most students become deeply involved and seem to enjoy the exercise very much.

Table 6 presents a list of evaluation criteria for the game, a short description of how the game applies to the criteria, and a word or two of evaluation.

COMPARATIVE EVALUATION OF SIMSOC

A gamed simulation—indeed, any learning exercise—can be evaluated most meaningfully in relation to other exercises. This section compares *SIMSOC* with five other prominent gamed simulations: *Generation Gap* (Schild and Boocock, 1969), *High School* (Coleman and Seidner, 1972), *Starpower* (Shirts, 1969), *Ghetto* (Toll, 1969), and *They Shoot Marbles, Don't They?* (Goodman, 1973). The data presented here were gathered as part of a larger project, some of the results of which appeared in Dukes (1976). Information on the games shows general trends only; it is not conclusive evidence. Ideally, the study should have many administrations of each game under experimental conditions.

TABLE 6 Summary and Evaluation of SIMSOC

Criteria	Description	Comments/Evaluation
Issues	Scarcity, order, quality of life	Issues are abstract and discovered by students. Issues covered are the really important ones. It is necessary to debrief on each of them separately. Excellent.
Content	Macrosocial concepts	Good focus.
Value orientation	Game allows the following value positions to emerge: Privileged groups seek to expand their resources, exploit other groups; deprived groups may have to take political action to gain a piece of the pie.	I'm convinced.
Model validity	Model is complex (see figures 1 and 3)	SIMSOC has a certain gut-level [face] validity
Objectives (of game)	To model an entire society. To get players to assess goal attainment in a simulated social system.	SIMSOC accomplishes its objectives.
Roles	Various roles as heads of groups (Table 1). Role performances are not prescribed. Roles are given resources to develop.	The roles modeled by the game are the necessary ones. They are well structured.
Skills	The game does not teach specific skills. It does give some feedback on which individual and group strategies tend to work, though which strategies work may be idiosyncratic to a particular play.	Perfectly okay for a *simulation* game.
Activities	Most basic activities of society are included.	Good.
Interactions	Game allows and defines many types. Others are emergent. Norms of the classroom limit range of behavior.	Good mix of defined and emergent interactions.
Chief advantage	Models an entire society.	One of the few simulations to accomplish this feat.
Chief limitation	Long time commitment for play.	Challenges teacher to make the time worthwhile.
Rules	SIMSOC has many of them.	They are well thought out, clearly written.
Resources/ Scoring	Being head of a group or agency, living in a privileged region, acquiring Simbucks. Scoring involves acquiring resources or attaining more intrinsic, personal goals.	The game has a good mix of objective-subjective dimensions.
Debriefing	Debriefing will take up entire last session. Suggestions appear in coordinator's manual.	Good.
Complexity	Game is quite complex.	Complexity seems necessary, and useful for student learning.
Flexibility	Game is flexible for number of players, length, and number of sessions. Game is somewhat flexible for room arrangement. Game is somewhat inflexible for total time commitment.	Good.
Completeness	It is as complete as can be.	Excellent.
Durability	Materials in book form; paper in third edition is of better quality. Pages seem easier to tear out more cleanly than in the last edition.	Good.
Overall quality	Game is well thought out and well put together.	Excellent.

Administration of the Games

All players were students in two introductory sociology courses offered in the spring and summer of 1974 at a public university in the western United States (see Table 7).

Data Gathering

The questionnaire consisted of 12 closed-ended items (presented in Table 8) based on published work by Elder (1973), Fletcher (1971), and Wentworth (1972). Players rated items from a low of zero points to a high of 100 points for each game played in the course. Students completed the questionnaires at the end of the course after all the games had been played. The instrument took players about ten minutes to complete.

TABLE 7 Numbers of Games, Administrations, and Players for Each Game

	Number of Games	Number of Administrations	Number of Players*
Generation Gap	36	2	70
High School	11	2	60
Starpower	2	2	53
Marbles	1	1	26
Ghetto	7	2	49
SIMSOC	2	2	67

*Discrepancies between the number of games and the number of players are due to missing responses. No student played any one game more than once.

Results

Table 8 presents the results. The numbers are the average scores for each item for each game. On items 1 through 3, *SIMSOC* scored near the lower end of the evaluated games. Its high complexity (item 6) may have stood in the way of students perceiving the exercise as fun and the rules as clear and easy to understand. I still consider *SIMSOC* to have clear, well-written (though lengthy) rules.

These student evaluations seem to indicate that for *SIMSOC,* at least, complexity has a payoff in terms of accuracy of representation (item 4), plausibility (item 8), and realism (item 9), areas in which *SIMSOC* scored the highest of any of the exercises.

Though *SIMSOC* takes the longest to play, students rate it only slightly longer than most of the other games in the time it takes to reach its objectives (item 5). *SIMSOC* does not appear to *waste* any time.

SIMSOC was near the top of the list as one of the most worthwhile (item 10), valuable (item 11), and relevant (item 12) exercises among an already select group of gamed simulations. My feeling is that the new edition would do even better!

TABLE 8 Average Scores of Student Evaluations

Items	*Generation Gap*	*High School*	*Starpower*	*Ghetto*	*They Shoot Marbles, Don't They?*	*SIMSOC*
1. To what extent was the exercise fun?	54	71	64	70	60	59
2. To what extent were the rules clear?	71	78	75	75	63	61
3. To what extent were the rules easy to understand?	67	78	77	73	65	62
4. To what extent does the erercise accurately represent the model of reality it is attempting to represent?	55	69	69	70	64	73
5. How long does it take the exercise to reach its objectives?	51	62	60	57	59	63
6. To what extent is the exercise complex?	32	47	48	53	61	64
7. To what extent is the exercise appropriate for this class?	56	66	70	74	70	69
8. To what extent is the exercise plausible?	55	64	65	69	64	71
9. To what extent is the exercise realistic?	50	64	64	68	63	72
10. To what extent is the exercise worthwhile?	52	66	68	72	68	71
11. To what extent is the exercise valuable?	54	63	68	72	67	71
12. To what extent is the exercise relevant?	55	64	69	73	68	71

NOTE: If players and plays of each game represented a population of players and plays, a difference of 10 scale points would be a statistically significant one at the .05 level.

Notes

1. The exact percentage of agencies that are located in each region is determined by the total number of players in the game (see Gamson, 1978a: 48).

2. These letter/number designations refer to the rows and columns in Figure 3.

Notes to Figure 3: "Functional Interdependencies in SIMSOC"

1. BANK is conducted by the game administrator.

- It sells private transportation tickets (at 25 Simbucks, good for unlimited travel), luxury living endowments (at 25 Simbucks, good for staying alive, permanently employed, and 5 Munchie tickets per round), luxury living packages (at 15 Simbucks, good for one round of staying alive, remaining employed, and 5 Munchie tickets), and Munchie tickets (at 1 Simbuck for 4 tickets, good in exchange for food and beverages–imported luxuries–at the Munchie Bazaar).
- The income rates the BANK pays to POP, SOP, JUDCO, and EMPIN are determined after round 1 by the National Indicators and the number of members/subscriptions the group has.
- The interest rates the BANK pays on industrial funds it holds varies directly with the lowest National Indicator.

2. BASIN (Basic Extractive Industry) buys written passages in which workers must identify how many times each vowel occurs. Each error reduces payment by 10% of purchase price (no payment for a job with 5 or more errors). A perfect job yields a 125% return on the investment.

3. RETSIN (Retail Sales Industry) buys 6- to 8-letter anagrams, from which workers must make up to five 5-letter words. Each correct word returns 30% of purchase price. A five-word solution yields a 150% return on the investment.

4. SUBSISTENCE: One ticket keeps one player alive for one round. After one round without subsistence, a player becomes unemployed. After a second round without subsistence, the player dies. The agent sets the cost of a ticket.

5. TRAVEL: One ticket buys one player a round trip on public transportation between home and one other region. The agent sets the cost.

6. NATIONAL INDICATORS must all remain above zero; if one falls below the society ends.

7. ECONOMIC INDICATORS (food and energy supply, standard of living) decline through depletion of natural resources. Investments in Research and Conservation increase them.

8. SOCIAL INDICATORS (public commitment, social cohesion) are adversely affected by worker discontent with what is to be produced and by the number of players unemployed, arrested, rioting, dead, or absent from class. Investments in Welfare Services increase them.

9. STANDARD OF LIVING is increased by investments in Welfare Services and in Research and Conservation. It declines if there is worker discontent in BASIN and RETSIN over production. It increases if production is successful.

10. MASMED (Mass Media) is the only institution that can see all National Indicators and can communicate with all regions simultaneously. Its income depends on how many subscriptions it gets.

11. POP, SOP, EMPIN, and JUDCO may receive contributions/dues

from many sources, and they distribute them as they see fit.

12. WELFARE SERVICES: Investments increase the standard of living by 10%, public commitment and social cohesion by 20% each, of the Simbucks invested.

13. RESEARCH AND CONSERVATION: Investments increase the food and energy supply by 40% and the standard of living by 10% of the Simbucks invested.

References

COLEMAN, J. S. and C. J. SEIDNER (1972) *High School.* (unpublished) Rules available from second author c/o School of Education, Boston University, Boston, MA 02215.

COPPARD, L. and F. L. GOODMAN [eds.] (1977) "SIMSOC," pp. 275-289 in Urban Gaming/Simulation '77. Ann Arbor: University of Michigan, School of Education.

DUKES, R. L. (1976) "Toward a general evaluation for simulation games: GEM." Simulation & Games 7 (March); 75-96.

--- and C. J. SEIDNER (1978) Learing with Simulations and Games. Beverly Hills, CA: Sage.

ELDER, C. D. (1973) "Problems in the structure and use of educational simulation." Sociology of Education (summer): 335-354.

FLETCHER, J. L. (1971) "Evaluation of learning in two social studies simulation games." Simulation & Games 2 (September): 425-454.

GAMSON, W. A. (1978a) SIMSOC, Simulated Society: Coordinator's Manual. New York: Free Press.

--- (1978b) SIMSOC, Simulated Society: Participant's Manual. New York: Free Press.

--- (1972) "SIMSOC, establishing social order in a simulated society." Simulation & Games 2 (September): 287-308.

--- and R. J. STAMBAUGH (1978) "The model underlying SIMSOC." Simulation & Games 9 (June): 131-157.

GLANDON, N. D. (1973) "Sexism: rigging SIMSOC to make the point." Simulation/Gaming News (September): 1-5.

GOODMAN, F. L. (1973) *They Shoot Marbles, Don't They?* University, Alabama: Institute of Higher Educational Research and Services.

INKELES, A. (1969) "Society, social structure and child socialization," pp. 73-129 in J. A. Clausen (ed.) Socialization and Society. Boston: Little, Brown.

ROSEN, B. et al. (1973) "Effects of participation in a simulated society on attitudes of business students." J. of Applied Psychology: 355-357.

SCHILD, E. O. and S. S. BOOCOCK (1969) *Generation Gap.* Indianapolis: Bobbs-Merrill.

SHIRTS, R. G. (1969) *Starpower.* Del Mar, CA: Simile II.

SILVER, B. R. (1973) "Social mobility and intergroup antagonism: a simulation." J. of Conflict Resolution 17 (December): 605-623.

STADSKLEV, R. (1974) "Little MAC (Measuring Analytical Components)," pp. 88-91 in R. Stadsklev (ed.) Handbook of Simulation Games in Social Education (Part I: textbook). University, Alabama: Institute of Higher Educational Research and Services.

TOLL, D. (1969) *Ghetto.* Indianapolis: Bobbs-Merrill.

WENTWORTH, D. (1973) "Reviewer checklist," in D. Zuckerman and R. Horn (eds.) The Guide to Simulations/Games for Education and Training. Lexington, MA: Information Resources, Inc.

WOLF, C. P. (1972) "SIMCANSOC: Simulated Canadian Society." Simulation & Games 3 (March): 53-77.

SIMULATIONS OF DEVELOPING A SOCIETY
An Evaluation

by Frank P. Diulus

INTRODUCTION

If you are a teacher or student of the social sciences, you will certainly be interested in the simulations presented in this essay. If you are a person who is interested in communication between people, the development of ideas and actions, or the typical patterns of groups and institutions, you are also in luck. These simulation games deal with how groups, social organizations, institutions, and societies develop. Students of society are also fortunate in that three of the seven simulation games we discuss here are renowned classics.

Definition

What are simulations of the development of a society? The simulation games included in this essay are dynamic models of social realities. They can be employed as learning opportunities, educational resources, or group dynamics exercises in studying or promoting social interactions and cooperative decision-making.

These simulations offer models of how societies are created, developed, maintained, and altered. This includes analogs of microsocieties, such as a specific family, the tenth-grade class, the local Boy Scout group, the Friday night Poker Club, and so forth. The simulation games we shall discuss begin with little prior definition of a society or institution; within the parameters of minimal prestructuring through rules, the participants themselves develop the processes and structures of a soceity.

Sociologists explore how societies typically begin with group experiences that quickly lead to a group identity. To belong to a specific group, an individual is expected to behave according to the expectations and norms of this group. There are typical developmental processes snd structures in all societies; but societies are differentiated by different contexts, relations, and ideas. To understand the various cognitive aspects of basic sociology would require extensive study, but direct and active involvement simulating a developing society can quickly enable participants to see, feel, and direct many of the critical social variables. To experience the multiple realities of a society goes beyond the study of abstract and anonymous theory to a direct grasp of social processes in the here and now and can lead, furthermore, to reflection and interpretations of social forces and human responses. Simulation of developing a society helps participants both think about, and act on, the social construction of realities. It is in this respect that some scholars regard simulation games as a new language of the social sciences; for simulation games can offer a holistic view of an active model.

TABLE 1 Simulations of Developing a Society

Simulation Game	Summary Description
Comco	Players living in different regions try to satisfy their personal basic needs by developing complex social structures.
Humanus	The only known survivors of a world-wide catastrophe must make several collective decisions about their new society.
Micro-Community	After students study some important aspects of American history, they simulate their own social processes of government, business, etc.
Micro-Economy	A structured micro-society develops in the classroom with pupils sharing in the control of the economy.
Simsoc	Players residing in different regions and working for one of seven basic groups seek to create and maintain a society through the delicate balancing of personal interests, group needs, and allocation of scarce resources.
Starpower	Participants trade chips of different values to establish a stratified society of three social classes.
They Shoot Marbles, Don't They?	Players construct a society through political processes, law making, and enforcement; and they compete for rewards of marbles.

Editor's Note: Listings for all simulations discussed in this essay are in the social studies section, except for *They Shoot Marbles, Don't They?*, which is in the frame games section.

A BRIEF OVERVIEW

This essay will describe, compare, and critically analyze seven simulation games. Table 1 offers a summary description of the central theme of each of the simulation games under consideration.

Playing Data

There are several preliminary matters to consider when trying to choose an appropriate simulation game. The playing data summarized in Table 2 may indicate whether a particular simulation of developing a society is appropriate for a specific group or setting. Table 3 identifies information for ordering the simulation games.

The space requirements for playing a simulation game can be an important selection criterion. All seven of these simulations can be played in a single room. However, although both *Simsoc* and *Comco* can be played in one large room with the space divided, it is preferable to have four rooms to house the different regions.

Preparation

The amount of time and energy that a game director invests before playing should yield related dividends. In this respect, two classic simulation games, *Starpower* and *They Shoot Marbles, Don't They?* are remarkable bargains. With under thirty minutes devoted to reading before the first playing, a game director can easily lead a group through a unique and rich educational experience. Both *Starpower* and *They Shoot Marbles, Don't They?* are complex in the many aspects of society simulated; but both are masterpieces of modeling, reducing complexity to the simplicity of directing and playing.

Of the four simulations that require more than two hours of preparatory reading and organizing (see table 2) *Simsoc* is the third classic, far out-distancing the other three (*Comco, Micro-Community,* and *Micro-Economy*) for dividends returned on the investment of preparation time. Among the seven simulations under study, *Simsoc* simulates the greatest quantity and complexity of social system variables; and *Simsoc* could comprise the majority of a semester of class time. Therefore, the several hours of initial preparation for *Simsoc* pay a rich dividend.

Finally, playing *Humanus* is a wise investment of limited time. After five minutes devoted to an overview of the simulation, the game director is replaced by an audio-tape cassette. While *Humanus* does not have the depth of issues or realism of three simulations above, the ease of beginning and using it, together with excellent educational results, rates *Humanus* high in this category of preparation in relation to results.

ISSUES AND ACTIVITIES

In selecting a simulation game it is important to know key aspects of both the content and the methods of a simulation. All seven simulation games under study deal with the development of a society, but each emphasizes different social processes and problems. What follows is a brief description of each

TABLE 2 Basic Playing Data

Simulation	Age Level	No. of Players	Playing Time	Preparation Time
Comco	high school – adult	15-60	5-10 sessions of 45-90 min.	3 hours
Humanus	grade 7-adult	5 or more	1-1½ hours	5 min.
Micro-Community	grades 7-12	15-40	1 month – 1 year	3 hours
Micro-Economy	grades 5-12	whole class	1 month – 1 year	2 hours
Simsoc	high school – adult	15-60	5-10 sessions of 45-90 min.	3 hours
Starpower	grade 5 – adult	15-35	1-1½ hours	20 min.
They Shoot Marbles, Don't They?	grade 5 – adult	12-50	1-2 or more hours	20 min.

TABLE 3 Ordering Information

Simulation	Cost	Author	Copyright	Producer & Address
Comco	$5.95	Susan Cumming, Elizabeth Manera, Gerald Moulton	1973	Arizona Training Labs, P.O. Box 26660, Tempe, AZ 85282
Humanus	$11.50	Paul Twelker, Kent Layden	1973	Simile II, P.O. P.O. Box 910, Del Mar, CA 92014
Micro-Community	$49.50	William B. Jarvis	1971	Classroom Dynamics, 231 O'Connor Drive, San Jose, CA 95128
Micro-Economy	$39.00	George Richmond	1973	The Center for Curriculum Design, Harcourt Brace Jovanovich, 757 Third Avenue, New York, NY 10017
Simsoc	participant's Manual – $5.95; instructor's manual-free	William A. Gamson	third edition 1978	The Free Press, 866 Third Avenue, New York NY 10022
Starpower	do-it-yourself kit $3.00; kit for 18-35 students $30.00	R. Garry Shirts	second edition 1978	Simile II, P.O. Box 910, Del Mar, CA 92014
They Shoot Marbles Don't They?	$40.00	Frederick L. Goodman	1973	Institute of Higher Education Research and Services, Box 6293, University, AL 35486

simulation's main issues, interactions, and sequence of activities.

Comco

Players attempt to handle such aspects of society as food, jobs, money, travel, and crime, while they simultaneously focus on satisfying five basic human needs: physiological, safety, love, esteem, and self-actualization. This simulation game imitates most of the format, contents, and activities of *Simsoc,* which is described below.

Humanus

This simulation of a survival group consists of a values clarification exercise concerning interpersonal relationships, the development of democratic laws, and methods for change. The group is asked by a survival computer to consider such issues as their basic needs for survival, getting along with others, problems of loneliness, and relating to other potential groups.

Micro-Community

There are several simulations contained within curriculum units that can be used to study American history. Simulations of developing a government, managing business investments, and solving pollution problems enable pupils to become active participants in American history.

Micro-Economy

Students can increase skills of practical economics by earning play money for completing assignments, getting good grades, doing school chores, working on civic projects, and achieving successes in the ongoing class simulation. Participants learn first about money, banking, and real estate. Then they apply and extend their knowledge in a monopoly type of simulation game.

Simsoc

Several sociological perspectives, processes, and decisions are simulated. Players actively engage in conflict, protest, deviance, control, change, and issues of legitimacy. Participants work for one of seven different groups and live in one of four different locales. This simulation usually progresses through three phases: problems of scarcity, problems of power and authority, and problems of prosperity.

Starpower

Participants create social situations that express many of the uses and abuses of power, wealth, and prestige. Three groups, which represent social class divisions of the rich, middle, and poor, attempt to bargain and exchange different color chips worth different values. In one type of round, participants can mingle freely to trade chips with anybody; in another, participants strategically divide bonus chips among their reference group. The wealthiest group is suddenly given power to change the rules of the game, and their typical self-serving rule changes stir a revolt.

They Shoot Marbles, Don't They?

Players experiment with social, economic, and political processes, and coalitions vie for control of rules and rewards. Five marble shooters are engaged in competitive shooting at marbles on the game board, in bargaining for bonus marbles, and in proclaiming their problems and interests to the other players. The rest of the players, who are assigned key roles in a society, are simultaneously making laws, enforcing them, building businesses and profits, developing an opposition party, and so on. Continuous changes of power, norms, rewards, and activities affect all of the groups in the society.

Each of these simulations highlights certain issues and interactions. Thus, participants are offered a laboratory in which to experiment with certain social processes. Involvement in any one of these simulation games provides concentrated opportunities to practice certain skills. Table 4 summarizes the major competencies that can be developed through playing each simulation game.

ROLE-PLAYING

All of the simulations of developing a society involve role-playing; therefore, various aspects of the role-playing in each simulation game can offer important evaluative perspectives. Potential users can use this section in considering such things as the kinds of roles participants are expected to play, the interest and motivation offered by the scenario, the degree of involvement by the participants, and the learning potential offered by the role-playing dimensions of a specific simulation game.

Comco

There are nine major groups for individual players to role-play—two industries, health and housing, police, recreation, education, communication, court, and philanthropy. Some participants also serve as food and travel agents. The role-playing in *Comco* can be as active and educational as in *Simsoc,* its precursor. However, *Comco* lacks the detail and clarity of *Simsoc's* role descriptions, so that the interest, involvement, and learning potential of *Comco* tend to be

TABLE 4 Competencies

Simulation Game	Skill Development In . . .
Comco	achieving personal needs via social control
Humanus	personal values clarification and group problem solving
Micro-Community	implementing democratic principles
Micro-Economy	managing personal and practical economics
Simsoc	establishing and maintaining social order
Starpower	the uses of power and authority
They Shoot Marbles, Don't They?	coalition building and lawmaking

Humanus

All players are assigned the same role of projecting themselves into a situation of survival and meeting basic individual and group needs. The scenario demands intense communication and decision-making. Each participant is asked to choose value priorities on crucial issues.

Micro-Community

Pupils represent themselves within the development of classroom government, business, ecology, and so on. Since the simulations relate to a student's interests and social world, they promote involvement and learning.

Micro-Economy

Students assume new roles in their real world, and they also take on roles in the simulation game such as banker, business leader, member of a protection group, and so forth. Players develop organizational skills and economic understanding through their high degree of involvement, decision-making, and problem-solving.

Simsoc

The role descriptions clearly depict the basic reference groups of industry, unions, politics, mass communication, and the judicial council. There are also travel and subsistence agents. The complex and interdependent needs for survival, power, and ownership highly involve all of the players in planning strategies, developing coalitions, and accomplishing tasks.

Starpower

Each participant is a member of one of three groups—the squares, the circles, and the triangles. Negotiations between individuals and groups are rapid and intense. Role players directly experience a social rewards system of haves versus have-nots.

They Shoot Marbles, Don't They?

The roles for the game consist of marble shooters, rule makers, rule enforcers, and a judge; several other optional roles in a society are suggested. However, the flexibility of this scenario permits a game director to substitute different institutions, roles, and rules for different audiences and instructional objectives. The interaction of typical social roles is highly realistic, enjoyable, and educational.

DIRECTOR'S MANUAL

Rules

Directions for games have historically been poor. Many games are difficult to play because the directions are confusing and incomplete. Table 5 shows an assessment of the rules for the seven simulations of developing a society.

TABLE 5 Rules

Simulation Game	Overall Rating for Clarity, Organization & Completeness		
	Outstanding	Good	Adequate
Comco			X
Humanus	X		
Micro-Community			X
Micro-Economy		X	
Simsoc	X		
Starpower	X		
They Shoot Marbles, Don't They?	X		

Debriefing

Post-game activities are another important part of a simulation game because they provide an opportunity to reflect upon, reconstruct, clarify, and reinforce the learning experiences. The director's manual should provide discussion topics or questions or suggest some other kinds of follow-up activities. Table 6 indicates ratings of the debriefing guidelines offered by the seven simulation games.

User's Reports

Based upon user's reports, we can offer game directors some suggestions concerning the directions for the simulations of developing a society. After several play experiences with a simulation game, some critical perspectives on the directions emerge, and game directors will frequently add some rules or change some emphases.

They Shoot Marbles, Don't They? is a good illustration of flexibility. Although game directors playing this simulation for the first time will not be disappointed by following the director's manual closely, they are encouraged by the game's developer, Fred Goodman, to modify the rules and roles to suit their instructional purposes. For example, a class in teacher education could simulate a school in which the pupils are marble shooters, the principal and vice-principal are law makers, the senior faculty are law enforcers, and the teacher's union is the opposition party.

The publication of revised editions also testifies to the fact that rule changes are necessary. The third (1978) edition of *Simsoc* includes simplified and more precise rules, new personal consumption options, and some much needed improvements in the participants' experiences of directly controlling the national indicators. There is also a second edition of *Starpower,* which includes a new debriefing section. An important rule change is that players are told at the beginning of

TABLE 6 Debriefing Guidelines

Simulation Game	Good	Adequate	Inadequate
Comco	X		
Humanus	X		
Micro-Community			X
Micro-Economy			X
Simsoc	X		
Starpower	X		
They Shoot Marbles, Don't They?	X		

the first trading session that the most successful traders will draw new chips from an enriched bag stacked with the most valuable gold chips.

Users of *Humanus* suggest that game directors improvise additional introductory directions and stage-setting both to clarify the scenario and to emphasize the importance of controlling laughter in response to role players' statements.

Teachers who have used *Micro-Economy* suggest an extensive mingling of real life interests and issues with the board game, using actual classroom curriculum, space, and furniture as a more real and life-related content for the simulation. Game directors of both *Micro-Economy* and *Micro-Community* will need to impose more organization on the playing manuals, clarify the directions, and carefully select aspects that relate to instructional objectives. Both of these simulation games contain elements of token economies, which could frustrate and embarrass low achievers. Some teachers have successfully used *Micro-Economy* and *Micro-Community* while de-emphasizing the play money rewards.

Users of *Comco* suggest several additions to the director's manual. The large amounts of money and food stamps available to players seem to be unrealistic because prosperity is too easily accomplished. The explanations for investing in stock should be increased. The aspect of experiencing personal need levels is weak, and the realities of affecting needs and the means of computing need levels both require considerable clarification. Finally, several of the group role descriptions could be expanded.

CONCLUSION

We have described and evaluated seven simulations of developing a society. They have many strengths and some weaknesses. After reviewing the critical comments, the following final ratings clearly emerge.

Four of the simulation games, *Humanus, Simsoc, Starpower,* and *They Shoot Marbles, Don't They?* are outstanding simulations that excel on most evaluative perspectives. The major liabilities of *Micro-Economy* are its weaknesses in the director's manual, its lack of debriefing guides, and its high cost. Although *Comco* and *Micro-Community* are rated only adequate, they can be used effectively by a persevering game director. Potential users of a simulation of developing a society should weigh their own purposes and constraints against the multiple and varying characteristics, strengths and limitations of the seven simulation games to make a suitable match. All of the simulation games can provide memorable and enjoyable educational experiences.

TABLE 7 Final Ratings

Simulation Game	Outstanding	Good	Adequate
Comco			X
Humanus	X		
Micro-Community			X
Micro-Economy		X	
Simsoc	X		
Starpower	X		
They Shoot Marbles, Don't They?	X		

SOCIAL STUDIES GAMES AND SIMULATIONS
An Evaluation

by Michael J. Rockler

INTRODUCTION

In the third edition of the *Guide,* I described new developments in social science education. These included changes from traditional history, civics, and name-place geography courses to more contemporary ones emphasizing conceptualization, inquiry, and values education. In general, these innovations required examination of important phenomena; they were all improvements. This essay advocates additional changes in the continuing evolution of the discipline.

The social studies ought to focus upon the use of education for expanding human potential, ways to facilitate creative behavior, interdisciplinary studies, and future studies. Simulation games enhance the study of all these topics.

Education for increasing human potential derives directly from John Dewey's concern for student-centered teaching. It is an updated, more sophisticated "life adjustment" curriculum. Rapidly changing circumstances and the increased complexity of our lives make it both necessary and appropriate.

Problem solving, based primarily on analytical and conceptual convergent thinking, remains an important aspect of social studies education. Problem solving also demands divergent thinking as an additional strategy. The use of theories espoused by W.J.J. Gordon, E. Paul Torrance, Sidney Parnes, and others accomplishes this objective by facilitating creative behavior.

The social studies have often emphasized traditional social science disciplines, which form the basis of many current programs. However, these methods of organizing knowledge will someday become obsolete, to be replaced by interdisciplinary studies using broader, more useful concepts such as system, model, and decision making. Concepts like these will increasingly be used to organize knowledge.

The rapid rate of change occurring in social institutions demonstrates the increasing need for emphasizing future studies. This need has been recognized by the recently formed Future Studies Special Interest Group of the National Council for the Social Studies.

Simulation gaming in the social studies promotes understanding of these new trends in education: for increasing human potential, for facilitating creative behavior, for interdisciplinary studies, and for future studies.

The process of gaming contains many important human development attributes. Those that involve role playing activity can be viewed as simulated psychodramas allowing players to work out their problems in nonthreatening environments. Human behavior can be understood from a gaming perspective; the increased use of the technique in the social studies helps develop more effective strategies for living, enriching education for human potential.

Divergent thinking—facilitated by open-endedness, by the need to discover new solutions to problems, by the juxtaposition of experiences not previously viewed as related, by an accepting-respecting attitude—leads to increased creative behavior. A simulation gaming experience contains all of these characteristics. Furthermore, it minimizes the required risk of creativity in a safe, nonthreatening, simulated setting. Social studies students who regularly participate in simulation games will improve their creative behavior.

Because of its basic nature, the simulation gaming process enhances interdisciplinary studies. The concepts noted above—system, model, and decision making—are all components of it: any simulation is a system; a model is a simulation; all games used in the classroom demand some form of decision making.

Richard Duke calls games "future's language." Games allow us to predict, understand, and adjust to rapid change. The use of games prepares social studies students for the future even as they simultaneously accomplish other purposes.

This essay reviews several miscellaneous social studies games that do not fit into the other essays; no theme connects them. They represent a variety of objectives, uses, formats, and topics.

Table 1 indicates those included in this review.

Editor's Note: *Committee* is listed in the domestic politics section, *Expressway* (*Environmental Simulations*) in ecology, *Fail-Safe* in the social studies subsection of the computer section, and *Inner City Housing* in community issues. The rest of the simulations discussed in this essay are listed in the social studies section.

TABLE 1 Simulation Games, Publishers, Packaging, Cost

Simulation Game	Publisher	Kind of Packaging	Cost
Committee	Interact	Interact pamphlet	$14.00
Expressway	Scholastic-Tab Publications, Canada (in *Environmental Simulations* by Ben Vass)	Unit (teaching guide, 30 student pamphlets, wall poster)	$24.00
Fail-Safe	(1) Minnesota Educational Computing Consortium	(1) instruction pamphlet	(1) $1.25
	(2) Bruce Tipple, Social Science Education Consortium, 855 Broadway, Boulder, CO 80302	(2) instructions and software	(2) not yet set
Flight	Interact	Interact pamphlet	$10.00
Inner City Housing	Free Press (in Samuel A. Livingston and Clarice Stasz Stoll, *Simulation Games)*	book	$4.95
The Propaganda Game	Wff. n Proof	kit	$11.00
Rip-Off	Interact	Interact pamphlet	$14.00
Shipwreck and Other Government Games	Scholastic Book Services	pamphlet	$9.95
Simulating Social Conflict	Allyn and Bacon	pamphlet	$10.20 set of 10 $3.04 instructor's guide

USE IN COURSES

One way to organize these games is to indicate in what particular kind of course they might be used.

Committee fits best in a government course at the senior high school level. Because it teaches decision-making skills, it could be used in some other courses as well.

Expressway also emphasizes decision making; it teaches geography, and it could also aid instruction in government classes.

Fail-Safe uses a computer to simulate the book by Eugene Burdick. It emphasizes decision making and applies best to courses in government. History classes examining the cold war will also find it beneficial. It can help demonstrate the role of the computer in decision making, education, and simulation gaming.

Flight simulates a cross-continent air race; it develops decision-making and map-reading skills and serves as an alternative to the usual ways of teaching maps in a geography course.

Inner City Housing examines tenant-landlord relations. Teachers of social problems and sociology courses will find it helpful.

Shipwreck and Other Government Games consists of four games developed by the Scholastic Book Company. All are intended for junior high government classes. The games are *Shipwreck, On Trial, You be the Mayor*, and *Curfew.*

Sociological Resources for the Secondary Schools developed *Simulating Social Conflict;* it is primarily a sociology game useful also for social problems courses. It teaches decision making as well.

Table 2 indicates how the games relate to subjects in the social studies and offers a brief description of each.

TABLE 2 Subject Area and Description

Simulation Game	Social Studies Subject	Summary Description
Committee	government	A simulation of the committee process by which a bill becomes a law. Involves negotiation and decision-making skills. Leads to an understanding of the interaction of the presidency, Congress, and lobbyists.
Expressway	geography	Demonstrates the competing demands that exist when ever decisions must be made about the environment. Players decide whether to complete an expressway through a city or to develop mass transit.
Fail-Safe	government history	Players assume the roles of the characters in the Eugene Burdick novel of the same name. They use data from a computer terminal and try to avoid disaster.
Flight	geography	Played in teams consisting of pilot, copilot, and navigator. Uses an air race to teach map-reading and decision-making skills.
Inner City Housing	sociology social problems	Players assume the roles of tenants and landlords and attempt to secure housing or make a profit. Demonstrates the frustrations of each in an even-handed way.
The Propaganda Game	social problems, most other subjects in the social studies.	Players compete to be the most effective in identifying a variety of propaganda techniques.
Rip-Off	sociology, social problems government as it relates to law	Players are to engage in shoplifting (arranged for by instructor), after which they discuss the problem in a debriefing experience that also examines juvenile justice.
Shipwreck and Other Government Games	government civics, political science	This Scholastic publication contains four games. Each attempts to examine an aspect of government; written for average and below-average junior high readers.
Simulating Social Conflict	sociology social problems	Contains two related simulations, *Dilemma of the Tribes* and *Resources and Arms.* In this modified prisoner's dilemma game, players compete or cooperate: they gain or lose depending on the patterns of interaction they establish.

BASIC CONSIDERATIONS

Whether to use one game or another sometimes depends on circumstances. A simulation may produce excellent results in one situation and be less useful (and even fail) in another. Results sometimes depend on what has come before and what will follow. Each person makes these judgments independently. The data described in this section may help you choose. The age level, playing time, number of players, and number of groups are important information specified by the designers, including their conceptions of the most effective conditions for playing the game. Another basic consideration involves a game's adaptability to settings other than those specified. I have successfully used simulations designed for secondary students with graduate students (*Expressway* is one example).

Table 3 describes basic considerations for the simulation games reviewed here.

COMPLEXITY AND MODEL VALIDITY

Meaningful simulation games require complexity; they must contain enough variables to yield an accurate reflection of the simulated events. However, too many variables can lead to an unmanageable experience. A game requiring several pages of rules may create boredom before it ever starts. One that is too simple may be badly distorted, limiting the potential of learning outcomes.

Table 4 indicates the complexity of the games on a continuum from easy to complex.

The issue of complexity also relates to the question of model validity. The concepts of theory, model, and simulation interrelate. A theory is a hypothesis or generalization about an event in the real world that is descriptive, predictive, and has high credibility. A model is a subclass of a theory. It can be a physical one or a mental abstraction; its central feature is that it is an analogy for some real-world event or thing. As model is a subclass of theory, so a simulation is a subclass of a model. A simulation game represents a real-world event in a reduced and compressed form that is dynamic, safe, and efficient.

Because of its relationship to model, it is appropriate to ask the extent to which a simulation accurately portrays the reality it mirrors. Determining model validity means examining issues like distortion and simplification. It means looking at the relationship of the simulation to reality in terms of what has been included and omitted. The difficulty of the issue of validity resides in the conflicting needs for some simplification (so that the event is simulated and not real) and the desire to avoid oversimplification and distortion.

The following comments evaluate model validity for the games considered in this essay.

In general, *Committee* is a valid representation. It is an accurate analogy of the way committees function in the legislative process, containing a reasonable balance between simplification and distortion. The complexity of this aspect of law-making is maintained to demonstrate the procedures with sensible fidelity.

TABLE 4 Complexity

Simple	Easy	Moderate	Complex
Shipwreck and Other Government Games	*Expressway*	*Committee*	*The Propaganda Game*
	Flight	*Fail-Safe*	*Rip-Off*
		Inner City Housing	*Simulating Social Conflict*

TABLE 3 Basic Considerations

Simulation Game	Age Level	Playing Time (Hours)	No. of Players	No. of Groups	Adaptability
Committee	senior high college	4	35	4	Limited
Expressway	junior high senior high college	2	35	4	High
Fail-Safe	junior high senior high college	2	12-24	3	Moderate
Flight	junior high senior high	10	Any number in teams of three	Any number	Limited
Inner City Housing	senior high college	4	20	2	High
The Propaganda Game	senior high college	1	4	1	High
Rip-Off	senior high	8	12-25	2	Moderate
Shipwreck and Other Government Games	junior high senior high	4	35	3-4	Moderate
Simulating Social Conflict	senior high college	7	27	2	High

Expressway also has acceptable model validity. The scenario describes a meeting of a committee in a legislative body. Instead of reflecting the institutional structure (as *Committee* does), this one is an analogy about the nature of the competing forces involved in decision making. Both theoretically and practically, this game is a valid model of legislative process.

Fail-Safe employs a computer to simulate the process of choosing alternative courses of action. The human actors, however, can only respond to the limited options presented in the computer program. These choices are based on Burdick's novel, from which the game was created. Because of this, distortion and oversimplification are introduced, weakening the model's relationship to reality. *Fail-Safe* has many strengths, but model validity is not one of them.

Flight teaches map-reading skills; the question of model validity is not appropriate in this case.

Inner City Housing, creating empathy and teaching about the difficulty in relations between tenants and landlords, considers important problems. However, some distortion and oversimplification occur. The game fails to consider the role of the absentee landlord. It leaves the impression that dealings between those who own property and those who rent it are based on face-to-face negotiation. The simulation also has a slight bias in favor of tenants. These factors weaken the model validity of the exercise.

Because *The Propaganda Game* is a skills exercise, the issue of model validity is not applicable to it.

Rip-Off blurs the distinction between reality and simulation in the version that requires actual stealing. Different circumstances may occur each time it is played (for example, players may be caught in a situation in which a mistake has been made about the location of a "simulated" theft). Then, too, the different versions (actual and pseudo-stealing) confuse the issue. This one distorts and oversimplifies, resulting in questionable model validity.

The four exercises in *Shipwreck and Other Government Games* lack depth to the extent that oversimplification must occur. This leads to distortion and inaccurate modeling. The developers may believe that simplicity is necessary to reach the intended audience (average and below average junior high students), but this is hard to accept. These materials lack validity as they model government organization, trial procedures, city problem solving, and community action.

The remaining game, *Simulating Social Conflict,* is interesting for the question of model validity. It attempts to simulate the process of social conflict on a large scale. This demands oversimplification, but because it examines the dynamics of large-group interaction, the simulation remains valid despite its limitations. Table 5 rates model validity.

TABLE 5 Model Validity

Model Generally	Model Valid with Qualifications	Model not Valid	Not a Relevant Issue
Committee	*Inner City Housing*	*Fail-Safe*	*Flight*
Expressway	*Simulating Social Conflict*	*Rip-Off*	*The Propaganda Games*
		Shipwreck and Other Government Games	

ISSUES

After a presentation I once made on creativity at an annual meeting of the National Council for the Social Studies, a speaker from the floor rose to argue against developing creative behavior in children. Such behavior, he asserted, might cause intolerable problems for them, leading to frustration because they would have no outlet for their divergent patterns of thinking. This had never occurred to me. I describe this incident as an introduction to the examination of issues. Most simulation games contain a variety of overt issues. Many contain covert ones as well. Before using a game, one must attempt to identify all of the issues, both overt and covert.

I identify five different kinds of issues. First, in any exercise, the designer describes specific issues. Second, any game contains topical issues (current questions that may soon fade from the scene) and, third, relevant issues (those that are important and will continue over some time). Fourth are issues that have endured for a long time and will continue. And finally, there are issues relating to the nature of social studies education itself.

Table 6 describes the kinds of issues examined in the various games.

CONTENT

The issue of content focuses on the amount of detail as opposed to breadth of coverage. This question is important for any social studies material. Superficiality results when a small amount of information tries to cover too many topics. An overabundance of information in games, however, transforms the process from one based on experience to one based on reading.

The following descriptions, as well as Table 7, analyze the content of these games.

The specific content given to players in *Committee*, combined with the suggested research project, leads to an overemphasis on content.

Expressway provides balanced content. Participants receive sufficient information for playing without being overwhelmed. It can be played in as little as one class period while successfully examining all the issues.

Fail-Safe provides balanced content. Requiring students to read the original book would expand it, but this is not necessary. Players receive enough information to make the game exciting.

Flight is another balanced one. Players make flight plans, learning to read maps. This is a commendable use of content, the nature of which could lead easily to overemphasis on data.

Inner City Housing is generally balanced. The materials contain needed information about costs and roles. The most important outcomes emerge from the experience rather than from the content.

TABLE 6 Issues by Type

Games	Specific Issues	Topical Issues	Relevant Issues	Enduring Issues	Social Studies Issues
Committee	Mine safety Education Pollution		Mine safety Education Pollution	Legislative process	
Expressway	Should the express-way be built?	The energy crisis crisis	Pollution Role of the automobile	Individual vs. group needs	The influence of group pressure on decision making
Fail-Safe	How should the U.S. respond to military accidents?	Détente	The nature of Pres-idential decision making	Civilian control of the military	The role of the com-puter in the social studies
Flight	Airplane safety	Air traffic control	Control of airlines		The role of skills in the social studies
Inner City Housing	Rights of tenants vs. rights of landlords	Rent control	Fair profit and fair treatment	Individual vs. group rights	
The Propaganda Game	Awareness of propaganda	Control of the media	Should critical think-ing be encouraged?	The right of free speech	Basic education and critical thinking
Rip-Off	The cost of shoplifting		The status of juvenile justice	Individual vs. societal rights	What kind of tech-niques are acceptable in the social studies?
Shipwreck and Other Government Games	The need for government	Participation vs. representation	The nature of gov-ernment in society	Nature of demo-cratic government	
Simulating Social Conflict	Self-interest vs. the common good	Priorities of resources	Societal cost of war	Conflict vs. cooperation	

The Propaganda Game includes considerable information for students to analyze. It requires knowledge about the categories of propaganda. Despite these problems, the game remains balanced between breadth and depth.

Rip-Off simulates shoplifting and the juvenile hearing process. The game contains two different experiences and is still balanced. Players receive information but not burdensome and unnecessary detail.

Shipwreck and Other Government Games is a pamphlet that contains four games. Players create a society, learn courtroom procedures, learn about problem solving in the city, and study community action. Everything comes in twenty-eight pages, amply illustrated, with several pages of dialogue. These games lack depth. (The Scholastic Book Company often operates on the assumption that brevity is simpler and easier. I am not sure that this is better, nor am I sure that the assumption is correct.)

Simulating Social Conflict makes excellent use of sociological data in a simulation. Players are often less familiar with this kind of content than any other in the social studies. Some information is needed; this episode provides it in an appropriately balanced manner.

VALUE ORIENTATION

The value orientation of simulation games for the social studies is particularly significant. Value education remains an important part of the curriculum. Three identifiable levels of value education exist in the simulation process. The lowest is recognition, leading to an understanding of the value positions in a particular games. The next level is clarification, through which players learn to identify their values. At the highest level, players analyze their values in a rational way. Most of the simulations examined here are effective at one or more of these levels.

TABLE 7 Content

Too Much Breadth	Balanced	Too Little Depth
Committee	*Expressway* *Fail-Safe* *Flight* *Inner City Housing* *The Propaganda Game* *Rip-Off* *Simulating Social Conflict*	*Shipwreck and Other Government Games*

Committee develops value recognition. Students examine the orientations of each issue considered. The game focuses on the committee process leading to legislation. Players do not advance beyond the first level.

All three levels are present in *Expressway*. Students recognize the value conflicts involved in building expressways; they clarify their own stands; they analyze the consequences of their value orientations.

Fail-Safe contains no explicit value orientation issues. However, it examines the nature of a balance-of-terror strategy for preventing nuclear confrontation. Implicit value considerations exist that could be examined in debriefing (for instance, the issue of military preparedness at the current enormous cost can be compared with other options, such as a strategic arms limitation treaty). In this way, value clarification and analysis can be achieved in this simulation.

Flight does not facilitate value education. Designed as an aid in the development of particular skills, it has little involvement with the process of valuing.

The balanced value orientation of *Inner City Housing* leads to value analysis. Players most often begin the experience from a position favoring either the landlord or the tenant. By the end of the game, the players recognize the difficulties of both positions; this forces them to reconsider their judgment and therefore analyze their values.

Like *Flight, The Propaganda Game* emphasizes skills—particularly the skill of critical thinking. Players learn that critical thinking is preferable to being uncritical. This represents both value clarification and recognition.

The value orientation of *Rip-Off* opposes shoplifting and favors more equitable juvenile hearing procedures. It requires students to recognize these values and to clarify their stands on stealing.

Shipwreck and Other Government Games requires players to recognize the value orientations of the various issues they consider. It also demands value clarification as the exercises proceed. Specific issues include questions of war and peace, the nature of social sanctions against law breakers, the importance of participation in social change, and problems of urban renewal.

Simulating Social Conflict helps participants recognize the difference between conflict and cooperation. They clarify their own stands with regard to these behaviors. They must also analyze the consequences of both types of action for society.

Table 8 summarizes the various levels of value education in the games.

TABLE 8 Value Levels

Value Recognition	Value Clarification	Value Analysis
Committee	*Expressway*	*Expressway*
Expressway	*Inner City Housing*	*Inner City Housing*
Fail-Safe	*Rip-Off*	*Simulating Social Conflict*
Inner City Housing	*Shipwreck and Other Government Games*	*Fail-Safe*
The Propaganda Game	*Fail-Safe*	
Shipwreck		
Simulating Social Conflict		

ROLE PLAYING

In the introduction, I suggested education for increasing human potential as a new direction for the social studies. Many simulation games contain all the necessary elements of a simulated psychodrama except for a protaganist. Psychodrama, based on the use of role playing, aids human growth; games that contain role playing facilitate education for human potential.

I do not wish to imply that simulations and games lacking this kind of process are necessarily less valuable. There are many useful board games and successful simulations in which players act as themselves. None the less, role-playing activity adds an important dimension to the process.

As noted elsewhere in the *Guide,* three kinds of roles exist: roles assumed by individuals (a player takes the role of President Granger in *Fail-Safe*); roles played by groups (tenants in *Inner City Housing*); and functional roles (the banker in *Inner City Housing*). All three kinds can be included in a single game; players can take individual and group roles simultaneously. The richer the role-playing experience, the better the game as education for human potential.

In *Committee,* players assume roles of members of the House of Representatives, Senators, the President, and lobbyists. The roles are fully described and are effective for providing meaningful experience because of the depth of the roles and the nature of their participation in the process.

Participants in *Expressway* play the parts of experts on one side or the other of the expressway controversy, or they act as members of the legislative committee. The roles are exceptionally well defined.

Fail-Safe provides rich role-playing opportunity. It contains individual, group, and functional roles, and they encompass a wide range of political positions. Players are divided into two groups. One group consists of the President and several advisors. The other contains military advisors at Strategic Air Command headquarters in Omaha, Nebraska. The President receives advice from each group (balanced among hawks, doves, and moderates) but acts on his or her best judgment and does not have to listen to any group.

Teams of three take part in *Flight.* Each group consists of a pilot, co pilot, and navigator. Role-playing possibility is limited.

Inner City Housing facilitates the process of role playing exceptionally well. Participants become tenants and landlords and have specific roles within these groups. Participants change sides during the activity; this adds depth to the experience and helps develop empathy. During each round, tenants must negotiate the cost of housing with a potential landlord. Chance cards introduce concerns such as vandalism, garbage disposal, and maintainance. The game requires both sides to arrive at a mutually acceptable level of rent and profit. Sometimes this is very difficult, but vacant premises or lack of housing result in the highest cost of all.

Designed to teach critical thinking, *The Propaganda Game* contains no role-play activity.

The second part of *Rip-Off* is a simulation of a juvenile hearing process and involves role playing. The shoplifting sequence, however, overshadows this part of the activity.

Shipwreck and Other Government Games fail to provide role-playing experiences. The materials contain parts for the students to act through dialogue provided by the designers. Therefore, little possibility for spontaneity exists, and this part of the activity loses most of its value.

In *Simulating Social Conflict,* participants belong to one tribe or another. This leads to limited role-play activity as some players become tribal chieftains and others become referees. However, these roles are game administration aids and lack much importance in and of themselves.

Tables 9 and 10 provide further information about this kind of activity.

TABLE 9 Quality of the Role-Playing Activity

Rich in Role-Playing Activity of High Quality	Limited Role-Playing Experience of High Quality	No Role-Playing Activity
Committee	*Flight*	*The Propaganda Game*
Expressway	*Shipwreck and Other Government Games*	
Fail-Safe	*Rip-Off*	
Inner City Housing	*Simulating Social Conflict*	

TABLE 10 Type of Role-Playing Activity

Individual Roles	Group Roles	Both	Functional Roles
Rip-Off	*Simulating Social Conflict*	*Committee*	*Fail-Safe*
Shipwreck and Other Government Games		*Expressway*	*Flight*
		Fail-Safe	*Inner City Housing*
		Inner City Housing	

TABLE 11 Rules

Simulation Game	Number of pages devoted to rules	Overall Rating for Clarity, Organization and Completeness		
		Adequate	Good	Outstanding
Committee	5		X	
Expressway	1			X
Fail-Safe	6			X
Flight	6		X	
Inner City Housing	4	X		
The Propaganda Game	12		X	
Rip-Off	8		X	
Shipwreck and Other Government Games	8	X		
Simulating Social Conflict	8	X		

RULES

Players may lose interest in a simulation game that requires the discussion of an overwhelming number of rules before it begins. They may become frustrated because of confusing rules that do not work (I used a game once in which no one could agree on what the rules meant; I no longer use it or review it). Rules should be flexible and adaptable; the possibility should exist for using a game for different educational objectives by changing the rules.

In evaluating the simulation games for this essay, I found that none contained lists of rules that were too long or complex. The rules for *Inner City Housing* and *The Propaganda Game* demand careful reading, probably more than once. These two, as well as *Simulating Social Conflict,* have complex rules and procedures. *Fail-Safe* is of average complexity, even though it is done with a computer. Both *Expressway* and *Shipwreck and Other Government Games* have easily understood rules and procedures.

Of the games reviewed here, I believe that *Expressway, Inner City Housing* and *Rip-Off* contain adaptable rules. *Fail-Safe* (because it is done with a computer program), *The Propaganda Game,* and *Simulating Social Conflict* are the least flexible. The remaining games are moderately adaptable. Table 11 provides further information about rules.

SKILLS

Social studies education requires many important skills, and successful participation in simulation gaming demands the acquisition of still others. Often, for example, participants must be able to negotiate. Another useful competency is the ability to plan strategy. Participants must learn to anticipate the plans of others and coordinate responses. As in poker, recall sometimes helps.

Many of the games reviewed in this essay develop a wide variety of skills. Table 12 describes them, and table 13 rates them.

DEBRIEFING

A common criticism of simulation gaming is that it lacks value because of its use for fun with inadequate debriefing. So we have all argued for better debriefing techniques. But I think we have overemphasized the problem, overreacted, and become too defensive. I have argued, in papers given at meetings of the North American Simulation and Gaming Association, that participation itself makes a positive experience. It facilitates creativity and aids healthy mental development. These advantages exist even when no debriefing occurs. A good simulation achieves important outcomes, and a poor one with excellent debriefing may accomplish nothing. Nonetheless, we must consider the quality of debriefing in the evaluation process.

Table 14 describes the debriefing of these games with reference to the number of questions provided for debriefing. The quality of postgame analysis is also an issue.

When the designer provides no questions for discussion at the end of the game, the quality of debriefing depends wholly on the person using the game, and that cannot be evaluated here. This is the situation with *Expressway, Inner City Housing,* and *Shipwreck and Other Government Games.*

Committee provides few debriefing questions; those that exist are factual. This reinforces the tendency of the game to be didactic rather than heuristic. *Flight* needs little debriefing since it teaches map skills; the learning occurs throughout the exercise. This is also true of *The Propaganda Game,* but this one provides some good questions that enhance its value. *Rip-Off* offers few postgame questions, and they are of limited value. The developers suggest that those using the simulation should structure the debriefing, making its quality dependent on the skills of the particular facilitator. The designer of *Fail-Safe* apparently assumes that the impact of the experience requires limited debriefing. This is partly true, but issues exist that could be examined in greater depth (see the earlier discussion of values).

Of all the games reviewed in this section, only *Simulating Social Conflict* contains a debriefing procedure that is both thorough and of high quality.

TABLE 12 Summary of Skills

Simulation Game	Social Studies Skills Present	Simulation/Gaming Skills Present
Committee	Ability to analyze and interpret data Ability to defend a position Ability to work with others Ability to write research	Ability to make decisions Ability to negotiate Ability to persuade Ability to play rules
Expressway	Ability to analyze and interpret data Ability to construct and use logical arguments Ability to defend a position Ability to think critically	Ability to make decisions Ability to megotiate Ability to persuade Ability to play roles
Fail-Safe	Ability to analyze and interpret data Ability to use a computer Ability to work in groups	Ability to negotiate Ability to persuade Ability to play roles
Flight	Ability to use maps Ability to work in groups Ability to write clearly	Ability to cooperate
Inner City Housing	Few skills – cognitive outcomes stressed	Ability to negotiate Ability to play roles
The Propaganda Game	Ability to think critically	Does not develop simulation/gaming behavior beyond critical thinking
Rip-Off	Ability to analyze and interpret data	Ability to play roles
Shipwreck and Other Government Games	Ability to analyze and interpret data Ability to work in groups	Simulation/gaming behaviors not developed
Simulating Social Conflict	Ability to analyze and interpret data Ability to draw conclusions on the basis of experience Ability to hypothesize	Ability to develop strategies

TABLE 13 Rating of Skill Development

	Quality of the Development of Social Studies Skills			Quality of the Development of Simulation Gaming Skills		
Game	Low	Average	Good	Low	Average	Good
Committee			X		X	
Expressway			X		X	
Fail-Safe			X			X
Flight			X		X	
Inner City Housing	X				X	
The Propaganda Game			X	X		
Rip-Off	X			X		
Shipwreck and Other Government Games		X		X		
Simulating Social Conflict			X			X

TABLE 14 Debriefing Guidelines

Totally Inadequate	Inadequate	Adequate	Thorough
(No debriefing questions	(few questions)	(some questions, 8-12)	(Several pages of questions and guidelines
Expressway *Inner City Housing* *Shipwreck and Government Games*	*Committee* *Flight* *Rip-Off*	*Fail-Safe* *The Propaganda Game*	*Simulating Social Conflict*

OVERALL EVALUATION OF THE TEN GAMES

Committee

Committee effectively teaches the use of committees in the legislative process. However, I do not like it as well as some others that do the same thing. My favorite remains *The Game of Democracy* (published by Bobbs-Merrill; out of print). Coleman's game describes legislative behavior in a simpler way without the excessive details of this one. *Committee* takes a long time to play, and the extra time does not increase understanding of the legislative process. I am wary of attempts to teach students the boring steps by which a bill becomes a law (usually misrepresented anyway). *Committee* comes close to replicating these steps.

Expressway

Expressway has become one of my favorites of all time. It takes a short time to play (two class periods, including extensive debriefing); nonetheless, it teaches much about geography and social issues. I have seen it played with junior high school students and with graduate students. It works at all levels, successfully examining a wide variety of issues.

Fail-Safe

Fail-Safe generates much interest. The use of the computer and the speed of play add to the excitement, and players become totally involved in their roles. This results in a genuine sense of anguish over the impending disaster that is simulated. (My older daughter once observed *Fail-Safe* being played when she was seven years old. She was convinced by the intensity of the participants that the crisis was real.) At the end of each game, players feel elated and relieved. I have no reservation about this game and recommend it highly.

Flight

I really like *Flight*. Richard Duke's idea of a future's language fits both simulation gaming and cartography. *Flight* represents both. Students need the map-reading skills this teaches, but they are often taught in a dull way that causes loss of interest. The approach taken here makes the process exciting while it accomplishes important goals at the same time.

Inner City Housing

Inner City Housing achieves many useful purposes. I rate this sociology game above average because it develops empathy and understanding while offering opportunity for meaningful role play. Players quickly become enmeshed in the roles; they learn to use appropriate strategies. I view the fact that players change sides (from tenant to landlord and vice versa) as a very positive aspect of the experience. The biggest problem with *Inner City Housing* is the confusion occurring at the beginning when players are examining the rules, but this diminishes in importance once the game is under way.

The Game of Propaganda

Propaganda teaches people to resist irrational action based on propaganda. Since it was first published in 1966 (with several subsequent revisions), the need for critical thinking has increased. Its only problem lies in its sophistication, which requires substantial preparation in terminology, particularly with younger players, before it can be played.

Rip-Off

I dislike the game of *Rip-Off*. It can be a dramatic one, teaching the players the consequences of shoplifting. However, I have many reservations about its effectiveness for an instructor who tries it with a large number of students in a twelfth-grade class in which juvenile theft is already a problem. The difference between the version in which students actually steal and the one in which they pretend to steal is great. The simulation of shoplifting lacks impact, but the real act seems unwise except in limited circumstances.

Shipwreck and Other Government Games

I have mixed feelings about this set of games. On the one hand, the games are simplistic and lack depth. On the other hand, they have been written for average (or perhaps below average) junior high school students and are probably effective with that population. They have some value because they will work for some teachers and some students in settings where other more sophisticated games will not. Still, I find them so superficial that I cannot recommend them highly.

Simulating Social Conflict

I like the material in *Simulating Social Conflict* very much. It contains two related simulation games– *The Dilemma of the Tribes* and *Resources and Arms,* modified versions of the classic "Prisoner's Dilemma." The players learn how conflict can be tempered with cooperation; they learn the consequences of both. This results in an exciting experience along with exceptionally good debriefing procedures.

Table 15 provides an overall rating of the games.

TABLE 15 Rating of Overall Quality

Simulation Game	Excellent	Good	Average	Low
Committee			X	
Expressway	X			
Fail-Safe	X			
Flight	X			
Inner City Housing		X		
The Propaganda Game		X		
Rip-Off				X
Shipwreck and Other Government Games				X
Simulating Social Conflict		X		

CONCLUSION

In the introduction, I indicated new directions for social studies education, which included education for increasing human potential, facilitating creative behavior, interdisciplinary studies, and future studies. I will conclude by examining the extent to which the games reviewed in this essay relate to that framework. Because I take the position that all simulation games achieve these goals to some degree, I will comment on how each relates to the new directions *beyond* what would be true of all of them.

Committee seems not to fit into this new framework. It applies a new technique to an old social studies topic—how a bill becomes a law. I have discussed its success with this problem, but clearly it fails as a harbinger of change in the social studies.

Expressway emphasizes an interdisciplinary perspective. The problem it considers cannot be solved from the perspective of any single orientation. Because it also helps students focus on new modes of transportation, it is a future study as well.

Fail-Safe contains an interdisciplinary perspective (focusing on decision making) as well as a future studies issue (because it introduces players to the use of the computer). The intensity of the role play facilitates education for human potential.

Limited to teaching some specific skills, *Flight* lacks an interdisciplinary nature. Map reading is part of cartography, which is a language of the future. *Flight* has some problem-solving elements that help it facilitate creative behavior.

Because it develops empathy, *Inner City Housing* aids in education for increasing human potential. It has an interdisciplinary focus. Through its problem-solving aspects, it helps develop creativity.

The Propaganda Game teaches critical thinking, which aids the development of all goals described in this section.

Rip-Off, although unique in the way it works, consists of a traditional lesson about honesty and court procedures. Under certain circumstances (the arrest of a player for shoplifting) the game may have great ability to effect education for human potential, but this is not the intention of its designer.

Shipwreck and Other Government Games also applies new techniques to old social studies topics. The attempt is commendable, but its shallowness does little to change the nature of the discipline.

Simulating Social Conflict potentially develops at least three of the advocated goals: it facilitates education for increasing human potential, it has interdisciplinary application, and it aids creative behavior through emphasis on problem solving.

It will be interesting to watch the development of games in the social studies through further editions of the *Guide.* I hope these new perspectives become more prevalent as emphasis changes in response to an ever-changing and more complex global society.

TOTAL ENTERPRISE BUSINESS GAMES
An Evaluation

by J. Bernard Keys

INTRODUCTION

Definition

Total enterprise is a term we will use to refer to games that include all of the main functions of business as decision inputs–marketing, production, and finance. Such games are most commonly used in college undergraduate and graduate business policy courses. They are also frequently used in management and executive development programs by both universities and industry. The less complex games are often used in introduction to business or basic management courses. Hand-scored total enterprise games are used by high schools, management development programs in industry, and occasionally by smaller colleges that do not have computer facilities.

Purpose

You will find that total enterprise games are quite flexible and therefore their purpose varies greatly. They are most often used as integrative teaching techniques to bring together all of the functions of business and allow participants to view them as a whole. In management development and executive programs games often serve to break down the perceptual walls of narrow specialized thinking and cause businessmen to better appreciate the difficulties of their counterparts employed in other business functions.

Business Game Play

How does one play a total enterprise business game? Usually participants operate as members of teams representing the upper level of management of a business firm, submitting decisions periodically, and receiving financial statements printed by a computer much like those that would be received from an actual business data-processing department. Successful decision-making requires careful analysis of feedback, marketing, financial, production, and engineering data.

The Purpose of this Evaluation

We will describe, compare, contrast, and evaluate nine of the best total enterprise computer-scored business games. In addition, information is provided on sources for hand-scored total enterprise games. We will consider the decision inputs of the games, along with their complexity, balance, intended purpose, and practical matters such as readability and supporting information.

Criteria for Selection

What were our criteria for selecting the nine best total enterprise games? These games have been played and evaluated many times, are commercially available from reliable publishers, and all are priced in a manner to make them available for classroom use. All nine games have been used in university-level courses and in executive development programs. Furthermore, eight of the nine chosen games have instructor's manuals available, and the ninth game contains a manual of information within the game itself. All of the games either furnish a computer card deck or program free to adopters or sell one at a nominal price.

GAME COMPLEXITY

Computerized business games, especially in the total enterprise category, are difficult to categorize. Because the experience and background of participants becomes a part of the game input, it is not uncommon to learn of the same game being used as an orientation device for freshmen college students and as a major laboratory for analysis by graduate MBA students. To some extent, time requirements or qualitative reports can be traded for game model complexity.

In discussing the complexity of a game, two dimensions are important–game variable complexity and computer model complexity. An efficiently programmed game may have numerous decision inputs and yet require very little computer memory capacity. The number of individual decision inputs per round of game play (a decision set) is perhaps the best measure of game complexity.

Generally speaking, the range of game complexity could be scaled as follows:

Computerized Total Enterprise Games

	Number of Decision Inputs
Simple games	Less than 15 decisions
Moderately complex	15–30 decisions
Complex	30 decisions or more

Caution is needed in this interpretation since many games, such as *The Management Game* (Harvard), require as many as 50 decision inputs by requiring the same decision variables numerous times for each of several products. Making the same decision *more* times is not usually as fruitful a learning experience as making different decisions.

Computer Scoring

Computer complexity is more difficult to define. Potential game users who are not programmers should contact the school data-processing department before adopting a game; that department probably will require such information as the language the game is written in (FORTRAN, BASIC, and so forth) the memory capacity required which is given in thousands of bites (8K, 10K, for example) and the quantity or lines of printing to be processed for each decision set. Note that computer complexity and game complexity are not necessarily parallel. Many of the moderately complex games listed can be run on a small computer. Complex games may require as much as 100K. Computers without sufficient K often can use larger programs by "segmenting" the program.

Some game authors offer "time-sharing" services for schools with a teletype terminal but no computer.

Operating Time

The operating time for a game can be understood better by examining a cycle of game play, shown in Table 1. Initial preparation time for students usually will require double the time in item 1.

The adaptation of simple or moderately complex games in the computer area often will require very little processing except for moderate changes in the control cards. This is especially true if you can find a game running on the same type computer as yours. However, you should allow from three to five weeks for computer adaptation and setup time since data-processing managers sometimes find it difficult to schedule "new" projects.

A game should never be used with students until a trial run is made, unless the students are aware that the game is being debugged. The best way to learn how to administer a game is to play one.

While many game descriptions indicate no limit to the number of teams or players, there usually is a limit of from five to seven players per team and often a limit of teams per industry. "Industry" is used to define the number of teams competing directly with one another and whose decisions affect each other in the computer model. Additional industries usually double the time for items 4, 5, and 6 shown in Table 1.

TABLE 1 Time (In Hours) for Cycle of Game Play

	Simple	Moderately Complex	Complex
1. Students read and study manual.	2	3-5	5 or more
2. An orientation session is held.	1	1-2	1-2
3. A decision set is made.	1/2-1	1-3	2-5
4. Decisions are key punched.	1/2	1/2-1	1 or more
5. Computer run is made (includes printing time).	1/2	1/2	1/2-1
6. Print-outs proofed, processed and returned.	1	1-2	2-3
7. Discussion of feedback held. Depends on number of teams and extent of personal counseling.	1	1-3	2-5

Price

The prices of student manuals for business games range from five to ten dollars. Given a certain number of decisions, quality may vary inversely with the number of pages. Publishers often charge for the game program, which may come in card deck or paper tape form. Some games, such as the *Decision Making Exercise,* offer a yearly rate—often in thousands of dollars, with no limit to student input and manual reproduction rights. Some offer an outright purchase price, as does this same publisher. *The Carnegie Mellon Game* offers the use of this very complex program to schools for a $100 copying charge.

The Nine Games Described

Table 2 shows a listing of each game along with bibliographical information. You will want to refer to Tables 3, 4, and 5 as you read the comparative descriptions of the nine total enterprise games. We will consider the game decision inputs and variables in the sequence with which corporate planning usually occurs: marketing, production, and finance.

MARKETING VARIABLES

Product

One of the most distinguishing factors you will want to consider in your choice of a game is the type of product the game portrays. Many of the games use a generic product description, defining the product only in terms of price range. The rationale for such a description is provided for the Gidget in *Tempomatic IV,* "This vagueness is intentional and precludes the participant's basing his decision on the known actions of any real company" (p. 4). Other games describe the product by industry or standard product classifications. For example, *The Executive Simulation* (3) uses a general product description: Product A consists of a branded consumer good sold to retailers, while Product B consists of an unfinished good sold directly to industrial customers. You will find a third approach used by *Executive Decision Making Through*

TABLE 2 Nine Computerized Total Enterprise Games

Simulation Game Number	Bibliographical Information
1	Henshaw, Richard C., and James R. Jackson, *The Executive Game* (third edition). Homewood, IL: Richard D. Irwin, Inc., 1978. 189 pages. Fortran II Program. Instructor's information included in student manual.
2	Smith, W. Nye, Elmer E. Estey, and Ellsworth F. Vines, *Integrated Simulation* (second edition). Cincinnati, OH: South-Western Publishing Co., 1974. 56 pages. Fortran II and IV Program. Instructor's manual.
3	Keys, Bernard, and Howard Leftwich, *The Executive Simulation* (second edition). Dubuque, IA: Kendall-Hunt Publishing Co., 1977. 173 pages. Fortran IV Computer Program available in card-deck form for most systems. Instructor's manual.
4	Barton, Richard F., *Imaginit Management Game.* Lubbock, TX: Active Learning Publishers, 1973. 304 pages. Computer programs in card deck or tape form available for many systems. Administrator's manual.
5	Scott, Charles R., and Alonzo J. Strickland, *Tempomatic IV.* Boston, MA: Houghton Mifflin Co., 1974. 97 pages. Fortran IV Computer Program available. Instructor's manual.
6	McFarlan, F. Warren, James L. McKenney, and John A. Seiler, *The Management Game.* New York: The Macmillan Co., 1970. 153 pages. Instructor's manual.
7	Jensen, Ronald L., and David J. Cherrington, *The Business Management Laboratory* (revised edition). Dallas, TX: Business Publications, 1973. Fortran IV Computer Program available. Instructor's manual.
8	Cone, Paul R., Douglas Basil, Marshall J. Burak, and John E. Megley, *Executive Decision Making Through Simulation* (second edition). Columbus, OH: Charles E. Merrill Publishing Co., 1971. 264 pages. Fortran Computer Program available. Instructor's manual.
9	Thorelli, Hans B., Robert L. Graves, and Lloyd T. Howells, *INTOP (International Operations Simulation).* New York: The Free Press, MacMillan Co., 1964. Fortran II and IV Computer Program tapes available. Instructor's book. Supplementary for instructors: Thorelli and Graves, *International Operations Simulation—With Comments on Design and Use of Management Games.*

Simulation (8), which uses as its products automobile tires, complete with historical data from *Business Week* and other industrial sources. You will find an index of product descriptions in Table 3.

The games using more generic products allow your students to improve their skills in analysis of data, without preconceived notions. On the other hand, games with actual products defined encourage marketing research and the analysis of industry data. There is some danger that in the latter case students will simply try to mimic the more successful firms without relying on their own analyses.

Defining specific products such as stainless steel flatware or cookware, as does *The Business Management Laboratory* (7), greatly enriches the marketing area and the potential for library and industry research. It also complicates the game considerably.

Imaginit (4) incorporates the most versatile product description, allowing product parameters in the computer to be changed by the game administrator for each game play. Numerous industry parameters are available—such as that of breakfast cereals or home computers—or the administrator may create a product description.

Most games incorporate some means of increasing product quality or attractiveness on a competitive basis. *Integrated Simulation* (2) uses a *general* quality variable that allows participants to increase the product attractiveness on a scale similar to their advertising function by investing money in this variable. Other games tie in product quality and market areas. For example, *The Management Game* (6) defines five market segments from which a product may be merchandized—one is most influenced by price, a second by the price-to-quality ratio, and a third by quality and product development. The other segments are variations on these three. *Executive Decision Making Through Simulation* (8) allows companies to differentiate their tires and to enter the bias belted and radial ply markets. Richer qualitative variables, such as this game's variable, teaches students the complexity of most American markets, but it may be a type of learning more difficult to generalize to other industries. It is debatable whether this much complexity is necessary at the policy level (total enterprise game). Perhaps it would be more desirable in a specialized marketing course. Certainly, the handling of so many variables in a one-semester or one-quarter course is likely to be more suitable for graduate students.

Price

The price variable described in *The Executive Simulation* (3) is a typical one for moderately complex games. The game incorporates a price/elasticity, meaning that with other decisions constant, more revenue may be generated by lowering price in the elastic range and by raising price in the inelastic range. The game is also price competitive, as described in the following paragraph:

> The finished goods market is somewhat price competitive, but the price chosen should be weighed against advertising, R & D number of salesmen, sales commissions, and number of distribution centers. Any of these variables could possibly offset a price differential. [The Executive Simulation, p. 5]

In this particular game the market for the industrial product is much more price competitive, as is true in real life. Some games allow price changes only in lump sums; others, such as *Business Management Laboratory* (7), allow any price to be set for game products.

You probably will want to choose a game that is both price elastic and price competitive, as is true of all the games listed in Table 2. Most of the games reviewed incorporate prices in

TABLE 3 Marketing Variables

	1	2	3	4	5	6	7	8	9
	The Executive Game	*Integrated Simulation*	*The Executive Simulation*	*Imaginit*	*Tempomatic IV*	*The Management Game*	*The Business Management Laboratory*	*Executive Decision Making Through Simulation*	*Intop*
FORECASTING:	x	x	x	x	x	x	x	x	x
PRODUCTS:									
Research and Development	x	x	x	By product	x	By product	By product		By product
Number of products	1	1	2	3	1	3	2	3	2
Product Description Generic (G)	G	G	By customer	Many choices	G	G	Flatware/ cookware	Auto tires	By grade
Licensing									By product
PRICE:	x	x	By product	By product	x	By product	By product/ area	By product Bids	By product/ grade/ nation
PLACE:									
Distribution centers/ Regional offices			Any number						By product
Marketing Areas					3	3	2	5	Int'l/ product
International									By product
PROMOTION:									
Salesmen:			Product A	By product					
Number		x	x		x		By area	By region	
Trained			x		x		x	By region	
Salaries			x				x	By region	
Commission			x				By product		
Advertising:			Product A	By product		By product	By prod./ area	By region	By prod./ nation
Local					By area				
National					By area				
Other promotion	General	General				x			
MARKETING RESEARCH:									
Purchase of Mktg. information			Instructor Controls		8 types		3 types/ product		4 types

the $10-$100 range, probably because they have chosen as their product small consumer goods with which participants are most familiar. Using expensive products complicates the bookkeeping function and often takes one out of the ordinary range of product.

Place

The simple games such as *Integrated Simulation* (2) and *The Executive Game* (1) do not define specific market areas for their products. Others define marketing areas generically in terms of price sensitivity, product quality sensitivity, and so on. *The Management Game* (see [6]. *INTOP* (9) allows participants to market products in America, Europe, or Brazil with different channels of distribution in different nations. Price elasticity, advertising effectiveness, and the rate of growth of the markets vary from area to area. The *Business Management Laboratory* (7) defines areas, but only for purposes of delivery of merchandise.

PRODUCTION

Plant Capacity

All of the games reviewed allow participants to expand capacity (see Table 4). Some of the more complex games include provision for adding new plants, new stages of production, or locating plants in other nations (games 5, 7, and

9). Others allow the addition of warehouses or warehousing facilities in distribution centers (games 3 and 8). Plant capacity additions presume the building of new fixed facilities; therefore you may want to choose a game that treats this as a long-run variable. Long-run variables are similated by a lag effect; for example, a decision to add capacity to plant may require one or more quarters lag before production may be scheduled. All nine games in Table 4 have one or more lag decisions in order to require long-run planning and goal-setting. The more complex games require two or three quarters of lag decisions (game 8) and thus demand a longer period of play or a more extensive college course to allow enough frequency of decisions within the lag period.

Production Scheduling

Production scheduling is a basic variable and should be included in any total enterprise game. The moderately complex and complex games include scheduling by multiple products on the same line, often by stages of production. Only games with such variables generate enough data to use quantitative decision techniques, such as linear programming or computer simulation, in decision-making. It is desirable to have some limit on the amount of production that can be scheduled each quarter. This is realistically done by an overtime variable that makes overscheduling more expensive and penalizes one for not planning ahead in capacity. Two of the more complex games allow subcontracting for production units, an emergency variable to be used sparingly (games 7 and 8). Others allow the transfer of inventory from plant to plant, greatly complicating scheduling, but enriching options (games 4, 5, and 9).

TABLE 4 Production Varibles

	1	2	3	4	5	6	7	8	9
	The Executive Game	*Integrated Simulation*	*The Executive Simulation*	*Imaginit*	*Tempomatic IV*	*The Management Game*	*The Business Management Laboratory*	*Executive Decision Making Through Simulation*	*Intop*
PLANT CAPACITY:									
Multiple plant	x	x	x	x	By areas	x	By stages	x	By nation
Warehouses								x	x
PRODUCTION SCHEDULING:	x	x	By product	By product	By areas	By stages	By stages	By area/ product	
Multiple shifts	Model 2		x				By product		
Stages						2	2		
Subcontracting							x	x	
Inventory transfers				x	By areas				By company
MATERIALS PURCHASE:	x			by product	x	x		By quality	
Materials choice				By product	3	2	2	3	
PRODUCTION WORKERS:									
Hired/Discharged					x	By Product/ stages		x	
Scheduled							By stages		
Training					x				
Hourly wage								Union bargain	
Overtime Pay	x		x		x	x	x		
Fringe benefits				x				Union bargain	
OPERATIONS RESEARCH:			x	x					By plant
Engineering							x	x	
Quality control		x			x		By product	x	
Pollution control								x	
Automation								x	
MAINTENANCE	x						By stages	x	
WAREHOUSING								4 areas	
INVENTORY COSTS	x	x	x	x	x	x	x	x	
Accounting choice								x	
Insurance								x	

Materials Purchase

Including a materials purchase variable in a game adds to the sequential planning process of production planning, but it is not appreciably different from the type of decision-making required of scheduling within plant capacity, unless choices of different quality materials are available. This choice is provided by all of the complex games reviewed (games 5, 6, 7, 8, and 9).

Production Workers

It is unwise to unduly complicate games with additional generic numbers (such as undefined production workers) unless some human factor or wage problem is to be simulated. Furthermore, the concept of overtime pay and stretching fixed facilities over two shifts may be added by dealing with labor as a lump sum. In other words, doing additional numerical computations is not in itself a management or business learning concept. Game 8 extends the production workers variable to include union bargaining for wages and fringe benefits, and thus enriches the game with a totally new dimension. Games 5, 6, and 7 complicate numerical decision-making by adding production workers.

Operations Research

Three of the variables listed under operations research—engineering, quality control, and automation—allow participants to reduce unit costs through continued investments. The more realistic games such as the nine reviewed here provide increasing returns to sustained efforts and use "S-shaped" functions. Such a function presumes that first dollars spent provide few results until "breakthrough" points are reached. After some point beyond "breakthrough," further dollars spent provide diminishing returns in lowered costs per unit.

Other Production Variables

Pollution control is a unique variable incorporated only in game 8. Most of the nine games include a built-in depreciation variable. Depreciation expenses are then automatically charged against plants each quarter to maintain the plant capacity at its current operating level. Game 3 charges 2-1/2 percent of book value per quarter (10 percent per year) to depreciation. Two games add a more realistic maintenance variable (games 1 and 8). In game 8 relatively small cost reductions in plant operation can result from a vigorous R & D program, while inadequate maintenance budgets can lead to significantly increased manufacturing costs. Game 1 teaches an additional principle by experience: "If you permit the factory to deteriorate, it may be difficult to get production costs back to normal" (*The Executive Game*, p. 16).

The rich production function described in game 8 allows the addition of warehousing (for automobile tires) and additional inventory variables. In addition to carrying costs for inventory, which the other eight games incorporate, it also allows a choice of inventory accounting (Lifo and Fifo) and the purchase of insurance on inventory.

Summary of Production

In summary, games 1, 2, 3, 4, and 9 provide well-rounded coverage of the basic production variables and thus serve as reasonable exercises in policy making in this area. Games 5, 6, 7, and 8 distinguish themselves by providing very rich production functions. You would want to use them only with students who had taken a college-level production course.

FINANCE

All nine games offer a reasonable number of financial decision options, with the exception of 1 and 2 (see table 5). Game 1 allows no financial decision inputs and game 2 allows only short-term loans. This is a serious deficiency in both games. In game 1 an automatic loan is provided when net cash assets fall below zero. This is a good variable and allows game play to continue even when students fail to provide sufficient cash. However, it does not allow participants to engage in long-run financial planning, and even encourages complacency in this function. For this reason the game is probably best used in lower-level introductory courses or in a policy course as a course introduction, to be enriched by financial cases. Game 2 allows the use of credit for which students are charged an interest rate. No further description of the loan is provided. An overt decision is required of participants.

Games 3, 4, 5, 6, and 9 allow short-term and long-term financing, bond financing, and sale and purchase of stock. Thus, they provide a choice of debt and equity financing and allow the experiential teaching of leverage concepts. A realistic finance function is also necessary to keep overly optimistic marketing departments in check and to prevent run-away expenditures in production capacity and development.

Games 7 and 8 provide almost as many finance options as those available to many corporations; they allow the teaching of many principles about finance options, such as the use of preferred versus common stock and the choice of investments.

Games 1, 2, 3, and 4 suffer from an occasional weakness that occurs when the industry is placed in a declining economy but when price competition or other marketing interactions allow firms to develop a flush cash position. In this stage of the economy firms may need to dispose of cash through outside investments and thus increase stock price and lower the invested base. This option is unavailable to participants in these four games.

As was the case with the production function, game 8 includes several financial options that are not available to participants in any other games—insurance purchases and officer salary plans. These are relatively specialized decisions and their value in a policy game is questionable.

A LEARNING MODEL

A good definition of learning potential in business games is one that provides "a good opportunity for the student to organize reality." Most total enterprise games have implied a model for learning that accomplishes this by providing three phases.

TABLE 5 Financial Variables

	1	2	3	4	5	6	7	8	9
	The Executive Game	*Integrated Simulation*	*The Executive Simulation*	*Imaginit*	*Tempomatic IV*	*The Management Game*	*The Business Management Laboratory*	*Executive Decision Making Through Simulation*	*Intop*
FINANCE:									
Short-term loans		x		x	x	x		6 kinds	Nation by Inter-company
Notes payable			x						
Factoring							x		
Accounts payable							x		
Special Loans			x			x	x		
Long-term loans						x	x		
Loan negotiation							x		
Transfer of cash									x
Bonds									
Purchase/sale				x	x		x	x	
Scheduled					x				
Stock									
Purchased/sale			x	x	x	x	x	x	
Preferred/common								Preferred x Common	
Divident payout	x	x	x	x	x	x	x	x	
Investments							Choices		
Choices					x	x	x		
Insurances purchases								4 kinds	
Officer salary plans								x	

(1) *Experience:* This phase of learning is provided by game play, decision inputs, and team interaction.

(2) *Content:* This phase includes dissemination of new ideas, principles, or concepts regarding business, management, and organizational practices.

(3) *Feedback:* This phase includes feedback in the form of financial statements, comparative team standings, and participant and team critiques by the professor or game administrator.

The effective learning situation is created when a proper balance is achieved among these three factors. "Balance" here denotes balance in terms of the particular participant group being considered.

A group of MBA students in the final course presumably would be already acquainted with many business principles necessary to provide content in a moderately complex game. For college seniors, some review of business concepts is usually necessary, either within the game or by supplementary readings and lectures within the same course.

Content is often provided in the form of game-planning sheets such as allow students to perform a funds flow forecast by completing blanks.

COMPARATIVE EVALUATION: EXPERIENCE

The qualitative evaluation of the kind of decision inputs available to you in total enterprise games is covered in tables 3, 4, and 5. Table 6 summarizes these decisions and will be used for reference in usage suggestions.

According to overall complexity using the criteria cited earlier, simple games are *The Executive Game* (1) and *Integrated Simulation* (2); moderately complex games are *The Executive Simulation* (3) and *Imaginit* (4); and complex games are *Tempomatic IV* (5), *The Management Game* (6), *The*

TABLE 6 Approximate Number of Possible Decision Inputs per Decision Period

Game	Marketing	Production	Finance	Total
1. *The Executive Game*	5	6	1	11
2. *Integrated Simulation*	6	3	2	11
3. *The Executive Simulation*	13	6	4	23
4. *Imaginit Management Game*	12	13	4	29
5. *Tempomatic IV*	15	17	6	38
6. *The Management Game*	14	15	6	35
7. *The Business Management Laboratory*	19	21	10	50
8. *Executive Decision Making Through Simulation*	29	36	8	73
9. *INTOP*	40	10	4	54

Business Management Laboratory (7), *Executive Decision Making Through Simulation* (8) and *INTOP* (9).

Although you will want to evaluate the games cited in Table 6 according to their specific decision inputs, the guidelines listed in Table 7 will prove useful.

The Management Game (6) is a two-level hierarchy business game, allowing one team to operate a corporate headquarters office that allocates financial resources to two divisions (other teams) and approves capital decisions. This arrangement is a good one if you have opportunities for some of your students to participate in the game twice—once as division managers and again as corporate managers. The Harvard game (6) has been used in this manner as a first- and second-year MBA experience.

The Business Management Laboratory (7) requires that a prospectus for any new securities be completed ahead of time. This gives participants a complete picture of the process of financing a corporate operation that is often quite vague when students merely complete a decision form.

Executive Decision Making Through Simulation (8) and *INTOP* (9) provide unique experiences. With the growing emphasis on international operations, the international markets provided by *INTOP* should be of even more interest. (Business Publications is scheduling a 1980 publication of such a game, which includes international operation, *The Multi-National Management Game.*)

Executive Decision Making Through Simulation is the only game of the nine that builds heavily on actual industry data (the rubber industry). In addition, it includes supplementary case studies from this industry. Research indicates that for use in a business policy course all the games should be supplemented by cases.

Complexity in games often can be incorporated in ways that enhance rather than compete with learning. For example, *Imaginit* (4) introduces game complexity gradually as students play round after round of the game. Game variables are first introduced in simplified form, and more and more is revealed about them as game play progresses. This is one of the most effective game decision introductions among the nine games reviewed. *The Executive Game* (1) uses a variation of this approach by introducing Model 2.

The Executive Simulation (3) uses another approach to handling complexity. All game decisions and variables are introduced at the beginning of game play, but four planning sheets are provided with detailed step-by-step instructions so participants are *led* through relevant planning cycles in the first decision set, even though they are unfamiliar with the game.

TABLE 7 Use of Business Games

	Recommended Uses
Simple games	High school courses Introductory college courses Orientation or stimulation sessions in senior policy courses One-day sessions in management development programs
Moderately complex games	Senior college courses for 1/3-1/2 of semester course, supplemented by cases Graduate policy courses, as 1/3 or less of course time, with business cases Management development programs in three to five-day sessions
Complex games	Senior college courses as major pedagogy (2/3 to total course) Graduate level policy courses (1/2 to total semester courses)

COMPARATIVE EVALUATION: CONTENT

Most games leave the content provision entirely in the hands of the administrator, who can provide such by assigned readings, a complementary text, or lecture assignments. If the required research and assignment of support materials is developed adequately by the administrator, this can be an effective method of disseminating content information.

Two games reviewed include considerable amounts of content information within the manuals. *Imaginit* (4) integrates content principles and concepts throughout the game instructions. *The Executive Simulation* (3) includes readings on business policy and strategies excerpted from classic books and articles.

COMPARATIVE EVALUATION: FEEDBACK

Much of the learning process in simulation games comes through reflection by participants as they compare their conceptual models of business activities with the results of their experiences in the simulation games. But competition and activity in a simulation game are so stimulating that reflection seldom occurs unless it is guided or required by the game or the administrator. All nine games reviewed provide feedback of a computerized income statement and balance sheet, although the one furnished by Integrated Simulation is quite abbreviated. In addition, several of the games require structured experiences that force reflection and learning.

Executive Decision Making Through Simulation (8) provides a chapter on understanding behavior patterns in the simulation teams. It includes six questionnaires that require students to reflect on their decision-making style, their objectives in the game, their organization, attributes of their team (cohesiveness, among others), and their perceptions of themselves as team players. These questionnaires have both a pre- and postgame requirement to force before and after game play reflection. Many games overlook the rich behavioral and organizational learning opportunities such as those in this total enterprise game.

One of the major weaknesses of total enterprise games has been the overemphasis on decision inputs and game complexity at the expense of supportive content and feedback. In other words, the time span for game play is often too short to complete the learning cycle. For this reason, games like *The Executive Game,* with very few decision inputs, can be quite useful in one-day sessions or when heavily supported by other materials.

Two types of participant planning sheets are provided by the simulations reviewed. *The Executive Game* (1), *The Executive Simulation* (3), *Imaginit* (4), and *The Business Management Laboratory* (7) provide "tear-out" sheets for computing such things as cash flow analysis. Providing lined sheets and headings will encourage much more reflective work by participants than a mere suggestion. *Tempomatic IV* (5), a more complex game, has a computerized routine that prints out a cash flow analysis, a production analysis, and several similar feedback items. This saves time and allows participants to handle large amounts of data in a short game cycle, but it does not provide as much involvement (and perhaps learning) as actually completing the sheets. Here is a prime example of how complexity and sophistication may be traded for learning. The Harvard *Management Game* (6) does the best job of remedying this dilemma. It provides several terminal-based planning models already programmed so participants can use computerized programs to organize their data, but they must *choose* the relevant input data and punch it in. This is also a good experience at using computer-assisted planning.

MANUALLY SCORED TOTAL ENTERPRISE GAMES

Hand-scored games free one from the need of computer assistance, but often offer several disadvantages. First, many scoring errors are likely to occur, especially if the game is moderately complex. This demoralizes the team. Second, even the simplest games are likely to require at least 45 minutes of instructor time per decision set, using an electronic calculator. If frequent repeat game sessions are to be held with a total enterprise game, it is advisable to get access to a computer and obtain a computer-scored game.

It appears that most schools using manually-scored games have been "trading up" to computer-scored games since the last edition of the *Guide.* Two of the more prominent manual games, *Executive Action Simulation* (by Herron, Prentice-Hall) and *Management Decision Simulation* (by Vance, McGraw-Hill) have gone out of print. None of the other manual games meet the criteria established for the nine games above. However, various games are available from the centers of gaming and publication suggested in Table 8.

TABLE 8 Manually Scored Total Enterprise Games

Use	Recommended Source
High School	The Avalon Hill Company 4517 Harford Road Baltimore, Maryland 21214 Several simple to moderately complex games.
Management Development	Didactic Systems, Inc. Erwin Rausch Box 457 Cranford, New Jersey 07016 Various published and customized games for management development.
General Usage	Book: *Dynamic Management Education,* by Allen Zoll, 1969. Addison-Wesley Publishing Co. Jacob Way Reading, Massachusetts 01867 Numerous hand-scored management development games suitable for high school and college. Also instructions on designing your own game.

URBAN GAMING SIMULATIONS
An Evaluation

by Mary Joyce Hasell

INTRODUCTION

Are you looking for a way to introduce your students, organization, or community group to a systems view of urban issues? Or would you like your group to experience the pressure faced by a decision maker in a large city? Then perhaps a gaming/simulation can help you by allowing hands-on experimentation with an urban gaming model.

Whether or not you have had previous experience either playing or directing a gaming/simulation, you can probably use some information about the wide variety of urban games available. For example, how do you know which game is most suitable for the age level and size of your group, or for the playing time you have available? Which gaming model is most similar to the real world problems you want to explore?

Begin by checking the gaming/simulation descriptions themselves, and then start narrowing your choices. There are many games from which to choose, and each one has something special to offer.

The Purpose of this Essay

In this essay, I shall examine, describe, and evaluate eighteen of the most accessible urban gaming/simulations according to their objectives, content, processes, and suitability for use. Because each gaming/simulation reviewed here was designed (1) with *specific communication objectives,* (2) with a *specific audience context,* (3) about a *specific problem context,* and (4) with the objective of *improving options* in real life for players, it is neither possible nor desirable to compare and rate the games one with the other. Rather, it is for you to make a careful self-evaluation of your goals and needs in relation to the four points enumerated above so you will choose the gaming/simulation appropriate for your own circumstances.

To help you in this selection, I shall begin by looking at some preliminary considerations such as playing time, complexity for various age groups, and number of players. I shall go on to consider the following components of the gaming/simulations: issues, scenarios, value orientations, roles, rules, skills, activities, interactions, the model, and debriefing.

Editor's Note: Listings for all simulations discussed in this essay are in the urban section, except for *Trilogy* (*At-Issue, The Conceptual Mapping Game,* and *Impasse*), which is in the frame games section.

Definition

Urban gaming/simulations began to appear in the 1960s with Allen Feldt's *Clug* and Richard D. Duke's *Metropolis* (see description). These two games have had a great influence on urban games since then, and many of the games we review here are similar in character to these games. Lately, urban games that diverge from these two basic prototypes have begun to appear. These divergences will become apparent as I describe the games. However, all of the urban games I review here have some of the following commonalities:

(1) *Scenario* (or setting the stage): an urban problem statement that allows for various ways of exploring city life.
(2) *Roles:* either *functional* roles, which become defined through the players' actions in the game, or *assigned roles,* which are given to players at the start of the game. These vary in nature from urban public sector decision makers to citizens with a vested interest in the city's development.
(3) *Steps of Play:* the sequence followed in the game, which is similar to a process followed in urban functioning and decision making. Decisions can vary in scale from building one's own house to building a regional transportation system.
(4) *Accounting System:* the distribution of resources that are available to urban dwellers and the means by which these resources are used in the system. These could be personal income, coalition, influence, regional tax base, public financial investments, and so forth.
(5) *Rules:* two basic types that are important in urban settings are natural rules (e.g., rivers flow downstream) and man-made rules (e.g., zoning). The gaming/simulations usually include both types of rules in their structure.
(6) *Events or Pulses:* exogenous factors or unexpected events in urban settings, when one does not know exactly what will happen next, that vary from political coalition formation to receiving federal grants. They always influence the direction of the gaming/simulation.
(7) *Symbols:* artifacts used in the game to represent land use, buildings (commercial, industrial, and residential),

production of resources, influence, and other elements commonly thought of in relation to urban areas.

(8) *Models:* the systems and subsystems that are abstracted, and simplified, from the reality of an urban area and linked together in a playable form.

As players of urban gaming/simulation, your group will have the opportunity to develop skills of communication, decision making, writing and promoting proposals, persuading, negotiating, forming coalitions, and creating new design solutions for urban spaces. Players will be involved in groups of various sizes, and they may play roles that are just the opposite of their own value system. They will have a go at influencing the course of events through hands-on experience in these games, and the insights they gain will facilitate their understanding of real-world urban systems.

Table briefly describes the eighteen gaming/simulations reviewed in this article.

Using Gaming/Simulations*

The specific situation for which you want to use a gaming/simulation should determine what game or games will be most appropriate. You can ascertain this by answering the following three questions.

(1) *What is the audience for which you want to use a gaming/simulation?* The audience matrix shown as Table 2 gives a wide range of audience types among which you will be able to find or include in your group. To be sure of a successful run, carefully match your own audience requirements with the audiences with which the games have been tested.

(2) *What specific communication purposes do you have that can best be served by a gaming/simulation?*

(A) *Motivation/sensitivity* to help participants develop empathy by experiencing the roles of those with value orientations and constraints different from their own.

(B) *Teaching/dissemination* to pass on information to participants through an active hands-on experience.

(C) *Information gathering/survey research* to query participants for their pragmatic reactions to nonquantifiable data.

(D) *Research design/problem formation* to share information about a problem among members of a research team.

(E) *Technology assessment/impact evaluation* to explore the secondary and tertiary outcomes that might be potential areas of impact.

(F) *Strategic planning/alternative formation* to develop new models for urban systems or institutions and to test new planning options and scenarios for the future.

(G) *Training* to develop participants' skills in decision making, problem solving, proposal writing, program development, the use of political influence, bargaining, and coalition formation.

Once you have determined your communication purposes, again check the gaming/simulation with this criterion in mind. If the game you choose is mostly process oriented (training) and you are interested in content (teaching/dissemination), you will not be pleased with your choice.

(3) *What is the specific problem context in which you have*

TABLE 1 Urban Gaming/Simulation Descriptions

Gaming/Simulation	Summary Description
Acres	A computer model simulation game that illustrates the impact of development on the following variables of a community: environment, business, building, and government. Players, through the use of functional roles, gain some insight into the dynamic interactions of a developing urban area.
Blackberry Falls	A computer model simulation that illustrates small-town formal decision making. Players learn how to handle problems of a rapidly growing population, land development, industrial and commercial growth, and environmental regulations.
Blight	Emphasizes idea generation, divergent thinking and creative inquiry around planning an urban area in decline. Players in the roles of citizen, planner, politician, industry, and business are organized around community sectors each representing a set of urban values.
City Model	A computer model based simulation game that illustrates the ecomonic, political, and social impacts of decision making on an urban system. Players work with the relationships among the various subsystems of the model.
Clug	Illustrates the economics of urban and regional growth. Players build a city considering the connections among industry, housing, municipal services, commercial and transportation needs, and their physical location.
Eco-Acres	Players work to create an "ideal" community, gaining insights into planning that will help them examine their own community more critically and objectively.
The Grand Frame	Illustrates a design process model for community planning. Players explore alternative design solutions and make trade-offs between the ever-growing numbers of environmental constraints, human actors, and the cost of satisfying both.
Hexagon	Resource allocation through human settlement planning in a developing country. Three concurrent, interdependent games at national, regional, and local levels introduce players to the problems of communication and decision making.
Inhabs	Illustrates through the use of a plausible physical model how housing is designed and developed. Players adopt roles related to real-life ones (house builder, developer, housing association) and experiment within them.

*The criteria listed in this section were developed for a conference on Global Interactions and gaming/Simulation held in Nijmegen, the Netherlands, in July 1977. For further information, see Appendix A in *Gaming: The Future's Language,* by Richard D. Duke (Beverly Hills: Sage, 1974).

TABLE 1 Urban Gaming/Simulation Descriptions (Cont)

Gaming/ Simulation	Summary Description
Metro	Computer model-based simulation game of city operation using a medium-sized U.S. midwestern city (Lansing, Michigan) as a prototype. Players in private and public sectors roles make decisions whose cross impacts are felt by the entire metropolitan area.
Metro-Apex	To the original model, METRO, an air pollution control compnent, water quality, solid waste, and transportation considerations.
Metropolis	Players in the roles of politicians, business, school board, and planners, make decisions about capital improvement budgeting and the tax rate in a medium-sized U.S. midwestern urban area. Cross impacts of these interactions affect the total system and are felt by players through rewards and punishments.
Mini-Apex	Essentially a simulation of Metro-Apex. The model can be operated for several cycles by programming individual roles as separate entities, using the standard output. The game director effects much of the role interaction and performs the calculations required to go to the next cycle.
New Town	Players build a new community with the goal of making it a pleasant place to live and work. Competition is fierce and the winner is the player who becomes the wealthiest.
Systems	Illustrates that there is no "right" plan for an area and that rightness varies according to the values and goals of those who plan the site. Players use general systems approach in combination with an acetate overlay system of decision making to make design decisions.
Trilogy	*Impasse, Conceptual Mapping,* and *At Issue* are three simple, expedient, and flexible communication tools designed to improve citizen participation in public policy discussions through planning, evaluation, and dialogue. They can also be used in an academic context as a teaching tool, or as a data collection instrument for recording players' views on the impact of specific public policies through a systems approach.
U-Dig	Illustrates the residential real estate investment process in an urban neighborhood. Shows the relationship between residential form and certain financial variables in the development process.
Urban Dynamics	Basic structures and interlocking systems in the growth and development of a northern U. S. metropolitan area, with emphases on the varied socioeconomic groups who are/were the major actors, 1920 and forward.

an interest? Each game we review here deals with an urban system or subsystem in a different manner. Make sure that the game or games you choose address your interest and purpose.

TABLE 2 View of the World Audience Matrix

Scale	Local	Regional	National
Types of Users			
Policy Makers	Mayor Council Chief	Council Governor	Ministers Presidents Military
Staff	Business Persons Planners Human Settlement Managers	Organizational Leaders Scientists Business Persons Academics	Secretaries Business Persons
Special Interests	Social Service Agencies Community Action or Planning Agencies	Farm Bureau	NAACP AAAS NSF Sierra NAS
Students Teachers	Grade School High School	Universities Colleges Technical Institutes	Universities

PRELIMINARY CONSIDERATIONS

Criteria for Selection: Background

The main criterion for selecting these gaming/simulations was that they had been played and evaluated at least ten times. With the exception of *Inhabs,* all of the game are available in the United States. Some are available through commercial sources and others only through the designers. Several of the computer gaming/simulations are very expensive to run—over $300—but most are available for under $100. In some cases it is possible to construct your own set of game materials by using the instructions in the manual. For further information, see Table 3.

Intended Audience, Group Size, Playing Time

One of your major considerations in choosing a gaming/simulation is how it suits the age and ability level, size, and playing time available for your class or community group. Table 4 should help you identify games that are appropriate for your situation in this area.

Complexity

We rank the gaming/simulations in Table 5 according to their complexity. The simplest games require little reading, no computer skills, and are easily understood by secondary school classes or groups that have a limited amount of time to play.

ISSUES

Determining the range of issues to be presented and resolved in the gaming/simulation you want to use is one of your most important considerations. In urban gaming/simulations there are usually numerous *issues* that are part of a basic unifying scenario. This stems from the fact that no urban problem can be looked at as a single issue but must be analyzed rather as a system of variables or as many sub-

TABLE 3 Preliminary Considerations: Background

Gaming/ Simulations	Designer	Date	Producer	Cost and Packaging	Are There Additional Research Articles?
Acres	George Pidot	1973	Project Compute Dartmouth College Hanover, N.H. 03755	instructor's manual $3.00 user's manual $4.00	yes
Blackberry Falls	Edmund Jansen Anne Knight Jerry Warren	1975	Computer Services Univ. of New Hampshire Kingsberry Hall Durham, N.H. 03824	player's and game master's manual and role and reference appendices $17.00	?
Blight	Ronald G. Klietsch	1971	System's Factors, Inc. 1940 Woodland Ave. Duluth, Minn. 55803	kit $64.50 + postage	?
City Model	Peter W. House	1970	John G. Symons, Urban & Reg. Studies Inst., Mankato State University, Mankato, Minn. 56001	very expensive in computer time and human input	yes (numerous)
Clug	Allan G. Feldt, Datson, Monroe, Sawicki	1966	(1) The Free Press Dept. F, Riverside, N.J. 08075 (2) Inst. of Higher Ed. Research and Services, Box 6293, University, Ala. 35486	(1) manual $6.95 (2) complete kit $75.00	yes (numerous)
Eco-Acres	Grayce Papps Eton Churchill Eric Van de Bogart	?	Eric Van de Bogart Maine Public Broadcasting Network, Alumni Hall, Univ. of Maine, Orono, Me. 04473	$1.00	yes
Grand Frame	Jo Hasell Webb Nancy Stieber Pat Miller	1976	Jo Webb 2707 Lohr Road Ann Arbor, Mich. 48104	manual $5.00 kit $300.00	yes
Hexagon	Richard D. Duke and 680 Game Design Seminar, 1975, U. of Mich.	1976	Multilogue 321 Parklake Avenue Ann Arbor, Mich. 48103	manual $5.00 kit $300.00	yes
Inhabs	Cedric W.B. Green	1971	Dept. of Architecture, Univ. of Sheffield, The Arts Tower, Sheffield S10 2TN, U.K.	£1	yes
Metro	Richard D. Duke Environmental Simulation Lab. Univ. of Mich.	1965	Multilogue 321 Parklake Avenue Ann Arbor, Mich. 48103	kit $5.00	yes (numerous)
Metro-Apex	Richard D. Duke COMEX, Univ. of Southern Calif.	1967 1964	Mark James, Director of Computing Services, COMEX, Davidson Center, Univ. of S. Calif, Los Angeles, Calif. 90007	$500 a run for computer average cost manual $7.00	yes
Metropolis	Richard D. Duke	1964	Gamed Simulations Inc. 10 West 66th Street New York, N.Y. 10023	complete kit $50.00 leader's manual $10.00 participant manual $5.00 charts $15.00	yes (numerous)
Mini-Apex	Theodore H. Rider	1974	Theodore H. Rider, Managing Director, NECEP, Boston College, Weston Observatory, Weston, Mass. 02193	kit $70.00	?
New Town	Barry R. Lawson	1969 1971 1975	Harwell Associates (1) Box 95, Convent Station; N.J. 07961 (2) Pinnacle Rd., Harvard, Mass. 01451	(1) educational kits $18.00, $30.00 (2) planner's set $85.00	yes
Systems	Carl Steinitz Peter Rogers	1977	Carl Steinitz, Graduate School of Design, Gund Hall, Harvard University; Cambridge, Mass. 02138	$150.00-$200.00; write for information	yes
Trilogy	Richard D. Duke Cathy Greenblat	1973	Sage Publishers, Inc., 275 S. Beverly Dr., Beverly Hills, Calif.	under $10.00	yes

TABLE 3 Preliminary Considerations: Background (Cont)

Gaming/ Simulations	Designer	Date	Producer	Cost and Packaging	Are There Additional Research Articles?
U-Dig	Erwin J. Bell	1969	Erwin J. Bell 1460 Moss Rock Place Boulder, Colo. 80302	instructions, tables, base $30.00-$40.00; user must buy Lego blocks	yes
Urban Dynamics	Loel A. Callahan Dwight A. Caswell Larry A. McClelan	1970	Inst. of Higher Ed. Research and Services, Box 6293, University, Ala. 35486	complete kit $95	yes

TABLE 4 Intended Audience, Group Size, Playing Time

Gaming/ Simulations	Age Level	Playing Time	Can be Played Over Several Sessions	Number of Players Minimum	Number of Players Maximum	Number of Groups
Acres	high school, university, professionals, citizen groups	10 hrs.-semester	yes	6	30	9 teams
Blackberry Falls	high school, college, citizen groups	12-18 hours	yes	15	40	dependent on game run
Blight	high school, college, citizen groups	7-9 hours	yes	23	40	6+
City Model	college, graduate school, urban education, professionals, citizen groups	10-12 4-hour sessions	yes	15	100	–––
Clug	high school, college graduate school, professionals, citizen groups	4-9 hours	yes, after 2 hours	3	25	3-5 teams
Eco-Acres	5th grade to college (with variations)	2 hours	no	5	15	–––
Grand Frame	high school, college, professionals, citizen groups	6-8 hours	yes	10	25	5 teams
Hexagon	high school to adult, professional planners, government bureaucrats	2-3 hours	no	17	24	–––
Inhabs	university, professionals, citizen groups	Sprawl—2 days Spiral—2 days Squat—1 day	yes	20+		–––
Metro	university, graduate school, professionals, citizen groups	3-9 hours in 4-hr. periods	yes	10	40	3 teams
Metro-Apex	university, graduate school, professionals, citizen groups	3-10 8-hr. sessions	yes	25	150	numerous
Mini-Apex	university, professionals, citizen groups	3 8-hr. sessions	yes	20	50	numerous
Metropolis	high school to unversity, professionals, citizen groups	6-8 hours	yes	3 Can run several games at once	45	3-4 teams
New Town	junior high to college, professionals, citizen groups (depending on version)	1-9 hours	yes	12	20	4 groups
Systems	high school science; university engineering, landscape, soils; citizen groups	3½ hours	no	9	21	3 teams

TABLE 4 Intended Audience, Group Size, Playing Time (Cont)

Gaming/ Simulations	Age Level	Playing Time	Can be Played Over Several Sessions	Number of Players		Number of Groups
				Minimum	Maximum	
Trilogy	high school to college, professionals, citizen groups	2 hours	no	3	175	———
U-Dig	college, graduate school, professionals, citizen groups	4-7 hours	yes	4	16	4 groups
Urban Dynamics	10th grade and up, college, graduate school, professionals	4-6 hours	yes	3-5 per group		4 groups

TABLE 5 Complexity

Simple	Easy	Moderate	Complex
Eco-Acres	*Hexagon*	*Grand Frame*	*Metro*
New Town	*Trilogy*	*Metropolis*	*Metro-Apex*
	Urban Dynamics	*Mini-Apex*	*City Model*
	Blight	*Systems*	*Blackberry Falls*
		Acres	
		Clug	
		Inhabs	
		U-Dig	

TABLE 6 Scenarios

Urban/ Land Developing Issues	Urban/ Social Issues	Management Issues	Design or Urban Components	Insert Your Own Issue
Acres	*Blight*	*Eco-Acres*	*Inhabs*	*Trilogy*
City Model	*Urban Dynamics*	*Hexagon*	*Systems*	*Grand Frame*
Clug		*Metropolis*	*Grand Frame*	*Inhabs*
New Town		*Metro*		
U-Dig		*Metro-Apex*		
		Mini-Apex		
		Blackberry Falls		

systems. As you begin to observe a particular urban system, you realize that a change in one part of the system (that is, the resolution of one issue) results in changes in other parts.

Table 7 lists the various issues that are included in the gaming/simulations and that make up numerous urban systems. Tied in with these issues are the roles through which all the games are played (see Table 11). By playing the parts of actors in a system, participants will experience dealing with the numerous issues in that system. The division of the gaming/simulations into five scenario categories in Table 6 may help you identify the basic scenario around which each was designed.

After you check the scenario type, refer to table 7 to find the issues that concern you. Some games are more complex than others and include many issues, so make your selection carefully. If you want to adapt a game to the specific needs and interests of your community or to a special-interest group, choose a gaming/simulation that is flexible enough to let you develop and insert your own interests—*Trilogy, Grand Frame, Inhabs.*

CONTENT

Another of your concerns in choosing a gaming/simulation is the scope and breadth of the content material included in the simulation/game itself or, perhaps, in the manual, and the availability of this information to players. With the exception of *New Town,* the urban games reviewed here are concerned with communicating principles rather than facts.

The major significance of urban gaming/simulation lies in communicating holistic understanding of problems. Thus, it is not as important to remember facts about a particular game as it is to understand a process or a set of principles or some insight into a system's functioning. When a player can transfer a process or principle learned in a game to a real-world problem, then the game has been successful.

As a help in judging the scope and breadth of content of a gaming/simulation, Richard D. Duke has expressed very simply the idea of complexity and multiplicity of gaming/simulation in a model he calls a cone of abstraction. At the top of the cone you can picture gaming/simulations that have little depth or breadth of content, and consequently limited content. As you move down the cone, you increase the number of variables and also the depth of inquiry into a specific problem. With this increased depth there is a great need for information to supplement the symbols employed in the gaming/simulation itself.

The gaming/simulations we review here are likewise distributed along this cone: some are highly abstract, have little content, few details, few symbols, and usually take a short time to play; some are very detailed, have heavily loaded symbols, and require a great deal of additional information to supplement the gaming/simulation and make it meaningful. The very complex gaming/simulations sometimes consume an entire semester of work and are supplemented with many additional sources of information (lectures, books, seminars, and so forth). The simple games usually require a couple of hours to play and serve as a motivator for players, who may want to do additional research on their own about the game topic or a related subject.

Table 8 is meant to help you determine the gaming/simulations that are appropriate to your needs by classifying them according to the scope of their content. It is not meant as a

TABLE 7 Issues

Issues	*Acres*	*Blackberry Falls*	*Blight*	*City Model*	*Clug*	*Eco-Acres*	*Grand Frame*	*Hexagon*	*Inhabs*	*Metro*	*Metro Apex*	*Mini-Apex*	*Metropolis*	*New Town*	*Systems*	*Trilogy*	*U-Dig*	*Urban Dynamics*
Urban Issues																		
population	S	S	S	S	S			S		S	S	S	S			F		S
migration	S	S	S	S	S			S		S	S	S	S			F		S
zoning/land use regulation	S	O	S	S	S	S	S-F	S	S	S	S	S	S	S	S	F	S	S
industry	S	O	S	S	S	S		S	S	S	S	S	S	S	S	F	S	S
housing	S	S	S	S	S	S	S-F	S	S	S	S	S	S	S	S	F		S
commercial	S	O	S	S	S	S	S-F		S	S	S	S	S	S	S	F		S
transportation	S	O	S	S	S	S	S-F		S	S	S	S	S	S	S	F		S
municipal services	S	O	S	S	S	S		S	S	S	S	S	S		S	F		S
agricultural development	O	O		S	O			S							S	F		
renewal/rehab			S														S	
Social Issues																		
employment	S		S	S	S					S	S	S	S	S		F		S
education	S	S	S	S	S	S				S	S	S	S	S	S	F		S
recreation	S	O	S	S	S	S	S-F			S	S	S	S	S	S	F		S
social disaster (fires, riot, crime)							S-F			S	S	S	S	S		F		
historic preservation							S-F								S	F		
cultural enrichment			S			S	S-F								S	F		
segregation										S						F		S
racial confrontation																F		S
block busting																F		S
redlining																F		
Equity Issues																		
elderly			S				S-F									F		
minority groups			S				S-F			S	S	S	S			F		S
women							S-F									F		
equal opportunity			S							S	S		S			F		S
low/middle/high income groups differences			S				S-F		S	S	S	S	S		S	F	S	S
Environmental Issues																		
aesthetics		O									S		S					
air quality									S		S		S			F		
water quality		S		S	O						S		S			F		
wildlife							S-F				S		S		S	F		
conservation			S								S		S		S	F		
solid waste		S		S							S		S			F		
transportation				S							S		S			F		
agricultural quality											S		S			F		
pollution standards											S		S	S		F		
environmental disaster														S				
Budgeting Issues																		
capital improvements	S		S	S	S					S	S	S	S		S	F		
municipal services/utilities	S		S	S	S		S-F			S	S	S	S		S	F		
discretionary funds	S			S	S					S	S	S	S			F		
tax base	S	S	S	S	S					S	S	S	S			F		
public institutions			S						S	S	S	S	S		S	F		
development cost constraints	S		S		S		S-F		S			S		S		F	S	
Experiencing																		
values clarification	S	S	S	S	S	S	S-F	S	S	S	S	S	S	S	S	F	S	
making trade-offs	S	S	S	S	S	S	S-F	S	S	S	S	S	S	S	S	F	S	
citizen participation		S	S	S		S	S-F			S	S	S	S			F		
group decision making	S	S	S	S	S	S	S-F	S	S	S	S	S	S	S	S	F	S	
policy making	S	S	S	S	S	S	S-F	S		S	S	S	S		S	F		S
political action	S	S	S	S	S	S		S		S	S	S	S			F		S
small town decision making		S				S										F		
rapid change	S	S		S	S					S	S	S	S			F		
compromising	S	S	S	S	S	S	S-F	S	S	S	S	S	S	S	S	F	S	S
developing	S	S	S	S	S	S	S-F	S	S	S	SS	S	S	S		F	S	S
speculating land	S	S		S	- S					S	S	S	S	S		F	S	S

TABLE 7 Issues (Cont)

Issues	*Acres*	*Blackberry Falls*	*Blight*	*City Model*	*Clug*	*Eco-Acres*	*Grand Frame*	*Hexagon*	*Inhabs*	*Metro*	*Metro Apex*	*Mini-Apex*	*Metropolis*	*New Town*	*Systems*	*Trilogy*	*U-Dig*	*Urban Dynamics*
Process Issues																		
design process							S-F		S						S		S	
real estate investment process	S			S	S				S	S	S	S	S				S	
planning process	S	S	S	S	S	S	S-F	S	S	S	S	S	S	S	S			S
political process	S	S	S	S	S	S	S-F	S		S	S	S	S	S				S
Third World Development																		
barter economy								S								F		
regional trading				S	O			S								F		
international trade								S								F		
technology transfer				S	O			S								F		
international exchange rates								S								F		
war								S								F		
famine								S								F		
U.N. grants								S								F		

Key: S = Issue significant
F = Frame game allowing players to design and insert their own issues
O = Issue optional and may be included in variations of the simulation game

Note: Some gaming/simulations explore a few specific issues. In others, the players' interests and values determine which issues will be explored. Still others encourage players to determine the issue they want to insert in the framework of the gaming/simulation.

comparative tool for evaluating whether one game is better than another. Only you can determine that.

With the exception of *Inhabs*, which was designed specifically for use by architecture and urban planning students, all of these gaming/simulations are interdisciplinary. Although the nature of their basic content is urban, the vastness of the topic allows many players of various interests to approach the simulations and gain insights into such areas as politics, law, or ecology.

VALUE ORIENTATION

There are certain underlying messages in each gaming/simulation that will be discovered by its players. Each gaming/simulation has its biases according to the designers' view of the "real world" and how it operates. Nevertheless, the participants have a tremendous amount of influence on the value orientation in all the urban games reviewed here. Although there are set rules as well as roles in the games, there is much latitude in interpreting both, especially the roles. So, depending on the interests of the participants and their willingness to experiment with "what if" questions, these games are amenable to variation in value orientation.

Table 9 lists the basic value orientation of each gaming/simulation. You may disagree with this orientation and still think there is worth in the gaming/simulation. You may make use of the experience as a contrast with some other orientation. In some gaming/simulations you can insert your own value orientation.

ROLES

Another important consideration is the kind of role-playing ability your students or participants have. It is necessary to realize that the roles in gaming/simulations are an integral part of the structure and have been chosen by each game designer to maximize the learning experience for players.

If the participants for whom you plan to run the gaming/simulation have not had previous experience with role playing, you may want to begin with a game that provides either group or team role descriptions. These roles, which are usually

TABLE 8 Scope of Content

Wide Scope: Multivariable with much additional information in manual and bibliography)	Medium Scope: Additional information in manual and bibliography	Narrow Scope: Some information needed to supplement simulation	Limited Scope: No additional information needed to supplement gaming/simulation
Metro	*Clug*	*Grand Frame*	*Trilogy*
Metro-Apex	*Acres*	*Hexagon*	*New Town*
City Model	*Mini-Apex*	*Inhabs*	*Eco-Acres*
	Metropolis	*Systems*	*Blight*
	Blackberry Falls	*Urban Dynamics*	
	U-Dig		

TABLE 9 Value Orientation

Gaming/ Simulations	Value Orientation
Acres	Western culture's ideas about planning.
Blackberry Falls	Democratic process in a small town.
Blight	Citizen participation as policy makers with emphasis on democracy.
City Model	Systematic planning process.
Clug	Enlightened self-interest through cooperation.
Eco-Acres	Setting priorities on needs and values for good of all.
Grand Frame	Users' participation in design decision.
Hexagon	Open for political interpretation with emphasis on self-sufficiency.
Inhabs	Homeowners in a community share local political and economic interest.
Metro	Open to alternative value orientation.
Metro-Apex	Open to alternative value orientation.
Mini-Apex	Intelligent cooperation leads to gains for all.
Metropolis	Intelligent cooperation leads to gains for all.
New Town	Capitalism, self-interest, and the need for cooperation.
Systems	Values and goals of designers determine design solutions.
Trilogy	Communication between citizens and planners.
U-Dig	U.S. real estate investment process.
Urban Dynamics	Political and economic action in cities with an emphasis on equity for all social classes and races.

played by three or more participants, minimize the risk participants may feel when they role play for the first time. Shared roles also allow inexperienced players to make decisions carefully and with due consideration.

Another good choice for inexperienced role players is a gaming/simulation in which the roles are described functionally through the game procedures. For example, *Hexagon* and *Clug* depend on the rules of the game and the steps of play to direct the participants' behavior. Participants interject their own personality traits, goals, and values.

A third type is the assigned role, which describes specific information, including age, employment, education, goals, values, and attitudes, in some detail. These roles are usually played by an individual and require little previous experience.

A fourth type of role playing in gaming/simulations recognizes that people in the "real world" often have more than one role (for instance, home owner and planner) and includes both roles for one player in the same game. This requires a bit more experience and a greater time commitment from game participants.

There are two other types of roles that you will find in games but that are not acted out by participants. The first is a simulated role, which is necessary to the game's smooth operation. It is a structural function built into the game events that causes things to happen at certain times. The second is a pseudo role, one that the game operator might assume extemporaneously or might assign to someone else. (See Richard B. Duke, Gaming: The Future's Language, Beverly Hills: Sage, 1974, p. 121.)

Table 10 indicates gaming/simulations that have functional and assigned roles, as well as individual or group roles. The

TABLE 10 Kinds of Roles

	Individual Role Descriptions	Group Role Descriptions	Individual & Group Role Descriptions
Functional Roles	*City Model*	*Acres* *Clug* *Metropolis* *New Town* *U-Dig* *Urban Dynamics*	*Hexagon* *Inhabs* *Metro* *Grand Frame*
Assigned Roles	*Eco-Acres*	*Systems*	*Blackberry Falls* *Blight*
Functional Assigned Roles	*Metro-Apex* *Mini-Apex*		

individual roles are usually played by one participant but can be shared by three or more. The group roles are usually played by three or more participants but can be played by one. This choice is determined by the game operator in response to the participants' gaming experience and the number of participants available for playing the game.

The *Trilogy* game series can either have role descriptions or not. Because they are not necessary for the game to function, the game operator makes the choice as appropriate.

Each game has specific types of roles (mayor, citizen, special-interest group), with specific objectives. Refer to table 11 to select the game or games that make use of roles that interest you.

Role Description Details

If one of your objectives is to provide your students with role playing of the sort that will help them develop empathy for a particular personality or a specific set of goals and values, then you may want to use a game with very structured and lengthy role descriptions. If you prefer that roles be open to interpretation, then you should choose a game with brief and generalized role descriptions. Table 12 indicates the length and specificity of the role descriptions given for each gaming/simulation.

Flexibility of Roles

If you are interested in adapting a gaming/simulation to roles that exist in your community or to roles that relate to specific objectives, you will find that *Grand Frame, Trilogy, Systems,* and *Inhabs* are all very malleable. These gaming/simulations are "frame" games in that new background data can be loaded into the game to fit a particular situation.

Although some of the games reviewed here are more flexible and open to interpretation than others, you will find that in all of them there is room for players to interject their own personalities and value systems. The only game that seems to be difficult for players is *City Model.* Its game structure is so complex and the cycles are so long that players often lose interest in their roles. However, those participants who do understand the model and their roles feel very positive about the game's value. *Metro-Apex*, too, is a very complex gaming/

TABLE 11 Roles: Types and Objectives

Gaming/ Simulations	Role Types	Role Objectives
Acres	Participants role play actors in the business, residential, and governmental sectors.	To enhance the town's well-being by making political, social, and economic decisions in relation to property development.
Blackberry Falls	Participants role play town officials (mayor, council members, school board, and so on and groups from private sectors (e.g., business, developers).	To illustrate the various value and goal orientations of members of small town as the population increases rapidly.
Blight	Participants role play members of community interest groups (taxpayers, homeowners' association, central community council, commerce and industry board, environmental council, metropolitan planners' guild, citizens' league) and task forces (economic development, environmental quality, municipal services, land use and development, human resources, community concerns), city council and mayor.	To analyze, bargain, and negotiate with other roles to achieve a task force position most favorable to their group's own interest.
City Model	Participants role play actors in economic, social or government sectors. There are up to 25 separate decision-making roles in both economic and social sectors. The government sector has seven roles with the capacity for additions.	None specified—open to participant development.
Clug	Participants are assigned to 3 to 5 teams designated by color. Specific role criteria develop funcitoning through game.	None specified—open to participant development.
Eco-Acres	Participants choose a role from the following: industrial contractor, member Angler's Association, vice-president Homebuilders Association, Trustee YMCA, President Snowmobile Association, President National Rifle Association, Garden Club Trustee, PTA President, President Audobon Society, Kiwanis President, Hospital Trustee, President Retired Citizens Council, Member Fraternal Order of Police, Bank President, President Automobile Association.	To explore the possibilities of both satisfying the needs of an ideal community and achieving the goals of each role.
Grand Frame	Participants assume roles of user, planner, landscape architect, developer, and city council member through the game functioning. No specific roles assigned. (Users include young low-income and mixed ethnic families, middle age and medium-income families and high-income and elderly families.)	To synthesize the information needed to develop design solutions, making trade-offs among different viewpoints represented.
Hexagon	Participants assume roles of national government (president, minister of foreign trade, and so on), regional government (e.g., banker, planner, engineer, governor), local people (marketeer, religious leader, fisher, farmer, entrepreneur, medical person, labor leader, teacher, beggar, and so on) and banker. Roles are developed through the game functioning.	To balance resource allocations in a third world country through policy making and to understand the difficulty of communication about dilemmas that arise in this context.
Inhabs	Participants role play individual home builders and establish partnerships as developers, or housing associations, or find customers for their buildings. The role of planner and banker also included.	To provide yourself with a house is primary. A secondary goal is to provide satisfactory housing for the community. Other goals are open to interpretation and development.
Metro	Participants role play major decision makers in the metropolitan area in categories of politicians, land developers, school people, and planners.	To make decisions about budgeting, political longevity, and land development that will benefit the metropolitan area.
Metro-Apex	Participants role play the following key decision makers: environmental quality agency with departments of air pollution, water pollution, and solid waste; politicians, planners, and administrative officers from a central city and a county; land developers and industrialists from the private sector; representatives from news media and pressure groups. Additional roles can be added as needed.	To make decisions through cooperation and/or competition with the other players to promote a particular strategy.
Mini-Apex	Same role as Metro-Apex.	Same as in Metro-Apex.
Metropolis	Participants role play teams of politicians, planners, or speculators. The specific roles develop through game functioning.	To improve the metropolitan area and maximize the rewards of each role.
New Town	Participants role play a public planner, or private developers in three to five teams. A council of representatives elected from the teams makes public decisions. There is also a banker role.	To maximize the developers' returns on property investments and to manage and regulate this development through the role of planner.
Systems	Participants role play one of the following teams: new town developers, status quo developers, and an environmental development group.	To allocate different land uses (housing, industrial, and commercial) within a given area, meeting their own goal requirements.

TABLE 11 Roles: Types and Objectives (Cont)

Gaming/ Simulations	Role Types	Role Objectives
Trilogy	Participants in groups of three can have a specific role or point of view in these games, but it is not necessary.	None necessary.
U-Dig	Participants are assigned to three to five teams, each given equal capital. Teams function as real estate investment brokers.	To increase the teams' capital through real estate investment.
Urban Dynamics	Participants are assigned to one of four teams and given resources that determine whether they will be aristrocacy, middle-class white, middle-class black, or poor. They become aware of resource disparity through playing the game.	Open to interpretation by the individual players.

TABLE 12 Role Description Details

No Role Descriptions (develop through play)	Brief Role Descriptions (open to interpretations)	Average Role Descriptions (several sentences)	Lengthy Role Descriptions (more closed than open)
City Model	*Hexagon*	*Blackberry Falls*	*Metropolis*
Clug	*Eco-Acres*	*Systems*	*Metro-Apex*
Trilogy	*New Town*	*Inhabs*	*Metro*
Grand Frame	*Urban Dynamics*		*Mini-Apex*
Acres	*U-Dig*		*Blight*

simulation, and it sometimes takes players a couple of cycles to feel comfortable with their roles.

RULES

The amount of time you need to prepare a simulation game and introduce it to a group of participants depends to a great extent upon the clarity, organization, and completeness of the rules. If an excessive amount of time is spent explaining and trying to understand the rules, participants' interest and motivation wanes. Some games, like *Metropolis, Acres, Blight,* and *Clug,* have a player's manual that can be given to participants several days before the game run so they will be familiar with rules, roles, and forms when play begins. Other games, like *Eco-Acres, Hexagon, Systems,* and Inhabs, have very brief sets of rules that can be explained and understood very quickly at the game introduction.

We find the rules of most of these eighteen gaming/simulations to be outstanding in clarity, completeness, and organization. These games represent the best and most often played urban gaming/simulations available, so this is to be expected. The exceptions are noted below.

The *City Model* game designed by Peter House has received both praise and criticism. It has served as a laboratory in which participants have learned to manipulate a very complex model with many variables, but this complexity has often kept participants from becoming involved in the model. It has been referred to as a "coarse simulation" rather than a gaming/simulation.

Mini-Apex, which is a simplified version of *Metro-Apex,* has a poorly organized manual. Although a great deal of the game's number crunching has been taken over by the operator in this version, there is very little explanation of this process. Though the information is there, it is presented in a very technical manner that may be hard for a person who is not familiar with the *Metro-Apex* game to understand. (*Mini-Apex* is being redesigned; these difficulties may not be present in the new version.)

U-Dig has a player's manual of rules that is very sketchy and difficult to follow. The steps of play for the players are displayed as a flow chart, which might present difficulty for a number of players.

Table 13 indicates the number of pages devoted to presenting the rules of the game and an overall rating for clarity, completeness, and organization.

SKILLS AND ACTIVITIES

Your next step in choosing a gaming/simulation should be determining which skills and activities are in accord with your objectives. Table 14 lists skills and activities down the left side and the gaming/simulations across the top. Match your objectives with the skills and activities list; then see what games incorporate them.

These gaming/simulations vary in the type of skills they employ and the extent to which these skills are emphasized.

TABLE 13 Rules

Gaming/ Simulations	Number of Pages Devoted to Rules	Overall Rating for Clarity, Organization, and Completeness		
		Adequate	Good	Outstanding
Acres	70			x
Blackberry Falls	100+			x
Blight	16			x
City Model	100+	x		
Clug	12			x
Eco-Acres	6		x	
Grand Frame	2			x
Hexagon	1			x
Inhabs	12			x
Metro	100+			x
Metro-Apex	100+			x
Mini-Apex	100+	x		
Metropolis	60			x
New Town	10			x
Systems	2			x
Trilogy	1-10			x
U-Dig	10+	x		
Urban Dynamics	26			x

TABLE 14 Skills and Activities

Skills and Activities	Gaming/Simulations																	
	Acres	*Blackberry Falls*	*Blight*	*City Model*	*Clug*	*Eco-Acres*	*Grand Frame*	*Hexagon*	*Inhabs*	*Metro*	*Metro Apex*	*Mini-Apex*	*Metropolis*	*New Town*	*Systems*	*Trilogy*	*U-Dig*	*Urban Dynamics*
Information Processing																		
analysis	vs	vs				vs	vs	vs		vs	vs	vs	vs		vs	vs	vs	
synthesis	vs	vs		vs		vs	vs	vs	vs	vs	vs	vs	vs		vs		vs	
information gathering	vs	vs	s	vs	vs	s	vs			vs	vs	vs	vs	vs	vs			s
rapid thinking	vs		s	s	vs	vs		vs	vs	vs	vs	vs	vs	vs				vs
planning		vs		vs	vs	vs	vs	vs	vs	vs	vs	vs	vs	vs	vs			vs
researching															vs			
problem solving		vs			vs	vs	vs		vs						vs		vs	
questioning		vs				vs	s											
brainstorming		vs																
innovating		vs				vs	vs											
divergent thinking			vs				s											
value clarification			vs	s	s		vs			vs	vs	vs	vs		vs	vs	vs	vs
Development of Skills																		
individual decision making	vs			s	s	vs			vs	vs	vs	vs		vs				
differentiating fact/opinion		vs								s	s	s	s					
memorizing																		
evaluating		vs	s				vs			vs	vs	vs				vs		
reading																		
writing																		
preparing reports																		
designing							vs							vs	vs			
calculating mathematically					s		vs			s	s	s	s					
manual construction																		
public speaking							s											
writing and presenting propsals			vs							vs	vs	vs	vs					
trading			vs		s			vs							vs			
budgeting: personal								vs	vs								vs	
community			vs				vs	vs	vs	vs	vs	vs	vs	vs				
town		vs						vs						vs				
city	vs			s	vs			vs		vs	vs	vs	vs		vs			
region				s				vs		vs	vs	vs						
national								vs										
managing a barter economy								vs										
allocating physical resources					vs	vs	vs	vs	vs	vs	vs	vs	vs	vs	vs		vs	
managing production and distribution of goods and services					vs			vs										
marketing					vs			vs									vs	
public finance								s		vs	vs	vs	vs					
Group Activities																		
role playing	m	vs	s	s	m	vs	s	s	vs	vs	vs	vs	vs	s	vs	m		vs
challenging others before a group		vs			s	vs												
debate		vs				vs												
parliamentary procedure						vs												
small group discussion		vs	vs		vs		vs			vs	vs	vs	vs					
class discussion															vs			
group decision making	s	vs	vs	s	vs	vs	s	s	vs	vs	vs	vs	vs	vs	vs	vs		vs
political maneuvering																		
Human Relations																		
coalition formation		vs	vs	s		vs		s	vs	vs	vs	vs	vs	vs			vs	
competition			vs	vs	vs			vs			vs	vs	vs	vs			vs	
compromising	vs	vs	vs	vs	vs	vs	vs	vs		vs	vs	vs	vs	vs		vs	vs	vs
negotiating	vs	vs	vs			vs		vs		vs	vs	vs	vs	vs		vs	vs	vs
bargaining	vs		vs	vs	vs			vs		vs	vs	vs	vs	vs				
persuading						vs										vs		
arbitrating																		
voting		vs	vs			vs	s											
organizing others				vs									vs	vs				
alert observation	s	vs						vs		vs	vs	vs	vs	vs				
skillful listening																		

Key: vs = very significant
s = signficant
m = minor

TABLE 15 Kinds of Interactions

Primarily Small Groups	Primarily Small Groups Interacting With Each Other	Primarily Large Groups	Primarily Large Groups Interacting With Each Other	Flexible Groups*
Grand Frame	*Inhabs*	*Eco-Acres*	*Metro*	*Hexagon*
Trilogy	*Acres*		*Metro-Apex*	*Blackberry Falls*
Systems	*Blight*		*Mini-Apex*	
	Clug		*City Model*	
	Metropolis			
	New Town			
	U-Dig			
	Urban Dynamics			

*Groups form and disband continually throughout play.

Games with a great deal of depth and breadth as well as many roles will not provide each participant with the opportunity to experience all the skills and activities during the game run. However, all players will gain an overview and a framework to fit these skills and activities into, and they will have the opportunity to study a system as they develop new skills and try out new experiences.

Interactions

If you are concerned with the development of skills in the affective domain as well as those in the cognitive domain, you should note carefully the types of group activities and human relations in Table 14. Table 15 shows the kinds of interactions that occur in the gaming/simulations.

Activities

With the exception of *Grand Frame, Trilogy,* and Systems, all the gaming/simulations repeat themselves in a discernible pattern of rounds or cycles. During the first two cycles of a game, the players are learning the boundaries of their roles and activities. By the third round, they begin to experiment within the roles until the point at which the gaming/simulation's model has been exhausted and there is little left to explore. The *Grand Frame* game and *Systems* differ in that an evolving, changing process with certain defined steps is the means for disclosing the model. *Trilogy* does not cycle at all and is an exercise for focusing discussion and communication.

Table 16 gives an overview of the sequence of activities and interactions that occur in the gaming/simulations.

RESOURCES AND SCORING

Another concern in choosing a gaming/simulation is the scoring procedures. Some games are highly competitive; others are based on cooperation. Most of the games reviewed here are a subtle combination of both. This is reflected in the resources that are distributed at the start of the game as well as in the procedures for totaling the accumulated resources at the end of the game. In order to determine which game best fits your criteria, refer to Table 17. Next refer to table 18, which summarizes the kinds of resources and scoring procedures used in these urban games.

TABLE 16 Sequence of Activities

Gaming/ Simulations	Activities
Acres	(1) Make decisions about price, investment, employment, transportion, recreation and pollution of an urban area, (2) Enter these into a computer, (3) Receive feedback, and make alterations in playing board, (4) Continue to round two.
Blackberry Falls	(1) Players choose roles. (2) Choose residences. (3) Form interest groups. (4) Receive town newspapers. (5) Examine news for issues that influence them. (6) Game operator provides action issues. (7) Interaction between groups. (8) Perform town functions. (9) File reports with computer terminal and game director. (10) Repeat steps 5-10 each session.
Blight	(1) Interest groups of community, city council, and task forces meet sequentially according to specified times. State priorities. Present proposals. (2) Council votes on proposals.
City Model	(1) Participants check past rounds to determine what actions occurred as a result of decisions. (2) Interaction as participants begin to plan. (3) Town meeting agenda setting and lobbying. (4) Town meeting. (5) Turn in decision to computer. (6) Repeat steps 1-5.
Clug	(1) Buy land. (2) Provide utilities. (3) Construct buildings. (4) Designate employment. (5) Set prices. (6) Receive income. (7) Pay employees. (8) Pay stores, offices, commercial, and transportation costs. (9) Pay taxes and vote. (10) Repeat 1-9.
Eco-Acres	(1) Receive community development pieces. (2) Draw role cards. (3) Create an ideal community through town meetings.
Grand Frame	(1) Conceptual planning. (2) Information processing: natural planning and development constraints. (3) Design phase. (4) Evaluation of designs. (5) Planning commission session.
Hexagon	*Local*: (1) Production. (2) Receive event cards. (3) Pay taxes. (4) Trade. (5) Form policy and make requests. (6) Fill out accounting forms. (7) Report on cycle. (8) Repeat 1-7. *Regional*: (1) Receive event cards. (2) Collect taxes. (3) Trade. (4) Form policy, make requests, allocate resources. (5) Collect accounting forms. (6) Report on cycle. *National*: (1) Receive event cards. (2) Receive

TABLE 16 Sequence of Activities (Cont)

Gaming/ Simulations	Activities
	taxes. (3) Discuss policy. (4) Receive requests, international trade, allocate resources. (5) Report on cycle.
Inhabs	(1) Build a family house with a certain amenity level. (2) Choose a community role. (3) Consider a development plan. (4) Buy land. (5) Design and build roads and paths. (6) Build houses. (7) Landscape, buy, and sell houses. (8) Repay loans. (9) Deduct penalities. (10) Repeat 1-9.
Metro	(1) Receive and review newspapers and current cycle output. (2) Complete decision forms for roles about an urban area. (3) Form political coalitions, discussion groups, and so on. (4) Planning, development, and the like. (5) Resolution of conflict.
Metro-Apex	Same as Metro.
Mini-Apex	Same as Metro. However it is possible to play for only two or three rounds, and cross impacts do not occur between activities.
Metropolis	(1) Distribute newspaper. (2) Complete public opinion poll. (3) Distribute and complete decision forms. (4) Compute budget for year. (5) End-of-cycle calculations. (6) Election. (7) End-of-cycle critique. (8) Repeat 1-7.
New Town	(1) Team assignments, distribute resources. (2) Bid for land. (3) Land development. (4) Record development and note bonuses for premium choices. (5) Calculate income and loss for the round. (6) Repeat steps 1-5 at least six times.
Systems	(1) Assignment to teams. (2) Begin planning area, considering numerous national variables. (3) Teams set criteria. (4) Design residential, industrial, and commercial sites. (5) Evaluation.
Trilogy	*Impasse*: (1) Each team has Descriptor Wheel on which to record assessments. (2) Discuss impact of central issue on each variable and record opinions. (3) Reverse the smaller wheel and examine experts' opinions; begin discussion. *At Issue*: (1) Players identify role (group constituency) they wish to represent and form groups of 3 ± 2 players. (2) Identify value priorities. (3) Select a strongly felt issue from list. (4) Evaluate cross impacts between this issue and the others on list. (5) Assess impact of position on each factor on a variable Impact Wheel. (6) Assess overall impact of this issue by taking your value score times the impact score from the variable identification wheel to obtain the "felt impact." (7) Discussion of issue should follow, using generated answers as basis. *Conceptual Mapping*: (1) Play the primer game At Issue. (2) Reveal the real world role. One is normally identified with. (3) Divide intro groups corresponding to the sectors of the conceptual map. (4) Review variables. (5) Select a specific issue for review. (6) Assess impact of issue on each of the variables by assigning a number value. (7) Transfer to master conceptual map for debriefing and discussion.
U-Dig	(1) Teams bid on property and finance loans. (2) Teams can sell property, raze old buildings, or construct new ones. (3) Evaluate investment strategy. (4) Round two: repeat 1-3.
Urban Dynamics	(1) Place population units. (2) Educate selected population units. (3) Place factories and corporations. (4) Negotiate loans. (5) Employ population units in factories and corporations. (6) Industry receives income. (7) Pay employees. (8) Pay rent. (9) Welfare paid to unemployed. (10) Pay transportation cost. (11) Pay taxes. (12) City council meeting.

TABLE 17 Kinds of Scoring

Mostly Cooperation	Mostly Competition	Both Competition and Cooperation
Blight	*New Town*	*Acres*
Eco-Acres	*Trilogy*	*Blackberry Falls*
Grand Frame	*U-Dig*	*City Model*
	Urban Dynamics	*Clug*
		Hexagon
		Inhabs
		Metropolis
		Metro
		Metro-Apex
		Mini-Apex
		Systems

TABLE 18 Resources and Scoring

Gaming/ Simulations	Resources	Scoring
Acres	Cash and some property are unequally distributed at start of game.	A complete set of environmental, business building, and government accounts that summarize the total impact of development.
Blackberry Falls	Budgets specific to roles.	Community as a whole win or loses as the town grows and changes.
Blight	No identifiable resources.	Optional: points may be awarded for achieving a match between a team's goals and priorities and the proposals adopted by city council.
City Model	Players begin with specific kinds and amounts of land, buildings, and official positions.	Scoring process embraces 25-50 variables that players may choose among to decide on relative score. These variables always posted and made available.
Clug	Play money equally distributed among teams.	Scoring or keeping track of who owns what is complex task that the accoun-

TABLE 18 Resources and Scoring (Cont)

Gaming/ Simulations	Resources	Scoring
		tant does all through the game. Teams keep their own financial records and playing board serves as physical reference of who owns what.
Eco-Acres	Development pieces: waterways, roads, institutional items, services and commerce, dwellings, and industry.	Community as a whole scores for approaching or reaching consensus.
Grand Frame	Groups receive budget according to family income level.	Players must present designs to planning commission for approval. Must be within the budget.
Hexagon	Players receive unequal survival resources according to settlement characteristics.	Scoring is a composite of how well the whole country provides resources for the entire population (i.e., national stability.)
Inhabs	Money distributed according to family size.	There are penalties for failing to buy a house and for not providing amenities.
Metro	Budget for specific roles.	Accounts of population growth, public revenue, property assessment, and rewards and penalties for each role kept by computer.
Metro-Apex	Budget for specific roles.	Composite accountability for entire urban area from all role decisions kept by computer.
Mini-Apex	Same as Metro-Apex.	Same as Metro-Apex.
Metropolis	Budget for specific roles.	Accounts of population growth, public revenue, property assessment, and rewards and penalties for each role kept by computer.
New Town	Play money and buildings are equally distributed to teams.	The player who amasses the greatest fortune is winner.
Systems	No identifiable resources.	Relative evaluation based on how well plans meet stated values and goals.
Trilogy	No identifiable resources.	None appropriate.
U-Dig	Equal amounts of capital.	Property holdings and cash flow recorded and profits determined.
Urban Dynamics	Financial means distributed unequally according to team role.	Community wins or loses according to relative assessment of the urban development.

MODEL

If you intend to use a gaming/simulation to introduce your students or coworkers to the actual workings of an institution, social process, or system, you will want to check the validity of the model it uses. To evaluate a gaming/simulation model for its validity and realism, you will want first to read the descriptions of the models in Table 19, checking for what you think are aspects of reality the model includes. Next ask yourself if these are the essential aspects of the model and what, if anything, has been left out. Table 20 shows the order in which we have rated the models for validity, but you should make your own judgments.

TABLE 19 Types of Models

Gaming/ Simulations	Model
Acres	The early version was based on *Clug* but soon changed to become an interactive computer game. The model is based on economic concepts of supply and demand in the land market, environmental science, and governmental budgeting. Educational quality has a direct impact on productivity and wages in the game, thus requiring the governmental sector to make decisions keeping this in mind. The highway system or transportation plays a heavy part in the game.
Blackberry Falls	*Apex* was used as a beginning model for this game. However, the form of government and roles had to be changed significantly to better fit New England small town growth. Now there is emphasis on local town government rather than on county government (e.g., town planning board and conservation commission). Also citizens, opportunity to influence and participate in the town's decision-making process is increased.
Blight	Social development, political participation, and cultural expansion provide three dimensions for viewing urban decline and growth. *Blight* is based on a model in which: (1) functionally defined task leaders, (2) task forces, and (3) competency-based representatives are all part of the city decision-making process and policy making.
City Model	*City I* was based on *Clug*, using the computer to do the bookkeeping. *City II* began to take over some of the decision making players had previously done in *Clug*. *City III* was basically an attempt to simulate five cities that varied in size from 40,000 to 25,000,000 people. *River Basin Model* attempted to look at water resources on a regional basis and integrate this element with the economic, political, and social subsystem. It highlights environmental issues, especially pollution and quality of life. All models are based on a systematic view of a metropolitical area.
Clug	Clug seeks to convey Brian Berry's mathematical description of the basic elements of urban growth and change. Cities are supported by "basic" activities, whose locations are determined exogenously to the city by comparative advantage in regional, national, and international economic systems. Such basic activities universally include a central business district, the focus not only of the city but also of its tributary region. Various specialized activities will also be pres-

TABLE 19 Types of Models (Cont)

Gaming/ Simulations	Model
	ent . . . . Locations of the basic activities, plus a transport system, provide the skeletal features of the urban pattern. This pattern is filled out in part by the residences of workers in the basic activities, and is given a dynamic quality by the daily ebb and flow of commuters and, from beyond the city's limits, of goods and customers to and from the sites of the basic activities . . . .
Clug (Cont)	Further patterning is provided by the orientation of business services to the basic activities and by tertiary activities to the consuming workers and their families. Shopping trips create yet another ebb and flow. Then appear all the "second-round" effects; locations of the residences of workers in the "nonbasic" activities, additional commuting, more demand for teritary activities, and so forth, in an increasingly complex chain of multipler effects (*Clug Player's Manual*, p. 7)
Eco-Acres	Although highly simplified, the basic model involves location theory and land use planning. Environmental perception and ecological use of land area are also important considerations. When you place people with differing values and incomes in this model, you find conflict over which solutions are best. A democratic process is used to make decisions about planning an ideal community.
Grand Frame	The model on which the game is based is a design process that includes the following steps: (1) conceptual planning—user constraints, (2) information processing—natural constraints, (3) information processing—planning constraints, (4) information processing—developer constraints, (5) design phase—housing development, (6) evaluation, (7) planning commission session. This process is one that has been taught by design teachers for a number of years to design students, but the symbols and structure used to illustrate the process are new.
Hexagon	*Hexagon* is based rather loosely on a model of central place theory and a generalized model of an organizational theory of western government (non-centrally planned). The land use is based loosely on geographic data from Ghana. Although the game was based on literature, research, and verbal information from the client, and although there was an attempt to match the basic structure of Ghana, the game itself is a generalized, abstracted model of the information.
Inhabs	The individual creative process is embedded in a matrix of decisions and regulating constraints made by a great many people (other than the designer), whose actions cannot be either predicted or ignored. The simplest case is the designer who must take into consideration the client, the users of the product, and the makers of it; all of whom will have different roles and objectives affecting the final design. If you add to that the specialists contributing to the design, people with financial interests that diverge from

TABLE 19 Types of Models (Cont)

Gaming/ Simulations	Model
	the client's, a variety of users, competing producers, planning or standards controllers, then the designer's individual contribution shrinks in proportion, and the relevance of working in isolation becomes less. The individual creative synthesis is vital to the whole process, and it must be developed to the context of the forces that act upon it, or at least, if that is not possible in design education, in a simulated context. Designing considered as a form of complex decision making can be practiced in a role-playing game that simulates the socio-economic context in which it operates. Inhabs was designed to simulate that area of designing that most obviously draws its validity from its social context housing [*Inhabs Game Manual,* 1971: 1].
Metro	The major models in the computer simulation are the GROW Model which contains exogenous industrial and bureaucratic employment in the county; TOMM controls the population changes as well as subsidiary commercial and industrial growth, the VOTER RESPONSE Model handles the simulated community's response to politicians' requests on bond and special millage issues; the CANDIDATE ELECTION Model allocates votes to the politician on the basis of previous budgeting decision; the AIR Model generates pollution indices for the county on the basis of gamed and simulated industrial output. Other models are used to handle solid waste and water pollution generated in the community. All of these models are linked together. Several subroutines process land transfers and financial transactions between players and the simulated community. Through these models and subroutines the dynamics of an urban area are mirrored in the simulation (Larry C. Coppard and Frederick L. Goodman (eds.), *Urban Gaming/Simulation—'77,* University of Michigan.)
Metro-Apex	Same as *Metro*, with the addition of air, water, and solid waste models.
Mini-Apex	Simulation of *Metro-Apex*.
Metropolis	*Metropolis* is an urban financial management game that focuses on the capital improvement program (CIP) aspect of local government patterns. It is an abstraction of an urban area with a 215,000 person population. The three major roles—administration, political, and speculation—are intertwined with the fate of the city. Pressure is applied to these groups through news items that list priorities of various interest groups in the city.
New Town	*New Town* is based on spacial models from geography. Neighborhood development, intra-urban development, and adjacency are important in the model. Concentric rings, multiple nuclear eye, and sector theory are represented. Focus is on the strategies of the private developer and the planner in an attempt to maximize financial returns, the distribution of community services, and the shaping of urban development.

TABLE 19 Types of Models (Cont)

Gaming/ Simulations	Model
Systems	*Systems* uses 9 land use models—public expenditures, conversation, recreation, public institutions, schools, and utilities. There are 15 land use and land type classifications: water and wetlands, forest, agriculture and open space, conservation and recreation, utilities, transportation, residential, commercial, industrial, commercial grounds, isolated business, highway related business, neighborhood centers, community centers, regional centers. There is also information on the effects of urbanization, glaciation, transportation patterns, historic sites and old settlement patterns, on soil characteristics, water tables, slopes, wildlife, and more. All this is on data maps for use by teams in designing the physical layout of a site that contains residences, industry, and commercial development. People obtain their goals by using information available.
Trilogy	General systems theory is the basis for the model for these three games. Systems are set of components and linkages that work together for the overall objective of the whole. *Impasse, At Issue,* and *Conceptual Mapping* are frameworks designed to investigate a problem considering the components that are an integral part of the system.
U-Dig	*U-Dig* is modeled on the real estate investment process—a complex affair that involves many specialists.
Urban Dynamics	This game's model began as a *Clug* game with modifications. The model contains an urban system; it has the city council becoming the focus of decision making. It also was modeled on the designer's own experiences and feelings about working in neighborhoods in Chicago. It includes dramatic racial and economic differences and allows inclusion of models for the dynamic experiences of blockbusting, racial confrontation, segregation, and the extreme differences between central city and suburban developments. The game further incorporates a sense that people come to the large cities for many reasons besides industry and employment.

TABLE 20 Model Validity (Completeness)

Simple	Adequate	High
Inhabs	*Hexagon*	*Clug*
New Town	*Eco-Acres*	*Grand Frame*
	Blight	*Trilogy*
		Metropolis
		Urban Dynamics
		Acres
		U-Dig
		Systems
		Metro
		Metro-Apex
		Mini-Apex
		Blackberry Falls
		City Model

Although most of the models are rated high in validity and realism, a few are rated adequate or simple. You should interpret this rating very carefully. It indicates that many variables in *Hexagon, Blight,* and *Eco-Acres* have been generalized and simplified, not that the models are incorrect or invalid. However, in the *Inhabs* game, although there is a good deal of reality in the model, several constraints are unrealistic; for instance, everyone has the role of a house builder and buyer in the same community. The main criticism of the *New Town* model is that the game neglects to deal with social and aesthetic values of people and political groups. It does afford players an opportunity to learn when a consideration of these social and aesthetic values plus economic values is relevant in the development process.

DEBRIEFING

The most important part of any gaming/simulation is the discussion that follows immediately after the playing has stopped. In fact, you should think of the game as a communication tool that prepares the participants for this debriefing session. Whether or not you are well versed in the subject matter of your simulation or have had experience leading simulations before, a good set of debriefing questions will make it easier to lead this discussion. Some of the gaming/simulations provide very thorough guides to debriefing, whereas others suggest a few questions, and still others provide no guidance at all. Table 21 indicates the scope of the debriefing guidelines included in the manuals of the various gaming/simulations.

If you are using a game for which there are few or no debriefing guidelines, you will find that the following three-step procedure works well for a large variety of gaming/simulations:

(1) Participants usually have strong feelings about the actions in the game. Therefore, you as the game operator need to help them express these reactions by asking questions like: What was your most frustrating experience in the game? What did you dislike about the game?

(2) Next, you want to shift to a different level of questioning and ask: Did you observe any similarities between the real world and the game?

(3) The final stage is to leave the game itself and shift to the real world problem or problems simulated in the game. By this time, everyone in the group has a common game experience and a common game language. It is now possible for the participants to transfer learning

TABLE 21 Debriefing Guidelines

No Guidelines	Few Guidelines	Adequate Guidelines	Thorough Guidelines
City Model	*Hexagon*	*Clug*	*Grand Frame*
Acres	*Inhabs*	*Metropolis*	*Trilogy*
U-Dig	*Eco-Acres*	*Systems*	*New Town*
Mini-Apex	*Blackberry Falls*	*Urban Dynamics*	
		Blight	
		Metro	
		Metro-Apex	

to a situation that is meaningful to them. Because this is your main objective in playing the game, allow a good deal of time for this discussion.

A game operator needs to keep these stages of debriefing in mind during the game run and jot down notes on participants' actions. These notes will help start the discussion and keep it going in the right direction. We usually allow from one-fourth to one-third of the entire game time (playing plus debriefing) for this debriefing session. Thus, if we have two hours of game time, we spend one and one-fourth to one and one-half hours on playing and thirty to forty-five minutes on debriefing.

CONCLUSION

At this point, having considered numerous criteria for selecting an urban gaming/simulation, you must make a decision about which game or games most closely match your objectives, your purposes, audience context, your time limits, and so forth. If you have defined your criteria carefully, then selecting a gaming/simulation will be fairly simple. However, remember that the best way to judge a game (and many experienced gamers say the *only* way) is to play it. The gaming model comes alive only with participants.

REFERENCE

Green, C.W.B. (1971) INHABS 3–Instructional Housing and Building Simulation: Operating Manual. Cheltenham Papers No. 5. Gloucestershire College of Art and Design.

Part II
ACADEMIC LISTINGS

A NOTE TO OUR READERS

In addition to the Listings, divided by subject, which follow, games and simulations in the three categories below may be found in the sections indicated.

NATIVE AMERICANS

See Collision DOMESTIC POLITICS
Indian Reservation SOCIAL STUDIES

Indians View Americans, Americans View Indians HISTORY
Mahopa SOCIAL STUDIES

NATIVE AMERICANS

Collision DOMESTIC POLITICS
Indian Reservation SOCIAL STUDIES
Indians View Americans, Americans View Indians HISTORY
Mahopa SOCIAL STUDIES
Opening the Deck SOCIAL STUDIES
Potlatch Game SOCIAL STUDIES

THIRD WORLD

See AFASLAPOL INTERNATIONAL RELATIONS
The Aid Committee Game ECONOMICS
Baldicer SOCIAL STUDIES
The Coffee Game RELIGION
Country Development Economics and Finance Game ECONOMICS
Diplomatic Practices INTERNATIONAL RELATIONS
Formento ECONOMICS
The Grain Drain ECONOMICS
Hexagon URBAN
The Hunting Game SOCIAL STUDIES
Imperialism INTERNATIONAL RELATIONS
Independence SOCIAL STUDIES
Kama SOCIAL STUDIES
The Poultry Game SOCIAL STUDIES
The Poverty Game SOCIAL STUDIES
Sanga SOCIAL STUDIES
Traders Arrive on the African Scene SOCIAL STUDIES
Uhuru SOCIAL STUDIES
Up Caste Down Caste SOCIAL STUDIES

WOMEN'S ISSUES

See Access SOCIAL STUDIES
Becoming a Person SELF-DEVELOPMENT
Encapsulation COMMUNICATION
Female Images SELF-DEVELOPMENT
Herstory SOCIAL STUDIES
Job Crunch EDUCATION
The Marriage Game SELF-DEVELOPMENT
Profair BUSINESS: PERSONNEL DEVELOPMENT
Pro's and Con's FRAME GAMES
Seneca Falls HISTORY
Women in Management BUSINESS: PERSONNEL DEVELOPMENT
Women's Liberation DOMESTIC POLITICS

ADDICTIONS

CHOOSE

Playing Data
Copyright: 1972
Age Level: junior high
Number of Players: 6
Packaging: professionally packaged board game

Description: Choose is a simple board game that encourages team cooperation while alerting students to the dangers of drugs. Players divide into two teams of three. Each player advances a token according to the throw of the die, landing on red, yellow, or green squares. The green squares are "safe" and are labeled with messages like "roller skating." The yellow, or "possible trouble," squares require players to throw an odd number with the die before they can advance. The red squares are labeled with the name of a drug such as marijuana or cocaine. A player who lands on one of these squares must decide, with the agreement of the team, whether to select a "choose" card and follow its instructions, or wait to throw a one on the die. The winning team is the first whose members complete one circuit of the board. (DCD)

Cost: $6.00

Producer: Spoken Arts Inc., 310 North Avenue, New Rochelle, NY. 10801

COMMUNITY TARGET: ALCOHOL ABUSE

Harry Silas and Jim Spears

Playing Data
Copyright: 1975
Age Level: grade 9-adult
Number of Players: 15-35
Playing Time: 4-5 hours
Preparation Time: 1 hour
Packaging: 35-page photocopied booklet

Description: This game is designed, according to the authors, to re-create "the various attitudes that people hold about alcohol abuse and how these attitudes can help or hinder efforts for solution." Players assume the roles of one of five "Advisory Committee" members, or members of two citizen action groups. The descriptions for each of these 35 roles include summaries of, and rationales for, each person's attitude toward alcohol abuse. The game is played in five rounds. During the first two rounds, participants discuss the excessive drinking in an unknown town as defined by a report, and all players have the opportunity to propose responses to the problem. In the third round, all players, except those on the advisory committee, are assigned to one of two groups which must prepare sets of recommendations for committee consideration. In the fourth round, the Advisory Commitee drafts a comprehensive plan for coping with problem drinking in the town. During the final round, the teacher leads a discussion of the successes and failures of the game. (DCD)

Comment: This simulation requires a knowledgeable instructor/discussion leader. It is oriented toward the California situation but could serve as a basis for most other North American communities. It is suitable for use within classroom constraints. (AEZ)

Note: For more on this game, see the essay on health and health care simulations by Amy E. Zelmer and A. C. Lynn Zelmer.

Cost: $6.00

Producer: Center for Health Games and Simulations, Department of Health Science and Safety, San Diego State University, San Diego, CA. 92182

DRUG ATTACK

Robert L. Nelson

Playing Data
Copyright: 1971
Age Level: grades 7-12, families
Number of Players: 3-5
Playing Time: 30 minutes
Packaging: professionally designed box, with folding game board, tokens, chance and player cards, play money, and instructions

Description: "Your community is about to come under a drug attack!" read the instructions for this game. "On the outskirts, criminal drug pushers are watching and waiting for the chance to strike. Their target is the community's drug users. Their goal is the community's money—and they will stop at nothing to get it.

"What can you do about it? As the mayor, Agent, or Health Officer, you must detect their attack, stop it, and treat the victims. If you fail, your community may become overrun by drug pushers and users, and collapse in a drug disaster!"

Drug Attack is based on the concept in the above quote. Players take the roles of mayor, narcotics agent, or health officer. Drug users and pushers are represented by plastic markers. No players assume these roles. Play begins when the mayor draws a chance card. Most of these cards instruct the mayor to move a pusher zero to ten spaces toward the nearest user, and some instruct her to distribute community funds. Agents and health officers play in turn after the mayor. These roles are required to apprehend pushers or rescue users, who may be rescued or busted only if a player can identify the drug the pusher or user possesses. For example, the printed answer to the question of what causes psychotic reactions and brain damage is marijuana. Agents and health officers compete to be the first to remove all, or the most, pushers and users from the board. Play ends when this is accomplished or there is a "drug disaster" (all community funds are lost to drug costs). (DCD)

Cost: $8.00

Producer: R. Frederick, P.O. Box 472, Girard, OH. 44420

TO DRINK OR NOT TO DRINK

Judith Platt, Ray Glazier, Games Central

Playing Data
Copyright: 1972
Age Level: grades 7-12, adults
Prerequisite Skills: grade 5 reading, grade 1 math
Number of Players: 5 to 16 maximum in 0 to 5 teams
Preparation Time: 15 minutes

Description: This is a board game in which players progress first along a Teenage track, deciding whether to drink, and if so how much, whenever the opportunity to do so comes up. Players then move onto the Young Adult track in either a blue-collar or white-collar life-style. Chance determines the life-style track as well as the life events that may or may not present drinking options.

The purpose of the game is to give players vicarious opportunities to consume alcohol and experience its consequences with the intent of alleviating pressure for real-life experimentation in extreme situations. (DCD)

Cost: $30.00

Producer: Games Central, Abt Publications, 55 Wheeler Street, Cambridge, MA. 02138

TRIP OR TRAP BINGO

Playing Data
Copyright: 1971
Age Level: grade 4-adult
Number of Players: 15-40
Playing Time: 10-60 minutes
Packaging: professionally designed box with bingo cards, markers, call pieces, instructions, and "drug abuse lesson guide"

Description: This game is identical to *Bingo* except that each square contains a picture of an illicit drug as well as a number. The "caller" calls out the name of a drug as well as a number. The object of the game is to cover a row of drugs with markers. The winner is the first to yell "trip or trap bingo" after covering a row of five squares. The object of the game, according to the producer, is for players to "learn to identify names, appearances, and actions of the drugs as they are called out in the course of the game." (DCD)

Cost: $14.50

Producer: Spenco Corporation, P.O. Box 8113, Waco, TX. 76710

COMMUNICATION

AGENCY

John Wesley

Playing Data:
Copyright: 1977
Age Level: grades 7-12
Number of Players: 14-48 in groups of 6-8
Playing Time: 15 one-hour sessions
Preparation Time: 3 hours (est.)
Special Equipment: video tape recorder
Packaging: 30-page instruction booklet

Description: *Agency* is designed to simulate the activity of competing advertising agencies. Players begin by examining and analyzing magazine and television advertisements to understand advertising techniques. Next, participants form agencies of six to eight players each, with six typical advertising agency jobs such as copywriter, artist, producer, and model. The game director then introduces five new accounts for pizza, toothpaste, gum, automobiles, and a new line of skateboards. The agencies must decide which account they wish to represent. Since each account will select only one agency, each agency in secret prepares a complete campaign (including radio, television, print, and outdoor advertisement). At the conclusion of the simulation each agency presents its complete campaign to all of the other participants. (DCD)

Cost: $14.00

Producer: Interact, Box 262, Lakeside, CA 92040

THE ANNUAL HANDBOOK FOR GROUP FACILITATORS

J. William Pfeiffer and John E. Jones, Editors

Playing Data
Copyright: 1972-1979

Age Level: college, professional, adult
Packaging: approximately 290 pages each, looseleaf notebook or paperbound

Description: The exercises in this series of eight annual handbooks are designed for use by group facilitators in personal growth, humanistic education, leadership and management training, and organization development. The publisher describes the design of each volume thus: "The Structured Experiences section includes step-by-step activities designed to focus on individual behavior, constructive feedback, processing, and psychological integration. The Instrumentation section contains questionnaires, scales, inventories, and measurements useful for data generation, teaching, training, personnel selection, organizational diagnosis, and research. The Lecturettes section is designed to provide brief statements of principles, models, and theoretical positions. The Theory and Practice Section is intended to provide a forum for emerging ideas in the human relations field. The Resources section offers material for reading and for further training and provides the facilitator with access to additional tools, books, background, bibliographies, and lists of resources." (AC)

Cost: looseleaf notebook $29.50 each, paperbound $12.50 each; *Reference Guide to Handbooks and Annuals* (3rd edition, 1979) $6.00

Producer: University Associates, Inc., 7596 Eads Avenue, La Jolla, CA 92037

CAN OF SQUIRMS

Arthur L. Zapel, Arthur Merriwether, Inc.

Playing Data
Copyright: various, 1968 through 1974
Age Level: primary-adult, depending on version
Number of Players: 2-40 in 2 teams
Playing Time: 5 minutes-2 hours in flexible time periods
Preparation Time: 10 minutes

Description: There are eleven Cans of Squirms. They are Primary, Intermediate, Junior High, High School, College, Adult, Generation Gap, Teenage Sex Education, American History, and Old and New Testaments (see religion section). Users can also write their own from a prototype. Two teams are chosen. A role-play dilemma is read aloud. Two or three people from one team assume roles and try to resolve the dilemma in three minutes, after which the group participates in discussion of the resolution. The Squirms are intended for use in introducing and defining certain dilemmas that otherwise might be difficult or awkward to introduce for group discussion. While it is the role of the players to be antagonists and protagonists of a specific viewpoint within a situation, it is the function of the teams to explore alternatives within the dilemma. Generally, the decisions are largely moral choices. If expediency is chosen over moral conviction, the players are scored accordingly. Discussion following game determines which decisions were major and which were not. (AC)

Cost: $7.95, quantity discount.

Producer: Contemporary Drama Service, Arthur Merriwether, Inc., Box 457, Downer's Grove, IL 60515

DILEMMA

Playing Data
Copyright: 1975
Age Level: adult
Number of Players: 5-14
Playing Time: 1-2 hours
Preparation Time: 30 minutes (est.)
Packaging: 7 professionally produced participants' packets and instructions

Description: According to the producer, *Dilemma* is designed to set "a dramatic stage where insights can emerge" and highlight "the ethics of how decisions are made." The situation the game dramatizes is one in which a deadly accident has occurred. Six innocent people have been exposed to a fatal virus, serum is available to treat only one, and time is critical. Players assume the roles of members of a special committee that must decide within 30 minutes which one is to be saved. During debriefing, players are asked to consider how small groups make decisions, what gives them the right to make them, and if some ways of making decisions are more ethical than others. (DCD)

Cost: $19.95

Producer: Creative Learning Systems, Inc., 936 C Street, San Diego, CA 92101

ENCAPSULATION

Playing Data
Copyright: 1972
Age Level: high school-adult
Number of Players: 6-20
Playing Time: 1-2 hours
Special Equipment: cassette tape player
Packaging: professionally packaged game set includes coordinator's guide, role folders, observer forms, casette tape recording, bias boards, and situation cards.

Description: Encapsulation is a conflict management simulation with five possible scenarios including (1) Labor versus Management, (2) Black versus White, (3) Affluent versus Deprived, (4) Career versus Homemaking, and (5) Parent versus Child. A "master kit" includes all game materials plus cards for one scenario. "Adaptation kits" to adapt any master set to another version consist of two "Bias Board Insert Cards," a "Situation Card," and a "Facilitator's Card."

In the game, regardless of scenario, "two adversaries with a social problem and a set of biased perceptions of self and other confront each other. Peers sit behind them and pass them notes that urge loyalty to their 'own kind.' A facilitator tries to encourage open and honest communication in spite of group pressures. The debriefing that follows play reveals the dynamics common to many inter-group differences." Each version provides roles for two adversaries and the peer groups they represent, for a facilitator, a "helpful observer," and two "non-participant observers."

"Labor versus Management," for example, depicts a situation in which the management of a manufacturing company has recently introduced a new work procedure. The affected workers assert that the new procedure is inefficient and needlessly complicates their jobs. Management, on the other hand—noting that the new procedure has not, as expected, increased the efficiency of the department—has accused the workers of deliberately slowing down production. Two players representing these two sides sit across a table from one another. In front of each is a plastic stand containing six assumptions, or attributions, each side makes to itself and its adversary. (For example, while labor's spokesman labels herself as a "producer" and her adversary as a "manager," management's spokesman thinks of herself as an "executive" and her adversary as an "employee.") These assumptions are masked by paper strips and may be unmasked at the direction of the facilitator or the discretion of either negotiator. Each negotiator receives advice from peers, from a "helpful observer," and from the facilitator. The negotiators are assigned the tasks of listening to this advice and negotiating a resolution to the conflict.

The two nonparticipant observers take notes on the intervention strategy of the game facilitator. Work sheets help these players decide if the facilitator (1) promotes his or her own solution, (2) takes sides, (3) promotes a win-lose mode, (4) smoothes things over, (5) promotes the clarification of ideas, (6) emphasizes interpersonal dynamics, and/or (7) deals with feelings.

The game concludes with a debriefing during which all players discuss what they observed, thought, and felt during play and what they learned from the simulation. (DCD)

Note: For more on this simulation, see the essay on frame games by Sivasailam Thiagarajan and Harold Stolovitch.

Cost: master set (frame game plus one scenario) $35.00; adaptation kit $7.95; all materials $66.60

Producer: Creative Learning Systems, Inc., 936 C Street, San Diego, CA 92101

ERNSTSPIEL KIT

MUST (Management Utilizing Staff Training)

Playing Data
Age Level: management and other adult groups
Number of Players: 5 to 30 or more in 3 to 15 or more teams
Playing Time: 1 to 1-1/2 hours for each of the 8 kits
Preparation Time: none for players, some for coordinator
Special Equipment: overhead projector

Description: The *Ernstspiel Kit* contains eight packets, each focusing on a different communication concept or skill. The first six packets deal with the single concepts of one-way and two-way, tacit, nonverbal, by-pass, overload, and written communication. The last two packages require participants to incorporate skills learned in the first six exercises to play the remaining two: group norms and individual versus group norms.

Each package in the kit is a self-contained exercise that is enjoyable to play. Meanwhile, the exercises create a climate that resembles a slice of life, which can be taken seriously and played with zest, but which carries no penalty if a participant fails to win or achieve some expected outcome. Each package is short enough to be used after staff meetings, but can also be used in combination with other packages in a workshop. This kit is particularly helpful in improving communication in ongoing groups. (Authors)

Cost: $94.75 (limited supply)

Producer: MUST, Research Division, Center for Educational Policy and Management, 1472 Kincaid Street, Eugene, OR 97403

FAMILY

Jim Deacove

Playing Data
Copyright: 1971
Age Level: grade 7-adult
Number of Players: 1-19
Playing Time: 30 minutes (est.)
Packaging: professionally packaged card game with two instruction booklets

Description: There are nine different cards in the 68-card deck used to play this collection of cooperative games. These cards are marked with the symbols U for Universe, F for natural forces, M for man, W for woman, A for animal, V for vegetable, S for soil or the earth, T for the passing of time, and Mu (which is short for Mumba, or the freezer card).

Instructions for 12 variations are included with the deck, but none differs significantly from the basic game described here. One player deals out 7 piles of 7 cards face down. The top card of each pile is exposed and any Ts revealed must be placed in a "clock area." The Ts represent the passing of time, and when all 15 Ts in the deck have been turned up the game ends.

As in all variations of solitaire (upon which *Family* seems to have been modeled) the players arrange cards taken from the 7 piles into 6 identical arrays. Each array must feature the same symbol cards arranged in the same order. Players may work on any number of arrays simultaneously. Whenever a Mu (or Freezer) card is turned up in one of the 7 piles that pile is frozen and no more cards may be removed from it until one of the other piles is exhausted, when the Mu card may be moved into that empty space. When the players can no longer extend any of the arrays of symbol cards by drawing cards from the piles, the dealer passes out the remaining 19 cards in the deck, distributing them as equally as possible among all players. The players then use these to try to complete the arrays (or families) of cards.

The idea of the game is for the players to work cooperatively to complete at least one family (or array) that contains all 7 of the symbol cards (except T and Mu). (DCD)

Cost: about $4.50. Send for current price list.

Producer: Family Pastimes, R.R. 4, Perth, Ontario, Canada

THE FAMILY CONTRACT GAME

Elaine A. Blechman, Yale University

Playing Data
Copyright: 1974
Age Level: all ages, intended for families
Number of Players: 2 or more
Playing Time: 40-60 minutes for two games
Special Equipment: optional videotape equipment
Packaging: game box

Description: The purpose of *The Family Contract Game* is to aid families in resolving their interpersonal problems without therapy. Families begin by listing their major problems on individual cards to make up a problem deck, and by listing rewards for acceptable behavior on cards which make up the reward deck. The 14 squares of the game board are divided into four basic components of problem-solving. Each time the game is played, one player is the target of a complaint while the other writes the contract for change. Each play of the game is set for 15 minutes, during which time players progress around the board, selecting a problem, agreeing on replacement behavior, determining how to reinforce and record it, and writing a contract. Failure to complete a step of the process results in fines of play money and the selection of risk cards which may, for example, require the penalized party to move back one space. Bonus and reward cards are awarded for successful transactions.

This game has been researched extensively, and studies describing its use have been published in David Olson's *Treating Relationships* (Graphic Publishing Company); *Behavior Therapy* 7, 1976; and *The Journal of Consulting and Clinical Psychology* volume 44, #3, 1976. The designer recommends that families play it first with a trainer and then attempt to use it on their own to resolve their interpersonal problems. (TM)

Cost: $40.00; 3 kits, $100.00

Producer: ASIP, Inc., P.O. Box 389, Madison, CT 06443

A HANDBOOK OF STRUCTURED EXPERIENCES FOR HUMAN RELATIONS TRAINING

J. William Pfeiffer and John E. Jones, Editors

Playing Data
Copyright: 1973-1979
Age Level: college, professional adult
Packaging: paperbound books

Description: This is a seven-volume series of books each describing 24 structured experiences, or what are sometimes called games in human relations training. Their major uses are in communications classes, personal awareness training, and organizational development. It would be too much to print even the table of contents of these exercises. Suffice it to say that many of the exercises which are frequently used in these sorts of classes are described in these books. They are most appropriate for college-age and adult participants. Some of the exercises border on simulations in that they ask people to attempt to do things like build a building in a team context so that the elements of cooperation, types of cooperation, and types of roles people play in teams can be observed.

The books are indexed with short exercises in the following categories: Ice Breakers, Awareness Expansion, Interpersonal Communication, Intergroup Communication, Personal Feedback, Dyads, Leadership, Group Process, Group Problem Solving, Competition, and Organizational Development. (REH)

Cost: each volume $6.00; boxed set of volumes 1-7 plus free Reference Guide to Handbooks and Annuals (3rd edition, 1979) $42.00

Producer: University Associates, Inc., 7596 Eads Avenue, La Jolla, CA 92037

HUMAN COMMUNICATION HANDBOOK

Brent D. Ruben and Richard W. Budd, Rutgers University

Playing Data
Copyright: 1975 volume one, 1978 volume two
Age Level: high school, college, adult
Number of Players: variable
Playing Time: variable
Packaging: paperback book

Description: Volume One contains 49 games concerned with personal communication, social communication, and communication systems, and 16 guides and forms for recording observations and insights. Most activities involve dividing a group into dyads, two teams, or five small groups, and are less than an hour in length. However, a number of the communication systems games are complete simulations and can be expanded from a week activity to an entire semester.

One communication systems game, *Hypothetica,* is geared for a group of 12-60, with a time span of 2-5 hours. Students receive a map of a fictional country, Hypothetica, and divide into groups representing five districts which differ in their density, level of development, and economic activities. After each group elects a chairman and all groups together elect a prime minister, groups meet individually to determine their position and strategies of negotiation in regard to the imminent national conference, where issues related to the discovery of rich mineral deposits in one district of the country will be discussed. Problems which may or may not be considered and resolved at the conference concern what action should be taken in regard to the mineral deposits, how they should be transported, whether a contract with a foreign mining company should be negotiated, and the determination of a long-range plan for national development.

Another simulation, the *Communication Systems Simulation,* involves 25-150 persons, and provides them with experience with the dynamics of mass communication in organizations. Students begin by researching a topic such as a state election, and then choose their roles as freelancers who prepare stories and try to sell them and as editor/managers who plan, prepare and distribute the best publication possible. Three enterprises of editor/managers contact an equal number of freelancers who sign a contract with them and attempt to cooperate within the guidelines determined by the editor/managers in their media enterprise charters. All players evaluate and score articles and publications at the end of the simulation.

In addition to such mass communication simulations as described above, this handbook includes a large variety of communication games, such as: *Eye Contact,* in which one teams avoids eye contact with another team who is trying to influence them; *Prediction and Interpersonal Perception,* in which students in dyads share assumptions they have made about each other and attempt to guess each other's interests; and *Cooperative and Competitive Communication,* in which five groups of students must construct five equal squares without talking, and are free either to help or not to help each other. (TM)

Volume Two contains fifty-eight exercises that "demonstrate processes such as self-perception, inference, stereotyping, trust, nonverbal communication, and intergroup organization."

Exercises include "A Communication Inventory" in which participants fill out a questionnaire on their interpersonal contact and media use; "Experience, Attention, Memory," an exercise that requires players to compile a list of objects scattered on a table that is briefly shown to the players; and a "Message Design and Style" exercise that has players design an advertisement. This volume also includes an essay, "The Structure and Function of Experience-Based Learning Environments." (DCD)

Cost: Volume One $6.95, Volume Two $8.95

Producer: Hayden Book Company, Inc., 50 Essex Street, Rochelle Park, NJ 07662

INTERACT

Bruce Moritz, Research for Educational Programs

Playing Data
Copyright: 1970
Age Level: college
Number of Players: 10-40
Playing Time: 1-1-1/2 hours
Preparation Time: 15 minutes

Description: Interact is a simulation of a task-oriented exercise with timed competition between groups who reconstruct a model shown them by the leader. They may use up to 45 minutes to decide how the model was constructed. Once actual construction of their reproductions are begun, discussion is discontinued.

At the end of the time limit, extensive discussion occurs with a strong attempt to involve each member. The discussion leader is encouraged not to become involved except in situations where clarification is needed. Normally, feelings of anxiety, tension, and competition are expressed by participants based upon their experiences in the exercise. At the conclusion of the discussion, the group is supposed to have a better perspective concerning the danger of being autocratic in any interpersonal activity. (Thomas E. Harris, Rutgers University)

Cost: $1.00 for administrative manual (plus postage)

Producer: Bruce A. Moritz, 1485 Ferguson Way, San Jose, CA 95129

INTERACT II

Brent D. Ruben, Rutgers University

Playing Data
Copyright: 1977
Age Level: college
Number of Players: 20-200
Playing Time: 10 hours-1 school year
Preparation Time: 15 hours (est.)
Packaging: 99-page book

Description: Interact II describes a highly refined format within which communication education can take place. It provides, in the designer's words, "the basic structure for a learning environment in which the theoretical and operational aspects of human communications merge," and also "provides the basic framework for a system which involves the participants in the creation and maintanenace of a society. The Interact Society is much like the larger society of which we're all a part, but it is more flexible, and the communications problems and opportunities facing its members are lessened in risk and consequence. And, participants need not wait weeks, months, or years to see the results of their decisions. The result is a community of communicators, in which participants are involved directly in the challenges and frustrations of human communication. This affords the opportunity and creates the necessity for participants to experience first hand, in an instructional context, the processes of individual, group organization, societal, and mass communication.

"The objective is accomplished by providing minimal structure and operating rules, beyond which participants are encouraged to creatively define their own structures, roles, rules, and goals—their own society."

Interact II has no beginning, middle, or end. It defines no individual roles, prescribes general rather than specific tasks, and has no content. What this simulation does provide are four institutions which share an industry. The first is the "communication organization." Various communication organizations operate within the reality of the game and "the fundamental task of these organizations is to conceive, plan, produce, and distribute mass communications products to Consumers." The second institution is the "agency." Agencies may be established by any player or group of players "to fulfill any identified or anticipated communication product or service need. The more obvious examples of Agencies include advertising, public relations, research, and management consulting, but no options are precluded." The third institution is "The Executive Council," which "is charged with performing regulatory, legislative, judicial, and service functions necessary for the society. The Executive Council monitors the inter-personal and inter-organization interactions within the community, striving to foster and insure the highest possible standards." The last institution is the consumer. All participants serve as consumers "and regularly evaluate the products of the various Communications Organizations. In much the same fashion as members of society at large, they make decisions to allocate time and money among the available mass communications products, each consumer within the laboratory community awards points to available products based on their quality and appeal." (DCD)

Cost: $5.95

Producer: Avery Publishing Group, Inc., 89 Baldwin Terrace, Wayne, NJ 07470

NOTHING NEVER HAPPENS

Kenneth G. Johnson, John J. Senatore, Mark C. Liebig, and Gene Minor

Playing Data
Copyright: 1974
Age Level: high school, college, adult
Number of Players: 6 or more
Playing Time: 1/2 hour-10 hours
Packaging: 320-page paperback book

Description: Nothing Never Happens is a collection of 33 communication exercises in which participants learn about their own and others' patterns of communication and develop their communication skills. The manual includes numerous feedback forms for use with each exercise, as well as 120 pages of related readings. The exercises focus on several areas: growing acquainted, semantics in action, group interaction and leadership, orientations, and encounters. Some examples of exercises are: a debate between groups in which each group must clearly restate the feelings and attitudes of the other group before presenting its own position; a self-disclosure inventory, after which students discuss how comfortable they are with self-disclosure and what they are and are not willing to reveal; and an exercise involving groups of three in which two make statements about a third and become aware of assumptions they make which are based upon their own imaginations. (TM)

Cost: $10.95

Producer: Glencoe Press, Encino, CA

THE O.K. GAME

Playing Data
Copyright: 1974
Age Level: grade 5 to adult
Number of Players: 2-6
Playing Time: 1-1½ hours
Packaging: professional game box

Description: The O.K. Game introduces players to the concepts of parent, adult, and child ego states as used in transactional analysis, and gives them experience in identifying these states within themselves. As players move around the board by throwing the die, they land on different colored spaces which require them to select a card and respond to the instructions printed on it. The cards either request that one answer a question such as "What basic message did you receive from your father?" or that one respond to a picture or a statement such as "Why did you do that?" If the space on which players landed is marked P, C, or A, they must respond from the ego state specified. Some spaces are connected to others by lines which indicate that the player, after following instructions, must engage in a transaction with another player, who is determined by the throw of the die. Players are encouraged to discuss the ego states they expressed in the course of these transactions. (TM)

Cost: $9.99

Producer: Simco Enterprises, 3012 Samoa Place, Costa Mesa, CA 92626

PREDICTION

William I. Gorden, Kent State University; Richard Goodman, SUNY at Fredonia

Playing Data
Copyright: 1972, 1976
Age Level: grade 9-college
Number of Players: 3-6
Playing Time: 30-90 minutes

Description: "Prediction is an academic game designed to teach people some relationships between communication and impression formation. In this game a player looks at a picture or statement and then predicts how another player-designate will respond to it. For example, one card reads, "Predict which of these terms the player on your right feels best describes himself or herself (a) efficient (b) determined (c) creative." The player writes down his prediction and reads the card aloud or shows it, as the case may be. The player-designate writes down his response. If the responses agree, both move forward; it not, both move backward. Since three to six people can play at one time, four or five games are usually enough for a beginning section of communication . . .

"Interest is high. Initial interest is focused on winning, but soon shifts to why predictions were made and why beliefs are held. A good feature is that items that do not seem to generate discussion can be discarded and new items can be created.

"In an hour period there is normally time to introduce the game, play it, and draw some conclusions from it.

"I have found *Prediction* to be of value in teaching how meaning is created, how impressions are formed, the limitations of stereotypes, and the beneficial effects of increasing feedback" (Charles G. Waugh, University of Maine at Augusta, in The Speech Teacher, 1973, 22: p. 265).

Cost: $9.00 (includes postage and handling)

Producer: WEGO Games, P. O. Box 212, Dunkirk, NY 14048

PRIORITY

Playing Data
Copyright: 1972
Age Level: adult
Number of Players: 5-15
Playing Time: 1-2 hours
Special Equipment: cassette tape recorder

Packaging: professionally packaged game materials include cassette tape and instructions

Description: This game is designed to help players become aware of some of the forces which affect group decision-making. Players assume the roles of tourists on a sight-seeing boat in the Okefenokee swamp. When the elderly captain of the boat has a heart attack, it runs aground and the players must "abandon ship." Twelve items including an air horn, a first aid kit, matches, and a rubber raft are salvaged. Each player must rank the items according to their importance to the group's survival. The entire group must then agree on a ranking. (DCD)

Cost: $15.95

Producer: Creative Learning Systems, Inc., 936 C Street, San Diego, CA 92101

THE ROAD GAME

Barbara Ellis Long and Thomas E. Linehan

Playing Data
Copyright: 1974
Age Level: grade 4-adult
Playing Time: 2 hours in 2-3 sessions
Number of Players: 16-32 in 4 groups
Special Equipment: water-based poster paint and four brushes
Packaging: professionally produced 44-page booklet

Description: The Road Game is a game of competition and cooperation, and of conflict and conflict resolution, with application to the behavior of communities and nations. The game is divided into three parts—road building, judicial review, and debriefing. During the road-building period, four groups of students each with a leader/negotiator attempt to draw as many roads as possible from their territory to the perimeter of the map, which consists of four differently colored sheets of construction paper each representing one group's territory. After the students have chosen teams and leaders, leaders begin by negotiating with each other about building roads, for they must obtain another group's permission to enter its territory and to cross its roads. They also must get unanimous permission from their own team in order to build a road. When the 20 minutes for road building are up, a hearing is held during which each group voices complaints about the legitimacy of one or more specific roads build by other teams, and the rest of the class votes on whether or not to count the road. Note that the students are advised to build as many roads as possible; they are not told to compete with other teams. How much they cooperate or compete is decided by them. During the debriefing session they discuss their behavior as well as feelings and attitudes which developed during the game and the parallels between the game and the actions of communities and nations. Alternative ways of playing *The Road Game* are described in the text—ways involving unequal distribution of power, specific directions to compete, and the establishment of four cultural/group identities before the start of the game. (TM)

Cost: $1.50

Producer: Global Perspectives in Education, Inc., 218 East 18th Street, New York, NY 10003

ROLL-A-ROLE

Playing Data
Copyright: 1976
Age Level: age 8 and up
Number of Players: 4-16
Playing Time: no set playing time
Packaging: professionally packaged game includes character cubes, topic cards, "Where it Happens" chart, hourglass timer, and instructions.

Description: This is a very simple, improvisational, role-playing game. It is played with two large plastic cubes (colored red and blue). Each side of each cube contains a number and a role (such as "4, Little girl;" "5, Red's father;" "6, person in wheelchair"). Participants roll one cube to determine the roles they will play. The two exposed numbers are added to determine one of eleven settings (such as "stranded in an elevator," "in the park," "at a laundromat,") for the encounter between Red and Blue. A "Talk Topics" card provides the topic for conversation between the two. (For example, Red has invented a gadget which she is trying to sell to Blue, who doesn't go in for "new fangled" ideas.) The players must act out a scene for three minutes. (DCD)

Comment: This is a terrific device for introducing role play. The roles, settings, and premises are inoffensive and easy to have fun with. (DCD)

Cost: $9.95

Producer: The Ungame Company, 1440 S. State College Blvd., Building 2-D, Anaheim, CA 92806

SOCIAL SECURITY

Playing Data
Copyright: 1976
Age Level: age 6-adult
Number of Players: 3-6
Playing Time: 1 hour
Preparation Time: 15 minutes
Packaging: professionally packaged board game

Description: The players of this board game meander pawns over a playing track (pictured as a circuit of interconnected stone pathways) according to the roll of a die. The little stones come in different colors that correspond to seven "cutely" named features on the playing surface—Pleasure Place, Feelings Fruitstand, Parking Lot, Values Market, Dynamite Solutions Juice Bar, Changing Stage, Sharing Table, and Be A Winner. A player who lands on "Be A Winner" must say something positive to each of the other players. At "Pleasure Place" they must twirl a spinner and answer a question or follow a direction (such as, "Share a funny thing you saw lately") printed on the spinner dial. For each of the other six categories, players must respond to a randomly drawn card which may require them, for example, to "Do a commercial for your favorite ice cream flavor" or "Show how you might sit when you are sad," or practice ingenuousness in some other way. (DCD)

Comment: This is a fine, warm game for friends and families, but potential users may find themselves uncomfortable with the degree of intimacy a successful playing requires the participants to share. (DCD)

Cost: $8.95; additional Christian packet $2.00

Producer: The Ungame Company, 1440 South State College Blvd., Building 2-D, Anaheim, CA 92806

SOURCEBOOK OF EXPERIENTIAL EXERCISES: INTERPERSONAL SKILLS

Samual C. Certo, Indiana State University

Playing Data
Copyright: 1976
Age Level: adult
Number of Players: approximately 6-24, varying with each exercise
Playing Time: 1-3 hours per exercise
Packaging: book

Description: This source book contains 16 task-oriented exercises designed to improve the interpersonal skills of managers, teachers, and administrators. Exercises focus on such issues as goal-setting, team-building, developing helping skills, revealing hidden assumptions, using participative management theory, and firing employees. In one exercise players as teachers decide how to cope with several dozen children who are delayed after school while a broken bus is being repaired. Players, according to their roles, express different attitudes about the situation at hand. In another exercise groups build castles while following the instructions of their assigned manager and the dictates of participative management theory. The last exercise in the book involves players in dyads interviewing each other about the self-awareness they have gained throughout the preceding exercises, and then expressing their partners' insights to the whole group. (TM)

Cost: $3.95 (plus $.50 postage/handling)

Producer: Clearinghouse for Experiential Exercises, Bureau of Business Research, School of Business, Indiana State University, Terra Haute, IN 47809

TAG: THE TRANSACTIONAL AWARENESS GAME

Thomas C. Oden, Drew University

Playing Data
Copyright: 1976
Age Level: high school-adult
Number of Players: 1-6
Playing Time: 1 hour or more
Preparation Time: 1/2 hour or more
Packaging: professional game box including theoretical book, instruction manual, poster playing board, and cards

Description: According to the back cover of the TAG book, "More than an explanation of an exciting new game based on Transactional Awareness, TAG is a revelation of the constant interplay of human interaction at all levels and moments. Accepting acknowledged insights into the nature of interpersonal behavior, TAG identifies and summarizes the 16 basic interactional styles, from the authoritarian 'Director' to the acquiescent 'Follower' to the headstrong 'Individualist.' By understanding how every person's style interacts with the styles of others, the reader can better understand the roots of empathy or antagonism experienced in relationships. Through TAG, the reader discovers more complementary transactions and avoids stifling ploys and fruitless struggles."

Four games are described in the TAG kit—beginning, advanced, solitaire, and analogy TAG. The playing board for the games consists of 16 descriptions, divided into two complementary groups of life positions or roles that people take in interactions with others. "Guide" and "Follower," for example, are complementary positions, as are "Judge" and "Dissenter." Each position is clearly elaborated in the text and described on the game board according to its positive and negative manifestations, as well as the needs and mottos associated with it. Seven sets of 16 cards, keyed to the 16 positions used in the game include impression, preference, request, admiration, vice, invitation, and contract cards. An example preference card, linked to the "Individualist" role reads, "Often what you seem to be wanting from me is that I recognize and applaud your uniqueness, strength and independence." An example vice card, linked to the "Director" role reads, "One quality which I sometimes dislike in you is your strong need to be in control of every situation."

In the beginning game, players in turn choose cards and either discard or read them (with alterations and explanations), giving them to the player to whom they are appropriate. When a player receives two cards for the same position, he places an object on that position on the game board. When two players have placed objects on complementary positions, they form a TAG, which is essentially a tendency toward complementary interactions. The game ends when everyone has made at least one TAG. Players are encouraged to give feedback and explain their statements throughout the game; after the game, players study the roles represented by the cards they have received to better understand how they come across to others.

In advanced *TAG*, players move their own and each other's tokens around the board as they identify impressions, wants, channels of communication, virtues, and vices in themselves and each other. In analogy *TAG*, players use news stories, movies and actual work situations as the basis for simulating interactions on the playing board. (TM)

Comment: TAG is not merely a game, but an entire construct for a field of psychology related to transactional analysis. It is not superficial or overly simplistic like many social psychology games, for it has the capacity to be a powerful instrument of self-awareness and awareness of others. The book and game manual are comprehensive and clearly written, and the multicolored poster playing board is a work of art as well as a detailed, graphic summary of transactional awareness theory. (TM)

Cost: $15.00

Producer: Harper and Row, Publishers, Inc., Keystone Industrial Park, Scranton, PA 18512

TRUST

Jacob Boulogne

Playing Data
Copyright: 1976
Age Level: junior and senior high school
Number of Players: 20
Playing Time: 1 hour
Packaging: professionally designed box, tokens, payoff charts, and instructions

Description: Trust is a simulation designed, according to the author, "to illuminate a simple but profound truth of morality: the crucial role of trust in human relationships." The group director is instructed by the author not to tell the participants the name or the purpose of the game.

The game is played with tokens marked "H" on one side and "T" on the other. Each player gets 15 tokens and is assigned to one of ten pairs. Each pair also gets a "bank" of 20 tokens. Players simultaneously expose either an H or a T and the winner is determined by the combination that results. Two Ts mean that both win, while two Hs mean that both lose. An H always wins over a T. Winners take tokens from the bank, and losers must surrender tokens. Players may change partners at any time. The game continues until the players "get the point" that the most successful strategy is for both members of a pair to always lay down T so that they both win.

During debriefing, the leader asks questions such as, "What would happen if you consistently choose H?" and "How would your opponent react?" or "If you are moral (i.e., choose T) but your opponent is not, what course of action should you take?" (DCD)

Cost: $8.75

Producer: Jacob Boulogne, 13965 64 Avenue, Surrey, B.C., Canada V3W 1Y7

THE UNGAME

Rhea Zakich

Playing Data
Copyright: 1972
Age Level: grade 3-adult
Number of Players: 2 to 6 (30 with Class Group Pak)
Playing Time: set by players
Preparation Time: none

Description: The components of *The Ungame* are a board, playing pieces, a die, and a deck of cards. There is a deck for use with adolescents and adults and one for use with children. As players move around the board, they pick cards from the "light" or "heavy" piles and answer the questions on them as honestly as possible. "The questions are worded so that a person only has to answer as deeply as he wants to. Players can be as superficial as they want to, so there's no feeling of vulnerability," the author says.

The Ungame began as a family game and has since been used as well in classrooms, teacher training classes, parent education workshops, drug education, a marriage and the family course, with children on probation and their parents, and in church groups. (AH)

Cost: $8.95; $2.00 each for additional game card sets: student, married couples, families, Christian, Spanish language

Producer: The Ungame Company, 1440 South State College Blvd., Building 2-D, Anaheim, CA 92806

YIN YANG

Jim Deacove

Playing Date
Copyright: 1976
Age Level: junior high-adult
Number of Players: approximately 2-10
Playing Time: 1-2 hours
Packaging: cloth bag with tokens, playing map, charts, and guidebook

Description: Yin Yang is a cooperative game which simulates the ways people plan their lives and influence other people's lives. Players begin with ten randomly chosen tokens—red tokens indicate positive energy, blue indicate negative energy, and yellow indicate balanced energy. The object is to get as many yellow tokens on the world map as possible. Players each have a life map upon which they can secretly plot the moves they will make on the world map, a chart divided into squares representing divisions of astrological signs. As players choose the squares upon which to place one or more of their markers each turn, they plan configurations which will enable them to marry and have children, and thus be able to place more yellow tokens on the world map. However, as they play the game, they discover, for example, that too many children cause problems in play. Throughout the game, they also learn how to function within the limits of such laws as the law of marriage, law of harmony, law of conflict, law of death, and law of dynamic change. For example, if four red or four blue markers are together on the map, all markers touching them must be removed. The group then must plan together how to institute the law of harmony by surrounding the conflicting markers with either markers of a different color. The game ends when all the spaces on the world map are filled with markers; players then score themselves as a group by counting the number of yellow markers they have been able to place on the world map. (TM)

Cost: about $10. Send for current price list.

Producer: Family Pastimes, R.R. 4, Perth, Ontario, Canada K7H 3C6

See also
Becoming Aware BUSINESS: PERSONNEL DEVELOPMENT
Dialogues on What Could Be FUTURES
Futuribles FUTURES
Nuclear Site Negotiation COMMUNITY ISSUES
Praise and Criticism RELIGION
Settle or Strike ECONOMICS
SELF-DEVELOPMENT
Essay on Communication games (by Brent D. Ruben)

COMMUNITY ISSUES

THE BLACK COMMUNITY GAME

Playing Data
Copyright: 1976
Age Level: age 8-adult
Number of Players: 2-7
Playing Time: 90 minutes
Preparation Time: 30 minutes (est.)
Packaging: Professionally produced board game

Description: This game was designed to simulate, in the words of the producer, "the lifestyle of black people," and to demonstrate that cooperation is an effective strategy for improving the general welfare in the community portrayed in the game.

The game is played on a circular track on a playing board. Most of the squares represent locations, such as the "Job Center," the "Country Club," and the "Baptist Church." There is a square marked "Payday" where players collect an income, a square marked "Pay Rent," and two squares marked "Situation." To begin, players select an income card at random. Incomes vary from $40 to $150 per round. Participants next select a place to live (inner city, outer city, or suburbs) and play begins. By following directions in the squares they land on, players not only accumulate income, but can also acquire black, white, and red poker chips symbolic of power and influence in the black and white communities and political power. A player who lands on a "Situation" square, must respond to that situation. There are three basic types of situations. The first involves a personal dilemma, such as being robbed, which reduces the player's total of money or chips. The second involves a windfall such as an income tax refund, or allows that player to do a favor for another player, such as getting the other player a better-paying job. A third type of situation requires cooperation among several players, such as pooling money or chips to stop a community-wide transit strike that would hurt everyone.

The game ends after two polls have been taken. These occur either when a "poll card" is drawn from the situation deck or when a majority of players vote to take a poll. During a poll, each player's black (power in the black community) and red (political power) chips are counted and recorded. After the second poll both counts are combined, and the player with the highest total wins. (DCD)

Comment: This is an unusually clever and honest board game. Recommended. (DCD)

Cost: $9.50

Producer: Motherland, Inc., 45 Wellington Hill Street, Boston, MA 02126

BUILDING A RAPID TRANSIT SYSTEM

Playing Data
Copyright: 1975
Age Level: grades 9-12
Number of Players: 30-45 (est.)
Playing time: 4 45-60-minute sessions
Preparation time: 1 hour (est.)
Packaging: 27-page photocopied booklet

Description: Participants are divided into six teams, each of which has the identical task of planning and building a public transportation system to service the whole metropolitan area of Urbantown. Each team is told that full funding for the project has been guaranteed and that they have an emergency fund of three million dollars to pay unexpected costs. No currency is used in the game. Each team draws the route of its proposed transit system on a map of metropolitan Urbantown in seven segments. After each segment has been drawn, the teacher evaluates the cost efficiency of that segment for each team. Certain routes involve additional expenses, which must be deducted from that team's emergency fund. These costs include the purchase of rights-of-way and construction of additional roads, tunnels, and parking facilities. The circumstances in which these additional expenses are incurred are thoroughly defined in the players' instructions, and each team must try to minimize these costs. An additional inflation factor is incorporated into the game so that, for instance, a tunnel will cost more to build at the end of the game than at the beginning. The goals of each planning team are to draw a route for the transit system and minimize emergency expenses. The team that has spent the least of its three million dollars at the end of the game wins. (DCD)

Cost: $3.00

Producer: Edu-Game, P.O. Box 1144, Sun Valley, CA 91352

COMMUNITY X

Playing Data
Copyright: 1971
Age Level: college, continuing education, community groups
Number of Players: 12-100 (10-24 best)
Playing Time: 3 hours
Preparation Time: 2-4 hours for first play

Description: After the director introduces this decision-making game, the participants draw cards that specify their roles in Community X according to age, sex, financial status, and place in the community. One-third belong to the Improvement Committee, and the rest are community members. Both groups receive the same prepared materials on current community issues, but the first group has an already-established structure to facilitate decision-making, while the second is unorganized and the information is scattered among individuals. The game consists of two simultaneous gatherings. The committee usually has a more or less straightforward decision-making meeting. The citizens go through a more complicated, less direct route in the process of identifying problems and defining possible solutions. They tend to make decisions as individuals rather than as groups, they tend to leave issues unfinished, and they may come under the sway of a particular individual or small group. The director stays in the room with the citizens to observe and help keep the game moving. The committee joins the citizens after its own meeting ends, when it probably will seek to impose its own ways of proceeding on the larger group. At this point, the director ends the game and initiates an hour-long debriefing discussion to analyze the processes, outcomes, and associated feelings of the decision-making efforts. (AC)

Cost: microfiche copy of "Operational Guide to Community X" $2.00

Producer: International Communications Institute, P.O. Box 8268, Station F, Edmonton, Alberta, Canada T6H 4P1

COMPACTS

Armand Lauffer, University of Michigan

Playing Data
Copyright: 1975
Age Level: advanced high school, college, adult education, community groups
Number of Players: 20-60 (32 ideal)
Prerequisite Skills: understanding of community dynamics
Playing Time: 2½ hours-3 days in periods of at least 2 hours
Preparation Time: 2 hours
Packaging: kit including 45-page manual, Lego blocks, graphs, forms, etc.

Description: Players of *Compacts* are community workers; community influentials; agency administrators; consumer representatives; and funders concerned with developing, promoting, and opposing proposals for community change. Players receive individual role cards describing who they are and their resources, objectives, strategies, and scoring procedures. There are six kinds of resources in *Compacts*—money, expertise, popularity, energy, social standing, and legitimacy.

During each 45-minute round, players, organized in groups based on their interests and resources, (a) choose an issue card which describes one of 15 issues (such as mental health, welfare, slum housing, or employment) and the number and kinds of resources needed for a demonstration project or permanent program related to that issue; (b) write proposals according to a carefully delineated format explained in the manual; (c) form coalitions to promote or oppose proposals; and (d) stack the required number of resources (Lego blocks) necessary to pass their proposal. Some players, particularly funders, are concerned with maintaining existing services and limiting expenditures; others ardently support widespread changes. Agencies must avoid antagonizing funders, whose money is affected by obligations to other agencies as well as by news and chance events which are posted periodically. A Community Directory and a chart describing players' dispositions to act on the various issues facilitate the interactions. At the end of each round, proposals and issue boards are examined, and resources expended and gained are tallied. Each Player is allowed to reward or punish five other players by taking away or giving resource units of popularity and social standing. The manual provides several pages of debriefing questions relating to proposal writing, resources, work models, coalitions, and strategies. The manual also includes suggestions for 12 variations of the game. (TM)

Cost: kit $125.00

Producer: Gamed Simulations, Inc., Suite 4H, 10 West 66th Street, New York, NY 10023

CRISIS IN MIDDLETOWN

David J. Boin and Robert Sillman

Playing Data
Copyright: 1972
Age Level: grades 7-12
Number of Players: class size
Playing Time: 3 to 4 class periods

Description: The class represents different elements of a small coast town where workers in the only industry, an electronics firm, are on strike. Students divide into different community groups: striking workers, nonstriking workers, management, and citizens. Character sketches are provided. The teacher presents a list of the proposals for negotiation to the community representatives. Each community group must decide whether the proposal is in the best interest of the group. By the outcome of their votes, these community groups will earn predetermined influence points. The group with the greatest increase points by the end wins the game. The simulation is intended to develop an awareness of the complexity of the labor-management relationship in a technological society and of the relationships between industry, labor, and the community, and to provide some insight into the difficulties of compromise when special interests are at stake. (REH)

Comment: This is a standard Edu-Game simulation with a few pages of teacher's manual and locally reproducible student material. (REH)

Cost: $3.00

Producer: Edu-Game, P.O. Box 1144, Sun Valley, CA 91352

DECISIONMAKERS

Julia A. Jitkoff and Edward Doty, American Friends Service Committee

Playing Data
Copyright: 1972
Age Level: grade 9-adult
Number of Players: 20-50 in 2 groups, each with 5 subgroups
Playing Time: 1½-2 hours, 1 or 2 periods
Packaging: complete kit including 13-page moderator's script, 28-page manual, 4-page fact sheets for each player, posters, i.d. buttons, forms, cards

Description: In *Decisionmakers*, players as change agents experience influencing other players who represent community decision makers. Any social change issue can be used in the game, which gives as an example the introduction of the course "Challenges of Peace Building" into a high school curriculum. The manual gives guidelines for adapting *Decisionmakers* to other communities and other issues.

The decision makers are students, parents, teachers, administrators, and community leaders, each with predetermined attitudes and goals. They control the community's resources–seals of approval, money, and letters of public opinion. The manual specifies the number and kinds of resources necessary for a course lasting six weeks, six months, or a year.

At the beginning of the game, decision makers and change agents meet separately to plan their strategies. Each group receives a list of questions decision makers may wish to ask (Parents: How will the unit help my child prepare for college?). At a joint meeting, decision makers then question change agents about their proposal. They next decide how many tokens to give the change agents, who meanwhile are planning their next strategy. A second joint meeting takes place before the decision makers make their final decision. To convince the decision makers of the viability of their proposal, change agents must know their project thoroughly, understand the concerns and the influence of the decision makers, and know how to use their own personalities and resources to accomplish their aims. (TM)

Cost: $5.50

Producer: American Friends Service Committee, N.Y. Metropolitan Regional Office, 15 Rutherford Place, New York, NY 10003

DIGNITY

Kenneth Christiansen

Playing Data
Copyright: 1969
Age Level: grade 7-college
Prerequisite Skills: reading, grade 8; math, grade 4
Number of Players: 2-5, in 3-4 teams (teams optional)
Playing Time: 45-90 minutes
Preparation Time: 10-30 minutes

Description: Dignity is a board game that intends to simulate life in the black ghetto. Players change roles frequently; in the course of the game one player may take the roles of a welfare recipient, public housing tenant, street gang member, community worker, suburbanite, building inspector, and high school dropout. All teams begin in the Start square of the track and try to reach a square marked Dignity. Chance and strategic competition determine who will reach Dignity. Movement of playing pieces is determined by the draw of situation cards. These cards specify a player's role and situation and direct the player to move the team's playing piece forward or away from Dignity. The first team to reach Dignity wins. (DCD)

Comment: Dignity is designed to "help middle class persons enter into the experience of others, especially that of ghetto blacks." Results of outside evaluation and of field tests are available. A community training organization with a goal of improved intergroup relations said *Dignity* was the "best available tool" to open up the issues and develop awareness, and purchased 600 copies. (DZ)

Cost: $6.95

Producer: Friendship Press, P.O. Box 37844, Cincinnati, OH 45237

GHETTO

Dove Toll

Playing Data
Copyrightt: 1970
Age Level: grade 7-college
Prerequisite Skills: reading, grade 7.5-math, grade 3.5
Number of Players: 7-10
Playing Time: 2-4 hours in 20-minute periods
Preparation Time: 15 minutes

Description: " 'Bout the only things left to do are hustling and welfare. When you got five kids, not much education, and no job, you don't have much time for anything else. If I get a job I lose my welfare reward points, plus have to pay for a baby sitter. The best I can do is invest one chip in welfare, two in hustling, and one in relaxation. At least at the end of the year, I will get 65 points for welfare, 30 for hustling and one for relaxation. Even if I get caught for hustling, I will only be in jail for a year and having a police record won't hurt. I could not get a job so it does not matter."

These statements are characteristic of the person who plays Liza in the Ghetto Game. Liza starts the game as an eight-grade drop-out, 35 years old, with five children (ages 12, 10, 5, 4 and 2), separated, and on welfare. She has four hour-chips.

Ghetto is a simulation game that starts where the political process ends and illustrates the conditions that lead to policies and decisions. The purpose of the game is to sensitize the players to the emotional, physical, political, and social context of poor people. The players are confronted face to face with the pressures and economic forces that drive ghetto dwellers into crime, welfare, and sometimes community action. The designers have identified activities in the ghetto in three major areas: (1) community, economic and family responsibilities; (2) slum education; and (3) the rewards and risks of illegal activities. Each player has to deal with these areas in the yearly life cycle of a ghetto dweller.

The game can be played successfully with high school students. A heterogeneous age and sex group from diverse social and economic backgrounds enriches the game. Much would be lost if the game were played with players having similar life experiences. From a dissimilar group different levels of awareness to inner city lives are exchanged and the players become increasingly sensitive to the needs and chances of ghetto livelihood.

There are ten descriptive profiles of ghetto dwellers in the game. Each of the profiles can be played by one or two players. The maximum number of participants for the game is 20 (two players for each ten profiles); the minimum for a fairly complete game is 7 (one player for seven profiles). The object of the game is to improve the ghetto dweller's life in terms of present and future rewards, starting from the profile sheet's description. The present rewards are measured by the number of points one acquires for activities in which the dweller participates. Future rewards are determined by advancing one's educational and employment status.

Each player starts with a set number of hour-chips. An hour-chip is equivalent to one hour a day for making a living and improving one's living standard. The minimum is twelve. A woman (Liza) with children and an eighth-grade education and family responsibilities begins the game with four hour-chips. Maria, a single woman with no dependents and an eleventh-grade education starts with twelve.

At the start of each round or year, the ghetto dwellers invest their chips in an activity or activities with expectations of receiving a number of reward points or/and self-improvement. A player may invest in any of the following activities: school (trade, high school and college); work (unskilled, semiskilled, Skilled I, Skilled II, semiprofessional and professional); hustling (small and big time); relaxation and recreation; welfare; and neighborhood conditions (housing, education, recreation, and safety). The activities may or may not have specific requirements before players are able to participate. For example, there are no requirements for going to high school full-time with exception of investing eight hour-chips. They also receive self-improvement and a year of high school; they must complete high school in order to get a skilled job. Another example to illustrate this aspect of the game is investing eight hour-chips in a semiskilled job after completing the requirement for one year of trade school.

Reward points for activities are divided into three major categories. Work, welfare, and hustling is the first category and offers the highest rewards or points. This group provides the dweller with direct financial rewards, plus self-satisfaction. The second category is school, relaxation, and recreation. These are self-satisfying, but have a small reward value in points. Neighborhood conditions is the third category; there are no direct rewards points for this group. Self-satisfaction for increasing the level of one of the neighborhood conditions may result in the dweller receiving some indirect rewards such as (at level C in education) parent involvement and increases in parent/teacher relations resulting in students investing one hour-chip less for school, (at level G in recreation) people are afraid to go out at night for recreation activities, or mothers losing one hour-chip. The levels for neighborhood conditions range from highest-A, to lowest-H. Each level describes some condition relating to the community as illustrated in the above two examples.

Just when it looks like the player going to graduate from a second-rate high school, he is involved in a severe accident on the way to school. He has to take all his hour-chips and put them in the hospital. All the expected rewards that year are lost and the player has to start back to school next year in the same grade. These actions are the result of a school chance card the player drew. Much of the ghetto life is based on chance rather than control over one's own life. Ghetto introduces this element through the use of chance cards, dice, and a dial spinner. The results of these chance elements can be negative or positive adjustments to the expected reward points.

One year the dweller decided to go to school full time and hustle small time to make it through his final year. The neighborhood safety condition is at level B, which means there is a 75 percent probability that he will get caught hustling. Or it could be that the dweller got caught for hustling the year before and lost all reward points because he had to go to jail. The next year, because of his record, he has to spin the spinner to determine if he can get a job. Still another example of chance, he may not have any negative adjustments thus far in the year, but near the end of the year he is a victim of some criminal activity. Victims loose 50 points. And, finally, if the player is a female in the ghetto, she may discover that she will have a baby next year. All females must role a die and if they throw the number of the die of "one"—guess what?

Ghetto is not a game that highlights or deals with major political and governmental concepts or ideals. One aspect of the game that does interject concepts of politics is the neighborhood conditions. In order to change the level of neighborhood conditions, there must be an investment of five hour-chips per level. Very few dwellers have five free chips to make the investment.

Not everyone receives positive benefits from improving neighborhood conditions, particularly the dwellers with extra chips. Thus, to improve the neighborhood conditions there almost has to be a community effort. A community effort must be initiated by a dweller presenting the need for improving the condition to get others to invest in it. Others may resist the community effort.

One player may have been a victim of some criminal activity the last year. This year he decides that safety conditions must be increased. A number of the other dwellers are hustling this year. The number of victims increases with the increase of three hour-chips in hustling. But when an increase in the safety level increases the odds for hustlers getting caught—the bargaining, negotiating, and alliance formation becomes an active and emotional political process. This aspect of the game produces ad hoc pressure groups or citizen action. Chance is brought into this aspect, adding more frustrations to the lack of control of the dwellers. Once a group has succeeded in raising the five chips for changing the safety level, however, there is no assurance that the level will go up. The spinner must be spun to ensure that the level will go up.

The process of responding from one's own point of view, values, and experiences in a ghetto environment is the major impact on individual players. The game emphasizes behavior through feelings, emotions, and pressures of living in the ghetto. It enables a nonghetto person to come to terms with some of the struggles of ghetto living. The everyday economic and political pressures that ghetto residents live under are brought into intellectual reach. The insights of why certain conditions are found in a ghetto are valuable learning experiences for players.

When it is played at least ten rounds, *Ghetto* achieves its objective (to sensitize the players to the emotional, physical, political, and social living conditions of being poor in an inner-city ghetto). The experiences are shallow when less than ten rounds are played. In order to play ten rounds without preparation takes over four hours. Players must allow this amount of time to get the most out of it.

Editor's Comment: Preparation before playing the game is a must! Someone who has played the game or has become very familiar with the rules should lead the preparation. A well-prepared group will get the most out of the game. The most effective way to prepare is to allow each player to read materials on the rules and description of the game in the coordinator's manual.

Following the game, there should be a discussion session. This is a significant part of the game and can be the most productive aspect. On the other hand, it could be a waste of time. If time has not been scheduled or the game runs over the alloted time, the session will not be productive. It is very easy not to plan to utilize time for discussion.

A shortcoming of the game is the neighborhood conditions aspect. The designers recommend that neighborhood conditions are not added to the game until the third year (round). The suggestion is a very good one because it usually takes about three rounds to come to terms with the game. Adding neighborhood conditions late in the game, however, may hinder its effectiveness. Many dwellers by the third round are locked into self-interest activities and do not make use of neighborhood conditions even when, in the long run, it is more beneficial. Even though self-interest is accurately represented, it may be a shortcoming in that it does not serve as a useful tool to illustrate the political process.

In playing the game, one may identify other shortcomings, however, these are not major and can be dealt with. For example, the victims of criminal acts are drawn from the names of all the players that are being used in the game. This seems unfair, because it is not realistic for residents to be continuously victimized, but instructors are free to add extra victim cards.

The game has a great deal of potential for expansion when a group has played it before or has become very familiar with it. An experienced group brings the game to life. The manual includes a number of suggestions for making additions to the game. *Ghetto* is an excellent base on which to develop creative and original components and provides more real-life learning experiences. The game comes with a list of teaching and reading materials that can be helpful before and after playing the game.

Ghetto is a simulation of inner-city chance. It is not a game that provides an opportunity to learn and deal with political concepts and ideals. The game provides experiences in organizing interest groups and offers insights into the powerlessness and lack of political influence of inner-city dwellers. *Ghetto* is only a game, but it succeeds in helping players understand what ghetto residents face. (from the Robert A. Taft Institute of Government Study on Games and Simulations in Government, Politics, and Economics by William Hayden, Jr., University of North Carolina at Greensboro)

This game was designed to give people who do not live in a ghetto some understanding of the problems people have to face when they do live in one. Players must decide how to divide their time to their benefit with four available time uses: work, school, leisure, and hustling. Their individual decisions also can be accompanied by collective actions to improve their neighborhood in various sectors of life, including more protection against crime.

According to L. Warren Nelson (*Simulation and Games*, Vol. I, No. 3, Sept. 1970, pp. 341-345) the game can be profitably played in inner-city schools as well as in the suburbs.

A session played recently with young researchers and students showed that even with only six players there might still be dynamism. During this condensed game, most players took female roles and the conditions for women in the inner city were stressed. (GLC)

Cost: $28.00

Producer: The Bobbs-Merrill Company, Inc., Education Division, 4300 West 62nd Street, Indianapolis, IN 46268

GREENHAM DISTRICT COUNCIL

Gordon Cooper

Playing Data
Copyright: 1974
Age Level: high school
Number of Players: 8-15
Playing Time: 2-3 hours
Packaging: unbound packet with role sheets, briefing sheets, and instructions

Description: Greenham District Council is an English simulation of a town or city council meeting. There are roles for seven councilors and an independent secretary who keeps the minutes of the meeting. Councilors may be assisted by one other person with whom they can discuss role and tactics. The council discusses and votes on five agenda items concerning the extension of shopping facilities, a new road and new housing, a new sports complex, and the disposition of an encampment of gypsies on the north side of the town. Before the simulated meeting, players discuss the agenda, the geography of the town, and read the role sheets of the law makers, who vary from Ted Briggs, Chair of the Council, retired school master, and lifelong resident of Greenham to Susan Chambers, a newcomer to Greenham and the Council and the wife of a foreman in the local oil refinery. As the Council debates and votes on the agenda items, the players have the opportunity to experience the various conflicts and compromises inherent in the political process. The game ends when the agenda has been completed. DCD)

Cost: £1.50 (plus 15% postage)

Producer: Community Service Volunteers, 237 Pentonville Road, London N1 9NG, England

GREENHAM GYPSY SITE

Gordon Cooper

Playing Data
Copyright: 1975
Age Level: high school
Number of Players: 7
Playing Time: 2 hours
Packaging: unbound packet with role sheets, briefing sheets, and instructions

Description: Greenham Gypsy Site is an English simulation of a town council meeting called to decide the main features of a gypsy site in the imaginary town of Greenham. There are roles for five councilors and two nonvoting advisors. Before the meeting, players discuss the agenda and the culture of gypsies and read the role sheets for the law makers. The Council debates and votes on the various agenda items concerning the location of the site, the services that will be provided, and the hiring of a Gypsy Warden. The game ends when the agenda has been completed. (DCD)

Cost: £1.50 (plus 15% postage)

Producer: Community Service Volunteers, 237 Pentonville Road, London N1 9NG, England

HALF WAY HOUSE

Mary Simpson Furlong and Louise Weinberg Jacobsen

Playing Data
Copyright: 1978
Age Level: high school
Number of Players: 35
Playing Time: 3 hours
Preparation Time: 1 hour
Packaging: professionally packaged game includes role cards, observer forms, and instructions.

Description: According to the producer, "This simulation is designed to pose the following questions: 1) Should citizens of a community support changes which will contribute to the reduction of crime, even if those changes may have an adverse effect on their lives?; 2) should facilities for helping ex-offenders be located in residential neighborhoods so that people who have been released from prison can participate normally in a community?; 3) does the 'average citizen' have any responsibility to help an ex-offender?"

The game simulates a city council meeting in which citizens request a permit for a zoning variance to allow locating a rehabilitation center for ex-convicts in the middle-class neighborhood of "Pacific View." Seven students assume the roles of city councilors, 22 represent interested citizens (half supporting and half opposing the variance), and six portray journalists. The role cards for the residents supply a name and a statement that epitomizes that person's attitude toward the halfway house. (These are statements such as "A rule is a rule. We moved to Pacific View so our children would be brought up in a decent neighborhood. Exceptions should not be made to the zoning regulations." Or, "The people of Halfway House are providing a real community service as a rehabilitation center. Laws should take into consideration human concerns.")

The game is played in three rounds. During the first round opponents and proponents of the house hold strategy meetings. The City Council session happens in the second round. During the third round, the City Councilors caucus to make their decision and the journalists critique the other players on the authenticity with which they played their roles. (DCD)

Cost: $8.95

Producer: Zenger Publications, Inc.; distributed by Social Studies School Service, 10,000 Culver Blvd. Culver City, CA 90230

THE INNER CITY HOUSING GAME

Samuel A. Livingston, Clarice Stasz Stoll; Center for Social Organization of Schools, Johns Hopkins University

Playing Data
Copyright: 1973
Age Level: grades 7-12
Number of Players: 12-20
Playing Time: 2 hours
Preparation Time: 1 hour
Packaging: included in paperback book

Description: This game is intended to make players familiar with the housing problems of the inner city. Participants assume the roles of banker, landlords, or inner-city tenants. Play of the game simulates the conflicts among these parties in the real world. The most significant general conditions and relations that affect play are the tenant's ability to pay rent, the high cost of building maintenance, and the slow and ineffective enforcement of building codes. (MJR)

User Report: I played the game with three groups of ninth graders, and found the game fairly effective in demonstrating that tenants and landlords have serious problems. Both competitive and cooperative interaction are required of players; the students enjoyed this. Overall, however, the game was unsatisfactory. The rules are too complicated, especially since the basic movements, once mastered, are very simple. Students must deal with a mass of cards and money and this is often confusing. Only a few students were able to delve deeply into the city inspection and housing court aspects of the game. Scoring is also complicated. The game works best with small groups (James Havelka, Papillion High School, Papillion, Nebraska).

Comment: This is a useful simulation game for demonstrating the difficulty of being a tenant or a landlord in the inner city. It simulates the conflict of a situation in which no one wins except at someone else's expense. *Inner City Housing* can further the development of concepts and inquiry skills and can serve as a means for examining value conflict. (MJR)

Note: For more on this game, see the essay on social studies simulations by Michael J. Rockler.

Cost: in *Simulation Games: An Introduction for the Social Studies Teacher* $4.95

Producer: The Free Press, 866 Third Avenue, New York, NY 10022

LOW INCOME HOUSING PROJECT

David J. Boin and Robert Sillman

Playing Data
Copyright: 1972
Age Level: grades 7-12
Number of Players: 1-40
Playing Time: 3 class periods

Description: Students are divided into committees representing various property owners, school, or church board members in the Town of Middleville, which has been given a federal grant for a low-income housing project to be occupied by minority group families. They have to decide where the housing project is to be built. After small committee meetings, a town meeting is held and a vote taken. (REH)

Comment: This is a planning exercise in a standard Edu-Game format which incorporates a few pages of teacher's manual and locally reproducible student material. (REH)

Cost: $3.00

Producer: Edu-Game, P.O. Box 1144, Sun Valley, CA 91352

MAKING A CHANGE

Paul A. Twelker and Kent T. Layden, Simulation Systems

Playing Data
Copyright: 1974
Age Level: high school-college
Number of Players: 6 or more in 2 or more teams
Playing Time: 4 to 6 hours in periods of 2 hours

Description: Participants are divided into planning teams of about five members each. Teams are numbered and manuals (to be reproduced from the coordinator's guide) are distributed. Students are briefed, then work through a 16-step problem, identification, and solving process focused on some aspect of technological development in their community. They must identify and clarify the problem, identify encouraging and resisting forces, identify appropriate courses of action to begin change and determine resources for carrying out the action, develop a comprehensive plan of action, and doing it! (author)

Note: For more on this game, see the essay on frame games by Sivasailam Thiagarajan and Harold Stolovitch.

Cost: $4.00

Producer: Simulation Systems, Box 46, Black Butte Ranch, OR 97759

MICROVILLE

John C. Snider, Colorado State University, and Wayne L. Schroeder; Florida State University

Playing Data
Copyright: 1970
Age Level: adult education
Number of Players: 5 or more in groups of 5-8
Playing Time: 3 hours (est.)
Preparation Time: 8 hours (est.)
Packaging: 28-page staple-bound photocopied booklet and suitcase-sized Microville model

Description: "The purpose of this endeavor," according to the authors, "is to develop and utilize a training device for teaching community-wide program development in adult and continuing education . . . specifically . . . whose purpose is to acquaint adult education leaders with the basic decision-making processes involved in program development.

"Of theoretical as well as practical importance is the testing and validation of the simulation-gaming device which is designed as a model for securing substantial changes in management behavior of adult education leaders and in their interpersonal perceptions of their professional roles as change agents, facilitators, coordinators, and energizers."

Groups of five to eight participants represent the community council of the fictitious community of Microville. Each council must accomplish five things: (1) develop a philosophy and a set of policies with which to organize its subsequent actions; (2) identify the needs and wants of the community; (3) write a set of realistic and attainable objectives; (4) implement those objectives; and (5) evaluate that implementation. The model of Microville is constructed of three 24" by 30" plywood shells hinged so they close like a suitcase. A model town is painted inside the shells with portions representing lower, lower- middle, middle, and upper-class neighborhoods and 27 institutions and agencies that are interested or involved in adult education programs. Each residential area and each institution has a sleeve to store index cards, which supply information about who the people represented by each card are, what they do, what they can do, and what they want. The case also contains a set of position cards, which assign individual players occupational positions to hold as the game progresses. Dice are used for determining the number of data cards each player may select. The game kit includes a community newspaper which contains general community information. (DCD)

Cost: $650.00

Producer: John C. Snider, Center for Continuing Education, Colorado State University, Fort Collins, CO 80523

MUCH ADO ABOUT MARBLES

Frederick Goodman (with Armand Lauffer), University of Michigan

Playing Data
Copyright: 1973
Age Level: grade 9-graduate school, community groups. Some versions can be played by grade and junior high school children
Number of Players: 60
Playing Time: 2-8 or 12 hours in periods of 2-5 hours
Preparation Time: 1-1/2-2 hours

Description: This social simulation contains roles for a multitude of community sectors including business and industry, social control agencies (police, courts), socialization agencies (schools, welfare), service and public utilities, government, and natural forces. The point of the game is to assemble a social system, maintain it, and, for the individual player, to keep at least enough marbles to stay in it. All players start with an equal number of marbles. An effective social system, in terms of this game, would be one that ensured an effective system of distributing and collecting marbles so that as many players as possible have marbles or so that all the players who should have marbles do have marbles. Marbles are the ultimate symbol of power, simultaneously representing status, wealth, and influence. Success depends upon each player's skill at shooting marbles, salesmanship, negotiation, and chance (dice rolls repeatedly occur, and marble shooting represents partial chance, partial choice). Inevitably, in the course of play, some players acquire more marbles than others, and it is this inevitability that encourages players to try out various political, economic, and social systems within the format of the game. (DZ)

Comment: *Much Ado About Marbles* is an open-ended simulation which can be modified with extreme ease by participants of almost any skill level.

The "message" of the game experience is that the social system can work better or less well depending upon how social, political, and economic forces are managed. The game demonstrates how events, built up from participant interactions and activity, conspire to create social situations that overwhelm some participants, while significantly enhancing the fortunes of others. It can be played many times with good effect, each under a different system of government or rule making. (DZ)

Much Ado About Marbles is a variant of Frederick Goodman's *They Shoot Marbles, Don't They*. It differs in that it includes an additional game dealing with the welfare or human service element in the community. Players representing the Welfare Department, vocational rehabilitation counselors, job trainers and placement officers, and others create a side game in which the players without marbles find themselves. Some like the welfare game, with its relatively well-defined rules and limited objectives, remaining there throughout the course of play. Others itch to get back into the action of the general economy. (Armand Lauffer and Thomas Morton, Continuing Education in the Human Services, University of Michigan)

Cost: $125.00

Producer: Gamed Simulations, Inc., Suite 4H, 10 West 66th Street, New York, NY 10023

MULBERRY

Dorothy Dodge, Jimmie Powell, Jr., Macalester College

Playing Data
Copyright: 1970
Age Level: grade 9-junior college
Number of Players: 30-6 teams
Playing Time: 5-8 hours in time periods of 1 hour
Preparation Time: 1 or 2 hours
Packaging: 35 participant manuals, set of roles, 7 sector maps, score forms

Description: This manual game deals with an urban renewal project within a 15-block area of the hypothetical city of Greenbriar. Players take on the roles of citizens, city officials, and professional planners who interact with one another through the institutions of the city.

The City Council has five members; a mayor who is elected at-large and four councilmen from different wards of the city. The council provides and administers various services for the community. The Community Development Authority is a five-member agency specially created for carrying out the renewal project. This agency must draft the workable programs (codes and ordinances, comprehensive community plan, neighborhood analysis, financing, housing the displaced, and community participation) and supervise the land clearance, site improvement, and resale of land to private developers.

Four private sector organizations lobby to protect the private interests. The Downtown Real Estate Association (four members) is interested in buying the cleared land and constructing new land uses. The Industrial Development Project (two members) is composed of industries that wish to build central offices or demonstration projects in the renewal area. The King's University Expansion Committee is represented by Vice President for Financial Affairs, who looks at the renewal area as a potential spot for future university expansion. The Mulberry Business Association (four members) has as its goal the promotion of the interests of the small businessmen within the neighborhood.

Six individual citizens represent the residents of Mulberry and several other individuals represent other individual decision makers who affect Mulberry neighborhood. The latter include the municipal judge who will hear all cases involving the acquisition of land, the editor of the city newspaper, a political columnist, a reporter, and the director acting as the Federal Government representative.

Points are scored by sectors and by individual for achieving goals, receiving approval from the press, holding land at the end of play, and increasing personal wealth. Play progresses for ten periods, with each period representing three months of time. Point total can be used to evaluate player success. (The State-of-the-Art in Urban Gaming Models)

Cost: $57.50

Producer: Paul S. Amidon & Associates, Inc., 166 Benson Ave., St. Paul, MN 55116

NEW TOWN

Barry Ross Lawson

Playing Data
Copyright: 1968, 1970, 1971, 1975
Age Level: 3 versions span grade 4 through urban planners and city officials
Number of Players: 2-8; 4-20
Playing Time: 45 minutes-2 hours; 2-10 hours
Preparation Time: 30 minutes; 3 hours est.)
Packaging: all versions are professionally packaged board games

Description: There are three *New Town* Games of increasing sophistication. All are played by two, three, or four players or teams who compete for profits while trying to build or develop a new town. The tension that results from opposing greed to intelligent planning is exploited to simulate the realities of the planning process in direct correlation with the complexity of the game version. The action in *New Town Family Game,* for example, is described by the producer as follows: "First you create your new town–and then you try to save it. You and your family find yourselves making speeches at Town Meetings. While you plan and build the ideal community in which to live and work–money-hungry opponents plot against you." The most sophisticated version, on the other hand, is described as "a gaming model of urban development and environmental planning which entails strategic thinking and decision-making. The roles included are those of an urban developer, public planner and the city council. The dual objective of private developers is the maximization of economic return and the accumulation of environmental points. . . . More important . . . however, are the processes of decision-making inherent in the *Planner's Set Model.*"

All three versions share similar procedures. The first part concerns the construction of the town. A town clerk gives players money with which to bid for land. Various types of buildings–including factories, apartment buildings, and stores–may then be built on any vacant lot on the game board. A dice roll determines which type of building will be constructed, and certain dice rolls are designated as "Happenings." These include disasters, pressure for urban redevelopment, and federal intervention to stop pollution. After the town is established, players may make or approve additional real estate transactions. During a public meeting phase, players may debate and vote on zoning regulations and the construction of public facilities. They may take advantage of all these events to make money or earn "bonus points." In various versions, either all players or the players who represent public officials earn bonus points for clustering department stores, for providing for resident parking, for maintaining pollution-free air and water, and for putting dwelling units–rather than stores or factories–in scenic locations. Players with the most money and the most bonus points at the end of the game are the winners.

The family game is designed for 2 to 8 players from grades four through eight, and for the parlor. The educational kit (revised in 1975) is for 4 to 20 players from junior high through college, and for civic groups. The planner's set, which is described separately in the urban section, is for players from college through graduate school, civic groups, urban planners, and local officials. (DCD)

Note: For more on this game, see the essay on urban gaming simulation by Mary Joyce Hasell.

Cost: (1) family game $12.00; educational kit for up to 10 players $18.00, for up to 20 players $30.00; (2) planner's set $85.00

Producer: (1) Harwell Associates, Box 95, Convent Station, NJ 07961; (2) Harwell Associates, Pinnacle Road, Harvard, MA 01451

NUCLEAR ENERGY GAME

Playing Data
Age Level: grade 9-college, adult
Number of Players: 25-35 in 11 teams
Playing Time: 10 rounds of 1 hour each

Description: The Nuclear Energy Game is a group simulation dealing with a community land use dispute. The conflict involves the proposed location of a nuclear power plant in the beautiful residential village of Oak Harbor.

The model involves the organization of eleven interested parties to the conflict and their participation in the Atomic Energy Commission hearings for the licensing of the plant. Ten proposals are suggested for the resolution of the conflict, and the participant's ability to influence passage or defeat of the proposals are the play of the game.

The game is divided into rounds in which teams organize, identify groups with like interests, secure witnesses, conduct an A.E.C. hearing, influence public opinion, debate the issues, and resolve the dispute.

A round of play requires one hour and represents six months of real time. The game ends when the proposals have been passed or defeated.

The simulation is modeled after an actual case in the northeastern United States. Play is possible with small groups by combining teams. The game is also available as a 5- to 11-player board game. (producer)

Cost: $45.00

Producer: Simulation Learning Institute, P.O. Box 240, Roosevelt Island Station, New York, NY 10044

NUCLEAR SITE NEGOTIATION

Ruth Bass, Bernard M. Bass, Zur Shapira

Playing Data
Copyright: 1976
Age Level: advanced high school, college, adult
Number of Players: 8 or more in groups of 4-6
Playing Time: 2 hours
Packaging: booklet

Description: Participants in this negotiation exercise learn about the issues and groups of people involved in determining the location of a nuclear power plant, and gain experience as negotiators involved in making such a decision. After completing questionnaires and reading about nuclear energy development, players choose roles either as utility representatives, generally in favor of a nuclear plant, or as concerned citizens, generally opposed to the plant. Each group discusses the cost-benefits, safety factors, and social and ecological issues involved in choosing a site and determines and ranks preferences for the site location or opposes the plant construction altogether. After groups have met individually for 30 minutes, players divided again into negotiation teams, each consisting of two utility representatives and two citizens, and attempt to reach an agreement. The game concludes with individual and group analyses of the process of negotiation and the decisions, a re-taking of the questionnaire, and a discussion of the game's application to reality. (TM)

Comment: This exercise was designed primarily as a negotiation experience not necessarily for participants who are dealing with this issue as a concern, but the issue of nuclear power plants was chosen as the vehicle for the exercise precisely because it is highly charged for many people. However, because nuclear power has finally emerged as a recognized, critical social and community issue, we've put it here. (AC)

Cost: $4.50

Producer: Transnational Programs Corporation, 54 Main Street, Scottsville, NY 14546

PLANNING THE CITY OF GREENVILLE

David J. Boin and Robert Sillman

Playing Data
Copyright: 1972
Age Level: grades 7-12
Number of Players: 10-40
Playing Time: 2 class periods

Description: The class is divided into teams of 2 to 4 students each, who are instructed that they are a planning team for developing a new urban suburban community in a new town. They are given a simplified map of the area chosen and a list of types of land use which they must incorporate into the plan. (REH)

Comment: This is a standard Edu-Game format with a short teacher's manual and locally reproducible student material. (REH)

Cost: $3.00

Producer: Edu-Game, P.O. Box 1144, Sun Valley, CA 91352

PLANNING AN INNER-CITY HIGH SCHOOL

David J. Boin and Robert Sillman

Playing Data
Copyright: 1972
Age Level: grades 7-12
Number of Players: 10-40
Playing Time: 2 class periods

Description: Students are divided into teams of four members, each of whom is a teacher, a parent, a school administrator, or a student. They are to plan the organization of a new inner-city school which has a large minority population. They are to make recommendations in this planning exercise on selection of teachers and administrators, curriculum and educational goals, political action groups on campus, special courses in ethnic studies, and the community advisory board. (REH)

Comment: This is a standard Edu-Game format with a teacher's manual locally reproducible student material. (REH)

Cost: $3.00

Producer: Edu-Game, P.O. Box 1144, Sun Valley, CA 91352

PSYCH CITY

Robert Cohen, John McManus, Davis Fox, and Connie Kastelnik; Institute for Community Development, Syracuse, N.Y.

Playing Data
Copyright: 1973
Age Level: high school
Playing Time: 8-40 hours in 1½-hour periods
Number of Players: 10-50
Preparation Time: 3-5 hours
Special Equipment: ditto machine (optional)
Packaging: 330-page paperback book

Description: *Psych City* provides students with over 100 roles (government officials, school administrators and personnel, civic leaders, political and social activists, citizens, students) and 10 issues (such as school integration, low-income housing, equal opportunity for women, land use, welfare system improvement) to solve in the simulated community of Psych City described in the text. A class may use few or all of the roles and work on one or more of the issues. They may even add their own within the format of the simulation.

Essentially, *Psych City* consists of three related activites—reading extensively in the literature, theory, and research included in the text; holding town meetings (eight per issue) to solve a chosen community problem; and completing assignments designed to clarify and elaborate on issues raised in the readings and at town meetings. The role profiles in the text explain the economic background, affiliation, attitudes, beliefs, and interests of each citizen. At the first town meeting, all citizens determine the decision and voting processes they will use throughout the eight-day period. Each session begins and ends with a town meeting, although a large portion of each day consists of subgroup caucuses in which players discuss the issues, develop proposals, and plan strategies to promote their proposals. Usually, students make presentations, debate, persuade, and negotiate with each other during this time. Between sessions, or before or after a particular simulation, students may read selections from the eight chapters of the book concerned with role expectancies, interpersonal perception, communication, norms and reference groups, social power, group decision making, individual change, and group and community change. Examples of exercises and assignments in these chapters are: interviewing a person who fits the description of a role in the simulation, visiting a public government meeting, and exploring the differences of opinion of several public figures on the issue at hand. After each simulation, with its appropriate exercises, students hold an evaluation and debriefing session. (TM)

Cost: $8.25 per book

Producer: Pergamon Press, Inc., Maxwell House, Fairview Park, Elmsford, NY 10523

SOUTH STREET HOSTEL REFORM

Playing Data
Copyright: 1974
Age Level: high school
Number of Players: 24
Playing Time: 1 hour
Packaging: 6-page pamphlet

Description: *South Street Hostel Reform* is a simulated community meeting of citizens concerned about the introduction of a hostel for mentally handicapped young people within the town of Biswick. Upon the results of this meeting rests the future of the plans for the hostel. Each participant is given a role card. About half are for persons who live in the affected area, and the guidelines for these roles are very sketchy. Mrs. Roger, for instance, believes "People like that should be kept out of sight. Don't we pay rates to have them looked after properly in institutions?" Another character, Mr. Golding, is described as "Afraid they will come to his shop and frighten away customers." (All quotes from role slips.) After the meeting, students draft a letter to the local newspaper to express the views of the community. (DCD)

Cost: 40 pence (plus 20% postage)

Producer: Community Service Volunteers, 237 Pentonville Road, London N1 9NG, England

SPRING GREEN MOTORWAY

Community Service Volunteers

Playing Data
Age Level: grades 7-12
Number of Players: 23-48
Playing Time: several class periods
Preparation Time: 15 minutes
Packaging: unbound instructions and role sheets

Description: This English game simulates a public meeting held to discuss the construction of a new highway near the fictitious village of Spring Green. The locals are divided on the issue, and their points of view are briefly described on role cards. The exercise follows a simple process, consisting of role assignments by the teacher, factual research on the construction of highways, and the hearing. (DCD)

Comment: *Spring Green* is more a suggested pedagogical strategy than a game. However, Community Service Volunteers' wit and charm make their publications a joy to examine. (DCD)

Cost: 40 pence (plus 20% postage)

Producer: Community Service Volunteers, 237 Pentonville Road, London, N1 9NG, England

SUBURBAN OPERATIONS SIMULATION

Henry W. Clark, Robin Louin, Jon Gosser; Community Change, Inc.

Playing Data
Copyright: 1970
Age Level: adult, community groups
Number of Players: 16-30 in 5 teams
Playing Time: one evening and next day
Special Equipment: P.A. system, mailboxes

Description: This is a planning game that emphasizes the effects of racism and poverty on the growth of the imaginary suburban town of Waterchester. (The producers suscribe to the view that a "white noose" of affluent suburbs is choking the life from America's increasingly black and impoverished cities.) The game is played on a four foot by six foot playing surface that represents Waterchester in its metropolitan context.

Participants are divided into five teams. Four represent socioeconomic constituencies within the town. The fifth represents governmental, religious, business, and other interest groups and institutions outside the community. The simulation is played through a series of four-year cycles. As Waterchester's population grows during each cycle, participants must rezone land, build schools, raise taxes, and cope with an increasingly acute housing shortage. The natural interests of the various groups, according to the authors, "become sources of conflict." Public issues are decided at town meetings. The simulation ends after four to six rounds. (DCD)

Comment: The producers write, "Community Change believes this exercise builds awareness of how systems and institutions operate to oppress not only the minority poor, but many other people as well. It illustrates the interface between suburbs and city and the obligations that suburban people have to themselves and to the larger metropolitan area. It provides participants with ideas on how and where to begin dealing with racism in their own communities. It exposes people to conflict management, and requires a "systems" approach to thinking about racism—to understand that the "white problem" is more than just personal prejudice."

S.O.S. requires a weekend of play. It has been used at several universities and theological schools and in training teams of community groups. (AC)

Cost: $600.00; sliding scale based on ability to pay, length of time.

Producer: Community Change, Inc., P.O. Box 146, Reading, MA 01867

SUNSHINE

David Yount and Paul DeKock; El Capitan High School, Lakeside, California

Playing Data
Copyright: 1968
Age Level: grades 9-12, average to above average ability. Grades 7-8, above average
Prerequisite Skills: reading, grade 11-math, grade 3
Number of Players: 20 to 35 in 4-6 teams
Playing Time: 16-22 class periods
Preparation Time: several hours

Description: *Sunshine* is a simulation of current racial problems in a typical American city. Originally written in 1966, it has been extensively revised for the 1970s. Each student pulls an ID tag to wear throughout the simulation. Tags are either white, tan, brown, or black and show education, job, income, street address, and pressure group (White Integrationist, White Gradualist, White Segregationist, Black Integrationist, Black Gradualist, Black Segregationist). Classroom area simulates Sunshine city's six neighborhoods with their varying segregation and integration patterns in housing and schooling. Pressure Cards (simulated radio newscasts or newspaper articles) summarize actions of individuals and groups within the community. These actions are measured in good or bad IMPS effects (IMPS represent "self-image" points; the effects differ, depending on the action). Students also earn IMPS for completing various reading, writing, and speaking assignments. An annotated bibliography specifically related to 19 historical generalizations helps teacher and student organize this work. Following investigation of the roots of racial problems, student citizens face two community crises. One involves schooling, with its social and housing problems; the other, crime, with its job, welfare, and civil rights problems. Then students try to resolve these crises by using the democratic process: organizing pressure groups, electing school board and city council members, politicking for pressure groups' viewpoints before public agencies. During this process students first role play the specific identities they drew; then during a debriefing session following the role-playing, they reenact the same situation as themselves. In addition to this interaction activity, the simulation includes an objective test, an essay evaluation, and pre- and postattitude tests. (The Teacher Guide also includes a shortened version for church and club groups.) (producer)

Cost: $14.00

Producer: Interact, P.O. Box 262, Lakeside, CA 92040

TENEMENT

Gordon Cooper

Playing Data
Copyright: 1972
Age Level: grades 9-12
Number of Players: 14-28
Playing Time: 1 1/2-2 1/2 hours
Preparation Time: 1 hour

Description: *Tenement* is a simulation concerned with the problems of families living in a multifamily housing unit in a large city (in Great Britain). Participants assume the roles of members of one of seven families living in such a house. Each family has different resources and different problems. The chief task of participants is to seek help for their problems or to solve their problems themselves. Personality characteristics within the roles are specified, but the main aim is not to introduce personality conflict too much as this would over-complicate the game, which is designed primarily to make participants aware of the housing crisis in Great Britain and its human consequences. (AC)

Cost: £1.50

Producer: Shelter, 157 Waterloo Road, London SE1 8UU, England

URBAN AMERICA

John A. Koppel

Playing Data
Copyright: 1972
Age Level: grades 7-12
Number of Players: 20-36
Playing Time: 2 weeks
Packaging: 44-page manual

Description: In *Urban America*, students acting as residents of four neighborhoods, ranging from low to high income, meet in caucuses and at town meeting to solve the problems (freeway, housing, education, crime, industry, budget, the environment) of a city of 21,000. To begin, participants get familiar with their role slips, which explain occupation, income, education, family, taxes, and property. Students elect a mayor and one member from each neighborhood to serve as the city council. Each session begins with a neighborhood gathering during which neighborhoods read the information in the manual about each problem and its proposed solutions, and then write out their own proposals. Each session ends with a town meeting during which councilors present their proposals, which may or may not reflect the views of their neighborhoods, and vote. If there is a difference in opinion within a neighborhood, nembers may ask to present their views before the council. One problem is dealt with each session. An example is the freeway problem: The city council must decide which of three proposed routes to use for a new freeway, each of which would involve a substantial cost and the destruction of homes in some neighborhoods. The manual includes considerable background information on the city and its problems, a number of maps, and a copy of the city charter. There is no scoring; success depends upon the resolution of each problem. After the simulation, students hold a debriefing session. The teacher may then want to quiz them on arguments for and against each proposal. (TM)

Cost: $21.50

Producer: Classroom Dynamics Publishing Company, 231 O'Connor Drive, Suite B, San Jose, CA 95128

WHIPP (WHY HOUSING IS A PROBLEM AND A PRIORITY)

Barbara Steinwachs, Extension Gaming Service

Playing Data
Copyright: 1972
Age Level: grade 10-graduate school, community groups
Number of Players: 15-40 in 5 teams
Playing Time: 2 hours, including discussion
Preparation Time: about 1 1/2 hours once game is mastered

Description: In an attempt to explore the causes of housing problems in a metropolitan area in as simple a way as possible, the game puts participants into five teams representing diverse socioeconomic groups. The teams then endeavor through negotiation to pass or defeat a number of proposals related to the housing situation of their imaginary metropolis. Role integrity is maintained by a feedback system which continually adjusts each team's influence level. The teams negotiate and then allocate influence for one or more proposals in each round. Three rounds are usually adequate. Debriefing is important. (author)

Comment: *WHIPP* is an introduction to problems behind inadequate housing for low and moderate income people. It is especially suitable for use as such by a group which will be exploring and working further together on problem areas in the real world, which the short game experience is designed to help participants identify. It was inspired by Simile II's *Sitte* (now out of print).

Modification of the game by use of local data in the structure of the game is encouraged. The author recommends that there be two or three game operators to keep the feedback system functioning best. (AC)

Cost: $5.00 (prepaid only)

Producer: Barbara Steinwachs, 1640 Argonne Place, N.W., Washington, DC 20009

YOUR COMMUNITY'S ECONOMIC DEVELOPMENT

Playing Data
Age Level: grade 9-adult
Number of Players: 15-25
Playing Time: 2½-3½ hours
Preparation Time: 1 hour (est.)
Packaging: professionally produced game includes instructions, forms, cards, and playing surface

Description: The purpose of this exercise, according to the producer, "is to emphasize and apply forty basic components of an effective plan of economic development." These components can be categorized as pertaining to Community Development, Action Groups, Labor, Fiscal Management, and Industrial Sites. Teams of participants must sort "essential component cards" into three categories: (1) elements that are already part of their community's plan; (2) elements that are needed to improve the community's plan; and (3) elements that are either impossible to get or are not needed. For example, one "Community Development Essential Component Card" proposes that the community "work on the elimination of substandard housing units." Each team of players would have to decide the importance of this goal in their town's overall economic development. After setting priorities on the forty "Essential Components," the players score their choices according to the probable result (which is stated on an "Economic Opportunity Card"). Finally, the players discuss their town's economic development plan. (DCD)

Cost: complete game designed for use by community groups (not available for review) $600.00; set for classroom use $15.00

Producer: Mississippi Research and Development Center, Post Office Drawer 2470, Jackson, MI 39205

See also
Change Agent COMPUTER/SOCIAL STUDIES
Dissent and Protest HISTORY
Equality SOCIAL STUDIES
Planning Exercises FRAME
Policy Negotiations FRAME
Settle or Strike ECONOMICS
Simpolis DOMESTIC POLITICS
They Shoot Marbles, Don't They? FRAME
HUMAN SERVICES
URBAN

COMPUTER SIMULATIONS

Contents

Domestic Politics	326
Ecology	328
Economics	331
Education	333
International Relations	334
Practical Economics	335
Policy	336
Science	337
Social Studies	342
Unclassified	345

COMPUTER SIMULATIONS

DOMESTIC POLITICS

ELECT

D. Klassen, J. McGrath, J. Eder, and L. Kaufman; Huntington Two Computer Project

Playing Data
Copyright: 1972
Age Level: grades 7-12
Special Equipment: computer; BASIC

Description: The *ELECT* package contains two separate simulation programs, both focusing on campaign decision-making and electoral politics.

The computer programs Elect 1-2 contain simulated voter attitudes for each of 14 past presidential elections. (The elections of 1828, 1840, 1844, 1868, 1876, 1884, 1896, 1920, 1928, 1932, 1948, 1952, 1960, and 1968 have been included.) Information about voter attitudes toward the candidates, the parties, and the issues is stored for each of these elections. The basic question facing the students using Elect 1-2 is how each candidate should allocate his political resources among the three areas of voter attitudes. In other words, how much emphasis should the candidate place on his image, on the party, and on the issues? Once these strategies have been determined, they are entered and the computer then indicates how the election would have turned out if these student-developed strategies had, in fact, been adopted by the candidates. It then gives the actual election results so that students can compare the two and attempt to explain the differences. For each election, a brief description of the political climate of the country prior to the election has been included to help students make their strategy decisions. Also included is a brief explanation of how the election actually turned out and why.

Elect 3 is based on the same model as is Elect 1-2; that is, the program simulates voter-attitudes toward candidates, parties and issues. Elect 3 has been designed as a role-playing game which can be used in the classroom to simulate a campaign and election. Students playing the roles of campaign managers, media specialists, candidates, speech writers, etc. make decisions which change the attitudes of the electorate. The class may be divided into two camps, each representing a candidate and his campaign staff. The basic objective of the campaign is to use resources in such a way as to increase the candidate's chances of winning on election day. The campaign is conceived as a series of actions taking place along a time continuum that begins several months before the election and culminates in the aggregate decision of the voters on the day of the election. Each group is given periodic poll results and information regarding the success of their campaign strategies to aid them in planning future campaign strategy.

Attempts to reach and change the attitudes of voters within the electorate is hampered by message and media distortion. Finally, a turnout rate, which can be influenced by the candidates, helps to determine the final outcome of the election. (Huntington Two)

Cost: individual packet (student, teacher, resource manuals and paper tape) $5.00; classroom packet $23.00. Quantity discount.

Producer: Huntington Two Simulation Package, Software Distribution Center 1-2, Digital Equipment Corp., 146 Main Street, Maynard, MA 01754

ENERGY (United States Energy, Environment and Economics Problems: A Public Policy)

Barry Hughes, American Political Science Association

Playing Data
Copyright: 1975
Age Level: college
Special Equipment: computer; FORTRAN
Packaging: 52-page student manual; software (batch FORTRAN program, 605 lines of code)

Description: With this simulation, students can study the formation of public policy through the interaction of the U.S. economy, energy supply and demand, and the physical environment (air, water and land quality). The model of public policy used in this simulation is composed of the following elements:

1. societal, group and individual values and goals;
2. governmental structures and processes; and
3. the non-political environment: economy, physical environment and energy system.

The simulation program contains data modeling the third element, thus allowing the student to examine various models representing the first and second elements and their implications in the area of environmental issues. (CONDUIT Pipeline)

Note: For more on this simulation, see the essay on computer simulation games by Ronald E. Anderson.

Cost: $10.00; additional student manual $4.00, 2 or more $3.50 each

Producer: CONDUIT, Box 388, Iowa City, IA 52240

MASPAR

Huntington Two Computer Project

Playing Data
Age Level: grades 9-12
Special Equipment: computer; BASIC

Description: This program contains a model that illustrates the relationship that exists in a society between social status and organizational involvement on the one hand, and mass political participation on the other.

MASPAR allows the student to explore (through the use of three variables) the relationship that exists between the pattern of mass political participation in a democratic society and the social class structure and level of organizational membership in the society. The three variables that the student manipulates are: (1) the social class structure of the society the student is studying; (2) the percentage of individuals in the society who belong to organizations whose membership is voluntary; and (3) the relationship that exists between social class structure and variable 2. (Huntington Two)

Cost: individual packet (student, teacher, resource manuals and paper tape) $3.00; without tape $2.00; classroom packet $23.00. Quantity discount.

Producer: Huntington Two Simulation Package, Software Distribution Center 1-2, Digital Equipment Corp., 146 Main Street, Maynard, MA 01754

POLICY

D. Klassen, J. McGrath, S. Hollander, and L. Oberlander; Huntington Two Project

Playing Data
Copyright: 1972
Age Level: grades 9-12
Number of Players: 6-48 in 6 teams
Playing Time: 5-10 hours in periods of 40 minutes
Preparation Time: 2 hours
Special Equipment: computer; BASIC

Description: This program simulates the interactive process of six interest groups and their impact on the kind of public policy the national government enacts.

The game data is stored in a file for continuing game periods. Before running the program, the user must open a file as follows: OPEN-PLCY1,1.

The "1" in the name PLCY1 corresponds to the game number. Thus, if you are playing game 7, the file name is PLCY7. A maximum of 9 game files, or 9 different games, may be played with their information stored at one time.

To use this program, six interest groups must be established: Business, Labor, Civil Rights, Military, Internationalist, and Nationalist. The effectiveness of the simulation depends directly on how well the members of the interest groups play their roles. Each group strives to allocate their 100 resources points per round to either pass or block any of the 14 possible policy proposals (maximum of 50 per policy and 50 negative).

Input data is in the form of two numbers, a policy number and the number of allocated resource points, separated by commas. (0,0 or a sum of 100 resource points terminates the input for a group). After the six groups have input their data, the program outputs which, if any, of the policies have been passed and a chart of eighteen socioeconomic indicators which reflect the actions of the various interest groups. (TIES)

Cost: individual packet (student, teacher, and resource manuals and paper tape) $3.00; without tape $2.00; classroom packet $23.00. Quantity discount.

Producer: Huntington Two Simulation Package, Software Distribution Center 1-2, Digital Equipment Corp., 146 Main Street, Maynard, MA 01754

POLSYS

R. Mazze, J. McGrath, D. Sobin, and K. Moy; Huntington Two Project

Playing Data
Copyright: 1971
Age Level: grades 7-12
Number of Players: 15-45 in 2 teams
Playing Time: 4-5 time periods of 40 minutes each

Preparation Time: 1 hour
Special Equipment: computer; BASIC

Description: This political simulation investigates the operation of two role-playing groups, pro and con, in their dealings with an elective governmental board on an issue.

Students should be divided into two groups which play roles of various community members. The groups determine the allocation of their resources in order to influence the vote of the board. Each side expends its resource points in three general areas; publicity, support, and knowledge. The game is essentially three rounds with the final outcome of passage or rejection being indicated.

A file is used to store the data from one round to the next. Thus, before running the game, the user must open a file to one record as follows: OPEN-POL7,1. The 7 in the file name corresponds to the game number. This number may vary from 1 to 9. Thus, nine games may be run simultaneously. (TIES)

Cost: individual packet (student, teacher, and resource manuals and paper tape) $3.00, without tape $2.00; classroom packet $23.00. Quantity discount.

Producer: Huntington Two Simulation Package, Software Distribution Center 1-2, Digital Equipment Corp., 146 Main Street, Maynard, MA 01754

PUBLIC ADMINISTRATIVE BUREAUCRATIC LABORATORY FOR UPPER MANAGEMENT—PABLUM

Edwin L. Heard, Indiana University

Playing Data
Copyright: 1974
Age Level: college, postgraduate
Number of Players: 4-20
Playing Time: 3 or more 1-hour rounds
Special Equipment: automatic data processing; BASIC

Description: This "gaming experience" is designed, according to the author, to provide the participant with "1) understanding of theoretical constructs derived from the opportunity to implement and observe the effects of normative concepts as well as verification of predicted impacts from descriptive concepts; 2) an opportunity to hone decision-making and improve human relations skills through cooperative and competitive interaction with his peers; and 3) heightened awareness of his own comparative strengths and of areas in which he needs further improvement."

PABLUM is played in cycles corresponding to fiscal years and quarters. Teams of players must prepare an agency budget beginning in the second quarter of each year. Funds are allotted quarterly. Teams must specify the "Operating" and "Personal" expense for each service project they wish to initiate, as well as for "internal administration, public relations, development and testing and fund seeking."

Results, in terms of funds approved and clients served, are generated by computer. The software for this simulation is written in time-sharing BASIC. (DCD)

Comment: This clever satire masterfully replicates the tone, syntax, and vocabulary of the modern American paper shuffler. Bureaucrats, of course, are just folks like you and me with a Ma who loves them and rent to pay, and there is a method to their hocus-pocus, the result of the well-founded and deep-seated fear that if other people knew what they really did their agencies would be denied funding and their lives would be ruined. Consequently, an educated guess becomes a "predicted impact from a descriptive concept." Should be useful to any serious student of government. (DCD)

Cost: unknown

Producer: Edwin L. Heard, Indiana University.

COMPUTER SIMULATIONS

ECOLOGY

BUFLO

L. Braun, R. L. Siegel, and E. A. Williamson; Huntington Two Computer Project

Playing Data
Copyright: 1972
Age Level: grades 9-12
Number of Players: 1-class unit
Preparation Time: 1-2 hours
Special Equipment: computer; BASIC

Description: This program simulates the natural lifecycle and population patterns of the buffalo, and allows the user to explore the effects of various harvesting policies on a buffalo herd.

Instructions are contained in the program. The user chooses between two policies of herd management: (1) total freedom of control or (2) automatic safeguards for herd population. The user also determines the years of observation, the year interval for changing harvesting policy, and the year interval for obtaining a herd population report. After the user inputs the carrying capacity and the starting year, a choice is given on how the herd will be initialized. If the herd is initialized by breakdown, the user should be prepared to input the number of adult males, adult females, yearling males, yearling females, calf males, and calf females. If the herd is initialized by total herd, the user will then input the total number of buffalo in the herd and the program will automatically break down the herd. The program will then ask for the harvesting quota for each of the herd subgroups. Program output is in the form of a table which lists the year; total herd size; the population breakdown for the subgroups; and the percentage of original population for adults, yearlings, and calves.

Variables such as mortality rates, birth rates, percentage of sexes at birth, and the critical herd size may be changed in the model to examine other factors controlling herd management or to investigate a different species. (TIES)

Cost: individual packet (student, teacher, resource manuals and paper tape) $3.00; without tape $2.00; classroom packet $23.00. Quantity discount.

Producer: Huntington Two Simulation Package, Software Distribution Center 1-2, Digital Equipment Corp., 146 Main Street, Maynard, MA 01754

COAL1

Roger F. Naill

Playing Data
Age Level: college
Special Equipment: computer; DYNAMO compiler

Description: This policy simulation is designed to allow the student to experiment with solutions to the energy crisis by inputting various parameters and discovering their effect on the model. A reference run uses historical data in energy production and demand up to 1970. A second run demonstrates the future of the energy system through the year 2010. The student can experiment with policy decisions which could affect this future energy system. (RA)

Cost: $34.55

Producer: Resource Policy Center, Attn: Publications Office, Thayer School of Engineering, Hanover, NH 03755

COEXIST

P. J. Murphy, Chelsea Science Simulation Project, University of London

Playing Data
Copyright: 1975
Age Level: college
Special Equipment: computer; BASIC
Packaging: 5 copies of 15-page student notes; 15-page teacher guide; software (interactive BASIC program, 280 lines of code)

Description: This unit simulates two biological situations. In the first, up to three populations are modeled to grow independently on identically limited food resources. The student can then investigate the situation in which organisms compete only with members of their own species. In the second situation, two populations in competition with each other for the same limited resources are simulated. In each situation, the student controls a number of parameters (initial population and generation time) which influence the outcome of species competition. (CONDUIT Pipeline)

Cost: $20.00; additional student notes 25 for $15.00; additional teacher guide $1.50

Producer: CONDUIT, Box 388, Iowa City, IA 52240

ECOLOGICAL MODELING

William Reiners, William Glanz, and Stanley Cornish; Dartmouth College

Playing Data
Copyright: 1973
Age Level: college
Special Equipment: computer; BASIC
Packaging: 3 copies of 57-page student manual; software (8 interactive BASIC programs ranging from 34 to 200 lines of code)

Description: By deciding on input and studying results from various computer-simulated models in population ecology, students learn about the exponential growth of a population and how additional factors can affect incremental growth. These eight programs present models of graduated complexity, designed for use in an undergraduate ecology course.

The models are EPOP (an exponential growth model), EXPOP (another exponential growth model), SIGPOP (a sigmoid growth model), RAND-K (a logistic growth model with density-independent effects), COMPET-1 (an interspecific competition model), COMPET-2 (a program to graph two-species interaction), COMPET-5 (interspecific competition with random environmental fluctuations and migration), and TUNDRA (Arctic tundra ecosystem simulation). (RA)

Cost: $50.00; additional student manual $5.00; 2 or more $4.50 each

Producer: CONDUIT, Box 388, Iowa City, IA 52240

GRAZE

Michael Chester

Playing Data
Copyright: 1973
Age Level: grades 7-12
Number of Players: 1-small group
Playing Time: ½ hour or more
Preparation Time: ½ hour

Description: This program deals with the ecology of a grasslands region and allows the student to vary animal population in order to attain an ecological balance. The flora and fauna balance is quantitatively compared by regulating the population of cattle, songbirds, and hawks which indirectly affect the number of grasshoppers and rodents per acre in the three-square-mile sample.

The user must enter values for the three controllable variables: cattle, songbird, and hawk populations. A chart is then printed listing the year, in half-year increments, and populations of cattle, songbirds, hawks, rodents per acre, and grasshoppers per acre. After two years, a relative score is printed which on a scale of 0 to 99 indicates quantitatively if an ecological balance has been attained. Then a complete 15-year chart may be printed or just the summary results of the study. (TIES)

Cost: student text (HP 5951-5653) $4.00; teacher's guide (HP 5951-5654) $4.00; Paper tape of program (optional) $10.00. Quantity discounts.

Producer: The Scientific Press, The Stanford Barn, Palo Alto, CA 94304

POLUT

L. Braun, T. Liao, D. Pessel, C. Losik, and E. A. Williamson; Huntington Two Computer Project

Playing Data
Copyright: 1971
Age Level: grades 7-12
Number of Players: 1-class unit
Playing Time: 15 minutes per run
Preparation Time: 1-2 hours
Special Equipment: computer; BASIC. Special language feature: TAB

Description: The interaction between water and waste is simulated on the computer, providing a context within which the user can control specific variables which affect the quality of a water resource. The user can control the following parameter: (1) type of body of water (large pond, large lake, slow-moving river, or fast-moving river); (2) water temperature (in degrees Fahrenheit); (3) rate of dumping of waste (in parts per million per day); (4) type of waste released into water (industrial or sewage); (5) type of waste treatment (none, primary, or secondary).

After the variables have been entered, the computer will determine the oxygen content and the waste content of the water for each simulated day until the system reaches equilibrium; for example, until the oxygen content and the waste content are constant. You can choose to have the computer print out this data in the form of a table, a graph, or both. (Huntington Two)

Cost: individual packet (student, teacher, and resource manuals and paper tape) \$3.00; without tape \$2.00; classroom packet \$23.00. Quantity discount.

Producer: Huntington Two Simulation Package, Software Distribution Center 1-2, Digital Equipment Corp., 146 Main Street, Maynard, MA 01754

POP

L. Braun, J. Friedland, S. Hollander, and K. Moy; Huntington Two Computer Project

Playing Data
Copyright: 1973
Age Level: grades 10-12
Number of Players: 1-class unit
Special Equipment: computer; BASIC

Description: The *POP* program consists of three population growth models which students may explore. Student exercises revolve around studies of the growth of a gypsy moth population.

The three models in the POP program are: POP1–simple exponential growth (population explosion); POP2–logistic model (environmental limiting factor); POP3–logistic model with a low density modification.

In each of the models student input includes: (1) P(O)–the starting population; (2) reproduction rate–the average number of offspring each individual will contribute to the next generation; (3) time unit per generation–the time necessary (in years) for a generation to produce its own offspring; (4) number of generations–the number of generations into the future for which students want the population projected. Input for models 2 and 3 includes: (5) Carrying capacity–the size of the population that uses up the limiting factor as fast as it becomes available. Model 3 requires one additional input: (6) At what population are low density effects first noted–what is the size of the population necessary for easy mate location? Output may be in the form of a chart, graph or both. (TIES)

Comment: The designers note that each of the programs is general enough to be used to model other plant and animal populations. (AC)

Cost: individual packet (student, teacher, and resource manuals and paper tape) \$3.00; without tape \$2.00; classroom packet \$23.00. Quantity discount.

Producer: Huntington Two Simulation Package, Software Distribution Center 1-2, Digital Equipment Corp., 146 Main Street, Maynard, MA 01754

RATS

Huntington Two Computer Project

Playing Data
Age Level: grades 10-12
Number of Players: 1-class unit
Special Equipment: computer; BASIC

Description: This program simulates the growth pattern of a rat population in either a city or an apartment house. The user controls the conditions of growth and sets the time of report. The program prints a daily count of the rat population for any specified period of time.

The user inputs the type of environment, whether the computer or the user will determine the initial number of rats, the date for the report, the garbage level of the area, and the type and quantity of poison to be used for rat control. At this point, the program has the information necessary to determine the growth pattern of the rat population. For the specified period of time, a report is printed showing the total population and the distribution of the population by age, the number of births and deaths, and the count of emigration and immigration. Upon termination of the run, a report is produced which lists the accumulated births, deaths, emigrations and immigrations for the entire run, the dollar values of the damage done by the rats, the cost of rat poison, and the amount and type of poison left uneaten. (TIES)

Cost: individual packet (student, teacher, resource manuals and paper tape) \$3.00; without tape \$2.00; classroom packet \$23.00. Quantity discount.

Producer: Huntington Two Simulation Package, Software Distribution Center 1-2, Digital Equipment Corp., 146 Main Street, Maynard, MA 01754

RIVER DOSE (RVRDOS) MODEL

James A. Martin, Jr., Charles Robbins, Christopher B. Nelson, Robert D. Cousins, Jr., and Mary Anne Culliton

Playing Data
Copyright: 1976
Age Level: adult
Packaging: professionally produced, 86-page booklet

Description: The River Dose computer model was developed to assess the consequences to the general population from the consumption of drinking water and fish from waters which receive radioactive liquid effluents.

A computer program RVRDOS has been developed to calculate population doses due to releases of radionuclides into flowing streams. Concentrations of the radionuclides downstream take into account dilution, decay, and the ingrowth of a daughter product. Population doses to four organs are calculated for drinking water and fish ingestion pathways. Individual doses due to swimming may also be estimated. A program manual for RVRDOS is included in the manual.

RVRDOS has been used to calculate population doses due to releases from nuclear power reactors on the Mississippi River Basin during 1973. The data base for these calculations and a summary of the calculations are discussed. (producer)

Cost: unknown

Producer: U.S. Environmental Protection Agency, Office of Radiation Programs, Environmental Analysis Division, Washington, DC 20460

STERL

A. Frishman, L. Braun, C. Losik, and K. Moy; Huntington Two Computer Project

Playing Data
Copyright: 1971
Age Level: grades 7-12
Number of Players: 1-class unit
Playing Time: 15 minutes per run
Preparation Time: 1-2 hours
Special Equipment: computer; BASIC

Description: This environmental program enables the user to explore the effects of application of pesticides and sterilization in controlling the growth of the screwworm fly population.

An instruction option is included at the start of each run. The goal is to eliminate the one million male flies in a 10,000-square-mile area, thus destroying the total fly population. Pesticides or sterilization techniques may be used either separately or together in controlling the population. In either case the user specifies the days of the 75-day investigation period in which the eradication procedures are to be used.

Results of the input strategies for control are given by the computer in graphical form (time in days versus normal adult male fly population in millions). Also projected is the cost of the fly control and estimated damage. The needed time for a single run is approximately ten minutes. (TIES)

Cost: individual packet (student, teacher, and resource manuals and paper tape) $3.00; without tape $2.00; classroom packet $23.00. Quantity discount.

Producer: Huntington Two Simulation Package, Software Distribution Center 1-2, Digital Equipment Corp., 146 Main Street, Maynard, MA 01754

TAG

J. Friedland, S. Hollander, and G. J. Smith II; Huntington Two Computer Project

Playing Data
Copyright: 1973
Age Level: grades 9-11
Number of Players: 1-3, adaptable for class
Special Equipment: computer; BASIC. Special language feature: RANDOMIZE

Description: The *TAG* program provides the user with an opportunity to investigate the size of a wildlife population through the technique of tagging and recovery.

TAG uses the large-mouth bass population of a simulated farm pond as the study species. User inputs include: the number of tagged bass released from the hatchery; the number of days after the release of the bass from the center of the pond that the sampling should be made; the X and Y coordinates of the site for the sampling; and the number of fish which should be sampled. The output includes the number of tagged and untagged fish caught at a site and, after several samplings and calculations, an evaluation of the accuracy of the student's results. (TIES)

Cost: individual packet (student, teacher, resource manuals and paper tape) $3.00; without tape $2.00; classroom packet $23.00. Quantity discount.

Producer: Huntington Two Simulation Package, Software Distribution Center 1-2, Digital Equipment Corp., 146 Main Street, Maynard, MA 01754

COMPUTER SIMULATIONS

ECONOMICS

BOARD OF DIRECTORS

Richard F. Loomis, President, Flying Buffalo Computer Conflict-Simulation, Inc.

Playing Data
Copyright: 1973
Age Level: grade 7-12
Prerequisite Skills: ability to read and do basic mathematical calculations. Fifth-grade students have played the game successfully.
Number of Players: 3-100 in teams of 3-5
Playing Time: played by mail; moves require one hour; 4-8 days between moves.
Preparation Time: 1-2 hours initially, less per move

Description: The class receives the rules and initial resources. Teams are assigned. The teams make the decisions of the game, and give them to the teacher. The teacher fills out the turn sheet and mails it to us. We return the results in the form of financial statements for each team. The teams consult these and make the same decisions again under the altered circumstances.

The game includes only a few of the aspects of the real situation (but this is intended in order to make the game easy enough for junior high and high school students with no business experience or limited business experience). It is possible for the players to make wild or ridiculous decisions just to "see what happens," but this doesn't usually happen in team play unless one member of the team dominates, and the other players allow him to make the decisions.

Some of the things that the players do not have to handle are: government intervention, raw materials acquisition, personnel administration, consumer reaction, pollution, and quality control. (author)

Comment: This game is very similar to the first total enterprise computer games designed by the American Management Association and UCLA. It has been uniquely tailored to make it suitable for junior high and high school students and offers a minimal amount of initial preparation, very simplified financial statements, and is one of the few games with a computer service option; in this case decisions returned by mail. Costs seem very reasonable. The author does not state how many computer print-out copies are returned for each team. Some copying may be necessary in order to facilitate each student's analysis. (BK)

Cost: $38.00 for 10 moves for 10 teams. Other options available.

Producer: Flying Buffalo Computer Conflict-Simulation, Inc., P.O. Box 1467, Scottsdale, AZ 85252

COMPUTER SIMULATION POLICY GAMES IN MACROECONOMICS

Richard Attiyeh, William Brainard, and F. Trenery Dolbear; University of California, La Jolla

Playing Data
Copyright: 1976
Age Level: college, graduate school
Number of Players: 1 and up in 1 or more teams
Playing Time: 1 week or more
Preparation Time: 10 hours
Special Equipment: computer; BASIC, FORTRAN

Packaging: 28-page student manual; 159-page instructor's manual; software (5 interactive BASIC programs, ranging from 140 to 340 lines of code, and one interactive FORTRAN program, 665 lines of code)

Description: These six games attempt to convey to students an understanding of the aggregate behavior of the economy by casting them into the roles of economist and policy maker for a simulated, real-world economy. Each game uses a time series model or "history" of the economy to inform students of past behavior of the model. In choosing values for the policy variables, students learn the value of economic analysis; in trying to identify and understand the structure of the model they apply economic theory; and in playing the game they appreciate the usefulness of economic theory in understanding economic events. Each of the following six games vary in the complexity of the hypothetical economy, the goals provided the student, and the aspects of macroeconomics being emphasized.

MPG-I: In this game the student manages fiscal policy in a simple model of the product market with the goal of keeping aggregate demand at or near potential output. (Program MICRO1, 170 lines)

MPG-II: In this game the goal is to minimize the difference between aggregate demand and potential output by student control of the policy variables–government expenditures and marginal tax rate. (Program MICRO2, 170 lines)

MPG-III: This game incorporates a money market and a factor market with the product market and involves the student in simultaneous decisions about several policy variables in pursuit of multiple targets defined by the welfare function. (Program MPG3, 210 lines)

MPG-IV: The structure of the model for this economy is identical to that in MPG-III except for the inclusion of a foreign sector so that policy decisions affect both the balance of trade and the balance of payments. (Program MPG-IV, 180 lines)

MPG-V: The student is provided with increasingly complex goals, beginning with a social welfare function in which the only argument is the deviation between actual and target income and concluding with a function involving deviations between actual and target rates of unemployment, inflation, and interest. (Program MPGV, 310 lines)

MPG-VI: This version is basically the same as *MPG-V.* It includes the nature and sequencing of policy goals, types of fiscal policy decisions available to the student, and the fixed interest rate regime; however, a more realistic monetary sector is incorporated, teaching the standard theory of money supply and demand. (Program MPGVI, 365 lines)

Each of these models casts the student or a student team in the role of policy maker whose decisions are based upon historical information about an economy and some stated objective. Decisions are entered interactively; the computer develops the problem and requires student response, evaluates reasonableness of answers, and finally reports on successfulness of student decisions. These models can be used with introductory, intermediate, or advanced students, although the complicated models are better suited to students who have nearly completed an introductory macroeconomics course or are taking an intermediate course. (Trinka Dunnagan, CONDUIT)

Cost: $50.00; additional student manual $4.00; 2 or more $3.50 each; additional instructor manual $5.00

Producer: CONDUIT, Box 388, Iowa City, IA 52240

MA 501

William Davisson

Playing Data
Age Level: college
Special Equipment: computer; CRT interactive equipment preferred

Description: This simulation allows the student to learn about the Keynesian system by looking at the impact of policy changes on the economy. The model is a simple Keynesian expression with nine equations and a built-in, Phillips-type relationship. The student plays the role of an economic policy maker trying to achieve satisfactory levels of inflation and unemployment by manipulating the following policy variables: (1) money supply, (2) government spending, (3) taxes, and (4) transfer payments. Once the appropriate policy or set of policies is determined, the student goes on to the second period, then the third, and so on until the problem in each 10-year period is alleviated. (RA)

Cost: $25.00 for MA501-4

Producer: Prof. William Davisson, Department of Economics, University of Notre Dame, Notre Dame, IN 46556

MANAGE

N. Thompson, TIES

Playing Data
Age Level: grades 11-12
Number of Players: class in three teams
Special Equipment: computer

Description: This business management simulation allows a class to be divided into three subgroups acting as teams of executives making quarterly decisions for their firms. All teams decide on selling price, advertising, production, research and development, and plant improvement based on a quarterly financial report giving the results of orders, sales, and cash flow for their companies.

A detailed description of the simulation model, operating instructions, instructional materials, and sample runs may be ordered from TIES. A bibliography which correlates this program to a variety of related texts and other educational materials is also available. (TIES)

Cost: $.40 for program documentation

Producer: Minnesota School Districts, Data Processing Joint Board, 1925 West County Road, B2, St. Paul, MN 55113

MARKET

S. Finkelstein, I. Staw, J. McGrath, D. Sobin, and E. A. Williamson; Huntington Two Computer Project

Playing Data
Copyright: 1972
Age Level: grades 10-12
Number of Players: 2-32 in 2 teams
Playing Time: 1/2 hour or more
Preparation Time: 1 hour
Special Equipment: computer; BASIC

Description: The computer program in this unit is a game which allows the users to play the roles of two companies which are competing for the market for a particular product.

At the start of the game, the computer informs the players of the fixed and the variable production costs involved in the marketing of their goods, and it assigns to each company initial values for inventory (stock on hand), cash on hand, and total assets. Each player is allowed to make marketing decisions quarterly and can determine the production level, the advertising budget, and the unit price of the product for his company.

After these decisions are made for a given quarter, the computer reports the results of the decision with respect to performance on the market. For each company, the report lists: (a) profit, (b) percentage share of the market, (c) cash on hand, (d) number of units sold, (e) number of units in stock (inventory), and (f) total assets. Players make their decisions for the following quarter on the basis of these reports. Contingencies which may require changes in strategy occur randomly throughout the game. The game ends when one company goes bankrupt or attains 12 million dollars in total assets. (Huntington Two)

Comment: For a bibliography correlating this program to related texts and other materials, contact Minnesota School Districts, Processing Joint Board, 1925 West County Road, B2, St. Paul, MN 55113 (AC)

Cost: individual packet (student, teacher, and resource manuals and paper tape) $3.00; without tape $2.00; classroom packet $23.00. Quantity discount.

Producer: Huntington Two Simulation Package, Software Distribution Center 1-2, Digital Equipment Corp., 146 Main Street, Maynard, MA 01754

COMPUTER SIMULATIONS

EDUCATION

CRITICAL INCIDENTS IN EDUCATION

Vincent N. Lunetta, CONDUIT

Playing Data
Copyright: 1975
Age Level: teacher training
Special Equipment: computer; BASIC
Packaging: 30-page instructor's guide, 5 copies of 10-page student's guide, software (26 BASIC programs, ranging from 500 to 700 lines of code)

Description: This package provides twenty-two simulations of classroom situations that allow prospective teachers to examine human interactions in schools. Each simulation begins with a description of a realistic incident confronting a teacher. The prospective teacher is then given a number of possible responses, and must select one which fits his or her teaching style. Responses are generally classified as extreme authority, ignore, moderate authority, and shared authority. The problem continues to evolve, based on the responses selected. During the simulation, feedback is provided in the form of reactions from principals, students, parents, and other teachers in much the same way as in a real school setting. Critical commentary and suggestions are presented to the student at the end of the simulation (CONDUIT Pipeline)

Note: for more on these simulations, see the essay on computer simulation games by Ronald E. Anderson.

Cost: $60.00; additional instructor manual $4.00; additional student manuals 25 for $15.00

Producer: CONDUIT, Box 388, Iowa City, IA 52240

EDUCOM FINANCIAL PLANNING MODEL

Daniel A. Updegrove

Playing Data
Copyright: 1978
Age Level: college, professional
Packaging: 75-page photocopied booklet

Description: The *EDUCOM Financial Planning Model (EFPM)* is an interactive system for generalized budgeting and financial planning which has been designed primarily for use in higher education. The objective of *EFPM* is to simplify and reduce the cost of answering such questions as: (1) Will the budget be in balance next year given the prevailing growth rates for income and expense categories? (2) What are the budgetary implications for this and succeeding years of optimistic and pessimistic estimates of utility rates, labor costs, and endowment growth? (3) What are the explicit trade-offs between such primary planning variables as tuition charges, faculty salaries, new construction, and special programs?

The model currently runs on the IBM 370/168 computer with the VM/370 (CMS) operating system. It requires one 3330 cylinder of disc storage for user input data and a 700-k byte virtual region. All usage of the system is by dial-up or through Telenet or TYMNET to the computer at Cornell University in Ithaca, New York. (Author)

Cost: $1250 annual subscription fee ($1000 for members of EDUNET); $1750 one-time consulting fee ($1500 for members of EDUCOM), plus run charges of $7 to $20 per hour

Producer: EDUCOM, P.O. Box 364, Princeton, NJ 08540

COMPUTER SIMULATIONS

INTERNATIONAL RELATIONS

INS2: INTER-NATION SIMULATION

Bahram Farzanegan and Ronald J. Parker, North Carolina Educational Computing Service

Playing Data
Copyright: 1978
Age Level: high school, college
Playing Time: 4-8 1-hour periods
Special Equipment: computer; BASIC or FORTRAN
Packaging: 57-page SRA participant's manual; 100-page CONDUIT instructor's manual; 120-page CONDUIT programmer's guide; software (12 interacting BASIC programs, ranging from 200 to 400 lines of code, and 12 data files, each 550 numeric data elements)

Description: *INS2, Inter-National Simulation* in Time-Sharing BASIC, is based on the *Inter-Nation Simulation* by Guetzkow and Cherryholmes (Science Research Associates, Inc., 1966). It improves on the original kit by eliminating the many manual calculations required of students and instructors. Time-consuming calculations are performed by the computer, leaving students free to concentrate on making decisions, formulating strategy, and implementing overall policy. Computer programs are provided for students to test their decisions on trade, budget, and war, and to project the consequences of their decisions during the simulation.

Likewise, *INS2* relieves the instructor of many hours of tedious calculations, leaving him free to concentrate on the development of the simulation. The programs for the instructor provide summary sheets for each nation's characteristics, determine office holding, and provide a master statistics report and an intelligence report.

In the simulation students assume government positions in hypothetical nations and make decisions concerning budget, trade, military force, and war. The simulation format consists of a briefing, a trial run, several decision periods focusing on one problem (a scenario chosen by the instructor), and a debriefing. Twelve scenarios are included with the package. The simulation is generally run for four to eight periods, each lasting about one hour. Several additional hours are needed for preparation and post-game analysis, or debriefing.

By replicating the international system through the creation of a cluster of prototype nations, students are exposed to the complexities of decision-making in foreign policy and the problems of trade, diplomacy, and war. *INS2* also permits research into areas otherwise not open to research by conventional methods. (CONDUIT Pipeline)

Note: For more on this simulation, see the essays on international relations simulations by Leonard Suransky and on computer simulation games by Ronald E. Anderson.

Cost: $95.00; additional instructor manual $12.00; additional student manual $3.00, 2 or more $2.50 each

Producer: CONDUIT, Box 388, Iowa City, IA 52240

POLIS NETWORK

Robert C. Noel, University of California, Santa Barbara

Playing Data
Age Level: college, graduate school
Number of Players: varies
Playing Time: varies
Special Equipment: teletype terminal
Packaging: manual, operations guide, scenario

Description: The *Polis* network is a computer-controlled communications network designed to facilitate political science gaming and simulation. Messages (or moves) are sent to other teams over the *POLIS Network*. Each message is scanned by a control team to evaluate message content. Messages are then either forwarded to their designated recipient or rejected and returned to the sender with some explanatory comments. The process is then repeated. Rejected messages may be revised or totally rewritten and then sent again.

When one thinks of simulation and gaming exercises, one tends to think of a model United Nations or of international conferences in which students engage in role playing. The *POLIS Network* exercise is a more analytic and less dramatic mode of participation. The scenario is independent of the computer software and network so that different scenarios may be substituted from time to time. The present scenario focuses on international relations in a multipolar world and has a global character rather than focusing on a narrowly defined problem or crisis. Players are expected to simulate the roles of foreign office officials, but are given considerable latitude to depart from current pattern.

This simulation tends to be excessively rational in that "chance happenings" do not occur. What is unique about the whole concept of *POLIS* is that it allows different schools, separated by long distances, to interact in a simulation. This exercise also permits different levels of work by participants and has been highly successful in motivating students to do outside research. (REH)

Note: For more about this simulation, see the article on international relations simulations by Leonard Suransky.

Cost: $100 per participating team (plus phone costs)

Producer: William D. Hyder, Polis Network Exercise, Polis Lab, Dept. of Political Science, University of Calif., Santa Barbara, CA 93106

PSW-1/SW POLITICAL SIMULATION

John Parker, IBM; Clifford N. Smith, Northern Illinois University; Marshall H. Whithed, Temple University

Playing Data
Age Level: grade 11-graduate school
Number of Players: 21-300 in 7-100 teams

Playing Time: 4 hours minimum
Preparation Time: several hours
Special Equipment: keypunch machined IBM 360/40 or IBM 360/90; PL/1, or CDC 6400, time share interactive

Description: *PSW-1/SW Political Simulation* engages participants in a large-scale, internation simulation as they represent heads of state, directors of business enterprises, and citizens (other roles may be added). They seek to increase their living standards, their business prosperity, and their governmental effectiveness. While the particular decisions involved in play vary according to the political, economic, and social parameters of each country, they generally concern allocating funds to various functions of government, conducting foreign relations, domestic "politicking," and preparing national budgets. (AC)

Cost: for information on materials, transfer capabilities, and costs, contact producer or Marshall H. Whithed, School of Community Services, Virginia Commonwealth Univ., Richmond, VA 23284

Producer: Westland Publications, P.O. Box 117, McNeal, AZ 85617

PRINCE, A PROGRAMMED INTERNATIONAL COMPUTER ENVIRONMENT

William D. Coplin, Stephen L. Mills, and Michael K. O'Leary; Syracuse University

Playing Data
Copyright: 1971
Age Level: college, graduate, adult
Number of Players: 1-100 in 1-20 teams
Playing Time: 3 hours minimum to an indefinite maximum in 2-hour periods
Supplementary Material: topical and theoretical material relating to foreign policy decision-making may be used in any of various uses of the general *PRINCE* model.
Special Equipment: computer; FORTRAN IV

Description: Players represent foreign policy planners of the United States (roles with the decision power of the President and the Secretary of State combined). Teams are an ad hoc adjunct to the game. The same foreign policy decisions must be made, whether by a single player or a team with self-assigned roles.

PRINCE may be used interactively or noninteractively by either individuals or teams. An interactive exercise of about two hours' duration is usually used to introduce potential users to the *PRINCE* Political Accounting System, the core of the program. The System, in turn, can be used apart from the computer if desired, and may be applied to any political situation.

Play involves strategic thinking, coalition formation, bargaining, and compromising. Players begin with equal resources and strive to promote successful policies. These policies are/are not successful (quantitative outcomes), and are not subject to chance determination.

PRINCE generates interest in and provides a framework for understanding the role of the United States in contemporary world affairs. It may, in addition, be used for research purposes. It should be emphasized that *PRINCE* is a general-purpose model of the real world programmed to provide both international and domestic consequences to any foreign policy decision. What we have here, then, is not simply a simulation game, but the basis for any number of simulation games.

PRINCE may be used repeatedly by the same players as they work toward an almost unlimited variety of educational, analytical, and research goals. (DZ)

Note: For more on this simulation, see the essay on international relations simulations by Leonard Suransky.

Cost: (1) participant's guide plus *PRINCE* model $5.00 (bulk order); (2) $6.95

Producer: (1) International Relations Program, Syracuse University, 752 Comstock Avenue, Syracuse, NY 13210; (2) Duxbury Press (1976).

PRACTICAL ECONOMICS

CHARGE (PERSONAL FINANCES)

Northwestern National Bank of Minneapolis and Minneapolis Public Schools

Playing Data
Copyright: 1972
Age Level: high school
Number of Players: groups of 10 or fewer
Playing Time: 2-4 hours
Preparation Time: 1 hour
Special Equipment: computer
Packaging: 32-page teacher's guide and complete set of documentation

Description: Students playing *Charge* gain experience with budgeting, using credit, and calculating finance charges. Each player is assigned one of six computer-related roles, five representing young people on limited budgets and the sixth allowing for the players to be themselves. During each of twelve turns representing the months of one year, students consult their assigned monthly income and savings goals and choose items (clothes, recreation, transportation, etc.) to buy on cash or credit according to item selection sheets which state the cash price or monthly payments for each item. They also decide how much money to save each month. After filling out their activity sheets each turn, they feed their selections into the computer, which evaluates their choices according to a satisfaction index, awards them points, adjusts their savings figures, and informs them of special events which may alter their choice of expenditures during their next turn. At the end of the simulated year, students total their savings and points earned and determine their comparative success at selecting items, budgeting, and saving. (TM)

Comment: The designers of *Charge* recommend a trial run of the game involving 3 or 4 months before beginning the simulated year; they also suggest replaying the entire game at least once. Since only a limited number of students can use the computer at any one time, teachers should plan accordingly. (TM)

Cost: $17.25

Producer: Paul S. Amidon and Associates, Inc., 166 Benson Ave., St. Paul, MN 55116

SHELTER

Minneapolis Public Schools and Northwestern National Bank of Minneapolis

Playing Data
Age Level: high school
Special Equipment: computer
Packaging: classroom set consisting of teacher guide, 20 student booklets, 3 each of 8 roles, 3 each of 14 Shelter sheets, supply of Activity sheets

Description: This is a simulation designed to provide students with experience in purchasing and maintaining a home successfully over a period of one year.

The users make decisions about their type of shelter at twelve different intervals (monthly). Decisions include buying or renting, type of shelter, and user financial conditions. Also included are fixed and variable expenses. The student may choose one of 14 different options to obtain a shelter and one of 8 different assigned roles. The computer furnishes the amounts needed for varying expenses, such as utilities and maintenance. The computer also prints out any unexpected events that may occur each month. (TIES)

Cost: $36.00

Producer: Paul S. Amidon & Associates, Inc., 1966 Benson Ave., St. Paul, MN 55116

WHEELS

Minneapolis Public Schools and Northwest National Bank of Minneapolis

Playing Data
Age Level: high school, junior college
Number of Players: 2-35
Playing Time: 4-5 40-minute periods
Special Equipment: computer; BASIC
Packaging: classroom set consists of teacher's manual, 35 participant manuals, 3 sets of student roles, activity sheet tablet

Description: This simulation is designed to provide students with experience in purchasing and maintaining an automobile successfully over a one-year time period.

The users make decisions about their car at twelve different intervals (monthly). Decisions include purchase of a car, method of financing, type of insurance, and plan of expenditures, giving careful consideration to fixed and variable expenses. The student may choose one of 14 different assigned roles or define a self-role. The computer randomly generates unexpected events such as accidents, repairs, additional expenses, and increased income. Students must maintain an activity sheet to record the financial conditions for each round. (TIES)

Note: For more about this simulation, see the essay on computer simulations by Ronald E. Anderson.

Cost: $46.00; specify with or without computer

Producer: Paul S. Amidon and Associates, 166 Benson Ave., St. Paul, MN 55116. Program available through Honeywell time-sharing (EDINET) centers. Check local area sources for availability.

COMPUTER SIMULATIONS

POLICY

IDIOM: A DISAGGREGATED POLICY-IMPACT MODEL OF THE U.S. ECONOMY

Stephen P. Dresch and Daniel A. Updegrove

Playing Data
Copyright: 1978
Age Level: postgraduate, professional
Number of Players: 1 or more

Description: This computer simulation has been designed and used as a federal policy evaluation model. Specifically, *IDIOM* (an acronym for Income Determination Input Output Model) was designed "for the assessment of multifaceted economic effects of large scale changes in fiscal structure (tax, transfer and expenditure policy) and of other exogenous economic developments. The central objective of this research effort has been the adaptation and the development of techniques of analysis capable of identifying these effects at a relatively high degree of disaggregation, e.g., by region, industry and occupation."

The national component is driven by predetermined final demands such as gross investment, government purchases of goods and services, gross exports, and transferred imports. The production required to fulfill these final demands generates labor and nonlabor income. These incomes serve to determine endogenous consumption final demands via consumption functions.

The regional component represents the total outputs from the national model distributed over regions according to predetermined distribution matrices indicating the share of each region in the total ouput or output change of each national industry.

The designer notes that although *IDIOM* "appears to be a potentially useful tool for teaching regional economics and policy analysis, no such use has yet been made."

IDIOM currently runs on the IBM 370/168 computer with the VM/370 (CMS) operating system. It requires two 3330 cylinders of disc storage for user input files and a 1000-k byte virtual region. All usage of the system is by dial-up or through Telenet or TYMNET to the computer at Cornell University in Ithaca, New York. (DCD)

Cost: $10 to $40 per hour depending on the complexity of the policy evaluations specified

Producer: Institute for Demographic and Economic Studies, Inc., 115 Whitney Avenue, New Haven, CT 06510

COMPUTER SIMULATIONS

SCIENCE

BOLA, BOMBARDMENT OF THE LIGHT ATOMS

Michael Chester, P.O. Box 985, Sunnyvale, CA 94088

Playing Data
Copyright: 1974
Age Level: grade 11-college
Number of Players: 1-2
Playing Time: 30 minutes-a few hours
Preparation Time: 1 hour
Special Equipment: time-shared system using BASIC. Hewlett-Packard 2000 or equivalent system.

Description: *BOLA* is a nuclear physics game. The computer displays the name of an atomic isotope, such as Li 7. The player chooses a bombarding particle. The computer displays the result of the bombardment (the new isotope formed–and, in some game modes, the energy of the reaction). Then it is the player's turn to choose the next bombardment. The computer also provides a periodic display of isotopes generated since the start of the game and updated player scores.

The player's text includes a gameboard derived from a chart of isotopes. The computer displays track of the player's motions on this gameboard, periodically offering the display of the entire gameboard, including a history of past moves. Players' options are also multiple, including descriptive "tours," analytical "toying," and one-person or two-person game modes. (Author)

Cost: $3.00

Producer: The Scientific Press, The Stanford Barn, Palo Alto, CA 94304

CHARGE (PHYSICS)

D. Scarl, A. Caggiano, C. Losik, and K. Moy; Huntington Two Computer Project

Playing Data
Copyright: 1971
Age Level: grades 11-12
Number of Players: 1-class unit
Special Equipment: computer; BASIC. Special language feature: RANDOMIZE

Description: This is a physics simulation of the Millikan Oil Drop Experiment designed to demonstrate the existence of a discrete unit of electrical charge. The program allows a student to perform the original experiment without the actual equipment, in much less time, and to isolate the data for drawing inferences.

Four charged latex spheres are dropped into an electric field with random velocities. The user then inputs a voltage between −1000 and 1000, attempting to make the velocity as close to zero as possible. An input voltage of 2000 signals a request for the calculations of the charge on a stopped drop. An input voltage of 3000 signals a request for four new spheres to be dropped. An input voltage of 4000 terminates the program.

A user may repeat results by answering yes to the question "Do you want repeatable results?" If yes, the number entered must be the same number used in the run to be duplicated. (TIES)

Cost: individual packet (student, teacher, resource manuals and paper tape) $3.00; without tape $2.00; classroom packet $23.00. Quantity discount.

Producer: Huntington Two Simulation Package, Software Distribution Center 1-2, Digital Equipment Corp., 146 Main Street, Maynard, MA 01754

COMPETE

M. E. Leveridge, Chelsea Science Simulation Project, London University

Playing Data
Copyright: 1975
Age Level: college
Special Equipment: computer; BASIC
Packaging: 5 copies of 15-page student notes, 11-page teacher guide; software (interactive BASIC program, 237 lines of code)

Description: The computer simulation *COMPETE* enables students to plan and carry out an investigation without the long delay usually associated with growth experiments. The *COMPETE* unit includes investigations with both real and simulated plants and other relevant data in the form of graphs, tables, and descriptions. The six investigations presented in the Students' Notes are: (2) Effects of crowding on plant growth (real experiment); (2) measurement of growth (second-hand data); (3) simulated growth (monoculture) (computer simulation); (4) interaction between clover varieties (real experiment and second-hand data); (5) simulated growth (mixture) (computer simulation); (6) interaction below the ground (second-hand data); and (7) direct plant interaction (second-hand data).

The real experiments need to be set up well in advance of the time when the results are required, but the others can be carried out in one or two hours. The Students' Notes describe the investigations in a continuous sequence, but if time is short, some of them may be omitted. No prior knowledge about plant competition is assumed, but it is desirable that the students have a general background knowledge of the development of flowering plants and of the resources which plants must obtain from their environment. (CONDUIT Pipeline)

Cost: $25.00; additional student notes 25 for $15.00; additional instructor manual $1.50

Producer: CONDUIT, Box 388, Iowa City, IA 52240

COMPUTERS IN THE BIOLOGY CURRICULUM

J. Denham, M. E. Leveridge, J. A. Tranter, and J. Pluck; Schools Council Project, Chelsea College, London

Playing Data
Copyright: 1978
Age Level: high school, college introductory
Special Equipment: computer; BASIC
Packaging: 162-page manual of student and instructor documentation; software (12 BASIC programs written in Level O BASIC, ranging from 175 to 400 lines of code)

Description: The computer programs included in this package provide the facility for simulation gaming, model building, computation, and data retrieval. The documentation is divided into eight chapters. The first provides instructors with an overview of instructional computing in biology with numerous examples. Later chapters deal with the specific curriculum topics. The corresponding computer programs are listed below.

(1) Inheritance: part one simulates inheritance of characters for fruit files, mice, and human beings; part two simulates multifactorial inheritance.
(2) Predator-Prey Relationships: a simple model.
(3) Pond Ecology: based on a computer simulation for a freshwater community consisting of three trophic levels–phytoplankton, herbivores, and fish. Fishing may be permitted, introducing man as a fourth trophic level.
(4) Transpiration: covers simulation of water loss by leaves.
(5) Countercurrent Systems: simulates two types of countercurrent systems found in the bodies of animals–exchangers and multipliers; discusses both the simplified mathematical models and their biological counterparts.
(6) Human Energy Expenditure: allows exploration of human energy requirements in relation to activity, sex, and body mass.
(7) Statistics for Biologists: computes simple statistics. (drawn from CONDUIT Pipeline)

Cost: $60.00 (manual not available separately)

Producer: CONDUIT, Box 388, Iowa City, IA 52240

ENZKIN

M. T. Heydeman, Chelsea Science Simulation Project, University of London

Playing Data
Copyright: 1976
Age Level: college
Playing Time: 1½ hours at terminal, 3 hours to write up results
Special Equipment: computer; BASIC
Packaging: 5 copies of 15-page student notes; 15-page teacher guide; software (interactive BASIC program, 283 lines of code)

Description: This unit on enzyme kinetics permits the student to obtain realistic laboratory-type results very rapidly, using a computer program to simulate enzyme catalyzed reactions. *ENZKIN* allows many cycles of experimental research–planning, experimentation, and interpretation–over a short time. The purposes of this package are: (1) to enable the student to answer fundamental questions of enzyme kinetics, (2) to provide experience in dealing with initially unknown systems, (3) to provide experience in various forms of data interpretation, and (4) to introduce students to some complex situations in enzyme kinetics. Six enzymes with different properties are simulated in the computer program so that each student or group of students can obtain different results. (RA)

Cost: $25; additional student notes 25 for $15.00; additional teacher guide $1.50

Producer: CONDUIT, Box 388, Iowa City, IA 52240

EVOLUT

S. McCormick, Chelsea Science Simulation Project, University of London

Playing Data
Copyright: 1975
Age Level: college
Special Equipment: computer; BASIC
Packaging: 5 copies of 19-page student notes; 11-page teacher guide; software (interactive BASIC program, 260 lines of code)

Description: EVOLUT is an introductory unit on evolution and population genetics, intended to teach (1) mechanisms generating variation and the selective process leading to adaptations; (2) adaptation to environmental conditions in relation to survival value; (3) manipulation of models of selection acting on populations; and (4) investigation of the power of selection in producing certain frequencies of alleles in a given environment and relation of adaptation to survival. An elementary knowledge of genetics is required for effective student use. The computer program allows the student to test the hypothesis that inherited variations showing a small positive survival value are sufficient for micro-evolution. Students select various parameters in the model, including zygote type, percentage of selection, size of population, initial percentage of green alleles, and a number of generations; they also observe the simulated process of natural selection and evolution. (CONDUIT Pipeline)

Cost: $20.00; additional student notes 25 for $15.00; additional teacher guide $1.50

Producer: CONDUIT, Box 388, Iowa City, IA 52240

HARDY

Huntington Two Computer Project

Playing Data
Age Level: grades 10-12
Number of Players: 1-class unit
Special Equipment: computer; BASIC. Special language feature: TAB

Description: This program leads the user through the formulation of the Hardy-Weinberg Principle and assists the user in determining the proportion of individuals within a population that are homozygous dominant, heterozygous dominant, or homozygous recessive with regard to a certain trait.

During the first run of the program, the user receives a brief explanation of the Hardy-Weinberg Principle and a sample rat population to work with. The user specifies the number of rats to be sampled, and the computer will generate the number of rats showing the dominant trait and the number showing the recessive trait. The computer will also calculate and print out the proportion of rats showing the recessive trait, the frequency of the recessive allele, and the proportion and number of rats that are homozygous dominant, heterozygous recessive. On subsequent runs of the program, users have the option of either using their own data or working with either of two rat populations generated by the computer. In each case, the computer will calculate and print out the proportion in the sample showing the recessive trait, the frequency of the recessive allele, and the genotypic ratios for the population. (TIES)

Cost: individual packet (student, teacher, resource manuals and paper tape) $3.00; without tape $2.00; classroom packet $23.00. Quantity discount.

Producer: Huntington Two Simulation Package, Software Distribution Center 1-2, Digital Equipment Corp., 146 Main Street, Maynard, MA 01754

HABER

R. Edens and K. Shaw, Chelsea College, London

Playing Data
Copyright: 1978
Age Level: college
Special Equipment: computer; BASIC
Packaging: 5 copies of 11-page student notes; 15-page teacher guide; software (interactive BASIC program, 400 lines of code)

Description: The production of ammonia by the Haber process is an important process in chemical industry, and yet one not easily performed by students in conventional laboratory investigations. The *Haber* simulation provides students with the opportunity to study the Haber process and how the various conditions (temperature, pressure, catalyst and reactant concentration ratios) influence the course of the reaction (i.e., the time required to reach equilibrium and the equilibrium yield of ammonia). In Investigation 1, Properties of a System at Equilibrium, the student specifies the constant molar ratio, the inital temperature, the increase in temperature, and the constant pressure, and selects to vary the pressure, temperature or initial hydrogen-to-nitrogen ratio. The program then calculates and displays the percentage yield of ammonia against the varying parameter for the conditions chosen by the student.

In Investigation 2, The Haber Process, the student specifies which catalyst to use (none, osmium, tungsten, molybdenum, iron or manganese dioxide), and then chooses a temperature and pressure for the investigation. At time intervals set by the student, the simulation calculates the amount of ammonia formed, expressed as a percentage of ammonia in the equilibrium mixture.

Before using this unit, students should be introduced to the concepts of the equilibrium law, Le Chatelier's Principle of Equilibrium, the rate law from kinetics, the ammonia synthesis mechanism, the effect of temperature on the rate constant, and the effects of catalysts. (CONDUIT Pipeline)

Cost: $25.00; additional student notes 25 for $15.00; additional teacher guide $1.50

Producer: CONDUIT, Box 388, Iowa City, IA 52240

IDGAME

Fred Hornack, Nancy Hetzel, and Molly Hepler; North Carolina Educational Computing Service

Playing Data
Copyright: 1975
Age Level: college
Special Equipment: computer; FORTRAN
Packaging: 52-page user's guide; software (interactive or batch FORTRAN program, 1500 lines of code and 3 data files, 220, 249, and 65 lines of code)

Description: IDGAME is a qualitative organic identification game which uses a data base of 20 organic compounds to teach strategies for analyzing organic compounds. The purposes of *IDGAME* include: (1) teaching qualitative analysis in the "dry lab" setting; (2) reducing the number of unknowns to be analyzed in a regular lab; (3) exposing the student to optimizing techniques through a gaming situation; and (4) establishing the concept of industrial costs related to chemical analysis. To use this program effectively, students should have a background similar to that required for beginning laboratory classes, but no computer background is necessary.

The instructor selects one or more unknowns from the data base and assigns them to one or more students, who then pursue the identification of the unknown by making computer runs to get simulated laboratory test results. The compound name itself is not available as output from the game; only the analytical results of the 41 tests can be requested. (CONDUIT Pipeline)

Note: For more on this game see the essay on computer simulation games by Ronald E. Anderson.

Cost: $65.00; additional student manual $3.00, 2 or more $2.50 each; additional instructor manual $5.00

Producer: CONDUIT, Box 388, Iowa City, IA 52240

INTERP

John Harris, Chelsea College, London

Playing Data
Copyright: 1976
Age Level: college
Special Equipment: computer; BASIC
Packaging: 5 copies of 11-page student notes; 15-page teacher guide; software (interactive BASIC program, 284 lines of code)

Description: This unit on wave superposition is designed to improve students' understanding of the use of models in physics. It pays particular attention to the wave theory of light and to how effective the theory is in explaining observed phenomena.

The Students' Notes provide information that guides the student through three investigations of interference and diffraction phenomena using the computer program. The simple model in the program calculates the intensity due to the superposition of radiation from two sources, or two slits, each having two secondary sources. The complex model in the program allows students to investigate the effects of the number of secondary sources in each slit.

This unit is not intended to replace the usual teaching of interference and diffraction phenomena. It cannot in any way replace actual observation of these patterns, which are, after all, easily enough seen using simple laboratory equipment. The aim is to focus students' attention on the physical model used to "explain" the observations in a way that, because of the mathematics involved, is not simple to do without a computer. The work is intended to encourage a more critical attitude toward the use of models in physics in general, by emphasizing and investigating some of the assumptions (nearly always made but hardly ever mentioned) in this one example. (CONDUIT Pipeline)

Note: For more on these exercises, see the essay on computer simulation games by Ronald E. Anderson.

Cost: $25.00; additional student notes 25 for $15.00; additional teacher guide $1.50

Producer: CONDUIT, Box 388, Iowa City, IA 52240

KSIMS

Dwight Tardy, H. Warren Smith, Joseph R. Denk, and Nancy Hetzel

Playing Data
Copyright: 1978

Age Level: college
Special Equipment: computer; FORTRAN
Packaging: 5 copies of 10-page student manual; 5-page instructor's guide; software (interactive FORTRAN program, 1700 lines of code including 200 comment lines)

Description (The primary objectives of *KSIMS* are to introduce the element of experimental design into the undergraduate study of chemical kinetics and to allow the student to interpret raw kinetic data from an experiment (also designed by the student) so that the balanced chemical equation, rate law, rate constant, activation energy, and pre-exponential factor can be obtained.

The student "performs" kinetic experiments using *KSIMS* by setting the experimental parameters of temperature, initial concentration of reactants, and which substances are to be analyzed at predetermined times. The student has the option of introducing experimental errors (standard deviations) associated with each of the above variables so that a "real" experiment can be simulated.

Twenty unknowns (to the student) can be explored using *KSIMS*. All of the reactions are symbolically represented. The reactions are either first or second order overall; equilibrium reactions (those involving back reactions) are not treated. Any reaction (first or second order rate laws) with new Arrhenius variables can be added to *KSIMS* repertoire.

To use *KSIMS,* the student should be familiar with the general principles of chemical kinetics. (drawn from CONDUIT Pipeline)

Cost: $65.00; additional student manuals 10 for $15.00; additional instructor's guide $1.50

Producer: CONDUIT, Box 388, Iowa City, IA 52240

LINKOVER

P. J. Murphy, Chelsea Science Simulation Project, University of London

Playing Data
Age Level: college
Special Equipment: computer; BASIC
Packaging: 5 copies of 11-page student notes; 15-page teacher guide; software (interactive BASIC program, 380 lines of code)

Description: This program simulates genetic mapping based on Morgan's theory. Through investigation, the arrangement of the ten genes within the chromosome is established.

In this program the objective is to establish the arrangement of the ten genes. The user chooses 3 of the 10 linked genes in the chromosome. Automatically, a strain of the organism which is homozygous recessive for these particular genes will cross with a strain which is homozygous dominant. The results of 100 cross-tests will be printed for these genes. These results can then be used to map the chromosome. After the results, the user may choose to continue or stop. (TIES)

The Teachers' Guide includes instructions to modify the model to simulate more closely the natural situation by: altering the number of offspring; randomizing the number of offspring; altering a gene symbol; altering the locus of a gene; adding and removing genes from the linkage group. (CONDUIT Pipeline)

Note: For more on this simulation, see the essay on computer simulation games by Ronald E. Anderson.

Cost: $20.00; additional student notes $25.00 for 15; additional teacher guide $1.50

Producer: CONDUIT, Box 388, Iowa City, IA 52240

LOCKEY

J. Friedland, B. Rosen, and D. Sobin; Huntington Two Computer Project

Playing Data
Copyright: 1971
Age Level: grades 10-12
Number of Players: 1-class unit
Special Equipment: computer; BASIC

Description: This biochemical program investigates the lock and key model of enzyme action. The user can test various hypotheses by regulating the amount of enzyme and the type and amount of inhibitor.

An instruction listing options is contained in the program. The enzyme to be investigated is acetylcholinesterase, with the codes for the inhibitors being: 1–Ammonium, 2–Dimethylamine, 3–Methylamine, 4–Prostigmine, 5–Trimethylamine, 0–No inhibitor.

The user inputs the following information: (1) the amount of enzyme (0 to 3 millimoles), (2) code number of the inhbitor to be investigated, (3) the amount of inhibitor, and (4) type of ouput (1 = chart, 2 = graph). Thus, this program can be used in an experimental setting of regulating variables and controls. (TIES)

Cost: individual packet (student, teacher, resource manuals and paper tape) $3.00; without tape $2.00; classroom packet $23.00. Quantity discount.

Producer: Huntington Two Simulation Package, Software Distribution Center 1-2, Digital Equipment Corp., 146 Main Street, Maynard, MA 01754

NEUTRON ACTIVATION ANALYSIS

Ted Hopkins, CONDUIT

Playing Data
Copyright: 1976
Age Level: college
Special Equipment: computer; FORTRAN
Packaging: 21-page student manual; 32-page teachers guide; software (interactive FORTRAN program, 873 lines of code)

Description: This program was inspired by a program RADIO included in the text *Numerical Methods in Chemistry,* K. Jeffrey Johnson, University of Pittsburgh, and an experiment conducted at the Radiation Center, Oregon State University. In this experiment, a short length of wire (an alloy of aluminum and indium) is activated by neutron bombardment in the TRIGA reactor. The radioisotopes which are produced, 2813 Al and 11649 In, each decay with different half-lives. The total decay curves for sample wires of different compositions are simulated by the program. The half-lives and initial activities of the various samples can be determined by analyzing the decay curves. The student chooses a sample number, how soon to begin counting the sample after ejection from the reactor, and details about the counting of the sample (e.g., counting times, intervals, duration). The total count rate calculated by the program is given an appropriate scatter through the use of a random number subroutine. The data are then presented in tabular form, followed by a log plot of activity versus time for the same data. The log plot enables the student to immediately decide whether proper decisions were made with regard to the counting parameters for the sample. The experiment can then be redone if changes are needed.

The material is suitable for students in beginning courses in physical chemistry or nuclear chemistry. First-year general chemistry students should have studied course work related to radioactivity before using the program. The program should be particularly useful as supplemental work for laboratory courses in physical, nuclear and general chemistry. (CONDUIT Pipeline)

Cost: $35.00; additional teacher's guide $3.00; additional student manual $2.00, 2 or more $1.50 each

Producer: CONDUIT, Box 388, Iowa City, IA 52240

NEWTON

J. Harris, Chelsea Science Simulation Project, University of London

Playing Data
Age Level: college
Special Equipment: computer; BASIC
Packaging: 5 copies of 10-page student notes; 11-page teacher guide; software (interactive BASIC program, 122 lines of code)

Description: This unit is designed to help the student achieve an appreciation of how the application of Newton's Second Law and Law of Gravitation leads to the prediction of satellite orbits. The computer program on which this unit is based uses an iterative method to calculate the path of a projectile launched horizontally. The student is instructed to find the initial velocity needed for the minimum (circular) orbit. (CONDUIT Pipeline)

Cost: $20.00; additional student notes 25 for $15.00; additional teacher guide $1.50

Producer: CONDUIT, Box 388, Iowa City, IA 52240

PH

Huntington Two Computer Project

Playing Data
Age Level: grades 10-12
Number of Players: 1-class unit
Special Equipment: computer; BASIC. Special language feature: TAB

Description: This program consists of three laboratory investigations dealing with the PH specificity of enzymes. The first investigation deals with why many enzymes exhibit a bell shaped curve when their activity is plotted against different PH values. The second investigation is an introduction to the importance of structure in an enzyme, and the third exercise deals with a specific enzyme, acetylcholinesterase.

Instructions are contained in the program. The user sets the number and type of ionizable amino acids present, and whether or not each of the amino acids is in a charged condition. The user also sets the PH range for the experiment. Output for each run is in the form of a graph with the specified PH range plotted against the enzyme activity level. (TIES)

Cost: individual packet (student, teacher, resource manuals and paper tape) $3.00; without tape $2.00; classroom packet $23.00. Quantity discount.

Producer: Huntington Two Simulation Package, Software Distribution Center 1-2, Digital Equipment Corp., 146 Main Street, Maynard, MA 01754

RKINET

A.W.B. Aylmer-Kelly, Chelsea Science Simulation Project, University of London

Playing Data
Age Level: college
Special Equipment: computer; BASIC
Packaging: 5 copies of 10-page student notes; 10-page teacher guide; software (interactive BASIC program, 170 lines of code)

Description: This BASIC simulation is intended to (1) extend students' laboratory experience and understanding of reaction kinetics by enabling them to carry out a wider range of investigations and (2) help students understand the relationship between a mathematical model and reality. The simulation model, which is based on data from real experiments, will broaden students' knowledge of first- and second-order reactions, rate constants, concentrations, and the effect of variation of temperature on reaction rate. (CONDUIT Pipeline)

Cost: $20.00; additional student notes 25 for $15.00; additional teacher guide $1.50

Producer: CONDUIT, Box 388, Iowa City, IA 52240

SCATTER

J. Harris, Chelsea Science Simulation Project, University of London

Playing Data
Copyright: 1975
Age Level: college
Special Equipment: computer; BASIC
Packaging: 5 copies of 15-page student notes; 17-page teacher guide; software (3 interactive BASIC programs, ranging from 132 to 159 lines of code)

Description: This package contains three models simulating experiments for student investigation of particle scattering. The experiments are:

(1) "Marbles" are scattered by hard, massive objects of regular shape. From the scattering produced, students can infer the shape and size of the object.
(2) The particles are scattered by a hard object or by an "inverse-square scatterer." In the latter case the particle's energy has an effect on the scattering produced and the "size" of the scattering object depends on the energy of the probing particles. Once again students are asked what the scatterer is and to estimate its "size."
(3) A simulation of scattering of alpha particles by a thin foil, using a simple nuclear model. Students can vary parameters (metal, foil thickness, energy of alpha particles) and are asked to decide, on the basis of their own "experiments," whether a hard sphere or inverse square scattering model of the atoms in the foil was used, and also to check predictions of this model against actual experimental results.

The shape and size of the objects in the first two programs are determined by random numbers generated by the computer. Thus, they are unknown to the student (and the teacher), neither of whom knows the "right" answer. (CONDUIT Pipeline)

Cost: $20.00; additional student notes 25 for $15.00; additional teacher guide $1.50

Producer: CONDUIT, Box 388, Iowa City, IA 52240

SCATR1/SCATR 2/SCATR 3

Huntington Two Computer Project

Playing Data
Age Level: grades 10-12
Special Equipment: computer; BASIC

Description: *Scatr* simulates alpha particle scattering as demonstrated in the laboratory and according to three theoretical models of the atom: the hard sphere, the Thomson, and the Rutherford or nuclear model.

The scattering package includes three separate programs:: SCATR1 contains a simulation of Rutherford's experiment using a thin gold foil and the 5 MeV alpha particles emitted by radioactive polonium. This program permits students to gather information concerning the scattering of alpha particles from the gold foil. SCATR2 produces theoretical angular distribution graphs for three atomic models to allow students to compare "actual" and theoretical results. The theoretical models include the hard-sphere model of kinetic theory, the plum-pudding model of J. J. Thomson, and Rutherford's nuclear model. SCATR3 calculates and plots trajectories for alpha particles scattered from individual Rutherford-type atoms. (Huntington Two)

Cost: individual packet (student, teacher, resource manuals and paper tape) $3.00; without tape $2.00; classroom packet $23.00. Quantity discount.

Producer: Huntington Two Simulation Package, Software Distribution Center 1-2, Digital Equipment Corp., 146 Main Street, Maynard, MA 01754

SLITS

A. Caggiano and E. A. Williamson, Huntington Two Computer Project

Playing Data
Copyright: 1971
Age Level: grades 11-12
Special Equipment: computer; BASIC. Special language features: RANDOMIZE, TAB

Description: The computer program in this unit simulates Young's double-slit experiment. The program allows the user to manipulate three of the key variables in the experiment, one at a time. The three variables are: (1) W, the wavelength (in angstroms) of the light source; (2) D, the distance (in millimeters) between slits A and B; (3) L, the distance (in meters) between the double-slit screen and the viewing screen.

Each time the user specifies a value for W, D, or L, the computer will plot a graph of relative light intensity on the viewing screen versus distance from the center of the viewing screen. On the computer's graph, the vertical axis is the distance axis and the horizontal axis is the intensity axis. A "self test" for users has been built into the program which helps them to determine their degree of understanding of the relationships among L, W, and D. The computer randomly selects a wavelength W for the light source which is not given to the user. The user must try to determine the selected wavelength by varying L and D. An estimate within 10% of the actual value is an acceptable approximation. (Huntington Two)

Cost: individual packet (student, teacher, resource manuals and paper tape) $3.00; without tape $2.00; classroom packet $23.00. Quantity discount.

Producer: Huntington Two Computer Package, Software Distribution Center 1-2, Digital Equipment Corp., 146 Main Street, Maynard, MA 01754

TITRATION

R. W. Collins, K. J. Johnson, and C. T. Furse

Playing Data
Copyright: 1977
Age Level: college
Special Equipment: computer; BASIC or FORTRAN
Packaging: 100-page manual; software (12 interactive BASIC or FORTRAN programs, ranging from 59 to 355 lines of code)

Description: This package covers a variety of topics within the general area of titration phenomena and ionic equilibria, including: aqueous equilibria for mono-, di-, and tripotic acids; metal ion-EDTA titrations; metal ion complexation; the potential of metal/metal ion electrodes versus pH in aqueous ammonia; solubility as a function of pH; processing titration data by the derivative method; and simulating a variety of titration curves. The majority of the programs produce typical laboratory data through simulation of a chemical process. Three of the programs produce student exercises, complete with answer keys for the instructor. These problem sets can be used as lecture assignments or as prelaboratory experience for the students. Another program allows students to process their own pH or potentiometric titration data. The instructor's use of the package is aided by a "usage flowchart" which helps the instructor decide which program(s) suit the needs of the course. By answering questions about the concepts presented in the course, the instructor "steps through" the sequence of programs, and how these programs may supplement classroom and textbook instruction. A thorough presentation of the concept in lecture or through a standard textbook is still necessary; the programs do not stand alone. (CONDUIT Pipeline)

Cost: $75.00; additional user's guide $6.50, 2 or more $6.00 each; no charge for instructor's guide

Producer: CONDUIT, Box 388, Iowa City, IA 52240

COMPUTER SIMULATIONS

SOCIAL STUDIES

CHANGE AGENT

Charles Weinberg, Stanford University

Playing Data
Copyright: 1973
Age Level: high school, college
Special Equipment: computer; BASIC
Packaging: 11-page student manual; 12-page instructor's manual; software (interactive BASIC program 376 lines of code)

Description: *CHANGE AGENT* is a simulation game intended to help students understand the role of a change agent and the strategies typically used by the change agent to accomplish diffusion of innovation. The student assumes the role of a change agent and develops and implements (via the computer program) a strategy for influencing change in a hypothetical farm community. (The scenario may be easily changed for any situation.) The student "purchases" information about the villagers, such as the percentage of villagers who read newspapers or listen to the radio. A total of seven types of information are available. For each purchase of information, a certain number of days is deducted from the student's total number of days given to achieve adoption (e.g., 50% adoption in 250 days). Based on this information, the student then selects from seven diffusion strategies, such as using newspapers or the radio to create knowledge of the innovation. After each diffusion strategy the program informs the student of the resulting number of adopters. Chance events which affect the number of adopters occur randomly during play of the game, and serve to enhance the reality of the simulation. (CONDUIT Pipeline)

Note: For more on this simulation, see the essay on computer simulation games by Ronald E. Anderson.

Cost: $30.00; additional instructor manual $3.00; additional student manual $2.00, 2 or more $1.50 each

Producer: CONDUIT, Box 388, Iowa City, IA 52240

CHEBO

James Sakoda

Playing Data
Age Level: college
Special Equipment: computer; FORTRAN

Description: *CHEBO*, which stands for "checkerboard," operates a simple model of social interaction based upon some elementary principles of field theory and attitude/attraction theory. In brief, actors move closer or apart depending upon their current interpersonal distances, their attitudes toward members of their own group, and their attitudes toward members of another group. For example, one group, the social workers, is represented by open circles and the other group, lost souls, is represented by solid circles. At cycle 0, they are randomly dispersed around the checkerboard space, but by cycle 15, the final structure, the social workers are huddled in the center and the lost souls are spread around the outer boundary. This occurred largely because the initial starting conditions defined the social workers as having positive attitudes toward themselves as well as toward the outgroup, while the lost souls (solid) held negative attitudes both toward themselves and toward the outgroup. The model is interesting because it demonstrates a variety of intersocial structures evolving from a simple social process. (RA)

Note: For more on this model, see the essay on computer simulation games by Ronald E. Anderson.

Cost: $15.00

Producer: James Sakoda, Department of Sociology, Brown University, Providence, RI 02912

COGNITIVE PSYCHOLOGY

William Bewley, Project COMPUTE, Dartmouth College

Playing Data
Age Level: college
Special Equipment: computer; BASIC
Packaging: 102-page student manual; 96-page instructor's manual; software (6 interactive BASIC programs, ranging from 150 to 290 lines of code)

Description: *Cognitive Psychology* was written for students in cognitive phychology or experimental psychology. It consists of a series of separate programs to simulate live experiments in each of these areas: pattern recognition, short-term memory, long-term memory, discrimination-net learning, concept learning, problem-solving, and decision-making in mixed-motive games. The package is modular, and an instructor can select out a single simulation program for a specific laboratory exercise. (RA)

Note: For more on these exercises, see the essay on computer simulation games by Ronald E. Anderson.

Cost: $50.00; additional instructor manual $2.00; additional student manual $4.50, 2 or more $4.00 each

Producer: CONDUIT, Box 388, Iowa City, IA 52240

FAIL-SAFE

Bruce Tipple

Playing Data
Copyright: 1976
Age Level: grade 7 and up
Number of Players: 12
Playing Time: 1-2 hours
Preparation Time: 2 hours (est.)
Special Equipment: 12 K computer

Description: This simulation of the edge of the end is based on the 1962 bestseller of the same name. The situation is as follows. Somehow, six American B-52s armed with hydrogen bombs on a routine mission have passed the point of recall, or Fail-Safe, and appear to be proceeding toward an attack on Moscow. The crews have standing orders not to receive radio transmissions after they have passed the fail-safe point, and there is no way to call them back. The Soviets know. Players assume the roles of the President of the United States and eleven advisors, and they must decide what to do. At fifteen-minute intervals the President must choose one of several options presented by the computer. For example, upon first learning of the situation, the President must decide whether to try to regain radio contact with the group, launch a full-scale attack, pursue the bombers with American fighters, or do nothing. After each decision is made, the computer responds with an analysis of the probable consequences. The game ends when the crisis is resolved or the nuclear holocaust begins.

Note: For more on this simulation, see the essay on social studies simulations by Michael J. Rockler.

Cost: (1) time-share system offers 17-page instruction booklet $1.25 (plus 33% out-of-state surcharge); (2) software under development

Producer: (1) Minnesota Educational Computing Consortium, 2520 Broadway Drive, Lauderdale, MN 55113; (2) for information contact Bruce Tipple, c/o Social Science Education Consortium, 855 Broadway, Boulder, CO 80302

GHETTO

Judy Edwards, University of Iowa

Playing Data
Copyright: 1977
Age Level: high school, college, teacher training
Special Equipment: computer; BASIC
Packaging: 20-page manual; software (3 interactive BASIC programs, ranging from 200 to 900 lines of code)

Description: The *Ghetto* simulation is designed to sensitize its players to the emotional, physical, and social world the disadvantaged inhabit. Players assume the role of one or more of ten "ghetto" residents, and experience vicariously the economic pressures that drive people into crime, welfare, and community action. The players attempt to improve their lives by investing hours in various activities and collecting as many reward points as possible. Each ghetto resident is described by sex, age, marital status, level of education, number of children, and number of hours to invest. Investments may be made in school, work, hustling (illegal activity), relaxation, and welfare. As capricious events occur which negate the player's investments and reduce or eliminate reward points, the player experiences frustration akin to the frustrations of the disadvantaged.

Ghetto was originally designed for an HP2000 system. Therefore, some difficulty may be experienced in implementing this package. In particular, *Ghetto* makes extensive use of random access data files and the CHAIN statement. (CONDUIT Pipeline)

Comment: While the computerized version of *Ghetto* may be played by individuals, the major benefits of the game are derived through post-game discussion where players can share experiences, compare strategies, and determine if what they have experienced is realistic.

Based on the CONDUIT review of *Ghetto* in preservice and inservice training of professional educators, the package offers potential for use in professional education courses, from introductory courses through inservice training activities. While *Ghetto* does not reflect an accepted formal social theory of life in the ghetto, it does offer potential for use in support of the teaching process. Its particular uses in education range from an example of a nonmathematical simulation in courses on computers in education, to general methodology courses and courses on teaching the disadvantaged. (CONDUIT)

Note: For more on this simulation, see the essay on computer simulation games by Ronald E. Anderson.

Cost: $50.00

Producer: CONDUIT, Box 388, Iowa City, IA 52240

LIMITS

Huntington Two Computer Project

Playing Data
Age Level: high school, college
Special Equipment: computer; BASIC

Description: *Limits* is a simulation which explores the need for long-term planning for the world's future. Students work with a world model to get an extrapolation of present trends in areas of population dynamics, industrial growth, consumption of resources, environmental pollution, and agricultural production.

The *Limits* simulation is made up of five parts or subsystems that are considered crucial to the survival of our world community. Population (P) is the total number of people in the world. The 1970 figure is estimated at 3.6 billion, with a growth rate of 2% a year. Food Supply (F) is the world's annual production of all foods averaged among the world's population in units of Calories per person per day. Natural Resources (R) is the total supply of nonrenewable materials presently used for industrial and agricultural production. Industrial Output (O) is the world production of manufactured goods and services. Pollution (X) is all industrial, agricultural, and human non-recycled wastes.

At the start of each run, users are asked if they would like a standard run, which produces a printout in which present trends in world growth are allowed to continue unchanged until 2100. Non-standard runs allow users to change any or all of the following variables: B–Birth Rate, D–Death Rate, F–Food Supply, R–Resource Usage Rate, O–Industrial Output Growth Rate, X–Pollution Generation Rate. In standard and nonstandard runs, users may request the printout in the form of a graph, table, or both. (TIES)

Note: For more on this simulation, see the essay on computer simulations by Ronald E. Anderson.

Cost: individual packet (student, teacher, resource manuals and paper tape) $3.00; without tape $2.00; classroom packet $23.00. Quantity discount.

Producer: Huntington Two Simulation Package, Software Distribution Center 1-2, Digital Equipment Corp., 146 Main Street, Maynard, MA 01754

MALAR

J. Friedland, A. Frishman, E. A. Williamson, and S. Hollander; Huntington Two Computer Project

Playing Data
Copyright: 1973
Age Level: grades 7-12
Number of Players: 1 or more
Special Equipment: computer; BASIC; Special language features: RANDOMIZE, FNR

Description: A computer program which allows the user to attempt to control a malaria epidemic provides a context within which to study the biological, economic, social, political, and ecological aspects of a classic world health problem.

This unit focuses on malaria eradication as a classic example of a world health problem. Although the materials in the unit consider the biological and medical aspects of the problem, they by no means stop there. The problem is put into economic, social, political, and ecological perspectives, and is treated as a representative case study of a health problem which involves a multiplicity of concerns for its solution.

The program *MALAR* simulates the attack phase of a malaria eradication plan. It allows the user to attempt to eradicate malaria from a given area within five years. The user can try to do this with or without a budget limitation. (Huntington Two)

Cost: individual packet (student, teacher, resource manuals and paper tape) $3.00; without tape $2.00; classroom packet $23.00. Quantity discount.

Producer: Huntington Two Simulation Package, Software Distribution Center 1-2, Digital Equipment Corp., 146 Main Street, Maynard, MA 01754

SIMSEARCH

Stephen Wietling, University of Iowa

Playing Data
Age Level: college
Special Equipment: computer; BASIC

Description: *SIMSEARCH* provides hypothetical scenarios for sequential decision-making. *SIMSEARCH* poses problems of a sociological nature and then cycles students through a number of decisions which allow them to construct a full-scale research design. The package is particularly useful in demonstrating the heirarchical nature of design decisions; that is, the way decisions sequentially reduce the range of the remaining choices. The importance of careful research planning is stressed, as is the close link between theory and method. The six design exercises deal with research problems in the areas of deviant behavior, family, sex role socialization, sociology of sport, minority group relations, and the sociology of youth. *SIMSEARCH* simulates research in each of these areas by making students feel as though they are managing a research project and constructing design plans. (RA)

Note: For more on this package, see the essay on computer simulations by Ronald E. Anderson.

Cost: request price from producer

Producer: CONDUIT, Box 388, Iowa City, IA 52240

USPOP

James Friedland and Stuart Hollander, Huntington Two Computer Project

Playing Data
Copyright: 1973
Age Level: grades 7-12
Special Equipment: computer; BASIC. Special language feature: TAB

Description: *Uspop* is a highly flexible human population model. In order to make student use easier, the model has been oriented toward investigation of the United States population projections. *Uspop* uses simulation techniques as a stimulus to learning in the teaching of many key demographic concepts involving population growth and age distribution. Students play the role of demographers projecting future population trends.

The student can investigate the effects of fertility, age of mother at birth of child, sex ratio of the offspring, and age-dependent mortality on population size and structure. Through use of 1970 census data, held in DATA statements, the student need enter only a few of the required inputs. (Huntington Two)

Cost: individual packet (student, teacher, resource manuals and paper tape) $3.00; without tape $2.00; classroom packet $23.00. Quantity discount.

Producer: Huntington Two Simulation Package, Software Distribution Center 1-2, Digital Equipment Corp., 146 Main Street, Maynard, MA 01754

UNCLASSIFIED

DATACALL: A COMPUTER-BASED GAME FOR TEACHING STRATEGY

Richard R. Johnson, Exxon Education Foundation

Playing Data
Copyright: 1970
Age Level: grade 10-graduate school
Prerequisite Skills: some understanding of statistical techniques
Number of Players: 1 or more in one or more teams; maximum players and teams limited only by materials and computer access
Playing Time: 1-2 hours or more
Preparation Time: I am assuming that the teacher or administration of the *DATACALL* game is already familiar with statistical techniques of data analysis and has some familiarity with simple forms of experimental design in research. In terms of catching on to the general way that the materials are used and being able to hand out student materials and explain them to a group, the preparation time should be relatively brief. However, my experience with instructors suggests that most will want to produce their own simulations and develop a game structure which will fit their goals in teaching the subject matter. Under these conditions the preparation time is significantly longer depending upon how sophisticated the instructor wants to be in designing simulations and developing games.
Supplementary Material: students may work along with a number of different books on experimental design and scientific research.
Special Equipment: IBM 1130: FORTRAN IV (others could be used, but some program changes would be necessary)

Description: *DATACALL* is a label for a group of data-generating simulation programs that are used in a particular fashion within a game format to simulate experimental research. At the present time, there are about ten different simulations which have been developed for this purpose, and any of these could be used within about a half-dozen different ways to simulate experimental research via a game structure.

Designed to teach certain basic strategies of experimental research, *DATACALL* can also serve to make relevant the use of statistical tests to analyze data and to provide a medium in which students can get a great deal of practice with statistical analysis.

Each player designs a computer-ready experiment, deciding on which variables to control, the numbers and arrangements of experimental groups, and the number of individual pieces of data to collect. Having run the simulated experiment, the player interprets the data and, in light of any new information gained, constructs the next round of the experiment.

Upon receiving the computer output for each round of play, the student receives a payoff in points in relation to the information gained. Students may play individually to try to build up point totals, they may play against a set of norms for the points they ought to achieve, or they may play competitively for the highest score. The games that have been experimented with so far have tended to emphasize cooperation, with students sharing information and at times working together as research teams in attacking the computer simulation problem. The kind of game that is developed depends on what the instructor wants to produce, and this can be adjusted by changing the rules or the cost-payoff structure of the game. (DZ)

Comment: One of the strengths of this approach is that there is a great deal of flexibility in the types of simulations which can be developed (besides psychology, simulations have been developed in chemistry, physics, and sociology), and any of these simulations can be used in one of a number of different game formats.

The use of the computer to simulate some problem situation is certainly not unique. However, imbedding computer simulations of research problems in a game format allows the development of relatively open-ended exercises by which students can learn research strategy. The game format ensures continuous feedback to allow students to gauge to what extent they are using the appropriate strategy. The game format also provides an excellent motivational device by which students come to see and search for the kinds of things they need to learn in order to master the game.

Results of field tests and of outside evaluation are available from Dr. Dana Main, Division of Behavioral Studies, West Virginia College of Graduate Studies, Institute, WV 25112. (DZ)

Cost: $4.00 kit of example materials including student materials for an individual game, program list, and a discussion of the rationale behind the game and some hints about how to build simulations to meet the purchaser's uses.

Producer: Ms. Norma Berry, Earlham College, Richmond, IN 47374

DECISION MATHEMATICS OPERATIONAL GAME

Geoffrey Churchill and Rachel Elliott Churchill

Playing Data
Copyright: 1974
Age Level: entry-level, business-oriented college mathematics
Number of Players: class
Playing Time: about one-fourth of course classroom time
Special Equipment: computer; BASIC (time-sharing system required) or FORTRAN (batch)

Description: This game was designed to be an integral part of a college entry-level mathematics course oriented to business decision-making that puts mathematical concepts and skills "into a goal seeking, model building, situational problem solving context." No previous experience in business, economics, or accounting is necessary. Most assignments use matrix algebra; some involve calculus, linear programming, or free-form modeling. A table in the instructor's guide lists the 25 assignments by decision topic and methodology; for example: forecast average demand for product 2 (quadratic curve fit), optimize product mix (linear programming), make or buy subassembly 3 (general modeling; calculus of extrema).

The instructor's guide describes the decision, demand, production, and financial structures in the model, discusses planning and using the game in the classroom and implementing and operating the BASIC and the FORTRAN versions. The game has been extensively tested and has gone through several revisions. (AC)

Cost: moderate, depending on volume ordered

Producer: Department of Quantitative Methods, Georgia State University, University Plaza, Atlanta, GA 30303

EXPER SIM (MESS) MODEL BUILDER'S KIT

Dana Main, University of Michigan; Robert Stout, CONDUIT

Playing Data
Copyright: 1978
Age Level: college, university
Special Equipment: computer; FORTRAN
Packaging: 50-page readings on *EXPER SIM*; 125-page model builder's manual; 125-page programmer's manual, software (interactive or batch FORTRAN, programmer version *MESS* driver and required subroutines [5,200 lines of code], student version *MESS* driver and required subroutines [4,100 lines of code]–about half of lines in drivers are comments, 24 model-building subroutines ranging from 30 to 230 lines of code, a random number table, and sample model subroutines)

Description: *EXPER SIM* is a set of computer programs and associated instructional techniques using simulation to teach research methodology in a variety of disciplines. While the approach was originally introduced in psychology, it has spread to applications in many other disciplines. The computer programs simulate a limited set of phenomena. The student, by interacting with the computer programs, learns about the particular phenomenon, and develops skills in asking research questions.

The student specifies the particular research by indicating the kinds of data and the conditions of collecting the data. Thus, each student in a class can conduct a separate, unique experiment. Since the computer generates the data, specialized research equipment is not needed. Computer generation of the data also eliminates the time-consuming data collection step of real experiments, thus permitting students to design and analyze complex studies in a short period of time. In response to each design, the computer returns sets of realistic data to represent pointer readings, counts, scores on tests, and other types of measurement. Students then analyze the data, draw conclusions, and design new experiments. As they compare the conclusions from several experiments, they learn which types of research design apply to a given inquiry.

This Model Builder's Kit is intended for instructors who wish to create their own *EXPER SIM* model. If you are interested in using the *EXPER SIM* models current available from CONDUIT, see *Imprinting* and *Schizophrenia.* (CONDUIT Pipeline)

Comment: The *EXPER SIM* approach is in use by about 50 colleges and universities so far, where individual instructors have developed numerous variations and applications. For further information, contact Dr. Dana Main at the Division of Behavioral Studies, West Virginia College of Graduate Studies, Institute, WV 25112. (AC)

Cost: model kit $85.00
Producer: CONDUIT, Box 388, Iowa City, IA 52240

IMPRINTING

D. W. Rajecki, University of Michigan

Playing Data
Copyright: 1978
Age Level: college
Special Equipment: computer; FORTRAN
Packaging: 5 copies of 5-page student guide; 20-page instructor's guide; software (interactive or batch FORTRAN, *EXPER SIM (MESS)* student version driver [5,100 lines of code], and model subroutine [60 lines of code])

Description: This model operates under the pedagogy and programs of *EXPER SIM (MESS)*. The unit is designed to provide students with the opportunity to investigate, through the application of research design, the theories of imprinting–in particular, imprinting in young precocial birds. The student studies the behavior of chicks as a function of target type, rearing conditions, age, arousal level and method, and number of tests.

In the typical application, the student first reads a scenario (provided with the unit) which describes the problem and includes some published research related to the problem of imprinting. Then the student designs an experiment by specifying the values of the independent variables available in the model. The *EXPER SIM* program then calculates the appropriate value of the dependent variable for the number of subjects specified by the student. The model algorithm, which computes the individual subject's score, provides for sampling error so that the scores derived from replications of an experimental design differ according to a distribution function specified by the computer model. Summary statistics are provided, and the student evaluates the results according to the original hypothesis. Based on these findings, the student proceeds to design additional experiments which, hopefully, follow from previous ones. Finally, the student prepares a report of his or her research written in APA format, or makes a presentation of the findings in class. The instructor can shape the use of this model by directing students' attention to specific independent variables. (CONDUIT Pipeline)

Cost: $35.00; additional student guides 10 for $10.00; additional instructor's guide $2.50

Producer: CONDUIT, Box 388, Iowa City, IA 52240

METRO-CHP

Playing Data
Copyright: 1973
Age Level: college, professional
Number of Players: 30-120
Playing Time: 4-10 4-hour rounds
Special Equipment: IBM 360-50 or larger

Description: *METRO-CHP* is designed to illustrate community response to a comprehensively planned system for the delivery of medical services. The simulation is intended for use by professional medical practitioners, administrators, and students. Participants "assume roles of various community decision-makers and are faced with decisions analogous to those made by their real-life counterparts. These decisions are then processed in a computer program which models the interactions and produces the results and effects on the simulated community." Specifically, these roles include health planners, politicians, spokesmen for interest groups, businessmen, hospital administrators, land developers, public planners, and representatives of public health departments. Issues raised in the game include the effects of Social Security and General Revenue Sharing, regional health plan development, the coordination of the delivery of services, and facilities expansion. (DCD)

Cost: Not commercially available but may be played by contacting the producer

Producer: COMEX, Davidson Conference Center, University of Southern California, Los Angeles, CA 90007

OREGON

Don Rawitsch, Minnesota Educational Computing Consortium

Playing Data
Age Level: elementary, high school
Special Equipment: computer; BASIC

Description: *Oregon* simulates a trip over the Oregon Trail from Missouri during the mid-1850s. The main challenge is to survive the full 2000 miles by spending money wisely, by typing "hunt" words, and, to a certain extent, by chance. The simulation is designed for elementary and secondary history or social studies courses and is intended to give students a better "feeling" for what the westward journey was like for the rugged individuals who attempted it. *Oregon* is very popular, and the secret to its success in the market place is that students learn something about early American history while they are having fun. (RA)

Note: For more on this simulation, see the essay on computer simulation games by Ronald E. Anderson.

Cost: Write for information on documentation and price

Producer: Minnesota Educational Computing Consortium, 2520 Broadway Drive, Lauderdale, MN 55113

QUANTITATIVE EXPERIMENTAL ANALYSIS

Richard Richardson and David Stones, University of Texas, Austin

Playing Data
Copyright: 1978
Age Level: college
Special Equipment: computer; FORTRAN
Packaging: 100-page student manual; 80-page instructor's manual; software (7 modules in ANS FORTRAN, up to about 1000 lines each)

Description: Without the traditional requirement of calculus, the materials present the theoretical foundations of statistical, analysis, estimation, and experimental design. Students empirically determine the relationships of bias, sample size, assumptions, and parameter values to probability functions and tests of significance. Comparisons are made among theoretical and among descriptive mathematical models. In learning these relationships, students learn ways to empirically determine the accuracy, precision, testability, and generality of predictions of models. These procedures are discussed in the context of scientific methodology and philosophy. (CONDUIT Pipeline)

Note: For more on these simulations, see the essay on computer simulation games by Ronald E. Anderson.

Cost: $90.00; additional student manual $4.00, 2 or more $3.50 each; additional instructor's manual $4.00

Producer: CONDUIT, Box 388, Iowa City, IA 52240

SCHIZOPHRENIA

David Malin, University of Michigan

Playing Data
Copyright: 1978
Age Level: college
Special Equipment: computer; FORTRAN
Packaging: 5 copies of 5-page student guide; 20-page instructor's guide; software (interactive or batch FORTRAN, *EXPER SIM [MESS]* student version driver (5,100 lines of code) and model subroutine (60 lines of code))

Description: This model operates under the pedagogy and programs of *EXPER SIM (MESS)*. The unit is designed to provide students with the opportunity to investigate, through the application of research design, theories of the incidence of schizophrenia in genetic and adopted relatives of people with or without known diagnosis of schizophrenia.

In the typical application, students first read a scenario (provided with the unit) which describes the problem and includes some published research related to the problem of schizophrenia. Then they design an experiment by specifying the values of the independent variables available in the model—diagnosis of subject, familial relationship, nature of relationship (adopted or biological), type of diagnosis, method of diagnosis, and sex of subjects. The *EXPER SIM* program then calculates the appropriate value of the dependent variable for the number of subjects specified by the student. The model algorithm, which computes the individual subject's score, provides for sampling error so that the score derived from replications of an experimental design differ according to a distribution function specified by the computer model. Summary statistics are provided, and the student evaluates the results according to the original hypothesis. Based on these findings, students proceed to design additional experiments which, hopefully, follow from previous ones. Finally, students prepare a report of their research written in APA format, or make a presentation of the findings in class. The instructor can shape the use of this model by directing students' attention to specific independent variables. (CONDUIT Pipeline)

Cost: $35.00; additional student guides 10 for $10.00; additional instructor guide $2.50

Producer: CONDUIT, Box 388, Iowa City, IA 52240

See also
Woodbury Political Simulation DOMESTIC POLITICS
Acres URBAN
Apex URBAN
Limits COMPUTER SIMULATIONS/SOCIAL STUDIES
Confrontation INTERNATIONAL RELATIONS
The Middle East Conflict Simulation Game (MESG) INTERNATIONAL RELATIONS
DOMESTIC POLITICS
ECOLOGY
INTERNATIONAL RELATIONS

DOMESTIC POLITICS

ALIEN SPACE SHIP

Playing Data
Copyright: 1975
Age Level: grades 7-12
Number of Players: 18-45 in 6 groups
Playing Time: 2 1-hour periods
Preparation Time: 1 hour
Packaging: 13-page mimeo

Description: This game, simulating the governmental decision-making process after an alien space ship appears over Washington D.C., is intended to help students appreciate the influence of pressure groups and advisors on presidential decisions and recognize the personal, nonrational attitudes that affect these decisions. The author states that the perspectives gained by students in playing the game will help them "develop skepticism in regard to the decisions of political leaders." To begin, one student is chosen as "President" and the remainder are divided into six groups—the Armed Forces Council, Counter Intelligence Agency, Industrial Advisors, Internal Security Division, National Civil Liberties Board, and National Security Council. The President and groups get briefing sheets defining their roles, and each group elects a spokesperson. The Industrial Advisors group, for instance, is told that it represents the largest industrialists in the country and that "if the alien space ship can be destroyed before it lands it might be possible that a scare could be created that would result in renewed contracts for war goods" and that they should use "whatever information or arguments that are necessary to convince the Chief of State to destroy the alien space ship before it lands." The National Civil Liberties Board, on the other hand, has just the opposite objective, "to convince the Chief of State to not destroy the alien space ship but instead to welcome it," using "whatever information or arguments that are necessary." After the groups have read their briefing sheets, the teacher tells them that the "fate of the nation and possibly the world" is in their hands; the teacher then begins to read news bulletins about the progress of the aliens at three-minute intervals. At any point the President may order the visiting craft destroyed. The activity ends when the alien vehicle either has been destroyed or has landed and made its intentions known. The teacher then leads the class in a discussion about how and why the President made the ultimate decision. (DCD)

Cost: $3.00

Producer: Edu-Game, P.O. Box 1144, Sun Valley, CA 91352

AMERICAN GOVERNMENT SIMULATION SERIES

Leonard Stitelman and William Coplin

Playing Data
Copyright: 1969
Age Level: grade 9-college
Number of Players: 32-49

Description: William Coplin and Leonard Stitelman have constructed a series of simulation exercises designed to provide students with laboratory experiences in the field of American politics. The five exercises in the series may be used individually or as a unit of instructions. They involve: (1) a simulated *Constitutional Convention* attempting to present students with some of the political issues confronting the American founding fathers; (2) *Congressmen At Work*, modeling a number of complex issues which a member of the U.S. House of Representatives might be called upon to handle while also meeting time pressures and responding to constituents' demands, party concerns, and maintenance of position or status with other members of the House; (3) *Presidential Election Campaigning*, creating an environment for an election campaign and the factors or strategy choices a candidate must consider in seeking election to office; (4) *Budgetary Politics* and *Presidential Decision-Making*, attempting to introduce the student to the general steps involved in producing the federal budget, the political pressures entering the budgetary decision processes, the role of special interests, and the series of considerations raised for presidential decision-making; and (5) *Decision-Making By Congressional Committees* illustrating the committee system of Congress, the variety of pressures entering into the process, and suggesting some of the possible behavioral patterns of the congressional members.

In constructing this series of exercises, the modelers warn that the student should be made aware of model limitations. The models are intended as suggestive of various aspects of American politics and are merely simulated political environments that contain only partial representations of American political life. The modelers point to their difficulty in building accurate models, since political science as a discipline has developed only partially validated hypotheses to be employed by those attempting to model political behavior. Conflicting interpretations of the data found in the discipline prevent a model based upon one set of tested or reliable conclusions concerning the nature of political reality according to the modelers. Most students of political science would recognize the difficulty of constructing a reliable model of the political system; therefore, the series of simulations should be employed instructionally as an illustration of certain aspects of the political process, rather than as isomorphic models of the various decision-making processes included in the series of exercises.

The exercises are intended to encourage critical analysis by the student of a number of concepts significant in the field of politics. Among these concepts are political role behavior, interest groups pressure and cross-pressure, management of resources, and decision-making dilemmas arising between the so-called "necessities" and the "ideals." The exercises are designed for the high school student or introductory college freshman course and are generally inappropriate as written for more advanced levels of students.

Procedures: The exercise series employs a number of instructional techniques. Structural role-playing is included in the *Constitutional Convention* and *Congressional Committee* exercises. A second technique used in the presidential campaign and budgetary exercises involves a scoring mechanism based on numerical tables. The table operates individually rather than in an environment of social interaction. While attempting to handle and evaluate the factors involved in their decisions, instructors use the numerical tables to score their decisions in relation to political "reality" as represented by the scoring mechanism provided. *The Congressman at Work* exercise also involves individual work requiring evaluation of a series of factors but not pitting the student against a scoring mechanism. A third technique employed in the series of models is the provision of background materials to provide the student with a context for decision-making. In some cases the materials employ historical situations, such as the constitutional convention, but in the majority of exercises the materials are artificial and only partially resemble historical or current political situations. The exercises generally are not intended to focus on political history, but to model factors of the political process. Use of the unit in the classroom will provide students with insights into political process, rather than factual information on past political events.

The exercises do not require elaborate instruction of students or computation or administration by the instructor. Student manuals with complete instructions and forms are provided, as well as an instructor's manual detailing all steps to be taken in conducting the exercise. Optional exercises are also included for the instructor, revising the exercises to permit students to work as teams, to program more structural discussion frameworks, or to include materials centering more on current issues than on a fictional environment. These changes in exercise format may be used to fit the unit to instructional style of a given class.

If the series of exercises is used as an instructional unit, students will be exposed both to individual homework units illustrating decision-making strategies and dilemmas and to structured role-playing problems involving interaction. The exercises should encourage discussion of the political process and the pressures and cross-pressures involved in decision choices. The exercises do not raise community issues as such. The focus is more on the "How it gets done" than on the "what is done."

Comment: These exercises have been used widely in high school and freshman-level college courses. Instructors' reactions generally have been favorable to their value as a means to encourage discussion and critical analysis of political processes. However, the instructor should provide class time for discussion of the models and their limitations in relation to "political reality." Since the modelers warn about their inability to construct isomorphic models of the political process, discussion should center on what factors are included and excluded and whether the models bear any similarity to observed political behavior in the national or local community or to historical data. Such discussion provides an opportunity for exploring additional reading materials and differing views of "political reality" found in the discipline of political science. (From the Robert A. Taft Institute of Government Study on Games and Simulations in Government, Politics, and Economics by Dorothy Dodge, Macalester College)

Note: For more on the series, see the essay on the uses of simulation in the social sciences by Dorothy R. Dodge. For more on *The American Constitutional Convention*, see the essay on historical simulations by Bruce E. Bigelow.

Cost: student handbook for each unit $1.95 list, $1.55 school; instructor's guide for each unit $1.35 list, $1.08 school

Producer: Science Research Associates, Inc., 155 North Wacker Drive, Chicago, IL 60606

AMNESTY

Charles L. Kennedy

Playing Data
Copyright: 1974
Age Level: High School
Number of Players: 15-40
Playing Time: 4-5 class periods or more
Packaging: 8-page foldout student guide, 2-page student guide

Description: This simulation deals with the problem of amnesty following the Vietnam war. Three pages of the student guide deal with an account of American military involvement in Vietnam and protest in the United States and a short history of amnesty in this and other nations. These accounts are followed by a bibliography.

Students score themselves on a value orientation activity of 10 items from strongly agree to strongly disagree ("Any individual, citizen or alien, should have the right to criticize or oppose any government policy or official without fear of penalty or restraint."). Then they assume one of these identities: opponents of amnesty, members of AMEX (organization of exiles), members of the President's Clemency Board, members of the Joint Alternative Service Board, a U.S. attorney, lawyers, hypothetical cases—which include people in the categories of convicted deserters, unconvicted deserters, convicted evaders, unconvicted evaders, undesirable dischargees. Succinct profiles are provided for those who are involved as partisans for or against amnesty.

Once they have their roles, students research for one or two days and prepare their positions and presentations.

The amnesty hearings will run for two or three periods. A detailed procedure is included in the student guide. When the board's chairperson has called the meeting to order, lawyers submit the list of cases to appear. The chairperson decides on the order in which the cases will be heard and calls the first one. The lawyer presents an oral argument. Board members question the individual. This is followed by an executive session at which only board members may speak. They discuss the case and determine the extent of alternative service to be required. This is repeated for each case.

Those who must earn reentry are physically separated in some way from the rest of the participants. Advocate groups supporting and opposing amnesty will be producing newsletters for distribution during lulls in the hearings and may also communicate with lawyers and board members by notes.

The teacher's guide sets out a time sequence for the simulation and some options for teachers of history, English, and government. (AC)

Note: For more on this simulation, see the essay on the uses of simulation in the social sciences by Dorothy R. Dodge.

Cost: $10.00

Producer: Interact, Box 262, Lakeside, CA 92040

BUDGET

Charles L. Kennedy, Gannon College

Playing Data
Copyright: 1973
Age Level: grades 7-12, average to above average ability
Number of Players: 25-40

Description: ***Budget*** is a simulation of the formation of the national budget. Students first receive one of 40 specific identities to role-play (President, cabinet member, Senator, congressperson, lobbyist); four budgetary goals for this identity; and the amount of votes this identity controls in the House of Representatives and the Senate. Next, students either research programs of the specific faction they represent (e.g., the Defense Department, HEW, FCC, ICC, Common Cause, NAACP, AFL-CIO, et al.), or they act as the President and his advisors and develop a national budget for Congressional consideration, or they sit on Congressional Committees and hold hearings on the President's budget. Since all roles control differing amounts of votes in the House and Senate and Since all roles have differing, conflicting budgetary goals, students are compelled to interact politically. Private conferences result in "blocs" supporting or attacking particular budget recommendations; every student also becomes a Congressperson or Senator when the Whole House or Senate formally meets. In this way, students learn that a national budget is formed only after much give and take by various factions, after considerable soul-searching by individuals holding great power, and as a result of complex forces generated by the national economy and past legislation. (producer)

Note: For more on this simulation, see the essay on the uses of simulation in the social sciences by Dorothy R. Dodge.

Cost: $14.00

Producer: Interact, P.O. Box 262, Lakeside, CA 92040

CAMPAIGN

Playing Data
Copyright: 1970
Age Level: grade 7-college, graduate school, civic groups
Number of Players: 16-32
Playing Time: 4-8 hours in 1-2 hour periods
Preparation Time: time to read materials plus 15-minute start-up time

Description: Campaign is a sophisticated and highly realistic simulation game that deals with the American political campaign system. This game is designed to allow senior high school students and adults to participate in decision-making involving the tactics and strategies necessary to produce a winning political campaign. Precinct workers, pressure groups, nominating conventions, political platforms, speech-making, vote-switching, and news coverage are among the elements of the political process that are simulated. The game is complex and challenging, but can be an exciting and effective instructional device.

Campaign is designed for the following purposes and to illustrate certain political ideas and relations:

(1) The procedures which occur before the selection and nomination of a candidate.
(2) The role of party resolutions in affecting political position.
(3) The role of the news media and public sectors in "getting messages and needs" across to the public sectors.
(4) The use of resources such as party organization, dollars, issues, time, party strength, candidate appeal, and news media to secure the election of a candidate.
(5) The effects of immediate past behavior of the political parties in creating opportunities and choices for the public sectors and voters.
(6) The relationship between issues and voting behavior between interest groups and party resources factors, between the amount of time a candidate spends in direct voter contact and the response of the public sectors, and between the party appeal and the news media in voting patterns.
(7) The role of election analysts in predicting strategies, responses, revisions of strategy, and/or adherence to predetermined plans of the political parties.

Campaign covers the following behavioral models:

(1) Planning and strategy development;
(2) resource allocation: human and capital;
(3) evaluation and assessment of past behavior: voting patterns;
(4) policy management and policy application: party decision-making;
(5) influence and bargaining: public sectors, parties, and news media;
(6) social control: campaign fair practices committee;
(7) decision-making at various sociopolitical areas and levels;
(8) problem formulation and processing of information;
(9) media relations and the role of the "free press"; and
(10) collective behavior and persuasion.

Comment: The developer suggests 10 hours as a minimum time allotment for use of the game from start to finish. Experience has shown a minimum of 15 hours to be more realistic, and the game could easily be used for longer periods of time. The game may be played with students more than once; each time it is played, the players develop strategies.

Instructors need to spend a number of hours in preparation before introducing the game to students. They should study approximately 20 different forms and become familiar with the processes involved with each. Students, however, only need to become familiar with the few forms that apply to their roles in the game. The individuals who assume roles of analyst and pollster should not be averse to working along and performing a large number of computations. The game should be played in an area that is spacious enough to provide the political parties with opportunities to develop political strategies in private.

Teachers should plan to spend a minimum of two weeks preparing to use the *Campaign* simulation game. Among the useful books that should be consulted to enlarge the teachers own background are:

Agranoff, Robert	*The New Style in Eelction Campaign*
Sorauf, Frank	*Party Politics in America*

(From the Robert A. Taft Institute of Government Study on Games and Simulations in Government, Politics, and Economics by D. Duane Cummins, Oklahoma City University)

Cost: $162.50

Producer: System's Factors, Inc., 1940 Woodland Ave., Duluth, MN 55803

CITY HALL

Judith A. Gillespie, Indiana University
Playing Data
Copyright: 1972
Age Level: grade 9-college
Number of Players: 14-48
Playing Time: 4-7 45-minute periods
Preparation Time: 3-4 hours

Description: Players assume the roles of voters, candidates, party or group leaders, and radio commentators, all involved in a mayoral election. On the first day of play, students are introduced to the rules of play and their roles, which specify party affiliation, socioeconomic status, and racial or ethnic identity. On the second day, students begin developing newspaper articles and radio programs about the election, bargain over issues, plan campaign strategy, and hold rallies for the candidates. On the third day, public opinion on campaign issues and on candidate preferences is sampled. The radio programs and newspaper articles prepared the day before are presented. Before the election, candidates must make decisions about whether to support or oppose an open-housing law, fund-raising for air pollution control, and expansion of the police force. Other players must analyze the candidates position papers on these issues and decide which candidate to support. An election is held to determine the winning candidate, after which the teacher ends play and leads the class in debriefing. (DCD)

Note: For more on this simulation, see the essay on the uses of simulation in the social sciences by Dorothy R. Dodge.

Cost: $10.85 (supply limited)

Producer: Ginn and Company, Xerox Education Center, P.O. Box 2649, Columbus, OH 43216

COALITION

Playing Data
Copyright: 1972
Age Level: grades 9-12
Number of Players: 16 or more in 10-12 teams
Preparation Time: 1 hour (est.)
Special Equipment: copier

Description: Players assume the roles of one of four presidential candidates, their campaign managers and staff members, political reporters, Congresspersons, and special interest lobbyists. Candidates try to gain the support of the interest groups and win the election. Interest groups try to influence the candidates' positions on the issues, and reporters keep the voters informed. During the first session of the game, the candidates and the eight special interest groups prepare and present a platform, and each interest group presents a resource card to the candidate who most strongly supports its platform. Subsequent game stages simulate the political campaign, the general election, voting by the Electoral College, and, in the event that no candidate wins a majority, election by the House of Representatives. (DCD)

Cost: $19.95

Producer: Changing Times Education Service, Division EMC Corp., 180 East Sixth Street, St. Paul, MN 55101

COLLISION

Rick Reid

Playing Data
Copyright: 1977
Age Level: grades 7-12
Number of Players: 20-35
Playing Time: 12 1-hour sessions
Preparation Time: 3 hours (est.)
Packaging: 32-page booklet

Description: Collision, according to the producer, is a "simulation of the conflicts between Native Americans and the U.S. Government." The simulation consists of three phases. During the three sessions of phase one, participants portray members of tribes that "are slowly being forced off their land and onto reservations and Indian Agents of the U.S. Government." The action of this phase consists of a series of provocations by the Agents (unfair treaties, etc.) and the Native Americans' reactions to these provocations (fighting and rebellion or acquiescence.) Phase two involves the division of the participants into groups. Each group studies specific tribes and their problems and examines one of five case studies of Indian Land Claims Cases that have been ajudicated within recent years. During the final phase, the players as members of Congress formulate policy and legislation to Indian Affairs. (DCD)

Cost: $14.00

Producer: Interact, Box 262, Lakeside, CA 92040

COMMITTEE

Charles Kennedy

Playing Data
Copyright: 1976
Age Level: high school
Number of Players: 25-35
Playing Time: 12 50-minute periods
Preparation Time: 1 hour
Supplementary Materials: reserve shelf of books on congressional procedures
Packaging: 1-page student guide and 35-page teacher guide

Description: Students playing *Committee* learn about the procedures of congressional committees and the process by which a bill becomes law as they role-play Congresspersons, lobbyists, and the President acting upon legislation. To begin, students read about three bills—the Federal Mine Safety Act, the Quality Education Act, and the Clean Air Act—each with three provisions and three alternate clauses. Then, in their assigned roles, each with specific goals and differing amounts of money and votes to use for bargaining, they meet in Democratic and Republican caucuses and at a lobbyists' convention, and begin politicking. Group meetings are followed by three days of subcommittee hearings and two days of full House and Senate judiciary committee meetings involving the final drafting of the three bills. During this time, students are also working on short research papers about actual Congresspersons, U.S. domestic problems, or the congressional system. The teacher's guide provides explicit directions and many forms related to goals and awards information for each player, resource allocation, scoring procedures, the text of the bills, congressional procedures, subcommittee reports, the final draft of the bill, and post-simulation examinations. On the last three days of the simulation, the President approves or vetoes the bills and Congress either sustains or overrides the vetoes; students also hold a debriefing session and take essay and objective examinations on the congressional system. (TM)

Note: For more on this simulation, see the essay on social studies simulations by Michael J. Rockler.

Cost: $14.00

Producer: Interact, Box 262, Lakeside, CA 92040

CONGRESS

Barry R. Lawson and Marjorie S. Jeffries

Playing Data
Copyright: 1976
Age Level: high school
Number of Players: 10-40
Playing Time: 1-2 hours
Preparation Time: 1 hour
Packaging: professionally packaged game including paper playing surface, role cards, chips, dice, instructions, news bulletin cards, scoring sheets, labels, and a bell.

Description: This game is designed to teach students how the United States Congress works and how laws are enacted. Students assume the roles of Senators and Representatives and take part in the passage of four separate bills affecting the environment, agriculture, war, and urban development. The roles specifically state the Congressperson's age, party, profession, seniority, margin of victory in last election, previous voting record, and several vital statistics about his or her home state or district. Eighteen students take additional roles as presiding officer of one of the chambers, minority or majority leader, committee chair, or bill sponsor. At the beginning of the game all of players get 12 influence points; those with a double role get three extra points.

Play begins in the House. The four bill sponsors roll dice to see who goes first. The first bill to be acted upon is read aloud and placed on the playing surface. Rolls of the dice affect the bill's movement between committees, and the sponsor of each bill must shepherd its progress through each chamber. Bills may be amended. Points are awarded to the sponsor of a bill which passes. The game ends when all four bills have been defeated or passed into law. (DCD)

Cost: $22.00

Producer: Harwell Associates, Box 95, Convent Station, NJ 07961

CONSTITUTION

Charles L. Kennedy

Playing Data
Copyright: 1974
Age Level: grades 10-12, average to above average ability
Number of Players: 25-40
Playing Time: 15 hours in 1-hour periods
Supplementary Materials: reading matter suggested in teacher's manual

Description: *Constitution* is a simulation of a convention called to revise the United States Constitution. Students role-play delegates at a modern political convention called to examine whether the Constitution of 1787 is hindering American attempts to cope with current political, socioeconomic, and foreign policy problems. Students' first responsibility is representing their states whose votes they control during the convention; their second responsibility is role-playing members of one of six political factions: Radicals, Moderates, Conservatives, Minorities, the Anti-President bloc, and the Big States bloc. Since states and factions differ on which amendments they feel should be given priority, students find themselves continually caught between conflicting pressures. A total of 35 amendments are considered by the Convention Rules Committee, but due to time pressure, the committee selects only a certain number for research and debate. From these amendments only a few are sent to state ratifying conventions, during which students continue the debates and have the opportunity to ratify these key amendments. Sample amendments from the 35: establishing minimum and maximum age requirements for Congresspersons; ending the seniority system in Congress; limiting presidential power; limiting the Supreme Court's power; and, most controversial of all, decreasing the number of states from 50 to 20 in order to form a more logical union based upon location of resources, population, transportation, and economic and social arrangements. Besides researching various amendments and writing Amendment Justification outlines, students have several opportunities to debate one another and practice political bargaining. Final debriefing discussion questions as well as essay and objective examinations contained in the simulation help students share what they have learned about how our Constitution works and how it might be changed. (TM)

Cost: $14.00

Producer: Interact, P.O. Box 262, Lakeside, CA 92040

COUNCIL

Playing Data
Copyright: 1974
Age Level: grades 5-8 (average to above average); grades 10-12 (below average ability)

Description: *Council* first has students examine positions in local government: mayor, council members, city planners, and others. Once they have this background, interested students file a petition as candidate for one of the five council positions. After candidates prepare and carry out a brief campaign, an election is held. The elected council then picks a major or chairperson and other needed government officials. The remaining students draw identities of community citizens, whose identities are correlated to 20 different ordinances that can be presented at council meetings. Normally, teachers select only eight ordinances which fit the needs and interests of their students (e.g., pollution controls, freeway extensions, raises for public employees, construction of a new council hall, better housing, and dog control). Each ordinance consists of the written ordinance and seven community identities who are directly concerned, either positively or negatively, with this ordinance. The community keeps a chart showing each council member's vote, and students use this chart at election time to determine if a council member's voting record warrants their support. After council members have deliberated on four ordinances, all must stand for reelection before a second cycle of deliberation considers four other ordinances. A council member's chances for reelection are based on his or her voting record and on Fate Cards which reveal both the wise and foolish things a politician sometimes does (being chairperson for a charity drive or being involved, either directly or indirectly, in accepting a bribe). (Producer)

Cost: $14.00

Producer: Interact, P.O. Box 262, Lakeside, CA 92040

DEFENSE

David Rosser

Playing Data
Copyright: 1975
Age Level: high school
Number of Players: class
Playing Time: 6 class periods
Packaging: 8-page foldout student guide, 2-page teacher guide

Description: This simulation takes students through the process of putting together a defense budget. They form six groups: the conventional army group, the defensive missile group, the modern navy group, the offensive missile group, the air power group, and the Armed Forces Committee. The position of each of these is described. Students spend a day or two doing research and then begin preparing their group presentations, getting their figures straight, deciding who will speak on what points, and so on. After the reports are presented to the Armed Forces Committee, it conducts its deliberations–in public, then draws up a budget bill to present to Congress.

Retaining their former interests, participants become members of Congress and hold a session in proper form. After bargaining and debate, congress votes on the bill and its goes to the President. The teacher has the good fortune to be this personage and, after analyzing it, addresses Congress concerning its strengths and weaknesses. The teacher then announces whether the bill will be signed or vetoed. This signals debriefing.

The teacher's guide sets forth a schedule and offers some suggestions for supporting materials. (AC)

Cost: $10.00

Producer: Interact, Box 262, Lakeside, CA 92040

DE KALB POLITICAL SIMULATION

William Harader, Indiana State University; Clifford N. Smith, Northern Illinois University; Marshall H. Whithed, Temple University

Playing Data
Copyright: 1969
Age Level: grade 10-college
Number of Players: 25-75 in 7 teams
Playing Time: 6-8 hours in 15-minute periods
Preparation Time: 2 hours (est.)
Special Equipment: team rooms; director and up to 6-8 umpires; hand calculators, duplicator (stencil, ditto, photocopier); typewriters.

Description: Participants in *De Kalb Political Simulation* assume the roles of political candidates, their campaign organizations, various political pressure groups, and representatives of mass media, all engaged in a political campaign. Different roles have varying degrees of wealth and political influence; some have specified personality characteristics. The decisions participants must make and the behavior they engage in around issues and coalition formation approximate those of an actual political campaign. (AC)

Cost: $5.00 per player manual and umpire manual (1 per participant). Quantity discount.

Producer: Westland Publications, P.O. Box 117, McNeal, AZ 85617

DELEGATE

Charles L. Kennedy

Playing Data
Copyright: 1976
Age Level: grade 10-college
Number of Players: class
Playing Time: 11 class periods
Packaging: 29 printed pages; student material reproducible

Description: After the presentation of some background material on the U.S. convention system, students take and score themselves on a liberal-conservative orientation. On the basis of this, the students form a literal political spectrum and are divided into five groups–radical, liberal, moderate, conservative, and reactionary–of equal size. Each faction gets a basic beliefs sheet and a list of role goals (hoped-for Cabinet positions). Groups begin preparing their strategies, and participants elect a Credentials Committee, which rules on three disputed delegates. Groups prepare testimony for presentation to the Resolutions Committee on seven issues and designate three members each to try for the nine-member committee. After the election, the chairperson of the Resolutions Committee conducts the meeting to consider platform proposals. After presentations are heard, a platform is developed and factions begin preparations for the nomination convention, which wheels and deals through two periods and ends with the election of a Presidential candidate. The following period is devoted to debriefing and the last to objective testing and an essay. (AC)

Cost: $14.00

Producer: Interact, Box 262, Lakeside, CA 92040

ELECT A PRESIDENT

Playing Data
Copyright: 1977
Age Level: grades 9-12
Number of Players: 30-42
Playing Time: 4 50-minute sessions
Preparation Time: 1 hour (est.)
Packaging: 19-page photocopied booklet

Description: In *Elect a President*, students become delegates from seven fictitious states and attend a national convention. There are three political parties, each with a candidate and a platform. The winners at the end of the game are the person elected President, the President's campaign advisor, and the state that was promised the most in the winner's campaign.

Each state has different problems and different interests. The candidates must choose one of three specified options for each of ten separate issues, varying from the use of federal money to develop state parks to the relative allocation of federal funds for welfare. The game structure consists of ten strategy sessions followed by ten voting periods for each issue. Delegates vote after each candidate has taken a stand on each issue. (DCD)

Cost: $3.00

Producer: Edu-Game, P.O. Box 1144, Sun Valley, CA 91352

ELECTION

Joseph Young and Marlene Joan Young

Playing Data
Copyright: 1974 (revised edition)
Age Level: grades 6-12, adult education
Number of Players: 4 or class
Playing Time: 30-60 minutes
Preparation Time: 30 minutes (est.)
Packaging: game board, dice, opportunity cards, tokens, spinner, score sheets, and teacher manual

Description: "The basic idea of this simulation," according to the designers, "is to actively pursue and achieve a career related to public service and political life." Players share a game board divided into four identical sections. Tracks of squares in each section are arranged to form a step pyramid. Sections of these tracks represent, in turn, the pursuit of career opportunities in town, city, county, state, and national government; the Presidential primary states; and the Electoral College.

To play the basic board game, players advance tokens through the career squares according to directions on a spinner. Each career square represents offices that may be held and events that can occur in the course of a political career. All offices are assigned point values, which vary from five for a Justice of the Peace to 100 for a Supreme Court Justice. Each of twelve primary states has a value of 50 points. Players record the value of the squares they land on, and after all four have progressed through the primary section, these scores are totaled. The two players with the highest scores at this point become the nominees for President of the United States. The other two players are chosen by the presidential nominees to be candidates for Vice-President. The presidential nominee with the highest score chooses first.

Both teams then simultaneously proceed through a track of squares symbolizing the Electoral College. They cast dice for each state; the team that rolls the highest number wins the votes assigned to each state. The team with the highest total is elected.

The teacher's guide makes suggestions for using the game in group or class play, having participants seek to influence and gain votes through developing issues, taking positions, and establishing platforms. (DCD)

User Report: Having used Election, I find it to be an illuminating way of presenting "basic" politics to students. The game was used with two classes of low-ability eighth graders. After the first play, lessons on politics, elections, and offices were introduced with time periods spaced between lessons for students to apply what they had learned to the game. Through this game my students are able to understand local government, state government, etc. and the role of each office and candidate. A problem many citizens have with our form of electing officials is understanding why we have primary elections. This game helps the teacher help the students gain this understanding.

The only area in which I would question the validity of this game as a simulation would be in rolling dice for primary and presidential elections state by state. In reality we do not depend on dice to win a state's electoral vote. However, this can be overcome with a little imagination and the use of issues and the class for voting or combined activity with other classes.

Once all terms have been defined and the students have had some background (through group discussion etc.) on local politics, members of the class then play for their careers at the local level. Because of the way that the game is set up for a "career" in politics and because of the simple, concise directions given with the game, students seem to have no difficulty in understanding the operation of town, city, and county governments. (Robert J. McNally, Half Hollow Schools, North Bellmore, N.Y.)

Note: For more on this simulation, see the essay on the uses of simulation in the social sciences by Dorothy R. Dodge.

Cost: $9.95

Producer: Educational Games Co., Box 363, Peekskill, NY 10566

ELECTION U.S.A.

Playing Data
Copyright: 1968
Age Level: grade 8-adult
Prerequisite Skills: familiarity with U.S. Constitution
Number of Players: 2, 4, or more with class as audience
Playing time: ½ hour
Packaging: professional game box

Description: Students playing *Election U.S.A.* answer legislative, executive, judicial, and general questions about the Constitution as they compete for electoral votes in a simulated campaign for President. At each turn, the Republican or Democratic player spins the spinner to determine which kind of question he or she must answer, draws the appropriate card, and answers the question (i.e., "What is the minimum age qualification for President?" "How many representatives constitute a quorum?"). Correct answers to general questions are worth 10 electoral points; correct answers to legislative, executive, and judicial questions are worth 20. At every turn players may "win" states by trading the correct number of electoral points for the electoral votes assigned to that state. The first player to get 270 electoral votes wins.

Comment: In a trial run of *Election U.S.A.*, both players tied 269-269, a highly unlikely occurrence. The last few minutes of the game simulated the last few weeks before an election, during which the strategy in choosing a state for campaigning greatly influences the electoral outcome, and one small mistake can have dire consequences. Generally, this is a highly informative, enjoyable game. Questions about the Constitution range from elementary to mildly difficult, but there are not many that a player with a high school history background would not be able to answer. (TM)

Cost: $14.95

Producer: Civic Educational Aids, 513 Holly, Crookston, MN

THE GOOD FEDERALISM GAME

Rodger M. Govea and George G. Wolohojian, Syracuse University

Playing Data
Copyright: 1975
Age Level: college, graduate school, government training programs
Number of Players: approximately 33 in 10 groups
Playing time: 2 consecutive 2-hour sessions, 2 weeks for follow-up activities
Supplementary Materials: books on library reading list
Packaging: instructor and student manual

Description: The purpose of *The Good Federalism Game* is to give participants an overview of intergovernmental relations and of the issues and problems associated with fiscal transfer. As central authorities, regional authorities, and local authorities of a hypothetical nation, players attempt to ensure their reelection and to provide residents with necessary services. Authorities are required to formulate policies in regard to the allocation of BRUs (Basic Revenue Units) and fiscal transfers, and to complete forms detailing their decisions and actions. During each game cycle, authorities learn about the tax resources available and the political satisfaction of the residents of the region(s) they represent. Residents belong to one or more of ten policy interest groups, such as labor unions or a homeowner's association. By the end of each cycle, authorities must make and justify their decisions. On the basis of their decisions, they receive political satisfaction scores at the beginning of the next cycle. The game guidebook includes four optional activities involving papers on different aspects of intergovernmental relations, as well as blank forms, facsimiles of completed forms, a bibliography, and a glossary. (TM)

Cost: $3.00

Producer: Learning Resources in International Studies, 60 East 42nd Street, New York, NY 10017

GOVERNMENT REPARATIONS FOR MINORITY GROUPS

David J. Boin and Robert Sillman

Playing Data
Copyright: 1972
Age Level: grades 7-12
Number of Players: class size
Playing Time: 3-5 class periods

Description: This is a simulation of a Senate subcommittee hearing. Students investigate whether or not the government owes reparations to certain minority groups. The class is six groups: the Senate committee, American Indians, Chicanos, Blacks, Taxpayers, and the Love It or Leave It Associates of America. Teams collect evidence and present it to the Senate committee, who may question them. At the completion, the Senate Fact-Finding Committee will present its report written after class or at home. (REH)

Cost: $3.00

Producer: Edu-Game, P.O. Box 1144, Sun Valley, CA 91352

HAT IN THE RING

Paul A. Theis and Donald M. Zahn, Changing Times Education Service

Playing Data
Copyright: 1971
Age Level: grades 5-12 and adults
Prerequisite Skills: reading, grade 5; math, grade 0
Number of Players: 3-27 in 3-8 teams
Playing Time: 2-4 hours in 40-50-minute periods
Preparation Time: 1/2 hour

Description: *Hat in the Ring* provides students with a fun and non-stressful way of learning the process of nominating a person to run for the office of President of the United States. Since the game is easy and uncomplicated to play, it can best serve children in their last years of grammar school.

The simulation shows in a simplified way how hopeful candidates travel from state to state to win the approval of the different people they encounter. Students learn some campaign techniques—that is, the more wholesome ones, such as holding square dances, movie parties, and picnics—to raise funds and win delegates. They learn this through direct experience.

What this simulation shows best is the amount of money spent just trying to become nominated. Based on the amount of money held, players makes their decisions on the type of campaign they will conduct: economy, full-scale, blitz, or fund-raising. This is not a common-sense type of decision; it is a gamble, since no one knows what reaction (chosen from a deck of cards) will be produced by the decision. Also, it must always be kept in mind that fund-raising cannot be done during Stage II: The Convention.

The simulation splits the players into competitive individuals, or groups if managers are used. If campaign managers are used, then candidates must discuss strategies with their managers. Reasoning comes into play when the students must decide which state to campaign in. The rules of the game affect this decision, since primary states must be won in succession and players cannot drop out of a primary once they are in. Also, if circumstances permit play into a second stage—The Convention—then the candidates and their managers can make deals with the other players.

Comment: The simulation gives the individual student knowledge of the steps taken to become nominated. It teaches participants the competitiveness of the candidates to *win* the nomination. The simulation shows in small part how democracy works.

To the best of my knowledge this simulation does not touch upon the relationship between teacher and student, or upon the life of the family.

Since this game takes a long time to play and younger children can become tired during play—as we older children did—I suggest the different stages be played on different days or between a lunch break.

I played this simulation twice with college students, first in a large class where we were able to use campaign managers, then in a smaller class where we only had enough students to be candidates. If campaign managers are used, the teacher can observe the decision-making process between the candidate and the campaign manager. With single players opposing other single players, more gambles were taken to liven up the game; with teams of two the discussion within teams livened it up. The competitiveness of the game was brought out more with the teams of two since it was played more seriously.

Stage II: The Convention is a big disappointment. The rules under the first ballot state: "The leading candidate on the first ballot immediately wins all uncommited delegates, plus a 'bandwagon' bonus of 40 delegates from each candidate." After this was done we had a winner in both classes tested. Stage II did not present itself as "the climax of the game." The problem is in the rules of the game. Perhaps the leading candidate should be required to play for the uncommitted delegates or win them during the Convention. Also, the bandwagon bonus should be given during the Convention, not during the balloting, to stimulate more excitement and enthusiasm in the players.

In summation, *Hat in the Ring* is an enjoyable educational simulation game which teaches students the democratic process of nominating candidates to run for the office of the President of the United States without relying upon any rote method of learning. (From the Robert A. Taft Institute of Government Study on Games and Simulations in Government, Politics, and Economics by Linda G. Quest, Pace University)

Note: For more on this simulation, see the essay on the uses of simulation in the social sciences by Dorothy R. Dodge.
Cost: $9.95

Producer: Changing Times Education Service, Division EMC Corp., 180 East Sixth Street, St. Paul, MN 55101

LEGISIM

Marshall Whithed, Temple University; William Harader, Indiana State University

Playing Data
Copyright: 1972
Age Level: college
Number of Players: 20-30
Playing Time: 3 hours in periods of 50 minutes-1 hour
Preparation Time: 1 hour

Description: *Legisim* is designed to simulate the American legislative process. The operation of the game follows the general procedural lines of the Congress and most state legislatures. Players are assigned roles as Senators and Congresspersons with partisan and regional identification, and the legislature is divided along partisan lines. It is also organized into committees. Bills or other pieces of legislation are introduced into the House by individual legislators and then referred to committee by the Speaker, where the bill may or may not be considered, amended, or reported. If the bill is reported out of committee, it is put on the calender for House consideration. In the House the bill may be amended, passed, not passed, or recommitted. (DCD)

Cost: $2.50

Producer: Center for the Study of Federalism, Temple University, Philadelphia, PA 19122

LOBBYING GAME

David Williams and Stanley Blostein, edited by Armand Lauffer

Playing Data
Copyright: 1973
Age Level: grade 10-graduate school
Number of Players: 20-50
Playing Time: 2 1/2-6 hours in periods of 2 1/2-3 hours
Preparation Time: 2-3hours

Description: Players assume the roles of lobbyists and legislators with varying degrees of power and influence. They aim to build their power and prestige and to influence legislative outcomes. Play is competitive, but participants must form temporary coalitions to achieve their intended objectives as they decide what bills to push, whom to make trade-offs with, and what to trade off. Success is reflected in whether legislators get reelected and in what bills pass. (AC)

Comment: The "message" of the game-reality is that the democratic process depends on informed opinion and the willingness of citizens to commit themselves to the available process. While the web of intrigue often obscures the actual democratic process taking place players learn that they *can* have an impact on the legislative process; that legislators need help and frequently welcome the inputs of lobbyists and concerned citizens. The game is designed to be played only once, and has been used to prepare people for legislative lobbying. From the descriptive material supplied, I can in no way distinguish *Lobbying Game* from *Democracy*. (DZ)

Cost: $125.00

Producer: Gamed Simulations, Inc., Suite 4H, 10 West 66th Street, New York, NY 10023

METRO GOVERNMENT

Rex Vogel

Playing Data
Copyright: 1972
Age Level: grades 5-12
Number of Players: 5-13
Playing Time: 45-90 minutes in 30-45-minute periods
Preparation Time: 2-3 hours

Description: *Metro Government* is a set of five games in which players assume the roles of aldermen seeking reelection. In the first game, all players begin with equal resources, but games two and three add various specific role characteristics such as age, sex, race, marital status, income, occupation, and organizational affiliations. Chance determines the attitudes of constituents, according to the deal of cards. These attitudes determine whether the aldermen are reelected, as satisfied constituents yield positive points and dissatisfied constituents yield negative points. (AC)

Cost: $40.00

Producer: Canadian Social Sciences Services, P.O. Box 7095, Postal Station "M", Edmonton, Alberta, Canada

METROPOLITICS

R. Garry Shirts

Playing Data
Copyright: 1970
Age Level: grades 8 (accelerated) - 12, adult
Prerequisite Skills: reading, grade 11.5; math, grade 1.5
Number of Players: 18-35
Playing Time: 1-2 hours

Description: *Metropolitics* is a game which is aimed at promoting understanding of the politics of governmental reorganization in metropolitan areas. The game is set in a hypothetical location called Skelter County which contains 16 incorporated municipalities encompassing approximately 50% of the county's land area. About half of the unincorporated area is populated. The municipalities cover a wide range of types. The largest is Skelter City, basically a middle-class and lower-middle-class community. Others include Pine Hills, an upper-middle-class community, and Rican City, a lower-middle-class town. Students play the roles of citizens of Skelter County who are voting on proposals for governmental reorganization in the county. There are six proposals before the voters:

(1) a single, unified government;
(2) a two-level (Dade County) approach;

(3) the special district approach;
(4) establishment of a municipal income tax for Skelter City;
(5) establishment of a blue ribbon committee to study governmental and fiscal affairs; and
(6) establishment of neighborhood governments.

Procedures: At the start of the game, players draw envelopes which contain role cards telling them something about their marital status, number of children, income, social class, place of residence, social problems which affect them, and attitudes they hold on matters pertinent to the issues before the voters. Each envelope also contains a number of chips which represent the influence individuals have in their simulated community. Finally, the envelope contains a Pressure Group card which players may use later to form alliances with other players. After roles are drawn, players are asked to announce whether they intend to form pressure groups. Those who do are given an opportunity to announce their position in the community, their position on the issues, and the reasons for their proposed action. Players are then given an opportunity to form pressure groups and receive bonus chips signifying additional influence if they are successful in doing so.

Following the period of group formation, a period of debate is conducted in which one or two brief speeches for and against each proposal are permitted. A referendum is then conducted in which players correspond to the six propositions. Votes are counted, and the proposition receiving the most votes is placed on the ballot for a final yes or no plebiscite. Prior to the final vote, the process of group formation, debate, etc., is gone through again, each player having once again received the original number of chips following the first ballot.

Objectives: Metropolitics both conveys information about various structures of government in metropolitan areas and creates in the classroom a model of the politics of referenda. The main impact of the game is on the cognitive understandings of the players. It promotes knowledge of metropolitan government, the politics of reform, the role of groups in referenda, and the positions of various types of people on reform proposals. Its main value is educational, but it does, of course, contribute to program diversification and may incidentally develop skills at bargaining, negotiation, and strategic decision-making.

This simulation is easy to administer and achieves its rather modest objectives. It is appropriate for use from junior high school (advanced groups) through adult education. The game is for 20 to 35 players and takes only one to two hours to complete. It is an easy game for an instructor to prepare for; preparation consists essentially of reading the operator's manual and the player's manual and organizing the game materials. It is also a simple game for students to play and not much time is required to explain it.

Weaknesses: The reason it may be difficult for junior high school students (rather than advanced groups) has to do with the game's concepts rather than with the complexity of the rules of the game. There are some minor imperfections in the game. For example, there is one point which the author neglects to make in the directions: If one proposition receives a majority on the first referendum, it would seem unnecessary to conduct the second vote.

One feature of the game with which some may disagree is the influence assigned to Good Government Advocates. They initially receive more chips (influence) than other players when they join a pressure group. Many people (perhaps especially good government advocates) may feel that Good Government Advocates are not this powerful in real life. The powerful position of the Good Government Advocates is intended to make them attractive as members of pressure groups and thus to cause those seeking to form pressure groups to seek out the Good Government Advocates and to try to convince them to join the group being formed. In one game, however this extra influence led to evidence which supports Lord Acton's dictum that power corrupts. The Good Government Advocates all got together and formed their own single pressure group. They did this before deciding what proposition to back. Once having established themselves as the pivotal force in the hypothetical metropolitan area because of their large bloc of votes, they then negotiated with other pressure groups and among themselves in order to decide which proposition it was most expedient to back under the circumstances. This was all done in the best power-broker style! The Good Government Advocates may have failed a bit in living up to their names in this instance.

Comments: This game may be related to real life very readily if any proposals for reorganization of government are pending, or have been made in the past, in your area. If not, places such as Dade County, Nashville, Los Angeles County, and others may be used to illustrate the proposals in the game. The two works cited in the director's manual are very useful additional reading to supplement the game: *The Metropolis: Its People, Politics and Economic Life* by Bollens and Schmandt, and *Forces Affecting Voter Reactions to Governmental Reorganization in Metropolitan Areas* by the Advisory Commission on Intergovernmental Relations.

Debriefing sessions for this game may profitably focus on such topics as the following:

(1) Who gains and who loses under each of the proposed reforms?
(2) Are Good Government Advocates really as influential as they are in the game?
(3) Does the formation of pressure groups really result in as much added influence as it does in the game?
(4) What proposals were most attractive to suburbanites? Why?
(5) What proposals were most attractive to city dwellers? Why?
(6) What proposals were favored by minority groups? Why?
(7) Which proposal would you like to see put into effect in your area?
(8) What groups or areas in your metropolitan area would favor and oppose each of the six proposals for reorganization dealt with in the game?

Overall, *Metropolitics* is a good, brief, well-designed game. It is relatively inexpensive and fairly easy to administer. If you are dealing with metropolitan governmental reform, it is worth using. (From the Robert A. Taft Institute of Government Study on Games and Simulations in Government, Politics, and Economics by Dayrl R. Fair, Trenton State College)

Note: For more on this simulation, see the essay on the uses of simulation in the social sciences by Dorothy R. Dodge.

Cost: $25.00, sample set $3.00

Producer: Simile II, P.O. Box 910, Del Mar, CA 92014

NAPOLI (NATIONAL POLITICS)

Developed by Western Behavioral Sciences Institute

Playing Data

Copyright: 1966
Age Level: grade 7-college
Prerequisite Skills: reading, grade 11.5; math, grade 4.5 (with instructor help in polling)
Number of Players: 9-33
Playing Time: 3 periods of 50 minutes
Preparation Time: 1 period of 50 minutes

Description: *NAPOLI (National Politics)* is a simulation which holds promise to every teacher who has felt the frustrations of trying to help students understand the nature of the legislative process. Those who are involved in politics (politicians, legislative assistants, lobbyists, concerned citizens) find it meaningful and fascinating. Our job as teachers is to help youngsters understand the process and feel its fascination. Properly used, *NAPOLI* can be one means of achieving these goals.

NAPOLI is a model of a legislature in action. Each legislator represents one of eight states: Garu, Agra, Coro, Efra, Baha, Inda, Hela and Dami. Each is also a member of either the Napoli Conservative Party (NCP) or the Napoli Liberal Party (NLP). Eleven bills are on the agenda:

(1) Limit on national debt;
(2) voting age bill;
(3) Electoral College amendment;
(4) eliminate extreme poverty via direct subsidies–$22 billion;
(5) Napoli should withdraw from the United Nations;
(6) enlarge civil defense by $500 million per year;
(7) eliminate oil and mineral depletion allowances;
(8) pollution tax bill;
(9) Federal Center Cities Project–$25 billion;
(10) centralize Napoli Geological Survey in Dami–$75 million;
(11) develop manned, orbiting, military space satellite–$4 billion.

Each bill is listed as sponsored by the Conservative Party, Liberal Party, or it carries bipartisan endorsement. The task facing every legislator is to vote in such a way on each bill as to ensure reelection. Three meetings are held. During the first meeting the legislators meet in party caucuses to debate strategy. They may decide whom they will support as speaker, which bills will be supported, and the order in which they will be introduced. Following this, state caucuses are held to develop strategies to protect state interests. Legislators are provided with sheets detailing the sponsoring party for each bill and the results of a poll taken in their states. These sessions are in preparation for the legislative session during which action will be taken on proposed measures which can spell victory or defeat at the next election. After four bills have been adopted, defeated, or tabled, the Legislature recesses.

A person serving as Calculator hands out forms to the legislators indicating their tentative chances for reelection. This cycle of state caucus, party caucus, and legislative session is repeated two more times, after which legislators are told whether or not they will be reelected.

Crucial to the successful operation of *Napoli* is selection of a Speaker. That person may be either elected at the first legislative session or appointed by the teacher. As well as possessing some understanding of parliamentary procedure, the speaker has to be a mature person who demonstrates leadership qualities and genuinely enjoys exercising them. On the college and adult level, it may be well to have a speaker selected by the group; otherwise this selection should be up to the teacher. A Director and Calculator must also be chosen. The Director's function is to see that each session moves along, while the Calculator uses a somewhat complicated formula to determine whether legislators' actions cause them to remain in the good graces of their constituents and party. These functions may be given to one person, or the teacher may assume the role of Director.

Legislators are involved in making countless decisions as they weigh the impact of each of their votes on their relations with their party, their constituents, and their consciences. In an effort to stay in office, legislators will make deals with members of their own party and members of the opposition; for, occasionally, the positions taken by their party clash with their constituents' wishes. Clearly, it is a lonely life. They won't bat a thousand, but they will strive to be successful to the point that their party and constituents will appreciate them. Thus, the activity that goes on in the caucus sessions becomes extremely important.

NAPOLI can be used to teach many things. Among these are:

(A) Understandings
  (1) Parliamentary procedure or rules are essential to the legislative process.
  (2) A major function of a legislative body is to decide which bills become law.
  (3) Politics is the art of compromise.
  (4) Political parties are essential to the proper function of government.
  (5) A strong two-party system enables democracy to work.
  (6) Concerned citizens can make a difference.
  (7) Political philosophies play an important role in politics.
(B) Concepts
  Caucus
  Parliamentary procedure
  Bipartisanism
  Public opinion
  Log-rolling
  Conservative
  Liberal
  Political party
  Speaker
  Legislative committee
(C) Skills
  Critical thinking or decision-making
  The art of negotiating or compromise
  Leadership
  Parliamentary debate
(D) Attitudes
  (1) Political decisions affect the welfare of everyone.
  (2) Compromise is often the key to success in politics.

Comments: NAPOLI can be a successful teaching tool. However, certain deficiences should be noted. The scoring is awkward and contrived. Some bills (the proposals to change the voting age and to enlarge civil defense by $500 million per year) are either outdated or lacking in interest. The complicated nature of the simulation (understanding parliamentary procedure, the substance of each bill, and how a legislature operates) poses problems for teachers who wish to use it with younger and slow students. Furthermore, it is expensive, considering that many of its items can be used only once. The publishers recognize some of these drawbacks and encourage teachers to make changes to suit their particular needs. Certainly, the proposed pieces of legislation could be updated to correspond with today's issues. Students will fight harder for laws they believe in. The press could be given a role to play, as could the Supreme Court, members of the President's cabinet, and the President himself.

NAPOLI should not be used without extensive preparation. Students should have some understanding of how a legislature works–the role of a Speaker, committees, caucus, log-rolling. To avoid vacuous debate, the bills to be proposed should be studied beforeheand.

Students, particularly average or slow ones, should write the points they wish to make in the legislative session on cards. Some means of introducing *NAPOLI* should be used. Perhaps one could discuss the pros and cons of a proposal for a national law to register and limit firearms. A discussion could follow concerning the roles of public opinion, the press, lobbyists, Congress, the President, and the Supreme Court. With the issue defined and the process to follow clearly understood, the students should be better able to conduct *NAPOLI.*

Given the nature of this simulation, *NAPOLI* is better used at the end rather than the beginning of a unit, at which point students can apply concepts and skills previously taught. Nevertheless, it should serve as a springboard for further learning. Basic to this is a discussion of the model itself. Among the questions to be discussed should be: (1) Is it realistic? (2) What changes would you make? (3) Why did you make the decisions you did? (4) What would you do as a legislator if your conscience conflicted with the wishes of your constituents? (5) What is the place of morality in politics? (6) What is a political party? (7) Why is it important to have a strong two-party system? Students could investigate other issues and write letters to their Congressmen. Congressmen could be invited to the school. Local and state legislative bodies could be studied. A skillful and imaginative teacher could find countless ways to adapt and use NAPOLI for his or her purposes. (From the Robert A. Taft Institute of Government Study on Games and simulations in Government, Politics, and Economics by Joseph Eulie, State University of New York at New Paltz.)

Cost: 35-student set $50.00, specimen set $3.00, additional consumable forms $2.50

Producer: Simile II, P.O. Box 910, Del Mar, CA 92014

THE NATIONAL POLICY GAME

John L. Foster, Thomas A. Henderson, and Daniel G. Barbee

Playing Data
Copyright: 1975
Age Level: College
Number of Players: 54
Playing Time: 12 hours (est.)
Preparation Time: 20 hours (est.)
Packaging: professionally produced 108-page softbound book

Description: This game, subtitled "A simulation of the American Political Process," attempts to simulate the federal budget process as players try to develop a budget that will allot money to solve the problems of American cities. All players fill specific roles in one of five groups: a hypothetical Department of Community Development, the Office for Management and Budget, a House committee that reviews the OMB's budget request, special interest groups, and the press. Players win or lose points in simulation of the points won or lost in Washington. For instance, department heads gain or lose points depending on whether their budgets are cut or increased. Players fill out forms to simulate the steps in the budget process. These steps are OMB recommendation, department request, agency request, OMB review, and congressional and presidential approval. The game is played in three-hour rounds, each of which corresponds to one fiscal year, and may be played for an indefinite length of time. (DCD)

Cost: $6.50

Producer: John Wiley and Sons, Inc., 605 Third Avenue, New York, NY 10016

NATIONAL PRIORITIES

David J. Boin and Robert Sillman

Playing Data
Copyright: 1972
Age Level: grades 7-12
Number of Players: class size
Playing Time: 1-2 class periods

Description: The class assumes roles as members of a political party that is making plans for an upcoming election. They are a committee to represent the "youth vote," and their job is to decide which problem should be the government's top priority and the party's number one issue. The issues are provided (unemployment, pollution, organized crime, etc.). Students are divided into area groups to discuss the issues, and the whole class then convenes as a national youth meeting to attempt to adopt a single list by two-thirds majority vote. (REH)

Comment: This simulation roughly approximates the process of deciding on a platform for a national political party, but the process is rough. I would characterize this as a decision-making exercise which enables students to think about, make decisions about, and debate national political issues. It is presented in the standard Edu-Game format; a few pages of teacher's manual and locally reproducible student material. (REH)

Cost: $3.00

Producer: Edu-Game, P.O. Box 1144, Sun Valley, CA 91352

OLD CITIES–NEW POLITICS

Faith Dunne

Playing Data
Copyright: 1971
Age Level: junior and senior high school
Number of Players: 15-35
Playing Time: 1-3 weeks

Description: *Old Cities–New Politics* is a multimedia teaching unit concerned with urban politics. The material focuses on a comparison of the city of 80 years ago with the city of today. It examines the urban poor in both settings and analyzes the political impact of the cities on them and their responses to it. The material contained in the teaching kit deals in some depth with the key political science concepts of "politics" and "influence" and the sociological concept of "relative deprivation."

The unit is designed for one to three weeks of classroom use, but may be expanded to five or six weeks if optional strategies are used. *Old Cities–New Politics* does *not* analyze the structure of municipal governments, nor does it survey American urban history. An integral part of the unit is a simulation game entitled *Parksburg*, which may be used within the context of *Old Cities–New Politics* or by itself.

The unit and game are both easy to administer, the supporting material is skillfully developed, and both student and teacher preparation is minimal. The subject matter and activities can be easily handled by average students as low as the seventh grade. Through inclusion of optional materials and more probing questions, we have found that the material will hold the interest of upper-level high school students as well.

Old Cities–New Politics comes in a compact boxed form that includes two short filmstrips, a teacher's manual and thirty copies of City Politics, a 31-page booklet of readings for students and a 33 1/3 rpm record. In addition, the kit includes five spirit masters for use in the simulation and the other activities included in *Old Cities–New Politics.* Finally, the kit contains two wall posters, one of which is a clear graphic map of Parksburg, the city which is the subject of the simulation. The teacher's manual is very well organized and enables the teacher to efficiently prepare for the various activities included in the unit.

As stated earlier, *Old Cities–New Politics* was designed to be used over a one- to three-week period or, if optional materials are included, up to six weeks. If the simulation alone is used, three to five days should be sufficient to properly complete it. The organization of materials and activities in *Old Cities–New Politics* is flexible enough to use in almost any fashion desired by the teacher.

The teacher needs to spend little time in preparation for the unit. The material is organized so that all work can be completed by students during class time, if the teacher does not wish to assign homework. The amount of reading required by students is minimal–never more than two pages per activity, the major pedagogical thrust being based on inference and discussion. Through use of the record, it is possible for students to skip reading any of the printed materials, which are interestingly presented on the record. The student readings are generally very short and highly readable, but thought-provoking. The filmstrips are shown a few frames at a time and are intended to trigger problem-framing as well as class discussion. The optional materials generally call for some outside student preparation.

Old Cities–New Politics is divided into five sections. The first section, "What is Politics?" defines it and stimulates students to think about how political they are in their everyday relations. The second section, "Old Politics: The Immigrant and the Machine," leads students to examine how the cultural background of European peasant immigrants reacted with social factors in America to produce the urban political machine. "New Times, New Voters" looks at changes that have produced a new population of urban poor. In this section students are introduced to the concept of relative deprivation, and they relate it to their own lives and attitudes. "The New Politics" analyzes the social, economic, and technological factors that have given impetus to new political tactics of great potency. In comparing old and new political action, students see how a new urban constituency can use new tactics to gain significant power. The fifth section, *Parksburg: Garbage and Politics,* is a simulation game. By assuming roles of residents and politicians in a medium-sized city, students deal with some of the variables that enter into political decision-making. This section ties together the material presented in the first four sections by focusing on the major concepts of "politics," "influence" and "relative deprivation."

The five sections outlined above form a thematic sequence. Each section may be used independently, but the designer advises that if the game is used, it is helpful to first use sections 1, 2, and 4. Materials and strategies from each section may be selected to fit different classroom situations.

The teaching guide allows maximum flexibility in planning class lessons. Each section starts with a thematic preview, a list of materials to be used, and a brief perspective for teacher use. The contents of each section are outlined in the teacher's manual.

In the first section, "What is Politics?", students are given a questionnaire to fill out. This questionnaire is designed to score students on the question "How Political are You" and is then used as a basis for discussion of what being "political" means, as well as what "politics" is. Students are then introduced to three different political tactics through reading the student manual and/or a short recording. These tactics are then applied and elaborated through a filmstrip segment entitled "A New School for Eastside" and further readings.

The next section, "Old Politics: The Immigrant And The Machine," leads students into an examination of American industrial life as a political arena. The focus is on the immigrant and the political machine. Recordings and/or readings dealing with the nature of politics and political machines during the nineteenth century are presented in order to explore the relationships between ethnic politics and political bosses during this period. The readings and recordings are varied as well as interesting; they lead to a clear understanding of the meaning of "Old Politics" and reasons why city political machines were so powerful during this period. The selections are well balanced, presenting both the good and bad sides of the old ward leader system. One of the highlights of this section is a role-playing exercise dealing with citizen protest and the response of the political "boss." Another highlight of this section is an interview with George Washington Plunkitt, a powerful political boss of the 1890s. This section is provocative and interesting to students, capturing the flavor of machine politics during this period of American history.

The next section is "New Times: New Voters." Recordings, brief readings, and short filmstrip excerpts are used to show how the changing constituency of the city from European peasants to Southern blacks leads to the eventual crumbling of the old-time political machines. The new expectations and needs of this later group are compared with the European immigrants of the nineteenth century and the Southern black and Puerto Rican "immigrants" in the twentieth century. The increasing political turmoil in the cities is partially explained through the introduction of the concept of relative deprivation, which is introduced by a little game appropriately entitled "The Relative-Deprivation Game." This game effectively and quickly gets the concept across to students. In this section, material relating to the suburban outmigration of the middle class is presented, largely through the use of a filmstrip.

The fourth section of *Old Cities–New Politics* deals with the so-called "New Politics", which is broadly defined by the author as a set of tactics used by the politically weak to influence a new majority–that part of the middle class which is prosperous, socially conscious, and powerfully effective in shaping contemporary public opinion in the United States. Once again using readings, recordings, and filmstrips, students are led into an analysis of the strategies of the new politics. They define social, economic, and technological factors that give these tactics such extraordinary power today. At the end of this section, a test is distributed to help evaluate the degree to which students understand and can recognize the two streams of political style operative in America today–the "Old" and the "New."

Finally, students are asked to bring to bear what they have learned in a simulation of a rather mundane, but very crucial problem–garbage. The problem is service, who is to be advantaged and disadvantaged by the political process. Students are divided into groups representing four neighborhoods of Parksburg, a medium-sized American city. The object of the simulation is for each neighborhood to establish its position on a bill before the city council and then try to get enough votes of council members to swing the decision in its favor. The setting for the simulation is effectively established through the use of a large poster map of Parksburg and a short filmstrip presentation. The simulation is an excellent vehicle for making students aware of political decision-making processes and coalition politics.

The teacher's manual is so well organized that the teacher's task in organizing each of the five sections in *Old Cities–New Politics* is made simple. The only real decisions that the teacher has to be involved with, if any, are optional materials to add to the basic materials or what units to exclude in order to meet individual teaching objectives.

The simulation is easy to set up and administer. Once the exercise begins, the teacher's role becomes simply one of circulating and answering questions. In most situations few will arise. Debriefing the simulation is, of course, important, but the suggestions contained in the manual make this phase of the game relatively easy.

Comments: *Old Cities–New Politics* is an effective multimedia device for teaching the basic concepts of politics, influence, and relative deprivation. The material is somewhat oversimplified in places, but this can be remedied by including supplementary materials adjusted to the level of student sophistication.

The simulation game, *Parksburg,* while carefully meshed with the preceeding four units, can easily be used by itself. This writer has done so on several occasions with some success, especially for purposes of teaching concepts of bargaining and coalition formation in a democratic political setting. With slight alterations, the game can be used to teach the basic dynamics of the two-party system in the United States.

On the whole, *Old Cities–New Politics* is an excellent tool for teaching the concepts intended by the designer. The only reservation held by this reviewer concerns the need to creatively supplement the basic materials in the unit for students above the junior high school level. The materials included establish the framework–it is up to the teacher to intelligently "beef up" the presentations for above-average junior and senior high school students. This reservation does not, however, apply to the simulation itself. It will stand alone as an instructional device, as students will naturally relate to it at their existing levels of political sophistication. (From the Robert A. Taft Institute of Government Study on Games and Simulations in Government, Politics, and Economics by Gerald Thorpe, Indiana University of Pennsylvania.)

Cost: $82.00 list, $75.00 school

Producer: Olcott-Forward, Division of Educational Audio-Visual, Inc., Pleasantville, NY 10570

PARKSBURG: GARBAGE AND POLITICS

Faith Dunne

Playing Data
Copyright: 1971
Age Level: junior and senior high school
Number of Players: 15-35
Playing Time: about 3 class periods

Description: *Parksburg* is a simulation of a political crisis in a middle-sized midwestern city. In attempting to resolve the crisis, the different socio-economic groups have different and unequal resources which determine the strategies and tactics available to them. The emphasis of this simulation is more on political processes and strategies than on particular solutions to Parksburg's problem.

The crisis which shatters this previously peaceful city has to do with garbage. During a heat wave in August, several garbage trucks broke down, and there was no money to buy new ones. Players learn that a bond issue for more equipment for the sanitation department was recently voted down. Because this is an election year, the issue will not be reopened. Garbage pick-up in the city's poorest area, Riverside–already somewhat below healthy levels–was cut back even further. For four weeks, no pick-ups were made in Riverside. Rats were drawn to the rotting garbage, and finally a two-year-old girl was killed by a rat bite. The residents of Riverside were aroused from their customary resignation and demanded more frequent garbage removal. Their councilman drafted a bill to equalize the frequency of garbage pick-ups in all four neighborhoods. The two wealthier neighborhoods, Hillcrest and Mill Park, would have their service cut back, and service would improve in Southside and Riverside. In the scenario, it is explained that there is no money to increase sanitation services, so the only alternative to the present system is assigning some trucks to different areas.

The students are assigned roles: neighborhood resident, mayor, and councilman. Hillcrest, Mill Park, Southside, and Riverside are stereotypical neighborhoods, each with characteristic social and economic status. The mayor lives in Hillcrest, along with other wealthy, highly respected people. The people in Mill Park are lawyers, doctors, and store-owners. The residents of Southside are ethnically diverse and working class. The city's main industry is a cannery, located in the local slum, Riverside. The residents there, if they have jobs at all, probably work in the cannery.

The city's political structure is also typical. One councilman is elected from each neighborhood; two more are elected at-large. Together with the mayor, they make up the seven-member council. (The mayor may only vote to break a tie.) The council members must be responsive to the demands of their constituents in order to keep their jobs and fulfill their political ambitions. One vote on the proposed bill to equalize the frequency of garbage pickups in all the neighborhoods has already taken place. The councilmen from each area voted according to their own areas' interests, the councilman-at-large who lives in Southwide voted for the bill, and the one who lives in Hillcrest voted against it. The mayor, instead of breaking the tie, decided to table the bill for two weeks and consider the problem very carefully.

The students are assigned brief scenarios and role profiles to read. The class as a whole discusses the situation and the characteristics of the town and its residents as an introduction to the roles they are about to assume. The players plan their strategies in groups (neighborhoods) and separately (council members), and fill out strategy sheets; the strategies are put into action in negotiating sessions, including meetings of citizens and councilmen, two or more councilmen, and citizens from different neighborhoods. The players are given an opportunity to alter and complete their strategies. They negotiate again, and finally they vote. (GR)

Comment: In a test session, the students (high school seniors) were enthusiastic about their roles, although some found them contrary to their own natures. In fact, they often adhered to the roles too strongly, and held out where a "real" person would probably have been able to compromise.

The game provided the students with valuable insights into the practice of politics, especially log-rolling, bargaining, negotiating, compromise, strategic thinking, decision-making, and the relationship between morality and politics. With respect to morality, a group of students said that they thought everyone was really in favor of passing the bill, but some of the councilmen could not vote for it without committing political suicide. When there was a contest between the two, morality usually took the back seat.

There was one major complaint about the game. The players felt that it was extremely unrealistic to set up a severe crisis and simultaneously rule out the possibility of acquiring more money for sanitation, or even diverting it from other, less pressing, programs. They thought that if the crisis were really that critical, priorities would have been changed to reflect the crisis. While this criticism is valid, altering the game to incorporate these other potential options would require detailed budget information and a history of the city's priorities; at very least, they would need to include several other items on the budget and give some idea as to which are less important to the citizens and therefore easier to cut back on. If all this were done, *Parksburg* would be a slice of life lasting a couple of weeks instead of a simulation lasting three days.

This game is also weak in terms of political concepts; however, these might have been taken care of in the social studies unit of which the game is actually a part.

Overall, this reviewer thinks that *Parksburg* is a valuable tool for bridging the gap between political theory and political practice, and for developing political skills. (From the Robert A. Taft Institute of Government Study on Games and Simulations in Government, Politics, and Economics by Alina Kurkowski and George Richmond, Educational Development Center.)

Note: For more on this simulation, see the essay on the uses of simulation in the social sciences by Dorothy R. Dodge.

Cost: unavailable apart from complete 5-part teaching unit *Old Cities–New Politics* (A2AK 0568) $82.00 list, $75.00 school

Producer: Olcott-Forward, Division of Educational Audio Visual, Inc., Pleasantville, NY 10570

POWER POLITICS

Russell Durham and Virginia Durham with Paul A. Twelker

Playing Data
Copyright: 1974
Age Level: high school, college
Number of Players: 20-40
Playing Time: 2-3 hours
Preparation Time: 2 hours (est.)
Special Equipment: overhead projector
Packaging: 78-page photocopied teacher's manual

Description: *Power Politics* is described by its designers as "a simulation exercise developed for the specific intent of showing certain facets of a political campaign, especially those that include the interactions between candidates and special interest groups." Players assume the roles of two candidates for the United States Senate; their campaign managers and press agents; reporters; and spokesmen and lobbyists for big business, real estate development, organized labor, the educational establishment, agriculture, and the Citizen's League For Clean Water, as well as homeowners and naturalists. (Each group has a different "power point" value which approximates its supposed power in the real world. Big business, for example, has 20 power points, while the naturalists have only 2.) Both Senatorial candidates must "walk a tightrope with each of the interest groups without losing the support of the others." Play proceeds through four distinct phases, and each is punctuated by a news broadcast. In the first phase the candidates formulate a stand on the primary issues of the campaign. (These issues are only generally specified in the game materials.) During the campaign phase, candidates present their positions to the various pressure groups. During the third phase, candidates defend their positions on the issues to the press. The game concludes with a last-minute flurry of campaigning, restatement, and speech-making as the various pressure groups pledge their power points to one candidate or the other. (DCD)

Comment: *Power Politics* presents a vision of an America of, by, and for the special interest group. It is wickedly fun to play, as practically any game about a political campaign has to be, but it is not about power or how to get or use power so much as it is about the worst sort of manipulative excess–what Lincoln would have described as fooling all of the people at election time. (DCD)

Cost: $14.00

Producer: Simulation Systems, Box 46, Black Butte Ranch, OR 97759

POWER POLITICS

Playing Data
Copyright: 1977
Age Level: grades 9-12
Number of Players: 30-50
Playing Time: 3 50-minute sessions
Preparation Time: 1 hour (est.)
Packaging: 14-page photocopied booklet

Description: As *Power Politics* begins, players are divided into groups representing legislative delegations from seven fictitious states. Their goal is to pass legislation favorable to their own states while working for the good of the nation. The game structure involves deciding how to vote and voting on 30 legislative proposals in ten categories (such as government purchase of surplus farm products, tariff on foreign-made autos, and federal aid to education). States differ markedly from one another and will be inclined to vote differently on each issue, so participants must use strategy and compromise to accomplish their objectives. States gain or lose up to three points in each of the three rounds depending on whether favorable legislation passes or fails. Winning or losing the game depends on how much favorable legislation passes, and the idea of the game is for participants to play and vote in such a way that not more than three of the seven states finish with negative total scores. (DCD)

Cost: $3.00

Producer: Edu-Game, P.O. Box 1144, Sun Valley, CA 91352

PRESSURE

David Rosser, Council Rock High School, Newtown, PA

Playing Data
Copyright: 1972
Age Level: grades 7-12, average to above average ability
Number of Players: 25-40 in 6 teams
Playing Time: 15-20 hours in 1-hour periods
Preparation Time: 3-5 hours

Description: *Pressure* is a simulation of decision-making in local government. Students become citizens in a community beset by the big problems facing local government today: zoning versus personal rights; cultural and ecological preservation versus economic gain; school modernization in plant and curriculum versus tradition and taxpayers. Students become involved in these crisis situations by forming six pressure groups found in almost all communities: a taxpayers' association, a chamber of commerce, a developers' association, a farmers' co-operative, an ecology club, and a historical society. Some students are elected to a community council (simulating a city or town council or county board of supervisors) and a school board. Others are appointed to a planning commission, a zoning board, and a water-sewer authority. Consequently, students learn the elements and processes of local government by actually participating in its simulated functions. During the simulation, students work on assignments which teach them how to work effectively within the local control system of most school districts. In addition, they learn to utilize group pressure to influence a public agency's decision. *Pressure* helps students realize that all citizens have the opportunity to participate in decision-making affecting their community's life-style, that apathy—or nonparticipation—is the greatest enemy of a democracy. (producer)

Comment: According to the producer, the message of this simulation for participants is, "Since some faction(s) will achieve goals by applying "pressure," the faction(s) I represent had better apply pressure too! That's democracy." Pressure may be adapted to suit localities where it is played. (AC)

Cost: $14.00

Producer: Interact, Box 262, Lakeside, CA 92040

REAPPORTIONMENT

Thomas P. Ryan

Playing Data
Copyright: 1973
Age Level: secondary, college
Number of Players: 20-30 maximum
Playing Time: 2 hours minimum
Preparation Time: 2-3 hours minimum

Description: This game is highly flexible, with most procedures determined by the instructor. Players may work individually or in small groups. The purpose of *Reapportionment* is to familiarize students with the problems involved in reapportioning a state legislature. A state (Northumberland) is mapped, and background information pertinent to the game is provided. Each county is also mapped with data on each. A panel of three judges is selected to act as the court deciding on the fate of the state's reapportionment. Students act in group committees representing the majority party putting together a recommendation for reapportionment. Students are concerned with preserving the strength of their party throughout their state, but they must also make certain that the "one man, one vote" guidelines are followed. If these guidelines are not followed in the recommendation to the panel, the court will order that "assemblymen be elected on an at-large basis in the next election," which will represent a defeat for the Majority Party. Students therefore must carefully follow recommendations provided in the manual to draw up the best recommendations possible. (WHR)

Cost: $1.95

Producer: SAGA, RR#2, Greentree Road, Lebanon, OH 45036

SIMPOLIS

Ray Glazier

Playing Data
Copyright: 1970, 1974
Age Level: grade 7-adult
Number of Players: 23-50
Playing Time: 3 hours and 45 minutes or 5 class periods
Preparation Time: 2 hours (est.)
Special Equipment: public address system
Packaging: professionally produced and packaged game includes role profiles and name tags for 50 players and 16-page teacher's manual.

Description: This simulation of urban political crisis is played in a rectangular grid arrangement of clustered desks and chairs which represent city blocks and streets. The name of the city is Simpolis, and the mayor—the first Repocrat to hold that office in ten years—is fighting for reelection against his critical Demopub opponent, the Police Commissioner. Up to 48 other players represent officials, labor leaders, gangsters, rabble-rousers, and concerned citizens in three classes (upper, middle, lower) and three racial-ethnic groups (black, white, Hispanic) spread throughout the city's sixteen neighborhoods.

Most of these roles have voting power in the mayoral election, and all have specific, authentic political goals and ideas about how to realize them. (The leader of "The Young Princes," the strongest street gang in the city, for example, is a self-appointed voice for Hispanic interests who demands to be heard. In the past he and his gang took over one of the city's schools and recently he has been thinking of kidnapping someone important. The president of PIG, the Policemen's Interest Group, on the other hand, is a tough ex-cop who considers the mayor soft on street crime and who would like to pressure the mayor into outlawing the Young Princes by starting a wave of "Blue Flu.")

Each role is carefully interrelated and includes enemies, allies, and built-in delusions. (The Sanitation Commissioner dreads a garbage collector's strike and is eager to play the two candidates against one another. The president of the garbage collector's union doesn't like anybody in city government very much. He has concluded that unless his workers get a raise they will all have to move into the city's projects with the blacks, and before he allows that to happen the trash can pile up on every street in Simpolis. In the meantime, the town's affluent are outraged that the Sanitation Department is dumping garbage into the Swill River. "Too bad about that name" the head of the Union replies to critics.) Every role has an ax to grind, from the president of OUT (the Organization of Unemployed Tramps—which started as a flophouse joke but has started to take itself seriously at election time) to the Spokesperson for FEM (the Female Equality Movement).

The game addresses seven major problem-areas—transportation, education, housing, civil rights, poverty, crime, and pollution. A public address system is used to represent the city's only news outlet, and access to it is as capriciously allowed as the game facilitator will tolerate. The interaction of the players is interrupted seven times during the campaign when the teacher reads "Crisis Bulletins" over the loudspeaker. The players may react to these crises (such as a race riot, a teacher strike, and an outbreak of hepatitis directly attributable to the pollution of the Swill) by making speeches, negotiating, forming coalitions, or conducting nonviolent protests. The game concludes with the election of a new mayor. (DCD)

Comment: The simulations for schools from Abt's Games Central are such a joy to behold that it is impossible not to rave about them. *Simpolis* is about a place that could be Cleveland, or—in less recent years—Philadelphia. *Simpolis* is at once brilliant satire, fun to play, and educational. It meets Picasso's definition of art as "a lie that makes you see the truth" and it has a kind of Randy Newman zaniness (burn on Swill River, burn on) about it. It is straightforwardly written, easy to set up, and among the very best of simulations and games for education and training.

The one drawback to the game is that it might prove too full of life for some potential users. Game sessions are noisy. The original version was first played at a "Design-In" in May 1967 and, according to *Newsweek,* was "the Design-In's most ingenious demonstration." Potential users who are interested in knowing more about what a play of the game is like may refer to page 26 of the May 27, 1967 issue of *The New Yorker.* (DCD)

Cost: $39.00

Producer: Games Central, Abt Publications, 55 Wheeler Street, Cambridge, MA 02138

STATEHOOD

Steve Denny

Playing Data
Copyright: 1977
Age Level: junior and senior high school
Number of Players: 8 groups of 4-5 players
Playing Time: 15 1-hour sessions
Special Equipment: duplicating machine
Packaging: 24-page teacher guide and 1-page student guide

Description: *Statehood* uses role-playing, political elections, and a variety of problem-solving activities to introduce students to the history of their state and some of its current political problems. It also allows them to explore future possibilities for resolving these problems. The simulation is divided into four phases described in the teacher guide as follows:

"PHASE I: History Research Project. The class is divided into eight regional groups of 4 or 5 students. A map of the state is similarly divided into 8 regions and each regional group of students is assigned one interest/investigative area. Each regional group has a geographer/environmentalist, an economist-businessperson or farmer, a culturalist-educator, and a politician-legislator. Students research their region in relationship to the history of the 4 areas above. Each regional group prepares a PRESENTATION FORM and gives an illustrated report to the entire class. . . .

"PHASE II: Problem-Solving. Three problems involving land use, taxation, and promotion of education-art have been developed in a case-study format. Students break into OCCUPATIONAL GROUPS . . . to consider the 3 problems. Representatives from each occupational group appear before a committee . . . that makes a decision on each problem.

"PHASE III: Political Campaign for Governor. Each politican (8 total) runs for governor, 4 from each political party. . . . A primary election is then held after other students are allowed to attend one or the other of the party's "conventions." . . . The two winners of the primary run on an Issues Platform based on views about PHASE II's problems and possible solutions. . . .

"PHASE IV: Future. Each occupational group creates an 'idea' situation for the state's future, the creation based on its own frame of reference. . . . Each student reports back to his/her regional group and tries to get the regional group to accept the occupational group's recommendations. The winner (if any) is the occupational group receiving the greatest regional support."(TM)

Cost: $14.00

Producer: Interact, Box 262, Lakeside, CA 92040

TAXES

David Rosser

Playing Data
Copyright: 1974
Age Level: high school
Number of Players: class
Playing Time: 6 periods
Packaging: 8-page foldout student guide, 2-page teacher's guide

Description: Students read the purpose and overview of this simulation of how major taxes affect individual and community decision. Then they draw their identities (which range from a very wealthy doctor through mechanics, salesmen, teachers, typists, artisians, to retired, disabled, and unemployed people). Besides occupation, identity categories include basic median salary, additional income from second jobs or investments, total bills per month, real property value, net worth, welfare, and rent. After studying definitions of four basic taxes, they fill out tax identity profiles.

Players are then confronted with a tax crisis: they have to increase tax revenues by 20%. They form factions to protect/assert their interests and put up candidates for Community Council. After a campaign and secret ballot election, the Council holds a public hearing. After the hearing, it deliberates in the presence of the other players, reaches a decision, and faces another election. Debriefing includes an Appeals Court for those who feel wronged.

Pressure is provided by the option (recommended by the designer) of having student grades reflect the degree of their success in keeping their taxes down while helping the community increase its tax revenues.

The teacher's guide consists of directions and suggestions for conducting the 6 periods of play, many of which will be rather heated. (AC)

Note: For more on this simulation, see the essay on the uses of simulation in the social sciences by Dorothy R. Dodge.

Cost: $10.00

Producer: Interact, Box 262, Lakeside, CA 92040

VOTES

Charles L. Kennedy, Gannon College

Playing Data
Copyright: 1973
Age Level: grades 7-12, average to above average ability, adult
Number of Players: 25-40

Description: Votes requires students to organize four political party campaign committees (two major parties, two minor parties) and to choose their candidates. The four parties' committees then simulate a three-month campaign for office. (The simulation materials may be used by students to conduct a presidential, congressional, gubernatorial, or local political campaign.) After each party's campaign committee receives a financial and volunteer worker base, it must decide how to allocate these basic resources in order to create and carry out these activities: TV and radio commercials, newspaper ads, speeches, rallies, public poll-taking, and research on the issues. Eight "citizen representatives" evaluate these activities. These evaluations are translated into electoral and popular votes for a presidential campaign, but only into popular votes for congressional, gubernatorial, or local elections.

Votes simulates, in three phases, the three months prior to the election. The first two phases end in "straw votes" which give parties an indication of their voting strength at those moments; the third phase has the final casting of votes that decides the election. With minor modifications, adult groups can use it as a training exercise in practical politics. (producer)

Comment: "Votes may be utilized as a national campaign for President, Senator, Congressperson, a state campaign for Governor, or a city campaign for Mayor. It simulates the problems and pressures the candidates, the campaign staff, and voters encounter during a political campaign. Students, intimately involving themselves in a political campaign, become vividly aware of the many problems and frustrations involved in developing strategy and tactics, allocating resources, coordinating the many diverse elements on the staff, and preparing position papers on the issues. In addition, students realize the importance of developing a campaign blueprint that will appeal to the wide variety of voting blocs in the United States. They learn to overcome early mistakes and inadequate decisions and make the necessary strategic and tactical adjustments in order to win the most votes. Finally, the simulation does more than merely help students understand the structures and functions of a political campaign, it also increases their insight into how politicians, their classmates, and they, themselves, react under the stress and strain of trying to win an election through group effort." (Teacher's Guide)

According to the manual, *Votes* stresses the operations of a political campaign, introduces students to typical political issues on which candidates run, acquaints them with the complexities that are associated with a candidate's attempts to appeal to a broad segment of the voting population, and exposes them to the operation of the Electoral College System. (From the Robert A. Taft Institute of Government Study on Games and Simulations in Government, Politics, and Economics by George Richmond, Educational Development Center.)

Cost: $14.00

Producer: Interact, P.O. Box 262, Lakeside, CA 92040

WOMEN'S LIBERATION

David J. Boin, Robert Sillman

Playing Data
Number of Players: class size
Playing Time: 3 class periods

Description: Students assume the roles of members of the six largest women's liberation organizations who are called to a convention to decide on the three highest priorities to be presented to Congress for legislation on the liberation of women. (REH)

Note: For more on this simulation, see the essay on the uses of simulation in the social sciences by Dorothy R. Dodge.

Cost: $3.00

Producer: Edu-Game, P.O. Box 1144, Sun Valley, CA 91352

WOODBURY POLITICAL SIMULATION

Bradbury Seasholes, Tufts University; Marshall H. Whithed, Temple University; H. Roberts Coward, Case Western Reserve University

Playing Data
Copyright: 1969
Age Level: college
Number of Players: 25-70 in 8 teams
Playing Time: 8-9 hours
Preparation Time: several hours
Special Equipment: team rooms; director and up to 6-8 umpires; hand calculators, duplicator (stencil, ditto, photocopier), typewriters

Description: One of the more unexpected, yet telling, adjectives one might attribute to the game *Woodbury* is the color purple. Players use written communications to make moves in a mayoral election situation. These communications are reproduced on hectograph masters and distributed to the player group. Purple finger and handprints appear on door jambs and wall edges. Students race arround the room talking, persuading, bargaining, testifying to the compelling nature of the exercise and of their involvement in it. *Woodbury* catches the imagination of students, even if the clean-up duties leave the janitorial staff less enthusiastic.

Woodbury has a dual purpose: (1) To illuminate both the components and the dynamics of a two-party election and (2) to demonstrate the linkages between urban problems and the electoral process. The game encourages political science analysis along with an appreciation for the way a campaign evolves over time. Background documents and a map of a hypothetical city in the midwest provide an overall view of the city and its two suburbs, and in so doing generate enough information and numerical data to allow a simulated election to occur. The game supplies data in credible form: excerpts from a purported Census, newspaper articles, etc. The election engages players with varying stakes, resources, strategies, and skills in making strategy against a backdrop of demographic and economic facts. These produce basic party affiliations. Care has been taken to create data which give either party an equal chance of winning the mayoral election. Indeed, Woodbury's City Council has a roughly even distribution of Republicans and Democrats, a necessary deviation from the likely reality of cities of this size, location, and economic character.

In the course of the game, students learn to measure the actions of candidates and the party organizations against a common criterion: potential votes temporarily gained or lost among segments of the eligible electorate. To do this, they must grasp the political significance of the population mix and its geographic dispersion. By design, Woodbury's racial, ethnic, and income subpopulations are somewhat more simplistically locatable by ward than is the case in real cities.

The outcome of Woodbury's mayoral elections depends not just on social and economic facts, but also on the structure of the situation. Winning the mayoral race is highly desirable: the salary is high, and the patronage appointments are substantial. The two political parties have ward and precinct organization. The mayoral candidates are chosen by primary.

Woodbury has the typical set of urban problems. Some or all of them have the potential for becoming campaign issues. The city has a typical assortment of interest groups–unions, League of Woman Voters, Chamber of Commerce, and others–which may become factors in the evolving campaign. The city has a large number of mass media outlets, including some that cater to ethnic preferences. Thus, the *Woodbury* simulation taps into demography, economic base, governmental/political structure, and the problems crosscutting all these. In addition to these variables, two others become factors. Once play begins–individual skill and "fate" (exterior events) can be decisive.

Student players are assigned to various groups. Each constitutes a single playing unit in the game. Republican and Democratic organizations, interest groups (some affiliated with one or the other party), and the media are represented. Within each group, individual roles can develop. The Republican organization may assign the Republican mayoral candidate one particular individual member to help in the campaign, or, if preferred, the mayoral candidate may develop his own group of supporters.

Moves take place in written form and are submitted "daily" (a "day" may be on the order of ten to fifteen minutes in the game's time frame). Each move is then examined by a panel of judges ("Umpires") for its impact on the eligible electorate. Since this impact is measured in potential votes temporarily won or lost, periodic "Gallup Polls" can be published by the media, providing the players with feedback on what their actions are accomplishing. True to the real world, these polls only report net changes (by ward) resulting from all moves to date; students cannot learn the specific effect on any one move until the postgame debriefing. The final poll is, of course, the election result itself.

Woodbury depends heavily on the abilities of the "Umpires." They must make continual judgments about votes gained or lost. One judge for each ward is recommended, although alternate bases for division of labor could be employed. However accomplished, the main lesson from use of *Woodbury* is that a knowledgeable team of referees is virtually essential.

Woodbury, then, is a game that should be undertaken by a *group* of teachers who feel at home with fundamental voting analysis. The game was originally designed for college freshman or sophomores, with a professor plus graduate or advanced undergraduate assistants doing the umpiring. Reasonably detailed instructions, with supportive forms, are provided for the umpires, but they are no substitute for a thorough familiarity with voting and political party behavior. The umpires must use their own judgment in determining the parameters of voter shifts in response to ongoing campaign events. The demands on the student participants for informational background are fairly heavy too, but are consistent with high school capabilities reached after exposure, for example, to a six-week unit on voting and parties. *Woodbury* should never be used to introduce such a unit, but it can be an exciting and integrative culmination to a study unit.

Comments: *Weaknesses*

(1) The prime weakness is the necessary presence of a number of judges who are both knowledgeable and willing to devote themselves to advance preparation and execution of the game.

(2) Under most circumstances, the game makes space demands which may be hard to meet: separate rooms for each of the playing groups, plus a central area for the umpires.

(3) Materials and equipment use is of sufficient magnitude to be of possible concern to some: perhaps 1,000 hectograph masters, 20 reams of paper, allied amounts of masking tape, and hectograph fluid.

(4) Substantively, some aspects of *Woodbury* are dated: a few of the urban problems, the stated voter registration requirements, reference to blacks as Negroes.

(5) As with many other games, players tend to be exaggerative in their actions, attracted more to noticeable effects than to verisimilitude.

(6) As an extension of the above point, exaggeration tends to escalate (often to the point of silliness) as this and similar games approach their ends. Unfortunately, students sense that, unlike the real world, the game's world has no tomorrow, and that the longer-term consequences of extreme actions taken for present advantage need not be taken into consideration.

(7) *Woodbury* lacks a financial dimension. Campaign moves implicitly involving expenditure can be made without reference to total financial reserves or to the typical timing of the inflow of campaign contributions. While players are forced to come to understandings with many kinds of groups and persons in the course of the game, they are not forced to reconcile themselves to financial backers.

Strengths

(1) It is well suited to the usual class size of 25 to 30 students. Subgroups of from 3 to 6 students encourage group work and avoid underengagement of marginal students that sometimes occurs in large groups.

(2) *Woodbury's* time allocation is flexible—once the minimum commitment of seven hours total lapsed time is recognized. The game is usually played intensively, during one day or over two half-days. But it can also be played on a one-to-one basis, with "daily" moves submitted daily.

(3) A major advantage of this game over many others (especially those in the international relations tradition) is its numerical standard. Actions are converted into a common currency of votes, and as a result, periodic progress readings as well as a definitive statement of a final result are useful.

(4) Because the game involves an election, there is a specifically known and accepted finale: election day. Again, in contrast to many international relations games, this adds substantially to the students' sense of planning, accumulation, and completion.

(5) Because of the inbuilt dynamic of the game (and of real elections), there is, empirically, a noticeable heightening of student engagement, excitement, and activity as the game moves toward its climax.

(6) The mechanics of game moves—chronicled with hectograph—produces a reasonably full record of what has happened and can support valuable postgame debriefing discussion.

(7) While, as mentioned, a few of the materials are dated, the great preponderance reflects the more stable and undatable aspects of electoral politics. *Woodbury* rests on a solid foundation of voting research and theory, as its continuing relevance over nearly 15 years demonstrates. (From the Robert A. Taft Institute of Government Study on Games and Simulations in Government, Politics, and Economics by Bradbury Seasholes, Tufts University.)

Note: Users interested in the computer-assisted version of *Woodbury,* which is being mounted on a CDC time-sharing system, should contact Westland Publications or Marshall Whithed, School of Community Services, Virginia Commonwealth University, Richmond, VA 23284, to inquire about availability, materials, transfer capabilities, and costs.

Cost: manual version, $5.00 per player manual and umpire manual (1 per participant). Quantity discount.

Producer: Westland Publications, P.O. Box 117, McNeal, AZ 85617

See also
The American Constitutional Convention HISTORY
Blackberry Falls URBAN
Bombs or Bread INTERNATIONAL RELATIONS
Build URBAN
Buyer Beware PRACTICAL ECONOMICS
Espionage HISTORY
Federalists versus Republicans HISTORY
The Good Society Exercise INTERNATIONAL RELATIONS
Guns or Butter INTERNATIONAL RELATIONS
Inflation ECONOMICS
Involvement HISTORY
Japanese American Relocation, 1942 HISTORY
Power SOCIAL STUDIES
The Progressive Era HISTORY
Protection PRACTICAL ECONOMICS
Settle or Strike ECONOMICS
Watergate HISTORY
COMMUNITY ISSUES
COMPUTER SIMULATIONS/DOMESTIC POLITICS

ECOLOGY

ALGONQUIN PARK

David Dagg

Playing Data
Copyright: 1973, revised 1978
Age Level: grades 7-12
Number of Players: 5-35 in 5 teams
Playing Time: 7-8 50-minutes periods
Preparation Time: 1 hour
Packaging: large area map and 5 small map/playing boards, resource sheets, role cards, and teacher's materials, including debriefing strategies

Description: Players fill the roles of a committee on land use, a forestry management company, land developers and planners, lawyers for cottage owners, and the mining company. They simulate a real-life situation that involves the multiple use or conflicting use of wilderness. The committee on land use must make a policy on land use for a sixty-square-mile tract.

Players receive resources and role cards and then begin consulting and negotiating pacts, deals, and so on. When each group has prepared its basic plan and finalized intergroup deals, it begins preparing its oral report with supporting materials and defense for its plans. The government committee sets the schedule for reports and then hears presentations from the four interest groups. It forms a policy on land use, hands it down, and then entertains group reactions to its policy. The last session begins with a secret ballot election of the whole government committee. Assessment of the fate of the other groups measured against the objectives stated on their role cards is led by the committee on land use chairman. A debriefing follows, and *Algonquin Park* ends with an announcement of election results. (AC)

Note: For more on this simulation, see the essay on ecology/land use/population games by Larry Schaefer et al. (AC)

Cost: $15; $10 if prepaid

Producer: David Dagg, 420 Isabella Street, Pembroke, Ontario, Canada K8A 5T5

BALANCE

David Yount and Paul Dekock

Playing Data
Copyright: 1970
Age Level: grades 7-12, adult ecology groups
Prerequisite Skills: reading, grade 8; math, grade 6
Number of Players: 18-35
Playing Time: 15-20 days in periods of 40-50 minutes—also there is a 3-hour version
Preparation Time: 2-4 hours
Supplementary Materials: listed in administrator's manual
Special Equipment: duplicating equipment, overhead projector

Description: The first hour introduces the concept of ecosystem by simulating the last one hundred fifty year history of an American geographical area: fifteen students are animals (mountain lions, beavers, jays, trout, and the like); four are Indians who live in harmony with their physical environment; the remainder are settlers who dominate and kill the animals, wipe out the Indians, and subdue the wilderness while their population soars to over 100,000. Survivors start *Balance* with a bonus in GASPS (goal and satisfaction points), the scoring system used throughout the simulation. Students are then divided into families of four members each living in Ecopolis, their burgeoning city with many ecological problems. Interviewing parents and adults plus reading about American air and water pollution, land usage, and population problems culminate in confrontations over whether each problem necessitates social action. Within their class families, students role-play four different identities on four different occasions: father, mother, young adult, and adolescent. Each family then fills out a family decision form which tests its ability to balance short-range economic-hedonistic goals with long-range environmental goals. Before an essay evaluation ends *Balance,* students conduct an ecological survey of their real community and hold a one-hour forum in which they argue about the ecological balance of their own environment. (Producer)

Note: For more on this simulation, see the essay on ecology/land use/population games by Larry Schaefer et al.

Cost: $14.00

Producer: Interact, Box 262, Lakeside, CA 92040

THE DEAD RIVER

E. Nelson Swinerton, University of Wisconsin, Green Bay

Playing Data
Copyright: 1973
Age Level: 14 and up
Number of Players: 10-32; for example, 20 players in 4 teams or 5 players each
Playing Time: 4 hours or 5 50-minute class periods
Supplementary Materials: appropriate slides, film strips, cassettes of films on water pollution
Special Equipment Needed: projectors, screens, cassette players, and so on
Packaging: instructor's guide and team guides included in one box

Description: One person is selected as moderator (either test instructor or anyone with experience in simulations). Players are divided into five teams. Players prepare by reading the "Hearings of the Interstate Water Pollution Control Commission." Additional audio-visual aids plus individual research into relevant topics may be necessary especially for young players. An "envirogram" from the governor (not present) is read to start the game. Each team "works through their exercises, arrives at a position, and prepares to argue their recommendations before the Regional Council." A team performance sheet is kept to record both team performance and compromises and adjustments. The objectives are to formulate an overall water-quality standard, develop a budget (in millions of dollars) for reclaimation of the "dead river," and to project in millions of dollars how the reclamation would effectively benefit the region served. Players are provided with five basic objectives for water quality that vary from excellent quality costing $5,000,000 to industrial/navigation quality costing $1,000,000. The five teams—the Taxpayers Association, the Valley Recreational Development Association, the Regional Industry team, the State Eco-Action Agency, and the Federal Enviro-Policy Agency—develop their own recommendations based on the above standards and costs (65 minutes). Each team completes its own work then chooses one spokesman or a small committee, which delivers a five-minute recommendation before the regional council (twenty-five minutes). During a recess of the council, each team compares its position to the others, revises its position, if necessary, and adapts a final regional position. In addition, supplementary funding guidelines are considered and a total budget is developed with proposed costs and benefits (twenty minutes). As the council reopens, the moderator seeks a motion from any team to adopt a single objective. The objective may contain amendments. Each team has one vote, and a majority of votes on an objective carries the council. The moderator presents the final motion and closes the meeting. Each team prepares a performance score sheet. (TM)

Note: For more on this simulation, see the essay on energy/environmental quality simulations by Charles A. Bottinelli.

Cost: $13.00

Producer: Union Printing Company, 17 W. Washington Street, Athens, OH. 45701

ECOPOLIS

John Wesley

Playing Data
Copyright: 1971
Age Level: grades 5-8, average ability, grades 9-12 below-average ability
Number of Players: 18-35
Playing Time: 12-15 40-50 minute periods
Preparation Time: 2-4 hours
Special Equipment: duplicating equipment, overhead projector

Description: *Ecopolis* is an elementary version of *Balance* (see above). It is designed to develop knowledge about the concept of ecosystem and causes of and solutions to pollution problems and overpopulation; to show how special interest groups interact with community ecologists to achieve their respective goals; and to develop skills in outlining, note taking, using data to support positions, and defending viewpoints. It makes use of several short simulation exercises. First is an ecosystem game tracing 150 years of land use, pollution, and population growth in the city of Ecopolis. There are also two crisis simulations. In the first, half the players assume assigned roles in attempting to resolve a proposed county park crisis. The remaining players act as observers and evaluate the role performance. For the other crisis involving population control, the activities of the two groups are reversed. The unit includes bonus projects that involve students in applying their learning to their own community. (HWM)

Note: For more on this simulation, see the essay on ecology/land use/population games by Larry Schaefer et al.

Cost: $14.00

Producer: Interact, Box 262, Lakeside, CA 92040

THE ENERGY-ENVIRONMENT GAME

Raymond A. Montgomery, Jr. and Toby H. Levine

Playing Data

Copyright: 1973
Age Level: grades 10-12, college, adult
Number of Players: 20-40
Playing Time: 5-7 hours
Preparation Time: 3-4 hours
Packaging: 28-page teacher's guide, 32 21-page player's guides, wall map, 2 sets of 32 role profiles, 1 filmstrip and cassette, 1 set of site selection information (4 spirit masters)

Description: This is a role-play simulation which deals with the escalating energy demands of a hypothetical region. The objective is to locate a new power-generating facility in one of three proposed locations. Twenty to forty students assume the roles of environmentalists, utility officials, business professionals, and lay citizens as they provide input to the "Governor's Commission on Energy and the Environment." During a series of task force meetings and commission hearings, players are exposed to the gamut of viewpoints and attitudes related to energy consumption, production, and environmental degradation. The necessity of arriving at viable trade-offs is a basic theme of the simulation. (CB)

Comment: Multidisciplinary in character, *The Energy-Environment Game* can be of special educational value when incorporated in traditional science, social studies, environmental/energy education lessons concerning the energy crisis. (CB)

Note: For more on this game, see the essay on energy/environmental quality games by Charles A. Bottinelli.

Cost: presently out of print, revision scheduled for 1979 (last edition was $20.00 per package)

Producer: Edison Electric Institute, 90 Park Avenue, New York, NY 10016

ENERGY X

Norman S. Warns, Jr.

Playing Data
Copyright: 1974

Age Level: junior and senior high school
Number of Players: approximately 30 in 9 small groups
Playing Time: 3-4 class sessions
Special Equipment: mimeograph machine, tape recorder, filmstrip projector
Packaging: professional game box including 8-page manual, cassette, filmstrip, many charts and mimeographs

Description: *Energy X* is based on the make-believe situation of a meteorite which lands in the United States and provides a pollution-free new energy source available only to a limited area of the country. Students representing eight regions spend two days preparing presentations in which they try to convince the project control group that their regions should be one of the three entitled to use *Energy X*. In addition to a filmstrip and tape in the kit, numerous maps, charts, and data sheets available to all students help them complete their regional analysis worksheets and forms for organizing their presentations. The simulation game ends when the project control group decides on the three groups, the other groups appeal the decision, and a final decision is made. Students then discuss their experience and complete self-evaluation forms. (TM)

Note: For more on this simulation, see the essay on energy/environmental quality simulations by Charles A. Bottinelli.

Cost: $19.95 (#3057)

Producer: Ideal School Supply Company, 1100 South Lavergne, Oak Lawn, IL 60453

ENVIRONMENTAL SIMULATIONS

Ben Vass

Playing Data
Copyright: 1976
Age Level: grades 7-12
Number of Players: 15-45
Playing Time: 40-60 minutes for each simulation
Preparation Time: 1 hour (est.)
Packaging: 28-page teaching guide, 30 student booklets, wall poster

Description: The four simulations in this collection are designed "to reflect the real problems of the modern world," according to the designer. Each is a period-long activity that can be used to supplement a course of study.

In *Expressway*, the students discuss the advantages and disadvantages of urban expressway systems. Six players assume the roles of informed advocates for or against the construction of an expressway in an unnamed city. The advocates present their views to the rest of the class. In *Algonquin*, twenty students represent participants in a public hearing called to examine a proposal to run part of a superhighway system through a national park. The players assume diverse points of view and at the end of the discussion the entire class votes on a recommendation to build the highway. *Airport* follows the same format as *Expressway*, except that the debate concerns, according to the author, "the construction of a new airport." In the final simulation, *Moscow*, the players learn about the city by pretending to tour it, and then compare that city to their own community. (DCD)

Note: For more on *Expressway*, see the essay on social studies simulations by Michael J. Rockler.

Cost: complete unit $24.00; replacements: teaching guide $2.00, student booklet $.80, poster $.80 (Canadian currency)

Producer: Scholastic TAB Publications Ltd., 123 Newkirk Road, Richmond Hill, Ontario L4C 3G5 Canada

EXPLOSION

Dan Guida, Roger Henke, and Dennis Porter

Playing Data
Copyright: 1975
Age Level: grades 7-12
Number of Players: 25-40
Playing Time: 5-25 1-hour sessions
Preparation Time: 2 hours (est.)
Packaging: 33-page instruction booklet and student handbook

Description: To quote the designers, "*Explosion* is a simulation which allows the student to experience the traumas of the population crisis and its social, environmental, and political impact on the spaceship earth—a finite system." A hypothetical nation called Scioto is a model of the United States, with the problems this nation may have between 1980 and 2015. It is comprised of the states of Egon, Nauwark, Aksarben, Agrion, Indio, and Zarion, which correspond to regions of North America. Nauwerk, for instance, represents the East Coast strip city from Boston to Norfolk, and Egon represents the Pacific Northwest. Various resources (pieces of colored paper) and population (number of players) are unevenly divided among these states. The game is composed of three phases which require five, eight, and ten hours, respectively, to play, and which may be played independently.

Phase I is designed to help players appreciate the social pressures of a population explosion in Scioto. States (groups of students) complete "research assignments" to earn food coupons. While they work on these assignments these groups experience "resource shortages (not enough colored paper to complete an assignment), unemployment (too many students for the tasks assigned), crowding (being restricted to a small area), a rising 'crime' rate (some may resort to 'stealing' resources from other states), hunger (if an assigned task is not completed no food coupon is earned), and many other feelings experienced by individuals living in such a society (anger, hopelessness, frustration, despair, etc.)."

In phase II, players assume the roles of members of planning agencies and special interest groups who recommend plans for the future use of Scioto's resources. State planning groups meet with representatives of the special interest groups to make plans in the best interests of their states. Representatives of each group debate and approve or reject these state plans before compiling a final set of recommendations for the future management of the nation's resources.

During phase III, students play the roles of concerned citizens from the six states. Citizens from each state must design a law which would solve that state's population-related problems but which is federal in its implications. After these proposed laws are designed, the players assume the roles of Senators and Representatives from those states who must act on the bills legislating those laws in Scioto's Congress.

The game ends with a two-hour debriefing, during which the players relate the problems of Scioto to the problems of the real world. (DCD)

Cost: $22.00

Producer: Interact, P.O. Box 262, Lakeside, CA 92040

GOMSTON: A POLLUTED CITY

Norman S. Warns, Jr.

Playing Data
Copyright: 1973
Age Level: junior and senior high school
Number of Players: 25-50 in 8 groups
Playing Time: 8-15 class periods
Special Equipment: overhead projector, tape recorder, filmstrip projector
Packaging: professional game kit including cassette, filmstrip, 40 copies each of half a dozen forms, transparencies, posters, role cards, and 24-page manual

Description: *Gomston* attempts to alert students to the problems of a degraded environment and to possible solutions. As representatives of city and state government, agriculture, local industry, the Chamber of Commerce, the news media, and an antipollution group, each with specific goals, students attempt to solve the environmental problems of a make-believe polluted city, Gomston. The first two days, the class discusses pollution problems and gets acquainted with Gomston through a filmstrip, maps, and transparencies. After they choose roles, elect their mayor, and meet in small groups, the city council (consisting of all students) convenes. The structure of the remainder of the simulation is up to the teacher. Basically, students plan, discuss, and vote on proposed solutions to the pollution problems while trying to accomplish their own groups' goals. While the news media record events, other students map and chart the plans and progress of the city council. Response forms for recording ideas, environmental analysis sheets, and other important data about Gomston are available to each student. The teacher's guide includes additional data on the financial situation of Gomston, a list of possible solutions, and a complete bibliography of resource materials. The simulation ends unexpectedly, at a time decided by the teacher, who also role plays the local geographer. There are no winners. The evaluation process includes class discussion, a quiz, and self-evaluation forms. (TM)

Note: For more on this simulation, see the essay on energy/environmental quality simulations by Charles A. Bottinelli.

Cost: $26.50 (3055)

Producer: Ideal School Supply Co., 11000 South Lavergne, Oak Lawn, IL 60453

INDIAN VALLEY

Norman C. Thomas, Springfield Schools, Oregon

Playing Data
Copyright: 1973
Age Level: junior and senior high school
Number of Players: 18-40
Playing Time: 2-3 hours
Supplementary Equipment: overhead projector, duplicating machine
Packaging: 14 printed pages

Description: The purpose of *Indian Valley* is to teach students basic principles of forest land management so that they can judge land-use issues more effectively. Five teams representing five different land-use interests (water resources, parks and recreation, fire protection, fish and game reserve, timber reserve) analyze the master map and the five or six goals they are given and draw up a plan for three projects to present to the multiple-use committee. Because some of the goals of the various groups overlap (e.g., salvage sick trees, develop a road system, build look-out towers), and because students receive ten points for each project approved by the committee for two points when another team presents the same project, teams will want to negotiate with each other to convince other teams to accept their chosen projects. At the end of the game, the multiple-use committee evaluates the projects according to their master evaluation sheet and chooses ten projects to be implemented. Teams then score themselves. (TM)

User Report: A real advantage of this game is the age range of the players with whom it may be used. It is simple enough that primary school children are able to get some useful land management ideas in short periods of play, whereas older students can arrive at more refined solutions to these land-use issues by doing outside research.

When the game is extended for several days, students often become so involved that they may continue planning and consolidating their projects outside of class time.

Indian Valley fits nicely into the science, social studies, or even language arts curriculum. (Gene Franks, McGuffey Laboratory School, Oxford, Ohio)

Cost: none

Producer:: American Forest Institute, 1619 Massachusetts Ave., N.W., Washington, DC 20036

LAND USE

Playing Data
Copyright: 1973
Age Level: grades 7-12
Number of Players: 20-50
Playing Time: 2-15 hours
Preparation Time: 3 hours (est.)
Packaging: professionally produced 69-page booklet

Description: This collection of environmental activities was developed by the Forest Service of the U.S. Department of Agriculture to help teachers examine different aspects of the environment. To quote the materials, "[they] provide a structure for learning in which one activity builds on others and leads to some generalizations about the environment. These generalizations, in turn, can provide a basis for a better understanding of environmental problems and their possible solutions. Even though the investigations are 'structured', they allow the student freedom to observe, collect, record, and interpret data at his own pace and level of understanding. The lessons also are designed to elicit a maximum of student response and involvement through the use of discussions and questioning techniques. In many instances, charts, tables, and other aids are included to help the student interpret the data he has collected." There are seven activities in this collection. They address land use planning, water quality criteria, environmental habitats, forest environments, urban environments, the comparison of environments, and include the *Land Use Simulation Game.*

Players of the game assume the roles of representatives of industry, recreational groups, utilities and real estate developers who must draft proposals for the use of six hundred forty acres of farmland four miles northeast of "Centerplace City" that have recently been incorporated. Groups of players speculate about uses for the land and then choose one idea to present to the county board of commissioners. Each group must then develop a strategy for presenting their proposed plan to the county board. Finally, the commission must select one plan and justify its choice. (DCD)

Note: For more on the game, see the essay on ecology/land use/population games by Larry Schaefer et al.

Cost: $.95

Producer: Superintendent of Documents, U.S. Government Printing Office, Washington, DC 20402

MAKE YOUR OWN WORLD

William R. Brown

Playing Data
Age Level: grades 4-12
Number of Players: 20-50 players in 11 groups
Playing Time: 50 minutes to 3 hours
Packaging: 13-page photocopies instruction booklet

Description: *Make Your Own World* is a kit designed to enable teachers to make their own materials based on the Coca-Cola Company's *Man in His Environment.* This kit includes mimeographed materials for making the following: one "Make Your Own World" chart; ten project proposal cards; cards simulating population, sewage disposal plants, a dump; and eleven team identification cards.

At the beginning of the game each student is assigned a role. The roles include three basic types: human, organism, or natural resource. The students are presented a chart representing the natural area for their own world. The game proceeds by presenting a series of proposals about adding man-made projects to their world. Each project is analyzed according to the student's roles, and each proposal has a list of items to consider concerning the impact of the project. After the discussion, a vote is taken to determine whether it should be added. If the project *is* added, a given number of population squares are also added to represent the population increase brought about by the project. The game ends when all the projects have been considered. (Author)

Cost: postage stamps to equal $.53

Producer: Science Education Center, School of Education, Old Dominion University, Norfolk, VA 23508

NO DAM ACTION

R. G. Klietsch

Playing Data
Age Level: grade 7-college, management, ecology groups
Special Prerequisites: curriculum unit treating water resources
Number of Players: 20-40
Playing Time: 6-20 time periods of 30-40 minutes
Preparation Time: 1-3 hours, unless quest activities are used

Description: This game focuses on the interplay between interpersonal and ecological factors. Players represent county residents–from lakeshore owner to business persons, county officials, and so on. Teams represent political factions and agencies in the county. Players address eighteen major problems, having a broad base of alternatives. Problems range from recreational water use to water purity standards, and each of these problems requires personal, group, and intergroup endorsement.

The game is designed to create contexts of paradoxes and dilemmas involving ecological and human factors that necessitate resolution in terms of values, to establish priorities and feasibility criteria relating to water resources and their use, and to illustrate cross-purposes and conflicts of interest in water resources planning and use.

This simulation emphasizes the need for cooperation in the face of inherently conflicting situations. Play involves all aspects of planning and governing water resources within a specified area. These include planning, prediction, communication, group skills, measurement, and willingness to compromise. Policy decisions with respect to county funding of water resource concerns, electrical energy potential, and flood-control costs and benefits can all be measured quantitatively. In addition, each special interest group has its own quantifiable objectives. (DCD)

Note: For more on this simulation, see the essay on energy/environmental quality simulations by Charles A. Bottinelli.

Cost: $145.00

Producer: System's Factors, Inc., 1940 Woodland Ave., Duluth, MN 55803

THE PLANET MANAGEMENT GAME

Victor Showalter, Educational Research Council of America

Playing Data
Copyright: 1971
Age Level: grade 7-college
Number of Players: 2-10 in groups
Playing Time: 1-5 45-minute periods
Preparation Time: 30-60 minutes
Special Equipment: overhead projector optional

Description: Players assume the roles of managers of the imaginary planet Clarion, and set out to improve living conditions there. Clarion is Earth-like and presents problems similar to those that would be encountered on Earth.

Students deal with contemporary ecological problems such as the population explosion, famine, and the quality of the environment, and they discover that changes in one area have an effect on other areas. For example, a project that would increase income may also cause pollution and, thereby, lower the environmental quality. Starting with equal resources, players attempt to regulate the food, population, income, and environmental levels while spending an alloted sum of money. The accuracy of their foresight determines success. The winning group, based on its self-evaluation and that of the other players, will have the most desirable combination of population, food income, and environmental levels. Students may wish to play the game more than once–to accomplish the same results at a lower cost or to improve one or more of the levels by spending money differently. (DCD)

Note: For more on this simulation, see the essay on ecology/land use/population simulations by Larry Schaefer et al.

Cost: $22.56

Producer: Houghton Mifflin Company, One Beacon Street, Boston, MA 02107

POLITICAL POLLUTION

David J. Boin and Robert Sillman

Playing Data
Copyright: 1972
Age Level: grades 7-12
Number of Players: 10-40
Playing Time: 5 class periods

Description: Students are members of Congress from three states, members of the Ecology Society of America, the Department of Defense, and a special fact-finding committee. They have to come up with an ecology antipollution bill that will directly affect the state's major industry, which is polluting a beautiful lake. The conflict is produced by another state, whose major industry is fishing in that lake. After several days of discussions, the students produce a bill and different elements of the society are given points, depending on the form of the bill. According to the producer, the instructional objectives of the simulation are "to help students understand some aspects of the ecology problem, the dynamics of political compromise, and the term 'special interests' and how they may affect legislation, and to help students organize ideas and develop logical as well as emotional arguments." (REH)

Comment: This is in the standard Edu-Game format with a few pages of teacher's manual and locally reproducible student materials. (REH)

Cost: $3.00

Producer: Edu-Game, P.O. Box 1144, Sun Valley, CA 91352

POLLUTION

Ron Faber and Judith Platt, Abt Associates

Playing Data
Copyright: 1973
Age Level: grades 5-12
Number of Players: 12-16 in 4 teams
Playing Time: 1/2-1-1/2 hours in periods of 30 minutes
Preparation Time: 20 minutes

Description: Each team (two community representatives and two factory owners) gets a standard amount of money at the beginning of each round. Each team also has the same amount of air, noise, and water pollution to contend with at the beginning. Each team must decide if and which abatement program they want to purchase and how to divide the cost. Chance cards are then picked which may alter the pollution levels. At the end of each round the consequences of the pollution level of each town is read. When these consequences result in a financial loss, the teams must decide how to pay.

Players are only told that they are either community representatives or factory owners. No further instructions are given but the method of winning builds in greed to offset the idealism generally found in young children.

The game demonstrates the long-term disaster of continuing unabated pollution and the large outlays necessary to *undertake* pollution control, which then pay off in long-term effectiveness. (Producer)

Note: For more on this simulation, see the essay on energy/environmental quality simulations by Charles A. Bottinelli.

Cost: $26.00

Producer: Games Central, Abt Publications, 55 Wheeler Street, Cambridge, MA 02138

THE POLLUTION GAME

Educational Research Council of America

Playing Data
Copyright: 1968, 1971, 1979
Age Level: grades 7-9
Number of Players: 4-5
Playing Time: 45 minutes to 1-1/2 hours in class periods
Preparation Time: 45 minutes

Description: Each player begins this board game with $3,500 and two election cards. Players' progress is determined by dice throws. Players buy property, pay and collect rent, collect $500 each time they pass start, and occasionally land on take a chance. As various businesses profit in the course of the game, air and water pollution indexes increase. If these levels pass the lethal limit everyone loses. A player may call for an election to vote on one of a number of possible proposals to reduce pollution. Each player may call for only two elections. The individual who ends up with the most money wins, whereas the team with the lowest pollution indexes in the class wins.

Note: For more on this game, see the essay on energy/environmental quality simulations by Charles A. Bottinelli.

Cost: not established

Producer: Carolina Biological Supply Company, 2700 York Road, Burlington, NC 27215

POLLUTION: NEGOTIATING A CLEAN ENVIRONMENT

Paul A. Twelker

Playing Data
Copyright: 1971
Age Level: grade 7-adult
Number of Players: 4-8
Playing Time: 1-3 hours
Special Equipment: overhead projector
Packaging: professionally packaged game includes all materials for four players

Description: This game was designed, according to the designer, to help players "understand some of the problems of air, water, land and visual pollution, as well as the principle interests at stake and their relationships to each other." The players assume one of four roles which are intended to epitomize the attitudes and clout of business, government, the one of four ways to any issue. All groups may support or not support an example of pollution. Conservationists may also petition or lobby against polluters, private citizens may write to their legislators or leave town, government has the options of ignoring or stopping pollution by court order, and business may, if it cannot afford to stop polluting, close down.

Pollution is played in five-minute rounds. During each round, participants examine and discuss one example of pollution. For example, one issue involves tearing down billboards within two hundred twenty yards of interstate highways. In another, the players must decide what to do about a small farmer's supply company which has been prevented from burning its rubbish and so has begun to throw the trash into a small stream. In the course of play, each representative group negotiates "with other players to increase the chances of each player realizing his own personal or corporate goals." Issues are decided by vote. Business and government have more voting power than do the other two groups, and depending on the outcome of each vote the players move tokens up or down "quality of life" and "goal satisfaction" columns on a playing surface. The game is interrupted when a trend toward either an increasing or decreasing quality of life becomes apparent. (DCD)

Note: For more on this simulation, see the essay on energy/environmental quality simulations by Charles A. Bottinelli.

Cost: $27.00

Producer: Simulation Systems, Box 46, Black Butte Ranch, OR 97759

POSSUM CREEK VALLEY

American Forest Institute

Playing Data
Age Level: grade 6-college
Number of Players: 12-40 in 6 teams
Playing Time: 1-4 hours
Special Equipment: Overhead projector, spirit duplicator or photocopier

Description: Each team of players represents a group of land-management specialists in a given discipline, exercising principles of that discipline. Their objective is to develop an integrated management plan for a forested valley. Teams are formed, one each specializing in water resources, recreation, timber, fish and game, fire protection. The last team is a panel of judges who will draft a multiple-use plan. Each team drafts a plan for its specialty, based on principles given, then determines projects for management it will present to the multiple-use team. A final integrated management plan is evolved. (Producer)

Comment: This game is designed to give students an introduction to broad principles of forest land management. The manual suggests making a story for *Possum Creek Valley* before play with younger students, and with older students discussing the increasing pressures of an increasing population on water, recreation facilities, and timber. This is the southern version of *Indian Creek.* (AC)

Cost: $1.50

Producer: American Forest Institute, 1619 Massachusetts Ave., N.W., Washington, DC 20036

THE REDWOOD CONTROVERSY

Educational Research Council of America

Playing Data
Copyright: 1971
Age Level: Grades 7-9
Number of Players: 21
Playing Time: 2-3 class periods

Description: *The Redwood Controversy* is based, with fictional roles, on an actual dispute over the appropriation of California coastal redwoods, owned mostly by lumber companies, for a national park. Some elements of the conflict have been omitted by the designers for the sake of simplicity; the teacher is informed of what those elements are.

Players take the roles of senators who must vote to establish a large, medium, or small national park, or none at all, and witnesses who represent conservation and lumbering interests and who attempt to persuade the senators how to vote. This process involves negotiation, bargaining, and compromise and exposes players to conflicting ecological, sociological, and political elements that come into play over such issues. (AC)

Note: For more on this simulation, see the essay on ecology/land use/population simulations by Larry Schaefer et al.

Cost: $14.25

Producer: Houghton Mifflin Company, One Beacon Street, Boston, MA 02107

WHALES

Thomas G. Cleaver, University of Louisville

Playing Data
Copyright: 1978
Age Level: grade 9-adult
Number of Players: 4-20
Playing Time: 1-2 hours
Preparation Time: 90 minutes (est.)
Special Equipment: bingo chips
Packaging: 30-page mimeo

Description: The designer writes, "In *Whales* the problems of whale population control and resource management are dealt with by allowing players to assume the roles of the heads of state of commercial whaling nations. They may seek quick profits by heavy fishing, or may adopt more patient policies which limit fishing. Other players represent the International Whaling Commission, environmental groups and consumers of whale products. In the play of the game, whalers conduct simulated fishing operations for the various species of whales while the other players interact with them to attempt to influence whaling policies. Overfishing may ultimately result in the extinction of species and reduction of profits, but sound management and cooperation among the players should bring about the stabilization of whale populations and a plentiful yearly whale harvest."

Players represent one of seven groups: The United States, the Soviet Union, Japan, Norway, the United Kingdom, a bloc of nonwhaling nations, and the International Whaling Commission. Each group receives a playing board on which whale harvests and economic actions may be recorded, and the five whaling nations receive factory ships. The "sea" in this game is represented by a box and the populations of four species of whales are symbolized by bingo chips as follows: blue chips for blue whales, green chips for finbacks, pale blue chips for humpbacks, and yellow chips for Sei whales. Orange chips are used to indicate unsuccessful whaling expeditions. Chips have a relative value, according to their color, that is equivalent to a unit of measure called the blue whale unit. Under the regulations of international whaling, blue whales are considered to be more precious than other species. This value, in terms of the chips used in the game, means that each blue chip is worth two green, two and a half pale blue, and six yellow chips.

The principal game events are the harvesting of whales and the initiation of sanctions against whaling nations. In each turn, each whaling nation must declare the species of whale it will hunt and then draw from the box a number of chips equal to the number of factory ships in its fleet. Orange chips must be returned to the box immediately, as must all but one chip representing any other than the declared species to be harvested. Each round, after the hunt, the moderator restocks the sea with a number of chips shown on a chart. Finally, in each round, all nations may impose economic sanctions against any whaling nation for any reason (for instance, because of the efforts of environmentalists in discouraging whaling or because a country appears to be overharvesting). These sanctions result in the loss to the target country of up to ten thousand blue whale units or one blue chip. Any country may increase or decrease its harvesting capacity during play by buying or selling factory ships. New ships may be built for the cost of one blue chip and a ship in service may be sold for whatever the market will bear or scrapped for a return of one yellow marker. The game continues for seven rounds. (DCD)

Cost: $1.00 for postage and handling

Producer: Thomas G. Cleaver, Department of Electrical Engineering, University of Louisville, Louisville, KY 40208

See also
Acres URBAN
Apex URBAN
Baldicer SOCIAL STUDIES
Eco-Acres URBAN
Extinction SCIENCE
Fire LEGAL SYSTEM
The Future Game FUTURES
Limits COMPUTER SIMULATIONS/SOCIAL STUDIES
Metroplex URBAN
New Town COMMUNITY ISSUES
Nuclear Energy Game COMMUNITY ISSUES
Nuclear Site Negotiation COMMUNITY ISSUES
Predator SCIENCE
Pro's and Con's FRAME GAMES
River Dose (RVDOS) Model COMPUTER SIMULATIONS/ECOLOGY
The Territorial Sea INTERNATIONAL RELATIONS
Wildlife SCIENCE
COMPUTER/ECOLOGY

ECONOMICS

THE AID COMMITTEE GAME

Playing Data
Age Level: high school
Number of Players: 5-50 in groups of not more than 10
Playing Time: 4 or more class periods
Supplementary Materials: background papers on countries under study; slides
Packaging: 14 mimeographed pages

Description: This game is designed for use by a group studying development problems. Participants study one developing country from background materials which are available for Guatemala, Upper Volta, and India. After one class session devoted to introducing the country under study, participant groups assume the role of aid committees charged with allocating limited funds for development projects in the country. On the basis of the background papers, each committee decides what they think that country's development priorities are. Then the members decide which projects (each has a budget) to fund, at what level, and why, seeking to support those that will be most productive for the country. This process takes about two class periods. The last session is devoted to a debriefing discussion. The materials suggest several ways the leader might choose to vary the game. (AC)

Cost: 20 pence; slides £2 deposit refundable

Producer: OXFAM Education Department, 274 Banbury Road, Oxford 0X2 7DZ England

THE ARTIFICIAL SOCIETY

James H. Campbell, University of Alabama; Dan L. Costley, New Mexico State University; Sandra L. Cook, University of Missouri at St. Louis

Playing Data
Age Level: college
Number of Players: up to 500 or more in teams of 6-12
Playing Time: 1 semester

Description: Artificial Society is a semester-long course which focuses on the principles of organization and management and the questions of dealing with people in organizations. Players are either members of government or of companies. Companies are formed by presidents that have been selected by a rules committee at the beginning of the semester. The president hires members of the company and the companies then bid on government-sponsored research. They must negotiate contracts to perform particular research and prepare reports. People can be fired from companies, go on unemployment, join other companies or become independent operators. Each company is ranked on each product it produces by all of the companies involved. Rankings are worth points and are translated into individual points. Points are then paid to students at the end of the semester and are translated into grades. These grade points are also combined with other assignments that are given during the course. In subsequent sessions, companies may issue stock. (REH)

Comment: The rules for this game are well worked out. Many contingencies have been explored. The importance of this simulation lies in the fact that it shows how a college course can be organized around a particular economic model. The economic model here is a case of consulting organizations and the government, one which is certainly important to a lot of graduate students in their future lives. However, the intricate grading system, based on performance of the company plus individual deals made with the group and with the government, throws into bold relief the simplistic grading systems that are normally used to keep score in normal college classrooms. Economics is simply keeping score in society as a whole. This simulation suggests that we could all take a look at the organization of our keeping score systems in our classrooms. Clearly, this group-plus-individual-responsibility model works well. What other models are there? Could we model, instead of a competing society, a series of interdependent cooperative groups plus individual point accumulation? This calls for a lot of exploration from all of us in the field. (REH)

Cost: none

Producer: Dr. James H. Campbell, Room 306A, MEB, University of Alabama School of Medicine, Birmingham, AL 35294

BUSINESS STRATEGY

Playing Data
Copyright: 1973
Age Level: high school-adult
Number of Players: 2-4
Playing Time: 1 hour or more
Packaging: professionally designed boxed package

Description: To begin, the business climate cards are shuffled and placed face down on the playing board. Each player receives $10,000 in play money, four pawns of the same color, two standard factories, four raw material units, and two finished inventory units. Play continues for twelve turns, each of which represents one month. Each turn has six phases. In phase one, each player pays fixed expenses which always total $4,200. In phase two, the top business climate card is turned face up. This card specifies the number of raw material units that are available for purchase and the minimum price for which they may be purchased. Each player writes down a bid for one or more of these units and then all simultaneously disclose these bids. The highest bid(s) may purchase these units. In phase three, all players pay production costs to the bank as stated on the game board. Phase four is identical to phase two except that players bid on the maximum market value for their products (as stated on the business climate card), and this time the low bid wins. In phase five, players may borrow money, using the factories they own as collateral, and in phase six they may use the money they have borrowed to build new factories. For instance, a player may purchase three raw material units in phase two for $800 per unit. She would pay the production costs on these units in phase three, and could then sell her finished inventory of nine units in for $6,000 per unit in phase four. In phases five and six she may raise more capital with which to build more factories to process more raw material units to make more money. The winner is the player with the most cash on hand at the end of the last turn. The length of the game varies from one year to several years. The fundamental game is called the *Family Game.* More involved simulations are available using the same format. These are *Basic Business* and the *Corporate Game.* The classroom version allows teams to operate each business. Such teams could elect a chairman, board of directors, officers, etc., to administer each phase of operations. (DCD)

Cost: $12.00

Producer: Avalon Hill, 4517 Harford Road, Baltimore, MD 21214

BUY AND SELL

Playing Data
Copyright: 1976
Age Level: grades 9-12
Number of Players: 30-42
Playing Time: 3 50-minute sessions
Preparation Time: 2 hours (estimated)
Packaging: 36-page photocopies booklet

Description: Buy and Sell, according to the producer, "is a consumer economics game" in which "the classroom is transformed into a business community. Banks, factories, wholesale distributors, and resale stores are set up. Students assume the role of industrial executives, loan officers, sales persons, and consumers. Businesses compete for the most profits while consumers try to get the best buy. At the conclusion of the game the class has the opportunity of assuming the role of legislators in an attempt to determine what kinds of laws, if any, are necessary for the business and consumer community."

After participants are divided into groups representing consumers and business institutions as listed above, play consists of six buy-and-sell sessions. Half are for consumers and retailers, half exclude consumers. The process consists of banks capitalizing to businesses, manufacturers selling to wholesalers, who sell to retailers, who sell to consumers. The game provides for five winners: the bank, manufacturer, distributor, and retailer who profit the most, and the consumer who bought the most products for the least money. (DCD)

Cost: $3.00

Producer: Edu-Game, P.O. Box 1144, Sun Valley, CA 91352

CHANCELLOR

Centre for Business Simulation

Playing Data
Age Level: college-adults
Number of Players: 2-8
Playing Time: 2-4 hours

Description: Chancellor is a macro-economic exercise simulating the type of budget decisions which the finance minister of an economically developed and democratic country would have to take with the objective of maximizing its gross domestic product.

The starting point of the exercise is that all wealth is created by industrial and other production (labor, capital, and enterprise). The effect upon economic performance of the allocation of that wealth between various competing claims within the public and private sectors is observed.

All decisions and their impact are recorded in a budget statement, which includes the gross national product, government income and expenditure, the balance of payments and various key economic indicators. (Producer)

Cost: £36

Producer: Management Games, Ltd., 63B George Street, Maulden, Bedford MK45 2DD, United Kingdom

COUNTRY DEVELOPMENT ECONOMICS AND FINANCE GAME (CDEFG)

Harold W. Adams, SUNY at Albany

Playing Data
Number of Players: 10-25
Age Level: graduate school, business executives

Description: This is a sophisticated board game designed to simulate a national economy of a developing country in transition from traditional to modern agriculture with an emerging industrial urban sector. There are five teams. Four represent private sector entrepreneur groups which seek to develop agricultural and industrial properties for profit. The other team, the government, provides development assistance and seeks to direct the country's economic development with a focus on a balanced economy.

The game seeks to develop the player's ability in investment analysis. To win it is to prosper. To prosper, teams must develop a clear overview of the entire economic situation and make decisions not only on their own economic capabilities but in accord with governmental policy and the world economic reality.

The board game involves developing economic units on a 576-square board. Each square represents a plot of land, an industrial site, or an urban property which can be developed somewhat along the lines of Monopoly. There is a twenty-page guide for accounting procedures and a twelve-page description of development potentials of the different property units. Teams are divided into team captain, who is responsible for what to buy and where to build or develop; team negotiator, who is responsible for getting government approval for development plans; and the team accountant, who maintains property records, balance sheets, and is responsible for keeping a general economic overview of the country. The game is run by a game controller who is responsible for the initial set-up of the economic situation and for introducing outside economic factors such as natural disasters and changes in the world economic situation which effect export markets. (JG)

Comment: This game is a complete and well thought out simulation of the possibilities of economic growth in a developing country. It is complex and requires effort to master the rules, accounting system, and procedures of play; but the results of the effort look rewarding. Participants who play through this game should leave it with a basic knowledge of the economic and political parameters involved in the development of a traditional economy into a modern economic state. (JG)

Cost: free instruction manual and advice on use

Producer: Harold W. Adams, Graduate School of Public Affairs, State University of New York at Albany, ULB96, 1400 Washington Avenue, Albany, NY 12222

ECONOMIC DECISION GAMES

Erwin Rausch

Playing Data
Copyright: 1968
Age Level: grade 12, college freshmen
Prerequisite Skills: reading, grade 10; math, grade 4
Number of Players: teams of 6, any number of teams
Playing Time: two to four hours per game
Preparation Time: 1 hour per game, first time only

Description: This series consists of eight games.

BANKING

This game simulates banking conditions in an area containing three banks, each represented by a player whose task is to make profits while stimulating the area's economy. Players must determine what interest rate to pay for deposits, which investment opportunities to accept, and which loans to make. Each opportunity is offered to all banks, and several may make the same transactions. The results of all transactions are figured from tables and recorded on worksheets. The object of the game is to win as many points as possible from profits and investments. (AC)

COLLECTIVE BARGAINING

This game simulates the atmosphere, the problems, and the attitudes that influence the behavior of participants in a typical but simplified collective bargaining situation. It assumes negotiations in a single company where employees are represented by one union negotiating a one-year contract. The game asks participants to assume that people's preference can be measured and expressed in numerical terms. Only basic bread-and-butter items of wages and fringe benefits are discussed in order to maintain simplicity. Furthermore, the game assumes that all employees earn the same wages and will receive the same increases. (Author)

THE COMMUNITY

Participants try to create an attractive, progressive community. Success is measured by the number of public improvements and the prosperity of the three local industries. Players make decisions in a team that functions as an employer when setting wages, as the government when setting the tax rate, and as taxpayers and officials when spending tax revenues. Points are received for increases in wages and profits and for new community improvements. Points are lost if wages or profits decline, if industries go out of business, or if community improvements are discontinued. Tables are used to determine the relationships between industry tax payments and the tax rate, between profits and rate of return, between the increase in welfare needs and delayed tax payments, between profits and wage levels, and the costs and rewards of civic improvements. The tables have been calculated based on a maximum play of eight rounds. The game director can influence play by altering the general business conditions or the minimum rate of return industries need to continue operations. (DCD)

THE FIRM

Participants play in teams of six, each representing one of six owners of a store in the following roles: the president, who fills out the balance sheet; the operations manager, who maintains the income statement; the buyer, who computes the cost of merchandise; the controller, who maintains a loan worksheet; an assistant controller, who maintains the weekly cash accounts; and the public accountant, who check the other owners' calculations. The chief action of the game consists of filling out forms. For instance, in one round the owners might decide to sell their merchandise for $9.75 per piece. The instructor then tells players what the business conditions for that week are. By checking a table, players can see how many pieces are likely to be sold at a given price under given business conditions. All calculations in the game are made in this manner. (DCD)

INTERNATIONAL TRADE

Two trading countries (the United States and the Common Market) and two products (skyhooks and bushbats) are represented. Players assume roles as importer, exporter, and banker for each country. They must decide how much to produce, where to sell the products, and how much to charge. Players make their decisions after examining a set of tables and record their decisions on worksheets. Each round has four steps.

In step one, the exporters and importers in each country decide on production levels. Chart one lists five alternative decisions specifying the production of the two products in a particular ratio. In step two, players calculate the numbers that can be sold in their own countries by consulting a graph that illustrates "satisfaction levels." If the level that corresponds to a set of production figures is low, players must trade on a foreign market, and vice versa. In step three, traders use another chart to calculate prices available for various amounts of traded items for both nations, and the two parties agree on a trade. In step four, the bankers calculate the trade deficit, if any, and recalculate the satisfaction levels for both countries. The inevitable increase in satisfaction levels from trade is each country's score for the round. The game ends after two to five rounds. (DCD)

THE MARKET

This game seeks to demonstrate how price is established through changing supply and demand and to illustrate the concept of diminishing returns. Each player represents one hundred college students deciding how much allowance to spend for books, records, sweaters, and how much to save. All scoring is in "satisfaction points" that quantify satisfaction from possessing the items or from banking money. Players determine their scores by referring to a table.

The game is played in four steps. In step one, each player decides how to spend the personal allowance assigned by the game director, using the *anticipated* price for each item to draft this budget. In step two, after all players have completed their budgets and reported what they intend to buy, the total demand for each item is calculated from a chart that converts the total number of desired items into an equivalent price per unit. In step three, after the actual prices have been set, players calculate their actual expenditures. In step four, they consult another chart to determine how many "utility points" they have earned with their purchases. (DCD)

THE NATIONAL ECONOMY

Players imagine that they "live in a small country where important economic decisions are made only by businessmen; government is not involved in the economy or in the decisions that affect it. . . . Once each year a top-level business council meets to discuss the health of the economy and to choose actions that will influence its course during the next twelve months. Representatives of three industry groups sit on the council. The groups are (1) consumer goods industries, which produce the necessities for modern living; (2) luxury goods industries, which make those things that people with higher incomes can afford; and (3) producer goods industries, which build the plants, offices, stores, and equipment that luxury and consumer goods industries need to maintain and expand production." Players represent members of this council.

The game is played in five steps. In step one, players are told that the value of that year's consumer goods production will be $288 million, and of luxury goods production, $65 million. These figures are entered onto worksheets. In step two, the council looks at a chart to find the optimal producer goods purchase for a year in which consumer production has a value of $288 million. According to the chart, the figure in such a situation is $430 million. Since the existing capital base is $397 million, the players must invest an additional $33 million in producer goods. In step three, players look at their charts to find the profits for each industry that result from this investment. In step four, they consult charts to determine the income needed for full employment. In step five, they consult their charts to determine whether their economic decision has caused an excess demand for or an excess supply of the goods they manufacture. (DCD)

SCARCITY AND ALLOCATION

Players represent survivors of a shipwreck who must try to use their work time as efficiently as possible to accumulate leisure time. The game is played in twelve rounds, each representing ten days. At the beginning, it is assumed that each player has twelve hours a day to perform useful work and that gathering food takes eleven hours a day per player. Therefore, at the end of the first round each player has accumulated ten hours of leisure. Players can reduce the time it takes to get food by buying tools with the time they save. A plow, for example, can be "bought" for thirty hours of accumulated time, and players with a plow can get the food they need in just six hours a day. Players who find and domesticate an animal (doing so "costs" seventy hours) and who also own a plow can sustain themselves with just two hours of effort each day. The player with the most accumulated hours and the shortest workday after twelve rounds wins. (DCD)

Note: For more on *Collective Bargaining, The Firm, The National Economy,* and *Scarcity and Allocation,* see the essay on economic simulations by Victor Pascale.

Cost: package of materials for 30 players (any one title) $13.30; sample set (materials for 6 players for all 8 games, plus teaching guide) $24.20; "A Guide to Teaching" (must accompany first orders for any one or combination of games) $3.85

Producer: Didactic Systems, Inc., Box 457, Cranford, NJ 07016

ECONOMIC SYSTEM

James S. Coleman and T. Robert Harris, Academic Games Associates

Playing Data
Copyright: 1969
Age Level: grade 7-college
Prerequisite Skills: reading, grade 11; math, grade 6
Number of Players: 7 to 13 in 1 to 3 teams
Playing Time: 2-4 hours
Preparation Time: 2-1/2 hours

Description: This game is designed to illustrate some important concepts about the operation of economic systems, including the interdependence among the parts of such systems, ways group demands can cause individuals to change their behavior, and how people can use their power—economic and otherwise—to ensure that their interests influence group demands and collective goals.

Players assume the roles of manufacturers, workers, farmers, and mine owners who want to profit and maintain a high standard of living. They make decisions about how much to buy, sell, consume, and at what prices, and how best to use their productive potential. At higher levels, players learn about problems of international trade and about problems of taxation and the provision of public goods. (AC)

Comment: Economic System is intriguingly complex. If you are doubtful about the effects of the complexity, ask for the results of outside evaluation. I like the fact that it calls for individual, rather than team, play. (DZ)

Note: For more on this game, see the essay on economic simulations by Victor Pascale.

Cost: $25.00; replacements available

Producer: The Bobbs-Merrill Co., Education Division, 4300 West 62nd Street, Indianapolis, Ind. 46268

THE ECONOMY GAME

Gottfried Krummacher

Playing Data
Copyright: 1975
Age Level: junior high-adult
Number of Players: 2-6
Playing Time: 1-2 hours
Packaging: professional game box

Description: The purpose of *The Economy Game* is to demonstrate the realities of the business of economics. Players begin with $11,000 each and proceed around the playing board by throw of the die. When they land on a consumer square, they must buy goods, the cost of which is determined by their availability, which changes throughout the game. When they land on decision squares, they must make one of nine decisions: (1) to pay for further training, resulting in a salary increase; (2) to invest in a company, whose success will depend upon one's future actions and other players' actions; (3) to invest in savings at 10 percent interest; (4) to pay production costs for a company, which will depend upon the number of employees and their training, as well as the cost of raw materials; (5) to sell goods, at a price which depends upon the quantity of goods available; (6) to raise cash by borrowing money or cashing in on savings; (7) to repay debts; (8) to sell holdings; (9) to take no action. Risk and situation squares require players to select cards and follow the printed instructions such as to pay a sales tax; stock exchange squares indicate that players may sell, buy, or trade stocks with each other. When the game ends, at a prearranged time, players total their assets. The winner is the person with the highest cash balance. (TM)

Cost: about $8.00 at retail outlets

Producer: Cadaco, Inc., 310 W. Polk St., Chicago, IL 60607

EDUCATION FOR JUSTICE

Thomas P. Fenton

Playing Data
Copyright: 1975
Age Level: high school, college, adult groups
Number of Players: 9-14
Playing time: 2 hours
Packaging: 122 page workbook

Description: Education for Justice is a participant workbook containing two didactic simulations: *The Coffee Game* and *The Money Game.* The focus of the book is national and international justice.

The Coffee Game simulates the international coffee trade from mid-1971 to mid-1973. Participants play the roles of a coffee grower from one of five Latin American countries, an import or export agent, a U.S. foreign aid official, or the chairperson of one of two conferences. Coversheets and worksheets are provided for these roles. Winners and losers are already programmed into the game since the actual statistics for the coffee trade for these years are used. As the game progresses, players may sell exports, pay taxes, deposit money in Swiss bank accounts, and/or attend conferences depending on the worksheet instructions for their roles. The game, according to the instructions, is designed to dramatize and illustrate "the realities of trade" where "the winners use 'the rules of the game' to their own advantage and to the disadvantage of the losers."

The Money Game simulates a meeting of industrialized and developing nations to determine how to spend $330 million for development. Players take the role of one of fourteen countries through four conference rounds. In the first round, one dollar is collected from each player and money is passed between countries to illustrate how funds flow in world markets. Rounds two and three involve setting national goals and forming international coalitions. Round four is used for discussion and debriefing. (DCD)

Cost: workbook $3.95, resource manual $7.95

Producer: Orbis Books, Maryknoll, NY 10545

ENTERPRISE

Davis Rosser and Donald Ernsberger, Council Rock High School, Newton, PA; David Yount, El Capitan High School, Lakeside, CA

Playing Data
Copyright: 1972
Age Level: grades 9-12, average to above average ability; grades 7-9 above average ability
Number of Players: 25 to 40 in 6 teams
Playing Time: 17 to 30 hours in 1-hour periods
Preparation Time: 3 to 5 hours
Supplementary Material: helps to have a general economics text available

Description: Players begin the simulation by being grouped into bankers, businessmen, brokers, consumers, welfare poor, politicians, and lobbyists. These groups interact with one another, buying and selling labor; buying and selling capital; and, generally, engaging in all the economic activities capitalism requires to function. In addition, each group uses its lobbyists to influence Congress and the President to enact legislation or take executive action favorable to the group's particular interests. Enterprise is not an easy simulation to conduct—but then economics is an extremely difficult subject to teach interestingly. It will help students understand how labor, capital, and governmental regulations and pressures all interrelate to produce the American economy. (Producer)

Cost: $14.00

Producer: Interact, P.O. Box 262, Lakeside, CA 92040

EQUITY AND EFFICIENCY IN PUBLIC POLICY

Robert Buchele and Howard Cohen

Playing Data
Copyright: 1978
Age Level: high school, college
Number of Players: 15-30
Playing Time: 1 hour (est.)
Packaging: 30-page photocopied booklet

Description: The players of this game try to establish criteria by which the wages paid to the members of a society are set. The class of students is divided into six groups which represent occupational classifications: twenty percent will be office workers, twenty percent will be skilled manual workers, and ten percent represent the unemployed. Players are told that the total wages paid to the class will be the number of participants multiplied by $10,000. Each player thinks about the kinds of jobs that fall into the categories and then allocates a portion of the total available wages to each occupational group. For instance, a player may decide that professionals, such as teachers, should make $15,000 a year and the unemployed should be given $5,000. Consequently, in a group of thirty players, the three professionals would be paid $45,000 and the three unemployed would be paid $15,000 out of a total of $300,000 in available wages. Next, players are randomly assigned to occupational groups and must, as a group, agree on a scheme of wages. (DCD)

Comment: This game explores the idea of the distribution of wealth by merit but seems to ignore the idea of the distribution of wealth by need. (DCD)

Cost: $3.00

Producer: Policy Studies Associates, P.O. Box 337, Croton-on-Hudson, NY 10520

EXCHANGE

David J. Boin and Robert Sillman

Playing Data
Copyright: 1972
Age Level: grades 7-12
Number of Players: class size
Playing Time: about 5 class periods

Description: Students are divided into stock clubs and given a hypothetical $5,000 with which to buy stocks. Each time the group plays, they decide to buy or sell particular stocks, following stock activity in the newspaper. The instructional objectives of Exchange are, according to the producer, "to provide students with a basic understanding of stock exchange transactions as reported in newspapers, with an opportunity to experience market fluctuation through buying and selling stock, and with an appreciation of wise investment and an awareness of risks involved in buying securities." (REH)

Comment: This exercise follows the standard Edu-Game format with a few pages of teacher's manual and locally reproducible student material. Although such things as broker's fees are not simulated in the game, they may be added by the teacher. (REH)

Cost: $3.00

Producer: Edu-Game, P.O. Box 1144, Sun Valley, CA 91352

EXCHANGE

Andrea Jane Richardson Taylor, Burley Middle School, Albemarle County, VA

Playing Data
Copyright: 1976
Age Level: high school
Prerequisite Skills: basic understanding of economics and the stock market
Number of Players: approximately 30-36
Playing Time: 5-6 1-hour periods
Packaging: 8-page teacher's manual, 8-page student manual

Description: Students playing *Exchange* gain an understanding of stock market mechanics as they role play foundation directors, mutual fund managers, brokerage firm executives, exchange accountants, and individual buyers. The first day of the simulation they read in their manual about stocks and mutual funds and get familiar with their roles, the foundations and brokerages, the stocks and mutual funds, and the trading procedures of the game. After groups have met to develop strategies and investors have initially planned their buying, trading begins and continues throughout the second, third, and fourth day. Each day the teacher sets margin requirements and periodically circulates rumor cards (e.g., a possible war in the Middle East) and crisis cards (e.g., a decrease in the prime lending rate) while brokers compete to sell services, mutual fund managers sell and invest mutual funds, foundation directors strive to maintain conservative investments, and individual buyers choose to buy or sell through brokerage firms or mutual funds. Throughout the game, all participants also write daily portfolio entries in which they express the viewpoints and concerns of their roles and describe the events of each day. The final day of the simulation, players calculate role performance points and discuss their role-playing experience and investment strategies. (TM)

Cost: $10.00

Producer: Interact, Box 262, Lakeside, CA 92040

EXECUTIVE DECISION

Sidney Sackson

Playing Data
Copyright: 1971
Age Level: grade 7-adult
Number of Players: 2-6

Playing Time: 30 minutes-1 hour
Packaging: professionally packaged game includes play money, decision forms, material cards, and playing surface

Description: This game is designed to simulate profits and losses in a free-market, industrial economy. Players represent corporate managers who buy, make, and sell materials and goods to make a profit. Each participant starts with three to six hundred dollars in capital (depending on the number of players). The game is played for twelve rounds, each of which corresponds to one month. During each round players attempt to buy raw materials, make finished goods, and sell them for a profit.

Players purchase raw materials by bidding for them. Raw materials come in three grades, standard, fine, and X-fine. A base price for each grade can be determined by checking a chart on the playing surface. The base price varies according to the demand. For example, if one player orders six units of X-fine, the base price per unit would be $36. However, if two players both ordered six units, creating a total demand for twelve units, the base price per unit would rise to $42. When they bid, the players try to guess the demand, and the consequent price, for units of raw material. Any bid which equals or *exceeds* the actual market price is successful. The players then secretly decide how many units of finished goods to make from their raw materials. Finished goods may be made in three qualities (A, B, and C) and it always takes three units of raw materials to make one unit of finished goods. For example, a player must use two units of X-fine and one unit of fine raw materials to make one unit of grade A finished goods. The sale of these manufactured goods is then arranged through a bidding process identical to the method used for purchasing raw materials, except that to arrange a sale the player's bid price for sale must be equal to or *lower* than the price calculated from the chart on the playing surface. The winner is the player with the most money after twelve rounds. (DCD)

Note: For more on this game, see the essay on economic simulations by Victor Pascale.

Cost: $12.00

Producer: Avalon Hill, 4517 Harford Road, Baltimore, MD 21214

FORMENTO

Ernest R. Alexander and Richard L. Meier

Playing Data
Age Level: graduate school
Number of Players: 3-24
Playing Time: 6-8 hours in 3 rounds
Preparation Time: 2 weeks
Packaging: 30-page mimeo

Description: This simulation attempts to provide development planning students with a highly abstracted, completely structured tool with which to apply alternative strategies for regional and national development. In *FORMENTO* the player is given an experiential introduction to the complexities of the development process. This process is hypothesized as one of generating human organizations (ORGS) such as firms, agencies, cooperatives, or clubs, and consists of three phases: design, evaluation, and resource allocation. Teams represent national and state or metro governments, the local private sector, and the development agency. Government teams are assigned specific ideological interests which reflect the political parties in power. The private sector is given the simple criterion of profit maximization by which to select ORGS for funding. The development agency (a *formento* in Latin America) must stimulate as rapid and successful development as possible. The simulation is particularly designed to reflect the developmental process in developing countries and postindustrial societies.

For instance, at the beginning of a game session, after receiving general instructions, a player may be assigned to a *formento* team for the city of Saigon. Since this development agency team is the only one with the pure interest of the city at heart, on the success of which its own survival depends, it will probably use its own funds to found ORGS that will plug gaps left by the other teams' special interests. It may also exchange any land that it has for money, favoring land-hungry ORGS such as housing and transportation. A player may function as an information gatherer, bargainer, evaluator, or decision maker. His or her only final knowledge of the selection process comes at budget time at the end of the round, when all teams announce the ORGS they are funding. In subsequent rounds the player may switch teams, suggest the creation of new ORGS, learn to create agency strategies, and discover the effects of his or her own decisions.

The game continues until the players are able to compare their own experience with the parameters built into the game, and in the authors' words "amend them—or accept them and modify their own preconceptions. When players reach this stage, the game has achieved its purpose—they are seeing development planning as a continuing political process which can be directed toward collective goals by a series of small scale decisions." (DCD)

Cost: free

Producer: Ernest R. Alexander, University of Wisconsin–Milwaukee, School of Architecture and Urban Planning, Milwaukee, WI 53201

THE GRAIN DRAIN

Brian Wren

Playing Data
Copyright: 1976
Age Level: high school-adult
Number of Players: 5-9
Playing Time: 1 hour
Preparation Time: 30 minutes (est.)
Packaging: playing surface, tokens, chance cards, play money, background paper on the world food crisis, and instructions

Description: This game, according to the producer, "is designed to bring home to the players in a lively and stimulating way some of the basic truths about the trading and commercial relationships of the rich and poor countries of the world." It illustrates the problems many Third World countries face in buying food on the world market, where high bidding can force prices up.

Before the game begins one person assumes the role of banker and the other players divide themselves, by throwing dice, into the rich and the poor. The game is played on two intertwining tracks that intersect at trading squares. The rich move their tokens over an orange track and the poor move over a yellow track. The orange squares offer more opportunities for high earnings than do the yellow squares. In addition, the rich earn a regular game income (each time they complete one circuit of the track) twice that of the poor. Players move their tokens according to the roll of a die.

Trading is the central event of the game. Whenever a player lands on a trading square the banker offers one grain token for bidding. All players, rich and poor, may bid for these tokens, and the token is awarded to the highest bidder. Beef tokens may be acquired from the banker in return for two grain tokens.

Each player in the game has a primary objective and a secondary objective. The primary objective for the rich is to collect three beef tokens and three grain tokens. The primary objective for the poor is to collect five grain tokens. Players unable to realize their primary objective have the option of surviving until the end of the game. Survival is defined as always having at least one grain token. The game ends when there is a winner, who is the first player to achieve his or her primary objective. (DCD)

Cost: 85 pence

Producer: OXFAM, 274 Banbury Road, Oxford OX2 7DZ, United Kingdom

INFLATION

Bruce E. Tipple and William E. Miller

Playing Data
Copyright: 1972
Age Level: grades 10-12
Number of Players: 30-45 (est.)
Playing Time: 7 1-hour rounds
Preparation Time: 1 hour (est.)
Packaging: 14-page instruction booklet, 28 pages of player instructions and supplementary materials in loose-leaf binder

Description: This simulation is designed to allow high school students insight into the way the American government makes economic decisions. The players are assigned specific roles as members of the federal government, as representatives of big business, organized labor, or special interest groups, as consumers, or as journalists.

The simulation consists of seven sequential phases. During the introductory phase, students are told what inflation is, how it affects the stock market, national unemployment, and banking, and the measures the federal government has taken to fight inflation in the past. During the second and third phases, each of the first five groups listed above meets to discuss how inflation affects it and selects representatives to meet with members of the legislative and executive branches of government. During the fourth phase, these members of government must recommend specific proposals to decrease inflation in a meeting held before the rest of the class, and the government as a whole must formulate a policy of action. This policy is then discussed by the whole class in phase five, and news reports describing the newly formulated policy and the reaction to it are prepared and critiqued in phase six. Phase seven consists of a teacher-led discussion which summarizes the entire simulation. (DCD)

Cost: $23.00

Producer: Paul S. Amidon and Associates, Inc., 166 Benson Ave., St. Paul, MN 55116

INVEST

Playing Data
Age Level: grade 9-junior college
Number of Players: 5 or more
Playing Time: 3 to 5 hours
Preparation Time: 1 hour
Special Equipment Required: overhead projector, copier

Description: Players represent investors, brokers, and bankers. Investors decide where to invest, whether to borrow money, and whether to sell for higher yielding investments in seeking to make the most lucrative investments possible. Brokers advise investors. Bankers set interest levels on passbook savings and find ways to encourage investors to save, invest, and borrow at the bank. While bankers and brokers are assisting their clients, they also seek to increase their own assets. (AC)

Comment: In this game, players learn about investment portfolio management, capital gains, liquidity investment, income and yield, diversification, risk, fixed yield and varying yield investments. They actually experience making or losing money. (HL)

Cost: $9.95

Producer: Changing Times Education Service, Division EMC Corp., 180 East Sixth Street, St. Paul, MN 55101

ISLAND

Colin Proudman, The College of Emmanuel and St. Chad, Saskatoon, Saskatchewan

Playing Data
Copyright: 1973
Age Level: grade 9-college, adult church groups
Number of Players: 12 to 24 in 4 to 6 teams
Preparation Time: 2 hours

Description: Island simulates the economic relationships between a subtropical island and outside business interests. Participants take the roles of nationals, government, foreign banking interests, a fruit company, Granite Mining Company, and Paradise Company—a tourist concern. After orientation, team representatives approach the government board to purchase land, and other team representatives negotiate and conduct business. Play continues for seven rounds and is followed by evaluation and discussion. (AC)

Comment: Teams do have added interests besides profit which are significant for the players. In the scenario they are directed to develop a poor, undeveloped island nation, and they are trying to survive.

When conducted properly *Island* can be counted on to illustrate dramatically the following points: (1) Undeveloped nations are almost helpless without capital and technology. (2) Inexperienced government officials, no matter how well intentioned, will make errors of judgment that will be interpreted by others as corruption. (3) There are limits to how much development can be done in a period of time. (4) Undeveloped nations are terribly vulnerable to strong outside interests if the welfare of their people is a prime concern. (5) Well-meaning church mission projects often prove to be cruel straws of hope for hungry and desperate people when they promise much and actually yield little. (6) The process of development which calls for capital investment, stable government, technological assistance and improving the condition of the people of the country is one that has thousands of possibilities for misunderstanding. (7) Developing a country is a rather brutal experience.

This simulation is based on the format of land development games that have been used extensively for research and education in the United States. It is a good adaptation and is a great deal of fun to play once the basic structure of the simulation is understood. It is my opinion that *Island* really does require an experienced game director or one that is willing to take a great deal of time to understand the basic operations of the simulation. The game directions seem to be written with the "bias" that potential users will have some understanding of development games such as *City Land Use Game* or *New Town*. If this is not true then the potential user of *Island* will probably find some directions confusing or obscure.

When adequately led with a group of interested participants, a great deal can be learned about developing nations. *Island* is an interesting and often exciting simulation. It takes a rather long time to play but rounds can be continued. This means that you can run the game an hour and stop, and continue at another time. Indeed, this may even be a desired approach because it gives participants times to work on different strategies. (GMM)

Cost: $10.00

Producer: Friendship Press Distribution Office, P.O. Box 37844, Cincinnati, OH 45237

MONOPOLY

Charles W. Darrow

Playing Data
Copyright: 1935
Age Level: grade 4-adult
Number of Players: 2 to 8 in an optional number of teams
Playing Time: 30 minutes to hours
Preparation Time: 5 to 15 minutes

Description: *Monopoly* has long been the most popular and commercially successful simulation ever invented. In fact, its almost universal distribution makes reviewing it virtually unnecessary.

Monopoly simulates a real estate industry. The materials include: player pieces, dice, chance and community cards, rules and money in several different denominations. The game is ideally suited for home play, but can be used in school. It takes a minimum of three players and can accommodate as many as seven. Play can take as few as two hours or as many as a dozen.

Comment: The commercial version of the game has more entertainment than educational value, but the game mechanism lends itself to any number of additions with cognitive payoffs. For example: (1) Teachers can revise the currency so that the bills have random values. This will force children to do some mighty heavy calculating every time they transact business. (2) If students make a move once a week instead of playing the game all at once, the educator can program in architecture, city-planning, model building, and various units on measurement. Teachers may even devise ways to recognize essays as improvements on property. (See Micro-Community and Micro-Economy.) (3) The older the players, the more impetus there will be for programming political experience into the simulation. One can do this simply by adding one rule to the commercial game: allow a majority of the players to change any rule in the game at the end of every full turn. Within minutes the weaker players will organize against the stronger.

Monopoly is a great game. The challenge is to attach educational agendas to it. Most teachers will have little difficulty in making the most of a good thing. (From the Robert A. Taft Institute of Government Study on Games and Simulations in Government, Politics, and Economics by George Richmond, Educational Development Center.)

Comment: Aside from being a whale of a lot of fun to play, *Monopoly* offers a game structure which is easily adaptable to additional rules, roles, or events. It *has* been used to teach economics concepts through such additions, and quite effectively according to the information I received. As you know, the basic game is a competitive game greatly influenced by chance, which calls for decision-making and bargaining by individual players and leads to quantitative (zero sum) outcomes. Beyond such basics, the options are limited only by your ingenuity. (DZ)

Cost: Varies

Producer: Parker Brothers, 190 Bridge St., Salem, MA 01970

ON STRIKE AND OTHER ECONOMIC GAMES

Editors of Scholastic Search Magazine

Playing Data
Copyright: 1974
Age Level: grades 5-9 (est.)
Number of Players: class
Playing Time: 1-1/2 to 3 hours each (est.)
Packaging: 28-page manual written for students

Description: This is a simply written manual for children describing four games—*On Strike, Profit and Loss, The Battle of Ripple Creek,* and *Madison Avenue.* The games involve role-playing, small group discussion and decision-making, and class discussion as well as reading dialogues and answering questions in the text.

On Strike is concerned with a strike in a blue jeans factory. After six players read the dialogue, the class divides into two groups—union members and management. Union members decide which of their eight demands are most important while management decides which demands they are willing to grant. Union players then make picket signs while management players write a newspaper ad soliciting town support. Finally, all students read a dialogue expressing townspeoples' attitudes toward the strike, decide how the strike will affect six businesses described in the text, read and interpret a graph concerning the factory's business operations, and finally role play negotiations with an arbitrator, spending approximately 10 minutes discussing each of the union's demands.

Profit and Loss, a game of big business competition, involves three teams representing 3 bakeries as well as a 5-member consumer board. Each bakery must privately make 6 decisions: What shall they name their bakery? What kind of bread should they bake? What size and at what price? From whom should they buy flour? How should they solve a profit loss problem? Should they merge with another bakery? After each decision is made, the consumer board chooses which bakery made the best decision and awards them winning points; they also judge which of the TV commercials presented by each group is the best commercial. Information and dialogues relating to each decision are included in the text. The winning team is the one with the most points at the end of the game.

The Battle of Ripple Creek is an ecology economics game involving a conflict between townspeople concerned with conserving land being torn up for strip mining and workers concerned with keeping their jobs. Students read aloud the attitudes of the people involved in the controversy, interpret graphs, and attempt to differentiate facts from opinions. Then a meeting is held in which students role play the chairman, conservationists, and mining workers, making up their own responses where the text indicates. A vote is taken at the end of the game.

Madison Avenue, an advertising game, involves several written exercises in which students compare three ads, interpret a commercial, differentiate facts from phony claims, and determine how to distinguish bargains. Then, after reading aloud a dialogue involving advertising agency workers attempting to sell soap, they decide a name for the soap and draw and write their own ads.

Cost: $9.95) (#382)

Producer: Scholastic Book Services, 50 West 44th Street, New York, NY

PINK PEBBLES

Craig M. Pearson and Joseph R. Marfuggi, Education Ventures, Inc.

Playing Data
Copyright: 1972
Age Level: grades 4-10
Number of Players: up to 36 in up to 6 teams
Playing Time: 20 to 45 minutes
Preparation Time: time to read teaching guide

Description: If you have children interested in learning about the impact of money on human society, introduce them to *Pink Pebbles.* In this simulation six individuals per team (one to six teams) may play at once. Each individual represents the head of a household in a tribe which is making the transition from a nomadic to an agricultural society, about 30 centuries ago. They discover that survival can be ensured by raising cows, sheep, and wheat. Players begin with nothing, on row #1, the first of seven tiers, and proceed to acquire cows. This acquisition is determined by chance: the role of a die tells the player which square he lands on, and he follows the printed instructions, e.g., "Cow falls off cliff and dies," "You claim one cow that wanders onto your field," "Two calves born." One square on each of the first three rows awards the player pink pebbles instead of a good or product. These are a unit of currency, of no use or worth except as a medium of exchange. All acquisition of goods or wealth is in the form of a unit token: 1 cow, 1 sheep, 25 bushels of wheat, 100 pink pebbles, etc. When or if misfortune causes a reduction in a player's stock, he returns the token to a common reserve.

Players stay on the first tier (the cow tier) until they have 3 cows. With appetites whetted for possessions, they then move up to the second tier where they begin raising sheep. They are locked in this row until they have 3 sheep. Once a player has done this, he continues his climb to the third tier where he raises wheat to add to his collection of goods. Players remain on the third tier until they have harvested 4 units of wheat, at 25 bushels per unit. These amounts of cattle, sheep, and wheat must be maintained through the next stage of the game in order to ensure the survival of the household. The penalty for falling below this minimum requirement is not to leave the game, but to return to the appropriate row until one's stock is replenished.

In the test situations, players were distributed rather evenly between the first three tiers by the tenth move. By the twelfth move, one player had succeeded in reaching the fourth of the seven tiers. By the 17th move, all players had been mobile enough to get off cow row, several still found themselves preoccupied with sheep, or wheat, and two players were in row four.

In contrast to the first three tiers, the fourth introduces a variety of new products together with the concept of leisure, which allows players to make goods beyond their requirements for survival. Unlike the first three tiers, there is no quota of possessions to attain. In a direct sense, the plows, woolen coats, dishes, and sandals available on the fourth (and subsequent) tier represent luxuries. One leaves the fourth tier for the fifth by chance: one must land on a square which directs him to move up. Thus, it may happen that a player spends little or virtually no time on the fourth row, and thereby fails to acquire a stock of luxury items available there. The reverse may also happen, so that another player hoards sandals, plows, coats, and dishes.

The fifth and sixth rows instruct players to "start trading" and "start saving," respectively. But the impetus to trade, save, or specialize doesn't make itself felt until players begin considering the requirement for winning: creating two stores, each consisting of seven or eight units of a single good or product. Once someone has achieved this, each player gets three more turns. In case two or more players have two stores, the winner is the one with more pink pebbles than anyone else with two stores.

By the time a player reaches the seventh tier, he has more than likely acquired a variety of goods. To obtain his objective, the formation of two stores, it becomes advantageous or even necessary for him to trade his surplus items for goods he needs to fulfill his goals. Other players in similar positions have reciprocal interests. Players who have at least small surpluses, either of goods they are especially saving, or goods they are collecting for trading purposes, discover the benefits of maintaining a surplus. By definition, this is saving.

Up to this point the fortunes of the players have largely been dependent on the roll of a die. But once trading begins in earnest, bargaining, persuasion, strategic thinking, alliance-making, and other political and economic skills come to take on a major importance. For example, Amy had five sheep and wanted a sixth. Bob was willing to trade—if she would give him a cow and two wooden plows for one sheep. Nancy (who had seven cows and needed eight for a store) interfered: "Don't trade with *him,* he's a *boy*!" Amy replied, "But if I trade with you, will you have a store?" "No, I only have six cows" (a lie). Nancy got her store two moves later, in order to dispel suspicion.

The "value" of a particular token—what it was worth in trade—fluctuated wildly throughout the game, depending on conditions of supply and demand. If only one player had surplus cows, and therefore monopolized the supply, he could charge the maximum his cotrader would pay. One player in this monopolistic situation asked for amounts so outrageous that no one could trade with him. They got their revenge; later, when he wanted to trade, they shunned him.

Moreover, when one reaches the seventh tier a few occasions arise in which players find themselves losing rather than gaining the three survival-oriented goods (cows, sheep, wheat). If a player lands on one of these squares and drops below the minimum, he has two options: either he can (1) trade with another player to regain his previous level of the essential item, or (2) if this fails, he must return to the appropriate row and stay there until he builds up his stock again. It is harder to go back to one of the first three rows once one has completed a store; his survival requirements drop to one unit each of cows, sheep, and wheat. This is the new level he must fall below before he may go back.

An interesting dynamic evolved in the game with respect to collecting enough of one item to form a store. First, players usually reached the fourth row with a bare minimum of cows, sheep, and wheat. At most, a player might have a surplus of one or two of a particular item. Second, the direction of progress is defined as upwards, tier seven being perceived as "better" or at least more advanced than tier six. Therefore, going back to one of the first three rows was seen as a punishment for falling below minimum levels, instead of as an extra opportunity for acquiring goods in short supply. The discovery of the latter interpretation of "going back" can have a powerful influence on the outcome of the game. In the second test game, Linda, to the amazement of the other players deliberately traded away one cow. Her trading partner, though he needed a cow badly, was compelled to warn her: "But you'll have to go back to the beginning!" She smiled, "Then I'll just have to go back." To the surprise of less innovative players her strategy paid off; she was the first to create a store. Before the others could successfully imitate her example, she had won.

Economic Understanding: Pink Pebbles provides children with excellent access to certain kinds of economic experience, concepts, and information. The trading component of the game coupled with the resource distribution mechanisms provides fairly realistic supply and demand situations. These situations, in turn, make it possible for students to study and learn concepts like comparative advantage, optimization of resources, capital accumulation, and pricing.

From the economic point of view, *Pink Pebbles* provides strong insights into the operation of a market place for goods, but provides rather limited insights (sometimes none at all) into the markets for labor and capital. Likewise, it offers little information about the production, distribution, or consumption processes. Overall, *Pink Pebbles* helps players to conceptualize a market for goods, and gives them a reasonable introduction to the characteristics and purposes of currency. (In response to the question, "Can anything be used as money?" one fifth-grader replied, "Yes, as long as the other people will accept it.")

Political Skills: Alliance formation and cooperation: Three kinds of alliances were made: (1) "Don't trade with him; he's a *boy*!" (2) "I'm your friend. Doesn't that count for anything?" (3) Two players saving different goods, and therefore not in direct competition with each other, joined forces and through wise trades, each helped the other to complete a store while blocking the nonaligned players.

Rapid thinking and decision-making: To trade or not to trade—and with whom?

Bargaining, compromise and decision making: The players learned quickly that trading involves compromise; it's a choice between accepting less than one wants, and not getting anything at all.

Optimizing resources and decision-making: If Amy offers two sheep for Nancy's cow, and Bob is willing to give her one wooden plow and two woolen coats, Nancy must decide which is the better deal, taking into account her other holdings.

Strategic thinking: (1) What to save? (2) Should I try to make a store by trading (interactive) or by going back to the beginning (individualistic)? (3) How many pink pebbles should I keep on hand?

Deception: The students were deceitful only toward the end of the game, when they were close to completing stores. Their lies all involved understating the number of units of a good they had, so that their competitors would think them farther away from completion than they actually were. Deception has other potential applications, though, such as falsely pleading poverty in hopes of getting a better deal.

Though *Pink Pebbles* has a rather broad usefulness in the teaching of political skills, it offers little toward the teaching of political concepts. Conceptually, *Pink Pebbles* emphasizes economics, although it is certainly possible for a teacher or parent to guide the players in an exploration of the conceptual basis surrounding those political skills. Bearing in mind that the game is designed for upper elementary level students, it seems that it has fulfilled the promise of its title. Furthermore, the interest of the players was sustained for the duration of the game. Indeed, one player commented at the end, "Boy, this is better than Monopoly!" (From the Robert A. Taft Institute of Government Study on Games and simulations in Government, Politics, and Economics by George Richmond, Educational Development Center.)

Cost: $10.00

Producer: Cardinal Printers, Inc., Wesleyan Station, Wesleyan University, Middletown, CT

RAILROAD GAME

Fred M. Newmann and Donald W. Oliver, Harvard Graduate School of Education

Playing Data
Copyright: 1967
Age Level: high school, college
Number of Players: 15-24 in 4 groups
Playing Time: 45-135 minutes
Preparation Time: 30 minutes (est.)
Packaging: professionally produced teacher's guide and student's manual

Description: The *Railroad Game* is part of a teaching unit called *The Railroad Era* adapted by the authors from a Harvard University Social Studies Project. The game seeks to simulate the bitter competition between American railroads during the 1870s. The players are divided into four groups representing the management of small railroads that compete for the business of carrying iron ore from Oretown to the mills in Steeltown. The game is played in rounds that represent one business day. Each railroad has a break-even point determined by its fixed and variable costs. They make money by successfully bidding for business. The teacher is the mineowner and must ship between twenty and thirty-five loads of ore every seven rounds. During each round, the mineowner posts the number of loads she or he wants to ship and asks for bids. After examining the bids the owner announces how many loads she or he will ship on each line, but need not declare the amount of the winning bids. The railroads use their earnings to pay their expenses. Any line that cannot meet its expenses goes bankrupt, and those players must leave the game. The mineowner may lie about bids to the players and the players may conspire to fix prices. (DCD)

Note: For more on this simulation, see the essay on economic simulations by Victor Pascale.

Cost: $1.25

Producer: Xerox Education Publications, 245 Long Hill Road, Middletown, CT 06457

SETTLE OR STRIKE

Ray Glazier

Playing Data
Copyright: 1974, 1977
Age Level: grade 9-adult
Number of Players: 6-32 in teams of 3-4
Playing Time: 3 hours
Preparation Time: 2 hours (est.)
Packaging: professionally produced and packaged game includes 18-page teacher's manual, role profiles, issue/options cards, and all other materials for 32 players

Description: This is an amended edition, for use in schools, of a simulation originally devised by the same producer for use in the Communication Workers of America's School for Local Negotiators. Groups of three or four represent bargaining teams for either the Lastik Plastik Company or the recently formed Lastik Plastik Unit of CWA Local 0001. The players must reason out the initial collective bargaining agreement for the workers at this plant. There are five issues in the negotiations as the union demands (1) an immediate, across the board raise of seventy cents per hour of labor for all employees; (2) at least one week's paid vacation for all employees who have been on the job more than six months; (3) the mandatory inclusion of all hourly employees in the bargaining unit (union shop); (4) a plantwide seniority scheme that would discourage the layoff of older employees; and (5) that the life of the contract be one year. Six bargaining positions, or "options," are specified and printed on cards that are color coded by demand. Each option is numbered so that number one always represents the union's initial demand and number six represents Lastik Plastik's initial offer. (For example, on card one of the wage set the union demands a seventy cent per hour raise and card six represents the company's offer of a thirty-five cent per hour raise. Cards two through five represent settlements on a wage increase of more than thirty-five and less than seventy cents an hour.) The simulation is played in three one-hour rounds. During the first forty minutes of each round the representatives of the two sides meet and negotiate. During the remainder of each round each side caucuses and prepares for the next round. The game ends after the third bargaining session. (DCD)

Comment: This outstanding simulation is very highly recommended for use by teachers of economics and civics courses for two reasons. In the first place, it is an extremely realistic portrayal of the collective bargaining process. The attitudes, aspirations, and limitations of the characters represented in the six major roles accurately reflect those of people who must sit down and hammer out union contracts. The demands and counterdemands reflect those that are made in real life and the players are given the information that the characters they must represent would have. The local chairman knows who is in the bargaining unit, what they will accept and what they will reject, knows what the effects of a strike would be on the membership and on the company, and knows what the international expects of him or her. The chief negotiator for the company knows just what each concession will cost, what other companies are giving, how much sales are expected to increase next year, and just what the company's president will and will not tolerate. Second, this game is refreshingly free of pro- or anti-union dogma. (DCD)

Cost: $45.00

Producer: Games Central, Abt Publications, 55 Wheeler Street, Cambridge, MA 02138

SPIRAL

David Rosser

Playing Data
Copyright: 1974
Age Level: high school
Number of Players: 20-40 in 6 groups
Playing Time: 6 periods
Supplementary Materials: a resource shelf of books on economics
Packaging: 2-page teacher's guide, 7-page students' guide

Description: Students playing Spiral are concerned with limiting inflation to less than 3% a year and with protecting their group (labor lobby, government employees, businessmen, farmers, consumers, and banker/investors) from financial disaster. During the first session, students read their student guide, choose groups, group leaders, a president, and an economic adviser. The next two days, they read about the economic situation and conduct group research projects which involve looking for facts, incidents, and quotes to explain or support general statements given about the economic views of each group (e.g., businessmen: the cost of borrowing money for expansion is too high). Meanwhile, the president and economic adviser have their own research and fill out a group proposal form listing its solutions or combination of solutions (as described in the manual) for controlling inflation and protecting their own interests. The last two days, the president conducts the economic advisory council, during which groups make proposals; then he or she develops a bill which he or she introduces at the congressional joint session. At this time, a final vote on the proposals is taken. Students are evaluated on how much of their group's proposal is adopted or on an optional postsimulation activity on the debriefing questions described in the manual. (TM)

Cost: $10.00

Producer: Interact Co., Box 262, Lakeside, CA 92040

STOCK MARKET GAME

The Avalon Hill Company

Playing Data

Copyright: 1970
Age Level: general
Number of Players: 1-6
Playing Time: 3/4-2 hours
Preparation Time: 10 minutes

Description: This board game simulates the action of the stock market. Players are individual investors who must accumulate wealth by buying and selling securities. The winner is the player with the greatest net worth at the end of the game. Players can reenact the crash of 1929. (AC)

Note: For more on this game, see the essay on economic simulations by Victor Pascale.

Cost: $12.00

Producer: Avalon Hill, 4517 Harford Rd., Baltimore, MD 21214

STOCKS AND BONDS

Mr. and Mrs. W. Stanley Hooper, Mr. and Mrs. A. Brooks Naffziger, Mr. and Mrs. Richard P. Hoffman

Playing Data
Copyright: 1964
Age Level: grades 7-12 and general
Number of Players: 2-8, can be adapted to more players
Playing Time: 1/2-1 hour
Preparation Time: 10 minutes

Description: Players of this stock market game assume the roles of individual investors who must decide what securities to buy and sell in order to profit. Chance determines market trends and prices. The winner is the player who has made the most money at the end of the game by successfully investing in the game's ten securities over a ten-year period. (DCD)

Comment: *Stocks and Bonds* may have some educational value, but as a simulation it seems to be on about the same level as *Monopoly*. (DZ)

Note: For more on this game, see the essay on economic simulations by Victor Pascale.

Cost: $12.00

Producer: Avalon Hill, 4517 Harford Rd., Baltimore, MD 21214

TIGHTROPE

Larry Baskind, Ann Buddington, Melvin Erickson, Ala Kay Hill, and Geraldine Murphy, all of El Paso Public Schools

Playing Data
Copyright: 1969
Age Level: grade 10-college
Number of Players: 3 or more in groups of 3
Playing Time: 4 20-minute rounds
Preparation Time: 30 minutes
Special Equipment: overhead projector
Packaging: 24-page photocopied booklet

Description: The designers describe *Tightrope* as "a simulation of maintaining the economic stability and growth of a country. A group of students called the Economic Advisory Council makes decisions regarding the fiscal and monetary policies of a nation." The game is played in four rounds. During each round each group of students must do the following: (1) determine the economic condition of the nation by consulting an economic indicator chart in the game materials; (2) improve the economic condition of the nation by taking a certain specific action, such as raising or lowering corporate income taxes, and record their action on a decision form; and (3) analyze their decision in terms of its overall effect on the economy by studying the economic indicator charts. Each round ends with a class discussion. (DCD)

Comment: *Tightrope* is based on Keynesian principles of economic policy and on the concept of government regulations; participants who do not accept these assumptions would find play unrewarding. (DZ)

Note: For more on this simulation, see the essay on economic simulations by Victor Pascale.

Cost: $2.00

Producer: El Paso Public Schools, Purchasing Agent, Box 1710, El Paso, TX 79999

TRADE

Playing Data
Copyright: 1974
Age Level: grades 7-8, above average; grades 9-12 average to above
Number of Players:
Playing Time:

Description: *Trade* is a simulation of the pressures and procedures of international commerce. It directly involves students in exchanging goods and in dealing in currency that together form the basis of international trade. The basic organization of the simulation–large industrial countries, medium-sized countries, small raw materials-producing countries, and some nonnational business organizations–forms a fairly accurate microcosm of the real world. Students, acting as the decision-makers of these countries and organizations, face typical real trade problems of our times. All simulated nations are placed in environmental situations which force their leaders to interact with leaders of the other nations to maintain the precarious balance of trade that symbolizes our modern world. Students, using research to aid their decision-making, try to survive as a nation without resorting to war. (The Teacher Guide contains a "refresher review" of world economic principles.) (Producer)

Cost: $14.00

Producer: Interact, P.O. Box 262, Lakeside, CA 92040

TRANSACTION

John R. Tusson, Study Craft

Playing Data
Copyright: 1968
Age Level: grade 7 and up, management, stockbrokers
Number of Players: 25
Playing Time: 1 hour or more
Preparation Time: 1/2 hour

Description: This stock market simulation exists in several versions. Depending on the version, players represent stock brokers, chairmen of mutual funds, specialists, advisers, members of investment clubs, treasurers, governors, senators, and so on. Teams represent corporations, mutual funds, brokerage firms, stock exchanges, and political groups. A manual data-processing unit provided with the game releases a flow of information that represents authentic past events in the actual stock market. On the basis of this information, in the basic game, players make decisions to buy, hold, sell short, or cover stock; place stop loss points; interpret chart movements; decide on buy and sell signals developed as a result of stock movements. In other versions, they may make decisions about forming and developing a new company; may decide to become brokers, specialists, advisers, members of the New York Stock Exchange, mutual fund managers, and so on. In the political version, some may decide to seek political office, find necessary votes among other players, exercise political power. (AC)

Comment: *Transaction* has three unique qualities that place it above competitors in its area: first, it is an open-ended simulation, allowing participants to develop the game further; second, it is programmed with authentic data (rather than die throws, etc.) which determine the direction a stock will move; third, it comes with a manual data-processing unit which can "stimulate thought on the part of the user as a tool for use in other simulations." It can be played repeatedly by the same players as they try to improve their decision-making skill by trying different approaches to the problems faced.

Letters from satisfied purchasers are available from the producers. *Business Week* has said, "Of all the stock market games, Transaction most resembles the market." (DZ)

Cost: $20.00 plus postage and handling

Producer: Study-Craft Educational Products, 2000 Woodland Highway, Belle Chasse, LA 70037

YOU'RE THE BANKER

Office of Public Information, Federal Reserve Bank of Minneapolis

Playing Data
Copyright: 1973
Age Level: grade 7-college
Number of Players: 2 or 3 to 30 in 1-5 teams
Playing Time: 3 or 4 periods of 50-60 minutes
Preparation Time: several hours

Description: *You're the Banker* (formerly called *Mr. Banker*) gives players experience in making the kinds of decisions an actual banker might make. The players begin with a stack of cards, an accounting sheet, currency, community resource cards, a set of unforeseen circumstances, and a game board. The cards are the meat of the game; they describe loan applications, and give detailed instructions about changing the account records depending on whether the loan is accepted or rejected. Players pace themselves by working quickly or slowly through the transaction cards. For each of three rounds there is a different set of cards; the cards cover several different economic states, including inflation, recession, and economic stability.

A team of two or three students represents one bank. Teams may compete with each other for the highest net earnings, or one team may play by itself just for the decision-making experience. The teams do not interact with each other during the game at all, so it does not make too much difference either way. The first few cards detail the founding of the bank, and give a couple of examples of accepted loan applications. The team takes it from there, and reads the next transaction card in round one. Throughout the game, each bank keeps a running record of bank earnings, currency, checking account totals, and loan totals. They begin with $10,000 in cash and no earnings and $100,000 in checking accounts. As they grant loans, the investment is reflected in the community in the form of new factories, new jobs, more machinery, or an increase in goods and services. These are all represented by cards for community resources. Bankers keep track of community resources in two ways: first, they pile up the community resource cards, and second, they record the undifferentiated total of resources by placing a token on the appropriate square of a track around the outside of the board.

Pretty soon in the first round, the teams will encounter an unforeseen circumstances transaction card. Then they must roll the dice and look up the number they have rolled, and finally follow the instructions they find. This is the only element of chance in the game, and is included realistically to remind the bankers that things happen to them which are beyond their control. After they have made decisions about each loan request, considering the security of the loan as well as the use to which the money will be put, the students read the outcomes of all the transactions. If they have passed up a good loan, they find that out here, and if they have made a bad one, they must pay the consequences. This feedback enables them to define for themselves the range of loans they should accept. The first round is extremely inflationary; loan totals, checking account totals, and bank earnings have swelled. Community resources have also increased, and it is becoming apparent that there is a relationship between how well the bank does and how well the community does—namely, the bank with more money has more money to invest.

The second round ushers in a recessionary period. There are several requests for loans in cash, and the banks' cash reserves are quickly depleted. Several customers cancel their loans and withdraw money from their checking accounts. Often they must "borrow" cash from their bank earnings in order to pay their customers. The students complain long and bitterly during this round, and wonder why this is happening to *them*, especially after such a great first round. They have no choice but to keep on subtracting from their loan totals, currency, earnings, and checking account totals. In the outcome section of the second round, things cheer up a little—no banks went broke during the second round, though several came close.

At this point the game is interrupted, and the teacher explains about the Federal Reserve Board to the students. The Fed offers discounting to its member banks, where they can turn over a loan to the Fed at a low interest rate, and receive the money from the Fed. The bank's earnings are reduced by the amount it pays in interest to the Fed. Member banks can also write bank drafts, to transfer money from their earnings to their currency reserve. The Fed controls both of these, especially the former. It sets discount rates, and issues policy statements suggesting speedups or slowdowns of investing. If banks ignore these policy statements, the Fed can refuse to discount any of their loans. The teacher, for the remainder of the game, is the Federal Reserve Board.

Besides the institution of the Federal Reserve Board, the third round has an important difference from the other two: feedback (the outcome of transactions) is immediate rather than delayed. The combination of the two makes the third round economically stable—the ups and downs are smaller and more frequent, but nothing drastic happens. Unless, of course, a bank happens to ignore the Fed's policies. For instance, toward the end of the third round, the Fed raises the discount rate, and suggests that it would be good to accept fewer loans. The Fed has, more directly, stated that banks should not grant loans to a certain builder, Bruisey Thumbnail. One time, Bruisey defaulted on his loan, and the bank went broke. Of course, even with the Fed, banks can theoretically still go broke, but the warnings and policies of the Fed are generally reasonable.

User Report: In each of the test sessions, at least one student asked me, "If the bank only has $12,000 (in earnings), how can it grant a $60,000 loan?" The idea that the banks use other people's money instead of its own to grant loans, was novel, challenging, and hard to accept for most students. They were immediately aware of the risk involved: What if everybody withdraws his money? This led to a discussion of savings versus checking accounts, a distinction neglected in the game per se, and to cash reserve requirements imposed on member banks by the Fed. They grasped, most for the first time, how money actually grows in a bank compared to under a mattress.

The players also learned to hone their skills in making quick judgments of situations, in this case loan requests. If they didn't learn to distinguish between a good loan and a bad one, they would soon go broke. Before the game, several players found it hard to believe that a bank, as a source of money, could go broke. They ended up with a sense of a bank's powers and its limitations. Instead of a source of infinite wealth, a bank merely lends out other people's money at a higher interest rate than it pays its depositors. That is quite a severe limit to its investing capacity; it is always dependent on its depositors.

The students had some difficulty grasping the nature and duties of the Federal Reserve Board, but most were relieved by the stability of the third round compared to the extremes of the first and second.

You're the Banker provides a solid introduction to the details of banking. It is a fairly simple game, but has been simplified in a way that is not misleading and actually paves the way for further learning. This is very rare in introductory, simplified games. A good game for demystifying the U.S. banking system. (From the Robert A. Taft Institute of Government Study on Games and simulations in Government, Politics, and Economics by Alina Kurkowski and George Richmond, Educational Development Center.)

Cost: $15.00

Producer: Office of Public Information, Federal Reserve Bank of Minneapolis, 250 Marquette, Minneapolis, MN 55480

TREBIDES ISLAND: THE ECONOMICS OF MONOPOLY MANAGEMENT

Playing Data
Copyright: 1977
Age Level: high school, college
Number of Players: 6-28 in 6-9 groups
Playing Time: 50 minutes
Supplementary Materials: cassette tape recorder
Packaging: professionally designed box with tape recording, map, instructions, discussion guide, and bid forms

Description: This game was designed to introduce students to such economic concepts as demand theory and monopolies. The scenario is a sovereign island that is about to receive phone service for the first time. At the start of the game, three students are selected to represent the island's homeowners, investors, and businesses, and the rest of the players are divided into groups representing the companies bidding on the right to operate telephone service on the island. The homeowners, businesses, and investors have conflicting interests which influence their responses to the bids the companies make. They must agree on one bid. When they do this, the game ends and the players discuss the economic concepts they have learned (these being natural and artificial monopolies, natural and artificial barriers to entry, average revenue, total revenue, marginal revenue, economies of scale, legal and illegal price discrimination, the central idea of demand theory, a demand schedule, a demand curve, elasticity of demand, complexities of economic transaction, and the theory of the invisible hand). (DCD)

Cost: $19.50

Producer: Simile II, P.O. Box 910, Del Mar, CA 92014

See also
Baldicer SOCIAL STUDIES
Big Business HISTORY
Bombs or Bread INTERNATIONAL RELATIONS
Boxcars GEOGRAPHY
Building a Rapid Transit System COMMUNITY ISSUES
Canada's Prairie Wheat Game SOCIAL STUDIES
Class Struggle SOCIAL STUDIES
The Coffee Game RELIGION
The Covert Farm Game SOCIAL STUDIES
Crisis in Middletown COMMUNITY ISSUES
Empire HISTORY
Guns or Butter INTERNATIONAL RELATIONS
IDIOM COMPUTER SIMULATIONS/SOCIAL STUDIES
Manchester HISTORY
Merchant HISTORY
Micro-Economy SOCIAL STUDIES
New City Telephone Company SOCIAL STUDIES
New Town COMMUNITY ISSUES
North Sea Exploration GEOGRAPHY
Panic HISTORY
The Poultry Game SOCIAL STUDIES
The Poverty Game SOCIAL STUDIES
Railway Pioneers GEOGRAPHY
Rainbow Game SOCIAL STUDIES
Relocation BUSINESS: PERSONNEL DEVELOPMENT
Serfdom SOCIAL STUDIES
Starpower SOCIAL STUDIES
Strike HISTORY
The Territorial Sea INTERNATIONAL RELATIONS
Trade-Off URBAN
Triangle Trade HISTORY
The Twenty-First Year SOCIAL STUDIES
The Uses of the Sea INTERNATIONAL RELATIONS
War Time INTERNATIONAL RELATIONS
Your Community's Economic Development COMMUNITY ISSUES
BUSINESS FOR SCHOOL
COMPUTER SIMULATIONS/ECOLOGY
ECOLOGY
URBAN

EDUCATION

ACHIEVING CLASSROOM COMMUNICATION THROUGH SELF-ANALYSIS

Theodore W. Parsons

Playing Data
Copyright: 1974, 1975
Age Level: teachers
Number of Players: 1-10
Playing Time: 18-20 hours in 10 sessions
Preparation Time: 2 hours (est.)
Packaging: professionally packaged game includes participant's manual, group leader's manual, cassette tape recording, and instructions

Description: This package of training materials for teachers is designed, according to the author, to help participants: "identify the overall pattern of communication of their classroom" by determining "the relative amounts of both teacher talk and pupil talk"; assess their "success in encouraging each pupil to develop his ideas verbally"; "identify the pupils who talk the most, those pupils who speak infrequently, and those who make no voluntary verbal communications"; and "increase pupil participation by reducing the amount of "teacher's talk and by encouraging and enabling all . . . pupils to express themselves verbally."

These materials are available in two versions. The basic version is self-administered, and the more recent version provides a format by which the materials may be administered in ten in-service training group sessions. In both versions the training materials are divided into four parts. In the first part the players analyze the patterns of communication (who talks to whom about what how often) among the students in their classes. In part two, the teachers pay attention to their own classroom talk and analyze how it may affect the classroom participation of their students. In part three, the teachers examine their responses to student participation and determine whether their talk aids or hinders class discussion. In the fourth and final part, the teachers practice ways to stimulate individual students and whole classes. (DCD)

Cost: leader's kit $21.50; additional participant manuals $8.95 each

Producer: Science Research Associates, 1540 Page Mill Road, Palo Alto, CA 94304

ACTIONALYSIS

H. Harvey Mette, Jr.

Playing Data
Copyright: 1971
Age Level: college, professional
Number of Players: 3-7
Playing Time: 1 hour or more
Packaging: professionally packaged game includes instructions, observation forms, role cards, die, and timer

Description: This game uses some of the elementary ideas of transactional analysis social psychology in such a way that players may observe and experience simulated student-teacher confrontations. The game is played in three-minute rounds. In any round one player assumes the role of teacher, one represents a student, and all others act as observers. One observer assumes the duties of "game master." The two players draw role cards. The student role card describes both an attitude, such as passive, and a situation, such as " 'Dirty' words and pictures are written on the blackboard as the teacher enters the room." The teacher role card specifies only an attitude, such as "sarcastic." The two players must dramatize this situation for three minutes before exchanging roles and continuing for another three minutes. Observers judge the players' authenticity and try to guess the attitudes that they are trying to portray. (DCD)

Cost: $25.00

Producer: H. Harvey Mette, 10 Broadview Drive, Huntington, NY 11743

ADAMS SCHOOL SYSTEM SIMULATION (ADSIM)

Edward Hickox, Nancy Howes, and Nicholas Nash

Playing Data
Copyright: 1978
Age Level: postgraduate
Number of Players: 25
Playing Time: 45-90 hours (est.) for complete simulation
Preparation Time: 40 hours (est.)

Description: ADSIM is the most recent addition to this producer's list of school system simulations. This extensive package of curricular materials attempts to describe comprehensively and dramatize the administration of a school system with 8,800 students and 500 staff in eleven schools. Numerous in-basket exercises and case studies, and occasional motion pictures, filmstrips, role-playing situations, and tape recordings are employed to assist the players, in the authors' words, "plan, decide, implement, and evaluate," while assuming the roles of superintendent, assistant superintendent for business services, assistant superintendent for instruction, secondary school principal, and elementary school principal. The full use of all of these materials should occupy at least a semester. Individual components of *ADSIM* may be played in as little as an hour.

Background information about and specific administrative problems is provided for the following areas: declining enrollments, affirmative action, union negotiations, special education, de-funding, school records' confidentiality, and administration-school board relations.

The simulation may be purchased complete, by component, or by individual item. (DCD)

Cost: $4,510 complete; contact producer for component prices

Producer: University Council for Educational Administration, 29 West Woodruff Avenue, Columbus, OH 43210

THE CHANGING HIGH SCHOOL

Frederick P. Venditti, University of Tennessee

Playing Data
Copyright: 1971
Age Level: high school teachers or trainees
Number of Players: variable
Playing Time: 3-30 hours in 5- or 6-hour periods
Preparation Time: none
Special Equipment Required: film projector

Description: A black teacher stops at a white colleague's desk with a plea for help. Ever since she asked one of her students to verify an excuse in the school office, the youngster has been "impudent" and increasingly difficult. She asks her colleague, "What do you think I should do?" How would you evaluate the problem? What factors do you hold to influence your suggestion?

This simulation is oriented to schools with a 16mm film equipment. Role plays with simulated problems are also presented on film and via the paper and pencil mode. Participants react individually to problems on incident response sheets. Then the participants discuss in small groups various ways of dealing with the problem. The simulation seeks to stimulate divergent thinking regarding problems of desegregated schools. (REH)

Cost: Preview/rental: $15.00 for 48 hours, $35.00 for one week, deductible from purchase price, otherwise nonrefundable. Purchase: $175.00 film only, $290.00 film and publications

Producer: The Anti-Defamation League of B'nai B'rith, 315 Lexington Ave., New York, NY 10016

CHOICE

Playing Data
Copyright: 1972
Age Level: adult
Number of Players: 12-15
Playing Time: 1 hour
Special Equipment: cassette tape recorder
Packaging: professionally packaged game includes player's pad, intervention card, and debriefing card

Description: Choice is designed, in the producer's words, to "highlight the effects on goal accomplishment of . . . behaviors typical of participants in small task oriented groups . . . and the hidden agendas that accompany them."

Players assume the roles of teachers assigned to a committee which must decide how a former classroom should be used during the next school year. Each player is instructed to advocate a particular opinion—that the committee should not make the decision, that the room should be used as a media center, or as a conference center, and so forth. After thirty minutes players are asked to stop advocating their assigned opinions and start discussing their feelings about doing so. They are also asked to assess the effects their actions had on the other members of the group and how their experiences in this group remind them of experiences in other groups. (DCD)

Cost: $15.95

Producer: Creative Learning Systems, Inc., 936 C Street, San Diego, CA 92101

EDPLAN

Clark Abt, Ray Glazier, Ezra Gottheil, Peter Miller, and Games Central, Abt Associates Inc.

Playing Data
Copyright: 1970
Age Level: grade 10 to college, adult groups (teachers, PTA, and the like)
Prerequisite Skills: grade 7 reading; elementary math
Number of Players: 29-36
Playing Time: unspecified
Preparation Time: 15 minutes

Description: Players begin with role profiles. They exchange views on education programs and budget allocation. The school board budget is submitted to the city council for federal aid representation for funds. These groups in turn take action on the request for funds. There are then primaries and elections for the school board and city council. Personal history and opinions related to educational issues are given in the text of the role profiles and the electorate is represented only in voting blocks, not in individual voters. Players are asked to act within the descriptions given by the role profiles except on issues where their opinions are not already stated. The game shows the necessity of mustering political support and potential funding for educational program decisions. (REH)

Note: For more on this simulation, see the essay on future-oriented simulations by Charles M. Plummer.

Cost: $32.00

Producer: Games Central, 55-C Wheeler Street, Cambridge, MA 02138.

EDVENTURE

Ezra Gottheim, Ray Glazier, and Ken Fischer

Playing Data
Copyright: 1974
Number of Players: 30-45 (or up to 300 sharing roles with add-on packages)
Age Level: grade 10-college, adult
Playing Time: 2-10 hours
Preparation Time: about 4 hours

Description: Edventure is an investigation of educational demands and options in the world of 1981 to 2000. Fifteen different learner personae present a wide range of education seekers, from the retired businessman who would like to develop a new (possibly lucrative) field of interest to the young divorcee trying to establish herself in a career—all of them engaged in a common pursuit, life-long learning. These learners interact with fifteen educational institutions, ranging from Omnimedia Network to the Jack of All Trades School to Ivy University. A learner may decide to work and not to study at all; otherwise, he selects a course and seeks admission at an appropriate institution. Once admitted, and only if he is able to pay the tuition (either with cash or federal education vouchers), the learner may pass or fail, according to his "learning type" and the roll of a die. During the simulation, learners can accumulate surplus income and satisfaction points. Winning at *Edventure* can be a number of things, depending on how you define the session. One winner, for example, could be the richest learner; another, the most satisfied learner; yet another, the institution which filled its maximum enrollment while accommodating the learning style of the greatest number of learners. (Producer)

User Report: Exhilaration, frustration, insight, impatience. These are just a few of the experiences which players of *Edventure* are likely to have as they plan and live 20 years of educational and work life in the short space of an hour. Each player begins a life role which specifies family background, social and economic advantages (or disadvantages), a learning type (ranging from abstract to concrete), and some tentative suggestions about life goals and ambitions.

A player enters the educational marketplace of institutions representing fifteen archetypes of postsecondary alternatives available to the life-long learner, a truly bewildering array of educational options.

The player considers his role, selects a course, and waits in line to register for it, simulating the experiences of students at any registration process, with all its attendant emotions. Will the course be open? (The availability of courses is geared to the number of persons playing so that demand always exceeds supply; and faced with the inability to get one's first choice consistently players must often rethink their life plan.) Can I afford it and do I have the prerequisites? (The reality of socioeconomic advantages and disadvantages is brought home, as is the availability of hitherto unexpected options in the learning society.) Having paid for the service, will I pass and will the benefits be worth the cost? (Built into *Edventure* are simulations of cognitive style which, when matched with course requirements, increase the probability of success, in terms of increasing capacity to derive wealth and satisfaction from life.)

Outcomes: The learning derived from the game is of several kinds. First, there is for most players a general raising of consciousness about the fact that post-secondary education includes a vast array of institutions beyond what has been traditionally considered "higher" education; and that a person with a background and interests ill-suited to traditional academic curricula has a plethora of opportunities and potential directions. Second, players develop awareness of the process of educational life planning, the ways in which a life-long learner's goals and conceptions vary with time and with success or failure. Third, and this is particularly eye-opening for administrators, players get a sense of what students feel as they are put through a course registration process with its long lines, its interminable discussions of course credits, requirements, and chances of passing; and the inevitable frustration (carefully built into the game) of finding just the right course at just the right time—closed! Finally, there is opportunity in *Edventure's* debriefing session to discuss more general issues such as the competition among postsecondary institutions for students and societal support, the feasibility of such innovations as voucher credits and credit for work experience, and issues such as future roles for faculty, and the relation of federal and state governments to postsecondary institutions.

Reactions: Edventure was an integral part of some thirty fall regional conferences of the American Association for Higher Education (which commissioned its development); and was quite successful in generating for most participants the learnings mentioned above.

Problems associated with it are instructive for anyone wishing to use it in similar or different settings (to which it is readily adaptable): (1) The game is complex and initially rather difficult to grasp. Particularly so is its scoring system, a problem for some players who often make mistakes. Preliminary briefing by someone thoroughly familiar with *Edventure* and with how learners approach its scoring system is crucial.

(2) There is a fine line between frustration which is consciousness-raising and frustration which creates defensiveness. Having too few institutional administrators or administrators who are not thoroughly briefed can turn the built-in frustrations (the waiting-in-line, the mix-ups, and the missed opportunities) into destructive rather than constructive feelings.

Moderate inconvenience opens a player's eyes about real-life registration problems and fulfills the goals of the simulation. If that inconvenience exceeds tolerable levels, attention is focused on the structure of the game itself with rather negative results.

(3) Finally, because the debriefing is of the utmost importance in tying together the game's learnings, a competent discussion leader who understands the purposes and pitfalls of simulation, who knows *Edventure* thoroughly, and who can both guide and facilitate a constructive discussion may mean the difference between a successful and an unsuccessful game.

In sum, my assessment of *Edventure* is that it is a difficult simulation to organize and carry out effectively; but if enough time and care are given to preparing properly for it, it can pay rich dividends in raising consciousness, generating discussion, and concretizing many important issues and problems of life-long learning. (John Bruce Francis, State University of New York at Buffalo.)

Cost: $60.00, add-on packages and replacements available separately in open stock

Producer: Games Central, 55-C Wheeler Street, Cambridge, MA 02138

EDVENTURE II

Ray Glazier

Playing Data
Copyright: 1974
Age Level: grades 9-12, adult
Number of Players: 1-15
Playing Time: up to 15 45-minute rounds
Preparation Time: 1 hour

Description: This is an adaptation of *Edventure* in which players take only the learner roles under the supervision of a guidance counselor. (AC)

Note: For more on this exercise, see the essay on future-oriented simulations by Charles M. Plummer.

Cost: $32.00, replacements available separately in open stock

Producer: Games Central, Abt Publications, 55 Wheeler St., Cambridge, MA 02138

FIXIT

Alice Kaplan Gordon

Playing Data
Copyright: 1970
Age Level: adult
Number of Players: 10 or more
Playing Time: 1-2 hours
Preparation Time: 2 hours (est.)
Packaging: professionaly produced 206-page softbound book

Description: FIXIT (Fostering, or Fighting, Innovations and Experiments in Teaching) is found in the appendix of an excellent and comprehensive introduction to educational simulations called *Games For Growth.* The book begins with a cogent rationale for using educational games and devotes several chapters to the administration and criteria for the use, design, and adaptation of simulations. A portion of *Games For Growth* describes the content and play of numerous games including the well-known *Grand Strategy, Propaganda,* and *Empire.*

FIXIT is designed to explore the process by which decisions are made as well as reinforce the player's (reader's) newly acquired knowledge of educational games. *FIXIT* focuses on the problems sometimes encountered in integrating educational innovations, such as games and simulations, into school systems. Ten players assume the roles of the teachers, parents, and administrators who comprise the budget committee of a public school. One of the items this committee must discuss is a request for funds to introduce experimentally game materials into the school's curriculum. After a negotiating session during which each player has the opportunity to talk to every other player, the committee must vote for, or against, the new program. After the vote the players discuss the roles they portrayed. (DCD)

Cost: $5.95

Producer: Science Research Associates, 1540 Page Mill Road, Palo Alto, CA 94304

FRESHMAN YEAR

L. Wayne Bryan

Playing Data
Copyright: 1973
Age Level: high school seniors, (entering) college freshmen
Number of Players: 30 or more
Playing Time: 40 minutes in 20-minute periods
Preparation Time: 1 hour
Supplementary Material: collection of books for simulated library
Special Equipment: tape recorder

Description: This is a simulation for high school seniors and entering college freshmen during orientation week. At the beginning of the simulation, players choose goals for themselves in these areas: academic, social, personal, career, and political. They then guide themselves through a series of experiences where they earn points in these areas. They grade themselves and score their own points after each experience. After two 20-minute semesters, they total the points to see if they have reached their goals.

Here is a sample list of goals that can be chosen: for an academic A average, one needs to earn 19 points on the simulated activities; a B average requires 15 points; C 11 points; and D 7 points. Failing is less than 7 points.

Play is broken for some students by an "emergency" such as being sick and missing school. All players are invited to celebrations such as homecoming and graduation. The major decision is how to use the time available to attain the goals one has set. (REH)

Comment: The game provides a "college milieu" of pressures, interactions, choices, and a system which helps players know what is waiting for them. The game appears to be a little unlike reality in that there is, for example, in the social goals, a goal of wanting to be everybody's friend for which you must earn 9 points; many good friends, 7; close friends, 5; and don't care much about friends, 3 points. In our estimation, "want to be everybody's friend" should have a rating of at least 20-50 points, compared with the other items on the scale. However, despite such wrinkles, the freshman year seems to be a basically good idea. It is, however, a simulation which emphasizes the frenetic rather than one which might emphasize careful planning for a week or a month of time and the making of realistic choices in regard to these time allocations. (REH)

Cost: $12.50 administrator's guide. Game administrator prepares materials

Producer: UCCM Games, c/o L. W. Bryan, 4215 Bethel Church Road, F-1, Columbia, SC 29206

IN-BASKET SIMULATION EXERCISES

Donald F. Musella and H. Donald Joyce

Playing Data
Copyright: 1973
Age Level: postgraduate
Number of Players: 1
Playing Time: 2-10 hours per simulation
Packaging: professionally produced participant's manual and 82-page leader's manual.

Description: These exercises are designed, according to the producer, "to simulate the problems encountered by administrators at various levels of the educational system. By placing the reader in a hypothetical situation, the authors challenge the reader's judgment in determining how best to deal with the administrator's daily problems." Playing materials include a leader's manual and participants' manuals for the eight roles of elementary school principal, elementary school consultant, intermediate school principal, secondary school division chairman, secondary school principal, area superintendent, director of education, and school board trustee.

Each exercise simulates the problems of a member of the educational system of imaginary Richland County, Ontario, and includes background information about the county, the town of Fleetwood, and the organization of the school system, as well as between twenty and thirty in-basket items including memoranda, correspondence, telephone messages, and "interruptions" (unexpected telephone calls or visits). The secondary school principal, for example, must immediately respond to a minor automobile accident in the school parking lot, whereas the school board trustee must immediately consider the request of a parent who wishes to withdraw his son from school to educate him at home.

The leader's manual includes the same background information as the player's manuals and several essays which discuss alternative procedures for using the simulations. (DCD)

Cost: leader's manual $4.65, participant's manuals $3.30-$5.25 each

Producer: Ontario Institute for Studies in Education, 252 Bloor Street West, Toronto, Ontario M5S 1V6, Canada

INNER-CITY SIMULATION LABORATORY

Donald R. Cruickshank

Playing Data
Copyright: 1969
Age Level: student teachers, elementary level

Description: The participants assume the role of Pat Taylor, a new sixth-grade teacher at Edison School in a socially and economically depressed section in the City of Urban.

A long-playing record and two filmstrips entitled "Orientation to Edison School" and "Orientation to Urban Public Schools" introduce the participants to the school and the public school system. They also receive the faculty handbook, introduction letters, comparative statistics on Gardner Park, and the cumulative record folder of each student in their class.

The thirty-four critical teaching incidents that occur in Pat Taylor's classroom were identified in a survey of twelve major cities as the most frequent and severe problems a teacher must face in the inner city.

Fourteen of the incidents are presented on sound and color movies. Pat Taylor views the incident through the lens of the camera and Pat's voice is subtitled.

The written incidents are presented as scripts, in-box memos, scene settings, course and committee assignments, and playlets to be acted out by two participants.

Five of the incidents are: Phyllis Smith Asleep in Class (child coming to school without proper food or sleep), movie; Sidney Sams Strikes Out (helping a child with social adjustment problems), movie; Marsha Wright Has an Excuse (child refusing or otherwise finding ways to get out of classwork), movie; Phyllis Smith's Hearing Problem (getting parents to take an interest in their child's health), playlet; Wesley Briggs Breaks Bardley Livesay's Watch (dealing with children who are destructive of others' property), movie.

Cost: Complete Director's Unit (including 14 16mm films) $875.00 (#13-1100). Components may be ordered individually; write for prices

Producer: Science Research Associates, College Division, 1540 Page Mill Road, Palo Alto, CA 94304

INTERACTION

John Wesley and John Montgomery Middle School, El Cajon, California

Playing Data
Copyright: 1974
Age Level: teacher education and in-service training
Number of Players: 9-25 or 30
Playing Time: about 2 hours

Description: Interaction is a simulation concerned with ecological problems, written in response to requests from principals, department chairpersons and college professors who have the responsibility for conducting in-service training and who want to introduce teachers or other interested persons to the world of education simulations. It can be adapted to almost any workshop situation. The simulation's content is concerned with a local governmental commission's problem: Should it accept a master plan that would fundamentally change a county park's ecological balance? Participants role-play members of this commission and various citizens whose personal convictions or vocations are directly related to the commission's decision. (Producer)

Cost: $5.00

Producer: Interact, P.O. Box 262, Lakeside, CA 92040

JOB CRUNCH

Joan Straumanis

Playing Data
Copyright: 1977
Age Level: college, graduate school, adult
Number of Players: 25-30
Playing Time: 2 hours
Preparation Time: 1 hour (est.)
Packaging: 28-page photocopied booklet

Description: Job Crunch is a learning experience designed to demonstrate a model for using simulation in teaching issue-oriented courses such as women's studies. The format of this game was adapted from that of *Psych City,* a simulation of community conflict which was designed by Robert and Nancy Cohen of the Institute for Community Development of Syracuse, New York.

The purpose of *Job Crunch* is to assist people in learning how an academic community might deliberate in a hiring situation involving the conflicting claims of candidates who are members of minority and majority groups. The simulation operates as a hypothetical community, in this case, a small liberal arts college, with its own people, resources, and problems. Participants are asked to play roles in this community and through these roles to examine the factors which influence the decision-making process.

The scenario in *Job Crunch* is the following. Three candidates of diverse training and background are being interviewed for an opening in the English department of the college. Members of the faculty and the administration and invited students meet the candidates in a "cocktail party" setting, which functions as a mixer. Next the various constituencies caucus briefly to plan strategies for the following phase, an open hearing to discuss the merits of the candidates and the obligations of the college under affirmative action guidelines. The issues of discrimination or "reverse discrimination" are likely to be raised at this stage. The hearing is conducted as a town meeting where participants try to come to consensus. They may present ideas, facts, plans, attempt to persuade others, establish procedures and rules, caucus, and vote on issues and motions. Finally, participants decide by vote who is to be hired and the game ends. (author)

Note: For more on this simulation, see the essay on sex roles games by Cathy Greenblat and Joan Baily.

Cost: $5.00 plus $.50 postage/handling

Producer: Denison Simulation Center, Denison University, Granville, OH 43023

THE PRINCIPAL GAME

Ralph J. Gohring and Leigh Chiarelott

Playing Data
Copyright: 1978
Age Level: school administrators
Number of Players: 4-20
Playing Time: 1½-3 hours
Preparation Time: 30 minutes
Packaging: professionally designed and packaged game includes instructions, playing boards, stickers, decision cards, and die

Description: *The Principal Game* is designed to help its players become aware of their own administrative styles. The game is played on 8½" X 11" single-use individual playing boards with dual, ten-space playing tracks. The back of each playing board contains scoring information and a chart to help each player analyze his or her administrative style.

Players answer questions in four categories concerning typical decisions a school principal has to make in regard to students, staff, the central office, and parents. On each turn a player answers a question drawn from one of these four numbered categories, determined by rolling a four-sided die. A card is then drawn with a multiple-choice question printed on it. Each answer has "task" score and a "people" score of 0, 1, or 2. After making the decision the player unfolds the card to learn his or her task and people scores, marking them on the separate tracks of the playing board. After answering ten questions each, players total their task and people scores and match the resultant number with an index which equates each possible score with one of five administrative styles. These styles can be briefly summarized as timid, task-oriented, people-oriented, passive, or task- and people-oriented. (DCD)

Cost: unknown

Producer: MESA Publications, 3445 Executive Center Drive, Suite 205, Austin, TX 78731

LABEL GAME

Robert A. Zuckerman

Playing Data
Copyright: 1975
Age Level: college, professional
Number of Players: 4 or more in groups of 4-8
Playing Time: 2 hours (est.)
Preparation Time: 1 hour (est.)
Packaging: self-published board game includes all materials needed for play by eight persons

Description: The author describes this game as being "designed and constructed to simulate the labelling process" (that is to say, labeling a person or group of persons as, for example, mentally retarded, emotionally disturbed, learning disabled, deaf, blind, or hyperactive), "and the effects of that process in terms of the consequences on those individuals who have been labelled."

Four to eight players navigate tokens through a maze of tracks on a game board according to the roll of a die and must follow any printed instructions in the square where they land. These instructions may require a player to move to another square or to draw a card. All cards "label" the player, who must then wear an orange ring, called a "Stigma ring," and who may move only on squares marked "Special Class," "T.M.R. Class," "Institution," or "Sheltered Workshop." Some, but not all, labels are "congenital." When a labeled player whose label is not congenital enters the "School," "Job Training," or "Work" spaces because of instructions on the game board or a drawn card he or she may place a blue "Normal" ring over the Stigma ring and proceed through the track. Players move their tokens around the game board once. The first to reach the squares marked "Parent" wins. (DCD)

Cost: not commercially available; contact designer

Producer: Robert A. Zuckerman, Department of Special Education, Kent State University, Kent, OH 44242

MADISON SIMULATIONS

Thurston Atkins, Donald Anderson, Hugh D. Laughlin, Glenn Immegart, Walter Hack, and Ben Harris

Playing Data
Copyright: 1966
Age Level: college undergraduate and postgraduate
Number of Players: 25
Playing Time: 90-120 hours (est.) for complete series of simulations
Preparation Time: 40-80 hours
Packaging: all materials are professionally produced and packaged

Description: The *Madison Simulations* are the earliest attempt of the University Council for Educational Administration to develop an interrelated set of curricular materials for school administrators in training. Six component simulations describe the school district and the duties of five key or typical administrators.

The background materials consist of three filmstrips which describe the district and school personnel, a set of instructor guides for all the remaining components, a bibliography, study guides, forms, transparency masters, and a 197-page document titled "The Madison School Community." This last item is a profile of the imaginary Madison School District "written by" two imaginary professors of education at imaginary "Lafayette State University" in imaginary Lake City, Lafayette. The book is very thorough and contains a fictional rationale, history of the area (including the names of imaginary Indian chiefs), demographic data, and a profile of the prevailing religious and social attitudes of the 30,000 persons who live in the suburban school district of Madison.

Edison Elementary Principalship includes a filmstrip and a ten-minute black and white motion picture describing that school, a packet of background material which describes and evaluates the staff and curriculum at Edison, a tape recording in which a former principal of the school describes each member of the teaching staff, two in-basket exercises which simulate the administrative duties of the principal, and an "interruption" tape recording for use with the in-basket exercises. As with all of this producer's materials, each item of the component may be ordered individually. *Madison Secondary Principalship* contains a black and white, twelve-minute film of the school's teachers in action, a filmstrip description of the school, two in-basket exercises with an interruption tape recording, and background materials including a course offerings booklet, bulletins for teachers and students, a teacher's manual, and an intern's report describing the school. *Assistant Superintendent for Instructional Services* includes a filmstrip with information about that position's role in the system, two in-basket exercises which simulate the duties of the position, and a packet of background information that includes, in the producer's words, "(1) Perspective—a brief look at how the role incumbent moved into this position, (2) a sociological study of the central office, (3) a news item announcing the appointment of the Assistant Superintendent for Instructional Service, (4) a letter from the superintendent concerning appointment to the position, (5) a file of important matters left by a former assistant superintendent, (6) a Budget, and (7) a Teacher's Handbook." *Assistant Superintendent for Business Management* includes two in-basket exercises, a film strip, and an interruption tape recording, in addition to such background materials as an annual budget, a plan for reorganizing the district, and a survey of property and building needs. *Madison Superintendency* also includes two tape recordings, two in-basket exercises, "Special Problems Material" with five brief role-playing episodes, and a packet of background information that describes who the incumbent superintendent is, how he got the job, and a budget.

All of these components are packaged with materials for at least twenty-five students, and all include an instructor's manual. Individual items or components may be purchased separately or as part of the total *Madison* package. (DCD)

Cost: contact producer for current price list

Producer: University Council for Educational Administration, 29 West Woodruff Avenue, Columbus, OH 43210

MONROE CITY SIMULATIONS

Jack Culbertson and others

Playing Data
Copyright: 1971
Age Level: postgraduate
Number of Players: 25
Playing Time: 90-270 hours (est.) for complete simulation
Preparation Time: 40-80 hours (est.)
Packaging: all materials are professionally produced and packaged

Description: The *Monroe City Simulations* are the University Council for Educational Administration's most extensive collection of interrelated curricular materials for school administrators in training. Taken together, they comprehensively describe the administration of a small, midwestern school system. The complete set of materials includes fifteen booklets, a thirty-minute long motion picture, a filmstrip and audio tape recording, and an instructor's manual which describes the community of Monroe City; a series of multimedia (but predominantly in-basket) exercises which simulate the problems and duties of six key administrators in the system; twelve "Interpretive Content Series" booklets; one in-basket simulation which addresses the problems inherent in "Curricular Decision Making"; two simulations concerning specific problems in high school administration; and two simulations pertinent to collective bargaining negotiations within the Monroe City system.

The complete set of background materials about Monroe City includes: (1) The Monroe City School System and its Environment: An Overview; (2) Monroe City: Its Setting and Demography; (3) The Political Environment of the Monroe City School System; (4) The Economic Environment of the Monroe City School System; (5) Monroe City's Mass Media; (6) Patterns of Influence in Monroe City; (7) Inter-Agency Relations in Monroe City; (8) Community Organizations in Monroe City and Their Demands upon the School System; (9) Monroe City's Board of Education; (10) Internal Organization and Decision Making in the School System; (11) Monroe City's Educational Program; (12) The School System's Professional Staff; (13) Monroe City's Public Schools: Professional Negotiations; (14) Perceived Challenges to Educational Leadership in Monroe City; and (15) Monroe City's Students. The motion picture, filmstrip, and tape recording all provide an introduction to, and an overview of, Monroe City.

The currently available simulations of the duties of Monroe City administrators include the following components. The *Abraham Lincoln Elementary School Principalship Simulation* contains seven films, a faculty handbook, a background information packet containing reports describing the school and its surrounding neighborhood, and five in-basket exercises designed to replicate the duties of the principal. The *Janus Junior High School Principalship Simulation* contains two color slide presentations that describe the school and its attendance area, a student handbook, a "data bank" of information about the school and its staff, two tape recordings (of student interruptions and interviews with local community workers), two motion pictures portraying student "incidents" and "problem situations," and two in-basket simulations. The *Wilson Senior High Principalship Simulation* includes student and faculty handbooks, a slide presentation about the school, a file of background information about the school and its students and faculty, six case studies describing intraschool and school-community relations problems, two in-basket simulations of the principal's duties, and a role-playing exercise in which the principal must stop the use of objectionable language by students. The *School Psychologist Simulation* contains two role-playing exercises, an in-basket simulation of the psychologist's administrative duties, a simulated PTA meeting, a "Resource Bank," a "Department of Child Study Handbook," and a filmstrip and tape recorded material which describe the programs administered by the psychologist for the city's schools. The *Special Education Administrator* contains eight tape recordings and/or filmstrips, three case studies, background files, and a booklet describing the duties and problems of that position, in addition to five in-basket planning items. The *Monroe City Superintendency Simulation* uses films entitled "Minority Group Employment Practices in Monroe City," "On the Spot with the Press," and "Sex Discrimination in Monroe City." This exercise also includes two general in-basket simulations of the superintendent's duties, and in-basket items pertaining to the hiring of an assistant superintendent and the disposition of a first-year teacher who refuses to sign a loyalty oath.

The "Interpretive Content Series" items are designed, according to the producer, "to facilitate the linking of theory and practice in simulation experiences. The papers set forth concepts for students and instructors and provide examples for those preparing their own interpretations." This series includes essays about the "relationships between organizations and their external environments within the Monroe City setting," analyses of several "types of organizations conspicuous by their presence or absence in the Monroe City school environment," and "behavioristic analysis" of incidents portrayed in the films.

Administration of Curricular Decision Making "focuses on a demand for a districtwide black studies program, and the factors impinging upon the administration's decision." Three in-basket items comprise this component.

Two parts of the package address problems specific to Wilson Senior High School. *Problem Sensing and Selection in Wilson Senior High* "enables participants to assume the role of staff members involved in consensual decision making." The core of this exercise is a document which "reviews the process of consensual decision making and group dynamics. It offers guidelines for formulating problem statements, combining and ranking problems, and problem choice and resolution." The *Site Budgeting Simulation,* "based on data from the Wilson High School, aids student's understanding of school site budgeting as an aspect of program leadership."

The *Negotiations Simulation* and *Impasse Simulation* are both in-basket exercises in which the participants must review sixteen items to be negotiated between the Monroe City Education Association and the Monroe City Board and then use these materials in completing strategy worksheets and agreement forms. Among the negotiated items are salary, pupil-staff ratio, benefits, and instructional materials.

All of these components are packaged with materials for at least twenty-five students and all include an instructor's manual. Each component may be purchased separately or as part of the total Monroe City package. (DCD)

Cost: contact producer for current price list

Producer: University Council for Educational Administration, 29 West Woodruff Avenue, Columbus, OH 43210

NAKED MONSTERS

Sivasailam Thiagarajan

Playing Data
Copyright: 1972
Age Level: teacher trainees
Number of Players: 2-15
Playing Time: 5 minutes to 3 hours
Preparation Time: none
Special Equipment Required: special deck of 52 cards

Description: *Naked Monsters* is a series of thirty-five games, none of which take more than five minutes to play, designed to aid teacher trainees to teach concepts. "The Naked Monsters are 52 delightful little creatures which share some attributes and differ in others. Some of the attributes are systematically varied while others are left to the artist's whims. As a result, these monsters can be classified into different concept categories. The cards were originally designed to simulate a subject-matter domain at the elementary level. When we started out designing these games on concept teaching, we used actual concepts and examples from the regular curriculum, but the teacher trainees' previous experiences with them kept interfering with their learning of new skills. Familiarity bred contempt, and nobody could really visualize the difficulties of a naive learner in acquiring a concept. We then went to the other extreme and used concept classes of fat, blue cubes and then, red cylinders. But these smacked of laboratory experimentation and were labeled artificial and irrelevant by the players. Finally we ended up with our monsters. We like them because the players like them. Also to a great extent they enable us to create concepts at the elementary school level. No teacher is an existing expert on any of these concepts. Unlike attribute tiles, the total universe of the monsters cannot be figured out by logic" (from the teacher's manual).

Here are the names of some of the games and their instructional objectives.

Bare Facts—to rapidly analyze a given example into its attributes.

Any Fool Can See—to identify salient attributes of an example.

Blind Spots—to identify nonsalient attributes.

Krazy Kinship—to analyze a set of examples into critical attributes.

Unique Uncles—to analyze a set of examples and identify irrelevant attributes.

Striking Similarities—to identify obvious critical attributes.

Drastic Differences—to identify noisy irrelevant attributes.

Subtle Samenesses—to locate subtle critical attributes.

Dim Distinctions—to locate subtle irrelevant attributes.

Monster Massacre—to rapidly identify examples and nonexamples of a concept class.

Example Hunt—to identify examples of a given concept.

Clear Example—to choose the best instructional example from a set of available examples.

A Rare Eg—to choose a difficult test example from a set of available examples.

Diverge—to choose an example which is divergent from a given example.

Eg Pairs—to choose a pair of divergent examples of a given concept.

Negatives—to identify nonexamples of a given concept.

Close In—to choose the most close-in nonexample for a given example.

Tough Test—to choose a subtle nonexample to be used for testing the mastery of a concept.

Matched Pairs—to choose an example-nonexample pair in which the example is matched to the nonexample as closely as possible.

Emphasis—to redraw the picture of a given monster to exaggerate its critical attributes and to de-emphasize irrelevant ones.

Viewpoints—to analyze a conjunctive, disjunctive, or relational concept.

Malc—to present a series of examples and nonexamples in the most effective sequence to teach a concept.

Jabberwock—to identify deficiencies in the learner's repertoire of subconcepts and teach appropriately.

Remediate—to diagnose type of learner error (undergeneralization, overgeneralization, or confusion) and provide the most efficient remedial instruction.

Confusion—to analyze the cause of various misconceptions in terms of misperceived attributes. (REH)

Comment: These games are a beautiful example of gaming as a teaching strategy. They teach some of the most important skills that teachers need to know—how to teach discriminations and generalizations. They can be reused many times and there are many variations of the games which are easy and inexpensive. The games are intrinsically motivating, so much so that they have been used as party games. There are some possible misconceptions about teaching concepts stemming from the limited range of concepts used. Nevertheless, I recommend this game to teacher trainees and teachers. (REH)

Cost: $4.00

Producer: Dissemination Unit, Center for Innovation in Teaching the Handicapped, 2853 East Tenth Street, Bloomington, IN 47401

QUESTIONEZE

Lewis B. Smith, University of Idaho

Playing Data
Copyright: 1972

Description: *Questioneze* is based on Bloom's taxonomy of cognitive objectives. The purpose is to help teachers vary the cognitive level of their oral and written questions. Results of field testing indicate that teachers tend to compose questions at more varied levels after use.

In different parts of the several games included, teachers play a card game where they must classify questions, and a much harder one in which they are given a variety of pictures or diagrams and they must then ask questions at the various levels in the Bloom taxonomy: knowledge, comprehension, application, synthesis, and evaluation. (REH)

Cost: $4.95

Producer: Charles E. Merrill Publishing Co., Columbus, OH 43216

REGULAR MEETING OF THE WHEATVILLE BOARD OF EDUCATION

Kenneth McIntyre, University of Texas at Austin

Playing Data
Age Level: school boards
Number of Players: 1-20
Playing Time: 2 hours (est.)
Preparation Time: 1 hour (est.)
Packaging: photocopied game materials include playing manual, agenda, and reaction forms

Description: The purpose of this simulation is to provide an educational experience for school board members. Agenda items are real ones which include high school marriages, selection of athletic directors and coaches, controversial library books, and the like. (producer)

Cost: $2.75 for UCEA members, $3.10 for nonmembers

Producer: University Council for Educational Administration, 29 West Woodruff Avenue, Columbus, OH 43210

ROLE-PLAYING METHODS IN THE CLASSROOM

Mark Chesler and Robert Fox

Playing Data
Copyright: 1966
Age Level: grades 6-12
Number of Players: 2 or more
Playing Time: 5-15 minutes per exercise
Preparation Time: 2 hours (est.)
Packaging: professionally produced 86-page soft bound book

Description: This general introduction to role-playing activities is designed to show, according to the producer, "how role-playing techniques can be used with students to identify classroom problems and help students understand the problems of social interaction." The book presents a theory of role behavior and role playing and includes several practical examples and case studies from classroom use. In addition, a three-stage, nine-step procedure for administering role-playing activities in a classroom is suggested. This procedure begins with a period of teacher instruction followed by the role-playing exercise and its debriefing, and concludes with the teacher's evaluation. An annotated bibliography includes a list of 126 "warm-ups, two-man situations, group situations, and problem stories" that a teacher may use to introduce role playing to students.

The warm-up exercises require minimal emotional investment by the participants. The simplest of these (such as having a student pretend to walk through snow drifts) are elementary and pantomimic, whereas others (such as asking a student to pretend that someone has just invited him to a party or to pretend that he sees two children beating up a smaller boy) allow the participants to practice complex role-playing skills. The most advanced scenarios listed in the book require participants to act out classroom problems or sensitive personal and community issues (such as the prosecution of a fourteen-year-old who has stolen food for his family). These scenarios may be played out by one to twenty players and may require up to one class period to complete. (DCD)

Cost: $2.95

Producer: Science Research Associates, 1540 Page Mill Road, Palo Alto, CA 94304

SCHOOL PLANNING GAME

Richard Lavin, Merrimack Education Center

Playing Data
Age Level: architects, students, teachers, administrators, community (school board members)
Number of Players: variable, in teams of 3 or 4
Playing Time: Variable

Description: Players gather around a magnetic board. They may move markers representing walls, windows, storage/unit equipment, students, teachers, administrators, and so forth. Playing procedures require the use of this aid to objectify different arrangements of space, pupils, and staff. The rules make it necessary for a team to relate education program objectives to space, to relate form and function. (REH)

Comment: As we do not have a set of the rules and equipment, it is difficult to evaluate this game from the articles and descriptions. However, it appears to be a good idea and one which could be implemented very easily on a local level, or on one which could be played with the sole use of the authors, and appears to be an excellent way to involve a group in decision making and planning. This is something which groups at all levels in our society need to do. We might just as well begin with a school. In fact, of the beginning of every school year, principal, the teachers, representatives from the community and—in the upper grades—representatives from the student body might very well work together to play the *School Planning Game.* And it might not be a bad game to play two or three times a year. Of course, one wouldn't have quite the flexibility of space that you have in designing a totally new school. But there's a lot more flexibility in education than we generally use, and regretably we often perform the same experiments over and over again each year. (REH)

Cost: $195.00

Producer: Merrimack Education Center—IGM, 101 Mill Road, Chelmsford, MA 01824

SEATS: SPECIAL EDUCATION ADMINISTRATION TASK SIMULATION

Daniel D. Sage

Playing Data
Copyright: 1970, 1972, 1973
Age Level: adults in graduate school with administrative/management orientation
Number of Players: no absolute limit
Playing Time: 1-5 periods of 5 hours each

Preparation Time: 30-60 minutes
Special Equipment: tape recorder, slide projector, 16mm movie projector, background data bank (provided)
Packaging: kit includes instructor's manual, 25 participant booklets, 25 sets of data bank

Description: The central role in this simulation is the Administrator/Director of Special Education, a newly created position in an average-sized suburban school district. Utilizing an interactive mode of both individual and team participation and relying heavily on role playing, this simulation makes participants familiar with key issues in administering educational programs for the handicapped. The simulation takes the participants through a series of in-basket situations that require responses and generate feedback. Some of the major decisions include what actions to recommend in emotionally charged situations, how to handle ambiguous problems such as the disposition of a child's placement, and how to report to the superintendent and other appropriate personnel. (Alan H. Kraushaar)

Comment: The most unique and outstanding aspect of this simulation is that it provides a high degree of realism and close correspondence to prototype decision-making situations from real life. The major goals of *SEATS* are to present relatively inexperienced educators with simulations of important issues which require considerable understanding and rapid decision making. In this way the participants can learn more about dealing with ambiguity under high emotional pressure and become more familiar with both their own problem-solving styles and different perceptions of real-life situations.

One apparent shortcoming of the simulation is that it tends to generate a good deal of frustration when participants find that their "good" decisions do not make an appreciable impact in struggling with succeeding problems. However, to a considerable extent, this may not be far removed from the realities of the school environment. But the realities of life situations are such that, to a large extent, incomplete and inadequate information are factors with which administrators must be able to function in everyday encounters. Thus, the participant in the key role comes to learn that even though he or she is an authority in a school system, his/her decisions need not always be acted upon.

This instrument has been used primarily as part of preservice training for special education administrators. However, it has also been used with other auxiliary and support school personnel to expose them to issues in related areas of responsibility with which they need to become more familiar.

Although this simulation costs $325.00 for a complete kit for 25 participants, with the increasing thrust of "mainstreaming" people with handicaps into society, it could be effectively used by many school systems in lieu of more expensive orientations, in-service meetings, and publication costs. (Alan Kraushaar, Clinical Services, Division of Mental Retardation, Massachusetts Dept. of Mental Health)

Cost: $325.00

Producer: Syracuse University Press, Box 10, Syracuse, NY

SEASIM

W. Michael Martin, University of Colorado; James R. Yates, University Council for Educational Administration, General Editors

Playing Data
Age Level: graduate and postgraduate, professional
Number of Players: 10-25
Playing Time: 45-120 hours
Preparation Time: 40 hours (est.)
Packaging: 104-page, photocopied monograph

Description: *SEASIM (The Special Education Administration Simulation in Monroe City)* is a complex series of overlapping exercises and simulations prepared by twenty-eight authors representing nine different universities for the University Council for Educational Administration, General Special Education Administration Consortium. These materials are designed, according to the authors, "to prepare urban school district administrators" (and administrators in training) "to deal with the handicapped student."

"The various components of *SEASIM* are composed of numerous stimulus items which may be used in total or in part by the instructor, depending upon the goals and objectives for the course or activity. It probably will not be possible for *SEASIM* instructors to utilize optimally all of the available stimulus items contained in *SEASIM* unless the game is adopted as the total curriculum for a minimum of one course (equaling about 45 instructional hours). It could require as much as 90 to 120 instructional hours to cover extensively all of the more likely considerations made available through the stimulus items contained in the game." "*SEASIM* utilizes the theoretical approach to administration developed by Getzels and Guba (1955), i.e., a transactional mode of interacting between individuals within an organization and individuals without the organization. Additionally, it adopts a Theory Y approach (McGregor, 1960) in dealing with individuals."

Simply stated, *SEASIM* is an in-basket simulation (with occasional role playing) in which the participants assume the role of Mare Grady, a person recently appointed Director of Special Education in Monroe City. (Monroe City is the subject of sixteen booklets published by the University Council for Educational Administration. See the descriptive entry in this section.)

The game has five components. The *Continuum of Services Component* simulates the relationships of special and regular education personnel. The *Identification and Placement Component* elaborates many of the problems presented in the *Continuum Component* pertaining to eligibility and responsibility as players define what services will be provided by whom and how they will be provided. The *Curriculum Component* examines the relationship between special and regular education curricula; players must determine a community definition of curricula and evaluate the curriculum in general and in reference to minority students. Play of the *Finance Component* requires students to draft and understand a special education department budget. In the *Evaluation Component* the players must evaluate the special education program in Monroe City. (DCD)

Cost: contact producer for current price

Producer: University Council for Educational Administration, 29 West Woodruff Avenue, Columbus, OH 43210

SOLVING MULTI-ETHNIC PROBLEMS

Frederick P. Venditti, University of Tennessee

Playing Data
Copyright: 1970
Age Level: teachers and trainees
Number of Players: variable
Playing Time: 3-30 hours
Special Equipment: film projector

Description: *Valleybrook Elementary School* and *Lakemont High School* are two series of films, simulated discussions, and role plays similar to *The Changing High School* (described earlier in this section). Each series focuses especially on problems of desegregated public schools, elementary and secondary, respectively. (AC)

Cost: Preview/rental for either series is $15.00 for 48 hours, $35.00 for one week, deductable from purchase price, otherwise nonrefundable. Purchase price for either film is $150.00; total Valleybrook package $225.00; total Lakemont package $318.50

Producer: The Anti-Defamation League of B'nai B'rith, 315 Lexington Avenue, New York, NY 10016

TEACHER EDUCATION CAN OF SQUIRMS

Playing Data
Copyright: 1978

Age Level: college, teachers
Number of Players: 2 or more
Playing Time: 10 minutes or more
Preparation Time: 30 minutes (est.)
Packaging: professionally packaged game includes role slips, leader's guide, and discussion questions

Description: This is the thirteenth in the series of *Can of Squirms* games. Pairs of players assume the roles of teachers and students, administrators, or paraprofessionals to act out typical classroom confrontations. The events prompt these encounters are printed on slips of paper called "Squirms" (because they describe situations in which people may want to cringe and squirm), which are drawn from a can. Typical of the twenty "Squirms" in this set are the following, in which a "new teacher tries to settle the question of classroom discipline with another teacher who has seniority," or a "teacher tries to help the 'loner' student," the "class clown," a "negative student," and a "student who may have symptoms related to drug abuse," or a "teacher must defend the purchase of library books instead of football uniforms at a parent-teacher association meeting despite parental criticism."

This game may be played by any number of participants for any length of time in five- to fifteen-minute rounds. This set of Squirms is specifically designed for use by practicing and prospective high school teachers. A "Can of Squirms" portraying elementary school situations may be purchased separately. (DCD)

Cost: $10.00

Producer: Contemporary Drama Service, Box 457, Downer's Grove, IL 60515

TEACHERS' LOUNGE

S. Joseph Levine and James W. Fleming, Michigan State University

Playing Data
Copyright: 1972
Age Level: college, graduate school, continuing education
Prerequisite Skills: a basic understanding of teaching
Number of Players: 1-8
Playing Time: 15-40 minutes
Preparation Time: 15 minutes

Description: *Teacher's Lounge* is designed to provide a sampling of diverse learning problems which can be encountered in the classroom. By analyzing these problems, the players should become better able to analyze problems in their own classrooms.

Players take turns selecting a game card and analyzing a learning problem, selecting a diagnosis, and receiving feedback on their selection. Players move their tokens according to whether they are correct or not. The first player to arrive at the finish line wins. (Chance cards are drawn if a player lands on a designated space.) [AC]

Comment: The game demonstrates the diversity of learning problems that exist, the ambiguity of problems, and the difficulty of making speedy diagnoses. According to the authors, the game has been used successfully with parent groups who have no teaching experience. The game makes them more aware of typical learning problems. A review copy was not provided, so we are unable to make further statements about this game. (REH)

Cost: $4.95

Producer: Mafex Associates, Inc., 90 Cherry Street, Johnstown, PA 15902

TEACHER TRAINING SIMULATION

Dale L. Bolton, University of Washington

Playing Data
Copyright: 1970
Age Level: undergraduate and postgraduate college, professional
Number of Players: 25
Playing time: 8-40 hours
Preparation Time: 6 hours (est.)
Packaging: professionaly packaged and produced 63-page instructor guide and 118-page game manual (25 copies), plus multi-media material not available for review

Description: This material is suitable for workshops or courses dealing with general decision making, or more specifically, with personnel selection problems. It is designed to provide experience at various stages of the teacher selection process. The simulation includes a description of a school situation (using slides, tapes, and written materials), a set of fictitious applicants (described in written documents and television films or interviews), and response devices that require decisions and provide for analysis and feedback. (Producer)

Cost: $595.00 for UCEA members, $695.00 for nonmembers

Producer: University Council for Educational Administration, 29 West Woodruff Avenue, Columbus, OH 43210

TEACHING PROBLEMS LABORATORY

Donald R. Cruickshank, Frank W. Broadbent, Duke University, and Rolf L. Bubb, State University College, Brockport, NY

Playing Data
Copyright: 1967
Age Level: college, graduate school, working teachers
Number of Players: 35-40
Playing Time: 8-1/2 days to 3 hours per week for a semester
Preparation Time: 5-10 minutes before each problem
Special Equipment: film projector, display system, record player

Description: *Teaching Problems Laboratory* introduces participants to the teaching problems encountered in every classroom—problems that can become an obstacle to imaginative and effective teaching. The participants discover approaches to solving these problems and can develop a familiarity in dealing with them that can help improve their teaching.

The teaching problems take place in a fifth grade classroom of Longacre School in the town of Madison, a suburb of Elton. The school is typical of thousands across the country, and the problems that occur here are representative of the problems that occur daily in most schools. As Pat Taylor, a new fifth-grade teacher, the participants solve the teaching problems that occur in their Longacre classroom.

Two filmstrips help create the environment Pat Taylor will teach in. Cumulative record folders, sociograms, and reading progress reports introduce the 31 students in the class—students with varied backgrounds, interests, and achievements; and students who will create the same problems for Pat Taylor that real students create for teachers in actual classrooms.

The thirty-one critical teaching problems that the participants are asked to solve cover a wide range of teaching experience—problems in student behavior, parent relations, curriculum planning, teaching methodology, classroom management, and evaluation of learning.

Comment: *Teaching Problems Laboratory* appears to be the most complete training experience available to the prospective teacher, short of OJT. It is, however, noninteractive, which may limit its utility. Problems involving other humans—teaching problems, for example—rarely have the clarity or easy solubility of their printed descriptions. Play is by individuals who are called on to employ decision-making and problem-solving skills. (DZ)

Cost: $795.00 (#13-0420)

Producer: Science Research Associates, College Division, 1540 Page Mill Road, Palo Alto, CA 94304

TEACHING STRATEGY

John E. Washburn, in consultation with staff of National Teacher Education Project

Playing Data
Copyright: 1971
Age Level: in most cases played with professional and volunteer teachers; however, can be played with teacher and class (elementary-adult)
Special Prerequisites: some understanding of concept development and behavioral objectives
Number of Players: 2-12 optimum; 18-24 maximum, in 1-8 teams
Playing Time: 20-30 minutes
Preparation Time: 1 hour
Special Equipment: suggest overhead projector or chalkboard

Description: According to the producer, *Teaching Strategy* is designed "to help players inquire into the wide variety of teaching styles." Players form teams of two teachers and agree on a concept and the age level of students they want to teach. Each team is then assigned the task of formulating a coherent teaching strategy with cards drawn from six different stacks, paying for each card with a bingo chip. Each team must first buy a "Start of Teaching Strategy" card, which will have a message like "Teacher and students view a filmstrip." Teams then may draw up to fourteen more cards from the piles of "Resources" (such as "Old Magazines"), "Presentation" (such as "Teacher asks a question"), "Teacher-Initiated Student Activity" (such as "Teacher directs students in role-playing episode"), "Student-Initiated Activity" (such as "Student raises question"), and "Conclusion/Summary" (such as "Teacher makes summarizing statement"). Teams have thirty minutes to assemble teaching strategies from these statements. After formulating their strategies they explain them, in turn, to the other teams. (DCD)

Cost: $9.75

Producer: National Teacher Education Project, 6947 East MacDonald Drive, Scottsdale, AR 85253

TENURE

Frank Bazeli

Playing Data
Copyright: 1977
Age Level: college, professional
Number of Players: 8-32
Playing Time: 2½ hours
Preparation Time: 2 hours (est.)
Packaging: professionally packaged game

Description: *Tenure* simulates the experiences of eight probationary teachers in an urban integrated high school over a period of two years. Each player is assigned a specific role, and the goal of each is to get tenure. Players get tenure by investing poker chips (symbolic of each role's capacity to invest time and energy in teaching and related activities, such as going to graduate school) and by answering questions about how they would react when confronted by situations at school. For instance, a player may be asked how she or he would handle an unruly student in a particular situation, and depending on the answer, the player could be awarded up to fifty points. The facilitator keeps score during the game's eight rounds. To get tenure, a player must score at least twelve hundred points and raise the instructional level of Homeroom 302 to a level "conducive to learning." (DCD)

Cost: $24.50

Producer: Educator's Workshop, P.O. Box 345, DeKalb, IL 60115

TRIP TO MARS

Playing Data
Copyright: 1972
Age Level: adult
Number of Players: 12-15
Playing Time: 1 hour
Special Equipment: cassette tape recorder
Packaging: professionally packaged game includes player's pad, instructions, and debriefing card

Description: *Trip to Mars* is designed to help players develop group decision-making skills. Participants assume the roles of members of a team assigned to pick ten educational items to send to a Martian colony to establish a school. Players must, first, agree on ten items, and then as a team, defend their choices. Finally, players are told to adjust their list so all ten items consume no more than forty cubic feet of storage space on the spacecraft.

During the debriefing players are asked what clashes in values among team members affected their choices and what assumptions about children conditioned the team's decisions. (DCD)

Cost: $15.95

Producer: Creative Learning Systems, Inc., 936 C Street, San Diego, CA 92101

WEB

Dannie L. Smith, Memphis State University

Playing Data
Copyright: 1973
Age Level: education students
Number of Players: class size
Playing Time: 12-15 hours

Description: Web is a simulation of working in an educational bureaucracy. It teaches how a typical educational bureaucracy works by having students assume various roles and their corresponding responsibilities and by requiring student role players to make decisions in simulated crisis situations that arise in any school system. The learning experiences *Web* gives participants are: taking pre- and post-educational attitude surveys, analyzing and outlining their own educational philosophy in relationship to specific school district problems (e.g., employee dress and grooming, salary and promotion, controversial topics, ethnic ratio, sex education, student grouping, objectives and evaluation, innovation), making decisions in a present or future educational role, evaluating subordinate decisions, speaking informally to fellow professionals in small group meetings, writing a letter of application, participating in an employment interview, organizing others with similar views into an effective political faction, participating in a problem-solving process discussion in order to define group goals, and speaking formally to elected citizens' representatives.

Comment: Every frontline professional teacher should know something about the workings of the administration that surrounds him—and sometimes makes him feel as if it is pushing him down. I know no better way of learning than by simulating the workings of the policy-making parts of the educational bureaucracy. *Web* appears to give a solid survey of these issues without the usual boring lectures that often accompany courses or lags in educational administration. This departs from the usual Interact package in that it is nearly 90 pages long and comes in a paperback book format. (REH)

Cost: $4.95

Producer: Interact, P.O. Box 262, Lakeside, CA 92040

See also
Auction SELF-DEVELOPMENT
College Game SELF-DEVELOPMENT
Dynamic Modeling of Alternative Futures FUTURES
EDUCOM Financial Planning Model COMPUTER SIMULATIONS
Encapsulation COMMUNICATION
Everybody Counts HUMAN SERVICES
Ghetto COMPUTER SIMULATIONS/SOCIAL STUDIES
Microville COMMUNITY ISSUES
Pipeline HUMAN SERVICES
Probe SELF-DEVELOPMENT

FRAME GAMES

AT-ISSUE

Richard D. Duke and Cathy S. Greenblat

Playing Data
Copyright: 1974
Age Level: grade 9-graduate school, citizen groups
Number of Players: 15-135 in multiples of 3
Playing Time: 1 hour per segment (6 optional segments are presented)
Preparation Time: 1-4 hours

Description: *At-Issue* (from the *Trilogy* series) is designed with a series of six components which the operator can mix and match to suit the need at hand. The operator may elect to prepare subject content in advance or have the participants provide the content.

In either case, a normal use of the exercise would proceed as follows: The audience is broken into groups of three or five. These teams select or are assigned "roles" which are color coded. An "issue" or problem is selected for discussion, using the procedure provided. The issue is checked against other related issues to establish cross impacts. Variables related to this "issue" or problem are determined, and weights assigned to indicate their significance to each team. The impact of the issue is assessed by each team against the variables. Each team's results are scored, the long-term impact of the decision on a few of the most critical variables is decided, and a critique is held.

It is not necessary to elect all steps or to play them in sequence. *At-Issue* is designed for on-going and recurrent use by a group exploring serious problems. (Publisher)

Note: For more on the *Trilogy of Games,* see the essay on urban simulation games by Jo Webb.

Cost: under $10.00 for *Trilogy of Games*

Producer: Sage Publications, Inc., 275 South Beverly Drive, Beverly Hills, CA 90212

CONCEPTUAL MAPPING GAME

Richard D. Duke and Cathy S. Greenblat

Playing Data
Copyright: 1974
Age Level: grade 9-graduate school, citizen groups
Number of Players: 3-21
Playing Time: units of 3 hours (may repeat as required)
Preparation Time: the first 3-hour session

Description: The *Conceptual Mapping Game* might best be described as a game-like exercise. In any event, it represents a procedure that can be readily adjusted to different group requirements. It is intended for mature audiences who grapple with serious problems. The objective of the exercise is to help the group organize their thoughts and to communicate their ideas to others.

There are two styles of use. It may be prepared in advance or it may be developed by the players while in use. The procedure is quite simple. Individuals (or teams, depending on group size) select or are assigned roles. These are normally their real-world role. Using the procedure provided, a problem ("issue") is selected for consideration. The teams evaluate the impact of this decision on a large number of variables displayed on a "wheel." This can be done independently, by sectors, or jointly. A summary of each team's conclusions is presented, discussion of the issue is continued while productive, and a vote is taken to determine how the issue would be resolved.

This procedure is in use by different groups and is used periodically to examine new issues that emerge. (Publisher)

Note: For more on the *Trilogy of Games,* see the essay on urban simulation games by J. Webb.
Cost: under $10.00 for *Trilogy of Games*

Producer: Sage Publications, Inc., 275 South Beverly Drive, Beverly Hills, CA 90212

CONFRONTATION

Harold D. Stolovitch

Playing Data
Copyright: 1979
Age Level: adult
Number of Players: 3 or multiples of 3
Playing Time: 45-90 minutes
Preparation Time: varies, depending on the complexity of the content being loaded on the frame
Packaging: printed manual

Description: During each round, players receive a card which specifies a confrontation situation and assigns roles. Two players are in adversary roles and the third is a mediator. Adversary players choose one of five possible positions on the issue and compare their choices. Adversaries discuss/debate their positions and with the help of the mediator reach a common ground. They exchange poker chips depending on the shift from their initial positions. (ST)

Note: For more on this game, see the essay on frame games by Sivasailam Thiagarajan and Harold Stolovich.

Cost: $2.00

Producer: Instructional Alternatives, 4423 East Trailridge Road, Bloomington, IN 47401

FACTS IN FIVE

R. A. Onanian

Playing Data
Copyright: 1966
Age Level: grade 7 and up
Number of Players: 1 to any number
Preparation Time: 5 minutes
Playing Time: 45 minutes

Description: Players test their knowledge, competing to score points by filling in their playcard blanks with words or phrases that fit the class/category in the allotted time. (AC)

Note: For more on this game, see the essay on frame games by Sivasailam Thiagarajan and Harold Stolovitch.

Cost: $12.00

Producer: Avalon Hill, 4517 Harford Road, Baltimore, MD 21214

GAMEgame

Sivasailam Thiagarajan

Playing Data
Copyright: 1972
Age Level: in-service and pre-service teachers
Number of Players: 9-70 in 3-7 teams
Playing Time: 7-1/2 hours to one semester
Preparation Time: 2 hours

Description: *GAMEgame* is a simulation/game which teaches various aspects of designing, adapting, and evaluating instructional games and simulations.

GAMEgame has a nested format: One set of players designs a simulation/game which requires that another set of players design another simulation/game. The first set of players, called game wardens, works from an outline of the *GAMEgame,* modifies it to suit local needs and runs the game. This involves dividing up the remaining larger group of players into teams and giving each team a simulation/gaming task and a "grant" to work with. Team members enroll in mini-courses on game design, receive consultative help and resource materials (for a charge), construct their own games and test them on real or simulated subjects. In the final phases on *GAMEgame,* all players cooperatively evaluate the simulations/games, determine the winning team and debrief themselves.

GAMEgame is made up of four sessions, each lasting for about two hours. The first session is for the game wardens only. The other three sessions are for all players.

GAMEgame is available in three versions: The teacher-proof maxi-version has all decisions, rules, forms and courses ready made along with a rigid script for the game wardens. The midi-version, designed for flexible teachers, offers a model game which requires local finishing. Finally, there is a miniversion, designed for creative teachers, in which game wardens are merely given a brief synopsis of GAMEgame and asked to take off from there. (Author)

Objectives: Upon the completion of the simulation/game, the game wardens should be able to demonstrate the following competencies: (1) modify an off-the-shelf simulation/game to suit local players and their instructional needs; (2) plan and conduct a simulation/gaming session; (3) evaluate simulations/games from different aspects of cost-effectiveness; and (4) debrief players at the end of a simulation game. The other players should be able to demonstrate these competencies: (1) design a simulation/game to help a specified group of target students reach specified instructional objectives; (2) modify the simulation/game on the basis of expert appraisal and player feedback; and (3) evaluate instructional simulations/games from different aspects of cost-effectiveness. (Author)

Cost: $2.00

Producer: Dissemination Unit, Center for Innovation in Teaching the Handicapped, 2853 E. 10th St., Bloomington, IN 47401

GAMEgame II

Sivasailam Thiagarajan

Description: *GAMEgame II* provides participants with a chance to practice distinguishing among "simulation games, nonsimulation games, nongame simulations, instructional games, noninstructional games, instructional simulation games, etc." Participants study a diagram of definitions. An item is then picked randomly and each participant writes down the classification he thinks it comes under. Responses are then checked with the classification assigned to the item by "experts." Each player scores one point for being correct and another point for each error made by other players in that round (thus incorporating a reflection of difficulty of discrimination). (AC)

Cost: reprint from *Phi Delta Kappan,* (1974) 55(7) free from designer.

Producer: Sivasailam Thiagarajan, Instructional Alternatives, 4423 East Trailridge, Bloomington, IN 47401

GAMEgame IV

Sivasailam Thiagarajan

Playing Data
Copyright: 1976
Age Level: adult
Number of Players: 3-30 players in 3-6 teams
Playing Time: 30-60 minutes
Preparation Time: none
Packaging: printed directions

Description: Teams create lists of five important items related to a selected theme. The game leader creates a common list. In secret, teams record their top choice from the common list and are rewarded for achieving consensus. This is repeated until the top five items are identified. (Author)

Note: For more on this game, see the essay on frame games by Sivasailam Thiagarajan and Harold Stolovitch.

Cost: $2.00

Producer: Instructional Alternatives, 4423 East Trailridge Road, Bloomington, IN 47401

GAMEgame VI

Sivasailam Thiagarajan

Playing Data
Copyright: 1978
Age Level: adult
Number of Players: 10-60
Playing Time: 30-60 minutes
Preparation Time: 20 minutes
Packaging: printed directions with contents for game cards

Description: The game leader prepares cards with individual opinions on a topic or issue. Players write four personal opinions on blank cards. These cards are randomly distributed to all players. Players exchange opinion cards at a discard table and with each other, then form coalitions of like-minded players and reduce their total number of cards to five. Each group writes a summary statement of its stance and selects an appropriate name for itself. (Author)

Note: For more on this game, see the essay on frame games by Sivasailam Thiagarajan and Harold Stolovitch.

Cost: $2.00

Producer: Instructional Alternatives, 4423 East Trailridge Road, Bloomington, IN 47401.

IMPASSE (The IMPact ASSEsment Game)

Richard D. Duke and Cathy S. Greenblat

Playing Data
Copyright: 1973
Age Level: grade 9-graduate school, citizen groups
Number of Players: 3-99 in multiples of 3
Playing Time: 1 hour
Preparation Time: 10 minutes for existing game (load new game 1-4 hours)

Description: *IMPASSE* (from the *Trilogy* series) was designed as a quick exercise to get a group to focus on some of the central dimensions of a problem.

Specifically designed for classroom use (the 50-minutes hour) it can also be readily adapted for use in various public forums. It can be used in either of two formats: preloaded with content by the game operator; or loaded by the participants at time of play.

As participants arrive they are broken into groups of three (or, alternatively, groups of five). They are assigned a role, or permitted to select a role. An issue or problem is described, and the teams must reach consensus on its impact on a list of related variables. Teams may elect to resolve the issue either pro or con. They are asked to estimate the impact at some future time, perhaps five or ten years hence.

IMPASSE is a good ice-breaker, a device to launch a group into thoughtful and organized discussion of a problem or to introduce facts or "expert" opinion. It can also be used as a questionnaire to record player opinion. It is simple to use and evaluate. (Publisher)

Note: For more on the *Trilogy of Games,* see the essay on urban simulation games by Jo Webb.

Cost: under $10.00 for *Trilogy of Games*

Producer: Sage Publications, Inc., 275 South Beverly Drive, Beverly Hills, CA 90212

NEXUS

R.H.R. Armstrong and Margaret Hobson, University of Birmingham, England

Playing Data
Copyright: 1970
Age Level: graduate school, professional
Number of Players: 1-125 in flexible number of teams
Playing Time: 3-30 hours in periods of 1½-2 hours
Preparation Time: 1 day
Supplementary Material: briefing material explaining approach and equipment
Special Equipment: display system, overhead projector and transparencies, video equipment (optional)

Description: An open simulation that can be applied to many specific contexts, *Nexus* is intended to develop appreciation of patterns of interaction and dependence between organizations in any given situation, time-scale consequences of decisions, and the range of alternatives available in a given situation. Players represent their own or other people's decision-making roles (individual or organizational). They begin with equal resources (finances, manpower, and so on) and work to make changes in the total situation through tradeoffs with respect to specific factors. Successful play evolves from the capacity of participants to appreciate complex display systems and to integrate this with administrative experience and skills.

As *Nexus* provides a framework into which concrete information concerning complex situations can be fitted relatively easily and as it can be used in working situations by individuals or groups, it is intended for repeated use so that examination of many alternative courses of action can be undertaken.

Nexus may also be employed as an accounting device by the operational umpiring group in a large-scale gaming-simulation, when its function is that of a "computer." Already developed variants include *Sin Minus* which is an enjoyable and amusing introduction to the more complex *Nexus* framework, *English Local Government,, Developing Countries, Goals for Dallas, Management By Objectives,* and *City of Liverpool, England.* (AC)

Cost: materials cost $80.00-$100.00 per game; construction costs are one to three person-months

Producer: Margaret Hobson, Water Industry Management Centre, University of Birmingham, P.O. Box 363, Birmingham B15 2TT, England

PLANNING EXERCISE

Paul A. Twelker

Playing Data
Copyright: 1971
Age Level: grade 7-adult
Number of Players: 16-36 in 3-7 teams
Playing Time: 3-5 hours; exercise may be broken into segments of about 1 hour each
Preparation Time: 15 minutes or less
Supplementary Material: additional curricular material optional with the instructor

Description: The group is divided into teams and presented with a problem. One team functions as evaluators. The other teams create solutions to the problem, while the evaluation team develops and shares its criteria for judging solutions. Teams present their solutions in turn, with the evaluation team providing feedback. Each team has an opportunity, during the rebuttal phase, to clarify any misunderstandings and summarizes the strengths and weaknesses of the solution. A superplan is created from all the strong points.

In addition to presenting this generic format, the manual includes five appendices presenting sample problem statements and background material. (AC)

User Report: Twelker begins his manual by defining a planning exercise as a technique which "expedites the examination of complex problems and the proposing of alternative solutions, especially where a number of divergent views, often coupled with strong emotional overtones, may be brought into the situation." It is "a mix of one part debate, two parts competition (or gaming), one part cooperation, and a measure of simulation. When mixed thoroughly with a precise statement of a goal to be strived for, either . . . an issue to be resolved or a situation to be improved, the effect . . . [is] an intense, maximally productive experience which . . . integrates cognitive learning of facts or principles with the application of ingenuity and skill in order to extend an individual's thinking and perception of a problem and its solution."

Skeptically, I moved beyond the opening definition to examine a typical planning exercise format which proved to be in sufficient detail for me to begin to see what it was all about. All phases of the exercises were carefully discussed with an eye to the kinds of concerns which crossed my mind. Specific instructor guidelines of the "do this now, and this next" variety were presented each step of the way. Finally, six sample exercises were presented. They convinced me and I selected *Bridging the Communications Gap* to try with my students. It worked as promised. I have now run a variety of planning exercises—other samples and some of my own design—with audiences ranging from seventh-grade social studies students to graduate students in high education.

If you desire group examination and solution of specific problems of concern, planning exercises are unexcelled. A *Manual For Conducting Planning Exercises* makes their design and use easy. It deserves wide reading. (Kent Layden, Assistant Professor of Educational Research, United States International University)

Note: For more on this exercise, see the essay on frame games by Sivasailam Thiagarajan and Harold Stolovitch.

Cost: $3.00

Producer: Simulation Systems, Box 46, Black Butte Ranch, OR 97759

POLICY NEGOTIATIONS

Frederick L. Goodman, University of Michigan

Playing Data
Copyright: 1974
Age Level: high school-adult
Number of Players: 10-25
Playing Time: 5 hours

Description: *Policy Negotiations* is a "frame" game which allows the user to adapt the game to consider the policy decision-making process in almost any subject area. Typical use includes: (1) playing a priming game in order (2) to learn its formal structure so that (3) players can build their own game relevant to a topic of their choice. Five priming games are included with this kit. One involves a policy negotiations situation between teachers and a school board. The school board and the teachers negotiate over some twenty-four possible issues ranging from a raise in pay to increased clerical staff to reduced class size. The four other priming games concern community issues, rapid transit, industrial park development, and a regional shopping center. Once players have played one of the priming games, they are ready to redesign the game around an issue of local importance. As the player-designer redesigns, he must be able to analyze the issue under consideration in order to identify key decision-makers, levels of influence and prestige, appropriate issues, and the interrelationship between issues and roles. Once completed, the players have built a totally new game which can be used with the designer group or a new group. (Producer)

Note: For more on this game, see the in-depth essay on *Policy Negotiations* by Barbara Steinwachs and the essay on frame games by Sivasailam Thiagarajan and Harold Stolovitch.

Cost: do-it-yourself kit $12.00

Producer: Institute of Higher Education Research and Services, Box 6293, University, AL 35486

PRESS CONFERENCE

Sivasailam Thiagarajan

Playing Data:
Age Level: adult
Number of Players: 10-30
Playing Time: 90 minutes-3 hours
Preparation Time: none

Description: Procedures for this game appear in full in the essay on frame games by Sivasailam Thiagarajan and Harold Stolovitch. (AC)

PROS AND CONS

Creative Learning Systems, Inc.

Playing Data
Copyright: 1976
Age Level: adult
Number of Players: 7-16
Playing Time: 1-2 hours
Preparation Time: none
Packaging: Box with 12 color-coded sets of flip cards specifying "pro," "con," or neutral stance to be taken with each of eight different issues; 12 copies of issue cards with two different issues; 4 copies of observer's guide; vote tally sheets (1 pad); 3-page coordinators manual.

Description: *Pro's and Con's* is a structured discussion exercise that exists so far in two versions—sex role options, and energy and the environment. Each player votes to agree, disagree, or remain neutral on a set of issues. The votes are recorded. Then the group arranges the issues to be discussed in order of priority. They spend ten to fifteen minutes on each one, taking "pro," "con," or neutral (facilitative) stances as specified on their flipcard packets. After discussion, players' votes of pro, con, or neutral stances are again recorded. The exercise ends with a debriefing session. (AC)

Note: for more on this game, see the essay on frame games by Sivasailam Thiagarajan and Harold Stolovitch.

Cost: each version $19.95

Producer: Creative Learning Systems, Inc., 936 C Street, San Diego, CA 92101

PX-190

Richard F. Tombaugh and Joseph L. Davis

Playing Data
Copyright: 1973
Age Level: grade 9-adult
Number of Players: 4-48
Playing Time: 6-8 hours in 4-6 rounds
Preparation Time: 1 hour
Packaging: 26-page photocopied instruction booklet

Description: The designers describe *PX-190* as "a systems model designed to show the interrelationship among discrete decisions about elements of that system. By making these decisions participants get a visual representation of their effect elsewhere and can map their progress towards stated goals and record the 'costs' of pursuing these goals." It "is a general decision-making model that can be adapted to many problems. Thus far the following models have been created: Family Financial Planning; New Directions in Rural Ministries; New Directions in Campus Ministries." For example, in Family Financial Planning, the participants make decisions about the family's finances based upon their life goals and the time available from the husband and the wife. The model then predicts the effects of those decisions over an entire lifetime. A "Life Event Card" represents unplanned events for the family, and newspaper headlines are read aloud each round to simulate external events which effect the family's finances. (DCD)

Cost: by individual contract

Producer: The Center for Simulation Studies, 736 De Mun Avenue, Clayton, MO 63105

SCIFI

Diane Dormant

Playing Data
Copyright: 1976
Age Level: adult
Number of Players: 3-30 (up to 8 individuals; after that 3-6 equal teams)
Playing Time: 30-90 minutes
Preparation Time: 10-20 minutes
Packaging: printed instructions

Description: *Sci Fi* is a problem-solving game for teams. Each receives an envelope on which a problem is written. The team writes a solution on an index card, puts the card in the envelope, and passes the envelope to the next team. The envelopes continue the circuit until they return to their original teams, who remove the solutions and rank them from best to worst. The winner is the team that suggested the most "best" solutions. (AC)

Note: For more on this game, see the essay on frame games by Sivasailam Thiagarajan and Harold Stolovitch.

Cost: 75¢

Producer: Instructional Alternatives, 4423 East Trailridge Road, Bloomington, IN 47401

SYSTEM 1

Instructional Simulations, Inc.

Playing Data
Copyright: 1969
Age Level: grade 4-adult
Number of Players: 2-8 per game
Playing Time: 20-40 minutes
Packaging: professionally packaged kit contains 4 two-sided game boards, 6 sheets of adhesive-backed paper for printing data used on plastic tiles, 160 plastic tiles (both sides usable), a timer, and an instructor's manual

Description: *System 1* is a game of classification and concept development. It is played on a grid, the two axes of which represent two dimensions in any chosen subject area. Information on the subject appears on plastic tiles that can fit into this grid. Players try to place their tiles in appropriate places on the board. *System 1* can be used in mathematics, social science, history, language arts, science religious education, and so on. (AC)

Note: for more on this game, see the essay on frame games by Sivasailam Thiagarajan and Harold Stolovitch.

Cost: $9.95

Producer: Griggs Educational Service, 1731 Barcelona Street, Livermore, CA 94550

TEAMS-GAMES-TOURNAMENT

David L. DeVries and Keith J. Edwards

Playing Data
Copyright: 1973
Number of Players: 12-60

Description: *Teams-Games-Tournament* has the purpose of creating a classroom environment in which all students are actively involved in the teaching-learning process and consistently receive positive reinforcement for successful performance. It alters the social organization of the classroom in that it creates interdependency among students and makes it possible for all students, despite differential learning rates, to have an equal chance to succeed at academic tasks. Students are assigned to four-member teams. Within each team there are students from all achievement levels. The teams remain intact throughout a semester. Each team has two practice sessions per week in which teammates can assist each other to learn particular tasks. Any commercially available games may be used. A list of suggested ones is given in the back of one of the publications. Teachers or students can also design their own games over a period of time (a semester seems to be a good chunk). The teams play the game at least once a week. However, here is a new twist: the teams do not compete as teams but rather each team member is assigned to a tournament table where he competes against two other students, each representing a different team. So that at any given tournament table the three students are roughly comparable in achievement level. The game scores are converted into points with a fixed number of points assigned to top scores, middle scores and low scores at various tables. Players can be bumped up and down the hierarchy and their points are summed with points of other members of the teams to compute a team score. (REH)

Note: For more on this game, see the essay on frame games by Sivasailam Thiagarajan and Harold Stolovitch.

Cost: manual, *Using Teams-Games-Tournament in the Classroom,* $3.00

Producer: Center for Social Organization of Schools, Johns Hopkins University, 3505 N. Charles Street, Baltimore, MD 21218

THEY SHOOT MARBLES, DON'T THEY?

Frederick L. Goodman, The University of Michigan

Playing Data
Age Level: junior high-adult
Number of Players: 15-30
Playing Time: 2-5 hours

Description: Originally designed for a police department, as a police-community relations game, "*Marbles*" has evolved into a very free unstructured learning tool with no single lesson in mind. There is no one particular way to play the game; rather the idea is to experiment with many different variations and interpretations. The overall goal is to encourage experiments in rule governed behavior, governmental structures, law enforcement policies and problems of wealth distribution during a process of developing an urban community.

The game begins with minimum structure and two rules, on the basis of which the players define and redefine the rules and goals of the game. The beginning scenario includes five players (A,B,C,D, and E) sitting around a board who represent private citizens involved in their daily routine of social interaction and survival. Marbles are the medium of exchange and may represent money, power, mobility, status, skill, employment, whatever the players perceive as important to survival.

Each round consists of a bargaining session and a shooting session. In the bargaining session the five players bargain for a pot of marbles (the number of marbles being determined by the roll of five dice) with the relative bargaining power of each player determined by the number of marbles he holds in his possession from previous rounds. Players have three minutes in which to strike a bargain between two or more players for the pot and to build a tower out of wooden dowels which symbolizes that bargain. Once established the tower must stand in the middle of the board through the shooting round which follows. In the shooting round all players have the option of toppling the bargain or shooting at job marbles which are scattered around the board and which, if hit, return an income. If the tower is toppled the bargain is nullified and no one receives the pot.

The shooting round is governed by the only two rules, or laws, which exist at the beginning of the game. One rule designates the position from which each player may shoot and the other restricts players from hitting the trouble marbles which are also placed on the board. To enforce these rules there is a police force. In addition to the citizens and the police there is a government which is chosen at the outset of the game. Once chosen, the government is solely responsible for making all laws of the game, free to make any laws they wish, to establish any system of law enforcement and justice they choose. The police force is responsible to the government.

Once the first round has been played and the players have assumed their roles the direction of the game is entirely up to the players. The only function of the game director thereafter is to keep the game moving, to insure that the sequence of play is correctly performed.

What has just been described is only the skeleton of the game. Many variations and experiments can and have been tried, including those involving the teaching of social studies as well as math and English. Within social studies there are countless experiments involving different forms of government, agricultural versus industrial societies, different economic structures, competition versus cooperation, and any other institutional framework which one wants to explore. The game was designed to be redesigned by the user. (Author)

Note: For more on this game, see the essays on frame games by Sivasailam Thiagarajan and Harold Stolovitch and on simulations of developing a society by Frank P. Diulus.

Cost: complete game kit $40.00

Producer: Institute of Higher Education Research and Services, Box 6293, University, AL 35486

See also

Decisionmakers COMMUNITY ISSUES
Encapsulation COMMUNICATION
Kathal SELF-DEVELOPMENT
Making a Change COMMUNITY ISSUES

FUTURES

COPE

Jerry K. Ward, Wilde Lake High School, Columbia, Maryland

Playing Data
Copyright: 1972
Age Level: grades 7-12
Number of Players: 20-40
Playing Time: 20-30 hours in 1-hour periods
Preparation Time: 3-5 hours
Special Equipment: tape recorder
Supplementary Material: anything in science fiction, or nonfiction on change, future shock, or the future could be helpful additional reading

Description: This simulation of change and the future focuses attention on questions such as: What are "good" and "bad" futures? Can we "cause" the kind of future we prefer? Can we adapt quickly enough to accelerating change? First, students read and discuss an article called "Coping With Change." The classroom then becomes a think-tank called Technopolis, in which students live through five future time periods, 2000 to 2040 A.D. TIME PERIOD 1: The city is a leisurely intellectual community whose citizens research what the future will be like. TIME PERIOD 2: A complex computer called COMCON helps with information and material problems and asks citizens to provide human input for problems that must be solved. TIME PERIOD 3: COMCON, having assimilated all international and intergalactic computer systems, directs all human activity and requires citizens to learn computer forms and to drastically increase their efficiency and productivity. TIME PERIOD 4: Citizens are forced to learn a new language called FUTURESPEAK and must compete with COMCON to create ever newer and more sophisticated technology. TIME PERIOD 5: COMCON has grown impatient with human inefficiency and tells citizens that human beings are apparently obsolete. COMCON asks citizens to choose between a life of uncaring bliss (COMCON would shoulder all responsibility and work) and a life of constant struggle (COMCON would continue constant change and demand increasing output). After this DECISION POINT the simulation ends, and students analyze their ability to adapt to the differing roles they filled during the five time periods. Debriefing centers on the implications of students' findings about change and the future. (Producer)

Comment: *Cope* involves cooperation as players work in teams; competition as teams vie for Creative Work Units (CWUs), and conflict as players confront their own internal stress in order to deal with the increasing demands of a simulated computer called COMCON.

The quantitative outcome is measured in CWUs earned for each team. The student guide indicates, "Your CWUs will increase dramatically if you follow directions and complete the tasks assigned to you," which sounds more like training for an industrial society rather than a postindustrial society; however, additional CWUs can be earned for "devising ways to cope with the many and rapid changes built into the simulation."

The author lists "confidence that man can fully use his vast technology while not losing control of it" as a qualitative outcome. Yet the game ends when the students decide to let COMCON do everything for them or to continue playing until they are unable to "cope." Considerable debriefing (which is built into the simulation) may be necessary to restore this confidence. Some futurists believe that creativity will no longer be necessary once technology works for man and the separation of humanists and technologists in Stage I may be the first step toward this type of future. Whether *Cope* simulates a dystopian or a utopian future may be one of the most important qualitative outcomes.

Cope is a self-contained teaching unit lasting for as long as four weeks. It includes daily lesson plans, pre- and posttests, reading lists, and good advice for teachers concerning room arrangement and hints for successful play of the simulation. This format attracts teachers unaccustomed to simulation but has caused some futuristics educators who pride themselves on being flexible to ignore this worthwhile simulation. In spite of its structured format, *Cope* can be adapted to any classroom situation. A special section assists English teachers in using it in science fiction classes; and it is particularly valuable to teachers who have added urban study units or who wish to expand a traditional unit on work.

Cope provides an excellent medium for students to experience future shock. The world view presented in this simulation is best expressed as "perhaps the ideal solution is not to stop change, but to cope with it." HUMELMCNB EFFIC!RPT:HUMELMCNB–EFFIC! (BBF)

Note: For more on this simulation, see the essay on future-oriented simulations, by Charles M. Plummer.

Cost: $14.00

Producer: Interact, P.O. Box 262, Lakeside, CA 92040

DIALOGUES ON WHAT COULD BE

Joe Falk, The Future Associates

Playing Data
Copyright: 1973
Age Level: grade 9-adult
Number of Players: 2 to unlimited small groups
Playing Time: 10 minutes to 2 hours

Description: Players of this game learn to orient their thinking toward the future, to communicate evocatively and to cooperate, and increase their self-confidence through association with others.

There are six dialogues: (1) living together in cooperative communities; (2) communicating evocatively as well as educatively; (3) working together within the system; (4) learning from each other; (5) planning our society together; and (6) changing human nature and attitudes. Each dialogue proceeds as follows: A player reads an evocative question card ("Are you more interested in yourself or others and why do you feel that way?") and then tries to involve the rest of the group in a dialogue. Another player reads another card, and the process continues to repeat itself. Players briefly summarize the responses to the cards they read. (DCD)

Comment: *Dialogues* is an ideal game for exposing people to life processes. In a family setting, the children experience participation in very important aspects of life. As the questions are worded to elicit the personal opinions of the participants, the child gets the experience of revealing his thoughts. Since the evoker controls the discussion, the danger of rebuke is avoided or at least lessened. Since the role of evoker moves from one group member to the next, each person also gets the opportunity to experience the role of supervisor or chairman and gains a feel for the responsibility the role entails.

Dialogues is especially appropriate in teaching courses dealing with politics and related disciplines. In addition to highlighting the importance of and the effects of an individual's personal feelings about a particular subject, the game allows the participant to experience confrontation in a controlled environment. Students, who ordinarily would not get the opportunity to lead or experience the anxieties or responsibilities of the role, could learn to control, manipulate, or develop a facility for handling discussion groups, committees, or other types of gatherings. (DL and BH)

Cost: $1.00 per book, quantity discounts

Producer: The Future Associates, P.O. Box 912, Shawnee Mission, KS 66201

DYNAMIC MODELING OF ALTERNATIVE FUTURES

Charles M. Plummer, John R. Rader, Harold D. Stolovitch, Sivasailam Thiagarajan, Robert A. Zuckerman

Playing Data
Copyright: 1976
Age Level: high school, college, professional

Description: This is a collection of five future-oriented simulations. *Focus on the Future*, by Harold Stolovitch, is a frame game that follows a modified delphi procedure for assessing needs and encourages participants to share their views of the possibilities of various events. *Alternative Futures Analysis and Review (AFAR)*, by Sivasailam Thiagarajan, is an operational game for generating participant insight into and awareness of their importance to others concerning multiple alternative desirable futures. John R. Rader's *Simulation on the Identification of the Gifted and Talented* is a model that merges an in-basket exercise with a futuristic scenario (enhancing creativity and genius in the year 2000). *The Label Game*, by Robert A. Zuckerman, based on a model of long-term social changes, shows how self-fulfilling educational prophecies develop and decline. *Toward Walden Two*, by Charles M. Plummer, simulates a social system based on operant conditioning to demonstrate a technique for accelerating transition from possible to probable. (AC)

Note: For more on this collection, see the essay on future-oriented simulations, by Charles M. Plummer.

Cost: VIEWPOINTS, Bulletin of the School of Education, Vol. 52, No. 2, March 1976 $2.00

Producer: Publications Office, Room 109, School of Education, Indiana University, Bloomington, IN 47401

EUROPEAN ENVIRONMENT: 1975-2000

Michael Bassey, Conservation Trust

Playing Data
Copyright: 1972
Age Level: grade 9-graduate school, management, conservation groups
Number of Players: 1 to 10 teams of 4 to 12
Playing Time: 100 minutes or more
Preparation Time: about 20 minutes to sort papers and organize the room
Special Equipment: one tape recorder; one random decision maker (can be pack of cards or a bottle with 3 white balls and 1 black ball)

Description: This in-basket simulation is designed to enhance its players' understanding of foreseeable environmental problems that will trouble Western Europe during the last quarter of this century. It stresses the urgent need to solve those problems now. Each player represents a member of the "Advisory Council of the Commission for the Environment of Western Europe." The Commission must respond to 25 reports that detail the year-by-year population growth, energy consumption, and use of land and other resources in Europe from 1975 to 2000. Each yearly report recommends the adoption of one or more proposals (such as establishing a "Water Resources Authority," enlarging the Strait of Gibraltar–as the Mediterranean becomes a great, stinking cesspool around 1995, or committing funds for gerontological research). The body of each report contains supporting arguments for the adoption of each proposal, and each player must simply either support or oppose the measure described. (DCD)

User Report: I have now run the simulation 12 times, involving 37 teams and a total of approximately 300 participants. Most of these have been managers–junior, middle, and senior. . . . In every case there has been a high degree of participation and the simulation experience has led to a stimulating discussion touching on a wide range of issues.

The news broadcasts have been recorded on tape, together with the announcements of each year, and this paces the simulation.

The "control" file for recording the responses from each Advisory Council is a standard layout. The work materials are photocopied from the printed book (we purchased the rights to do this from The Conservation Trust), so that each report can be handed out at the appropriate date. Careful preparation of these materials avoids confusion during the simulation when time is short.

The speed of events is such that a fairly brief discussion has to be followed by a quick decision, and emphasis is placed on content of the reports rather than on the processes of decision making. The fact that a decision has to be made means that there is high concentration on appreciating information and the issues. The four-minute year seems the optimum period.

Running the simulation is not difficult but does require general understanding of the issues contained in the reports. Occasionally Advisory Councils will ask questions or make suggestions which need a ready and plausible response if the credibility of the simulation is to be maintained. Some of the things raised by Advisory Councils include dispatching radioactive wastes by rocket to the sun; more information on the research studies and on the ecological effects, for example, of building the Bering Straits Dam. . . .

Up to 4 teams of 8 people have participated at one time, and this is all right provided that the "controller" works with an assistant, preferably someone who can type (Colin Hutchinson, Training and Development Advisor, British American Tobacco Co.)

Comment: The author states that the projections are "surprise free" (no eco-catastrophies have been built into the simulation), which is not surprising since one of the recommended sources is *The Year 2000* by Kahn and Wiener. The simulation's "either/or" format suggests that the author was more influenced by the negative extrapolations of the MIT team of Forrester and Meadows than the positive extrapolations of Kahn and Wiener–particularly since the quantitative outcome–The Erlich Award–is based on the number of "correct" eco-decisions.

Positive extrapolists would question the limited role played by the market economy, since technology does not come to the rescue in solving environmental problems in this simulation.

The chief qualitative outcome is an increased awareness of the kinds of environmental dilemmas confronting humans today and the urgent need to think in terms of the long-range consequences of the decisions made today for tomorrow. To ensure this, when players receive statistical reports, they find that their third decision influenced the outcome and they now must learn to live with a future of their own creation.

Rules are flexible enough so that participants can challenge and even reject the simple solutions. They soon realize that time is a problem and decisions must be made on the basis of inadequate and outdated information, but "That's life," writes the author.

Futuristics educators who accept the concept of Spaceship Earth and who have a pelagian view of human nature will appreciate this simulation as humans are viewed as "intelligent animals able or not able to solve problems of their own making." This simulation is based on the view that if the "decision makers in our society (politicians, senior management) have a better understanding of the social and environmental problems facing the world, they can through the democratic processes safeguard the quality of life for all people on earth." (BBF)

Cost: 40 pence U.K. plus postage for single copies. Permission to reproduce copies can be obtained from The Conservation Trust

Producer: Conservation Trust, 19, Criftin Road, Burton Joyce, Nottingham, England

FUTURE DECISIONS: THE IQ GAME

Betty Barclay Franks

Playing Data
Copyright: 1975 second edition
Age Level: grade 9-adult
Number of Players: 10-400
Playing Time: 3 hours (est.)
Preparation Time: 30 minutes
Special Equipment: cassette tape player
Packaging: 39-page booklet and cassette tape recording

Description: This is a decision-making game with a science fiction premise. The players assume that they live in the United States at a time when advances in obstetrics have made it possible for physicians to accurately predict both the sex and intelligence quotients of unborn children. Now, the scenario goes, a chemical has been discovered which can be given to pregnant women to guarantee that their children will be born with genius IQs. Unfortunately, the cost of administering this new chemical is very high and existing supplies are so meager that the public demand outstrips its availability. The burden of deciding who may get the new "IQ Drug" (as it is called in the game) falls on the four hospitals authorized by the Food and Drug Administration to use it. One hospital's response is to form a committee.

In the first phase of this game players represent members of a committee that must decide which of fifteen women who have requested treatment with the chemical will receive it. Players meet in groups of six to eight. Each committee receives materials that describe the game scenario and the age, race, general physical condition, and occupation of fifteen sets of prospective parents and the sex and IQ of each unborn child. Players who decline committee membership become observers. The players who choose to serve on a committee elect a chairperson and then go about the business of disqualifying some of the couples who have requested the chemical treatment and rank ordering those they do not disqualify. When this task is finished the committee members and observers discuss why they did or did not wish to serve on the committee and the validity of the criteria each committee used to make its decisions.

In phase two, participants form triads and listen to a tape recording of news items. After each one, players discuss the implications of what they have heard. For example, the first news item on the tape announces that a child is suing her parents because they failed to obtain treatment with the "IQ Drug" before she was born even though they were wealthy and could easily afford it. Consequently the plaintiff was doomed, she claims, to go through life with an IQ of only ninety. The players are then asked to invent as many headlines as they can "think of on the consequences of the IQ treatment for society." The players hear and respond to four more such announcements. (DCD)

Comment: Through the IQ Game participants interact to decide how an IQ drug will be administered. Even though their first decision is intentionally couched in an either/or format, all players do not have to be involved in the same way. Some will choose to serve on the board which selects the applicants whose unborn children will receive the drug while others will opt to observe the board in action. Those who protest during debriefing that they had no other choice but to serve or observe learn that alternatives are always available to those who feel strongly enough to seek them, they could have left the room to explore other alternatives and to develop a plan to be presented to the group later in the simulation.

Flexible rules permit board members to ignore the demands for rapid thinking by the hospital administrator (the game director), but this rarely happens in our authority-conditioned society. The chairperson of the board generally insists that members reach their decisions in haste often in spite of their protests.

One of the qualitative outcomes is that players realize that reality is often perceived through their own assumptions. For example, board members frequently indicate that the applicants described in the left-hand column are male since they often have the more prestigious occupation. This poses a problem when board members discover that two women have chosen to have a child through artificial insemination. Conventional couples are included in addition to a marriage corporation and a single male adopting a child of a different race which forces board members to grapple with the question of whether the future will be radically different or whether they should not make a conscious effort to preserve certain aspects of our present society.

Values clarification occurs during debriefing as the observers tell what they were doing and interact with the board members. Participants bring much of themselves to this game—corporation employees viewing the children as products; nurses seeking to gain new medical knowledge; teachers questioning the value of IQ in creating happy individuals, and low-ability students fearing that they would be unable to control children with high IQs.

To this point, this simulation is a decision-making exercise which involves a good deal of valuing; however, the tape recorded debriefing confronts the participants with the long-range consequences of the decisions they have been making. To prevent this simulation from becoming just an exploration of the unknown, current research on IQ improvement is presented and the participants are left with the question: Is the future already here?

The chief criticism of this simulation is that the author assumes that futures can be prevented, altered, and/or invented and that humans can indeed create their own futures. (BBF)

Cost: $5.95 for a "do-it-yourself" booklet containing bibliography and suggested activities; $6.95 with the cassette tape

Producer: SAGA Publications, 4833 Greentree Rd., Lebanon, OH 45036, Attention: Michael Raymond

THE FUTURE GAME

John Wexo

Playing Data
Copyright: 1973
Age Level: continuing education
Number of Players: 1 or 2

Description: *The Future Game* is a board game designed to foster an awareness of the need to manage resources with a future-oriented perspective. To that end, players have the task of determining the best balance for society in the near and far future. Each player begins with 100 ponts which must be invested in a way that minimizes penalty points and maximizes opportunities for earning resource points. The board has three "Horizon areas," each of which represents a stage in the future development of society. While moving around Horizon I, the player finds that resources can be exploited without heavy penalty, opportunities are plentiful, and the future can be ignored. As the player moves into Horizons II and III, the penalties become more frequent and severe, and the number of opportunities diminishes substantially. However, a player does not have to move into a new horizon until she or he has accumulated enough resource points to assure future balance. (BBF)

Comment: The chief qualitative outcome of this game is that players realize the necessity of future planning. However, if one player loses all points by mismanaging resources, the other (if two are playing) is declared the winner. This may not be the case on Spaceship Earth!

This game, combined with the readings in the unit of which it is part, is an excellent way to introduce individuals to the need for futures planning. (BBF)

Cost: complete kit (including game, book of readings, and study guide for program "America and the Future of Man") $10.00; game not available separately

Producer: Courses by Newspaper, University of California, San Diego, X-002, La Jolla, CA 92093

FUTURE PLANNING GAMES

David L. Bender, Gary E. McKuen and Dewey Hinderman

Playing Data
Copyright: 1972, 1973, 1974, 1975, 1976
Age Level: high school
Number of Players: groups of 4-6
Playing Time: approximately 1/2 hour per exercise
Packaging: ten oversize sheets

Description: The Future Planning Games kit contains thirteen separate game brochures, each with brief descriptions of four or five exercises. The titles are: *Constructing a Life Philosophy, Constructing a Political Philosophy, Dealing with Death, Determining America's Role in the World, Determining Economic Values, Determining Family and Sexual Roles, Examining American Values, Facing the Ecology Crisis, Planning American Policy in Developing Nations, Planning Tomorrow's Prisons, Planning Tomorrow's Society, Preventing Crime and Violence,* and *Protecting Minority Rights.* In most of the exercises, students in small groups examine information and attempt to reach a decision. Four of the games will be described here.

In *Constructing a Life Philosophy,* students in small groups examine six lifestyles, determine the viewpoints persons with each lifestyle would be likely to have on major social issues, and decide upon the one they believe best. Then after examining the lifestyles of several famous people, they attempt as a group to develop a lifestyle they all can accept. A third exercise involves making moral decisions about cloning, and a fourth involves students as presidential commission members recommending legislation of moral issues of the future. In the last exercise, groups attempt to reach a consensus on their viewpoints about a particular future city.

Determining Economic Values includes four exercises. First, students in groups examine and determine the best of three possible retirement programs. Second, as political candidates, they campaign on the issue of bribery in foreign countries. Third, they read hypothetical data about twenty-first-century Detroit and as a group distinguish between fact and opinion. Finally, they discuss competitive and non-competitive economic systems which might exist in the future, and determine which they prefer.

Facing the Ecology Crisis involves similar exercises. Students in groups decide the best method for limiting China's population. They determine the priorities for the use of $200 million in the near future from 21 funding requests. They examine viewpoints on population growth, decide which of a list of citizens should survive a future ecological crisis, and differentiate fact from opinion in descriptive statements given for two possible future cities.

A fourth example of the Future Planning Games is *Protecting Minority Rights.* The five exercises in this game include discussing solutions for future school integration problems, distinguishing truth from falsehood in statements about school integration, deciding and supporting a recommendation for a governor to take on an equal rights amendment, and determining from a list of fifteen the five priorities for social programs in the future. (TM)

Note: For more on these games, see the essay on future-oriented simulations, by Charles M. Plummer.

Cost: 13 games at $.95 each

Producer: Greenhaven Press, 577 Shoreview Park Road, St. Paul, MN 55112

FUTURIBLES

George E. Koehler, United Methodist Church, Board of Discipleship

Playing Data
Copyright: 1973
Age Level: grade 9-graduate school, continuing education, management, church planners
Prerequisite Skills: average reading comprehension
Number of Players: minimum 1-maximum hundreds, in groups of 6 to 8
Playing Time: 15 minutes to 4 hours
Preparation Time: 30 minutes

Description: *Futuribles* consists of a deck of 288 cards, each citing a projection of some future probability in such areas as communication, energy, food, human experience, learning, natural resources, population, religion, transportation, etc. The cards may be used in innumerable ways, including games and simulations invented by the user group. The leader's guide suggests that players on successive rounds pick from a hand of seven cards one representing the most likely to occur, the least likely to occur, the one toward which they feel most negative, and most positive, the one most consistent with their values, and inconsistent, etc. Also suggested are rounds whereby players create cause-and-effect chains of cards, describe antecedents or consequences of cards, write scenarios for given cards, combine cards so as to create new inventions, and the like. (Author)

Comment: *Futuribles* is an informal gaming situation for people who want to learn about the future, for people who want to plan for the future, and for people who want to focus on particular futures. No special expertise is required in playing the game. Players take with them into the game only their background and experiences. Since no special expertise is required, players may find themselves defending their theories, biases, and unsupported but preconceived notions about many areas of life. The game provides an interesting way of gaining insight into one's own ideas about the future and into the need for interplay and discussion in developing policies for the future (DL and BH)

Note: For more on this game, see the essay on future-oriented simulations, by Charles M. Plummer.

Cost: $9.45

Producer: Book Service, World Future Society, 4916 St. Elmo Avenue, Washington, DC 20014

GLOBAL FUTURES GAME

Bill Bruck

Playing Data
Copyright: 1974 (revised 1975)
Age Level: high school, college, continuing education
Number of Players: 8 minimum; 48 maximum (divided into 8 teams)
Playing Time: 2 hours minimum; 3 hours maximum (less if broken up into 2 or 3 periods)
Preparation Time: 30 to 60 minutes
Packaging: kit containing one set of 8 scoresheets, rules, facilitator manual

Description: The *Global Futures Game* involves participants in a cooperative effort to ensure that Spaceship Earth does in fact have a future.

Players engage in limited role play as they represent eight regions: South America, Japan, Africa, China, North America, USSR, Western Europe, India. Teams are formed in larger groups.

Players begin the game with unequal resources with regard to population, food, and technology which are bartered in ten-minute trading rounds simulating five years of real time. Players must try to predict and control their future development in each of the three areas. If they decide to invest in education, the fourth variable, the effect is postponed for two rounds.

The quantitative outcome is measured in terms of world destruct points. All teams begin with five world destruct points. If the players compete without forethought, they soon find that this strategy brings them closer to the fifteen points which will end the game and simulate destruction of the world. (BBF)

Comment: The *Global Futures Game* is patterned on the *World Game*, a computer-based simulation developed by R. Buckminster Fuller. Its chief weakness is its inability to present an accurate model of the world with only four variables, but its chief strength is the qualitative outcome of permitting the players to view the world from new perspectives, the first being the dymaxion map which appears on the cover of the simulation. Participants become familiar with the concept of Spaceship Earth and realize (in Fuller style) that education has cumulative effects—you can even learn from your mistakes! Its holistic approach to problem-solving, especially the use of math, drives home the point that we need input from all disciplines when thinking about the future.

Political scientists and regional experts have difficulty playing this simulation, as its assumption is that common sense, not our preconceived political notion about nationalism and power politics, will save Spaceship Earth from oblivion. It often upsets the "experts" when the generalists fail at role playing (which is not mandated in this simulation), and succeed in acting on behalf of humanity.

The 1975 revision corrected the main problems in this simulation. The rules have been rewritten and the method of keeping score has been revised. Also included in the new version are questions and ideas which will be extremely useful to the facilitator during debriefing. (BBF)

Cost: complete kit $17.75 postpaid

Producer: Earthrise, P.O. Box 120, Annex Station, Providence, RI 02901

HUMAN SURVIVAL—2025

David J. Boin and Robert Sillman

Playing Data
Age Level: grades 7-12
Number of Players: class size
Playing Time: 3-4 class periods

Description: Earth has been destroyed and the members of the class have escaped on a spacecraft. The passengers must organize and set up their society. Students are divided into six committees dealing with: rules, leadership, human rights, the economy, enforcing the rules, and what to do with rule breakers. They are also provided a form which lists the previous occupations of the passengers and a current supply list of equipment and materials aboard the spaceship. According to the designers, the instructional objectives of the exercise are "to have students appreciate some of the difficulties our forefathers faced when settling the land, to show the need for cooperation in survival, to have students understand that democracy is weakened when people are constantly concerned with day-to-day survival, and that priorities change when the need for survival increases, to provide an atmosphere for purposeful debate in which students can learn the give and take of compromise, and to provide students with some understanding of the dynamics of group pressure and how it might influence voting behavior." (REH)

Cost: $3.00

Producer: Edu-Game, P.O. Box 1144, Sun Valley, CA 91352

THE HYBRID-DELPHI GAME

Ron Saroff

Playing Data
Copyright: 1974 (revised 1976)
Age Level: high school, college, continuing education
Number of Players: 6 to 36 (optimum number 15-21)
Playing Time: 2-1/2 to 3 hours
Preparation Time: 1/2 hour; longer if modifying the instrument to a particular topic
Components: instructions, Hybrid-Delphi Instrument
Supplementary Material: newsprint, marking pens or chalkboard

Description: *The Hybrid-Delphi Game* is an *interactive* simulation involving participants in a *cooperative* effort to identify the 15 most desirable futures from a list of 90 for the year 1996.

During round one, each participant indicates those futures which have a 10% desirability, a 50% desirability, and a 90% desirability. The facilitator then records the responses of all participants on newsprint or a chalkboard for immediate feedback.

During round two, groups of three use consensus as the method for arriving at the 15 most desirable futures. The entire group must identify the 15 most desirable futures during round three. No instructions are given as to how this should be accomplished as the hidden agenda of this simulation is group process, and this plays an important part in debriefing.

This is a "hybrid" delphi because it does not involve experts; it brings about direct confrontation of values and priorities; and the participants are known to each other. Since the instructions do not contain complete information on recording round one, those unfamiliar with delphi studies should read the chapter on group opinion in *Teaching the Future* by Draper Kauffman, Jr. (Palm Springs, Calif.: ETC Publications, 1976).

Round one generally requires 30 minutes; round two 1 hour; and round three up to 1-1/2 hours.

This simulation can be used to help groups establish goals and priorities and can be made a part of a planning and goal-setting workshop. The qualitative outcome is that participants have an opportunity to examine divergent values and opinions. Through *The Hybrid-Delphi Game* participants learn that alternative futures do exist and that conflicting values and opinions must be reconciled in order to prevent, invent, or alter futures.

Note: For more on this simulation, see the essay on future-oriented simulations, by Charles M. Plummer.

Cost: $5.50 plus postage
Producer: Ron Saroff, 6246 S. W. 37 Avenue, Portland, OR 97221

SIMULATING THE VALUES OF THE FUTURE

Olaf Helmer

Playing Data
Copyright: 1971
Age Level: adult
Number of Players: 38 in 10 groups
Playing Time: 5 hours
Preparation Time: 2 hours (est.)
Packaging: this exercise is found on pages 193-214 of *Values and the Future,* edited by Kurt Baier and Nicholas Rescher

Description: Participants must define and then evaluate two alternative futures. This game is organized into four phases during which groups of players make decisions that, if implemented, would affect the character of our environment, and then estimate the societal consequences of those decisions and evaluate their desirability.

In the first phase, two planning groups of five participants each allocate points (which represent a concerted effort by the federal government) according to their judgment of the desirability of twenty potential developments such as fertility control, an increased life span, or the availability of personality control drugs. One group is instructed to support developments that would "raise the gross national product to the highest expected level" while the other aims "to create a world in which values prevail, that, if anything, are better than those of today; in particular, individual liberties should be curtailed as little as possible." As a result of this activity, each group defines three to six developments that will, for the remainder of the simulation, be assumed as actualities by the year 2000.

In the second phase, two groups (corresponding to the planning groups) predict the social consequences of these developments and then assign a percentage probability for each development that its potential consequences will actually occur.

In the third phase, six teams of evaluators (three participants each, and representing teenagers, housewives, the employed middle class, persons over 65, the cultural elite, and the poor) must evaluate the two futures predicted by the planning and social consequences groups in terms of what each future would be like for them.

In the final phase, each participant judges the importance of the opinion of each evaluating group and all players make final determination about which of the two future worlds is better. (DCD)

Comment: This is an interesting game with two major defects: The directions are poorly written and the content is outdated. For example, since 1966, when the game was first played, it has become possible to doubt that a class of persons called "housewives" will even exist in the year 2000. Other examples are to be found in the list of possible future developments, such as the development of personality control drugs. It is now common knowledge that the U.S. Central Intelligence Agency had tested both LSD-25 and psilocybin during the 1960s to develop just such a drug, and that various exotic mood and perception altering substances have been created for psychiatric and research purposes since then. (DCD)

Note: For more on this game, see the essay on future-oriented simulations, by Charles M. Plummer.

Cost: $4.95 in paperback

Producer: Free Press, 866 Third Avenue, New York, NY 10022

SPACE COLONY

Janice Johnson

Playing Data
Copyright: 1977
Age Level: age 8 and up
Number of Players: 2-6
Playing Time: 90 minutes (est.)
Preparation Time: 30 minutes (est.)
Packaging: professionally packaged board game with playing board; colony, experiment, accident, and expedition cards; oxygen, food, and water coupons; module markers, playing tokens, dice, and 25-page instruction booklet including glossary and bibliography

Description: This board game is designed to introduce players, in general terms, to the exigencies of, and the tasks that would be performed by, a colony on the outermost terrestrial planet. The playing surface consists of a U.S. Geological Survey Office topographic map of Mars surrounded by a track of squares, and participants play on both sections of the playing surface simultaneously. Each player assumes the role of station chief of a Martian colony and places one marker on the first square of the track (called "Supply Day") and an identical marker in one of the grid squares of the map to indicate the location of his or her colony. All players must ensure the survival of their colonies and try to make their Mars bases self-sustaining.

The game events which occur as a result of moving markers around the track are intended to simulate the colonies' normal consumption of food, water, and oxygen, as well as accidents and good luck. The markers are moved according to the roll of dice, and one cycle of the track represents the passage of 90 Earth days–the interval of regular resupply from Earth. Each player starts with 150 units of food and equal amounts of water and oxygen. Each time a player's token advances past supply day the player receives an additional 300 units. These units may be used to acquire food, water, air, or components of a self-sustaining colony such as a greenhouse, water recycler, or solar generator. To be self-sustaining a colony needs at least one of all of the above plus living and recreation modules, a library, and a medical center. Any time a player's supply of food, water, *or* oxygen is exhausted, the player must leave the game and that colony is assumed to be abandoned. Any remaining supplies or colony component modules are left intact at the colony site.

Abandoned supplies may be retrieved by surviving colonies if they choose to mount an expedition. Expeditions, which are carried out on the grid square map, may be attempted for other reasons (to retrieve needed components from a Mariner craft, for instance, or to look for frozen water at Alba Patera or primitive life at the bottom of the Valles Mareneris). Any expedition requires expending reserves of food, water, and air at a standard rate for each grid square of distance traveled from the colony, in addition to the depletion of supplies as a result of activity at the base. Chance determines the degree of each expedition's success. A successful expedition may result in the salvage or acquisition of 200 units or more of needed supplies. A less successful expedition may result in the acquisition of barely enough supplies to cover the cost of the trek, or, in some cases, no rewards at all. Care must be taken at all times to ensure that no one supply is acquired at the foolish expense of others, as food, water, and oxygen are all needed if the colony is to survive.

The winner is the first player to assemble a self-sustaining Martian colony. (DCD)

Comment: *Space Colony* is an impressive game. For one thing, it seems to be an absolutely painless way to learn a great deal about life support systems, about terrestrial planets in general, and about Mars in particular. For another, it is fun to play. It is a rarity among games, being both meticulously accurate in regard to content and at the same time romantic enough in its conception to capture anyone's imagination. *Space Colony* is a do-it-yourself science fiction novel.

Cost: $13.95

Producer: Teaching Aids Company, 925 South 300 West, Salt Lake City, UT 84101

SPACE PATROL

Michael Scott Kurtick and Rockland Russo

Playing Data
Copyright: 1977
Age Level: age 12-adult
Number of Players: 2 or more
Playing Time: 4 hours (est.) per scenario
Preparation Time: 2 hours (est.)
Packaging: professionally produced 25-page booklet

Description: This science fiction/fantasy game may be played with any one of the following five scenarios. In *Interstellar Police,* according to the authors, "Crime runs rampant in the spaceways. The police party may be searching a star ship, city, or building for contraband. They may be trying to trap the Stainless Steel Rat or merely rousting the natives." *Landing Party* is described as having "a sort of Star Trek format. Here the players are making contact with a new world." The drama of *Space Salvage* is played out on an apparent space derelict. "The mission may be to salvage the ship, for use in the gallant rebellion, or merely the curiosity of alien architecture and drives." The players of *Soldier* can pretend to "be one of Robert Heinlein's *Starship Troopers*" or "Joe Haldeman's *Forever War* troopers." In *Hero At Large* the "hero may be a character who has found himself in several of the above situations and survived." The players use these scenarios and a complex outline to create the characters, society, and physical laws of the reality of the game. Once the antagonists and protagonists are defined, the game is played in "basic game moves" and "combat turns." The moves represent the distinct actions of each of the game's fictional creations in opposition to one another, and the results of these moves are judged subjectively by one player who is appointed "game master." (DCD)

Comment: This is more a do-it-yourself science fiction story than a game. It can be grand fun, but sophisticates should be forewarned that this is BEM SF. (DCD)

Note: For more on this game, see the essay on future-oriented simulations, by Charles M. Plummer.

Cost: $5.00; inquire for price of polyhedra dice needed

Producer: Gamescience, 7604 Newton Drive, Biloxi, MS 39532

TEACHING THE FUTURE

Draper L. Kauffman, Jr.

Playing Data
Copyright: 1976
Age Level: variable
Number of Players: variable, in groups of 3-6
Playing Time: 5 or more hours
Packaging: 296-page paperback

Description: The book *Teaching the Future* contains extensive reading material on future studies, 22 short exercises adaptable to any level on planning for the future, two longer exercises for high school students, and a broad outline of a simulation game entitled the *President's Select Commission.* This simulation game requires the teacher or class first to select a potential future problem and a future date at which this problem might reach crisis proportions, and to write a scenario of events leading up to the crisis, including the names and roles of the key persons involved. Then, as the *President's Select Commission,* the class divides into subcommittees or task forces, each researching a specific aspect of the problem and determining their specific recommendations for solving it. The class as a whole then meets as a commission to discuss, recommend, and vote on the most economically and socially feasible proposals. A sample scenario included in the book has been used with teacher in-service workshops concerned with educational reform in the future. (TM)

Cost: $7.95

Producer: ETC Publications, Dept. 1627-A, Palm Springs, CA 92262

2000 A.D. FUTURA CITY

Playing Data
Copyright: 1973
Age Level: grades 7-12
Number of Players: 20-30
Playing Time: 10 50-minute sessions
Preparation Time: 10 hours (est.); the simulation is the concluding activity in the Newsweek Current Affairs Case Study 2000 A.D.
Packaging: professionally packaged game

Description: The scenario of this game projects many of the constituent issues of what, in the 1960s, was called the urban crisis onto a far and future place called Central City. The solution the game offers to these problems is based on Alvin Toffler's idealized "Constituent Assembly" (to vitalize city government). Players are asked to pretend that it is 2000. The game begins with the announcement "of a sudden windfall for Central City; the availability of two adjacent land tracts for use by the city."

The players assume the roles of consultants or spokesmen for special interest groups called commerce, comm/trol, develop, integrate, ethnic, ecol, and life/plan, or members of the Mayor's Planning Council.

Together they form the Constituent Assembly. Each interest group defines its group objectives and develops a comprehensive plan for the redevelopment of the core city and the development of the new tracts. The ecol group, for example, "will be primarily committed to allocating funds for the creation of new parklands, the preservation of open space, and the lowering of pollution levels." The plans are presented to, and debated before, the Constituent Assembly which must agree upon one scheme. The winning team is that one which has "outlined priorities and objectives which are closest to those adopted by the Assembly in the final urban plan." (DCD)

Note: For more on this game, see the essay on future-oriented simulations, by Charles M. Plummer.

Cost: $47.00 for multi-media kit with record, $49.95 with cassette

Producer: Newsweek Educational Division, 444 Madison Avenue, New York, NY 10022

TWO YEARS OF HORROR

J. M. Kirman

Playing Data
Age Level: high school, junior college, adult groups
Number of Players: 6-18 in 2-3 groups
Playing Time: 3-4 hours
Packaging: 90 pages of game materials and 7 pages of instructions in a looseleaf notebook

Description: According to the author, "this simulation is designed to make people more aware of the impact of science and technology on society." Participants are asked to pretend that they are members of a group 100 years in the future named the "Science-Society Evaluation Agency." This agency was formed 50 years ago (game time) after the "two years of horror" when a man-made virus almost destroyed the world. The agency's job is to officially sanction or suppress the release or manufacture of new hardware and computer software, surgical procedures, chemicals, drugs, etc. All members of each group read a card which describes the innovation in question, discuss the social repercussions of the release or manufacture of this new technology, and decide whether to release the item unconditionally, with restrictions or changes, or to refuse to release it.

The game presents eighteen situations, which may be discussed in any sequence. For instance, there is a ray that, when directed into the brain, "neutralizes" tendencies toward violence; a computer parent training program for training and licensing prospective parents; an auto-grow process for regenerating human organs; Mentol, a chemical that blanks out normal personal control of the thinking portion of the brain; a beam that transports animate or inanimate material with the speed of light. Each situation is explored through a set of five questions each, which potentially extend discussion beyond the hypothetical. (DCD)

Cost: $20.00

Producer: Puckrin's Production House, 35 Mill Drive, St. Albert, Alberta, Canada T8N 1J5

UTOPIA

John Hildebrand and David Yount

Playing Data
Copyright: 1975
Age Level: high school
Number of Players: 16 or 24 in groups of 8
Playing Time: 13 or more hour periods
Supplementary Materials: a reserve shelf on Utopian literature
Packaging: 8 page teaching manual, 8 page student guide

Description: *Utopia* involves the development of the political, technological, economic and moral systems of an ideal society. The game is divided into 3 phases of 3-5 periods each. During the first phase, students read the Sunrise Cabin Experiment, which concerns an isolated community and how it deals with its problems. In small groups of 8, students discuss 5 problems which occur at Sunrise (e.g., snowbound in late winter) and individually write solutions. During phase two, each group begins to create its own ideal society by dividing up research tasks on political, technological, economic and moral systems of a society. Each student utilizes the information and questions included in the manual to aid him in filling out his general systems research form describing 3-5 main ideas of the system he has chosen to research. When his research is completed, he reports to the group which evaluates his work and prepares a general systems group report to present to the entire class. The class, in conclave and using parliamentary procedure, then decides which ideas for which system should be a part of its ideal society. Phase three begins with each student using his subsystem research form to determine the subsystems his system will have. For example, a student researching the political system may be concerned with lawmaking, law execution, elections, the penal system, the judicial system, foreign relations, and procedures for social change. Questions regarding each subsystem are listed in the student guide. When students have completed their research, they report to the group which in turn reports to the conclave. The conclave determines which proposals to adopt and writes a declaration of utopian commitment. Throughout the game students use evaluation forms to evaluate each other's contributions. (TM)

Note: For more on this simulation, see the essay on future-oriented simulations, by Charles M. Plummer.

Cost: $10.00
Producer: Interact, P.O. Box 262, Lakeside, CA 92040

WORLD GAME

Franciscan Communications Center

Playing Data
Copyright: 1972
Age Level: high school-adult
Number of Players: 25-50 in 5 groups
Playing Time: 2 hours
Packaging: professionally produced materials unbound; including 33 rpm phonograph record, role and situation sheets, name tags, posters, and instructions

Description: Much of the effect of the *World Game* lies in the ability of its presenters to surprise the players. Immediately after the group has been assembled, the presenter plays a recording of "Message from Planet X," in which an announcer declares that his planet, Planet X, with a population of eight million, must soon be evacuated because its sun is dying. The announcer states that he has taken over all communication media on Earth, and that his voice is being heard all over the world. He goes on to explain that the total population of his world will appear on Earth in one year, and that the people of Earth have six months in which to make preparations for the aliens' arrival. Three subsequent messages, on the same record, from other inhabitants of the alien planet explain that Planet X is much more advanced than Earth, and that Earth has been chosen by the aliens for their new home because of its similarity to their own world. Immediately after the record has finished the presenters quickly divide all the players into five groups. Each group represents a distinct world power, and individuals within each group take defined roles. The country called a "Socialist Republic," for instance, has as its members a doctor, a village head, a student, a diplomat, and a journalist. After the players in all groups study their roles, each group decides on a national policy concerning Planet X. The five countries then assemble for a meeting of a World Council, at which they must reach consensus about how to respond to the messages from Planet X. After this, the presenters lead a discussion of what happened during the game. (DCD)

Cost: $14.95

Producer: Teleketics, 1229 S. Santee St., Los Angeles, CA 90015

See also ECOLOGY

GEOGRAPHY

ADAPT

Jerry Lipetzky and John Hildebrand

Playing Data
Copyright: 1975
Age Level: high school, junior high
Number of Players: 28-35 students in groups of 5
Playing Time: 5-7 hours per cycle, 1-5 cycles
Special Equipment: overhead projector, movie projector, the film *The Hunters* for the first cycle
Supplementary Materials: a resource shelf in the library or classroom
Packaging: a 34-page looseleaf manual

Description: *Adapt* is a research-oriented game which explores the effect of physical environment on possible hunting-gathering, agricultural, preindustrial, industrial, and future societies. The game may be played in 1 or more of 5 cycles, and may therefore be concerned with 1 or more of the above kinds of societies. A description of the first cycle follows; the others are similar, and maps, fact sheets, and research sheets for all cycles are included in the manual. The first cycle begins with students reading their student guide, discussing the importance of various environmental factors on a hunting-gathering society, and determining on their comprehensive maps of Schunkland, a newly discovered continent, an ideal place for a society. Then, in groups of 5 they share their ideas for the location of the society and agree upon one location. Each member of the group then chooses a research task for homework (or for an additional work period)—climate, vegetation, landforms, wildlife, and minerals and receives a detailed map, research sheet, and fact sheet pertaining to this project. The next day, when researchers present their findings to their group, the group awards them research investigation points. Each group then pools its research by filling out a Symposium map and copying it on a transparency. During the next two class periods, the chairperson of each group presents his group's research and map and is cross-examined by the entire class, which awards research evaluation points to each group. The final day, the teacher shows a film on a real hunting-gathering society and students fill out evaluation sheets. (TM)

Cost: $14.00

Producer: Interact, Box 262, Lakeside, CA 92040

BOXCARS

Joan Steffy

Playing Data
Copyright: 1975
Age Level: grades 8-11
Number of Players: class
Playing Time: 4-5 class periods
Packaging: 8-page foldout student guide, 8-page teacher's guide

Description: *Boxcars* is a simulation of competing companies in Western Europe. In playing it, students learn locations (and native names) for 12 countries, their capitals, their various imports and exports, and deal with the way supply-and-demand economy works. The process for players involves decision making in a partnership, map reading, addition, subtraction, and percentages, and record keeping.

Students pair off, name their own shipping companies, and plan a strategy. (This is preceded by explanations and class discussion as the teacher judges it is needed.) The object of the strategy is maximum profit during a 15-"day" period of moving boxcars across Western Europe.

In secret they select a capital where they will buy exports for trade in other cities and plan a trade route. Then, weighing travel time and costs against their selling profits, they travel around Europe buying and selling. The company that makes the highest profit, as that which best understands the area's economic geography, wins.

The final period is devoted to discussion and evaluation.

The teacher's guide includes several options teachers might take in modifying the simulation to meet their own goals.

Cost: $10.00

Producer: Interact, Box 262, Lakeside, CA 92040

THE CARGELLIGO TOPOGRAPHIC MAP GAME

Harry Irwin

Playing Data
Copyright: 1978
Age Level: grades 7-12
Number of Players: 4
Playing Time: 35 minutes
Preparation Time: 1 hour (est.)
Packaging: professionally produced and packaged board game includes map playing surface, tokens, die, and instructions

Description: This game is designed to introduce students to elementary map-reading skills. It is played on a topographic, grid square map of the countryside surrounding the North South West Australian town of Cargelligo. Each player pretends to be a photographer hired by the Cargelligo Chamber of Commerce to take photographs of twelve geographic features for a tourist brochure. The features, such as a farmhouse or a hill over 1200 feet in elevation, are listed on a card. Players advance tokens, one grid square at a time, around a circuit printed on the map. Each time players recognize a feature on their list in the grid square they occupy, they place a plain white card over it and proceed. Some features are circled by a green line, and a player who lands in a square with one of these must answer a question from a quiz book. These questions are designed to test the student's map-reading skills (for example, "This square is coloured like many others on the map. What does this colour mean?"). The first player to "photograph" the 12 features and return to town wins. (DCD)

Cost: $19.95 (Australian)

Producer: Cassell Australia Limited, P.O. Box 52, Camperdown, N.S.W. 2050, Australia

LONGMAN GEOGRAPHY GAMES

Stephen Cotterell, Geoffrey Dinkele, and Rex Walford

Description: *Beat the Bell* (grades 5-8, 1-2 periods): This game is based on the familiar problem of beating traffic delays on journeys to and from school. A hypothetical route map is provided, but teachers may prefer to develop their own map using local landmarks and experience. The game shows the essential elements, flows, and blockages of an urban transport system and allows students to make decisions to use the system to best advantage. (Producer)

Beef Cattle in Northern Australia (grades 9-11, 4 periods): This game represents the situation in northern Australia in the early 1960s and requires players to make decisions about where to send beef cattle to market. The unpredictability of weather and such factors as animal diseases, plant pests, and the influence of government grants affect the simulation. Follow-up work takes account of recent changes caused by improved road networks, the introduction of large trucks, and the establishment of abattoirs in the area. (Producer)

Bobtree Moves into Western Europe (high school, 10-12 periods): Bobtree, a food-mixer manufacturer, is thinking of developing a new factory in a western European location. Students role play a director of the company as they work through seven related exercises which embody geographical skills ranging from map drawing to graph plotting and the analysis and presentation of statistics. The critical phases of the appreciation cover the whole of the European Economic Community region and lead to close study of two regions, Nordheim Westfalen and southern Italy. (Producer)

Bread Line (high school, 10-12 periods): This exercise reveals the problems of rural life in a Third World village. Chance factors determine the year-to-year shift above or below the "bread line." The simulation is worked out in three stages: the first considers the general problem, the second looks at the effect of foreign aid, and the third deals with the implications of introducing large-scale commercial farming. (Producer)

Caribbean Fisherman (grades 5-8, 1-2 periods): This game brings out the hazardous nature of fishing as an occupation and the game element in the fisherman's constant struggle against the weather. Levels of rule development can be increased for use by higher age/ability groups. (Producer)

Developort (high school, 4-6 periods): The first part of the simulation is a six-round sequence intended to reveal common elements in port development and an understanding that geographical advantages alter with time. This part also acts as a starting point for individual port study, and by showing the fundamentals of growth at a number of ports leads into a comparison of world ports which forms the second part of the activity. (Producer)

Honshu (high school, 4 periods): This is a four-part activity particularly concerned with factors influencing industrial development in Japan. The simulation is of the mathematical "Monte Carlo" type in which an area is evaluated for environmental advantages and disadvantages and then industrially developed (or colonized) by reference to this evaluation and to chance factors introduced through random numbers. (Producer)

Motorway (high school, 4 periods): This is a map-based unit in two parts which can be linked or used separately. The first part is a problem-solving exercise in which students develop their map-reading skills as they plan the route of a new motorway. The second part is a simulation in which students role play planners and other groups with an interest in the route and fight to exert influence. The emphasis is on the identification of conflicting interests and on the need for reconciliation if the motorway is to proceed. (Producer)

Noigeren (grades 6-9, 3-5 periods): This simulation is a variant on the theme of the changing location of the iron and steel industry. The reversed names and maps disguise the location of the exercise, which is the northeastern region of the United States. At the end of the game, classroom-evolved patterns can be compared quickly with reality. Players are given an appreciation of industrial location factors and the idea that they are relative to technology and economic circumstances, rather than absolute. (Producer)

Plant Succession (high school, 4-6 periods): This mathematical simulation demonstrates the processes of plant succession. Based on the heathlands of West Surrey and northeastern Hampshire, it can be adapted for use with local data if this is known. The simulation is designed to help students understand the principles of colonization and plant succession and to show some of the factors which affect the processes. (Producer)

The Power Game (high school, 6-8 periods): This is a programmed decision-making simulation linked to other exercises and intended to help students understand the problems of electricity generation and varying demand. It is based on the northwestern England and northern Wales region, where the variety of power stations in the area enables players to become familiar with the characteristics of different types of generating plant. (Producer)

Super-Port (grades 11 and 12, about 6 periods): This is an extensive simulation of the problems involved in setting up a superport and providing a supporting communications and industrial corridor based on Roberts Bank, Vancouver, Canada. Students represent various groups and interests, including the government, the media, mineral corporations, railways, and local interest groups. They define their own objectives from role biographies and work toward the resolution of conflicts through negotiation and discussion. (Producer)

Tea Clipper Race (grades 6-9, 1-2 periods): This is an operational game based on one of the most famous race situations in history—the 8000-mile journey from China to London in the competition to sell early tea crops in the 1850s and 60s. The game introduces some basic world geography and players, as clipper captains, choose their routes by evaluating the respective risks and disadvantages associated with patterns of prevailing oceanic weather. (Producer)

Urbanisation (grades 11 and 12, 6 periods): This is a complex game concerned with urban pressures on a rural environment. It has proved an effective lead-in to field work. The game aims to reveal some of the processes involved in the urbanization of a British rural landscape. Players assume a variety of roles which interact through different viewpoints and values about the undeveloped environment. The possibility of controlled and well-organized development can be pursued and the planning policies of the "local council" are a key factor in this situation. (Producer)

Weather Forecasting (high school, 2-3 periods): Dice, or a set of chance cards, provide the unpredictable element in the simulation of the easterly movement of a depression across the British Isles. Players become familiar with the characteristics of weather surrounding depressions and study the weather-forecasting process. (Producer)

Cost: boxed reference set (1 copy of each game) £6.00 plus postage; contact producer for other options and quantity discount schedule

Producer: Longman Group Ltd., Resources Unit, 33/35 Tanner Row, York, England

MERIDIAN 36

Lawrence Ralston

Playing Data
Copyright: 1972
Age Level: junior and senior high school
Number of Players: 2-4 individuals or teams
Playing Time: 30-60 minutes
Packaging: professionally designed package

Description: There are three games possible on the single game-board format. The board is a clear plastic peg-board with a 36-square grid, plus magnetic strips top and bottom. These attach to compatible metal strips on one of two underlays. The first of these is a world continental outline map. The second is a world political area map. *Population* is played on the continental map, *Meridian 36* and *World* are played on the political area map. Each player or team is given a set of pegs, either red, yellow, green, or blue. Moves on the board and the number of pegs used in the move are indicated on orange playing cards. Six cards are played in a single move; when all players have moved once, one entire round ends and another begins. There are three rounds in one game. The object of the game is to enclose the greatest number of grid squares. This is done by placing pegs in contiguous lines (e.g., no Ts or Xs) according to directions on the cards. A typical direction will read B-5-2, which directs the player to grid square B (down) and 5 (across), where 2 pegs may be placed along any side of the square, but they must link up with any other pegs played in an adjacent square in the same hand. Thus if the player held cards for A-4, B-4, B-5, C-6, C-7, and D-8, it would be possible to string a whole contiguous line of pegs along the borders of those squares. The moves usually involve more strategy, however, and less chance. Provisions are made for unusual conditions such as holding directions to move through captured areas on the board. *Population* includes the use of a supplementary scoring table. *Meridian 36* and *World* use a one square = one point scoring system. In addition, *World* features a question-and-answer element which allows for earning one point per correct answer to geographical questions.

Meridian 36 is the basic game of the three. It is "based on competition to capture 36 squares," thus players seek to dominate the world map. After 6 playing cards are dealt to each player or team, the entire hand of 6 cards is played as one turn. Pegs must be cleverly aligned within the bounds of the instructions so that the maximum area is captured per turn. A "round" is one turn for all players–6 cards played. Six more complete hands are dealt with each new round. Three rounds are one game. Scoring is one point for each square captured.

World adds the element of geography to the basic format. After a team captures a square, the corresponding territory is identified, and additional points may be awarded for correct answers. A limit on questions may be used when time is short.

Population adds the more sophisticated element of timed population growth to the basic format. Each new round of turns on the board doubles or triples the population of a global area beneath each square. Each square carries a number that indicated how many millions of persons (e.g., continental United States 200) are living there. A growth table lists starting populations for round one, 50-year projections for round two, and 100-year projections for round three. Population growth is pitted against a "Life Sustaining Elements" factor of 400 units per square. One "LSE" unit supports one million people. By the end of the game, players discover whether or not the areas they have captured will support the popualtions they contain. Population projections are derived from the work of Ehrlich and others. (WHR)

Comment: A verbose and confusing instruction booklet obscures an otherwise interesting idea. The publisher claims that one game board will support a variety of underlays forthcoming. These are to provide modular input into the basic apparatus. Such areas as vocabulary, mathematics, science, and economics are put forth as possibilities. The claim of reinforcement for class and text work remains to be seen. Basically, *Meridian 36* is a peg-board game that demands little skill. It is more a clever organizational technique than a new concept. (WHR)

Cost: $14.95

Producer: Meridian House, Inc., 21 Charles Street, Westport, CT 06880

NORTH SEA EXPLORATION

Rex Walford

Playing Data
Copyright: 1973
Age Level: grades 9-12
Number of Players: 12-24 in groups of 6
Preparation Time: 1 hour (est.)
Packaging: professionally packaged and produced game includes teacher's manual, deed cards, rig cards, weather cards, game map, markers, and role cards

Description: The designer describes this as "a game about the exploration for gas and oil which has taken place in the North Sea since the mid 1960's. The players who take part represent the boards of various (fictitious) companies similar to those involved in real life. They face the problems of deciding how to go about exploring for oil and gas and what equipment to use."

Students assume one of six roles within four companies. These roles are Chairman (who conducts the team, or company, meetings), Geophysicist (who recommends drill sites), Treasurer (who supervises company spending), Surveyor (who maps discoveries and wells), Rig Manager (who selects the rig for each drill site), and the Engineer (who moves the markers representing rigs, platforms, and plugs around the playing surface). Through each of the 10 rounds the players bid for sites, buy rigs, and drill for crude and gas. The results of their operations are determined by a master resource list and are announced by the teacher. The effects of weather and current political events are simulated by chance cards. The company that has operated most profitably after 10 rounds wins. (DCD)

Cost: £4.50

Producer: SGS Associates (Education) Limited, 8 New Row, London WC2N 4LH, United Kingdom

RAILWAY PIONEERS

Rex Walford

Playing Data
Copyright: 1972
Age Level: grades 9-12
Number of Players: 16-36 in teams of 4-6
Playing Time: 10-minute rounds
Preparation Time: 1 hour (est.)
Packaging: professionally packaged board game includes film strip, two maps, and instruction booklet

Description: This English exercise, which includes an introductory filmstrip, is designed to introduce students to the problems that hindered the construction of the railroads in the western United States during the nineteenth century and to improve the players' knowledge of the geography of the American West. Teams assume the roles of corporate officers of 4-6 railroad companies. Each company is assigned a capital base of 22-25 "units" and guaranteed an income of 15 units per round. In the course of play each company moves tokens across a map of the central and western United States. Numbers on the map represent the unit cost of laying track through an area of land. For example, a short move through the Great Plains might cost 2 units, while a move of the same distance through the Colorado Rockies costs 6-10 units. Companies earn additional profits by connecting with railheads such as Wichita, Denver, and Salt Lake. River crossing cost an additional 3 units. The game continues for 10 rounds or for 2 rounds after any company has reached a city on the Pacific Coast. The winner is the company with the greatest assets at the end of play. (DCD)

Cost: £4.50

Producer: SGS Associates (Education) Limited, 8 New Row, London WC2N 4LH, United Kingdom

REMOTE ISLAND

Richard A. Schusler, University of Kansas

Playing Data
Copyright: 1971
Age Level: grades 4-8
Prerequisite Skills: reading map recognition, writing
Number of Players: groups of 6 players
Playing Time: 3 hours to 6 days in periods of 1 hour

Description: Players divide into groups of 6 persons, each representing an inhabitant of a remote island in the Pacific (pop. 700,000 in 1970). In the first phase players collect data about the island from a transparency shown by the teacher and make a master map together. In the second phase they have the task of locating three large cities and a number of small cities, appropriate roads, and railroads. In the third phase they role play a problem given in the worksheet—a wealthy American wants to start an iron smelting plant. Students must consider the benefits of more jobs and a more even balance of trade for the people versus pollution, possible overpopulation, political corruption, and the loss of a way of life integral to a peaceful, beautiful place. (AC)

Comment: This is a relatively simple simulated learning situation of 15 Xeroxed pages. It provides only a framework for a class situation to build on. It is an example, also, of how a teacher can make and use his own simple simulations with a class. (REH)

Cost: $5.00

Producer: Dr. Richard B. Cohen, Dept. of Elementary and Early Childhood Education, Univ. of Arkansas, 33rd and University, Little Rock, AR 72204

See also
Euro-Card SOCIAL STUDIES
Flight SOCIAL STUDIES
Heritage HISTORY
Le Pays de France LANGUAGE
Spatial Marketing Simulation COMPUTER SIMULATIONS

HEALTH

BLOOD MONEY

Cathy Stein Greenblat, Rutgers University and John H. Gagnon, State University of New York at Stonybrook

Playing Data
Copyright: 1975
Age Level: college, continuing education
Number of Players: 20-35
Preparation Time: 3 hours (est.)
Playing Time: 2-1/2 hours, 1-2 classes
Supplementary Materials: recommended reading: Caroline Roth's *Hemophilia, Hemophiliacs and The Health Care System*

Description: Blood Money is a simulation game which gives players the experience of role-playing hemophiliacs and the helpers who care for them. The game makes players aware of the medical problems of hemophiliacs, their need for money, treatment, and blood and the difficulties obtaining all three, as well as the rigidity and red tape of the work world with its discrimination against the handicapped. Players begin by studying role sheets for their assigned roles as operators, employers, welfare agents, blood bankers, medical personnel and (sixteen) hemophiliac citizens; they also familiarize themselves with the treatment and cost charts and records appropriate to their roles. The room is divided into two sectors—a work area and a helping area. The game itself is composed of two thirty-five to forty-five minute rounds each beginning with a five-minute hiring period during which employers choose workers who demonstrate the most skill at dart throwing. Citizens wear tags which indicate the degree of their disability and thus the disadvantaged position (sitting down, using the left hand) from which they must throw darts. During each round they throw darts in order to win white chips (money) which they will later exchange for blue chips representing medical care and red chips representing blood. They must also choose cards indicating whether or not they had an attack and the degree of its severity. Those citizens who had an attack must leave work to report to the operator and then choose whether or not to treat the attack by paying doctors and blood bankers for their services. Only the operators (who supervise the employment area and give out attack cards) and medical personnel know the length of required treatment for each attack, the cost to the patient, and the expected consequences. Meanwhile, welfare agents, doctors, blood bankers, and employers are all trying to earn as many white chips as possible. Welfare agents are also distributing white chips to those citizens whom they feel are most in need. After the final round, players are given the opportunity to express feelings, discuss their roles, and relate the game to the reality of the hemophiliac's world. (TM)

Note: For more on this simulation, see the essay on health and health care simulations by Amy E. Zelmer and A. C. Lynn Zelmer.

Cost: (1) manual, DHEW Publication No. (NIH) 76-1082, no charge (2) complete kit $125.00

Producer:
(1) Superintendent of Documents, U.S. Government Printing Office, Washington, DC 20402
(2) Gamed Simulations, Inc., Suite 4H, 10 West 66th Street, New York, NY 10023

THE CALORIE GAME

Hazel Taylor Spitze

Playing Data
Copyright: 1972
Age Level: general
Number of Players: 1 or 2-6 players or teams
Playing Time: flexible
Preparation Time: 2 minutes
Packaging: box containing: instruction booklet; packet containing 6 player pieces, set of dice, 23 cards; playing board; set of calorie nutritive, score and nutrient award cards

Description: "The object of the Calorie Game is to obtain 100% of the RDA's of eight specific nutrients before running out of calories. In the process of doing this, players learn something of the relation between calories, nutrients, energy, and activity." How the game is played is somewhat flexible, with alternative rules described in the instruction manual. Players in sequential turns move their pieces around the playing board (number of spaces per move determined by throw of dice) landing on spaces which offer opportunities to buy (in calories) specific foods. Without looking up the nutritional value of the food, the player decides whether to purchase the food item. With purchases, players build up toward 100% RDA of each nutrient. Upon obtaining 100% RDA a nutrient award is given, and the first player to win all eight of the nutrient awards wins the game. (WHR)

Cost: price on request

Producer: Graphics Company, P.O. 331, Urbana, IL 61801

CLINICAL SIMULATIONS

The College Committee on Student Appraisal, University of Illinois College of Medicine; Christine H. McGuire, Lawrence M. Solomon, and Phillip M. Formon, editors

Playing Data
Copyright: 1976, 2nd edition
Age Level: medical students
Prerequisite Skills: knowledge of assessment, diagnosis, and medical management of the problems selected
Number of Players: 1
Playing Time: 15-60 minutes per exercise
Preparation Time: none
Packaging: softcover book on newsprint paper

Description: *Clinical Simulations: Selected Problems in Patient Management* is a set of twenty common patient problems. In each case, the student takes the role of the attending physician, takes appropriate diagnostic or treatment actions, and receives feedback on the basis of the choices made. The appendix provides a comment and recommended management. (AEZ)

Comment: The exercise might be enhanced for less experienced students by having them work in pairs. The materials can be used only once because the answers are permanently visible once disclosed. (AEZ)

Note: For more on this collection, see the essay on health and health care simulations by Amy E. Zelmer and A. C. Lynn Zelmer.

Cost: $23.50 plus handling

Producer: Prentice-Hall, Englewood Cliffs, NJ 07632

CLOT

Playing Data
Copyright: 1977
Age Level: adult (medical laboratory personnel)
Number of Players: 1-4 or 2-4 teams of up to 4
Prerequisite Skills: knowledge of hemostasis
Playing Time: 30 minutes (est.)
Packaging: professionally packaged board game including two softbound reference texts on coagulation

Description: *Clot* is designed for medical professionals, to teach and reinforce knowledge of hemostatis and the blood coagulation process. The game may be played by teams or individuals. To begin, each player puts a marker at the start of each of two tracks (inner and outer) on the playing board. The squares on both tracks are of various colors. A token can advance on the inner track only when the player's token on the outer track is on a similarly colored square, and after the player correctly answers a question about the process of blood coagulation. All questions and their answers are included in the instructions. The winner is the first player to complete one circuit of the inner track. (DCD)

Cost: $15.00

Producer: Ortho Diagnostics, Inc., Raritan, NJ 08869

CONTRACEPT

David E. Corbin and David A. Sleet

Playing Data
Copyright: 1978
Age Level: high school-adult
Number of Players: 2-8
Playing Time: 10 minutes-1 hour
Preparation Time: 15 minutes
Packaging: professionally produced board game

Description: This game is designed to introduce its players to the way women become pregnant and various methods for preventing this. Players advance sperm and ovum markers (according to the action of a spinner) over a double-tracked representation of a woman's reproductive organs. After penetrating the womb each player must choose one of eight birth control methods (such as withdrawal and rhythm) and attempt to advance the sperm marker to the end of one of the fallopian tubes. (Pregnancy results from an encounter between a sperm and an egg in the same tube.) Progress into a tube can be accomplished only after the player has answered a question about birth control such as, "name two disadvantages of withdrawal." A player who becomes pregnant must trade the fertilized egg for a baby. Any player who gets pregnant twice must stop playing. The first player to go all the way to the end of one of the tubes wins. (DCD)

Comment: A previous reviewer (a woman) found this game to be "offensively glib." She particularly objected to playing spaces labeled "Kama Sutra cum laude," "love it or leave it," "all shapes and sizes," and "jam up and jelly tight." We have, subsequently, been informed that many groups involved in planned parenthood find this game to be a good "icebreaker" for introducing a discussion of contraceptive techniques. Some potential users may find this game to be in poor taste. This reviewer (a man) can't really picture an audience of potential users who would both appreciate the cloying double entendres rampant in the game materials and be totally ignorant of contraceptive techniques. We know the game is well intentioned and does contain a considerable amount of good information about birth control. (DCD)

Cost: $24.00

Producer: Center for Health Games and Simulations, San Diego State University, San Diego, CA 92182

DISTRICT NUTRITION GAME

Playing Data
Copyright: 1972
Age Level: elementary or basic literacy level adult
Prerequisite Skills: counting; minimal reading
Number of Players: 2-4
Playing Time: 15 minutes or more
Preparation Time: 1 hour to make board
Special Equipment: game board, dice or spinner, markers
Packaging: description is in book; user must make own materials

Description: *District Nutrition Game* is a board game of the "snakes and ladders" type. Players roll dice and move individual pieces in sequence along the board. They make extra advances if they land on a square that indicates a good food habit and must move back if they land on a square with a poor food habit. (AEZ)

Comment: The game requires no skill; there is a large chance element. It is, however, useful for introducing or reinforcing basic concepts. (AEZ)

Note: For more on this game, see the essay on health and health care simulations by Amy E. Zelmer and A. C. Lynn Zelmer.

Cost: $10.00 for paperback book, *Nutrition for Developing Countries,* Maurice King, editor. Game costs are negligible as it can be made from readily available materials

Producer: Oxford University Press, 200 Madison Avenue, New York, NY 10016

GOOD LOSER

Dietor Self-Instruction Systems, Didactron, Inc.

Playing Data
Copyright: 1972
Age Level: elementary school, junior high, high school
Number of Players: 2-6
Playing Time: one hour
Packaging: game box including playing board, playing pieces, tokens, fat chance, and opinion cards

Description: The winner of *Good Loser* is the first player to lose twenty pounds. In the course of the game players learn about weight control, calories, and exercise. Taking turns throwing the dice and moving around the board, they attempt to gain willpower tokens in order to avoid weight-gaining circuits, and to lose overweight tokens by correctly answering opinion card questions about weight control (e.g. sour cream is a better dressing for weight watchers than mayonnaise). Fat chance cards and the crash diet programs on the playing board also give players opportunities to lose pounds and gain willpower. Variations of the game are also described in the manual. (TM)

Cost: $14.00

Producer: National Health Systems, Box 1501, Ann Arbor, MI 48106

HANDLING CONFLICT IN HOSPITAL MANAGEMENT (CONFLICT AMONG PEERS)

Erwin Rausch and Wallace Wohlking, adapted for hospital use by Grace E. Phelan, R.N.

Playing Data
Copyright: 1973
Age Level: college, graduate school, management
Number of Players: 3 or more in 1 or more teams
Playing Time: approximately 2-1/2 to 3 hours
Preparation Time: 1/2 to 1 hour

Description: This is a noninteractive workbook exercise in which the players assume the role of nursing unit supervisors who must minimize personal conflicts and achieve the best possible performance for their own departments. The exercise contains seven sections: Facing the Conflict Situation, Approaching the Extended Care Supervisor, Recognizing Emotional Reactions, Anticipating Emotional Reactions, Opening Communications, De-escalating the Conflict, and Establishing an Open-Communications Climate.

In each section, the players must choose the best of several possible responses to a hypothetical situation. For example, at the beginning of the exercise, the player is "clearly annoyed" at the extended care unit supervisor for instituting a new, and time-consuming, procedure. The player has the option of approaching the supervisor in several ways (by telephone, in person, or by memorandum), of approaching the director of nursing, or of not being annoyed with what appears to be an accomplished fact. Each alternative has been assigned different point values and the players try to choose the response that will promote the most constructive work atmosphere and earn the most points. The player with the most points at the end of the game wins. (DCD)

Cost: 5-participant set $26.50; 2 or more sets $19.25 each; meeting leader's guide $.50.

Producer: Didactic Systems, Inc., Box 457, Cranford, NJ 07016

HANDLING CONFLICT IN HOSPITAL MANAGEMENT (SUPERIOR/SUBORDINATE CONFLICT)

Erwin Rausch and Wallace Wohlking, adapted for hospital use by Grace E. Phelan, R.N.

Playing Data
Copyright: 1974
Age Level: college, graduate school, management
Number of Players: 3 or more in 1 or more teams
Playing Time: approximately 2-1/2-3 hours
Preparation Time: 1/2-1 hour

Description: This is a noninteractive workbook exercise in which the players assume the role of a hospital administrator. This administrator has assigned a subordinate the job of drafting and implementing a new hospital procedure. Having received no word on the subordinate's progress for several weeks, the administrator concludes that the subordinate is avoiding the job. This exercise delineates the administrator's alternatives for getting the subordinate to finish the job. The superior must notify the subordinate of his or her dissatisfaction with the lack of progress, meet with the subordinate, analyze the subordinate's reaction to this meeting, cope with the subordinate's defensiveness, evasiveness, and hostility, and prepare for disciplinary action against the employee. Through each step of the exercise the player is asked to choose from among several possible responses to each hypothetical situation. For instance, when first aware that the subordinate is not making progress in drafting the new procedure, the administrator may take one of six actions that require deciding whether to involve his or her own boss, whether to call the employee or send a memo, and whether to discuss the matter immediately or make an appointment to discuss the matter at some future date. Points are awarded for the best responses and the player tries to pick the alternatives that will promote the most constructive work atmosphere and earn the most points. The player with the most points at the end of the game wins. (DCD)

Cost: 5-participant set $26.50; 2 or more sets $19.25 each; meeting leader's guide $.50

Producer: Didactic Systems, Inc., Box 457, Cranford, NJ 07016

HEALTH GAMES STUDENTS PLAY

Ruth C. Engs, Indiana University; S. Eugene Barnes, University of Southern Mississippi; Molly Wantz, Ball State University

Playing Data
Copyright: 1975
Age Level: grades 6-13
Number of Players: 10-30
Playing Time: 10-100 minutes
Preparation Time: 3 hours (est.)
Packaging: 166-page softcover book

Description: This is a collection of eighty-four structured experiences, simulation games, and experiential exercises organized into thirteen categories and designed, in the authors' words, "to help stimulate the teacher of health education to create an assortment of activities which will be interesting and meaningful to students. The topics of the games included here are Getting Acquainted and Mental Health; Drugs, Alcohol, and Tobacco; Family Life and Human Sexuality; Death; Personal Health and the Body; Aging; Dental Health; Nutrition; Chronic Diseases; Communicable Diseases; Consumer and Community Health; First Aid and Safety; and Ecology and the Environment.

Most activities are what the authors call "structured experiences"—that is, "techniques or activities which are usually led by the teacher" and which "may last from a few minutes to a whole class period." *Live Sculpture, Smoke,* and *Dentopoly* are representative of these. In this last game, players earn points by answering questions about dental health (for example, true or false: Most dental disease is inevitable) and the player who answers the most questions correctly wins. *Smoke* is a game like Bingo, with the cards covered with terms relating to smoking and their definitions. In *Live Sculpture,* a device for getting students to portray their concepts about mental health, players are divided into teams of five. Each player writes down his or her definition of mental health. One of the players on each team must then assume a posture which represents and incorporates all of the ideas of the group.

Trial: Legalizing Marijuana is typical of the several simulation games in this collection. In this exercise the legalization of marijuana is placed on trial and students assume the roles of judge, prosecution and defense attorneys, expert witnesses, and members of a jury. Students do research and prepare for their roles outside of class, and a class session is devoted to conducting the mock trial.

Typical of what the authors classify as "experiential exercises" in this collection are those which have students write their own epitaphs, speculate about the causes of their deaths, talk about chronically ill people they have known, and record all the food they consume in a week. (DCD)

Cost: $5.95

Producer: Kendall/Hunt Publishing Company, 2460 Kerner Blvd., Dubuque, IA 52001

HOSPITEX: HOSPITAL WARD MANAGEMENT EXERCISE

Management Games Limited

Playing Data
Age Level: hospital management courses, student nurses
Number of Players: 6-30, in teams of 2
Playing Time: 3-5 hours
Preparation Time: 1-2 hours

Description: This combination in-basket exercise and case study is designed, according to the producer, "to put students and others into a simulated ward situation and then to pose problems which require a 'managerial' and/or technical response." The wards for which versions of *HOSPITEX* exist are surgical, medical, psychiatric, geriatric, midwifery, and pediatric.

Each player assumes the role of a "charge nurse" who is fifteen minutes late for work the first day back from vacation. Each player is given a roster specifying for each patient diagnostic details, any special treatment required, and general condition. The player must then develop a routine plan for the morning.

Each participant is also confronted in the course of play with "abnormal decisions" that result from unexpected events on the ward and must record on a form how each event is responded to, how long it would take to respond to each, and who would be delegated to resolve each problem. (DCD)

Comment: The authors suggest that the nursing profession's primary task is patient care, and rightly so. To this overriding objective they add that patient care in a ward is, to some extent, dependent upon the effectiveness of the management function as carried out by the head nurse. Furthermore, the authors suggest that this game has been designed to put participants into a simulated ward situation where, under considerable pressures, they must make clinical and managerial decisions.

The exercise can be played by individual participants. It is designed primarily for first-line management courses in a hospital, and for student nurses who have had to take charge of a ward as a head nurse. (HL)

Cost: basic set for 20 students in any one ward (specify which) (1) $60.00; (2) £36.00

Producer: (1) Didactic Systems, Inc., Box 457, Cranford, NJ 07016; (2) Management Games, Ltd., 63B George Street, Maulden, Bedford MK45 2DD, United Kingdom

KEEP QUIET

Scottusa Company

Playing Data
Copyright: pending

Description: *Keep Quiet* is a game for people who can talk but wish to learn sign language used by the deaf. Players play a series of crossword games using small wooden cubes that have the various deaf hand sign language, called finger spelling, on them. (REH)

Cost: $7.00

Producer: Kopptronix Company, Box 361, Stanhope, NJ 07874

MYTH-INFORMATION (ABOUT HUMAN SEXUALITY)

Jackie Reubens

Playing Data
Copyright: 1972
Age Level: high school-adult
Number of Players: 4-100
Playing Time: 30 minutes-3 hours
Preparation Time: 15-20 minutes
Packaging: 17-page game manual with instructions and explanations of each topic described on the playing cards, answer cards, 100-card deck, some buttons for winners

Description: According to the designer, "The goal of *Myth-Information* is to simulate thought and discussion and to acquaint people with the myths and truths that exist around the area of human sexuality and, hopefully, to dispel some of the myths." The class or group is divided into teams and the deck of cards evenly distributed to all the players. To begin, one player reads the statement written on one of his or her cards and determines whether the answer is myth or fact. Other team members may provide information to the player, but the final decision is the individual's. Each player reads a card in turn, and scores on correctness. After all cards have been read, teams exchange cards and repeat the process. The team with the highest number of correct responses wins the sexpert buttons. (WHR)

Cost: $10.00 plus postage/handling

Producer: Jackie Reubens, 34 Andrew Drive, Tiburon, CA 94920

NOURISH

Camille Freed Pfeifer and Mary Shaw Snaith

Playing Data
Copyright: 1975
Age Level: general
Number of Players: 1-20
Playing Time: 20 minutes
Packaging: 144-card deck and 30-page instruction booklet

Description: *Nourish* is a deck of cards that is intended to create an interest in and awareness of the nutritional values of various foods among those who play with it. The game provides instructions for fourteen separate card games including Rummy for the Tummy, Crazy Calcium (similar to Crazy Eights), Old Hen (like Old Maid), and Fifty-One (modeled after Twenty-One). The deck is divided into different colored suits. Each suit represents a different nutrient like protein, thiamine, or calcium; each card represents a different food. According to the designers, "The foods represented have been selected to appeal to a wide variety of people. They are 'Basic Foods' whose nutritional values have been calculated from accepted scientific sources." One of the games that can be played with the deck (the one the authors suggest that beginners play first) is called Daily Tally. Each player is dealt eight cards and each, in turn, picks and discards until one player has a hand with eight different colored cards. This player then scores the total number of calories on the cards he or she holds. All other players score zero. Play continues until one player has scored a predetermined number (say 2500) calories, at which time that player wins. (DCD)

Cost: $9.00

Producer: Fun With Food, P.O. Box 954, Belmont, CA 94002

NURSING CROSSWORD AND OTHER WORD GAMES

Sheryll Dempsey

Playing Data
Copyright: 1973
Age Level: students in health related fields
Prerequisite Skills: some familiarity with medical terminology
Number of Players: One
Playing Time: 10-30 minutes per puzzle
Packaging: in softcover book

Description: This book includes thirty-two crossword and similar word puzzles using medical terms. Answers are given. (AEZ)

Comment: Useful for recall practice of terminology. Materials can only be used once per student. (AEZ)

Note: For more on this collection, see the essay on health and health care simulations by Amy E. Zelmer and A. C. Lynn Zelmer.

Cost: $5.95

Producer: Trainex Press, P.O. Box 116, Garden Grove, CA 92642

THE NUTRITION GAME

Hazel Taylor Spitze

Playing Data
Copyright: 1972
Age Level: children and adults
Number of Players: one or more (2-4 players or teams are best)
Playing Time: determined by players (i.e., minutes, hours, number of turns)
Preparation Time: 5-10 minutes

Description: *The Nutrition Game* is a simulation in which players in rotating sequence each buy foods to obtain their recommended daily allowances of calories, protein, minerals, and vitamins. The game needs no instructor. There are eight options which have different goals, and each option can be treated as a separate game. The purpose is to familiarize players with the daily nutritional requirements and the nutritional value of specific foods purchased in the game. Being highly flexible, the time of play and options followed are determined by the players. After repeated playing, the players eventually become more aware of the relationships between food costs, nutritional values, and health. (WHR)

Cost: price on request

Producer: Graphics Company, P.O. 331, Urbana, IL 61801

PLANAFAM II

Katherine Finseth, Harvard Center for Population Studies

Playing Data
Copyright: 1972
Age Level: junior high school-adult
Number of Players: 3-10 in one group role
Playing Time: 1-2 hours
Preparation Time: 1 hour for initial materials preparation
Special Equipment: 3 sets double-six dominoes, 2 packs playing cards, 6 different colors of felt pens, colored markers (buttons), large sheet cardboard or newsprint
Packaging: print description only

Description: The players work through one female's reproductive life cycle, making decisions regarding life style and contraception which would affect reproductive events. Probabilities of events are based on 1960 U.S. census data. (AEZ)

Comment: Some updating would be helpful, but this still remains a useful tool for introducing the idea of choice and chance affecting reproductive behavior. (AEZ)

Note: For more about this simulation, see the essay on health care simulations by Amy E. Zelmer and A. C. Lynn Zelmer.

Cost: Eric Document ED064 228, hard copy $3.29, microfiche $.69; game materials readily available and can be assembled for under $10.00

Producer: ERIC, 855 Broadway, Boulder, CO 80302

POPPIN' SWAP (NUTRITIONAL GAME)

Playing Data
Copyright: 1973
Age Level: 10 years-adult
Number of Players: 4-7 per deck of cards
Preparation Time: 1-2 minutes
Packaging: classroom size: 5 deck package of playing cards and pre- and post-tests

Description: This is a game to teach players the nutritional value of various foods. Each card in the deck has a specific food pictured on it with its primary nutritional value (protein, calcium, vitamins, niacin, etc). Players (four-seven per deck) are dealt the entire deck, and on the signal of the dealer, they simultaneously exchange cards (verbally advertising what they want to trade) with one another to form sets of three or more cards of the same nutrients and/or one set of two cards from the same nutrient group. The first person to place his hand face-up on the table is the winner. Points are accumulated with each hand played. (WHR)

Cost: single deck $6.00; 5-deck pack $27.50

Producer: Poppin' Swap Game, The Pillsbury Company, Box 60-090, Dept. 377, Minneapolis, MN 55460

PSYCHIATRIC NURSE-PATIENT RELATIONSHIP GAME

Carolyn Chambers Clark

Playing Data
Copyright: 1977
Age Level: nursing students
Number of Players: 2
Playing Time: approximately 1 hour
Preparation Time: about 30-60 minutes to become familiar with material
Packaging: mimeographed materials and playing board in cloth bag

Description: This game is a programmed learning system guiding the players through the orientation, working and termination phases of the nurse-patient relationship. The game calls for role playing to simulate realistic situations and poses direct questions to integrate concepts and to test understanding and knowledge. (Author)

Comment: Provides a guided learning experience with quite an elaborate scoring system. Materials are rather flimsy and will not stand repeated use. (AEZ)

Note: For more on this game, see the essay on health and health care simulations by Amy E. Zelmer and A. C. Lynn Zelmer.

Cost: $15.95 plus postage

Producer: Carolyn Chambers Clark, P.O. Box 132, Sloatsburg, NY 10974

RESUSCI-ANN

Playing Data
Age Level: high school-health professionals
Number of Players: 1 or 2
Playing Time: basic resuscitation courses usually allow up to 8 hours of teaching time for students to achieve skill mastery
Preparation Time: 10-20 minutes
Special Equipment: Resusci-Ann model recording tapes
Packaging: model comes in carrying case; must be assembled for use each time

Description: Resusci-Ann is a life-size human model used for demonstration and practice of cardiopulmonary resuscitation. Depending upon the model selected, students may receive feedback via a light system or recording tape so student and instructor can determine if CPR techniques are being carried out effectively. (AEZ)

Comment: A qualified instructor is required. (AEZ)

Note: For more on Resusci-Ann, see the essay on health and health care simulations by Amy E. Zelmer and A. C. Lynn Zelmer.

Cost: $1,000 or more for the model; recording tapes approximately $3.00

Producer: check local safety supply companies or heart foundation for nearest agent

SOUP'S ON

Didactron, Inc.

Playing Data
Copyright: 1970
Age Level: elementary school to high school
Number of Players: 4-40
Playing Time: 15-30 minutes
Packaging: professionally designed box and playing pieces

Description: *Soup's On* is a bingo game made for introductory learning of nutritional concepts. A "dietician" is chosen to spin the master selector which provides a selection of eighty-four different foods, divided into six basic categories: protein, vegetable, fat, fruit, milk, and starch. The dietician spins the pointer, announced the outcome to the players, and records the item on a register sheet, should anyone have it on his or her playing board. Each player receives a card on which three "balanced" meals are listed from left to right. Six columns of food divide meals into the basic six parts. One of each kind of food makes up one balanced meal.

Instructions advise that while these meals may not seem so appetizing, there are many attractive ways to serve them. These are discussed later. When the dietician announces a food item, the player who recognizes it on his card then "calls" it. Successful calls are marked with blue chips on the playing card.

To win, a player must complete a "balanced" meal, marked by a line of six blue chips, and call out "soup's on." Rules are provided for ties and variations of play. A "review session" is to follow playing time, during which players familiar with certain foods and their preparation may share that experience. (TM)

Cost: $14.00

Producer: National Health Systems, P.O. Box 1501, Ann Arbor, MI 48106

SUPER SANDWICH

Barry Zaid

Playing Data
Copyright: 1973
Age Level: grades 5-8
Number of Players: 2 to 4
Playing Time: 25-45 minutes
Preparation Time: 10 minutes to explain rules

Description: Super Sandwich explains the concepts of nutrition and a balanced diet.

Players go to breakfast, lunch, and dinner and buy foods as they try to concoct the "super sandwich" (a sandwich composed of all the foods necessary for a balanced daily diet). Each food is broken down into its nutritional content, which is recorded onto laminated score cards. There are fattening snacks and nutritious between meal snacks as well as trips to the "Gym" to work off excess calories. The first player to acquire sufficient foods to complete his nutritional requirements and remain within allowable calorie levels is the winner of the game. (Publisher)

Cost: $14.95

Producer: Teaching Concepts, Inc., P.O. Box 2507, New York, NY 10017

VITAMINS

Tiff E. Cook

Playing Data
Copyright: 1975
Age Level: grade 5-adult

Number of Players: 2-6
Playing Time: 45 minutes (est.)
Packaging: professionally packaged card game

Description: This game is designed to teach its players the vitamins necessary for their good health. According to the action of a spinner, players draw cards from one of three piles. Piles of cards include vitamin cards (e.g., riboflavin or ascorbic acid), source cards (e.g., citrus fruits) and function cards (e.g., cell repair). Players must collect cards that form vitamin chains, which each consist of a vitamin, source and function card for one vitamin. The first player to form three vitamin chains wins. (DCD)

Cost: $9.50

Producer: Union Printing Co., Inc., 17 West Washington Street, Athens, OH 45701

WHEELS

Didactron, Inc., Dietor Systems

Playing Data
Copyright: 1972
Age Level: junior high, senior high
Number of Players: 5-30 minutes
Packaging: complete game box including spinner, cards, chips and nutrition booklet

Description: Wheels is a simple game designed to teach students about the food sources of vitamins and minerals, which function as "wheels" propelling the vehicle of the body. Each player receives a card portraying a good nutrition truck with eight wheels listing eight essential vitamins and minerals. Also on each card are twenty-one different foods; each card has a different selection of foods. Throughout the game, the leader spins the spinner and calls out a vitamin or mineral and related food such as "calcium-milk." Players cover these items on their cards, as well as any other vitamin or mineral contained in that particular food. Not only luck but also how much each player knows about nutrition will determine whether or not she will be the first to cover all the wheels on her card and declare herself winner. A booklet reviewing nutrition is included in the game kit and may be studied by students before or after the game is played. (TM)

Cost: $14.00

Producer: Didactron, Inc., Dietor Systems, Box 1501, Ann Arbor, MI 48106

See also
MALAR COMPUTER SIMULATIONS/SOCIAL STUDIES
METRO-CHP COMPUTER SIMULATIONS
ADDICTIONS
HUMAN SERVICES

HISTORY

AFTER THE WAR AND THE ELECTION OF 2020

Martha Doerr Toppin

Playing Data
Copyright: 1976
Age Level: high school
Number of Players: 10-40
Playing Time: 2-5 hours
Packaging: 10-page stapled booklet

Description: *After the War* simulates the conditions in Germany in 1918-1932 and allows students to experience the frustration and powerlessness of the German people and the factors contributing to their support of Hitler during the election of 1932. Students, however, are not aware during the simulation that they are learning about Nazi Germany; they are involved in studying and discussing the fictitious history of America in the early twenty-first century, drawing up a list of programs for action, and dividing into groups to prepare speeches for four candidates running for office. One of the candidates, unknown to the students, is the equivalent of Hitler in 1932. After students present their speeches and hold a mock election, they discuss the factors influencing their voting decisions as well as the parallels between the simulation and Hitler's rise to power. (TM)

Cost: $4.00 (California residents add sales tax.)

Producer: MDT, 11 Conrad Court, Oakland, CA 94611

THE ALPHA CRISIS GAME

William A. Nesbitt

Playing Data
Copyright: 1973
Age Level: grades 9-12
Number of Players: 21-40 in 7 teams

Playing time: 4 50-minute periods
Preparation Time: 2 hours (est.)
Packaging: 115-page photocopied game manual and instructions

Description: "The Alpha Crisis Game," its author explains, "is intended as an introduction to the set of materials also available" from the same producer and "titled 'The July 1914 Crisis: A Case Study in Misperception and Escalation'." Teams of players represent hypothetical nations with Greek letter names. Although the students are not told that they are playing a simulation of the origins of World War I, the seven teams represent Serbia, Germany, Austria-Hungary, Russia, France, England, and Italy. The players are given background information about their nations describing the geography of each as well as its military and economic resources. In addition, players are each given a description of a precipitating crisis, a description of the web of alliances linking the countries, a common set of rules and procedures, and individual role descriptions. The game begins with the announcement that the "heir to the throne of Omega was assassinated in Botnia, and each country must decide what to do." The Omegans blame the Alphans for the assassination and issue an ultimatum. The teams try to resolve the crisis and the game ends when the crisis has been resolved, there is a stalemate, or war breaks out. (DCD)

Note: For more on this game, see the essay on history simulations by Bruce E. Bigelow.

Cost: $2.00 (checks to Regents Research Fund)

Producer: Center for International Programs and Comparative Studies, University of the State of New York, Albany, NY 12210

THE AMERICAN CONSTITUTIONAL CONVENTION

Leonard Stitelman and William Coplin

Playing Data

Copyright: 1969
Age Level: high school, college
Number of Players: 32-49 in 13 teams
Playing Time: 2-6 hours in 1 hour periods

Description: Students are assigned the roles of members of the Constitutional Convention. By reading the student handbook they become familiarized with the position of each delegate to the convention. They are required to adopt the position, philosophy, and point of view of the delegate they are assigned. The issues to be debated arise in one of the following broad areas: (1) the Federal Executive Branch; (2) the Federal Legislative Branch; (3) the Federal Judiciary; (4) Federal-State Relations; (5) Procedure for Amendment. Proposals relevant to each of these broad areas are presented in the following form:

> The executive power of the United States shall be invested in ______ a) a single person; b) three persons; c) a chief executive or president and such vice-presidents as Congress shall from time to time establish [Student Handbook, p. 30].

The teacher and/or class select the specific issues to be discussed. The convention passes through a five-step procedure in dealing with a given proposal: (1) state delegations caucus and determine the stand that the delegation will take; (2) the convention as a whole meets. At this time the chairman, George Washington, (played by the instructor) recognizes each delegation and they indicate the position they have taken; (3) the convention recesses and students are free to mingle with other delegates and work toward the development of a group consensus; (4) the state delegations again meet separately and take a final stand; (5) the convention approves one proposal.

It is important to note that this process is required for each proposal. Thus two proposals on different subjects are never handled at the same time so that trade-offs do not develop between them.

Objectives: The objectives of this simulation (in the evaluator's opinion) are these: (1) to provide students with an understanding of the process that the Founding Fathers utilized; (2) to enchance a student's appreciation for the genuine differences of opinion that existed among the Founding Fathers; (3) to enhance a student's knowledge of the role and philosophy of each member of the Constitutional Convention; (4) to develop an understanding of the importance of negotiation and compromise; (5) to enhance a student's appreciation of the Constitution.

Main impact: (1) The student will increase his cognitive understanding of the major issues that confronted the Founding Fathers. (2) The student will increase his cognitive understanding of the various ideological disagreements that existed among the Founding Fathers. (3) The student's motivation to study an important aspect of America's past would be enhanced. (4) The simulation helps the Constitutional Convention come alive.

Comments: Most simulations do very little to increase a student's cognitive knowledge of a subject. *The American Constitutional Convention* has a strong cognitive orientation. Students are required to read and understand an excellent essay on the Constitution, to examine the forces that surrounded it, and to study the men who fashioned a powerful government on the document. The simulation makes the topic "fun" and acquaints students with the facts simultaneously.

The major weakness of the game is that at least 32 students are needed before it can be played. In addition, the simulation is not deterministic. There are no winners or losers. Thus, little incentive exists for students to act out their roles in an appropriate manner. Finally, the game tends to be slow because only one issue is to be dealt with during the play period.

This evaluator is very favorably impressed with the game. It would be a worthwhile learning experience that would have a positive impact on both the cognitive and the affective domain. (From the Robert A. Taft Institute of Government Study on Games and simulations in Government, Politics, and Economics by John Beasley, Southern Illinois University, Carbondale.

Note: For more on this simulation, see the essay on history simulations by Bruce E. Bigelow.

Cost: student handbook $1.95 list, $1.55 school; instructor's guide $1.35 list, $1.08 school

Producer: Science Research Associates, Inc., 155 North Wacker Drive, Chicago, IL 60606

AMERICAN HISTORY GAMES

Alice K. Gordon et al., Abt Associates

Playing Data

Copyright: 1970
Age Level: grades 7-12
Number of Players: 15-40
Playing Time: 2-3 hours
Packaging: professionally packaged set of board games

Description: Each of the six games in this collection concerns a dominant issue during some significant period in American History. *Colony* allows students to assume the roles of colonial merchants, bankers, and members of the English government so they may examine the trade relations between England and her American colonies. In *Frontier,* players represent settlers in the Northwest and Southwest and bankers between 1815 and 1830. The idea in *Frontier* is to grow cotton or grain to sell at a profit and to establish schools and roads. In *Reconstruction,* two teams representing landowners and freedmen compete to increase their wealth. *Promotion* is designed to illustrate the development of industrial urban society in America. Players representing railroad officials, industrialists, farmer, bankers, and politicians work together to try to join the country by railroad. *Intervention* is about American foreign policy following the Spanish-American War. Players assume the roles of businessmen, politicians, diplomats, soldiers, and bankers who try to promote the stability and independence of other countries. *Development,* the final game in the series, is designed, in the producer's words, to illustrate "the relationship of major world powers and developing countries of the world." In this game four teams representing developing countries compete to increase their wealth. (DCD)

Comment: This set is clearly designed so that game outcomes are *not* an accurate reflection of historical realities. I have read little discussion of the advantages/disadvantages of such efforts to "spare the feelings" of student participants. Obviously it makes each game more fun as a game, and less valid as a simulation. I wonder whether the decision in this case was based on good classroom psychology, or whether it is just a case of "playing it safe" by SRA. (DZ)

Cost: $456.00; schools $342.00

Producer: Science Research Associates, Inc., 155 N. Wacker Drive, Chicago, IL 60606

ARMADA

Education Development Center

Playing Data
Copyright: 1973
Age Level: grades 8-12
Number of Players: 5-30 in groups of 5
Playing Time: 45-90 minutes
Packaging: professionally packaged board game

Description: Players assume the roles of Spanish and English admirals in 1588 who must decide how to arrange their fleets to accomplish or prevent a Spanish invasion of the British Isles. In each group, two players represent an English admiral, two represent the Spanish, and one assumes the duties of referee, who adjudicates disputes and keeps time. The game is played in ten six-minute rounds. During each round each side has three minutes to maneuver or engage their squadrons and fire on the enemy. The results of these naval bombardments is determined by a spinner. Players win points for effectively bombarding their opponents. The player with the most points after ten rounds wins. After the completion of play the participants discuss their winning or losing strategies and compare the events that occurred during play with the events that occurred during the actual invasion attempt of 1588. (DCD)

Cost: $69.00 #82205

Producer: Denoyer-Geppert Company, 5235 Ravenswood Avenue, Chicago, IL 60640

BIG BUSINESS

Jack Needham and David Dal Porto, Mt. Pleasant High School, San Jose

Playing Data
Copyright: 1976
Age Level: high school
Number of Players: 18-50
Playing Time: 4-50-minute periods
Packaging: 26-page manual

Description: *Big Business* is a simulation about the unregulated growth of U.S. business during the late nineteenth century. Its aim is to teach students the reasons for the rapid expansion of business during this period, the advantages of consolidating business concerns, and the role of transportation and natural resources in developing businesses. Students as bankers, railroad owners, oil company owners, coal mine owners and iron ore mine owners compete to make the most profit during eight rounds of negotiations for mergers, agreements, loans, and bids. Each role description includes six suggestions for possible actions during the game, such as considering a merger or checking on shipping costs. Students must keep a record of all transactions, income, and expenditures, and they must also complete such forms as bid sheets, merger agreements, and railroad shipping agreements whenever a negotiation is completed. The manual includes all role slips and forms, a price and quantity pruchasing guide, and an evaluation sheet. (TM)

Cost: $15.00

Producer: History Simulations, P.O. Box 2775, Santa Clara, CA 95051

BRINKMANSHIP: HOLOCAUST OR COMPROMISE

David Dal Porto, Mt. Pleasant High School

Playing Data
Copyright: 1970
Age Level: grades 7-12
Number of Players: 20-45 in 2 teams
Playing Time: 3-5 periods of 50 minutes each
Preparation Time: 2 weeks background material on cold war
Supplementary Material: cold war readings

Description: The game of *Brinkmanship* is a simulation which involves two super powers, the U.S. and the USSR in the background of several Cold War crises such as the Berlin Blockade, the Hungarian Revolution, the Korean War, Quemoy and Matsu, the Cuban Missile Crisis, and the Middle East War of 1967. The teacher may add any up-to-date crises he feels are more important.

As members of two teams, the U.S. and the USSR, students must decide how to deal with each crisis. After receiving the information about the crisis in the "background session," the students must decide what their goals and objective's are during the "policy session." The next step is the "negotiations session" whereby each side sends three negotiators to deal in a confrontation situation with the other side. Time limits which must be followed if war is to be avoided force the students to develop proposals for compromise. If a compromise is reached, the negotiators "return home" for approval or rejection of the compromise by the other students on their team. Finally, the settlement is evaluated either by a student panel or the teacher in the "judge's evaluation session." This evaluation determines which side got the most from the settlement. (Author)

Cost: $8.50

Producer: History Simulations, P.O. Box 2775, Santa Clara, CA 95051

THE CENTRALIZED POWER GAME

Playing Data
Age Level: grades 5-12
Number of Players: 5-25 in teams of 1-5
Playing Time: 2 hours or more in 45-minute periods

Description: This simulation guides participants in the discovery of the power structure of the times from feudalism to absolutism. The participants, taking the roles of king, nobles, clergy, merchants, and serfs, debate a series of decrees which, if enacted, will enhance the position of the king. Historical-events cards move players through time. These unplanned events cause players to revise strategy as their position changes. A round of play consists of team meetings, the king's court hearing, the decision period, and the selection of the historical events cards. (Producer)

Cost: $35.00

Producer: Simulation Learning Institute, P.O. Box 240, Roosevelt Island Station, New York, NY 10044

THE CH'ING GAME

Robert B. Oxnam

Playing Data
Copyright: 1972

Age Level: high school, college
Number of Players: 30; instructions for modifying for more or less
Playing Time: 4-6 or 8 hours in 1 to 1 1/2 hour periods
Preparation Time: 1-2 weeks of reading

Description: *The Ch'ing Game* is an historical simulation exercise which focuses on the political and social life of the Chinese upper classes in the eighteenth century (during the middle of the Ch'ing or Manchu Dynasty which lasted from 1644 to 1912). The Ch'ien-lung Emperor who ruled from 1736-1796 was a powerful monarch who maintained careful control over the bureaucracy, commanded the armies so that the borders remained secure, and patronized the arts and letters.

The Ch'ing Game reproduces the basic structure and systems of mid-Ch'ing government and sets forth various roles (from emperor to merchant) which are similar to those of eighteenth-century society. One of the key roles in traditional China (and a key rule in the game) is that one must seek prestige and power "within established routes": success in the civil service examinations, extraordinary service in his local community, efficient operation as a bureaucrat, contacts through marriage, display of sophistication through art and poetry, suitable gifts to superiors, charitable activities, and the like. To try to work outside the established system, the "break out of one's role," was not acceptable in the Ch'ing period and is not acceptable in *The Ch'ing Game.* Players who, in their quest for greater recognition, use unrecognized techniques (such as rebellion) will be severely punished and perhaps even eliminated from the game.

A successful player is one who has carefully prepared for his role, who understands Ch'ing government and society, and then who performs sensitively and imaginatively in the game. There is no winner in *The Ch'ing Game*; instead each player is evaluated (by the game supervisors and his fellow players) in a discussion after the game is over.

The Ch'ing Game consists of four or five rounds of one to two hours each, which can be played consecutively (allowing fifteen minutes for breaks between each round) or over several days. Each round corresponds to the passage of three years. At the beginning of the game each player will start by performing his assigned role. As the game proceeds, some players will shift their activities to other functions depending on their success or failure in their initial roles.

A wide variety of activities occur during the game. Much of the time will be consumed in regular functions ("systems") which take place each round: paying and collecting taxes, buying and selling land, arranging marriages, sending memorials to the emperor, obeying imperial edicts, evaluating officials, preparing for and taking the civil service examinations. In addition, certain players will be given "round cards" at the beginning of rounds which indicate special activities in which they will become involved (emperor's tour, a legal dispute, a tax case, and the like.) (Author)

Editor's Comment: The first chapter includes a discussion of history and simulation and descriptions of experiences and problems that occurred during development of the game. For instance, "We had intended the game to be arduous and wanted a tiring routine to develop, but we failed to anticipate the true exhaustion that occurred." The next chapter is a 30-page Handbook for Students, intended for careful study. The supervisor's chapter is 25 pages long and discusses who should play, how to prepare for and set up the game, and what modifications are possible.

The designer clearly appreciated the game's intricacy, which is intentional as a reflection of the reality simulated, and gives careful attention to preparing the supervisor and participants to master that intricacy for what looks like an involving and illuminating simulation experience. (AC)

Note: For more on this simulation, see the essay on history simulations by Bruce E. Bigelow.

Cost: $2.00

Producer: Learning Resources in International Studies, 60 East 42nd Street, Suite 123, New York, NY 10017

CONGRESS OF VIENNA

B. Barker and R. Boden

Playing Data
Copyright: 1973
Age Level: high school, college
Number of Players: 10-30 in 5 groups
Playing Time: 3 hours (est.)
Preparation Time: 2 hours (est.)
Packaging: professionally produced 17-page booklet

Description: *Congress of Vienna* is a simulation of that 1814-1815 historical event. Teams of players represent the nations bargaining at the Congress. All five share the same vital interest in creating a balance of power in Europe and all have national aims which conform to those of the actual participants in the conference. The exercise is structured by an agenda which consists of the following four items: (1) the territorial settlement of Saxony and the Grand Duchy of Warsaw; (2) the territorial settlement of north and west Germany; (3) the territorial settlement of Italy and the Adriatic; and (4) the future form of the Germanic Confederation. Each agenda item is disposed of through a strict procedure of formal and informal sessions within which the consensus needed for a resolution is specified. For example, a Russian motion on agenda item one may be adopted only if Russia, Prussia, and Austria all vote for it and either France or England remains neutral. After each item has been resolved the teacher announces a score for each team that reflects the comparative advantage or disadvantage to each nation of the agreement they have made. The highest any team can score is 120 points. The winner is the team with the most points at the end of the game. (DCD)

Note: For more on this simulation, see the essay on history simulations by Bruce E. Bigelow

Cost: £6.75. Quantity discounts

Producer: Longman Group Ltd., Resources Unit, 9-11 The Shambles, York, England

CZAR POWER

R. G. Klietsch

Playing Data
Copyright: 1971
Age Level: grades 9-14
Number of Players: 20-35
Playing Time: 3-5 50-minute rounds
Preparation Time: 2 hours (est.)
Packaging: professionally packaged game includes role sheets, scoring tables, problem cards, and instructions.

Description: *Czar Power* attempts to illustrate how feudal autocracies self-destruct. The author states that "Czar Power is designed to replicate the relations within and between social classes, including the obligations, positions on pressing issues, and alternatives. Each problem is an historical statement of the conditions which surround the problem, its probable causes, effects and social implications."

Players assume the roles of fictional (except for the czar) persons in six social classes– royalty, nobility, clergy, civil service, merchants, and serfs. There are thirty specific roles and each has a personal history, a set of resources, a specific class goal, and a personal want. Czar Alexander II, for example, wants to preserve the dynasty and rule well, and he has five thousand estates and nine million rubles a year with which to do this as well as the power to award titles and honors. A fictional cloth merchant on the other hand, although moderately wealthy, can only petition those with influence to promote certain tax and trade reforms, and a serf named Ivan Kirov, who has nothing, has nothing to lose by supporting revolution.

During each round the czar picks four issues from a list of fourteen to address. Some players in all the social classes have an interest in these issues, but not all players have an interest in all issues. A set of charts specifies how each of the czar's decisions enriches or impoverishes each player in the game. The game ends after three rounds. (DCD)

Note: For more on this simulation, see the essay on history simulations by Bruce E. Bigelow.

Cost: $68.50

Producer: System's Factors, Inc., 1940 Woodland Ave., Duluth, MN 55803

CZECH-MATE

Daniel R. Place

Playing Data
Copyright: 1976
Age Level: high school
Number of Players: 20-50 in 7 teams
Playing Time: 10 1-hour periods
Supplementary Materials: reserve shelf in library
Packaging: 29-page teacher guide, 1-page student guide

Description: Students divide into eight teams representing the eight nations involved in the Munich Crisis of 1938. The first three days each team researches its own country and the events leading to Hitler's attempt to take over Czechoslovakia and writes a policy outline of goals and methods for resolving the crisis and satisfying its national interest. The manual includes detailed information on the chronology of pre-1938 events, the state of the world in 1938, military alliances and relative military strengths, and profiles and maps of the eight nations. After the initial research period, teams negotiate through message forms, ambassadors sent to other teams, and international conferences. Students also take a matching test and complete daily progress reports, both included in the manual. The simulation ends after five days of negotiations or earlier if war breaks out. The last two days teams rate themselves according to their military success, and students write essays about what they learned during the simulation. (TM)

Note: For more on this simulation, see the essay on history simulations by Bruce E. Bigelow.

Cost: $14.00

Producer: Interact, P.O. Box 262, Lakeside, CA 92040

DESTINY

Paul DeKock and David Yount, El Capitan High School

Playing Data
Copyright: 1969
Age Level: grades 7-12, average to above-average ability, college
Prerequisite Skills: reading, grade 12
Number of Players: 25-40 in 6 teams
Playing Time: 15-30 hours in 1-hour periods
Preparation Time: 5 hours

Description: Destiny is a simulation of American foreign policy during the Cuban Crisis of 1898. Divided into six factions present in America at the time, students use inquiry skills in order to determine the positions of their factions: Spanish Diplomats, the Cuban junta, newspapermen, businessmen, Imperialists, and Anti-Imperialists. Pressure cards containing information about actual historical events from fall 1897 to April 1898 increase or decrease each faction's PAP's (presidential advice points used as grade points during the simulation). While competing factions debate one another in front of President McKinley during three crises, students practice the social studies skill of differentiating fact, inference, and judgment as they pressure President McKinley to ask Congress either to declare war or not declare war on Spain. During McKinley's April 1898 speech, all students become Congresspersons and either accept or reject his requests. Pre- and post-tests show both teacher and class how the experience has changed student attitudes toward and knowledge about how presidential decisions affect America's "destiny." (Publisher)

Note: For more on this simulation, see the essay on history simulations by Bruce E. Bigelow.

Cost: $14.00

Producer: Interact, P.O. Box 262, Lakeside, CA 92040

DISCOVERY

John Wesley, Chase Ave. School, El Cajon, California

Playing Data
Copyright: 1972
Age Level: grades 5-8, average ability; grades 9-12, below-average ability
Number of Players: 20-40 in teams
Playing Time: 15-30 hours in periods of 1 hour
Preparation Time: 5 hours
Supplementary Material: encyclopedia, geography and history texts

Description: After a brief study of maps, natural resources, and geography, the class is divided into groups that face problems seventeenth-century American colonists faced. Assigned a role, each student must fulfill certain responsibilities throughout the simulation. Right away he helps his group make crucial first decisions. Once these decisions have been made and recorded, the ships set sail. After crossing the sea, each colonizing group selects its site, establishes its colony, and makes daily economic, political, and military decisions. Fate cards also affect each colony's success or failure. For example, if a severe storm batters a coastal colony dependent on fishing, the colonists must find other food. Interaction among the colonies results in trading goods, forming alliances, and sometimes killing one another. Thus individual students playing different roles are put under daily pressure to make decisions that will help save their colony or insure greater economic success. Once 20 rounds of colonial activity have been completed, each colony totals its wealth and one colony wins. A debriefing discussion helps students evaluate the degree to which the simulation experiences made textbook descriptions of colonization come alive. (Publisher)

Cost: $14.00

Producer: Interact, P.O. Box 262, Lakeside, CA 92040

DISSENT AND PROTEST— THE MONTGOMERY BUS BOYCOTT

David J. Boin and Robert Sillman

Playing Data
Copyright: 1973
Age Level: grades 7-12
Prerequisite Skills: reading, grade 7
Number of Players: class size
Playing Time: 3-4 class periods

Description: This is a reenactment of the Montgomery, Alabama bus boycott of December 5, 1955. The class is divided into five to ten action groups. Each one must decide on a list of proposed demands in the light of these considerations, quoted from the teacher's manual for presentation to students: (a) Is the demand too weak? Will it result in only minor improvements? (b) Is the demand too strong? Is it unrealistic and certain to prolong the boycott? (c) Is the demand necessary? What will it do to change the present situation? (d) Will the demand result in lasting change or will it solve the problem only temporarily? (e) Will the demand benefit the entire black community or will it help only a selected few? At various times during the simulation, newspaper bulletins that may influence discussions and decisions are read. (REH)

Comment: This appears in our history section although it seems almost too recent to some of us to be history. It is presented in the usual Edu-Game format with a few pages of teacher's manual and locally reproducible student material. (REH)

Cost: $3.00

Producer: Edu-Game, P.O. Box 1144, Sun Valley, CA 91352

DISUNIA

David Yount and Paul DeKock, El Capitan High School, Lakeside, California

Playing Data
Copyright: 1968
Age Level: grades 10-12, average to above-average ability; grades 7-9, above-average ability
Prerequisite Skills: reading, grade 10.5
Number of Players: 20-35 in 7 teams
Playing Time: 2-4 hours

Description: Divided into states on a new planet in 2087, students assume state offices, create a state flag and motto, and then struggle with problems paralleling American problems during 1781-1789. The governor binds the state to official decisions with his signature; the economic analyst supervises trading; the political analyst levies tariffs and forms alliances; the military analyst develops and disperses the state's military forces; the congressperson writes and presents legislation in Eisenhowerton, the national capital. Students complete assignments (tests on the American philosophy of government in the Declaration of Independence and the problems Americans faced in the 1780s; study and discussion of Madison's Federalist Paper No. 10) and fulfill state duties by filling out and giving report and recommendations forms to their governors. For completing all such assignments and state responsibilities, citizens receive blue (mercantile-industrial) or green (agricultural) CGSs (Consumer Goods and Services), which their states trade in attempts to balance their overly industrial-mercantile or agricultural economies. The jealousy of the separate sovereign states undermines attempts to solve their economic problem (a plague of insects devouring crops); their moral problem (an expedition returns through the Cumberley Gap after capturing Applegreens near the Ohayoo River in "The Great Valley"); their military problem (taxing students' CGSs in order to raise an army capable of turning back border attacks by Grittbrayton and Spaynoriana). Each day the governor compiles all economic, political, and military data and fills out a state activities summary sheet and a state power chart. These forms reveal the states' economic and military ranking. Eventually an astute "founding father" sees the parallel between 2087 and 1787 and calls a constitutional convention to revise the Articles of Confederation of Disunia. If the convention is successful, its members write a new federal Constitution based on the Constitution of 1787. Students who have experienced *Disunia* understand why we have the federal constitution we do today because they have experienced the confederation problems that brought about the "miracle at Philadelphia" in 1787. (Publisher)

Comments: *Disunia* appears to be an effective device for teaching concepts relating to American political philosophy as enunciated in the Declaration of Independence and the Constitution. It also appears to be strong on the concept of federalism. It helps make vivid and understandable the period of American political history on which it focuses.

The simulation is complicated for the teacher to prepare for and administer but proves fairly easy for students to play. Although the simulation supplies few readings for students, this reviewer believes it necessary to supplement the game by extensive use of the library and intensive study of the confederation and constitutional period of American history. Writing and research activities are integral to proper utilization of the game which is a strong plus in its favor, but the teacher must stay on top of this game to keep from being sidetracked. Thorough organization is an absolute must.

Only teachers who are willing to put the time and energy into this exercise should attempt *Disunia*. The investment pays off, however. Below-average high school students learn and remember the political intricacies of the American federal system they encounter in the simulation.

The authors of the simulation point out that student abilities, the decisions they make in the game, the amount of time each teacher feels the simulation warrants, and the contest within which the game is used, determines the length of the play period. The teacher's manual contains a detailed sample unit time chart which supports a three-week utilization of the simulation. The experience this reviewer has had with the sample sequence has been positive. Ample material is provided, both to utilize the simulation fully and to lead students through an evaluation of the process. The same sequence calls for an assignment and for an evaluation of an essay based on game experiences. There are many advantages to be gleaned from making the essay an integral part of the simulation experience, even for average or below-average ninth-grade students. (From the Robert A. Taft Institute of Government Study on Games and Simulation in Government, Politics, and Economics by Gerald Thorpe, Indiana University.)

Cost: $14.00

Producer: Interact, P.O. Box 262, Lakeside, CA 92040

DIVISION

Paul DeKock and David Yount, El Capitan High School, Lakeside, California

Playing Data
Copyright: 1968
Age Level: grades 8-12, average to above-average ability
Prerequisite Skills: reading, grade 10.5; math, grade 3
Number of Players: 16-35 in 6 teams
Playing Time: 5-15 days in 1-hour periods
Preparation Time: 2-4 hours
Supplementary Material: specified in teacher's manual

Description: *Division* is a simulation of the divisive issues of the 1850s and the crisis election of 1860. Students first join one of six factions: Western Republicans, Eastern Republicans, Abolitionists, Southern Democrats, Northern Democrats, or Constitutional Unionists. They then research and prepare position papers on 14 issues dividing Americans during the 1850s; extension of slavery into the territories, the Dred Scott Decision, the protective tariff, perpetuity of the union, and the like. The four political parties of 1860 then caucus and choose students to role-play Lincoln, Douglas, Breckinridge, and Bell. Campaign speeches and debates between the candidates take place prior to the final two days when factions use their IPS (influence points) to pressure, bargain, and block one another during the last six 'weeks" of the election of 1860. Students who have successfully researched all six factions' stands on all 14 issues have the best chance of electing their candidate president of the United States. (Publisher)

Cost: $14.00

Producer: Interact, P.O. Box 262, Lakeside, CA 92040

EMPIRE

Education Development Center

Playing Data
Copyright: 1973
Age Level: grades 7-12
Number of Players: 15-36 in 7 teams
Playing Time: 5-6 40-50-minute rounds
Packaging: professionally packaged board game

Description: Teams represent interests groups in the British Empire's

Atlantic trading community around 1735. These are London merchants, New England merchants, European merchants, colonial farmers, southern planters, and British West Indian planters. The interests of each group is, in some way, intrinsically opposed to the interests of the others, and each must negotiate with the others to buy, sell, and ship goods. Each must increase its own wealth at the other's expense. The game is played on a large map and proceeds through trading and shipping periods. At the end of each round the teams tabulate their profits and losses. The winning team is the one that achieves the greatest percentage increase in wealth during play. (DCD)

Cost: $83.00 (82223)

Producer: Denoyer-Geppert Company, 5235 Ravenswood Avenue, Chicago, IL 60640

ESPIONAGE

William Lacey, Fountain Valley High School, California

Playing Data
Copyright: 1974
Age level: high school
Number of Players: 29-35
Playing Time: 3-5, 50-minute periods
Packaging: 28-page teacher's manual and 7-page student guide

Description: *Espionage* is a simulation game of the Rosenberg trial of 1951. While playing *Espionage,* students enact the trial of Julius and Ethel Rosenberg and attempt to understand the issues which made the trial a national controversy. On the first day of the game, they complete an attitude survey on communism and espionage, read in their student guide about the Cold War and the trial, and choose roles. The teacher's guide contains 1-page role descriptions (to be photocopied) on each of the 17 witnesses, attorneys, and court staff involved in the case, as well as the 12 jurors. On the second day students are tested on their reading under controlled espionage conditions—they are to be given group rather than individual grades, some students receive correctly marked answer sheets, and the teacher leaves the room during the test. When she returns, the tests are graded and students discuss their experience. The third day, after the trial procedure and role-playing sheets are reviewed, the trial begins. Witnesses must be true to their roles but may interpret them freely when questioned by the attorneys. The trial continues throughout the fourth day when the rest of the testimonies are given, the jury deliberates and reaches a verdict, and the judge passes sentence. The last day of the game students discuss debriefing questions included in the manual (e.g., Is the testimony of a blood relative acceptable? What is treason? How should we judge people accused of espionage?), retake the attitude survey, and read follow-up material about the consequences of the Rosenberg trial. (TM)

Cost: $14.00

Producer: Interact, Box 262, Lakeside, CA 92040

EXPLORING THE NEW WORLD

David J. Boin and Robert Sillman

Playing Data
Copyright: 1972
Age Level: grades 5-12
Number of Players: class size
Playing Time: 5 class periods

Description: The class is divided into crews for different ships exploring the new world in 1550. Their job is to bring back information which is realistic and accurate. The crews each elect a captain. They must gather and do research, by finding books, maps, articles, diaries, and any other kind of information. Different fixed points of departure and areas of departure are assigned. The class must create various diaries and journals of the trip. (REH)

Comment: This has the bare simple elements of an exercise. Most of it must be filled in by the teacher and the students. There is not much more to it than what has been described above. However, it would appear to be an excellent activity for a history class, particularly an advanced one. This is the standard Edu-Game format with a few pages of teacher's manual and locally reproducible student material. (REH)

Cost: $3.00

Producer: Edu-Game, P.O. Box 1144, Sun Valley, CA 91352

FEDERALISTS VERSUS REPUBLICANS

Eric Rothschild

Playing Data
Copyright: 1970
Age Level: grades 9-12
Number of Players: 5-28
Playing Time: 10-20 hours
Preparation Time: 2 hours (est.)
Packaging: Professionally packaged multimedia unit includes long-playing record, filmstrip, 28 student's booklets, and 96-page teaching guide

Description: This multimedia kit assembles a variety of resources for a participatory student exercise about the thinking and actions of the leaders of the Federalists and Republicans between 1789 and 1815. The thinking of Hamilton and Jefferson, spokesmen for the opposing parties, is closely followed. Through recordings and a filmstrip as well as simulations involving dramatizations, role-playing scripts, newspaper headlines, cartoons, letters, and memorabilia, students experience the timeless problems of decision-making under the Constitution and come to realize how, in a representative government, points of view and even firm convictions must often be modified. (Producer)

Cost: $82.00

Producer: Educational Audio Visual, Inc., Pleasantville, NY 10570

FIFTIES

William Lacey

Playing Data
Copyright: 1977
Age Level: grades 7-12
Number of Players: 28-36
Playing Time: 30 1-hour sessions
Preparation Time: 3 hours (est.)
Packaging: 67-page instruction booklet

Description: According to the producer, *Fifties* intends to simulate "the events, personalities, life-styles, and culture of the 1950's." The simulation is divided into five phases. In Phase One, which lasts for four sessions, participants write historical summaries of a dozen events that occurred in the United States during the 1950s. During the ten sessions of Phase Two, participants simulate the "life-style" and "creatively rework dialogue for Peanuts cartoons, survey people on conformity, compose Beatnik poetry, write jokes as Shelly Berman or Mort Sahl might have done, draw doodles, and as a finale schedule an ideal family weekend in the Fifties." The six sessions of Phase Three are intended to simulate "mass media impact," and are spent writing and performing "scenarios" from "What's My Line?," "Ozzie and Harriet," and "Dragnet," composing pop tunes of the *Fifties,* and silently acting out "Rebel Without A Cause." During Phase Four, participants devote eight sessions to "a) interviewing people who grew up in the era, and b) researching yearbooks of the 1950's, continually generalizing and hypothesizing." During the two-session debriefing participants assess the positive and negative characteristics of the 1950s and take the final exam included in the simulation." (DCD)

Cost: $14.00

Producer: Interact, P.O. Box 262, Lakeside, CA 92040

GATEWAY

Jay Mack

Playing Data
Copyright: 1974
Age Level: grades 7-12 average to above-average ability
Number of Players: 25-40
Playing Time: 16 hours or more in 1-Hour periods
Supplementary Materials: reading matter suggested in teacher's guide

Description: *Gateway* is a simulation of immigration issues in past and present America. During the simulation's four phases students assume various identities and live through crises dealing with immigrational and "Americanism" issues throughout our history. Phase I: As either Scandinavians, Poles, Germans, Italians, or Russians, students simulate travel across the Atlantic Ocean in the late 1890s. While on board ship they write autobiographies based on skeletal identities furnished them. Scores on three performance tests on immigrational history plus crossing bulletins determine their ships' rate of progress. Phase II: The three ships' immigrants who travelled the most miles in Phase I become processors at Ellis Island during Phase II; the other three ships' immigrants must be processed through the Ellis Island of 1900. Each immigrant carries a detailed immigrant checklist; clearance officials ask questions and write responses on this checklist as the immigrant progresses through the background, character, vocation, and health stations. Most immigrants gain admission to the United States and repeat the loyalty oath, but some are deported because of character or health flaws. Phase III: A congressional committee holds a hearing on various immigration bills presented by five different factions who have varying ideas of what kinds of persons should be allowed to enter the United States. Phase IV: In an American secondary school the social studies department is being pressured by Black, Chicano, Oriental and Native American students to offer separate Black studies, Chicano studies, Oriental studies, and Native American studies courses in place of the traditional American history class. Students role-play members of one of these factions as well as WASP students, the social studies department members, a citizen's patriotic group, and school board members. In a school board confrontation, five issues are debated before the board members make a decision about what the content of the secondary school's history courses should be. Debriefing: The simulation ends with students reflecting on how America has been a "gateway" for millions of persons. Students also exchange their definitions of "American" and their conclusions about whether our pluralistic society's ideal should be to become a "melting-pot" or a "mosaic." (Publisher)

Note: For more on this simulation, see the essay on history simulations by Bruce E. Bigelow.

Cost: $14.00

Producer: Interact, P.O. Box 262, Lakeside, CA 92040

GOLD RUSH

Myron Flindt

Playing Data
Copyright: 1978
Age Level: grades 7-12
Number of Players: 24-48 in 8 teams
Playing Time: 13 1-hour sessions
Preparation time: 2 hours (est.)
Packaging: 24-page player's manual, 36-page teacher's guide.

Description: Students assume the roles of fortune seekers eager to go west to Golden Gulch, where "the gold is so plentiful that it is just waiting for you to pick it up off the ground, and rumors have it that the main deposits have yet to be discovered!" The players are told that they must accumulate as many golden nuggets as possible and that their first task is to form mining companies and find a way to get to the goldfields. They are offered five ways (by ship around Cape Horn, by ship via a Panama portage, or by one of three wagon trains), but even the safest route is risky, and many of the would-be miners must try over and over to get to this fabled place. When they do get to Golden Gulch, they must stake claims, purchase supplies, and begin their search for gold.

Students collect gold when they correctly answer questions about its value, about nineteenth-century mining techniques, or about life in the mining camps of the old west. Their task is complicated by claim jumpers and fate. Claims are jumped when a mining team that has already staked a claim cannot correctly answer a question and another team can. Fate takes the form of random events (such as illness, theft, or the arrival of the first woman the miners have seen in fourteen months) printed on cards the teams draw. Teams (mining companies) of players make a total of four decisions that affect the amount of gold they find or lose (how best to get to the mining camp, what to do about a hypothetical group of miners that builds a dam upstream from their claim, what to do about a captured bandit, and whether to continue to search or quit and get a job in town). Miners may collect additional gold by writing a research paper, preparing a Saturday night activity (similar to those that amused the forty-niners), and by maintaining a log.

This highly structured game ends after thirteen one-hour sessions and the activities prescribed to fill them have been completed. The winner is the player with the most gold at the end of the game. (DCD)

Cost: $22.00

Producer: Interact, P.O. Box 262, Lakeside, CA 92040

GOLD RUSH DAYS

Playing Data
Copyright: 1975
Age Level: grades 5-12
Number of Players: 20-45
Playing Time: 5 50-minute sessions
Preparation Time: 90 minutes
Packaging: 13-page photocopied instruction booklet

Description: Participants assume the roles of groups involved in the California Gold Rush of 1849. These groups represent the people who went to California by way of Panama, by way of the Great Plains, the miners, and the inhabitants of the mining camps. Each member of each group is given a one-page profile of the situation of the people that group is to role play. These groups then have three sessions to produce a dramatization of their profile. The four dramatizations are presented during the fourth session. During the fifth session, the players are asked to discuss such questions as "How realistic was the average forty-niner?" and "If you were living during the gold rush days would you have been one of the forty-niners?" (DCD)

Cost: $3.00

Producer: Edu-Game, P.O. Box 1144, Sun Valley, CA 91352

GRAND STRATEGY

Clark C. Abt and Ray Glazier

Playing Data
Copyright: 1970, 1975

Age Level: grades 9-12
Number of Players: 10-30 in 10 teams
Playing Time: 3-4 50-minute periods
Preparation Time: 1 hour (est.)
Packaging: professionally produced and packaged simulation includes map of Europe, 30 copies of rules, role profiles, and 23-page teacher's manual

Description: The authors state that the objective of this simulation of the origins of World War I "is that the process of international diplomacy be experienced, witnessed, comprehended." Students, or teams of students, represent diplomats for the nations principally involved in the beginning of the conflict: Austria-Hungary, Belgium, France, Germany, Italy, Russia, Serbia, Turkey, the United Kingdom, and the United States. The game is played in alternating Conferring Sessions and Declaring Sessions. "Conferring Sessions are for the purpose of nation team caucus, negotiations between nations via emissaries, planning strategies, etc. The purpose of Declaring Sessions is to make public moves, such as announcements to the world press, mobilization and deployment of troops and/or ships, attacks and defensive military actions, and conduct of international meetings convened by players or by neutral Switzerland." The authors also stress that "1) this is a diplomacy game and not a war game, and 2) the events in the game need in no way conform to the historical facts following the end of June, 1914." (DCD)

Note: For more on this simulation, see the essays on history simulations by Bruce E. Bigelow and on international relations simulations by Leonard Suransky.

Cost: $39.00

Producer: Games Central, Abt Publications, 55 Wheeler Street, Cambridge, MA 02138

GREAT PLAINS GAME

Lowell Thompson and Charles Kaiser, University of North Dakota

Playing Data
Copyright: 1972
Age Level: grades 5-12
Number of Players: 2-6
Playing Time: 1-4 hours in 1-hour periods
Preparation Time: 2-3 hours

Description: This is a board game where players advance in the usual fashion of board games. It lacks reality in that all of the confrontation involves an exchange of money which is really not characteristic of confrontations in the wild west as any TV viewer knows. Primarily players need to decide how to invest their wealth, when to negotiate with other players, and when to provoke confrontation and how to deal with nature. The rules of the game are kept simple and rather vague in keeping with the legal structure of the nineteenth-century West. (REH)

Editor's Comment: Despite my prejudice against pure and simple board games, I have to indicate that the various moves and chance cards and instruction cards in this simulation are much better than average. So I can believe the designers of the game will accomplish some of their objectives which are to get over to students what the tools, culture, and life of the Great Plains was like. It is neither difficult nor expensive to change the game to suit individual teacher or player circumstances and players are encouraged to do that. (REH)

Cost: $2.95

Producer: Northern School Supply, P.O. Box 2627, Fargo, ND 58201

GUERILLA WARFARE

Northern Vision Services

Playing Data
Copyright: 1974
Age Level: high school, adults
Number of Players: 8-24 in 4 teams
Playing Time: 2 1-hour periods
Packaging: 8 page manual

Description: Players of *Guerilla Warfare* gain experience with the leadership, decision making, and strategies necessary for successful guerilla warfare. After the game is explained and players familiarize themselves with the map, group profile sheets, achievement points chart, casualty records chart, and rules sheet included in the manual, they divide up into four groups, the Vietcong and the North Vietnamese infiltrators which consult together, and the ARVN (South Vietnamese troops), and the U.S. troops, which also work together. Each group has a different number of military strength points and achievement points as determined by its number of troops and its military supplies. After groups have chosen leaders (who will make all final decisions) and seconds-in-command, they discuss their initial strategy and determine secretly their initial placement of troops on the map. Troops may be divided into a number of units, and units may all be moved separately each turn. Every round, after troops have been placed by each group, the trainer reveals their locations. If enemy units are within two squares of each other, a battle takes place and the winning unit is the one with the greatest number of points. The number of points lost or gained as a result of battle are determined by difference in the strength of the units engaged in the battle and whether or not a unit is the victor or the defeated. The game may end when the trainer determines or when one force is annihilated. Players then discuss the application of the game, their development of strategies, and the skills necessary for effective leadership. (TM)

Cost: $4.00

Producer: Northern Vision Services, 77 Swanwick Ave., Toronto, Canada

THE HAYMARKET CASE

David DalPorto, Mount Pleasant High School

Playing Data
Copyright: 1969, 1972
Age Level: grades 7-12
Prerequisite Skills: reading, grade 9; math, grade 0
Number of Players: 25-55
Playing Time: 5-6 periods of 50-minutes each
Preparation Time: 1-week study of court producer

Description: *The Haymarket Case* is a simulation of the famous court trial of 1886. Students assume the roles of attorneys, legal aides, witnesses, defendants, jurors, and court personnel. The case begins with selection of the jury, continues with rulings by the judge, statements by the attorneys, presentation of cases by each side, including witnesses and evidence, and ends with the jury's verdict.

During the simulation, students must discuss such vital issues as the definition and limits of free speech. Courtroom dialogue brings out the major issues that divided labor and business during the late 1800s. Students must prepare for their part in the simulation by knowing the information given in the fact sheet and in their role sheets. Students who play the part of defendants and witnesses must elaborate on the information contained in their role sheets. This requires good use of their imagination and careful cooperation with the attorneys. (Author)

Note: For more on this simulation, see the essay on history simulations by Bruce E. Bigelow.

Cost: $15.00

Producer: History Simulations, P.O. Box 2775, Santa Clara, CA 95051

HERITAGE

Diane Wesley, Fuerte School, El Cajon, California

Playing Data
Copyright: 1976
Age Level: grades 5-8
Number of Players: variable, in teams of 3
Playing Time: 14 1-hour sessions
Special Equipment: overhead projector
Packaging: 8-page student guide, 22-page teacher guide

Description: Students playing *Heritage* study road maps and U.S. geography and make themselves familiar with historical landmarks as they participate in a cross-country road race. During the first two days, they take a pretest; choose teams consisting of a leader, driver, and navigator; discuss the forty-five historic sites; identify the fifty states; and plot their travel course. Each team then begins work on 500-word research reports on one historic site. During each day of the race, which lasts about ten days, each team draws a fate card (e.g., Bridge out, Detour, Lose 80 miles), calculates the consequences, keeps a diary and travel log, and works on and presents its research project. The roles within each team change daily. Teams score their progress in miles and in points for excellence in their written work. The game ends with the announcement of the winning team, a test, and an evaluation. (TM)

Cost: $14.00

Producer: Interact, P.O. Box 262, Lakeside, CA 92040

HOMEFRONT

Arthur Peterson

Playing Data
Copyright: 1977
Age Level: high school
Number of Players: 4 teams of 8 players
Playing Time: 20 one-hour sessions
Packaging: 55-page bound manual

Description: The purpose of *Homefront is to examine the U.S.* homefront during World War II in depth, encouraging dialogue between adults who remember the period and students who do not, and to introduce students to historical skills. In the words of the game manual, "In Homefront students work as teams, study pairs, and individual. They perform a number of activities which increase their understanding of American life during the World War II years and their knowledge of how historians work. The unit has five overlapping phases that require approximately 20 days to complete. In phase I, teams of students "create headlines for important dates in World War II," and "study pairs search out facts to support general statements about the causes of the war." During phase II, students are encouraged to seek information that many Americans born before 1927 can provide. "Students conduct interviews, seek out memorabilia, learn and/or tape record and interpret certain significant popular songs of the period." Each student works as a specialist in one of four areas—entertainment, work, family life and education, social problems. During phase III students examine the homefront in depth as they "practice the skills of an historian." Activities include identifying various homefront persons and phenomena, generalizing from facts, interpreting documents, forming hypotheses, and making decisions. In phase IV students are encouraged to do special projects such as group and individual reports. Phase V includes special group presentations and debriefing. (DCD)

Cost: $14.00

Producer: Interact, P.O. Box 262, Lakeside, CA 92040

HOMESTEAD

Jay Reese

Playing Data
Copyright: 1974
Age Level: grades 5-8 average ability; grades 9-12, below average
Number of Players: 25-40
Playing Time: 10 class periods

Description: Students draw family identities as early pioneer homesteaders out on the American prairie after the Civil War. Then students read a brief history of the Homestead Act, discuss what homesteading was, who homesteaders were, where people homesteaded and why such things as sections and townships will be important to them as settlers. After gaining this background knowledge, students examine a map of the land to be homesteaded and pick their own quarter sections of land. They can select either valley land, prairie land, grazing land or timber land. During the following five class periods, students simulate five years of life on their homesteads. They plant crops, raise livestock and cut timber in an attempt to become prosperous farmers. By the end of these five periods, some homesteaders realize that they are not cut out to be farmers, whereas others begin to see the need for special services such as general stores, banks, blacksmiths, and doctors. From this natural need, they form a community and their frontier town grows. During the simulation's next four periods, the homesteaders hold elections, pay taxes, and build roads and railroads. Added to all of these activities are fate cards which simulate such things as dry wells, lame horses, extra crops, and neighbors' help. (Producer)

Cost: $22.00

Producer: Interact, P.O. Box 262, Lakeside, CA 92040

INDEPENDENCE

Charles Kennedy and Paul DeKock

Playing Data
Copyright: 1975
Age Level: high school
Number of Players: 12-35
Playing Time: 13-15 1-hour periods
Packaging: 39-page manual

Description: Independence is a simulation game based on the events leading to the American Revolution. Students each participate in three groups—a political group (Loyalists, Patriots, and Neutralists), a section group (New England, Middle, and Southern Colonies), and an area group (rural and urban). The first few days of the game involve reading bulletins about historical events, doing research, and meeting in political groups to decide what pressure actions to use. Lists of possible pressure actions (e.g., hanging someone in effigy) are included in the manual. Each group has a number of power units and pressure points which they may spend in taking pressure actions against another group or may lose when a group takes an action against them. During political meetings, pressure cards (e.g., your children are harassed in school. Lose 3 power units) and historical bulletins are read, both of which result in loss or gain of pressure points to one or more groups. The fifth and sixth days of the simulation, the stamp act Congress convenes and a test is given on the readings. Students also gain points for their participation in debates during the various congresses, their scores on tests, and their completion of various research projects (e.g., research the Albany Plan of Union) or challenge projects (e.g., draw a series of cartoons satirizing a revolutionary event). The seventh and eighth days, area and section meetings are held. The remainder of the game involves group discussion, debating during the First and Second Continental Congresses, political meetings, and several short tests on readings assigned in the manual. The game ends with a group decision about declaring independence, and a debriefing period. Considerable background information, discussion questions, scoresheets, assignment suggestions, and a bibliography of film and books are included. (TM)

Cost: $14.00

Producer: Interact, Box 262, Lakeside, CA 92040

INDEPENDENCE '76

Charles L. Kennedy

Playing Data
Copyright: 1975
Age Level: 8-adult
Number of Players: 2-8
Playing Time: ½-1 hour
Packaging: professional game box

Description: Players throw the die each turn and move revolutionary leader markers on the playing board from 1763 through historical events to July 4, 1776. As they land on historical events, they gain or lose "patriot" pieces based on the outcome of these events. When they land on stockade or redcoat spaces, they are delayed in jail. Pressure and destiny spaces indicate that players must pick cards that require further actions, such as a move or a gain or loss of patriots. A player who loses all his or her patriots and goes into debt must lose one turn for every ten patriots owed. The game ends when a player lands exactly on July 4, 1776. (TM)

Comment: Independence '76 is based entirely on chance and should be played only for fun; it teaches very little American history. (TM)

Cost: $14.95

Producer: Civic Educational Aids, 513 Holly, Crookston, MI 56716

INDIANS VIEW AMERICANS, AMERICANS VIEW INDIANS

Rachel Reese Sady and Daniel C. Smith

Playing Data
Copyright: 1970
Age Level: grades 9-12
Number of Players: 5-25
Playing Time: 10-20 hours
Preparation time: 2 hours (est.)
Packaging: professionally packaged multimedia unit contains long-play phonograph record, 2 filmstrips, 25 student's manuals, and 82-page teaching manual

Description: This multimedia unit stimulates students to analyze their own cultural biases through studying Indian-white relations during the Colonial and immediate post-Colonial period.

The unit is divided into three sections. *Red Men and White Men* examines how new and native Americans regarded each other and identifies some of the cultural components in human responses to peoples and situations. *The Black Hawk War and Cherokee Removal* studies these two events with particular emphasis on the viewpoint of the Indian groups. This section is presented in the form of case studies and includes simulation exercises. *The Sun Dance and Ghost Dance* demonstrates how customs and institutions can be understood only in their cultural context. Students are called on to see seemingly barbaric and irreligious ceremonies from the Indian point of view. (Producer)

Cost: $82.00

Producer: Educational Audio Visual, Inc., Pleasantville, NY 10570

INTERACTION—A BALANCE OF POWER IN COLONIAL AMERICA

David J. Boin and Robert Sillman

Playing Data
Copyright: 1972
Age Level: grades 7-12
Number of Players: class size
Playing Time: 3-4 class periods

Description: Students are divided into groups of six to nine members representing frontiersmen, bankers, merchants, and small farmers to represent the society of Colonial America. They are presented with ten bills proposed by the Colonial government to discuss, negotiate, and vote on. This simulation emphasizes the social and economic differences in the colonies just before the American Revolution. (REH)

Editor's Comment: This is a standard Edu-Game format with a few pages of teacher's manual and locally reproducible student material. (REH)

Cost: $3.00

Producer: Edu-Game, PO Box 1144, Sun Valley, CA 91352

INVOLVEMENT

William Krause and David Sischo

Playing Data
Copyright: 1971, 1974
Age Level: high school
Number of Players: 15-45
Playing Time: 50-100 minutes per exercise
Preparation Time: 1 hour (est.) per exercise
Packaging: four staple-bound booklets, each containing materials and instructions for six activities

Description: This series contains twenty-four activities, games, and simulations pertaining to U.S. history and government. The first volume contains activities for high school classes studying the period that concludes with the Civil War. The second contains activities about modern American history from reconstruction to the end of World War II. The third contains decision-making activities that reflect themes in American history from 1620 to 1972. The final volume, entitled "American Inquiry," addresses modern national concerns such as the presidency, Black progress, and inflation.

"Federalist or Anti-Federalist" typifies the activities in Volume I. Groups of four or five students read the biographies of fictional Americans who might have lived during the 1780s. The groups must decide whether each person was a Federalist or anti-Federalist and defend his decisions.

In "Stock Market Game" (Volume II), participants buy and sell eight different stocks during seven rounds that represent 1920 to 1929. The stock prices fluctuate from year to year, and these fluctuations represent the appreciation and depreciation of several common stocks during the 1920s. After the final round, students are encouraged to explain how they made or lost money and what they think caused the stock prices to rise and fall.

Volume III contains exercises in which the players found an English colony in America in 1620, recreate Revolutionary War and Civil War battles, and simulate the Dred Scott trial and a trial resulting from the Pullman Strike.

Typical of the activities in Volume IV is a research project in which the students compare and contrast the Black and White populations of the United States. This exercise includes information students can use to compare the life expectancies, median incomes, occupations, unemployment rates, median years of education completed, and housing conditions of the two races. (DCD)

Cost: $10.00 per volume, $40.00 per set

Producer: Involvement, 3521 E. Flint Way, Fresno, CA 93726

JAPANESE AMERICAN RELOCATION, 1942

Rachel Reese Sady and Victor Leviation

Playing Data
Copyright: 1970
Age Level: grades 9-12
Number of Players: 5-24
Playing Time: 10-20 hours
Preparation Time: 2 hours (est.)
Packaging: professionally produced multimedia unit includes long-playing phonograph record, 2 filmstrips, 24 copies of student booklet, and 84-page teaching guide

Description: This multimedia unit actively involves students in a case study of one of the most extraordinary episodes in American history. It explores the causes, consequences, and implications of the evacuation and detention by the U.S. government during World War II of 110,000 residents–citizens and aliens–on the basis of their ancestry. Students are given speeches, articles, interviews, Supreme Court opinions, and other official documents, along with poignant written and recorded texts and a variety of visual materials. From these sources, they learn the underlying motives for the relocation and the patterns of hate and suspicion involved. Finally, through a simulation entitled *Nisei Scouts,* students actually experience the problems faced after the war by Japanese-Americans who had to decide whether to return to their former homes. (Producer)

Cost: $82.00

Producer: Educational Audio Visual, Inc., Pleasantville, NY 10570

JUDGMENT

Jonathan Harris

Playing Data
Copyright: 1977
Age Level: grades 9-12
Number of Players: 28-36
Playing Time: 10 1-hour sessions
Preparation Time: 3 hours (est.)
Packaging: 41-page instruction booklet

Description: *Judgment* simulates the trial of President Truman for his decision to drop the first two atomic bombs. The leader assigns participants the roles of Truman, his defense attorneys, the prosecuting attorneys, bailiff, court reporter, a panel of judges, members of a jury, or one of fourteen witnesses including Generals Marshall and Eisenhower. The simulation is divided into pretrial, trial, and posttrial phases. Before the trial participants must examine and reveal their opinions about the bombings, make themselves familiar with the events of World War II and court procedure, and learn their roles. During the trial, participants must play their specified parts and the jury must decide if Truman was guilty of crimes against humanity. After the trial the participants again examine their opinions about the bombings. (DCD)

Cost: $14.00

Producer: Interact, P.O. Box 262, Lakeside, CA 92040

KING TUT'S GAME

Peter A. Piccione

Playing Data
Copyright: 1977
Age Level: age 10 and up
Number of Players: 2
Playing Time: 1 hour
Packaging: professionally packaged boards game with playing surface, dice, tokens, and instructions

Description: King Tut's Game is a modern version of an ancient Egyptian game called Senet (which means "passing"). Although scholars have long been aware of the game, there was no certainty until recently about how it was played. A graduate student at the University of Chicago's Oriental Institute studied similarities between certain elements in religious papyri and the game and reconstructed what he estimates to be 90% of the original game.

The game is played on a T-shaped track of squares. Each player starts at opposite ends of the top and tries to move five tokens to the bottom of the T and back to the starting point. The first to do so wins. Rules governing movement of the tokens give play a strategic nature. (DCD)

Cost: about $5.00 retail

Producer: Cadaco, Inc., 310 West Polk Street, Chicago, IL 60607

LIBERTÉ

Sister Marleen Brasefield, University High School, San Diego, California

Playing Data
Copyright: 1970
Age Level: grades 7-12, average to above-average ability
Prerequisite Skills: reading, grade 11; math, grade 7
Number of Players: 25-40 in 5 teams
Playing Time: 14-25 hours in 1-hour periods
Preparation Time: 3-5 hours

Description: Liberté simulates the social stratification, economic conditions, and political process leading to the French Revolution of 1789. Students are divided into five socioeconomic groups in the troubled France of 1789: royalty, clergy, nobles, bourgeoisie, peasants. RIPS (revolutionary influence points) are gained and lost through researching role identities and historical background information and through daily taxation, tithing, title selling, trading (for which forms are provided), and fate (which is represented by historical bulletins that change financial, social, and political status and push factions toward revolution). After this opening economic phase, politics enters with the opening of the Estates-General and the introduction of the National Assembly, before which students debate and decide twelve of the gravest issues of the revolution. Then after researching historical indictments against their king, students bring Louis XVI to trial. Suddenly Robespierre introduces his Reign of Terror. Finally, an evaluation discusses the nature of revolution, past and present. (REH)

Note: For more on this simulation, see the essay on history simulations by Bruce E. Bigelow.

Cost: $14.00

Producer: Interact, P.O. Box 262, Lakeside, CA 92040

LIFE IN THE COLONIES

Patricia D. Flowers (with Ray Glazier)

Playing Data
Copyright: 1977
Age Level: grades 7-10
Number of Players: 6-8
Playing Time: 45-50 minutes
Preparation Time: 1 hour (est.)
Packaging: professionally packaged board game includes role profiles, chance cards, die, playing surface, and 38-page teacher's manual

Description: Life in the Colonies, according to the designers, is designed to simulate "daily colonial living including family responsibilities, relationships with the church, and even occasional violence, with an emphasis on the political and economic factors of life." The players' roles represent typical colonial farmers, merchants, ministers, planters, printers, royal officials, slaves, and widows. The action of the game occurs on a folded, paper playing surface on which is printed a star-shaped track of squares. The track includes blank and "historical events" squares, which are used to announce the player's participation in some event between the French and Indian Wars and the Declaration of Independence. Players advance tokens according to the roll of a die. Each role specifies a point goal that that player must try to achieve. Until the skirmish at Lexington, points are gained or lost according to how the historical event players land on would have affected the real-life counterparts of their roles.

All players must rest their tokens at the Battle of Lexington and Concord and declare themselves either Patriots or Tories. Subsequently, players gain or lose points according to their declared loyalties. The number of points gained or lost because of the effect of some historical actuality is included with a rationale in an "Outcome Booklet." For example, the farmer starts the game with fifty points and must finish with sixty. Before the Battle of Lexington and Concord, the farmer would gain points by landing on the repeal of the Townshend Acts and lose points for landing on the decree of the Quartering Act. After the start of the war a Patriot farmer gains points when the British evacuate Boston, and a Tory farmer loses points by landing on Bunker Hill. The game materials include handouts describing colonial life and a thorough bibliography for teachers and students. (DCD)

Cost: $26.00 plus $1.50 shipping/handling

Producer: Abt Publications, 55 Wheeler Street, Cambridge, MA 02138

LONGMAN HISTORY GAMES

B. Barker, D. Birt, R. Boden, and J. Nichol

Playing Data
Age Level: grades 9-12
Number of Players: class

Description: This is a collection of games which are flexible in use and suit most syllabuses and teaching methods. They span 400 AD to 1926 and although the accent is on British history, games on North America and Africa are included. Each game contains illustrations and documentary materials.

The Norman Conquest: Five-part simulation showing course of action open to Duke William of Normandy in the year 1066.

The Development of the Medieval Town: Flexible, five-part simulation guiding pupils to make development decisions and trace the growth of a settlement between the Saxon invasions and the twelfth century.

Trade and Discovery: Two-part map-based simulation highlighting the factors affecting English overseas development through the activities of Elizabethan and Jacobean merchants and adventures.

Frontier: Map-based simulation emphasizing the importance of the frontier in American history and the problems facing British colonial settlement in North America 1690-1782.

Ironmaster: Based on the Darby family's Coalbrookdale Company, guides students through developments in the eighteenth-century iron industry.

Canals: Players form committees to represent canal companies and enterprising landowners competing to develop a fictional industrial district. Simulates the engineering and business problems confronting entrepreneurs after 1760.

Congress of Vienna: Players are divided among the different national delegations and sent to Vienna to conclude a general European settlement following the defeat of Napoleon.

Harvest Politics. Simulates the conflict and compromises which dominated English political life in the decade and a half following the Great Reform Bill.

Railway Mania: Recreates the situation in Britain in the 1830s and 1840s when, as a result of furious speculation, tens of millions of pounds were poured into railway companies.

Village Enclosure: Set in the early nineteenth century shows the transition from "open fields" to hedged fields and scattered farmhouses of today. In four parts, graded for difficulty.

The Scramble for Africa: Reproduces the balance of interests and conflicts involved in the colonization of Africa from the Treaty of Berlin (1884) to the end of the Moroccan Crisis (1911).

General Strike: A simulation of the political and industrial conflicts culminating in the General Strike of 1926. (Producer)

Note: For more about *Congress of Vienna* and the *Scramble for Africa,* see the essay on history simulations by Bruce E. Bigelow. Both simulations also have separate descriptions in these listings.

Cost: boxed reference set (1 copy of each game) £ 5.50 plus postage; contact producer for other options and quantity discount schedule.

Producer: Longman Group Ltd., Resources Unit, 9-11 The Shambles, York, England

MANCHESTER

John Blaxwell (revised by Thomas K. Dorman with Ray Glazier)

Playing Data
Copyright: 1977
Age Level: grades 7-12
Number of Players: up to 8
Playing Time: 5-10 10-minute rounds
Preparation Time: 3 hours (est.)
Packaging: professionally produced and packaged board game and 36-page teacher's manual

Description: Manchester is about the Industrial Revolution in England from 1820-1840. The players assume the roles of squire, farmers, mill owners, and families of workers. Their probability of success at fullfilling the object of the game, to make as much money as possible, exactly parallels the chances for success that their historical counterparts enjoyed. A squire's wealth can be expected to increase slightly in a short game and substantially in the long run. Farmers may expect to break even or, possibly, accumulate a small amount of land or money. Mill owners should increase their profits, productive capacity, and net worth as the game progresses. The workers must struggle hard to stay out of the poorhouse and survive.

Each player starts with assets, fixed expenses, and instructions on how she or he may make money. Farmers, for example, may buy or lease land from the squire, who enjoys the option of hiring labor and farming the land himself. Both earn income by buying and selling grain. Mill owners must buy looms and hire workers if they are to make money. All three roles must try to extend their profits and maintain their standard of living by paying their workers as little as possible. Each worker family is responsible for keeping ten workers gainfully employed. Each worker must pay seven pounds a year for living expenses but can expect to earn only five pounds a year for home weaving or six pounds a year for working in the mills. Worse, before a worker can even get a job in a mill she or he must pay a one-pound travelling fee.

A multicolor game board represents the city of Manchester and the surrounding countryside. Players put tokens on the board to represent the movement of workers, the ownership of land, and the location and ownership of looms. The market price for cloth and grain is announced by drawing and reading a chance card. Players compute their income (unless they are workers who need only count their wages) by multiplying the market price by the number of looms or plots they own or rent and the number of workers in their employ. Expenses for living, traveling, paying wages, and buying or renting land or looms is then deducted from income each year, or ten-minute round. Players who cannot meet their expenses must go to the poorhouse and work for someone else, with all of their wages going to the bank, until their debts are paid. The net worth of any player is calculated by adding his or her cash assets and the cash value of any property or equipment that he or she owns. The winner is the player who achieves the greatest percentage increase in net worth during the play of the game. (DCD)

Cost: $35.00 plus $1.50 shipping/handling

Producer: Games Central, Abt Publications, 55 Wheeler Street, Cambridge, MA 02138

MERCHANT

Jay Reese

Playing Data
Copyright: 1974
Age Level: grades 5-8 above average; grades 9-12 below average ability
Number of Players: about 32
Playing Time: 15 class periods

Description: *Merchant* begins when students are paired as owners of sixteen competing general stores located in four different western communities at the turn of the century. During the next fourteen hours students work in an environment of change. Because their community is gradually emerging from the nineteenth century, they must change their stores from simple general stores to shops and stores which meet the ever-growing needs and demands of their consumers. At the beginning of the simulation, the merchants' first job is to select an initial inventory from a master list of merchandise. Then decisions must be made concerning store hours, work schedules, and employee pay rates. After they have made these decisions, the store owners must begin to create merchandise displays, advertising and a store floor plan. Once they have set up their stores, the merchants must begin making decisions concerning the running of their stores. Each pair of store owners makes decisions concerning the pricing, special buying, advertising, handling, salesmen, merchandise arrangement, and holding sales. In this way the stores compete with one another as they strive to gain larger percentages of the available market. All these experiences produce a simulation which gives students insight into how a small business operates and what life must have been like during this hectic period of change and progress. (Producer)

Cost: $22.00

Producer: Interact, P.O. Box 262, Lakeside, CA 92040

MR. PRESIDENT

Louis Alpern and Robert W. Allen

Playing Data
Copyright: 1970
Age Level: junior and senior high school
Number of Players: 3-6
Playing Time: 20 mins. 1 hour
Packaging: professionally designed box with cards, chart, and instructions

Description: *Mr. President* consists of a set of cards, each with a picture of one of the first thirty-seven presidents on the front, and facts about and quotations from that president on the back. A chart pictures each president, and under each picture appears the president's birth and death dates, years of office, alma mater, and political party. The authors suggest six game variations to play with these materials. In the first, one player holds up a picture of one president and the other players identify that president's number. In the second, the players must identify the president by name. In the third variation, the player holding the cards must announce a fact to be identified about the president, such as political party or year of birth, before holding up that president's picture. In another variation, the players must identify a president when the dealer reads a fact about him from the back of a card. (For example, who said "whenever we have tried to purchase peace at any price, the price has always been an installment payment on a bigger war"? Answer: Richard Nixon.) In a fifth version, players may wager cards on their answers. The sixth variation has the players act out clues about a president. If none of the other players can determine the president, the actor loses a card. The winner is the player with the most cards after all have been played. (DCD)

Cost: $10.00

Producer: Wff 'n Proof, 1490 South Boulevard, Ann Arbor, MI 48104

MISSION

David E. Yount and Paul DeKock, El Capitan High School, Lakeside, California

Playing Data
Copyright: 1969
Age Level: grades 7-12, average to above-average ability
Prerequisite Skills: reading, grade 11; math, grade 0
Playing Time: 12-20 hours in time periods of 1 hour
Preparation Time: 5 hours

Description: As members of various factions that arose in America during our involvement in Vietnam, students research, then argue the viewpoints of Hawks, Doves, or Moderates. While pressure groups strive to gain major influence over the president's foreign policy decisions, everyone faces the possibility of being drafted, wounded, and killed in Vietnam (or having one of these happen to a relative). Communication barriers, draft protests, prestige factors, popularity polls, and a national presidential election, all coalesce during a crisis carrying America to the brink of World War III. Pre- and post surveys show teacher and students how the experience has changed what students know and believe about the complexities of America's "mission" in Southeast Asia. (Publisher)

Cost: $14.00

Producer: Interact, P.O. Box 262, Lakeside, CA 92040

MUMMY'S MESSAGE

Tony Maggio

Playing Data
Copyright: 1978
Age Level: grades 7-12
Number of Players: 18-42 in six teams
Playing Time: 4-10 1-hour sessions
Preparation Time: 2 hours (est.)
Packaging: professionally packaged game includes 3 pyramid maps, 35 student guides, question, problem, and hieroglyph cards, discovery cards, and 55-page instruction booklet.

Description: Mummy's Message introduces students to classical Egypt by simulating the excavation of a pyramid. The class is divided into three groups which play the game simultaneously. Each group is divided into two teams that race to be the first to excavate the pyramid and piece together a "Mummy's Message."

The game is played on a flat surface that represents passageways in a pyramid. The competing teams enter the structure through different doors and progress, one square at a time, by correctly answering questions about the people, places, history, and customs of ancient Egypt (such as, "True or False–The Nile Delta is referred to as Lower Egypt"). Problem cards and hieroglyphic cards are placed on various squares in the interconnected passageways. Problem cards recount problems faced in actual excavations (such as blocked passageways and physical hardships) and cause teams to lose one or more turns. Hieroglyphic cards contain parts of a message. Players must explore the pyramid and piece together the Mummy's Message. The first teams in the three groups to do this are the winners. (DCD)

Cost: $22.00
Producer: Interact, P.O. Box 262, Lakeside, CA 92040

NORTH VS. SOUTH

David N. DalPorto

Playing Data
Copyright: 1971
Age Level: grades 7-12
Prerequisite Skills: reading, grade 9

Number of Players: 20-80 in 6-8 teams
Playing Time: 4-6 periods of 50-minutes each
Preparation Time: variable
Supplementary Material: additional curricular material

Description: *North vs. South* is a simulation of the pre-Civil War period where students must deal with the issues that divided the North and South. The students are divided into six teams that represent different positions on the slavery issue. The three northern teams are the Abolitionists, Free Soilers, and Moderates whereas the three southern teams are the Rebels, Realists, and Moderates. These teams must consider three major case problems while playing the role of Senators. The three major case problems are the Compromise of 1850, the Kansas-Nebraska Act, and John Brown's raid.

Finally, after the election of Lincoln in 1860, those that represent the South must decide whether or not to secede whereas the Northerners must decide whether to fight a war to "save the union" or to let the South go "peacefully." (Author)

Cost: $12.50

Producer: History Simulations, P.O. Box 2775, Santa Clara, CA 95051

NUREMBERG

Arthur Pegas, El Capitan High School, Lakeside, California

Playing Data
Copyright: 1971
Age Level: grades 7-12, average to above-average ability
Prerequisite Skills: reading, grade 9; math, grade 0
Number of Players: 25-40
Playing Time: 15-25 hours in 1-hour periods
Preparation Time: 3-5 hours
Supplementary Material: materials on German history, 1933-1945

Description: Nuremberg is a simulation of the International Military Tribunal of 1945-1946. Few events in the history of mankind have the dramatic impact and the legal significance of the International Military Tribunal at Nuremberg 1945. The tribunal tried to answer questions that have confounded men for centuries: What should an individual do in a society that has gone beserk? What should an individual do when his government orders him to commit an act contrary to his beliefs and to the beliefs of civilization? Under what laws or what moral code may a defeated nation be tried? Can the leaders of victorious nations try leaders of defeated nations when the actions of the victors are also questionable? The precedents for these questions and their inherent moral-legal dilemmas comprise this simulation. Every student in class portrays one or more of the following roles: four judges (French, British, Russian, American); prosecution and defense attorneys; prosecution and defense witnesses; several defendants (Hermann Goering, Rudolph Hess, Joachim Von Ribbentrop, Alfred Jodl, Julius Streicher and others). The simulation kit includes a detailed bibliography for older, more capable students doing most of their own research; already researched information for younger students; and step-by-step instructions for each role. Students who have simulated Nuremberg will never forget the horrible issues raised by Hitler's Germany of 1933-1945. They will also see how these same issues continue to haunt us in a world still practicing war. (Producer)

Note: For more on this simulation, see the essay on history simulations by Bruce E. Bigelow.

Cost: $14.00

Producer: Interact, P.O. Box 262, Lakeside, CA 92040

ORIGINS OF WORLD WAR II

The Avalon Hill Company

Playing Data
Copyright: 1971
Age Level: grade 7-college
Number of Players: 2-5
Playing Time: 45 minutes to 2 hours
Preparation Time: 30 minutes

Description: As the U.S. president, what are you going to do about Germany? Can you stop France and Britain from making a pact with Germany? What will Russia do?

The simulation *Origins of World War II* is a five-player game which recreates the opportunities for diplomatic alliances and conflicts of the 1930s between the United States, Russia, Britain, France, and Germany. Each of the five players represents a country. Each country has its own objectives, political strengths, and military power. Germany, as with the countries, can only win the game by making one or more alliances. The players' ability to negotiate a favorable alliance offers them an opportunity to win the game.

Winning the game is determined by the player having the most number of points after six turns. If the player representing either Germany or Russia wins and obtains fifteen or more points, it is assumed that World War II has begun.

A mapboard of 1939 Europe showing thirteen countries including the areas of Alsace-Lorraine and the Rhineland is the playing board for the game. Each player receives counter points called political factors to negotiate and attack other countries. There are understanding counterpoints and control counterpoints each worth five political factors. The use of an understanding counter shows that there is a diplomatic agreement with a country. A control counter shows that one country has taken over another country.

Counter cards are allotted to each player on the basis of a political factor allocation chart. A player's counter points are increased or reduced by diplomatic conflict (attacking another country) in which a diplomatic conflict table and a roll of the die are used to determine the outcome.

The game kit includes a map board of Europe 1939, five sets of counters for Germany, France, Britain, Russia, and the United States, a set of five historical objective cards, die, charts, and an era of diplomacy booklet. Although there are optional rules and suggestions for playing the game, the game generally takes forty to sixty minutes to play. The game is a six-player game; however, teams can be used to increase the number of participants.

Objectives: The game provides experiences in critical thinking as well as opportunities to learn social studies concepts. These include:

- I. Critical Thinking Skills
  - A. Comparisons
    1. How does this fulfill my national interests?
    2. What does France get out of an alliance with Germany?
  - B. Decision Making
    1. What is my strategy?
    2. Should I take a chance?
  - C. Interpretation
    1. What does Russia's actions mean for the United States?
    2. What are Germany's national priorities?
- II. Social Studies Concepts
  - A. Alliances
    1. Causes, effects, and obligations of alliances
    2. Forces creating alliances
  - B. Compromise
    1. The use of compromise as a means of winning
    2. When to compromise
  - C. Winning
    1. A negotiated strategy
    2. Methods of winning
  - D. Bargaining
    1. Self- and common interests and concerns
    2. Strategies for bargaining
  - E. National interests and commitments
    1. Nationhood
    2. Economic and geographical interests
  - F. Historical content

1. Chronology
2. Sequence
3. Cause and effect

G. Internationalism
1. Differences and commalities of countries
2. Conflict resolution
3. Power

Comment: The primary strength of *Origins of World War II* is its historical setting and the need for negotiations in order to win. The country's objectives and the options to fulfill these objectives allows players to experience and exhibit their reactions and methods for achieving the objectives established for each country. Such class discussion topics as nationalism, political geography, and alliances can be pursued by students. Contemporary foreign policy can be easily initiated and examined in terms of cultural ties, location of resources, military security, and international compacts. (From the Robert A. Taft Institute of Government Study on Games and Simulations in Government, Politics, and Economics by Harry Miller, Southern Illinois University, Carbondale.)

Note: For more on this simulation, see the essay on history simulations by Bruce E. Bigelow.

Cost: $12.00

Producer: The Avalon Hill Company, 4517 Harford Road, Baltimore, MD 21214

PANIC

David Yount and Paul DeKock, El Capitan High School, Lakeside, California

Playing Data
Copyright: 1968
Age Level: grades 7-12, average to above-average ability, easily adaptable to various ages
Prerequisite Skills: reading, grade 11; math, grade 5
Number of Players: 25-36 in 6 teams
Playing Time: 23 class days
Preparation Time: several hours, to familiarize teacher with sequence
Supplementary Material: as outlined in teacher's manual
Special Equipment: duplicating equipment

Description: By chance students pull ID tags showing their economic pressure group, 1920-1940: bankers, businessmen, laborers, farmers, women, and social critics. One-sixth have blue tags (upper class); one-half, green (middle class); one-third, red (lower class). Students take a brief political attitudes survey which groups them as Republicans or Democrats during the 1920s. While studying the prosperity of the 1920s, students earn fantastic quantities of WPS (Wealth Points) during a bullish stock market that allows them to buy on a 10% margin. Suddenly the market crashes, the banks in which students have saved their WPS start closing, and the nation plunges into the Great Depression of 1929. Many students whose grades had ballooned to As now stare at Fs on their WPS balance sheets. The simulation culminates in Senate committees and a Congress which considers actual New Deal bills each pressure group researches and presents: WPA, AAA, NRA, Social Security, FDIC-SEC, the Wagner Act. The most fascinating debate of all concerns whether the legislation should be funded by taxing all students an equal number of WPS or by taxing progressively according to ability to pay. (Publisher)

Cost: $14.00

Producer: Interact, P.O. Box 262, Lakeside, CA 92040

PARTY CENTRAL

R. G. Klietsch

Playing Data
Copyright: 1972
Age Level: grades 9-14
Number of Players: 30
Playing Time: 12 rounds totaling 8 hours, 45 minutes
Preparation Time: 3 hours (est.)
Packaging: professionally packaged game includes role sheets, party guidelines, and instructions

Description: Party Central is an "active case study" which recounts the rise to power in Germany of Hitler and the Nazis from 1925 through 1933. Players assume the roles of historical persons (such as Hitler, Hindenberg, and Thaelmann) in one of six political parties or party coalitions. The action of the simulation consists of party, cabinet, and Reichstag meetings, and one presidential election. The issues discussed during these thirty-six meetings and the events that occur independent of the players' actions are specified for each round on a summary sheet. (These are historical actualities.) In addressing issues, players rely on the philosophies of their parties. Scoring is by party rather than by individual players.

In 1925, for instance, players conduct party meetings, a closed cabinet meeting, and a Reichstag assembly, followed by a presidential election and another Reichstag assumbly. Each party has a suggested position on each issue (for 1925 the issues are the tariff bill, national military organization, the revaluation of state loans, trade treaties, the amnesty bill, and the Locarno Pact), and each party tries either to gain endorsement of its position by the Reichstag or to use these issues to create or test coalitions.

Although the events in this game replicate those in Germany during the years when the Nazis were coming to power, a Nazi takeover of the political apparatus is not an inevitable outcome of play. (DCD)

Cost: $68.50

Producer: System's Factors, Inc., 1940 Woodland Ave., Duluth, MN 55803

PEACE

Arthur Peterson

Playing Data
Age Level: grades 7-12
Prerequisite Skills: reading, grade 11; math, grade 0
Number of Players: 25-40 in 5 teams
Playing Time: 14-20 hours in periods of 1 hour
Preparation Time: 3-5 hours

Description: August 1914, war has erupted in Europe. Students are grouped into five factions: Anglophiles, Francophiles, Germanophiles, idealists, realists. All give President Wilson advice based on what their research tells them is their best interests. New international crises pressure the factions to offer the President specific recommendations. By the time the 1918 armistice has ended the war, the factions have informed views as to how Wilson should represent their beliefs at the Versailles Conference. Factions next work hard publishing separate newspapers that propagandize their differing positions. Interaction intensifies as each staff examines the other newspapers for logical fallacies. When Wilson returns to the United States, the scene shifts to the Senate, where students group into new factions: Wilsonians, Irreconcilables, Loyalists, Strong Reservationists, and Mild Reservationists. In these new roles students debate whether the United States should accept or reject the Versailles Treaty. (Producer)

Cost: $14.00

Producer: Interact, P.O. Box 262, Lakeside, CA 92040

PIONEERS

Interact

Playing Data
Copyright: 1974
Age Level: grades 5-8 average to above; grades 9-12 below-average ability
Number of Players: 20-40
Playing Time: 10-15 hours in 1-hour periods

Description: As members of four simulated wagon trains heading for Oregon in 1846, students face many challenges that force them to make many of the same decisions made by early pioneers. After drawing identities as farmers, merchants, doctors, lawyers, blacksmiths, and craftsmen—or their wives—students begin the decision-making process which will help them cross the continent. The first major decision involves selecting supplies and goods to carry west in their prairie schooners. Once they have decided between items such as salt and family heirlooms, they write the first of three diary entries which will chronicle some of the highlights of their trip west. By completing class assignments, writing diary entries, making intelligent decisions, writing a short research report, and doing extra-credit projects, the students earn points that move their wagon trains across the simulation's detailed map of the route to Oregon. Working first as individuals and then together with the other members of their wagon train, the students make four major decisions and a variety of minor decisions regarding their trip. For example, the students reach a fork in the trail. Having been given information concerning each branch and several actions which might be taken, they must analyze each action and then select the best way. No matter what they decide, fate cards simulate other events such as Indians, weather, snakes, broken wagon wheels, and various obstacles in nature. Mixed in with such simulation activity is instruction in listening skills and simple research techniques, using high interest material from Western history. (Producer)

Cost: $22.00

Producer: Interact, P.O. Box 262, Lakeside, CA 92040

PREVENTING THE CIVIL WAR

David J. Boin and Robert Sillman

Playing Data
Copyright: 1972
Age Level: grades 7-12
Number of Players: class size
Playing Time: 3-4 class periods

Description: It is always fascinating to assume that wars might have been avoided. In this simulation the class holds a special convention for the purpose of amending the Constitution to avoid a very costly war. Students are divided into three main groups: northern, southern, and border states. Different Constitutional amendments are presented, each with a point value attached to it. Students must learn to negotiate and compromise if war is to be avoided.

Editor's Comment: This is another one of the Edu-Game "What If" series on history. It does not actually simulate any real situation but brings students together in a mock format so that they can get at the issues that were prevalent just before the outbreak of the Civil War. It is in the standard Edu-Game format with a few pages of teacher's manual and locally reproducible student material. (REH)

Cost: $3.00

Producer: Edu-Game, P.O. Box 1144, Sun Valley, CA 91352

THE PROGRESSIVE ERA

David N. DalPorto

Playing Data
Copyright: 1975
Age Level: junior high and high school
Number of Players: 13-37 in 6 teams
Playing Time: 4 50-minute periods
Preparation Time: a number of class periods
Packaging: 26-page manual with instructions, background materials, charts, sample articles, role sheets, record sheets, problems, and evaluation sheet

Description: This simulation will make students familiar with the Progressive Era of American history and introduce them to the processes by which reform programs are drawn up and adopted through legislative action. This means that students will need to learn about and "understand the problems that necessitated the Progressive movement in the United States history, . . . become aware of the great similarities between the problems that existed in the early 1900s and those of today, . . . be able to see the strong inter-relationship between various problems created by growth of industry in the U.S.," learn about the complexities of domestic problems, and, finally, understand the need for practical solutions in reform programs if they are to succeed.

The class is divided into six teams, each of which selects its own "progressive program" with the intention of having as much of its program as possible be adopted as the final one. (WHR)

Cost: $15.00

Producer: History Simulations, P.O. Box 2775, Santa Clara, CA 95051

PUZZLE

John McLure, University of Iowa

Playing Data
Age Level: grades 7-12, average to above average ability
Number of Players: 25-40 in 4 teams
Playing Time: 13-20 hours in 1-hour periods
Preparation Time: 3-5 hours

Description: In *Puzzle* students perform the tasks of a biographer-historian, learning to make inferences and judgments and to write biographical sketches. Before the simulation begins, the teacher hides documents (old letters, newspaper clippings, journal entries, bills of sale, family geneologies) from the lives of Amos Tuttle, an eighteenth-century Virginia planter, and Molly Andrews, his talented slave. Four teams search for simulated documents "planted" in the school library and try to discover which faculty members and which citizens in the community represent Tuttle's and Andrews's descendants. Then students talk these "descendants" into allowing their team to use the "family heirloom" documents (given the "descendants" by the teacher). Once a team has found sufficient documents to piece together its subject's life, team members divide tasks: some work on a chronology; some negotiate with other teams to trade, buy, and sell documents; some create original documents that logically could have been part of their subject's life. After each team member writes a brief biographical sketch on either Tuttle or Andrews, each team selects its most skilled biographer to write its OTB (Official Team Biography). Then in a challenge-defense session, the teams challenge and defend the biographical inferences and judgments in the four OTBs. A panel of five judges sustains or overrules each challenge. From all *Puzzle* activities teams receive BIOPS (biography points), and a winning team is declared. A presurvey, a midunit test, a post essay, and a debriefing discussion help teacher and student realize how knowledge of and attitudes toward biography have changed during the simulation. (*Puzzle* contains two sets of handwritten documents.) (Producer)

Editor's Comment: We have here an example of not only an excellent subsimulation task which will convey to students some of the excitement of the historical quest and the historian's job, but also some very clever pedagogy. Distributing the documents among other members of the school community and even in the libraries is really a brilliant twist, having the Galaxy Encyclopedia Company seeking articles from the students-as-historians, and finally the challenge and defense session all in my estimation put this together as an exceptional package. It comes in the usual Interact form, a 37-page booklet containing teacher's directions, student directions, and about 40 pages of original handwritten documents printed on heavy brown paper. (REH)

Cost: $14.00

Producer: Interact, P.O. Box 262, Lakeside, CA 92040

RADICALS VS. TORIES

David Dal Porto, Mount Pleasant High School

Playing Data
Copyright: 1969, 1972
Age Level: grades 7-12
Prerequisite Skills: reading, grade 9; math, grade 0
Number of Players: 24-100
Playing Time: 3 periods of 50-minutes each
Preparation Time: 1-week background on the American Revolution

Description: *Radicals vs. Tories* is a simulation of the struggle between those favoring close ties with Great Britain and those favoring independence during the American Revolution. All students assume one of three roles, Radicals, who favor independence; Tories, who favor loyalty; and Moderates who have not as yet made a commitment. During the first two days of the simulation, the Radicals and Tories try to persuade as many Moderates as possible to accept their arguments. Students meet as delegates to colonial section meetings of the northern, middle, midsouth, and southern colonies. At the end of each meeting the Moderates may vote for independence or loyalty or they may abstain.

On the last day of the simulation all the delegates meet together to hear formal presentations from the leading spokesmen for the Radicals and Tories. At the conclusion of these formal arguments, the 13 colonies vote on either independence or loyalty. (Author)

Cost: $12.50

Producer: History Simulations, P.O. Box 2775, Santa Clara, CA 95051

THE REVOLUTIONS GAME

Playing Data
Age Level: grades 5-12
Number of Players: 8-40
Playing Time: about 45 minutes per revolution

Description: As players work their way through the issues of America 1776, France 1789, Mexico 1911, Russia 1917, and China 1949, they discover, through the eyes of the major protagonists, the similarities and differences in these great events.

Players gain knowledge about the rigid social structures, economic inequalities, the inability of the ruling classes to change, the role of revolutionary leaders and propagandists, the influence of ideology and outside factors, and how these contribute to the course of human history. (Producer)

Cost: $35.00

Producer: Simulation Learning Institute, P.O. Box 240, Roosevelt Island Station, New York, NY 10044

ROARING CAMP

Jay Reese

Playing Data
Copyright: 1972
Age Level: grades 4-8
Number of Players: 18-35
Playing Time: 10-15 30-minute rounds
Preparation Time: 20 minutes
Packaging: professionally produced and packaged game includes 11-page teacher's manual, play money, and two maps.

Description: In *Roaring Camp* students are given a $600 grubstake to file a mining claim and try their luck as prospectors. The first year the student must pay $400 to the "Roaring Camp Mercantile Company" (RCMC) to pay for mining tools, supplies and equipment to get him started, and $200 each year thereafter to keep him going. Two large maps marked off in sections and plots are placed on the wall. He selects a plot to mine by putting his initials on one of the plots and filing a claim form. When all of the students have selected their plots, the teacher consults a hidden "Master Map of the Gold Field" and tells the students whether they have struck it rich, and if so, for how much. A student may hit pay dirt the first time out and become an instant millionaire or he may vainly seek elusive Lady Luck year after year and end up thousands of dollars in debt. (Author)

Cost: $10.00

Producer: Simile II, P.O. Box 910, Del Mar, CA 92014

THE SCRAMBLE FOR AFRICA

B. Barker and R. Boden

Playing Data
Copyright: 1973
Age Level: high school, college
Number of Players: 15
Playing Time: 2-3½ hours in 6-10 20-minute rounds
Preparation Time: 2 hours (est.)
Packaging: professionally produced 17-page booklet

Description: The authors describe this game as designed "to recreate the circumstances which led to the acquisition of African Colonial Empires by the major European powers in the thirty years following the British occupation of Egypt (1882)." Teams of players represent English, French, and German diplomats, soldiers, and colonists. Each twenty-minute round consists of three stages. The first is negotiation, when diplomats, army commanders, and the leaders of civilian missions meet informally to discuss the situation and make agreements. Any player may negotiate with any other player, but only agreements between diplomats are binding. "Negotiations," the authors explain, "may involve offers of cooperation, non-intervention, friendship, support formal alliance; also threats, the spread of rumours, blackmail and so forth." In the second stage, players write orders. The diplomatic, military, and commercial missions of each country may stand at, or move from, any city or port to an adjacent city or port and attempt to bring it under their control. Diplomats may also formalize treaties. These orders are publicized during the last stage of each round, and their results are announced by the teacher. Generally, the first government to occupy any town holds that town until a larger military force compels it to withdraw. Points are awarded for control of various African cities, and the government with the most points at the end of the game wins. (DCD)

Note: For more on this simulation, see the essay on history simulations by Bruce E. Bigelow.

Cost: (1) contact producer (2) $3.95 U.S.

Producer: (1) Longman Group, Ltd., Resources Unit, 33/35 Tanner Row, York, United Kingdom (2) Longman Inc., 19 West 44th Street, New York, NY 10036

SENECA FALLS

Paul DeKock, El Capitan High School, Lakeside, California

Playing Data
Copyright: 1974
Age Level: grades 7-12 average to above-average ability
Number of Players: 25-40
Playing Time: 3-4 hours in 1-hour periods
Supplementary Materials: reading matter suggested in teacher's guide

Description: *Seneca Falls* is a simulation of the first women's rights convention in July 1848. It introduces students to nineteenth-century conceptions of woman's nature and her intended place in the world. The teacher first selects capable students to role-play Elizabeth Cady Stanton, Lucretia Mott, John Mott, and Frederick Douglass, all of whom were present at Seneca Falls. The remaining boys and girls go to separate rooms (or separate corners of the same room) in order to prepare for the convention. Girls are told to choose identities from several clusters–some to be married, some widows or spinsters, some girls 14-18 years old. Other suggestions tell the girls how to dress and how to prepare their arguments to meet the boys' likely arguments. Boys are given various identities such as ministers, merchants, and farmers; they are told where to find Biblical and societal justification for the "truths" that women are physically and mentally inferior but morally superior to men. After overnight or one class period's research, the convention begins. Participants debate and vote on four motions growing out of the Declaration of Sentiments that the ladies patterned after the Declaration of Independence of 1776. Motion 1 reaffirms the natural rights doctrine, stressing that it should be applied to all human beings, not just men. Motion 2 encourages women to "leave the jail of the nursery and the kitchen" in order to get out into the world and find other worthy professions. Motion 3 recommends abolishing the double standard of morality for men and women. Motion 4 urges the adoption of a constitutional amendment giving women the right to vote. During a debriefing session students discuss how alive the nineteenth-century conceptions of male and female roles are and reveal their real convictions about these conceptions. (Publisher)

Comment: Although *Seneca Falls* is available as a self-contained package, it also comes as part of a larger exercise, Herstory, which is described in the social studies section. (AC)

Note: For more on this simulation, see the essays on sex-role simulations by Cathy Greenblat and Joan Gaskill Baily and on history simulations by Bruce E. Bigelow.

Cost: $10.00

Producer: Interact, P.O. Box 262, Lakeside, CA 92040

1787

Eric Rothschild, Scarsdale High School, Scarsdale, New York

Playing Data
Copyright: 1970
Age Level: grades 7-12
Special Prerequisites: general knowledge of American history into the period of the Articles of Confederation. Reading, grade 9; math, grade 0
Number of Players: 20-36
Playing Time: 3-10 periods of 45 minutes
Preparation Time: 2 hours
Special Equipment: record player

Description: This game is designed to illustrate the conflicting interests that had to be reconciled during the drafting of the U.S. Constitution. Players assume the roles of Franklin, Washington, and Madison, a reporter, and fictional delegates to a mock constitutional convention. Delegates represent one of four conflicting interests (North versus South, Federalists versus states righters, big states versus small, or landowners versus commerical interests) and must represent their views as strongly as possible while the group creates a constitution that will be acceptable to everyone.

The mock convention follows a seven-item agenda pertaining to the establishment of legislative, executive, and judicial branches of a federal government, powers granted to those branches and powers denied to the states, and the establishment of procedures for ratifying the Constitution. For each item a delegate must move the adoption of any alternatives listed in the agenda. (For example, in deciding the nature of the legislative branch of government, delegates must decide if there shall be one house or two; if the members of the legislature shall be chosen by the people, the state legislatures, or the state or federal executive; if the legislature shall represent states or numbers of citizens; and if the terms of office of these legislators shall be one, two, three, four, five, six, or more years.) Debate follows any motion and the question is moved after debate. If a motion is carried, the next item on the agenda is considered. If a motion is defeated, an alternative proposal, either one listed on the agenda *or* one invented by a delegate, must be debated and moved to a vote. This procedure continues until a complete constitution is drafted. The acceptance of this draft is voted on and then compared with the actual one. (DCD)

Cost: $36.00 (#2EO 565)

Producer: Olcott Forward, Division of Educational Audio Visual, Inc., Pleasantville, NY 10570

SKINS

William Lacey, Ft. Valley High School

Playing Data
Copyright: 1975
Age Level: high school
Playing Time: 7-10 periods
Number of Players: 12-40
Supplementary Materials: reserve shelf on the Rocky Mountain fur trade
Special Equipment: egg timer, dice, beef jerky
Packaging: 50-page manual

Description: *Skins* is a simulation game concerned with the economics and history of the Rocky Mountain fur trade in the early nineteenth century. Students role play the Head Booshway, Little Booshway, and trappers of three competing fur companies, all trying to trap beavers and survive under arduous conditions. The first two days of the game, they read background information on the fur trade, form companies, write short essays, and take a pretest on their readings. Members of each group add their scores to obtain the amount of capital or number of plew points each group has to spend or lose. The next five days of the game represent the twelve-year span of the Rocky Mountain fur trade, including the last two years of disaster and collapse. Every three minutes during each period plew (beaver skin) cards are drawn which indicate an event that has occurred (e.g., beaver skins are hard to find. Lose 300 plew points) or a task requested of each group member (e.g., Write a 3 page summary of the Lewis and Clark expedition and describe its significance. Reward: 400 plew points. Or write a dialogue between 2 trappers at a rendezvous. Reward: 300 plew points). Students work in class and at home on these tasks, but the value of each task changes if it is handed in the next day, for at the beginning of each period new profit sheets are posted. Each period also, plew points are tallied and competition is spurred among groups. The *Skins* manual includes background information, pretest, ledgers, price and profit sheets, and twelve pages of research "cues" on the mountain men. Some of these "cues" describe the men's diet, their speech and vocabulary, and their savagery. During the final debriefing, students eat beef jerky and discuss their roles, how their companies profited, and why the fur trade declined. (TM)

Cost: $14.00

Producer: Interact, Box 262, Lakeside CA 92040

SLAVE AUCTION

Mary Simpson Furlong and Louise Weinberg Jacobsen

Playing Data
Copyright: 1976
Age Level: junior, senior high school
Number of Players: 36
Playing Time: 1-2 class periods
Preparation Time: 1 hour (est.)
Packaging: professionally packaged game includes facsimile specie, role cards, and instructions.

Description: This simulation recreates a slave auction in the Virginia of 1860. Thirty-six roles are briefly described. Twelve players take the part of the slaves and seventeen the parts of the prospective buyers. Two others who take the parts of the auctioneer and his assistant have no name, age, home, or attitudes; their part is merely functional. Five players assume the roles of abolitionist observers. The role descriptions for the slave buyers give each a name, an age, and a marital status. The number of slaves the customer already owns, what he is looking to buy, and the amount of money he has to spend is also given. John C. Wallace, for instance, is a forty-eight-year-old married gentleman from Virginia who runs his place with the help of five slaves and has come to the auction with seven hundred dollars with which to buy another. The slaves are known only by what the auctioneer has to say about them. They include the Turner family, Tom, Sarah, and their thirteen-year-old daughter Melissa, a young unmarried woman named Saphne, and Thaddeus, who is hawked as the "fastest cotton picker around."

The auctioneer and his assistant have the job of bringing the slaves to the block and getting the best possible price for them. The abolitionists have the task of evaluating the realism of the other player's performances. In the course of play each slave is brought to the block and sold. After the auction the teacher leads the class in discussing their feelings during the auction by asking questions such as, "Is it man's nature to be prejudiced?" and "What are some of the consequences of treating human beings as property?" (DCD)

Cost: $12.95

Producer: Zenger Publications; distributed by Social Studies School Service, 10000 Culver Blvd., Culver City, CA 90230

SLAVE COAST GAME

Ray Glazier, Games Central

Playing Data
Copyright: 1973
Age Level: grades 7-12
Prerequisite Skills: grade 6 reading, elementary math
Number of Players: 26-30
Playing Time: 45 minutes to 2 hours in 45-minute periods
Supplementary Material: "Life of the Egba Yoruba" (30 copies in Afro-City package), filmstrip, tape cassette, 6 student projects

Description: *Slave Coast Game* is part of a unit called Afro-City, an anthropological and historical approach to the New World by the chains of the slave trade. In two classroom weeks of varied activities, students learn about an African society from different aspects and the history of the tribe from which most Black Americans are descended. (Publisher)

In the game, students take the roles of chiefs, kings, traders and missionaries and so forth in one of four Yoruba city-states in 1850. One city-state (or more) will invariably attack another as the British, missionaries, and some local residents try to stop the slave trades while other locals and slave traders try to continue it. Warfare may result in slave captives. Characteristics of role players are specified and they include: position, background, goals, opinions, and action suggestions. The major transactions are loan and sale of guns, sale of goods and slaves, exchanges of intelligence data. (REH)

Editor's Comment: This game shows the strong preying on the weak and efforts to protect the weak by others of conscience. It appears to be a well-researched, historically accurate simulation and the only one we know of available on the West African slave trade. The simulation parameter is limited to Africa and does not deal with the sea-crossing or slavery in America. The complete package of Afro-City includes a student text of 30 pages which is a simplified but complete ethnography of history of the tribe and a history of Abeokta, capital city of the province. There is also a film strip with narrated sound track that covers the local geography. (REH)

Cost: $44.00 if purchased apart from Afro-City Unit package; less as part of package, which is $75.00 total

Producer: Games Central, 55-C Wheeler Street, Cambridge, MA 02138

SPANISH-AMERICAN WAR

David J. Boin and Robert Sillman

Playing Data
Copyright: 1973
Age Level: grades 7-12
Number of Players: class size
Playing Time: 3-5 class periods

Description: Students form a congressional fact-finding committee and a variety of other politically influential groups in the country in 1898. The congressional fact-finding committee has the job of finding out whether the United States should have gone to war over Cuba. Simplified role statements are provided for each side. However, students may do considerably more research and are encouraged to do so in this hypothetical situation. There are nearly always congressional fact-finding committees following wars and other crisis events. It is a good format for students to work in. It provides a game-like structure to channel research and the results of research. (REH)

Editor's Comment: This is in the standard Edu-Game format with a few pages of teacher's manual and locally reproducible student material. (REH)

Cost: $3.00

Producer: Edu-Game, P.O. Box 1144, Sun Valley, CA 91352

THE SPANISH AMERICAN WAR

Ronald Lundstedt and David DalPorto, Mt. Pleasant High School, San Jose, California

Playing Data
Copyright: 1969, 1972
Age Level: grades 7-12
Prerequisite Skills: reading, grade 9; math, grade 0
Number of Players: 32-105
Playing Time: 5 periods of 50 minutes each
Preparation Time: 1 week
Supplementary Material: background readings on the Spanish American War period

Description: *The Spanish American War* is a simulation of the period prior to our entry into war with Spain in 1898. Students represent either a pressure group favoring or opposing war with Spain or assume the role of members of the U.S. government (congressmen or members of the executive branch—the president and his cabinet). The pressure groups, Imperialists, The Yellow Press, Cuban Rebels, Business-Arms Industry, Wall Street, Anti-Imperialists and the Spanish Ambassadors, through a series of hearings, try to bring the president, members of the cabinet, and key congressional committees to their view of the situation.

Members of the government are asked to express their vote on seven different occasions. Their decisions are also influenced by the fact that they are members of a political party. The two party leaders in the simulation have opposing viewpoints on going to war with Spain.

The many events of the period, stories from Cuba, the DeLome Letter and the sinking of the *Maine* are all introduced at various key points throughout the simulation. On the final day the president goes before Congress to either recommend or oppose war with Spain. Finally, Congress has a roll call vote on the issue. A running total of votes throughout the simulation determines the final outcome. (Author)

Cost: $15.00

Producer: History Simulations, P.O. Box 2775, Santa Clara, CA 95051

STRIKE

David Yount and Paul DeKock, El Capitan High School, Lakeside, California

Playing Data
Copyright: 1970
Age Level: grades 9-12 average to above-average ability; grades 7-12 above average
Prerequisite Skills: reading, grade 11.5; math, grade 7
Number of Players: 25-40
Playing Time: 12-20 hours in 1-hour periods
Preparation Time: 3-5 hours
Supplementary Material: U.S. history texts

Description: *Strike* is a simulation of the history of American labor-management relations. The classroom first becomes a late nineteenth-century environment with two towns: one having a steel mill, one having a coal mine. Students have roles with conflictg goals: two are owners; two are managers; four are foremen; ten-twenty are workers; the remainder are unemployed (immigrants, Negro migrants, labor union organizers, socialists, anarchists). Horatio Alger-type effort and chance can help students climb the golden ladder of economic opportunity. Each two-day cycle simulates a month: The owners buy raw materials (assignments and tests) and make many capitalistic decisions regarding wages, sales, and research and development; workers labor in mine or mill at menial jobs (researching and taking tests on 100 questions about labor-management history, 1800 to the present day); the unemployed try to get hired, organize a union, or start a revolution. Each identity has to earn GPS (grade points) to meet varying cost of living expenses, but only the rich are able to buy luxuries such as pillows and secretarial help. The simulation is heated up with two kinds of bulletins: historical bulletins, which make workers aware of labor's plight throughout America; fate regarding those bulletins, killed or injured (e.g., molten steel burns bodies or mine shafts collapse). Eventually, phase I culminates in either a strike, a lockout, or fumbling attempts at collective bargaining. During phase II students realize how much labor-management relations have changed during the last 75 years when they participate in collective bargaining sessions on a contemporary case study. (Publisher)

Editor's Comment: *Strike* follows the usual Interact format. It is a 42-page attractively printed booklet containing the teacher's manual and locally reproducible material. The comparison of the classroom to a factory should not go unnoticed. In fact, it is one of the subtler indictments of American education. One wonders if the simulation produces any real strikes in high schools after the simulated strikes in the history classrooms. (REH)

Cost: $14.00

Producer: Interact, P.O. Box 262, Lakeside, CA 92040

TRADE-OFF AT YALTA

Daniel C. Smith, Woodland High School, Greenburgh, N.Y.

Playing Data
Copyright: 1972
Age Level: grade 9-college
Prerequisite Skills: some knowledge of World War II
Number of Players: 12-40 in 3 teams
Playing Time: 2 hours in 1-hour periods
Preparation Time: 1-2 hours
Special Equipment: record player

Description: *Trade-off at Yalta* is a replay of the summit conference between Roosevelt, Churchill, and Stalin in February 1945. By participating in the five rounds of play, students acquire a working knowledge of at least fifteen of the problems that had to be dealt with at the end of World War II. At the beginning of each round, teams form their strategies and secret planning sessions. The major issues are the Polish borders, keeping the postwar peace (structure of the UN), how the Far East war will be completed, who shall get what, how shall Germany be divided, and reparations shall be made. (REH)

Comment: This is a well worked out, carefully put together simulation in which the game director still has options. There are five rounds, each with three specific problems to be solved. The director may choose among the problems to be dealt with and may also complicate play by creating new problems. Unlike some historical simulations, which require the student to dig into original source materials and other source books, much of the material for play of this simulation is provided in the form of advisor's briefcases, profile cards, and newspapers. The package also includes a phonograph record with music from the time, maps, a debriefing chart, and so forth. (REH)

Note: For more on this simulation, see the essay on history simulations by Bruce E. Bigelow.

Cost: $35.00

Producer: Prentice-Hall Media, 150 White Plains Road, Tarrytown, NY 10591

THE TRIAL OF GEORGE WASHINGTON

David J. Boin and Robert Sillman

Playing Data
Copyright: 1972
Age Level: grades 7-12
Number of Players: class size
Playing Time: 4-5 class periods

Description: This simulation assumes that the British won the American Revolution. The scene is a trial in London and the defendant is George Washington. The charge is treason. Students are prosecutors, defenders, and judges of George Washington, as well as witnesses and newspaper and editorial writers from various parts of the world. (REH)

Comment: The materials suggest only the bare bones of the situation. Students must do all the research to prepare for the trial on their own. This simulation opens up a vast "what-if" realm. Almost any historical event could be assumed to have gone the other way and be debated and played out in that way. However, this does get at some important issues and could represent a significant activity in a history class. It is presented in the standard Edu-Game format of a few pages of teacher's manual and locally reproducible student material. (REH)

Cost: $3.00

Producer: Edu-Game, P.O. Box 1144, Sun Valley, CA 91352

THE TRIAL OF HARRY S. TRUMAN

David J. Boin and Robert Sillman

Playing Data
Copyright: 1972

Age Level: grades 7-12
Number of Players: class size
Playing Time: 4-5 class periods

Description: This is a hypothetical situation assuming that the Allies lost World War II and that Harry Truman was ordered to stand trial for crimes against humanity, for dropping two atomic bombs on Japan. The class represents judges, defense and prosecuting attorneys, and newspaper editorial writers from various countries.

Editor's Comment: The simulation material only provides the framework for this simulation. Students must do all the research to assume their roles. It is in the standard Edu-Game format with a few pages of teacher's manual and locally reproducible student material. (REH)

Cost: $3.00

Producer: Edu-Game, P.O. Box 1144, Sun Valley, CA 91352

THE TRIAL OF JEFFERSON DAVIS

David J. Boin and Robert Sillman

Playing Data
Copyright: 1972
Age Level: grades 7-12
Number of Players: class size
Playing Time: 4-5 class periods

Description: *The Trial of Jefferson Davis* is an activity which involves group work, research, and the interaction of ideas and interpretations of the secession question. The culmination is the actual trial followed by a general discussion and analysis. (Publisher)

Editor's Comment: Students must do all of the research required for the different positions which they occupy. This is a standard Edu-Game simulation with a number of pages of teacher's manual and locally reproducible student material. (REH)

Cost: $3.00

Producer: Edu-Game, P.O. Box 1144, Sun Valley, CA 91352

TRIANGLE TRADE

Russ Durham and Jack Crawford

Playing Data
Copyright: 1969
Age Level: grades 7-12
Number of Players: 20-44
Playing Time: 2-3 hours
Preparation Time: 3 hours
Special equipment: overhead projector
Packaging: 116-page staple-bound photocopied teacher's manual

Description: This simple game is designed to simulate the triangular trade in the eighteenth century between England, its Atlantic colonies, Africa, and the West Indies. It also attempts to illustrate the relationship between the English mercantile system and the economic development of New England, including the conflicts inherent in this system. Each player assumes one of eighteen different roles including Boston bankers, English abolitionists, African slave traders, and West Indian planters, in addition to shipmasters and mates. Each player starts with finances represented by slips of paper and tries to profit from the trade. (An exception to this is the abolitionists, who want the trade in human beings to stop.) The simulation is not designed to arrive at any climax. It continues until the players become familiar with the mechanics of the triangular trade. (DCD)

Cost: $16.00

Producer: Simulation Systems, Box 46, Black Butte Ranch, OR 97759

THE UNION DIVIDES

Eric Rothschild, Joan Platt, and Daniel C. Smith

Playing Data
Copyright: 1971
Age Level: grades 9-12
Prerequisite Skills: reading, grade 9; math, grade 0
Number of Players: 30
Playing Time: 5-10 class periods

Description: Players represent governors of the thirty states of the Union before 1850. They understand their interests from a profile card. Participants divide into blocks of Abolitionists, Moderates, Southern Extremists, and Southern Moderates and present their positions in seven rounds of conferences and meetings that represent the progress of events through the decade preceeding the Civil War. The events involved are the admission of California to statehood, the Kansas-Nebraska Act, the Dred Scott decision, John Brown's raid, the Black abolitionist movement, the secession of South Carolina, and the Union loss of Fort Sumter. (REH)

Comment: A phonograph record gives a kind of "You Are There" quality to different events that have just happened before each round. This simulation seems to be well worked out and carefully done. (REH)

Cost: $40.00 (#A9E0566)

Producer: Olcott Forward, Division of Educational Audio Visual, Inc., Pleasantville, NY 10570

VIRTUE

David Yount and Paul DeKock, El Capitan High School, Lakeside, California

Playing Data
Copyright: 1975
Age Level: high school
Number of Players: class
Playing Time: 2 1/2 weeks
Supplementary Materials: reading *The Crucible* is recommended
Packaging: 34-page loose-leaf manual

Description: *Virtue* is a simulation of a 1980 Alaskan wilderness community with a moral code similar to that of a seventeenth-century Puritan village. While participating in the simulation, students gain a deeper understanding of Puritanism and clarify their own moral attitudes in personal journals. The first two days are introductory—background materials on the history of Oakania are read, roles are chosen (three elders, judge, minister, teacher, and family members), the Oakania Compact is signed, and students familiarize themselves with the moral enforcement of the community. These include keeping their own private morals improvement journal, charting their moral progress and decline, and writing moral improvement slips on classmates' behavior. The game itself occurs in three cycles of three periods each, each cycle including a simulation phase and a reality phase. During the simulation phase, students study the ten virtues of Oakania in a simulated school, hold family, community, and fellowship meetings, and witness the judge, who has examined the moral-improvement slips, give out public punishment. The teacher also reads a community crisis bulletin which implicates an Oakanian for some vice or crime. Punishments include a grade point fine or the wearing of a scarlet letter. During the reality phase of each cycle, students read Puritan literature in their manual and answer questions, participate in activity groups (e.g., a discussion of the relevance of the Ten Commandments to their community), and complete a follow-up chart or quiz. They are also required to write several entries in their journal, examining their own behavior from two viewpoints—as themselves and as an Oakanian. Throughout the game they record the work done and the income earned (grade points) on a balance sheet included in the manual. When the three cycles are completed, a debriefing session follows during which students review Puritanism and relate Puritan beliefs and morals to those of American youth. All materials necessary for the simulation are included in the manual and may be photocopied for student use. (TM)

Cost: $14.00

Producer: Interact Company, Box 262, Lakeside, CA 92040

WAGING NEUTRALITY

Russ Durham and Virginia Durham

Playing Data
Copyright: 1970
Age Level: grades 9-12
Number of Players: 15-33
Playing Time: 4 50-minute periods
Preparation Time: 1 hour
Special equipment: overhead projector
Packaging: 90-page staple-bound photocopied playing manual

Description: The designers describe this game as concerning "the significant events occurring between 1914 and the United States entry into World War I in 1917. Specifically, it revolves around the economic circumstances which played an important part in our eventual military involvement." The "economic circumstances" were the trade the neutral United States carried on with both principal combatants during the first three years of the war. Players assume the roles of J. P. Morgan and House of Morgan executives, the British and German ambassadors, various industrial and farm brokers, and the purchasing agents for the antagonists, New York and Atlantic shipping firm executives and sea captains, and the captains of German submarines and English destroyers. Play is structured around three activities. During "Commercial Action," cargo is bought, sold, and shipped off in the direction of Europe. The fate of these ships is determined by drawing chance cards. The House of Morgan, the English and German ambassadors, and the shipping firms hold "Operations Meetings" during which the players must respond to the actual events of 1914 to 1917 as the teacher reads them. During the "Open Forum" phase, representatives from each operations meeting make public announcements and defend their countries' actions and policies to all participants. (DCD)

Note: For more on this simulation, see the essay on history simulations by Bruce E. Bigelow.

Cost: $14.00

Producer: Simulation Systems, Box 46, Black Butte Ranch, OR 97759

THE WAR CRIMES TRIALS

David Dal Porto and John Koppel, Mt. Pleasant High School, San Jose, California

Playing Data
Copyright: 1974
Age Level: grades 9-12
Number of Players: 25-40

Description: *The War Crimes Trials* is a simulation of the numerous trials that have been held in Germany since the end of World War II. Four individuals, a high-ranking Nazi official, a field commander, a camp commandant, and a camp guard, are placed on trial. The degree of guilt for crimes of aggression, crimes of war, and crimes against humanity is examined in the background of Nazi Germany and World War II. The simulation is divided into two separate trials where teams of attorneys select groups of students to play certain already established witnesses. Attorneys may also develop surprise witnesses on their own. The major issue that develops during the trial is where responsibility should lie for the inhuman treatment of people during World War II. Students examine the makeup of a totalitarian society. They also try to decide what limits should be placed on actions taken during war and whether these limits can be enforced. A panel of "Allied Judges" decides the final verdict for the four defendants. (Author)

Cost: $17.50

Producer: History Simulations, P.O. Box 2775, Santa Clara, CA 95051

WATERGATE: THE WATERLOO OF A PRESIDENT

Richard W. Hostrop

Playing Data
Copyright: 1975
Age Level: junior, senior high
Number of Players: 19-50
Playing Time: 3 hours
Preparation Time: 2 days
Packaging: 31-page bound booklet

Description: The purpose of *Watergate* is to teach students about presidential impeachment procedures and the events leading up to Nixon's resignation. The manual lists fifty-six specific steps for reenacting Watergate-related events from July 1971 to August 1974—the Ellsberg trial; the break-in itself; the sentencing of the burglars; the resignations of Haldeman, Ehrlichman, and Dean; the discovery of the tapes; the firing of Cox; the Ervin Committee; Nixon's resignation, and so on. Brief newspaper cards are duplicated and passed out periodically to inform students of the changing events. Portraying nineteen of the major figures involved in these events (the manual suggests thirty additional roles for larger classes), students must research their roles and play them as accurately and dramatically as possible, guided by role cards that summarize the background, knowledge, feelings, and some public statements of each figure. The manual includes a bibliography as well as fourteen debriefing questions relating to the impeachment procedure, the Watergate period, and its significance in U.S. history. (TM)

Cost: $4.95

Producer: Etc. Publications, P.O. Drawer 1627-A, Palm Springs, CA 92262

WORLD WAR I

David J. Boin and Robert Sillman

Playing Data
Copyright: 1973
Age Level: grades 7-12
Number of Players: class size
Playing Time: 3-5 class periods

Description: World War I is a simulation of congressional fact-finding committee hearings. Students take parts of the congressional committee and pressure groups of that time. These groups present evidence, and the Committee determines what happened and whether the United States should have gone to war. (REH)

Cost: $3.00

Producer: Edu-Game, P.O. Box 1144, Sun Valley, CA 91352

See also
Amnesty DOMESTIC POLITICS
City Planning URBAN
Collision DOMESTIC POLITICS
Confrontation INTERNATIONAL RELATIONS
Equality SOCIAL STUDIES
First Amendment Freedoms LEGAL SYSTEM
Gestapo RELIGION
The Hunting Game SOCIAL STUDIES
Mahopa SOCIAL STUDIES
Micro-Community SOCIAL STUDIES
Newscast LANGUAGE
On the Move: Soviet Jewry RELIGION
Opening the Deck SOCIAL STUDIES
Oregon COMPUTER SIMULATIONS
Potlatch Game SOCIAL STUDIES
Railroad Game ECONOMICS
Railway Pioneers GEOGRAPHY
Small Town SOCIAL STUDIES
Traders Arrive on the African Science SOCIAL STUDIES
Trebides Island ECONOMICS
Urban Dybamics URBAN
MILITARY HISTORY

HUMAN SERVICES

ALL FOR THE CAUSE AND THE CAUSE OF EACH

Adrienne Ahlgren Haeuser and the faculty of University of Wisconsin

Playing Data
Age Level: college, graduate school, social service agencies
Number of Players: 9-20, 15 preferred
Playing Time: 1½ hours
Preparation Time: 1/2 hour
Packaging: 42-page manual

Description: The purpose of this simulation is "to facilitate a community team approach to child abuse and neglect" which allows paraprofessionals and trained volunteers to help with child abuse problems. Six players, as pediatrician, planning federation director, mental health association president, protective services supervisor, and judge, role play members of a task force on child abuse which has thirty to forty-five minutes to determine their recommendations for funding. Their personalities, biases, and recommendations, clearly delineated on their role sheets, reflect the acknowledged biases of the authors, such as that doctors are very concerned with maintaining their status, tend to give orders rather than cooperate on an equal basis, and view child abuse problems primarily as physical rather than as social problems. Players who do not participate in the role playing are assigned to evaluate those who do, by answering questions regarding the exhibited values, interests, and cooperativeness of task force members to whom they are assigned. (TM)

Cost: $2.50

Producer: Center for Advanced Studies in Human Services, School of Social Welfare, P.O. Box 786, University of Wisconsin-Milwaukee, Milwaukee, WI 53201

BROOKSIDE MANOR

Dorothy H. Coons and Justine Bykowski, Institute of Gerontology, The University of Michigan–Wayne State University

Playing Data
Copyright: 1975
Age Level: adult; staff of institutions
Number of Players: 15-60 in two groups
Playing Time: approx. 2 hours
Preparation Time: 30-60 minutes
Special Equipment: duplicated forms for each participant, large newsprint or overhead transparency guides for discussion, felt pens and flip chart or chalk and chalkboard, *two* meeting rooms
Packaging: sufficient forms for one play of the simulation plus leader's manual in paper folder

Description: Brookside Manor presents the problem of what personal belongings are to be brought into a retirement home by prospective elderly residents. The simulation requires dividing the participants into two groups; one group assumes the roles of prospective residents, and the other the staff. Each group is then engaged, separately and simultaneously, first in rank ordering items from a list of 23 and then in eliminating those which are considered non-essential.

When the two groups—the "staff" and "residents"—have completed the rank ordering and elimination steps, they come together to compare their priorities. Following this, "staff" and "residents" meet together in small groups. "Residents" have this opportunity to plead their respective cases and to negotiate with staff regarding their personal possessions. Similarly, staff have an opportunity to explain the rationale for their decisions and the problems they anticipate. (author)

Comment: This deceptively simple structure can be used to open up a number of issues (such as safety, independence, socialization, rights, union agreements) that arise in designing and operating group facilities. Issues related to aging in general can also be approached. (AEZ)

Note: For more on this simulation, see the essay on health and health care simulations by Amy E. Zelmer and A. C. Lynn Zelmer.

Cost: $5.00

Producer: Institute of Gerontology, The University of Michigan–Wayne State University, 520 East Liberty, Ann Arbor, MI 48109

THE BROTHERS AND SISTERS GAME

Donna Sweedler and Alan Kraushaar, Mass. Dept. of Mental Health

Playing Data
Copyright: 1973
Age Level: siblings age 7-16 or young children with mental retardation
Prerequisite Skills: oral and written comprehension, some practical desicion-making ability
Number of Players: minimum 3-7 maximum; optimim, 6 with leader
Playing Time: minimum 45 minutes to 2 hours maximum or more
Preparation Time: 15-30 minutes

Description: Players of the *Brothers and Sisters Game* move their pieces around a game board which represents them in their neighborhoods—for example, at home with their family, in school, and at their local shopping center. On these journeys into their communities, their pieces, which move according to a roll of a die, sometimes land on special squares, some with writing on them. The player then moves his piece according to written directions. These squares represent situations in which the player finds himself regardless of whether or not he has a sibling who is retarded.

When he lands on a color-coded square, he picks up different "situation cards." Each of these cards presents a brief, pertinent, challenging situation stemming from a sibling's retardation (that they are involved in or can identify with) and four different ways of responding to it. The situation cards are read out loud for the consideration of the other players, and then the spinner is used to determine which of the four alternatives is chosen. After this selection has been determined, the person to the player's right considers whether he feels it is an appropriate response to the particular situation, and rates it by physically moving the player's piece one or two spaces forward, or backward on the board, or by leaving the piece where it is. (Moving the piece in a forward direction indicates a feeling that the choice is a good one. On the other hand, moving it back indicates disapproval of this alternative. Not moving the piece indicates that he feels neutral about the alternative selected).

The movement (or nonmovement) of the piece and the thinking behind it can then be challenged or discussed by the other players. After a challenge, a majority vote then decides where the player's piece should be placed, and the next player takes his turn by rolling the die.

In this way, a format is provided to open up sharing about these important issues in a game-like, nonthreatening atmosphere which can be promoted by the adult game facilitator-leader. (authors)

User Report: The *Brothers and Sisters Game* presents everyday situations and concerns, with a range of responses to the players, allowing the participants an opportunity to discuss the responses in an objective, nonjudgmental atmosphere.

I have played the game with the siblings of retarded children for a series of three sessions. Using the game as the focus for group interaction and discussion, children were able to openly share experiences and concerns. The game structure allowed children to comfortably terminate a discussion that became too anxiety provoking and also allowed the leader to help the players with reality testing and information seeking.

Parents whose children were involved in the sessions not only approved of their youngsters' participation but encouraged this type of group interaction. The parents were aware that even though their children did not voice questions, feelings, and concerns about the retarded sibling at home, they did need an opportunity to express themselves within a structured, supervised setting.

This game's value in stimulating discussions around specific topics is almost unlimited. Some attention should be exercised that the game per se does not become so competitive that conversation and interaction around the issues and feelings raised are excluded. Second, the leaders should review the "situation cards" prior to meeting with the group to determine their appropriateness to the age group involved and to the age of the retarded child. The game is nicely designed so that "situation cards" can be added or deleted without changing the concept.

This game is highly recommended for siblings of retarded children. With adaptations it has been successfully used with mildly retarded persons. It has great potential for use with any group of persons who are in contact with children with special needs and could profit from a structured group experience. (Barbara S. Greenglass, Associate Director, South Shore Mental Health Center)

Cost: $4.75

Producer: United Community Planning Corp., 87 Kilby Street, Boston, MA 02109

EVERYBODY COUNTS!

Harry Dahl

Playing Data
Copyright: 1976
Age Level: adult; health professionals or trainees
Number of Players: up to 60 (25-40 is optimal)
Playing Time: Individual activities 30 minutes or more in day-long workshop
Preparation Time: considerable; a great deal of special equipment must be assembled or adapted
Special Equipment: a great deal is required if all activities are to be carried out (wheelchairs, special goggles, symbol charts, plastic letters and symbols, plaster casts)
Supplementary Materials: 16 mm film "Hello Up There," audio cassette "Blind Cindy," record "An Unfair Hearing Test"
Packaging: workshop manual

Description: *Everybody Counts! A Workshop Manual to Increase Awareness of Handicapped People* outlines a program of activities to provide participants with the experience of functioning with various types of handicapping conditions (e.g., mobility restrictions, visual and hearing impairments). (AEZ)

Comment: This is an elaborate program to organize, but it can provide a learning experience for those who work with the handicapped that cannot be achieved in any other way. (AEZ)

Note: For more on this program, see the essay on health and health care simulations by Amy E. Zelmer and A. C. Lynn Zelmer.

Cost: $10.00-$15.00 range

Producer: The Council for Exceptional Children, 1920 Association Drive, Reston, VA 22091

EL BARRIO

R. L. Meier (adapted from Fred Goodman)

Playing Data
Copyright: 1973
Age Level: college, graduate school, continuing education, human service workers
Prerequisite Skills: interest in urban immigrants' problems
Number of Players: 7-15
Playing Time: 2 to 3 or 4 hours
Preparation Time: 1-2 hours

Description: The game embodies the forces affecting the Latin immigrant to the big city in North America. He must acquire a skill, build a social network, learn how to use a vehicle, and decide whether to collaborate with the System or fight it. Players may aim to become leaders in the Latin Community, join the System and graduate to integrated suburbs, or move up and out still contesting.

This is an outsider's analysis of the unique social processes and the distinctive political behavior found in *la colonia*; it is designed to illustrate to people who are providing human services in the cities and to students that their preconceptions about Latins are often faulty. Although most suited to extension courses, short courses, and encounter groups, this gaming situation can be fitted into sectional meetings of large college courses.

The game itself evolved from Fred Goodman's original series of "Marbles Games." He set out to discover how inner city young people created strategies for action in their street behavior. Rules and "environment" for one of the original marbles games are also provided. Community structure existing among the players is revealed to themselves and the gaming director by the events of the game and the situations it generates. (author)

Cost: $25.00 plus shipping charge

Producer: Institute of Urban and Regional Development, University of California, Berkeley, CA 94720

THE END OF THE LINE

Frederick L. Goodman, University of Michigan

Playing Data
Copyright: 1975
Age Level: high school, college, professional
Number of Players: 16-60, 30 optimal
Playing Time: 1½-2 hours to play, 1+ hours to debrief
Preparation Time: 1 hour

Description: This game is designed to give participants experiences in growing old and in obtaining services as an elderly person. Most of the participants represent "citizens" in the process of growing old. There are two direct service agencies, X and Y, and a funding agency, Z, which represents the government–the source of funds and funding criteria. Citizens begin with artifacts that represent life resources–including physical mobility, money, the ability to remember, the ability to form relationships. They are tied to chairs and may move to engage in various activities–including a "Game of Life" through which they may seek to gain more resources necessary to meet the requirements for staying alive–only as far as the cords allow. The Game Overall Director (note the initials) runs the Game of Life, through which citizens increase or suffer decreases in relationships. Another director, the Grim Reaper, moves through the community making citizens pick cards that may result in their losing artifacts for relationships, money, mobility, and even, eventually, the pencil (representing memory) without which they cannot keep track of changes in the large, posted matrix on which the Game of Life is based. The helping agencies seek to help the elderly in any way they can. (AC)

Note: For an in-depth description and discussion of this simulation, see the essay on the End of the Line by Andrew W. Washburn and Maurice E. Bisheff.

Cost: (1) instruction manual $5.50; (2) complete kit with instructions $75.00

Producer: (1) Publications, institute of Gerontology, University of Michigan, 520 East Liberty, Ann Arbor, MI 48109; (2) Institute of Higher Education Research and Services, Box 6293, University, AL 35486

HORATIO ALGER (A WELFARE SIMULATION GAME)

Ann Kraemer, Bob Preuss and Helen Howe of Citizens for Welfare Reform and People Acting for Change Together

Playing Data
Copyright: 1971
Age Level: high school, college, community and church groups
Number of Players: 13-32
Playing Time: 2 hours
Preparation Time: 1/2 hour
Special Equipment: 1 tinker toy kit for every 4 citizens, chance, and work cards (written up according to manual)
Packaging: 12-page xeroxed pamphlet including description of all cards used in game

Description: *Horatio Alger* is a welfare simulation game designed to introduce players to the realities of the welfare system: that the cards are stacked against the poor, that racism is common, that training programs are often useless and irrelevant, and that organizing is the only solution. At the beginning of the game, players are assigned roles as governor, administrator, case-workers, citizens, and policeman. Citizens draw work chance cards to find out if they work and those wearing red-striped citizens tags usually learn that they will remain unemployed (e.g. You have been evicted from your apartment; drop out of the job training program to look for a new home). At this point the administrator announces the work project for the year and citizens under caseworker supervision must build with tinker toys for a five-minute period equivalent to one year. At the end of each round, the administrator distributes tokens (money) to the caseworkers according to the number of their clients who were working and the caseworkers choose how many tokens they want themselves and how many will be distributed to the workers and the unemployed. Each person in the game must accumulate enough tokens to maintain his or her status, for at the end of the game those players who are short of tokens are demoted and the poor who cannot maintain themselves must go to jail. The policeman may also send citizens to jail if he catches them cheating or misbehaving. Several times during the game the governor and administrator draw chance cards which describe conditions affecting the amount of tokens the welfare department will have to distribute during the year. The governor may demote the administrator if he feels that the administrator is handling the funds inadequately. When the game is completed, players discuss the parallels between the game and reality, their reactions to their roles, their effects on each other, and their ideas about changing the welfare system. (TM)

Note: For more on this simulation, see the essay on community issues simulations by Tracy Marks.

Cost: $3.00

Producer: Connie Soma, Citizens for Welfare Reform, 305 Michigan Ave., Detroit, MI 48226

THE LIBRARIAN'S GAME

Barbara F. Weaver and Barry R. Lawson

Playing Data
Copyright: 1977

Age Level: professional
Number of Players: 5-25
Playing Time: 6 hours
Packaging: professionally packaged game including instructions, role manuals, and forms.

Description: The game is designed to simulate program planning and budgeting for up to ten libraries. All of these libraries are located in adjacent communities and must compete for the same federal and state funds. The game procedure consists of preparing budgets. Participants assume the roles of administrators of various hypothetical libraries. During each round they are confronted with client requests. In many cases, these requests elicit a referral to a library program at the first or a second library. For example, a local chamber group may request the use of a library for rehearsals; a librarian may agree or refuse. Or, to cite another example, a client may request information on available social services for the town's elderly. If that library or another has a program specifically designed to educate senior citizens about such services, the client may be referred to that program, and if no such program exists the library may consider establishing one. When they prepare their budgets, players consider the incidence of these requests, the number of participants in their own programs, and the availability of funds. (DCD)

Cost: $27.50

Producer: Harwell Associates, P.O. Box 95, Convent Station, NJ 07961

LIFE-CYCLE

G. Maureen Chaisson

Playing Data
Copyright: 1977
Age Level: adult, staff of institutions
Number of Players: 4-8
Playing Time: 3 hours
Preparation Time: 2 hours (est.)
Special Equipment: audio or video tape recorder
Packaging: professionally packaged game includes playing surface, feel wheels, dice, player tokens, grease pencils, critical incident cards, chips, and directions

Description: *Life-Cycle,* according to the designer, is designed to offer "players the opportunity to learn about old age by trying on the roles of old persons and their significant others (people who play a significant part in the old person's daily life) in interactions that are problematic and closely resemble the real life dilemmas of the elderly."

At the beginning of the game players are divided into two groups of equal size. One group assumes the roles of senior citizens and the other plays significant others (clergymen, family members, doctors, nurses). The elderly roll dice and move tokens around the game board.

This board has three rings of playing spaces and a suicide circle. An elderly person who throws a double with the dice or simply wants to quit the game moves his or her marker to the suicide circle. There is also a withdrawal circle which allows the senior citizen to withdraw from the game for one play to avoid an encounter. But most of the game revolves around the Play Circle. The spaces in this ring are marked (by the game director before play begins) with the name of specific significant others (such as a minister or a son) and when the senior citizen lands on one of these spaces she or he role-plays an incident with that significant other. These role-plays each center on one of the eighteen "Critical Incidents" supplied with the game (but not available for review). Each encounter is limited to five minutes and is video or audio taped by the game director.

Players are encouraged to comment on these encounters after they occur, and to award red or blue chips to role players who they think acted in a positive or negative way. "Cop out chips" are awarded to any player whose role was perceived to be inauthentic, evasive, or overly manipulative.

Four role-play encounters occur in each game with twenty to thirty minutes of discussion between each. During the general discussion which follows the game, the game director plays back the audio or video tape recordings of the role-play encounters. (DCD)

Cost: $195.00

Producer: Academy of Health Professions, 2220 Holmes Street, Kansas City, MO 64108

NURSING HOME CARE AS A PUBLIC POLICY ISSUE

Mary W. Bednarski and Sandra Florczyk

Playing Data
Copyright: 1978
Age Level: college
Number of Players: 3 or more
Playing Time: 1-3 weeks
Preparation Time: 10 hours (est.)
Packaging: 62-page photocopied booklet

Description: This package of curricular materials, according to the producer, "describes the policy environment for nursing home care, the key actors, possible legislative and other policy changes, and alternatives to nursing home placement. Through a series of exercises students learn from actual experience how to conduct interviews and gather information in other ways on nursing conditions and patient loads and how to evaluate public policy proposals for dealing with nursing home problems and alternatives to nursing home placement."

The package contains five exercises. In the first, participants find and write about "newspaper or magazine articles dealing in some way with nursing homes." In the second exercise, students contact one resident of a nursing home and try to determine whether or not that person is receiving adequate care. In exercise three, the participants use the information they gathered in the first two exercises to "identify all of the actors relevant to the analysis of the public policy surrounding nursing homes." In the fourth activity, students diagram the laws and administrative procedures affecting nursing homes. In the final exercise, students write a five-page paper examining one public policy "prescription" that "would affect both the quality and the cost of nursing home care." (DCD)

Cost: $3.00

Producer: Policy Studies Associates, P.O. Box 337, Croton-on-Hudson, NY 10520

PIPELINE: AN EMPLOYMENT AND TRAINING SITUATION

Jean L. Easterly and David P. Meyer

Playing Data
Copyright: 1978
Age Level: college, adult
Number of Players: 25-30
Playing Time: 90 minutes
Preparation Time: 2 hours (est.)
Supplementary Materials: cards, dice, poker chips, play money, gloves, magazines, ten shoe boxes, glue, twenty pairs of scissors
Packaging: 30-page photocopied monograph

Description: *Pipeline,* according to the authors, "is designed to represent the main features of current employability programs through environmental constraints and a step by step progression through the program phases by a majority of the players." The game attempts to simulate the entire process by which clients are enrolled in, and graduate from, manpower programs (at this writing, primarily CETA) beginning with enrollment and ending with getting and keeping gainful employment in the economic mainstream.

Before the game begins, seven participants are assigned administrative roles in the bureaucracy of the game (as one of three "Gatekeepers," or as "Pipeline Supervisor," "College Master," "Head of

Circle," or "Chance Master"). All other players are "Unemployed." The unemployed are allowed to play for twenty minutes for free, but for the rest of the game they must, at regular intervals, pay a "buck" to the game director or die. There are three ways for the unemployed to get money. The first, which all of the unemployed are encouraged to follow, consists of entering the "pipeline" (which corresponds to and simulates the programs of the various CETA titles). Another alternative is to opt for self-employment, which consists of playing cards or dice with the Chance Master. A third option for the unemployed is income maintenance (simulating unemployment compensation, welfare, or Social Security). Income Maintenance is paid by the Head of Circle, but the player who chooses this option must stand in line and wait until everyone in the pipeline is paid.

Players are admitted to the Pipeline by a Gatekeeper on the basis of the Gatekeeper's personal choice and the player's classification into one of several groups (corresponding to Vietnam Vets, heads of single-parent households, ex-offenders, members of minority groups, women). Players proceed through the pipeline according to their opinions about and ability to construct collages. After passing through the pipeline, players may, depending on their ability, and the availability of jobs, be given a job making collages which they may keep or lose. The game ends when at least one-third of the players have been employed as collage makers. (DCD)

Comment: This is a new and highly recommended game. (DCD)

Cost: (#PB 277 83/AS) $4.50

Producer: National Technical Information Service, 5285 Port Royal Road, Springfield, VA 22151

TERRY PARKER

Nigel Gann, Bosworth College

Playing Data
Copyright: 1975
Age Level: high school-adult
Number of Players: 17-29
Playing Time: 1 hour
Packaging: unbound 20-page packet with role sheets, background sheets, and instructions.

Description: This simulation is designed to allow participants to practice role playing and decision making as Terry Parker's probation officer, teacher, mother, headmaster, and girlfriend. Terry, who does not have a role in the game, is fifteen years old and was placed in Beechwood Children's Home a year ago when his divorced mother could no longer control him. He often skips school and fraternizes with an undesirable group. The role players must decide what recommendations the Children's Home should make about Terry's future. They will advise the juvenile court either to return Terry to his home or to place him under court supervision. After the role players are selected and briefed, the rest of the class is divided into groups of six. Each group interviews each role player. After the interviews, each group reports its recommendations on Terry's future. The instructions state that these reports "must be based on an understanding of the difference between punishment and rehabilitation" and "may lead on to a wide-ranging study on the family, on legal provisions for young offenders or take any other direction the teacher wishes." (DCD)

Cost: 70 pence + 20% postage

Producer: Community Service Volunteers, 237 Pentonville Road, London N1 9NG England

TRACY CONGDON

Nigel Gann, Bosworth College

Playing Data
Copyright: 1975
Age Level: high school-adult
Number of Players: 6-30
Playing Time: 2 hours
Packaging: unbound 20-page packet with role sheets, background sheets and instructions

Description: This simulation is intended to help students gain an understanding of the complexities of social problems. The situation concerns Tracy Congdon, who is eight years old, and who may be a battered child. Tracy lived with a foster family for six years but now has been returned to live with her mother, brother, sister, stepfather, and his three children. Six participants take the roles of Tracy's mother and stepfather, her foster parents, her teacher, and an overworked social worker. Tracy's neighbors report that she is often bruised and gossip about her supposed mistreatment. It is the task of the six characters listed above to determine whether Tracy should be removed from her home. The other participants are divided into groups of six persons each. Each group interviews each role player. After these interviews, each group reports its recommendations on Tracy's future, bearing in mind that it might be detrimental to remove Tracy from her home, that there is no positive evidence of her being ill-treated but that she may have been abused and that she is more familiar with her foster family than her own. Participants are instructed that their only concern should be for Tracy's welfare and that they must be able to justify their decision. (DCD)

Cost: 70 pence + 20% postage

Producer: Community Service Volunteers, 237 Pentonville Road, London N1 9NG England

WELFARE WEEK

Ann Marie Kraemer and Katherine Consori; edited by Armand Lauffer

Playing Data
Copyright: 1973
Age Level: grade 9-graduate school, church and civic groups
Number of Players: 5-500 in 2-100 teams
Playing Time: 2 evening sessions; rest of play is done at home for one week
Preparation Time: 3-5 hours

Description: Players assume the roles of welfare recipients and try to survive on a minimal food budget and to maintain normal family life during a simulated week of living on welfare. Chance enters this game as households must cope with a visit from their case worker and deal with what turns up in the crisis cards. Individual and team play emphasizes cooperation among players as they handle conflict with the system in which they operate. Play involves strategic thinking (above all else), good budget management, and lots of courage.

The game is designed so that players can develop an empathy with those on public relief and an understanding of how public welfare works and how it can be supportive or destructive of its clients. (DZ)

Cost: $75

Producer: Gamed Simulations, Inc., Suite 4H, 10 West 66th Street, New York, NY 10023

WORKSHOP MANAGEMENT TRAINING SIMULATION

William J. Wasmuth, Industrial & Labor Relations, Cornell University

Playing Data
Copyright: 1968
Age Level: management, supervisory and social nonprofit agency staff
Prerequisite Skills: experience as a manager or executive director
Number of Players: 15-20 in 3-4 teams
Playing Time: 4-5 days
Preparation Time: 3 days

Special Equipment: slide and overhead projectors

Description: This simulation is designed to foster an appreciation of the complex knowledge, attitudes, and skills involved in the management process and of the dynamic changes which occur in organizations over time, to explore the interrelated roles of managers and supporting staff in organizations; to provide knowledge of research-supported concepts and practices used in managing rehabilitation facilities and sheltered workshops; to provide for analysis and understanding of the relationships among such key functional areas as client admissions, staffing, subcontract procurement, wage policy, community relations, board of directors composition and function, financial record keeping, and so forth.

Workshop places each participant in the position of the executive or key staff person of a rehabilitation facility which is experiencing problems concerning its future admissions, staffing, financing, board of directors, and community relations. The simulation is played by month with twelve intervals requiring about two to three hours each. There are aspects of group problems solving, role playing, and in-basket techniques in this simulation. Conflict situations develop, resources become scarce, opportunities present themselves, tradeoffs between two or more factors are required, analysis is necessary, and a strategy needs to be hatched out within a team framework and them implemented. Problems include rapid rate of growth from about fifteen clients served on a daily basis to an average of fifty-six presently, and a deficit has grown, however, from $5,000 to over $37,000. Teams then begin by trying to formulate their management objectives. Each team communicates its decisions and requests in writing to a coordinator who represents the outside environment and who responds with the help of a thirty-page manual. (REH)

Cost: Special arrangements necessary

Producer: William J. Wasmuth, ILR School, Cornell University, Ithaca, NY 14850

See also
Compacts COMMUNITY ISSUES
Serfdom SOCIAL STUDIES
System SOCIAL STUDIES
ADDICTIONS

INTERNATIONAL RELATIONS

AFASLAPOL: THE GAME OF POLITICS IN A MODERNING NATION

Timothy Gamelin and Hank Kennedy

Playing Data
Copyright: 1972
Age Level: high school
Number of Players: 10-30
Playing Time: 6-12 sessions of approximately 50 minutes each
Preparation Time: 1-2 hours
Packaging: 42-page player manual includes extensive instructions, explanations, charts, and forms

Description: The purpose of this game is to simulate the politics of a nation-state in the midst of modernization and development. The simulation is based on the modernization of African, Asian, and Latin American nation-states. The focus of the simulation is on the struggle of "transitional" politics to increase political capabilities and performance. Phenomena that are simulated include social and economic mobilization, government taxation and allocation dilemmas, bureaucratic corruption, difficulties of gaining political legitimacy and of regulating a moderning society, irregular selection and turnover of leadership, and other problems which are found to varying degrees within the nations of the "Third-World." The class is divided into ten groups each with one leader who has a "distinct base of political support within the country"; the ten constituencies are: the Aristocracy, the Bureaucracy, the Industrialists, the Intellectuals, the Military, the Peasantry, the Policy, the Proletariat, the Traditionalists, and the Villagers. The goal of the game is "to maximize the economic and force potential of one's constituency and/or maximize one's status." Through the transactions between the ten sectors of the society, each constituency has opportunities to accumulate various forms of political and economic strength potential, or, in other words, increase their power. With a chart, a player is able to determine his relative success and convert it to a score, which may be compared with the scores of the other players. There are twelve cycles (each representing one simulated year in the nation's life), each approximately fifty minutes to one hour in length. (WHR)

Cost: $4.50

Producer: Gerald L. Thorpe, Department of Political Science, 101 Keith Hall Annex, Indiana University of Pennsylvania, Indiana, PA 15701

THE ARAB-ISRAELI CONFLICT

Karen Ann Feste

Playing Data

Copyright: 1977
Age Level: college
Number of Players: 18-60
Playing Time: 4-6 hours in 4-10 rounds
Preparation Time: 2 hours (est.)
Packaging: professionally produced, 66-page soft-bound book

Description: This is a role-playing game designed to model, in the author's words, "the chaotic complexity defining interrelationships among Middle Eastern Nations."

At the start of the game, players are assigned to one of ten teams representing Israel, Egypt, Syria, the Palestinians, the United States, the Soviet Union, the United Nations, the rest of the world, or the press. Each player on the first seven teams drafts a policy paper addressing a specific issue important to that nation. The United Nations team lists the ways it is currently active in the Middle East. The rest-of-the-world team writes an overview of the Middle East situation, and the press assumes the roles of *Pravda* and the *New York Times*.

The game then follows a general pattern of challenge and response moves. During each move period, each nation decides on problems to be addressed and strategies to be employed. Each country formulates a policy, negotiates with other countries, and announces its policy to the press, which then makes that policy public. This process continues for four to ten rounds, and then the players must evaluate and assess the importance of the changes that have taken place during the game in their originally formulated policies. (DCD)

Cost: $3.50

Producer: The American Political Science Association, 1527 New Hampshire Avenue, N.W., Washington, DC 20036

BOMBS OR BREAD: CONSUMER PRIORITIES VERSUS NATIONAL SECURITY

Playing Data

Copyright: 1978
Age Level: grades 9-12
Number of Players: 28-49 in seven groups
Playing Time: 4 50-minute rounds
Preparation Time: 2 hours (est.)
Packaging: 15-page photocopied booklet

Description: Players assume the roles of President, Secretary of National Security, Secretary of Trade, or Secretary of Production for one of seven countries named Antiem, Brunwick, Calizonia, Dromain, Europia, Freedonia, and Garbania. Each nation tries to achieve the highest possible standard of living for its citizens while defending them from attack by other nations. Each country's standard of living is indicated by its accumulation of "products," and its ability to defend itself is indicated by its stockpile of "bombs."

The game is played in four rounds, and there are three parts to each round. During the opening priority session, of each round, the decision makers of each country confer and decide how many products and how many bombs will be produced that round. A country may order a total of twenty items in any round and these items may be any combination of bombs and units of that country's single product. (For example, Antiem produces only airplanes, Calizonia produces beef, Europia produces oil. In the trade session of each round, the countries try to barter their products. Nations earn points for the number of *different* products they possess at the end of the game. (One set of all seven products, for example, earns the nation that owns it fifteen points, while one pair of two different products has a value of only one point.) Each round ends with an attack and defend session in which each nation must decide if it wants to attack any of its neighbors. In an attack, the country with the most bombs wins. The winning nation may keep all of its products and bombs while the losing nation must forfeit everything. The nation with the most points after four rounds wins. (DCD)

Cost: $3.00

Producer: Edu-Game, P.O. Box 1144, Sun Valley, CA 91352

CONFLICT IN THE MIDDLE EAST

Stephen M. Johnson, The Lincoln Filene Center for Citizenship and Public Affairs

Playing Data

Copyright: 1969
Age Level: grades 9-12
Number of Players: about 50
Playing Time: 3-4 hours

Description: Players take the roles of top leaders in nine nations concerned with Middle Eastern affairs. Each team is composed of five members; Head of State, Foreign Policy Advisor, Military Advisor, Political Advisor, and the Opposition Leader. All decisions and game actions must pass across the control desk to maintain the realism of the simulation. Students are supplied with a manual giving only the background of the Middle East situations. The first act of players is the preparation of a statement of goals. Two scenarios are provided which enable the policy simulation to begin. One assumes that Jordan and Israel independently decide the issues between them and set up a nation on the West Bank of the Jordan River. The second scenario assumes that the United States withdraws its past verbal commitments to Israel. This sets off a time of uncertainty and change. (REH)

Comment: This is a relatively open-ended simulation where most of the effort must be provided and filled in by the students who are using the simulation. Scenarios are brief and little more information is provided than to give players roles. (REH)

Cost: $1.50

Producer: The Lincoln Filene Center for Citizenship and Public Affairs, Tufts University, Medford, MA 02155.

CONFRONTATION: THE CUBAN MISSILE CRISIS

Ralph Meyers, Arthur Solin, and Gerald L. Thorpe

Playing Data

Copyright: 1978
Age Level: high school, college
Number of Players: 8-36
Playing Time: 50 minutes
Packaging: kit includes 18-page pressbook, 4 filmstrips, 4 cassettes, teacher's guide, simulation game

Description: According to the authors, "*Confrontation* is a decision-making exercise designed to place students in positions comparable to those occupied by Soviet and American decision-makers during the days of the Cuban Missile Crisis. Participants in the simulation puzzle over the same alternatives faced by the protagonists in the actual crisis as they make decisions and respond to counter-moves.

"The basic objective of the exercise is to illustrate the chronic dangers inherent in a world political system that depends upon countervailing military power to stabilize that system."

The exercise begins with the division of participants into two or four working groups with four to nine people in each. All of the work is done in these isolated groups. Half of the participants assume the roles of Soviet leaders and the other half pretend to be American policy makers. The process of the exercise consists of following the instructions of memoranda that are distributed to the groups. There are four separate memoranda for the Americans and four for the Soviets, detailing various decisions that can be made by either group and differing from one another in the tone of their belligerence. For instance, one American "advisor" may favor bombing Cuba, another may only consider that action as one option out of many, and a third may be completely opposed to bombing. The game follows the chronology of the historical crisis. After the players have studied their roles, the groups meet and decide on a course of action in response to each challenge from the other side. The Americans, for instance, must decide what action to take when they first discover Soviet missiles in Cuba, and the Soviets must decide what to do when the Americans discover the missiles. The only choices open to the players are those elaborated in the memoranda, and these are deliberately limited to reflect what, in the authors' opinion, were the political world views of the historical antagonists. The groups work independently, and after they have all resolved the crisis in some way, all participants from all the groups convene to discuss their experiences and decisions and to speculate about optional courses of action. A BASIC computer version by Gerald Thorpe, codesigner of *Confrontation,* has recently become available, and a FORTRAN version will follow. (DCD)

Note: For more on this simulation, see the essay on international relations simulations by Leonard Suransky.

Cost: (1) $96.00 (#CA 340); (2) $20.00 BASIC version

Producer: (1) Social Studies School Service, 10,000 Culver Blvd., Culver City, CA 90230 (2) Gerald L. Thorpe, Dept. of Political Science, 101 Keith Hall Annex, Indiana University of Pennsylvania, Indiana, PA 15701

CRISIS

Western Behavioral Sciences Institute

Playing Data
Copyright: 1966
Age Level: grade 6-graduate school
Prerequisite Skills: reading, grade 11.5; math, grade 6
Number of Players: 6-35 in 6 teams
Playing Time: 1 class period minimum, no maximum

Description: *Crisis* is a game designed to bring the child out of the man and cause even professional educators to threaten to blow up the earth. I have used it with junior and senior high school students, social studies teachers, nuns teaching in colleges in Nova Scotia, secondary-school and college teachers in England, and staid members of Her Majesty's Inspectorate of Education in Great Britain. In each occasion the participants were carried away by their involvement in the process and they behaved as nations often do when they are confronted with an international dilemma. To my amazement and horror one group of Nova Scotian participants threatened to blow up the earth. That was a *Crisis*!

A simulation is a model of a real-life situation. In this instance, *Crisis* involves six nations–"Axiom," "Burymore," "Cambot," "Dolchavit," "Ergosum," and "Fabuland." Each of these countries becomes involved because a rare element "Dermatuim" is mined in the "Rhu Day Valley" which borders the two small countries Ergosum and Fabuland.

According to the historical background provided in the simulation: "Until now an agreement held together by the Peaceheard Pact has allowed both countries to mine this element. Several times during the past few years each of the countries has disputed the right of the other to use the mines, but on every occasion a new agreement has been reached and fighting avoided.

"Dermatuim is a volatile element and thus is very dangerous to mine. Until recently it was used primarily in the production of skin cream. Last year, however, scientists discovered that by shooting Dermatuim through the dried digestive tracts of monkeys who had been fed iodized salt, a new synthetic substance, Balonium, could be created, one pound of which would contain energy sufficient to run all of man's present machines for the next 10,000 years; or, if used destructively, enough explosive power to destroy the world. This discovery obviously has increased the value of the Rhu Day Valley and is a major factor in the crisis which is now developing. . . ."*

Though the terms "Dermatuim" and "Balonium" are fictitious and strange, one sees encapsulated in this story the many world crises, past and present: the struggle between the Spanish and the English for the gold of the New World; the wars fought between the French and the English for the fur trade; the division of Africa by the European countries before World War II; and the oil of the Middle East today. Some simulations deal with an historical problem. Others are concerned with a contemporary issue. *Crisis* illustrates a perennial problem, albeit one which is becoming increasingly critical in a crowded planet with diminishing resources and more deadly weapons.

Crisis quickly thrusts the participants into the situation. Chief decision makers in each country meet. They delineate the problem from their perspective, draw up their options, make alliances, and meet in the World Organization. Students learn the give and take of decision-making. Instant feedback is provided as the results of their actions become evident. Total involvement leads the participants to develop a deeper understanding of the political process on national and global levels. The twin goals of critical thinking and understanding are achieved.

Simulations often require the teacher to motivate students beforehand. *Crisis* can stand on its own. One should move right into it. However, the instructions should be thoroughly understood by teacher and participants. It is essential to select a responsible person as the Director. If no one can be found, the teacher should assume the role. The teacher should also move about, unobtrusively, to see that all is proceeding smoothly. As with all simulations, feel free to improvise. Greater use could be made of the press. Create a newspaper headline center in one corner of the room, issue trial balloons, send out feelers. *Crisis* could be played in one long session or one session a week for several weeks. The latter approach will cause the participants to make alliances and deals outside the class.

Crisis may be used at the beginning or the end of a political unit. If it starts a unit, activities and understanding flow from it. If it is the culminating activity, knowledge previously acquired could be applied.

These are some of the understandings, concepts, skills, and attitudes which may be taught through *Crisis.*

(1) Understandings
 A nation's foreign policy should reflect its best interest.
 A nation's foreign policy is produced by a number of internal and external forces.
 We are part of an international community (a "global village").
 Events in any part of the world can affect us.

(2) Concepts
 Foreign policy
 Balance of power
 Collective security
 Logrolling
 Propaganda

(3) Skills
 Decision-making
 Parliamentary procedure
 Leadership

(4) Attitudes
 Appreciation of the complexity of foreign affairs problems which confront every President, Congress, and Secretary of State
 Appreciation of the need for the nations of the world to cooperatively seek solutions to global problems.

A discussion should follow *Crisis.* Participants should reflect on such questions as "Why did you make the decisions you did?" "Would you make them again?" "Is *Crisis* realistic?" "How may it be made more realistic?" "Where do we face similar problems today?" "Where, in the past, did similar problems occur?"*Crisis* also makes it easy to individualize instruction for participating in this simulation can lead students to do the following:

(1) Collect or create a series of cartoons and headlines regarding current world crises;
(2) Study the rise and fall of the League of Nations;
(3) Report on earlier efforts at international cooperation;
(4) Invite their congressman to speak on current foreign problems;
(5) Report on international cooperation in Antarctica, undersea exploration, and outer space. The origins of international law could be studied as well as its growth and development in the last two areas;
(6) Report on the Middle East and Cyprus;
(7) Write papers on international crises since 1945;
(8) Study the United Nations;
(9) Study the roles played by the president, congress and the State Department in making United States foreign policy;
(10) Study current United States relations with the Soviet Union and China.

Crisis can be used with secondary school students (grades 7-12), college students, and adults. It is thoroughly entertaining; even reluctant players can be drawn into the "wheeling and dealing" of politics, the excitement of decision-making, and the formation and dissolution of alliances. Human relations skills and critical thinking skills are developed. Participants also acquire a deeper meaning of the concepts, skills and attitudes previously noted. Crisis presents a realistic foreign policy dilemma, one which continually confronts the United States. But, no teaching technique operates on its own. Simulations, too, require an expert teacher, one who fully understands the objectives and mechanics of the simulation and who can fully exploit its potential in the matrix of total teaching. (from the Robert A. Taft Institute of Government Study on Games and Simulations in Government, Politics, and Economics by Joseph Eulie, State Univ. of N.Y. at New Paltz)

User Report: I am an enthusiastic advocate of simulations in the classroom, and I have used *Crisis* more often than any other simulation (twenty to thirty times) to illustrate the process because it does it so admirably. The participants introduced to simulation become excited about its possibilities and want to use *Crisis* in their own classrooms. It is a well designed simulation which involves the participants very quickly and maintains their interest and involvement–an involvement which becomes a springboard to profitable debriefing sessions.

Because of its good design and simplicity, it lends itself to many variations, allowing for different purposes in using the simulation, for different principles and concepts to be stressed, and for different levels of sophistication in different groups. Although students' interest in it usually wanes after three to five hours, it can be played again after a period of instruction, with new insights resulting.

As with most simulations, the director's manual for *Crisis* is less than adequate. It is difficult to run *Crisis* the first time from only the instructions, and it usually takes some help from someone familiar with the game or a dry run or two if the director has never been a participant in it.

Overall, however, it is an excellent simulation. (Jack W. Sutherland, Professor of Education and Social Science, California State University, San Jose)

Note: For more on this simulation, see the essay on international relations games by Leonard Suransky.

Cost: $25.00

Producer: Simile II, P.O. Box 910, Del Mar, CA 92014

*"World History," *Crisis*. Western Behavioral Sciences Institute, La Jo la, Ca., 1966.

DANGEROUS PARALLEL

Foreign Policy Association

Playing Data
Copyright: 1969
Age Level: grades 7-8
Number of Players: 18-36 in 6 teams
Playing Time: 5 hours in periods of 40 minutes

Description: This simulation presents a model situation in which conflict threatens world peace. Participants are divided into six imaginary nations. Each nation has a Chief Minister and 2 to 5 supporting ministers in his cabinet. During each round, students go from one nation to another negotiating for a peaceful solution to the crisis. The teacher acts as Control, interpreting the results of each round. Each Kit contains all needed materials. (Producer)

Cost: $72.00

Producer: Scott, Foresman and Company, 1900 E. Lake Ave., Glenview, IL 60025

THE DECISION MAKING MODEL

World Affairs Council of Philadelphia

Playing Data
Copyright: 1966
Age Level: high school
Number of Players: 25-42
Playing Time: 5 45-minute periods
Packaging: kit includes teacher manual; one set of nation profiles; tables and charts for five country groups

Description: This game attempts to simulate interaction between five disparate nations by focusing on the means by which each one makes domestic and international policy. Each nation has a government consisting of a budget director, a defense minister, and a head of state, and a population consisting of influential citizens.

The simulation begins as all players are assigned a nation and a role. They then make themselves familiar with the profile for their nation, which is included with the game materials. The nations range from very poor to very rich and are analogous to third world and European countries, the Soviet Union, and the United States. The leader of each nation must then, with the advice or consent of the rest of the government and the citizenry, define national goals. These goals concern the maintenance or improvement of the standard of living, avoiding or initiating war, international trade, and foreign aid. The head of state must then realize these goals within budget constraints, defense needs, and the goals and ambitions of the other nations.

After five game sessions, the participants discuss the political and economic efficiency of each of the governments. (DCD)

Cost: kit $3.50, single manuals $1.00 each, extra forms $2.50 per set

Producer: World Affairs Council of Philadelphia, 3rd Floor Gallery, The John Wanamaker Store, 13th and Market Streets, Philadelphia, PA 19107

DIPLOMATIC/FOREIGN POLICY GAME

Harvey Starr

Playing Data
Copyright: 1974
Age Level: college
Number of Players: class
Playing Time: 3 class periods

Description: This game is designed to be played toward end of a comparative foreign policy course. Players take four decision-making roles in each nation-team: executive, foreign policy, internal/domestic, and military. They decide on the relative voting weight of each role within a team according to their knowledge of the country they represent.

When the game scenario has been presented, each nation-team develops a policy, which guides activity through the first round, and a mid-game statement that guides policy through the second round. Teams have freedom of movement, of approach and refusal, and access to a "world newspaper"–which announces truth, lies, and propaganda, as it is told. There is also a presence named God, from whom no secret may be kept, whose announcements of the outcome of human enterprise are made when he or she so desires. Success or failure is determined by God's arbitrary decisions regarding relevant matters, which can range from assassination attempts to U.N. votes.

Postgame sessions deal with what foreign policy tools were used in the game, evaluations of the success of nation-teams, and the styles and influences players perceived at work during the game.

Cost: Not published, but can be played by contacting producer.

Producer: Harvey Starr, Dept. of Political Science, Woodburn Hall, Indiana University, Bloomington, IN 47401

DIPLOMATIC PRACTICES

Paul M. Kattenburg, University of South Carolina

Playing Data
Copyright: 1975
Age Level: college, adult
Number of Players: approximately 35, minimum 24
Playing Time: 6 class sessions, and 5 hours outside class
Preparation time: 3 hours
Packaging: 51-page manual

Description: Players of *Diplomatic Practices* learn about the effect of bureaucratic factors on policy formation, analyze diplomatic situations in foreign countries, and increase their understanding of, and experience with, negotiations. After reading about the political situation of Ecks, a fictitious, poverty-stricken, isolated country of 50,000 square miles, students write a political analysis of a border dispute and choose individual roles as American or Ecksian diplomats. The first day of the simulation, students divide into country teams of eight members each to discuss their analyses according to several criteria described in the manual and to make recommendations. The second and third session, students discuss the bureaucratic organization of Ecks, and prepare for, and conduct, meetings in interdepartmental groups, during which students express the views held by the diplomats they are role playing. For example, the U.S. Public Affairs Officer favors a "hands off" approach until more data is available on the Ecksian political situation; the Ecksian Minister of Economic Affairs supports the establishment of a U.S. military base operated by Ecksians to stimulate the economy. After the meetings, students write option papers about the pros and cons of continuing negotiations with Ecks and establishing a Navy/Air Force tracking system there. The last three days of the simulation examine bilateral negotiations. Students divide into at least two delegations of six members and up to six advisers and role-play negotiations between the U.S. and Ecks government over the terms of their agreement to establish a radar tracking base. Delegations meet separately to determine their tactics and then meet jointly for a maximum of eighty minutes. This final simulation exercise is followed by discussion and evaluation. (TM)

Cost: $2.00

Producer: Learning Resources in International Studies, 60 East 42nd Street, New York, NY 10017

FOREIGN POLICY DECISION MAKING

William D. Coplin and J. Martin Rochester

Playing Data
Copyright: 1971
Age Level: college
Number of Players: 15-60 in groups of 3-5
Playing Time: 3 hours including briefing and debriefing
Preparation Time: 1 hour (est.)
Packaging: 30-page photocopied booklet

Description: This simulation (with its introductory and postplay materials) is designed, in the authors' words, to introduce "concepts and propositions that scholars have employed in describing the process of foreign policy decision-making." The authors stress that foreign policy decisions are made by men (there are no women's roles) rather than by nations and that these decision makers are not blessed by the miracle of immaculate perception. That is to say, the information on which major decisions is based is sometimes inaccurate, often includes political and ideological considerations, and is always filtered through the mentation of the decision maker.

Players are assigned to one of three categories of decision-making groups. In *rational* groups, participants are not assigned roles and thus must make their decisions without regard to personal or political bias. In *political* groups, players assume the roles of hypothetical, subcabinet-level appointees whose political orientation varies from reactionary to liberal. They must make their policy decisions in consideration of the possible domestic and international responses that may result from their action. Participants in *bureaucratic* groups represent career diplomats and soldiers who make decisions using the same criteria as political groups.

Each group addresses three categories of problems. In the *crisis* problem, the groups must respond to a revolution in Greece. In the *decision* problem, they reevaluate American policy (as of 1971) pertaining to the People's Republic of China. In the *administrative* problem, each group must formulate an American policy to be pursued at a United Nations' Disarmament Conference. (DCD)

Cost: $2.00

Producer: Learning Resources in International Studies, 60 East 42nd Street, New York, NY 10017

THE GAME OF NATIONS

Dr. Henry Murray, Harvard University (the original game) Michael Hicks-Beach, et al.

Playing Data
Copyright: 1973
Age Level: high school, college, continuing education
Number of Players: 4
Packaging: 15-page manual, playing board, character pieces, and 3 packs of cards

Description: The *Game of Nations* is a simplified version of the *Game of Nations* used at the Peace Games Centre and by the CIA. It is based on the idea that political leaders play three interrelated games at one time–an international game, a domestic game, and a game of personal satisfaction. Four players, each with 4 home ports located near the eight oil-bearing countries of Kark portrayed on the board aim to secure revenue for their own country and to gain control of the countries of Kark. In order to move their leaders–kings, politicians, dictators, guerillas, and secret agents–players must pay a set revenue. On each turn players may make a number of moves–they may move a leader, a secret agent and tanker, may buy and sell tankers, and may place a pipeline. A traveling leader who attacks the capital city of a country then occupies it and seizes its tankers. Tankers moved to the home port and paired with derricks yield revenue. Secret agents may move 1 to 7 spaces on each turn and may pick up an espionage card with secret directions to help them accomplish their purposes. The game is won when all but one player is eliminated or when one player gains control of six countries. The winner always proves to be the one who has made his decisions in strict accordance with mathematical games theory rather than according to personal inclinations. During the course of the game, each player's leadership style is revealed. (TM)

Cost: $14.00

Producer: House of Games Corp. Ltd., P.O. Box 2138, Toronto Bramalea, Ontario, Canada

THE GOOD SOCIETY EXERCISE: PROBLEMS OF AUTHORITY

Steven Apter, Jon Bramnick, and William D. Coplin, International Relations Program, Syracuse University

Playing Data
Copyright: 1971, 1973
Age Level: grade 10-college, continuing education
Number of Players: 16-60
Playing Time: at least 1 hour, no top limit
Preparation Time: 30 minutes
Special Equipment: video equipment optional

Description: Stokely Carmichael has said: "The status quo persists because there are no ways up from the bottom. When improvements within the system have been made, they resulted from pressure–pressure from below. Nothing has been given away; governments don't hand out justice because it's a nice thing to do."

Barry Goldwater has written: "Maintaining internal order, keeping foreign foes at bay, administering justice, removing obstacles to the free interchange of goods–the exercise of these powers makes it possible for men to follow their chosen pursuits with maximum freedom. But note that the very instrument by which these desirable ends are achieved can be the instrument for achieving undesirable ends . . . Government can, instead of extending freedom, restrict freedom. . . . This is because of the corrupting influence of power, the natural tendency of men who possess some power to take unto themselves more power."

Elton McNeil has observed: "Once an animal has achieved a position of dominance over other animals, it will violently resist any effort made to lower its status . . . a human can adjust to subordinate status if he has known no other way of life, but, once having acquired a taste for being first in relationship with others, it is not abandoned gracefully."

The *Good Society Exercise* is designed to demonstrate the above propositions in a simulated setting. Participants in the *Good Society Exercise* gain insights into three major concepts in the social sciences–authority, justice and order. The game can be employed to introduce a unit concerned with these concepts or as a terminating exercise following a discussion of laws or governmental programs that have as their purpose the reallocation of rewards and privileges in society. The game is an abstraction of the relation between authority, justice, and order. Students make judgments about these concepts by distributing power among the teams. It can be played with as few as fifteen people and as many as sixty. It can extend over a period of six hours or as few as two hours depending on the number of participants and the amount of time. The rules of the game are relatively simple, thus making it suitable for students from the seventh through the twelfth grades.

The assumption underlying the *Good Society Exercise* is that the essence of politics in every society is related to the bargaining among groups over the distribution of power. In operation, the game creates an actual political process in the classroom. For the game to be most effective, it should be subordinate to an informational package in which authority, justice, and order are defined and explained by philosophers, politicians, and social scientists.

The rules for the game are not complex and are explained rather fully below. The Good Society consists of seven teams, six of which consist of two to five players each and are called groups (A, B, C, D, E, F). The other team consists of four to six players and is called Central Authority. The groups and Central Authority behave in the Good Society according to the following statement of principle: "This society guarantees equal opportunity and equal rights to every member. It advocates the need for stability, change, and justice. Inequality is as enslaving as disorder. Order and justice must prevail."

In spite of this statement of principle, however, the "goods" of the Good Society as measured by Power Units are not distributed equally. Moreover, the ability to make changes in the Good Society is not shared equally by the groups. The simulation exercise is designed to represent what happens in a society in which order and justice are its guidelines, yet where realities in the society run counter to the guidelines.

Power Weights (the units used to measure the wealth and influence of each team in the society) and Petition Weights (the weight of each team in the signing of a Petition) are distributed in the following manner among the seven teams of the *Good Society*:

Name of Team	*Power Weights*	*Petition Weights*
Central Authority	100	3
Group A	45	2
Group B	35	2
Group C	10	1
Group D	8	1
Group E	4	1
Group F	4	1
TOTAL	206	11

Changes in the distribution of Power Units as well as any other facet of life in the Good Society can occur in two ways: A petition can be signed by the teams representing at least six Petition Weights which call for any specific change. A war (the contest could be referred to as interest group action before a city council, state legislature or the Federal Congress) can be fought between any team in the society in which the more powerful (the larger number of Power Weights) gets all but one of the losing side's Power Weights.

In any redistribution of power points, no fractions may be used. Power points may be neither created nor destroyed.

The maintenance of order in the Good Society is the responsibility of all of the members, although the Central Authority has a special role, given the large number of power units at its disposal. It can engage in war against members that fail to live up to the principles of the society and can attempt to get Petitions signed. At the same time, it is up to the six groups to make sure that the Central Authority does not abuse its role. These groups can sign Petitions to limit the actions of the Central Authority, even to the point of redistributing the power. (It is important to note that any redistribution of power units restricts the use of fractions and restricts creating new power units or destroying power units.)

The main idea of the exercise is contained above. However, certain procedures and rules must be followed.

(1) The exercise can be run in one classroom by separating the groups. If possible, separating the groups into seven different classrooms is recommended, but not essential for a successful exercise.

(2) Each group will be issued travel passes. Groups A through E will be issued travel passes depending on the number of members in the group. If groups A through E have three or less members, only one travel pass will be issued per group. If those groups have four or more members, two travel passes will be issued. The Central Authority is issued half as many passes as number of group members. Simulation Control will distribute these passes at the beginning of the exercise.

(3) The instructor will act as Simulation Control. He will distribute all communications and answer any questions about the game. Simulation Control must prepare for the exercise in the following manner:

(a) Simulation Control should prepare signs, designating the letters of the groups, and post these signs in the respective classrooms. If played in one classroom, some method of identifying the groups is necessary.

(b) (Optional). A video tape on the Central Authority during he exercise provides an excellent teaching aid for debriefing discussion.

(c) (Optional). During the first fifteen minutes of the exercise restrictions can be placed on travelling between groups. This is a time for strategy meetings within groups. Following the fifteen-minute strategy meeting, a fifteen-minute negotiation period should follow. During this time, members with passes may travel between groups to discuss strategies and alternatives. Simulation Control or observers representing Simulation Control should notify their respective groups when negotiation and strategy periods begin and end. The fifteen-minute negotiation-strategy periods may continue throughout the game at the discretion of Simulation Control.

(4) In addition to personal discussions, communications may be sent through written messages or may be distributed as part of a Newsletter. A Newsletter, if it is published, will contain announcements by groups and Central Authority as well as editorial statements by reporters. The Newsletter will officially announce the results of all Wars and Petitions. If the game is played without written communications, Simulation Control must travel from room to room or group to group informing the groups of Wars and changes by Petitions.

(5) All decisions taken by the groups must be unanimous decisions. All decisions taken by Central Authority must be supported by a majority of its members.

(6) A Petition Form will be used for purposes of peaceful change. Any Group or the Central Authority can initiate a Petition. Once enough signatures of teams to constitute six Petition Weights has been achieved, the Petition should be submitted to Simulation Control. It goes into effect *as soon as Control officially announces the Petition.* It remains in effect until a different Petition is completed to alter the original one. No Petitions can go into effect between the first declaration of War and the end of the War ten minutes later.

(7) War occurs when one or more teams declare war against one or more other teams. A War Form is submitted to the Simulation Control, who will inform the other teams. The attacked team will have ten minutes to counter-attack. More than one team can attack and more than one can counter-attack. The side with the largest number of Power Units will win the war and will get all but one of the other team's Power Units. If more than two teams attack and win, the power units won will be distributed in accordance with the relative power units of the winning teams unless the winning teams unanimously agree to a different distribution within five minutes after the war. Any teams submitting a War Form may retract it before the ten-minute time limit is up.

The main impact of the game is an understanding of the innate drive for equality and justice among men, and the dilemma it poses for those in authority. For many students the game forcefully presents the moral dilemmas inherent in the striving for power by the "have nots" on one hand and the attempts by those in power to maintain a position of superiority on the other. The game requires students to collaborate to maintain a favored position or to replace other teams and share more fully in the Good Society. Deception is an important aspect of the game as various teams attempt to negotiate private advantage at the expense of other teams and most frequently at the expense of Central Authority. In practice the game centers on the strategies developed by various teams to upgrade their power units.

The game has a high entertainment value, especially if video tapes of strategy sessions and negotiations are played during the debriefing session. The video tapes provide the opportunity to review and analyze the plots and counter plots that played a part in the simulated decision-making process. The function of the instructor is to compare and contrast student behavior patterns with (1) statements made by philosophers, political commentators, and politicians; and (2) real world examples of political decision making. Many students gain an appreciation of factors other than the merits of a particular argument in political decision making, e.g., the desire of those who enjoy a privileged position to maintain it and the ingenuity demonstrated in rationalizing that position. Students have the opportunity to sharpen such skills as bargaining, negotiating, and log rolling. Emphasis is placed on maximizing a team's resources (power units) through alliance formation, cooperation, deception, image building, and effective written and oral communications. Ideas such as democracy, balance of power, and orderly change are usually referred to in the bargaining and negotiations that take place between and among the Good Society teams.

Although not necessarily a shortcoming, the *Good Society Exercise* is a very abstract version of the real world. Competition is for power units which do not relate directly to real world benefits and authority. The instructor or game manager must bridge the gap between the exercise and real world situations. Consequently, it must be used in connection with a teaching unit or learning package that focuses on the concepts of authority, justice and order. If used in isolation the game has a minimal impact on the educational process. As in most games and simulations, the instructor must effectively relate the game experience to the real world in the debriefing session. (From the Robert A. Taft Institute of Government Study on Games and simulations in Government, Politics, and Economics by Scheley R. Lyons, University of North Carolina at Charlotte.)

Cost: $2.00

Producer: Learning Resources in International Studies, 60 East 42nd Street, New York, NY 10017

GUNS OR BUTTER

William A. Nesbitt

Playing Data
Copyright: 1972
Age Level: grades 7-12
Number of Players: 25-30 in 5 teams
Playing Time: 6 45-50-minute rounds
Preparation Time: 90 minutes (est.)
Packaging: professionally produced and packaged game includes director and team manuals, buttons, and forms

Description: The leaders of this game assume the roles of leaders of five hypothetical nations. The leaders, in the author's words, must "advance the economic and social welfare of the people" in their countries "without endangering national security in the international system." During debriefing, the players "discuss the relative success of the various national decision makers. Among the factors that may be considered are: the human costs of actions in the game; the increase of resources in industry and agriculture over the start of the game; the distribution of resources among agriculture, industry and health education and welfare; and the improvement of health, education and welfare." In the course of play each nation-team determines goals and allocates resources from one sector of the economy to another (from agriculture to industry, for example, or from industry to weapons), negotiates trade and mutual defense agreements with other nations, and may attack other nations. The game ends after five or six fifty-minute sessions. (DCD)

Cost: $30.00

Producer: Simile II, P.O. Box 910, Del Mar, CA 92014

IMPERIALISM

Playing Data
Age Level: grades 5-12
Number of Players: 10-25
Playing Time: 2 hours or more; can be broken into smaller time segments

Description: A crisis between Panama and the United States over control of the Panama Canal serves as a model for the study of imperialism.

The play of the game is the current attempt to negotiate a mutually satisfying treaty concerning the control and operation of the Canal. The participants, in three teams representing the U.S., Panama, and other interested nations, meet in an atmosphere of nationalism, politics, economics, need, cultural differences, and historical mistrust.

The game is designed so the players can acquire an understanding of the frustration of less powerful countries in their dealings with powerful controlling nations, national sovereignty and how it conflicts with international interests, propaganda and the role of the press, and the problems involved in writing the treaty. (Producer)

Cost: $35.00

Producer: Simulation Learning Institute, P.O. Box 240, Roosevelt Island Station, New York, NY 10044

INTER-NATION SIMULATION KIT

Harold Guetzkow, Northwestern University; Cleo H. Cherryholmes, Michigan State University

Playing Data
Copyright: 1966
Age Level: grade 10 (advanced)-graduate school
Prerequisite Skills: reading, college; math, grade 8
Number of Players: 24-50 in 5-6 teams
Playing Time: 1 1/2 hours
Preparation Time: normal class preparation time

Description: "As the confusion and turmoil from the war subsided, the students in Political Science 203, Introduction to International Relations, were still in a state of shock as to how the war came out." These students had just completed playing *Inter-Nation Simulation,* a hypothetical model of the international system. INS, originally designed for research into behavioral patterns among nations, has become in recent years the most popular simulation in the teaching of international relations.

While several versions or adaptions of INS exist, the *Inter-Nation Simulation Kit* (SRA, Inc., 1966), which was developed by Harold Guetzkow and Cleo H. Cherryholmes of Northwestern University, is the most usable for international relations courses including those in high schools. The INS Kit, like its predecessor, involves five to seven nations, a structured decision period, and the utilization of some basic formulas to calculate the impact of certain decisions. Included in the package are enough copies of all the forms to run ten teams for ten decision periods (in most classes, there will be fewer than ten decision periods), but it does not include message forms. The instructor is able to make modifications to the kit including increasing the number of nations, adding guerilla groups, and developing world histories for each of the prototype nations.

One of the major purposes of the *Inter-Nation Simulation* is to give "participants the experience of making decisions in a miniature prototype of the complicated international world." What has been developed is a sophisticated yet manageable model. By participating in this simulation, the student's skills, talents, and understanding of international politics are put to a real test. At the disposal of the model nations are most of the resources possessed by their counterparts in the real world. To implement foreign and domestic policies, these nations are able to declare war, communicate and negotiate with other countries, establish or break diplomatic relations, and create new international organizations. One may also suffer revolutions, experience domestic economic chaos, and many other consequences, both good and bad, of one's decisions.

In this particular simulation, there is not a one-to-one correspondence between real world nations and simulated nations. There is, however, a representation of *types* of nations current in the contemporary world. The nations vary with respect to type of government, size, population, resources, and certain technical achievements. In addition, the time period which is simulated is not a specific historical period. The basic framework of international law is considered roughly to be that of the modern world. This type of generality was purposely developed to give the participant the necessary flexibility so that biases would not spoil the learning experience.

At the beginning of the simulation, participants receive from the simulation director a statistical report that contains key information about the economic status of their state as well as others. This report provides each nation with information regarding its form of government, defense position, population, and basic resources. The participants within each nation use this information to plan and determine their goals and strategies. At the beginning of each subsequent period, nations receive a new statistical report which records the results of the decisions of all actors in the previous period.

In addition to the five to seven nations involved in the INS Kit, there are two other functional bodies. The first is the International Organization, a body composed of representatives from all the participant nations. During each simulated period the International Organization meets and discusses issues which the various diplomats present for solution. The strength of this body and its role is highly determinant upon the diplomats themselves, as well as by the goals and policies of the various model nations.

In addition to the International Organization, some students are involved in a world press. At the beginning of each period, participants receive a copy of the World Newspaper which contains a summary of the important events of the previous period as well as some secret messages that were intercepted. (This feature of the press simulates real world spying and security leaks.) The paper may also contain "plants" from time to time and can be used for propaganda by the various actors. As with the International Organization, the World Press's success depends on the individual students involved in the production as well as the degree to which it is used by the various nations participating.

The collapsing of time is one of the most basic features of the simulation. The period of one hour (or one and one-half hours) is equivalent to a year, so one minute is about the same as a week in the real world. As in the real world, there will not be enough time to do everything one would like and thus it is necessary to plan carefully and work hard to make the best use of limited time. If possible, it is worthwhile to simulate about a decade of time (eight to ten periods) so one has a chance to develop policies over a significant period and experience the rewards or penalties of decision making.

After the decision has been made to use the INS Kit in the classroom, the teacher will divide the class into five to seven nations (each nation should have from four to seven students) with two or three students to assist in the production of the world press. Within each nation there are several key roles that need to be filled:

(1) The Head of State is the chief government official in each nation and concerned with developing both domestic and foreign policies as well as selecting the means to implement various policies and programs.
(2) The foreign policy advisor assists the Head of State in the overall planning of foreign policy.
(3) The official domestic advisor is responsible for internal affairs and plays a key role in the economic development of the nation.
(4) The foreign affairs diplomat serves as ambassador to either a nation or to the international organization. Within any given nation, there may be two or three individuals that act as diplomats involved in various types of negotiation.

(5) Within some nations there will be a domestic opposition leade: who opposes the way the current government conducts foreigr and domestic affairs.

The role that each participant plays varies with the size, type ol government, and domestic situation of each particular model nation.

Comment: Although the *Inter-Nation Simulation* has been used most often in the postsecondary educational experience, its use is becoming fairly widespread within high schools. While there are certain shortcomings and prerequisites for using this particular simulation (of which more will be said later), those of us who continue to use it have found it rewarding.

Students normally report feeling the tensions and strains that world leaders feel; they report having a better understanding of the complex interactions within the international system, as well as having a lot of fun attempting to implement strategies and accomplish goals that were stated at the onset of the exercise. Instructors desiring the detailed presentation of the simulations operations, forms, and resources should obtain a copy of the participant's manual for the *Inter-Nation Simulation Kit.* (From the Robert A. Taft Institute of Government Study on Games and Simulations in Government, Politics, and Economics by Earl L. Backman, Univ. of North Carolina at Charlotte)

Note: For more on this simulation, see the essay on international relations simulations by Leonard Suransky. To learn about *INS2,* the computer version in BASIC, see the listing in the computer section.

Cost: $82.00

Producer: Science Research Associates, Inc., 155 North Wacker Drive, Chicago, IL 60606

INTERNATIONAL CONFLICT

Anne Thompson Feraru, California State University, Fullerton

Playing Data
Copyright: 1974
Age Level: college, graduate school, adult
Number of Players: 5 or more in teams of 5, for the role-playing activity
Playing Time: 1 hour for role playing, 2-6 weeks for the learning package
Packaging: 70-page paperback

Description: *International Conflict* introduces students to problems of individual, societal, and intersocietal conflict and various means of conflict resolution. This learning package contains eleven activities involving the study and evaluation of conflict theories and situations, and one role-playing activity. The latter requires players to divide into groups of five that represent advisors to the President on issues of foreign policy and national security. Groups have approximately 40 minutes to discuss a situation involving Iran's termination of a security agreement with the U.S., and to determine three constructive courses of action the U.S. could take, as well as the most destructive course of action. Other activities in the learning package include recording and evaluating one's responses to conflict situations, studying Anatol Rappaport's theories, and responding to quotations of political leaders which express their views on national conflict. (TM)

Cost: $2.00

Producer: Learning Resources in International Studies, 60 East 42nd Street, New York, NY 10017

THE MIDDLE EAST CONFLICT SIMULATION GAME

Leonard Suransky and Edgar C. Taylor, Jr.

Playing Data
Copyright: 1979
Age Level: grade 10-college, continuing education
Number of Players: 50-60 in 7-10 teams
Playing Time: 12-20 hours over a weekend
Preparation Time: in context of 6-12 week college course
Special Equipment: video equipment, computer for communications (Both optional)
Packaging: simulation game Manual

Description: *The Middle East Conflict Simulation Game* (MESG) was designed to fit an existing course on the Arab-Israeli Conflict at the University of Michigan. As a "game-within-a-course" it cannot be easily used as a prepackaged game without considerable preparatory work by the facilitators. Rather, the essence of this simulation game is that it integrates with, and becomes the center of, a course.

Students are thrust into the play of the game from day one of the twelve week course, although outwardly the impression is of a conventional lecture/discussion group procedure. In the twenty-member discussion groups students choose their teams (Israel, Egypt, Palestine, Syria, USA, USSR). Then the three to eight team "cabinet" delegates its decision-making portfolios as they see fit (Head of State, Foreign Minister, Defense Minister, Opposition, and so forth).

Players now begin a gradual indentification with the role they are to play, and in turn become identified in the class with their country-team and role. This is aided by the systematic preparation by each player of (a) a character/personality sketch; (b) a portfolio position paper (e.g., the defense history and capacity of, say, Jordan, or the energy dilemma of the USA); (c) a personal, and (d) a team strategy outline.

In this sense it may be said that the game runs for an entire term, with a preparatory research and data collecting period which includes "fore-play" of a lobbying, posturing, and scheming nature. This reaches an experiential climax on the actual week-end of the game play, and culminates in the intensive post-game two weeks of debriefing and reflection on the experience.

Theatricality and the inherent drama of the conflict are emphasized. Students who feel comfortable dressing up and impersonating accents and personality traits are encouraged to do so. In some cases we have auditioned major role players with key positions in the actual conflict, to assure compatibility with the adopted personalities.

The country-teams are spread along a university/school corridor, often decorated with flags, maps, posters, and other national or cultural symbols. Couriers carry messages through Control on triplicated forms. Face-to-face meetings or conferences can be arranged. All messages or moves pass through Control, where their plausibility is validated. An (optional) WGOD-TV news broadcast, produced by Control, is shown after most game recesses, three times a day.

Communication can also be handled through a CONFER, computer conferencing system, designed by Robert Parnes of the University of Michigan especially for the MESG. The advantage is an easily retrievable and indexed computer printout of all the written game activities for debriefing or research purposes. The major disadvantage is the magnetic focus of the computer terminal, which intrudes upon the cabinet atmosphere of the room, and interrupts the flow of debate and decisionmaking.

Debriefing is stressed as a period for learning about the events of the game, and their relation to the real-world conflict, and to theories and practices of international politics and diplomacy. (LS)

Comment: The unusual aspect of the MESG is its emphasis on the theatrical and dramatic aspects of the game. This approach to role playing in simulation games is born of an assumption that the "simulated-yet-real" experience of the actor/players is an alternative route to understanding and appreciating the complex ramifications of the international politics of the Middle East, or other conflicts. The realism of the impersonations heighten rather than distort the insights gleaned. However, the role of Control to maintain the plausibility of the game is greater (a) in its validating role; and (b) in being sensitive to the rare instances in which a student identifies so intensely with the role impersonation, that interpersonal process skills are needed to help him or her come down from an emotional entanglement. (A *Southern*

Africa Conflict Simulation Game (SASG) has been similarly designed and pilot tested.) (LS)

Note: For more on this simulation, see the essay on international relations simulations by Leonard Suransky.

Cost: (1) manual $5.00; (2) videotape, "The Impact of a Simulation Game on a College Course" (Sigman, Taylor, Suransky) $75.00; (3) CONFER manual $2.00

Producer: (1) Simulectical Simprovisations, 1225 Prospect, Ann Arbor, Mich. 48104, (313)769-2063; (2) The University of Michigan TV Center, Univ. of Michigan, Ann Arbor, Mich. 48109; (3) Bob Parnes, Center for Research into Learning and Teaching, Univ. of Michigan, Ann Arbor, MI 48109

MIDDLE EAST CRISIS

David J. Boin and Robert Sillman

Playing Data
Copyright: 1972
Age Level: grades 7-12
Playing Time: 3 class periods

Description: Teams representing Egypt, Israel, the Soviet Union, and the United States work out a series of proposals to settle the Middle East conflict. The scoring system requires that at least three of the four countries agree to a proposal. Students are provided with demands of the countries and must do most of the research on the simulation themselves. (REH)

Comment: This is in the standard Edu-Game format with a few pages of teacher's manual and locally reproducible student material. (REH)

Cost: $3.00

Producer: Edu-Game, P.O. Box 1144, Sun Valley, CA 91352

NATIONALISM: WAR OR PEACE

David J. Boin and Robert Sillman

Playing Data
Copyright: 1973
Age Level: grades 7-12
Number of Players: 10 to 40
Playing Time: 5 class periods

Description: Students assume roles as President, Secretary of State, Treasurer, or Negotiators for seven fictitious countries. In successive trading, war, and peace rounds, they enter into trade agreements, defensive alliances, and offensive alliances, always faced with the necessity of accumulating wealth through trading and protecting themselves in peace time from the threat of war. Special formulas for making trade agreements are provided as well as simulated military strength of different countries. The purpose of the simulation is to help students understand the relation of economics to war, the meaning of "entangling alliances," and why nations enter into different types of alliances. (REH)

Comment: This appears to be a simple and very easily understood international relations game. It follows the standard Edu-Game format with a few pages of teacher's manual and locally reproducible student material. (REH)

Cost: $3.00

Producer: Edu-Game, P.O. Box 1144, Sun Valley, CA 91352

PHANTOM SUBMARINE

Richard A. Schusler and Richard B. Cohen

Playing Data
Copyright: 1974
Age Level: junior high
Number of Players: groups of 5
Packaging: 19 mimeographed pages

Description: The discovery of a "phantom" German submarine floating in the Gulf Stream under the surface creates a problem. Its ballast is elementary mercury, worth a fortune to a salvager, but salvage is risky because any accidental spill would be an ecological disaster. The submarine's continued drift presents the same danger, however, since it might break up on submerged mountains.

Students are formed into groups of five, representing a member of the U.N. Committee on World Pollution Control, a treasure-hunter, a U.S. politician, an ocean ecologist, and an admiral. Each has a role card explaining areas of interest and expertise and a data card with some physical characteristic of the floor of the Atlantic.

Each group begins by compiling information about the floor of the Atlantic. Students share the information on their data cards (ocean depth, route of Gulf Stream mid-Atlantic ridge, speed of Gulf Stream, bordering countries). On the basis of the information they all share, they discuss what is best to do about the submarine.

Then, players assume their roles and participate in a meeting called by the U.N. Committee on World Pollution Control in response to the crisis posed by the intent of a Mr. Greedy to begin salvage operations. Discussion continues until a group decision is reached. After the discussion, groups report to the whole class on what and why they decided. (AC)

Cost: $5.00

Producer: Dr. Richard B. Cohen, Department of Elementary and Early Childhood Education, University of Arkansas at Little Rock, 33rd & University, Little Rock, AR 72204

POLITICAL EXERCISE

Lincoln P. Bloomfield, Massachusetts Institute of Technology

Playing Data
Copyright: 1958
Age Level: college, professional
Number of Players: 8-50 in 1-5 teams
Playing Time: 16 hours
Preparation Time: 40 hours

Description: The *Political Exercise* is a general term used to describe successive generations of foreign policy games, including several versions of the games POLEX and CONEX. In each, a particular foreign policy crisis is stated as a "Scenario-Problem." This, as the author describes it, "is a document, up to a dozen pages in length, in which a hypothetical but plausible series of events—say, a revolution in a Central American country (which was the topic of CONEX I)—is depicted to the game players in some detail. This starts the game, and it remains the basis for the interaction between the teams and between the teams and the Control Group (and its subteams) for however long the game runs." The teams simulate making American foreign policy. (DCD)

Comment: Both because of its publication date and its unimpeachably establishment producers, PE must be ranked as the Big Daddy of International Relations Simulations. (DZ)

Cost: nominal cost of publication

Producer: MIT Center for International Studies, 30 Wadsworth Street, Cambridge, MA 02139

SALT III

Frank Hawke and Dan Caldwell

Playing Data
Copyright: 1976
Age Level: college, adult
Number of Players: 24 in 2 teams
Playing Time: 16-20 hours
Preparation Time: 40 hours (est.)
Packaging: 68-page photocopied booklet

Description: This simulation of the Strategic Arms Limitations Treaty negotiations between the United States and the Soviet Union has been offered as a five-credit, two-month course at Stanford University. At the start of play the participants are divided into two groups representing the parties to the negotiations, and are given specific information on the limitations on weapons and delivery systems that have already been agreed upon at the SALT talks. The sides prepare opening statements and negotiating positions which may reflect either what the two nations would, or should, propose. In previous plays of the game, negotiating sessions have been held twice a week for eight weeks and have culminated in drafting an arms limitations treaty. (DCD)

Note: For more on this simulation, see the essay on international relations simulations by Leonard Suransky.

Cost: $3.00

Producer: Prof. Dan Caldwell, Social Science Division, Pepperdyne University, Malibu, CA 90265

SECURITY

Charles E. Osgood, University of Illinois

Playing Data
Copyright: 1964
Age Level: grade 11-college
Number of Players: 2 or more
Playing Time: 1 to several hours
Preparation Time: 2 1/2 hours

Description: *Security* is a game in which all of the complexities of strategy in international relations can be simulated–with nothing more than a bridge table and an ordinary deck of cards. Each player represents the chief decision-maker of his own nation. His goal is to end each game with as much human welfare for his own nation as possible. Some leaders will prove to be suicidal, of course, and many worlds will end in shambles. But this game offers chance to learn from past mistakes, to learn how to create a better world, and to learn how to win real security. The end of each game (each world) is called Doomsday, and it is unpredictable–you never know exactly when it's going to come. The winner is that decision-maker who has built up the most human welfare for his people.

Welfare cards are heart and diamond head cards; warfare cards are club and spade head cards. Jacks represent manpower, queens represent manpower plus resources, kings represent manpower plus resources plus industry, and aces represent manpower plus resources plus industry plus science. The joker is the Doomsday Card. (Author)

Comment: *Security* has many of the complexities of strategy in international relations. Foreign aid, military aid, intelligence, secrecy, disarmament, arms control, deterrence, and credibility, all come into play in this simple yet very fascinating game.

We recommend this way of learning through everyday elements used in an outstandingly creative way. (REH)

Cost: none

Producer: Not published, but rules can be obtained from Charles E. Osgood, Institute of Communications Research, University of Illinois, Urbana, IL 61801

SIMULEX IV: INTER-NATION SIMULATION

David L. Larson

Playing Data
Copyright: 1968, 1970, 1975, 1978
Age Level: secondary, undergraduate, graduate
Number of Players: approximately 25
Playing Time: flexible, usually 3-6 hours to play, but recommended student participation in preparation (research and design) is extensive
Preparation Time: extensive
Packaging: 131-page manual

Description: *Simulex IV* consists of over 100 pages of material on international relations, a nine-page procedure guide for making an internation simulation, a list of resources, and a selected bibliography.

The chapters preceding the simulation discuss a definition and history of international relations; national policy and power; military power and policy (new chapter with this edition); international politics and economics (new); and international law, organization, and disarmament (new).

The simulation procedures discuss–in practical terms–the crucial role of preparation; choosing and establishing a crisis/referent system; assigning teams and roles; conducting research (writing position papers); setting up and running the simulation, including possible types of messages/moves; dealing with multination conferences, military action, national elections, disputed moves, the press, and central control; and, finally, how to organize debriefing and what points to cover. (AC)

Comment: This manual is an excellent introduction to international relations and to a simulation. (LS)

Note: For more on this simulation, see the essay on international relations simulations by Leonard Suransky.

Cost: $5.00

Producer: The New Hampshire Council on World Affairs, The University of New Hampshire, 11 Rosemary Lane, Durham, NH 03824

THE STATE SYSTEM EXERCISE

William D. Coplin, Syracuse University

Playing Data
Copyright: 1971
Age Level: college
Number of Players: 24 in 8 teams
Playing Time: 3-4 hours in periods of 50-75 minutes

Description: The purpose of *The State System Exercise* is to illustrate the dynamics of those factors affecting the stability of the international system. The simulation focuses on the accumulation of power and the maintenance of security in the international system. *The State System Exercise* has three cycles, each representing a historical stage in the development of the state system. The first cycle represents the classical state system of the eighteenth century. The second represents the modern state system of the late nineteenth and early twentieth centuries. The last cycle represents the contemporary system where nuclear weapons have changed many of the basic relations between states during the classical and modern periods.

Through the assumption of decision-making roles within each of these three cycles, the student should develop an awareness of some of the more important pressures originating in the international system that affect the behavior of nations. As they play each of the three cycles, the twofold desire to survive and to increase power will interact as participants attempt to guide their nations.

The author points out that it is not the purpose of *The State System Exercise* to simulate the international system as it has existed at any given point in time. Students' behavior need not be constrained by what they think is "realistic" in terms of the real world, but rather by what they think is reasonable given the rules of the game and their assumptions about the behavior of other players. The de-briefing of the exercise should allow the students to compare the behavior which occurred in the exercise to the behavior of real leaders and real states.

The State System Exercise takes from four to ten hours of classroom time, does not require extensive or elaborate materials, and involves a comparatively small amount of extra planning effort from the instructor. The instructor may, of course, supplement the exercise as desired, or break it into three sections corresponding to the three cycles and use each cycle as a separate exercise when chronologically appropriate to, for instance, a course in world history.

The simulation can be played in an average-sized classroom with students arranged in groups of eight. Each group represents a separate nation. Although not necessary, it is helpful to reproduce the "World Map" as a transparency and project it on the wall during game play.

Very little advanced preparation or reading is necessary and the rules are relatively uncomplicated. It is suggested, however, that the teacher assign readings or perhaps even use media to give students some background into each of the three international systems being studied. The degree to which *The State System Exercise* is successful depends upon the skill of the teacher in providing background material to students and conducting a meaningful de-briefing session. Although The State System Exercise is easy to play, the more able students in the class will get more out of it, simply because they will be more readily able to relate past learning and on-going assignments to the game experiences. Consequently, the teacher can help maximize learning by average and below-average students through carefully thought-out grouping practices. Spread your more able students out evenly among the eight groups. They will help the others and make the exercise more meaningful for all.

The general purpose of *The State System Exercise* is to stimulate and focus discussion on various aspects of international politics. More specifically, the author's objectives include:

(1) an awareness of the importance of information in defining situations and of time constraints in searching for alternatives in international affairs
(2) an awareness of the difficulties faced by "real world" foreign policy decision-makers in the pursuit of security maintenance in an unstable world
(3) a knowledge of the types of strategies followed by different states
(4) the ability to draw historical parallels between the strategies followed in the game and the strategies followed in the different historical periods
(5) an understanding of the terms "balance of Power," "Collective Security," and "Balance of Terror"
(6) an understanding of the historical development of the modern state system and shifting environmental constraints on decision makers operating within different historical time periods

It is completely up to the individual instructor to provide methods of reaching these objectives, as *The State System Exercise* does not include suggested de-briefing questions or assignments.

The State System Exercise actually contains three simulations, which may be clustered together for purposes of comparison and contrast or used in a mutually exclusive fashion. As mentioned earlier, there are three cycles, or separate simulations, contained in the exercise. Cycle I: "The Classical State System," has the following five characteristics:

(1) relatively even distribution of power among five of its members and three smaller states
(2) high degree of flexibility in the alliance structure
(3) prospect of high gain from victorious wars
(4) a large role played by geographic factors
(5) autocratic decision-making within states

These characteristics are evident in the design structure and processes of the simulation model.

Cycle II: "The Modern State System" has the following characteristics:

(1) Relatively even distribution of power among four of its members and four smaller states
(2) Less flexibility in alliance structure
(3) Prospect of only moderate gain from victorious wars
(4) Reduction of the effect of geography in international interaction
(5) Majority-rule decision-making within some of the nations

As with the "Classical State System," these characteristics are built into the simulation model.

Finally, Cycle III: "The Contemporary International System," has been designed to manifest the following characteristics:

(1) Highly uneven distribution of power
(2) Less flexibility in alliance structure
(3) Prospect of only small gain from war
(4) Elimination of the effects of geography on international interaction
(5) Majority rule decision-making in all states

The procedures for playing each of the three cycles are roughly the same. Three students are assigned to each nation represented in the simulation and decision-making roles distributed. Only the foreign minister from each nation may visit other nations. Students are provided with a map depicting the nations in the system. The spatial relations of the map influence the rules of warfare and alliance. Each nation is also assigned power units.

The author characterizes game procedure as involving five processes, aside from the decision-making processes, which are present within each nation. These five processes are negotiation, alliances, war, bargaining for peace and division of the spoils. The purpose of each nation is to survive and one survives by not being on the losing side in a war. Loss in a war means the elimination of all but one of the country's total power units. Victory results in the acquisition of the total power units, less one, of the losing nation. Given these basic rules, one can choose from at least three basic strategies—a defensive posture by joining an alliance which will protect your nation from any set of enemies, a neutral policy, or an offensive posture in which, by seeking to beat all other states, you build up the preponderant amount of power. Each strategy has its own risks.

The rules for conducting the five processes mentioned above are simple and easy to administer. Alliances and wars are undertaken by the filing of either Alliance or War forms with "Control" (the teacher). Control also makes any changes in power units at the end of each round of game play. At the end of the simulation, an extensive de-briefing session is held.

The State System Exercise can be a powerfully effective device for teaching basic concepts relating to the international political system. It is compact, easy for students to play, highly motivating, and simple to administer.

The fact, however, that it is so compact and streamlined (no elaborate forms, etc.) leads to some words of caution. It is important that the teacher have some solid background in the subject matter being dealt with in the simulation if one hopes to maximize its potential. There is little or no help provided to the teacher in the form of suggested supplementary activities and de-briefing questions. In this area the teacher is on his or her own and must have some clear ideas about what he or she wants students to get out of the experience and how to get it. The more subject matter background the teacher has the easier the task is. A few background reading suggestions are included in the participant's manual. I have found them too difficult for most high school students, but they are well chosen and it is strongly recommended that teachers read as many as possible before trying this simulation, unless they are already well versed in the history and theory of the international systems under consideration.

The author also emphasizes "Realpolitik" too much at times for this reviewer's personal taste. This lack of attention to ethical problems related to international politics is easily handled, however, if one is aware of the game's structural bias toward Machiavellianism built into *The State System Exercise*. A few well-thought-out questions directed toward the ethics of national behavior will add a new and interesting dimension to the de-briefing session. This reviewer has found that asking students to identify what the assumptions of the author of the simulation seem to be and then inquiring of them whether they agree with him will almost inevitably lead to the most lively and productive give-and-take discussion of the de-briefing period, as some students support the design assumptions and others take issue with these assumptions. (From the Robert A. Taft Institute of Government Study on Games and Simulations in Government, Politics, and Economics by Gerald Thorpe, Indiana Univ. of Pennsylvania)

Cost: $1.50

Producer: Learning Resources in International Studies, 60 East 42nd St., New York, NY 10017

SUMMIT

Playing Data
Copyright: 1976
Age Level: grades 9-12
Number of Players: 28-49 in 7 teams
Playing Time: 4 1-hour periods
Special Equipment: mimeograph machine
Packaging: looseleaf manual

Description: The purpose of *Summit: A Game of International Compromise* is to help students gain an understanding of contemporary international problems, of the meaning of "national interest," of the use of diplomacy and compromise, and of the difficulties involved in forming international agreements. Each of the seven groups representing a fictitious nation studies statistics about their country, and selects a president, two advisers, and one or more negotiators. During each of ten fifteen-minute rounds, groups determine their position on one of ten issues and send their negotiators to a summit meeting to try to influence other negotiators. Each round ends with a voting period followed by scoring. When a treaty is passed, the groups score according to benefits to their nation; if no treaty is passed, all groups lose a point. The ten issues discussed are national waters, size of army, price of lumber, access to ocean, size of navy, price of oil, air space, price of wheat, size of air force, and nuclear bombs. (TM)

Cost: $3.00

Producer: Edu-Game, P.O. Box 1144, Sun Valley, CA 91352

THE TERRITORIAL SEA

Playing Data
Age Level: grade 11-adult
Number of Players: 10-30 or more

Description: As nations feel the impact of food, mineral, oil, and gas shortages and the effects on their economies, the international arena springs to life over the abundance of raw materials and nutrition in the world's oceans.

The Territorial Sea is a model which brings together all the issues and the interested parties—the international community subdivided into six power blocks and the world business interests vying for the riches of the Territorial Sea.

The game is an attempt to introduce this conflict to the players as they assume the roles of the protagonists. The International Community, divided by national interests, attempts to write a treaty regulating the uses, distribution, control, finances, and environmental impact of the resources in the seas. Coincidentally, the World Business Interests are organizing individually, as national consortiums and cartels to reap the riches of the deep. Investments are arranged, research and development started, explorations undertaken, lobbying initiated and the race for control of the Territorial Sea begins.

A round of play consists of meetings, negotiations, strategy, decisions, and outcome, and represents one year of real world time. The model can be expanded by adding supporting teams. (Producer)

Cost: $75.00

Producer: Simulation Learning Institute, P.O. Box 240, Roosevelt Island Station, New York, NY 10044

THE USES OF THE SEA

John Gamble

Playing Data
Copyright: 1974
Age Level: college, graduate school
Number of Players: 16-32 in 8 teams
Preparation Time: 1 hour

Description: This game is intended to give participants a general overview of the economic and political factors shaping the attempts of nations to regulate the uses of the sea. Players assume the roles of policy makers from seven nations and an international organization. During each of the three one-hour cycles they work toward a more favorable defination of claims and uses of maritime territory. Their actions are chosen from a list of twenty possible actions on an agenda. Players must decide whether to support, oppose, or remain neutral on each of these actions. They receive feedback on the consequences of their actions and the actions of the world.

After playing the game students will be able to define territorial sea, continental shelf, exclusive fishery zone, and inclusive and exclusive claims; describe the role of domestic and transnational factors in the formation of a state's marine policy; state the types of marine and nonmarine characteristics used by the states in shaping the transnational rules governing the uses of the sea; and apply the concepts and relationships expressed in the simulation exercise to dynamics of the uses of the sea in the contemporary world. (AC)

Comment: This simulation, for the sake of simplicity, limits the types of claims under consideration to three: claims to territorial sea (fish in the water, ocean floor, and navigation), claims to the continental shelf (insofar as it extends beyond territorial sea limits), and claims to exclusive fisheries zones (when they extend beyond territorial sea claims).

Data for each nation involved is clearly and extensively provided in tabular form, with a brief paragraph setting the sociopolitical context. The hard facts participants must work with look impressive. Nonmarine characteristics given for each country are area, population, GNP, the ratio of GNP to population, amount of political competition, percentage of food produced domestically, and industrialization. Marine characteristics are coastal and distant water fish catches, the ratio of total catch to the GNP, total seaborne trade and its ratio to the GNP, expenditures in oceanographic research, research vessels working in foreign waters, the value of offshore oil and its ratio to the GNP, and the amount of coastal water pollution. There are tables showing the relation of activities between nations, such as how much fishing one country is doing near the coast of another country. For each country there are graphs of trends and projections for five kinds of marine activities.

Thus, while the claims to be considered are limited for the sake of simplicity, the material participants have to work with is not simplistic. The package includes four optional follow-up exercises, each of which represents one to two weeks of normal course work. (AC)

Note: For more on this simulation, see the essay on international relations simulations by Leonard Suransky.

Cost: $2.00

Producer: Learning Resources In International Studies, 60 East 42nd Street, New York, NY 10017

WAR TIME

David J. Boin and Robert Sillman

Playing Data
Copyright: 1973
Age Level: grade 7-12
Number of Players: 10-40
Playing Time: 5 class periods

Description: Students are divided into roles of government leaders, generals, industrialists, or workers of four fictitious countries. In short periods, the industrialists engage with each other in trading certain war products from the military-industrial complex. Then the generals seek to get enlistments from the workers to increase military strength until they can declare war on some other country. The purpose of this simulation is to help the students understand the relation of economics to war, the motivation of the military-industrial complex, the meaning of "war-time economy," and why many citizens support a war-time economy. (REH)

Comment: This is a relatively simple simulation, which may, in fact, represent some aspects of reality. It is only the workers who have to enlist and die in wars. It does not teach much about the intricacies of international relations, but it does provide an introductory opportunity to explore the relations between war and economy. It is in the standard Edu-Game format with a few pages of teacher's manual and locally reproducible student material. (REH)

Cost: $3.00

Producer: Edu-Game, P.O. Box 1144, Sun Valley, CA 91352

WORLD

Steve Denny

Playing Data
Copyright: 1975
Age Level: intermediate
Number of Players: 16-36
Playing Time: 25 hours for full simulation
Preparation Time: approximately 2-3 hours each for three phases

Description: "In *World,* students make decisions about a world they build. The simulation gives students problem-solving alternatives and invites interaction and creativity." The class is divided into four, five, or six groups which each form a nation. Each nation makes its own rules, establishes its own government. Nations appoint or elect their own spies, diplomats, leaders, and citizens who function as such within their own countries until "ultimately one country controls the world." Throughout the simulation, "alliances are broken, balances of power are disturbed and students gain insight into the organization and workings of world politics."

The simulation is divided in three phases as students develop their world. Students will have much opportunity in these phases to develop or reinforce skills in map and content areas to adequately perform projects. In Phase I (Power Point Buildup), the groups each develop their nation choosing their form of government and selecting leaders, spies, and diplomats. Each nation earns points (there are twenty-five ways to do so listed for each group) "by completing and having the teacher approve assigned projects."

In Phase II (War) "interest reaches a high point with a war simulation based on the number of points earned by each country during Phase I." By following a defined set of rules, one country is eventually victorious and becomes the world power. In Phase III (Peace Conference), "the victorious nation leads the peace conference as all nations struggle with this problem: What should be done with a world after national conflicts have resulted in war? The victorious nation decides what to do with the world." At the conclusion of Phase III, "Students fill out an evaluation form that helps them reflect upon how the development of their world strongly parallels the development of the world in which they live." (TM)

Cost: $22.00

Producer: Interact, Box 262, Lakeside, CA 92040

WORLD POLITICAL ORGANIZATION

Bruce Downing, James Green, and Bruce Riddle

Playing Data
Copyright: 1973
Age Level: college
Number of Players: 9-25
Playing Time: 55 minutes
Preparation Time: 1 hour (est.)
Packaging: 31-page staple-bound photocopied booklet

Description: This game is designed "to introduce students to 'The International Politics' framework and 'The World Policy' framework, two alternative models for understanding the structure of global policy-making." Each player is assigned to a lettered nation and to a specific group within that nation. The result of this assignation is that each participant has *two* loyalties, to Nation A and organized labor, for example, or to Nation C and the business community. During the eleven five-minute rounds the players must try to acquire as many satisfaction points as possible. In alternative rounds these points are distributed among the nations and then among the interest groups within each nation. Players' final scores are determined by their nation's total points, their group's total points, and their own total points. The player with the most points wins. (DCD)

Cost: $1.50

Producer: Learning Resources In International Studies, Suite 123, 60 East 42nd Street, New York, NY 10017

See also
The Alpha Crisis Game HISTORY
Brinkmanship HISTORY
Congress of Vienna HISTORY
Czech-Mate HISTORY
Diplomacy MILITARY HISTORY
The Road Game COMMUNICATION
Whales ECOLOGY

LANGUAGE SKILLS AND ARTS

CLASSROOM GAMES IN FRENCH

Maurie N. Taylor

Playing Data
Copyright: 1974
Age Level: high school
Number of Players: 10-40
Playing Time: 10-60 minutes
Packaging: 32-page, soft bound booklet

Description: This is a collection of twenty-five classroom games that have been designed, according to the producer, to be "amusing and competitive enough to appeal to students," and that feature "language learning skills in oral-aural practice." All have been classroom tested with students at an intermediate level in the study of the French language.

"Je pense a un mot," "Me Voila," and "Cherchez le mot" are representative of the games included in this collection. In "Je pense a un mot" one student thinks of a word and then chalks on the blackboard the correct number of dashes (as in a crostic) to indicate the number of letters in that word. The class guesses letters, and when they correctly guess one the player at the blackboard marks it over the appropriate dash. The class must deduce the word in ten guesses or less. "Me Voila" is mental hide-and-seek in French, in which one student picks a place in the classroom and the other students guess what that place is. "Cherchez le mot" is a conversational game; a student introduces a certain word or phrase into random conversation in such a way that it cannot be spotted by his opponent. Each student thinks of a "secret word" to whisper to the player on the right and then tries to guess the secret word given to the player on the left. (DCD)

Cost: $2.67 list, $2.00 school

Producer: National Textbook Company, 8259 Niles Center Road, Skokie, IL 60076

CODE

Catherine D. Johnson

Playing Data
Copyright: 1976
Age Level: grades 7-10
Number of Players: any number
Playing Time: 7-9 1-hour sessions
Packaging: 11-page teacher's guide, 2-page student guide

Description: In this language arts simulation game, students learn about the nature of language and design and decode codes. In groups of four (consisting of a leader, recorder, emissary, and chief recorder) they write limericks and translate the best in each group into a code they have designed. The next four days are spent decoding. Emissaries procure codes from other teams, and groups work on each code for as long as they choose, returning them to the teacher for scoring before getting another code. The final day, winners are announced, limericks shared, and the simulation evaluated. (TM)

Cost: $10.00

Producer: Interact, Box 262, Lakeside, CA 92040

INVENTING AND PLAYING GAMES IN THE ENGLISH CLASSROOM

Kenneth Davis and John Hollowell

Playing Data
Copyright: 1977
Age Level: high school
Number of Players: 3-30
Playing Time: 15 minutes to 12 hours
Preparation Time: 3 hours (est.)
Packaging: 160-page paperback book

Description: This collection is designed to serve as a handbook for teachers who have a minimum of previous experience playing, administering, designing, and evaluating games and classroom simulations. The text is divided into three sections. The first contains six essays that address the subject of games in general terms (titles include "Why Play Games," "How to Run the Game," "How to Design the Game," "The Classrom as Game," "Why *Not* to Play Games," and "Tomorrow's Games"). The last contains three appendices (an "Annotated Bibliography of Publications on Games and Simulations," a list of "Recommended Games for the English Classroom," and a "List of Academic Games Development Centers"). Most of the text, however, is devoted to eight "Examples of Teacher-Made Games." These games are all remarkably engaging and applicable to most high school English classrooms.

"Madison Avenue" represents the composition games in this collection. The players form teams and collaborate in writing a 200-word television commercial according to special instructions. Following the best traditions of American consumption, each team must imagine a market of people with money to spend, that has a specific "hang-up" (author's term), and that habitually watches some particular television show. When each team has its market in mind they brainstorm a product to sell and write the copy for a commercial. Once the ads are drafted, teams exchange their copy and each team judges the success of another team's ad. For an ad to be considered successful, the players who did not write it should be able to tell what is being sold to whom, what hang-up the budding copywriters are trying to exploit, and what television program is being used to gain access to this market.

Another game, titled "H. Z. Zilch" (and reminiscent of the fiction of Borges and Cortazar), illustrates the general quality and inventiveness that is to be found in this collection. For the most part these games are not fancy ways to teach specific skills or painless ways to swallow facts. They are, rather, ticklers of the fancy. Zilch, for example. "No one knows much about H. Z. Zilch" the author of the game tells us. "Some feel he was some kind of real loser. Some feel he was a person with extraordinary insight into human nature, the world, and whatever else philosophers tend to spend their time insighting about. All we have to go on are a few scraps of poetry which H. Z. wrote in 1915 and 1916, parts of a book H. Z. wrote some time before 1932, the diary of H. Z.'s next door neighbor Willa Sue Swink, and a collection of odd bits and pieces of information gathered by an unknown graduate student at the University of Michigan. Unfortunately we do not know anything else about this graduate student beyond the fact that 'someone' was reported several times 'lurking' in the third floor stacks of the Graduate Library and was known to have used a vacant study carrel on that floor. It was in this study carrel #321 that on January 14, 1973, then librarian Charlotte Pederson discovered the research and threw it out. What we have are the remnants of the research picked out of a trash compactor at the city dump of Ann Arbor by Charles Schwartz, a garbage collector with the Right Way Rubbish Service." All of the basic research sources are included with the game. Other game materials include newspaper clippings about Zilch and his alleged associates. His poems are written in two different styles and it is not known if all are authentic Zilch. (This is partially due to an event that took place in July, 1916, when Zilch put a large collection of poems into a large brown envelope, put the envelope over his heart, put the barrel of a gun to the envelope, and so forth. The upshot was that Zilch survived this suicide attempt in much better shape than did the thick sheaf of poetry. While he was recovering in the hospital, Zilch was questioned by a reporter from the Detroit Times, who later quoted Zilch as saying, of the poems that saved his life, "They're all mine . . . more or less.") The author suggests a number of uses for this whimsical apocrophia. All of these uses are research and creative writing projects that serve to introduce students to sophisticated concepts like internal criticism, and challenge them to consider of what stuff truth and poetry are made. (DCD)

Comment: Re:joyce reader! Smart, clever, funny, well-informed, down to earth folks have put together a book that can help you to become a better teacher. (DCD)

Cost: $6.50 nonmember, $5.00 member

Producer: National Council of Teachers of English, 1111 Kenyon Road, Urbana, IL 61801

LE PAYS DE FRANCE

Playing Data
Copyright: 1973
Age Level: high school, college
Number of Players: 5-30
Playing Time: 70-80 minutes
Packaging: professionally produced and packaged game includes instructions, film strip, wall map, and language cards

Description: This package of instructional materials for intermediate French students includes the *French Holiday Travel Game,* which simulates a seven to fourteen day automobile vacation in France. Groups of five or six students plan their vacation, name something they would appreciate about visiting each place they stop, and try to memorize something about the places they visit. A film strip examines the agriculture, industry, popular resorts, and the countryside of France. (DCD)

Cost: £ 9.00

Producer: SGS Associates (Education) Ltd., 8 New Row, London WC2N 4LH England

LET'S PLAY GAMES IN FRENCH

Bernard Crawshaw

Playing Data
Copyright: 1977
Age Level: age 8-adult
Packaging: 129-page paperbound book

Description: This is a collection of 146 games, skits, and teacher's aids, to assist in teaching French to beginning, intermediate, and advanced students. The authors suggestions include giving your students French names, conducting spelling bees in French, playing treasure hunt in French, singing songs in French, debating in French, making up limericks and puns in French, and so on. (DCD)

Cost: $7.24 list, $5.43 school

Producer: National Textbook Company, 8259 Niles Center Road, Skokie, IL 60076

LET'S PLAY GAMES IN GERMAN

Elisabeth Schmidt

Playing Data
Copyright: 1977
Age Level: age 8-adult
Packaging: 81-page paperbound book

Description: This is a collection of games and skits to aid teachers of German. Categories include spelling, article, verb, number, object identification, map, and story telling games. Some, such as the counting games, are fairly simple. For instance, the students may be asked to draw a tree with a number at the end of each branch, and then be asked to count all the numbers on the tree. Some of the story-telling games, on the other hand, require fluency. Students may be asked to build a story around specific objects, or tell a story in round robin. All of the exercises are arranged in order of difficulty. (DCD)

Cost: $7.24 list, $5.43 school

Producer: National Textbook Company, 8259 Niles Center Road, Skokie, IL 60076

NEWSCAST

Mike Beary, Gary Salvner, Bob Wesolowski

Playing Data
Copyright: 1977
Age Level: high school
Number of Players: 7-14 in teams of 7-14
Playing Time: 7 or more 1-hour sessions
Special Equipment: optional videotape machine, large room
Packaging: 80-page guide

Description: The teacher's guide describes the simulation as follows: "Students are organized into network news teams of 7-14 players. Each team includes an anchorperson, one or two ad writers, a comment writer, and 5-10 reporters who investigate stories and write news copy for the evening edition of the news. In completing this task, the reporters interview personalities important to their stories, consult newsroom files for background information, and receive periodic news updates which chronical late-breaking developments in their stories.

"After the reporters write their copy, they gather for a brief organizational meeting with their anchorperson, and finally, they appear before the camera or microphone to report the day's news. Later, the class views or listens to a tape of the broadcast and discusses its strengths and weaknesses.

"In an optional Phase II the news teams follow the procedures above, but their newscast focuses on a historical period or on literary characters. In the history option . . . a model unit organization is given for the Civil War; in the literature option, a model format centers on characters in literature who exhibit the "loss of innocence" theme."

The news stories included in the manual, ranging from national news to sports to education, are all fictitious and are largely humorous, such as the discovery of the cure for a disease that curls toes. The first option, using these stories, is suitable for classes in English, media, speech, or journalism; the other options are oriented specifically for literature and history classes. (TM)

Cost: $22.00

Producer: Interact, Box 262, Lakeside, CA 92040

QUERIES 'N THEORIES

Layman E. Allen, Peter Kugel, and Joan Ross

Playing Data
Copyright: 1970
Age Level: high school-adult
Number of Players: 2-4
Playing Time: 1 hour or more
Packaging: professionally produced game includes 54-page instruction booklet, chips, and playing mats.

Description: The authors describe this as "a series of games designed to introduce some basic ideas of modern linguistics." In the most basic version two players assume complimentary roles: One player assumes the role of "Native" and the other represents a Querist-Theorist or "QT." The Native privately defines a particular language which the QT must try to understand. The method the QT uses, as defined in the game, is called "Query-Theory." To gain information the QT constructs queries about the language–just as a scientist might construct an experiment. The outcomes of the queries constitute data. The QT must use this data to build theories about the Native's language. The winner is the first QT to understand the Native's language well enough to predict from his theory the answers to queries constructed by the Native. (DCD)

Cost: $13.00

Producer: WFF 'N PROOF, 1490-TZ South Boulevard, Ann Arbor, MI 48104

RIGHT IS WRITE

Gordon Russell, Jr.

Playing Data
Copyright: 1971
Age Level: grades 4-10
Number of Players: 20-50
Playing Time: 2 hours or more
Preparation Time: 15 minutes.

Description: The game focuses on the material written by the children that play the roles of writers. They write the stories; they compose the poems; they create the exercise in written thought. The activities of all other players are based on what these writers produce.

Writer's agents work for the writers. They assist writers with ideas and with editing. After the writing is completed, the agent will attempt to "sell" the material to the editor.

The editors are responsible for the mechanics of punctuation, spelling, and usage of the written word. They are expected to be selective and critical of the written work offered by the agents. After a selection is accepted by an editor, he will, in turn, offer it to a publisher.

Publishers attempt to purchase high quality, well edited material, acceptable to the audience for which they were prepared.

Right is Write is completely flexible and can serve to strengthen any writing skill. Specifically, it assists in making the task of editing second nature. The assessment of quality comes from within the class itself.

This experience has been classroom tested for more than five years. (Publisher)

User Report: It is by far one of the best simulation games I have ever seen, particularly in the area of language arts/reading. My class was so enthusiastic about writing for the purpose of publishing (with the members of the class acting as agents, editors, and publishers), that the entire task of "editing" has become a positive exercise. We have used it several times now and the growth of the quality of their writing has changed tremendously. For the first time I can honestly say that my class enjoys writing, but they also enjoy the thrill of seeing and doing and using the skills needed to make their writing special. (Barbara Simkin, 6th grade teacher, Highland School, Chelmsford, Mass.)

Cost: $19.95

Producer: Curriculum Associates, Inc., 8 Henshaw Street, Woburn, MA 01801

SATURATION

Ken Hogarty

Playing Data
Copyright: 1976
Age Level: high school
Number of Players: 15-40 in 3 groups
Playing Time: 15-24 1-hour sessions
Special Equipment: duplicating machine, playing cards, 700 poker chips
Packaging: 32-page manual

Description: *Saturation* provides students with the experience of being traditional journalistic reports or saturation reporters as they participate in a SuperBowl of Poker or in a Poetry Festival. According to the manual: "During each simulation the class is divided into three groups. Eight students become participants, role-playing poker players (with one student acting as host/dealer or host/judge-moderator in each simulation). The rest of the class is divided evenly to act as either traditional journalistic reporters or saturation reporters as they particivary throughout the simulations. Traditional reporters organize a 'cityroom' to meet daily assignment deadlines. They interview participants, rotate in 'pool coverage' of the event, and depend upon press conferences and formal interview sessions to gather facts for their stories. At the end of the activities of the simulation, all traditional reporters submit journalistic reports of the proceedings in a traditional journalism form that they have learned (good leads and bodies presented in an 'inverted pyramid approach). Saturation reports have much more freedom to attend events. They have to learn that this 'freedom' must be balanced by through note-taking and an alert eye for details. Receptions given by the hosts may be as meaningful to these journalists as the more 'artificial' press conferences. At the conclusion of the simulation, these reporters have to submit a saturation report that does justice to the Super Bowl or Festival."

Meanwhile, participants complete saturation reports, write press release biographies or role-play a poet of their own choice, and arrange and participate in press conferences and receptions. The simulation ends with students evaluating each others' work using journalistic analysis sheets, followed by a debriefing. The game manual is complete with information on the two schools of journalism, proofreading data, role profiles, and evaluation sheets. (TM)

Cost: $14.00

Producer: Interact, Box 262, Lakeside, CA 92040

SEARCH

Paul DeKock, El Capitan High School, Lakeside, California

Playing Data
Copyright: 1971
Age Level: grades 7-12
Prerequisite Skills: reading, grade 11; math, grade 0
Number of Players: 20 or 40 in 4 teams
Playing Time: 10-20 hours in time periods of 1 hour
Preparation Time: 3 hours

Description: After joining one of four teams named after four historical ages, team members assume historical names (e.g., Queen Elizabeth and Sir Francis Drake during the Elizabethan Age). An artistic team member designs name tags which include assumed historical name, motto, and coat-of-arms. Students then tour their school library. Next, individual team members start earning LIPS (Library Intelligence Points) by taking short quizzes and completing assignments on the following: library location, the Dewey Decimal System, the card catalog, "The Reader's Guide to Periodical Literature." Daily bonuses are awarded teams who come in first and second places in average totals on team members' LIPS Balance Sheets. Now understanding where to find materials in their school library, students face a more difficult challenge: they fill out detailed Search Forms which show how they created and solved problems requiring understanding of how to use all library resources. Daily competition continues as teams report their average LIPS totals and receive bonuses. In a final optional phase, team members divide up questions that necessitate delving into the historical age assigned their team. Answers to the questions are found only by performing all the unit's learning objectives (given in behavioral terms). Each team then uses its research to present 20 minute dialogues to the class, bringing the past's great men and women into the 1970's to comment on our way of life. (Publisher's brochure)

Cost: $14.00

Producer: Interact, P.O. Box 262, Lakeside, CA 92040

SHAKESPEARE

Playing Data
Copyright: 1966
Age Level: grade 7-adult
Prerequisite Skills: basic level, none; advanced tournament, familiarity with plays and sonnets
Number of Players: 2-4 in 2 teams
Playing Time: 15 minutes-1 hour
Preparation Time: none

Description: Players move around board edge as determined by dice roll. When they land on a square marked "pick card," they must answer a question which can be challenged by other players. Both challenger and challengee can be penalized for incorrect answers or challenges. (Producer)

Cost: $12.00 (10% educator's discount available)

Producer: The Avalon Hill Company, 4517 Harford Road, Baltimore, MD 21234

TAKING ACTION

Lynn Quitman Troyka and Jerrold Nudelman

Playing Data
Copyright: 1975
Age Level: grades 7-12
Number of Players: 6-35
Playing Time: 35-55 minutes
Preparation Time: 3 hours (est.)
Packaging: professionally produced 150-page game manual, 50-page teacher's manual

Description: The six simulations in this collection are designed for use in the English classroom. Each may be played within one class period and all share the same format. In each game about five minutes are budgeted for role selection, seven to fifteen minutes for planning strategy, eighteen to twenty-five minutes for negotiation, and the last five to ten minutes for decision making.

In "Uprising Behind Bars," players assume the roles of prisoners, correctional personnel, and mediators at the Medford Correctional Facility, a state prison for men. The prisoners are on strike to protest living conditions in the facility, and successful play requires the negotiation of a compromise that will end the strike. In "Taxis For Sale," players assume the roles of automobile salesmen, a cab driver, and an executive of a taxi company planning to purchase seventy-five new cabs. Players must negotiate this sale. In "Conservation Crisis," a group of residents must convince their town's major employer, a chemical company, to stop the polluting the town's water supply without closing down. "Women On Patrol" is about nontraditional roles for women. "Dollars In Demand" is about government spending. The final game, "Population Control," forecasts a time (A.D. 2204) when families must be issued permits to have children. Players must decide who can have children and be able to justify their decision. (DCD)

Cost: $5.50

Producer: Prentice-Hall, Inc., Englewood Cliffs, NJ 07632

TALKING ROCKS

Robert F. Vernon

Playing Data
Copyright: 1978
Age Level: grades 4-8
Number of Players: 6-60 in groups of 3-5
Preparation Time: ½ hour (est.)
Packaging: 24-page instruction manual

Description: The participants of this simulation are divided into groups of five players or less who represent bands of the "Eagle People," a tribe of neolithic shepherds who live in small groups and seasonally migrate in search of new pastures. These small groups communicate with one another by leaving messages near abandoned campsites. These messages contain information important to the survival of the other groups. (Such as where to find water or fresh grazing land and where there is danger.)

The groups of players are segregated into different parts of one large room and each group is given a printed "survival message" by the game director. Each band of the "Eagle People" must then translate this message into pictures, draw the pictures on paper, and leave this message for subsequent groups to find. No talking is allowed between groups, and when all of the survival messages have been translated into pictographs, the game director collects these printed messages and all of the groups move to new stations (or camp sites). There each band tries to translate the message that was left behind by the previous group. Those groups which fail die, and must leave the game.

This game is primarily designed to illustrate a variety of points (according to the interest and inclination of the teacher) about language development and the development of writing, and can also be used to illustrate various concepts about intergroup dependence, or the nature of primitive culture and art. The game ends when all of the messages have been comprehended. (DCD)

Cost: $5.00

Producer: Simile II, P.O. Box 910, Del Mar, CA 92014

WALK IN MY SHOES

Alan Teplitsky and Ronald Hyman

Playing Data
Copyright: 1976
Age Level: grades 7-12
Number of Players: 5-50
Playing Time: 30 hours (est.) per unit
Preparation Time: 5 hours (est.) per unit
Packaging: eight professionally produced units, each of which contains a teacher's guide, plus anthologies and "eyewitness journals" for students.

Description: *Walk in my Shoes* is a series of curricular materials for language study encompassing eight thematic units. These materials are designed to improve students' reading and writing skills, their ability to think logically, their understanding of language concepts, and their vocabulary, as well as to increase understanding of their own lives and the lives of others. This last goal is achieved by having "students walk in the shoes of [or role play] another person—a potential dropout, for example, or a social worker, advertising copy writer, Hollywood casting director, or teenage run-away. By trying out roles and attempting to solve real-world problems within a controlled or simulated setting, students learn the *uses* of language and begin to develop a feeling of mastery over their own lives."

The following units are available. In "Are You With Me?" students analyze the process and results of communication by learning to recognize the difference among fact, inference, and opinion. In "Fair/No Fair" players explore the idea of justice in everyday life. This unit contains thirty-five activities that range from very simple to complex. In a beginning exercise, for example, a group of students must define "fair." A more advanced exercise has students hold a mock trial to examine the role of evidence in argumentation and decision making, and problems that arise from viewing language as inflexible and unchanging introduce participants to ways of coping with injustice. In the "Ins and Outs" unit, students participate in role-playing exercises that examine various social attributions. In "Cities/USA" students use reference materials as tools to evaluate the credentials of "experts." In "Something of Value" players investigate values—what they are, how they are acquired, and what influences them. "It's a Free Country" is a series of exercises about freedom in communications. "To be Somebody" helps students explore their own self-image and their personal goals. The last unit, "Escape Routes," gives students practice in ways to cope with real problems. (DCD)

Cost: five components total about $40.00 per unit

Producer: Prentice-Hall, Inc., Englewood Cliffs, NJ 07632

See also
Puzzle HISTORY
COMMUNICATION
SELF-DEVELOPMENT
SOCIAL STUDIES

LEGAL SYSTEM

CONFRONTATION IN URBIA

Ronald Lundstedt and David Dal Porto

Playing Data
Copyright: 1972
Age Level: grades 7-12
Number of Players: 20-36
Playing Time: approximately 2 weeks
Preparation Time: reading time—teacher's book

Description: This trial simulation is designed to help students understand the legal consequences and ramifications of sociopolitical confrontation. Before beginning the simulation, players read a news account of a confrontation between a store owner and a group of young people, a second narrative description of the same incident, and a transcript of a preliminary hearing of the charges against the demonstrators. Players then assume the roles of presiding judge, attorneys, arresting officers, spectators, defendants, and plaintiff for the trial. Although the players are expected to act within the framework of rules and limitations that would govern their roles in real life, the decisions demanded by this simulation are not made for them. The trial is followed by a class discussion of the outcome, various alternatives, and implications. (DCD)

Comment: *Confrontation in Urbia* is designed to develop player's understanding of concepts embodied in the Bill of Rights. Although the trial simulation is highly structured, instructors have considerable latitude in the use of the article, story, and court transcripts to make connections between urban situations and application of constitutional concepts to them. (AC)

Cost: $21.50

Producer: Classroom Dynamics Publishing Company, 231 O'Connor Drive, Suite B, San Jose, CA 95218

COURT POLICY NEGOTIATIONS

Michael C. Vander Velde

Playing Data
Age Level: college, graduate school, continuing education, administrators
Number of Players: 15 or more
Playing Time: 1-2 hours
Preparation Time: none

Description: *Court Policy Negotiations* is a board game designed to acquaint players with the political complexities of administering a local judicial system. The main focus of the game is to come to decisions on agenda items of the judicial executive committee of a large California county judicial system. Policy issues range from buying a pool table for the judges' lounge and coordinating vacations to making recommendations to the state legislature on a new no-fault insurance plan and radical reorganization. The decision process is tempered by taking into account each of the committee members' consistency, his prestige, and his political leanings, i.e., establishment judge, avant-garde judge. In addition, external influences from the bar association, country clerk, district attorney, news media and the historical propensity of each issue are incorporated. The game requires each participant to develop the use of prestige, influence, media, and timing to achieve the desired result for the role assumed.

Cost: price on request

Producer: COMEX, Davidson Conference Center, University of Southern California, Los Angeles, CA 90007

THE DISPOSITION EXERCISE

Mary A. Harrison

Playing Data
Copyright: 1972
Age Level: professional
Number of Players: 12-24 in 2-4 teams
Playing Time: 5-8 hours
Preparation Time: 2 hours the first time

Description: This role-playing exercise is designed to help new and experienced juvenile officers improve their decision-making and interviewing abilities. Individual players and teams assume the roles of an arresting officer, a juvenile criminal offender, his parents, neighbors, teachers, and an alleged victim. The player(s) representing the arresting officer(s) must define a strategy for each case and decide what information to gather, what to do with the offender, and to which social service agency, if any, the case may be referred. After the role-playing exercise the participants analyze the factors that most influenced their decisions and discuss policies their own police departments might establish. (DCD)

Comment: To our knowledge this is the "only available training simulation for police officers (or other enforcement officers) who must make dispositions of juvenile offenders. It has been called an exceptional training tool by a police lieutenant who feels the exercise should be used to speed development of skills vital to juvenile officers. The trainer's manual contains a section which describes in detail how to adapt or create game materials for situation-specific uses; thus, agencies using the exercise can tailor it to their needs and goals." Results of field testing are available. (DZ)

Cost: $18.00 plus $2.50 postage and handling

Producer: Mary A. Harrison, 401 16th Street, Manhattan Beach, CA 90266

FIRE

John Wesley and Diane Wesley

Playing Data
Copyright: 1977
Age Level: grades 7-12
Number of Players: 30-40
Preparation Time: 2 hours (approx.)
Packaging: 38-page booklet

Description: This is a simulated trial of two teenage boys accused of negligence in starting a large and destructive forest fire. Players are assigned the roles of attorneys, court personnel, defendants, and witnesses. They prepare for the trial by studying role sheets, which include information that only they, and none of the other players, may know. During the trial, arguments must be made, testimony given, and a jury must reach a decision concerning the guilt of the accused. (DCD)

Cost: $14.00

Producer: Interact, Box 262, Lakeside, CA 92040

FIRST AMENDMENT FREEDOMS

David J. Boin and Robert Sillman

Playing Data
Copyright: 1973
Age Level: grades 7-12
Number of Players: class size
Playing Time: 5 class periods

Description: Students take the parts of judge, defense attorney, prosecuting attorney, witnesses, and jurors in various real cases having to do with First Amendment freedoms. One is the school prayer decision; the next, a freedom of speech issue, in which a rabble rouser is starting what looks like a riot. The third is a freedom of assembly issue. According to the authors, the simulation's instructional objectives are "to provide students with an opportunity to interpret basic constitutional issues, with some insight into the dynamic quality of constitutional law, with an opportunity to study and debate actual Supreme Court cases and to become familiar with courtroom procedures; and to help students develop a definition of individual freedom." (REH)

Comment: Succinct summaries of different witnesses are preprepared. This is a good model for the study of the Constitution and could well be replicated in a number of different areas. It is, nevertheless, short and simple, in the standard Edu-Game format with a few pages of teacher's manual and locally reproducible student material. (REH)

Cost: $3.00

Producer: Edu-Game, P.O. Box 1144, Sun Valley, CA 91352

INNOCENT UNTIL . . .

Peter Finn

Playing Data
Copyright: 1972
Age Level: grades 7-12

Prequisite Skills: reading, grade 6
Number of Players: 13-32

Description: This trial simulation is designed to increase players' knowledge of court procedures and prepare younger players for jury duty in later life. Players assume the roles of the judge, the defendant, his wife, two friends, his two lawyers, two prosecutors, two policemen, the court bailiff, four witnesses, and four to fourteen jury members. The innocense or guilt of the accused is determined by the deliberation of the jury, so players are encouraged to play their roles "to the hilt." (MJR)

Comment: This is an exciting game for studying court procedures. It is carefully structured with regard to situation and roles, but open-ended with regard to outcome. Students are provided with enough data to become interested but not so many clues to the outcome to make the simulation boring. The careful structuring and the open-endedness make this a stimulating, dramatic experience that is motivating and leads to player involvement.

According to the teacher's manual, "This simulation is especially useful for: promoting understanding of and empathy with different people, such as lawyers, policemen, witnesses and jurors, experimenting with occupations, roles and personalities students may be unable to experience in real life, confronting participants with the necessity for making decisions and attempting to influence people, and encouraging the development and expression of feelings about a variety of issues." (MJR)

Cost: $36.00

Producer: Games Central, 55-C Wheeler St., Cambridge, MA 02138

THE JAIL PUZZLE

Dave Wasserman, Sam DeBose, and Bernie DeKoven

Playing Data
Copyright: 1976
Age Level: high school-adult
Number of Players: 15-25
Playing Time: 1½-2 hours
Supplementary Materials: 2 decks of playing cards, pair of dice, poker chips, envelopes, $15,000 in play money, 2 checker boards.
Packaging: 24-page booklet and 9 role cards.

Description: This game is designed to simulate the experience of prisoners, social workers, and guards in a county jail. The game involves the playing of four simultaneous and interrelated "subgames," each of which, according to the authors, "models a component of the criminal justice system: (1) in the Arrest Arena, a game of 'I doubt you' is played; (2) in the Prison, a board game is played; (3) in the Community Resource Center, a game of 'Concentration' is played; and (4) in the Court, there is a game of chance."

Play begins with arrest. All players cut a deck of cards and those with the highest draw become, in order, judge, guards, social workers, or lawyers. The other players all become prisoners. They are then dealt all the cards and must, by turns, lay down their cards in order. (When one player puts down a nine, the next player must put down a ten, the player after that must put down a jack, and so on.) Players do not have to show their cards and are encouraged to bluff. They are also encouraged to challenge the other players. Any player who is unable to lay down the correct card is arrested and arraigned. Each player may take $20 from the bank when another player is arrested.

In arraignment court, a prisoner roles dice to determine his bail. A prisoner with enough money may pay bail and go free. If she or he does not have enough money, imprisonment follows.

In prison, the prisoners and the guards play a game like checkers in which guards must stop prisoners from moving their tokens to the guards' side of the board, and, the prisoners must avoid having their tokens on squares adjacent to the guards' tokens. When this happens, the guard collects $100 from the bank and the prisoner's token must return to the first rank. If a prisoner's token advances to the sixth rank or beyond, she or he may go to court with a social worker or lawyer.

Meanwhile the social workers and lawyers are playing a game called "concentration" with a deck of cards. The idea of this game is for players to pick matched pairs of cards at random. They may pick two cards for each prisoner in their caseload. Each time one of these players is successful he or she is awarded $50 from the bank and may use this money to bail out any prisoner. He or she may also solicit money from the guards for this purpose. Social workers or lawyers may visit prisoners only when the prisoner requests it. After arraignment, prisoners may return to court only by reaching rank six or better in the prison game, and when they do visit the court they must be accompanied by their social worker or lawyer.

A prisoner who is bailed out returns to the arrest arena to play the arrest game again.

The game continues for two hours. Debriefing and discussion follow play. (DCD)

Cost: $2.00

Producer: Dave Wasserman, c/o First Presbyterian Church, Ninth and Main Streets, Cedar Falls, IA 50613

JURIS

Gary Zarecky and William M. McCarty

Playing Data
Copyright: 1975
Age Level: high school
Number of Players: class
Packaging: 23-page printed student guide, 21-page teacher's guide with reproducible material

Description: The student handbook for *Juris* contains a history of the common law legal system, background material in contracts, torts, juvenile and criminal law (which has subdivisions for theft, alcohol and drugs, hitch-hiking, burglary, and murder), and a 4-page glossary of legal terms.

The teacher's guide, which includes case studies and questions (arguendo) for each legal area, describes three different teaching strategies. They are: to use the student handbook as a textbook, with the case studies and questions as assignments for discussion; to organize the class into five groups representing the elements of our adversary system of justice: plaintiff, defense, judge, jury, and appellate court. Given a legal controversy in one of the four areas, each group works to develop its particular frame of reference with regard to the case. Someone from each group reports its view at a meeting of all the groups. The class may then move to the next area of law. Individual work may also be pursued, to use the materials as a vehicle for independent study, with a student on a case, given appropriate school settings. (AC)

Cost: $22.00

Producer: Interact, Box 262, Lakeside, CA 92040

THE JURY GAME

Richard Weintraub, Richard Krieger, Constitutional Rights Foundation; George W. Echan, Jr., and Stephen Charles Taylor, Members of the State Bar of California

Playing Data
Copyright: 1974
Age Level: high school
Number of Players: 25-35 students
Playing Time: 2-3 50-minute sessions for *The Jury Game,* 2 50-minute sessions for the mock trial
Prerequisite Skills: class should have had a lesson on the background and history of jury selection
Preparation Time: 15 minutes

Supplementary Resources: the participation of a practicing attorney is helpful but not essential
Packaging: a professionally designed box including a 14-page manual, an 8-page pamphlet, pictures, question information sheets, tally, evaluation and role sheets

Description: *The Jury Game* includes two games, one involving the selection of jury members and the other a mock trial. In the first game, students select a jury of 6-12 members who will render a fair verdict on the civil or criminal case chosen from the four cases described in the manual. The criminal cases involve charges against a young woman for theft and embezzlement. The civil cases are concerned with claims made because of a fire caused by an appliance and because of a car accident. All students receive question-asking information sheets designed to help them develop their critical thinking and questioning abilities. As jurors, attorneys, judge, defendant, plaintiff, reporters, and observers, they familiarize themselves with their roles and tasks, and then act the jury selection process, with attorneys questioning, challenging, and dismissing jurors. After jurors are sworn in, the reporters report on the proceedings and observers evaluate the session. A de-briefing period follows during which students share their discoveries about the nature of prejudice and their opinions of the jury system. In the mock trial game, a follow-up to the first game, the chosen case is enacted. Attorneys make opening and closing statements and question witnesses, and the jury deliberates in front of the class. This dual game kit includes a 2-page description of each case and its jurors, background information on recent court decisions, a glossary, a complete bibliography, and multiple copies of information, tally, evaluation and role sheets. (TM)

Cost: $15.00; extended version with 35 copies of 8-page player's pamphlet $22.00

Producer: Zenger Publications; distributed by Social Studies School Service, 10000 Culver Blvd., Culver City, CA 90230

THE JUSTICE GAME

Alan S. Engel, Miami University

Playing Data
Copyright: 1974
Age Level: 7 years-adult, college, continuing education
Number of Players: 20 or more
Playing Time: 1 hour sessions, five recommended
Preparation Time: 2 hours

Description: An introductory-level game designed to simulate the functioning of a typical criminal justice system. The focus is on process of criminal suspect from time of arrest through release from prison. The game seeks to familiarize the player with the system's complexity. The effect of extra-legal influences and the cross pressures, frustration and alienation, are involved in the realities of the system. Stress is on the effect of time and cost consequences of various decisions at different junctures in the process.

The game is designed for classroom play and simulates, in a scaled-down version, the nature of the criminal justice system. There is a legal case with 18 laws, 48 indictable offenses. Every 2 minutes a suspect is produced who must be taken to police headquarters, assigned a public defender or obtain private counsel, be interviewed by the prosecutor, indicted, arraigned, tried, and, if guilty, sentenced and imprisoned. At each of these junctures various decisions must be made in handling the case. Decisions are influenced by quota pressure, the media, cost, and time requirements. (JG)

Cost: player's manual $5.95; instructor's manual $4.95

Producer: Glencoe Press, 17337 Ventura Blvd., Encino, CA 91316

KIDS IN CRISIS

Constitutional Rights Foundation

Playing Data
Copyright: 1973
Age Level: high school
Number of Players: 25-41
Playing Time: 1½-2 hours
Supplementary Materials: optional use of *Law: You, the Police and Justice* (Scholastic Magazine)
Special Equipment: duplicating machine
Packaging: packet containing role description folders, probation reports, role tags, 32-page Bill of Rights Newsletter, 16-page teacher guide

Description: The purpose of *Kids in Crisis* is to help students understand some of the serious problems of young people and how the juvenile and adult courts deal with these problems. Students divide into five groups, each with a different case to enact, and as defendants, lawyers, parents, judge, and observers role play either a juvenile disposition hearing or an adult sentencing hearing. Role descriptions and information describing the procedures for both hearings are available to all. One case concerns Angel, a seventeen-year-old Mexican-American who has been arrested over a dozen times for petty theft and shoplifting. Another involves Ronald, a white thirteen-year-old, who has been beaten repeatedly by his stepfather; the court must decide what to do with him. After each group completes its hearing and discusses it, the class convenes to share experiences and to complete observer rating sheets. (TM)

Cost: $32.50

Producer: Zenger Publications; distributed by Social Studies School Service, 10000 Culver Blvd., Culver City, CA 90230

MOOT

Gary Zarecky

Playing Data
Copyright: 1972
Age Level: grades 7-12
Prerequisite Skills: reading, grade 9
Number of Players: 20-40
Playing Time: 15 weeks-1 semester, periods of 1 hour

Description: *Moot,* according to the producer, "was designed to stimulate young people's interest in and understanding of our legal system." Players learn about crime in the United States, the need for law, due process, and courtroom procedures.

Students face three legal issues and attempt to solve the problems they present during mock trials. The first and most comprehensive of these trials involves a criminal drug case. Players assume the roles of drug sellers, buyers, prosecutors, defense attorneys, informants, and courtroom personnel. After simulating the crime, students turn the classroom into a courtroom where they apply the theories and practices derived from juvenile and adult law.

The other two trials concern juvenile court procedure and contract law. Several situations for more complicated trials (murder, assault and battery, school law violations, draft evasion, and robbery) are outlined in the teacher guide. (DCD)

Cost: $14.00

Producer: Interact, P.O. Box 262, Lakeside, CA 92040

PLEA: A GAME OF CRIMINAL JUSTICE

Ethan Katsh, Ronald M. Pipkin, and Beverly Schwartz Katsh

Playing Data
Age Level: grade 11 and up

Number of Players: 22-35
Preparation Time: 1 hour
Playing Time: 3 hours (more or less)
Packaging: professionally designed box
Special Equipment Required: 4 or 8 long tables (for 1 or 2 courts), 22-35 chairs, blackboard

Description: The actual setting of *Plea* is the Hall of Justice Building of Grand City, a large metropolitan area in the United States. The roles of judge, district attorneys, public defenders, defendants, and spectators are assigned and prepared. Preparation materials include printed case notes for each role, a suggested arrangement for the room employed, guidelines for determining the sequence of the actions, and copies of the Grand City Criminal Code.

The judge posts the order of the hearings, the public defenders meet with defendants, and district attorneys confer on charges, sentences, and acceptable means for plea bargaining. After 30 minutes, the judge opens court and calls the first case. Play continues for the next 90 minutes. At the end of this period, all defendants pleading guilty are to be sentenced and all defendants pleading a final "not guilty" are to be released on bail or returned to jail while their cases are "continued for trial." (TM)

Cost: 18-student kit $17.50, 35-student kit $25.00

Producer: Simile II, P.O. Box 910, Del Mar, CA 92014

POINT OF LAW

Michael Lipman, J.D.

Playing Data
Copyright: 1972
Age Level: grade 9-adult
Number of Players: 2-6
Playing Time: 1 hour
Packaging: professionally produced and packaged game includes 263-page case book

Description: The players of *Point of Law* randomly pick 10 of the 100 legal cases presented in the case book. Each case is a one-page condensation of a civil or criminal complaint that is distinguished for the broad principles of law which apply to it. A moderator reads the cases one by one, and as they are read each player selects one of the four possible rulings listed after the presentation of facts. One of these is the actual decision made by the court hearing the case and a second is, in the author's view, a *fair* decision. The remaining two decisions are considered *poor*. Players score four points for identifying the actual decision and two points for selecting the fair decision. The winner is the player with the highest score after ten rounds. (DCD)

Comment: Potential users who are committed to expunging sexism from American thought and culture should be forewarned that the female witness in the cover illustration of the case book appears to be testifying in her lingerie. (DCD)

Cost: $12.00

Producer: Avalon Hill, 4517 Harford Road, Baltimore, MD 21234

REVOLT AT STATE PRISON

David J. Boin and Robert Sillman

Playing Data
Copyright: 1974
Age Level: grades 9-12
Number of Players: class size
Playing Time: 4-5 class periods

Description: The class is divided into cell block groups of six members in each group. Groups must discuss and vote on certain predetermined issues and proposals from the warden. There is also a predetermined penalty and reward scale depending on their decisions. Finally, all of the prisoners vote on each of the issues. According to the authors, the instructional objectives are: "to provide students with some insight into the problems of the U.S. prison system, to provide an opportunity to debate the issue of prison reform, to discuss the purpose of prisons in a democratic society, and to make suggestions as to how prisons can best serve society in the future." (REH)

Comment: This is a neat way of bringing some current event topics immediately into the classroom. The simulation adequately represents a cross-section of prisoner issues and it can serve as an introduction to a study of the whole penal system in the United States. It is in the standard Edu-Game format with a few pages of teacher's manual and locally reproducible student material. (REH)

Cost: $3.00

Producer: Edu-Game, P.O. Box 1144, Sun Valley, CA 91352

RIP-OFF

Gary Zarecky

Playing Data
Copyright: 1976
Age Level: high school
Number of Players: 12-25
Playing Time: 8-10 50-minute periods
Preparation Time: 5 hours
Packaging: 3-page student guide, 25-page teacher guide

Description: The purpose of *Rip-Off* is to make students more aware of and concerned about teenage theft in America by introducing them to the problems and laws related to theft and to the juvenile hearing process. The teacher's guide suggests three options—an actual prearranged shoplifting in a local business followed by a simulated juvenile court hearing, an actual school theft followed by a hearing, or a make-believe theft—but it describes only the steps involved in the first option. To plan the theft, which provides the necessary background for further activities, the teacher must carefully arrange with a local business, school administrators, and students' parents for the theft to be performed and supervised in such a way that all stolen goods are returned and no legal difficulties result. After this lengthy preparation is concluded and students have discussed, analyzed and role played two case studies of teenage shoplifting, the planned theft takes place outside school and the class as a whole discusses the apprehension. At this point, students are assigned roles such as judge, probation officer, police officer, prosecuting and defense attorneys, and students prepare for the simulated juvenile court hearings that occur on the fifth and sixth days of the simulation. The simulation ends with an analysis of a third case study, a discussion of what can be done about shoplifting and an evaluation and debriefing. (TM)

Comment: Apart from the obvious difficulties involved in arranging a theft at a local business, with all the legal repercussions that might result if one or more students proves to be dishonest, the teacher should be aware that the actual shoplifting experience involved in *Rip-Off* may generate such enthusiasm that students may actually be eager to try shoplifting themselves. *Rip-Off* may have disastrous consequences if played with untrustworthy students. (TM)

Note: For more on this simulation, see the essay on social studies simulations by Michael J. Rockler.

Cost: $14.00

Producer: Interact, Box 262, Lakeside, CA 92040

VOWS

Gary Zarecky and William M. McCarty

Playing Data
Copyright: 1977
Age Level: high school
Number of Players: 15 or more, in 3 groups
Playing Time: 8-13 1-hour sessions
Packaging: 7-page teacher guide and 27-page student guide

Description: The purpose of *Vows* is to introduce students to the legal aspects of dating, engagement, marriage, and the three ways marriage ends—annulment, divorce, and death. According to the teacher guide: "Vows has three phases: Phase I, the legal background of dating and engagement; Phase II, the legal background of marriage; Phase III, the legal background of possible termination of marriage. During each phase the class is divided into three groups, with each group receiving a different case study to analyze. Each group first carefully examines its case study's facts; then it searches through the law section in the student handbook to find legal principles dealing with this fact situation. The group then splits itself into two or three sections, with each section being assigned to a different person or persons in the case study. Each section's responsibility is to define as clearly as possible this person's legal position in relationship to the others.

"All sections then use this preparation as the basis of a presentation made to one of three judges, who are students randomly picked by the instructor. After each case study has been role-played in front of the judges and the class, the judges write separate opinions on what should be done while the remainder of the class prepares answers to DISCUSSION QUESTIONS found at the end of each case study. This preparation then is used the next day as the basis of an EVALUATION SESSION in which any confusion can be cleared up or identified as an area needing clarification by further research. Following all three phases, an OBJECTIVE TEST contained in this Teacher's Guide can be duplicated and given."

Example case studies used in the simulation include situations involving premarital pregnancy, return of gifts after an engagement is broken, circumventing age requirements by marrying in a foreign country, and determining who has custody of children affected by divorce proceedings. (TM)

Cost: $22.00

Producer: Interact, Box 262, Lakeside, CA 92040

See also
Halfway House COMMUNITY ISSUES
The Haymarket Case HISTORY
Nuremberg HISTORY

MATHEMATICS

BIG DEAL

Playing Data
Copyright: 1976
Age Level: junior high, high school
Number of Players: 2-4
Playing Time: 30-60 minutes
Packaging: professionaly packaged game with game board, play money, property deeds, pawns, and chips

Description: *Big Deal* is a board game in which, according to the instructions, "players invest in property and engage in various transactions, many involving percentage calculations, to amass large estates."

Each player starts with $10,000 and moves according to the spinner. When players lands on unowned property they may buy it at a percentage discount of its face value by using the discount spinner. For instance, if a player lands on the pawn shop (which has a face value of

Cost: $4.00

Producer: Simile II, P.O. Box 910, Del Mar, CA 92014

BINARY

Michael Abrams

Playing Data
Copyright: 1971
Age Level: grades 7-9
Number of Players: 2
Playing Time: 15 minutes (est.)
Packaging: professionally produced game includes playing surface with bingo chips and paper dots.

Description: This simple math game is played with small plastic chips that can be arranged in rows on a 130-square grid. Both players use 25 chips. Fifteen of these represent the number one, and ten (on which paper dots are glued) represent zero. The idea of the game is for the players to try to place four or more chips in a row anywhere on the playing surface. Players who do this score a number of points equal to the binary number formed by their row of chips. (For example, a player may lay down 1111 and, consequently, score fifteen points.) The game ends when all of the chips have been played. The player with the highest score wins. (DCD)

$4,800) and wants to buy it, the player spins the discount spinner to see what percentage of the face value to be paid. If the spinner stops at 20%, for instance, the player could purchase the pawn shop for 20% of its face value, or $960. Players also use a spinner to determine the rent they must pay when they land on a property owned by someone else and in determining their income tax. These also involve percentage calculations. Each time players pass payday, they collect $1,900. The first player to amass $20,000 wins. (DCD)

Cost: $7.95

Producer: Creative Teaching Associates, P.O. Box 7714, Fresno, CA 93727

CAL Q LATE

Cecile L. Hurley

Playing Data
Copyright: 1971
Age Level: grades 7-10, special education
Prerequisite Skills: computational skills involving addition, subtraction, multiplication
Number of Players: 1-4
Playing Time: 15 minutes (if desired) or indefinite
Preparation Time: minimal

Description: Game components include four playing boards for linear computation, four sets of number cards 1-9, and four sets of operation cards (+, −, x). Players place their sets of operation cards in alternate spaces across their playing boards in an order determined by chance. Three number cards, drawn at random, supply an "answer." Players then arrange their sets of number cards so that by computing in order across the board, the "answer" is obtained. (producer)

Cost: $4.59 school price (#02162-9 71)

Producer: Scott, Foresman and Company, 1900 East Lake, Chicago, IL 60025

CONFIGURATIONS

Harold L. Dorwort

Playing Data
Copyright: 1967
Age Level: high school to adult
Number of Players: 1
Playing Time: 30 minutes-3 hours
Packaging: professionally produced and packaged game includes 17 puzzle boards, 5 sets of plastic numbers, and 28-page instruction booklet.

Description: This puzzle game consists of a progressively difficult series of four problems. All four require the player to figure out the mathematical relationship between arrays of numbers (each with three rows and seven to ten columns) and geometric configurations. (A "configuration" is a geometrical figure consisting of seven points and seven lines, with each line containing exactly three points and with exactly three lines of the configuration passing through each point.) In the first puzzle, which describes the Fano 7/3 configuration, for example, the player must first align three sets of the numerals 1, 2, 3, 4, 5, 6, and 7 in seven columns of three rows in such a way that the three numerals in any column are different and so that the same pair of numerals does not occur in two different columns. The player must next transfer the numerals from the array onto the points of a diagram of a configuration in such a way that every column of numbers will correspond to one of the lines on the diagram. (DCD)

Cost: $6.75

Producer: WFF 'N PROOF, 1490-TZ South Boulevard, Ann Arbor, MI 48104

EQUATIONS

Layman E. Allen

Playing Data
Copyright: 1969, 1971
Age Level: elementary school to adult
Number of Players: 2-8
Playing Time: 20 minutes-1 hour
Preparation Time: 1 hour
Packaging: professionally produced and packaged game includes dice, timer, playing surface, and 55-page instruction booklet; also available is a series of Instructional Math Play (IMP) kits for solitary play

Description: This collection of games is designed, according to the author, "to provide a stimulating and entertaining occasion for learning some of the elementary operations of mathematics–addition, subtraction, multiplication, division, exponentiation, and root." *Equations* is similar to the game *WFF 'N PROOF* and *On-Sets* by the same producer.

In all variations, two or more players must efficiently balance a mathematical equation. The equation has two parts, a "goal," which is the result of some operation, and a "solution," which is an operation that would result in the goal. The game begins with the generation of resources for that play. These resources are dice marked with arabic numerals and the symbols for the operations of addition, multiplication, division, square root, and exponention. After the dice have been rolled, one player chooses up to five numbers or symbols to represent a goal. The remaining players take turns moving one symbol at a time in an effort to balance the equation. Players earn points by catching the mistakes of the other participants. The game ends when the equation is balanced.

A series of "Instructional Math Play" booklets for solitary play is also available for use with this game. In each, the instructions set a goal and explain various ways in which a solution may be developed from given resources.

There is also a "snuffing version" of *Equations* that "provides incentive for each player in the course of play to reveal to the other players mathematical ideas that he is thinking about in seeking to win." A photocopied set of rules for this version is available from Layman E. Allen at the Mental Health Research Institute, The University of Michigan, Ann Arbor 48109. (DCD)

Cost: Equations $10.00; IMP kits $1.25 each

Producer: WFF 'N PROOF, 1490-TZ South Boulevard, Ann Arbor, MI 48104

FOUR IN A ROW

Larry L. Whitworth, Community College of Allegheny County, Pittsburgh, PA.

Playing Data
Copyright: 1971
Age Level: grades 4-9
Number of Players: 2-any number in 2 teams
Playing Time: 5-15 minutes
Preparation Time: minimal
Special Equipment: overhead projector (optional)
Description: *Four-in-a-Row* gives students practice in both addition and multiplication facts and in plotting points on a grid. A coordinate grid on an overhead transparency provides the gameboard. Students take turns either adding or multiplying the coordinate numbers on the grid, trying to place four of their marks in a row. (producer)

Cost: $3.72 school price (#02340-0 71)

Producer: Scott, Foresman and Co., 1900 East Lake, Glenview, IL 60025

GEOMETRIC PLAYTHINGS

Playing Data
Age Level: grades 3-8
Prerequisite Skills: be able to read and follow directions
Number of Players: 1 or more
Playing Time: variable
Preparation Time: minimal
Versatility: activity, puzzles, model

Description: Students can take this book into a corner, each one take a page, cut it out, fold it, or color it and learn some aspect of geometry and afterwards trade information. (RI)

Related Mathematics: This is the best kind of book, one which is consumed while being used. A wide variety of geometric topics can be explored, including strange things like moebius strips, flexagons, and deltahedra. (RI)

Cost: $2.50

Producer: Troubador Press, 126 Folsom St., San Francisco, CA 94105

HERE TO THERE

Wayne H. Peterson and Shirley Ringo, Seattle Public Schools

Playing Data
Copyright: 1971
Age Level: grades 4-9
Number of Players: 2-4
Playing Time: 15-45 minutes
Preparation Time: minimal

Description: On a gameboard listing gas stations, free parking zones, and jail, students travel the road from "Here" to "There" by adding and subtracting mixed numbers. The fractions all have a common denominator of 12. Components include an 18" x 26" vinyl gameboard, a deck of fraction cards, plastic markers, and an instruction booklet. (producer)

Cost: school price $7.47 (#02114-9 71)

Producer: Scott, Foresman and Company, 1900 East Lake, Glenview, IL 60025

IN ORDER

Kenneth Kidd

Playing Data
Copyright: 1973
Age Level: grades 3-12
Number of Players: 2-52 in groups 2-4
Playing Time: 10-30 minutes (est.)
Preparation Time: 1 hour (est.)
Packaging: cardboard sheets from which 13 decks of cards and 36 card racks may be cut, and instructions

Description: *In Order* is a math drill game that can be played with any of 13 decks, which display the counting numbers, multiples of four, addition and subtraction problems, multiplication and division problems, area problems, angles, decimals and percentages, base five numerals, fractions, scientific notation, temperatures, integral exponents, and fractional exponents.

Groups of two to four players select a deck. Six cards are dealt to each of the players who place them in a rack in the order in which they were dealt. Any cards not dealt go face down in a "blind pile." The top card in the blind pile is turned face up to start a discard pile. The idea of the game is for the players to remove cards from the rack and replace them with cards drawn from either of the two piles so that the numbers represented on the cards in their racks are in increasing order from front to rear. The first player to do this says "In Order" and gets 30 points. The game may continue for several hands. The player with the highest score at the end wins. (DCD)

Cost: $12.95

Producer: Midwest Publications Company, Inc., P.O. Box 129, Troy, MI 48084

INTERCEPTOR

Ralph McMullin

Playing Data
Copyright: 1977
Age Level: up to grade 10
Number of Players: 2
Playing Time: 30 minutes (est.)
Packaging: professionally packaged game includes playing board, playing pieces, problem cards, instructions, and answer sheet

Description: The players of this simple game try to remove one another's playing pieces from a checkerboard playing surface. This is done, as in checkers or chess, by moving a token onto a square already occupied by an opponent's piece. Each player starts with five pieces called interceptors.

Before moving, a player must solve a problem in addition, subtraction, multiplication, or division that appears on a card. These problems use numbers up to five digits and consequently require a successful player to possess a solid comprehension of basic arithmetic. A player may move a piece to "intercept" an opponent's piece from two to seven spaces, as indicated at the bottom of the various problem cards. The player who captures all the opponent's pieces wins. (DCD)

Cost: $5.00

Producer: Teaching Aids Company, 925 South 300 West, Salt Lake City, UT 84101

METRICAT10N

Playing Data
Copyright: 1973
Age Level: grade 5 and up
Number of Players: 2-4
Playing Time: 80 minutes
Packaging: professionally packaged board game includes playing board, tokens, dice, ownership chips, play money, diplomas, "Metricat10n" cards, and instructions

Description: This complex, well thought out board game emphasizes metric prefixes. Players move tokens over a long, convoluted track of squares according to the roll of dice. Road signs, commands, and "schools" on this playing track require or allow players to move extra spaces; to give or take money from and to other players, the bank, or a "Metric Holding Company"; to draw "Metricat10n" cards (on which are printed commands similar to the commands on Monopoly "Chance" cards); to exchange denominations of currency; and to buy property and subsequently collect rent from other players. The currency consists of bills named MILLI, CENTI, DECI, METER, DEKA, HECTO, and KILO. At the beginning of the game each player is given nine bills of each denomination. The object of the game is for players to collect ten of each bill. This task is complicated by a rule which requires all transactions to be made with exact change. In fact, the only time players may exchange one kind of bill for another is when they land on one of the several "Bill Exchange" squares.

The producers are obviously enthusiastic about conversion from English to metric measurement. The "Metricat10n" cards, for example, bear such messages as: "The English system of measurement wastes time, lose one turn;" "Using the English system of measurement retards progress, go back six spaces;" and "Want to get ahead! GO METRIC!

Take another turn." The English system appears on an alternative section of track called "the Old English Trail" which is crooked, heavily penalized, and terminates with a dead end sign. The number 10 appears throughout the game, both obviously and subtly, to reinforce the basics of the metric system. The producer points out that "1) no player can bankrupt himself out of competition, 2) there is no emphasis on using conversion factors, 3) players need not know anything about the metric system to play, and 4) there is an allowance for a time limit on the game in that a winner can be declared at any point necessary."

The game ends when a predetermined amount of time has expired or when one player has collected at least ten of each type of metric bill. If the game ends without one of the players achieving this last objective, the winner is the player who holds the most KILO bills. (DCD)

Cost: $10.00

Producer: Metrix Corporation, P.O. Box 19101, Orlando, FL 32814

METRIC POKER

M. Hamsher

Playing Data
Copyright: 1975
Age Level: high school-adult
Number of Players: 2-5
Playing time: 1 hour
Packaging: deck of cards, metric conversion cards, and instructions in professionally designed box

Description: *Metric Poker* is designed to help students learn conversion and differentiation within the metric system. At the beginning of a hand, players are dealt five cards each and arrange their cards according to the three suits–length, mass, and volume. In turn, players draw one card from the center pile or the discard pile and set down any sets of cards that total one thousand metric units (grams, milliliters, millimeters, or cubic centimeters). A hand ends when one player can lay down all of his or her cards. Play continues until a player reaches 25 points. (DCD)

Cost: $3.75; set of 5 decks $16.00

Producer: Environmental Education Center, 800 Dixwell Avenue, New Haven, CT 06511

METRI-MAGIC

Michael Shwarger

Playing Data
Copyright: 1975
Age Level: 8-adult
Number of Players: 1-20
Playing Time: 20 minutes-1 hour
Packaging: deck of 48 cards with instruction booklet

Description: The purpose of this deck of cards is to teach, in an enjoyable format, most of the basic concepts of the metric system which we may soon join the world in adopting. The deck is arranged in six suits, each representing a common metric unit: meter, square meter, cubic meter, liter, gram, and watt. The eight ranks in each suit, listed from lowest to highest, are: milli, centi, deci, one, deka, heco, kilo, and mega. Included as well are six reference cards which table the values of each unit represented in the cards. The accompanying pamphlet describes the deck in detail and outlines approximately 30 games which can be played with it, varying from solitaire to classroom bingo, and with learning goals ranging from the introduction of metric terminology and sequence of values to working with ratio and scientific notation. (JJ)

Cost: $6.40

Producer: LaPine Scientific Co., 6001 S. Knox Avenue, Chicago, IL 60629

MULTIFACTO/PRODUCTO

Wayne H. Peterson, Seattle Public Schools

Playing Data
Copyright: 1971
Age Level: grades 4-9
Prerequisite Skills: computation with factors
Number of Players: 2-any number
Preparation Time: minimal
Playing Time: 10-30 minutes

Description: *Multifacto* is a skill-building card game that gives students practice with both multiples and factors. After hands are dealt, one card is turned face up. The first player must play a card that is a multiple of the largest 2-10 factor of the number on the face-up card. Then the player names any 2-10 factor of the card played and the next player must play a multiple of that factor. The one to first get rid of all the cards wins.

Producto has bingo-like rules which make it fun to find factors. A duplicating master provides for each student a gameboard/grid, the squares of which are filled with different shapes. Numbers 1-10 are printed down the side of the board; across the top each player writes the numbers 1-10 in random order. A deck of cards supplies the call numbers. The caller reads a number, and players mark the spaces on their gameboards that are the meeting of the factors of that number. Winner is the one to first fill in all the squares that contain the same shape. (producer)

Cost: School Price $4.80 (#02339-7 71)

Producer: Scott, Foresman and Co., 1900 East Lake, Glenview, IL 60025

NUMBER LINE

Playing Data
Age Level: elementary or high school
Number of Players: depends on activity
Versatility: game board, model, teaching device

Description: This number line is plastic coated, comes in 3-foot sections, runs from −20 to +100, and can be written on. There are many activities and games which can be played with number lines. Though they are easy to make, it's nice to have one you can write on and wipe clean. (RI)

Related Mathematics: The ideas of ordering, addition and subtraction, distance, negative numbers, and operations with negative numbers are some of the elementary topics which can be taught with number lines. For high school students studying algebra, transformations on the line are a natural use. (RI)
Cost: package of 12 $2.95 (#7802)

Producer: Ideal School Supply, 11000 Lavergne Ave., Oak Lawn, IL 60453

OH-WAH-REE

Edward Ross

Playing Data
Copyright: 1972
Age Level: grades 1-8
Prerequisite Skills: none
Number of Players: 2

Preparation Time: 15 minutes for the basic rules
Versatility: Basic use as a game (offers some use in logical thinking–predictions of outcomes)

Description: *Oh-Wah-Ree* consists of a playing board with two rows of 6 pits, each row facing a player, with small, irregularly shaped stones that are moved from one pit to another. This is a very fast-moving game, once learned. Winning is determined on a "how many captures" basis. Opponents move a group of stones from one of their pits, dropping one at a time into consecutive pits, while trying to predict where the last stone will go–hopefully capturing one or two of their opponent's stones. At the same time they must try to anticipate their opponent's moves which might lead to a loss. Basic tactics are early discoverable; more subtle questions of overall strategy are there, but not nearly so visible. The game can be played on different levels by widely varying persons. And the variations to the game enlarge even this broad base. (RI)

Related Mathematics: General skills include visualization of many changing numbers at once, predictions, setting up traps, and avoiding them yourself (feedback). No particular arithmetic skills are involved aside from counting, which makes it a good game for small children. However, as the game moves so quickly, an estimation of numbers of pieces rather than exact counts is often used. (RI)

Note: This game also goes by the names of *Mancala* and *Kalah*. In Africa, where this game originated, children play in the spaces between their fingers using pebbles from the streets. (RI)

Cost: $10.00

Producer: The Avalon Hill Company, 4517 Harford Road, Baltimore, MD 21214

ON-SETS

Layman E. Allen, Peter Kugel, and Martin F. Owens

Playing Data
Copyright: 1969
Age Level: high school-adult
Number of Players: 2-6
Playing Time: 30 minutes (est.)
Preparation Time: 1 hour
Packaging: professionaly packaged game includes cards, timer, playing surface, and symbol dice

Description: *On-Sets* teaches some of the basic ideas of set theory. The game uses four symbolic devices to accomplish this: (1) cards picturing one, two, or three red, green, or blue dots; (2) color cubes, dice with red, green, blue, or orange dots on their sides; (3) symbol cubes, dice that feature the symbols for set-union ("cup"), set-intersection ("cap"), set-differentiation ("minus"), and set-complimentation ("prime"); and (4) number cubes, dice with a number from one to five on each side. At the start of play a "universe" of six randomly selected cards is dealt. These cards represent the resources available to the players. The sets named during play are always sets of cards from that universe. Color, symbol, and number dice are rolled after the universe has been established. For example, a die with a red dot exposed represents the set of cards in the universe that picture red dots, and the same relationship exists between green dots on dice and red dots on cards. Sets may overlap. The symbols, colors, and numbers revealed become the resources available to the players. One player sets a "goal," which is expressed by one or more of the number cubes. The cubes representing "2" and "3," for instance, may be used to describe the number 5 (since 2 + 3 = 5). The players then arrange symbol and number cubes to represent the number designated as the goal. The winner is the first player to do this, or the first player to recognize that a solution can be made in one more move. (DCD)

Cost: $10.00

Producer: WFF 'N PROOF, 1490-TZ South Boulevard, Ann Arbor, MI 48104

ORBITING THE EARTH

L. L. Whitworth, Community College, Allegheny Co.; Harry B. Cohen, University of Pittsburgh

Playing Data
Copyright: 1970
Age Level: grades 1-8
Number of Players: 2-6 or teams
Playing Time: indefinite
Preparation Time: minimal

Description: After tossing numeral blocks and correctly stating the basic fact called for, students may move markers from space station to space station and win the game. There are four separate games: Addition, Subtraction, Multiplication, and Division. Each game includes a 26″ square vinyl playing field, numeral blocks, disc markers, and complete instructions. (producer)

Cost: Addition $6.39 (#02169-6 70); Subtraction $6.39 (# 2170-X 70); Multiplication $6.93 (#02199-8 69); Division $8.76 (#02157-2 69)

Producer: Scott, Foresman and Company, 1900 East Lake, Glenview, IL 60025

POLYHEDRON RUMMY

Wayne Peterson, Seattle Public Schools

Playing Data
Copyright: 1971
Age Level: grades 7-10
Prerequisite Skills: knowledge of geometric plane figures and 3-dimensional figures
Number of Players: 2-4
Playing Time: 12-30 minutes

Description: This fast-moving card game–with rules based on aspects of rummy and double solitaire–improves students' ability to visualize dimensional figures. After hands are dealt, four cards are turned up in the center. Each player in turn draws a card and plays one on any center card to begin forming a polyhedron. The player who completes the set of faces forming a polyhedron receives a score based on the number of faces. Wild cards make the game exciting; display cards showing polyhedrons to be built make it easy. (publisher)

Cost: School price $3.00 (#02113-0 71)

Producer: Scott, Foresman and Company, 1900 East Lake Ave., Glenview, IL 60025

PICTURE PATTERNS

Playing Data
Age Level: grades 3-8
Prerequisite Skills: none
Number of Players: 1 per book
Playing Time: variable
Preparation Time: 15-25 minutes
Versatility: activity, construction

Description: *Picture Patterns* is an activity book which explains how to construct interesting and creative geometrical designs using a template of different geometric shapes (called a geo-graph). It also includes instructions and colored paper. Good "project" activity for kids. (RI)

Related Mathematics: Shapes, design construction, etc. Geo-Graph templates include squares, triangles, hexagons, trapezoids, parallelograms, and extreme parallelograms. Essentially the same shapes as Pattern Blocks. (RI)

Cost: *Picture Patterns* $5.50, supplementary kit $1.95, Geo-Graph $1.25. Quantity discount.

Producer: Stokes Publishing Co., P.O. Box 415, Palo Alto, CA 94302

PSYCH-OUT

Playing Data
Age Level: age 6-adult
Prerequisite Skills: none
Number of Players: 2
Playing Time: a few minutes
Preparation Time: 1 minute
Versatility: game

Description: The game consists of a plastic board with 3 rows of flippable tabs (3 in the first row, 5 in the second and 7 in the third). The name of the game belies the fact that if players analyze it completely they can decide, on the first move, whether or not they will win. It is one of those games which is deterministic, but does not appear so. *Psych-Out* is usually called *Nim* among mathematicians and game players. The greatest pleasure of this game is that of deducing the winning strategy. Players who know how to win can wreak havoc on the self-confidence of others who do not. (RI)

Related Mathematics: The game can be approached as a manipulative puzzle by youngsters, or the starting point for an extensive, complicated binary analysis by a mathematician. Computers can easily be programmed to play *Nim*. If it is difficult to get the game or if the reader wishes to start right away, here's how:

Reach in your ashtray and take out a bunch of matches, or get some pebbles out of your driveway, or use poker chips—any bunch of little things will do. Arrange them into 3 piles with 3 matches in pile 1, 5 matches in pile 2, and 7 matches in pile 3. Players take turns by taking any amount from any pile. The player who takes the last match wins (or loses, if you want to vary the game). Actually you can start with any number of matches in any number of piles; there's always a winning strategy; each separate situation has a complete mathematical analysis, and a generalized analysis is possible. Try writing down the outcomes of simple starting situations and see if you can discover an analysis. (RI)

Cost: $2.00

Producer: Mag. Mf., Inc., P.O. Box 69, Mentor, OH 44060

QUBIC (3 DIMENSIONAL TIC-TAC-TOE)

Playing Data
Age Level: grade 1 through college and beyond
Prerequisite Skills: knowledge of regular tic-tac-toe helpful
Number of Players: 2-3
Playing Time: 5-30 minutes
Preparation Time: 2 minutes for basic game
Versatility: game, conceptual model for a 3-dimensional coordinate system; 4 of the games can be a model for a 4-dimensional cube in hyper space

Description: The game is a stack of four platform game boards with 4 x 4 squares or cells on each platform—altogether making a cube of 64 squares. Several different colored poker chips are used as markers instead of Os and Xs (noughts and crosses). For the teacher and those inclined to play and analyze the game mathematically, it is good to make strips of paper with numerals 1, 2, 3, and 4 on them to number or "give coordinates to" each row, column, and level of the game.

The game is best played by two persons, the winner being the first to get a straight line of four chips in any direction. It requires a person to think and reason spatially in three dimensions and is much more challenging than the traditional two-dimensional game. It is possible, though rare, to have a draw. With chance lacking (except perhaps in how you choose your opponent), five- and six-move winning strategies can be worked out, making chess-like analysis possible. Playing with coordinates and writing moves down allow play by mail and postgame analysis. Really ambitious players can combine four games to play in the fourth dimension—coordinates being row, column, level and game.

Students are often quite stimulated by the game, and teachers love it even more because it provides a model for problems from so many topics. Tournaments, as in chess, can be organized, as can school championships. *Qubic* is generally a two-person game, so it's better used in decentralized math classrooms or at times when students have extra time. But it's rich enough to be a topic in the regular math curriculum. *Qubic* is a must for every math lab. (RI)

Related Mathematics: *Qubic* provides a model for problems from a variety of mathematical topis, including: geometry, coordinate geometry, graphing, algebra, topology, heuristics, number theory, and logic. It provides, for many, the much-needed experience of visualizing and thinking spatially. Heuristically speaking, it provides the opportunity to discover a general formula for the coordinates of any straight line combination of four chips (formula for all wins!). Topologically speaking, it is a model for Euler's formula, and four games together satisfy his formula for a four-dimensional cube. Four-dimensional space and four-dimensional coordinates can be explored this way—even with high school students. For students and teachers interested in computers, the game can be programmed so you can test yourself against a machine. Higher dimension games can be explored and general theorems deduced; the formula for winning coordinates can be extended to any dimension game. (RI)

Cost: $4.75 approximate retail

Producer: Parker Brothers, 190 Bridge St., Salem, MA 01970

THE REAL NUMBERS GAME

Layman E. Allen

Playing Data
Copyright: 1966
Age Level: elementary-adult
Number of Players: 2 or more
Packaging: professionally produced and packaged game includes pen and five symbol dice

Description: The object of this game is to detect and write down more numbers (of a specified kind) than other players detect and write down. The game is played with five symbol dice which are rolled to determine the digits and mathematical symbols that can be used to construct numbers. The numbers may be natural or negative integers, non-integer rational fractions, or irrational numbers. For example, the dice might reveal the symbols 7, 3, 9 +, and –. The players might then decide to form as many natural numbers as possible from combining these symbols. The player with the longest list at the end of a predetermined period of time wins. (DCD)

Cost: $2.25

Producer: WFF 'N PROOF, 1490-TZ South Boulevard, Ann Arbor, MI 48104

RECKON

Playing Data
Copyright: 1976
Age Level: kindergarten-grade 12
Prerequisite Skills: knowledge of numbers
Number of Players: variable

Playing Time: 15 minutes per exercise
Packaging: deck of cards and instructions

Description: *Reckon* consists of a deck of numbered cards and instructions for playing over a dozen mathematical learning games involving addition, subtraction, multiplication, and division, as well as factoring, identifying prime numbers, and other nonelementary skills covered in first-year algebra. Games involve such activities as playing a mathematical variation of bingo, competing to form equations equaling a number identified on a playing card, and capturing from other players cards that are factors of another card. The instructions suggest many variations for each game as well as a number of additional drills involving the cards. (TM)

Cost: $2.00

Producer: Creative Publications, Inc., P.O. Box 10328, Palo Alto, CA 94303

SOMA CUBE

Edward Ross

Playing Data
Age Level: upper elementary and up
Prerequisite Skills: none, except knowledge of what a cube is, although this puzzle might give someone just that
Number of Players: 1
Playing Time: literally any amount of time one might wish to put into it
Preparation Time: minimal; for use in free play, none at all
Versatility: puzzle, conceptual model (for surface and volume)

Description/Dynamics: *Soma Cube* is aesthetically a pleasure to handle and fool around with. It is a set of 7 pieces that can be put together to form a cube; hence, loss of a piece is critical. The puzzle can be easily made by gluing small wooden cubes together in the same shape of the puzzle pieces. This might be advisable, since the commercial plastic pieces have an annoying habit of coming apart. The puzzle is absolutely fascinating, though. Aside from the cube itself, many intriguing things can be built—some of which are illustrated in the accompanying booklet. The advantage is in the range of versatility with which one person can play alone or a group of people can make up their own activities. It is a 3-D analog to the tangram puzzle, always a guaranteed interest provoker. (RI)

Related Mathematics: Many questions concerning measuring of surface areas and volumes of built shapes can quite naturally be asked. So, for making these topics enjoyable and understandable, this puzzle is highly recommended. Activities of one person duplicating another person's shape increase perceptual abilities, while just the act of describing how you constructed a shape you built will get you thinking of the relation of words to things. (RI)

Cost: $3.50 approximate retail

Producer: Parker Brothers, 190 Bridge St., Salem, MA 01970

TAC-TICKLE

Harry D. Ruderman

Playing Data
Copyright: 1967
Age Level: age 6-adult
Number of Players: 2
Playing Time: 15 minutes-1 hour
Packaging: professionally packaged game

Description: The basic moves of this pure strategy game, according to its author, "strongly suggest 'doing' and 'undoing' which correspond to an operation and its inverse," and "the notion of commutativity, a property of many binary operations." Moves in a variation of the game correspond "to taking reflections in a line, a very important type of mapping in geometry."

In the basic game, two players symmetrically arrange four red and four blue tokens in eight of the twenty holes cut in a foam playing mat. The players take turns moving either red or blue playing pieces until one of them manages to align (vertically, horizontally, or diagonally) three tokens of the same color to win. (DCD)

Cost: $1.75

Producer: WFF 'N PROOF, 1490-TZ South Boulevard, Ann Arbor, MI 48104

TRI-NIM

Bruce L. Hicks, University of Illinois; Harvey C. Hicks, Carnegie-Mellon University

Playing Data
Copyright: 1968
Age Level: grades 7-12
Prerequisite Skills: beginner's game: reading, grade 3; math, grade 2
Number of Players: 2 or 3
Playing Time: 10-20 minutes

Description: *Tri-Nim* contains 45 counters: 36 red, 3 blue, 3 green, and 3 yellow. The red are playing counters and the others are marking counters. The game can be played with any number of red counters, but 10 to 20 is recommended for all but the most expert players. These counters are placed on a triangular board sectioned off into small triangular spaces, and marked with numerals of successively higher value. The basic game is played by two players who move counters from the center of zero spaces on the *Tri-Nim* board to other spaces at higher levels and finally off the board into a goal. Points are scored when a player "wins" a corner, but only after moving within certain restrictions, one of which requires that the counters must be moved upward numerically in one or more steps between adjoining spaces. (producer)

Cost: $5.75

Producer: WFF 'N PROOF, 1490-TZ South Boulevard, Ann Arbor, MI 48104

TRI SCORE

Playing Data
Age Level: elementary, junior high
Prerequisite Skills: knowledge of basic operations with whole numbers
Number of Players: 2-4
Playing Time: 30 minutes
Preparation Time: 15 minutes
Versatility: game

Description: This is a collection of novel games played with tiles on a board which give students practice in basic operations and combinations of operations. The novelty of the games can keep older students who have difficulty with their basic math interested. For example, "they can be doing remedial work without feeling like it." (RI)

Related Mathematics: The games provide practice and reinforce the knowledge of basic operations of addition, subtraction, multiplication, and division. They also provide practice in the combinations of the operations. New fraction number tiles could be made by the teacher and injected into the game after students get the hang of playing. (RI)

Cost: $3.95

Producer: Creative Teaching Associates, P.O. Box 293, Fresno, CA 93708

TUF

Joan Brett and Peter Brett

Playing Data
Age Level: grade 3 or 4 through college
Prerequisite Skills: know what +, –, x, –, = mean
Number of Players: 1-4
Playing Time: 1-8 minutes per round; a full game can run 30-60 minutes
Preparation Time: 5-10 minutes
Versatility: game, model

Description: The game of *Tuf* comes with an instruction booklet, number cubes, fraction cubes, blank cubes, operator cubes, special symbol cubes, and = cubes. A whole course in algebra can be taught using this game. It is one of the best equations-algebra games available. It can be used competitively or noncompetitively, as a game for fun or as a rich source of math problems in a progressive study of arithmetic and algebra. In the game itself players roll their cubes and then try to use as many as they can to form a valid equation. Three timers are provided (3 minutes, 2 minutes, and 1 minute) so that the player who first forms an equation yells, "Tuf," and wins after three minutes if no one has yelled "Tuffer" while making a longer equation. The "Tuffer" player wins after two minutes unless someone yells, "Tuffest," giving everyone one minute after that. Scoring is 1 point for each cube used (2 points for each fraction cube) in a player's equation at the end of each round. Everyone has the right to inspect each other's equations for invalidating mistakes (penalty could then be a negative score). Sometimes it is very easy to make equations and at other times it is difficult. The roll of the cubes determines what a player has to work with; skill and ingenuity determine the rest. Many times students don't like to play competitively as much as just rolling the cubes and seeing what they can make with no pressure. This kind of practice is necessary for being good (quick) at the game. The game is excellent for the lab approach to teaching mathematics and decentralized and small group approaches. (RI)

Related Mathematics: The genius of this game is the simple, straightforward, evolutionary instruction book by Joan and Peter Brett which takes a player from game 1 (which anyone can master in a few minutes) through more difficult games, step by step, adding only one or two new rules or concepts at a time. There's not the frustration of having to "spend hours reading complicated instructions," which, unfortunately, seems to be a drawback found in many of the Wff 'N Proof games. The impatient, low-attention-span student can easily read and understand the rules and be playing, in action, in minutes. As soon as a game becomes "boring" or its possibilities have been exhausted, players add a new cube or read a new rule, and it's a whole new game.

The simplest game involves just whole numbers, =, and the operations: +, –, x, and ÷ (without the use of parentheses), and the content of the games range through order of operations, distributive law with parentheses, fractions, decimals, exponents, roots, different number bases, ratios, percentage, logarithms, imaginary numbers, tangent function, pi, radians, factorial, equations in one unknown or two unknowns, and simultaneous equations. The booklet gives numerous examples with each new game showing how each new concept and rule works. It really is a "math course in a game." All math classrooms and math labs should have several sets. (RI)

Cost: $10.00

Producer: Avalon Hill Co., 4517 Harford Rd., Baltimore, MD 21214

TWIXT

Playing Data
Age Level: upper elementary and up
Prerequisite Skills: none
Number of Players: 2
Playing Time: 30 minutes (varies, of course)
Preparation Time: 15 minutes reading, 2 minutes telling
Versatility: game, source of logical problems, like chess problems

Description: In this game players place pegs and links on a large, square peg-board and try to form connected, continuous chains from one side to the other; the first to do so is the winner. As in chess, winning depends entirely upon the thinking of the player and the mistakes of the opponent. Books could be written on openings, strategies, traps, etc., as in chess and Go, although there are none we know of. One feature is that a successful defense against an opponent's strategy has the property of being, at the same time, an offense that the opponent must in turn try to counter. It's exciting and absorbing to play, simple to learn, yet it can get quite complicated. (RI)

Related Mathematics: Logical thinking, planning ahead, and visualization are some of the mental skills developed by the game. Concepts of graphing and shapes are inherent. It involves many of the thinking processes of chess. The game can be played by coordinates. Graph paper variations are possible. (RI)

Cost: $10.00

Producer: Avalon Hill, 4517 Harford Road, Baltimore, MD 21214

WFF: THE BEGINNER'S GAME OF MODERN LOGIC

Layman E. Allen

Playing Data
Copyright: 1963, 1970
Age Level: elementary-junior high school
Number of Players: 2
Playing Time: 10 minutes-1 hour
Preparation Time: 10 minutes
Packaging: professionally packaged and produced game includes 12 symbol dice, instruction booklet, and playing surface cards

Description: WFF (an acronym for well-formed formula, pronounced "woof") is a pair of logic games about WFFs that are also included in the twenty-one-game collection *WFF 'N PROOF*. In the first game, designed to provide practice in recognizing well-formed formulae, two players must construct a WFF from a random assortment of letters and operations symbols printed on tossed dice. The second game is an advanced version of the first, in which fewer dice are tossed so that the resources available for constructing a WFF are, logically, diminished. (DCD)

Cost: $2.25

Producer: WFF 'N PROOF, 1490-TZ South Boulevard, Ann Arbor, MI 48104

WFF 'N PROOF: THE GAME OF MODERN LOGIC

Layman E. Allen

Playing Data
Copyright: 1962
Age Level: elementary-adult
Number of Players: 2-6
Playing Time: 10 minutes-3 hours
Preparation Time: 1 hour
Packaging: professionally produced and packaged game includes 168-page instruction manual

Description: The current (1969) version of this well-known logic game (originally developed to help law students master logic) consists of a progressively difficult series of twenty-one variations. These variations are arranged in the instructions so each incorporates the ideas of all the preceding games and introduces one new idea.

WFF 'N PROOF is a game in the sense that Norbert Weiner described language as "a joint game between speaker and listener against the forces of confusion." In each variation the players must make a predetermined number of random elements (dice marked with symbols) articulate together in a meaningful way, that is, to form logically correct, mathematical statements. In all variations, the first player to exhaust his or her stockpile of resources (dice) constructing valid WFFs or PROOFs wins. (DCD)

Cost: $13.00

Producer: WFF 'N PROOF, 1490-TZ South Boulevard, Ann Arbor, MI 48104

See also
Home Economy Kit PRACTICAL ECONOMICS

MILITARY HISTORY

Ancient History

Troy
Ancient Conquest
Chariot
Spartan
Trireme
Thermopylae
The Conquerors
Alexander the Great
Raphia
The Punic Wars
Hannibal
Legion
Pharsalus
The Siege of Jerusalem
Decline and Fall

Medieval European History

Warlord
Viking
Siege
Kingmaker
Siege of Constantinople

Renaissance to 18th Century

Conquistador
Yeoman
Musket and Pike
Mercenary
A Mighty Fortress
Thirty Years War:
- Freiburg
- Lützen
- Nordlingen
- Rocroi

Breitenfeld
Cromwell
English Civil War
En Garde!
Frederick the Great
Torgau

The Age of Sail

Frigate
Wooden Ships and Iron Men

The American Revolution

American Revolution
1776

Napoleon

Grenadier
Micro-Napoleonics
Napoleon at War:
- Marengo
- Jena-Auerstadt
- Wagram
- The Nations: Leipzig

La Grande Armée
Austerlitz
Borodino
La Bataille de la Moskowa
Waterloo
Napoleon at Waterloo
1815: The Waterloo Campaign
Wellington's Victory
Napoleon's Last Battles:
- Wavre
- La Belle Alliance
- Quatre Bras
- Ligny

The 19th Century

Veracruz
Crimea

Four Battles from the Crimean War:
- Alma
- Balaklava
- Inkerman
- Tchernaya

Königgrätz
The Battle of Roark's Drift [sic]

The American Civil War

War Between the States
The American Civil War
Blue and Gray:
- Chickamauga
- Shiloh
- Antietam
- Cemetery Hill [Gettysburg]

Blue and Gray II:
- Fredericksburg
- Hooker and Lee [Chancellorsville]
- Chattanooga
- Battle of the Wilderness

Manassas
Stonewall
Road to Richmond
Lee Moves North
Gettysburg
Terrible Swift Sword [Gettysburg]
Battle of Chickamauga

The American Indian Wars

Custer's Last Stand
7th Cavalry

The 20th Century

Patrol!
Dreadnought
The Russo-Japanese War:
- Port Arthur
- Tsushima

Red Sun Rising
Mukden 1905
Viva!
Raid!
Up Scope!

World War I

First World War
World War I
Soldiers
Fight in the Skies
Richthofen's War
Tannenburg [sic]
Tannenberg
Verdun

The 1920s and 1930s

Russian Civil War

World War II

Global War
Rise and Decline of the Third Reich
War in Europe
World War II
Squad Leader
Sniper!
Tank!
Case White
Winter War
Battle for Germany

World War II in Western Europe and the Atlantic

War at Sea
Victory at Sea
Rhein Bung
Narvik
Eagle Day
Battle of Britain
Spitfire
Air Force
Seelowe
Submarine
Wolfpack
U-Boat
MTB
Luftwaffe
D-Day
1944
Overlord
Cobra
Caen 1944
Panzer Leader
Atlantic Wall
Panzer '44
Westwall:
- Hürtgen Forest
- Bastogne
- Remagen
- Arnhem

Highway to the Reich
The Ardennes Offensive
Battle of the Bulge
Wacht am Rhein

World War II in Southern Europe and North Africa

Descent on Crete
PanzerArmee Afrika
Afrika Korps
Four Battles in North Africa:
- Cauldron
- Crusader
- Kasserine
- Supercharge [El Alamein]

Sidi Rezegh 1941
Battles for Tobruk
Fall of Tobruk
Tobruk
Desert War
Kasserine Pass
Avalanche
Cassino 1944

World War II in Russia

Barbarossa
The Russian Campaign
East Front
War in Europe: War in the East
Drang Nach Osten! and Umentschieden
PanzerBlitz
Panzergruppe Guderian
Crimea 1941

The Moscow Campaign
Kharkov
Drive on Stalingrad
Turning Point: Stalingrad
Kursk

World War II in Asia and the Pacific

Victory in the Pacific
USN
"CA"
Fast Carriers
Flying Tigers II
Coral Sea
Battle for Midway
Midway
Solomons Campaign
Island War:
Bloody Ridge [Guadalcanal]
Saipan
Leyte
Okinawa
Burma
Operation Olympic

Since World War II

Arms Race
Air War
Foxbat and Phantom
Korea
Yalu
Citadel
Fast Carriers
Bay of Pigs
Jerusalem!
Sinai
Bar-Lev
Arab-Israeli Wars
October War
Modern Battles:
Chinese Farm
Golan
Modern Battles II:
The Battle for Jerusalem

Imaginary Contemporary and Future Wars

Strikeforce One
Firefight
World War III
Oil War
SSN
Missile Crisis
Fortress Rhodesia
South Africa
Fulda Gap
NATO
Revolt in the East
Sixth Fleet
Red Star/White Star
Mech War '77
Modern Battles:
Wurzburg
Mukden
Modern Battles II:
Yugoslavia
Bundeswehr
DMZ
The Next War
Invasion: America
Objective: Moscow
Minuteman
After the Holocaust

Semi-Historical and Imaginary World Games

Mission Aloft
Diplomacy
Tactics II
Blitzkrieg
Imperialism
Ballistic Missile

AFRIKA KORPS

Thomas N. Shaw

Playing Data
Copyright: 1964
Age Level: grade 9 and up
Number of Players: 2
Playing Time: 2-4 hours
Preparation Time: 1 hour

Description: Each move in this, the original North African game, represents one-half month. The game goes from April 1941 to November 1942. Counters represent the fighting units, supply units and Rommel, who causes units with him to move faster. Axis supplies arrive with the throw of a favorable number on the die. (MC)

Comment: *Afrika Korps* is an easy game, but it has little claim to historical accuracy. It does reflect the uncertainty of Axis supply but probably too drastically. (MC)

Cost: $12.00

Producer: Avalon Hill, 4517 Harford Road, Baltimore, MD 21214

AFTER THE HOLOCAUST

Redmond A. Simonsen and Irad B. Hardy

Playing Data
Copyright: 1977
Age Level: grade 12 and up
Number of Players: 1-4
Playing Time: 10 hours
Preparation Time: 3 hours

Description: This game posits the destruction of the United States in a nuclear war followed by a recovery starting from four different centers in the old U.S. As the game begins, the four centers are starting to move out to attach new areas and they are therefore in danger of running afoul of each other and getting into a new war. The game, however, is mainly about the economic management of these four developing economies, and if the players opt for war, they will discover that it is economically very expensive. It is also possible, even necessary because of the unequal distribution of resources, to trade with the other players. The economies are divided into five sectors: farm, metal, fuel, industrial, and transportation. The players assign population and mechanization to these. They can build more mechanization, more plant sites, consumer points, or military units, while population increases 10% every four turns (years). One wins the game by raising his area's standard of living ("social state"). Optional rules allow for the creation of a bank and for research and development. (MC)

Comment: This game is a serious attempt at a simulation of an economy, and as such it is very complex. I have proved, with a small group of volunteers from our class, "The Future as History," that it can be played by ordinary college students, but it takes a long time to get the game started. It is a fascinating study of the relationship between economic power and military might in developing countries. The countries in this game do not have outside big powers coming in to arm everybody to the teeth, so they are poorer militarily than developing countries in the world today. (MC)

Cost: $12.00

Producer: Simulations Publications, Inc., 44 East 23rd Street, New York, NY 10010

AIR FORCE

S. Craig Taylor, Jr.

Playing Data
Copyright: 1976
Age Level: grade 11 and up
Number of Players: 2 or more
Playing Time: 2 or more hours
Preparation Time: 2 hours

Description: The players fly specific British, American, and German aircraft in various realistic missions from the European Theater of World War II. The differences among the aircraft are numerous and subtle. Performance depends on speed and altitude. Planes can perform a variety of maneuvers and have a variety of different weapons. Movement is simultaneous with all moves written down before enactment. If a plane ends its move pointed at the enemy at the right altitude, it may fire. Damage is taken in various sections of the aircraft. Optional rules allow the use of surface terrain which gets in the way of low altitude combat. They also allow bombing and strafing of a number of kinds of surface targets, and spotting problems involving visibility, sun, and clouds. A few scenarios are included, but players are encouraged, by the presence of many extra counters if nothing else, to make their own. The game system is intricate and hard to learn, but it plays very easily once the players get started. (MC)

Cost: $11.00

Producer: Battleline Publications, Inc., P.O. Box 1064, Douglasville, GA 30134

AIR WAR

David C. Isby, Greg Costikyan, and Redmond A. Simonsen

Playing Data
Copyright: 1977
Age Level: high school and up
Number of Players: 2 or more
Playing Time: 1-10 hours
Preparation Time: 2 hours

Description: Air War is a plane-to-plane simulation of past, present, and future air combat since about 1950. Plane counters vary from the MIG 15 to the B1. There is even a counter for a flying dragon with deadly breath. This is an extremely complex game. The forty-eight page rules booklet goes into great detail of modern air combat. Such rules include maneuvers, missile combat with both radar and heat seeking missiles, radar, radar countering measures, ejecting, and pilot experience (pilots vary from turkeys to superhonchos). Included with the game are various charts showing throttle and energy points, acceleration, altitude, turning degrees, flight indicators, and weapons. The game is played on eight map sections. Of the 600 counters only 140 are air units; the rest include damage, missile chaff, radar, ground units. There are many scenarios ranging from simple to very complex. The rules booklet also explains tactics, the reasoning for rules and aircraft maneuvers. (Richard J. Rydzel)

Comment: This game is far too complex for the usual gamer and cannot be used in a normal classroom situation. It is a very thorough and interesting game but due to its complexity it will be played only by air war enthusiasts. (RJR)

Cost: $11.00, $12.00 with box

Producer: Simulations Publications, Inc., 44 East 23rd Street, New York, NY 10010

ALEXANDER THE GREAT

Gary E. Gygax and Donald J. Greenwood

Playing Data
Copyright: 1971, 1974
Age Level: high school-college
Number of Players: 2
Playing Time: 3 hours
Preparation Time: 1 hour
Materials: mounted mapboard (with morale track), rules booklet, combat results chart card, die-cut counters, die

Description: The game is a simulation of the decisive Battle of Arbela which was fought in 33 B.C. between Alexander, the Macedonian invader, and Darius, the last emperor of the first Persian Empire. The mapboard shows the level plain on which the battle was fought. The counters mainly represent various sizes of units with both hand-to-hand and missile arms. Some counters represent the army and wing commanders. The counters can be set up historically or according to the players' own plans. The most distinctive feature of this game is the keeping for each side of a changing morale index, which controls the results of combat and determines the winner and loser. There are four general morale states and four combat results tables to match. Both sides start at level 1. The side that reaches level 4 has lost almost all ability to damage the enemy. (MC)

Comment: The game is a fair representation of an ancient battle and has a larger number than usual of intriguing and innovative design features. This edition of the game requires, or nearly so, that the purchaser write unit information on the backs of the counters. I understand that a third edition, in which the counters will be printed on both sides, is being prepared. (MC)

Cost: $12.00

Producer: Avalon Hill, 4517 Harford Road, Baltimore, MD 21214

THE AMERICAN CIVIL WAR

James F. Dunnigan

Playing Data
Copyright: 1974
Age Level: grade 9 and up
Number of Players: 2
Playing Time: 4-8 hours
Preparation Time: 2 hours

Description: The game covers the whole war on a strategic level. The map shows the main area of the conflict. Each turn represents three months. The counters represent abstract points worth of armies and fleets, or fortresses, or leaders. Both sides draw reinforcements each turn. A command control rule interferes with the movement of units each turn and is harder on the Union than on the Confederacy. The leadership rule simulates the search for competent leadership on both sides. The Union wins by dividing the Confederacy into nonviable sections. The Confederacy wins by surviving to the end. (MC)

Comment: The game is an excellent simulation of many of the strategic problems of both sides in the war. The game shows the importance of railroads, rivers, and Union seapower. It would be particularly good for team play since there are several fronts to look after. See the introductory essay to this section for more on this game. (MC)

Cost: $10.00

Producer: Simulations Publications, Inc., 44 East 23rd Street, New York, NY 10010

AMERICAN REVOLUTION

James F. Dunnigan

Playing Data
Copyright: 1972
Age Level: grade 9 and up
Number of Players: 2
Playing Time: 4-6 hours
Preparation Time: 1 hour

Description: The game map shows the thirteen states and Canada divided into geographical areas and sub areas. Each move represents three months. The object of the British player is to control a variable number of points of territory by driving out American forces and leaving a British garrison. The counters represent British regulars, Tories, Continentals, militia, French, forts, and fleets. The American player gets varying reinforcements depending on the luck of the die and the number of uncontrolled areas. British movement is restricted by the die so that the British player cannot always do what he plans to do. The French arrival depends on the Americans winning a major victory. In addition to the historical game, there are twelve "what-if" possibilities—events that might have happened and might have changed the course of the war. (MC)

Comment: The game as designed is a fair representation of the strategic problems of the war. On the tactical level, the game is quite abstract. I used the game in a small class with slightly changed rules. I eliminated the ordinary restrictions on British movement and divided the British command among six students, one representing the King's government in England and the others generals in America. Then I prevented them from communicating with each other except through writing notes which I delivered, and I told each of them only about the enemy situation on their immediate area. The Americans operated under fewer restrictions. The result was a good representation of the confusion of the British command during the war. One plaintive note from England to one of the generals ended with the question, "By the way, where are you?" (MC)

Cost: $11.00, $12.00 with box

Producer: Simulations Publications, Inc., 44 East 23rd Street, New York, NY 10010

ANCIENT CONQUEST

Playing Data
Copyright: 1975
Age Level: high school-college
Number of Players: 4
Playing Time: 4-5 hours
Preparation Time: 1 hour
Material: mapboard, order of appearance combat results card, four score cards, die-cut counters, message tablet pad, die

Description: The game show the movements, rises and falls of peoples and empires in the ancient Near East. Each of the four players controls (or attempts to) the destinies of several peoples who enter the board in various places at various times during the fifteen turns of the game (each turn representing about sixty years from 1500 B.C. to 600 B.C.). Each player can achieve up to forty points by performing various actions with his peoples. The first player, for example, controls the Egyptian Empire, the Kingdom of Media, the Kingdom of Urartu, and the Aramean peoples. They each get points by occupying cities and "smiting" specific numbers of their foes. The Medes will get two points if they "smite 12 factors or more of Cimmerians." Rules for movement and combat are ordinary. Communication among the players is limited to exchanging one written message with one other player per turn. (MC)

Comment: This game is somewhat overpriced but it is on a unique subject. From a game point of view, the objectives of each people in the game seem to be rather arbitrary but they keep the action directed along roughly historical channels. The first playing of the game is more than ordinarily confusing for a beginning player because of the lack of any logical connections between his different peoples. (MC)

Cost: $9.95

Producer: Excalibre Games, Inc., P.O. Box 29171A, Brooklyn Center, MN 55429

THE ARAB-ISRAELI WARS

Seth Carus, Russell Vane, Richard Hamblen, and Randall C. Reed

Playing Data
Copyright: 1977
Age Level: grade 9 and up
Number of Players: 2
Playing Time: 2-5 hours
Preparation Time: 1 hour

Description: This is a tactical game on armored warfare as practiced in the Middle East from 1956 to 1973. Like other tactical games, it includes a large number of unit counters and uses various groupings of these in a number of scenarios (twenty-four in this case) on a mapboard showing typical terrain. Arab-Israeli Wars includes four small boards, one of which contains a section of the Suez Canal for the 1973 scenarios, all of which can be put together in various ways. The units, platoons for the most part, include many kinds of tanks, infantry, engineer, antitank gun, antitank missile, trucks, armored personnel carriers, helicopters, aircraft, and artillery. The scale is 250 meters per hex. Some of the conditions reflected in the rules are the effect of indirect artillery fire, the use of fortifications, building bridges, the influence of the varied morale in the wars, and the effect of terrain. (MC)

Comment: The tactical system used in Arab-Israeli Wars is an improvement from the point of view of realism over the system used in its ancestors, *PanzerBlitz* and *Panzer Leader*. The system is still somewhat abstract, but it gives a convincing representation of reality. In a classroom game, a chain-of-command system can be used. The commander-in-chief on each side should be provided with a separate map and not allowed to look at the actual playing field. Moves on each side should be timed, and players on one side should not be allowed to view the board while the other side is moving. (MC)

Cost: $12.00

Producer: Avalon Hill, 4517 Harford Road, Baltimore, MD 21214

THE ARDENNES OFFENSIVE

James F. Dunnigan

Playing Data
Copyright: 1973
Age Level: grade 9 and up
Number of Players: 2
Playing Time: 10-14 hours
Preparation Time: 1 1/2 hours

Description: *Ardennes Offensive* uses the same map as *Bastogne* but with somewhat simplified details. In all other respects, *Ardennes Offen-*

sive is a simplified version of *Bastogne.* There are no separate artillery units or supply units, for example, and units may not change size. There are still options available in the reinforcement rates for both sides. (MC)

Comment: The game still shows, like its predecessor, that traffic was a headache for the Germans. It shows further how impossible the aims of the historical offensive were. (MC)

Cost: $9.00, $10.00 with box

Producer: Simulations Publications, Inc., 44 East 23rd Street, New York, NY 10010

ARMS RACE

Playing Data
Copyright: unknown
Age Level: grade 8 and up
Number of Players: 2-3
Playing Time: 6-12 hours
Preparation Time: 1 hour

Description: The combat system for *Arms Race* is extremely simplified since the emphasis is on building arms and the problem is to build the right thing at the right time. Players may choose between conventional or nuclear forces, and between land, sea, or air forces. They can also choose to invest in political subversion, guerrilla forces, spying, transportation systems, or economic aid to contested countries. The game may be played for a single decade or for the whole period 1950 to 2001. There are four moves per year. The three players are the U.S., the U.S.S.R., and China. A hypothetical four-person game includes a Nazi Germany that has somehow survived World War II. (MC)

Comment: *Arms Race* is a very good idea but somewhat flawed in the execution. The rules are sometimes more suggestive than explanatory, the counters are poorly printed and impossible to punch out, and the map is hard to read and aggressively ugly. If you want to use it (and I probably will sometime) be prepared to make up some rules as you go along. Also, it will be better to plan on making your own map. (Mc)

Cost: $7.98

Producer: Attack International Wargaming Association, 314 Edgely Avenue, Glenside, PA 19038

ATLANTIC WALL

Joseph M. Balkoski and Steven Ross

Playing Data
Copyright: 1978
Age Level: high school and up
Number of Players: 2
Playing Time: 4-100 hours
Preparation Time: 1 hour

Description: *Atlantic Wall* is a simulation of the invasion of Normandy, June 6, 1944, and the ensuing battles in the bocage up to July 1, 1944. Each hex represents one kilometer. The 2000 playing pieces represent companies, battalions, air wings, and individual ships. Each turn represents 4 1/2 hours of daylight and 10 1/2 hours for night turns. The five-section map covers the Normandy area from Cherbourg to Coutances and Caen. The rules include special armor benefits on combat, strategic movement, naval and air support, battalion integrity, replacement, step reduction, dispersal and demoralization, and limited supply capacity for allied divisions. Several short scenarios as well as a 104-turn campaign game can be played. (Richard J. Rydzel)

Comment: This game is a very accurate simulation of the Normandy campaign. It is easily played for its seemingly complex handling of the tactics of that campaign. Of special interest is the ability of battalions to break down into companies. The short invasion scenarios are exciting but are usually just a matter of die rolling for a few hours. The campaign is a long and tedious but fun exercise on how many different ways the German player can be beaten. (RJR)

Cost: $26.00, $28.00 with box

Producer: Simulations Publications, Inc., 44 East 23rd Street, New York, NY 10010

AUSTERLITZ

John M. Young

Playing Data
Copyright: 1973
Age Level: grade 9 and up
Number of Players: 2
Playing Time: 2-3 hours
Preparation Time: 30 minutes

Description: The mapboard shows the area of the historic battle of 1805. The scale is 400 meters to the hexagon. Each of the thirteen turns represents one hour. Counters show the historic units by brigades and divisions. There are cavalry, infantry, and artillery units. Artillery units can fight at a range of two hexagons. All other units must be adjacent to the enemy in order to fight. (MC)

Comment: With any kind of competent players, the French will always win. The question is, by how much? This is a simple game and can be used to introduce people and classes to the idea of gaming. The mechanics are not very realistic but the action tends to follow the original battle. (MC)

Cost: $11.00, $12.00 with box

Producer: Simulations Publications, Inc., 44 East 23rd, Street, New York, NY 10010

AVALANCHE

Frank Chadwick, Rich Banner, Doug Poe, and John Harshman

Playing Data
Copyright: 1976
Age Level: grade 11 and up
Number of Players: 2
Playing Time: 14 hours
Preparation Time: 1 1/2 hours
Materials: include two 22" x 27 1/2" mapsheets

Description: This game deals with the nine-day battle of the Salerno beachhead at three turns per day on a scale of 1300 yards per hexagon. The map shows the terrain around Salerno in six different elevations separated by contour lines. The game has a strong tactical flavor. The units are companies and platoons, and there are many kinds, including engineers, heavy weapons companies, artillery units, and different kinds of armored units. the game system gives the players the feel of tactical problems, like the desirability of coordinating artillery with the other arms, and the necessity of armor-infantry cooperation. Naval gunfire and tactical air attacks play an important part. The Allied objective is to establish a viable beachhead without excessive casualties, while the German objective is to eliminate the beachhead or cause excessive casualties. (MC)

Comment: Because this is a large game on a small subject, I will probably not use it, but it is good in its way. The rules are inexact at times and would have to be supplemented extemporaneously from time to time. (MC)

Cost: $12.75

Producer: Game Designers' Workshop, 203 North Street, Normal, IL 61761

BALLISTIC MISSILE

Playing Data
Copyright: 1975
Age Level: grade 11 and up
Number of Players: 2
Playing Time: 1-2 hours
Preparation Time: 30 minutes
Materials: include two cards of cut-out playing pieces

Description: The plot of the game is that one country in the 1970s is attempting to make a preemptive nuclear strike against another. One side deploys three cities and assorted defenses on a table or desk top while the attacker launches bombers, ICBMs, etc., from a small area of the playing surface. The rules are intricate but easily understood. (MC)

Cost: $2.50

Producer: Tabletop Games, 92 Acton Road, Arnold, Nottingham, United Kingdom

BARBAROSSA: THE RUSSO-GERMAN WAR, 1941-1945

James F. Dunnigan

Playing Data
Copyright: 1971
Age Level: grade 9 and up
Number of Players: 2
Playing Time: 3-14 hours
Preparation Time: 1-1 1/2 hours

Description: The map is a small scale map of European Russia and its western neighbors. Each game turn represents one month. The units represent armies and corps (which keeps the number of units manageable), and supply units. Several scenarios are available based on the action of different years: 1941 (eight moves), 1942 (eight moves), 1943 (nine moves), and 1944 (six moves); or the players can use the Campaign Game of forty-five moves which goes from June 1941 to February 1945. (MC)

Comment: Barbarossa is an excellent simulation of the strategic problems of the war in Russia. The movement system is fairly complex. The game shows that the power of the German offensive lay in mobility, and that Russia's ability to survive depended on mobilizing large numbers and throwing them in to barely halt the Germans. It shows that Russian supply was weak, even in victory, and that Russian command and organization improved in the course of the war. See the introductory essay to this section for more on this game. (MC)

Cost: $9.00, $10.00 with box

Producer: Simulations Publications, Inc., 44 East 23rd Street, New York, NY 10010

BAR-LEV

John Hill, Frank Chadwick, Doug Poe, and Simon Ellberger

Playing Data
Copyright: 2nd edition, 1977
Age Level: grade 11 and up
Number of Players: 2-4
Playing Time: 6-24 hours
Preparation Time: 1-2 hours

Description: *Bar-Lev* is a simulation of the Arab-Israeli War of 1973, played on two separate mapboards, one showing the Egyptian front around the Suez Canal, the other showing the Syrian front. Leaving out the inactive area in between allows the active area to be shown in larger scale, six kilometers per hexagon for the Egyptian front and three kilometers for the Syrian front (which means that movement and ranges are doubled in terms of hexes on the Syrian front). The game may be played with different sets of rules and boards. Players may leave one front out of the game, or, even more drastically, leave the air rules out. The game lasts up to twenty-eight turns, each representing one day. The game features ranged artillery, amphibious units, Arab commandos, rebuilding destroyed units, loss of morale, and full air war rules. (MC)

Comment: This is a thorough reworking of the first edition. The rules are in pretty good shape now, though still not perfect. The game is good on its subject and features larger than normal unit counters and map hexagons, which are boons for classroom use. A teacher using this game might consider coupling it with a U.N. role-playing simulation on the political crisis which accompanied the war. (MC)

Cost: $10.00

Producer: Game Designers' Workshop, 203 North Street, Normal, IL 61761

BATTLE FOR GERMANY

James F. Dunnigan

Playing Data
Copyright: 1975
Age Level: high school-college
Number of Players: 2-4
Playing Time: 2-3 hours
Preparation Time: 40 minutes
Materials: mapboard (with chart and track), rules folder, randomizer chits (die substitute)

Description: The game deals with the last six months of the land war in Europe, December 1944 to May 1945. The map shows Germany and parts of surrounding countries. The main competition in the game is not between the Germans and the Allies but between the Russians and the Western Allies. In the two-player game the Western player also controls those Germans opposing the Russian advance while the Russian player controls those Germans opposing the Western advance. Odd, but it works. Victory goes to the player who takes the most points in cities. The three-player game gives the Germans a unified command, while the four-player game divides the Germans again and allows for two winners. An imaginary two-player scenario allows for Russians and Americans to fight it out after the German collapse. The rules for fighting are fairly standard. (MC)

Cost: $4.00

Producer: Simulations Publications, Inc., 44 East 23rd Street, New York, NY 10010

BATTLE FOR MIDWAY

Marc William Miller

Playing Data
Copyright: 1976
Age Level: grade 9 up
Number of Players: 2 or more
Playing Time: 8-10 hours
Preparation Time: 1 hour

Description: When the two map boards of this game are placed on the table, Tokyo is in the upper left-hand corner of the complete map and French Frigate Shoals is in the lower right-hand corner, with Midway in the middle of the right-hand map. All but a few hexagons are colored blue for Pacific Ocean. The map is on the scale of thirty-three nautical miles to the hexagon and the game moves at four turns per day to the total of forty-eight turns. The counters represent individual carriers, battleships, cruisers, destroyers, submarines, abstract land forces, and groups of two to six of aircraft. Moving cloud markers designate squalls and overcast areas. Movement is secret, but a task force must give clues to its own whereabouts to search for the enemy. The players must find the enemy before they can attack and ready aircraft before they can fly. Combat is fairly abstract. (MC)

Comment: This is a good strategic simulation, and it is not very detailed tactically. I believe it could be made a little more accurate by lessening the ability of antiaircraft guns to shoot aircraft down. In a classroom game, a fully secret movement and search system under the direction of the teacher/referee could be used, as suggested in the multiplayer rules. Since all the hexagons are numbered, and since the game already uses written moves, multiple sets of the game are not actually required as the rules indicate to implement secrecy rules. Commanders of separate task forces on the same side should not be allowed to talk to each other unless they break radio silence and reveal their locations to everyone, friend and foe. (MC)

Cost: $10.00

Producer: Game Designers' Workshop, 203 North Street, Normal, IL 61761

THE BATTLE OF BRITAIN

Louis B. Zocchi

Playing Data
Copyright: 1968
Age Level: grade 9 and up
Number of Players: 2
Playing Time: 6-16 hours
Preparation Time: 2 hours

Description: The mapboard, which is overlaid by a staggered square grid, shows England, the English Channel, the North Sea, and the Channel coast of France. English cities and English and French airfields are located on the map. Each airfield has a number indicating the number of aircraft it can service. The English "radar line," or place where the radar can pick up attacking aircraft, is also shown. The original game contains a beginner's game (of no value) a basic game and an advanced game. The object of the German player is to defeat the British by bombing cities containing aircraft factories. One week's bombing, by the required number of bombers, destroys the factories and British production suffers. Each turn represents ten minutes and each sequence of thirty turns represents one week of combat. The whole game lasts five weeks. The revision adds a primary game, which lasts only one sequence of thirty turns. The revision also adds "The Ultimate Game," which adds another week and several new rules. (MC)

Comment: *The Battle of Britain* contains much information about the battle, but its total impression of the battle is erroneous. Particularly, on the mapboard and in the order of battle, it forces the German player to select targets that never occurred to the German command, allows the British player to abandon the defense of the area where the battle was actually fought, and allows the German player to sustain casualities that are totally unrealistic. (MC)

Cost: $10.00

Producer: Lou Zocchi, President, Games Science, 7604 Newton Drive, Biloxi, MS 39532

BATTLE OF CHICKAMAUGA

David Tomasi

Playing Data
Copyright: 1974
Age Level: grade 9 and up
Number of Players: 2
Playing Time: 2-3 hours
Preparation Time: 30 minutes

Description: This game is a simple representation of the 1863 Civil War battle with brigade-sized units. It covers two days with twenty turns, each representing an hour and a half of daylight or a half a night. The map is drawn on a scale of one-fourth mile to the hexagon. (MC)

Comment: The game has a plain map, plain counters, and plainly written rules. It would make a good beginning game. (MC)

Cost: $5.00

Producer: Flying Buffalo, Inc., P.O. Box 1467, Scottsdale, AZ 85252

THE BATTLE OF ROARK'S DRIFT

Herbert O. Barents and Thomas B. Webster

Playing Data
Copyright: 1978
Age Level: grade 9 and up
Number of Players: 2
Playing Time: 3-4 hours
Preparation Time: 1 hour

Description: *The Battle of Rorke's Drift* (the game title is misspelled) was the action of the Zulu War on which the movie, "Zulu," was based. The British are represented by individual men, with the heroes in the film given appropriate combat values. The Zulus are represented by combat groups. The post is shown on a large but unspecified scale which makes the hospital eight hexagons long. (NC)

Comment: The appeal of the game to anyone who has seen and enjoyed the movie is overwhelming. If the British follow correct tactics the game will tend to turn out as the battle did, but there is plenty of room for error on the British port and only a few errors might prove fatal. The game is good as a demonstration of the value of firepower over mere valor and of the reasons why vast empires in the nineteenth century could be controlled by a few European soldiers. (MC)

Cost: $9.95

Producer: Historical Alternatives Game Co., 1142 South 96th, Zeeland, MI 49464

BATTLE OF THE BULGE

Lawrence Pinsky

Playing Data
Copyright: 1965
Age Level: grade 9 and up
Number of Players: 2
Playing Time: 4-6 hours
Preparation Time: 1 hour

Description: The game map shows the area of the battle. Each turn represents one half day. The counters represent regiments, brigades, and combat commands. The Germans have immense power and always exceed the historical results. The only way the allies can win is not by confronting them head on but by attacking the flanks of the Bulge in hopes of cutting off supply to the powerful units in front. (MC)

Comment: The balance of forces has been totally distorted in this game with the effect of making the German attack look like a sane military operation with reasonable objectives. (MC)

Cost: $12.00

Producer: Avalon Hill, 4517 Harford Road, Baltimore, MD 21214

BATTLES FOR TOBRUK

George Munson, Dan Hoffbauer, and Bill Comito

Playing Data
Copyright: 1975

Age Level: college
Number of Players: 2
Playing Time: 5 hours
Preparation Time: 1 hour

Description: This game includes four scenarios lasting ten turns (2 1/2 days) each, and is played on a map of the Tobruk/Egyptian border area on a scale of three miles to the hexagon. The units are battalions and regiments. (MC)

Comment: A convoluted movement/combat system makes this game more difficult to play than its level of simulation warrants. It does have several good rules and ideas that could be used in other games. Its system could also be simplified to use the map, counters, and situations. (MC)

Cost: $5.00

Producer: Balboa Game Co., Box 989, Bellflower, CA 90706

BAY OF PIGS

Jim Bumpas, Seth Fine, Barry Eynon, and Steve Goodman

Playing Data
Copyright: 1976
Age Level: grade 11 and up
Number of Players: 2
Playing Time: 4 hours
Preparation Time: 1 hour

Description: The game pits the forlorn Cuban exile invasion of Cuba against the massive Castro response on a scale of one turn for six hours and 1000 meters of terrain per hexagon. The task of the exiles is to seize as much of the board as they can and then to hang on. Rules include sea transport, amphibious landing, airborne landing, and air combat. Forces include artillery, bombers, and tanks. Supplies for the invaders are represented by separate markers. (MC)

Comment: The rules are rather sketchy, so, if you are interested, be prepared to make extemporaneous additions as needed. (MC)

Cost: $5.00

Producer: Jim Bumpas, 948 Lorraine Avenue, Los Altos, CA 94022

BLITZKRIEG

Lawrence Pinsky

Playing Data
Copyright: 1965
Age Level: grade 9 and up
Number of Players: 2
Playing Time: 6-10 hours
Preparation Time: 2 hours

Description: The board shows a hypothetical part of a continent, where there are two large powers separated by several neutral countries. The war begins immediately and the two powers overrun the neutrals and meet in the middle. The counters represent land and air forces. Sea forces are not directly represented but land forces and air forces may operate amphibiously. The rules and forces are roughly based on the armies of World War II. (MC)

Comment: *Blitzkrieg* was not designed as a historical game, although it was inspired by history. Its result is more like that of World War I than World War II because the two opponents are equal in power and weapons. So it is a lengthy game of attrition which is quite likely to be given up before it is concluded. (MC)

Cost: $12.00

Producer: Avalon Hill, 4517 Harford Road, Baltimore, MD 21214

BLUE AND GRAY: CHICKAMAUGA, SHILOH, ANTIETAM, CEMETERY HILL

Irad B. Hardy, III, Christopher G. Allen, Thomas Walczyk, and Edward Curran

Playing Data
Copyright: 1975
Age Level: high school-college
Number of Players: 2
Playing Time: 2-3 hours
Preparation Time: 30 minutes
Materials: four mapboards, two general rules leaflets, four specific rules leaflets, four sets of die-cut counters, die

Description: In this set of four games, most of the rules are the same for all games. The maps, the orders of battle, victory conditions and a few rules are peculiar to each game. The rules are simple. The scale is 400 meters to the hexagon with counters representing formations of 500 to 2000 men. The time scale varies from four moves per day to ten moves per day. Units represent infantry and cavalry which function identically or artillery which is different and vitally important to the play of the game. A unique attack effectiveness rule plays havoc with the offensive by rendering units that have been thrown back once in an attack ineffective for further attacks the same day. (MC)

Comment: The games are simple. The combat system does not pretend to any detailed accuracy but does tend to produce convincing general results. The attack effectiveness rule is a worthwhile innovation but perhaps too drastic. See the introductory essay to this section for more on *Chickamauga.* (MC)

Cost: $11.00 ($3.00 apiece separately), $12.00 with box

Producer: Simulations Publications, Inc., 44 East 23rd Street, New York, NY 10010

BLUE AND GRAY II: (FREDERICKSBURG, HOOKER AND LEE [CHANCELLORSVILLE], CHATTANOOGA, BATTLE OF THE WILDERNESS)

Joe Angiollilo, Richard Berg, Frederick Georgian, and Linda Mosca

Playing Data
Copyright: 1975
Age Level: high school-college
Number of Players: 2
Playing Time: 2-3 hours
Preparation Time: 30 minutes
Materials: four mapboards, two general rule leaflets, four specific rules leaflets, four sets die-cut counters, die

Description: This does the same thing as the first *Blue and Gray* for four more Civil War Battles. (MC)

Cost: $11.00 ($3.00 apiece separately), $12.00 with box

Producer: Simulations Publications, Inc., 44 East 23rd Street, New York, NY 10010

BORODINO

John M. Young

Playing Data
Copyright: 1972
Age Level: grade 9 and up
Number of Players: 2
Playing Time: 3-6 hours
Preparation Time: 1 hour

Description: The map shows the historic battlefield on a scale of 400 meters to the hexagon. The units are the historic cavalry and infantry divisions and artillery. Each day turn represents one hour. Night turns represent about five hours each. Players can play eight turn, fourteen turn, or forty-two turn games. The games system is the same as for Austerlitz above. (MC)

Comment: This is a simple game but it can be a lengthy one. The mechanics are not very realistic but the game looks very authentic and the problems of the game are analogous to the problems of the actual battle. (MC)

Cost: $9.00, $10.00 with box

Producer: Simulations Publications, Inc., 44 East 23rd Street, New York, NY 10010

BREITENFELD

J. A. Nelson, and Brad E. Hessel

Playing Data
Copyright: 1976
Age Level: grade 9 and up
Number of Players: 2
Playing Time: 2-3 hours
Preparation Time: 30 minutes

Description: The game uses the same system as developed for the Thirty Years' War set of games (see below). This game lasts seventeen turns. It allows the Imperialists to make short work of the Saxons as was done historically. The importance of leadership in history is represented by the importance of leader counters in the game. (MC)

Comment: Since the mapboard of this game contains a nonexistent marshy stream, anyone who plays the game ought to simply ignore that terrain feature. (MC)

Cost: $4.00

Producer: Simulations Publications, Inc., 44 East 23rd Street, New York, NY 10010

BURMA

Bob Fowler and Marc Miller

Playing Data
Copyright: 1976
Age Level: grade 9 and up
Number of Players: 2
Playing Time: 6-8 hours
Preparation Time: 1 hour

Description: The game is a strategic simulation of the Burmese campaign from the period just after the original Japanese conquest (December 1942) until the end of the war, by which time the allies historically had conquered it back again. Each of the twenty-six turns represents one month in good weather or two months in the monsoon season, when nothing much can happen. The infantry units are regiments and brigades. Other counters represent supplies and air transport capability. As is appropriate to the subject, the game stresses supply problems and the use of air transport power and eventually amphibious assaults by the Allies. The Allies can build roads and will usually try to complete the Burma Road as a means of winning. The Allies also may create long-range penetration groups or parachute units by training infantry units off the board for four turns. (MC)

Comment: This game has both sketchy rules and overly intricate rules, but it is a valuable simulation of a very unusual campaign. I understand that it has been used successfully in some college classes. (MC)

Cost: $8.00

Producer: Game Designers' Workshop, 203 North Street, Normal, IL 61761

"CA"

James F. Dunnigan

Playing Data
Copyright: 1973
Age Level: grade 9 and up
Number of Players: 2
Playing Time: 1-4 hours
Preparation Time: 45 minutes

Description: The game map shows Guadalcanal, Savo Island, and the "slot" in three different positions for different scenarios or the land shown on the map may be ignored for other scenarios. *"CA"* is a simplified tactical game. Each move represents six minutes of a battle. Each hexagon represents an area about 930 meters across. Each counter represents a single ship, battleship, cruiser, or destroyer. There are ten scenarios based on real and hypothetical battles from the Pacific war. All battles assume the absence of airplanes, which were absent from the real battles depicted because they were fought at night. (MC)

Comment: Some of the aspects of this game will undoubtedly appear to be oversimplified to any player. Nevertheless the result is remarkably faithful to the action of a naval engagement. (MC)

Cost: $11.00, $12.00 with box

Producer: Simulations Publications, Inc., 44 East 23rd Street, New York, NY 10010

CAEN 1944

R. J. Hlavnicka, Dennis P. O'Leary, and Michael A. Trdan

Playing Data
Copyright: 1977
Age Level: grade 9 and up
Number of Players: 2
Playing Time: 4 hours
Preparation Time: 1/2 hour

Description: Caen deals with Montgomery's half of the offensive in which the Allied armies broke out of the Normandy beachhead. Time and distance scales are not given. Unit markers represent infantry, armored divisions, and assorted kinds of artillery units. The game lasts fifteen turns. To win, the British player must take Caen and reach Falaise to close the pocket that I presume exists off the map by then. (MC)

Comment: This is a fairly short game but not a simple one. The rules are in poor shape. The artillery rules are quite foggy. I believe that players are supposed to get from the rules that German artillery and Allied naval fire is used defensively just as artillery is used defensively in Simulations Publications' *Westwall* and similar series, but players who have not played the other games would have a hard time figuring out what is meant, if indeed that is meant. (MC)

Cost: $6.00

Producer: Excalibre Games, Inc., Box 29171, Brooklyn Center, MN 55429

CASE WHITE

Frank Alan Chadwick, Paul Richard Banner, and Marc W. Miller

Playing Data
Copyright: 1977
Age Level: grade 9 and up
Number of Players: 2
Playing Time: 3-6 hours
Preparation Time: 1 hour

Description: Case White was the German invasion of Poland in 1939.

The game is part of a larger series, not yet completed, which will show all of World War II in Europe at the same scale. The turns in this game, which is complete in itself, represent three days each. The map scale is sixteen miles to the hexagon. The unit scale is basically one division to the counter but with many other nondivisional units represented separately. The air battle is represented on about the same level of abstraction as the land battle. Soviet intervention is a strong possibility. An abstract Western attack on the Siegfried Line may occur (with die rolls) if the Germans do not destroy Polish resistance quickly enough. (MC)

Comment: This is a well thought out game, perhaps a little too complex for use in any but a fairly skilled group. However, the fact that the map was printed as part of a longer series has made the game very awkward to assemble or play in a small area. The three map sheets come together on the Polish battlefield and leave lots of unused territory to get in the way of the players of this game. (MC)

Cost: $11.00

Producer: Game Designers' Workshop, 203 North Street, Normal, IL 61761

CASSINO 1944

R. J. Hlavnicka, Dennis P. O'Leary, and Michael A. Trdan

Playing Data
Copyright: 1977
Age Level: grade 9 and up
Number of Players: 2
Playing Time: 3-4 hours
Preparation Time: 30 minutes

Description: This game deals with the assault on the German Gustav Line, with its center at Cassino, which coincided with the Anzio landing, which is handled abstractly in the game. The game emphasizes the Allied problem of German fortifications and rough terrain, compensated for somewhat by massive artillery and air support. Unit counters represent armored and infantry divisions. Time and distance scales are unknown. (MC)

Comment: Like Caen (see above), this game can be played more easily if you are familiar with the rules of similar games. (MC)

Cost: $6.00

Producer: Excalibre Games, Inc., Box 29171, Brooklyn Center, MN 55429

CHARIOT

Playing Data
Copyright: 1975
Age Level: high school/college
Number of Players: 2
Playing Time: 2-4 hours
Preparation Time: 1 hour
Materials: mapboard (with charts and tracks), standard rules booklet, special rules sheet, die-cut counters, die

Description: This is a revision of *Armageddon.* The rules have been changed little, except that a morale rule has been added by which an army routs when losses have reached a certain point, defined by the scenario. The scale has changed and the range of bows has been reduced severely. An optional rule now provides for simultaneous movement. (MC)

Cost: $9.00, $10.00 with box

Producer: Simulations Publications, Inc., 44 East 23rd Street, New York, NY 10010

CITADEL

Frank Alan Chadwick and Doug Poe

Playing Data
Copyright: 1977
Age Level: grade 11 and up
Number of Players: 2
Playing Time: 1-50 hours
Preparation Time: 1-2 hours

Description: Citadel is a battalion- and company-level simulation of the siege of Dien Bien Phu in which the Vietminh destroyed French hopes of maintaining themselves in Vietnam. The two-sheet map is on a scale of 200 yeards to the hexagon, and each turn represents one of the fifty-seven days of the fighting. Scenarios are provided for one, four, five, seven, or fifty-five day scenarios. The players use mortars, howitzers, antiaircraft guns, and the French player has tanks and aircraft, both transport and attack. The map shows French wire-enclosed strongpoints, and the Vietminh can build trench and tunnel complexes. Advanced rules covering supply and airpower are necessary for complete realism. (MC)

Comment: This is an intricate and long game, but it is also a fascinating and convincing study of a crucial battle in contemporary history. (MC)

Cost: $10.00

Producer: Game Designers' Workshop, 203 North Street, Normal, IL 61761

COBRA

B. E. Hessel and Dave Werden

Playing Data
Copyright: 1977
Age Level: high school and up
Number of Players: 2
Playing Time: 7 hours
Preparation Time: 1 hour

Description: *Cobra* is a simulation of the allied breakout from Normandy during World War II. Counters represent divisions, regiments, and brigades. Each hexagon represents an area 3.2 kilometers wide, and each turn represents three days. The map covers the Normandy province of France from St. Lo to Alençon (the area Patton's troops took in the actual campaign). Set-up is facilitated by having the initial placement hex printed on each counter. Reinforcement time is also printed on each counter. Much of the map is bocage country, which helps the defense and slows any advance after combat. Headquarters units help in attack and defense for the Germans and in attack only for the allies. (Patton's is the only headquarters unit represented.) Monty apparently does not rate in this game. The allies are limited in attacks due to supply problems, and the Germans are limited on movement due to Allied air interference. Poor weather (determined by die roll) could ruin any allied long-range plans. Once per game the allies may carpet bomb a hexagon, automatically making a six to one attack on any German unit. (Richard J. Rydzel)

Comment: Once the fine points of this game are understood, it makes for exciting and realistic play. Due to some new concepts such as truck counters to mechanize some infantry units and H.Q. units helping attacks, some time must be spent understanding each rule. The slightest mistake on either side could mean defeat. The problems of the bocage and the city of Caen to the Allies are readily apparent when they are attacked. This is a very good simulation of World War II on the western front. (RJR)

Cost: $9.00, $10.00 with box

Producer: Simulations Publications, Inc., 44 East 23rd Street, New York, NY 10010

THE CONQUERORS: THE MACEDONIANS AND THE ROMANS

Richard H. Berg, Redmond A. Simonson, and Frank Davis

Playing Data
Copyright: 1977
Age Level: grade 9 and up
Number of Players: 2-3
Playing Time: 4-20 hours
Preparation Time: 1 hour

Description: The Conquerors is a set of two related games; one on Alexander's conquest of the western Persian empire and the other on two of the Roman wars in Greece and Asia Minor—the second Macedonian War of 200-197 B.C. and the Syrian War of 192-189 B.C. Both games use approximately the same system, which is a combination of strategic and tactical simulation. The armies and fleets maneuver on a strategic map with a scale of twenty miles to the hexagon. When opposing armies arrive on the same hexagon, they may fight it out on an abstract tactical board. Sieges, naval battles, and skirmishes are settled more simply by a single die roll. Larger battles may also be settled by dice rolling, but this destroys much of the attraction of the game. Both the strategic and tactical games emphasize the value and necessity of leadership by using leader counters and making them necessary to any movement or attack. The Romans game uses economic/production and diplomatic as well as military rules. Winners of a siege can profit from sacking the city. Other funds come from a taxation die roll at the end of the year. Funds are used to raise armies and fleets, to supply and maintain them, and to buy allies or suborn those of the enemy. The three-player Syrian War with the Romans, Seleucids (Syrians), and Macedonians as the major players is probably the most interesting game in the set. (MC)

Comment: This is one of the best wargames available, although I have a few doubts about some of the details, especially about the tactical game. The important thing about the tactical simulation is that it is a detailed game in itself on an appropriate scale. For classroom use, *The Conquerors,* especially the Roman versions, gives many chances for conversion to multiplayer versions. For more information, see the introductory essay to this section. (MC)

Cost: $14.00, $15.00 with box

Producer: Simulations Publications, Inc., 44 East 23rd Street, New York, NY 10010

CONQUISTADOR

Richard Berg and Greg Costykian

Playing Data
Copyright: 1976
Age Level: grade 11 and up
Nmber of Players: 2-5
Playing Time: 10-14 hours
Preparation Time: 2 hours

Description: The map is a small-scale map of most of North America and all of South America divided into areas such as "The Aztec Empire" and "the Caribbean." The players represent Spain, England, France, Portugal, or the German bankers. Each of the twenty-one turns represents five-year periods between 1495 and 1600. The colonizing players manipulate counters representing soldiers, colonists, ships, and gold on the map and attempt to settle areas, discover gold, fight Indians (represented as an abstract "native label"), and survive. Separate counters represent the historical individual Spanish conquistadors, English privateers, and explorers of all nations. The optional German banker wins by making money from the efforts of the colonizing nations. (MC)

Comment: The game is often more complex than I would ideally want it to be, and it is guilty of hurrying French and English colonization along to make a more competitive game, but it is fascinating and the subject is big enough to justify the expense in time and effort. See the introductory essay to this section for more on this game. (MC)

Cost: $8.00, $9.00 with box

Producer: Simulations Publications, Inc., 44 East 23rd Street, New York, NY 10010

CORAL SEA

Marc William Miller

Playing Data
Copyright: 1976
Age Level: grade 9 up
Number of Players: 2
Playing Time: 6-8 hours
Preparation Time: 1 hour

Description: The mapboard shows eastern New Guinea, a corner of Australia, the Solomon Islands, a bit of New Caledonia and the surrounding seas. The game system is the same as that of Battle for Midway (see above), but there are no battleships, no submarines, and no overcasts. (MC)

Comment: As in Battle of Midway, the antiaircraft ought to be toned down and a fully secret movement and search system used. (MC)

Cost: $8.00

Producer: Game Designers' Workshop, 203 North Street, Normal, IL 61761

CRIMEA

Frank Alan Chadwick

Playing Data
Copyright: 1975
Age Level: grade 11 and up
Number of Players: 2
Playing Time: 2-30 hours
Preparation Time: 1-2 hours

Description: Crimea is a strategic game about the Crimean War of 1854, in which battles are handled in much more detail than is normal with strategic games. To allow for greater detail, the board, instead of showing the area of the campaign as a whole, shows five areas in detail, and the five maps are connected during the game by off-board movement. The largest units are divisions which can be broken down into smaller units. Each strategic move represents one month, but it may contain a large number of battle moves when the occasion arises. Rules provide for building fortifications, using siege guns, morale problems, supply, naval warfare, and varied approaches to the strategic problems. Several battle scenarios allow the players to set up and fight historical battles instead of making their own in the course of a campaign. (MC)

Comment: This is an intricate game. Several of the rules, especially those having to do with battles, are poorly thought out or poorly presented. The way I played it, it seemed faithful to the tactical and strategic conditions of the war. (MC)

Cost: $8.75

Producer: Game Designers' Workshop, 203 North Street, Normal, IL 61761

CRIMEA 1941

R. J. Hlavnicka, Dennis P. O'Leary, and Michael A. Trdan

Playing Data
Copyright: 1977

Age Level: grade 8 and up
Number of Players: 2
Playing Time: 3 hours
Preparation Time: 1 hour

Description: This is a short simple treatment of the German conquest of the Crimea in 1941-1942. The units are infantry, cavalry, mountain, and panzer divisions, Russian armored brigades, and miscellaneous artillery units. The armor is stronger but does not function differently than infantry. Mountain divisions and engineer units have special powers. Russian units can make amphibious assaults. Artillery units are different, having a range of two hexagons. German units can take losses and still exist while Russian units go all at once—a provision that reflects the greater brittleness of the Russian army in this period. Supply and air power are reflected abstractly. (MC)

Comment: *Crimea* 1941 is a good game for take-home assignments since it has few complexities to confuse beginning players and is at least suggestive of the actual campaign. It is overpriced by about $1.00 compared with similar games. (MC)

Cost: $6.00

Producer: Excalibre Games, Inc., Box 29171, Brooklyn Center, MN 55429

CROMWELL

Leonard Kanterman, Doug Bonforte, and Dana F. Lombardy

Playing Data
Copyright: 1976
Age Level: grade 9 and up
Number of Players: 2 players or teams
Playing Time: 2-10 hours
Preparation Time: 1 hour

Description: *Cromwell* is a grand strategy game on the English Civil War. Each of the forty-six moves represents one month of the campaigning season or two or three winter months. The map shows England and Wales and a little bit of Scotland divided into areas corresponding to the counties and major cities. The game can be shortened by confining it to one eight- or ten-turn year. The rules can be followed at various levels of difficulty. Players manipulate musketeers, pikemen, cavalry, militia, ships, and leaders. In advanced versions, players pick out, within certain limits, the armies they will lead. Some battles are resolved on a separate battleboard. (MC)

Comment: The game seems to be faithful to the strategic problems of the war. The producer, Simulations Design Corporation, has gone out of business, but the game is still available through retail outlets for under $10.00. (MC)

CUSTER'S LAST STAND

Richard Zalud and R. Patrick Mirk

Playing Data
Copyright: 1976
Age Level: grade 9 and up
Number of Players: 2
Playing Time: 4-6 hours
Preparation Time: 30 minutes

Description: The game covers what might have happened at Little Big Horn, unless the army player insists on being as rash as Custer, in which case the game may see history repeated. The main combat units are Indian war parties and Army companies. There are also leaders, Indian camps, army artillery, pack trains, and entrenchments. The time scale is indeterminate. The map shows the location of the Indian camps and a large area for maneuver. The Custer player tries to spot the Indian village so he can send for reinforcements and then he tries to last until the reinforcements arrive. The Indians are hampered by rules which require them to leave their camps gradually. Victory is decided by the accumulation of points for various activities including eliminating enemy units. (MC)

Comment: The game is faithful, not to what happened, but to what should have happened. A more historical problem could be created by requiring Custer to be at least moderately aggressive. (MC)

Cost: $9.00

Producer: Battleline Publications, Inc., P.O. Box 1064, Douglasville, GA 30134

D-DAY

Charles Roberts

Playing Data
Copyright: 1965
Age Level: grade 9 and up
Number of Players: 2
Playing Time: 2-6 hours
Preparation Time: 1

Description: The mapboard shows all of France, the low Countries and the Western edge of Germany in a more abstract fashion than usual. There are fifty turns, each representing one week. The counters show armored, infantry and parachute divisions. The board shows several invasion areas with data on the number of units that can invade there and be supplied from there. The German player sets up his defense and the Allied player then selects an invasion area. He must breach the Rhine line by the end of the game to win. (MC)

Comment: The game shows the hazards of an amphibious invasion and the necessity of attacking against weakness. It also shows the importance of supply, although the supply rule is too arbitrary and hard to work. Combat and movement are not very realistically handled and the balance of forces on both sides is erroneous. The game contains an abstract air power rule which is particularly fallacious. (MC)

Cost: $12.00

Producer: Avalon Hill, 4517 Harford Road, Baltimore, MD 21214

DECLINE AND FALL

Terrence P. Donnelly

Playing Data
Copyright: 1972
Age Level: grade 9 and up
Number of Players: 4
Playing Time: 6-8 hours
Preparation Time: 30 minutes

Description: The board shows most of Europe and the Mediterranean lands of the Roman Empire. The players represent the Romans, the Huns, the Goths, and the Vandals. The Goths and Vandals actually represent numerous other German tribes. There are fifteen turns, each representing a period of five years. "Conflict," the designer tells us, "is not just combat. It represents all aspects of competition between peoples and armies: economic, military diplomatic, etc." However, it is played just like combat in other games. The game is won on points. The Romans get points for saving cities in the Empire. The Goths and Vandals get points for looting Roman cities and for the number of spaces they manage to settle on by the end of the game. The nasty Huns get points for destroying the units of the others and for looting cities. (MC)

Comment: I have used this game in several classes and it has always proven fascinating and suggestive. I have used it with individuals or with teams representing the various sides. The game gives a good idea of the chaos of the period represented (375 A.D. - 450 A.D.) The multiplayer aspect of the game makes each player come out quite different from the others. The game shows the division of the Empire, and something of the effect of leadership. See the introductory essay to this section for more information on this game. (MC)

Cost: £4.75 (about $9.50 U.S.)

Producer: Philmar, Ltd., 47-53 Dace Road, London E3 2NG, United Kingdon. Inquire about U.S. distributors.

DESCENT ON CRETE

Eric Goldberg

Playing Data
Copyright: 1978
Age Level: grade 10 and up
Number of Players: 2
Playing Time: 5-90 hours
Preparation Time: 1-2 hours

Description: This game simulates the German landing on Crete and the battle for the island in World War II. It is a very detailed game. The hexes on the two map sheets represent 640 meters. Units are batteries and companies. The full game is very long. Shorter games last only a few turns, each of which represents a period of two to four hours. (MC)

Comment: This is a very interesting intricate game but hopelessly big and complex for most classroom situations. (MC)

Cost: $14.00, $15.00 with box

Producer: Simulations Publications, Inc., 44 East 23rd Street, New York, NY 10010

DESERT WAR

James F. Dunnigan

Playing Data
Copyright: 1973
Age Level: grade 9 and up
Number of Players: 2
Playing Time: 3-6 hours
Preparation Time: 1 hour

Description: The map shows a piece of land with very open desert terrain on a scale of one hundred meters to the hexagon. Each move represents three minutes and forty seconds. The counters represent the infantry, guns, and, most important, the tanks of the Italians, Germans and British in the North African desert from 1941 to 1945. Movement and firing is simultaneous according to written orders, but the orders are canceled for some units by the terms of a so-called panic rule. (MC)

Comment: *PanzerBlitz, Combat Command, Kampfpanzer* and *Desert War* constitute a series of games (arranged there in the order of their design) in which Mr. Dunnigan tries to depict small scale action, especially armored action, in World War II. All of them have proven popular in classes. *PanzerBlitz* has been around the longest and is the most popular among hobbyists but it is the most complex and has several quirks in its rules which make it unsatisfactory. Furthermore, its board is too crowded with rough terrain to allow for much maneuvering. *Combat Command* was an attempt to improve the system. Mr. Dunnigan did eliminate some of the quirks in the rules and loosened up the board but he changed the scale in an unhappy direction, and made *Combat Command* more abstract that its parent. Also, *Combat Command* comes with a set or scenarios which are nearly all predetermined. *Kampfpanzer* and *Desert War* have the same rules; only the setting and the available units make then different from each other. They have the advantage of a workable simultaneous movement system, and a change of scale to bring the player closer to the action. However, the panic rule is frustrating and has the effect often of fixing the results of the scenarios. In *Desert War* it is nearly impossible for any British force to beat a German force, while the Italians are just as easily defeated by the British. Also both games show that infantry is worthless and helpless against tanks and that antitank guns are easily spotted and destroyed. I do not believe that these are correct lessons. So far the last two are the best for class use but they ought to be used cautiously. (MC)

Cost: $9.00, $10.00 with box

Producer: Simulations Publications Inc., 44 East 23rd Street, New York, NY 10010

DIEPPE

Stephen M. Newberg

Playing Data
Copyright: 1977
Number of Players: 2

Description: A classic "raid in force." Units are company, platoon, and section sized. Turns represent one hour and each hex is 500 meters across. Six scenarios. (Publisher)

Cost: $9.75

Producer: Simulations Canada, P.O. Box 221, Elmsdale, Nova Scotia, Canada, BON 1MO

DIPLOMACY

Alan B. Calhamer

Playing Data
Copyright: 1971
Age Level: grade 9 and up
Number of Players: 5 or 7
Playing Time: 8-16 hours
Preparation Time: 1-2 hours

Description: The map shows Europe in about 1913. It is divided into land and sea spaces for movement. Thirty-four of the land spaces are also "supply centers," which means they can support an army or fleet. There are seven players, each representing one of the major powers: England, France, Germany, Italy, Austria-Hungary, Russia, and Turkey. Players make agreements and may help or attack one another. They do not have to abide by any agreements and movement is simulataneous according to secretly written orders, so violations of agreements may come as a shock to the victims. Players start out equal and new armies and fleets come from the conquest of new supply centers. There are a number of neutral supply centers that can be easily taken but after the third move or so, countries cannot grow except at each other's expense. Victory consists of capturing eighteen of the thirty-four supply centers, or, for a shorter game, of growing more than the others. No power can win without betraying some trust or other. (MC)

Comment: Diplomacy is unhistorical and unrealistic in several ways: the power structure, with all countries equal, is wrong for any historical time; the game requires all countries to conquer or die; even the process of conquest is not very well simulated. However, the game does simulate some elements in international and European power politics. It shows, for example, that small countries cannot exercise much influence; it shows that it might be hard to demonstrate to the leader of another country the rationality of a course of action which is perfectly obvious to you. It shows something about the geography of power: why England has a historic interest in Belgium and why England is automatically the enemy of a continental power that builds a large fleet; why Russia might be interested in the Dardanelles. It shows something of the psychology of foreign relations: the necessity and the precariousness of trust; the power of emotional factors, like a desire for revenge. This leads to the last comment. Diplomacy embodies the paradox of the amoral diplomat: a man as head of state or diplomat does things that he would never do as a private person. So it is in the game. Lying and betrayal are essential to it, and the best liars and most ruthless betrayers are people who would never lie or betray in real life. See the introductory essay to this section for more on this game. (MC)

Cost: $15.00

Producer: Avalon Hill, 4517 Harford Road, Baltimore, MD 21214

DRANG NACH OSTEN! AND UNENTSCHIEDEN

Paul R. Banner and Frank Chadwick

Playing Data
Copyright: 1973
Age Level: grade 9 and up
Number of Players: 2
Playing Time: 100-200 hours
Preparation Time: 2-3 hours
Materials: DNO: two map sheets, unit counters, rules folder, five order of battle sheets, seven charts, errata sheets, die. Unent: 4 mapsheets, counters, 13 order of battle sheets and charts

Description: *D.N.O.* alone shows the war between Russia and Germany up to March, 1942. *Unentschieden* takes it from there to the end and also makes some important changes in the first game.

The nine mapsheets form a large map of the area of the war. The counters represent divisions and lower land units, air units (with specified types of airplanes) and a few naval units. There are no shorter scenarios included. A large number of special rules deal with special events and specific operations. There is a counter representing one man who was a deadly Russian tank destroyer with his Stuka. *D.N.O.* can be used by itself. *Unentschieden* cannot be played without *D.N.O.* (MC)

Comment: This set shows a lot about the war but at a terrific cost in time and trouble. (MC)

Cost: $14.75 for *D.N.O.*, $13.85 for *Unentschieden*

Producer: Game Designers Workshop, P.O. Box 582, Bloomington, IL 61701

DREADNOUGHT

John Michael Young and Irad B. Hardy

Playing Data
Copyright: 1975
Age Level: high school-college
Number of Players: 2
Playing Time: 2 hours plus
Preparation Time: 1 hour
Materials: six-part mapboard, two tables, cards, rules booklet, die-cut counters, one SiMove pad, die

Description: The game deals with campaigns and battles involving fleets of battleships, with accompanying cruisers and destroyers, from 1906 and 1945. The counters include separate counters for each battleship built during that period. For ships that were refit, during the period, two or three counters are included. The game includes five historical and two hypothetical scenarios. Then, because the history of dreadnoughts is actually rather dull, the designers have given us a hypothetical campaign game and extended campaign game, which allow the players to cause their own naval battles by building fleets and attempting to perform missions with them. The battle rules are not very difficult. They include simultaneous movement, torpedo attacks by the cruisers and destroyers, and (optionally) smoke and radar rules. (MC)

Comment: The game does not provide for multiple players but there is a lot of room to introduce more than one on a side. (MC)

Cost: $8.00, $9.00 with box

Producer: Simulations Publications, Inc., 44 East 23rd Street, New York, NY 10010

DRIVE ON STALINGRAD

B. E. Hessel and Greg Costikyan

Playing Data
Copyright: 1977
Age Level: grade 9 and up
Number of Players: 2
Playing Time: 13 hours
Preparation Time: 1 hour

Description: *Drive on Stalingrad* is a simulation of the 1942 German offensive in southern Russia. The two-section map shows the Kursk, Stalingrad, and Caucasus Mountains area. Unit counters are divisions, but German mechanized units are regiments. Russian units are usually untried and their strength is not known until they get into combat. Supply for Russian and Axis allied units must be channeled through headquarters units. German units may be air supplied temporarily. The Axis player may also establish a supply line with truck counters, but each time supply is drawn from a truck unit the supply system is used up and must start over again. German movement is severely restricted by the Hitler directive table. This table dictates how many units may be in a given area and a time limit to capture certain cities. There are severe penalties if either directive is not obeyed. Other rules include extra reinforcements, Soviet strategic withdrawal, major rivers, and overruns. (Richard J. Rydzel)

Comment: This is an interesting and playable game but only fairly historical. The Axis player can win only against an incompetent player. The Don River is a fortress that cannot be breached. Stalingrad never sees any action. In the Caucasus region, a good Russian player can easily trap many Axis units if they go much past Rostov. This game does not fit well into historical learning situations. (RJR)

Since the above was written, SPI has recognized the complaints and attempted to fix the problems by amending the rules. I do not know how successful they have been. (MC)

Cost: $11.00, $12.00 with box

Producer: Simulations Publications, Inc., 44 East 23rd Street, New York, NY 10010

EAGLE DAY: THE BATTLE OF BRITAIN—1940

Laurence J. Rusiecki

Playing Data
Copyright: 1973
Age Level: grade 9 and up
Number of Players: 2
Playing Time: 3-8 hours
Preparation Time: 1 hour

Description: The map shows the southeastern corner of England and the English Channel. The map shows the location of London, English airfields, and aircraft factories. Each turn represents ten minutes. Each sequence of twenty turns represents one week. There are three weeks in the ordinary game. Counters represent units of specific kinds of aircraft. The Germans win the game by bombing British airfields to gain victory points. An optional rule allows the game to be continued for two more weeks. (MC)

Cost: $6.00

Producer: Histo Games, 34 Sharon Street, Brooklyn, NY 11211

EAST FRONT

John Wilson, Stephen G. Bettum, and Christopher Allen

Playing Data
Copyright: 1976
Age Level: grade 9 and up
Number of Players: 2
Playing Time: 6-60 hours
Preparation Time: 1 hour

Description: The two-sheet map shows all of European Russia and all of Poland, Slovakia, and Scandinavia, with large parts of Germany, Turkey, Hungary, Rumania, and Bulgaria. The game proceeds at one turn per week and covers either one five- to fifteen-turn season or all 101 turns of the entire Russo-German War of 1941-1945. The units are German corps and Russian armies, both represented as somewhat abstract units, with all units of the same kind being identical. There is a major exception: the Russians get better experienced units as the game progresses, while the Germans tend to get more and more depleted units. Supply rules are simple to operate yet representative of the campaign's problems. The weather automatically produces mud or winter on a schedule. The game shows the Russian army as slow, methodical, and tough, while the German army is flashy in the summer and very handicapped in the winter, especially the first. Russian production varies according to how much Russian territory is overrun. (MC)

Comment: This is a splendid simulation and a valuable compromise between the abstraction of *Barbarossa* (q.v.) and the excessive detail of *War in the East* (q.v.). This game might displace *Barbarossa* (which is in the introductory essay) as my favorite classroom game on its subject. So far I have not had the opportunity to use it in a classroom and I am still puzzling out a few rules problems. (MC)

Cost: $15.00

Producer: Control Box, Inc., 627 68th Street, Brooklyn, NY 11220

1815: THE WATERLOO CAMPAIGN

Frank Chadwick

Playing Data
Copyright: 1975
Age Level: grade 9 and up
Number of Players: 2
Playing Time: 3 hours
Preparation Time: 45 minutes
Materials:

Description: This game covers the three days of battles in the Waterloo campaign. The map scale is 600 yards per hexagon. The time scale is one daylight hour or two nighttime hours per turn. The rules are fairly simple, but the game nevertheless makes realistic distinctions in the way it handles infantry, cavalry, and artillery. Rules also cover loss of and recovery of morale, and contain an assortment of optional forces. An optional rule allows movement off the board by one or both players. (MC)

Comment: Headquarters units provided with the game are not functional but could be easily made so in a multiplayer classroom game. The map is hard to read because of too little difference between two kinds of road. (MC)

Cost: $8.75

Producer: Game Designers' Workshop, 203 North Street, Normal, IL 61761

EN GARDE!

Darryl Hany and Frank Chadwick

Playing Data
Copyright: 1975
Age Level: high school, college
Number of Players: any number
Playing Time: any amount
Preparation Time: 1 hour
Materials: rule booklet, paper and dice—no board is used

Description: This is a tongue-in-cheek simulation of the kind of life lived in The Three Musketeers and other historical adventures. The players represent swashbucklers of some place like Paris. Each move represents a week. The object for the players is to acquire greater and greater status. The dice define the lineage, wealth and abilities of each players' character and other supporting characters before the game begins. Then the characters engage in assorted peaceful activities like carousing, gambling, whoring, and dueling—the dueling system was the first part of the game designed and is usable alone. Characters can also join the army and go on campaigns. Then an abstract war has to be conducted to see if any of the heroes are affected by it. No criteria for ending the game are included in the rules. (MC)

Cost: $4.00

Producer: Games Designers' Workshop, P.O. Box 582, Bloomington, IL 61701

ENGLISH CIVIL WAR

Roger Sandell, Ken Clark, and Hartley Patterson

Playing Data
Copyright: 1975
Age Level: grade 7 and up
Number of Players: 2
Playing Time: 4 hours
Preparation Time: 1 hour

Description: As the game begins, the rival armies are gathering to their standards. England and Wales are divided into areas, some of which contain important cities which can be controlled separately. Enemy armies in the same area can try to bring on a battle or attempt to refuse a battle. The army that wants to fight never has a greater than 66 2/3 percent chance of forcing the issue. When battles do occur, the armies of infantry, cavalry, and leaders are put on a battle board. Flank battles between the cavalry are decided first, and the victorious cavalry then see if they will join in the main battle or just run around the countryside. Each move is a month. During the winter, the armies recruit new men based on the number of areas they control. Later in the game, the Scots intervene in Parliament's favor. (MC)

Comment: The game is a superb simplification of the strategy and even the tactics of the English Civil War. For a classroom, each student ought to be given a leader to control and communications between leaders ought to be inhibited unless they are in the same area.

Cost: £5

Producer: Philmar, Ltd., 47-53 Dace Road, London E3 2NG, United Kingdom. Inquire for U.S. distributors.

FALL OF TOBRUK

Frank Alan Chadwick

Playing Data
Copyright: 1975
Age Level: grade 9 and up
Number of Players: 2
Playing Time: 3-4 hours
Preparation Time: 1 hour
Materials:

Description: In this twenty-four-move game, each move represents one day of the Battle of Gazala and the German-Italian capture of Tobruk. In the game, the Axis must do it again or lose. The counters generally represent battalions. The rules have several unusual provisions which help to recreate the ebb and flow of desert warfare. The minefields that restricted the battle field are printed on the board. Much of the emphasis appropriately is on tank warfare. The tank units are identical partly by a silhouette of the tank type that was used by the unit. The combat rules show the differences between types of tanks, and the German superiority in tank recovery, which went a long way to assuring a German victory, is reflected by a set of recovery rules. (MC)

Comment: See the introductory essay to this section for more on this game. (MC)

Cost: $11.98

Producer: Conflict Game Company, P.O. Box 432, Normal, IL 61761

FAST CARRIERS

James F. Dunnigan

Playing Data
Copyright: 1975
Age Level: high school, college
Numbers of Players: 2
Playing Time: 3-12 hours
Preparation Time: 1 hour
Materials: map sheet (with 6 separate maps plus charts, tables, tracks), 2 sheets of die-cut counters, 8 U.S. and 8 Japanese display charts, 2 search templates, rules booklet, die

Description: *Fast Carriers* deals with operations of carriers and their aircraft from 1941 in World War II through Korea and Vietnam to a hypothetical action in the North Sea between NATO and Soviet forces in 1977. The ship counters represent individual carriers, and also battleships, cruisers, and destroyers used as escorts, some auxiliary vessels used mainly as targets, and submarines for the North Atlantic, 1977, scenario. The aircraft counters represent units of six or three aircraft each. They include fighters, dive bombers and torpedo planes for the World War II scenarios. Five of the maps are strategic on the scale of ninety miles to the hexagon showing the Guadalcanal-Coral Sea area, and the North Sea area, the Midway Islands area, Korea area, and Vietnam area. The sixth map is a hex-divided, but otherwise featureless, board on which is fought the aircraft versus carrier battles of the World War II scenarios. The battles of the other scenarios, and the World War II battles if desired, and any surface battles that occur, are all fought abstractly without maneuvering. The rules for manipulating all of this equipment are fairly complex but the whole game is rendered much easier than usual, for the complexity of the rules, by a set of players' aids (multiple turn tracks and the operational displays) which help direct the players' movements at every step. The game system requires the players to make many small decisions and leaves a lot of room for assigning multiple players to each side. (MC)

Comment: This is a fast-moving, generally accurate game, which shows particularly the devastating effect of the use of aircraft and carriers in World War II. See the introductory essay to this section for more on this game. (MC)

Cost: $11.00, $12.00 with box

Producer: Simulations Publications, Inc. 44 East 23rd Street, New York, NY 10010

FIGHT IN THE SKIES

Mike Carr

Playing Data
Copyright: 1971
Age Level: grade 9 and up
Number of Players: 2-10 (more could be unwieldy)
Playing Time: 1-10 or more hours
Preparation Time: 1-2 hours

Description: The mapboard, featureless except for a square grid which distorts movement slightly, provides an area for dogfights between World War I aircraft. Each player pilots one plane in three dimensions. Each plane is located two-dimensionally on the map and its height is noted separately. The cards describe fifteen allied and thirteen German aircraft. The aircraft differ in their speed, climbing, diving, and turning abilities and in their abilities to take punishment. The planes may move ordinarily or may perform various maneuvers (Immelmann, barrel roll, etc.) The situations have to be constructed by the players. (MC)

Cost: $10.00

Producer: TSR, Inc., Box 756, Lake Geneva, WI 53147

FIREFIGHT

James F. Dunnigan, Irad B. Hardy, Redmond A. Simonsen, and Frederick Georgian

Playing Data
Copyright: 1976
Age Level: grade 9 and up
Number of Players: 2
Playing Time: 2-6 hours
Preparation Time: 1 hour

Description: *Firefight* is a game of small-unit contemporary tactics. Each counter represents a two to four man fire- or weapons-team or a single tank, armored personnel carrier, or other vehicle. The map is a two-piece representation of typical terrain at a scale of fifty meters to the hexagon. The opposing forces are U.S. and Soviet and consist of actual weapons deployed in Europe today with a few possible future weapons. Well-developed combat charts make combat easier to figure out than is normal in games at this scale. (MC)

Comment: *Firefight* was developed by SPI in cooperation with the U.S. Army, which then bought a large number of sets as training aids. The game was designed to teach current army tactical doctrine. At the moment, the Army is no longer using the first edition of the game because they are in the throes of developing a second edition which will carry the message in a somewhat simpler fashion. Since it is not a historical game, my interest in it is slight. (MC)

Cost: $11.00, $12.00 with box

Producer: Simulations Publications, Inc., 44 East 23rd Street, New York, NY 10010

FIRST WORLD WAR

Frank Davis and Mark Herman

Playing Data
Copyright: 1977
Age Level: grade 9 and up
Number of Players: 2
Playing Time: 10-45 hours
Preparation Time: 2 hours

Description: The game map for this game is the same nine-part one designed for *War in Europe*. *First World War* uses four or six of those parts for a division/corps level simulation of the whole war on a time scale of three moves per month. The counters distinguish between active corps, reserve corps, cavalry corps, and shock divisions, which come in late in the war. Fortifications are present on the map when the game begins, and trenches can be built later on. Divisions are the same for all armies, so the main difference is in morale, which changes as the game progresses. Surface fleet warfare and submarine warfare are handled very abstractly. Every three months, both sides get a new supply of resources points, which are used to build new units and to supply combat on the part of the old. New nations enter the war either as they are invaded or according to a historical time table. A demoralization-level chart is kept for each nation and as it loses troops, ships, and territory, its demoralization level rises; when certain numbers of points are accumulated, its national morale falls. When it falls to zero, the nation is out of the war. (MC)

Comment: *First World War* has many good ideas, but it does not work as a simulation of the war. I am not sure what is wrong, but it appears that the main problems are an unrealistic assignment of resource points, a combat system that makes offensives even more impossible than they were historically, a supply system that prevents operations in areas where they occured historically, and assorted other rules. (MC)

Cost: $14.00 with no map or box, $15.00 with box; $20.00 extra for *War in Europe* map set

Producer: Simulations Publications, Inc., 44 East 23rd Street, New York, NY 10010

FLYING TIGERS II

Louis B. Zocchi

Playing Data
Copyright: 1973
Age Level: grade 9 and up
Number of Players: 2
Playing Time: 1 hour

Description: The map shows Burma with parts of China, India, Siam, and Indo-China. The time is that of the Japanese advance into the area in early 1942. The counters represent groups of Japanese, American, and British aircraft. The number of planes represented by each counter is recorded on the record sheet. The game is played in sequences of fifteen turns each. The airplanes fly and fight and the Japanese planes bomb allied airfields during these fifteen turns. Each of the seven sequences of turns represents one month. When an airfield has been bombed adequately, it comes under Japanese control and may be used by the Japanese the next month. (MC)

Comment: *Flying Tigers* is a strange mixture. The game system of movement and combat reflects the tactics of air-to-air combat, while the game itself tries to show the advance of the Japanese across Burma, during which all the important battles were fought on land. The game is not at all faithful to the campaign, but its tactical aspect is a fair simulation. Nearly half of the over one hundred counters were printed with erroneous information and have to be corrected by hand–a job taking ten or fifteen minutes. (MC)

Cost: $6.00

Producer: Lou Zocchi, President, Games Science, 7604 Newton Drive, Biloxi, MS 39532

FORTRESS RHODESIA

Michael J. Raymond

Playing Data
Copyright: 1976
Age Level: grade 9 up
Number of Players: 2
Playing Time: 2-4 hours
Preparation Time: 1 hour

Description: This is a representation of the guerrilla war being fought in Rhodesia between the white-dominated government and guerrillas based on surrounding countries. As one might expect, the game does not deal much with the details of the situation, but there is a variety of Rhodesian units: infantry, engineers, artillery, parachute, and police. The guerrilla player manipulates upside-down dummy units as well as upside-down real units–to confuse the Rhodesians. The game lasts twenty-four turns and simulates a full year of the struggle. (MC)

Comment: The game is very simple to learn and play. Unfortunately, for a game on a guerrilla war, it includes no simulation of political structure. (MC)

Cost: $3.50

Producer: SAGA, 4833 Greentree Road, Lebanon, OH 45036

FOUR BATTLES FROM THE CRIMEAN WAR: ALMA, BALACLAVA, INKERMAN, TCHERNAYA RIVER

Steven Ross, Redmond A. Simonsen, J. Matisse Enzer, Thomas Gould, and Martin Goldberger

Playing Data
Copyright: 1978
Age Level: grade 9 and up
Number of Players: 2
Playing Time: 4 hours
Preparation Time: 1 hour

Description: This game set covers four of the field battles of the Crimean War that centered on the Siege of Sevastopol and Russian efforts to lift the siege. Time and distance scales differ. The games last from twelve to twenty-one turns. Unit counters represent infantry, artillery, and cavalry, each of which has different powers. Standard Rules apply to all four games, while Exclusive Rules apply to each game separately. The Standard Rules deal with fire, movement, disruption, rallying, and terrain. The exclusive Rules deal with special terrain, engineers, tactical doctrine, special fatigue, night, fog, skirmishes, and other factors that enter into only one of the battles. (MC)

Comment: The four games in this set are more realistic than other SPI four-game sets, but the games are not significantly more complex. Like others in the series, these could be made the subjects of take-home assignments. The game set includes a sixteen-page pamphlet on the war, which is much more historical background information than is normally packaged with a game. (MC)

Cost: $13.00, $14.00 with box ($4.00 apiece separately)

Producer: Simulations Publications, Inc., 44 East 23rd Street, New York, NY 10010

FOUR BATTLES IN NORTH AFRICA: CRUSADER, CAULDRON, SUPERCHARGE, KASSERINE

David C. Isby, Frank Davis, Howard Barasch, and Greg Costikyan

Playing Data
Copyright: 1976
Age Level: grade 9 and up
Number of Players: 2
Playing Time: 4-6 hours
Preparation Time: 1 hour

Description: The game set deals with four North African battles, using a standard system with a few special rules for each game. The maps show the terrain, including minefields and fortifications built before the start of the battles. The units are mainly the brigades and regiments of the historical armies. The combat system is very abstract with a combat results table that gives few eliminations of unsurrounded units. (MC)

Comment: The games play well but are not very convincing as representations of World War II desert warfare. (MC)

Cost: $13.00, $14.00 with box ($4.00 apiece separately)

Producer: Simulations Publications, Inc., 44 East 23rd Street, New York, NY 10010

FOXBAT AND PHANTOM

James F. Dunnigan

Playing Data
Copyright: 1973
Age Level: grade 9 and up
Number of Players: 2
Playing Time: 2-8 hours
Preparation Time: 1-2 hours

Description: The game "map" represents a piece of open air on the scale of 1000 meters to the hexagon. Each move represents thirty seconds. The counters represent the two-dimensional location of the aircraft on the map and numbers printed on the aircraft counters show their height. One to four aircraft on each side can be used. The scenarios are written for three on each side. The aircraft control cards allow any of fourteen different types of modern fighters and fighter-bombers to be "flown" in the game. Also there is a set of counters representing a force of heavy bombers of no particular type. A set of scenarios define two basic missions and apply the game to the Middle East, Vietnam, Europe, and other real and potential battle areas. (MC)

Comment: *Foxbat and Phantom* is a very believable simulation of the problems of dogfighting with planes capable of supersonic flight. However, the missions are written in such a way that many of the planes in the game cannot win. Although the planes have three different kinds of weapons—radar-guided missiles, heat-seeking missiles, and cannon—only the first of these can ever shoot anything down. The game is quite difficult to learn although it moves quickly once it is learned. (MC)

Cost: $8.00, $9.00 with box

Producer: Simulations Publications, Inc., 44 East 23rd Street, New York, NY 10010

FREDERICK THE GREAT

Frank Davis and Edward Curran

Playing Data
Copyright: 1975
Age Level: high school-college
Number of Players: 2
Playing Time: 4-5 hours
Preparation Time: 1 1/2 hours
Materials: mapboard (with charts and tracks), die-cut counters, rules leaflet, die

Description: The map shows most of Germany, Bohemia, East Prussia, part of Poland, and a corner of France. Four scenarios deal separately with the campaign year from 1756 to 1759, the year when Frederick successfully held off increasingly stronger enemy forces attempting to advance from all directions. Much of the action is controlled by a leadership rule which forces the players to work with incompetent and pusillanimous generals and armies that, if defeated, will refuse to do anything more for the rest of the year. Movement is partially hidden—units may be stacked under a leader with the opponent unable to examine the stacks. Supply requirements are very important in preventing rapid marches into enemy territory. Battles are quite hazardous and most players will soon realize that sieges are more profitable. (MC)

Comment: The game is remarkably faithful to the spirit and tempo of eighteenth-century warfare. Although designed for two players, the sides can be easily split among more players, an action that will improve the simulation somewhat, especially if communication among allies is inhibited. See the introductory essay to this section for more on this game. (MC)

Cost: $11.00, $12.00 with box

Producer: Simulations Publications, Inc., 44 East 23rd Street, New York, NY 10010

FRIGATE

James F. Dunnigan

Playing Data
Copyright: 1974
Age Level: high school-college
Number of Players: 2
Playing Time: 2-8 hours
Preparation Time: 1 hour
Materials: six-part mapboard, die-cut counters, rules folder, scenarios folder, die

Description: The board consists of six identical featureless hexagon-gridded sheets that represent open water and can be shifted around to follow the action as the game progresses. The game deals with many fleet actions and a few single-ship actions from 1702 through 1805 (Trafalgar) to 1812 and with a hypothetical U.S.-British action dated 1825. In spite of the name of the game, frigates are not very important. The ship counters represent ships from ships of the line of 120 guns to corvettes of 14 merchantmen. Movement is simultaneous. Wind direction and velocity are, of course, variables of great importance. The combat system is fairly abstract and simple. Hits are assessed as either crew hits or mast hits. Range and field of fire are both taken into account. Much movement is by formations instead of by individual ships. (MC)

Comment: This is an excellent simulation on the fleet level. Less so for single ship action. It is sometimes very time consuming because of the necessity of writing down the move for each formation or ship before moving. There are many opportunities for turning the large scenarios into multiplayer games. (MC)

Cost: $8.00, $9.00 with box

Producer: Simulations Publication, Inc., 44 East 23rd Street, New York, NY 10010

FULDA GAP

James F. Dunnigan, Thomas Walczyk, and Redmond A. Simonsen

Playing Data
Copyright: 1977
Age Level: grade 9 and up
Number of Players: 2
Playing Time: 3-5 hours
Preparation Time: 1 hour

Description: *Fulda Gap* is described as "the first battle of the next war"—the war that might occur in Europe some day and be fought in Western Germany between NATO forces and invading Soviet forces The units are regiments and brigades. The game can be played on several levels of difficulty and with an assortment of different assumptions about the political and military conditions of the war. The rules reflect several aspects of warfare not usually dealt with in a wargame and do so without excessive difficulty. Units do not have a certain combat strength until they actually get into combat. Then the players roll dice to find out their true strength. Disengagement is dealt with in an unusually realistic fashion. The importance of supply and of organization, including organization by different nationalities, is made manifest. Basic rules include artillery and chemical warfare. Advanced rules include airpower and nuclear weapons. (MC)

Cost: $8.00, $9.00 with box

Producer: Simulations Publications, Inc., 44 East 23rd Street, New York, NY 10010

GETTYSBURG

Mick Uhl

Playing Data
Copyright: 1977
Age Level: grade 7 and up
Number of Players: 2
Playing Time: 2-10 hours
Preparation Time: 30 minutes-2 hours

Description: The game comes with three versions, introductory, intermediate, and advanced. The first is extremely simple, the last extremely complex. The intermediate game emphasizes the details of combat. The advanced game represents the details of combat formations more thoroughly than any other game and also emphasizes chain of command and the problems of control. In all of the games, terrain plays a smaller than normal part in the battle. In all versions, victory points decide the game. For the intermediate and advanced game, these victory calculations are quite complex. (MC)

Comment: This game is not to be confused with the game on the same subject published by the same company and carrying the same name that was included in previous issues of the Guide. That *Gettysburg*, one of the two oldest board wargames, is now out of print. This *Gettysburg* is not only different from its predecessor but is unusual among its contemporaries, especially for its introduction of battleline counters which allow a unit to spread out over several hexagons. The combat rules are probably more intricate than most teachers will have time for, but they might appreciate that the game contains plenty of opportunities to use multiple commanders. (MC)

Cost: $12.00

Producer: Avalon Hill, 4517 Harford Road, Baltimore, MD 21214

GLOBAL WAR

James F. Dunnigan

Playing Data
Copyright: 1975
Age Level: senior high-college
Number of Players: 2 or more
Playing Time: 6-12 hours
Preparation Time: 2 hours
Materials: two-part mapboards, two charts/scenario leaflets, two turn record production track cards, three sheets of die-cut counters, rules booklet, errata sheet, die

Description: The map shows the world, except for the polar areas, on a scale of 300 miles to the hexagon–in places. Although the smallness of the scale might have led to confusion on the map, this problem is eliminated by a system of coloring the hexagon and marking the sides according to the kinds of movement possible in or across them. The players read the hexagon pattern, not the map itself. Each highly complex move represents three months of time. The players dispose of not only infantry, mechanized, fortification, surface fleet, submarine fleet and aircraft strength points, but also merchant ships, amphibious assault fleets and, if they care to pay the heavy price, atomic bombs. The players must decide not only where to attack and when, but also how much production to devote to the various arms and to supporting merchant vessels, and on which sea lanes to keep their merchant vessels. The production task is complicated by the fact that it takes from two to nine turns for a production decision to be translated into a usable counter. (MC)

Comment: This is a very ambitious and valuable game, but it does not succeed in simulating World War II in a convincing fashion. The scale is too small, the naval rules are either too constricting or much too permissive. One major fault in the game is that there is no production cost for maintaining a unit on the board, only for building new ones. Therefore, the latter part of the game tends to bog down with an excess of units on both sides. In spite of all this, I recommend the game–not so much for playing as it is, but for use as a source of inspiration for designing variant or alternate games on the same subject. (MC)

Cost: $11.00, $12.00 with box

Producer: Simulations Publications, Inc., 44 East 23rd Street, New York, NY 10010

GRENADIER: COMPANY LEVEL COMBAT, 1700-1850

James F. Dunnigan

Playing Data
Copyright: 1971
Age Level: college
Number of Players: 2
Playing Time: 3-6 hours
Preparation Time: 2 hours

Description: The map shows a part of a typical battlefield about 1 1/2 x 2 kilometers in area. Each hexagon is fifty meters across and each move represents ten minutes. The counters represent eight kinds of infantry, skirmishers, three kinds of cavalry, artillerymen, artillery transport, and five sizes of artillery. There are sixteen scenarios, based on historical battles or parts of historical battles. (MC)

Comment: I have used this game with a small group of upper level students with some success, but the game is too complex and the mechanics are too abstract for it to be used very widely. The game shows something of the relationship between the various kinds of armies that fought during the French Revolutionary and Napoleonic Wars, but much of the interpretation of the tactical conditions of the period is questionable. (MC)

Cost: $11.00, $12.00 with box

Producer: Simulations Publications, Inc., 44 East 23rd Street, New York, NY 10010

HANNIBAL

Lawrence J. Rusiecki

Playing Data
Copyright: 1972
Age Level: grade 9 and up
Number of Players: 2
Playing Time: 4-8 hours
Preparation Time: 1 hour

Description: Game board shows Western Mediterranean, including Italy, northern Africa, Spain, and southern Gaul. Each counter represents one or more abstract unit of infantry, cavalry or ships. The game starts in 218 B.C., at the beginning of the Second Punic War. Each turn is one year, and there are a maximum of seventeen turns in the game. A record is kept of points gained and lost as a result of capturing or losing cities, controlling or losing territory, and winning or losing battles. As soon as one side has a twenty-five point superiority, that side wins. Battles are decided mainly by weight of numbers and lucky die rolling, so skillful players avoid battles, unless the stakes are high and they can afford the possible loss. (MC)

Comment: More of the rules in this game are a little ambiguous than is usual. However, it is an excellent simulation of the main strategic problems on both sides during the Second Punic War. (MC)

Cost: $7.00

Producer: Histo Games, 34 Sharon Street, Brooklyn, NY 11211

HIGHWAY TO THE REICH

Jay Nelson, Redmond A. Simonsen, and Irad B. Hardy

Playing Data
Copyright: 1977
Age Level: grade 11 and up
Number of Players: 2 or more in 2 teams
Playing Time: 3-30 hours
Preparation Time: 1-3 hours

Description This game is about the Battle of Arnhem—the abortive attempt of the Allies to break through into Northern Germany by using airborne units to capture bridges and relieving them with a ground assault column. The scale is large (600 meters to the hexagon, two or four hours to the turn, units are infantry companies and armored platoons) and the game is large (four map sheets) and detailed. The rules cover air, land, weather, destruction and repair of bridges, and various kinds of weapons and their relations to each other. The rules particularly emphasize the importance of organization and leadership. There are functional counters for generals and headquarters. Units that get mixed up in the course of the battle are penalized severely. Several one-map scenarios allow the play of a short game, but short games only involve a small part of the action. (MC)

Comment: This is a large game and probably too complex and on too limited a subject to consider using in a classroom. (MC)

Cost: $18.00, $20.00 with two boxes

Producer: Simulations Publications, Inc., 44 East 23rd Street, New York, NY 10010

IMPERIALISM

Richard Barr

Playing Data
Copyright: 1974
Age Level: high school-college
Number of Players: 2-8
Playing Time: 2-4 hours
Preparation Time: 30 minutes

Description: *Imperialism* is a game on an imaginary world composed of eight islands and the surrounding seas. Each player has a home country on the largest island and they send their armies and fleets out to explore, claim and fight over the territories of the other islands. The game is based loosely on European imperialism of the seventeenth and eighteenth centuries. The counters represent fleets, armies, units of wealth (extracted from the colonies), pirates, and restless natives. Cities, ports and forts are drawn on the board. The game ends when the end-of-game card is drawn and victory is decided by the point value of the players' possessions. (MC)

Comment: It's a fun, easy game with a tenuous but noticeable relationship to history. (MC)

Cost: $10.00

Producer: Flying Buffalo, Inc., P.O. Box 1467, Scottsdale, AZ 85252

INVASION: AMERICA

James F. Dunnigan

Playing Data
Copyright: 1976
Age Level: high school-college
Number of Players: 2-4
Playing Time: 4-20 hours
Preparation Time: 1 hour
Materials: two-part extra large (35" x 46") mapboard (with charts), rules book, die-cut counters, die

Description: This game deals with an imaginary invasion of the United States in 1997. The U.S. is being attacked by the forces of the European Socialist Coalition, the South American Union, and the Pan Asiatic League. Nuclear weapons are not used. The map shows all of North America and a corner of South America for the South Americans to start from. The U.S. has no offensive capability. The short scenarios last eight turns representing one month each. The long game lasts five years (sixty turns). The unit counters represent land, sea, and air units, militia, supply ports. The scenarios include invasion and partisan scenarios. (MC)

Cost: $11.00, $12.00 with box

Producer: Simulations Publications, Inc., 44 East 23rd Street, New York, NY 10010

ISLAND WAR (BLOODY RIDGE [GUADALCANAL], SAIPAN, LEYTE, OKINAWA)

Kevin Zucker, Kip Allen, Jay Nelson, and Larry Pinsky

Playing Data
Copyright: 1975
Age Level: high school-college Number of Players: 2
Playing Time: 3-4 hours
Preparation Time: 1 hour
Materials: four mapboards, rules folders, die-cut counters, die

Description: This four-game set uses a simple set of rules applying to all four games. Each of the games uses a different map, different order of battle and counters, and a few peculiar rules. The scale of the maps is 500 to 2000 yards to the hexagon. Each move is one or two days. The games come with two to five scenarios each. The counters represent regiments or battalions. Artillery functions in the game quite different from other units. It is particularly flexible and therefore important. A significant flavor is given by a rule allowing Japanese Banzai attacks. (MC)

Cost: $13.00 ($4.00 apiece separately), $14.00 with box

Producer: Simulations Publications, Inc., 44 East 23rd Street, New York, NY 10010

JERUSALEM!

John Hill

Playing Data
Copyright: 1975
Age Level: grade 9 and up
Number of Players: 2
Playing Time: 3-4 hours
Preparation Time: 1 hour

Description: The game centers on the battle for Jerusalem in the first Arab-Israeli War of 1948. The main problem for the Jews is the provisioning of Jerusalem by convoys, which have to find a way by an easily blocked system through a hostile countryside. The Counters represent leaders, arab guerrillas, Israeli bulldozers for building new roads, terroritsts, as well as many kinds of regular units. (MC)

Comment: Jerusalem! is a fast moving, accurate simulation which uses an unusually flexible game system to reflect the unusually fluid situation around Jerusalem in 1948. The original producer, Simulations Design Corporation, has gone out of business, but the game will no doubt become available through another publisher. (MC)

KASSERINE PASS

John Hill and Frank Chadwick

Playing Data
Copyright: 2nd edition, 1977
Age Level: grade 9 and up
Number of Players: 2
Playing Time: 3-4 hours
Preparation Time: 30 minutes

Description: This game is a brief treatment of the World War II battle in which American troops first met the Germans. Units are battalion size. The map is on the scale of 4 1/2 miles per hexagon. Each move represents a day. The German surprise is simulated with special rules applying only the first turn. (MC)

Comment: Kasserine Pass is a popular subject, and this is probably the better of the two games on it. The rules are plain enough that the game could be used for take-home assignments. (MC)

Cost: $11.98

Producer: Conflict Game Co., P.O. Box 432, Normal, IL 61761

KHARKOV

Stephen B. Patrick, Brad Hessel, and David Werden

Playing Data
Copyright: 1978
Age Level: grade 9 and up
Number of Players: 2
Playing Time: 4 hours
Preparation Time: 30 minutes

Description: Kharkov is a simulation of the Russian offensive against the Axis armies in Southern Russia in the spring of 1942. The map shows the area of Russia around the city of Kharkov. Each hexagon represents an area 6.9 kilometers across, and there are ten daily turns. Counters represent divisions and brigades. The rules are basically the same as Panzergruppe Guderian but are altered and expanded to show Russian inflexibility and German unpreparedness. Russian units can infiltrate German lines by moving from one zone of control to another for two game turns. Both sides are restricted from moving units on the south edge of the map. The Russian player has reserve armies but they are released only slowly and are limited in their movement. The Germans may supply units with air counters. (Richard J. Rydzel)

Comment: Kharkov is an easy-to-learn game and is an excellent example of Russian combat doctrines in World War II. The short preparation and playing time make this a good simulation for school use; however, both players should be prepared to lose lots of units and a Russian player who isn't careful may lose all units. (RJR)

Cost: $9.00; $10.00 with box

Producer: Simulations Publications, Inc., 44 East 23rd Street, New York, NY 10010

KINGMAKER

Andrew McNeil

Playing Data
Copyright: 1974
Age Level: high school-college
Number of Players: 2-22—recommended for no more than 12
Playing Time: 3-8 hours
Preparation Time: 30 minutes
Materials: mounted mapboard, 72-card Crown Pack, 80-card Event Pack, 38 cardboard counters

Description: The game deals with the War of the Roses. The crown pack contains cards designating nobles, titles, offices, cities, bishops, mercenaries, and ships. A number of these are dealt to each player to give him his beginning coalition—his forces will change considerably by the end of the game. The rest of the cards form a drawing deck which will probably be exhausted for half the game, although it is periodically replenished by deaths. Titles and offices which are not used go into another deck called "Chancery," which cards can be distributed by the player controlling the king if there is only one king. The kings and potential kings are represented by seven counters distributed around the board at the beginning of the game. These have no will of their own. They are siezed by the nobles that the players control and are then controlled too. One king is crowned at the start and another one from the other family can be crowned too. Then other claimants can be crowned as earlier ones are killed off. The player controlling the last royal piece wins the game. (MC)

Comments: Kingmaker is a fascinating game on a minor civil war. It teaches a great deal about English geography, especially place names. It shows something of the organization of a late medieval kingdom and of late medieval anarchy. Players need to exercise diplomatic skills as well as strategic skills in order to win the game. See the introductory essay to this section for more information on this game. (MC)

Cost: $12.00

Producer: Avalon Hill, 4517 Harford Road, Baltimore, MD 21214

KÖNIGGRÄTZ

James Gabel and Richard Spence

Playing Data
Copyright: n.d. (1977?)
Age Level: grade 9 and up
Number of Players: 2
Playing Time: 4 hours
Preparation Time: 1 hour

Description: This is a simulation of the culminating battle of the Seven Weeks' War between the Prussian and the Austro-Saxon armies. The units are infantry brigades, which are reduced to battalion size by casualties, and artillery battalions. The rules make a distinction between fire combat and melee combat. Appropriately, the strength of the Prussian army is in its infantry's firepower while that of the Austrian army is in its artillery's firepower. The scales are 500 yards per hexagon and one hour per move. The game lasts twelve moves. (MC)

Comment: This is an accurate portrayal of the tactical situation of the Battle of Königgrätz (or Sadowa), without the Austrian command difficulties that were responsible for their losing the battle decisively. In a classroom, the original situation could be more closely duplicated by dividing the commands and giving some of the Austrian generals somewhat different instructions than the rest. The rules for the game are a little confused in places, especially in the adjustments made for terrain, but I believe they can be figured out. (MC)

Cost: $5.50

Producer: S&G Games, 2105 Custer Ave., Bakersfield, CA 93304

KOREA

James F. Dunnigan

Playing Data
Copyright: 1971
Age Level: grade 9 and up
Number of Players: 2
Playing Time: 4-14 hours
Preparation Time: 1 1/2-2 hours

Description: The map shows Korea with roads, railroads, and cities. Each turn represents one week. The counters for most of the forces represent divisions and regiments. Units may be built up or broken down into their parts. The Chinese forces also have army counters.

Other counters represent U.S. naval forces, amphibious landing craft units, sea transport units, supply units and entrenchments. There are four ways of playing *Korea,* reflecting different times in the war: The *Invasion* game lasts seventeen turns; the *Chinese Intervention* game lasts nine turns, the *Stalemate* game lasts twenty-one turns, and the *Campaign* game lasts fifty-two turns. Variations to the *Invasion* and *Campaign* games can be made according to various "what if" rules. (MC)

Comment: Korea is an excellent simulation of the most important military problems of the campaign—supply, reinforcement, the strength of the defense, terrain, and amphibious operations. However, the game is very difficult to play because it has a large number of very specific rules. Every force moves and fights according to different rules. (MC)

Cost: $9.00, $10.00 with box

Producer: Simulations Publications, Inc., 44 East 23rd Street, New York, NY 10010

KURSK

Sterling S. Hart and James F. Dunnigan

Playing Data
Copyright: 1971
Age Level: grade 9 and up
Number of Players: 2
Playing Time: 3-5 hours
Preparation Time: 2 hours

Description: The mapboard shows the area around Kursk, including the Russian and German fortified lines, at the scale of ten miles to the hexagon. Each turn represents two days. The counters represent German divisions and Russian corps, German Kampfgruppen (remants of divisions), and air units. It is possible to begin the battle in May, June, or July (the historical month). (MC)

Comment: Kursk shows clearly the hopelessness of the historic German offensive and the growing but still very imperfect Russian offensive power in 1943. (MC)

Cost: $9.00, $10.00 with box

Producer: Simulations Publications, Inc., 44 East 23rd Street, New York, NY 10010

LA BATAILLE DE LA MOSKOWA

Lawrence A. Groves and John Harshman

Playing Data
Copyright: 1977
Age Level: grade 11 up
Number of Players: 2 +
Playing Time: 4-24 hours
Preparation Time: 1-4 hours

Description: This game is a microsimulation of what is usually called the Battle of Borodino—the big battle of Napoleon's invasion of Russia in 1812. Four map sheets show the battlefield on a scale of 100 meters to the hexagon. Some 1000 markers show infantry regiments, infantry battalions to replace the regiments, cavalry regiments, artillery companies, and officers and their aides. Other markers are used to represent casualties, formations, and to mark routed units. Various scenarios can be played, ranging from six to forty-eight turns, each turn representing twenty minutes. The game stresses the importance of leadership, the use of a variety of formations (e.g., skirmishing and squares), the combination of various arms. Multiplayer rules outline the use of the game by groups. (MC)

Comment: This is one of the best of the monster games. It allows the study of a major battle at close range. The availability of short scenarios and of rules for groups is a boon for a teacher. The game is big, but the rules are not particularly difficult. (MC)

Cost: $18.00

Producer: Game Designers' Workshop, 203 North Street, Normal, IL 61761

LA GRANDE ARMÉE

John Young, Arnold J. Hendrick, and Philip Orbanes

Playing Data
Copyright: 1972
Age Level: grade 9 and up
Number of Players: 2
Playing Time: 4-5 hours
Preparation Time: 1 hour

Description: The map shows the lower two-thirds of Germany, Austria, and Bohemia. The scale is about fifteen km. to the hexagon. Each move represents one-third month. The game covers Napoleon's campaigns of 1805, 1806, and 1809 in the area. The counters represent the French forces, the old Austrian army, the old Prussian army, the Russian army, and the reformed Austrian army of 1809. The kinds of units are infantry, cavalry, supply, depot, and leader. Victory in each of the three scenarios depends on the ratio of points achieved by each side. Points are given mainly for taking or besieging cities and fortresses. (MC)

Comment: The game shows a lot about the marching and concentration for battle of Napoleon's army, about the differences between his army and those of his opponents in these years, and about the superiority of French leadership. It is too complicated and abstract in its mechanics for most uses. I have used the 1806 scenario with some simplification of the rules, and with the players on both sides sitting at separate boards where they could not see their opponents' dispositions or actions. I acted as umpire and kept the actual situation on my board. The results was very realistic regarding the conditions of warfare at the time, although the French group in my class did not include any Napoleons, and so they lost rather badly. (MC)

Cost: $11.00; $12.00 with box

Producer: Simulations Publications, Inc., 44 East 23rd Street, New York, NY 10010

LEE MOVES NORTH

John Young

Playing Data
Copyright: 1973
Age Level: grade 9 and up
Number of Players: 2
Playing Time: 6-8 hours
Preparation Time: 1 hour

Description: The game map is on the scale of 4.5 miles to the hexagon. Each turn represents two days. The game deals with two campaigns: Lee's first invasion of the North from August to September, 1862, and the Gettysburg campaign of 1863. Each game is twenty turns. The counters represent army units, fortifications, depots, trains, and generals. Movement is semihidden, pieces upside down mixed with dummy units moved like the others. A unit moves with difficulty and cannot attack unless it is in range of "command control" exercised by one of the few generals. (MC)

Comment: Tricky game mechanics sometimes get in the way of the simulation. For class play, the instructor should exercise his option to change things. (MC)

Cost: $11.00, $12.00 with box

Producer: Simulations Publications, Inc., 44 East 23rd Street, New York, NY 10010

LEGION

Playing Data
Copyright: 1975
Age Level: high school-college
Number of Players: 2
Playing Time: 1 hour
Materials: mapboard (with charts and tables), general rules booklet, special rules folder, die-cut counters, die

Description: This game is a revision of *Centurion.* The map has been enlarged and changed considerably at double the scale. Two additional kinds of units are used. The scenarios have been improved (with better victory conditions and terrain rules) and increased in number (from seventeen to twenty). The rules have been simplified and generally improved, although some options have been lost in the process, mainly the ability to use field fortifications. A new optional rule allows simultaneous movement. (MC)

Comment: See Yeoman. (MC)

Cost: $9.00, $10.00 with box

Producer: Simulations Publications, Inc., 44 East 23rd Street, New York, NY 10010

LUFTWAFFE

Louis B. Zocchi

Playing Data
Copyright: 1971
Age Level: grade 9 and up
Number of Players: 2
Playing Time: 3-16 hours
Preparation Time: 1-2 hours

Description: The map shows Germany and neighboring occupied areas on the scale of twenty miles to the hexagon. The map shows major cities, air bases, aircraft factories, oil refineries, industrial complexes, and a few rail lines. Each counter represents about one hundred fifty aircraft. Each turn represents twenty minutes. Each sequence of twenty turns further represents three months of combat. Counters can be turned over to show another face to represent losses. Players can play a single quarter game (twenty turns) or campaign games of twenty sequences of twenty turns each. In the campaign games both sides get new aircraft counters each turn. German reinforcements can be lessened by bombing. The "advanced game" allows Germans to improve their production and acquire jet aircraft earlier than they did historically. (MC)

Comment: Luftwaffe contains some doubtful historical information—that bombing aircraft factories was highly effective, for example. The emphasis of the game is on bomber vs. fighter combat, not on the effects of the bombing itself. (MC)

Cost: $12.00

Producer: Avalon Hill, 4517 Harford Road, Baltimore, MD 21214

MANASSAS

Thomas Eller

Playing Data
Copyright: 2nd edition, 1976
Age Level: grade 9 and up
Number of Players: 2
Playing Time: 4 hours
Preparation Time: 1 hour

Description: This game about the First Battle of Bull Run (the Union name) covers four days of the campaign on a scale of about 300 yards to the hexagon and two daylight hours to the turn (with no night action). Its rules are quite intricate, featuring written orders and simultaneous movement and combat. The number of combat units (brigades mostly) is quite small (twenty Union and fifteen Confederate) so that the difficulty of the system can be handled. Rules cover combat formations, the effects of leadership, and Cavalry operations by J.E.B. Stuart. Optional rules allow for weather and artillery. The latter increases the difficulty of the game and nearly doubles the number of pieces. (MC)

Comment: This game could be used as the basis of an appropriately confusing multiplayer game in which the McDowell and Beauregard players would write to their subcommanders, while the latter would prepare separate written orders for their own units. (MC)

Cost: $8.00

Producer: Game Designers' Workshop, 203 North Street, Normal, IL 61761

MECH WAR '77

James F. Dunnigan

Playing Data
Copyright: 1975
Age Level: high school-college
Number of Players: 2
Playing Time: 2-4 hours
Preparation Time: 1 hour
Materials: mapboard, charts and tables sheet, die-cut counters, rules booklet, die

Description: Mech War '77 is a simulation of small battles, mainly battles of the hypothetical immediate future between U.S. and Russian forces in West Germany. Eight scenarios deal with this subject while one each is devoted to a small Arab-Israeli 1973 battle and a hypothetical Russian-Chinese battle. The counters include eleven kinds of U.S. units, mostly kinds of platoons, seven kinds of West German, four kinds of Israeli, six kinds of British, five kinds of Chinese, and eighteen kinds of Soviet units. The British and West German units are not used in any of the scenarios. Movement and fire are semisimultaneous. Off-board artillery and air strikes aid the units on the board. Weapons include many kinds of armored vehicles, mines, helicopters, and antitank guided missiles. A game turn represents one to six minutes while a hexagon represents a piece of ground two hundred miles across. Games last from five to thirty turns. (MC)

Comment: This was originally intended as a replacement for *Red Star/White Star,* a game on the subject, but turned out to be different enough in its approach that both games were kept in print. This game is a companion to *Panzer '44*—the two have nearly identical rules, the differences due mainly to the differences in weapons. (MC)

Cost: $8.00, $9.00 with box

Producer: Simulations Publications, Inc., 44 East 23rd Street, New York, NY 10010

MERCENARY

Chris Ruffle

Playing Data
Copyright: 1977
Age Level: grade 9 and up
Number of Players: 1 to 5

Playing Time: 5-10 hours
Preparation Time: 2 hours

Description: *Mercenary* deals with diplomacy and conquest in Europe from 1494 to 1560 in eleven complex turns, worth six years each. The players represent "merchants" who move around Europe controlling the destinies of one power after another, if they are not very lucky, or staying in one place if they are. Such extremely mobile adventures never existed on such a high level, but they function like royal or imperial ministers who did exist. At the beginning of the game, major powers that have no player will automatically follow a set plan of conquest. The map of Europe uses large area for movement. The counters represent appropriate kinds of arms for the time and change during the game to reflect the development of new arms (crossbowsmen are replaced by arquebusiers, for example). Diplomacy is a vital part of each move as the players use money, marriages, and conquests (each major power has two daughters and two sons to marry off) to influence small countries. (MC)

Comment: Mercenary is a great idea, but it has grave difficulties. The rules have to be described as suggestive rather than descriptive. It may be better suited to class play than to social play, since in a classroom the instructor can impose new rules as the need for them arises, while social players must laboriously negotiate the rules as they go along. I slightly prefer this game to the very similar *Mighty Fortress* (below). (MC)

Cost: $10.00, $12.00 with box

Producer: Fantasy Games Unlimited, P.O. Box 182, Roslyn, NY 11576

MICRO-NAPOLEONICS

Playing Data
Copyright: 1975
Age Level: grade 11 and up
Number of Players: 2 or more
Playing Time: 2 hours or more
Preparation Time: 1-1/2 hours
Materials: include unit markers and terrain features to be cut out. Other terrain features must be hand made to construct a table-top battlefield

Description: This game is played with cardboard markers in the miniature fashion—with terrain made up for the occasion and negotiated victory conditions. The set comes with markers for forces representing 45,000 French and 35,000 British troops. Other cards of troops can be purchased separately. The game does not, of course, simulate a particular battle but instead the tactical conditions of the Napoleonic period. The rules cover firing and morale in a detailed manner but are sketchier for movement. (MC)

Cost: $2.50

Producer: Tabletop Games, 92 Acton Road, Arnold, Nottingham, United Kingdom

MIDWAY

Lindsley Schutz, Thomas N. Shaw, and Lawrence Pinsky

Playing Data
Copyright: 1964
Age Level: grade 9 and up
Number of Players: 2
Playing Time: 5-7 hours
Preparation Time: 1-2 hours

Description: The two search boards show *Midway* and the surrounding ocean divided into two sets of square grids. One set of counters represents the search board location of U.S. and Japanese ships. The players sit so that they cannot see each other and are allowed to spot in a limited number of squares each turn. When they find each other they move to the battle board for either air-to-ship or ship-to-ship combat. The other sets of counters represent aircraft in various numbers and individual ships. (MC)

Comment: This is a very realistic game except perhaps for the excessively large number of casualties suffered by attacking aircraft. In a classroom situation, the awkward search board rules could be replaced by the activities of an umpire. (MC)

Cost: $12.00

Producer: Avalon Hill, 4517 Harford Road, Baltimore, MD 21214

A MIGHTY FORTRESS

Rudolph W. Heinze, Redmond A. Simonsen, and Richard Berg

Playing Data
Copyright: 1977
Age Level: grade 9 and up
Number of Players: 6
Playing Time: 10-12 hours
Preparation Time: 1 hour

Description: This game deals with European power politics and religion, supposedly 1532-1555, although some of the situations that arise are more characteristic of earlier or later years. Four of the players represent the great powers of the period: the Hapsburg Empire of Charles V, France, England, and the Ottoman Empire. The fifth player is the Pope and the sixth (called the Lutheran player) represents the political/military interests of the German Protestant states, as well as the religious interest of Protestantism everywhere. The game has twenty-four turns, each representing one year. Each player has a set of goals which will get points if they are achieved on the last turn of the game. Mostly the goals are the conquest of new territories and cities. Goals for the Lutheran and the Pope involve the conversion or reconversion of territories. Each turn is preceded by a round of freewheeling diplomacy. Then each player plays in turn. Play consists of the movement of fleets, armies, and missionaries, and combat between rival fleets and armies or "theological debate" (which looks a lot like combat) between rival missionaries. At the end of the turn, players collect their taxes and raise new armies and fleets, which will appear in two years. An intricate financial system makes warfare outrageously expensive and any military movement almost as expensive, so players will attempt to sit out the struggle several years in a row to save money for future campaigns. (MC)

Comment: *A Mighty Fortress* has a number of excellent ideas, and it ought to be a great game and a great simulation, but it is disappointing in both ways. It ought to show the intimate relationship between politics and religion in the period, but in fact, none of the players has any interest in the religious struggle except the Pope and the Lutherans. The game seems to be unbalanced in favor of the Lutheran, at least between beginning players, for the Pope has no religious power in the game until it is half over and even then his power is only equal to that of the Lutheran, while he must take the offensive to accomplish anything. The representation of military and naval combat is quite odd. Military forces cannot really be destroyed, nor can they really be increased very much, so the power structure has a peculiar frozen quality. Then, in an era when armies were small and marched and fought concentrated, this game shows them operating on a broad front like World War I and II armies. (MC)

Cost: $9.00, $10.00 with box

Producer: Simulations Publications, Inc., 44 East 23rd Street, New York, NY 10010

MINUTEMAN

James F. Dunnigan, Joseph Balkoski, Edward Curran, and Redmond A. Simonsen

Playing Data
Copyright: 1976
Age Level: grade 9 and up
Number of Players: 2-4
Playing Time: 6-10 hours
Preparation Time: 1 hour

Description: This is a game about a highly speculative future. The United States is under a military dictatorship—or is being occupied by foreign armies as a consequence of the events in *Invasion: America* (q.v.). Opposition grows and leads eventually to open rebellion. One player is the government, or occupation forces, and the other one, two, or three are rebels. The rebels' object is to overthrow the government and, in the multiplayer games, to establish a new government over the rival rebels. The counters represent rebel agents (Minutemen) and organizations, government counterintelligence agents and organization, and guerrilla and government military units. The action is mechanical with, for example, the rebel rolling dice on a "Sedition Combat Results Table" to see what success his or her Minutemen are having in fomenting riots in cities. The map shows the United States and parts of Canada and Mexico. (MC)

Comments: *Minuteman* is a lively and interesting game. The politics of insurgency is handled very abstractly, but this is the only game that covers the subject along with the military side at all. The multiplayer rules are vague in some important areas. (MC)

Cost: $11.00, $12.00 with box

Producer: Simulations Publications, Inc., 44 East 23rd Street, New York, NY 10010

MISSILE CRISIS

D. Gallagher and D. Casciano

Playing Data
Copyright: 1974, 1975
Age Level: grade 9 and up
Number of Players: 2
Playing Time: 3-4 hours
Preparation Time: 30 minutes

Description: This is a game not about the historical Cuban missile crisis but about the military campaign that presumably might have come from it. The U.S. player commands the forces that are to invade Cuba, prevent Soviet reinforcement by sea, and destroy the missile sites by force. (MC)

Comment: This is a quick game on a campaign that surely might have been fought. It ought to have some nonmilitary elements, but those could be added. Its main problem is that the garish map hurts the eyes of those who look at it very long. (MC)

Cost: $7.00

Producer: Attack International Wargaming Association, 314 Edgley Ave., Glenside, PA 19038

MISSION ALOFT

William Streifer

Playing Data
Copyright: 1976
Age Level: grade 12 and up
Number of Players: 2
Playing Time: 1-8 hours
Preparation Time:
Materials: include 20 small copies of the map (on 10 sheets of paper) to use in plotting positions and moves

Description: The game map shows an imaginary country. The ten scenarios are based on real and hypothetical air actions of the 1960s and 1970s. In each case, one side is the defender of the country and the other side is trying to fly over it on reconaissance or bombing missions. The simulation is quite detailed but well explained. Aircraft simulated include the F-86 Sabre, the F4 Phantom, and MiGs 17, 19, 21, 23, and 25. Weapons include air-to-air missiles, surface-to-air missiles (SAMs), guided bombs, ordinary bombs. Planes fly at either high or low altitude. (MC)

Cost: $6.00

Producer: Jim Bumpas, 948 Lorraine, Los Altos, CA 94022

MODERN BATTLES: WURZBURG, CHINESE FARM, GOLAN, MUKDEN

James F. Dunnigan, Howard Barasch, Irad B. Hardy, and David C. Isby

Playing Data
Copyright: 1975
Age Level: high school-college
Number of Players: 2
Playing Time: 2-4 hours
Preparation Time: 1 hour
Materials: four mapboards, two copies of standard rules, two copies of charts and tables sheet, four exclusive rules leaflets, four sets of die-cut counters, die

Description: The four games use basically the same set of rules with a few additions (*Wurzburg* has helicopters, *Chinese Farm* and *Golan* have SAMs, *Mukden* has guerrillas, paratroops and gunboats, for example). They all have armored units, infantry units, infantry units (most of them mechanized), and—very important—artillery. Artillery can fight at a distance and is very valuable on attack or defense. The game maps, in color, show actual terrain at the scale of about one mile per hexagon. Each turn equals twelve hours. Two of the games, *Chinese Farm* and *Golan*, come from the recent past of the Arab-Israeli War of 1973, while the other two are set in the immediate future—*Wurzburg* is a town which the Americans contest with advancing Russians while *Mukden* is a town which the Chinese contest with advancing Russians. Each of the games comes equipped with three scenarios. (MC)

Comment: The game rules are basically simple but the artillery rules tend to put a larger strain than usual on players' abilities to add numerous numbers in their heads and to consider several alternatives at the same time. (MC)

Cost: $11.00 ($4.00 apiece separately), $12.00 with box

Producer: Simulations Publications, Inc., 44 East 23rd Street, New York, NY 10010

MODERN BATTLES II: BATTLE FOR JERUSALEM '67; YUGOSLAVIA; BUNDESWEHR; DMZ

Mark Herman, James F. Dunnigan, Redmond A. Simonsen, Phil Kosnett, Virginia Mulholland, and Joseph M. Balkoski

Playing Data
Copyright: 1977
Age Level: grade 9 and up
Number of Players: 2 mostly: 3 in Yugoslavia scenario
Playing Time: 3-5 hours
Preparation Time: 1 hour

Description: This game set contains one historical game, *Jerusalem*, on the battle between Israeli and Jordanian forces around that city in 1967, and three speculative games: *Yugoslavia*, on the battles that might rage around the city of Zagreb following a Soviet invasion of that country; *Bundeswehr*, on the North German battles following a Soviet invasion of that area; and *DMZ*, on a renewal of the Korean War. The games use a common fund of rules on terrain, movement, short-range combat, artillery combat, and air power. In addition, each game has appropriate special rules on special terrain, special kinds of units, city combat, command problems, nuclear weapons, and, in all the games except *Jerusalem,* untried units, i.e., units whose strength is not known until the moment of first combat. The units in the games represent battalions and regiments. The time scale is twelve hours per turn, and the distance scale is about one mile per hexagon. Each game has three or four scenarios, each of which is eight to sixteen turns long. (MC)

Comment: The rules for these games are much like those for the first Modern Battles set, with untried units as the major addition. (MC)

Cost: $13.00, $14.00 with box ($4.00 apiece separately)

Producer: Simulations Publications, Inc., 44 East 23rd Street, New York, NY 10010

THE MOSCOW CAMPAIGN

James F. Dunnigan

Playing Data
Copyright: 1972
Age Level: grade 9 and up
Number of Players: 2
Playing Time: 3-9 hours
Preparation Time: 2 hours

Description: The game map shows central western Russia on the scale of six miles to the hexagon. Each turn represents two days. The counters show infantry, cavalry, and armor units—corps, divisions, and brigades, mostly divisions. There are thirty-one turns in the long game. The player may also play part of the campaign with four different time schedules (eight or sixteen turns each). There are an assortment of alternate history "what-if" scenarios included. (MC)

Comment: This is a realistic game but a little too complicated for the amount of realism achieved. The game shows that the task that Hitler set for his army was nearly impossible. It shows also the effect of Russian ability to pour men into the battle, the effect of Russian weather, and the effect of German inability to keep supplies moving as fast as the front. (MC)

Cost: $11.00, $12.00 with box

Producer: Simulations Publications, Inc., 44 East 23rd Street, New York, NY 10010

MUKDEN 1905

Richard Spence and James Gabel

Playing Data
Copyright: 1976
Age Level: grade 9 and up
Number of Players: 2
Playing Time: 3-4 hours
Preparation Time: 30 minutes

Description: The Battle of Mukden was the largest field battle of the Russo-Japanese War. This game simulates the battle with division and brigade counters on a scale of three miles to the hexagon and one day to the turn. In addition to infantry and Cavalry, the Japanese have artillery units to reflect their superiority in that arm, while the Russians have leader units, who are included for the purpose of being subjected to a command control rule which freezes portions of the Russian forces erratically. Victory in the game depends on the number of enemy units destroyed. (MC)

Comment: This is a simple but very helpful treatment of a little-known battle that should be better known. (MC)

Cost: $5.50

Producer: S&G Games, 2105 Custer Ave., Bakersfield, CA 93304

MUSKET AND PIKE: TACTICAL COMBAT, 1550-1680

John M. Young

Playing Data
Copyright: 1973
Age Level: grade 9 and up
Number of Players: 2
Playing Time: 2-4 hours
Preparation Time: 1 hour

Description: The map represents a piece of terrain about 1 x 1.3 kilometers. Each hexagon represents a space fifty meters across, and each move represents five minutes of real time. The counters represent pikemen, musketeers, cavalry, and artillery, each in several kinds. There are eighteen scenarios, each based on a historical battle, each lasting from ten to twenty turns. (MC)

Comment: I have used Musket and Pike in my class, "War in Western Civilization," playing it with somewhat altered rules as a simultaneous movement game and using the team system I described under *Armageddon* (above). Played ordinarily it gives a fair representation of the tactical problems of the time although the game mechanics are too abstract. (MC)

Cost: $11.00, $12.00 with box

Producer: Simulations Publications, Inc., 44 East 23rd St., New York, NY 10010

MTB

Playing Data
Age Level: grade 9 and up
Number of Players: 2
Playing Time: 1-2 hours
Preparation Time: 30 minutes
Materials: include 2 cards of cut-out ships (bird's-eye view), with speed measure, firing and turning gauges, depth charge, torpedo and fire markers, and a pad of "combat order charts"

Description: *MTB* has no board but uses any convenient table or desk. The game simulates a typical clash between motor torpedo boats and a convoy or between hostile boats in the English Channel in 1942-43. The rules are fairly direct. Orders are written out simultaneously and executed serially. (MC)

Comment: The instructions for setting up a game are a little confusing. (MC)

Cost: $2.50

Producer: Tabletop Games, 92 Acton Road, Arnold, Nottingham, United Kingdom

NAPOLEON AT WATERLOO

James F. Dunnigan

Playing Data
Copyright: 1971, 1972
Age Level: grade 9 and up
Number of Players: 2
Playing Time: 2-4 hours
Preparation Time: 30 minutes

Description: This game covers only the battlefield of Waterloo on a scale of 400 meters to the hexagon. Each move represents one hour.

There are two games. The standard game uses mostly division-sized counters. The advanced game uses mostly brigades. (MC)

Comment: Neither the advanced game nor the standard game are particularly realistic, but the standard game is useful as an introductory game. (MC)

Cost: $11.00, $12.00 with box

Producer: Simulations Publications, Inc., 44 East 23rd St., New York, NY 10010

NAPOLEON AT WAR: MARENGO, JENA-AUERSTADT, WAGROM, AND BATTLE OF THE NATIONS

David C. Isby, Thomas Walczyk, Irad B. Hardy, and Edward Curran

Playing Data

Copyright: 1975
Age Level: high school-college
Number of Players: 2
Playing Time: 2-5 hours
Preparation Time: 30 minutes
Materials: four mapboards (with turn tracks), two standard rules leaflets, four exclusive rules leaflets, four sets of die-cut counters, die

Description: The four Napoleonic battles, dating from 1800 to 1813, are handled in the same simple fashion as the other Simulations Publications, Inc. Napoleonic battles. The maps are on scales from 400 to 800 meters per hexagon. The daylight represent one to three hours (the one day of *Jena-Auerstadt* is represented by twelve turns; the three days of *the Nations* by only twenty turns). The unit counters represent cavalry and infantry, which serve the same function, and artillery (except for *the Nations*) which may bombard from a distance. (MC)

Comment: The games are fast-moving and suggestive of the strategic situation of the original battles. One tactical system is not very realistic but tends to give realistic general results. The game maps are excellent. The games are simple and make good introductory games. (MC)

Cost: $13.00 ($4.00 apiece separately), $14.00 with box

Producer: Simulations Publications, Inc., 44 East 23rd St., New York, NY 10010

NAPOLEON'S LAST BATTLES: LIGNY, QUATRE BRAS, WAVRE, LA BELLE ALLIANCE [WATERLOO]

Kevin Zucker, Redmond A. Simonsen, and J. A. Nelson

Playing Data

Copyright: 1976
Age Level: grade 8 and up
Number of Players: 2
Playing Time: 2-8 hours
Preparation Time: 1 hour

Description: The last battles of Napoleon are the four battles of the Waterloo campaign. This game set can be used to separately play any of the battles or to simulate the campaign from the point at which the armies made contact to the end of the third day following. The basic system used is the same as that used in the *Napoleon at War* set above, but the rules introduced for the campaign game provide for much more realism without much more trouble. The geographic scale is 480 meters per hexagon. Most of the units represented are brigades. Each daylight turn represents one hour. (MC)

Comment: The campaign game is large and takes a large area with its four-part 34"x 44" map, but it offers many opportunities for involving a lot of people in the command structures of the three armies represented in the campaign, and the campaign game rules are very informative on command structure and its problems. A shorter game with some of the advantages of the longer game might be made by adapting the campaign rules to one of the battle games. See the introductory essay to this section for more information on this game. (MC)

Cost: $11.00, $12.00 with box ($5.00 apiece separately)

Producer: Simulations Publications, Inc., 44 East 23rd Street, New York, NY 10010

NARVIK

Frank Chadwick and Paul Richard Banner

Playing Data

Copyright: 1974
Age Level: grade 9 and up
Number of Players: 2
Playing Time: 4-6 hours
Preparation Time: 1 hour

Description: Narvik deals with the German invasion of Norway in April 1940, with four-day turns on a map using a scale of sixteen miles to the hexagon. Most of the land unit counters represent regiments and battalions. Both sides also use air units and naval units, some representing individual ships and others more abstract. The game begins with the German landing and proceeds for fifteen turns. The game is won on victory points gained for destruction of enemy units and holding territory at the end of the game. The rules emphasize organizational problems, the difficulties of Norwegian mobilization, and German supply problems. (MC)

Comment: Narvik is difficult and not well presented for beginning players, but it is a convincing simulation of many aspects of the campaign, particularly its opening phases. It shows the sheer audacity of the German invasion and the great importance of German airpower. It is a remarkable game for its integration of land, sea, and air forces and actions. (MC)

Cost: $8.75

Producer: Game Designers' Workshop, 203 North Street, Normal, IL 61761

NATO: OPERATIONAL COMBAT IN EUROPE IN THE 1970's

James F. Dunnigan

Playing Data

Copyright: 1973
Age Level: grade 9 and up
Number of Players: 2
Playing Time: 4-8 hours
Preparation Time: 1 hour

Description: The map shows West Germany and parts of the countries on its borders on a scale of sixteen kilometers to the hexagon. Each turn represents two days of a hypothetical war between the Warsaw Pact and NATO. The counters represent assorted land units on the division and brigade level, supply units and nuclear contamination markers (to mark the use of "tactical" nuclear weapons). "To simulate the use of strategic nuclear weapons, simply soak the map with lighter fluid and apply a flame." It is assumed that the air forces on each side cancel each other out. There are a few political rules to complicate the military action. The game has two basic scenarios, one of twenty war turns and the other of ten turns. (MC)

Comment: The game involves a large number of units and is difficult for that reason. (MC)

Cost: $9.00, $10.00 with box

Producer: Simulations Publications, Inc., 44 East 23rd Street, New York, NY 10010

THE NEXT WAR

James F. Dunnigan and Mark Herman

Playing Data
Copyright: 1978
Age Level: high school and up
Number of Players: 2-8
Playing Time: 4-30 hours
Preparation Time: 2 hours

Description: *The Next War* is a simulation of Warsaw Pact countries invading western Europe in the near future. The three section map covers Europe from Denmark to Belgium to Italy and Yugoslavia. Each hex represents fourteen kilometers, each unit represents a division, battalion, air group, or naval group. Rules include electronic warfare, nuclear contamination, air attacks, S.A.M. units, fuel and ammo supply limits, and a special naval segment. Units attack individually during movement and expend various amounts of movement points for different types of movement and combat. Units may greatly extend their normal movement allowance but are then subject to fatigue. (Richard J. Rydzel)

Comment: This is an interesting but complex game. There are so many possible die modifications on attacks that each attack may require ten or fifteen minutes to compute the right odds. Also, keeping track of all the air and antiaircraft units is a burden and subtracts from the game's playability. However, if one wishes to spend the time and effort on this game, it is a good simulation that seems to cover most areas of such a conflict. (RJR)

Cost: $26.00, $28.00 with box

Producer: Simulations Publications, Inc., 44 East 23rd Street, New York, NY 10010

1944: THE INVASION OF FRANCE AND THE BATTLE OF GERMANY

Lawrence J. Rusiecki

Playing Data
Copyright: 1972
Age Level: grade 9 and up
Number of Players: 2
Playing Time: 3-6 hours
Preparation Time: 1-1/2 hours

Description: The map shows France, the Low Countries and Western Germany. Each of the ten turns represents one month. The counters represent infantry troops, armor and parachute divisions, supply units, and ports. The Allied player uses an abstract "tactical air power" in conducting his land attacks. The Allied player must invade, overrun France, and penetrate deep into Germany before the end of the game. It is a very difficult task. There are two short games (eight turns each) which allow the allies to attempt a 1942 or a 1943 invasion. (MC)

Comment: This game gives a very good idea of the strategy of the war in France and Germany in 1944-1945. The shorter games show that earlier invasion attempts would not have been good ideas. (MC)

Cost: $6.00

Producer: Histo Games, 34 Sharon Street, Brooklyn, NY 11211

OBJECTIVE MOSCOW

Joe Angiolillo, Jr. and Phil Kosnett

Playing Data
Copyright: 1978
Age Level: high school and up
Number of Players: 2 or 3
Playing Time: 2-24 hours
Preparation Time: 1-2 hours

Description: The four-section map shows most of Europe, all of Russia, part of China, and parts of many adjacent countries. There are over 1000 unit counters which represent divisions, air wings, naval units and headquarters. Each hex represents an area sixty kilometers wide. There are five scenarios, including one for future possibilities. Basically, each scenario pits Russia against China or the U.S. Nato alliance. Rules include a reaction phase that allows the nonturn player a chance to move units between combat and the mechanized movement phase. Other rules include nuclear combat, Japanese Samurai units, which come out of a resurgence of Japanese tradition in the 1990s, cruise missiles, space marines, and the collapse of the Warsaw Pact. (Richard J. Rydzel)

Comment: This is an interesting but unwieldy game. Air units for example must return to the exact base they left before combat. With a total of thirty to fifty plane counters, it is difficult to remember where each started. The game also seems based in favor of the U.S. alliance in the representations of strengths of the counters. (RJR)

Cost: $19.00, $20.00 with box

Producer: Simulations Publications Inc., 44 East 23rd Street, New York, NY 10010

OCTOBER WAR

Irad B. Hardy and Mark Herman

Playing Data
Copyright: 1977
Age Level: grade 9 and up
Number of Players: 2
Playing Time: 2-4 hours
Preparation Time: 30 minutes

Description: This game simulates small unit actions during the 1973 Arab-Israeli War. The units represent platoons; armored units represent three vehicles each. The map shows an area that does not exist but is "typical." Its scale is 200 meters to the hexagon. Each move represents two minutes, and the eight individual scenarios last fifteen to twenty turns. There are also two sets of linked scenarios. Three short battles are played as a set with forces available depending on what has been done or is being done in the other scenarios. (MC)

Comment: This game gives some feeling of realism without great rules complexity. It suggests the reasons for Israeli tactical superiority and shows other factors that influenced the fighting in the Middle East. The game equipment can easily be used to develop other scenarios on the war. (MC)

Cost: $9.00, $10.00 with box

Producer: Simulations Publications, Inc., 44 East 23rd St., New York, NY 10010

OIL WAR

James F. Dunnigan

Playing Data
Copyright: 1975
Age Level: high school-college
Number of Players: 2
Playing Time: 2-3 hours
Preparation Time: 1 hour
Materials: mapboard (with chart, tables and tracks), die-cut counters, rules leaflet, randomizer chits

Description: This game simulates an imaginary descent by American troops on the Arabic oil fields around the Persian Gulf. The unit counters represent air landing, infantry, mechanized and armored brigades, and several kinds of aircraft squadrons. Games last eight turns, each representing two days. There are three scenarios, one in which the United States is acting with Israeli help, one in which Israel is neutral and one in which the attacker is Iran while the United States helps the smaller Arab states. (MC)

Cost: $3.00

Producer: Simulations Publications, Inc., 44 East 23rd Street, New York, NY 10010

OPERACAO LITTORIO

D. H. Casciano

Playing Data
Copyright: 1977
Age Level: grade 9 and up
Number of Players: 2-4
Playing Time: 6 hours
Preparation Time: 30 minutes

Description: *Littorio* is a game covering a fictional war in which Brazil takes on the rest of South America. Counters represent armies but have no numbers printed on them. Argentina may have nuclear weapons, Russians may indirectly help Brazil, and units may conduct air drops and amphibious assaults. (Richard J. Rydzel)

Comment: This could be an interesting game but the rules are vague and incomplete. The map is also vague and, because of its garish coloring, difficult to focus on. Revision is necessary before this game can be played in a school. (RJR)

Cost: $6.99

Producer: Attack International Wargaming Association, 314 Edgley Avenue, Glenside, PA 19038

OPERATION OLYMPIC

James F. Dunnigan

Playing Data
Copyright: 1974
Age Level: high school-college
Number of Players: 1 or 2
Playing Time: 3-4 hours
Preparation Time: 1 hour
Materials: mapboard (with charts and tracks), rules sheet, die-cut counters, die

Description: This game deals with an imaginary but historically planned event, the invasion of Japan in World War II. The map shows the southern-most Japanese island, Kyushu, on a scale of 6.5 kilometers per hexagon. The counters represent specific regiments and brigades of land forces or generalized American air power. The American player must choose invasion sites, invade, and attempt to seize the southern half of the island in a limited time. For the solitaire version, the Japanese actions are governed by die rolls on the "Japanese Doctrine Chart." (MC)

Cost: $11.00, $12.00 with box

Producer: Simulations Publications, Inc., 44 East 23rd Street, New York, NY 10010

OVERLORD

John Hill, Frank Chadwick, and John Harshman

Playing Data
Copyright: 2nd edition, 1977
Age Level: grade 9 and up
Number of Players: 2
Playing Time: 10 hours
Preparation Time: 1 hour

Description: *Overlord* is a simulation of the breakout of the Allies from the Normandy beachhead in World War II. Each hexagon represents 3.5 miles, counters are basically regiments but also include many battalion-sized units. Turns are three-day periods. There are several scenarios of various lengths and possible nonhistorical set-ups like the Rommel plan. Play sequence is movement and combat. Units may move from one zone of control to another. Any possible movement after combat depends on a die roll. Both sides have naval fire for defense and offense near the coast. Weather, commandos, armor restrictions, and allied attack restrictions are also included. (Richard J. Rydzel)

Comment: This is an interesting game, but it leaves no room for tactics and maneuvers. It is basically a slugging match with the winner being the player who inflicts the most losses. There is an interesting order of battle, but the unit strengths seem to overrate the German counters. This is a quick and easy game to learn and play but there are better simulations available on this subject. (RJR)

Cost: $11.98

Producer: Conflict Games Co., P.O. Box 432, Normal, IL 61761

PANZERARMEE AFRIKA

James F. Dunnigan

Playing Data
Copyright: 1973
Age Level: grade 9 and up
Number of Players: 2
Playing Time: 3-4 hours
Preparation Time: 1 hour

Description: This game, like the three which follow, has a map of the Northern coastal areas of Africa from Eastern Libya to Egypt, and covers the war in that area from 1941 to 1942. The units represent the historic brigades, regiments and battalions, supplies and trucks. Each move is a month. There are 20 moves. The supply rule is unusually comprehensive and complex. The British have a superiority of numbers but are unable to use it effectively because of a "command control rule" which paralyzes some of their units every turn. (MC)

Comment: For a short game, *PanzerArmee Afrika* gives the feel of the North African campaign well. The supply rule is abstract and difficult to remember, but the rest of the game moves smoothly. The game shows many of the factors that were important in the campaign but it does not show that Axis supply was erratic, a fact which continually influenced Rommel's plans to attack. (MC)

Cost: $9.00, $10.00 with box

Producer: Simulations Publications, Inc., 44 East 23rd Street, New York, NY 10010

PANZERBLITZ

James F. Dunnigan

Playing Data
Copyright: 1970
Age Level: grade 9 and up
Number of Players: 2
Playing Time: 3-8 hours
Preparation Time: 1-1/2 hours

Description: The game map, which can be arranged in various patterns, shows a variety of supposedly typical Russian terrain at the scale of 250 meters to the hexagon. Each move represents a six-minute span. The counters represent Russian and German units, companies and platoons, from the period 1943-1944. The counters show a large number of specific tanks and other armored vehicles, and units armed with guns, howitzers, mortars, antitank guns, submachineguns, rifles, etc. There are twelve scenarios, and the background pamphlet gives a great deal of information helpful in creating more. (MC)

Cost: $12.00

Producer: Avalon Hill, 4517 Harford Road, Baltimore, MD 21214

PANZER '44

James F. Dunnigan

Playing Data
Copyright: 1975
Age Level: high school-college
Number of Players: 2
Playing Time: 3-4 hours
Preparation Time: 1 hour
Materials: mapboard, two charts and table sheets, die-cut counters, rules booklet, die

Description: The game map shows a piece of presumably representative northwestern European terrain on a scale of 200 meters to the hexagon. Each turn represènts one to six minutes depending on the scenario. There are fourteen scenarios each based on a historical battle in the period from June 1944 to March 1945. The unit counters represent twenty-seven kinds of German units, twenty-four kinds of American, and seven kinds of British. These units are infantry platoons, gun batteries, many kinds of tank platoons, etc. Movement is partially simultaneous—both players decide simultaneously which units are going to move, but the actual movement is sequential. (MC)

Cost: $9.00, $10.00 with box

Producer: Simulations Publications, 44 East 23rd Street, New York, NY 10010

PANZERGRUPPE GUDERIAN

James F. Dunnigan and Richard Berg

Playing Data
Copyright: 1976
Age Level: grade 9 and up
Number of Players: 2
Playing Time: 3 hours
Preparation Time: 1 hour

Description: This game deals with the battle of Smolensk in July 1941. The map shows a large area around Smolensk. The unit counters represent Russian infantry, armored and mechanized divisions, a German infantry division, and German Panzer and mechanized regiments, which can unite to form divisions. Other counters represent Russian generals (and their staffs) with varied leadership abilities. Something of the fog of war is introduced by having "untried" Russian units, which means they operate inverted until they get into a fight. Then they are turned over to reveal their strength. Strengths vary from very weak, even nothing, to very strong. Each of the twelve turns represents two days. The terrain has woods, rivers, swamps, roads, railroads, and towns. Both players dispose of a small amount of abstract airpower in addition to their land forces. (MC)

Comment: *Panzergruppe Guderian* is a superb representation of the German Blitzkrieg in Russia in 1941—if the players take advantage of the possibilities. The rules are understandable enough that I have been able to give the game to students to play on their own. (MC)

Cost: $9.00, $10.00 with box

Producer: Simulations Publication, Inc., 44 East 23rd Street, New York, NY 10010

PANZER LEADER

Dave Clark, Nick Smith, and Randall C. Reed

Playing Data
Copyright: 1974
Age Level: high school-college
Number of Players: 2
Playing Time: 2-5 hours
Preparation Time: 1 hour
Materials: four-part mounted mapboard, five scenario cards, terrain effects and other charts card, combat results tables card, rules booklet, die

Description: *Panzer Leader* is based on the long-time favorite wargame, *PanzerBlitz,* with a different locale. The scene is France and western Germany from June 1944 to March 1945. There are twenty scenarios in which the players can command counters representing platoons and batteries of infantry, tanks, and guns. There are forty-five kinds of Allied (British and United States) ground units, including twenty-four kinds of armored vehicles. The Germans have thirty-nine kinds of units including nineteen kinds of armored vehicles. The Allied player sometimes has airpower in the form of light planes for spotting or fighter-bombers for ground support. The rules show several improvements over those of *PanzerBlitz,* but are basically the same. The new rules can be used for *PanzerBlitz* and the counters can be interchanged. (MC)

Cost: $12.00

Producer: Avalon Hill, 4517 Harford Road, Baltimore, MD 21214

PATROL!

James F. Dunnigan

Playing Data
Copyright: 1974
Age Level: high school-college
Number of Players: 2
Playing Time: 2-4 hours
Preparation Time: 1 hour
Materials: six-piece mapboard, combat-table card, two terrain-chart cards, vehicle crew status chart card, rules booklet, die-cut counters, die

Description: The six-part map shows assorted terrain features, the values of which are changed according to the different "terrain modes," on a scale of five meters to the hexagon. The map parts may be fitted together in various ways. The counters represent individual infantrymen. Other markers show variable "terrain": barbed wire, mines, shell craters, smoke screens, pillboxes. Other counters represent individual tanks, personnel carriers, and horses. The game is a descendent of *Sniper* (q.v.) and simulates small unit actions with five to fifteen men on each side. The game gives nearly 200 scenarios with five kinds of situations (patrol, reconnaissance, raid, ambush, and assault) and forty-one different times and places from France in 1914 through the Russian Civil War, colonial wars, many from World War II, to various hypothetical battles of the 1970s. A small amount of artillery fire (from off the board) is included in many scenarios. "Panic" and "preservation" rules limit the control that the player has over his men. (MC)

Comment: Like *Sniper,* this is a convincing and gripping simulation of situations that are often simulated in the cinema and on television. (MC)

Cost: $9.00, $10.00 with box

Producer: Simulations Publications, Inc., 44 East 23rd Street, New York, NY 10010

THE PELOPONNESIAN WAR

Stephen M. Newberg

Playing Data
Copyright: 1977
Number of Players: 2 or 3

Description: Athens and Sparta clash for control of the Greek world. Turns represent one year. . . . Hexes are 36 kilometers across. (Publisher)

Cost: $9 75

Producer: Simulations Canada, P.O. Box 221, Elmsdale, Nova Scotia, Canada, BON 1MO

PHARSALUS

Loren K. Wiseman and John Harshman

Playing Data
Copyright: 1977
Age Level: grade 9 and up
Number of Players: 2
Playing Time: 3-4 hours
Preparation Time: 1 hour

Description: The Battle of Pharsalus decided that Caesar rather than Pompey should rule the Roman world. Little is known about it, and most of that little comes from the writings of the master-propagandist who won the battle. Quite a bit is known about how Romans fought in general, and some of this is incorporated into the game. Each legionary counter represents a cohort, nonlegionairies are light infantry, archers, slingers, cavalry and, on Caesar's side only, what the designer anachronistically calls "voltigeurs" for want of a handy historical designation. The map has a minimum amount of terrain which is historically correct. The most interesting thing about the game is the way the combat rules provide for the modification of combat power by "efficiency" (generally higher in Caesar's outnumbered army) and "fatigue," which comes from fighting, running, and suffering missile fire. (MC)

Comment: The game has a good tactical flavor, although it does not reflect leadership as a tactical consideration. *Pharsalus* is not the best battle to turn into a tactical game because the enemy armies are so much alike it tends to get dull. (MC)

Cost: $8.75

Producer: Game Designers' Workship, 203 North Street, Normal, IL 61761

PLOT TO ASSASSINATE HITLER

James F. Dunnigan, Fred Georgian, Greg Costikyan, and Redmond A. Simonsen

Playing Data
Copyright: 1976
Age Level: grade 9 and up
Number of Players: 2
Playing Time: 4 hours
Preparation Time: 1 hour

Description: This is a very abstract game which at first glance looks rather concrete. The players do not represent anyone in the situation but instead simply guide the fortunes of one side or the other. The board contains a group of adjacent areas labeled to represent places in occupied and neutral Europe. "Berlin" takes up more than half the allotted area and contains many hexagons, including some areas labeled as different kinds of headquarters. The counters for the map represent individuals who figured or might have figured in the plot. Some are SS men and protect Hitler, others are Abwehr (plus one civilian, Karl Goerdeler) and opposed to Hitler. Others are neutral when the game begins and can be persuaded to join one camp or the other or remain neutral. The game system used to represent this political activity is based on war gains. The active characters on the board attack each other on a combat results table called a "harassment table" or "attack" neutrals on a "recruitment table." The units move on the board and can maneuver to cut off each other's retreat from combat, but it is hard to say what such action represents. (MC)

Comment: The game is interesting but the representation of political infighting in mechanical fashion often gives grotesque results. (MC)

Cost: $12.00, $13.00 with box

Producer: Simulations Publications, Inc., 44 East 23rd Street, New York, NY 10010

THE PUNIC WARS

Irad B. Hardy

Playing Data
Copyright: 1975
Age Level: high school-college
Number of Players: 2
Playing Time: 3-6 hours
Preparation Time: 1 hour
Materials: mapboard (with charts, tables, and tracks, rules leaflet, die-cut counters, randomizer chits (die substitute)

Description: This game simulates all three Punic Wars. The first two are nearly fair contests; the third a foregone conclusion. None of the wars involves a time limit. The map shows the western Mediterranean, with north Africa, much of Spain, a little of Gaul and all Italy. Each move represents one year. The counters show land or naval forces. Combat and movement, especially sea movement, depend a great deal on the leadership factors of various leaders. Roman leaders are uneven with a mixture of 1s and 2s, while the Carthaginian leaders are professional 2s. Hannibal and Scipio Africanus are 3s. Players build up and maintain their forces each year according to the state of their treasuries which are replenished according to the number of provinces they control. (MC)

Comment: This is a sensitive rendition of the strategic problems of the three wars. Battles are excessively chancy perhaps but leadership advantages make them less so. The diplomacy rules in my experience do not seem to work but they are easily replaced or ignored. The second Punic War is the most complicated and most interesting of the scenarios but required the hand manufacture of more counters when the limited numbers supplied with the game were already on the board. (MC)

Cost: $4.00

Producer: Simulations Publications, Inc., 44 East 23rd Street, New York, NY 10010

RAID!

Mark Herman and Tony Merridy

Playing Data
Copyright: 1977
Age Level: grade 9 and up
Number of Players: 2
Playing Time: 2-4 hours
Preparation Time: 30 minutes

Description: *Raid!* deals with small-scale commando or special forces operations. Each unit counter represents a group of four men, or a weapon team. Other markers represent vehicles and helicopters (used for transport or support). Five of the eight scenarios are based on historical events from World War II and Vietnam and include a very uninteresting Entebbe raid. The other scenarios are "representative." The designer encourages most players to design their own scenarios. (MC)

Cost: $9.00, $10.00 with box

Producer: Simulations Publications, Inc., 44 East 23rd Street, New York, NY 10010

RAKETNY KREYSER

Stephen M. Newberg

Playing Data
Copyright: 1977
Number of Players: 2

Description: A broad spectrum survey of the naval and naval air units that would be engaged in any major world conflict from now until the turn of the century. (Publisher)

Cost: $9.75

Producer: Simulations Canada, P.O. Box 221, Elmsdale, Nova Scotia, Canada BON 1MO

RAPHIA, 217 B.C.

Marc Miller and Frank Chadwick

Playing Data
Copyright: 1977
Age Level: grade 8 up
Number of Players: 2
Playing Time: 2-3 hours
Preparation Time: 30 minutes

Description: This is a simple game but one that seems to accurately reflect the warfare of the period. It deals with a battle in 217 B.C. in the Sinai between the two Diadochi kingdoms of the Ptolemies and the Seleucids. The forces include heavy (phalanx) infantry, light infantry, cavalry, and elephants. The rules stress the importance of morale, the immobility of the phalanx, the difficulty of tackling elephants with cavalry, the unreliability of elephants, especially the African elephants of Ptolemy, and to a small degree the value of the leadership of the two kings who were present on the field. (MC)

Comment: The game is short partly because the rules end the battle before it comes to a conclusion by providing for a point count to decide the game. There is room for at least three people on a side, especially as the game requires a lot of dice rolling. See the introductory essay for more information. (MC)

Cost: $5.00

Producer: Games Designers' Workshop, 203 North Street, Normal, IL 61761

RED STAR/WHITE STAR: TACTICAL COMBAT IN EUROPE IN THE 1970's

James F. Dunnigan

Playing Data
Copyright: 1972
Age Level: grade 9 and up
Number of Players: 2
Playing Time: 3-6 hours
Preparation Time: 1 hour

Description: The game map shows a hypothetical, supposedly typical, piece of contemporary western Europe on the scale of 300 meters to the hexagon. Each turn represents six minutes and forty seconds. Most of the counters represent platoons, companies, and battalions of Russian, American, and West German soldiers. The ten scenarios pit these units against each other in hypothetical incidents from a hypothetical, mostly nonnuclear, war of the 1970s. The counters show specific types of armored units and artillery weapons. Some of the units are armed with wire-guided antitank missiles. The American counters include observation and armed helicopters, and both sides have abstract "air strike" counters. (MC)

Comment: It is essential that the revisions in the errata sheet be used when playing *Red Star/White Star*. This game is a part of a series of games (beginning with *PanzerBlitz*) that is discussed above. It came after *Combat Command*, before *Kampfpanzer* and *Desert War*. *Red Star/White Star* generates the illusion of accurately reflecting events that have not happened, except in one way. If the game is played without considering the unit size, everything seems all right. But when the player notices the unit size, he realizes that the game is saying that five M60A1 American tanks are superior to 10 T62 Russian tanks. And that is hard to believe. (MC)

Cost: $11.00, $12.00 with box

Producer: Simulations Publications, Inc., 44 East 23rd Street, New York, NY 10010

RED SUN RISING

Frank Davis and Mark Herman

Playing Data
Copyright: 1977
Age Level: grade 9 and up
Number of Players: 2
Playing Time: 8-12 hours
Preparation Time: 2 hours
Materials: the game package contains two pages of hints and designer's notes and a sixteen-page article on the military history of the war

Description: This game simulates the entire Russo-Japanese War, both naval and land aspects. The board shows a small map of the seas around Japan for naval movement and a large one of northern Korea, Manchuria, and southeastern Siberia for land movement and combat. Naval combat takes place on a separate, abstract "Tactical Naval Display." The strategic game, including all movement and land combat, is on the scale of one move per month. Naval combat allows as many moves on an indeterminate scale as are necessary to decide the battle. (MC)

Comment: This game has a few puzzling rules that would need to be solved, but its main problem for class use is that it has no short scenarios. (MC)

Cost: $11.00, $12.00 with box

Producer: Simulations Publications, Inc., 44 East 23rd Street, New York, NY 10010

REVOLT IN THE EAST

James F. Dunnigan, Christopher Allen, and Redmond A. Simonsen

Playing Data
Copyright: 1976
Age Level: grade 9 and up
Number of Players: 2
Playing Time: 3 hours
Preparation Time: 30 minutes

Description: This game posits the outbreak of revolts against Soviet domination in the eastern European countries followed by a NATO invasion of the area. The units represent ground, airborne, and air force corps and armies. Two scenarios are placed in futures, one of which involves Yugoslavia as well as Warsaw Pact countries. The other two scenarios are placed in the year of the Hungarian suppression, 1956, and the year of the Czechoslovakian suppression, 1968, with history changed to allow the conflicts to spread. (MC)

Cost: $4.00

Producer: Simulations Publications, Inc., 44 East 23rd Street, New York, NY 10010

RHEIN BUNG

Tony Morale

Playing Data
Copyright: 1971, 1976
Age Level: grade 9 and up
Number of Players: 2
Playing Time: 2-6 hours
Preparation Time: 30 minutes

Description: This is a semiboard game on the sortie of the German battleship Bismarck in World War II which led to its destruction, the sortie code named "Operation Rheinübung" by the Germans. The name is inelegantly altered by the publisher for what he describes as copyright reasons. The strategic part of the game is played on two boards (so that the opposing players cannot see each others' locations). If surface vessels locate each other, they proceed to settle matters using cardboard cut out ship plans on a convenient floor. (MC)

Comment: The rules are simple. For classroom use, the cumbersome naval search technique could be replaced by the services of an umpire, and different commands could be given to different students. (MC)

Cost: $7.00

Producer: Attack International Wargaming Association, 314 Edgley Avenue, Glenside, PA 19038

RICHTHOFEN'S WAR

Randall C. Reed

Playing Data
Copyright: 1972
Age Level: grade 9 and up
Number of Players: 2 or more
Playing Time: 1-20 hours
Preparation Time: 30 minutes

Description: The game map is a hexagon grid laid over a full color painting of an aerial photograph. There are twenty British, two French, and twelve German aircraft types available. As in the previous game, each plane should have one pilot, and the aircraft are significantly different in performance. There is an even more fully developed set of scenarios than Flying Circus, and a long game which covers the operations of one German and two allied squadrons for a week. (MC)

Comment: The games *Fight in the Skies* and *Richthofen's War* have the same subject—plane-to-plane combat in World War I. *Fight in the Skies* is the most difficult to learn and to play because it does more with smaller details of combat. On the whole, it is not very suitable for class plays. So if I wanted to use a game on this subject, I would use *Richthofen's War.* (MC)

Cost: $12.00

Producer: Avalon Hill, 4517 Harford Road, Baltimore, MD 21214

RISE AND DECLINE OF THE THIRD REICH

John Prados

Playing Data
Copyright: 1974
Age Level: high school-college
Number of Players: 2-5
Playing Time: 5-20 hours
Preparation Time: 2 hours
Materials: three-part mounted mapboards, five scenario cards, rules booklet, die-cut counters, die

Description: The map shows Europe, North Africa, and the New East on a small scale. Each move represents one season of three months. The game deals, at the players' option, either with the whole war in Europe from the conquest of Poland to the end, or with a period of two years starting in 1939, 1942, or 1944. In the five player version, four of the players represent Germany, Italy, Russia, Britain, and the last represents both France and the U.S. The players must decide not only military strategy but also production strategy and, of course, diplomatic strategy, particularly important in the multiplayer versions. Each country gets an income each year of "Basic Resource Points," which are used to create new units, to declare war, to conduct offensives, or to reinvest for future production. Counters represent land, sea, and air forces, at different levels of simulation: the sea and air forces are quite abstract while the land forces are less so since they are divided into infantry, armor, and airborne forces. There is a large number of rules to allow for many of the diplomatic and military events of the war. (MC)

Comment: *Third Reich,* as it is generally called for short, is an innovative and challenging game which allows a player to experience the complexity of events in a total war. Although the final goal is clearly defined by the victory conditions, it is seldom clear which of a set of competing immediate goals ought to be pursued. Opportunities often appear and then disappear unseized because the production plans of several turns before did not provide for the necessary resources. The game breaks down in many details however. It is impossible for the Germans to make quick work of Poland and France, for example, and this throws the whole beginning of the war off the historical track. (MC)

Cost: $12.00

Producer: Avalon Hill, 4517 Harford Road, Baltimore, MD 21214

ROAD TO RICHMOND

Joe Angiolillo

Playing Data
Copyright: 1977
Age Level: grade 8 to college
Number of Players: 2
Playing Time: 3 hours
Preparation Time: 45 minutes

Description: This game uses the same basic system as the *Blue and Gray* games to simulate the Seven Days' Battles, June 26–28, 1862. As the game starts, the larger Union army is mostly off the board. The smaller Confederate army is in a position where it might destroy the Union army in detail as the Union forces attempt to enter the board and withdraw in another direction without losing excessive casualties to the Confederates. There are twenty turns, two nights and eighteen days. The days turns represent about two hours each. The units involved are brigades. (MC)

Comment: The game uses the well-tried *Blue and Gray* (q.v.) system. (MC)

Cost: $4.00

Producer: Simulations Publications, Inc., 44 East 23rd Street, New York, NY 10010

THE RUSSIAN CAMPAIGN

John Edwards, Tom Oleson, and Donald Greenwood

Playing Data
Copyright: 1976
Age Level: grade 9 and up
Number of Players: 2
Playing Time: 3-10 hours
Preparation Time: 1 hour

Description: This game deals with the whole Russo-German War from the invasion of Russia to the capture of Berlin on a small scale. Each move represents a two-month period. Most unit counters represent corps-sized units. Alternate scenarios allow for various starting points. The complete game lasts twenty-five turns. The rules allow two impulse movements so armored forces can exploit holes in enemy lines. Other rules provide for sea movement, partisans, German Stukas, Russian paratroops, and supply. Victory is confirmed by the elimination of counters representing Hitler and Stalin. (MC)

Comment: This game can be compared to *Barbarossa* (q.v.) and *East Front* (q.v.). So far I am ranking this one third, but that is partly, I suspect, because I simply have not played it enough. (MC)

Cost: $12.00

Producer: Avalon Hill, 4517 Harford Road, Baltimore, MD 21214

RUSSIAN CIVIL WAR

James F. Dunnigan, Redmond A. Simonsen, and Frank Davis

Playing Data
Copyright: 1976
Age Level: grade 9 and up
Number of Players: 3 to 6
Playing Time: 3-6 hours
Preparation Time: 1 hour

Description: This game separates the players a little more from reality than most games, since the players in this game do not represent any particular group in the Russian Civil War. Instead, each player draws a "hand," so to speak, which may be composed of Reds, Whites, interventionists, and non-Russian nationalists in a variety of combinations. Then, as the game proceeds, the player has to decide whether to try to win as a Red or as a White.

The game board shows European Russia and its neighbors with areas for movement marked out. Asiatic Russia is represented by an abstract group of areas. The counters used on the map represent various fighting forces and leaders of the Reds and the Whites. The leaders are ranked as one-, two-, or three-point leaders. Lenin, Trotsky, Wrangel, and Deniken are three. The players are given control of the leaders and move them to get control of the troops. Two other counters represent the Czar and the Imperial gold. Off-board counters can be drawn and used to assassinate other player's leaders or to show political power in the "politburo," which can take Red leaders away from players. During the first few turns, the game appears to be interminable, but after the fifth turn, the replacement of combat units closes, and the war soon runs down. The Reds usually win, but any of the players may turn out to have more Red points than another. (MC)

Comment: The game has some interesting points, but the lack of identity that it forces on the player makes it more difficult than it should be to use in a serious educational application. (MC)

Cost: $12.00, $13.00 with box

Producer: Simulations Publications, Inc., 44 East 23rd Street, New York, NY 10010

THE RUSSO-JAPANESE WAR: PORT ARTHUR AND TSUSHIMA

Marc William Miller

Playing Data
Copyright: 1976
Age Level: grade 9 and up
Number of Players: 2
Playing Time: 3-10 hours
Preparation Time: 1 hour

Description: This game simulates either the whole Russo-Japanese War or just the naval portion (under the name *Tsushima*) or just the land part (under the name *Port Arthur*). If one game is played without the other, actions in the other area are represented by appropriate die rolls. The naval game, in which battles are decided on an abstract battle board, while strategic movement takes place on a partly abstract strategic map, has several scenarios, including two battle scenarios. The rules that connect the two games are short but adequate. Victory in the land and whole war games is decided by territorial control, or, failing a clear-cut victory for either side, by diplomacy. (MC)

Comment: Some of the rules need to be interpreted in the spirit of the game rather than by their letter, so active refereeing is necessary if you want to use this game. (MC)

Cost: $10.00; *Port Arthur* alone $6.00; *Tsushima* alone $6.00

Producer: Game Designers' Workshop, 203 North Street, Normal, IL 61761

SEELOWE

John M. Young

Playing Data
Copyright: 1974
Age Level: grade 9 and up
Number of Players: 2
Playing Time: 3-4 hours
Preparation Time: 1 hour

Description: *Seelowe* is about a hypothetical campaign in which the Germans invade Britain in 1940 after the defeat of the British air force and navy. The map shows southern Britain on a scale of five miles to the hexagon. Each of the fifteen turns represents two days. The counters represent British divisions and brigades, German regiments British partisans, air units, and German supplies. The British player sets up his units. Then the German player begins to land troops on the beaches. The victor is decided by a victory point formula based on the number of German-held ports and the strengths of the two armies at the end of the game. The major German problem is the weather, which interferes with land reinforcements and supplies. There are several alternate scenarios which may be used with the game. (MC)

Cost: $9.00, $10.00 with box

Producer: Simulations Publications, Inc., 44 East 23rd Street, New York, NY 10010

1776

Randall C. Reed

Playing Data
Copyright: 1974
Age Level: grade 9 and up
Number of Players: 2
Playing Time: 2-40 hours
Preparation Time: 2 hours

Description: The northern mapboard shows the Northern states and

Canada. The southern mapboard shows the Southern states. Each hexagon represents an area about "18.6 miles" across. Each turn represents a period of one month. The counters represent various points of soldiers. Each point is about 500 men. The counters represent British regulars, Tory "militia," Indians, Continentals, rebel militia, French, battle fleets, transport fleets, boats, forts, entrenchments, supply units, and magazines. The game comes with a simple basic game, four complex but short scenarios of five to nine moves, and three large games of eighteen, twenty-four, and sixty turns. Victory is decided generally on the basis of the control of "strategic towns." (MC)

Comment: I used the short games and the long games in my "War" course. The students were content with the limited accuracy of the short games but they severely criticized the long game as being too tricky and too unrealistic with its allowing perfect coordination on the British side. This problem could be solved with a divided command game as described in *American Revolution.* In general, however, the long game is too detailed and too long for class play. See the introductory essay to this section for more on this game. (MC)

Cost: $12.00

Producer: Avalon Hill, 4517 Harford Road, Baltimore, MD 21214

7TH CAVALRY

Playing Data
Copyright: 1976
Age Level: grade 9 and up
Number of Players: 2
Playing Time: 2-4 hours
Preparation Time: 30 minutes

Description: This is a tactical game with counters representing two to five men each. The counters represent Indians and Cavalry as armed with different weapons (from lances to Winchesters) and as mounted or dismounted. Also included are army Gatling guns and cannon. Ten scenarios represent various actions and include three on the Battle of Little Bighorn. (MC)

Comment: The game looks interesting, but the rules are rather sketchy. There is a suggestion for replaying the Battle of the Little Bighorn with various Indian orders of battle so Custer will once more be tempted to charge against great odds. This is particularly suggestive, but the instructions for making it work are incomplete. (MC)

Cost: $7.00

Producer: Attack International Wargaming Association, 314 Edgley Avenue, Glenside, PA 19038

Sidi Rezegh 1941

R.J. Hlavnicka, Dennis P. O'Leary, and Michael A. Trdan

Playing Data
Copyright: 1977
Age Level: grade 9 and up
Number of Players: 2
Playing Time: 3 hours
Preparation Time: 30 minutes

Descriptiion: This is a short, simple game on the battles around Tobruk in November 1941. The units are divisions, regiments, brigades, reconaissance units, and artillery units. Abstract air points help out the British. The scales of the game are uncertain, but it has fifteen moves. (MC)

Cost: $6.00

Producer: Excalibre Games, Inc., Box 29171, Brooklyn Center, MN 55429

SIEGE!

Richard Jordison

Playing Data
Copyright: 1974
Age Level: high school-college
Number of Players: 2
Playing Time: 2-4 hours
Preparation Time: 1 hour
Materials: mapboard, two supplementary map sections, two blank map sections, terrain effects/combat results chart sheet, die-cut counters

Description: The game boards shows a section of a medieval city. This can be covered by the extra map sections so as to replace the medieval walls with a Roman mile-castle or Saxon Burh. The scales are ten yards to the hexagon, five to twenty-five men to the counter and ten to fifteen minutes to the turn. The time scale is set to the assault part of a siege and it allows the preliminaries like bombardment to be gotten over rather too quickly. Countries represent several types of armed men and assorted siege engines and ladders. There are two scenarios, but they are sketchy and uninteresting. The game includes many pieces that can be used by players to set up their own scenarios. (MC)

Cost: $5.50

Producer: Fact and Fantasy Games, P.O. Box 1472, Maryland Heights, MO 63043

THE SIEGE OF CONSTANTINOPLE

Richard H. Berg and Eric Goldberg

Playing Data
Copyright: 1978
Age Level: grade 9 up
Number of Players: 2
Playing Time: 3-4 hours
Preparation Time: 1 hour

Description: The subject of the game is the siege of 1453 which resulted in the fall of Constantinople to the Turks. The map shows the whole city, but the game is played almost entirely in the outer wall area (one-third of the map at most). Each hexagon represents 200 yards except in the wall area, where the terrain has to be spread out a bit to allow room for the action. Each turn represents two days, but assault "turns" are subdivided into ten "impulses" of indeterminate length in which the assault is conducted. (MC)

Comment: The map is very attractive and gives the general idea of the city adequately. The game plays well in details but it is probably too difficult overall for the Turks, who will have a hard time pulling off a historical result. (MC)

Cost: $9.00, $10.00 with box

Producer: Simulations Publications, Inc., 44 East 23rd Street, New York, NY 10010

THE SIEGE OF JERUSALEM, 70 A.D.

Stephen F. Weiss and Fred Schachter

Playing Data
Copyright: 1976
Age Level: grade 8 and up
Number of Players: 2
Playing Time: 4-20 hours
Preparation Time: 1-2 hours

Description: This is a detailed simulation of the Roman conquest of Jerusalem under Titus. The geographical scale is fifty meters to the hexagon. The game proceeds by one to five assault periods, each of eight to about thirty turns. The counters represent a variety of forces and implements including legionary infantry, light infantry, archers, cavalry, siege engines, movable towers, and ladders. Ramps, wall damage, and breaches have appropriate markers. Alternate scenarios allow the simulation of the ejection of the Roman garrison or the first Roman attempt to retake the city of 66 A.D. Another alternate scenario simulates only the main Roman assault. (MC)

Comment: *The Siege of Jerusalem* has an exceptionally detailed and attractive map. Its size (29" x 45") makes it a little hard to store between sessions, but it is in four parts that can be stored separately. In a classroom, the command of the two sides could be easily divided into four legionary and one overall command on the Roman side and four faction and one overall command on the Jewish side. (MC)

Cost: $15.00

Producer: Historical Perspectives, Box 343, Flushing Station, Flushing, NY 11367

SINAI

James F. Dunnigan

Playing Data
Copyright: 1973
Age Level: grade 9 and up
Number of Players: 2
Playing Time: 4-6 hours
Preparation Time: 1½ hours

Description: The game map shows Israel, the Sinai peninsula, and parts of Lebanon, Syria, Jordan, Saudi Arabia, and Egypt on the scale of twelve kilometers to the hexagon. The land unit counters represent brigades and battalions. Other counters represent Israeli air strikes, supply units, trucks, and SAM sites. Four different games can be played: the 1956 war (sixteen turns), the 1967 war (twelve turns), the 1973 war (variable), or a hypothetical war placed in the late 1970s. Each turn represents one-half day in 1956 or 1967 and one day in the two 1970 games. There are large number of alternate history scenarios. Each game has an assortment of political as well as military rules. (MC)

Comment: Trying to fit the game to three different wars has resulted in many awkward rules switches. Also some rules, especially the air rules, are uncomfortably abstract. Sinai was designed in unusual circumstances. The staff of Simulations Publications, Inc. were in the process of developing the game, which then included a hypothetical war in the 1970s, when a real war in 1973 broke out. Immediately they began watching television and reading newspapers and redesigning the game on a day-to-day basis. Finally, after the cease fire they finished the game with three instead of two historical scenarios. (MC)

Cost: $8.00, $9.00 with box

Producer: Simulations Publications, Inc., 44 East 23rd Street, New York, NY 10010

SIXTH FLEET

James F. Dunnigan

Playing Data
Copyright: 1975
Age Level: high school-college
Number of Players: 2
Playing Time: 3-4 hours
Preparation Time: 1 hour
Materials: mapboard (with charts and tracks), rules leaflet, die-counters, die

Descriptioion: This game deals with an imaginary sea war between NATO and Soviet air and sea forces in the Mediterranean. The map shows the central and eastern Mediterranean and the surrounding land on a scale of about forty-five nautical miles to the hexagon. Each move represents one-third of a day. The scenarios provide for a ten-turn and a twenty-one-turn game. The latter allows the opposing Mediterranean fleets to be reinforced by a Franco-British fleet from the Atlantic and by a new Soviet fleet from the Black Sea, presumed to be released by the land conquest of Turkey. The game starts out with plenty of action and then tends to run down into a stalemate as the Soviet fleet is bottled up in the East while the NATO FORCES are excluded from the East. (MC)

Comment: *Sixth Fleet* is a lively game but the game system forces the opposing units into competing battle lines and makes their operations look more like land operations than sea-air operations. This gives players a strange feeling of unreality. (MC)

Cost: $9.00, $10.00 with box

Producer: Simulations Publications, Inc., 44 East 23rd Street, New York, NY 10010

SNIPER!

James F. Dunnigan

Playing Data
Copyright: 1973
Age Level: grade 9 and up
Number of Players: 2
Playing Time: 3-8 hours
Preparation Time: 1 hour

Description: The map shows a small section of an ideal city at the scale of two meters to the hexagon. The counters represent individual soldiers moving through the buildings of the city. They are armed with rifles, grenades, sub-machine guns, automatic rifles, machine guns, and sometimes flame throwers and antitank weapons. Special counters represent tanks and trucks in some situations. Movement and firing is simultaneous according to written orders. A panic rule erratically voids some of the movement and firing ordered. Each move represents two seconds of activity and thirty seconds of time passed. There are three basic scenarios: patrol, ambush, and block clearing. These are modified to fit city fighting situations in sixteen different times and places in World War II, from Leningrad, 1941, to Berlin, 1945. (MC)

Comment: This is a very accurate simulation of the danger and activity in a city-fighting situation. It is very gripping to students, who have watched the same thing on TV frequently and who identify readily with their roles in the game. (MC)

Cost: $8.00, $9.00 with box

Producer: Simulations Publications, Inc., 44 East 23rd Street, New York, NY 10010

SOLDIERS: WWI TACTICAL COMBAT, 1914-1915

David C. Isby

Playing Data
Copyright: 1972
Age Level: grade 9 and up
Number of Players: 2
Playing Time: 2-4 hours
Preparation Time: 1 hour

Description: The map represents a piece of land with a variety of terrain, including a lot of open terrain, at a scale of one hundred meters to the hexagon. One move represents ten minutes. There are five sets of die-cut counters, one for each army: British, French, German, Belgian, and Austro-Hungarian. The counters represent infantry companies, machine gun sections and companies, cavalry squadrons (mounted and dismounted), and field gun and howitzer batteries and platoons. Other counters show improved positions and trenches. There are fourteen scenarios which pit various sized forces against each other. (MC)

Comment: This is one of the best games I have used in my classes. The rules are fairly simple and mostly logical. The individual scenarios are short. The game shows vividly the tactical conditions of 1914, before the building of the great trench lines. It is still possible to advance, if the advance is carefully planned to take advantage of cover and if it is adequately supported by artillery, but any attempt to rush the defenders across open ground is punished severely. I generally have the students play Scenario 7, "Belgian Defensive Action on the Schipdonck Kanal (October 11th, 1914)," in which the Germans, with twenty-four infantry companies and accompanying artillery, cavalry, and machine guns, try to establish a bridgehead across the canal, defended by eleven Belgian infantry companies with their machine guns, artillery, and cavalry. German players who size up the situation correctly, who are not impatient, who use their artillery correctly, and who use the available cover in approaching the Belgian defenses, are successful. But if they make only a few mistakes, they will suffer so many losses as to make their attack impossible. If they lose, they will repeat history. This is an excellent game, because of its basic simplicity, to play with limited information. An umpire with a master board can easily coordinate the activities on the two players' boards and feed them limited information or even misinformation as they might be likely to discover it. Another kind of difficulty may be added to the lives of the players. They can be introduced to the game without being taught all the rules. Particularly, they can be kept in the dark on the subject of the combat rules, reproducing in some degree the confusion of most of the officers of 1914, who did not know what the odds were of successfully charging a concealed defender with a battalion of infantry across open terrain. Then the game becomes a good means to teach people why wargames are a lot better for them than the real thing. See the introductory essay to this chapter for more on this game. (MC)

Cost: $11.00, $12.00 with box

Producer: Simulations Publications, Inc., 44 East 23rd Street, New York, NY 10010

SOLOMONS CAMPAIGN

James F. Dunnigan

Playing Data
Copyright: 1973
Age Level: grade 9 and up
Number of Players: 2
Playing Time: 4-6 hours
Preparation Time: 2 1/2 hours

Description: The game map shows an abstract view of Guadalcanal and the area surrounding it. This "map" is divided into nineteen large hexagons for movement. Each of the sixteen moves represents one week of the campaign. The counters represent land, sea, and air forces. The land counters show abstract point values rather than specific units. The sea counters represent specific large ships or groups of smaller ships. The air counters represent land-based aircraft or naval aircraft, not specific units or types of planes. The game uses simultaneous movement, partly according to written plans made before each turn. Each turn is divided into a multitude of separate operations or possible operations. (MC)

Comment: *Solomons Campaign* is based on USN (below), but it is not as complex as USN because the options for each player are fewer and because in the transformation the scale has been stretched out to allow much less abstraction. *Solomons Campaign* is one of the few games that successfully combines land, sea, and air operations on approximately the same level of simulation. (MC)

Cost: $11.00, $12.00 with box

Producer: Simulations Publications, Inc., 44 East 23rd Street, New York, NY 10010

South Africa

Irad B. Hardy and Frank Davis

Playing Data
Copyright: 1977
Age Level: grade 9 and up
Number of Players: 2
Playing Time: 5-20 hours
Preparation Time: 1 hour

Description: *South Africa* posits an almost interminable war between the Black nations of southern Africa and the Republic of South Africa and Assumes that the former would win eventually. The scale is one week per turn and sixty kilometers per hexagon. Forces include infantry, reconnaissance, artillery, police, engineer, airborne, armored, guerrilla, helicopter, and air strike units. Rules cover various kinds of combat, supply, taxation, mobilization of new units, demobilization of existing units, manpower, and assorted international events that could alter the situation. (MC)

Cost: $9.00, $10.00 with box

Producer: Simulations Publications, Inc., 44 East 23rd Street, New York, NY 10010

SPARTAN

Playing Data
Copyright: 1975
Age Level: high school-college
Playing Time: 2-4 hours
Preparation Time: 1 hour
Materials: mapboard (with tracks and charts), general rules booklet, special rules folder, die-cut counters, die

Description: This is a revision of Phalanx. The scale of the map has been doubled to fifty meters per hexagon. Unfortunately the number of different kinds of units has been cut down and the name of the units changed to fit into the series. The scenarios have been improved with the addition of more careful victory conditions and with rules which tailor the terrain to the simulated battle. One scenario has been added and two substituted for older ones. The rules are much simplified, but now an optional rule allows simultaneous movement. (MC)

Comment: See Yeoman. (MC)

Cost: $9.00, $10.00 with box

Producer: Simulations Publications, Inc., 44 East 23rd Street, New York, NY 10010

SPITFIRE

James F. Dunnigan

Playing Data
Copyright: 1973
Age Level: grade 9 and up
Number of Players: 2
Playing Time: 1-3 hours
Preparation Time: 1 hour

Description: The aircraft cards describe the performance of fifteen German and Allied fighters and bombers from the period, 1939 to 1942, and serve as charts to mark their actions in the game. The counters mark their place on the board which is a featureless hexagon grid. Two scenarios allow the play of fighters against fighters or fighters against bombers. (MC)

Comment: The rules of *Spitfire* are excessively rigid and abstract and they cause the aircraft to behave in an unpleasant, wooden fashion. This game might profit from new rules supplied by the user. (MC)

Cost: $9.00, $10.00 with box

Producer: Simulations Publications, Inc., 44 East 23rd Street, New York, NY 10010

SQUAD LEADER

John Hill and Donald Greenwood

Playing Data

Copyright: 1977
Age Level: grade 9 and up
Number of Players: 2
Playing Time: 3-30 hours
Preparation Time: 1 hour

Description: At its fullest development, this game is a microsimulation of small-unit combat in World War II Europe. Counters represent squads, leaders (company and battalion as well as squad leaders), vehicles, and support weapons (machine guns, mortars, antitank guns). Assorted markers show fires, barbed wire, shell bursts, star shells, smoke, etc., or are used to show damage to counters. The unusually attractive mapboard shows a wide variety of terrain, including buildings, some scattered and some concentrated in a town. The game rules include some representation of almost everything that would affect tactics on this level, including morale, weapons breakdown, the devastating effect of artillery, and especially, the effect of leadership. The game covers Western and Eastern front actions and includes counters for Russians, Germans, and American infantry and vehicles. It comes with twelve scenarios and full instructions for creating an infinite number of others. (MC)

Comment: Although this is a complex game, it can be played at several levels. The instructions are arranged according to levels of difficulty and also according to levels of necessity, so it is easy to get a group started on the game. The game gives a great feeling of verismilitude. It would be best to introduce the game to a new class in its simplest form without a great deal of instruction on the rules. Then introduce new ideas and rules as it occurs to the players to want to do different things, or as appropriate problems arise. Of course, the teacher must be familiar with the game. (MC)

Cost: $12.00

Producer: Avalon Hill, 4517 Harford Road, Baltimore, MD 21214

SSN: MODERN TACTICAL SUBMARINE WARFARE

Stephen M. Newberg and Marc W. Miller

Playing Data

Copyright: 1975
Age Level: grade 9 and up
Number of Players: 2
Materials: includes simultaneous movement plotting pad

Description: The board shows a featureless ocean or sea on which pieces representing contemporary surface ships, helicopters, and submarines maneuver at a scale of 1000 yards per hexagon and six minutes per turn. The game comes with three peacetime scenarios based on the way Soviet and American ships play electronic tag with each other on the oceans and with twelve wartime scenarios. Most scenarios pit one submarine against a surface fleet of four to twelve ships. Some scenarios, for which an umpire is desirable, pit submarines against each other. (MC)

Comment: Each player plots movement and attacks secretly. Then the actions are performed simultaneously. Submarines move by hexagon number, not visibly on the board, unless they are detected. The movement and combat systems are not difficult but they seem to portray the most important elements of contemporary naval warfare if it were to occur. (MC)

Cost: $8.00

Producer: Game Designers' Workshop, 203 North Street, Normal, IL 61761

STONEWALL

Mark Herman and Thomas Walczyk

Playing Data

Copyright: 1978
Age Level: grade 9 and up
Number of Players: 2
Playing Time: 4 hours
Preparation Time: 30 minutes

Description: *Stonewall* is a simulation of the Battle of Kernstown, during the American Civil War. This game is based on the *Terrible Swift Sword* (q.v.) game system but has only one hundred counters. Units are brigade level with each strength point representing one hundred men or one cannon. Each turn represents twenty minutes and each hexagon one hundred twenty-five yards. Units may fire at various ranges, depending on what type of weapon the unit had historically. Leaders must be within command radius of units for them to be effective. Units must have the proper formation to move and engage in combat in the most effective manner. Other rules include routes, melee, cavalry charges, artillery overshoots, demoralization, and morale. (Richard J. Rydzel)

Comment: At first glance, this game seems very complex and too hard to learn but once the rules are read and the game has been played a couple of turns, it is fairly simple. The new concepts are easy to learn. It is also these new concepts that make this an historical and playable game that seems to accurately depict the tactics used during this period. This is a very good game to fit into American history classes. (RJR)

Cost: $9.00, $10.00 with box

Producer: Simulations Publications, Inc., 44 East 23rd Street, New York, NY 10010

STRIKEFORCE ONE

Redmond Simonsen and James F. Dunnigan

Playing Data

Copyright: 1975
Age Level: high school, college
Number of Players: 2
Playing Time: 5-10 minutes
Preparation Time: 20 minutes
Materials: rules leaflet with attached map-chart sheets, 10 die-cut counters, sample game leaflet, die

Description: This is a very basic game which pits six Soviet units against four American in an imaginary situation. The main point of the game is that it introduces some of the most common mechanics of wargames to the novice player. Even with the small number of pieces on the board, the game offers several choices of strategy. See the introductory essay to this section for more on this game. (MC)

Cost: Free

Producer: Simulations Publications, Inc., 44 East 23rd Street, New York, NY 10010

SUBMARINE: TACTICAL LEVEL SUBMARINE WARFARE, 1939-1945

J. Stephen Peek

Playing Data
Copyright: 1976
Age Level: grade 9 and up
Number of Players: 2 or more
Playing Time: 2-4 hours
Preparation Time: 1 hour

Description: This game pits a small number of submarines against a small number of escorts and a suitable target, such as a merchant convoy or aircraft carrier. Each counter represents an individual ship. Movement is plotted one or more turns in advance. An optional rule allows the submarines to use hidden movement. Escorts may use vision, sonar, radar, or star shells to spot the enemy. Torpedos, depth charges, and guns are used in the battles. Scenarios are drawn from the Atlantic War and the Pacific War. More counters than are needed for the prepared scenarios are available as an aid to making one's own. Each kind of ship and weapon is identified as available after a certain time in the war, so that a series of games could be played with increasingly more sophisticated equipment. (MC)

Cost: $12.00

Producer: Avalon Hill, 4517 Harford Road, Baltimore, MD 21214

TACTICS II

Charles S. Roberts

Playing Data
Copyright: 1961, 1973
Age Level: grade 7 and up
Number of Players: 2
Playing Time: 2-4 hours
Preparation Time: 30 minutes

Description: This game shows two imaginary countries at war with World War II and post-World War II formations and weapons. Only land forces are represented, but some of them can conduct amphibious assaults. Weather and nuclear weapons are important options. The forces start from immobilized positions and must move into wartime positions on the first moves. The game continues until one side suffers total defeat. Conceivably, this could lead to an interminable game, but it seldom does. (MC)

Comment: Tactics II, first published in 1958, was one of the two original commercial wargames, and many old-timers have a soft spot in their hearts for it. It has little historical interest because its representation of World War II conditions is so oversimplified. It is sometimes recommended as an introductory game and in fact has served that function for many war gamers, but it has a square grid that makes its geography somewhat different from that of the now-standard hexagon grid board. (MC)

Cost: $6.00

Producer: Avalon Hill, 4517 Harford Road, Baltimore, MD 21214

TANK!

James F. Dunnigan

Playing Data
Copyright: 1974
Age Level: high school-college
Number of Players: 2
Playing Time: 30 minutes-2 hours
Preparation Time: 45 minutes
Materials: mapboard, counters, rules folder, chart/scenario, card simultaneous movement chart pad, die

Description: This game simulates tank vs. tank and tank vs. infantry combat on a tactical level. Each counter represents a single tank, single gun, or a squad of infantry. The game uses a small number of tank and gun counters, none of which has specific information printed on it. By using the accompanying charts they can be made to represent specific kinds of weapons from 1937 to the present, from the French S-35 and the German DZKW II of the earliest period to the American M60A3 of the contemporary period. The game map is similarly versatile since it can be read to represent three different kinds of terrain. The rules for using these weapons use simultaneous movement, hidden placement for infantry and guns, a panic rule so that the players cannot count on their units completely, and a fairly involved terrain system. Advanced rules allow such things as the intervention of artillery and aircraft. (MC)

Comment: Compared to *Tobruk,* combat in *Tank!* is still quite generalized while compared to *PanzerBlitz* and similar games it is quite detailed. Tank offers a fairly simple approach to the tactical problems of armored warfare and to the comparisons of the weapons used in it. (MC)

Cost: $8.00, $9.00 with box

Producer: Simulations Publications, Inc., 44 East 23rd Street, New York, NY 10010

TANNENBERG

David C. Isby

Playing Data
Copyright: 1978
Age Level: grade 9 and up
Number of Players: 2-3
Playing Time: 3-7 hours
Preparation Time: 1 hour

Description: *Tannenberg* deals with events on the northern part of the World War I Eastern Front either in August and September 1914 or in February 1915. Players can simulate the earlier Russian offensive which led historically to disaster or the later successful German offensive. Each turn represents three days, each hexagon eight miles. Play of the game emphasizes the importance of leadership (there are ordinary headquarters units and heroic headquarters units), supply, rail movement, and tactical expertise (which the Germans have more of than the unfortunate Russians). (MC)

Comment: The game is intricate–maybe too intricate for many classes, for the German success at least would depend on the German's knowledge of the details of the rules. The three-player game would be a valuable class exercise. In it, the Germans are under one player, while the Russian forces are under two players who cannot communicate without the German listening in. (MC)

Cost: $9.00, $10.00 with box

Producer: Simulations Publications, Inc., 44 East 23rd Street, New York, NY 10010

TANNENBURG

Richard Spence and James Gabel

Playing Data
Copyright: n.d. (1977?)
Age Level: grade 8 and up
Number of Players: 2
Playing Time: 3-4 hours
Preparation Time: 1 hour

Description: The game covers the East Prussian campaign of 1914 which led historically to the destruction of one invading Russian army at Tannenberg and the total defeat of the other. In the game the Russians outnumber the Germans in number of units but have fewer combat points. So the Russians must attack with less force. The only thing they can use is their ability to surround German units. Or they can just enter East Prussia and then let the Germans try to throw them out. The rules give importance to supply, fortifications, and German rail movement. (MC)

Comment: The game seems to be an accurate simulation of the situation, which means it's a bad one for the Russians. There are many possibilities for splitting commands on both sides for classroom play. The name of the battle is misspelled in the title. (MC)

Cost: $5.50

Producer: S&G Games, 2105 Custer Ave., Bakersfield, CA 93304

TERRIBLE SWIFT SWORD

Richard Berg and Redmond A. Simonsen

Playing Data
Copyright: 1976
Age Level: grade 9 and up
Number of Players: 2 or more
Playing Time: 3-60 hours
Preparation Time: 1 1/2-4 hours

Description: This is a microsimulation of the Battle of Gettysburg, with each hexagon of the large map representing about 120 yards and each unit representing (usually) a regiment of 100 to 800 men. In addition to unit counters representing infantry, cavalry, and artillery units, there are counters representing individual officers from Meade and Lee down to the colonels and brigadiers who commanded brigades and divisions. The game reflects the differences among the various weapons of the period and rewards the discovery of realistic tactics for their use. Infantry units can form in columns, for speed; or lines, for fighting They can throw up field fortifications. Cavalry can charge, but any general who orders a charge against unshaken infantry and artillery is likely to regret the decision. The game includes an ammunition supply problem, a provision for units routing and then reforming under their leaders, and other leadership provisions. The long game covers the whole battle, but shorter games cover only one day or a part of a day. The shortest game covers the sharp fight for Little Round Top in the late afternoon of the second day. The rules incorporate a short history of the battle and an order of battle. The scenarios also include directions for representing Pickett's charge on the board. (MC)

Comment: The duration of the grand battle version of this game makes it unlikely that any class could play it in this way, but much can be learned from shorter playings. The game is very complex, but it has an advantage in that it can be played by teams of players who do not know very many of the rules, provided the referee knows them. You can teach the players how to move, how to change formations, and a few other things. Then get them started on a series of made-up small battles involving a few divisions on each side. Let them learn by doing, as the Civil War generals did. After each short battle give promotions and demotions on merit and start again. After three or four short preparatory battles representing the experience of the participants from 1861 to July 1863, give them one of the shorter scenarios of Gettysburg to play out. (MC)

Cost: $18.00, $20.00 with two boxes

Producer: Simulations Publications, Inc., 44 East 23rd Street, New York, NY 10010

THERMOPYLAE

Martin E. Sliva

Playing Data
Copyright: 1973
Age Level: high school, college
Number of Players: 2
Playing Time: 3 hours
Preparation Time: 45 minutes
Materials: two-part mapboard, counter sheet, battle results table card, rules folder, die

Description: The map shows Greece, its northern neighbors, the Hellespont, a slice of Asia Minor and the surrounding seas. The land is divided into small (16 mm.) hexagons while the sea is divided into large (32 mm.) hexagons. The game lasts twenty turns. The Persian player must seize Athens and Sparta or Corinth and Sparta to win. The Greek Player wins by keeping the Persian conquests to one of the cities. The movement and combat rules are standard. The interest is in the strategic situation with the Greeks defending several places in which the Persian superiority of numbers cannot be brought to bear.

Comment: Playing the game *does* require a small amount of assembly. The counter sheet must be peeled to expose its adhesive backing and fastened to a piece of medium weight cardboard. Then the counters must be cut out. The whole job will take perhaps fifteen minutes. The game is easy to learn and play, yet it reflects the historical situation very well.

Cost: $5.00

Producer: Martin E. Sliva, 620 Briarcliff Road, Bolingbrook, IL 60439

THIRTY YEARS WAR: LÜTZEN, NORDLINGEN, ROCROI, FREIBURG

Brad E. Hessel, Redmond A. Simonsen, Thomas Walczyk, Linda D. Mosca, and Stephen B. Patrick

Playing Data
Copyright: 1976
Age Level: grade 9 and up
Number of Players: 2
Playing Time: 2-4 hours
Preparation Time: 1/2 hour

Description: This set of games deals with four battles of the Thirty Years War using a common battle system varied with one or two special rules per game. The common system is simple but differentiates neatly between cavalry, infantry, and artillery. It simplifies but reflects the differences between the old-style armies of the imperialists and the Spanish and the new-style armies of the Swedes and the French. The importance of leadership is reflected by having leader counters add to the combat power of units in the same hexagon and by making leaders aid in rallying disrupted units ("disruption is a result of combat"). Special rules allow some cavalry to gain effectiveness by charging and define the punishment that various armies can take before they become demoralized. (MC)

Comment: *Rocroi* is analyzed in the introduction essay to this section. Either *Lutzen* or *Nordlingen* might have taken its place, but *Freiburg* is dull and confusing. (MC)

Cost: $13.00, $14.00 with box ($4.00 apiece separately)

Producer: Simulations Publications, Inc., 44 East 23rd Street, New York, NY 10010

TOBRUK

Harold E. Hock

Playing Data
Copyright: 1975
Age Level: high school-college

Number of Players: 2 or more
Playing Time: 2-12 hours
Preparation Time: 2 hours
Materials: three-part mounted mapboard, two hit probability/damage tables cards, gunfire and other tables card, roster pad, die-cut counters, rules booklet, two dice

Description: This is a very detailed examination of very small parts of battles that took place in the neighborhood of Tobruk in 1942. The mapboard is a featureless hexagon grid which represents the desert. Most counters represent *individual* tanks, guns, mortars, machine guns. Other counters represent infantry squads or headquarters groups. For these the individual men are kept track of on the roster sheet. The combat system is quite intricate. The dice are rolled for each round fired by tank or antitank guns, first to see if the round is a hit and then, if it is, to see where it hit. Then, if the target is an armored vehicle, a table is consulted to see whether that kind of gun can do any damage on that kind of tank at that range. This makes for a lot of dice rolling although the process is not as long as it sounds. The heart of the game is tank-to-tank and gun-to-tank combat, but rules for infantry and artillery, fortifications, minefields, smoke, aircraft (German dive-bombers)—all of these and more put the armored conflict into an authentic setting. The game comes with nine large scenarios arrayed in order of increasing difficulty for programmed learning, with ten small "firefights" and with suggestions for making more scenarios. (MC)

Comment: *Tobruk* is fascinating and, except for the lack of terrain, fairly accurate. It is a great deal of work, however, for the amount of warfare simulated and probably too difficult for most classrooms. (MC)

Cost: $12.00

Producer: Avalon Hill, 4517 Harford Road, Baltimore, MD 21214

TORGAU

Frank A. Chadwick

Playing Data
Copyright: 1974
Age Level: high school-college
Number of Players: 2
Playing Time: 6-10 hours
Preparation Time: 1 hour
Materials: mapboard, six charts, cards (2 identical), two sets of die-cut counters, rules booklet, die

Description: The game simulates the battles of 1760 between the Austrian army under Daun and the Prussian Army under Frederick the Great. The map shows the terrain at 200 yards to the hexagon. The counters represent units of 300 men or less. Each move represents 15 minutes of real time. The game lasts up to 56 turns (a full day from 8:00 a.m. to 10:00 p.m.). The combat system is detailed and intricate. Losses are often taken by substituting smaller factored units for larger, instead of by removing the whole unit. (MC)

Comment: The game is difficult but seems faithful to eighteenth-century warfare. There are many ways of dividing the sides up among several commanders. (MC)

Cost: $8.75

Producer: Game Designers' Workshop, P.O. Box 582, Bloomington, IN 61701

TRIREME

Ed P. Smith

Playing Data
Copyright: 1971
Age Level: grade 8 and up
Number of Players: 2-4
Playing Time: 3-8 hours
Preparation Time: 1 hour

Description: The board is blank except for a staggered square grid. Its appearance can be varied by adding the islands, reefs, and sandbars, which are provided. The battle is a detailed one with up to ten triremes on a side. Players keep track of hull damage, oar and rudder damage, and losses among the crewmen, bowmen, and "marines." Also they record the number of turns at top speed—the oarsmen are capable of delivering only four such turns. The game does not include any ideas for setting up any specific battles. (MC)

Comment: This is a good fast game. The game is detailed, but the players' charts make bookkeeping easy. (MC)

Cost: £5.25

Producer: Skytrex, Ltd., 28, Brook Street, Lymeswold, Leicestershire, United Kingdom

TROY

Donald A. Dupont

Playing Data
Copyright: 1977
Age Level: grade 8 and up
Number of Players: 2
Playing Time: 2-6 hours
Preparation Time: 30 minutes

Description: The map shows Troy at the size of one hexagon and the area around it, which includes mountains, rivers, valleys, and the coastline. The counters represent bodies of fighting men, infantry, cavalry, and chariots, and, more interestingly, the individual heroes who fought in Homer's Iliad. Other markers show the magical possessions of the heroes. The deities intervene from time to time through the medium of a deck of deity cards. One can play Homer's Trojan War on less fanciful scenarios based on the earlier history of Troy. (MC)

Comment: The game is lovingly crafted and is one of the most attractive around. Although it is not strictly historical, it is certainly based on a historically important document and reflects what was probably a real event. (MC)

Cost: $10.00

Producer: Donald A. Dupont, P.O. Box 6274, Albany, CA 94706

TURNING POINT

James F. Dunnigan

Playing Data
Copyright: 1972
Age Level: grade 9 and up
Number of Players: 2
Playing Time: 3-9 hours
Preparation Time: 2 hours

Description: *Turning Point* was formerly called *The Battle of Stalingrad* and it deals with the period of the Russian counteroffensive of 1942. The map shows the area of the battle. The counters represent land units (mostly divisions) and air units. Each turn represents two days. There are two short (seven-turn) basic scenarios as well as the long (twenty-one-turn) game. There are a large number of alternate history, "what-if" possibilities included in the scenarios. Special rules allow the Russians to achieve surprise and allow for Hitler's demand that the Stalingrad positions be held. (MC)

Comment: *Turning Point* is a lively, moderately difficult game on an important development. The game shows the importance of the irrational German dispositions at the beginning of the Russian attack, the hopelessness of the German army caught inside the trap, and the importance of the Russian supply difficulties that prevented an even greater victory. (MC)

Cost: $11.00, $12.00 with box

Producer: Simulations Publications, Inc., 44 East 23rd Street, New York, NY 10010

U-BOAT

Playing Data
Copyright: 1975
Age Level: grade 9 up
Number of Players: 2
Playing Time: 1-3 hours
Preparation Time: 1 hour
Materials: includes 2 cards with cut-outs representing ships, U-Boats, torpedos, turning gauge, ASDIC search arc, etc. Players need a tape measure, preferably metric, and two dice.

Description: The game covers a typical battle between a small Atlantic convoy of three to eight transports and tankers and one to four escort vessels and an attacker with one to four u-boats. The time is sometime in 1942. Orders for a fleet are written in advance. The convoy and escorts cannot change their routine until a U-boat is spotted. (MC)

Cost: $2.50

Producer: Tabletop Games, 92 Acton Road, Arnold, Nottingham, United Kingdom

UP SCOPE!

Joe Balkoski, Frank Davis, and Steve Ross

Playing Data
Copyright: 1977
Age Level: grade 9 and up
Number of Players: 2 or more
Playing Time: 1-10
Preparation Time: 30 minutes-1 hour

Description: This game deals with submarine and antisubmarine warfare from 1914 to the present, including the World Wars and hypothetical contests from the contemporary period. The game is about ship-to-ship battles; each counter represents an individual submarine, escort vessel, or potential victim (ranging from battleships and aircraft carriers to freighters). The simulation of submarine/destroyer tactics and weapons is executed in some detail. Torpedos come in assorted types and move through the water after launching, survivors of previous sinkings get in the way of the defenders and need rescuing. Oil slicks appear. Water conditions affect gunfire and detection. Undetected submarines engage in hidden movement. (MC)

Comment: The game is quite complex and, unlike many complex games, should not be played at all by players who do not know the game. On the other hand, the rules make sense because they correspond well to reality, and it is easy to adapt the game to several players. (MC)

Cost: $11.00, $12.00 with box

Producer: Simulations Publications, Inc., 44 East 23rd Street, New York, NY 10010

USN

James F. Dunnigan

Playing Data
Copyright: 1971
Age Level: grade 9 and up
Number of Players: 2, preferably 2 teams
Playing Time: 3-150 hours
Preparation Time: 3-4 hours

Description: The game map shows a small scale map of nearly the whole Pacific Ocean and Southeast Asia. Each game turn represents one week. There are eighty-one turns in the long campaign game. Or two short campaign games of twenty turns each may be played to show the first or second four-month period of the war. Or, three short battle games of two or three turns may be played. The counters represent land, sea and air forces. The land forces are divisions, regiments and battalions. The sea are "divisions" of ships or individual carriers. The air forces are points of either land-based or sea-based aircraft. A large number of different operations in all parts of the theater can be or must be performed by each player each turn. (MC)

Comment: To play this game, one ought to have several people who are thoroughly immersed in this kind of game, and, of course, unlimited time. *USN* does offer an excellent simulation of some of the problems of the war. However, the Japanese player is allowed to coordinate and plan his vast offensive much better than he ought so that the offensive never slows down because of indecision or lack of information, and the Japanese will achieve more than they should. (MC)

Cost: $9.00, $10.00 with box

Producer: Simulations Publications, Inc., 44 East 23rd Street, New York, NY 10010

VERACRUZ

Richard Berg and Joseph Balkoski

Playing Data
Copyright: 1977
Age Level: grade 9 and up
Number of Players: 2
Playing Time: 8 hours
Preparation Time: 1 hour

Description: This game simulates Scott's invasion of Mexico and the ensuing campaign that took him from Veracruz to Mexico City. The map shows this area at a scale of five miles to the hexagon. Each unit represents a regiment at the most. Each turn represents one week and the game lasts twenty-five turns, from March to September 1847. Rules emphasize supply and organizational problems and the importance of leadership, with leader counters representing the generals on each side. The rules also allow for the influence of disease, Mexican politics, American short-term volunteers leaving, and Mexican reluctance to see their cities destroyed. (MC)

Comment: This is the only simulation of this conflict that is important to American history but is seldom studied except as part of an introduction to the Civil War. The flavor of the campaign is well presented in the game. (MC)

Cost: $9.00, $10.00 with box

Producer: Simulations Publications, Inc., 44 East 23rd Street, New York, NY 10010

VERDUN

John Hill

Playing Data
Copyright: 1972, 2nd edition 1978
Age Level: grade 9 and up
Number of Players: 2
Playing Time: 3-4 hours
Preparation Time: 1 hour

Description: The second edition was not available to me. The first uses a map that shows the northeastern approaches to the city of Verdun over which the massive battle of 1915 was fought. The game shows the effect of trenches and forts, the importance of artillery, and the way the battle chewed up units on both sides. (MC)

Cost: $11.98 for second edition

Producer: Conflict Game Co., P.O. Box 432, Normal, IL 61761

VICTORY AT SEA

Playing Data
Copyright: 1971, 1977
Age Level: grade 9 and up
Number of Players: 2
Playing Time: 2 hours up
Preparation Time: 30 minutes up

Description: This little package contains a rule book, a set of charts, and some paper ship plans to be cut out to fight three small battles on a nearby floor. It also includes many figures and suggestions to allow players to devise their own battles with their own designed cut-outs. (MC)

Comment: These rules are not as complex as they look at first, but instructor using the game in a class must be thoroughly familiar with them. It is not necessary for the actual players to be so thoroughly used to them. (MC)

Cost: $5.49

Producer: Attack International Wargaming Association, 314 Edgley Avenue, Glenside, PA 19038

VICTORY IN THE PACIFIC

Richard Hamblen and Donald Greenwood

Playing Data
Copyright: 1977
Age Level: grade 7 and up
Number of Players: 2
Playing Time: 4-6 hours
Preparation Time: 1 hour

Description: This is a somewhat abstract representation of the war in the Pacific. Surface fleet units from battleship to cruiser are each represented by a single counter. Submarines are represented by one counter on each side; land-based aircraft and land forces are also represented by somewhat arbitrary units. The game map uses various sized areas as movement spaces, so scale is not important. The game system is one in which each side occupies various sea areas, conducts battles involving much dice rolling for areas that are in competition, and gathers points each turn for those areas they hold at the end of the turn. There are ordinarily eight turns, each representing one to six months, and the side with the most control points at the end wins. (MC)

Comment: This game is very abstract but still shows a lot about the strategy of the Pacific War. Players can learn about the relationship of sea, air, and land forces in the Pacific, about the importance and the limits of land-based aircraft, about the great mobility of carrier forces. Combat is sometimes a bit tedious. In a classroom, each side ought to be played by a small committee. The abstraction of the game would make it difficult to divide up the commands. (MC)

Cost: $12.00

Producer: Avalon Hill, 4517 Harford Road, Baltimore, MD 21214

VIKING

Playing Data
Copyright: 1975
Age Level: high school-college
Number of Players: 2
Playing Time: 2-4 hours
Preparation Time: 1 hour
Materials: mapboard (with charts and tables), general rules booklet, special rules folder, die-cut counter, die

Description: This is a revision of *Dark Ages* (tactical warfare, 700-1300 A.D.). There are four more scenarios (nineteen instead of fifteen). The rules are simplified. The counter mix includes one kind of counter less. The map has been changed considerably. The scenario instructions provide for modifications of the printed terrain in many battles. An optional rule allows for simultaneous movement. (MC)

Comment: See Yeoman. (MC)

Cost: $9.00, $10.00 with box

Producer: Simulations Publications, Inc., 44 East 23rd Street, New York, NY 10010

VIVA!

Russ Beland

Playing Data
Copyright: 1975
Age Level: grade 7 and up
Number of Players: 2
Playing Time: 3-6 hours
Preparation Time: 30 minutes

Description: The game deals with revolutionary war in Mexico with scenarios on a "typical" revolution of the 1800s, on the Carranza-Villa-Pershing War, and on the French invasion of Mexico. The map shows all of Mexico with smaller maps showing enlarged battle areas for Vera Cruz, Guadelajara, and Mexico City. The maps show swamps, mountains, rough terrain, lakes, and railroads, used only in later scenarios. The game is simple and concentrates almost exclusively on combat, with an "attrition" die roll every turn taking care of all other problems. Leader counters play an essential part in the game. Reinforcements enter most turns depending on the control of the places they enter. Victory is decided mostly by control of strategic cities and towns. (MC)

Comment: It would be well to have a good game on this subject; a game like this that ignores the politics of the situation is not very valuable. (MC)

Cost: $5.00

Producer: Flying Buffalo, Inc., P.O. Box 1467, Scottsdale, AZ 85252

WACHT AM RHEIN

James F. Dunnigan, Joseph Balkoski, and Jay A. Nelson

Playing Data
Copyright: 1977
Age Level: high school and up
Number of Players: 2-4
Playing Time: 60 hours
Preparation Time: 3 hours

Description: *Wacht am Rhein* is a complex but very playable game covering the battle of the Bulge in World War II. The four map sections cover the Ardennes area over which the battle occurred. Each hex represents one mile; each daylight turn represents 4 1/2 hours and each night turn fifteen hours. Unit counters are battalion or regiment; however, U.S. units may break down into companies. Artillery units are either in battery (firing) or out of battery (moving). Most other units are back-printed for step reduction to simulate losses. Artillery units may fire at nonadjacent hexes for offense or defense. Air point allocation may be used to reduce the supply length of German units. All units must trace supply to an appropriate H.Q. unit bearing the battalion's designation. Bridges may be destroyed or made by engineer units. Units may use an extra night turn to move and attack but are subject to fatigue. The proper combination of different types of units can bring combat advantages. Units may construct improved positions and entrenchments. The game includes 1600 counters, mostly combat units, but also air units, bridges, march mode markers, isolation markers, and others. A complete and accurate order of battle for both sides is given. (Richard J. Rydzel)

Comment: This is an excellent game for history and playability and gives the feel of real tactics through its rules. It is a very complex and long game but is worth the time spent playing it. Because of its size and length it is not well suited to school use. Though there are many new concepts, after a couple of turns the mechanics become a habit and the pace picks up. Some rules options add even greater realism and playability. (RJR)

Cost: $18.00, $20.00 with two boxes

Producer: Simulations Publications Inc., 44 East 23rd Street, New York, NY 10010

WAR AT SEA

John Edwards and Don Greenwood

Playing Data
Copyright: 1976
Age Level: grade 7 and up
Number of Players: 2
Playing Time: 2-4 hours
Preparation Time: 30 minutes

Description: This is a quick, abstract representation of the Battle of the Atlantic with the Germans given a boost by being allowed to complete their aircraft carrier, which historically was never operational, and by being allowed a generally greater level of activity. So the game changes the Battle of the Atlantic from a basically submarine/antisubmarine war to a conflict of surface ships. The action is quite abstract. Points are made by occupying areas with ships. Combat is decided very simply by lengthy die rolling. (MC)

Comment: In spite of its historical distortions and lack of realism, *War at Sea* is some help in seeing the situation in the Atlantic and shows why the British feared the small German Fleet. Although the game is simple, it should not be considered introductory (except to its sister game *Victory in the Pacific*), because its system is not even distantly related to the systems used in more complex games. (MC)

Cost: $6.00

Producer: Avalon Hill, 4517 Harford Road, Baltimore, MD 21214

WAR BETWEEN THE STATES

Irad B. Hardy and Redmond A. Simonsen

Playing Data
Copyright: 1977
Age Level: grade 11 and up
Number of Players: 2
Playing Time: 10-220 hours
Preparation Time: 1-3 hours

Description: Like The American Civil War, its predecessor, *War Between the States* covers the whole war, but at a very expanded scale with expanded detail. The area of the war is shown on three map sheets. Each unit is a division. Leaders are named, but in one version, players cannot see who their leaders are for a time to simulate the fact that some of the best and the worst generals were unknown quantities at the war's beginning. Short games deal with a single year's campaign in either the West (two map sheets) or the East (one). The game can also cover the whole war with production rules including delicate political decisions like whether or not to start conscription. (MC)

Comment: If I taught a course on the Civil War only, I would be tempted to run this game as the main or only classroom activity in the course. As it is, the game contains at least a reference to the most important political and economic effects on military operations and the effects of operations on political and economic conditions. It is eminently suited to multiple player use and will give everyone in a group of twenty-four something to do if the teams are organized correctly and if the production rules are used. Economic and political rules should be kept under the control of the instructor, who can use the published rules as guidelines rather than as absolutes. (MC)

Cost: $18.00, $20.00 with 2 boxes

Producer: Simulations Publications, Inc., 44 East 23rd Street, New York, NY 10010

WAR IN EUROPE (WAR IN THE WEST, WAR IN THE EAST)

James F. Dunnigan, Redmond A. Simonsen, Irad B. Hardy, Tom Walczyk, Edward Curran, and Steve Bettum

Playing Data
Copyright: 1976
Age Level: grade 9 and up
Number of Players: 3
Playing Time: 3 hours to near infinity
Preparation Time: 1-4

Description: This game—or rather, game set—can be operated at several levels. The basic combat system is straightforward, so that one can play the shortest scenario of War in the West, the conquest of Poland, in a relatively short time. The full game of War in the West goes to 302 game turns and allows variable production (especially for the Germans), strategic bombing, partisan warfare, variable entry of neutrals, and, of course, it allows the war to develop very different from history. In *War in the West,* the Russian campaign is treated abstractly and the German player must take most of his forces off the board in 1941 to abstractly fight it. *War in the East* treats the Russian campaign in detail with variable Russian production. The full *War in Europe* puts all the variables together. (MC)

Comment: *War in Europe* is very impressive. The rules are not flawless, but because of the length of development they have now reached a state where they are better than one would expect in such a big game. The only way to play the game properly in a classroom situation would be to devote almost all the classroom time for a semester to it. It would also require its own room, which could not be used by any other classes. If you had the time but not the room, much of the game could be played on paper—all actions and dispositions except on the fronts where battles were raging could be written down, and parts of the nine-part map could be set up as needed. The possibilities for role assignments are almost endless. Although the game is big, the roles are not complex—but there are a lot of them. With a variety of role assignments, individual rules could be taught only to those people whose roles required the knowledge. The German production chief, for example, would learn about production but would not have to be concerned with the rules for combat. This kind of game would require a great deal of activity on the part of the instructor/referee. The diplomatic rules could be demechanized and administered by the referee without telling the player exactly what is going on.

I have used only the short Polish invasion scenario. It works well but it is only a military situation, and using it makes me feel as though I am using a semitrailer tractor to pull a child's wagon. (MC)

Cost: $37.00, $40.00 with three boxes; *War in the West* $28.00, $30.00 with two boxes, *War in the East* $22.00, $24.00 with two boxes

Producer: Simulations Publications, Inc., 44 East 23rd Street, New York, NY 10010

THE WARLORD GAME

Robert B. Williams

Playing Data
Copyright: 1976
Age Level: grade 9 and up
Number of Players: 2-6
Playing Time: up to 20 hours
Preparation Time: 1 hour

Description: This game sets up a pseudo-medieval countryside and invites the players to conquer as much of it as they can. Each "fief" occupied or taken forcibly from another player gives a player new men to use in conquering other territories. Some accessions of territory also results in a player's promotion—so that he goes up from knight to duke to king to emperor. The game lasts for twenty-five years at four turns per year and allows the players to build castles, towns, bridges, and roads as well as troops. With multiple players, diplomatic skills are at least as important in playing the game as mechanical game-playing skills. A player also needs luck, especially to avoid disastrous rolls of the "fate" table. (MC)

Comment: The game is great fun and makes the same general points as any diplomatic/military game. It does not seem to make any particular points about medieval warfare, diplomacy, or society. (MC)

Cost: $16.00

Producer: Robert Williams Games, P.O. Box 22592, Robbinsdale, MN 55422

WATERLOO

Lindsley Schutz and Thomas N. Shaw

Playing Data
Copyright: 1961
Age Level: grade 9 and up
Number of Players: 2
Playing Time: 2-4 hours
Preparation Time: 1 hour

Description: The game covers the march to and the battle of Waterloo. The units are mostly divisions. (MC)

Comment: Among avid players of wargames, Waterloo is known as a "classic," which means that it has little claim to historical accuracy but that it is old and fun to play. (MC)

Cost: $12.00

Producer: Avalon Hill, 4517 Harford Road, Baltimore, MD 21214

WELLINGTON'S VICTORY

Frank Davis, Fred Georgian, Tom Kassel, Pete Bennett, Joe Balkoski, and Ron Toelke

Playing Data
Copyright: 1976
Age Level: grade 11 and up
Number of Players: 2
Playing Time: 10-50 hours
Preparation Time: 1-2 hours

Description: This game deals in great detail with the Battle of Waterloo. The scale is tactical with each hexagon representing one hundred yards, each move fiteen minutes, and each unit a company, battalion, or regiment. Other pieces represent the individual commanders and their staffs. The rules emphasize small unit tactics, firepower, morale, battlefield supply, and leadership. Combat strength can be lost gradually, and its loss is indicated by marking the units with combat strength markers. (MC)

Comment: This is an immense game. The complete four-sheet map occupies a space 44″ by 68″ although two of the scenarios can be played on only half the map. Like others of the gigantic games, this has rewards that might in some cases justify the large amount of time and the area devoted to playing it. It invites multiple player, limited communication, applications. It could give everybody in a class plenty to do. The rules themselves are not extremely difficult and could be modified, given a classroom teacher/referee, to make them easier to absorb. But the whole battle game remains very big and would necessarily take a very long time. (MC)

Cost: $20.00, $22.00 with two boxes

Producer: Simulations Publications, Inc., 44 East 23rd Street, New York, NY 10010

WESTWALL: ARNHEM, HURTGEN FOREST, BASTOGNE, REMAGEN

Jay Nelson, Christopher Allen, Howard Barasch, Larry Pinsky, and Stephen Patrick

Playing Data
Copyright: 1976
Age Level: grade 9 up
Number of Players: 2
Playing Time: 3-6 hours
Preparation Time: 1 hour

Description: Four games are done on four separate mapsheets with four separate sets of counters using some rules in common and some "exclusive rules" for each game. The games all deal with battles near the border of Germany in 1944-1945. (MC)

Comment: The games are fairly simple, and I have used Arnhem and Bastogne as take home assignments. Of the four, Arnhem is probably the one that comes closest to reflecting important aspects of the battle it presents. (MC)

Cost: $13.00, $14.00 with box ($4.00 apiece separately)

Producer: Simulations Publications, Inc., 44 East 23rd Street, New York, NY 10010

WINTER WAR

James F. Goff

Playing Data
Copyright: 1972
Age Level: grade 9 and up
Number of Players: 2
Playing Time: 3-4 hours
Preparation Time: 1 hour

Description: The map is of Finland and the northwestern arm of Leningrad to Murmansk on a scale of twenty kilometers to the hexagon. The game covers the Russo-Finnish War of 1939-1940. Each of the ten moves represents ten days. The Russian numerical superiority is overwhelming, but the Fins have the advantage of maneuverability, defense lines, weak Russian supply, and the fact that the Russians have to beat a timetable to win. (MC)

Comment: This is an excellent simulation of a minor war. It is easy to learn and it shows the importance of supply, the effect of small numbers in a favorable situation, and, conversely, the difficulty of making an impression with large numbers in an unfavorable situation. (MC)

Cost: $11.00, $12.00 with box

Producer: Simulations Publications, Inc., 44 East 23rd Street, New York, NY 10010

WOLFPACK

James F. Dunnigan

Playing Data
Copyright: 1974
Age Level: high school-college
Number of Players: 1
Playing Time: 3 hours
Preparation Time: 1 hour
Materials: mapboard (with time track and charts), rules booklet, counters, die

Description: This is a solitaire game which works. The player is in charge of U-boat operations in the North Atlantic for a half a month in 1943 at the rate of one turn per day. He chooses to play either the February, March, April, or May scenario. The last two are very difficult to win. The Allied forces (convoys, escorts and aircraft) move and attack according to a workable combination of chance and routine. Convoy markers are moved across the map, each hexagon of which represents forty-three miles, but the player must get one or more U-boats in Positions to search for the convoys, since a real convoy cannot be told from a fake convoy until the marker has been searched effectively. The U-boats have to be dispersed for search and protection from aircraft and have to concentrate for attacks. (MC)

Comment: The game is a fairly realistic challenge for a single player. It would be possible to turn it into a multiplayer game I suppose but I doubt it would be worth the effort. (MC)

Cost: $12.00, $13.00 with box

Producer: Simulations Publications, Inc. 44 East 23rd Street, New York, NY 10010

WOODEN SHIPS AND IRON MEN

S. Craig Taylor

Playing Data
Copyright: 2nd ed., 1975
Age Level: high school-college
Number of Players: 2 or more
Playing Time: 1-10 hours
Preparation Time: 1 hour
Materials: mounted mapboard, card of tables, die-cut counters, rules booklet, die, ship's log pad

Description: This game simulates a large number of naval actions of the period 1776 to 1814. The board contains a hexagon grid pattern and areas of various shades of blue and gray. The shaded areas are designated land areas for some of the scenarios. The counters, which occupy two hexagons each, represent individual ships. The counters have only a little of the required information, most of which is recorded for each scenario on the "ship's log". Thus many kinds of ships can be represented by relatively few counters. The bookkeeping, movement and combat rules are detailed, allowing, for example, for a differentiation between long-range guns and short-range but deadly cannonades. Hits affect the crew, the rigging (sailing ability) and the guns. Guns can fire different kinds of ammunition. Crews can leave the guns to form a boarding party. Crews are classified according to their skill. There are a large number of scenarios from single ship actions to fleet battles. A campaign game which strings together a series of five battles is included. (MC)

Comment: This game can be best compared to *Frigate*. They deal with approximately the same subject, but *Wooden Ships* does it in much more detail. The former is best used for fleet actions with two to four players. The latter is best used in single ship actions with two players or fleet actions with a lot of players to do the bookwork. (MC)

Cost: $12.00

Producer: Avalon Hill, 4517 Harford Road, Baltimore, MD 21214

WORLD WAR I

James F. Dunnigan

Playing Data
Copyright: 1975
Age Level: high school-college
Number of Players: 2
Playing Time: 3 hours
Preparation Time: 1 hour
Materials: mapboard (with charts), rules leaflet, die-cut counters, randomizer chits (die substitute)

Description: This is a fairly simple simulation of World War I in Europe, on a grand strategical level. The map shows central Europe from a peculiar and at first unsettling angle, including a large enough slice of France to get Paris in, enough of Russia to show Riga and Kiev, and all of the Balkans including European Turkey. The counters represent armies, a scale which gives Germany only eight units to start, while France and Russia dispose of seven and five respectively. New units are built with "combat resource points" (CRPs), acquired according to a regular and, except for Russia, invariable schedule. The CRPs, which can be loaned by the stronger to the weaker, are also used to satisfy some losses by the armies in the field. Each move in this very strategical game represents a half a year. The only sea activity in the game is the possible transport of allied troops from the Western front to the Mediterranean. The last part of the game, the Germans may build Stosstruppen armies which gives them a great tactical advantage if they have enough CRPs left to use it. Optionally, the French may build a tank army which may only be used on the last turn. (MC)

Comment: *World War I* is a convincing simulation, or near, of the grand strategical problems of World War I. There are, I believe, a few problems with the simulation. It should not be possible, for example, for generals to give up territory lightly for strategic advantage. The game does not reflect the moral impossibility, for example, of allowing the French to occupy sacred German territory merely to save CRPs for use elsewhere. See the introductory essay to this section for more on this game. (MC)

Cost: $4.00

Producer: Simulations Publications, Inc., 44 East 23rd Street, New York, NY 10010

WORLD WAR II

James F. Dunnigan

Playing Data
Copyright: 1973
Age Level: grade 9 and up
Number of Players: 2-3
Playing Time: 3-8 hours
Preparation Time: 1-2 hours

Description: The game consists of grand strategy covering the years 1939-1945 in Europe. On the map, each hexagon is one hundred twenty miles wide. Each turn represents three months. There are twenty-three turns in the long game. The game covers mostly land operations with only a suggestion of air operations and with sea forces present only in an abstract power to move troops and conduct invasions. There are numerous political rules. There are three situations, or times when the players can start: 1939, at the beginning; 1940, with the invasion of France; or 1941, with the invasion of Russia. In the two player game, the players are Allies or Axis. In the three player game, the third player in Russia. (MC)

Comment: *World War II* is a complex game, but it should be useful in a class because its complexity can be broken into many parts and parcelled out to the class members. Some of the rules are ambiguous and others too rigid, so there should be an impartial umpire available. The teacher should improvise to keep the game on a realistic track. The game shows a great deal about the strategic options open in World War II. It also shows clearly how the German invasion of Russia closed most of those options for Hitler and became the dominating fact in the war. The game contains a few errors in setting up the politico-military situation in 1939, fewer errors for the 1940 setup, and no important ones for the 1941 setup. See the introductory essay to this section for more on this game. (MC)

Cost: $9.00, $10.00 with box

Producer: Simulations Publications, Inc., 44 East 23rd Street, New York, NY 10010

WORLD WAR III

James F. Dunnigan

Playing Data
Copyright: 1975
Age Level: high school, college
Number of Players: 2-3
Playing Time: 3-15 hours
Preparation Time: 1 hour
Materials: mapboard (with charts and tracks), three charts and tables sheets, die-cut counters, rules booklet, die

Description: The game deals with hypothetical nuclear or, by some chance, nonnuclear world war in the immediate future. The map shows the world, except for the polar regions, on a scale, in the middle of the map, of 500 miles to the hexagon. Since the map is to be read as if it were the outside surface of a cylinder, there is considerable distortion in many places, but most people who are used to looking at maps rather than globes will not notice it. The counters represent land forces, surface fleets, submarine fleets, naval air fleets, coast defense emplacements, amphibious fleets, merchant fleets, ports, supply forces, ICBM missile bases, or long-range bombers. The game provides for scenarios lasting six, eight, or twenty turns, each representing a three-month period. The players not only manipulated their military, naval and support forces but also decide how to spend production points on building new units. (MC)

Comment: Like the similar, later game *Global War* (q.v.), *World War III* has some problems. As a game, the victory conditions give the game to the Western player if he follows the best strategy, while in the three player version the third player, China, has little to do and has no chance of winning. The simulation aspects of the game are not easily faulted since the events have not happened (and if they had, who'd be playing games anyway?), but there are things in the play of the game that look suspicious. However, the game is unique and susceptible to improvements. By the way, the nuclear rules are designed to show that any use of nuclear warfare will probably lead to total destruction of the world. (MC)

Cost: $8.00, $9.00 with box

Producer: Simulations Publications, Inc., 44 East 23rd Street, New York, NY 10010

See also Origins of World War II HISTORY

YALU

John Hill and John Harshman

Playing Data
Copyright: 1977
Age Level: grade 9 and up
Number of Players: 5 hours
Playing Time: 2
Preparation Time: 1 hour

Description: *Yalu* is a game of the Chinese and North Korean armies' offensive against the U.S. and South Korean armies in North Korea in 1950-1951. The hard-backed map includes most of North Korea and a part of South Korea to near Seoul. Turns are weekly, and each hexagon is ten miles from side to side. Units are basically divisions, but U.N. units can break down into battalions and regiments. U.N. units are also supported by air and naval units. Communist forces have special conscript units and Inmun Gun guerrillas. Other rules allow for the "bug-out" (retreat) of U.N. units, paradrops, and sea movement. (Richard J. Rydzel)

Comment: This is an interesting and simple game, but it requires a very good U.N. player to survive, let alone win. It is very easy to learn and play. (RJR)

Cost: $11.98

Producer: Conflict Games, P.O. Box 432, Normal, IL 61761

YEOMAN

Playing Data
Copyright: 1965
Age Level: high school-college
Number of Players: 2
Playing Time: 2-4 hours
Preparation Time: 1 hour
Materials: mapboard (with charts and tables), general rules booklet, exclusive rules folder, die-cut counters, die
Description: *Yeoman* is a revision of the *Renaissance of Infantry*, which was once called *Tac 14*. The rules are somewhat simplified although the standard rules for the series are considerably modified by the special rules for *Yeoman*. The scenarios, when kept, are improved. Seven old scenarios are eliminated and only five new ones added. (However, one of the old battles is now dealt with in another game of the series.) (MC)

Comments: *Yeoman*, with *Chariot, Spartan, Legion*, and *Viking*, makes up Simulations Publications, Inc.'s *PRESTAGS* (pre-Seventeenth Century Tactical Game System). Together the five games study the art of war from 3000 B.C. to 1550 A.D. The beauty of the system is its range and flexibility. As it stands, it is occasionally very convincing but mostly only moderately convincing in its simulation of the events of the battles. However, it is a wide-open system for amendments and for the inclusion of new battles. There are many opportunities for the use of multiple players on each side. One of my favorite ideas, never yet implemented, is to design a strategic game of ancient or medieval warfare in which the battles, when they occurred, would be fought out using one or more of the tactical games. The change to these games from their predecessor was generally a happy one, but sometimes some of the flavor of the individual periods was lost in the process. Simulations Publications, Inc.'s *Musket and Pike*, while not belonging to this series, is definitely related, uses a very similar system, and with a little effort could be made compatible so as to continue the story to 1680. (MC)

Cost: $11.00, $12.00 with box

Producer: Simulations Publications, Inc., 44 East 23rd Street, New York, NY 10010

PRACTICAL ECONOMICS

BANK ACCOUNT

Arthur Wiebe

Playing Data
Copyright: 1976
Age Level: junior high, high school
Number of Players: 2-5
Playing Time: 30-60 minutes
Packaging: professionally packaged game including game board, markers, play money, checks, deposit slips, and balance sheets

Description: *Bank Account* is a board game which, according to the designer, "provides players with extensive experience in handling a bank account by engaging in life-simulating transactions." Players begin with a bank balance of $2,500 and $500 in cash. The spaces around the board are labeled with various expenses and windfalls such as "Buy insurance–$145" or "Collect $125 from every player." Each time players complete one circuit of the board and pass "payday," they collect $371.75. Players are encouraged but not required to conduct many of their transactions by check. Player moves are determined with a spinner. The first player to achieve a bank balance of $5,000 wins. (DCD)

Cost: $7.95

Producer: Creative Teaching Associates, P.O. Box 7714, Fresno, CA 93727

BUDGET

Playing Data
Copyright: 1977
Age Level: junior high, high school
Number of Players: 2-6
Playing Time: 20-60 minutes
Packaging: professionally packaged game including playing board, play money, option and savings certificates, and instructions

Description: *Budget* is a board game which, according to the instructions, "exposes students to some of the realities of real-life economics. The value of home ownership, insurance, and investments is stressed. Careful budgeting, though subject to unexpected events, is rewarded on Pay Day."

Budget is played on a board with red, yellow, and blue spaces. Red spaces allow the player the option of making one purchase of insurance or making a down payment on a house. Yellow spaces require the player to make some payment such as "repair appliance–$35" or "go bowling–$10." Blue spaces allow the player to make some sort of collection. All players can receive the $650 Pay Day, the $285 commission, and the $115 tax refund. Only homeowners, however, can sell their homes for $1500, and only the owners of savings certificates may collect interest. Each player moves according to the action of a spinner. The first player to accumulate $5000 wins. (DCD)

Cost: $7.95

Producer: Creative Teaching Associates, P.O. Box 7766, Fresno, CA 93727

THE BUDGETING GAME

Staff, Changing Times Education Service

Playing Data
Copyright: 1970
Age Level: grades 7-12
Number of Players: 4-16 in 1-4 teams
Playing Time: 3-5 periods of 40-50 minutes
Preparation Time: 1 hour

Description: *The Budgeting Game* introduces students to the problem of matching actual expenses with those that are projected for a given time period. Each player begins play with $20,000 received in 12 equal monthly installments. Using a budget worksheet, players make a series of purchase decisions, i.e., they select the type of housing they want, the make of automobile, and one luxury item. With each purchase decision they incur certain financial obligations: a mortgage, an auto finance loan, an installment contract, and so on. In addition to these costs, consumers must program their finances to deal with financial contingencies the average family must face. These include income tax payments, contributions to social security, and unforeseen expenditures. (From the Robert A. Taft Institute of Government Study on Games and Simulations in Government, Politics and Economics by George Richmond Educational Development Center.)

Comment: From the struggle to make ends meet, students gain insight into the relationship of credit purchases to cash income, the impact of unforeseen financial calamities and cash reserves (savings), and the utility of a budget to cut down the uncertainties one encounters. In its present form, *The Budgeting Game* would appear to provide teachers, students, or even parents and their children with a useful device for learning the ins and outs of personal finance. The game with some minor alterations might even be used to show children the consequences of inflation on a fixed income. (GR)

Cost: $9.95

Producer: Changing Times Education Service, Division EMC Corp., 180 East Sixth Street, St. Paul, MN 55101

BUYER BEWARE

Playing Data
Copyright: 1975
Age Level: high school
Number of Players: 36-50
Playing Time: 1 hour
Preparation Time: 3 hours (est.)
Packaging: 29-page manual

Description: *Buyer Beware* is designed to make students aware of what to look for when shopping and to encourage discussion of possible consumer protection laws. Participants are divided into retailers and consumers. The 16 retailers are assigned to four department stores. Each store has a manager, accountant, and two salespersons. Each store is given $1650 in play money and each consumer is given $2000. Each store is also given a wholesale price list (included with the game) and each store manager sets the retail price for each item on the list. When these preparations are completed, a buy-and-sell session is held. Each consumer is instructed to seek the best possible value for his money and each accountant records the store's accumulated profit. Trading stops after 20 minutes, a record is made of the stores' profits to this point, and the players are told that two years of inflation and recession have passed. The wholesale price lists are collected and an inflationary price list is distributed to each store. A second buy-and-sell session is conducted and at its conclusion the winners are determined. The winning store is the one that has earned the most profit. The winning consumer is determined by a formula which considers the total amount of money each consumer has spent, the number of items purchased, and where they were purchased. (DCD)

Cost: $3.00

Producer: Edu-Game, P.O. Box 1144, Sun Valley, CA 91352

CONSUMER

Gerald Zaltman, Academic Games Associates

Playing Data
Copyright: 1969
Age Level: grades 7-12, adult education
Number of Players: 11-34
Playing Time: 90-150 minutes in 40-minute periods
Preparation Time: 150 minutes at first only

Description: Players assume the roles of consumers, credit agents, and store owners who do business with one another. Consumers compete to get maximum pleasure for their purchase and minimum credit charges; credit agents compete for the best terms to the most people. The game is designed to teach players how to calculate true interest rates, how to negotiate contracts with credit managers, and the problems and economics of budgeting and buying.

This is a highly competitive game with a joker in the deck: minor random events such as doctors' bills appear unexpectedly—as is their wont—to harass the individual player who strives too hard to maximize pleasure in consuming while riding the thin edge of credit. (DZ)

User Report: *Consumer* has been successfully used with both junior and senior high school students as well as with groups of adults. The simulated conditions and circumstances are so realistic that players are highly motivated and animated. Groups of teachers in game orientation sessions even play the simulation enthusiastically and refuse to stop until they have completed the eight-month cycle.

Lively discussions follow the simulation as participants are eager to compare installment loan contract terms and conditions. They exclaim about the rates of interest charged and the penalties paid for slow payment. The high scorer is always proud to detail his rationale for spending and borrowing and to argue the pros and cons of using credit.

Through follow-up studies and field testing of *Consumer*, Gerald Zaltman finds that game participants learn the benefits of budgeting income and allocating funds according to a scale of priorities. The game teaches that one must pay for the privilege of borrowing and that the amount of interest one pays varies from lender to lender. The value of comparative shopping is demonstrated, as well as the implications of contracts. Participants learn to calculate percentage rates and to look for legal protection. They have an experience base for analyzing the values they have demonstrated.

Extensive experience with *Consumer* supports the observation that learners participate enthusiastically in simulated processes that require thinking, planning, deciding, human relations, value assessment, negotiation, communication, calculation, and strategy development. These are some of the processes of daily living that all students need experience with and that require more of the learner than absorption of knowledge.

Using credit is one of the necessary evils, or one of the painful luxuries, of modern American life. *Consumer* is one fun way to learn to use credit intelligently without the harsh penalty of a real financial catastrophe. (C. Raymond Anderson, University of Maryland, Simulation/Gaming/News, March 1973, p. 10)

Cost: $30.00 f.o.b. Indianpolis. Replacements: 11 pads $7.00, manual $1.25.

Producer: The Bobbs-Merrill Company, Inc., Education Division, 4300 West 62nd Street, Indianapolis, IN 46268

CONSUMER DECISION

David J. Boin and Robert Sillman

Playing Data
Copyright: 1973
Age Level: grades 9-12
Number of Players: 25-40
Playing Time: 3-4 class periods

Description: Students are divided into family groups (of about four) with monthly take-home salary of $800. They are faced with the problem of budgeting a limited income. They make purchases, arrange for loans, keep a monthly financial statement, and handle financial emergencies as they occur. (REH)

Editor's Comment: This is a standard Edu-Game format with 10-page teacher's manual and locally reproducible student material. (REH)

Cost: $3.00

Producer: Edu-Game, P.O. Box 1144, Sun Valley, CA 91352

CONSUMER REDRESS

Staff, Changing Times Education Service

Playing Data
Copyright: 1971
Age Level: grades 9-12
Number of Players: 15 or more in 2-3 teams
Playing Time: 40-50 minutes
Preparation Time: 1 hour

Description: *Consumer Redress* begins with each player receiving an identical set of 50 grievances against sellers. From these, each team selects the grievances it wishes to have redressed and then chooses the channel he thinks is most likely to comply with his request. Altogether, there are seven channels for redress: business and trade association channels, legal channels, state and local government channels, federal government channels, and private consumer organization channels (public interest law firms). In the game, coordinators are chosen to administer each of the redress channels. There is a limbo keeper, a person whose main duty is to keep the consumer in limbo (without a response to his complaint). Another person acts as scorekeeper and maintains a record of each team's success in having its grievances redressed.

Within each channel there are from one to four subchannels. For example, the legal channel is composed of a state court, a small claims court, the legal aid society, and a lawyer. It would appear possible, in a more advanced version of the game, to ask players to select the correct subchannel as well as the main redress channel for their complaints.

The educational objectives of *Consumer Redress* appear to be: to familiarize students with the channels of appeal for specific consumer complaints; to help students learn and compare the roles different agencies play with respect to consumer appeals; and to aid students in distinguishing between consumer and supplier responsibilities in the marketplace. (From the Robert A. Taft Institute of Government, Politics and Economics by George Richmond, Educational Development Center.)

Comment: Although the materials provided by the publisher offer little in the way of moral direction or insight for either the teacher or the student, *Consumer Redress* would seem to provide both with ample opportunities from which to draw moral distinctions and lessons. The game would also seem to be useful in helping its intended audience distinguish between outright fraud and more inscrutable forms of information manipulation and seller persuasion. (GR)

Cost: $9.95

Producer: Changing Times Education Service, Division EMC Corp., 180 East Sixth Street, St. Paul, MN 55101

DESIGN

John Wesley, El Cajon, California

Playing Data
Copyright: 1974
Age Level: grades 5-12

Description: *Design* is a simulation of designing and furnishing a home. Students first examine types of housing common in the United States today. They look at the costs, relative sizes, advantages, and disadvantages of single-family homes, condominiums, apartments, and mobile homes. Next, each student draws a family identity which outlines the size, income, occupation, and loan amount for his family. (Family identities represent a cross-section of U.S. families capable of affording a home today.) Having acquired this background on housing and a family identity, each student makes a first decision: What type of home will best match my income while meeting my family's needs? Next, the student designer makes a series of preliminary design decisions such as these: How many bathrooms and bedrooms? Do I want a fireplace? Can I afford air conditioning or a swimming pool? After making such initial decisions, the student calculates the number of square feet he can afford to build and then draws a rough outline of his new house. When this is completed, he starts sketching in where the various rooms will go and discovers that determining room size and then placing them is no easy task. Once he has finished this rough draft of his house, the student refines it and then copies it neatly onto his Home Design Sheet. Having completed his house plan, the student must now turn it into a home, buying and arranging furniture and other home decorations. Each family identity has a budgeted amount to spend on furnishings and decorations. By using catalogues, newspapers, sales brochures, and/or visiting furniture stores, each student furnishes his own home. *Design* concludes with an evaluation period during which students and teacher examine what they have learned about the cost and aesthetics of good home design.

Comment: This is a well worked out series of exercises that should tie in very neatly with children's fantasies of how they would like to have their houses when they grow up. It then combines a variety of skills from social studies, math, home economics, and so forth in the execution of the various parts of the exercise. Students need to draw, calculate, and observe houses and do some consumer economics in furniture buying. It would be nice if, in the next edition of this simulation, the author incorporated the types of options that children will be needing to consider when they are grown up, such as solar energy usage and other quite fundamental energy-use concepts. These, of course, can be incorporated into the exercise by any teacher knowledgeable in the area. (REH)

Cost: $22.00

Producer: Interact, Box 262, Lakeside, CA 92040

F.L.I.P. 2/80

Ron Klietsch

Playing Data
Copyright: 1970, 1979
Age Level: grade 7-adult
Number of Players: 1-14
Playing Time: 2 hours (est.)
Preparation Time: 1 hour (est.)
Packaging: professionally produced game includes instructor's manual, players' manuals and background sheets for eight players or teams, scoring sheets, and record sheets

Description: *F.L.I.P. 2/80* is the second edition of a family budgeting game (*Family Life Income Planning*) and is intended to simulate budgeting in the late 1970s and the early 1980s. The game is designed to let players "learn budgeting by going through all the steps involved."

Players represent one of six families or two independent individuals. They must methodically determine how the person or family they represent should spend the money at their disposal to buy what they want to own. "Babs Grady," for example, is a 33-year-old divorced mother of two with a net income of $2000 a month. The player representing Babs determines how much she should spend for food, shelter, utilities, clothing, furniture, personal care and services, transportation, alcoholic beverages, recreation, reading and education, gifts and charity, and medical care and insurance. The player must also determine which purchases Ms. Grady should make from a list of items she wants. These include a microwave oven, season tickets for the Houston Oilers, a new stereo, and a new couch. Players determine the budgets for the characters they represent for 12 rounds. Each round corresponds to one month. (DCD)

User Report: *F.L.I.P. 2/80* is an invaluable tool in a premarital counseling situation or with married couples having budgeting problems. It teaches sound, practical, easily understood budgeting principles in a remarkably short period of time. I took finance in college five hours a week for one year and did not learn much more than *F.L.I.P. 2/80* teaches in two hours. But even more important, (if one does not rush the simulation), it gives a couple time to talk over what categories of expenses are necessary and which ones are incidental. Areas often assumed to be understood surface and are faced for the first time. For instance, one couple could not agree on whether education was necessary or not. Finally, the husband-to-be asked how he was going to be able to get his master's degree without setting money aside for education while they both worked. Amazed, his fiancee admitted that she never realized that he wanted to get his master's degree. Repeatedly we have seen these kinds of family communications break through in *F.L.I.P. 2/80*.

The game is easily administered. The administrator has a complete leader's guide that outlines procedure step by step. The complete game would last 12 hours *but* we have found that optimum learning and involvement happens during round one, which lasts about 90-120 minutes. The beauty of this game is that one can stop almost any time. Learning and insight come as one plays (postgame discussion is often not needed). When played in groups, it allows couples to compare values and priorities, and this has proven to be exciting and helpful. (George M. McFarland)

Comment: The characters represented in this simulation are all affluent. It is reasonable to wonder why people with so much money have to be concerned with budgeting at all. (DCD)

Cost: $68.50

Producer: System's Factors, Inc., 1940 Woodland Ave., Duluth, MN 55803

HOME ECONOMY KIT

Ray Glazier

Playing Data
Copyright: 1975
Age Level: grades 7-12
Number of Players: 10-30
Playing Time: 1 hour a day for 5-30 days
Special Equipment: receipt pads, adding machine
Packaging: large carton containing 8 collapsible display stands, 360 item cards, section food labels, play money, adding machine, grease pencil, 30 problem cards, 28-page teacher's guide

Description: The objective of *Home Economy* is to provide students with realistic experience in home problem solving. The game begins with 2 students chosen as staff deciding how to place the 360 item cards on the display stands. Either they set prices according to the 1974 price listing the guide, or the entire class as homework finds out current prices. The staff are responsible for stocking, changing prices, and advertising while the rest of the students individually choose problem cards requiring them, for example, to plan a party for 10 children for $5 or to choose items for less than $1 which will enable them to clean thirty-two windows. The designer gives many suggestions for varying the game–adding items, introducing brand competition, wholesale buying, and so on. In the course of the game, students gain practice with simple math and learn basic economic concepts such as unit pricing and the law of supply and demand. The teacher grades them according to their success at problem solving; the manual includes solutions to the problem cards. During the follow-up discussion, students may discuss a number of issues presented in the guide, such as the relative benefits of canned, fresh, and frozen green beans or of homemade rather than ready-made cupcakes. (TM)

Cost: $48.00 + $3 postage and handling

Producer: Games Central, Abt Publications, 55 Wheeler Street, Cambridge, MA 02138

THE HOUSE DESIGN GAME

Luis H. Summers

Playing Data
Age Level: graduate students, architects, and their clients
Number of Players: 2-6
Playing Time: 3 90-minute rounds

Description: This game is a didactic tool designed to demonstrate the constraints and possibilities in house design and ownership more effectively than such standard methods of illustration as salesmen, booklets, and lectures. The game is played in three rounds corresponding to three discrete stages: preplanning, planning, and operational. There are two roles: architect and client.

During the first round, the manual states, "the client(s) and the architect(s) explore the familiar time dependent spatial and functional needs and desired lifestyle. The rationale is that people are often confused in ascertaining exactly what they want or really need in a house." In this round the client's family size and income are determined as well as the total size and cost of the house; the client purchases rooms for the house, and the architect arranges the rooms as the client desires.

The object of round two is to interest the client in an affordable house that will meet his or her needs. In the final round, the operation of the house over a period of 30 years is evaluated. (DCD)

Cost: $25.00

Producer: Luis H. Summers, 101 Engineering Unit "A," Pennsylvania State University, University Park, PA 16802

THE HOUSEHOLD ENERGY GAME

Thomas W. Smith and John Jenkins

Playing Data
Copyright: 1974
Age Level: adult
Playing Time: 30-60 minutes
Number of Players: 1 or more
Packaging: 20-page manual

Description: *The Household Energy Game* is not a game but is rather an informative exercise in determining how much household energy a person uses and in helping him or her decide how to use less. First, "players" utilize tables in the manual and a tally sheet to determine how much energy, measured in units equivalent to 1% of the average family's use, they use in transportation, heating, air conditioning, lighting, and major and minor appliances. This information is transferred to a 100-point game grid based on energy units. Then, "players" read about energy-saving measures, record the number of energy points they are willing to save, figure out their final number of energy points, and redo their game grid in an attempt to substantially reduce their number of energy points. (TM)

Cost: free

Producer: Sea Grant College Program, Marine Studies Center, 1800 University Avenue, Madison, WI 53706

HOUSEHUNT

Playing Data
Copyright: 1974
Age Level: junior and senior high school
Number of Players: 10-32
Playing Time: 2-3 1-hour periods
Special Equipment: duplicating machine
Packaging: multimedia kit in cardboard cover

Description: *Househunt,* part of a multimedia resource kit for teaching housing issues, focuses on developing awareness and skills in selecting private housing. Students choose roles as househunters, who compete for housing which meets their particular needs and income levels; as real estate brokers, who try to earn high commissions; and as bankers, who attempt to make wise loans. The kit includes househunter profile cards, instructions and financing guides for brokers and bankers, classified ads and housing cards describing available housing options, and numerous forms which must be completed, such as loan agreements and leases. After housing deals are made and closed, househunters discover the outcome of their choices, such as an increase in property value or the deterioration of the neighborhood. (TM)

Cost: $9.95

Producer: Changing Times Educational Service, Division EMC Corp., 180 East Sixth Street, St. Paul, MN 55101

MANAGING YOUR MONEY

Cumis Insurance Society, Inc. Cuna Mutual Insurance Co.

Playing Data
Copyright: 1970
Age Level: grades 7-10 most often and educatable mentally retarded classes
Number of Players: 2-6

Playing Time: 30-60 minutes
Preparation Time: time to study manual and run through game
Packaging: game kit including playing board, money, credit passbooks, ledgers, and cards

Description: The object of *Managing Your Money* is to be the first to repay a $2000 loan and to save $2000. Players begin by spinning the dice and taking turns moving around the board. When they land on a vocation square they may pay a fee and select the vocation; thereafter, every time they pass the credit union square they collect a salary. Other squares on the board allow a player to upgrade his vocation and therefore increase his salary, to collect dividends, to buy insurance, to pay bills, and to draw unknown or unexpected cards which may require payments, give a bonus, or provide additional opportunities. Winning the game is more a factor of chance than it is of skill, but players nevertheless gain experience with determining financial priorities and making decisions about spending, saving, and borrowing money. (TM)

Cost: $2.50 plus shipping

Producer: Advertising Dept., CUNA Mutual Insurance Society, P.O. Box 391, Madison, WI 53705

MARKET

William Rader, Florida State University at Tallahassee

Playing Data

Copyright: 1971
Age Level: grades 6-9
Prerequisite Skills: reading, grade 8.5; math, grade 6
Number of Players: 18-40 in 9-20 teams
Playing Time: 2-3 hours in periods of 40 minutes
Preparation Time: 1 hour
Supplementary Material: the game is part of "Economic Man," a 21-week program; or "Economic Man in the Market," a 4-week program

Description: Market is part of an elementary economics program developed at the University of Chicago. The game supplements other curriculum materials for a total instructional program. *Market* has a simple focus. The game materials provide teacher and students with all they need to simulate a market in a classroom.

At the start of play, students divide into teams of consumers and into teams of retailers. The consumers receive a set of materials that include: a shopping board, a supply of money, a dinner menu sheet, and the rules for play and for scoring. Retailers, on the other hand, receive a list on which they publicly advertise the prices they are charging for their commodities, money with which they can make purchases from the wholesaler, order lists used to obtain additional supplies of commodities they offer for sale, and a copy of the rules with instructions for scoring. To begin with, the teacher prepares envelopes for each team containing a recommended distribution of products. Once these are passed out, students begin planning their purchase decisions. Retailers rent store locations in the classroom, and then purchase one of the envelopes that the teacher has prepared for the retailing group. Consumers, during the same interval, begin planning their menus for the week.

Consumers seek to achieve a balanced diet, represented in the game by a card distribution that includes: meat, potatoes, a vegetable, a beverage, and a dessert for each of six meals. Retailers price their goods to maximize their profits and minimize their losses.

Market is structured so as to produce conditions of competition between sellers and buyers. Buyers compete to find the lowest price for each category of goods. Sellers compete to sell their goods without taking a loss. During play, buyers circulate in the room looking for the best price for a given commodity. It would appear from my examination of the material that the drama of the game occurs as different sellers, starting with the same resources, assign different values to their commodities. In effect, the consumer is given the chance to pit one seller against the other in the hunt for the best price.

The scoring mechanism reinforces the playing sequence. Consumers are rewarded with 3 points for each food card they purchase, with 10 additional points for each meal they complete, with 1 point for each $.50 they save, and with a 150-point bonus for satisfying all meal requirements. Retail scoring falls along a different dimension. One's score is computed by a simple profit formula: Goods sold + inventory − cost of goods sold = profit. The retailer with the highest profit wins. (From the Robert A. Taft Institute of Government Study on Games and Simulations in Government, Politics, and Economics by George Richmond, Educational Development Center.)

Comment: While this game was designed to help teach the principles of supply and demand, a side-effect has been noticed: Students tend to change their behavior with regard to their style of shopping for goods and may attempt to influence their parents to change.

Results of field tests and evaluative research are available from the Industrial Relations Center, University of Chicago, 1225 E. 60th St., Chicago, IL 60637. (DZ)

Cost: $72.00, game only (#048587)

Producer: Benefic Press, 1900 North Narragansett, Chicago, IL 60639

MORTGAGE

Jack W. Reynolds

Playing Data
Copyright: 1976
Age Level: high school
Number of Players: 3-70 in 1-10 groups
Playing Time: 2 1-hour periods
Preparation Time: 1 hour
Packaging: 11-page manual

Description: This simulation is designed to acquaint students with the difficulties in purchasing a home and to encourage setting priorities in making a major purchase. Players also gain practical experience in using percentages. The participants assume the roles of a family member (each family has a husband, wife, and four children) or a real estate agent. All family members have input on the purchase of the home. They must consider the size, type, and location of the home they will buy, must decide whether to purchase mortgage insurance, must purchase homeowners insurance, and must figure out how to pay for a $35,000 home on an annual income of $10,000. Property taxes, maintenance costs, and living expenses must all be calculated. All the decisions presented in the game must be made by the players in 60 minutes. Matters are further complicated by the lassitude of the real estate agent who does not like his job and does not want to be rushed. During the debriefing, participants are asked to discuss the frustrations they experienced in buying a house. (DCD)

Cost: $1.95

Producer: SAGA, 4833 Greentree Road, Lebanon, OH 45036

PROTECTION

David Rosser

Playing Data
Copyright: 1976
Age Level: high school
Number of Players: 18-50 in 6 groups
Playing Time: 5 50-minute periods
Packaging: 8-page student guide, 2-page teacher's guide

Description: *Protection* simulates the federal government's efforts to protect consumers and business. Students discuss the concept of trade-off, take a survey of adult opinion on consumer problems, and then divide into six activity groups (consumers, consumers/labor, consumers/investors, manufacturers, dealers, advertisers) to complete a short reading and question-answering assignment and to compare their surveys. Groups are then introduced to the simulation problem-developing consumer protection standards for a microwave oven. Each group must complete individual proposal forms and reach a consensus on a specific group proposal concerning construction, warranty, inspection, sales, and advertising of microwave ovens. The proposals they choose are influenced by their group's goals and the "advantage points" and "cost points" of a proposal for a particular group, as outlined in the manual. The last two days of the simulation, students discuss and enact a Federal Commission Hearing. After each group presents its proposals, the commission votes and groups score themselves according to their success at getting their proposals adopted. Groups may choose to appeal the decisions of the commission, in which case a revote is taken. (TM)

Cost: $10.00

Producer: Interact, Box 262, Lakeside, CA 92040

RIP OFF

Paul F. Ploutz, Educational Games

Playing Data
Copyright: 1975
Age Level: 12 and up
Number of Players: 2-12
Playing Time: 30-60 minutes
Packaging: professionally designed box

Description: Each player in this game is a consumer involved in the day-to-day challenge to save money by identifying appeals of advertisers. This game tests a player's ability to spot 1 or 12 basic appeals commonly employed in marketing consumer products. The 120 products appear as playing cards which list the price, and the product name (Early American Reproductions @ $40 per item, the Regency Hotel @ $100 per night; Sun & Sea Health Clubs @ $180 per membership, and so on), plus the appeal to purchase self-gratifications, social status, sex, and such. Each player receives 5 product cards to start. Play is begun by the player holding the least expensive product. The goal is to get rid of product cards. The first player with no product cards wins. Play involves use of "Liquidation" cards which either add or subtract "Product" cards or direct the player to pass a "Product" card to the other player. "Odd-job" cards make or lose money for the player, based on situations printed on each card. One player's turn involves the following:

- draw a "Product" card
- Choose which of all "Product" cards held to use for identification
- identify the appeal aloud to the group
- the identification is checked against the key
- if correct, discard that product to the bottom of the "Product" pile, draw a "Liquidation" card, and follow the instructions
- if incorrect, "buy" the product (i.e., keep the card), draw an "odd-job" card, and follow the instructions.

Final scoring is based first upon elimination of all products, second upon money earned, third upon remaining products and their negative worth (liability). The idea of the entire game is to save money. (REH)

Editor's Comment: The situations and products depicted on the playing cards are mundane but occasionally humorous. The appeals are effectively presented such that a player could identify them in real-life situations. (REH)

Cost: $9.00, 5 for $38.00

Producer: Union Printing Company, 17 West Washington Street, Athens, OH 45701

SHARE THE RISK

Playing Data
Age Level: grade 9-junior college
Number of Players: 8-40 in 2-3 teams
Playing Time: 5-7 hours
Preparation Time: 1 hour
Materials: instructor manual
Special Equipment: copy machine

Description: Players represent heads of families and members of a share-the-risk pool. The objective of family heads is to protect and improve family security by increasing resources from year to year. The objective of risk-sharing team members is to see that the pool adequately protects personal security without taking too much money (that is, that they are neither over- nor underinsured). The team objective is to increase net worth.

The class is divided into two or three equal teams. "Risk" cards are distributed to each team member, and players assess what perils they face and decide what amount of coverage to buy, if any. At the end of every round, players draw risk cards that determine their fate during the round. (AC)

Cost: $9.95

Producer: Changing Times Education Service, Division EMC Corp., 180 East Sixth Street, St. Paul, MN 55101

SMART SPENDING

Playing Data
Copyright: 1971
Age Level: grades 7-12
Number of Players: 5-26
Playing Time: 10-20 hours
Preparation Time: 2 hours (est.)
Packaging: professionally packaged, multimedia unit includes long-playing record, 2 filmstrips, 3 role-playing scripts, 26 copies of student booklet, and 100-page teaching guide

Description: This series of 10 exercises in consumer education was developed in cooperation with The Better Business Bureau of Metropolitan New York. The unit was designed to help "students become knowledgeable and responsible buyers. Several kinds of buying situations are studied and simulated to sensitize the student to the pitfalls of business transactions." The units address an overview of consumer affairs; budgeting; buying a car; buying clothes, appliances, and food; shopping for services; renting an apartment; advertising practices; and credit.

The unit that is concerned with buying a car typifies the exercises in this collection. Here the students are told that they must purchase an automobile for $1,200 or less. They examine advertisements in their local newspaper, select cars that they would like to see, and then must justify their choices. A film strip and a record are used to simulate a trip to a used car lot. Then the students role play a used car salesman and a prospective customer. (DCD)

Cost: $82.00

Producer: Educational Audio Visual, Inc., Pleasantville, NY 10570

SWINDLE!

Staff, Changing Times Education Service

Playing Data
Age Level: grades 7-12
Number of Players: any number
Playing Time: 1 or 2 40-50 minute periods
Preparation Time: 1 hour

Description: *Swindle* simulates a small consumer marketplace. Each student begins play with a $1500 stake. He uses this sum to purchase a used car from one of two car dealers and then must purchase the services of one of two mechanics to correct defects in the car bought. Students make their purchase decisions in response to advertising. Many get swindled either because they accept seller statements at face value or because they fail to ask key questions about the product they are purchasing.

Altogether, there are nine separate roles in the game: a coordinator, 2 used car managers (salesmen), 2 auto repairmen, a money-making (investment) counselor, a bargain-market manager, a better business bureau representative, and a manager of a car clinic. The remaining participants form the consuming public.

According to the publisher, *Swindle!* helps students improve their buying and question-asking skills; they become conscious of a variety of techniques used either to gyp the public or to exaggerate product qualities. The winner of this game is the person who gets the most for his money, or the player who avoids being conned more than anyone else. (From The Robert A. Taft Institute of Government Study on Games and Simulations in Government, Politics and Economics by George Richmond, Educational Development Center.)

Cost: $9.95

Producer: Changing Times Education Service, Division EMC Corp., 180 East Sixth Street, St. Paul, MN 55101

See also
Buy and Sell ECONOMICS
PX-190 FRAME GAMES

RELIGION

THE BANNER-MAKING GAME

Paul Dietterich, Charles Ellzey

Playing Data
Copyright: 1972
Age Level: grade 5-adult
Number of Players: 6-30 players in 3-6 teams
Playing Time: 2 to 2-1/2 hours in periods of 1/2 hour, 1 hour and 1 hour
Preparation Time: 3/4 to 1 hour

Description: Players plan and negotiate on behalf of their teams to finish their own banners and to place symbol pieces in their color on other teams' banners. (AC)

Comment: This game is always interactive and usually culminates in encouraging cooperative behavior. Sometimes younger or highly competitive groups do not realize that they must cooperate in order to win. The game is somewhat misnamed in that it does not deal with the skill of banner-making.

The game is very nicely packaged and contains some useful concepts for teaching. It seems best suited for junior and senior highs. (GMM)

Cost: $10.50

Producer: Graded Press, 201 Eighth Avenue, South, Nashville, TN 37202

BIBLICAL SOCIETY IN JESUS' TIME

Jack Schaupp

Playing Data
Copyright: 1973-1974
Age Level: grade 5-adult
Number of Players: 10-30 in 5 teams
Playing Time: 1 to 2 hours in periods of 3-15 minutes
Preparation Time: 1+ hours

Description: Players are divided into groups of Romans, Zealots, Sanhedrin, Herodians, and Disenfranchised–the major groupings of Palestinian society at the time of Jesus. Each of these groups represents a council with different objectives. During successive periods, rumors are spread by the Game Director. Then, various crises begin to occur: a food shortage, rioting against the Roman Army in Egypt, and so forth. Scoring is based on the effectiveness of the job of each of the teams. (REH)

Comment: Participants work in groups to arrive at decisions most appropriate to the situation and their objectives. Individual and group interaction is kept to a minimum except for brief open communication periods. The total game time is brief. It lasts less than one hour, although proper preparation will take longer than the game director's instruction booklet indicates. Proper preparation considerably enhances the game experience.

One unique aspect of this is its flexibility in use with differing age groups. This is a learning experience that entire families may participate in together. The role-play requirements are minimal and preparation is not difficult if properly explained. Children younger than fourth graders have difficulty understanding the simulation. Designed to be used in church school situations in typical Griggs' style, this is a helpful teaching aid for both professional and volunteer teachers. (GMM)

Cost: $5.00

Producer: Griggs Educational Service, 1731 Barcelona Street, Livermore, CA 94550

CAN OF SQUIRMS (OLD AND NEW TESTAMENTS)

Playing Data
Copyright: 1973
Age Level: teenagers-adults
Number of Players: 2 or more
Playing Time: one hour or more
Preparation Time: 2 hours (est.)
Packaging: professionally packaged and produced set of two games includes role sheets and instructions in screw-top cans

Description: Pairs of players assume the roles of Biblical characters to act out personal religious, ethical, or moral dilemmas recorded in the Old or New Testaments. The situations players must act out are printed on slips of paper called "Squirms," drawn from a can. There are an equal number of men's and women's roles.

For example, in one Old Testament Squirm, two players assume the roles of Noah and his wife. The wife is loving and devoted, but she is certain that her husband has become a little crazy and is making fools of the entire family. First he began to speak to himself as if there was someone else with him, and then he became obsessed with the idea of building a huge ship in the back yard. Today he started to collect pairs of exotic animals, and the poor wife's patience is exhausted. She must confront her husband with the facts of his bizarre behavior and put a stop to it. In a New Testament Squirm, one player assumes the role of Herod's son, Agrippa, who is on an official visit to Caesarea with his sister, Bernice. Agrippa meets Paul, experiences the man as powerful and convincing, and finds himself fascinated with and terrified of his strange Nazarene cult. Bernice then begins to tease Agrippa, telling him that he is afraid to become a Christian, and Agrippa must answer her.

The game may be played for any length of time by any number of players. Sets of Old and New Testament Squirms must be purchased separately. (DCD)

Cost: $7.95 per set, quantity discount

Producer: Contemporary Drama Service, Arthur Meriwether, Inc., Downers Grove, IL 60515

CATALYZER

Neil Topliffe

Playing Data
Copyright: 1973
Age Level: grade 9 through adult
Number of Players: 16-28 in 4 teams
Playing Time: 3 or 4 time periods
Preparation Time: 1-2 hours

Description: *Catalyzer* is designed to give participants some understanding of the third world and the relation of church missions to third world societies. Teams of participants represent different socio-economic or interest groups whose objectives are either to survive, to make a profit, or to change the existing structure. The game shows participants how people act and feel as they assume positions of power, aggression, or change. (author)

Comment: The manual is *very* complete. The game experience is well thought out. Explicit and helpful directions are given for the game director. The game itself is interesting and raises a number of challenging questions concerning the third world and the nature of the church's mission to that world. The suggested procedure for postgame discussion is extremely complete and will add significantly to the overall teaching experience.

I believe the directions allow a bit too much time for playing the game. It suggests six rounds and minimum playing time of one and one-half hours. My experience with *Catalyzer* indicates five rounds to be enough and playing time closer to an hour maximum. The game is more dynamic and enjoyable when time limits are reduced 10% to 20% and strictly enforced.

The *Catalyzer* manual is a must for people who feel somewhat inadequate about leading simulations but want to increase their skill. Directions are extremely clear and suggestions for setting up the simulation are relevant and helpful. Use *Catalyzer* only if the participants are sincerely interested in understanding some dynamics of third world nations and the relation of the U.S. church to those nations. *Catalyzer* should be a great experience. It is a fine resource for a church wishing to become more informed about the nature of missions. (GMM)

Cost: $1.95

Producer: Friendship Press, Distribution Offices, P.O. Box 37844, Cincinnati, OH 45140

THE COFFEE GAME

Megan McKenna and Joanne McPortland

Playing Data
Copyright: 1972
Age Level: grades 7-12
Number of Players: 10-15
Playing Time: 50 minutes (est.)
Packaging: 4-page teacher's guide and 3 black-and-white posters

Description: *The Coffee Game* is a device for promoting the discussion of a complex ethical dilemma: What can or should anyone who grows, sells, distributes, or drinks coffee do to ameliorate the suffering of the Colombian peon who harvests the beans? Game materials include three striking posters that picture a peasant, an American housewife, and a group of middlemen, with short statements by each. Seven students assume the roles of the people in the posters to present their opinions to another group, which represents an International Coffee Commission. The commission must seek an answer to the ethical question posed above, and should they find an answer, the authors urge them to write to the appropriate committee of the U.S. House of Representatives. (DCD)

Cost: unknown

Producer: Teleketics, 1229 South Santee Street, Los Angeles, CA 90015

ETHICS

Art Fair

Playing Data
Copyright: 1971
Age Level: grade 7-adult
Number of Players: 3 or more
Playing Time: 30 minutes or more
Prerequisite Skills: reading, grade 11; math, grade 0
Preparation Time: 1 hour

Description: *Ethics* is a game of ethical choices. Players are projected into situations that range through the spectrum of human experience. Faced with difficult decisions, one must choose a course of action—defend it or modify it—and, finally, face the judgment of one's peers. In settings as different as a supermarket and a concentration camp, the participants explore their own concepts or right and wrong and those of other players. It's a game that could change perspectives.

An ideal number of players is from five to seven. If there are more than this, can get two, three, or more games going at the same time. It is a useful and interesting exercise to have different groups deal with the same issues and compare results. (GMM)

Comment: This is not a true simulation, but it is a teaching game, a clever way to gather data concerning a number of people's opinions and feelings about various ethical situations. It is extremely valuable because of its flexibility. It is easily administered to any number of people in any time period, and it can be stopped and started later. Postgame discussion can be done any time.

To play *Ethics* is not a life-changing event, but it does stimulate participation and provide a context in which participants can share insights and opinions about many issues people often are unwilling or unable to discuss. The game has a way of bringing issues that seem remote or intellectually sterile into the immediate here and now. (GMM)

Cost: $4.95, $59.40 a dozen

Producer: Art Fair, Inc. 18 West 18th Street, New York, NY 10011

EXODUS

Elissa Blaser

Playing Data
Copyright: 1974

Age Level: high school
Number of Players: 20
Playing Time: 30 minutes
Preparation Time: 30 minutes (est.)
Packaging: professionally produced, unbound game packet including instructions, discussion guide, role sheets, and documents

Description: This game, according to the author, "is designed to simulate the experience that Soviet Jews must face in trying to get out of Russia." The game has roles for 15 would-be emigrants, 3 Soviet officials, the Swiss Ambassador, and the head of a Zionist group. Each emigrant has a reason for wanting to leave the Soviet Union and must secure an exit visa, a security clearance, and the money for the exit tax. The Swiss Ambassador and the Zionist organization can offer them a limited amount of help, and the punctillious Soviet bureaucrats try to frustrate the Jews. The process of the game consists of the 15 would-be emigrants going from official to official filling out forms and standing in line. The game ends after 30 minutes of this activity, and the players are then asked to discuss how they felt about their plight. (DCD)

Cost: $6.95

Producer: Behrman House, Inc., 1261 Broadway, New York, NY 10001

EXPERIENTIAL EDUCATION: X-ED

John Hendrix and Lela Hendrix

Playing Data
Copyright: 1975
Age Level: all ages
Number of Players: 10 or more
Playing Time: variable
Packaging: paperback book

Description: *Experiential Education* contains dozens of short interpersonal exercises for church groups designed to develop awareness of group process, to elucidate the teachings of Jesus, and to plan future church activities. The range includes simple encounter techniques such as practicing giving feedback to other group members and holding two-minute discussions with each other in shifting double circles, activities that involve focusing on walking or eating together, and task-oriented activities such as designing a conference or developing a scenario for the future of the church. A variety of intergenerational games for getting acquainted and making contact are also described. (TM)

Cost: $6.50

Producer: Abingdon Press, Nashville, TN

GAMING

Dennis Benson

Playing Data
Copyright: 1971
Age Level: adults
Number of Players: 4-8
Playing Time: 1 hour (average) per exercise
Preparation Time: 2 hours (est.)
Special Equipment: record player
Packaging: professionally produced 64-page hardback book and record

Description: The author describes this as "a book that is created to enable you to play learning games with your people as your situation dictates. You will quickly gain confidence by realizing that you can create games as good (or better) than those created by the folk in the book. You will then be on your own. You won't need this book, nor will you need any boxed games. You'll be free to do something authentic on your own."

Blind, *Bread*, and *Flight 108* are typical of the seven games in this collection. *Blind* is designed to help players grasp the significance of the passage 1 John 2:7-11. Players are blindfolded and must spend 30 minutes in each of five different rooms. In the first room a maze, through which the players must crawl, is marked on the floor with masking tape. In the second room the blindfolded players must make caves from cardboard boxes. In the third room the players listen to the sound track of a motion picture and try to figure out what is happening in the film. In the fourth room the players try to divine the identity of a mystery woman by feeling her feet, and in the last room players put a personal object in a plastic bag and pass it around to the other members of the group. In *Bread*, which is designed to help the participants comprehend the immensity of the problem of hunger in the world, red and green balloons represent people and food. The only way people can produce more food balloons is by risking them. Every five minutes a controller throws darts at the food balloons that are risked. The players earn an extra balloon for each miss, but for each hit another balloon, representing people, must also be burst. In *Flight 108* players are shown the photographs of 12 people and told that they are on an airplane which is going to crash. Told that they have the power to save just 5 of the victims, players must then choose who will live and justify their decisions. (DCD)

Note: To learn about *Ralph*, another game in this book, see the essay on religions simulations by Lucien E. Coleman, Jr.

Cost: $9.95

Producer: Abingdon Press, Nashville, TN

GESTAPO: A GAME OF THE HOLOCAUST

Rabbi Raymond Zaverin, Audrey Friedman Marcus, Leonard Kramish

Playing Data
Copyright: 1976
Age Level: junior high to adult
Number of Players: 4-1000
Playing Time: 1-2 hours
Preparation Time: 1/2 hour
Packaging: professionally produced package includes instruction booklet, perforated cards, pad of value sheets

Description: The purpose of *Gestapo* is to teach about the events in Germany from 1933 to 1945 and to give students the opportunity to become aware of what values they would sacrifice in conflict situations. Students begin with three markers each on seven values on their value boards—the values of community, family life, pride, religion, house, civil liberties, and income/job. Each turn, they must all risk one marker. When an information card is read, presenting chronologically events related to the Holocaust, students lose their marker if it corresponds to the value associated with the information card. After every five plays, a trading session allows students to exchange markers to gain the values of their choice. Then, they are allowed a chance to gain additional markers. Students survive the simulated Holocaust if they possess one life marker at the end of the game. They are not told until afterward that the cards are so marked that their chance of survival depends to a large extent on their willingness to risk life markers at the beginning of the simulation. (TM)

Note: For more about this simulation, see the essay on religious simulations by Lucien E. Coleman, Jr.

Cost: $8.50

Producer: Alternatives in Religious Education, Inc., 3945 S. Oneida Street, Denver, CO 80237

GOING UP: THE ISRAEL GAME

Joel Lurie Grishaver

Playing Data
Copyright: 1978
Age Level: age 12 to adult
Number of Players: 2-6 players or teams
Playing Time: 1 hour (est.)
Preparation Time: 30 minutes (est.)
Packaging: professionally packaged board game with playing surface, dice, tokens, cards, and instructions

Description: *Going Up* (or *Aliyah*) is designed, according to the producer, "to facilitate learning about the modern state of Israel. The kind and degree of learning which takes place as a result of playing *Going Up* depends on many factors. These include: prior knowledge, proficiency in Hebrew, the number of times the game is played, the introduction of the game and the kind of debriefing, follow-up and additional activities which follow play."

The game is played on a track of squares on which are distributed "experience cards." Players move tokens around the track according to the roll of dice and follow the directions printed on the card or square on which they land. Experience cards, for instance, recount what can happen to people who visit or live in Israel (such as visiting the Dead Sea or Yad Vashen, hitchiking, or just hanging out) and require players to move forward or backward, gain or lose turns, gain or lose Nachas cards, or go to a Kibbutz, Milium, or a Makolet. Squares without Experience cards are the Kibbutz (where players must stay at least three turns), the Milium (compulsory military service), the Makolet (grocery store), the Ulpan (language school), and the Bus Stop. Landing on any of these squares requires a player to give the Hebrew names for certain common objects (such as pizza). A player who can do this earns Nachas cards. A player who cannot give the Hebrew names for these objects loses Nachas cards.

The winner is the first player to complete three circuits of the track and work on the Kibbutz at least once, do Milium at least once, and have at least five Nachas cards. (DCD)

Cost: $9.95

Producer: Alternatives in Religious Education, Inc., 3945 S. Oneida Street, Denver, CO 80237

GOSPEL GAME

John Washburn

Playing Data
Copyright: 1971
Age Level: grade 5-adult
Number of Players: 12-30 in 4-6 teams
Playing Time: 1-2 hours
Preparation Time: 2-3 hours

Description: Players are organized into teams who prepare a "Gospel" from a variety of printed materials. Teams interpret their "Gospel," individuals select the "Gospel" they like best, and discussion follows. (producer)

Comment: This game has been designed specifically for use of communicant classes, Bible study workshops, youth programs, and for use in church retreat experiences. It is a simulation type exercise about the fourth stage in the development of the Gospels. It deals with the final editing of the Gospels in the early church to communicate the best to the needs and concerns of its people.

The directions for the game are clearly composed and easy to follow. Once the game materials are gathered and organized, very little remains for the administrator to do. Preparation of the game does require an extensive amount of time. Game components include old magazines and newspapers, a dozen or more small bottles of glue, 50-60 blank 4 x 6 cards, a dozen or more sheets of 3 x 5 construction paper, 4 x 6 cards with Scriptures verses typed on them and 5 x 8 cards with typed instructions. All these items must be gathered and prepared by the administrator.

This is a typical John Washburn creation. It integrates some elements of group process, art, and simulated data into one overall experience. The directions suggest many options for the administrator. These options do make this simulation highly flexible and thereby usable in most any kind of Bible study situation. The one caution for this game is that the participants must be able to read. It does require some reading as well as a fair amount of creativity.

This game has proven to be especially helpful for communicant classes and for Bible study in a retreat kind of setting. It is most effective in about a two-hour slot. Attempts to abbreviate or compress *Gospel Game* have not been very successful. Participants, especially young people, have expressed enjoyment of the game.

A list of books, films, and records is included in the directions. These can be used in conjunction with *Gospel Game* either as introductory material or for follow-up. These resources combined with the game can provide a substantial introduction into how the Gospels came to be and thereby provide the participants with a greater understanding and appreciation of "The Good News." (GMM)

Cost: $3.00 prepaid; include self-addressed 9 x 12 envelope with order

Producer: Simulation Sharing Service, 4740 Shadow Wood Drive, Jackson, MS 39211

KIBBUTZ

Rabbi Fred Rubinger, Northern Hills Synagogue, Cincinnati, Ohio

Playing Data
Copyright: 1975
Age Level: high school, college
Number of Players: 15
Playing Time: 5-6 hour class periods
Packaging: 14-page pamphlet

Description: The objectives of *Kibbutz* are to increase students' understanding of the difficulties involved in establishing a kibbutz in Israel and to make them more aware of the influence of individual background on beliefs and values. Fifteen character descriptions delineating life history, feelings, and personal and political beliefs are included for the students to role play. Among them are descriptions of two leaders, a member of the World Zionist Organization who serves as a mediator and a representative of the Israeli government. There is also a chairman of the meeting. Students as kibbutz settlers have to decide first whether to settle on disputed land which will be financially guaranteed by the government and protected by the army or whether to choose an undisputed site which will not be funded or protected. They then have four other decisions to make regarding the economic base of the kibbutz, the kinds of religious observances, the methods of childrearing, and qualifications for membership. Periodically, the teacher interrupts each hour meeting with radio announcements of news relating to the issue being discussed. During the last session students discuss such issues as how they imagined Israelis and Americans might differ in making vital decisions about establishing a kibbutz. (TM)

Cost: $1.50

Producer: Fred Rubinger and SAGA, 4833 Greentree Road, Lebanon, OH 45036

LEELA

From the ancient Sanscrit text, with commentaries by Harish Johari

Playing Data
Copyright: 1975
Age Level: high school-adult
Number of Players: 1-6
Playing Time: 1-2 hours
Packaging: paperback book, with enclosed playing board

Description: The purpose of *Leela* is to detach oneself from one's own personality and to identify with cosmic forces as one progresses along the spiritual path toward cosmic consciousness. The playing board is divided into 8 rows of 9 squares each, each row representing a *chakra* or energy center and a level of existence. Each step or square symbolizes a stage of the Hindu spiritual path (such as envy, plane of violence, plane of joy, and positive intellect) and is accompanied by a lengthy commentary which explains its meaning. Players use rings, or other small items which they normally wear or carry, as playing pieces; they move forward by throwing the die. Landing on a snake's head means that, due to egoism, they must regress to the step which occurs at the tail of the snake; landing at the bottom of an arrow means that due to spiritual devotion they may progress to the arrow's head. The game ends when one or all players have reached cosmic consciousness. (TM)

Comment: As the text claims, a synchronous effect does seem to exist between the psychological experience of the players and the squares upon which they repeatedly land. Most players are aware of a strong connection between their game and the problems with which they are struggling at the moment, rather than with the problems or experiences which predominate throughout their entire life. This raises the question of whether or not the game might be useful if played on a regular basis to help persons to clarify their present issues and to gain perspective upon their own processes of development. (TM)

Note: For more about this game, see the essay on religious simulations by Lucien E. Coleman, Jr.

Cost: $4.95

Producer: Coward, McCann and Geoghegan, Inc., 200 Madison Avenue, New York, NY 10016

MISSION GAMES

Playing Data
Copyright: 1978
Age Level: high school, adult
Number of Players: 10-40
Playing Time: about 1 hour per game
Packaging: professionally produced and packaged materials for six games

Description: *Mission Games* contains six group activities. Each is intended to help clarify a different value for evangelical Christian missionaries and prospective missionaries. In the *Dot Exercise,* each player must draw four straight, connected lines in such a way that they transverse all of nine dots arrayed into a square, without lifting their pencils. In *Decisions, Decisions,* four teams of equal size vote to say "blue" or "gold." All groups must try for the same color. When all vote for the same color, all teams earn points. In *Tag—You're Who,* players wear tags of different geometric shapes in the colors red, white, and blue. Players must group and regroup according to their tags, first by shape and then by color. In *Missionary Roads,* teams of four players draw roads on a large sheet of white paper that has been marked with X's, which signify places in the world where there are Christians. Players score points by drawing "Missionary Roads" to blank places. When 14 roads have been drawn, the game ends and players consider the passage Matthew 25:20-21. *Confronting Culture* is a two-hour game in which the players are divided into Alpha, Beta, Omega, and Iota cultures. As in the well-known cross-cultural simulation *Bafa Bafa,* from which this exercise was adapted, members of each culture spend time visiting the other and then report to their own groups about what the visit and the other culture were like. The final exercise in this collection is *Formissia.* Here, participants represent foreign missionaries at an annual meeting of missions. (DCD)

Note: For more about *Formissia,* see the essay on religious simulations by Lucien E. Coleman, Jr.

Cost: $15.00

Producer: Brotherhood Commission, Southern Baptist Convention, 1548 Poplar Avenue, Memphis, TN 38104

MOSES AND THE EXODUS

Jack Schaupp and Donald L. Griggs

Playing Data
Copyright: 1974
Age Level: grade 5-adult
Number of Players: 15-30
Playing Time: 15-60 minutes
Packaging: professionally produced 96-page booklet

Description: This booklet contains four role-playing exercises relevant to the Jewish Exodus. In the first, *To Stay Or Go,* three groups of players represent the followers of Moses, the citizens of Egypt, and undecided Hebrews. The game follows a highly structured scenario in which the players hear Moses speak, learn of the plagues of Egypt, and debate whether to leave the country or not. The exercise is designed to illustrate, according to the authors, that "there were many complicating forces that influenced the Hebrews to want to go with Moses or to stay in Egypt." Similarly, *Wilderness Survival* encourages participants to "identify with the needs of the Hebrews as they traveled through the desert." Each player must rank a list of 14 items in the order of each item's importance to the group's survival in the desert. Then, groups of players must agree on a ranking of the items. In *Faith In Darkness,* players wander through a darkened room for about 10 minutes and then try to creatively express how they felt and how the Hebrews must have felt as they wandered. *The Call Of Moses* is the most loosely structured of all the exercises. The participants in this activity assume the roles of people who were involved with Moses in Midian and try to identify with those people. (DCD)

Note: For more about *To Stay or Go,* see the essay on religious simulations by Lucien E. Coleman, Jr.

Cost: $5.00

Producer: Griggs Educational Service, 1731 Barcelona Street, Livermore, CA 94550

NEXUS

Playing Data
Copyright: 1974
Age Level: adult
Number of Players: 8-20 (est.)
Playing Time: 4 75-minute rounds
Packaging: professionally packaged board game with instructions, playing surface, tokens, and worksheets

Description: This game was designed for congregations and ministries of recommitted Christians. There is no clear-cut outcome to the exercise. The principle activity of the players consists of placing tokens on the spaces of the *Nexus* circle or playing surface. This circle is divided into wedges that are labeled with terms that describe life tasks or gifts of recommitted Christians (prophecy, evangelism, worship).

The circle symbolizes the duties and obligations of Christians. Players ask themselves what their spiritual gifts are and what area of the circle best identifies their gifts. Players reflect on this for several moments and then place tokens in the area of the *Nexus* circle that best categorizes their thinking. After placing these tokens (symbolic of an act of faith) in the circle, players explain why they put their tokens where they did and discuss their relation to the placement of the other players' tokens. (DCD)

Note: For more on this game, see the essay on religious simulations by Lucien E. Coleman, Jr.

Cost: $7.75

Producer: Broadman, 127 Ninth Avenue, Nashville, TN 37234

ON THE MOVE: SOVIET JEWRY

Playing Data
Copyright: 1974
Age Level: junior and senior high
Number of Players: 5-20
Playing Time: approximately 1 hour
Packaging: paper cards and board

Description: The objective of *On the Move* is to teach students the history of Jews in Russia from prerevolutionary times to the present. Each turn, the teacher reads an information card which describes an event in Russian history. Students then place their risk markers one, two, or three spaces ahead of their move markers on the playing board, and one student answers a question about the information presented. If the student's answer is correct, all students advance their move markers to the position of their risk markers. If the first answer is wrong but a second student answers the question correctly, all players advance their move markers one space. Some spaces impose penalties, such as the loss of a turn. The winner is the first player to reach the last space—to Emigrate. (TM)

Cost: $7.00

Producer: Alternatives in Religious Education, Inc., 3945 South Oneida Street, Denver, CO 80237

PERSECUTION

Reverend James Buryska

Playing Data
Copyright: 1975
Age Level: junior high-adult
Number of Players: 25
Playing Time: 2-3 hours
Preparation Time: 1-2 hours (discussion)
Special Equipment: statue, incense, cane, candy, candle, cassette tape recorder, 5 rooms
Packaging: booklet with cassette and role cards

Description: Students playing *Persecution* learn firsthand about the price of early Christian commitment. As pagans, slaves, bishops, magistrates, Roman soldiers, and Christians, each with a carefully defined role, they enact a situation in which Christians are asked to offer incense to the image of a Roman leader or to risk questioning by authorities and, ultimately, imprisonment. A player who admits to being Christian remains in "prison" throughout the game and can interact only with fellow prisoners. After the simulation, students discuss the ways Christians avoided sacrifice or imprisonment, the experience of the prisoners, and the moral dilemmas raised during the game. (TM)

Note: For more on this simulation, see the essay on religious simulations by Lucien E. Coleman, Jr.

Cost: $9.95 plus postage

Producer: Contemporary Drama Service, Arthur Meriwether Inc., Box 457, Downers Grove, IL 60515

THE PLANNING GAME

Donald L. Griggs

Playing Data
Copyright: 1971
Age Level: teachers of all grade levels in church education
Number of Players: 2 or more
Playing Time: 1-3 hours

Description: Participants, who are teachers, practice the process of lesson planning and apply it to their own curriculum. They select main ideas, identify instructional objectives, and choose the teaching activities and resources they need to communicate the main ideas and achieve the objectives.

Each pair of participants needs one set of cards. These cards are available in four sets: Moses (elementary to adult), Parables of Jesus (elementary to adult), Acts of the Apostles (youth and adult), and I am Me (preschool and kindergarten). (AC)

Comment: This is a teacher training device that Don Griggs has used in a number of his teacher training workshops. It is designed to give direction to those who need help on how to select material and teach it as well as dealing with resources and objectives. The manual and game cards require preparation for the potential user. There are a number of educational concepts that should be clearly understood before using the game. There are now a number of different sets of cards that go with the manual. This means that a variety of lesson plans can be explored through *The Planning Game*. The focus is on Biblical study for all ages. (GMM)

Cost: manual $3.00, cards $1.50 per set

Producer: Griggs Educational Service, 1731 Barcelona Street, Livermore, CA 94550

PRAISE AND CRITICISM

Peg Oliver

Playing Data
Copyright: 1975
Age Level: high school-adult
Number of Players: 5 or more
Playing Time: 6-13 1-hour sessions
Packaging: folder with 7 copies of scripts and a 9-page guide

Description: *Praise and Criticism* contains 13 dialogues which may be integrated into a three-month church school program on family relationships. The dialogues are concerned with interpersonal conflict and are loosely based upon transactional analysis theory, since lines can easily be identified as parent, adult, or child. Two people enact each brief dialogue by reading their scripts in front of the group, and then the group as a whole uses the discussion questions listed in their manuals as a starting point for exploring and resolving the problem dramatized before them. In "We're Just Trying to Help You Grow Up Right," for example, a mother nags her son for not cleaning his room and he defends himself; the group then discusses the concepts of self which led to the misunderstanding and suggests ways the characters can help each other and themselves. Other themes for dialogue are: "I Praise You, Why Don't You Praise Me?," "I Just Don't Want To Disappoint You," and "You Really Spoiled It for Me." A few of the dialogues are concerned with situations which would only occur within a church setting. (TM)

Cost: $14.50

Producer: Arthur Meriwether Inc., Box 457, Downers Grove, IL 60515

PROPHETS AND THE EXILE

Jack Schaupp and Donald L. Griggs

Playing Data
Copyright: 1973
Age Level: grade 4-adult
Playing Time: 1 hour
Packaging: professionally produced 96-page booklet

Description: This booklet contains five role-playing exercises relevant to the preachings of the Old Testament prophets. In *Meet The Prophets,* participants pick one prophet they would like to know more about and then, after gathering information about the prophet, write two letters. The first is a letter of introduction recommending the prophet to God, and the second is a first-person letter in which the participant, in the voice of the prophet, introduces himself to a stranger. In *Called To Be God's Spokesman,* players imagine that God has called on them to be prophets. Each must make a list of the advantages and disadvantages of pursuing prophecy as a career and then compare his or her list with the other participants'. In *Amos of Tekoa,* players describe some of the personal characteristics and the historical significance of the prophet Amos, and then creatively interpret one aspect of Amos' life. Similarly, in *Jeremiah in Prison,* the participant must express the feelings Jeremiah might have had as a result of being a spokesman for God. In the last exercise, *Hebrews in Exile,* three groups of players represent Babylonians and Hebrews committed to or undecided about returning to Israel. They debate the question of going or staying in Babylon, and then vote to determine who will go and who will be left behind. (DCD)

Cost: $5.00

Producer: Griggs Educational Service, 1731 Barcelona Street, Livermore, CA 94550

REBIRTH: THE TIBETAN GAME OF LIBERATION

Mark Tatz and Jody Kent

Playing Data
Copyright: 1977
Age Level: high school-adult
Number of Players: 1-8
Playing Time: 30-120 minutes
Special equipment: markers, die
Packaging: paperback book with map gameboard

Description: Rebirth is a thirteenth century Tibetan game that reveals the Tibetan map of the universe and Tibetan ideas about transcending human existence. The board is divided into 104 squares corresponding to the various levels of rebirth in the worlds of desire, form, and formlessness—from square 1, vajra hell, to square 104, nirvana. Taking turns, players begin at square 24, the heavenly highway, and roll the die, consulting the text to determine their next move. For example, a player on square 24 who rolls one on the die goes to square 27, the Realm of the Four Kings; a player on square 24 who rolls four becomes an animal and goes to square 11. Commentaries in the book describe the meaning of each square. Supposedly, players' tendencies to land on certain squares on the board are suggestive of their actual paths from lifetime to lifetime. The text includes a glossary and background material on Tibetan Buddhism and the playing board. (TM)

Cost: $6.95

Producer: Anchor Press/Doubleday, Garden City, NY

SHANTIH

Playing Data
Copyright: 1975
Age Level: high school, adult
Number of Players: 2-4
Playing Time: approximately 1 hour
Packaging: draw-string pouch containing hand-silk screened playing cloth, cards, shells, and Indian markers consisting of models of spiritual adepts

Description: According to the authors, "the aim of *Shantih* is to move from 1) being a Student to 2) Master to 3) Bodhissatva. The chief strategy is to attain 12 different Karma Cards and break the Chain of Dependent Origination. This is accomplished by Cultivating the Dharma at Sacred Cities and New Age Sites, building Ashrams and Institutes, and giving away all of one's Natural Foods Currency. Money is received from the Divine Mother for developing these sites and paid out to other players who land on them in return for Service Work. This procedure is just the opposite of most other games."

Players begin with markers, two karma cards, and a selection of natural foods currency. They proceed around the playing mat by throwing the die. Different spaces allow one to give away currency, collect karma cards, and receive the seals of New Age Communities; others penalize players by requiring them to move to such places as the Holiday Inn or the Ice Hell, where one may lose several turns or receive additional unwanted currency. Still other spaces require one to select a dharma or vajra card and follow its instructions, which are often humorous (e.g., Go to the Holiday Inn for interrogation by the FBI about smuggling brownies to Tim Leary). Once a player receives the required number of ashram cards for a specific grouping, he may build ashrams and institutes which allow him to accumulate additional karma cards. When he gives away all his currency or obtains 9 different karma cards, he becomes a master and is entitled to special privileges; when he gives away all his currency and receives 12 different karma cards, he becomes a Bodhissatva and may move freely around the board, helping other players to become Bodhissatvas and thus end the game. The game kit includes complete instructions with a glossary, as well as a variety of handmade playing pieces which are truly works of art. (TM)

Cost: complete kit $12.95; without Indian markers or sewn pouch $8.00

Producer: Kanthaka Press, P.O. Box 696, Brookline Village, MA 02147

SIMULATION GAMES FOR RELIGIOUS EDUCATION

Richard Reichert, Diocese of Green Bay, Wisconsin

Playing Data
Copyright: 1975
Age Level: junior and senior high
Number of Players: variable, often in groups of six
Playing Time: 1-10 hours
Packaging: 106-page book

Description: The 16 games described in this book cover four main subject areas—basic theology, community, the nature of prayer, and moral problems. All are accompanied by suggestions for use and tips for effective discussion. Students playing *The Stretching Game,* for example, try to grab a suspended dollar bill with their teeth, and then must come to terms with their own limitations and the limitations of being human. *The Oatmeal Game* involves students in teams lapping bowls of oatmeal without using their hands and then discussing how they feel about doing something which does not seem "right" and how they feel about sinning. In *Go Away, Come Back,* the teacher secretly chooses one student in each group to be a troublemaker and asks each group to reject one player. Later, after the rejected players, usually the troublemakers, try to get back into their groups, students discuss how sin disrupts a community and share their feelings about rejection and reconciliation. *Do It Yourself* involves students in developing individual prayers, consolidating them into a group prayer, and then into a class prayer. There are 12 other games and exercises included in the book. (TM)

Cost: $4.50

Producer: St. Mary's College Press, Winona, MN 55987

TEACHING STYLES

Lucien E. Coleman, Jr.

Playing Data
Copyright: 1977
Age Level: adult

Number of Players: 4-30
Playing Time: 60-90 minutes
Packaging: professionally packaged game includes 120 statement cards, blank cards, and 16-page instruction booklet.

Description: *Teaching Styles,* according to the producer, is a "structured group experience (or game) for Bible teachers and teachers-in-training of adults and youth designed to bring them into lively discussion with one another about the things which bring about good and bad teaching."

Play begins with the game director randomly distributing five statement cards to each participant. Each card contains a statement about teaching (e.g., "A great deal is said about creating a 'spirit of freedom' in the classroom; but the learner needs guidance from an experienced teacher much more than he needs freedom to think for himself." "Learners should be trained to use critical judgment when engaged in Bible study.") Players examine their cards and each discards the one statement with which he or she most disagrees. Then the players try to exchange cards so each can "collect a set of cards which reflects his or her own philosophy of teaching and learning." Players then pair up around views on teaching. These pairs discard the two statements with which they most disagree so that each pair holds a total of six cards. Pairs of Players then join with pairs of similar attitudes, and these newly formed quartets each discard four statement cards.

At this point the game director reads some of the discarded statements and asks if any teams wish to trade cards they hold for any of these discards. If they do, these teams are asked to discuss why they disagree with the statements they want to discard and why they agree with the statements they want to acquire. When the subsequent general discussion of Bible teaching methods has run its course, the game director asks the players what they learned from the experience. (DC)

Note: For more on this simulation, see the essay on religions simulations by Lucien E. Coleman, Jr.

Cost: $5.95

Producer: Broadman, 127 Ninth Avenue, Nashville, TN 37234

TEN COMMANDMENTS BIBLE GAME

Playing Data
Copyright: 1966
Age Level: 10-adult
Number of Players: 2-6
Playing Time: 1 hour
Packaging: professionally designed box with playing board, journey wheel spinner, commandment tablets for travelers and steward, harvest basket, pieces of silver, playing pieces, and scroll cards

Description: Players are called travelers. One traveler is selected at the beginning of the game to act as steward, who watches over all game supplies and distributes them according to the rules of the game. The steward always moves first, and travelers move around the board according to the luck of a spinner. There are various categories of squares on the board; the most important are the Commandment, Phillistine, and Good Neighbor positions. Travelers who land on one of four Commandment positions receive a Commandment Card such as "Thou shall not commit adultery." When travelers land on a Phillistine square they must read aloud any Commandment Cards they have and the steward turns over the matching master tablet for this commandment. When travelers land on a Good Neighbor Square, they can contribute up to two baskets of grain to the poor. Before any traveler can win, baskets of grain equal to 12 times the total number of travelers in the game must be donated, and all the commandment tablets must have been turned over by the steward. The first traveler to collect all 10 commandments wins.(DCD)

Cost: $5.00

Producer: Cadaco, Inc., 310 W. Polk St., Chicago, IL 60607

THANATOS

Thomas E. Herrold
Playing Data
Copyright: 1978
Age Level: adults
Number of Players: 8-15
Playing Time: 3 hours (est.)
Preparation Time: 1 hour (est.)
Packaging: 25-page staple-bound photocopied booklet

Description: According to the author, *Thanatos* is a simulation game which deals with attitudes toward death and dying. It is designed to be used primarily in the Christian parish. Hopefully, through participation in this game, participants will be able to become aware of and to constructively examine their attitudes toward death and dying. Ultimately, in this experience members of the Christian community should be able to come to an understanding of their attitudes so they can initiate change through counseling and educational opportunities, if desired."

The game is played in three phases. In each phase the players consider some situation and then answer several questions about what they would do or think in that situation. The game director then reads the *correct* answers.

In the first phase, the players are asked to imagine that they "are in a funeral home for the service of a 45-year-old football coach who had a coronary at a game. His family and friends have come to pay their last respects. The minister selects I Corinthians 15:41-57 to read during this service." The game director reads the passage and asks the players to answer six questions about it. (For example, "Is it wrong for a Christian to fear death?" "Is the scripture saying that a person is never really dead?") After the players answer these questions, the game director tells them the correct answers.

The second phase is designed to help the players "evaluate and understand their personal views about death." The game director defines eight ways people may "deceive themselves" (which is to say, hold beliefs different from the author's) about death—by "rationalizing," for example, or by "oversimplifying." The game director then reads several dozen statements that people might make about death and players must correctly identify which "self-deception" is implicit in these statements.

In the third phase, players consider 11 dramatic situations and try to identify the correct response to make in each. These situations describe specific, hypothetical instances of death and dying. Again, in this phase, as in the others, there are right and wrong answers. (DCD)

Note: For more on this game, see the essay on simulation games by Harry Farra.

Cost: $3.00

Producer: Simulation Sharing Service, 4740 Shadow Wood Drive, Jackson, MS 39211

325 A.D.

John Washburn
Playing Data
Copyright: 1971
Age Level: grades 4-7
Number of Players: 12 or more
Playing Time: 45-90 minutes

Description: Teams attempt to create a creed that will be acceptable to themselves and the church. (GMM)

Comment: *325 A.D.* does not deal directly with the Council of Nicea of that year. It is rather a process that deals with the nature of creeds and how they came into existence. It is designed to acquaint the participants with a process such as the one that created the Nicene Creed. It provides them with an experience that culminates in the creation of their own personal and group confession of faith.

325 A.D. is a combination of some elements of group process, an art project, and a simulation structure. The administrator's directions are very clear and easy to understand. It has worked extremely well with confirmation classes and in youth programs. In addition to various art materials, the filmstrip, "Council on Nicea," is highly recommended, although not absolutely necessary. The filmstrip is available from the United Church of Christ, Office of Audio-Visuals, 1720 Chouteau Avenue, St. Louis, MO 63103.

The game itself is divided into three parts. If time is a problem, part one can be done in one time period of 20-40 minutes. The last two parts can be done later in a time slot of at least an hour up to 2.5 hours. Although it is best to do the entire game at one time, it is possible to divided it into two programs as long as the group does not add new participants to the second session. (GMM)

Cost: $3.00

Producer: Simulation Sharing Service, 4740 Shadow Wood Drive, Jackson, MS 39211

USING BIBLICAL SIMULATIONS

Donald E. Miller, Graydon F. Snyder, and Robert W. Neff

Playing Data
Copyright: 1973 volume one, 1975 volume two
Age Level: high school, junior college, religious education classes
Number of Players: 4-50
Playing Time: 1.5-6 hours per simulation
Supplementary Materials: class set of Bibles
Packaging: volume one, 224 pages; volume two, 222 pages

Description: Volume one of *Using Biblical Simulations* contains several chapters on the use of simulations in Bible classes and 11 complete simulation games, with perforated instruction and data sheets which may be punched out and distributed to each group involved in each simulation. The subject matter is derived from the following Bible chapters: Genesis 2-3, Numbers 14, 1 Samuel 8, 2 Kings 22-23, Jeremiah 26, Job, John 11, Acts 15, I Corinthians 12-14, Mark 8, and chapters concerning the Last Supper.

What Happened in the Garden?, for example, involves Adam and Eve on trial, with students as prosecution and defense presenting their interpretation of the temptation scene and the meaning of "the knowledge of good and evil." Four judges representing the God of love, the God of justice, the God of wrath, and the God of mercy determine the nature of the crime committed, its significance, and its punishment.

The simulation game on Job has four small groups (Job, the three comforters, Elihu, and God) dramatize the comforters' attempts to help Job while asserting their own convictions, Job's continued lamenting, and God's proclamations. Students study their roles, speak in their own words, and pay particular attention to how they are attempting to comfort a person in a distressing situation.

A final example, The Upper Room, aims to help students improve their understanding of the Last Supper and the events which preceded it. Four groups representing the Mark and Matthew gospels, Luke and the Acts, John, and the letters of Paul study only their respective texts; decide what food to serve, rituals to perform, and issues to discuss at the Last Supper; and determine how and when each aspect of the Last Supper will be conducted. In a council meeting, the groups present their plans and the class votes.

Volume two follows the same format as volume one and contains the same introductory material, with 11 more simulation games. The simulation games are concerned with the following chapters of the Bible: Genesis 27, Genesis 37-48, Genesis 4:7, Ruth, Jeremiah 27-28, Mathhew 2, Luke 4:14-30, the Gospels, Acts 19, and Philemon.

One simulation involves a priestly group, prophetic group, shepherds, and narrator using a lexicon to translate Genesis 4:7, determining their own interpretations and presenting their group translation to the class for debate. In another simulation, four groups representing Naomi's household, Boaz' household, the reapers, and Amminadab's household take a position on Boaz' request to marry Ruth. The groups present their positions to the council of elders, which moderates the debate and makes the final decision about the marriage. A third simulation has Herod presiding over a debate among the Pharisees, the priests, and the scribes who examine Biblical sources and defend their positions on the identity of the Messiah and where and when he will be born. A group representing the wise men asks questions and makes suggestions to Herod, who seeks to maintain his power and is the final authority on actions to be taken in regard to the birth of Jesus.

All the simulations in this volume, as in the previous volume, require a study of Biblical passages and make use of dicussion and debate to illuminate the diverse viewpoints expressed in the Bible and to reveal the ambiguity of Biblical interpretation. (TM)

Comment: This is one of the most outstanding collections of simulation games available. Most of the simulations are highly inventive, dramatize important theological issues and events in a way which stimulates further research and discussion, and provide for active involvement of all students. The authors include pertinent background data and references and give many hints for preparing, leading, and debriefing the simulations, as well as adapting them to small or large groups and to time periods of varying length. (TM)

Note: For more on What Happened in the Garden, see the essay on religions simulations by Lucien E. Coleman, Jr.

Cost: $5.95 each volume

Producer: Judson Press, Valley Forge, PA 19481

See also Education for Justice ECONOMICS
The Grain Drain ECONOMICS
Island ECONOMICS
PX-190 FRAME GAMES
COMMUNICATION
COMMUNITY ISSUES
FUTURES
HUMAN SERVICES
SELF-DEVELOPMENT
SOCIAL STUDIES

SCIENCE

AC/DC

Ampersand Press

Playing Data
Copyright: 1975
Age Level: high school
Number of Players: 2-6
Playing Time: 30 minutes
Packaging: professionally designed box containing 84-card playing deck, and instructions

Description: *AC/DC* is a game teaching the design and function of electric circuits. Its players lay down playing cards (which picture electric sources and users, straight and bent wires, wiring connections, switches, fuses, electric shocks, and short circuits) in a formation that depicts workable electric circuits. A workable circuit is defined as one which contains one electric source, uses a switch, and contains enough connecting wires to link the other elements into a closed circuit. To begin, each player is dealt five cards face down and the rest of the deck is placed between the players. Play proceeds as players draw cards from the deck, request cards from other players' hands, and lay down or discard cards, until one player constructs a workable circuit. (DCD)

Cost: $5.00

Producer: Ampersand Press, 2603 Grove Street, Oakland, CA 94612

BLOOD FLOW

Rosette Dawson

Playing Data
Copyright: 1974
Age Level: grade 7-adult
Number of Players: 2-6
Playing Time: 1 hour
Packaging: professionally produced board game

Description: This game is played on a four-color, oversized (2′ X 4′) diagram of the human circulatory system. Each player must move a red corpuscle token from heart to lungs (to pick up an oxygen token), from lungs to a specified capillary bed (in the head, arm, kidney, intestines, or leg, where the oxygen token is traded for a carbon dioxide token), and then return to the right atrium with carbon dioxide. The first to arrive wins. (DCD)

Comment: This is a very nifty game. It is easy to learn, simple to play, and painlessly informative. (DCD)

Cost: price on request (between $10.00 and $15.00)

Producer: Carolina Biological Supply Company, 2700 York Road, Burlington, NC 27215

THE CELL GAME

Playing Data
Copyright: 1979
Age Level: high school-adult
Playing Time: 1 hour (est.)

Description: This game is designed, according to the producer, to teach "basic cell structure and some aspects of cell function." Play involves collecting the following components: chromosomes, mitochondria, golgi apparatus, endoplasmic reticula, lysosomes, chloroplasts, protein, polysaccharide, fat, glucose, salt, water, oxygen, and carbon dioxide. Each participant starts with a single complex molecule and draws cell components from a "bank" in accordance with life processes such as photosynthesis, respiration, parasitism, autolysis, synthesis, and random cell events. To win, a player must complete a plant or an animal cell. (DCD)

Cost: price on request

Producer: Carolina Biological Supply Company, 2700 York Road, Burlington, NC 27215

CHEM BINGO

Robert L. Gang

Playing Data
Copyright: 1971
Age Level: high school
Number of Players: 5-40
Playing Time: 10 minutes (est.) per round
Packaging: eight call lists, four chem bingo pads, and instructions

Description: This variation of bingo is designed to motivate students to memorize the symbols of important elements and ions. Each player takes one Chem Bingo sheet. Printed in each of its twenty-five squares is a chemical symbol. Each Chem Bingo sheet is different. The teacher then reads from a list of elements and ions, while players think of the chemical symbols and see if they appear on their *Chem Bingo* sheets. Squares with the appropriate symbols are marked. The first player with a row of five marked squares wins. (DCD)

Cost: $6.00

Producer: Teaching Aids Co., 925 South 300 West, Salt Lake City, UT 84101

CHEM CHEX

Robert L. Gang

Playing Data
Copyright: 1971
Age Level: high school
Number of Players: 2-4
Playing Time: 30 minutes (est.)
Packaging: professionally produced board game includes playing board, checkers, ion labels, periodic table, score pad, formula chart, and instruction booklet.

Description: This game borrows checkers as a device with which to generate random, hypothetical chemical reactions which players must identify and explain to score points. There are five basic games of *Chem Chex,* and their objective is to "teach recognition and value of chemical formulas and equations."

In all five variations, two players (or two teams of two players each) face each other across a checkerboard. Each side takes twenty-five playing pieces (these are in two colors, representing metals and nonmetals, with "ion labels" on the bottom of each) without looking at the ion labels. Playing pieces are then draw at random to set up the board for a game of checkers. Whenever a piece is removed from the board, the "jumping" player may replace the lost piece with one drawn at random from his or her reserve pile. The ion symbols on each piece removed from the playing surface are revealed, and these ions are assumed to react with one another.

In the first variation, "jumping" players may score points by correctly naming the ions exposed. In the second variation, players score by writing the correct chemical equation for the reaction that would occur between the two ions. In the third variation, players score if they can name and write the formula for the compound formed by the reaction of the two ions. In variations four and five, players score by writing the chemical equations for the reactions between three or four ions and naming the resultant compounds. The game ends when all of one player's pieces have been removed from the board. The winner is the player with the highest total score. (DCD)

Cost: $9.95

Producer: Teaching Aids Company, 925 South 300 West, Salt Lake City, UT 84101

CHEM TRAK

Robert L. Gang

Playing Data
Copyright: 1975
Age Level: high school
Number of Players: 1-30
Playing Time: 10-60 minutes
Packaging: board and instructions

Description: *Chem Trak* can be used as part of a class exercise, as a game, or as a tutoring device. It consists of a card with a track of chemical formulae around its edges. Half are for ions, and half are for elements. There are wheels on the back of the card and windows at each end of the track. These wheels turn to reveal formulae for other elements or ions. The player reveals one ion or element at a time and seeks to match that symbol with a symbol on the track to construct the formula for the resulting compound. Players may also be required to name the compound formed and/or write the equation for the combination reaction. The answers may then be checked on a chart attached to the card. The producer suggests several racing games that students may play. (DCD)

Cost: $1.95 each, 25 for $40.00

Producer: Teaching Aids Company, 925 South 300 West, Salt Lake City, UT 84101

CHEMINOES

Chemical Teaching Aids

Playing Data
Age Level: high school
Number of Players: 2-6
Playing Time: 10-60 minutes
Packaging: 36 professionally packaged playing pieces and instructions.

Description: *Cheminoes* is a game played with 36 domino-like chips which are decorated with matching valences rather than spots. Play consists of matching the valences or radicals on the chips. The game ends when a player has played all his or her cheminoes or when all of the players are stumped. (DCD)

Cost: $2.75 postpaid

Producer: Ampersand Press, 2603 Grove Street, Oakland, CA 94612

CIRCULATION

Robert Grossman

Playing Data
Copyright: 1973
Age Level: upper elementary, junior high
Number of Players: 2-4
Playing Time: 20-40 minutes

Description: *Circulation* explains the workings of the human circulatory system. Each player receives a Plasma Tray and moves it through the blood stream, picking up and dropping off oxygen and food at appropriate places and dispensing with wastes. "Emergencies" occur which send the players off to perform specific tasks in order to control these situations. They are "Germ Attacks" which cost the players white blood cells, and trips to the marrow to get them back. The first player to supply the arms, legs, and head with food and oxygen is the winner. (Publisher)

Cost: $14.95

Producer: Teaching Concepts, Inc., P.O. Box 2507, New York, NY 10017

CLASSIFYING THE CHEMICAL ELEMENTS

Hugh Oliver and Menaheim Finegold

Playing Data
Copyright: 1976
Age Level: high school
Number of Players: 4-6
Playing Time: 40 minutes-2 hours
Packaging: professionally produced and packaged game includes element cards and 10-page instruction booklet

Description: The purpose of this "puzzle" based on the periodic table, according to its authors, "is to introduce high school students taking chemistry to some of the problems of classifying scientific data, and, as well, to acquaint them with the names of the chemical elements, some of their properties, and the advances in chemistry that led to the discovery of particular types of elements. The puzzle is so designed that students experience many of the same problems nineteenth-century chemists faced in their attempts to classify the elements."

The game materials consist of two packs totaling 104 cards, each of which represents a chemical element, with four "joker" cards which represent substances at one time thought to be elements but later proved to be compounds. The first pack contains cards representing elements discovered before Mendeleef's classification in 1870. The second pack contains cards representing elements discovered after 1870. During play, participants arrange these cards to form a periodic table, first separating elements by the chronology of their discovery, then trying to classify the elements. The game allows students to replicate the historical process of classifying the elements and ends when the periodic table has been devised and completed. (DCD)

Cost: $22.50

Producer: The Ontario Institute for Studies in Education, 252 Bloor Street West, Toronto, Ontario M5S 1V6, Canada

COMPOUNDS

Nelson Payne

Playing Data
Copyright: 1974
Age Level: junior high and up
Number of Players: 2-6
Playing Time: 30 minutes

Description: *Compounds* is a simple rummy-like card game, the purpose of which is to provide experience with the names of 21 common chemical elements and some of the compounds they make. Each player receives a nine-card hand to start. A draw pile and a discard pile are set up using the remaining cards. With each turn, players seek to produce one or more chemical formulas from their own cards, the discards, or the draw cards. All compounds are checked against a master list, which may be consulted during play, or reserved only for scoring, depending on the level of familiarity with the compounds. A special five-point bonus is awarded for using all nine cards in one turn. An interesting feature of the card deck is that each card is "marked" for easy reading by observant players. A color-coded rim on one edge reveals one of six natural states–e.g., heavy metal, liquid, gas–for the particular element. Thus, the learning process is multidimensional. (TM)

Cost: $7.00, 5 for $30.00

Producer: Union Printing Company, 17 West Washington Street, Athens, OH 45701

ELEMENTS

Paul F. Ploutz, Ohio University

Playing Data
Copyright: 1970
Age Level: grade 6-high school
Number of Players: 2-6
Playing Time: 20-60 minutes
Preparation Time: none

Description: This is a competitive board game designed to teach the Periodic Chart of Elements and the symbols and names of the 105 elements. There are rules for four games and eight variations. (AC)

Cost: $13.00, 5 for $54.00

Producer: Union Printing Company, 17 West Washington Street, Athens, OH 45701

EXTINCTION

Stephen P. Hubbell, University of Iowa

Playing Data
Copyright: 1978
Age Level: high school-adult
Number of Players: 2-4
Playing Time: 3 hours
Packaging: professionally produced and packaged board game includes 23 page instruction booklet.

Description: This board game simulates the natural and man-altered ecology of the imaginary island Darwinia. Each player represents an imaginary species with specific characteristics in seven genotypes. (These are reproduction rate; optimal habitat for reproduction; ability to migrate to optimal habitats; predator type–such as swift, nocturnal, crafty, or strong; vulnerability to specific predator types; capacity to cross natural and man-made boundaries such as rivers, mountains, and jetports; and tolerance for natural and man-made environmental aberrations such as unseasonable cold, drought, fire, or pollution. The "genes" for each species of imaginary beings are specified on "gene cards" drawn at random by the players. The proliferation and diminution of each species are represented by four sets of twenty colored dice. The number of dots showing on a die is the number of that species that a die represents. (Consequently, no species may include more than 120 individuals, and the total population of all creatures in Darwinia may never exceed 480.) The game is played on a map of the island marked into hexagons. Each hexagon represents the territory of one to six individuals of any one species and is color coded to indicate its habitat type. (Marshes are blue, for example, and fields are green.) In play, the events which affect Darwinia's ecology are generated with a spinner. Two events always occur simultaneously. The events are reproduction, environmental change for better or for worse, the placement of natural and man-made barriers, migration, predation, and evolution (during which each species may exchange as many as four old genes for new ones). All species compete with one another to multiply and linger. The game continues until only one species is left to win. (DCD)

Comment: This is an attractive and intelligent game. (DCD)

Cost: $15.95

Producer: Carolina Biological Supply Company, 2700 York Road, Burlington, NC 27215

E-Z-SCIENCE GAMES

John H. Woodburn

Playing Data
Copyright: 1973
Age Level: high school
Number of Players: 2-6
Playing Time: 50 minutes
Packaging: materials include paper playing surface, die, markers, 6 decks of question cards, and one deck of "application-of-principles" cards.

Description: There are three *E-Z-Science* games (*BIO-E-Z*, *CHEM-E-Z*, and *SCI-E-Z*), and each must be purchased separately. All three versions are essentially the same, differing only in content. Each game includes six sets of question cards. (For instance, in SCI–E-Z cards ask questions about physics, astronomy, biology, geology, chemistry, and science principles.) Only one deck is used in any one game. Participants move playing pieces around a track of squares printed on a playing surface. In BIO-E-Z these squares contain the names of taxonomic groups, in *BIO-E-Z* these squares contain the names of taxonomic groups, in *SCI-E-Z* the squares are labeled with the names of pieces of laboratory equipment, and in *CHEM-E-Z* the squares are named after chemical question drawn from the deck of question cards by another player. If no other player "owns" that square, the first player to land there and correctly answer a question becomes its owner and is awarded a number of points (from 35 to 100). Players who land on a square owned by another player must also answer a question. If they answer correctly, they are awarded points, but if they fail, the owner of the square is awarded these points. The winner is the player with the highest score at the end of the game, which ends when the time available to play it has expired. (DCD)

Cost: $5.00 each

Producer: E-Z-Science Games, 9208 Le Velle Drive, Washington, DC 20015

FORMULON

Chemical Teaching Aids

Playing Data
Age Level: junior and senior high school
Number of Players: 2-8
Playing Time: 20 minutes-1 hour
Packaging: professionally packaged card game with instructions

Description: Formulon is designed as an aid for students learning about how atoms combine to form molecules. It is played with a deck of one hundred atom, ion, or multiplier cards. At the beginning of each hand, players are dealt ten cards each and must, in turn, use the cards to construct the formulae for molecules. For instance, a player may hold two uncombined hydrogen and one uncombined oxygen molecules in her hand. On her turn, she may lay down these cards to construct one molecule of water. During subsequent hands, she may pick up additional cards and alter the formula she has already constructed for instance, by playing one sulfur and three more oxygen atoms to construct an acid molecule. When one player can lay down all of her cards, the round ends and the other players total the valences of the cards they hold and after several rounds, players tally their total scores, and the lowest wins. (DCD)

Cost: $6.50 postpaid

Producer: Ampersand Press, 2603 Grove Street, Oakland, CA 94612

GEOLOGIC TIME CHART GAME

Paul F. Ploutz, Ohio University

Playing Data
Copyright: 1972
Age Level: grades 7-12
Number of Players: 2-6
Playing Time: 30-60 minutes
Preparation Time: 5 minutes
Packaging: game kit including playing board, cards, chips, and dice

Description: This game teaches players about evolutionary eras as they move in two directions across a geological time chart—forward through time and "up" as they evolve from protozoa through various animal forms to human beings. The first player to reach finish (the present day), the first to become a monkey or a man, or the only player left after others have become extinct or have lost their population is the winner. At the beginning of the game, players as protozoa receive 15 chips representing their original population of 15 million. By throwing the dice, players move ahead across spaces which each represent 5 million years. Some result in loss or gain of population. Chance spaces require players to answer questions about the evolutionary eras explained on the board. Evolution and adaptation spaces both give conditions of adaptation for players' species. Players who land on an EVOLVE space must move up to a higher animal form, gaining population as a result. Not all evolutionary moves are positive: a player who remains a dinosaur for too long becomes extinct and drops out of the game. Players may pass up a turn in hopes of evolving or to avoid being eaten, which occurs when a higher animal form occupies a space immediately above a lower animal form. (TM)

Cost: $15.00, 5 for $62.00

Producer: Union Printing Company, 17 West Washington Street, Athens, OH 45701

THE GREAT PERIODIC TABLE RACE

Richard E. Rodin

Playing Data
Copyright: 1973
Age Level: grades 7-12
Number of Players: 2-4
Playing Time: 20 minutes
Packaging: periodic table game board, player pieces, discovery cards, dice, and instructions

Description: This simple board game is designed to help students become familiar with the chemical elements as they are arranged on the Periodic Table. Players roll dice and move a token the number of spaces (elements) indicated, starting with hydrogen and following the atomic numbers on the playing surface. Whenever a player lands on a space occupied by another player's token, he or she moves that token back one space. A player who lands on silver or gold moves ahead an additional ten spaces to lanthanum or actinum. A player who rolls doubles moves the indicated number of spaces and then draws a discovery card, which will require the player to miss a turn or to move ahead or back a certain number of spaces. ("Pour yourself ahead or back to the nearest element that is a liquid at room temperature.") The winner is the first player to get past uranium and into the man-made transuranium elements. (DCD)

Cost: $7.75

Producer: Science Kit, Inc., 777 East Park Drive, Tonawanda, NY 14150

IN-QUEST

Robert W. Allen, Leo E. Klapfer, and R. Lawrence Liss

Playing Data
Copyright: 1970
Age Level: grades 7-12
Number of Players: 2-4
Playing Time: 2-8 hours in 50 minute periods
Preparation Time: 20 minutes

Description: Players initially get practice in recognizing common and subtle errors which are made in the processes of scientific inquiry, such as observing, measuring, and interpreting. Later, players examine experimental evidence to determine into which of two models the evidence fits.

Cost: $10.00

Producer: Wff 'N Proof, 1490-TZ South Boulevard, Ann Arbor, MI 48104

LAB APPARATUS

Mary E. Hawkins, Paul F. Ploutz

Playing Data
Copyright: 1975
Age Level: 10-adult
Number of Players: 2-6 best; up to 10 can play
Packaging: professionally designed box, cards, spinner

Description: Players spin a spinner to white, green, or orange when in turn. For white, the player looks at one of 100 pictures of lab instruments (calipers, test tubes, micrometers). For orange, a player has to answer one of fifty safety questions ("Give three musts for fitting a thistle tube into a rubber stopper". For green, a player gets a chance card ("What instrument is used to measure blood pressure?"). Correct answers get points. Ten variations are included. Solitare is also possible. (REH)

User Report: We played this with a seventh grader. It produced some good discussion and some learning. However, there were some silly aspects to some of the cards (e.g., this safety card: "How should horseplay or tricks in the lab be treated?" Answer: "Severely").

Some cards are poorly written (e.g., "What should you check as you leave the lab?" Student response: "They didn't tell me to give five reasons!") We also felt that chance gains or losses in points were not justified in this kind of game (e.g., "CHANCE: You tried to focus a microscope directly to high power. You broke the slide and scarred the lens. Lose 10 points.") It's easy to sort them out and throw them away.

Some pictures were poor (e.g., on the Geiger counter picture there are no clues to tell what it is). Our student suggested a good rule change: If you can explain what it does, you get it right. Our overall evaluation: a pretty good game, useful for learning names and functions of lab instruments and some rudimentary safety rules. (REH)

Cost: $9.00

Producer: Union Printing Company, Inc., 17 West Washington Street, Athens, OH 45601

LONGMAN SCIENCE GAMES

Playing Data
Age Level: grades 7-12

Description: Part 1: Biology and General Science

The Great Blood Race: A competitive team game tracing the path of the circulatory system to the main organs of the body. The aim is to give an appreciation of the structure and functions of the blood and blood grouping.

Nutrition: A competitive team game covering the requirements of the balanced diet, the world food problem, and deficiency diseases. Suitable for use with a wide ability and age range.

The Digestive System: A competitive team game in which players attempt to get food from the mouth through the alimentary canal. The players learn how the organs connected with the digestive system function and which fluids and enzymes aid the digestive process.

Transport in Plants: A game which traces the flow of food and water in plants. Players investigate various theories concerning mass-flow systems. This competitive game includes information on such topics as transpiration, osmosis, and turgor pressure.

Microbes: A competitive game in which players take on the role of invading pathogenic organisms. The players compete against the body's defenses. Useful over a wide range of courses.

Classification: The objective of this competitive game is to understand the basic concepts of classification. The game involves the ordering of animate and inanimate materials and leads to a good understanding of the classification process. Suitable for all elementary general science courses.

The Water Cycle: This competitive game is based on the water cycle and covers topics such as evaporation, condensation, precipitation, and the importance of water in the lives of all living things. Suitable for all general science courses.

The Air About Us: This game stresses the importance of air in our lives. The game includes the composition, uses and pollution of air. Suitable for a wide ability range and most general science courses.

Science Sense: The game emphasizes safety in the laboratory. Players are made aware of the many hazards found in a science laboratory and the need to take precautions. This unit is particularly suitable for all pupils embarking on science courses.

Part 2: Physics and Chemistry

Chemical Families: This unit includes two competitive games on the classification of the elements. Players build up groups or families of elements of the periodic table. The purpose of the unit is to give an understanding of the periodic table and some historical knowledge of its development. Useful in all chemistry courses.

Atomic Structure and Bonding: This is a competitive game in which the players first build up the atomic structures of certain atoms and then go on to use these to interpret the bonding in simple molecules and ionic compounds.

Chemistry's Alphabet: A competitive circuit game based on the names, formulae, and properties of some common elements and their properties. This unit is aimed at the middle school pupil beginning a course in chemistry.

Competition Amongst the Metals: A game devised to study the simple properties and reactions of metals. Based on the reactivity series of metals, the game involves occurrence, extraction, reactions, displacement, corrosion, and so on.

Keeping Warm: A game devised to draw attention to the necessity for energy conservation. Pupils are first made aware of the ways in which heat is lost from a home and how it can be conserved. The game develops to illustrate the economics of heating a home in an age of dwindling energy resources.

The Solar System 1 and 2: An extensive and fairly complex game which, because of the quantity of material involved, is presented in two units, both of which are required to play the game. Each player takes the part of a flight director in charge of a fact-finding probe to one of the planets in the solar system. Players use basic laws of physics governing space flight, collect information on our solar system, and are made aware of the advanced technology involved in space exploration.

The Electric Circuit: A game played to illustrate the function and symbols of various electrical components and the flow of electricity in simple circuits. An aid to the teaching of elementary circuit electricity.

Energy Conversions: The conversion of energy into different forms is of fundamental importance in all branches of science. This game concentrates mainly on energy conversions applicable to physics. Suitable for use with all elementary physics courses. (Producer)

Cost: Set of one copy each unit Parts 1 and 2 £11.00 plus post and packing; or Part 1 or 2 for £5.50 plus post and packing. Quantity options available.

Producer: Science Games, Longman Group Ltd., Resources Unit, 35 Tanner Row, York, England

THE MOUSE IN THE MAZE

Victor Showalter, Educational Research Council of America

Playing Data
Copyright: 1971
Age Level: grade 7-college
Number of Players: 1-8
Playing Time: 20-100 minutes in periods of any length
Preparation Time: 15-30 minutes

Description: One choice you have in teaching how animals learn would be to get yourself a mouse and some cardboard and wood and build a simple T maze. Put the animal in one end and some food down at the end of one of the branches, and you can either time the animal or count the number of mistakes it makes. You then put the animal in again and see if it makes fewer mistakes. For more advanced tricks, buy one of B. F. Skinner's books. But mice are messy and some people do not like to handle them. Ah! But you can get a simulated mouse.

The mouse is simulated by cards. Colored cards with holes punched in them stand for experimental conditions. One can choose different conditions and change the number of Ts in the maze, the kinds of animals in it, and the number of trial experiments. Randomness in animal behavior is simulated by rolls of dice. Data cards provide the time or the mistakes different kinds of animals make in the simulation. The data on the cards is reliable and was collected using laboratory animals. (REH)

Comment: We think this is a cleverly made simulation and one which could be repeated with several other types of scientific experimental conditions where repetition of the experiment a number of times in order to collect data is important. (REH)

Cost: $12.21

Producer: Houghton Mifflin Company, One Beacon Street, Boston, MA 02107

P & S

Chemical Teaching Aids

Playing Data
Age Level: high school
Number of Players: 2-8
Playing Time: 10-60 minutes
Packaging: professionally packaged card game with instructions

Description: This is a card game for chemistry students who are learning to match chemical properties and chemical substances. The deck consists of pink substance cards (each having a numerical score suiting players' probable familiarity with the substance and the card's possible frequency of use in the game) and yellow property cards (describing properties of substances used in the game). Players are dealt ten cards each and they take turns laying down pink and yellow cards to match properties and substances. For instance, a player may lay down a yellow card saying "a gaseous element." The next player may lay down a pink flourine card next to it. The play of the flourine card counts seven points, which are added to that player's score. The game proceeds in this manner until one of the participants has laid down all of his or her cards. The player with the highest score wins. (DCD)

Cost: $6.50 postpaid

Producer: Ampersand Press, 2603 Grove Street, Oakland, CA 94612

THE POLLINATION GAME

Marie Miller Lowell

Playing Data
Copyright: 1977
Age Level: 7-adult
Number of Players: 2-6
Playing Time: 1 hour (est.)
Preparation Time: 30 minutes
Packaging: professionally packaged card game

Description: The author suggests four games that can be played with the seventy-two card pollination deck. The deck contains five kinds of cards. Flood, fire, rain, frost, and night cards are representative of the cards which indicate environmental conditions. Ants, beetles, and crab spiders are representative of insect pests. Cards picturing the Capri Fig and the Capri Moth illustrate a noteworthy example of a pollination symbiosis. Most of the cards in the deck picture "pollinators" (such as bees, flies, wasps, and butterflies) or plants described in terms of their pollinators (such as "bee flowers," "fly flowers," or "butterfly flowers").

The instructions briefly describe the playing procedures for *Pollination Bug, Pollination Rummy, Pollination Whist,* and *In the Garden.* In all four games the players collect "books" of cards—two cards for the same flower and the card for the appropriate pollinator. (DCD)

Cost: $6.50

Producer: Ampersand Press, 2603 Grove Street, Oakland, CA 94612

PREDATOR

Marie Lowell

Playing Data
Copyright: 1973
Age Level: high school
Number of Players: 2-6
Playing Time: 15 minutes-1 hour
Packaging: professionally designed box

Description: *Predator* is a card game designed as an aid in teaching food chain and food web concepts. The cards represent elements in the food chain of a temperate zone forest. These include plants, seeds, birds, and animals. To begin, all cards are dealt and the player to the left of the dealer starts play by asking any other player for a "showdown." At a given signal, each of these two players lays down a card and if one card "eats" the other, that player takes the eaten card. Play continues in this manner while the cards in each person's hand are revealed. A player who knows where a certain card is held can make a "challenge" on his or her turn, demanding a certain card and showing the card that can take it. The winner of a challenge is entitled to another turn. Play continues for a predetermined period of time. At the end of the game, the player with the most cards wins. (DCD)

Note: For more on this game, see the essay on ecology/land use/population games by Larry Schaefer et al.

Cost: $5.00

Producer: Ampersand Press, 2603 Grove Street, Oakland, CA 94612

SPACE HOP

Helmut Wimmer

Playing Data
Copyright: 1973
Age Level: upper elementary and junior high
Number of Players: 2-4
Playing Time: 15-30 minutes
Preparation Time: 5 minutes to explain rules
Packaging: professionally packaged full-color game board, cards, rocket ship markers, dice, decoder

Description: *Space Hop* illustrates the scientific facts of our solar system, its sun, moons, planets, comets, and asteroids. It was designed by Helmut Wimmer, resident artist at New York's Hayden Planetarium. Each participant selects a space craft, and draws a mission card. The mission card indicates the player's objective by describing various aspects of his destination without naming it. If the player's knowledge of the solar system is sufficient, he may roll the dice and proceed immediately. If not, he may consult the "Space Hop Decoder" to get the proper answer, but only at the cost of a turn. There are rules for space travel, hops through hyperspace, "Space Bumps," and various other maneuvers. (Producer)

Cost: $12.95

Producer: Teaching Concepts, Inc., P.O. Box 2507, New York, NY 10017

WILDLIFE

R. L. Meier and Jane Doyle

Playing Data
Copyright: 1973
Age Level: grade 7 (accelerated)-graduate school, continuing education
Number of Players: 2-6
Playing Time: 3-100 hours in periods of 20-40 minutes
Preparation Time: 1 hour

Description: About 1960, we discovered that it was very difficult to convey the concept of *community* in an ecosystem to students of almost any level. Therefore we set about creating a working model that allowed students to see predator-prey relationships, tophic levels, overgrazing, migration pressures, population explosions, carrying capacity, and many other related concepts evolving over time.

The simple moose-beaver-wolf-vegetation system of Isle Royale National Park was reduced to the dimensions of a chessboard so that its patterns could be comprehended with a single glance. A single year of natural history could be played out in twenty to forty minutes. One to four students play out a future for the community, recording births and deaths of animals, migration, and shifts in the overall quality of the environment.

An important feature of *Wildlife* is that it is a model of a real living system that has been photographed many times, written about repeatedly, and can be visited any summer. It has been greatly simplified, but the population dynamics fit historical data so well it can be used as a model for forecasting the impact of various insults and challenges to that community. It is not a smooth illustration of a hypothetical instance, but contains some of the awkwardness of real data and uncertainties about causes and consequences that biologists face as observers of nature. (authors)

Cost: $25.00

Producer: Institute for Urban and Regional Development, University of California at Berkeley 94720

See also
Queries 'n Theories LANGUAGE
Space Colony FUTURES
COMPUTER SIMULATIONS/SCIENCE

SELF-DEVELOPMENT

AUCTION

Playing Data

Copyright: 1975
Age Level: adult
Number of Players: 7-17
Playing Time: 1-2 hours
Packaging: professionally packaged game with instructions, players' manuals, and bills of sale

Description: There are four versions of *Auction* (*Lifestyle, Life Goals, Personal Preference,* and *Components of Schooling*), each of which can be purchased separately, but all of which share the same frame. Players bid competitively on "items" that may be, depending on the version, the components of a hypothetical school, accomplishments, or various lifestyle scenarios. In each game the players are encouraged to set priorities on choices, but they are not limited to a specific budget once the bidding begins.

The *Auction* games are designed, according to the producer, to "give participants a chance to make choices that reflect what their values are." In the School Components Auction, for example, the players have 2000 "units" with which to purchase such items as an "Olympic-Size Swimming Pool," a "Lot of 50 Behavior Modification Kits," or "10,000 Sq. Yds. Carpeting." There are twenty-five items designed to "represent a variety of educational goals."

After the auction, players discuss their feelings about the items they bid for and what they have learned about themselves and their society by participating. (DCD)

Cost: $19.95 for each version

Producer: Creative Learning Systems, Inc., 936 C Street, San Diego, CA 92101

BARRIERS IN THE MAZE

Twyla Wright and Robert L. Husband

Playing Data

Copyright: 1978
Age Level: high school, college
Number of Players: 15-18 in groups of 5-6
Playing Time: 45-90 minutes
Preparation Time: 30 minutes (est.)
Packaging: game includes instructions, playing surface, values inventory sheet, score sheets, and playing tokens

Description: This simple values game is played in three parts. At the beginning, each player lists five goals she or he wants to accomplish within five years. The players then evaluate these goals and select one they believe to be desirable, realistic, attainable, and measurable. This becomes their goal.

During the second part, groups of five or six record their goals on individual score sheets and move tokens through a maze playing surface. When players encounter a barrier in the maze they must draw a barrier card, which asks what they would or would not be willing to do to accomplish their goal (eat dinner with parents each Sunday for the next fifteen years, for example, or refrain from listening to any type of music). Each time a player draws a card she must record both the request and her response to it. A player who acquiesces to the request may continue through the maze, but if she declines the request as unreasonable she must stay at the barrier and continue to draw barrier cards on each of her turns. The first player to successfully navigate the maze wins.

During the third part of the game, the players discuss their strategies in traversing the maze, their methods in making decisions, and the values they would or would not sacrifice. (DCD)

Cost: $10.00

Producer: Twyla Wright, Route 4, Box 282N, Batesville, AR 72501

BECOMING A PERSON

Jan and Myron Chartier

Playing Data
Copyright: 1973
Playing Time: 1-1/2 hours
Number of Players: 2-36, in 2-6 teams
Age Level: junior high, high school
Packaging: game kit including instructions, fulfillment alternative sheets, selection sheets, and score sheets

Description: *Becoming a Person* is a game allowing players to experience how society discriminates against women. Players divide into teams of pink and blue, without being informed of the male/female symbolism of the teams or of the game's real objectives. In four rounds of approximately ten to fifteen minutes each, teams use fulfillment alternative sheets to determine, from a given set of choices, the most fulfilling profession, personality characteristic, behavior, and life value. The Professions Alternatives Sheet, for example, gives players five choices from five sets of professions; one set consists of barber, mail carrier, welfare worker, child care worker, and night club singer. Each round, after five choices have been made, teams receive scoresheets to determine the number of points they receive for their choices. Unknowingly, pink teams receive scoresheets which penalize them for making traditional male choices. As the game progresses, they will become angry or apathetic as they begin to suspect that the game is rigged. Meanwhile, the leader continually refers to the teams by color and urges the losing pink teams to figure out the strategy of the winning blue teams. When the rounds are completed, or when players become aware of the objectives of the game, the group as a whole discusses the feelings that have emerged, the game dynamics, its realism, and the group process. *Becoming a Person* may also be played with individuals or dyads as competitors. (TM)

Note: For more on this game, see the essay on self-development games by Harry Farra.

Cost: $7.95

Producer: Teleketics, Inc., Franciscan Communications Center, 1229 South Santee St., Los Angeles, CA 90015

CAREER DEVELOPMENT PROGRAM

Esther E. Diamond and Frederic Kuder

Playing Data
Copyright: 1975
Age Level: grade 9-10, adult education
Number of Players: variable
Playing Time: variable
Packaging: package containing program guide and planning notebook and package containing career development profile and career development inventory (may be used separately)

Description: The aim of the *Career Development Program* is to help students gain awareness of their values and abilities, to determine feasible career options that are extentions of themselves, and to explore these options. The fifty-eight activities in the program guide include a number of individual and group projects such as interviewing working people, researching and reporting on a particular career, and debating work-related issues, as well as approximately ten role-playing scenarios. Some examples of role-playing activities, which generally involve several students in front of the class, and which are described only briefly, are: (a) a girl defending her desire to take an auto shop class; (b) a parent and teenager arguing about the teenager's career choice; (c) two young people applying for jobs as garbage collectors, and exhibiting different attitudes; and (d) three students defending their career orientations, each trying to convince the others that he or she has the strongest argument. In one role-playing activity, students act out their fantasy of an actual job situation about which they have daydreamed and discuss the difference between their fantasies and reality. (TM)

Cost: $52.67 list, $39.50 school; quantity discounts

Producer: Science Research Associates, 155 North Wacker Drive, Chicago, IL 60606

THE CAREER GAME

E. N. Chapman

Playing Data
Copyright: 1970
Playing Time: varies with the individual
Preparation Time: 15-30 minutes to study the Teacher's Guide and preview materials and 15-30 minutes to preview the introductory filmstrip and tape
Special Equipment: tape player and filmstrip projector

Description: The Career Game is a programmed trip through approximately 400 cards, taking the individual on a career search. Beginning on one of three trails—professional, mechanical/technical, or service—the player makes a series of decisions under branching conditions and eventually discovers a career choice he could not necessarily have predicted. His decisions along the way have been based on career information and an educational price is attached to each choice. The decisions are also based on his own individual interests, aptitudes, variations and aspirations. Warning signals, forcing players to consider negative as well as positive career aspects, are frequent, and to win the game he must find and score himself on two or three career possibilities. Players are encouraged to repeat the game, trying other trails. The simulation is designed to be played with the Career Development Laboratory, which is a reference unit of taped interviews on various occupations. In addition, all career cards or occupations have an address or suggestion where a student can follow-up for more information out in the real world. (REH)

Note: For more on this game, see the essay on self-development games by Harry Farra.

Cost: (#20-31) $118.50

Producer: Educational Progress, A Division of Educational Development Corporation, P.O. Box 45663, Tulsa, OK 74145

CHOICES

Jim Deacove

Playing Data
Copyright: 1976
Age Level: age 10-adult
Number of Players: 2-8
Playing Time: 1 hour
Packaging: professionally packaged game with 45 choice cards and instructions

Description: Choices is a discussion game designed for use by families and other small groups. According to the author's instructions, the game is planned "to focus attention on the choices we make in life, choices large and small, that give opportunities for spiritual growth." The choice cards are arranged in four categories with questions representing those a person might have to answer in childhood, adolescence, maturity, and old age. Each player starts in childhood and makes three choices in each of the four age groups. A player may start by deciding about what to do about another child who bosses her around, go on to decide on a teenage marriage in her adolescence, later be asked by her own teenage daughter for permission to marry, and finally face the death of her husband in old age. Each player makes a total of twelve choices in twelve rounds of play. The players discuss their choices after each round, and make a final assessment of their "life" at the end of the game by considering their hardest choice and what they have learned. (DCD)

Cost: about $2.50. Send for current price list.

Producer: Family Pastimes, R.R. 4, Perth, Ontario, Canada, K7H 3C6

COLLEGE GAME

Ronald R. Short

Playing Data
Copyright: 1968
Age Level: grade 9-college
Number of Players: 3 or more
Playing Time: 2-2-1/2 hours
Preparation Time: 30 minutes

Description: This game is for college students and high school students considering college. They must decide how to successfully allocate time for each role, given the role's personal profile and the occurrence of chance cards, among various activities. They seek to balance academic, social, and personal elements and to identify the best representative of each activity. Debriefing discussion is an integral part of this game, which is intended to stimulate discussion about how to live through college. It can be especially effective as part of freshman orientation. (AC)

Cost: for 1 game (3 players); $10.00 for 1 set (15 players) $40.00

Producer: Ronald R. Short, Whitworth College, Spokane, WA 99218

CRUEL CRUEL WORLD VALUE GAME

Bert K. Simpson, Drug Abuse Preventive Education, Orange County, California

Playing Data
Age Level: grade 7-adult
Number of Players: 2-4 in 2-4 teams
Playing Time: 1/2-1 hour
Preparation Time: 15 minutes

Description: Players begin the game with neutral status in each value territory. They move around the board by rolling dice and drawing cards. Players have the option to move into each value area and work toward that value. For example, a player may choose to work for wealth or skill. Each player attempts to gain a balance of the eight value categories of wealth, respect, responsibility, well-being, enlightenment, skill, influence, and affection. Or a player may choose to emphasize a certain value and "accept the consequences." Players represent themselves in the game. (REH)

Comment: The idea for this game is rather good, but the execution seems rather weak. The major reason is that values develop in concrete situations rather than in abstraction. Thus, a person can get well-being or enlightenment points in the game without having any real comprehension, even in the simulated fashion of what has to be gone through to obtain these points. We give the designer an A in goals and a C– in execution. (REH)

Note: For more on this game, see the essay on self-development games by Harry Farra.

Cost: $14.95

Producer: Pennant Educational Materials, 8265 Commercial Street, Suite 14, La Mesa, CA 92041

CYCLE

David Yount and Paul Dekock, El Capitan High School, Lakeside, California

Playing Data
Copyright: 1973
Age Level: grades 8-12, average to above average ability

Description: Interaction helps students to learn the eight stages of Erik Erikson's human life cycle (I-Infancy; II-Early Childhood; III-Play Age; IV-School Age; V-Identity; VI-Young Adulthood; VII-Adulthood; VIII-Mature Age). In "problems" seminar groups on stages IV through VII, boy-girl study pairs analyze case studies of persons whom Erikson would label "healthy" or "unhealthy" (e.g. in IV, School Age, a "healthy" child achieves "industry"; an "unhealthy" child experiences "inferiority"). After contacting persons living in various life cycle stages and getting their opinions about the case studies, the study pairs join "solutions" seminar groups that role-play solutions for the human problems. Each student fills out a decision form helping him to understand how he values certain activities for persons living in differing life-cycle stages. Then in a "life-cycle stage" seminar group, each student chooses a contact project that leads him into the world to contact and analyze one person in depth. These contact projects concentrate on a person's attempts to achieve Erikson's definition of "health" (e.g., "industry" in stage IV, "identity" in stage V, "intimacy" in stage VI, "generativity" in stage VII, and "integrity" in stage VIII). After hearing contact projects reports, students evaluate other "life patterns" persons have used to explain a human being's journey from birth to death: the Mt. Peak Pattern; the Ever-Downward Pattern; the Rollercoaster Pattern. Finally, a life and death survey is used as a debriefing activity to help students understand how a person's conception of death influences how he lives his unique life cycle. The teacher's guide includes a detailed bibliography and an approach to literary analysis keyed to stages IV through VIII so that English teachers can use Erikson's life cycle to help their students meaningfully read, analyze, and discuss literature either within or after this unit. (Publisher)

Cost: $14.00

Producer: Interact, P.O. Box 262, Lakeside, CA 92040

DECISION MAKING FOR CAREER DEVELOPMENT

Don H. Parker, Shelby W. Parker, and William H. Fryback

Playing Data
Copyright: 1975
Age Level: junior and senior high, adult
Number of Players: 5-30
Playing Time: 1/2-1 hour per activity
Special Equipment: cassette tape recorder
Packaging: includes four cassettes, one program guide, and 30 student response booklets

Description: This package contains nine role-playing situations with follow-up activities, all geared to give students the experience of actual decision making about vocational issues. The cassettes describe the role-playing situations, which generally involve two or three students in front of the class, and the workbooks provide short exercises for evaluating them and preparing discussion. The role-playing activities include themes like determining whether to ask for a raise in salary; resolving the issue of whether to steal steaks for a friend from one's place of employment; deciding between a job and college; and handling the responsibilities arising from having a car accident on the job. Since the cassettes and program guide were not available for examination, it is impossible to describe here the extent to which each scenario is presented. (TM)

Cost: $86.67 list, $65.00 school

Producer: Science Research Associates, 155 North Wacker Drive, Chicago, IL 60606

DECISIONS

J. L. Easterly, Oakland University

Playing Data
Copyright: 1976
Age Level: junior high
Number of Players: 15-20 in groups of 5-6
Playing Time: 2 1-hour sessions
Packaging: professional game box

Description: The purpose of this game is to help students learn to make decisions consistent with their values and goals. Students begin by dividing into groups and selecting a banker who distributes eight different value cards and thirty-two bucks to each player. Each card is a different color, and lists a value such as having money, caring about people, having fun, and obeying parents. Students individually rank their eight values and assign them points in accordance with their own value system.

During the four rounds of the game, participants buy, sell, and trade value cards as they attempt to accumulate the number of cards equivalent to the points they assigned each value. After each round, they must each make four decisions on the basis of situations presented to them, such as whether to help a friend with homework or to make extra money working for a neighbor. Participants then receive extra bucks which correspond in color to the value card related to their decision. To buy the value cards they want, they need bucks of the same color. Thus, those who make decisions consistent with their values accumulate the cards they need, and those who make inconsistent decisions are continually frustrated in their attempts to obtain the proper cards. The winner is the first player to collect the correct kind and number of value cards, and to pay back the banker the original thirty-two bucks. At the end of the game, players score themselves and identify the values underlying the decisions they made. (TM)

Note: For more on this game, see the essay on self-development games by Harry Farra.

Cost: $8.00

Producer: Innovative Education, Inc., 15020 Woodruff Rd., Wayzata, MN 55391

EXERCISES IN PERSONAL AND CAREER DEVELOPMENT

Barrie Hopson and Patricia Hough

Playing Data
Copyright: 1973
Age Level: high school
Number of Players: 20-40
Playing Time: 15-60 minutes
Packaging: 147-page soft cover book

Description: This book, according to the producer, "sets out to provide source material for teachers concerned with the personal development of their students. The exercises have been designed to help the student develop an awareness of himself, of other people and of the roles that he wants to adopt in life." There are twenty-six exercises in this collection. Eighteen are identified by the authors as exercises in personal development, while the remaining eight address career development.

The first and largest group consists of activities similar to those used in sensitivity and personal growth groups in the United States. In "Touch and Know," for instance, the players close their eyes and fondle everyday objects such as crumpled sheets of paper, sponges, or pieces of cloth to notice, as for the first time, tactile sensations they would normally ignore. In subsequent exercises players are helped to become aware of the somatic components of their psychological defenses (Hello), their shared feelings (Me Too), their anger (When I'm Angry), their feelings about themselves (Names), the way they solve problems (Brainstorming), their interests (Famous People), beliefs (I Believe), and their strengths (My Strengths). There are also exercises to help students become cognizant of their activities and their attitudes and inclination toward group or solitary problem solving. The section concludes with an exercise (in which students must agree on a collection of objects they can use to survive on the moon) designed to give players practice at making consensus judgments.

The last series of exercises is designed to enhance vocational awareness. While the first group of exercises can readily be played out of the context of the entire collection, the second series is an integral part of a comprehensive teaching unit called SPEEDCOP. SPEEDCOP is designed to help students become aware of vocational opportunities, and realistically evaluate their ability to perform *and enjoy* performing these jobs. Consequently, students are asked to describe the jobs that they see performed around their school, to learn more about these jobs, to examine their own expectations about the job they want to have, to discuss what they perceive to be the difference, if any, between careers for men and for women, and finally to write their own obituary, to gain perspective on what they have accomplished and what they would like to accomplish with their lives. (DCD)

Cost: £5.40

Producer: Hobsons Press, Bateman Street, Cambridge, CB2 1BR, United Kingdom

EXPLORING CAREERS

Keith Ober, Vermont Department of Education; Kathryn Kearins, Abt Associates, Inc.

Playing Data
Copyright: 1973
Age Level: grades 7-12, vocational
Preparation Time: minimal
Packaging: 38-page teacher's manual and 52-page spirit master book of student instructions and worksheets

Description: Exploring Careers provides a series of sixteen exercises and role-playing simulations for occupational education designed to help students learn about career opportunities and how to approach them. Among the simulations or explorations are "Let's Build a House," which requires different people in the classroom to work in groups involved with different aspects of a house—the plumbing, insurance, electrical system, interior and exterior finishing, tax, mortgage, and so on. All groups finally put their observations together to design the house. In the course of their work, they do research in the community and in the library to learn what is needed. In "Brave New World," students research different occupational titles and then debate the reasons for being granted passage on the first boat, wagon, rocket, or whatever, to the site of a new society. "Costs of Living" steps students through the economic consequences of different clear goals they choose. (REH)

Cost: $35.00

Producer: Games Central, Abt Publications, 55 Wheeler Street, Cambridge, MA 02138

FEMALE IMAGES: A LIFE SKILLS EXERCISE

Ronald G. Klietsch and Amy E. Zelmer

Playing Data
Copyright: 1971
Age Level: grade 8-adult
Prerequisite Skills: reading at 8th grade level
Number of Players: 4-8
Playing Time: 2-3 hours
Preparation Time: 1-1/2 hours

Description: Female Images uses a group of four to eight to jointly perform the life skills exercise, which consists of five basic steps. First, the concept of identity is presented as a stimulus and is explored for meaning. Second, during the evocation step, participants relate the dimensions of identity to their own ideas and examine components of identities in graphic form. All information is retained by the participant and is not examined by the exercise leader or other persons. Third, a subjective and objective inquiry is made into criteria women use in selecting, maintaining, changing, or modifying an identity. The model of identity deals with female identities as managers, pace-setters, stylizers, and facilitators, plus many associated identity forms. Fourth, a problem-solving exercise examines options in identity presentation and problem alternatives. The exercise ends with an evaluation and appraisal activities. (AC)

Comment: The only element of simulation in this exercise is the final one which involves the person in role playing a chosen image in a short problem-solving or role-playing situation. The overall exercise is interesting in that it assumes that identities can be matters of choice rather than tradition. Although this may sound like a strong female liberation position, in actuality, most of the exercise material more or less favors traditional female images. (REH)

Cost: $35.00 plus postage and handling

Producer: System's Factors, Inc., 1940 Woodland Ave., Duluth, MN 55803

HANG-UP

W. J. Gordon and T. Poze, Synectics Education Systems

Playing Data
Copyright: 1969
Age Level: grade 7-adult, management, community groups
Prerequisite Skills: reading, grade 9; math, grade 0
Playing Time: 1 hour or more
Preparation Time: none

Description: The players of this board game in turn enact charades of people with psychological hang-ups in stress situations and try to understand the predicaments the actors are trying to express. Some of the situations are ridiculous, some are authentic. The producer notes "The stress situations and hang-ups bring out ordinarily dormant racism, especially in those who still associate racism with others while refusing to see their own prejudices. However, much of the game's impact depends upon the verbal interaction among the players; the game provides the setting and (hopefully) is catalytic—but the game experience is not complete except through the players' unprogrammed discussion of the decisions and hang-ups. The game also helps players become aware of their real hang-ups, though this is not a stated purpose. Humorous situations are built in (allowing the tension-release of laughter), as are rules protecting those who are more inhibited about acting out their feelings." (AC)

User Report: The game is very simple to learn and easier to teach. Students quickly learn to enjoy the game but if the teacher is unfamiliar with the kind of commitment required by simulation games, that teacher is wise to explain the game to the students and let them play it by themselves.

In its original form the game is designed to bring out racial stereotyping. However, after the game has been played several times, a general discussion of hang-ups and stress situations often leads to valuable changes that bring out the hang-ups in parent-child, teacher-student, and teenager-child relations. The game is one of a few avenues open that encourage empathy, introspection, or analytical thinking actively and openly. It is an enjoyable, easy game. I recommend it. (Bruce Camber)

Note: For more on this game, see the essay on self-development games by Harry Farra.

Cost: $18.75 each for 1-4 copies, $14.75 each for 5 or more

Producer: Synectics Education Systems, 121 Brattle St., Cambridge, MA 02138

JOB EXPERIENCE KIT

Science Research Associates

Playing Data
Copyright: 1970
Age Level: junior and senior high school
Number of Players: 1 or more
Playing Time: 1/2-1 hour each
Packaging: individual packets

Description: Twenty booklets present simulated work experiences in a variety of occupations, such as lawyer, auto mechanic, truck driver, X-ray technician, librarian, and beautician. Their purpose is to stimulate students to explore career possibilities further. Each booklet includes introductory information about an occupation, a simple explanation of a representative problem, and one or more sample problems for the student to solve. For example, the Accountant Kit presents material related to using and processing checks and provides a record of checks and sample endorsed checks so the student can look for errors such as discrepancies in amounts or forged signatures. After solving the problem, the students score themselves and are free either to research the career further or to try another job experience kit. (TM)

Cost: $265.34 list, $199.00 school

Producer: Science Research Associates, Inc., 155 North Wacker Drive, Chicago, IL 60606

KATHAL: A GAME OF CREATIVE INTIMACY

Sivasailam Thiagarajan and Diane Dormant

Playing Data
Age Level: adult
Number of Players: 2-4
Playing Time: 20 minutes
Packaging: playing pads and rules for 15 variations.

Description: Kathal is a word game which simulates the joys and sorrows of two people engaged in a common task—couples worried about their children, collaborators writing a novel, a team searching for the biological base for cancer, or a pair of figure skaters. The game is played on a grid divided into individual and mutual territories. Players take turns placing letters on the cells of the grid to create words. The score depends on how many letters of any word are in the individual or mutual territories. (Authors)

Cost: $6.95

Producer: Instructional Alternatives, 4423 East Trailridge Road, Bloomington, IN 47401

THE MARRIAGE GAME

Cathy Stein Greenblat, Rutgers University; Peter J. Stein, City University of New York; and Norman F. Washburne, Rutgers University

Playing Data
Copyright: 1974, 1977 (second edition)
Age Level: young adult
Number of Players: 10-100 in groups of 2
Playing Time: 8 or more 50-minute rounds
Preparation Time: 6 hours (est.)
Packaging: professionally produced, oversized paperback

Description: Although *The Marriage Game* may be played by any two or more people in a variety of settings its "major use," according to the authors, "is as a segment of courses or units on marriage and the family at the college level." Consequently, of the 124 pages of text, approximately one-third is devoted to playing procedures and the remainder consists of thirteen scholarly articles that may be assigned as collateral reading.

This game is based on an exchange theory model; the costs and rewards of certain life decisions are quantified, recorded on forms, and balanced. The stage of life simulated is early adulthood, the settling down time between twenty-one and thirty, and each round represents one year. While several pages in this edition are devoted to suggestions of ways the exercise can be modified to represent lower-class and under-class economic constraints, the game admittedly mirrors the values of, the opportunities available to, and the decisions that are likely to be made by college-educated members of the middle class. Players must decide whether to get married or stay single, whether to obtain a graduate degree, whether to work at a full-time or part-time job, and whether to have children, as well as what to consume, what to do in their spare time, and where to go on their vacations. There are no winners or losers. Rather, players calculate a score for each round that reflects "their success in maximizing those rewards" they "most value."

One play of the game usually comprises seven to nine rounds and during each round the players work through six steps. In step one players define their preferences for certain rewards they may earn by their actions. Players set goals by assigning a point value, from one to five, to each of six categories of rewards so that the six numbers total twenty. The rewards and how they may be earned follow. Financial Assets Points may be earned by accumulating savings and acquiring real property. Players earn Enjoyment Points by doing what they think they would enjoy doing, rather than what is most profitable in other ways, or by purchasing goods or services. Social Esteem Points are earned or lost in several ways. How players decide to structure their leisure time partly determines their reward in social esteem. (For example, a player may decide to spend some leisure time visiting art museums. Doing this costs $75 of expendable income, gives a reward of one Social Esteem Point, and–depending on whether the player loves or hates art museums–may cost or pay as many as four enjoyment points.) Social esteem also depends on the kind of car and house (if any) a player buys and where he or she vacations. (Visiting parents is cheap, for instance, but has no social esteem, while luxury travel abroad pays fifteen Social Esteem Points.) Sex Gratification Points are earned by exchanging sex gratification cards with members of the opposite sex. The score (no pun intended) for unmarried players equals the number of cards exchanged. For married players this score is the number of cards exchanged outside the marriage plus the number of cards exchanged between the partners and multiplied by a number (either ½, 1, or 1½) determined by a "quality of sex relations chance draw." Parental Status Points are awarded to players who decide to have children. All of the possible ways in which an adult may have responsibility for a dependent child are listed on a chart (included in the game materials) and assigned a value in parental status points as well as a cost to each adult in the relationship in terms of leisure time lost and social esteem gained. (For example, a child under the age of six pays both mother and father ten Social Esteem Points and 150 Parental Status Points at a cost of ten leisure time units plus living expenses.) Ego-Support Points are calculated by each player's spouse. Each player first evaluates his or her partner's competence as a spouse by means of a standard form. This rating is then modified by adding the total number of *strokes* and *zaps* (positive and negative units of recognition) each partner has given the other during that round.

After players determine what rewards they want to work for throughout the round, step two begins. Each player must decide what the major characteristics of his or her role will be for the remainder of the round. A player must decide whether to marry, divorce, or stay single, whether to have a child, and whether to work at one job full time, one job part time, or to work full time and part time. All these decisions affect the amount of disposable income and the amount of leisure time. At the start of the game, for instance, men who work full time are assured an income of $11,000 per year and a total of 100 leisure time units. A man may increase his income by $6500 a year by working at an additional part-time job at the cost of an additional twenty-five leisure time units, or he may choose to go to graduate school. After five years (rounds) of graduate school a player's yearly income is automatically increased by $2000.

During step three, players calculate their disposable income (gross income less taxes, living expenses, and debts), draw up budgets, and buy goods and services with which to earn Enjoyment and Social Esteem Points. Then each player draws an economic chance card which introduces random elements–chance costs and rewards (such as getting robbed, becoming sick, losing a job, finding a great apartment, or getting a gift of money).

In step four, the players interact and spend leisure time units. If the activities on which these units are spent require the participation of other people (such as playing cards or joining a personal growth group), these other people must be found among the participants. After these leisure activities have been performed each player goes on vacation and obtains sexual gratification.

Partner ratings are calculated in step five, and in step six, players figure their total score for that round. This score gives the players information about how well they suceeded in earning the rewards they intended to earn and what rewards they gained inadvertantly.

Players then decide what they want to accomplish next and redefine or reaffirm their preferences to start the next round. (DCD)

Note: For more on this game, see the essay on sex roles games by Cathy Greenblat and Joan Baily.

Cost: $6.95 per manual

Producer: Random House, Inc., The College Department, 457 Hahn Road, Westminster, MD 21157

MENTAL HEALTH

Tiff E. Cook and Paul F. Ploutz

Playing Data
Copyright: 1975
Age Level: age 12-adult
Number of Players: 2-6
Playing Time: 30 minutes (est.)
Packaging: professionally packaged card game with instructions

Description: This simple card game is designed to introduce players to one of the key ideas of psychologist Abraham Maslow: that people have certain needs in a hierarchy that must be satisfied in order, the basic needs taking precedence.

Each player starts with three need cards (I Physical, II Security, III Love) and each in turn twirls a spinner to generate a number from one to six and then draws that number of cards from four piles. These piles contain need, stress, coping, and anxiety cards. (These cards are color coded and match colors on the spinner.) Coping cards protect need cards from stress and anxiety. There are five different kinds of need cards, which include physical, security, love, self-esteem, and self-fulfillment needs. The first player to collect all five need cards and five coping cards wins. (DCD)

Cost: $9.00

Producer: Union Printing Co., 17 West Washington Street, Athens, OH 45701

MY CUP RUNNETH OVER

Bert K. Simpson

Playing Data
Copyright: 1973
Age Level: grades 7-12
Number of Players: 2-4
Playing Time: 1 hour (est.)

Packaging: professionally packaged game with four playing cups, four racks, and wealth chips

Description: According to the producer "the essential purpose of *My Cup Runneth Over* is to assist the player in communicating his values and interpreting the values communicated by others." The game recognizes eight values: affection, respect, skill, enlightenment, influence, wealth, well-being, and responsibility. These values are printed on plastic wealth chips. Half of the chips for each value are also marked with a symbol (+) indicating the enhancement of that value and half with a symbol (–) indicating the deprivation of that value. To begin, each player takes six chips, a small cup, and a masonite rack to hold the chips during play.

The first player selects one chip, places it face down in front of his or her rack, and then verbally or physically expresses the value on that chip. A value may be acted out (1) as one's own; (2) as a value one attributes to another player; or (3) as a value one thinks another player attributes to one. The other players try to interpret the actor's message. If they hold a chip they think matches the actor's value they place it face down in front of their rack. Otherwise, they "pass." At the end of each turn the chips are exposed and compared. Played chips which match the actor's go into the player's cups. The actor's chip goes into his or her cup only if another player correctly interprets the message. Additional chips are drawn to maintain six in each rack and the next player takes a turn.

Ten chips fill a cup. The game ends when any player collects eleven chips and the "cup runneth over." (DCD)

Cost: $14.95

Producer: Pennant Educational Materials, 8265 Commercial Street, Suite 14, La Mesa, CA 92041

O.K.

Patricia and Noel Phelan

Playing Data
Copyright: 1976
Number of Players: any size class
Playing Time: 4 phases of 5 days each, occurring in a row or spaced through the school year
Age Level: junior high, senior high
Packaging: 57-page manual including lectures and descriptions of all activities

Description: The purpose of *O.K.* is to teach students about transactional analysis and to help them develop awareness of their own parent, adult or child, and their own stroke and game patterns. *O.K.* is played in four phases, each a week long, and involving lectures, discussion, and individual and small group activities. During the first phase, ego states, the teacher gives three short lectures on the parent, adult and child in each of us, and students participate in six activities, including writing a paragraph analyzing ego states portrayed in a picture and describing their own ego states. The second phase, concerned with giving and receiving strokes, involves three lectures and five activities including analyzing strokes given by characters in a short story and individually identifying strokes given and received during the course of a day. This phase ends with a celebration which encourages students to give unlimited positive strokes to each other. The third phase, psychological games and discounts, involves four lectures on identifying games and discounts and stopping them and six activities including individual analysis of games, role-playing ways of stopping discounts, and a quiz. The fourth phase, staying O.K., may occur immediately after phase three or later during the year. During this last phase, the teacher gives two lectures and students participate in three activities, including inventing O.K. messages to cancel out not O.K. messages and an exercise for developing listening ability. The unit concludes with the presentation of personal projects which may be completed by students over an extended period of time and may involve such activities as summarizing one's own psychological games or carrying out a plan to treat another person as a "princess" rather than a "grog." Individuals and groups then evaluate themselves. The *O.K.* manual itself includes twelve one-page lectures, eleven diagrams, a glossary, and descriptions of dozens of required and optional activities. (TM)

Note: For more on this game, see the essay on self-development games by Harry Farra.

Cost: $14.00

Producer: Interact, Box 262, Lakeside, CA 92040

PERSONALYSIS

Playing Data
Copyright: 1965
Age Level: grade 9-adult, management
Prerequisite Skills: familiarity with the other players
Number of Players: 3-4
Playing Time: 1 hour
Preparation Time: 10 minutes

Description: This game involves players trying to judge how other players see them on each of twelve traits (including personality, naturalness, and appearance). Its purpose, according to the producer, is to help participants better understand themselves, isolate their strengths and weaknesses, plan changes to become more effective and better able to have satisfying relationships. The player who most accurately assesses how the others see him or her is the winner. Within this structure, the decisions of what to say or refrain from saying to one another (and how), appear to be the significant dynamics of the experience, while insight and awareness of how one comes across to others are the potentially significant outcomes. (AC)

Comment: I object to the normative pressure of *Personalysis* and to some of the traits considered significant (personality, naturalness, appearance–too midcult Miss America for my bent mind), but the process–pointing up the disequilibrium between self-perception and others' perceptions–should be growth-producing. (DZ)

Cost: $4.50 per set; quantity discounts

Producer: Administrative Research Associates, Irvine Town Center, Box 4211, Irvine, CA 92716

PLAY YOURSELF FREE

Virginia Anne Church

Playing Data
Copyright: 1974
Age Level: junior high-adult
Number of Players: 3 or more
Playing Time: 1-2 hours
Packaging: game box with playing board, instructions, cards

Description: *Play Yourself Free* is a therapeutic game based on Albert Ellis' rational-emotive therapy, which stresses that irrational beliefs underlie negative emotions. Each turn, a player chooses a card that describes a conflicting situation and spins the spinner to determine the negative emotion triggered by the situation. He then states the irrational belief he believes would lead to a person's being mired in this particular emotion. Other players may challenge this belief and present their own. At the end of his turn, all players vote on the winner–the player who expressed the most probable irrational belief operating under the chosen circumstances–and the winner advances ten kilometers up the Road to Reason on the playing board. Players take turns until one player reaches the end of the road. (TM)

Note: For more on this game, see the essay on self-development games by Harry Farra.

Cost: $15.00 plus $1.75 shipping

Producer: Institute for Rational Living, Inc., 45 East 65th Street, New York, NY 10021

POWERPLAY

George Peabody and Paul Dietterich

Playing Data
Copyright: 1973
Age Level: college, adult, management, community organizers, alcoholics, addicts, women seeking their liberation
Number of Players: 6-200, 3 or more teams
Playing Time: 4-10 hours, straight through or in 1-hour sessions
Preparation Time: 1-2 hours

Description: Participants represent themselves using power and seeking to use it more creatively and effectively. Their most important objective is to identify their real self-interest (deep values and goals) and learn various options for achieving those goals. Teams tend to represent the self-interests and goals identified by their members, and their objective is to help their members achieve their goals. These change as the game is played and new self-interests and goals emerge.

Participants begin in an odd number of teams. Individuals and then teams identify goals. After some reading and team planning periods, the action periods take place. Teams negotiate, collaborate, or use coercive tactics to gain power points. A team's power points must be committed sixty seconds before the end of a round. The winner takes all the power points committed for a round. Players have various options: win-win, win-lose, and so forth. (REH)

Comment: This is a wide-open power game; all of the power tactics we use are available. Persons have wide latitude to fill in with their own particular assumptions, and they presumably have the opportunity to become aware of what major strategies they picked, whether they were collaborative or not, and how successful they were. We were unable to determine from examination of the game how the scoring formula affects such an evaluation. In general, it looks like a pretty good simulation for people who want to introduce an examination of power into various groups. (REH)

Cost: $35.00, 10-player set, $85.50, 30-player set

Producer: Powerplay, Inc., Box 411, Naperville, IL 60540

PROBE

Florence Denmark, Bernice Baxter, and Ethel Shirk, Melnyk Associates

Playing Data
Copyright: 1972
Age Level: grade 12, college
Number of Players: groups of 2-5, or classes up to 40
Playing Time: 3-4 hours
Preparation Time: 1 hour

Description: Probe is a tool for educational guidance and career planning. It consists of a series of steps alternating between individual self-examination and group discussion of career, educational, and personal goals. As individuals, players gain some experience in reviewing their future, their attitudes toward it, and in constructing realistic plans. As group members, players engage in giving and receiving help, in listening to issues that concern others. *Probe* is a version of *Prospects* (see below) adapted for younger players. It should lead to the same results. (AC)

Cost: $4.50 per participant program; staff guide free

Producer: Transnational Programs Corp., 54 Main St., Scottsville, NY 14546

PROCESS

Judith Krusell, Franco Vicino, Ed Ryterband, Bernard Bass, David Landry

Playing Data
Copyright: 1972
Age Level: grade 12-graduate school, management, church groups
Number of Players: 5-9
Playing Time: 2-3 hours for each of 8 exercises
Preparation Time: 1 hour

Description: Each exercise deals with a specific personal or interpersonal issue and successive exercises build on principles developed in earlier ones as follows: (1) Participants share with others some of their self-perceptions and aspirations, and new patterns of personal and interpersonal behavior are explored; (2) Participants learn how others view them on certain traits of interpersonal behavior such as openness, ability to take risks and listening skills; (3) Emphasis is on accepting one's self and others and listening rather than judging; (4) Effective and rewarding membership is examined in this group and indirectly in other groups; (5) Members are helped to deal openly with and share many of the feelings they have developed about each other; (6) Males or females discuss how they see themselves as men and women and how this affects their relations with each other; (7) Members examine some of their values, look at the inconsistency of what they say versus what they do, and explore their willingness to change; (8) Members evaluate any change in their attitudes and behavior and discuss ways of transferring what they have learned in the group to other situations.

Distribution of *Process* is limited to qualified professionals who subscribe to the "Guidelines for Psychologists Conducting Growth Groups." American Psychologist, 1973, 28, 933. (Producer)

Cost: $12.75 per participant manual

Producer: Transnational Programs Corp., 54 Main St., Scottsville, NY 14546

PROSPECTS

John Miller and Franco Vicino, University of Rochester

Playing Data
Copyright: 1972
Age Level: management, professionals, engineers
Number of Players: any multiple of 3
Playing Time: 3-4 hours
Preparation Time: 1 hour

Description: *Prospects* assumes that: (1) Change is a source of concern for personnel. This makes it difficult to self-plan systematically for the future; (2) Given realistic thinking and guidelines for assistance, personnel can make reasonably accurate forecasts about their near future at work; (3) Personnel do have the necessary skills to plan; they usually use these skills, however, in planning for their organization's future or the development of their subordinates, and seldom consider using these skills for their own career plans; and (4) Personnel need and appreciate help from others, especially if they feel secure and not threatened by the offers of assistance.

Prospects consists of five phases: I. *My past and my present*: The work history of the participant is examined. Trends and experiences are rediscovered. The question "where are you going" is posed with respect to the next two to five years. Some goals are set. II. *Getting to know others*: The participant shares with two others his career history and goals. A working relationship is established in the trio. III. *Developing the future career profile* (FCP): The individual considers present skills, interests and goals to plan for the future. He develops a career profile (FCP). IV. *Some reality testing*: The plans developed in Phase III are shared in the trios. A greater commitment, a clearer vision and some reality testing are the goals of this phase. V. *Relevant others*: A follow-up session is planned which is to involve those individuals which would be most influenced by the participant's new plans. (Producer)

Comment: *Prospects* is self-administered and requires only marginal supervision. Results of field tests and of outside evaluation are available from the Management Research Center, University of Rochester, Rochester, NY 14627. (DZ)

Note: For more on this exercise, see the essay on future-oriented simulations by Charles M. Plummer.

Cost: $6.50 per participant program, staff guide free

Producer: Transnational Programs Corp., 54 Main St., Scottsville, NY 14546

THE SHARING GAME

Marvin Fine, University of Kansas

Playing Data
Copyright: 1976
Age Level: age 10-adult
Number of Players: 2-6
Playing Time: approximately 1 hour
Packaging: professional game box

Description: The purpose of *The Sharing Game* is to increase self-awareness. Players throw the die and move around the board, landing on spaces that require them to "risk," "think," "do," or make a decision and receive a payoff. Cards for each category call for activities ranging from the humorous (wiggling your toes) to the personally serious (discussing an action by someone you like that makes you angry). Throughout the game, players are encouraged to give feedback on each other's responses to the directives. Awards may be given at the end of the game capsulizing performances—such as the "Shadow Award" for the person who most avoided being open and sharing. (TM)

Cost: $11.50 (Kansas residents add $.30 tax)

Producer: Psych-Ed Associates, P.O. Box 2091, Lawrence, KS 66045

SHRINK

Playing Data
Copyright: 1971
Age Level: adult
Number of Players: 3-8
Playing Time: one hour (est.)
Packaging: professionally packaged deck of cards with instructions

Description: Shrink, according to the producer, "is designed to provide amusing and stimulating insights into the ways in which people see themselves." The producer suggests variations in the play of this game for two people and for larger groups, but all variations are similar to the basic game for three or more players described here.

The game is played with a deck which has nine sets of cards (Trains, Animals, Playthings, Substances, Clothing Styles, Homes, Amusements, Viewing, and Tools). Before the game begins one person is chosen to be the shrink. The shrink separates the deck into its sets and stacks all of the sets together. The players then examine each set, and from each set each player selects the object that he thinks he is most like, the object he thinks would be the worst to be, and the object he thinks would be the best to be. For example, the Train set consists of five cards (locomotive, coal car, passenger coach, club car, caboose). A player might decide that she is most like a locomotive, and that that is the best thing to be, while the worst thing to be would be a caboose. After the players have examined all nine sets of cards, the shrink reads the interpretations printed on the back of each card and in the instructions. (The player above, for instance, who likes locomotives, is, according to the game, a powerful, hard-driving person with a good self-image who is afraid of being timid or last.) (DCD)

Cost: $4.95

Producer: Art Fair, Inc., 18 West 18th Street, New York, NY 10011

STEADY JOB

Jean F. Sanzone

Playing Data
Copyright: 1971
Age Level: grades 7-12
Number of Players: 2-4
Playing Time: 30 minutes
Packaging: professionally packaged board game with die, place markers, and steady job cards.

Description: According to the producer, this game "is designed to reinforce the positive attitudes and behaviors essential to employability." Players move tokens, according to the roll of a die, once around a playing board. All start at an employment office. About half the spaces require a player to draw a "plus" or a "minus" card. These depict positive and negative work attitudes and require the player to back up or move forward additional spaces. For example, a negative card might be "Talked too much –3 spaces," "Fight with other worker –3 spaces," or "Sloppy clothes –2 spaces." Positive cards carry labels such as "Put tools away +3 spaces," and "Good attendance +8 spaces." All cards carry illustrations of the work attitudes. The first player to reach the end wins. (DCD)

Cost: $9.50

Producer: Mafex Associates, Inc., 90 Cherry Street, Box 519, Johnstown, PA 15907

TIMAO

Playing Data
Copyright: 1973
Age Level: high school
Number of Players: 3-6
Playing Time: 30 minutes (est.)
Packaging: professionally packaged game with Timao cards, event cards, and instructions

Description: Timao is a values game. It is designed, according to the producer, "to assist the player in understanding that events are multi-valued and any event may involve both the enhancement and deprivation of values. . . . The player will become aware that the ability to see and express values is important in both self-analysis and in understanding others."

The game begins with the dealing of eight *Timao* cards to each player. There are three each of sixteen different cards in the deck. Each card represents the enhancement (+) or deprivation (–) of the values influence, responsibility, well-being, wealth, enlightenment, affection, skill, and respect. Next an event card is drawn from a pile. These cards describe events such as "Read a book," "Paid a bill," or "Drove a car." Each player in turn places a *Timao* card face up near the event card, and describes how that event enhances or deprives the value printed on the *Timao* card. Eight different *Timao* cards are then played on the event card. The player of the eighth *Timao* card keeps the event card and returns the *Timao* cards to the deck. The deck is shuffled, a new event card is drawn, and play continues. The game ends when any player collects three event cards, plays his or her last *Timao* card, or at the end of a determined time period. (DCD)

Cost: $7.95

Producer: Pennant Educational Materials, 8265 Commercial Street, Suite 14, La Mesa, CA 92041

VALUE BINGO

Bert K. Simpson

Playing Data
Age Level: grade 4-adult

Number of Players: 2-40
Playing Time: 45 minutes
Preparation Time: none

Description: This is a variation of the familiar Bingo game which concerns the player's ability to identify value emphasis in statements. These statements may be categorized as negative or positive views regarding affection, respect, skill, enlightenment, influence, wealth, well-being, and responsibility. The purpose of the game is to sharpen a player's awareness and rapid recognition of underlying values in personal statements and actions. (FEJ)

Cost: $8.95

Producer: Pennant Educational Materials, 8265 Commercial Street, Suite 14, La Mesa, CA 92041

THE VALUE GAME

Thomas E. Linehan and William S. Irving

Playing Data
Copyright: 1970
Age Level: grade 9 and up
Number of Players: 5-35
Playing Time: 45 minutes-1-1/2 hours
Preparation Time: 30 minutes

Description: This game consists of ten to twenty situations which require players to make decisions or judgments on the actions of other individuals or groups. Designed to demonstrate the lack of ethical consensus among any randomly selected group of people, *The Value Game* demonstrates the inadequacy of a moral-ethical system which affirms an absolute right and an absolute wrong. (producer)

Cost: teacher resources .95¢, student project $1.95

Producer: Seabury Press, 815 Second Ave., New York, NY 10017

VALUES

Colin L. Proudman

Playing Data
Copyright: 1972
Age Level: grade 7-adult
Number of Players: 3-6
Playing Time: 2-4 hours
Preparation Time: 15 minutes

Description: *Values* is a board game. Players compete to reach a final exit point on the board. Moves are determined by reading and reacting to cards chosen in response to directions from a spinner. (Cards say, e.g., "The use of violence for social change," "Radical protests," "What I really believe in.") As the game proceeds, its purpose changes. The players attempt to improve the quality of their evaluative reasoning. They examine and defend their values. Winning the game becomes less important than the value examination which takes place. (MJR)

Comment: I played *Values* with a group of graduate students and found it quite effective. It was also effective in involving the class in the activity. My major concern with *Values* is that it could be a volatile experience for some groups who might not be able to handle its tendency to become a sensitivity session. (MJR)

See also
Dialogues on What Could Be FUTURES
Edventure EDUCATION
Edventure II EDUCATION
Ethics RELIGION
Leela RELIGION
Pipeline HUMAN SERVICES
Rebirth RELIGION
Shantih RELIGION
COMMUNICATION

Cost: $5.95

Producer: Friendship Press, P.O. Box 37844, Cincinnati, OH 45237

VALUE OPTIONS

L. R. Mobley and Henry Malcolm, Mobley and Associates

Playing Data
Copyright: 1971
Age Level: college, adult
Number of Players: 15-1,000
Playing Time: 2-4 hours in 1 or 2 sessions
Preparation Time: 2 hours

Description: This is a simulation for large groups. In the first phase, players receive four "value cards" that make moral statements designed to evoke positive or negative responses. In phase two, players bargain to trade cards containing value statements they do not believe in for ones that come closer to their own values. In phase three, players form small groups that share similar values. Finally, each player pairs off with someone who has opposing values, and they try to form a third, synthesized value, signalling that they have successfully negotiated between their opposing views.

Decks are available for the following areas: basics, learning, schooling, man/woman, sex/love, black/white, theology, religious beliefs, science/technology, medical issues, leadership styles power, and money. (AC)

Cost: $15.00 per deck, including instructions, plus $1.00 postage per deck

Producer: Mobley, Luciani Associates, Inc., 16 West 16th Street, New York, NY 10011

WHERE DO YOU DRAW THE LINE?

R. Garry Shirts

Playing Data
Copyright: 1977
Age Level: high school-adult
Number of Players: 5 groups of 5
Playing Time: 50 minutes
Packaging: professionally packaged game with instructions, decision forms and markers, situation and summary forms

Description: Five groups of participants make ethical judgments about the actions of people described in various activities. Each group records not only its own ethical judgment, but the judgment it assumes the business community and the public at large would make. For instance, a group may be asked to categorize the behavior of Adam, who steals $10 from Betty's purse at lunch break as acceptable, somewhat acceptable, somewhat unacceptable, or unacceptable. After all groups have judged the situations, their responses are summarized, and the game leader leads the participants in a discussion to identify the assumptions groups based their judgments on and to consider the implications of those assumptions. (DCD)

Cost: $19.50

Producer: Simile II, P.O. Box 910, Del Mar, CA 92014

SOCIAL STUDIES

ACCESS

Susan Ebel Arneaud and Jean L. Easterly

Playing Data
Copyright: 1978
Age Level: adult
Number of Players: 8-40
Playing Time: 90 minutes
Preparation Time: 1 hour (est.)
Special Equipment: jacks game, rubber ball, playing cards, bingo chips
Packaging: 39-page staple-bound booklet

Description: According to the designers, *Access* is designed to raise "a number of issues related to the roles of women in western society; for example, the extent to which women have access to status, the values which society places on the roles of women and men, and the advantages and disadvantages of partnership." The game was originally conceived as a device with which to sensitize male counselors to women's issues. Its administration requires a game director and two facilitators.

Players are separated into red and green teams and are told that they must get at least one red chip and one green chip by the end of each of three rounds. The red and green teams are assigned to opposite halves of a single large room. Lanes into each zone allow members of teams limited access to each other's zones. Players earn red chips by throwing a rubber ball into a wastepaper basket in the red zone and earn green chips by successfully picking up jacks from a table in the green zone. At the start of the game each player is also given a "game tip" by the director. These tips differ slightly for the reds and the greens and are intended to reflect differing male and female attitudes toward competition and cooperation in this society.

During play the participants realize that the green game, although initially seeming to be equal, is not as highly valued as the red game. Whereas both red and green chips are necessary for survival, only red chips can be exchanged for gold stars and subsequent access to an "inner circle" poker game. Various subtleties make it hard for greens to get red chips: greens get fewer turns and have less access to the red game than reds have to the green game. Players may form partnerships to exchange chips, but these partnerships require two players to be tied together with a rope or a string.

Debriefing begins after the third round. Players are encouraged to draw parallels between game events and life events, to discuss how they succeeded or failed during the game, and to suggest ways the game might be made more fair for greens. (DCD)

Note: For more on this simulation, see the essay on sex roles games by Cathy Greenblat and Joan Baily.

Cost: $5.00

Producer: Simile II, P.O. Box 910, Del Mar, CA 92014

BAFÁ BAFÁ

Garry Shirts

Playing Data
Copyright: 1973
Age Level: grade 7 and up
Playing Time: 1-1 1/2 hours
Preparation Time: 1/2-1 hour

Description: ***Bafá Bafá*** is a simulation of two different cultures, Alpha and Beta. Each culture has its own rules for interaction and behavior, the details of which are unknown to the other society. The object of the game is to develop skills for observing and interacting with people of an alien culture. The game is divided into three stages: learning and practicing the rules of the culture, visiting the other culture to figure out as much as possible about their rules, and debriefing discussion.

The Beta culture is aggressive, competitive, and individualistic. Status is earned rather than conferred or inherited. In terms of Beta standards, those who achieve more have a higher status in the society. Within this culture, success is measured in points, won by acquiring a sequence of cards from 1 to 7, all of one color. The successful Betan must be an acute bargainer, getting more than he gives in trade, in order to reach and maintain a high relative level of achievement. Each Betan begins with ten randomly drawn cards. Typically he will hold cards of four or five different colors (out of a total of six colors), with a number between 1 and 7 on each card. Throughout the culture, there is a severe shortage of 3s and 5s, so these two cards are the focal points of building a sequence. (If I have a green 3 and 5 and 1, 2, and 4 in pink, the sequence in green will probably be easier to build.) In order to obtain needed cards, the players must bargain with each other, in a language made up of gestures and syllables. Two-, three-, and four-to-one trades are possible.

In bargaining, the players have to use Beta language for describing the cards they need. A color is translated into Betan by taking its first two letters and adding a vowel. In Betan, "green" might be translated as "gra" or "gru." Numbers are designated by syllables also: two initials of the speaker's name, each followed by a vowel. Thus, Linda Black might translate "5" as "Lo Bo Lo Bo Lo." The number of syllables is what counts here. The players learn gestures for essentials such as "yes," "no," "please repeat."

The trading and language form the skeleton of the Beta culture; it is filled out by the norms of behavior. Behavioral norms indicate whether a particular action is flattering, gauche, or downright insulting. For instance, it is customary for Betans to blink unobtrusively three times before getting down to the business at hand, trading. To a Betan, this means that the partner in trade is a member of the bargaining community—he will drive a hard bargain, but will not resort to deception.

A person who does not blink, usually a visiting Alphan, is treated suspiciously, but suffers no other consequences. A more rigid rule forbids Betans to communicate in English. If a Betan breaks this rule, he is barred from further trade. If the culprit is an Alphan, he is treated more leniently, but he is still rebuffed.

Unlike the Betans, an important quality of the Alpha culture is friendliness. Age is respected, and the society is patriarchical. The "king" of Alpha is the oldest male in the group. Girls are considered collectively owned by the boys, especially the oldest. The other boys need his permission to initiate contact with a girl. Each member of the society starts off with three cards and ten pennies. Each card has a picture of a cultural symbol on it and everybody has the same three cards. The Alphan interaction is called matching cards: each of two people puts a card face down on a flat surface.

If a player gets a particular card, called a Stipper, it is a grave insult to the other player. The one who uses the Stipper always wins the transaction but irrevocably loses the friendship. After each transaction, the two players initial each other's card (a separate 3 x 5 card set aside for this purpose) if all the rules were kept. If some rules are broken, the card is signed with numbers instead of initials. Alphans are wary of matching cards with people who have broken too many rules.

Before and after a transaction, the two people involved must make small talk, in the interests of a friendly atmosphere. Touching is also an important part of the Alpha culture; it signifies warmth, trust, friendliness. Alphans must touch each other at least once during each interaction—and a handshake doesn't count, because it connotes distance rather than warmth.

The first stage of the game consists of learning and practicing the rules of the culture one is assigned to. In the second stage, small groups of visitors are exchanged. They are expected to participate in the alien culture as best they can, and to ascertain information about the other culture's rules without asking direct questions. Each stage is supposed to take approximately one class period, but it might be advisable to keep practicing for two periods. By then, the players would be more "fluent" in their cultures, and less prone to awkwardness and revealing their rules through mistakes in their transactions.

User Report: The reviewer worked with a class of about 25 seventh graders in the test session. They divided into two groups, and each culture was in a separate room for the remainder of the game. The game design emphasizes the second stage—visiting and learning about the other culture. Though this was explained to the students, they seemed to accord greater importance to the intraculture transactions. As a result of this, they spent most of their initial efforts in excelling within their particular culture. When they realized that the major part of the game is trying to learn about the other culture, they formulated plots to steal each other's rules, instead of concentrating on learning through participation and observation. The students seemed to take the game almost too seriously; they wanted to win, even though "winning" is hard to define in this simulation game.

In spite of these preoccupations, the players gained a profound understanding of the problems that arise between alien cultures. For almost all, *Bafá Bafá* was a novel experience that allowed them to discover that certain norms of behavior they thought natural were offensive to others—a handshake in the Alpha culture falls into this category. They realized that the two cultures are parodic abstractions of aspects of our American culture. In the debriefing session, the members of Alpha were asked to report on their conclusions about Beta, and vice versa. The members of each culture then explained their rules in their own terms. The Betans made an interesting value judgment of their culture: they characterized it in terms of being capitalistic and money-grubbing (even though there was no actual money involved), and they thought a real culture along these lines would be very alienating and unpleasant, because the competitive characteristics were not balanced by warmth or friendliness.

The Alphans generally liked their culture, but thought it could be greatly improved if it were egalitarian rather than patriarchal. They couldn't accept the systematic oppression of one group of people by another. One student commented that "we don't see the sense to a lot of things, but that doesn't mean that they are wrong."

Players of *Bafá Bafá* have to deal with several important political and social concepts. On the most general level, they are exposed to the idea of culture, and some of the features which differentiate between cultures. They experience two cultures with different structures, different kinds of behavior, and end the simulation with a greater tolerance for unfamiliar ways of behaving and interacting. They also use their own values, external to the two cultures, to decide which culture they prefer and why.

Comment: The students did not adhere to the assigned roles in this game perfectly, but they found these roles easier to deal with than some of the others encountered in other simulations. Perhaps this is because Bafá prescribes roles in terms of structure rather than content. When content is too restricted, students often experience conflicts between their own convictions and the attitudes they are obliged to assume for the sure of play.

Rapid thinking is an essential skill in this game. The students have only five minutes each to observe the alien culture and form some kind of conclusion about their modes of behavior. Within the Beta culture (in which all the Alphans participated as visitors) bargaining, decision making, and compromise are crucial skills for making wise trades.

For a simulation with such a wide and abstract scope as the differences between cultures and how these differences might be approached, *Báfá Bafa* gives the players a very sound basis for further thought exploration. (from the Robert A. Taft Institute of Government Study on Games and Simulations in Government, Politics, and Economics by Alina Kurkowski, and George Richmond, Educational Development Center)

Cost: kit for 18-36 students $35.00; directions for making your own kit $3.50

Producer: Simile II, P.O. Box 910, Del Mar, CA 92014

BALDICER

Georgeann Wilcoxson

Playing Data
Copyright: 1970
Age Level: grade 9-adult
Prerequisite Skills: reading, grade 9; math, grade 0
Number of Players: 10-20
Playing Time: 1 1/2-3 hours, preferably continuous, or 8 rounds of 20 minutes each
Preparation Time: 2-3 hours

Description: Players assume the roles of "food coordinators," each representing 120 million people, who must feed their populations each year. They begin with unequally distributed resources. One has 50 Baldicers (balanced diet certificates), one has 35, and the rest have only 5 each. These resources, representing distribution of wealth in the world today, are distributed by chance. Baldicers are earned by simulated labor multiplied by mechanization, and they are lost through natural disaster. Natural/social forces cards are drawn each round; typical are flood, discovery of oil, drought, earthquake, strike, and hurricane card. Players must decide how to invest their resources as efficiently as they can, balancing among subsistence, capital investment, population control, and readiness for natural disaster. In the course of the game, individuals sometimes form cooperatives, communes, or corporations. If participants fail to feed their people, the people die and they are out of the basic game. However, they then represent the World Conscience. In this role, they "haunt" the other players, picketing, writing posters, demonstrating, and arguing in an attempt to influence the choices of other players toward world population survival. (DZ and AC)

User Report: The game was played at the end of a day long UNICEF Seminar on Global Malnutrition. The entire group formed a cooperative strategy from the beginning so that no one starved. Like most participants we would recommend this game for junior high and up.

(What follows is a selection of some participants' sentence-long responses to evaluation questions.)

What were the strengths of this experience? The opening of people's thoughts to world unity and hunger needs. The togetherness of players—openness and necessity for honesty. An interesting experiment in the idea that it exemplified the possibility of cooperation between a group of strangers. Teaches point dramatically. People were drawn together in an atmosphere of trust, generosity, and productive thinking and doing.

What were the weaknesses? That real life pressures for money are absent. There was no follow-up simulation game taking into account more facets of life. The potential cost to the environment of human survival rather than the survival of all life forms on earth which support us may not have been apparent to us. Too simple to apply to the problem of world hunger, in more specific sense emphasis on rules, on competitiveness as world situation is now. (Wilf Allan, Learner Centre Project, University of Saskatchewan, and Pat Roy Mooney, D.E.A.P. Animateur, Edmonton, Alberta)

Comment: *Baldicer* is designed to give participants some experience of the economic interdependence of the world and to provide insight into the issues of population control, mechanization, colonialism, inflation, the unequal distribution of wealth, differing forms of social organization, and the issue of "backyard" ethical commitment.

As Paul Dietterich has pointed out, "One important insight from the game grows from the drudgery of writing on the world-yield inventory slates; no matter how hard the person works" (unless he cheats), "he can not greatly alter his basic food situation. This suggests that solutions to the hunger problem depend more on the major changes in the way the world food system operates than on the motivation of the worker."

The game structure creates an environment in which the players' personalities interact with the simulated situation to produce roles in which their own behavior is often surprising. "Liberals" often find themselves acting out the kinds of roles they oppose in real life, "independent" players are often surprised at their willingness to give up self-determination for food, and "socialists" often become capitalists during play. These occurrences result from the game structure rather than from role descriptions.

Virginia Buus has written in *Circle Magazine*, "In addition to raising the obvious questions, the players must come to grip with the role of capital in developing increased food production. They must also ask who shall control capital. Each player's emotional involvement provides extensive insight into his own behavioral pattern and the way he views the world."

While the long-range issue of depleting resources within a closed system will be the final determinant of our life or death, the immediate issue of food and survival illustrates the problem effectively. Players who experience Baldicer may be ready to take a more critical view of the world around them and of the way people operate with one another. (DZ)

Note: For more on this simulation, see the essay on ecology/land use/population simulations by Larry Schaefer et al.

Cost: $25.00

Producer: John Knox Press, 341 Ponce de Leon Avenue, N.E., Atlanta, GA 30308

CLASS STRUGGLE

Bertell Ollman

Playing Data
Copyright: 1978 Age Level: all ages
Number of Players: 2-6
Playing Time: 2 hours (est.)
Packaging: professionally produced and packaged board game includes chance cards, dice, play money, and instructions for beginning and advanced play.

Description: *Class Struggle* is played on a triple-tracked playing surface with eighty-four squares. Players represent classes. There are two major classes, Workers and Capitalists, and four minor classes, Farmers, Small Businessmen, Professionals, and Students. After throwing a "Luck-of-Birth" die at the start of play to determine their class, players progress in turn around the tracks a number of spaces equal to a random number generated by throwing dice. There are squares of five different colors on the board. Red Squares affect Workers by rewarding them with assets or penalizing them with debits; Blue Squares affect Capitalists in the same way; one Black Square symbolizes a nuclear war, and if a Capitalist lands there, the game automatically ends; six Gold Squares represent the opportunity for a confrontation between the Workers and the Capitalists; players who land on Pale Blue Squares draw a chance card. Players representing the four minor classes get assets or debits when they land on any Red or Blue Square until they form an alliance with one of the two major classes, and then they are rewarded or deprived as that class would be. The idea of the game is for Capitalists and Workers to acquire wealth and power by traveling around the board and by forming alliances until one or the other has a clear-cut advantage, can force a confrontation, and defeat the other. The class, including its allies, that has the most assets when a confrontation occurs, wins. (DCD)

Comment: First the self-evident. Everybody can take offense. There is a dogma implicit in this game which may make it dangerous for teachers in some school districts to even recommend its play. By the same token, Engels, who wrote eloquently about the irony of history, might, at best, smile sardonically at the thought of a company named Class Struggle, Inc. Next the less obvious. This is a *terrific* game to play. Just as *Monopoly* tickled a hungry nation's greed during the thirties, *Class Struggle* may just catch on in the eighties by tickling a cynical nation's anger. Class Struggle, Inc.? (DCD)

Cost: $11.95

Producer: Class Struggle, Inc., 487 Broadway, New York, NY 10013

COMCO (COMPETITION—COOPERATION)

Susan Cumming, Elizabeth Manera, and Gerald Moulton, Atlabs

Playing Data
Copyright: 1973
Age Level: grade 7-adult
Number of Players: 20-60
Playing Time: 6 hours or more in roughly 90-minute periods
Preparation Time: 1 hour

Description: *COMCO* is a complex social simulation which is designed, in its authors' words, to provide its players with "an opportunity to examine societal process, particulary in view of competition and cooperation" and in which the "effects of specific behaviors in terms of meeting individual and societal needs can be examined." "As it unfolds, the simulation forces all participants to operate at a physiological need level. As survival becomes insured, decisions of the players usually lead to successively higher need concerns in keeping with Maslow's theory."

The players, citizens of Comco, live in four regions; Cardboard Alley, Duplex Acres, Model Manors, and Executive Estates) and are served by nine major groups (Blue Chip Industries, the police, the board of education, the philanthropic foundation, Health and Housing, Independent Industries, the recreation department, the Supreme Court, and mass communications. There is also Syn, the criminal syndicate.) Each player must secure food stamps from food stamp agents to survive. This and all other transactions must be accomplished with "Kash." Kash, goods, and jobs are distributed by group "heads." Players who do not secure food stamps for any round become "unemployed." Players who do not have food stamps for two succeeding rounds die. During each round, players make investments and purchases, transact business, and make many decisions. Between each round, the game director calculates the "state of the nation," which affects the chances for the success of criminal acts, the unemployment rate, profits, and other variables for the next round. The game ends after four or more rounds. (DCD)

Note: For more on this simulation, see the essay on simulations of developing a society by Frank P. Diulus.

Cost: $5.95

Producer: American Training Laboratories, P.O. Box 26660, Tempe, AZ 85282

CANADA'S PRAIRIE WHEAT GAME

Joseph M. Kirman, University of Alberta

Playing Data
Copyright: 1973
Age Level: grades 4-8
Number of Players: 7
Playing Time: 90 minutes
Packaging: unbound game materials include farm land and bank account record sheets, playing squares, spinner chance cards, and instructions

Description: This game simulates the production of a wheat crop from the purchase of the land, seed, and fertilizer, through growing and harvesting, to sale. The game allows for seven players, one banker and six farmers, to participate. No money changes hands, and all of the transactions are paper transactions recorded by the banker.

Each farmer begins with a bank balance of $3,000 and a line of credit for up to $7,000, which may be borrowed at 5% interest. In the course of the game, farmers moves their playing pieces over four sets of squares. The first set of squares (one square per farmer) instructs them to purchase land for $40 to $90 an acre. Farmers buy as much land as they can afford. The second set of squares allows each farmer to buy fertilizer, seed, and insurance. The set of growing squares details growing conditions, potential problems, acreage lost, and insurance settlements. The last set of squares indicates wheat yield per acre and price per bushel. Farmers then determine their gross incomes and repay their debts. The winner is the farmer with the most money. (DCD)

Cost: about $3.00

Producer: Alberta School Book Branch, 10410-121 Street, Edmonton, Alberta, Canada

CANADIAN "CIVIL WAR"

Playing Data
Copyright: 1977
Age Level: high school-adult
Number of Players: 3-4
Playing Time: 1-3 hours
Packaging: professionally packaged board game with 28-page instruction booklet

Description: *Canadian "Civil War"* is, according to the producer, a "simulation of Canadian politics in the past, present, and future. Each player represents a general political outlook or school of thought. The object of the game is for a player to attain political ascendency for his particular element. In game terms, the players must gain control of 'Issues' to win. Each Issue in the game represents a concern that is more or less basic to Canadian politics today."

Players assume the roles of Federalist, Seperatist, Provincial Autonomist, or Provincial Moderate. They control government officials, interest groups, and constituencies who share their ideology. The game is played on an abstract representation of the Canadian government, with "administrative," "legislative," and "grassroots" sections for each ideology. The players define issues and attempt to force and win elections based on these issues. A player wins by defeating those players who openly rebel, or by collecting all of the required Issues in any of the multitude of prescribed scenarios for this game. The instruction booklet suggests several ways players can construct scenarios to create their own views of the Canadian political scene. (DCD)

Cost: $11.00; $12.00 with box

Producer: Simulations Publications, Inc., 44 East 23rd Street, New York, NY 10010

COMMUNITY

Jim Deacove

Playing Data
Copyright: 1972
Age Level: age ten-adult
Number of Players: 3-10
Playing Time: 1 hour
Packaging: professionally packaged game with playing board, resource chips, event and counter cards, and instructions

Description: The object of this board game is to form a community. The board has two tracks or rings of spaces. The outside track represents places in the community, such as the library and the lake, and the inside spaces symbolize the resources of the community, such as education and industry. The movement of players around the outside track is determined by drawing counter cards. Resource chips are distributed around this outer ring, and the purpose of going around the board is to collect them. To build a community, players must transfer these chips to blank spaces on the inner track. A player can place two chips in these inner spaces every time that she or he has collected a total of ten chips. The game ends when the inner track is filled and the community is established. (DCD)

Cost: about $13.00. Send for current price list.

Producer: Family Pastimes, R.R. 4, Perth, Ontario, Canada K7H 3C6

CONTRACTORS

Jay Reese

Playing Data
Copyright: 1976
Age Level: high school
Prerequisite Skills: elementary knowledge of construction terms, architecture
Number of Players: 20-50
Playing Time: 10 periods
Packaging: 15-page teacher's manual, 15-page student's manual

Description: Students in small teams that represent construction companies bid each round on construction projects (such as building a suspension bridge or a performing arts center) chosen from a list of fifty projects. If they bid the lowest on a particular project, they get a contract, and must complete a bonus project related to the project for which they have contracted. Example bonus projects include drawing a map, building a small-scale model, or investigating labor unions. If they do not get their bid, they complete mini-projects, such as drawing up a floor plan for a home, during the round. The simulation consists of seven rounds. During each round, players draw incident and accident cards that indicate strikes, weather, and other conditions influencing project completion and profit. Students record their team's bids, working hours, and profits on worksheets, bookkeeping forms, and completion forms as they attempt to get contracts for projects that will bring the most profit and to complete them efficiently in the shortest period of time. The teacher evaluates each day's work each night and returns the work to the students at the start of the next day or round. After seven rounds, each representing six months, students take a posttest and evaluate the simulation. (TM)

Cost: $22.00

Producer: Interact, P.O. Box 262, Lakeside, CA 92040

THE COVERT FARM GAME

Anita Covert

Playing Data
Copyright: 1977
Age Level: age 9-adult
Number of Players: 10-30
Playing Time: 1 hour
Packaging: 7-page instruction booklet including map of farm

Description: This game is designed as a framework in which people can actively participate in order to learn more about farming, patterns of agricultural land use, choice of crops, and market influence. The scenario is the administration of an actual farm in Leslie, Michigan. The farm has six fields, four wooded areas, and two marsh areas. The players can plant whatever crop they want in five of the fields and must grow alfalfa in the sixth.

The farm is administered by individuals or groups, and all of the farms in the game are identical and subject to the same climatic and market conditions. The climate and the market place are the chance variables in this game, and what they will be is determined by a throw of dice by the game facilitator. The price each farmer gets for his or her crop depends on the crop chosen, the weather, and the market for that crop. The game is played in three rounds, each simulating one growing season. At the conclusion of each round, players determine their profit or loss for that year. (DCD)

Cost: $2.00

Producer: Anita Covert, 3589 Tuttle Road, Leslie, MI 49251

CRISIS IN LAGIA

John Elliott, Center for Applied Research in Education

Playing Data
Copyright: 1974
Age Level: grades 9-12, continuing education
Number of Players: approximately 20
Playing Time: no limit
Preparation Time: 1 hour
Special Equipment: tape recorder, display system
Supplementary Material: preferably with Schools Council Humanities Curriculum Project Materials, "War and Society," Heinnemann Educational (London)

Description: *Crisis in Lagia* is divided into four case studies with cumulative fact sheets. The game is designed to explore the differences between nation, state, and revolutionary warfare and their implications for society. Players assume a variety of role positions in different fictional situations, which they must interpret. Successful play involves imaginativeness, a capacity to understand motivation, the ability to confront situations, and relations among groups. (REH)

Cost: $1.00

Producer: Center for Applied Research in Education, University of East Anglia, Norwich, England

CULTURE CONTACT II

Ray Glazier

Playing Data
Copyright: 1976
Age Level: grade 7 and up
Number of Players: 20-30
Playing Time: 5 45-50-minute sessions
Preparation Time: 2 hours (est.)
Packaging: professionally produced package includes ditto masters and all materials for thirty to play

Description: Participants in this cross-cultural simulation assume the roles of specific persons within one of two cultures. Elenians are a preindustrial people who live on an isolated island. Their society is horizontally structured by speciality. They maintain no economy but subsist by fishing and gathering. They worship sacred groves. The Grannisters are members of an early industrial culture with an economy based on importing raw materials for manufacture into finished goods for export. Their society is heirarchal and competitive. Players extend the fantasy that a trading ship from Grannes has just landed on the recently discovered, little-known island of Elena. The Grannisters have been sent to establish a trading post and to barter for a cargo of tall timbers suitable for ship's masts. The Elenians are fascinated with the Grannisters' tools and knowledge of how to make things but are uninterested in the goods they brought to trade.

An artificial language barrier is imposed between the two groups by allowing only three members of each culture, or team, to communicate only with their counterparts on the opposing team. The immediately apparent cultural differences between the two groups are intentionally superficial and are manifested by differences in etiquette. (The Elenians greet people by rubbing their tummies and consider people who avoid eye contact to be sneaky and rude. The Grannisters say hello by vigorously scratching their heads and consider direct eye contact menacing and arrogant.) However, the game is designed so that deep-seated cultural differences become apparent after the first or second session. Each player's role generally describes his or her character and specifically capsulizes the person's goals, wealth, and the ways in which he or she can get power over others. The place of each role within its culture is well defined. Among the Grannisters, there exists a built-in schism between the ship's company and the traders, as well as numerous personal differences and grudges. The Elenian's culture is such that any decision always has to be made by a pair of individuals, and each of these pairs is composed of an older, conservative person and a younger, more innovative one.

Actual play consists of alternating "planning" and "talking" sessions. The members of each culture caucus during the first session, and the two groups confront each other during the second. The Grannisters must trade and the Elenians must protect their sacred groves, although other trees may be traded. A full playing of the simulation may not reach a clear-cut win or loss for either side. The designer states that the winning team is that which achieves "its goals without conceding its 'birthright.' " There is also a scoring system which allows the awarding of points to each culture based on what it obtains from the other in trade. (DCD)

Comment: This is a fine cross-cultural simulation. In less interesting and less skillfully contrived games, the conflict that inevitably occurs when two groups of strangers struggle to co-exist tends to be blatantly prevented by the simple error of bringing the differences between the groups into the foreground of play. That does not happen in *Culture Contact II*. The game postulates plausible roles within a cultural context and builds in intracultural conditions that submerge the differences between the Elenians and Grannisters beneath personal ones and specific game events. The players have enough to do that they can easily accept the two artificial cultures as merely part of their roles. The devices within the game that compel them to relate with their cultural peers reinforce a process of polarization.

This is a brilliantly worked out game with almost everything–

including a mysterious, unsolved murder. Users will find game sessions hectic, quarrelsome, and dramatic. They can expect their students to have fun and learn something valuable about the ways human beings can and do misunderstand one another. (DCD)

Cost: $35.00 plus $1.50 shipping and handling

Producer: Games Central, Abt Publications, Inc., 55 Wheeler Street, Cambridge, MA 02138

DECISION POINT

Roland F. Moy

Playing Data
Copyright: 1974
Age Level: grade 8-college
Number of Players: 5-9

Description: *Decision Point* is a package of four games, each of which uses a different decision-making style. *Majority Rule Polyarchy* uses coalition formation. Possibilities for winners depend on their negotiating skills, the order of items on the agenda, and player resources. In *The Autocracy Game,* a minority can dominate majority. The first player get resources, tax authority, and interest pattern to lead a minority-rule coalition. There is room for negotiation in determining who will be part of the ruling minority. In *The Conflict Game,* agreement on a desired outcome is made improbable through the structure of the game model. Potential losers have the option of using resources in such a way that everyone will lose. The structure of *The Consensus Game* requires at least a minimum level of contribution to community efforts. The structure also requires that contributions will be unequal unless some players contribute more than necessary. As self-interest tends to affect levels of contribution, extended negotiation is involved in backing a consensus. Game outcomes depend entirely on game rules and the negotiating skills of the players. (AC)

Cost: $10.00

Producer: Roland F. Moy, Political Science Department, Appalachian State University, Boone, NC 28608

DIG

Jerry Lipetsky

Playing Data
Copyright: 1969
Age Level: grade 5-college
Prerequisite Skills: reading, 8.5; math, 0
Number of Players: 30-40 in 2 teams
Playing Time: 15-30 hours
Preparation Time: 5 hours

Description: Divided into two competing teams with the task of secretly creating two cultures, each team in the class first writes a description of its hypothetical civilization. This description stresses the interrelationship of cultural patterns: economics, government, family, language, religion, and recreation. After designing and then constructing artifacts which reflect their civilization's cultural patterns, team members carefully place these artifacts in the ground, according to the archeological principles learned during simulation. Then each team scientifically excavates, restores, analyzes, and reconstructs the other team's artifacts and culture. A final confrontation reveals how creatively and how perceptively each team has applied the archaeological principles they have experienced. In final discussion, students use what they have learned inductively about patterns of culture to analyze their own American civilization. (If time and situation do not allow an actual archaeological dig, the teacher may have students bury their artifacts in large sawdust-filled boxes in the classroom or simply exchange their artifacts.) (publisher)

Cost: $14.00

Producer: Interact, P.O. Box 262, Lakeside, CA 92040

EQUALITY

John Wesley, Chass Avenue School, El Cajon, California

Playing Data
Copyright: 1971
Age Level: grades 5-8, average ability; grades 9-12, below average ability
Number of Players: 20-40
Playing Time: 20-30 hours in periods of 1 hour
Preparation Time: 3-5 hours

Description: Students first write imaginary diaries of their lives as Uglies, slaves on the advanced planet of Fantasia. Inductively they discover what happens to free individuals' personalities when someone "owns" their bodies and therefore controls their lives. Next students draw colored ID tags (white, yellow, red, brown, or black signifying an ethnic background). These tags contain role information: age, general description, education, occupation, address. Then students move into classroom neighborhoods. Independence, their mythical city of 340,000, has 6 neighborhoods from a black ghetto to Tranquility Estates with $75,000 homes. Students study and discuss 5 short essays on the history of black Americans. The citizens of Independence roleplay certain incidents involving tension between minority groups. From daily studying and role-playing, students increase their IMPS (self-image points) and enter them on their IMPS Balance Sheets. However, Fate Cards are regularly handed out which give or take away IMPS. These cards also act as news bulletins relating community problems that stem from racial misunderstanding. A community crisis arises over whether or not to integrate the schools. After a community meeting, each neighborhood's school board member votes for or against integration. During a final debriefing session, students discuss their own beliefs and feelings about racial problems in urban America and evaluate the simulation as a learning strategy. (publisher)

Comment: Players are able to achieve an overview of race relations in the United States by using this intensive simulation. It can serve as a springboard for further study. It is a highly motivating simulation and interest and involvement should be high for junior high and upper elementary students.

Race relations are difficult to examine in American schools because of the tendency for students to say what the teacher expects to hear. *Equality* minimizes this problem as players learn to empathize with the problems of ethnic groups by taking the roles of members of ethnic groups. (MJR)

Cost: $14.00

Producer: Interact, P.O. Box 262, Lakeside, CA 92040

EURO-CARD

Robert W. Allen, John M. Clark, Nova University

Playing Data
Copyright: 1967
Age Level: high school
Playing Time: 1-5 hours
Packaging: professional game box with map and cards
Supplementary Materials: reference material about Europe

Description: *Euro-Card* consists of five games that aim to develop students' thinking and resource ability while teaching them about the nations of Europe. The object of the first game is to obtain nation cards (illustrated with pictures of European countries) by identifying the countries on the basis of their size and shape. Another game involves students in doing research on questions about European nations, challenging each other's answers, and voting on the superior answer. The final game in the series requires students to write multiple-choice questions about five nations; students then receive or lose points according to their ability to answer each each other's questions. (TM)

Cost: $9.95

Producer: National Academic Games Project, Box 214, Newhill, CA 91321

FAMILY TREE

Paul DeKock

Playing Data
Copyright: 1977
Age Level: grades 7-12
Number of Players: 15-50
Playing Time: 10-12 weeks
Packaging: 4-page teacher's guide, 23-page student guide, sample family tree.

Description: *Family Tree* provides a format in which participants can do personal genealogical research. It includes detailed instructions for contacting relatives, recording their remarks, making a genealogical chart, and writing an essay about their personal heritage. (DCD)

Cost: $22.00

Producer: Interact, P.O. Box 262, Lakeside, CA 92040

FLIGHT

Jay Reese

Playing Data
Copyright: 1976
Age Level: grades 6-9
Prerequisite Skills: some map-reading ability
Number of Players: any number, in teams of 3
Playing Time: 10 1-hour periods
Preparation Time: 1 hour
Special equipment: overhead projector
Packaging: 16-page teacher manual, 8-page student manual

Description: Students playing *Flight* develop their map-reading skills, learn about the hazards of flying lightweight planes, and practice small group decision making as they compete in a flying race over an imaginary continent. Each team consists of a navigator, who maps the route and fills out the daily flight plan; a copilot, who writes the diary entries; and the pilot, the chief decision maker, who supervises the work of the other two. The first four days of the simulation involve a pretest, review of map-reading skills, and participation in two map-reading games. The next four to five days the race takes place. Each day, after the teacher announces the weather conditions, the teams draw fate cards (which might require them, for example, to detour because of fog), plot their routes on individual maps and on the class map, and write their flight plans and diary entries. Often, teams must adjust their plans to conform to such limitations as weather conditions, geographical hazards, fuel allowance, and rules pertaining to maximum elevation and mileage. The simulation ends when a team completes the race; the winning team is determined by a combination of the distance flown and the quality of the diary entries. Complete map-reading information is included in the student manual, as well as sample flight plans and diary entries. Suggestions for varying the game, such as by using kilometers rather than miles or adapting the game to another continent, are also included. (TM)

Note: For more on this exercise, see the essay on social studies simulations by Michael J. Rockler.

Cost: $10.00

Producer: Interact, P.O. Box 262, Lakeside, CA 92040

HERSTORY

Paul DeKock, El Capitan High School, Lakeside, California

Playing Data
Copyright: 1972
Age Level: grades 9-12, average to above average ability; grades 7-8, above average ability
Number of Players: 20-40
Playing Time: 10-25 hours in 1-hour periods
Preparation Time: 3-5 hours
Supplementary Material: bibliography included which lists articles and books that will help with research activities

Description: *Herstory* is a simulation of male and female roles emphasizing the American woman's circumstances, past and present. Paired by chance, boy-girl study couples join seminar groups that study male-female role expectations. During separate three-day cycles, students examine different American marital relationships: traditional marriage, androgynous marriage, collective family, living together. Ceremonies attending such relationships are thoroughly analyzed (e.g., the religious service in traditional marriage and the written marriage contract in androgynous marriage). Students simulate aspects of marriage, such as deciding who does the domestic work and who makes key decisions. Scholarship is central to each cycle as seminar group members read and observe information substantiating or attacking HYPS (hypotheses) and then report their findings to their group. The forty-four HYPS are divided into four categories on the history and position of women: manners-courtship; marriage and divorce; jobs, achievements, reform; nature-nurture. During each cycle, all students also participate in and evaluate role playing of contemporary sexual problems. Other activities include 66 Sisters research into the contributions of American women, past and present; a two-day simulation of the first women's rights convention at Seneca Falls, New York, 1848; a Contact Project in which students examine sexual roles in the real world; a Future Forum in which groups discuss what they hope and expect American sexual roles will be in 2025; pre- and post-Male-Female Surveys to chart attitude changes. Herstory helps young persons crystallize sexual identities during this era of change and future shock. (Publisher)

Note: *Seneca Falls* is available as a self-contained package. A description appears in the history section.

Cost: $14.00

Producer: Interact, P.O. Box 262, Lakeside, CA 92040

HUMANUS

Paul A. Twelker and Kent Layden

Playing Data
Copyright: 1974
Age Level: grade 7-adult
Number of Players: 5 or more
Playing Time: 1-1 1/2 hours in 40-minute periods
Preparation Time: 10 minutes, for teacher or administrator

Description: Students participate as "survival cell" group members. They are the only known survivors of a worldwide catastrophy. They are enclosed in a small room and linked to the outside world by their computer, Humanus, which communicates to them by various "print outs." These are recorded on tape. Humanus requires that the group make certain decisions if it is to survive. They need to make such decisions as what resources they need for survival, what kind of government their survival cell is going to have, ways they have of helping people so that they are not lonely, how to counsel a citizen who wants to commit suicide, and whether to let a potentially dangerous person enter the survival cell. The educative purposes of *Humanus* are effective: to provide an intense, compressed communication experience about the values individual participants hold, an experience that is characterized by close interaction among participants both during and after the exercise. The game aims to highlight alternative views of the future through the examination of assumptions about the nature of man, man's relationship to his social and physical environment, the nature of societal change and the methods man employs to achieve change.

Editor's Comment: We have heard that this is an exciting one. It looks like a good one for getting a group started. (REH)

Note: For more on this simulation, see the essay on simulations of developing a society by Frank P. Divlus.

Cost: $11.50

Producer: Simile II, P.O. Box 910, Del Mar, CA 92014

THE HUNTING GAME

Donald Valdes

Playing Data
Copyright: 1975
Age Level: high school, college
Number of Players: 30
Playing Time: 65 minutes in three sessions
Packaging: 12-page photocopied instruction booklet

Description: This game attempts to simulate the subsistence of a band of bushmen over three years. To begin, participants are divided into three equal groups, which represent the children, adults, and elderly of the band. The adults are further divided into groups which represent men and women. All groups have different skills and nutritional needs (men hunt, women gather, and children eat less than adults), and each player's task is to insure the survival of the band. This is done by exchanging various tokens which represent the nutritional potential of the Kalahari eco-system and the capacity of various members of the band to exploit it. Adults are more efficient at this than children, and so the children's survival depends on them. A system of penalties is built into the game to punish overexploiting the environment by either hunting or gathering so that men and women must cooperate with each other. According to the author, "Actual play consists of individual and group decision making, bargaining for terms of reciprocity, and combining resources in order to obtain the designated amount and combination of calories, both collectively and individually." At the end of the three sessions, the welfare of the band and of its individuals is calculated. (DCD)

Cost: $2.25 plus $.50 postage/handling

Producer: Denison Simulation Center, Denison University, Granville, OH 43023

IDENTITY

Paul DeKock

Playing Data
Copyright: 1975
Age Level: high school
Number of Players: class
Playing Time: phase 1, one week; phase 2 (optional), no set time frame
Packaging: 6-page teacher's guide, 22 pages of reproducible student material
Supplementary Materials: high school literature (optional)

Description: *Identity* is designed for use in English and social studies classes to give students knowledge of Erikson's analysis of seven areas of identity and to teach them a process for clarifying those areas in which the students themselves may feel identity confusion.

Phase 1: Students read the student guide for an overview of the unit and a 3-page handout which sets forth the core information. They divide into 6 activity groups, choose a chairperson for the day, and pair up. In pairs, they analyse 6 short cases in the 7 areas of identity on a scale of 1-6. The activity groups then discuss the cases. Groups are given different lists of identity points. (One of these is: Time 2; Self-Images 2; Roles 3; Work 4; Sex 2; Involvement 2; Values 6.) There are three lists, and from each list an imaginary boy will be created by one group, and an imaginary girl by another. The creations are described to the whole class. The week ends with a quiz testing ability to identify the identity areas. Optional but strongly recommended by the author is a long-term identity journal, which the teacher would then present along with the evaluation process.

Phase 2: These are optional activities which can be used in applying the newly acquired knowledge to literature and life. Prime among them is the journal keeping. The others are called Clarifying Identity, Portrait, Understanding Time, Female and Male, Valuing, Loyalty, Role-Playing, Alternatives Search. Each activity has worked out materials for students and suggestions for teachers, who will be using them according to their own class situations. (AC)

Note: For more on this exercise, see the essay on self-development games by Harry Farra.

Cost: $14.00

Producer: Interact, P.O. Box 262, Lakeside, CA 92040

INDEPENDENCE

Jane Sterk

Playing Data
Age Level: high school
Number of Players: 24-34
Playing Time: 9 45-minute periods
Preparation Time: 2 hours (est.)
Packaging: professionally packaged game with play money, briefing booklets, and instructions.

Description: This game simulates the metamorphosis of an African colony of the great power Nacra into the independent nation Libra. This area is home to five peoples: the Faible, the Helix, the Litta, the Nerva, and the Kano. These groups share the colony's resources and envy one another. The game is designed so that before Libra can become free various national goals must be met, and before these goals can be met, a series of intermediary goals must be formulated and accomplished. The process involves dividing the participants into the interested factions, having them negotiate a common goal besides independence and then try to achieve it. The game ends when they achieve the goal. (DCD)

Cost: $20.00

Producer: Puckrin's Production House, 35 Mill Drive, St. Albert, Alberta, Canada T8N 1J5

INDIAN RESERVATION

Ron Stadsklev and Ron Wagner

Playing Data
Age Level: grade 7-adult
Number of Players: 12-32
Playing Time: 3-6 hours

Description: This simulation introduces students to the problems encountered by residents of Indian reservations. Participants role play members of four Indian families, each seeking education and employment while dealing with tribal politics, all within a structure created and administered by the Bureau of Indian Affairs. Because of differing ages, degrees of responsibility, and percentages of Indian blood accorded to each role, participants start the game with differing numbers of status points and time chips. Overcoming this disparity is one of the game tasks. (producer)

Cost: Do-it-yourself kit $12.00

Producer: Institute of Higher Education Research and Services, P.O. Box 6293, University, AL 35486

INTENTIONAL COMMUNITY

Harriet Linda Tamminga

Playing Data
Copyright: 1975
Age Level: high school-adult
Number of Players: 6-10
Playing Time: 1½-2 hours in 6 15-20-minute rounds
Preparation Time: 1 hour (est.)
Packaging: 54 unbound, photocopied pages

Description: This game simulates the experiences of the residents of an "intentional community" (defined as "a territorially localized type of social organization purposefully created by its members to provide for their residence, sustenance, and social needs in mutually shared and agreed upon ways") in the Southern Colorado Rockies during the first year of its existence. "Participants in the game," according to the author, "assume the role of communtarians who are searching for an antidote to alienation in the large, impersonal city and to the isolation of the modern American nuclear family. They hope to find this through the creation of a community where there is a sense of belonging, fellowship, and mutual concern, as well as opportunity for freedom and personal fullfillment." All of the players have the goal of sustaining the community for six rounds (each corresponding to two months). Each player also seeks "self-actualization." (The game uses Abraham Maslow's idea that human beings have a hierarchy of needs which must be satisfied in order of their place in that hierarchy.)

Players keep score of how well they are meeting their own and their community's needs each round. Scores are in resource points (or resources) and the score for each player each round is determined as follows: A player's resources are added to his or her personal production rate (a number from 1 to 4 assigned by the game director to simulate differing abilities) multiplied by another random number (generated by rolling a die). Then five points are subtracted to represent what a player would use to subsist. This score is modified again when a player draws a "contingency card," which describes an event assumed to have happened during the preceding two months. (For example, "The bill collector finally caught up with you–subtract six points," "Drinking from the stream has caused diarrhea–subtract twelve resources from the community to obtain a water purifier.) The score is changed again by adding or subtracting the points gained or lost through "Policy Decision;" the community as a whole must resolve some issue (such as whether to permit the use of illegal drugs in the community) during each round. If all players agree on a decision, everyone in the community gains a point. If one or two players dissent, no points are gained. If three or four dissent from the majority, a point is subtracted. Each player's score after all of these operations may be larger or smaller than five. A star may be bought for five resource points, and satisfies one level in the hierarchy of needs. In order to be self-actualized, then, a player must produce five surplus resource points in five of the six rounds. By the sixth round, the entire community must have acquired thirty surplus resource points with which to make a land payment.

The game ends after six rounds. (DCD)

Cost: $2.50

Producer: Harriet L. Tamminga, Department of Sociology, Search and Rescue Project, University of Denver, University Park, Denver, CO 80208

IN THE CHIPS

John Beasley

Playing Data
Number of Players: 12-30

Description: *In the Chips*, a game about power, illustrates a concept; teachers from many diverse disciplines can use this instructional technique profitably. A greater understanding of such issues as the plight of the oppressed, world trade, the sorry battleground of Latin American politics, or the unequal distribution of the world's resources, can be acquired. Further, playing the game will provide students with some insight into what it "feels" like to be powerless and subject to the whims of a ruling clique.

(1) To enhance a student's understanding of power.
(2) To illustrate the concept that power tends to perpetuate power.
(3) To develop an understanding of the feelings and strategies of the powerless.

In the Chips is an inexpensive, do-it-yourself operation, requiring few materials. One of the game's greatest advantages is that it may be used by anyone interested. It will be necessary to secure five or six sets of brightly colored identification tags: strips of crepe paper will work quite satisfactorily. In addition, one hundred fifty poker chips are needed. Break these down as follows: five gold chips at a value of fifty points each; twenty green chips, twenty-five points each; forty red chips, 15 points each; forty blue chips, 10 points each; forty white chips, five points each; and five yellow chips, fifty points each.

Prior to playing the game, the class should be set up in four or five groups. Each group should have between four to seven members. The identification tags should be distributed to each group on the basis of color, i.e., red tags to one group, blue to another, etc.

The game is simple to play. Each player draws five chips sight unseen at the beginning of each round. After chips are drawn, a round of trading follows. During the trading session each player should attempt to obtain as many high value chips as possible, while disposing of those chips that are of little or no value. While the students are mingling, there is to be no discussion during the trading period. However, there can and will be nonverbal communication. Chips are traded by looking your trading partner in the eye, shaking hands, and exchanging chips. Players should only trade with members of other groups. When a player is satisfied with the five chips in his possession, he should return to his chair.

When everyone has been seated, each group tallies the total number of points it has acquired by computing the value of the chips in its possession. After the total has been established, the group chairman goes to the board and records the group score. The team that has the highest score can then make any rule that it chooses. The entire process is then repeated as many times as desired.

The most important aspect of the game, of course, is that the winning team can make any rule it chooses. Inevitably, the rules proposed by the winning team are self-serving. For example, a team might state that their score is to be doubled. Or, they might rule that all gold chips must voluntarily be turned over to them in all subsequent rounds.

While the behavior of winning team members remains constant from one game to the next, the reaction of losing players is much less predictable. Some cheat by recording astronomical scores after the trading sessions. Other tactics designed to co-opt and subvert "the system" are also devised. For example, it is a common occurrence for losing teams to ban together and flaunt all rules, thus initiating a microrevolution. Others simply quit.

(1) The main impact of the simulation is to enhance a student's affective understanding of power.
(2) Students find themselves acting as a microcosm of society. Those who are losing the game adopt such strategies as rebellion or dropping out to cope with their situation.

It is usually during the debriefing session that players fully understand the game. They come to realize for example, that those in power often tend to extend their authority by establishing self-serving rules and regulations. Beyond this, they can better understand why the "outs" in society act as they do. The student that revolts can understand, on an effective level, revolution. The student that quits can understand society's drop-outs and those that are apathetic. Those that cheat can better understand lawlessness.

In the Chips can be a powerful learning experience. Students will have fun, they will be joyous and frustrated, and they will always be unpredictable.

Comment: The strengths are:

(1) The game is a powerful learning tool. It increases a student's interest in topics relevant to the concept of power.
(2) The game is inexpensive and easy to utilize in the classroom.
(3) It is extremely flexible and can be used to supplement a wide variety of topics (i.e., lobbying, Latin America, revolution, the behavior of the dispossessed, rules and regulations of the stock market)

The weaknesses are:

(1) Little direct cognitive growth will result from the game. Cognitive development must result from lectures and readings that surround the game.

(from the Robert A. Taft Institute of Government Study on games and simulations in Government, Politics, and Economics by John Beasley, Southern Illinois University at Carbondale)

Cost: $1.50

Producer: SAGA, 4833 Greentree Road, Lebanon, OH 45036

KAMA: A SIMULATION DEALING POST-COLONIAL AFRICA

Michael J. Raymond

Playing Data
Copyright: 1974
Age Level: grades 9-12
Prerequisite Skills: 8-9 grade reading ability (Flesh Scale of Readability) familiarity with the U.S. Constitution; teacher familiarity with Africa, particularly Uganda
Number of Players: approximately 32
Playing Time: 7 55-minute class periods
Special Equipment: 1 die
Packaging: 44 page pamphlet

Description: As generals, kings, bishops, delegates, and political party members of Kama, an imaginary new African country similar to Uganda, students are concerned with establishing the constitution, protecting the country from military threats, achieving their party's goals, and developing their own goals for the country. Before they can role play the leaders of Kama, they must familiarize themselves with the map, history, social situation, leaders, political parties, armed forces, and proposed constitution as included in the text. After an introductory session, each of the following five hours is divided into three segments—ten minutes for planning and party caucuses, a thirty-five-minute constitutional conference during which time delegates must agree on each of the articles of the Kaman constitution (similar to the U.S. Constitution) and eventually ratify it, and ten minutes for additional meetings. Students also create propaganda materials in order to convince other delegates of their political beliefs; they also learn and use parliamentary procedure as they participate in the conference. The five sessions of conferences are followed by a debriefing session. A bibliography on Africa and Uganda in particular is included in the game pamphlet as a teacher resource. (TM)

Cost: $3.95

Producer: SAGA, 4833 Greentree Road, Lebanon, OH 45036

MAHOPA

John Wesley, Chase Ave. School, El Cajon, CA

Playing Data
Copyright: 1972
Age Level: grades 5-8, average ability; grades 9-12, below average ability
Number of Players: 20-40 in 3 teams
Preparation Time: 3-5 hours

Description: The class is divided into three imaginary tribes. Students receive tribal identities along with identification necklaces. After a brief Native American quiz reveals how much they know about Indian life, they read and discuss two brief essays on Native American history. Next they analyze a large wall chart showing them how to contrast value systems of Indians and contemporary Americans. During the remainder of the simulation the tribes compete with one another as they discover several Indian values: family, child-training, property, religion, nature, time, work. Students write diary entries which reveal how their Indian identities react to an eclipse of the sun, the death of a brother, the absence of snow and rain. Each tribe presents a minidrama about a crisis or ceremony: a visit by Spanish missionaries; the choosing of a new war chief; a fall harvest festival. Working on three-dimensional projects deepens students' understanding of Indian culture: making pottery, jewelry, weapons; grinding grain; weaving fabrics or baskets; creating pictographs or sand-paintings. Competition between the tribes is maintained by daily rewarding of all activities with CIPs (cultural image points), which are entered on individual balance sheets. The simulation ends with a one-hour activity covering the last 400 years of North American Indian history. The classroom floor becomes North America, 1550, and the students become either Indians or settlers and witness the Native American's fate. Finally students retake the Native American quiz and engage in an evaluation and debriefing.

Comment: *Mahopa* provides for a variety of activities during the simulation. This variety adds to the effectiveness of the simulation by making it interesting and increasing the possibility of involvement.

Mahopa capitalizes on one of the advantages of simulation by fostering empathy. The simulation demands cooperation as well as competition and this gives it balance. *Mahopa* is an exciting simulation for the audience for which it is intended. The attainment of its objectives should not be difficult. (MJR)

Cost: $14.00

Producer: Interact, P.O. Box 262, Lakeside, CA 92040

MICRO-COMMUNITY

William B. Jarvis

Playing Data
Copyright: 1971
Age Level: grades 7-9
Number of Players: 15-40
Playing Time: school year

Description: *Micro-Community* is an outline for a junior high level course in U.S. history structured by a classroom economic and social system that the teacher kicks off during the first week by assigning "dollar" values to quiz grades, establishing desk rents and taxes, and defining communities within the classroom according to seating arrangements.

The society develops with the establishment of a government, constitution and laws, and community institutions. The economy grows and experiences various structural and historical phenomena corresponding with some elements of the U.S. economy. The teacher remains the originator of money and from this position engineers inflation, a stock market crash, introduces the costs of pollution, and so on. Money enters the economy as the teacher pays students according to their success at activities ranging from special endeavors (e.g., each member of the community whose Bill of Rights gets adopted receives $20) to papers, tests, and day-to-day contributions to discussions. Students thus run a private enterprise economy and community within parameters the teacher as ultimate authority grants them.

The materials include a book of sample forms used by the designer in his own classroom, a book of lesson plans for a school year, and a book of thirty-two worksheets that go from European explorers to the 1960s. (AC)

Note: For more on this simulation, see the essay on simulations of developing a society by Frank P. Diulus.

Cost: $49.50

Producer: Classroom Dynamics, 231 O'Connor Drive, San Jose, CA 95128

MICRO-ECONOMY

George Richmond, Harvard University

Playing Data
Copyright: 1973
Age Level: grades 5-12
Prerequisite Skills: some reading, and some math
Number of Players: entire class
Playing Time: 1 month-1 year

Description: *Micro-Economy* sets up an economic system in the classroom that pays students with micro-economy money for work in five possible areas: academic (such as payment for math tests, compositions, reading, homework, etc.), employment in the school (blackboard monitors, class officers, etc.), micro-economy jobs (real estate manager, auditor, banker, payroll secretary), professional service (model builders, architects, lawyers, newspaper reporters, etc.), and civic projects and recreation (to set up and run a foundation which will pay for various programs such as cultural programs, neighborhood cleanup drives, etc.). It is suggested that the teacher obtain real goods to back the micro-economy currency from parents, students, the school community, themselves or from merchants in town.

The game is then played in phases. There is a period of about two minutes for transportation in which students move to a new place on the playing board supplied in the game package. The next phase is called marketplace and takes about ten to twenty-five minutes. At this stage, landlords collect rent from tenants, students may form partnerships and buy property, etc. This is followed with five to ten minutes in which the Board of Assessors increases or decreases the value of property. Finally, there are three auctions for each transportation move. All unsold properties are auctioned off in the first. In the second, students who wish to sell property they own may sell it, and in the third, students who have other things to sell—houses, building materials, improvements, and so forth—may do so in a general auction. It presumably takes a considerably longer period for students to keep the records and do other aspects of the transactions that take place.

A series of lesson plans on basic monetary concepts, banking, economic theory, types of business, and organizing is also included in the teacher's manual. In addition, a series of four suggested salary matrices show how a teacher can vary payment for different tasks to stress one of the following four aspects: (1) academic skills, discipline, and routines, (2) development of good citizenship, (3) economic development and work achievement, (4) creativity and initiative. In each one of these salary matrices the amounts paid for different assigned tasks, extra labor and creative or social contributions differ. With the payoff being different students will presumably vary their time, activity, and effort accordingly. These matrices are illuminating and well worth a teacher's look to see how she awards points in the particular version of the game that is played in her classroom. Obviously these salary matrices are just suggestions and may be varied to suit a particular teacher's or indeed a particular individual's situation. (REH)

Comment: George Richmond has done it. He has found a way to bring *Monopoly* into the classroom. Richmond observed that the average classroom game was feudalism. The teacher owned everything. The kids owned nothing. He also saw the results of this in the fifth-grade classroom into which he was placed as a substitute teacher in New York City. The other game besides feudalism that was going on in the classroom was, as Richmond put it, war. He soon found that the fifth graders were winnning the war and he, George Richmond, was losing it. He recognized that he had to change the game and change it fast.

So he invented *Micro-Economy*. How did he use *Monopoly*? First, the board was the whole classroom; shelves and places around the room which could be purchased by students if they landed on them. The kids could build houses and hotels on this property. The kids had to make the houses and hotels in art class. Twice a week they had moves. This apparently was similar to *Monopoly* in that kids rolled the dice and moved their markers from place to place.

Richmond mimeographed money for the game and began to pay people with it. He hired students to do all sorts of jobs: take trash out, keep the blackboard clean, tutor other students, take care of the plants in the class, and he began to pay them in micro-economy money for compositions and other academic accomplishments. Why should kids do schoolwork for phony *Monopoly* money? Well, Richmond put a gold standard behind the money. He told the kids that at the end of the semester they would have an option where they could buy real things: toys, games, baseball bats, basketballs, etc., with the micro-economy money. This put real incentive behind the earning of money in the game. Since kids could build their own houses, there was an obvious need for an assessor. George hired an assessor from the class. Soon a bank came into existence with bankers, accountants, and the other jobs that go to making commercial enterprises. Other commercial enterprises as well developed: a building supplies company, and in one of the situations a whole corporation for buying land at opportune times. As disputes developed, new jobs were needed and paid for: lawyers, judges, and so forth.

What were the results of this experiment? Well, the entire nature of the classroom changed from a primitive warlike society to an orderly economic society in which kids were learning a great deal more than they had in the past. And they were learning a great deal about writing, reading and arithmetic from the game. They had to write leases and do all of the arithmetic involved with the money transactions. In addition, some of the norms in the classroom began to change. Brains rather than strength began to pay off. This, interestingly enough, transferred to the real world.

What is fascinating to me about what Richmond has done is that it is difficult to tell where the game stops and the reality of the classroom starts. The game has affected reality itself. Good old *Monopoly* has changed the classroom norms.

It is obvious that the game bears much resemblance to some of the token economies implemented by the behavior modifiers. People can, and probably will, throw up their arms in disgust at both the capitalistic and behavior modifier aspects of this classroom design and this game. However, this should point out very clearly that every classroom does have a design and a reward system. There is no classroom that does *not* incorporate aspects of behavior modification by definition. The major question is what behavior modification you use. What kind of game is being played in your classroom? As many people have pointed out, in most schools the game is prison. (REH)

Note: For more on this simulation, see the essay on simulations of developing a society by Frank P. Diulus.

Cost: program with teacher's manual $39.00, teacher's manual separately $6.00

Producer: The Center for Curriculum Design, Harcourt Brace Jovanovich, 757 Third Ave., New York, NY 10017

NEW CITY TELEPHONE COMPANY SIMULATION GAME

R. Garry Shirts

Playing Data

Copyright: 1974
Age Level: high school (possibly junior high)
Number of Players: up to 6 teams of 5 players each or up to 6 individuals
Playing Time: 90 minutes followed by minimum of 30 minutes discussion (up to 3 hours)
Preparation Time: 10 minutes
Special Equipment: cassette player and overhead projector
Packaging: all materials professionally designed and packaged

Description: This game reflects situations that a telephone company experiences in the course of doing business. Players listen to tape-recorded debates and comments about four typical telephone company management decisions. These concern the automation of switching facilities, the negotiation of a collective bargaining agreement, coping with inflation, and customer-employee relations. (Specifically, the last of these typifies the quandaries presented in the game. A surly employee, who has worked for the phone company for a long time, has begun to evoke numerous customer complaints, and the players must decide whether to fire, transfer, or reprimand him.) After listening to the tape-recorded discussion of each situation, players are asked to choose the best of several listed decisions in terms of the probable resulting customer and employee satisfaction, and the resulting company profit.

The teacher's manual includes an answer guide with a rationale for what the author considers the best decisions to make in each situation. (DCD)

Cost: $15.00

Producer: Simile II, P.O. Box 910, Del Mar, CA 92014

OPENING THE DECK

Ray Glazier, Games Central, Abt Associates, Inc.

Playing Data
Copyright: 1972
Age Level: grades 7-12
Prerequisite Skills: 6th grade reading
Number of Players: up to 36
Playing Time: 30-45 minutes
Preparation Time: 15 minutes
Supplementary Material: "Life of the Kwakiutl Indians" (30 copies Potlatch Package), Potlatch Game (kit included in Potlatch Pkg.)

Description: This game is part of an Indian anthropology unit, and though it can be played independently, it is designed to demonstrate materials in the text and to help students understand and identify with Kwakiutl Indians and the ways their society was organized. The unit is also intended to foster reflection on contemporary American society and economic practices.

Each player receives a role card with information about a Kwakiutl Indian person. As players group and regroup according to family, residence, age, social status, and the day's work activity, they learn about the social organization of the tribe. Players may elaborate on their role descriptions by inventing more data. See the *Potlatch Game* for more on the unit. (AC)

Cost: Potlatch package unit $65.00; game alone $14.00

Producer: Games Central, 55 Wheeler Street, Cambridge, MA 02138

POLICE STATE

Harley L. Sachs

Playing Data
Copyright: 1969
Age Level: 10-adult
Number of Players: 2-6
Playing Time: ½-1 hour
Packaging: professional game box

Description: *Police State* is a board game in which players attempt to improve their standard of living while functioning under the restrictions and secrecy of a Communist society. The goal is either to become premier, by attaining a four room apartment and the only state car, or to be the last player to be sent to Siberia. Players begin as peasants, all living in one room, and moving only along one side of the board. Gradually, through attaining wealth and power cards which enable them to rent larger apartments, players gain access to the other sides of the board as they become workers, party members, and party elite. Cards are kept face down and hidden from other players until one has enough to move to a higher status. Players may denounce each other for hoarding twice the number of power and wealth cards allowed for their status, but because some cards are phony (marked worthless) the accused may be able to prove himself innocent, in which case the accuser rather than the accused is demoted and loses all his or her power and wealth. A demoted peasant is sent to Siberia and is out of the game. A third kind of card, secret police cards, require players to gain or lose power and wealth cards, or promise them immunity from denunciation. However, three of the four immunity cards are phony and serve only to scare opponents. Because of the limited number of apartments on each status level, a player often will not be able to move up in status to the next level, and must wait to be able to skip a level, while risking being denounced for accumulating too much wealth and power. (TM)

Comment: *Police State* is a clever game, but is primarily based upon chance, as players can demonstrate strategic skill only in their choice of timing for denouncing other players or buying the state car. The game is *not* an accurate model of life in a police state or in Russia and should not be presented as such; as a propaganda tool to increase anti-Communism it could only further international misunderstanding. For maximum enjoyment two players should modify the rules so that they may return from Siberia after losing several turns; otherwise the game may repeatedly end after only five minutes of play. (TM)

Cost: $8.00

Producer: Idea Development Co., Inc., Houghton, MI 49931

POTLATCH GAME

Ray Glazier, Games Central, Abt Associates, Inc.

Playing Data
Copyright: 1972
Age Level: grades 7-12
Prerequisite Skills: 6th grade reading, simple math
Number of Players: up to 30 in two teams
Supplementary Material: "Life of the Kwakiutl Indians" (30 copies in Potlatch Package), Opening the Deck Game (kit in Potlatch Pkg.)

Description: The *Potlatch Game* follows the game of *Opening the Deck* (see above) in a two-week Indian anthropology unit. This game, too, may be played independently of the unit. In the *Potlatch Game,* players simulate this highly competitive social institution of display of wealth and power. Players get individual role profiles. They meet in two separate clans, and one invites the other to a potlatch, at which the host gives guests the most lavish possible gifts and may also destroy goods to demonstrate wealth and gain prestige. Guests are obligated to reciprocate gift-giving (with interest) and goods-destruction. (AC)

Cost: Potlatch package unit $65.00; game alone $40.00

Producer: Games Central, 55 Wheeler Street, Cambridge, MA 02138

THE POULTRY GAME

Playing Data
Age Level: high school

Number of Players: 10-18 in 5 or 6 groups
Playing Time: 1-2 hours
Packaging: game packet in plastic bag containing villager, chance, and insurance cards, cost and benefit sheets, and instructions.

Description: This didactic game aims to give an understanding of the difficulties of a farmer in a village in the third world by simulating decisions and frustrations. At the beginning of the game participants are divided into groups of two or three. These groups each represent one villager who has decided to enter the local poultry market. Each villager decides how many chickens to buy, how to house and feed them, and figures income and expenditures. Egg yield is based on figures from Zaire in 1972. The group leader introduces chance cards after the first round. For instance, a house may burn down, or half of a villager's chickens may die. The survivors of these calamities can purchase insurance for a small fee. The game continues in this manner for four or five rounds. In most cases, the villagers work independently of one another although they may form a cooperative. Participants are asked to discuss who did best and why during the debriefing. (DCD)

Cost: 95 pence

Producer: OXFAM, 274 Banbury Road, Oxford OX2 7D2, United Kingdom

THE POVERTY GAME

Adapted from the original by Jim Dunlop

Playing Data
Age Level: 14-adult
Number of Players: 8-30 in groups of 2-5
Playing Time: 1 hour
Packaging: plastic case

Description: The purpose of *The Poverty Game* is to teach students about the economic difficulties of a typical West African subsistence farmer, and the extent to which a farming village is dependent upon chance. Players divide into small groups representing an African village and study a crop chart to determine which crops to plant on their ten fields. Yams, for instance, yield 70 units of food in a wet year, and 20 in a dry one. Then, by throw of the die, they determine the weather condition for the year, which generally has a 40 percent probability of being favorable; they also choose a disaster card to determine what other factors (such as rodents and insect pests eating a maize field) affect their yield. Production is then calculated. If a village produces less than 450 units of food, it suffers malnutrition and loses workers for the next year due to disease; the number is determined by drawing a disease card. Villages producing less than 250 units in two years may draw a help card to see if other villages will help them; those producing less than 200 units die. Students repeat the same steps for six rounds, representing six years, calculating their production level each round, and rotating crops if they so choose. The authors admit that the game is an oversimplification of realtiy but believe that it nevertheless depicts the kinds of problems African farmers face, and their struggle for survival against overwhelming odds. (TM)

Cost: 95 pence

Producer: Oxfam Youth Department, 274 Banbury Road, Oxford, United Kingdom

POWER

Eric Edstrom

Playing Data
Copyright: 1972
Age Level: grades 7-8
Number of Players: 9-35
Playing Time: 4 or more 40-minute to 1-hour periods; five periods are preferred
Preparation Time: 1-2 hours

Description: *Power* presents four simulated exercises, each one based on a realistic situation in which the exercise and maintenance of power, or authority, depends upon the control of the communication system.

Each game is based on a different level of politics. Simulation 1 deals with international politics. In it, two subordinate nations question the authority of the dominating nation. In Simulation 2, considering national politics, a president attempts to gain support for his Supreme Court nominee. Simulation 3, about local politics, fits into any American history or American government course. The game centers around a big-city mayor whose authority is being challenged by a suburban leader. In Simulation 4, about corporation management, a company president is retiring and several people are vying for the office. A total of five class periods generally is needed for one simulation. One kit contains the materials necessary for one classroom to play all four games. (Producer)

Cost: $14.97 (school price)

Producer: Scott, Foresman and Co., 1900 E. Lake Avenue, Glenview, IL 60025

THE PROPAGANDA GAME

Robert W. Allen and Lorne Greene

Playing Data
Copyright: 1966, 1975
Age Level: high school to adult
Number of Players: 2-4
Playing Time: 90 minutes
Preparation Time: 1 hour (est.)
Packaging: professionally produced and packaged game includes cards, tokens, and instructions

Description: This game is designed to make its players familiar with various rhetorical and propagandistic devices. The authors have defined fifty-five such devices and classified these into six categories, techniques of self-deception, of language, of irrelevance, of exploitation, of form, of maneuver. Each play of the game concerns just one of these categories or "techniques," and proceeds as follows: All players begin with markers on the line of the playing surface. (This surface is ruled into twenty-five spaces which increase in value, from bottom to top, from minus five to plus twenty.) One player reads an example of propaganda and then all must identify the propagandistic device contained in that example. (For instance, in a game examining "techniques of language" each participant must decide if the specific device in an example is emotional terms, metaphor and simile, emphasis, quotation out of context, abstract terms, vagueness, ambiguity, or shift of meaning.) When a majority of the participants agree on the device, all players in that majority may advance their tokens one space. Players who dissent from the majority view may not advance their tokens without issuing a "bold challenge," which involves looking up the authors' opinion in an answer key. If the dissenters and the authors agree, the dissenters advance and the rest must move their tokens backwards. Dissenting players must retreat if their challenge fails. The winner is the first whose token reaches line twenty. (DCD)

Note: For more on this game, see the essay on social studies simulations by Michael J. Rockler.

Cost: $11.00

Producer: Wff 'N Proof, 1490-TZ South Boulevard, Ann Arbor, MI 48104

RAFA RAFA

R. Garry Shirts

Playing Data
Copyright: 1976
Age Level: grades 4-8
Number of Players: 10-50
Playing Time: 90 minutes
Preparation Time: 1 hour (est.)
Special Equipment: cassette tape recorder
Packaging: professionally packaged 31-page instruction booklet, cassette, and cards

Description: The purpose of this cross-cultural simulation, which is a simplified version of *Bafá Bafá,* is to help students become aware of value and behavior differences between different cultures, and to understand the importance of nonverbal communication in cross-cultural interaction. The participants are divided into Alpha and Beta cultures, and the two groups are segregated. Both culture groups are thoroughly briefed by the game director on the features of that culture; a cassette tape recording of these instructions is included in the game. People in the Alpha culture are easygoing, superstitious, and like to touch one another. Members of the Beta culture are hardworking, businesslike, and do not touch. Each culture has a detailed set of rules. Alphans, for instance, spend their time playing a "good luck" guessing game, while Betans structure their time by trading cards. During the game visitors and observers from each culture visit and observe each other. They may not ask questions about the rules of the culture which they observe but they must try to infer what these rules are. The game ends when every participant has had the opportunity to be an observer. After the game, participants discuss how they feel and what they thought about their experience. (DCD)

Cost: $15.00

Producer: Simile II, P.O. Box 910, Del Mar, CA 92014

RAINBOW GAME

Ray Glazier

Playing Data
Copyright: 1975, 1976
Age Level: grades 7-12
Prerequisite Skills: reading level grade 6, math level grade 7
Number of Players: 20-30
Playing Time: 5 45-50-minute periods
Preparation Time: 1 hour
Packaging: professionally produced and packaged game includes all materials necessary for thirty players

Description: This simulation creates a socially and ethnically stratified society, called Rainbow Union, within which the players must distribute government benefits and also gather and interpret demographic data.

Rainbow Island, the fictional location of the game's society, was discovered by "Blues" and "Oranges" two centuries ago. For the first hundred years the Oranges, although fewer than the Blues, dominated the society. In the early times only Oranges had guns. Oranges recruited "Purples" and "Greens" from abroad to work in the island's mines and help promote the island's culture. At the end of that first century the Blues, who by that time numbered sixty percent of the population, overthrew the rule of the Oranges, rewrote all the history books, and started the "Fun Stamp Program." At that time the various ethnic groups assumed their present places in the nation's socioeconomic hierarchy. The Blues became the class of managers and, as the majority, set the standards for society. The Oranges, proud and resentful, were relegated to the bottom of the social order. The Greens became dominant in the professional classes, and the Purples, who are unambitious and carefree, assumed a place in society just above the Oranges.

Each player is assigned to one of these groups and must wear an arm band of that color. All are given a "group profile" and each must complete a "do-it-yourself role profile." Students fill out this document with some variation of their own name and address and determine their age, sex, marital status, family size, years of education, occupation, and income by matching a random number (generated by a dice roll) to a key for their color group.

Play centers around two activities. The first is the distribution of "Fun Stamps" (which is stamping a player's hand with a rubber stamp that says "Fun Stamp"). Stamps are dispensed by a committee of seven Blues who may use any criteria they want for accepting stamp applicants, but they are prohibited from Fun Stamping more than half of the players. Participants who do not work at the Fun Stamp Center are divided into groups which must each survey the classroom community on one characteristic such as age, sex, or marital status as it is distributed across the four color groups. One group must also do a follow-up survey on the characteristics of Fun Stamp recipients. The game ends after half the players have been fun stamped and their characteristics as a group have been identified. (DCD)

Comment: This is the best existing simulation for promoting intelligent discussion by participants of all ages of who should get what in America, why, and how. Its play would be specially useful for adults who for the first time must interpret census or labor market information to establish goals and timetables for affirmative action, for example. (DCD)

Cost: $30.00 plus $1.50 shipping

Producer: Games Central, 55 Wheeler Street, Cambridge, MA 02138

REVOLUTIONARY SOCIETY (REVSOC) SIMULATION

Donald M. Snow

Playing Data
Copyright: 1974
Age Level: secondary
Number of Players: 7-35
Playing Time: up to 2 hours
Preparation Time: approximately 45 minutes
Supplementary Materials: pencils, printed copies of forms from manual, and a die
Special equipment: an electronic calculator
Packaging: 20-page manual

Description: "The primary objective of this simulation is to create a situation in which the student can gain a clearer understanding of the processes and perceptions that are involved in the decision to engage in revolutionary activity." The simulation takes place in a hypothetical country (Fidelia) about which a background description is given. Within Fidelia there are five social groups (descriptions of each are provided in the manual) which individually or in a coalition (determined by a majority of "Total Influence Points (TIPS)" form the government. "The object of the game is to maximize the number of Total Influence Points that one has or can retain." TIPS are initially allocated to the various groups; however, the groups are unsatisfied with their meager security and demand more. Throughout the "Reaction Period," dissatisfied groups struggle attempting to achieve their desired goals. There are a number of alternative ways in which dissatisfied groups may react. At the conclusion of each reaction, another cycle commences (a simulated year). Each succeeding cycle has two stages: Budget Allocation (fifteen minutes) and Reaction Period (fifteen minutes). The player with the highest number of TIPS after the last cycle is the winner. (WHR)

Cost: $4.50

Producer: Gerald L. Thorpe, Department of Political Science, 101 Keith Hall Annex, Indiana Univ. of PA, Indiana, PA 15701

SANGA

Patricia Ward and E. Craig Williams

Playing Data
Copyright: 1974
Age Level: grades 7-10
Number of Players: up to 36
Playing Time: 5 class periods
Packaging: 8-page foldout student guide, 34-page teacher's guide

Description: *Sanga* is a simulation of village life of the Dogon in Mali, West Africa. There are six families of mother, grandmother, older and younger daughters, father, grandfather, older and younger sons. The first day students are given roles with background material on their place in the family, data on their particular family, and a description of village life. Family data include names and identities for each member (the mother, Dankele, a potter), resources (land planted in maize, dancing skills of Danu and Beluko, fifty yards cloth, bird mask, one hen, and five Malian Franks money). Each family also has goals (rebuilding a house, paying the blacksmith, a wedding) which are explained at some length, and is daily touched by fate, represented by fortune cards.

Days two and three begin with drawing fortune cards. Then each family holds a planning session in which members function according to their roles. Then comes a time in which families may bargain, sell, buy, and consult the village's chief priest. The families are working toward their various goals.

Day four is market day. The villagers trade excess goods or finish trying to accomplish their family goals. Day five is devoted to debriefing.

The teacher's guide gives directions for initial preparations, suggest various options in using the simulation, and has bibliographies for background and further study, as well as daily lesson plans and fortune cards, resource sheets, and family score sheets. (AH)

Cost: $14.00

Producer: Interact, P.O. Box 262, Lakeside, CA 92040

SERFDOM: A SIMULATION IN CLASS ACHIEVEMENT

Pat Bidol, Ann Kraemer, Virginia Stewart, and Fr. James Trent; People Acting for Change Together

Playing Data
Copyright: 1970
Age Level: high school-adult
Prerequisite Skills: maturity to examine minimal functioning of organizations and classes and to work in groups
Number of Players: 21 or more in 3 teams
Playing Time: 2 1/2-3 hours
Preparation Time: set up time: 20 minutes-1/2 hour; explaining instructions: 20 minutes-1/2 hour; trial game is recommended for those unfamiliar with simulation
Packaging: mimeographed directions, review, and debriefing suggestions

Description: *Serfdom* is a simulation of organizational structuring which imposes roles and values upon the individuals operating within its system. Setting up a hierarchical society of leaders, middlemen, and workers, it pressures the middleman, who is responsible to his leader, to exercise an efficient and productive economy over his workers. The organization which operates most efficiently in bringing the greatest amounts of reward with the least amounts of expense will spell victory for its leader and the greatest chance of survival for its other participants. Inefficient organizations will provide survival for its leader but certain destruction for its participants. All mobility within the organization is determined by the needs of the one who heads the organization. The system itself divides its subordinates by having them compete with one another to provide benefits for the one heading the structure. This same role-goal competition establishes the control lever between the powerful and the powerless. (Teacher's manual)

Serfdom demonstrates how a hierarchical society or institution maintains itself by keeping the enemy invisible and how it imposes roles and goals on its people, dividing them against each other, and reducing some to powerlessness. Players begin by familiarizing themselves with the scenario and their roles as princes, lords, policemen, and serfs. Three princes head teams of three lords and a number of serfs who are required to build towers and fortresses with tinker toys during each of five ten-minute turns. The princes are competing for the best construction in order to be chosen by their father as the next king. They determine the nature of the construction and supervise the lords, who in turn choose serfs to work for them. The policeman fines offenders and disrupters. At the end of each turn the prince may demote inefficient lords and promote serfs to take their places. Some players are intentionally and unknowingly discriminated against by being fired and never being promoted to lord. Rewards in the form of tokens are given out after each turn according to place in the competition, the role of the player and the economic chance card. Each player must maintain his role by receiving a set number of tokens each turn or he is demoted. If a prince does not maintain his standard, he and all his workers are to be considered executed at the end of the game. When the five rounds are completed, the trainers choose as winner the prince who has accumulated the most tokens beyond his expected standard. In the debriefing session which follows, players then discuss the parallels between *Serfdom* and real life, how they felt about their roles, and the possibilities for change within a hierarchical system. (TM)

User Report: After a number of *Serfdom* experiences here in Morgantown and other places it has proven to at least open up conversation about how systems work. It certainly raises the question of how people go about building up trust and how fragile groups are when they must work competitively under time pressures.

Serfdom is a very involving simulation for nearly all participants. The various roles tend to involve the participants in different ways. Some people are very busy, some become alienated, others become frustrated. All kinds of subgames develop as the game progresses. Because of this, it is essential to have the participants discuss their game experiences with others in the postgame discussion. It's the only way to get an overview of what really happened.

Some suggestions for *Serfdom* not included in the instructions: (1) Discriminate by the color of eyes rather than red name tags. (2) Do not use chance cards—just arbitrarily vary rewards giving headlines that explain the economic conditions. (3) Keep the princes out of the playing area and in their own room (the castle). Allow only policemen and lords to go to the castle. (4) Make the police get their rewards from the princes. (5) Shorten the game to four rounds with maximum playing time being one hour. (6) Do not allow serfs to exchange their tokens until they have their own survival assured. (7) Make unemployed serfs walk only on their knees. (8) Give the princes one "on site" visit for any five minutes they choose during the course of the game.

Comment: The simulation creates an environment in which personal survival is apparently linked with organizational or institutional survival, and the predominant ethic is that which permits or promotes the survival of the organization or institution. It shows how we are born into organizations and institutions which impose their goals upon us which affect our personal, interpersonal, group, intergroup and intragroup behavior and communication. It is clear that this simulation has a strong political message. One of the authors has told us, "*Serfdom* is designed to raise consciousness upon reflection that institutions and organizations tend to perpetuate themselves at human expense. They protect themselves by competitive elimination of opposition. Competitive values serve the growth and maintenance of institutions, organizations and some individuals but not the total community. Institutional and organizational values are likely to be assumed, defended and not questioned by those who have most to gain by those values even if it cost them their other values such as freedom and integrity. These entrenched organizations and institutions can be successfully changed only by organized movement." The author also notes that the game has been played by all sorts of groups: industrial factories, office personnel, church hierarchies, religious congregations, local communities, unions, schools, school systems, police-community relations groups, and so forth. (REH)

Cost: $.75

Producer: Simulation Sharing Service, 4740 Shadow Wood Drive, Jackson, MS 39211

SHIPWRECK AND OTHER GOVERNMENT GAMES

The Editors of Scholastic SEARCH magazine

Playing Data
Copyright: 1974
Age Level: grades 5-9
Number of Players: class
Playing Time: each game 3-5 hours
Packaging: 28 page manual

Description: Shipwreck, You Be the Mayor, On Trial and *Curfew* are the four games included in this simply and clearly written manual for students. *Shipwreck* involves the building of a new society by players who imagine themselves shipwrecked on an island during a time of worldwide devastation. They have seven days or periods to make seven important decisions–whether or not to have a leader, what jobs must be done, how to divide land and property, what to do about money, how to deal with dissenters, what laws to make and enforce, and what to do about another group who arrives and wants half the island. Short dialogues in the text present each problem.

You Be the Mayor is a city problem-solving game. As mayor and townspeople students must decide whether or not to turn a busy shopping street into a mall, what to do about bus drivers on strike, and whether or not to spend fifteen million dollars on fixing the baseball stadium. Views of important townspeople are described in the text and some students are given opportunities to enact their roles. Students also meet in groups to decide on problem-solving plans. The final decision about each problem is made by the student or students playing the mayor.

On Trial is a courtroom simulation game designed to give students experience as judge, jurors, attorneys, plaintiff, landlord, witnesses, and neighbors involved in a civil case about an accident on unsafe stairs. a large portion of the trial and a description of the witnesses are included in the text. As students enact the entire trial including the jury's decision-making process, they make up their own parts where the manual suggests, and learn to distinguish opinions from facts in the testimony.

Curfew is a community action game involving teenagers who are angry about their city's curfew laws for young people. Playing the roles of these teenagers, some which are described, students read the dialogues in the text, write letters to the mayor, and write newspaper ads. They also role play attempts to convince people to sign their petition, put on a radio talk show, and enact a city council meeting in which teenagers, supporters of the law, and the six council members debate the issue. Eventually, the city council members vote about whether or not to abolish curfew. The game involves individual, limited participation, and group activities. (TM)

Note: For more on this collection, see the essay on social studies simulations by Michael J. Rockler.

Cost: $9.95 (#381)

Producer: Scholastic Book Services, 50 W 44th St., New York, NY

SIMSOC: SIMULATED SOCIETY

William A. Gamson, University of Michigan

Playing Data
Copyright: 1969, 1972, 1978
Age Level: grade 12, college
Number of Players: 15-60 (40 ideal) in 3 or 4 regions
Playing Time: 5-10 sessions (6 average for fruitful run) of 50-90 minutes (75-minute sessions ideal), with 1 session for debriefing
Preparation Time: several hours, usually 1 between each session

Description: According to the designer, "*SIMSOC* attempts to create a situation in which the student must actively question the nature of the social order and examine the processes of social conflict and social control." Now in its third, revised edition, *SIMSOC* incorporates economic, social, and political institutions and dynamics. Participants begin with individually defined goals, with no structure, with marked inequalities in regional resources and individual privilege, and with certain barriers to communication. From this starting point, they seek to create or perpetuate a structure that will further their goals, which may well go through changes in the process. (AC)

Note: For an in-depth description and discussion of this simulation, see the essay on *SIMSOC* by Richard L. Dukes.

Cost: participant manuals $5.95 each, instructor manual gratis with a set of participant manuals

Producer: Free Press, 866 Third Avenue, New York, NY 10022

SIMULATING SOCIAL CONFLICT

Sociological Resources for the Social Studies

Playing Data
Copyright: 1971
Age Level: high school
Number of Players: class in pairs
Playing Time: 7 class periods

Description: *Simulating Social Conflict* is a series of simulations based on variations of the Prisoner's Dilemma game. It is a seven-day instructional package, carefully worked out by a team sponsored by the American Sociological Association. The first variation of the game is called "The Dilemma of the Tribes," where the variations are win-lose, win-win, and lose-lose. Strategies can be discovered. The second half of the curriculum in this is called "Resources and Arms." Throughout, students play in pairs so that everybody in the class is involved in making his own decision with one other person. (REH)

Comment: The teacher's manual is excellent for these simulations, in that the focus is very careful on setting up these simulation episodes to enable students to draw their own conclusions about cooperation and conflict, to gather evidence as sociologists do, and to reason about evidence. In most situations, several versions of the games are played so that students can explore different choices and rules. (REH)

Note: For more on these simulations, see the essay on social studies simulations by Michael J. Rockler.

Cost: $10.20 set of ten, $3.04 instructor's guide

Producer: Allyn and Bacon, 470 Atlantic Avenue, Boston, MA 02210

SMALL TOWN

David Yount and Paul DeKock

Playing Data
Copyright: 1977
Age Level: grades 7-12
Number of Players: 25-35
Playing Time: 15 1-hour sessions
Preparation Time: 3 hours (est.)
Packaging: 24-page instruction booklet

Description: This simulation, according to the authors, "provides students with a literary experience that enables them to better understand the small town values system which is such an important part of our national character." A significant portion of time is spent dramatising portions of Edgar Lee Masters' *Spoon River Anthology* and Thorton Wilder's *Our Town.* The players are assigned roles as residents of an imaginary turn-of-the-century town named Gentryville. These assigned roles are very sketchy and are developed by the players in conjunction with their growing knowledge of the two works mentioned above. The simulation is climaxed by a Memorial Day ceremony during which each player reads the epitaph for the character he has invented (as with Masters' device in *Spoon River.*) The players debrief by holding panel discussions on the proposition that the federal government should deliberately break up large metropolitan areas and disperse the population to municipalities with populations of 20,000 persons or less. (DCD)

Cost: $14.00

Producer: Interact, P.O. Box 262, Lakeside, CA 92040

SOUTHERN MOUNTAINEER

Michael J. Raymond

Playing Data
Copyright: 1976
Age Level: grades 7-9
Prerequisite Skills: 8th grade reading level as measured by Flesch Scale
Number of Players: 22-40
Playing Time: five 55-minute periods
Preparation Time: 2 hours
Packaging: 83-page booklet and cards

Description: This simulation is designed to give students, through role playing, an understanding of and appreciation for rural family life in the southern mountains of the United States in the 1920s. Participants are introduced to the daily and seasonal tasks these people performed, as well as the songs, dances, and recreational activities that were an integral part of southern mountain life. The first of the five game sessions simulates family activity in the fall of 1927. The class is divided into two groups, each representing the Jason Fletcher family. The students are introduced to the roles of family members and given tasks to do according to their roles. These tasks are detailed on task cards included with the game, and include activities like tending the moonshine still, shucking corn, and making quilts. Also during this first day a banker offers to buy the family farm—an offer the family discusses. On days two and three, students change tasks to reflect the changing of the seasons and may change roles to vary the tasks with which they may become acquainted. On the third day two young lovers become engaged, and a new cabin must be built for them, and wedding invitations must be prepared. The wedding occurs during the fourth session and participants perform and sing period songs and dances. The wedding ends the role-playing phase. Day five is spent in debriefing. (DCD)

Cost: $5.95

Producer: SAGA, 4833 Greentree Road, Lebanon, OH 45036

SPECTRUM

Andrea Jane Richardson Taylor

Playing Data
Copyright: 1975
Age Level: high school
Number of Players: 30-36 in 3 groups
Playing Time: 5-6 class periods or one 3-hour seminar
Packaging: 6-page teacher guide and 7-page student guide

Description: *Spectrum* is a simulation game of a high school at which ethnic group conflict is occurring at the time of a student government election. Eleven areas of controversy are described in the manual, including discrimination in clubs, segregated restrooms, and the predominance of whites among hall monitors. The first day of the game, students read student guides and choose roles as Wasps, Jews, Blacks, Chicanos, City Whites, Native Americans, and Oriental Americans. The thirty-six role descriptions given indicate role in the election and political viewpoint—there are eight radicals advocating drastic change in school policies, twelve moderates, and sixteen undecided students at the beginning of the game. The second day, moderate and radical students hold party meetings to develop platforms and slogans while undecided students sit in, trying to make up their minds about their political affiliations and beliefs. Campaigning occurs the next two days, culminating in a rally. Meanwhile, at home each student has been keeping a journal from the point of view of his or her role, and in class, rumor cards (e.g., Rosa Ramirez is flunking English. How can we have a Chicano as secretary when Chicanos are illiterate?) have been periodically distributed to stir up the controversy. The fifth day, when campaigning ends, students participate in two elections, voting in one as they themselves would vote, and in the other as they believe their role would vote. The game ends with a debriefing session for students to discuss questions in the guide related to their feelings and prejudices, the realism of their roles, and the difference between the two election outcomes. This game may also be played as a three-hour workshop; this time sequence is described in the text. (TM)

Cost: $14.00

Producer: Interact, P.O. Box 262, Lakeside, CA 92040

STARPOWER

R. Garry Shirts

Playing Data
Copyright: 1969
Age Level: any age group with an adventurous teacher
Prerequisite Skills: reading, grade 0 if instructor explains; math, grade 0 if instructor explains
Number of Players: 18-36
Playing Time: 1-2 periods of 50 minutes
Preparation Time: 20 minutes

Description: *Starpower* creates a political process in a classroom. The primary goal of that process is to provide students with the opportunity to generate a small social system and then, having generated it, to explore and discover how a social system shapes human behavior and human decision-making. In the process of discovering some of the truths and half-truths of organizations and bureaucracies, students also discover things about themselves and their classmates.

Starpower begins with a distribution of resources. In the first round, the distribution takes place through a random mechanism; chips of five different values are placed in three paper bags. Each player selects five chips from one of the bags. The game kit includes a point total chart that establishes special values for the various colors and for certain color combinations. Typically, collecting a color group will lead to a rise in a player's point value total.

After the initial distribution, the *Starpower* scenario calls for a trading session. In this session students attempt to improve their scores by exchanging pieces with other players. The exchanges are governed by a trading rule that prohibits exchanges of equal value. The motive for negotiations grows from a desire to improve one's point total equivalent to improving other "financial" situations.

At the end of the negotiating session, players are segregated into three groups. Players with the highest point totals become squares. The middle third of the playing group become circles and the bottom third of the class become identified as triangles.

Once group identifications are made students are instructed by the game director to seat themselves in their groups. The physical change in the seating arrangement forces immediate group identification. The final step in the first round of the game sequence is called a bonus trading session. During the bonus session each team receives three bonus chips worth twenty points each. Since the recipient group usually has from seven to ten members, the bonus chips also present them with a distribution problem. The solution to these problems, like the initial trading session, is governed by certain rules. One of the rules states that the distribution of chips must be done through unanimous consent of the group. The alternative is the forfeiture of any chips that are not distributed by unanimous vote. As far as the game is concerned, the bonus distribution session operates as an introduction to group decision-making and to the problem of evolving a set of distribution principles that allow some to receive the premium and others to forego it. By itself, it is a model for a great display of human political activity.

The principles each group evolves vary, interestingly enough, with their designated place in the social system. In two out of three test simulations, squares distributed the bonus resources to members of their group with the lowest scores. The rationale was simple. Their response to the situation was automatic. They were top dogs. Several of their number were in danger of being dislodged by members of the circle group. The instinct of the group was to protect its own. In the third test group, a similar sentiment was expressed, but several very charismatic students prevailed on the group to distribute the bonus resources through a chance mechanism. Under time pressure, they finally decided on this solution to the distribution problem. In one group two of the three bonus chips were given to the players with the highest scores. The third chip was given to the player with the fewest points. Observation revealed that only two players in the group were within range of changing their group membership. Observation also revealed a certain miscalculation on the part of Circles who failed to take into account the defensive strategy Squares adopted in an effort to protect those with low point totals who belong to their group. The result in the first group was no change in social status for any players.

In the second test group the game progressed differently. Students in the Circle group managed to promote one of their number into the Squares. Their success in doing this produced two kinds of satisfactions: one for having turned a member of the high and mighty group out and another for having made one of their own group part of the elite group. None of the players seemed to realize that the satisfaction arose from a false result. Despite all the effort and energy expended, there had been merely an exchange of one individual between the elite and subordinate groups. There were still the same number of losers and winners. Efforts to improve their lot failed.

In a third test group some rather interesting events took place around the issue of mobility. Tommy, a boy with high charisma and someone who typically used informal networks among his peers to support a self-image the formal hierarchy of the school did not. During the first bonus session, he managed to convince several key members of the Squares group to "let him in." Essentially, he was using the network of friendships and alliances he had already formulated to circumvent the rules of the game. When the game director prohibited his upward mobility on this basis, Tommy returned to the Circles and negotiated a loan of two bonus chips so that he could enter the Squares legally. In one of the other test groups something on the same order occurred. Dillon, probably the brightest player, and certainly one of the most powerful students in the school, succeeded in badgering a classmate into giving a gold chip in exchange for a blue one. Equivalent to receiving a hundred dollars on the dollar, the exchange defies the game's logic and clearly indicates the possibilities of charisma.

Certain rather devilish adjustments are made in the *Starpower* sequence during the second round. The game director stacks the deck. He does this by enriching one of the three resource bags; i.e., he places all the gold chips in one bag and allows only the Squares to select their chips from the enriched source. The effect is obvious. The rich get richer relative to the other players while, at the same time, the poor groups have the illusion that they are improving their lot.

A second bonus session follows the second distribution session. In all three test groups, Circles discovered that the small chances for social mobility they enjoyed in the first go around had been revoked. The divisions between rich and poor had grown so wide that a twenty, forty or even sixty point bonus simply did not bridge the gap.

At the end of the second bonus point session, the game manual instructs the director to congratulate the Squares on their achievement, implying thereby that they deserve their Starpower. In fact, few, if any, students even in the Squares feel they have achieved their station because of ability, the exceptions being those students who have borrowed and badgered their way into the elite group. Most Squares recognize they are on top and simply wish to remain there.

To hasten the game to its point and to its conclusion, the game designers bestow political power on the Squares. That power comes in several forms. First, the Squares receive the right to legislate. They can change any rule in the game. Second, the director announces that the three students with the highest numbers will eventually be declared winners. A third rule permits a majority of any group to eliminate a minority by a simple vote. These two rules eventually force the elite group to turn on itself. By contrast, the first rule permits that group to prey on the subordinated Circles and Triangles.

In the one test group, the Squares changed the distribution rule for the bonus chips. Instead of three to a group, they ordered all bonus chips distributed to the Squares. The same test group also legislated against outcries from the Circles, but only produced shrugs and acceptance among the Triangles. The Triangles in all three test groups had thrown in the towel. They did not care. Unable to improve their condition, they became apathetic.

In a second test group, the Circles attempted to pool their resources bestowing on two of their number the point totals of all others. When the Squares learned of this arrangement, they immediately met to prohibit it. The resulting frustration experienced by the Circles produced a great deal of anger and animosity between the parties. Under this pressure, the class voted to end the game. Even the Squares who had made the laws supported the termination of play, trying thereby to transfer to the game blame for such a miserable outcome.

Similarly, in the other test situations, Squares used their power to maintain the status quo. In one game, the Squares legislated a direct tax on all transactions taking place among Circles and Triangles. In another, they attempted to broaden the number of winners from three to a majority of the Squares group. Since the group had four girls and four boys, the four girls joined and stripped the masculine minority not only of its membership in the group but its resources as well.

In one debriefing session after the game, one teacher likened their certain aspects of the *Starpower* experience to the bind the middle class now finds itself facing. Although a majority of the seventh graders did not understand the connection, several did. In general, those who were able to connect play with something in the real world reacted positively to the game. Others, generally those who saw *Starpower* disturbing the fragile balance of relations among students, regarded the experience more cryptically, "It was a waste of time." All, however, felt strongly about the experience, and all, even in the slowest group, wrote voluminously about their frustration, anger and disappointment. Many insisted quite religiously that the world was not organized around wealth and its distribution. (Robert Mundt)

Comment: On the whole, *Starpower* provides school age children with a rather extraordinary experience. But the experience is not without its drawbacks, some traceable to the game, others to the special environment of the school in which the game is played.

Starpower takes a minimum of five hours to play. It takes a minimum of ten class hours to reasonably capture the essential experience the simulation provides. With some alterations, the game could be redesigned for use in a semester or year-long course. Played at the rate of one to two hours a week over an extended period, students would be able to explore the nuances of human relations and the subtleties of social system dynamics on human behavior. In its present form, *Starpower* fits only very awkwardly into the forty-minute time segments permitted the test groups at the test site. Ideally, one might play a segment of each round each day and allow the time in between to be utilized for alliance building, trading, negotiating, strategic planning, and to coordinate cooperative efforts. Under most school conditions *Starpower* will necessarily seem somewhat forced. Conclusions will be reached too quickly and will engender experience with an accompanying loss in subtlety.

Because of time compression, what is actually learned from the game, to a large extent, depends on the training of the game director and especially on his ability to lead students to insights into the social process they generate.

If the game were played over an extended period, it might be possible to reduce this dependency and possible for the game mechanism itself to produce the desired depth in understanding. The difference in duration of play may at first appear to be stylistic issues, but this reviewer sees it as a substantial issue of power and dominance.

During several of the test situations, another weakness became apparent. As presently designed, *Starpower* moves too quickly into social class confrontation. Soon into the second round, group is pitted against group. In more realistic circumstances, this pitting of forces appears to happen gradually. The elite group in society has the opportunity to buy off the opposition. The present edition of *Starpower* makes no provision in its structure to teach the elite group the art of governing. Some future edition might improve in this direction.

Finally, this reviewer discovered another structural defect in the game. As presently designed, the recommended value system for the "financial" chips was felt to be oversimple and self-defeating. In every case, one person's loss was another's gain. In no case was it possible for both traders to improve their position relative to others by trading. In other words, students traded because they were supposed to. Real impetus to trade might be achieved by some rather minor alterations in the rules, by assigning higher point totals to chip color combinations, or by allowing students to enter their social class groups on some other basis than pure chance.

Notwithstanding these reservations, *Starpower* stands as one of the outstanding simulations in the field. (From the Robert A. Taft Institute of Government Study on Games and Simulations in Government, Politics, and Economics by Robert Mundt, University of North Carolina at Charlotte.)

Comment: There are two minor negative aspects to the game. To achieve full impact, the game should be played in a continuous session of at least eighty to one hundred minutes, excluding debriefing. Without this continuous session, some of the true feeling of the game is lost. Another point to be avoided is allowing any of the players to gain second-hand information about the simulation. This usually means you can use it with only one class at a time and possibly only once every year. One player with a little knowledge could destroy the impact for which the game strives.

This simulation lends itself extremely well to a variety of debriefing sessions. As the result of the high level of frustration felt by most participants there is seldom any difficulty in creating group interaction. The most usual pattern for the group is to question their own actions in light of the society created during the play. An interesting concept to pursue is getting the group to restructure the simulation in a manner that every participant would call fair. (Robert C. Bilek, Curriculum Associate in Social Studies, leader of teachers' workshops, Salinas, CA)

Note: For more on this simulation, see the essay on simulations of developing a society by Frank P. Diulus.

Cost: do-it-yourself kit $3.00, kit for 18-35 students $30.00

Producer: Simile II, P.O. Box 910, Del Mar, CA 92014

SYSTEM

L. Wayne Bryan

Playing Data
Copyright: 1972
Age level: grade 9-college, adult
Number of Players: 25 or more
Playing Time: 1-2 hours
Preparation Time: 1/2 hour
Supplementary Material: administrator makes trading cards and posters, buys M&Ms for rewards

Description: Everyone begins with four cards each of Control, Freedom, Dollars. They are told to trade these cards with each other. After first trading around, values become apparent. Those with most Controls make a rule for the next round, those with most Freedom are set free (you can do whatever you wish), everyone is given one M&M for each Dollar. Play proceeds through four such rounds, altered only by the rules given by the Control people. In rounds three and four, players may invest in Machines (cards), which can augment their wealth. After round four, everyone converts to Dollars and players with largest wealth are given Giant Rewards (candy bars). (Author)

Comment: *System* is designed to help players understand the use and abuse of power and the creative use of freedom, and to help players understand people who are disenfranchised, not in control of their own lives. Concerning the game experience, the author writes, "Players create their own history and have to live with it. They continue to balance the forces of personal and outside control on their lives. Each affects the lives of others."

The desired game outcomes for each player are self-chosen. Satisfaction is, of course, another matter. Some players get to eat lots of M&Ms, but the crux of the matter is in the interplay of tension between power and freedom. The author wants players through the game experience to see in themselves the possibility of being corrupted by power, the possibilities of creative use of freedom, the bind of powerlessness. Considerable postgame discussion is essential to deal with the strong feelings that play can generate and "to translate these feelings into learning."

System has been used to introduce discussions on power in philosophy and education classes and in church groups to focus on the situation of disenfranchised people. (AC)

Cost: $3.00 administrator's booklet

Producer: UCCM Games, c/o L. Wayne Bryan, 4215 Bethel Church Road, F-1, Columbia, SC 29206

TIME CAPSULE

Don Eells

Playing Data
Copyright: 1978
Age Level: grades 9-12
Number of Players: 25-45 in groups of 4-6
Playing Time: 19 1-hour sessions
Preparation Time: 2 hours (est.)
Packaging: 32-page instruction booklet

Description: Although *Time Capsule* is designed to occupy at least twenty-five students for nineteen class periods it, like all of this producer's games, is easily adaptable to larger or smaller groups and less liberal time allowances. If necessary, it can probably be played by as few as ten in no more than five one-hour sessions. Three students must assume the major roles of coordinator, interviewer, and photographer, and in the smallest groups the remainder of the students may simply be assigned to "cultural artifacts groups." With larger groups, players may also be assigned to the roles of geographer, burial engineer, pack supervisor, document chief, press agent, protocol officer, and investment broker. Players assigned to the cultural artifacts groups (Or "CAGs") may be assigned to the roles of cultural artifacts chief, manuscript editor, or futurist.

The procedures of this game involve selecting and/or manufacturing items to be placed in a time capsule. "Students begin," in the author's words, "by considering artifacts as clues to culture and lifestyle in the past. Then after broadly categorizing American culture, students volunteer to bring or prepare representative artifacts and documents for inclusion in a class-prepared Time Capsule. Interaction, packed with growing insight into life in America today, becomes more and more intense as students debate what items should be included or rejected for the Time Capsule. Eventually, the Time Capsule is filled, sealed, and buried with directions left for its discovery by another social studies class in the future."

Each CAG selects an item or items for inclusion in the capsule by considering the questions "What does this item tell us about American life?" and, "Does the artifact accurately depict part of Americana or is it misleading in some way?" A cultural artifacts chief supervises the activities of each CAG. A futurist in each group makes predictions about the future. The time capsule coordinator helps the teacher organize the CAGs. A student photographer photographs the work of the CAGs and takes and submits photographs that portray a cross-section of American culture. An interviewer is assigned to tape record interviews with members of the CAGs.

When all the work is done, the selected artifacts are put in plastic bags and stored in the capsule; the school administration is notified of these proceedings and given a map showing where the capsule is to be buried and a letter from the time capsule coordinator to the future students at that school, informing them of the existence of the capsule; and finally, the artifacts are buried. (DCD)

Cost: $14.00

Producer: Interact, P.O. Box 262, Lakeside, CA 92040

TOGETHER

Jim Deacove

Playing Data
Copyright: 1974
Age Level: junior high school
Number of Players: 2-36
Playing Time: 1 hour
Packaging: professionally packaged game with problem and solution cards, colored sticks, and instructions

Description: *Together* is a simple classroom discussion game that emphasizes cooperation. At the beginning, players draw sticks (like drawing lots) from the teacher's hand. The sticks are differentiated by colored tips, and are used to divide the class randomly into tribes, states, nations, or whatever designated groups the teacher chooses. Each player then takes either a problem or solution card, with about twice as many solution cards being distributed as problem cards. The groups must solve their problems by setting priorities for which problems to solve first and allocating solution cards for this purpose. A group with the problem of air pollution may allocate money and education cards to the solution of this problem. The group would then have to explain how money and education can solve the problem and must still have the resources to solve any other problems the group has. The game is designed so that the groups will often lack a solution to their problems, and in these cases the teacher is supposed to suggest cooperation among groups. The game ends when all the problems are solved. (DCD)

Cost: about $4.50. Send for current price list.

Producer: Family Pastimes, R.R. 4, Perth, Ontario, Canada K7H 3C6

TRADERS ARRIVE ON THE AFRICAN SCENE

Mary E. Haas

Playing Data
Copyright: 1976
Age Level: high school, college
Number of Players: 26-36
Playing Time: 3 phases of 2 hours (estimated) each
Preparation Time: 1 hour
Packaging: 36-page instruction booklet

Description: During the first stage, all participants are assigned one of thirty-one roles, as traders or as members of three distinct tribes. The players in each group discuss their roles, both as individuals and as members of a social unit. During phase one, the tribesmen also anticipate the arrival of the traders and discuss how they will react to them.

Stages two and three follow a set scenario of interaction between the tribes and the traders, and intertribal conflict. During these phases the traders have the goal of acquiring native products by trading western products for them.

Very much factual material is presented during the course of the game, and the focus of the game is the way in which the tribal societies were changed by their interaction with the Western traders. Although the action of the game is completely described in the instructions, players may, to some extent, affect events by their own actions (such as slowing or accelerating the amount of trading).

There are no winners or losers other than those previously described by historical events. The point of the game is for participants to experience the impact of those events. (DCD)

Cost: $12.00

Producer: Social Studies School Service, 10,000 Culver Blvd., Culver City, CA 90230

THE TWENTY-FIRST YEAR

Playing Data
Copyright: 1979
Age Level: high school-adult
Number of Players: 1 or more
Playing Time: 60 minutes
Packaging: professionally packaged, 12-page unbound folder of game materials and instructions is part of a "World Hunger/Global Development Education Kit For Schools"

Description: The producer categorizes this as an exercise "in clarifying priorities in global development." Working alone, one or more participants assume responsibility for choosing projects to meet basic human needs in a fictitious developing nation. The deadline for the completion of all projects is the year 2000, and the game is played in three rounds which each represent seven years. The players must choose development projects from a list which includes proposals for the construction of hospitals, a plan for the redistribution of land, and a scheme for the development of new agricultural products. A scoring mechanism is included with the game materials for judging the player's development decisions and the effect those decisions would have if implemented. The game ends when the third round ends. A set of discussion questions is included with the game materials. (DCD)

Cost: $5.50

Producer: American Friends Service Committee World Hunger/Global Development Project, 15 Rutherford Place, New York, NY 10003

UHURU

Northern Vision Services

Playing Data
Copyright: 1974
Age Level: high school, college, continuing education
Number of Players: 15 or more in 5 groups
Playing Time: 2 hours or more
Packaging: 12-page manual

Description: *Uhuru* is designed to help players understand the problems of a third world country and the power politics involved in stabilizing and developing it. *Uhuru* is a fictionalized African nation now in the hands of a large black majority after centuries of European rule. Students divide into five groups, each group trying to protect its own interest and secure as much wealth as possible. The Bantu and Kollani tribes, one communal and one aristocratic, are in serious conflict. The white minority, which possesses 80% of the wealth, is threatened by both tribes; and the two superpowers, Milk 'N Honey and Redperil, which possess equal power and resources, are intensely interested in gaining control of the strategic Wabenzi peninsula, as well as profiting from the uranium deposits of Uhuru. At the beginning of the game, each group reads its profile, chooses a leader, plans its initial strategy, writes and reads its Independence Day speech, and fills out a form listing its objectives. Then negotiations begin. Groups meet and attempt to reach agreements about the development and sale of uranium deposits, the use of the Wabenzi peninsula, the creation of a political system, and the formation of policies regarding the future development of Uhuru. When group objectives are reached, when open conflict breaks out, or when groups are obviously stuck and unable to reach an agreement, the game ends. Players then discuss their success with reaching their objectives, the strengths and weaknesses of their groups, and the realism of the game. (TM)

Cost: $4.00

Producer: Northern Vision Services, 77 Swanwick Ave., Toronto, Canada

UP CASTE DOWN CASTE

Mary E. Haas

Playing Data
Copyright: 1977
Age Level: grade 5-12
Number of Players: 2-8
Playing Time: 3-4 hours
Packaging: kit with playing map, cards, die, instructions

Description: The purpose of *Up Caste Down Caste* is to teach how a set of beliefs such as Hinduism influences a society. The object is to move upward through the caste system over a period of several lifetimes and to eventually attain union with Brahman. Players begin by throwing the dice to determine their caste. They then select the appropriate role cards, which define their jobs, salaries, and expenses. Moving around the board by throwing of the die, they attempt to keep out of debt and attain merits by landing on the right squares. giving birth to sons, marrying off their daughters, and fulfilling their religious duties. Different squares require them to follow instructions on different cards—family cards, special cards and debtor's cards, which may mean paying or receiving money, being banished from a caste, adding to their family, gaining merits. Each time players complete their paths around the board they recieve their salary and pay expenses determined by their debts and their number of children. When players "die," they determine their caste for their next lifetime on the basis of their merits, select a new role cards, and continue to play.

The author acknowledges that the game is an oversimplification of Hindu religion and society but believes that it nonetheless enables players to experience some of the problems encountered by Hindus in India. An examination of the special cards for each caste reveals one of the basic problems of Hinduism encountered in the game—the untouchables have the fewest opportunities to earn money and to obtain merits. (TM)

Cost: $15.00

Producer: Social Studies School Service, 10,000 Culver Blvd., Culver City, CA 90230

THE WEEKLY NEWS GAME

Donald Criswell

Playing Data
Age Level: grade 7 to adult
Number of Players: 10
Playing Time: 30 minutes

Description: This is current events game based on the format of "The College Bowl" television show. Two panels of students compete to answer the moderator's questions. The game comes in thirty-three weekly installments during the school year. Each edition asks one hundred questions taken from *Time, Newsweek, People, U.S. News and World Report* and *T.V. Guide.* An electric Quiz-a-matic game machine is available to facilitate the game. Sample question: "Poet Sylvia Plath wrote an autobiographical novel, 'The Bell Jar' published in 1971 after her 1963 suicide. Give the title of the book which her mother will be publishing in December as an answer to 'The Bell Jar.' " (JG)

Comment: Clean format and challenging questions make this an excellent instructional game for junior and senior high school current events class work. (JG)

Cost: $35 per year (33 issues)

Producer: Creative Educational Services, P.O. Box 30501, Santa Barbara, CA 93105

THE WORLD WITHOUT WAR GAME

Beverly Herbert, Anne Stadler, and The World Without War Council of Greater Seattle

Playing Data
Copyright: 1971
Age Level: high school, adults
Number of Players: 30-40 (6 staff members for retreat)
Playing Time: 2 days (weekend retreat) or 5-10 1-hour classes
Preparation Time: 5 hours
Special Equipment: tape recorder or record player, 16 mm projector or screen, movie such as *The War Game* or *Hiroshima-Nagasaki*
Packaging: game box including 138-page manual, the book *To End War* and an adaptation for high schools

Description: *The World Without War Game,* designed as a weekend retreat experience for adults but adaptable to high school settings, acquaints participants with the dangers of war, introduces them to alternative ways of resolving conflict, and motivates them to work for a world without war. The high school adaptation involves a one-act drama, the war game itself, and a critique—all included in the adult game. The adult weekend retreat version begins with a one-act drama about assumptions concerning war and peace and a group discussion. For the second session, participants divide into two teams, each with a common goal such as access to the front door, and set land boundaries. Supreme commanders, soldiers, negotiators, and guards are chosen. Teams formulate strategies, the director announces fate events, and soldiers fight by arm wrestling or stare-down until one team wins, a compromise is reached, or a team uses its "ultimate weapon." Other sessions of the weekend involve a film such as *The War Game* or the shorter and less expensive *Hiroshima-Nagasaki*; a three-part Huntley-Brinkley presentation and discussion about weapons and the threat of war, a script presentation on active peace making, and a project-planning session for determining what can be done by individuals to prevent war. All scripts are included in the manual. (TM)

Cost: $10.00, school version $10.25

Producer: World Without War Publications, 67 E. Madison, Suite 1417, Chicago, IL 60603

See also

Agency COMMUNICATION
City Planning URBAN
Cycle SELF-DEVELOPMENT
Equity and Efficiency in Public Policy DOMESTIC POLITICS
The Good Society Exericse INTERNATIONAL RELATIONS
The Grain Dráin ECONOMICS
Much Ado About Marbles COMMUNITY ISSUES
The Road Game COMMUNICATION
Talking Rocks LANGUAGE
They Shoot Marbles, Don't They? FRAME GAMES
COMMUNITY ISSUES
COMPUTER SIMULATIONS/SOCIAL STUDIES
ECONOMICS
FUTURES
GEOGRAPHY
HUMAN SERVICES
LEGAL SYSTEM
SELF-DEVELOPMENT
URBAN

URBAN

ACRES

George Pidot, Temple University; John Somer, University of Texas at Dallas

Playing Data
Copyright: 1973
Age Level: college, adult
Number of Players: 6-30 in teams of 3
Playing Time: 3-60 hours
Special Equipment: computer time-sharing system

Description: *ACRES: Area Community Real Estate Simulation* is a medium-size computer-based simulation designed to give participants insight into the dynamics of land development and environmental impact in an urban area. Teams representing business, residential, and government sectors play the game on a 15 x 15 grid of parcels.

Residence owners may build four levels of residential development and must choose where to shop and work, how to get recreation, and how to deal with pollution. Business teams, which may run industries, retail stores, service outlets, and offices, make decisions about purchases, pricing, sales volume, and hiring. These eight teams jointly choose a government, elected every two rounds, to zone and assess property, collect taxes and other fees voted on by other participants, and construct and maintain public facilities such as highways, utilities, schools, a dump, municipal buildings, parks, and so on. Industries pollute air and water (destroying recreational facilities in the process; possibly) and government policies may try to deal with this.

The computer program simulates the economy and environs. At the end of each round of play (one year), participants are given summary accounts of that round's developmental impacts on the environment, business, building, and government. The game manual includes 18 scenarios. (AC)

Note: For more on this simulation, see the essay on urban gaming simulation by Mary Joyce Hasell.

Cost: instructor's manual $6.00, user's manual $4.00

Producer: Project COMPUTE, Dartmouth College, Hanover, NH 03755

APEX

COMEX Research Project and the Environmental Simultation Laboratory

Playing Data
Copyright: 1969
Number of Players: 12-60 in 16-23 teams
Playing Time: 2 hours
Special Equipment Required: IBM 1130 computer

Description: This gaming simulation has been designed to serve as a training tool for those concerned with air pollution decision making in a simulated metropolitan environment.

Apex is an extensively modified version of *METRO*. The Lansing data base is still employed, but the number of analysis areas has been reduced from 44 to 29. The School Department is simulated in *APEX* (i.e., players no longer make school decisions), but the actions of six local industrialists and a county air pollution control officer (APCO) become player-controlled roles. The APCO is the only player concerned exclusively by the air pollution conditions of the simulated area. The other players are concerned with air pollution only insofar as the costs of controlling pollution fall on them, the impact of regulations effect them, or the simulated electorate become vocal in their opposition to undesirable pollution levels.

The jurisdictional divisions and player responsibilities have been modified so that in *APEX* there is a central city, a suburb, and two townships. The central city jurisdiction is operated by players and it provides all general government services usually associated with local government, except for public health and welfare and air pollution control which are county functions. The county includes the entire metropolitan area, and it provides some services and facilities to the suburb and townships, which are simulated jurisdictions (i.e., controlled by the computer). Five gamed politicians serve as the county decision-making body. One politician is elected from each of the three simulated jurisdictions and two are elected from the central city.

Each industrial team represents one firm in the area that is a major potential air polluter (power plant, cement plant, foundry, copper smelter, rendering plant, and pulp mill). The remainder of the industrial sector is computer simulated.

As in *METRO*, the players who are land developers buy and sell land, develop land, and sell developed land in response to a simulated market. Simulated developers take up any remaining demand which gamed developers do not meet.

The APCO role is similar to that of the planner, in that his success depends on his ability to persuade the other local decision makers of the worth of his programs. He may gain authority from the politician to exercise control over air pollution emissions. The APCO may pay to monitor air quality and specific emission sources and/or set and enforce emission standards. He must receive local funds to match federal money that is made available through the game director's actions.

In addition to the conventional concerns of the politician (reelection, budget, tax rate, and such), the effect of air pollution on residents, property, and property values, as well as any air pollution regulation effects on employment and tax base, become a concern to the politicians.

The industrialist's chief objective is to operate his plant as efficiently as possible, and he does this by setting his production level and sales price, by estimating costs and sales volume, and possibly purchasing more equipment and/or expanding his plant. As part of his production process, he may use fuels that may cause air pollution. A switch to more expensive fuels may reduce pollution or a pollution reduction might be brought about through an expenditure on special pollution control equipment. Industrialists may choose to fight the APCO by commissioning a consultant (i.e., computer) survey of his plant's emissions and taking the case to court. He may also try to influence the politicians through campaign contributions, a threat to close down his operations, or by cash transfers to other players who will oppose the APCO or put pressure on the politicians. (The state-of-the-art in urban gaming models.)

Cost: Unknown

Producer: COMEX, Davidson Conference Center, University of Southern California, Los Angeles, CA 90007

BLACKBERRY FALLS TOWN GOVERNMENT GAME

Jerry Warren, Edmund Jansen, and Anne Knight; University of New Hampshire

Playing Data
Copyright: 1975
Age Level: college, adult
Number of Players: 15-40
Playing Time: 12-18 hours
Preparation Time: 4 hours (est.)
Special Equipment: time-shared computer; BASIC
Packaging: photocopied game materials include player's manual, game master manual, role appendices, and reference appendix

Description: This computer-assisted game simulates the way decisions are made in the imaginary New England town of Blackberry Falls. The town is located at the intersection of the railroad, a river, and an interstate, and most of its 3,500 residents work in an industrial city north of Blackberry Falls. The town is snuggled in a pretty valley and, as with many pretty New Hampshire towns, tourism has replaced a deteriorating row of mills as the major industry.

Players assume the roles of Selectmen and members of the Planning Board, Budget Committee, Conservation Commission, and a local citizen's group, as well as a real estate developer. The simulation is played for a minimum of four three-hour sessions which represent one year of the political process of the town. Issues for discussion are raised first in the local newspaper and are generated either by the game administrator or the computer. The players may base their decisions on the basis of data or reference material supplied by the computer. According to the producer, "Computer programs will supply an analysis of alternative methods of solid waste disposal; provide cost impact figures of housing and industrial development in the town; project the return on investment in the purchasing of land or construction of a building, etc. Also the computer is used to store and retrieve information during and between game sessions." (DCD)

Note: For more on this game, see the essay on urban gaming simulation by Jo Hasell Webb.

Cost: \$17.00 (player's manual may be purchased separately for \$.50, as may the role appendices for \$3.00 and the reference appendix for \$1.00)

Producer: Computer Services, University of New Hampshire, Kingsbury Hall, Durham, NH 03824

BLIGHT

Ronald G. Klietsch

Playing Data
Copyright: 1972
Age Level: grade 9 to graduate school, professional, community groups
Number of Players: 25-40, ideally 30
Playing Time: 8-12 hours in periods or 15-30 minutes
Special Equipment: 1 large room and 4-6 smaller rooms, tape recorder optional
Packaging: instructor's guide, participant materials, inquiry and record forms

Description: *Blight* is a task-force simulation that explores priority-setting in a situation of urban deterioration and reclamation. Participants assume roles as members of six community interest groups (Environmental Council, Citizens' League, Taxpayers and Homeowner Association, Commerce and Industry Board, Central Community Council, Metropolitan Planner's Guild) or of six city-wide task forces (Land Use and Development, Human Resources, Community Concerns, Economic Development, Environmental Quality, Municipal Services).

At the start of the game, interest groups meet to determine the needs of the parts of the community they represent. These interest groups then try to influence one another as well as the various task forces, which are charged with establishing criteria based on values, setting priorities, and developing proposals to deal with urban blight in Zenith City (the game materials provide a list of 130 quality-of-life criteria to work from). The City Council meets to approve or reject these proposals, as the task forces and community groups argue their cases. This process of planning, lobbying, and seeking official sanction is repeated twice more.

Blight is designed to teach participants 23 urban concepts, how task forces operate, how a municipal advisory group functions, and it provides experience with an issue development model and change-action system, in making presentations, in planning issues, and in value management. (AC)

Comment: The important message of this simulation is that conflict is a product of issue creators, not urban problem resolvers and implementors. *Blight* uses the task-force model to illustrate how communities of interest can achieve issue objectives. This model does not stress the exact costs of different proposals, but estimated dollar costs can be added by the instructor or by modifications in the simulation. These factors, plus a fixed revenue-sharing amount, could be used to expand the economics of urban renewal aspect. (REH)

Note: For more on this simulation, see the essay on urban gaming simulation by Mary Joyce Hasell.

Cost: $64.50 plus shipping and handling

Producer: System's Factors, Inc., 1940 Woodland Ave., Duluth, MN 55803

BUILD

J. A. Orlando and A. J. Pennington, Decision Sciences Corp.

Playing Data
Copyright: 1969
Age Level: grade 9-graduate school, management, community groups
Number of Players: 15-25
Playing Time: 4-8 hours in 1-hour periods
Preparation Time: 1-8 hours
Special Equipment: computer: Time-Sharing System/Fortran: source program in Fortran IV presently available on the GE Mark II Time Sharing System

Description: This computer-based game is a general model that can reflect any small community or portion of a large metropolitan area by changing the data base that is loaded at the start of play. A grid board is used to locate some activities within the community. *Build* has roles that are assumed by the players: mayor, city planning and zoning, school, health and welfare, police, national business, local business, community planner, agitator, and the people.

The mayor sets tax rates on three bases, appropriates funds to four departments, borrows funds, and he may differentially tax the four types of national businesses (manufacturing, assembly, warehouse, and office). The City Planning and Zoning Department may build roads, construct terminals, purchase parks, and build playgrounds. The School Board hires teachers, sets salaries for the teachers, offers preschool programs, offers adult education programs, and constructs low, middle, or high level schools. The Health and Welfare Department sets the unemployment payment level, sets the child allowance level, determines the extent of the job-training program, builds health centers and/or health clinics. The police chief hires policemen, sets their salary level, and builds new police stations.

The owner of national businesses may finance a job-training program, specify the number of workers to be trained, hire a specified percent of the required labor force from the community, invest money inside the local community on manufacturing plants, assembly plants, warehouses, or office buildings. The owner of local businesses may invest only in the local community and this must be on either retail or service establishments.

Game inputs and outputs are shown on teletype output that is run on the G.E. Mark II time-sharing system and CDC 6400. Minimum use is made of game manuals and heavy reliance is placed on learning the game by interacting with the teletype which spells out the options available to each team. (The state-of-the-art in urban gaming models.)

Comment: *Build* represents a situation of deterioration of a variety of urban elements—housing, services, and economy—with an emphasis on political, social, psychological, and even cultural aspects and the preservation of community values. The game can be made specific to a given area and can be defined as a flexible role-playing computer game. *Build* is considered one of the easiest of all the computer-based models. It is oriented to allow maximum player interaction with the computer, without much prior computer experience. (GLeC)

Cost: Variable as a function of modifications required.

Producer: Marshall H. Whithed, School of Community Services, Virginia Commonwealth University, Richmond, VA 23284

CITY

D. Vandeportaele and D. Garside

Playing Data
Copyright: 1977
Age Level: college
Special Equipment. computer: Fortran IV
Packaging: Program is described and listed in *Computer Simulation and Modeling,* by Richard Lehman

Description: *City* is a computer program designed to produce the results of establishing priorities on city growth. New priorities and models of growth can be specified and the program rerun, however, the program is not highly interactive. The program simulates 4 10-year periods of urban growth. A grid map is printed showing 1 of up to 9 different land uses for each lot (cell) in the geographic display. Particular city land use patterns can be specified as starting values. Population size and residential density can also be varied as parameters in the model.

Note: This program is designed for individual student use; it should not be confused with a group decision game on urban political economy, also called *City.* (RA)

Cost: *Computer Simulation and Modeling,* by Richard Lehman $19.95

Producer: Halstead Press, Div. of John Wiley and Sons, 605 Third Avenue, New York, NY 10016

CITY I

Envirometrics

Playing Data
Copyright: 1973
Age Level: grade 11-college, graduate school, urban education
Number of Players: 12-36 in 15-20 teams
Playing Time: 2-5 days in time periods of 2 hours
Preparation Time: 1/2 week of training
Special Equipment: 1130 IBM; Fortran IV

Description: *City I* is a computer-assisted simulation game which allows participants to make economic and governmental decisions affecting a hypothetical metropolitan area. It is appropriate for use from high school through graduate school and in adult education. Because it is somewhat complicated, it may work best with more able high school students, but it has been played successfully with high school students from a variety of levels and with varying abilities. It may be played with as few as 9 participants and as many as 50, but it works best with 12 to 36 players. The objective of the game is to view urban decision making as part of a complex decision-making environment which influences the effectiveness of individual decision-making activities.

The players are divided into nine teams which are designated by letter, A through I. These teams are the economic decision makers of the metropolitan area. The teams begin the game with roughly equal economic resources but with varying political resources. The importance of the differences in political resources depends upon the way in which the game administrator chooses to structure elections. Each team is assigned a voting power which depends upon the number of residential units controlled by that team. This means, of course, that voting power will change absolutely and relatively during play of the game as population changes take place. The voting power figure may be used in the conduct of elections, but each team may also simply be assigned one vote by the game director. The latter method makes elections easier to conduct and as a result is often used. Its use, however, obviates the differences in political power among the teams.

The nine economic teams elect a chairman and two city councilmen. The chairman then appoints a finance director, school superintendent, highway commissioner, public works and safety director, and planning and zoning officer.

Following selection of the government sector, play of the round continues. A round represents one year in the development of the city. During the round, the economic teams make various decisions, e.g., whether to borrow or lend money; whether to purchase land from another team; whether to invest money in conservative or speculative securities; whether to bid on land owned by the outside economy; whether to renovate residences or businesses; whether to build new businesses or residences, or to expand ones they already own; whether to change salaries their businesses pay or prices they charge.

The public sector in the meantime must decide what appropriations to grant each department; what tax rate to set; whether to construct, renovate or increase the capacity of roads, terminals, schools, utility plants, or municipal service installations; whether to employ additional public servants, teachers, and such; whether to change salaries; whether to service additional parcels of land. These decisions, for both the public and private sectors, if undesirable results are to be averted, must be based on realistic estimates of the needs of the metropolitan area and on reasonably accurate projections concerning changes which will take place during the next round. If, for example, the government overestimates the tax base and sets tax rates too low, the result will be public indebtedness for the round. If a private decision maker overbuilds in heavy or light industry and cannot obtain workers from the local community, he will find himself employing outside labor at much higher salaries. If public service and public goods outlets are overbuilt, or if prices are set too low, the owner will find himself losing money. Thus players must pay close attention to what they are doing and must balance individual desires with individual and community capabilities.

The computer reports, both those which are posted for all players to see and those which are given to each team and government role only, provide information which helps players make the necessary projections and decide on the optimum levels of activity which they are seeking. For some things, the computer makes a projection for the next round which usually proves to be reasonably accurate; e.g., interest rates on loans from the outside economy, rate of return on conservative and speculative investments, ratio of industrial earnings to normal. On other things, the computer provides data from which estimates may be made, e.g., knowing the property tax base for the next round. For other data, decision makers must seek information from their fellow players; in order to estimate population growth for the next round, it is necessary to know how many residential units each team is planning to construct, and the way to learn this is to ask the teams their plans.

The computer performs several functions in the game. First, it is a bookkeeper which keeps track of all the transactions which are made, stores all relevant data about the metropolitan area, and updates these data in terms of the transactions which take place; it also prints out yearly reports for each team and government role as well as for the area as a whole. Second, the computer simulates the outside economy and outside decision makers. For instance, it sets interest rates on loans from the outside and is the source of required goods and services when the local economy is not able to supply them.

Third, the computer performs routine functions based on the assumption that players would wish to choose the option which is most beneficial to them economically. It assigns workers to jobs, buyers of goods and services to the commercial establishments that supply them, and children to public or private schools. Decisions must be coded into the proper format before being entered into the computer. While this process is fairly simple, it does provide ample opportunity for mistakes which can have both amusing and disastrous results. For instance, many dollar amounts must be rounded off to tens, hundreds, or thousands before being entered. On numerous occasions an economic team has entered prices in hundreds instead of tens and has found itself doing a huge volume of business but still losing money. This may be funny, but it has dire effects not only on the team making the error but on others as well; other teams will lose money, too, and the sales tax collected will decline as the total dollar volume of sales declines.

Comments: In one instance a college faculty member playing the role of highway commissioner incorrectly entered the location of a road he wanted to build and found in the next round that the city had acquired a new superhighway which described a large circle in the midst of an undeveloped portion of the metropolitan area. This problem of errors in coding may be dealt with by placing all responsibility for accuracy on the players, but that is not fair to everybody because many parties may be affected by the errors of one team. It is probably better to have the game director check all decisions carefully before they are entered into the computer. But even that does not provide an ironclad guarantee against mistakes both because game directors make mistakes, too, and because sometimes it is simply impossible to tell from the entry itself what the team intended to do. For instance, did they wish to borrow $50,000 or $500,000 from the outside economy? Short of asking them, the administrator will not be able to tell from the entry itself unless he requires each entry to be written out in real amounts as well as coded amounts and even that will not guard against errors in coordinates for land parcels since there is no way to identify these except by the coordinates. (Parcels are located by two numbers, one giving location on the vertical axis, the other on the horizontal—e.g., 90-26—similar to the way in which one finds the town of Podunk on a road map.)

One final point on the matter of preventing errors. It is probably better to input decisions by means of data cards rather than typing them in at the console of the computer—unless the game director plans to be there when they are typed in. Certain errors can be caught and corrected before the reports are run using either input procedure, but if data cards are used, the director need not be present when the data is fed to the machine.

The major impact of the game is on the cognitive understanding of the individual participants. By giving them direct experience with the dynamic process of urban development, *City I* gives them insights which it is difficult to obtain in any other way—even through field experience. The educational value of the game is high, as is its value as a program diversifier. Skill development may also result from the play of the game. For instance, participants may very well become more able analysts of data and makers of decisions because of the experience they gain in these activities during play of the game. Because of the sophistication of the game, as well as the length of time it takes to play, teachers might wish to structure an entire course, or a major segment of a course, around *City I*. Supplemented by appropriate readings such as Loewenstein's *Urban Studies,* Bollens and Schmandt's *The Metropolis,* or Danielson's *Metropolitan Politics, City I* provides an excellent nucleus for an urban studies course.

The electoral aspect of *City I* is one of the weaker features of the game. The designers do not specify clearly, or even recommend, a method or choice of methods for conducting elections. Because they do not explain a simple way to make use of the voting power differentials provided in the computer reports, game directors often take the easy way out and assign one vote to each team. The computer reports indicate for each team an average turnout and a standard deviation but provide no directions for using these data in conducting elections. A game director can devise a method for doing this, but it is somewhat laborious. If there is no easy way of using these data, there is no reason to provide them and confuse the issue. If there is a fairly simple method of basing elections on them, then the director's manual should explain what that method is.

If you have gotten the impression that *City I* is complicated, you are quite correct. While it is complex, it is also a very rewarding experience for participants when it goes well. Unfortunately, the game's designers have not done everything they could have to make sure that it does go well. The problem with the conduct of elections has been mentioned. In addition, the player's manual, which runs to over 100 pages, is not indexed and information is sometimes located at places where one would not expect it to be. This makes for frustrating loss of time in locating what one needs to know and leads players to guess at what to do, thus producing errors. However, the game is well worth the 4 to 10 hours required to play it. A round takes approximately 2 hours and one should run two to five rounds for players really to begin to see how the metropolitan area develops as a result of private and public decision making.

Because of its complexity, *City I* requires a great deal of preparation. A prospective director should, if at all possible, participate in a session as a player before attempting to direct a game session. The director should also familiarize himself thoroughly with the operator's manual prior to the game session. Once these things have been done, the players should be walked through a practice round for which more than the normal 2 hours has been allotted before play of the game is begun. The director might want to conduct a dummy run making all the decisions for all of the roles himself in order to learn the game thoroughly before directing a session. Once some sessions have been directed, videotapes of previous sessions may be used to induct new players into the game more quickly. During the conduct of a session, the director should be as helpful as possible in enabling players to do what they want to do, but should not make decisions for them. The director should also make and enforce time limits and not let players get behind schedule, particularly in early rounds of the game. During debriefing sessions, discussion may focus upon the role of public and private decisions in determining metropolitan growth. It is useful to take decisions made during the game (i.e., building a new highway) and to do case studies of the reasons for the decisions as well as their impact. The game can be related to real life by drawing parallels between the simulated metropolitan area and other such areas with which participants might be familiar. It might also be possible for a class to take their own metropolitan area and to try to represent it in the terms used in *City I*. If this can be done successfully, there is provision for inputing a new data base into the game. Thus it is possible in principle for students to direct the development of their own metropolitan area for several years through use of *City I*.

All in all, *City I* is difficult to obtain at the present time and taxing to direct, but a rewarding experience if all goes reasonably well. (From the Robert A. Taft Institute of Government Study on Games and Simulations in Government, Politics, and Economics by Daryl R. Fair, Trenton State College.)

Cost: Request price for paper copy or microfiche; orders must include the "COM" number. (1) 24-page operator's manual describing how to run the Central Processor relative to the game along with the special coding required to change and/or update the computer program and core mapping; assumes operator knows how to cold start the 1130, change carriage tapes, and load and clear the card reader: COM 74-10701-2; (2) 76-page director's manual describing details of administering the game from the director's point of view with examples of decision codes, formats, and general information: COM 74-10702-1; (3) 150-page player's manual describing player details for economic and government sectors with general information required for game play: COM 73-11191.

Producer: National Technical Information Service, 5285 Port Royal Road, Springfield, VA 22151

CITY IV

Envirometrics

Playing Data
Copyright: 1974
Age Level: college, graduate school, professionals
Special Equipment: IBM 360/70 computer; Fortran

Description: *City IV* is an operational simulation game in which participants make economic, government, and social decisions affecting a hypothetical metropolitan area. Each player is assigned to a team which shares an economic and governmental role. The round-by-round play gives the players experience in selecting the type of analysis to move them toward their objectives while the allocation of their time and game resources is a critical determinant of the success they hope to achieve. As the game progresses, players learn to increase their involvement in the management of the environment while learning more about the relationships between business and society. The aim is to encourage players to view the activities of the city as being closely related and interdependent (e.g., an unemployment problem will exacerbate a health problem, the loss of industry and jobs in the private sector will reduce the number and quality of services offered in the public sector through reduced tax revenues, and so on). The game also encourages players to use an interdisciplinary perspective when dealing with urban problems, that is, to look at the problem not only from the viewpoint of an economist but also from the perspective of a geographer, planner, political scientist, and such (from game introduction).

Cost: Request price for paper copy or microfiche; orders must include the "COM" number. (1) 56-page operator's manual describing necessary Fortran and JCL commands for computer operation along with test commands and sequences and complete computer operating instructions: COM 74-10702-3; (2) 286-page director's manual describing details of game administration from the director's point of view and presenting examples of decision codes, formats, and general information necessary to run the game: COM 74-10702-1; (3) 276-page player's manual describing player details for economic and government sectors along with general information required for play: COM 74-10702-2.

Producer: National Technical Information Service, 5285 Port Royal Road, Springfield, VA 22151

CITY MODEL

Envirometrics

Playing Data
Copyright: 1970
Age Level: college, graduate school, urban education, citizen groups
Number of Players: 20-100 in 20-40 teams
Playing Time: 3 days to 4 weeks in 2.5-hour periods
Preparation Time: 1 week of training
Special Equipment: Fortran IV–360/40 or larger, UNIVAC 1108 EXEC 8

Description: This model is an extensive evolution beyond the *City I* model. A larger computer is used and a social sector is added to the system. Also multiple jurisdictions are possible, and the transportation component is expanded to include auto, bus, and rapid rail travel.

This game has been developed as a laboratory tool for education, training, and research purposes. It is designed so that the participants become decision makers in charge of government, economic, and/or social resources in a holistic environment. Players individually and collectively set goals for themselves and for the community.

The computer program is general in the sense that any set of initial data can be loaded into the 25 x 25 playing board at the start of play. Up to 26 economic teams and/or social teams may be represented, and play may start with up to three distinct political jurisdictions. Present initial starting configurations range from a 10,000 population rural county to a 1.5 million population metropolitan area with a decaying central city and a vigorous suburb. The 625 parcels of land may represent either a square mile or a ninth of a square mile depending upon the chosen scale.

Economic teams usually begin play with some developed property and with certain amounts of cash and undeveloped property. In order to develop new parcels of land, zoning, utilities, and highway access must be secured from the government sector.

The 11 types of private land use which economic decision makers can develop on a parcel of land are: heavy industry, light industry, national service industries, business goods, business services, personal goods, personal services, construction industries, and single-family, garden apartment, and highrise residences.

Social decision makers make the decisions for the population units in the metropolitan area. Time allocation decisions (to spend extra time at work, education, politics, or recreation) and boycotting decisions (not to shop or work at specific locations) are made for the three major socio-economic classes of residents in the community (high, middle, and low income).

Governmental decision makers are elected by the social players or appointed by already elected officials to assume the duties of one of the governmental functions, which are performed simultaneously with the economic and social functions. The elected officials must satisfy the voters (social decision makers) in order to stay in office at election time. The chief elected official in each jurisdiction appoints others to execute the functions of the School, Municipal Services, Highway, Planning and Zoning, and Assessment Departments. The governmental departments build schools, provide municipal services, build and upgrade roads and terminals, maintain roads, buy parkland, zone land, and estimate revenues. Utilities, Bus, and Rapid Rail Departments may be operated publicly or privately based upon choices made by the players.

Players set their own objectives for both the public and private actions they undertake. Decisions are recorded for each cycle of play (approximately two hours in length) by a computer, which acts as an accountant and indicates the effects of decisions on one another and on the entire metropolitan area.

The participants in the model make most of the decisions for the local system. External effects on the local system (federal-state aid received, migration of people, and national business conditions) are simulated by the computer. The computer also performs a number of allocation processes (matching full and part-time workers to employers taking into account distances, modes of travel, and educational levels of the workers; matching P1's to housing based upon rent, quality of housing, crowding, and social preference criteria; matching buyers of goods and services to sellers taking into account transportation costs, prices, and boycotts; matching workers to mode of travel using dollar costs and time costs criteria; and assigning students to public or private schools using quality of education criteria).

The model has no formal scoring system. Players are encouraged to measure their performance and the status of their city using their own standards. (The state-of-the-art in urban gaming models.)

Sherry H. Olson et al. (1973; "Computer Modeling in a Course on Urban Decision-Making" in Simulation and Games 4: 440-453) showed a great satisfaction with the game as an element in a course that included undergrad and grad students. Students of both levels cooperated and it "provided some situations in which students could recognize their dependence on other people, their need to share information and to educate one another" (p. 445).

However, these authors have expressed some criticism. Its complexity is so great that it is not as much instructive as a "coarser simulation" and there is a "lack of dramatic conflict" (p. 448) and there are too many resources in each sector (p. 449).

These authors provide recommendations that "educational users" should consult before trying the simulation. (GLeC)

Comment: *City Model* is in use at the Urban and Regional Studies Institute, Mankato State University, Mankato, MN 56001, where persons with serious interest may contact John G. Symons for information. (AC)

Note: For more on this model, see the essay on urban gaming simulation by Mary Joyce Hasell.

Cost: Request price for paper copy or microfiche; orders must include the "COM" number. (1) director's guide: COM 72-11540; (2) player's manual: COM 72-11541

Producer: National Technical Information Service, 5285 Port Royal Road, Springfield, VA 22151

CITY PLANNING

Forrest Wilson, Department of Architecture and Planning, Catholic University

Playing Data
Copyright: 1975
Number of Players: 1 class
Age Level: grade 10-adult
Packaging: 96-page hardbound book

Description: This book contains a series of 11 games in which participants develop and play out different kinds of human settlements. The games are embedded in brief analytical descriptions of the historical, social, architectural circumstances of these social living forms. For each game, participants are given instructions on how to construct simple materials, rules, and the rationale and mechanism of scoring. All are played on boards representing actual space.

The *Hunting-Gathering Game*: Each player has a band of 8 hunters. Movement on the 36-square board is controlled by dice. Dice also represent the game (1 is rabbit, double 4 a hippopotamus). Players remain on the hunting ground for 7 rounds (days) and return to the square of origin. The hunting party which returns with the most points (game) and the most surviving hunters wins. Complicating factors exist in play and are set forth in the rules.

The African Village Game: This game is about how a village grows. Through trade and speculation in land and animals, two or more players try to build the largest and most prosperous compound. Play proceeds through dice-rolls; number combinations indicate events of the year (the fate factor). Players begin with equal resources. Fortunes develop according to luck and skill in bargaining.

The Game of Warring Castles: The object of this prechivalric game is command of the board. Three or more players represent the king, clergy, and nobles. Land is the power base. Play begins with a period of negotiation. Players can form alliances, pooling the power of their land resources, to attack another. The outcome of an attack is determined by dice, but odds (explained in the rules) favor a combatant with more land over one who has less. Thus, the church, which lacks armed men, can participate in an attack on the king or a noble by lending the strength of its land. Play involves guile, deceit, and force.

Medieval Town Game: This game of competition and strategic negotiation involves lords, who are competing with each other and exploiting merchants, and the merchants who make money and bide their time to seize the city from the lord who owns it. In the process, participants play through the forces of a developing land-independent merchant class, the dynamics of life inside and abutting on medieval walled cities in Europe. Negotiations precede each round of play; lords try to get merchants to live in their cities, and merchants locate. Players have a number of options for developing their power, and play incorporates tax-paying (protection money from merchants to lords), battles, the annexation of land, relocation of marketplaces. Outcomes of battles are decided by dice as in previous games.

New England Town Puzzle: Players represent townspeople. The town must find a way to organize its space in the best possible compromise between the competing needs for security (defensible stockades) and space and privacy. No house may be more than .5 miles from the meeting house. Livestock and outbuildings must also be considered in the community's strategy.

The Game of Industrialization: Play is carried on by four groups—the country squire, farmers, city millowners, and laborers. Having begun with resources suited to their respective stations in life, players try to maximize their wealth. Laborers, beginning with 5 credit points, have their labor to sell. Farmers, with 50 credit points, require labor to maintain the productivity of the land they rent from the squire. Millowners, starting with 50 credit points, require labor and capital to operate. The squire has no points but does have land. While they are conducting their business, players are also involved in meeting expenses of rent and travel between city and country village. The game makes an effort to deal with complex and changing economic relationships.

The Cattle Town Game: Players are a cattle agent, store keepers, ranchers, homesteaders, a banker, and an entrepreneur land speculator. The game plays out the conflict between the interests of cattlemen, who need free access to town across farm land for their cattle, and the farmers, and the consequences in the town's economy of this conflict. Game rules provide for economic resources, transactions, profits, expenses, and taxes.

The Growth of a City Game: This and the following two games are a series about the modern city. Given at this stage are an established city centered on a highway and sprawling haphazardly. Living space in the innercity is factory-grimed and decaying, along with warehouses, alongside the railroad tracks and river. Players are office building owners and factory owners, who all start with equal resources; a land speculator; and farmers, who have equal amounts of land adjacent to the city. Play involves negotiation, investment strategy, and—represented in 11 different combinations of dice—chance. The objective of all players is to maximize wealth.

The Game of People Working in a City: Given the city as shaped by previous play and the same rules, executives, office workers, and unskilled laborers try to survive in the city, improve their living conditions, and maximize wealth. Rules cover purchase of housing, rent levels for various incomes, who lives where, and how far it can be between home and job. (For instance, if workers live more than 20 squares away from their jobs, they can no longer go to work and must stay home on welfare at half their wages; businessmen are taxed for resources to support unemployed workers.) The game can be varied to make more jobs than workers, or more workers than jobs.

The Game of City Government: Coming into play with the businessmen, entrepreneur, and others is city government, which is represented by one or more players. Its aim is to try to make the city work. Participants have great latitude in devising ways and means for improving the city. But if *the* objective continues as it began—maximization of wealth—the city will eventually break down, period.

You Design a City: Introductory material for this exercise includes this directive: "Review the games of the book and find the rules that made life more difficult than it already was, from the time of prehistoric hunters to the modern city dweller, and then consider alternatives." The probably sadder but presumably wiser survivors of the history of human settlement make their own rules for their new city, but questions and suggestions exist to guide them in their effort. (AC)

Cost: $8.95

Producer: Van Nostrand Reinhold Company, 135 West 50th Street, New York, NY 10020

CLUG (COMMUNITY LAND USE GAME)

Allan G. Feldt, Cornell University

Playing Data
Copyright: 1966
Age Level: high school-adult
Prerequisite Skills: interest in urban economics; reading, college; math, grade 7
Number of Players: 12-25
Playing Time: 3-20 hours
Preparation Time: 5-7 hours
Special Equipment: two operators

Description: *CLUG* is designed to illustrate a number of basic factors that affect land use decisions in and around a community. The basic model is a framework on which more realistic models can be constructed.

The game is played on a grid by teams that begin with some amount of cash which they invest in the purchase and development of land. Play proceeds in rounds consisting of delineated steps that structure the complex interactions that evolve as the community develops. The manual contains suggestions for several variations on the basic model. (AC)

Comment: This game started as a project in an urban ecology course. Feldt wanted to provide a common framework for the presentation of urban land use patterns. In 1966, 200 or 300 hours of playing experience with the game had been completed at Cornell University with a number of different groups: undergrads, graduates, in informal and classroom situations, and with faculty members and professional planners. It had been used in several planning-related courses in several universities. There is no doubt that *CLUG* is currently the most widely used of all games that deal with urban matters. Envirometrics, Inc. (The State-of-the-Art of Urban Gaming Models, Washington, D.C., 1971) reports that "a number of State Cooperative Extension Agents and Community Resource Development Specialists are using the game with a variety of groups, professional and lay, in various sections of the state of New York." Similar uses are reported also in Iowa, Ohio, New Jersey, and Pennsylvania. The same report indicates that *CLUG* "had provided a base for discussion of future planning needs and has been a useful tool in teaching planning concepts."

The consensus among players is often that the basic version is an oversimplification of urban economy and land use. It should always be said that the basic version is a necessary step in order to play a more sophisticated version. *CLUG* is also a game where rules are not strict and dictated. It is well-known for its quality of flexibility, and can be easily adapted to the needs of the users. There is a lot of freedom to improvise; there is, for example, more than one way to start the game. In fact, various versions of the game now exist. It was applied to the case of Syracuse, NY and advanced modified versions were developed by G. Duggar and E. Foster at the Institute for Public and International Affairs, University of Pittsburgh.

The game can be used to stimulate ideas, to think about a specific area, and it should be played more than once. Most players who were involved in the game tend to show enough motivation to play it again.

This game may be computer-assisted for bookkeeping. (GLeC)

Note: For more on this simulation, see the essay on urban gaming simulation by Mary Joyce Hasell, as well as the in-depth essay on *CLUG* by Joseph L. Davis.

Cost: (1) player's manuals $6.95 each, instructor's manual gratis; (2) complete kit $75.00

Producer: (1) The Free Press, Department F, Riverside, NJ 08075; (2) Institute of Higher Education Research and Services, Box 6293, University, AL 35486

ECO-ACRES

Grayce Papps, Eton Churchill, Eric Van de Bogart

Playing Data
Age Level: grade 9-adult
Number of Players: 15-25
Playing Time: 2 hours (est.)
Packaging: professionally packaged board game

Description: The players of this game try to create an ideal community by drawing, or being dealt, "affiliation cards" (which describe each player's point of view, such as "Industrial Contractor" who favors the industrial use of land) and development cards (which indicate waterways, roads, institutions, businesses, dwellings, and industry). Play proceeds from left to right with each player placing a development card on the map-playing surface. If no player objects to the placement, the game continues uninterrupted. A player who does object, because of his or her affiliation, calls "Town Meeting" and must propose an alternative for debate. After the debate a vote is taken and the proposal with the most votes is adopted. Each proposal must be accepted or rejected by "Town Meeting." Play continues until all of the pieces have been either played and placed on the playing surface or rejected. At that time all participants reveal their affiliations to the other players and discuss whether they were able to achieve the goals on their affiliation cards *and* also satisfy the needs of the community. (DCD)

Comment: *Eco-Acres* was developed as part of a larger project, a television series called "The Land and Me," produced by the Maine Public Broadcasting Network, which included a viewer-active simulation by the same title. Anyone interested in this is welcome to request *Blueprint for A Television Environmental Simulation Project* from Eric Van de Bogart at the address below. (AC)

Note: For more about *Eco-Acres,* see the essay on urban gaming simulation by Mary Joyce Hasell.

Cost: $1.00

Producer: Att: Eric Van de Bogart, Maine Public Broadcasting Network, Alumni Hall, University of Maine, Orono, ME 04473

GAMBIT

Cedric W. B. Green, University of Sheffield

Playing Data
Copyright: 1977
Age Level: college, students-adults
Number of Players: 1-25
Playing Time: 8 hours
Preparation Time: 3 hours (est.)
Special Equipment: magnetic rubber color bands, magnetic rubber tiles, and 600mm square sheet of tinplate (ordering instructions for materials from an English manufacturer in game manual)
Packaging: professionally produced 53-page manual

Description: *GAMBIT* is the most sophisticated of several generations of architectural simulations including *INHABS* (see listing below). *GAMBIT,* in the designer's words, "is a building design simulation which integrates building shape, construction, structure, cost, and thermal design using a magnetic tile model manipulated by a group who take the roles of client, architect, quantity surveyor, and specialists." The game is intended to provide an overview of a building in its context as a system, "not just a mechanical system, but one which integrates social and human goals (represented by client needs and cost limits), its physical structure and performance (represented by the magnetic tile model and thermal calculations), the social context of its design production (represented by the roles and their interaction), and its aesthetic and symbolic expression in a particular context (represented by the designer's perspective and the optional process of obtaining planning consent)."

During play, each participant must obey rules applicable to her or his role. The objective for all players is to design a building of 1000 square meters or less in size that can meet certain stated cost or thermal performance goals. (DCD)

Cost: £1

Producer: Department of Architecture, University of Sheffield, The Arts Tower, Sheffield S10 2TN, United Kingdom

THE GRAND FRAME

Pat Miller, Nancy Stieber, and Mary Joyce Hasell

Playing Data
Copyright: 1976
Age Level: college, adult
Number of Players: 15-25 in 5 teams
Playing Time: 6-8 hours
Preparation Time: 4 hours (est.)
Packaging: photocopies 52-page operator's manual and game kit

Description: This planning game is based on a design process consisting of the following steps: conceptual planning, information processing of natural constraints, information processing of planning constraints, information processing of developer constraints, a design phase, the evaluation of that design, and its presentation at a planning commission meeting. The authors describe *The Graphic Analytic Didactics Frame Game* as adjustable "to numerous design problems" but specifically supply the content for a problem in which a housing development must be planned for a 25-acre tract.

In the first step of the example game, each team of three or five players is given a set of user constraint cards. The teams assume the roles of users and rank order these constraints according to their importance to the users. Then, using pins, chips, and spheres, the players construct a three-dimensional representation of the features of the housing development which includes all the essential components (housing, parking, community facilities, and recreational and open space) and linkage systems (transportation routes and buffers). Next, each team pieces together a two-dimensional map of the proposed development using plastic shapes and an acetate playing surface. The surface includes the natural constraints (such as vegetation, soil type, and drainage) of the site. The plan is then modified according to planning and developer constraints (such as cost of land, marketing, engineering, and utilities) printed on cards. The last step in the design process is an overall evaluation of the design to insure that all constraints have been accommodated before presenting the plan to a "planning commission" composed of students. (DCD)

Note: For more on this game, see the essay on urban gaming simulation by Mary Joyce Hasell.

Cost: complete kit $300.00, operator's manual $5.00

Producer: Gaming/Simulation Center, College of Architecture and Urban Planning, University of Michigan, Ann Arbor, MI 48109

GSPIA: AN URBAN REGIONAL DEVELOPMENT SYSTEM

Francis Hendricks, California State Polytechnic University; Clark Rogers, GSPIA, University of Pittsburgh

Playing Data
Copyright: 1966
Age Level: college, graduate school
Number of Players: 18-100
Playing Time: 60 hours in periods of 2-10 weeks
Preparation Time: 3 hours
Special Equipment: tape recorder and video equipment, optional; computer, 360-40 minimum or equivalent Fortran IV

Description: The University of Pittsburgh game has been developed as an educational tool for planning students and others. The game is a combination math simulation and role-playing model that can be loaded with data for any location.

One version of the *GSPIA* game simulates Allegheny County and the other simulates Westmoreland County (the predominantly rural county to the east of Allegheny County which contains Pittsburgh). [Editor's note: Other locations since used are San Francisco, San Luis Obispo, Berlin, and Puget Sound.] The Westmoreland County model contains 7300 grids (located by an "x" and "y" coordinate) and each grid is 90 acres in size. The land ownership is divided among five Development Teams. All teams in this model have one member. The Development Teams compete for share of the county development which is determined by a computer program. That is, the Developers may develop industrial, commercial, or residential land uses, but the total number of grids which can be developed each cycle is determined by the economic condition of the county in comparison to that of the nation. Residences can be of one income class and of several density levels.

There are five Councilmen which represent the five political jurisdictions in the county. There is a County Manager who handles current expenditures for the county, and a Planner who handles capital expenditures in the county.

There are several other planners who submit plans for zoning, highway development, and master plans to the Council and Manager. The proposed plans are also evaluated by HEW Administrators (concerned with education, health, and welfare facilities), Urban Renewal Planners (concerned with deteriorated housing), Federal-State Administrators (who disburse federal and state aid to the county for programs which they have approved). A community role is played by participants who represent manufacturing interests, citizens, planning consultants, county solicitors, a school board, and transportation planners. A private banking role and a stock market are also part of the private sector.

The Urban Renewal Planners try to get Developers to raise low residential land values. The number of roles in the game changes to fit the class size. The number of Developers and Planners can most easily be increased or decreased according to need.

Land is allocated to the Developers at the beginning of a play day according to a Poisson distribution; that is, the Developer who receives the first grid has the greatest chance of receiving the grid next to it. Therefore, the land ownership tends to be in clusters. All land is owned by the Developers and absentee owners whose land is held in trust by the banking player.

The Urban Renewal teams have a more meaningful role when a heavily urbanized county is used. In the game, issues such as zoning, highway construction, capital expenditures on the county infrastructure (utilities) are interdependent role decisions.

The computer program is general in nature, and the specific area simulated and the number of grids can be changed. The game director can influence development and play through a variety of instruments. For instance, he can change the attractiveness of grids for development by changing the likely rates of return on those parcels through increasing their transportation access. (The state-of-the-art in urban gaming models.)

Editor's Comment: The maker of this simulation says, "Initially, players feel the Game Director is pushing buttons with the computer program. Then they study the program to try to manipulate the cause-and-effect sequences. Then it becomes clear that the city is a complex, richly connected system and to manage and plan for it involves treating it as a totality rather than as separate parts." And of the political message, he says, "Growth is economically determined. Energy resources and human will organize activities that are antientropic and fill niches. These grow until limits are reached or decline entropically. All of these events produce disturbances that shift costs and benefits. Governmental activities provide essential inputs that balance and regulate to maximize benefits to the shifting clientele that control local government machinery. The image of man? "Continuous struggle to maintain and better the human condition. Greater sophistication, finer distinctions are more human than uniform rules applied to stereotypes." The game covers 10 to 15 years, thus students encounter a wide range of conditions, for example, economic boom and bust, political turnover, physical disasters. Among the weaknesses admitted by the author are, "It does not deal with multilevel government problems. The scale of development is still too large. Some will still have excessive clerical work." However, the author points out, "The meetings were twice mistaken by outsiders for real government public hearings." He also says that he was once accused of writing a "play" because observers could not believe the level of dialogue was spontaneous and part of the natural communication of the game. He claims it is far more realistic than *Metro-Apex* or *CLUG*. (REH)

Cost: Not published, but can be played by contacting the authors.

Producer: Francis Hendricks, School of Architecture and Environmental Design, Calif. Polytechnic State Univ., San Luis Obispo, CA 93407; and Clark Rogers, 803 Bruce Hall, GSPIA, Univ. of Pittsburgh, Pittsburgh, PA 15213

G.U.L.P.: THE GROWTH OF URBAN LAND AND POPULATION

Robert M. Sarly, Michael Safier, Michael Tyler and Lea MacDonald

Playing Data
Copyright: 1974
Number of Players: 19-25
Playing Time: 4-5 hours.

Description: *G.U.L.P.* is a study of processes of change in the capital city of a developing country undergoing rapid urbanization. The entire cycle of play represents one year. Before the cycle begins, each participant prepares a score sheet indicating important objectives and priorities. Participants enter into a one-hour planning period (representing about three months) in which each one plans a strategy for achieving his objectives. This process involves negotiation among roles. A development period (one hour representing three months) follows. Development decisions, investments, and economic commitments are made. In the evaluation period each participant views the state of affairs in the city in light of his own interests. After discussing the outcomes of their own role performances and strategies, participants consider the problem of evaluating urban development in general. (AC)

Cost: unknown

Producer: Development Planning Unit, Bartlett School of Architecture and Planning, University College London, 9-11 Endsleigh Gardens, London WC1H OED England

HEXAGON

Richard D. Duke and 680 Game Design Seminar, University of Michigan

Playing Data
Copyright: 1976
Age Level: high school-adult, professional planners, government bureaucrats
Number of Players: 17-24
Playing Time: 2-3 hours

Description: Participants in *Hexagon* explore issues and dynamics of resource allocation through human settlement planning in a developing country. The game consists of three interdependent, simultaneous games at three levels—local, regional, and national. It involves a wide array of urban issues, such as municipal services, housing, commerce, industry, land use, and population, as well as a number of issues pertaining to third world development, including a spectrum of economic activity that ranges from barter to international trade and exchange rates. (AC)

Note: For more on this game, see the essay on urban gaming simulation by Mary Joyce Hasell.

Cost: manual $5.00; complete kit $300.00

Producer: Multilogue, 321 Parklake Avenue, Ann Arbor, MI 48103

IM-CLUG

Joseph L. Davis and Richard F. Tombaugh

Playing Data
Copyright: 1978
Age Level: grade 9-adult
Number of Players: 60-180
Playing Time: 6-12 hours
Preparation Time: 4 hours
Packaging: 27-page, photocopied booklet

Description: *IM-CLUG (Intergovernmental Model-Community Land Use Game)* is a modification of *CLUG* to accommodate very large groups and to explore the interrelationship among neighboring cities in an economically interdependent region. The simulation essentially involves four simultaneous *Polis/Richland* hybrid games (see also) played with possible interrelations between games.

The object of *IM-CLUG* is to build a city; however, this city takes the form of a twentieth century metropolitan area complete with separate political and economic jurisdictions. (producer)

Cost: $5.00

Producer: The Center for Simulation Studies, 736 De Mun Avenue, Clayton, MO 63105

INHABS 3

Cedric W. B. Green

Playing Data
Copyright: 1971
Age Level: college
Number of Players: 20 or more
Playing Time: 2 days for Sprawl, 2 days for Spiral, 1 day for Squat
Special Equipment: users must make their own acetate playing surface
Packaging: professionally produced 28-page booklet

Description: *INHABS 3: Instructional Housing and Building Simulation,* "was designed to simulate that area of designing that most obviously draws its validity from its social context–housing." The simulation is comprised of three similar games–*Squat, Sprawl,* and *Spiral*–which use different building sites. *Sprawl* simulates suburban income levels, land costs, and planning practices (in the U.K.), *Spiral* reproduces conditions in central cities, and *Squat* represents a condition where unzoned land is occupied by low-income families. Land costs 100£ per plot in *Squat,* 1000£ per plot in *Sprawl,* and ten times that amount in *Spiral.*

At the beginning of any of the three games, the players are given personal record and construction costing sheets. Each participant represents a family of one to six persons, or a developer, contractor, planner, or banker. All of these parties try to develop their neighborhood for their mutual profit and well being. (Families, for example, must try to own at least one house that is adequate for their family size; the developers want to meet this need and earn a profit, and the planners must see that any proposed development of the neighborhood include plans for roads, footpaths, and open space.) A scoring mechanism is defined for each role. While the game is in play, the participants construct a model of the neighborhood using plastic models, and players may drop in and out of the game, playing a round at the beginning and one or more rounds later. Each round takes one to two hours to play and represents one year in the life of the neighborhood. (DCD)

Note: For more on this simulation, see the essay on urban gaming simulation by Mary Joyce Hasell.

Cost: 1£

Producer: Department of Architecture, University of Sheffield, The Arts Tower, Sheffield S10 2TN, United Kingdom

THE LIVING CITY/LA CITE VIVANTE

Amy Elliott Zelmer and A. C. Lynn Zelmer, International Communications Institute

Playing Data
Copyright: 1972
Age Level: college, community groups
Number of Players: 18 or more in teams of about 20
Playing Time: 8-10 hours in 2-hour periods
Preparation Time: about 8 hours for the first time
Special Equipment: video equipment, VTR playback for tape essential, access to cable system useful for large groups

Description: This game, which introduces players to some of the factors that need to be considered in community planning, is designed both to increase appreciation of the complexity of planning issues and to demonstrate that citizens should have a voice in community planning. Participants portray private citizens and members of special interest groups. They are responsible for zoning tracts of land, setting building code standards, and planning transportation systems. Each issue is decided by vote. (AC)

Cost: unknown

Producer: Urban Gaming/Simulation '78, School of Education, University of Michigan, Ann Arbor, MI 48109

MINI-APEX

Theodore H. Rider

Playing Data
Copyright: 1974
Age Level: university, professionals, citizen groups
Number of Players: 20-50
Playing Time: 3 8-hour periods

Description: *Mini-Apex* is basically a simulation of *Metro-Apex.* Participants are engaged with numerous urban, social, and environmental issues in planning and political processes. (AC)

Comment: At press time, *Mini-Apex* is in the process of evolving further and is being redesigned as a minicomputer package. (AC)

Note: For more on this simulation, see the essay on urban gaming simulation by Mary Joyce Hasell.

Cost: kit for original version $70.00; write for price of 1979 version

Producer: Theodore H. Rider, Director, NECEP, Boston College, Weston Observatory, Weston, MA 02193

METRO

Richard D. Duke, University of Michigan

Playing Data
Copyright: 1969
Age Level: college, graduate school, management
Number of Players: 10-40 in 3 teams
Playing Time: 3-9 hours in 4-hour periods
Preparation Time: 2 hours
Special Equipment: desk calculator, film projector, IBM 1130: FORTRAN

Description: This computer-based role-playing game evolved out of experience from the game of *Metropolis.* Like its predecessor, *METRO* is based upon the real data base for Lansing, Michigan. *METRO* was designed to be a training tool, a research tool, and, ultimately, a decision aid. In fact, *METRO* (Michigan Effectuation, Training, and Research Operation) was designed as an integral part of the overall regional planning program of the TriCounty Regional Planning Commission.

Players are assigned as members to two types of teams, a functional team (politician, planner, school board, or land developer) and a locational team (central city, suburbs, or urbanizing township). Therefore, each player has a role and a jurisdiction to represent. The computer is given the responsibility to simulate household, industrial, and commercial behavior, as well as the task of being the data bank and processing the inputs and generating outputs.

Locational decisions take place on a map of the Lansing metropolitan area comprised of 44 analysis areas. These areas are irregularly shaped combinations of census tracts that are identified by a number and not a pair of coordinate numbers.

The politicians remain in office only if reelected by the simulated population controlled by the computer. Other players affect the vote by their inputs to the elite opinion poll. The politician is in charge of the jurisdiction budget, some public land purchases, zoning, and carrying out specific capital improvement projects.

Land developers attempt to make successful land purchases and building decisions given the demand for growth that is generated each round by the simulated land users (industry, commerce, and households).

The school decision makers try to improve schools, and they must also be reelected in order to stay in power. They set school tax rates, purchase land, allocate a budget, and make capital improvements.

Planners work one year ahead of the politicians and try to plan future programs. They have little formal power beyond access to good data, so their task becomes one of trying to persuade the community to accept their suggestions.

Each team receives computer output that is specific to his team and also a copy of the computer-generated newspaper. The newspaper reports on national, state, and local news for the year. The local news indicates the issues that are of importance to the simulated population of the metropolitan area. Some general computer output is also generated each round and is available to all teams.

The three major computer operations in *METRO* are those for growth, growth distribution, and voter response. Growth is a macro simulation that provides (with director inputs) the overall amounts of population (by category), employment (exogenous and endogenous), income, and property value in the region. The distribution model is a modified version of the TOMM model developed as part of the Pittsburgh Community Renewal Program. This model allocates employment (endogenous), commercial (retail and service), and residential (by household types) land uses among the analysis areas. New exogenous manufacturing is allocated manually by the game director. The voter response model simulates the actions of groups of voters during an election to (1) vote, (2) vote on a particular issue, or (3) vote yes or no. Factors affecting the model are population size, median family income, the kinds of issues being presented, the kind of election (special, local, state, or national), the support of simulated pressure groups, the elite opinion poll, and the performance of players up for election.

The voter response model generates output for the politician showing voter attitudes (by five types of groups ranging from NAACP to Right Wingers). Other politician outputs include the jurisdiction's credit rating, the present budget, and a projected budget (assuming current spending levels). The school people receive the above information for their department, plus detailed information on the age and number of students and the quality of schools in each analysis area. One of the objectives of the *METRO* exercise is to acquaint the players with computer output and the potentials of sophisticated data bank usage. (The state-of-the-art in urban gaming models.)

Note: For more on this simulation, see the essay on urban gaming simulation by Mary Joyce Hasell.

Cost: kit $500.00

Producer: Multilogue, 321 Parklake Ave., Ann Arbor, MI 48103

METRO-APEX

Richard D. Duke, COMEX, University of Southern California

Playing Data
Copyright: 1967, 1974
Age Level: college, graduate school, management, policy makers
Number of Players: 15-120 in teams; about 50 is optimal
Playing Time: at least 5 cycles of 3-8 hours each
Preparation Time: 2 hours per cycle
Special Equipment: desk calculator (optional). Versions are available for an IBM 1130 System or an IBM 360/370 System with the following minimum configurations: IBM 1130–at least 8K core and one disk drive. Primary input is from a 1442-6 or 7 card read/punch and output to a 1403 line printer. We can supply versions configured for 2501 card reader and/or 1132 printer on request. IBM 360 or 370–version is for OS with a partition size of at least 120K. The program is written primarily in FORTRAN IV but includes many nonstandard IBM features. There are a few assembly language subroutines. (University of Michigan Extension Service.)

Description: *METRO-APEX* was devised to permit sophisticated, dynamic exploration of several of the major components or "systems" of a major American community. This is accomplished through linking a number of community decision makers (roles) to a computer simulation representative of the community. The roles include six separate business corporations, seven land developers, politicians for each of three jurisdictions, and several other roles. The computer simulates the housing market, the transportation system, the municipal financial system, infrastructure, employment base, and similar components.

Play is cyclical, with each cycle representing a year of growth for the community. A typical cycle of play focuses on a particular problem and goes through several major steps: Review of last cycle's output; current newspaper; decision on pending issues; interteam bargaining; individual team decisions, operator coding, and computer processing of results. A critique is held each cycle and at the end of the exercise.

METRO-APEX is designed as a tool for the serious exploration of community problems. It is adaptable to a full term of study and/or to an intensive seminar on community problems of a week's duration. While it requires some initiative to launch, the exercise is inevitably well received by players. (R. Duke)

Note: For more on this simulation, see the essay on urban gaming simulation by Mary Joyce Hasell.

Cost: $500.00 a run for computer, manual about $7.00

Producer: Mark James, Director of Computing Services, COMEX, Davidson Conference Center, University of Southern California, Los Angeles, CA 90007

METROPLEX

Richard F. Tombaugh and Joseph L. Davis

Playing Data
Copyright: 1972
Age Level: grade 9-adult
Number of Players: 15-50
Playing Time: 5-10 hours in 1-2 hour rounds
Preparation Time: 1 hour

Description: *Metroplex* is a four-part simulation of urban air pollution and interrelated issues in which the parameters of the final part of the simulation are fixed by the results of the first three parts.

The exercise begins with the players identifying their positions on the seriousness and causes of pollution as well as "the focal points of legal controls." Next, the participants must define the characteristics of a community in which the issues surrounding the control of air pollution can be raised and decide who in this community should be involved in decision making. Then, assuming the roles of those decision makers, the players "decide what type of decision-making process shall be used for deciding about control of air pollution." The concluding part of the simulation is played in rounds in which the players generate income and make decisions about the control of pollution. (DCD)

Cost: $100 within metropolitan St. Louis, $200 elsewhere

Producer: Not published but can be played by contacting The Center for Simulation Studies, 736 De Mun Avenue, Clayton, MO 63105

METROPOLIS

Richard D. Duke, University of Michigan

Playing Data
Copyright: 1962, 1974
Age Level: high school through adult
Number of Players: 3-30
Playing Time: 6-8 hours
Preparation Time: 2-3 hours

Description: *Metropolis* is a medium-size city in mid-America. It is composed, politically, of three wards: Inner City, Blue Collar, and

Elite. Players are assigned as teams of three to four roles: Politician, School Board, City Administration, and Land Developers.

Play proceeds through a series of cycles or rounds each representing a year of growth for the city. A typical cycle of play includes review of last cycle's results; a newspaper of current events; decisions on pending issues; concurrent individual team decisions; and calculation of results.

The game focuses on the allocation of a scarce resource (City Budget Funds) to the many demands for municipal services (e.g., schools, teacher salaries, fire and police, new construction projects). The administrator, equipped with the "Crystal Ball" edition of the cyclical newspaper, is required to estimate revenues and expenditures for *next* year. The private sector (Land Developers) lobbies for various public improvements that they seek, and the School Board and City Politicians compete for the citizen's tax dollar. (Author)

Comment: This game, which has exerted a great influence on many urban games, has been run with city officials throughout the United States, urban professionals, students, and citizens. Some experiments were made to measure its "effectiveness." "Although the kind of experience acquired in a gaming simulation is not readily measured by an objective test, the students' test scores were clearly superior to those of a control group learning about community process and structure in a normal fashion. This game has demonstrated its flexibility in permitting from 3 to 30 to play, and it can be operated so the intercycle computations are handled either manually or with a computer. Also, unusual crises that may possibly affect a metropolis of the size and structure modeled (an inland city without satellites in the 100,000-500,000 category) have, on occasion, been introduced and played out." (From Richard L. Meier and Richard D. Duke, "Gaming Simulation for Urban Planning," Journal of the American Institute of Planners, XXXII, January, 1966, pp. 3-17.) (GLeC)

Note: For more on this game, see the essay on urban gaming simulation by Mary Joyce Hasell.

Cost: $50.00 for kit (9 participant manuals, 1 leader's guide, set of wall charts, forms; by the piece: leader's manual $10.00, participant manual $5.00, wall charts $15.00

Producer: Gamed Simulations, Inc., 10 West 66th Street, New York, NY 10023

NEW TOWN PLANNER'S SET

Barry Ross Lawson, Harvard, Mass.

Playing Data
Copyright: 1969, revised 1971
Age Level: college, graduate school civic groups, local officials, urban planners
Number of Players: 2-5 in 2-5 teams
Playing Time: 3-10 hours
Preparation Time: 30 minutes to 4 hours

Description: This game has four versions (I through IV), with each version adding complexity to the preceding version. The board is an 11 x 14 grid that has some topographic features (a lake, a river, and several parcels that may not be developed) and railroad tracks that offer locational advantage. Each parcel is divided into four equal-size subparcels. Three zones on the board represent the downtown, the transitional area, and the suburbs. The land uses are HI (1-3), PG (1-3), and RA, RB, and RC, each of which consumes a given number of subparcels. All four versions begin with no private development on the board.

Version I has as a specified objective for the teams (up to four in number) the attainment of the highest ratio of total revenue to total land costs. Each round, teams bid on land and then roll dice to determine the type and density of development units they may place on owned or rented land that round. Bonuses in the form of increased incomes are provided for retail agglomerations, retail neighborhoods, industrial locations adjacent to the rail and/or the river, and residence locations adjacent to the lake.

Version II introduces (1) money as a medium of exchange among the teams (all teams start with the same amount of cash), (2) bidding for retail and industrial units, (3) taxation, and (4) redevelopment of property. The criteria for determining a winner in this version is the highest rate of return. Additional bonuses are awarded to the team with the largest amount of industrial income (representing the extra return to specialization) and with homogeneous land uses on a parcel.

Version III introduces the public sector and a planner's role. The planner supplies the system with needed public services in the form of utilities, school units which use one subparcel of land, and parks which consume a full parcel of space. He pays for these out of tax revenue or bonding. The objective of the developers is still to maximize their rate of return, whereas the planner's objectives are either set by the system (a balanced budget) or self-established (to allocate public development in such a way as to serve or shape future development). Bonuses are now affected by the placement of utilities, parks, and schools.

Version IV expands the range of government activity and the transformation of some of the economic bonuses into point bonuses to reflect sociological and aesthetic benefits. This allows for mixed player objectives in the tradeoff of money for points in some cases. The new public land uses include fire stations, health clinics, a town hall, sewerage plants, institutions, a civic center, a refuse disposal plant, and an airport. The demand for these facilities is generated by the actual community mix of housing, retail, and industry as compared with an "expected" mix. (The state-of-the-art in urban gaming models.)

User's Report: Educators and those familiar with urban planning may wish to change some of the rules after they have played the game once or twice. This is not because the original rules do not work; quite the contrary, they work very well. Rather, this is because the rules involve significant value judgments, which certainly are open to question and experimentation. This flexibility illustrates the essential soundness of the basic principles and operating concepts of the game; i.e., the game may be played a number of different ways to add diversity to the educational experience. By manipulating rules, the game can be played over and over under a number of rigidly controlled rule situations in order to develop or test theoretical propositions about the complex forces which lead to different patterns of urban growth.

In sum, *New Town* is an excellent game for everyone. (Joseph Zikmund)

Comment: *New Town* exists also in a family game version (elementary through high school and parlor play) and as an educational kit (junior high through college, and civics groups). See the community issues section for a general description. (AC)

Note: For more on this game, see the article on urban gaming simulation by Mary Joyce Hasell.

Cost: planner's set $85.00, each additional set $70.00; computerized supplement (cards, write-up, and instructions) $50.00; instructor's manual and playing board (only) $40.00.

Producer: Harwell Associates, Pinnacle Road, Harvard, MA 01451

PLANNING OPERATIONAL GAMING EXPERIMENT

Francis Hendricks

Playing Data
Copyright: 1959
Age Level: college, continuing education
Prerequisite Skills: knowledge of zoning
Number of Players: 2-10 in 2 teams
Playing Time: 6-10 hours in one day
Preparation Time: none

Description: There are 15 hearings per year. Players in the roles of town planning director (and staff) and real estate developer (and staff) select strategies based on land use and location. A matrix representing the commission determines outcomes. Strategies selected have variable costs. The number of hearings won determines the planner's budget and the developer's taxes and yield. The planner has five to seven strategies and limited resources; the developer has four to six strategies and ample resources. (author)

Cost: $1.50

Producer: Francis Hendricks, School of Architecture and Environmental Design, California Polytechnic State University, San Luis Obispo, CA 93407

POLIS

Richard F. Tombaugh

Playing Data
Copyright: 1971
Age Level: grade 9-adult
Number of Players: 15-25 in teams
Playing Time: 6-8 hours
Preparation Time: 2 hours
Packaging: 18-page photocopied booklet

Description: *POLIS* focuses on the politics of metropolitan development and growth. It is a new game based on *CLUG-alum*, developed by the staff of The Center for Simulation Studies. In *POLIS*, the players deal with the competitive and cooperative decisions in a metropolitan region which includes an old, struggling center city and booming suburbs.

As well as being actors and investors in the private sector, the players form the governing body of the metropolitan region and are responsible for the financing, distribution, and maintenance of public services and facilities, e.g., schools, parks, medical facilities. Team voting in the governing body is weighted to reflect relative economic strength and population. In their political capacity the players must also deal with both positive and negative external factors that affect the community. These outside influences are introduced into the game through a randomly selected series of headlines.

The metropolitan area of the game is derived by playing *CLUG-alum*. With that city and region as a base, *POLIS* begins by introducing some rule changes, new types of industries, and environmental pollution that reflect modernization and contemporary urban problems. (producer)

Cost: $5.00

Producer: The Center for Simulation Studies, 736 De Mun Avenue, Clayton, MO 63105

RICHLAND, USA

Richard F. Tombaugh and Earl S. Mulley

Playing Data
Copyright: 1971
Age Level: grade 9-graduate school
Number of Players: 15-25 in five teams
Playing Time: 6-8 hours
Preparation Time: 2 hours
Packaging: 21-page photocopied booklet

Description: *Richland, USA* is a modification of the *POLIS* game that begins with an existent community comprised of five socio-economic groups—old rich, new rich, mercantile professional, lower middle white, and underemployed poor.

The basic process and rules of *POLIS* apply, with certain new factors added—differentiated wages, strike and riot options, public housing, immigration, various federal programs, and a fluctuating external economy.

In the other games the playing of roles tends to be insignificant. In *Richland, USA*, the combination of role definition and model building adds new dimensions to the learning experience.

It is necessary that those who play this simulation game have previous experience in playing *CLUG-alum* and *POLIS*. (producer)

Cost: $5.00

Producer: The Center for Simulation Studies, 736 De Mun Avenue, Clayton, MO 63105

SUPRCLUG (CENTER REGION)

Luis H. Summers

Playing Data
Age Level: graduate school
Number of Players: 14-40 in 7-19 teams
Playing Time: 9-60 hours in 30-minute rounds
Preparation Time: 4 hours
Packaging: 50-page manual

Description: *SUPRCLUG* is a modification of *CLUG* designed to demonstrate how basic economic factors affecting land use shape the growth patterns of a region. Each team begins play with a specified amount of cash with which to buy land. Play takes place on a 22 x 22 game board that has been divided into three counties, each containing a city. These cities have adopted contrasting urban planning policies. "Free City" has no zoning policy, for instance, while "Planned City" has an administrator in control of a public corporation with the ability to purchase land on the open market or to condemn land not otherwise for sale. Each city has up to five developer teams and one banking team. Play consists of proceeding through an ordered set of transactions during each round. The factors involved in these transactions include the purchase, assessment, depreciation, and sale of real property and the construction and demolition of buildings to achieve compatible land use. The game also simulates situations concerning employment, inflation, transportation costs, taxes, and political change. Each team's success is measured after every round and at the conclusion of the game in terms of the team's net worth and each city's quality of life. (DCD)

Cost: $25.00 manual and construction instructions

Producer: Dr. Luis H. Summers, 101 Engineering "A," Pennsylvania State University, University Park, PA 16802

SYSTEMS

Carl Steinitz, Peter Rogers, et al.

Playing Data
Copyright: 1977
Age Level: graduate school, planners
Special Equipment: computer

Description: *Systems* is a set of 28 models that, taken together, describe "the pressures and consequences of suburban growth." In the project summary, the designers note that the models "may be operated separately to address specific tasks or linked in a variety of ways to respond to more complex questions beyond the scope of any single model. When linked, the models exchange information through a shared computer programming system and a common data base" (constructed of 75,600 grid cells of 1 hectare) which, with a land use and land type classification system, compose the technical framework for the models.

The models are of two types. The allocation models provide a set of decision rules assigning a land use to a location. The data files that describe the original conditions are updated with the results of user decisions. The evaluation models operate on the updated files, describing environmental, fiscal, and demographic impacts of the new land use allocations. The allocation models are: housing, industry, commerce, public expenditure, public institutions, conservation, recreation, and solid waste. The evaluation models are: three soil models, two vegetation models—one concerning wildlife habitats, water quantity, water quality, critical resources, visual quality, transportation, air quality, noise quality, historic resources, land value, public fiscal accounting, demographic descriptions, and legal implementation.

The operational modes of the system and models are: preanalysis, single-model analysis, project evaluation, plan evaluation, sensitivity analysis, gaming, optimizing, legal testing, and alternative strategy simulation (through which users can test the consequences of various strategies).

Many, but at this point not yet all, of the models have been published. A 46-page summary of the project and the system is available. (AC)

Note: For more on this model, see the essay on urban gaming simulations by Mary Joyce Hasell.

Cost: *Managing Suburban Growth: A Modeling Approach* (summary description) $6.00, individual models between $3.00 and $7.00 each

Producer: Dr. Carl Steinitz, Graduate School of Design, Harvard University, Cambridge, MA 02138

TELECLUG

Marshall H. Whithed

Playing Data
Copyright: 1970
Age Level: grade 7-graduate school, management
Number of Players: 12-15 to 30 in 3-5 teams
Playing Time: 2 hours to several days
Preparation Time: several hours
Special Equipment: computer terminal Time Share, CDC 6400 or access to commercial CDC Time Share Service

Description: Participants in *TeleCLUG* take the roles of real estate entrepreneurs and community leaders. Their objectives are self-defined and might include maximizing team profits and/or developing a viable community as they buy land, develop facilities, and plan community development. The simulation is designed to give participants an opportunity to explore city development and the economic influences on it, as well as to explore the conflict between entrepreneurial motivations and community development in terms of quality of life. (AC)

Comment: A 52-minute videotape of *TeleCLUG* for use in a classroom is available. It is entitled "The Implementation of a Political Simulation" and can be obtained by writing to George Cranford, Instructional TV Services, New York State University College at Plattsburgh, NY (GLeC)

Cost: for information on materials, transfer capabilities, and costs, contact producer or Marshall H. Whithed, School of Community Services, Virginia Commonwealth Univ., Richmond, VA 23284

Producer: Westland Publications, P.O. Box 117, McNeal, AZ 85617

TRADE-OFF

Jerry Berger

Playing Data
Copyright: unknown
Number of Players: exact number unknown, but around 20-50
Playing Time: several hours

Description: Trade-off is an aid for resident planning which attempts to involve the members of a community in the planning for that community. Essentially the players are stepped through the problems and procedures of urban planning and design. The game has been played indoors and outdoors. Indoors a grid system is laid out on a board representing nine city blocks. Flags are placed on each block which describe the physical and social conditions in that block. Outdoors a blocked-off street in the affected neighborhood is used as the board.

Players are first asked to build the best possible community assuming no dollar constraints. They build improvements out of cardboard, wood, and whatever material is around. Professional game-aids calculate the price of each improvement. Then players are asked to assign a point value to the benefits of each improvement. It is only at this point the players are told the dollar amount that they have expended and also, for the first time, a dollar figure representing the total dollars available for improvement in their neighborhood. They must then begin the process of "trade-off." They learn the process of cost-benefit evaluation on the local level. (REH)

Comment: Any solution to the problem of the city must first and foremost be a solution to the problems of citizen participation. The authors of this simulation have shown that simulation gaming can be a very effective tool in bringing a neighborhood together and improving the rational and efficient allocation of resources in response to self needs. (REH)

Cost: $25.00

Producer: Joint Council of Economic Education, 1212 Avenue of the Americas, New York, NY 10036

THE TRANSFER-OF-DEVELOPMENT RIGHTS GAME

George H. Nieswand, Teuvo M. Airola, and B. Budd Chavooshian; Cook College

Playing Data
Age Level: college, graduate school, government
Number of Players: 4 teams of 1-5 each; 2 leaders
Playing Time: 6-7 hours
Preparation Time: 1 hour
Special Equipment: duplicating machine, game board or overhead projector
Packaging: manual

Description: Transfer-of-development rights is a new concept in land use management which involves "the preservation of environmentally important areas with equitable compensation to owners at no cost to taxpayers." The first half of the manual explains the concept and how it was applied successfully in New Jersey; the second half contains the priming game and the *Transfer-of-Development Rights Game* (T.D.R.). In both games (the latter is an extension of the former), participants are concerned with developing the land of a hypothetical low-density community and with maximizing their team's net worth. Each game is played in rounds of 12 steps beginning with bids on property and including taking out loans from the bank, making investments, paying activity costs and interest charges, negotiating with other teams for land, and receiving interest and returns on investments. In the priming game, teams attempt to gain the title for land they want to develop, which must be located in the appropriate zone. After 1.5-2 hours, the leaders unexpectedly announce a transition to the *T.D.R. Game*, in which zoning is no longer in effect. A natural preservation district is created, and owners of land in this district are awarded titles to land in other areas which they may develop at a higher density level than may landowners who did not lose land to the natural preservation district. Teams must revise their strategies in the second game, as they gain experience with the practical applications of the T.D.R. concept. (DCD)

Cost: $2.00

Producer: Communications Department, College of Agriculture and Environmental Science, New Brunswick, NJ 08903

TRANSIT

R. G. Klietsch

Playing Data
Copyright: 1972
Age Level: grade 11-graduate school, management

Number of Players: 20-40; 30 is ideal
Playing Time: 7 or 8-10 or 12 hours in 1 hour periods
Preparation Time: reading time 2 hours, and start up
Special Equipment: tape recorder, optional

Description: In three rounds of play, participants try to resolve eight interlocking urban transportation problems—the location of an airport, warehouse, traffic terminal, and industrial park, the completion of a freeway with parking ramps, the expansion of a mass transit system, and the development of a satellite transit system. Players assume the roles of the Airport, Transit, Environmental Quality, and County Board Commissions, The Taxpayers Association, the Metropolitan Transportation Agency, and the Industrial Council.

The game is designed to teach 16 urban transportation concepts including queue system factors and operations, transport requirements, and stand-by systems. Players are introduced to eight major urban transportation problems, issue factors, and alternatives that have recently been proposed. While operating within the political context of transportation (including customary controls, resource sanctions and policies), they learn current problem parameters and consequences of transit options.

There are two alternative scoring systems. The first uses press, boycott, endorsement, and sanction point balance to cover the eight problems. In the second method, queue factors, volume/density factors, and process factors are quantitatively scored for the alternative adopted. (AC)

Comment: The author's words can probably best establish the context for the game of *Transit*: "Until recently, transportation took a subordinate position to production and consumption aspects of economic behavior. Now, distribution of goods, services, and people looms large as constraints to continued economic well-being and accustomed life-styles. Transit focuses on unexpected reserves of alternatives and the benefits of fundamental views of transportation as a queue. The idea is that distribution systems (and all varieties of human physical spatial movement) are queues. Queues operate by design; the problem is: people knowingly violate queue rules and bring about problems that require new and greater controls. Queues work, but human factors do not readily accommodate that structure or function. The moral is, 'Don't get out of line—unless you can pay for your own queue energy costs.' "

Transit may be modified by the addition or deletion of problems considered necessary by the group using it. For instance, the current gas and fuel oil shortage could be added as a factor. The author notes that *Transit* is specifically about urban (not global or national or state-based) transportation. (AC)

Cost: $62.50 plus shipping and handling costs

Producer: System's Factors, Inc., 1940 Woodland Ave., Duluth, MN 55803

U-DIG

Ervin J. Bell, University of Colorado

Playing Data
Age Level: college, graduate school, professionals, citizen groups
Number of Players: 4-16 in 3-5 teams
Playing Time: 4-7 hours
Preparation Time: 3 hours

Description: *U-DIG* is designed to model economic factors of real estate operations in U.S. cities and to give participants the opportunity to change exogenous factors and observe the effects relating to improvement of the environment. The game is set in a city neighborhood of six blocks. Teams, functioning as real estate investors, begin with equal capital and seek to maximize their profits and improve the physical quality of the neighborhood. There are two rounds, each representing one year. In each, teams bid on property and arrange financing, sell property and demolish or construct buildings (Lego blocks), and evaluate their investment strategy. (AC)

Note: For more on this game, see the essay on urban gaming simulation by Mary Joyce Hasell.

Cost: between $30.00 and $40.00 (without Lego blocks)

Producer: Ervin J. Bell, 1460 Moss Rock Place, Boulder, CO 80302

URBAN DYNAMICS

Loel A. Callahan, II, Dwight A. Caswell, Jr., Larry A. McClellan, Robert E. Mullen, William N. Savage; Urbandyne, Inc.

Playing Data
Copyright: 1970
Age Level: grade 10-adult
Number of Players: 12-20
Playing Time: 3-4 hours (can be played in 1-hour periods)
Preparation Time: 2 hours the first time
Special Equipment: cassette tape player, optional; 35mm slide projector, optional

Description: This game is played by four teams (Yellow, White, Blue, and Red) that begin play with different mixes of assets and are given the responsibility each round of locating a predetermined number of new assets on a grid playing board that is divided into a central city and suburban area. A round represents 10 years in the life of the metropolitan area represented by the model.

Residential units and population units are identical (R1 = P1). More than one R1 may be placed on a parcel. The two types of businesses (HI and NS) may only be placed on parcels that are zoned as "economic sector." Government land uses (MS) increase with total population size and require a work force from the local population.

Population units are classified as blue collar (PL) or white collar (PM) and are employed by HI, NS, and MS. Owners of land receive rents from the tenants on the land (rents may be raised only with City Council approval) and new land may be purchased in the suburbs.

The game starts in 1920 with each team and the government having prespecified land ownership and an allocation of land uses. In the 1930s the players allocate the new growth to parcels selected using locational, ownership, and bias criteria. The bias is that all new Blue and Red land uses must be placed on original Blue or Red squares (their "ports of entry"). One new MS is added to the local system and jobs are filled. Incomes are earned (fixed for HI and NS unless full employment is not achieved), $6000 per PM and $3000 per PL if owned by Yellow and White and $5000 per PM and $2000 per PL if owned by Yellow and White and $5000 per PM and $2000 per PL if owned by Red and Blue, and $1000 in welfare for unemployed P's. Rents are paid, transportation charges are deducted, and taxes are paid on improved parcels and for HI and NS.

Councilmen are elected from wards that have attained the required population. The council elects the Mayor who then decides which PL's are employed by the government. The council also establishes a school within the system that will begin to educate P1's during the 1940's (the next round of play). The size of the school will determine the amount of space available for P1's in the following round.

In the 1940's the restrictive covenant is removed and P1's may be placed on any parcel provided that the ownership of the parcel permits the PL to rent or buy. Blockbusting may be employed by Blue and/or Red to oust Yellow or White, but Yellow or White can only oust Blue or Red through urban renewal. Blue and Red may purchase land in the suburbs only through dealing with Yellow or White, whereas Yellow or White may purchase suburban land by purchasing it from the bank which represents the outside system.

PL's are placed in the school up to the capacity limit set by the council. Each PL sent to school stays out of the labor force for one round, becomes a PM in the next round, and imposes a cost of $5,000 on the government. During the 1940's costs of building a HI and NS increase tremendously and roads may be constructed into the suburbs. Most of the remainder of the round is the same as before.

During the 1950's, HI and NS units may be built in the suburbs and the City Council may receive federal-state aid on a probability basis. During the 1960's all previous rules apply, unless altered by a majority vote of the City Council. Blue and Red may boycott working for Yellow and/or White. The income from HI and NS units is reduced in proportion to the number of workers boycotting.

The second version of this game deals with future urban growth, but follows much the same sequence round by round. Four rounds (1960 through 1990) are played with teams once again given predetermined amount of new cash and new economic units to locate on the board. The purpose of the game is to illustrate some of the effects of racial prejudice on urban development. (The state-of-the-art in urban gaming models.)

Comment: *Urban Dynamics* is a complex game and might for that reason be more suited to advanced high school classes rather than a general secondary school audience. Because of its complexity, it offers insights into the dynamic process of urban growth and development which more simplified games cannot duplicate. The main impact of the game is on the individual, usually on his cognitive understandings. However, on occasion players get very involved in the game emotionally and find that their values and attitudes are affected by playing the game. Players on the Yellow and White teams often find themselves faced with moral difficulties as they sense a conflict between their own values and the position they feel appropriate to their role in the game. Blue and Red often experience feelings of extreme frustration because of the disadvantages they suffer in the system. The game recognizes that these factors are involved and provides for boycotts by Blue and Red in the game rules. The operator's manual cautions that sometimes "riots" have taken place during play of the game and indicates some steps the director might want to take in order to capitalize on such an event in terms of making it a learning experience. There are no "rules" as to how to stage a riot, however, for there are no rules pertaining to riots in real life. Thus, *Urban Dynamics* has educational value of several dimensions as well as program diversification value.

Because *Urban Dynamics* is a complex game, it is probably best not to "tinker" with it too much. Some games are easy to modify and some are not. This one is not. The operator's manual suggests several variations on the basic game which may be employed and these should be sufficient for the purposes of most people. Despite its complexity, this game is not difficult to administer. There is an audiovisual package available which helps introduce participants to the game. This package may be advisable for someone who has had no previous experience with the game and who is attempting to run it for the first time. (From the Robert A. Taft Institute of Government Study on Games and Simulations in Government, Politics, and Economics by Daryl R. Fair, Trenton State College.)

Note: For more on this simulation, see the essay on urban gaming simulation by Mary Joyce Hasell.

Cost: \$75.00. Average package for introduction to game procedure, \$10.00

Producer: Institute of Higher Education Research and Services, Box 6293, University, AL 35486

See also Formento ECONOMICS
Nexus FRAME GAMES
Policy Negotiations FRAME GAMES
Simpolis DOMESTIC POLITICS
COMMUNITY ISSUES

Part III
BUSINESS LISTINGS

TOTAL ENTERPRISE—COMPUTERIZED

THE BUSINESS MANAGEMENT LABORATORY

Ronald L. Jensen, Emory University; David Cherrington, Brigham Young University

Playing Data

Copyright: 1973, 1977
Age Level: college, graduate school, management
Number of Players: 9-40 in teams of 3-5 (may be increased)
Playing Time: 6 hours or more, about 1-2 hours per decision
Preparation Time: 2 days (not full-time)
Special Equipment: computer; any FORTRAN IV can be adapted

Description: This simulation models a flatware manufacturing industry of up to eight competing firms. Participants represent management teams that are replacing previous management. They start from essentially the same position (but the instructor can vary this) and compete for market share. Decisions, many of which are options that may be included or deleted by the instructor, include marketing and promotion plans (salesmen, advertising, budget), capital investment plans (plant), cash management, production (plant balancing, shift operations, materials purchasing, inventories management), budgeting, and dividends. Because the complexity can be varied by the administrator, the simulation is adaptable to a range of situations–introductory-level business and management courses, graduate-level courses, and executive training and development. (AC)

Note: For more on this simulation, see the essay on total enterprise business simulations by J. Bernard Keys.

Cost: player's manuals $7.50 each; request prices for other materials supplied to adopters

Producer: Business Publications, 13773 North Central Expressway, Suite 1121, Dallas, TX 75243

THE BUSINESS POLICY GAME

Richard V. Cotter, University of Nevada

Playing Data

Copyright: 1973
Age Level: college, graduate, management
Special Prerequisites: advanced college level or experience
Number of Players: 9 or more players in 3 or more teams
Playing Time: 16-60 hours in periods of 1-3 hours
Preparation Time: 2-10 hours
Special Equipment: computer; FORTRAN II and IV

Description: This management education game is designed to help players learn to work effectively as a management team. Players assume the roles of top managers of the manufacturer of a single imaginary product. Management teams must set goals for the company and formulate and implement business strategies and policies to maximize profits. The company markets its products in four geographical regions. Players must make pricing, advertising, sales force, research and development, marketing, production, and financial decisions for each region. (DCD)

Comment: This game was tested in classes and general colleges and 6 years of intercollegiate competition. (BK)

Cost: player's manuals $7.95 each; computer deck $15.00

Producer: Prentice-Hall, Englewood Cliffs, NJ 07632

BUSOP

Jerald R. Smith, University of Louisville

Playing Data

Copyright: 1974
Age Level: grade 9-college, management
Number of Players: 9-60 in teams of 3-5
Playing Time: 1/2-1 hour
Preparation Time: 2 hours
Special Equipment: computer, Hewlett-Packard 2000 A or equivalent; BASIC

Description: This game is designed to teach players how an independent manufacturing company operates, to help them learn to cooperate as managers with responsibilities, and to sensitize them to the socially responsible and ethical decisions which the managers of such a company must make. Players assume the roles of operating officers of a small, single-product company that makes and markets radios. Their objective is to operate the company ethically, efficiently, and profitably. Consequently, the players must make and report decisions concerning marketing, production engineering, research and development, plant size, capital procurement, dividends, and pricing. The consequences of the decisions are simulated by computer. (DCD)

Comment: Although this is an otherwise typical computerized business game, the author incorporates "off-line" qualitative decisions for each decision set. These include such things as business responsibility and ethical problems with several alternatives for each problem. The team's choice of an alternative is entered as a decision into the game with resultant consequences becoming output in succeeding periods. (BK)

Cost: player's manuals $3.00 each; request prices for time-sharing or mail service

Producer: Simtek, P.O. Box 109, Cambridge, MA 02139

DECISION MAKING EXERCISE

John E. Van Tassel, Boston College

Playing Data
Copyright: 1970, 1978
Age Level: college-management
Number of Players: 15-30 in teams of 3-5
Playing Time: 75-150 1-hour periods
Preparation Time: 15-20 hours
Special Equipment: At present, the game is operating on a DEC-STATION 78 Micro-Computer and in a time-sharing version.
Packaging: professionally packaged game includes players' manuals, forms, and computer program

Description: This game is designed to allow players to learn to apply and integrate the planning process into a dynamic and competitive environment in order to understand the system of managerial controls necessary to achieve business objectives. Teams must specify goals in terms of profit or market share.

The players form into management teams which are responsible for business decisions in a company which produces low-priced, durable consumer goods. Teams must project demand and sales for these products, develop new products, raise and invest capital, work within a balanced budget, and pay the production and sales forces.

Game outcomes include profit, stock price, share of market, cost of production. There are random variations in some of these, and there is uncertainty about such elements as rate of growth and the business cycle, which is designed to reflect reality. (DCD)

Comment: The simulation was designed as a coordinate part of a two-year Master's program in Business Administration as an integrating device in the final year following the first-year preparation in management processes and business operations. It is also used in a similar fashion in a six-week management development program. The objective is to emphasize the importance of planning and to demonstrate the necessity for coordination among the operations. (DZ)

Cost: $15,000 for outright purchase or $3,000 per year—both subject to 20% educational discount

Producer: Decision Associates, P.O. Box 392, Westwood, MA 02090

ELECTRONIC INDUSTRY GAME

Francisco James

Playing Data
Copyright: 1978
Age Level: graduate school, management
Number of Players: 4-42 in 4-14 teams
Playing Time: 6 decision periods, 2 hours each
Preparation Time: 2 hours
Supplementary material: available from Simtek or from Intercollegiate Case Clearing House. Electro Industries, Inc. (C) and a Note on the Boston Consulting Group Concept of Competitive Analysis and Corporate Strategy
Special Equipment: remote terminal optional

Description: This simulation is designed to teach the impact of portfolio management decisions on corporate success. Participants represent top management of an electronics instrumentation firm competing for market shares in six distinct product groups. Their objective is to execute a strategy that will not only insure survival of the firm in the long run but also place it in a leadership position in the market segmentation they choose. Participants can make up to twelve decisions; there is an option to eliminate up to three. The decisions include investment in plant and equipment, shares to be issued, shares to be repurchased, promotion for each of the six divisions, and three decisions on which divisions to sell or acquire. Participants can run their firms by liquidating their least profitable divisions, acquiring or merging with other divisions to gain economies of scale, or by diluting their earnings risk over a fully diversified product line. Several strategies can yield favorable results, but consistency in implementing a strategy is essential to success. (Producer)

Comment: The producer notes that this simulation was used in the First Annual National Management Tournament with participants from over twenty universities and over twenty-two corporations. (AC)

Cost: student manual $4.00; program deck rental $7,000/year; full service by mail (including student manual) $21/person. Quantity discounts available

Producer: SIMTEK, P.O. Box 109, Cambridge, MA 02139

EXECUTIVE DECISION MAKING THROUGH SIMULATION

Paul R. Cone, Douglas C. Basil, Marshall Burak, and John Megley

Playing Data
Copyright: 1971 (second edition)
Age Level: junior college-graduate school, management
Number of Players: 12-40 in teams of 3-8 (teacher with more students can start a second industry)
Special Prerequisites: quantitative methods, management courses
Playing Time: 6-30 in 1-hour periods
Preparation Time: 2 hours
Special Equipment: computer, preferably 360 model; FORTRAN

Description: This simulation allows students to adjust their strategy through several ongoing rounds. Individual players assume the roles of corporate officers in a tire company. Play is by teams, each of which represents a competing company, who have equal resources at the start of play. Each team must try to fulfill short- and long-range objectives in about eight different areas within the context of industry-wide competition and an always changing business climate. Generally, teams (companies) try to increase profits and cut costs. Winning is an individual matter based on accurate forecasting and the achievement of sales, profit, and financial objectives.

In the course of the game, players are called upon to make decisions concerning the following: share of market, rate of growth, market segments, market geographical areas, product mix, contract bids to auto companies, sales staffing, training, salaries, advertising (national and regional), pricing, warehousing, shipping, insurance, taxes, production levels, inventory control and balancing, staffing of plants, construction or leasing of plants, construction or transfer of lines, automation, line conversions, market researching, pollution expenditures, research and development, quality control, labor negotiations, safety, financial requirements, cash flow and planning, dividend policy, capital structure, short-term borrowing, long-term funds, executive compensation, and profit improvement. (DCD)

Comment: This is an unusually well-detailed business game based on a single company (Uniroyal, Inc., Tire Division) and forecast data from standard sources. The realistic, rather than theoretic, data base may have advantages. The basic game involves only top management and only one product; replay at Level II can involve top and middle management (headquarters and divisions) and diversified products.

With 35 different decision areas shared among five functional roles, the major challenge is making personal strategies compatible within each team; transactions are mostly of the team-coordination type (Marketing has to meet with Finance, Personnel with Marketing, Marketing with Logistics, and so on). The constraints demand interaction under stress, and thus the game provides an excellent laboratory for the study of group dynamics and the learning of interpersonal relations skills; a "behavior" section is contained in the administrator's manual.

Executive Decision has been widely adopted and has been in use for years. Results of this "field testing" are available from the authors or the publisher. (DZ)

Note: For more on this simulation, see the essay on total enterprise business games by J. Bernard Keys.

Cost: player manuals $7.95 each

Producer: Charles E. Merrill Publishing Co., 1300 Alum Creek Drive, Columbus, OH 43216

THE EXECUTIVE GAME

Richard C. Henshaw, Jr., Michigan State University; James R. Jackson, UCLA

Playing Data
Copyright: 1978
Age Level: college, graduate school, management
Number of Players: 9-45 in teams of 3-9
Playing Time: 4-20 hours, with an initial time period of 2 hours and time periods of 20 minutes-1 hour thereafter
Preparation Time: 8 hours
Special Equipment: computer (debugged programs available for several computers. FORTRAN makes conversion to most machines easy.)

Description: This game, according to the authors, "is a 'dynamic business case' in which the outcome is determined by interactions within and between the participating teams in the framework of the economic structure which is programmed into the computer." Two models at different levels of complexity suit the game to beginning or advanced students.

The computer simulates a small industry in which there are a few companies manufacturing and selling a single product. Players represent top management of companies in competition with each other. Each team makes quarterly decisions on price of product; marketing budget; research and development budget; maintenance budget; production volume scheduled; investment in plant and equipment; purchase of materials; and dividends declared. When these decisions have been made, the data they contain are fed to the computer which generates the following reports for each firm: information on competitors; market potential; sales volume; percent share of industry sales; production this quarter; inventory finished goods; plant capacity next quarter; income statement; cash flow statement; and balance sheet. (DCD)

Note: For more on this game, see the essay on total enterprise business games by J. Bernard Keys.

Cost: player's manuals $6.95 each

Producer: Richard D. Irwin, Inc., 1818 Ridge Road, Homewood, IL 60430

THE EXECUTIVE SIMULATION

Bernard Keys, Tennessee Technological University; Howard Leftwich, Oklahoma Christian College

Playing Data
Copyright: 1973, 1977
Number of Players: teams of 3-15, up to 8 teams per industry
Age Level: college, graduate school, management
Playing Time: 2 hours or more per decision set
Preparation Time: 3 hours minimum
Special Equipment: IBM 360 or 1130; FORTRAN IV

Description: Players represent top executives and middle operating management. Decisions may include, for marketing–price, advertising, number of salesmen, commissions, R&D, distribution centers, sales forecasting information; for production–capacity and production scheduling; for financing–notes payable, capital stock issued, dividends paid. Participants submit decisions to the computer and receive one-page financial statements as feedback after each round. The administrator controls economic indices, variable input, and starting positions (cash, inventory, production capacity, and financial statement variables). The player's manual includes readings on business policy and strategy. (AC)

Comment: *The Executive Simulation* has been extensively field tested by college undergraduates and graduate students and executives in several nationally known firms and has been adopted by twenty colleges and universities. (BK)

Note: For more on this simulation, see the essay on total enterprise business games by J. Bernard Keys.

Cost: (1) player's manuals $6.95 each; card decks furnished by publisher on adoption; (2) request prices of time-sharing or mail service

Producer: (1) Kendall/Hunt Publishing Company, 2640 Kerper Boulevard, Dubuque, IA 52001; (2) SIMTEK, P.O. Box 109, Cambridge, MA 02139

THE IMAGINIT MANAGEMENT GAME

Richard F. Barton, Texas Tech University

Playing Data
Copyright: 1971, 1973
Age Level: college, graduate school, management
Number of Players: minimum of 2 teams of 2, no maximim; can be played solitaire
Playing Time: 3 days-15 weeks or longer (6-15 or more decisions)
Preparation Time: varies with complexity, which is controlled by game administrator
Special Equipment: computer, FORTRAN; a segmented modification is available for small computers

Description: The designer notes:

> "*The Imaginit Management Game* is designed to serve a broad spectrum of educational and research purposes. In each use, it is highly specific to the needs and objectives of the teacher or researcher and his player participants. The game may be custom designed by the user and players, or a ready-made version of the game may be adopted. All game controls, industry parameters, starting positions, and qualitative indexes are accessible to users. These features provide generality and flexibility from play to play and also enable the game user to adapt the game to his needs and to player behavior after a play has begun. By experimental manipulation of controls and parameters between plays or during a play, the game can serve as a vehicle for research into managerial behavior."

Available industry versions are: breakfast cereals, early television, automobiles, encyclopedias and texts, home laundry, home climate, home computers, tires, post-World War II typewriters, and washing products. Products may be sold in two markets, and one may contain two competing products from the same company. A decision may be comprised of as many as twenty-eight elements; these include: price, dollars of materials inputs per unit, expenditures for salesmen, expenditures for advertising, investment in product research and development, dollars of materials to be ordered, number of units to be produced, employee fringe benefits per hour, dividends per share to be paid, expenditures for operations research, short-term loans to be made or repaid, bonds to be issued or redeemed, shares of stock to be offered, and dollars of factory capacity to be purchased or sold. (AC)

Note: For more on this game, see the essay on total enterprise business games by J. Bernard Keys.

Cost: For course credit for a degree, there are no charges for use of computer program if player's manual is adopted. Higher education users are asked to sign a no-charge license agreement. Minitape and mailing costs may be incurred. For other uses, write for charges.

Producer: Active Learning, P.O. Box 16382, Lubbock, TX 79490

INTEGRATED SIMULATION

W. Nye Smith, Elmer E. Estey, and Ellsworth F. Vines, Clarkson Institute of Technology

Playing Data
Copyright: 1974 (second edition)
Age Level: college, graduate school, management
Number of Players: 4-96 in 4-12 teams
Playing Time: 4-17 hours in 30-minute periods
Preparation Time: 1 hour
Special Equipment: computer; FORTRAN II and IV Referee decks available for IBM 1130 and IBM 360/370 40

Description: *Integrated Simulation* requires students to act as the officers of an industrial corporation who must make key decisions affecting the operation of their corporation during the next quarter of a fiscal year. Successful play requires students to integrate the basic disciplines of accounting, finance, marketing, management, and data processing. Several companies (all producing the same product, selling to the same market, and all of the same size at the start of the simulation) compete with one another. Record keeping is done by computer. (DCD)

Note: For more on this simulation, see the essay on total enterprise business games by J. Bernard Keys.

Cost: (1) player's manuals $5.40 each; computer deck $20.00; (2) request prices of time-sharing or mail service

Producer: (1) South-Western Publishing Company, 5101 Madison Road, Cincinnati, OH 45227; (2) SIMTEK, P.O. Box 109, Cambridge, MA 02139

INTOP (INTERNATIONAL OPERATIONS SIMULATION)

Hans B. Thorelli, Robert L. Graves, and Lloyd T. Howells

Playing Data
Copyright: 1964
Age Level: college, graduate school, management
Number of Players: 12-175 in 4-27 teams
Playing Time: 7-15 hours in 40-75-minute periods
Preparation Time: 3 hours
Special Equipment: IBM 7040, 7044, 7094; FORTRAN II, IV. IBM 360 OS version, Control Data 6600, IBM 370; FORTRAN IV
Supplementary Material: Thorelli and Graves, *International Operations Simulation–with Comments on Design and Use of Management Games* (for instructors)

Description: *INTOP* provides for three to fifteen company teams, each consisting of four to seven executives, whose internal organization may be oriented to product, to function, or to geographical area. The simulation is designed to focus the player's attention on the problems of planning and coordinating far-flung, decentralized operations. The game is played in rounds (or decision periods) which represent three months.

The companies may operate in one, two, or all three of the following areas: Brazil, the United States, or the European Economic Community. Each area has its own characteristics in terms of size, demand, and production functions, as well as economic climate and governmental policies with regard to corporate taxation and international trade. Each company may manufacture and/or market either or both of two different products, and there are transfer costs between areas. Companies may also hold patents, sell two grades of each product in each area each quarter, invest profits to improve plant efficiency, market products in several ways, and conduct marketing research. Several devices for absorbing surplus liquidity are incorporated into the simulation. The data for decision making (a balance sheet, income statement, marketing research survey, and ancillary data) are generated by computer. (DCD)

Note: For more on this simulation, see the essay on total enterprise business games by J. Bernard Keys.

Cost: manuals $5.95 each; instructor book $15.00

Producer: Free Press, Macmillan Company, 866 Third Avenue, New York, NY 10022; for program tapes, contact Hans B. Thorelli, School of Business, Indiana University, Bloomington, IN 47401

THE MANAGEMENT GAME

F. Warren McFarlan, James L. McKenney, and John A. Seiler

Playing Data
Copyright: 1970
Age Level: graduate school, management
Number of Players: 20 or more in 5 or more teams
Playing Time: 3-10 days
Preparation Time: 1 month minimum
Special Equipment: 360/50 or larger computer; time sharing like GE 235 optional

Description: Participants assume the roles of managers of manufacturing firms selling generically defined products. In each decision period, participants are responsible for about fifty decisions in the areas of marketing, finance, and production. Marketing decisions concern forecasting, product research and development, pricing, marketing areas, advertising, and other promotion. Financial decisions concern the purchase and sale of stock, payment of dividends, investments, and loans–long-term, short-term, and special. Production decisions concern scheduling, what materials and how much to purchase, hiring and scheduling of workers, inventory costs, and plant capacity. (AC)

User Report: *The Management Game* is a very complex simulation. It requires players to analyze, integrate, and balance the business functions of marketing, production, and finance in a market determined by changing economic and competitive factors. The game offers a realistic and motivational mix of predictable and unpredictable events. A large quantity of data output is generated, which requires discerning analysis by players. Since the game offers poor initial conditions for each firm along with a propitious atmosphere for recovery, the players must contend with the diverse problems associated with both very poor and very good business performance. Players report increased awareness of the functional interrelationships in business, and increased awareness of the problems of adaptability and stability in organizational management. (Ralph M. Roberts, Faculty of Management, University of West Florida)

Note: For more on this simulation, see the essay on total enterprise business games by J. Bernard Keys.

Cost: manuals $8.95 each

Producer: The Macmillan Company, 866 Third Avenue, New York, NY 10022

MADS-BEE MANAGING A DYNAMIC SMALL BUSINESS

James E. Estes, University of South Carolina; and C. Brian Honess, University of South Carolina

Playing Data
Copyright: 1973
Age Level: college, graduate school, management
Number of Players: any number of teams or players
Playing Time: variable
Preparation Time: variable
Special Equipment: computer: moderate size or subroutine on small computer. FORTRAN IV. 30 K. IBM 360/65G compiler or WATFIV compiler

Description: Players assume the roles of the operations managers of a small company which manufactures from one type of raw material a single product used in building construction. Players must maximize the company's profit while making twelve decisions (including marketing, production, and finance) during each quarter of the fiscal year.

The game is designed for junior level introductory management courses, business policy courses, or executive development programs, to vitalize concepts applied to short-term and long-range planning, budgeting, and business strategy and policy, and to humanize the student's comprehension of business firms by simulating a small company that the participant can comprehend and in which the employees can be thought of as individuals rather than as the "workforce" of a "macro" approach. (DCD)

Comment: This simulation has been an integral part of the University of South Carolina's Executive Development Program for many years. It is a moderately complex game allowing administrator control of price elasticity and market indices. The seasonal variation is typical of a small product in the production industry. Optimal one- or two-product operation can be introduced, posing the challenge of new product obsolescence. Companies can bid for contracts to supply the original products. The game is played by individuals or teams. (BK)

Cost: player's manual $4.95 each; instructor's manual and program deck $25.00

Producer: Wentworth Corporation, 615 Meeting Street, West Columbia, SC 29169

POCKET CALCULATOR BOOM

Francisco James

Playing Data
Copyright: 1978
Age Level: graduate school, management
Number of Players: 4-40 in 4-20 teams (usually 1-2 per team)
Playing Time: 5 decision periods, 2 hours each
Preparation Time: 2 hours
Special Equipment: remote terminal optional
Supplementary Material: available from SIMTEK or from Intercollegiate Case Clearing House: Experience and Cost: Some Implications for Manufacturing Policy; also, Bowmar Instruments Corporation

Description: This simulation is designed to teach the impact of the Learning Curve Effect on manufacturing policy. Participants represent top management of a pocket calculator firm that leads the industry in the production and marketing of its product, and their objective is to survive the dramatic industry shakeout with a dominant market share. They can run their firms individually or in pairs to face an exponential growth rate (usually 100%) and dramatic drop in costs (around 50%). They have a total of six decisions: production (units), price, promotion budget, shares issued, shares repurchased, and investment in plant and equipment. Several strategies (high quality, low cost, me-too) can prove successful, but consistency in strategy implementation is essential to success. Game parameters were extracted from the real industry during the booming early 1970s. (producer)

Comment: The designer notes that this simulation requires more than cursory understanding of marketing, accounting, forecasting, finance, and competitive decision making, and that it has been used with favorable comments at a Fortune 100 computer company and a well-known eastern business school. (AC)

Cost: student manual $4; program deck annual rental $5,000; full service (including manual) by mail $16/person. Quantity discounts available.

Producer: SIMTEK, P.O. Box 109, Cambridge, MA 02139.

SIMQ

Arthur C. Nichols and Brian Schott, Georgia State University

Playing Data
Copyright: 1972, 1973, 1975
Age Level: college, management
Number of Players: 8 or more in 4 or more teams
Playing Time: 12 hours (est.) in 4 rounds
Special Equipment: computer with 48 K capacity
Packaging: professionally produced game materials include 48-page game manual

Description: The objectives of this business simulation game are (1) to provide a competitive, dynamic, and probabilistic environment in which students will have the opportunity to apply a number of quantitative decision tools; (2) to give students a better understanding of quantitative models, including the subjective aspects of model use and design, and of how these models assist in business problem-solving and decision-making; and (3) to encourage students to view quantitative techniques realistically, allowing them not only to see the advantages of the use of quantitative methods in business decision-making but also to see the limitations and disadvantages of their use.

In the game, each team will serve as a management team. There will be several management teams, each team managing a different company in the industry. The business simulation does not represent any particular industry but includes certain features of the industrial world in general. Each company produces and sells three products, Product X, Product Y, and Product Z. The companies sell the products to common markets, and are faced with the same operating and manufacturing expenses. In addition, the companies have similar historical backgrounds and start the year with identical financial statements. (Author)

Comment: This game is very similar in its basic format to many of the general management aggregate variable games on the market. It is an imaginary multi-product total enterprise game in which players represent top management teams. It is an oligopolistic type of game requiring quarterly decision inputs by team per hypothetical quarter. However, the game is much richer in the production and material scheduling areas than most general management games. Its unique and most outstanding feature is the teaching package of classroom/short/course quantitative assignments available to accompany the game. It is likely that *SIMQ* will be used frequently in courses known as "quantitative methods" by seniors and graduate students. (BK)

Cost: (1) player's manuals $3.25 each; computer program $25.00; (2) request prices of time-sharing or mail service

Producer: (1) Kendall/Hunt Publishing Company, 2640 Kerper Boulevard, Dubuque, IA 52001; (2) SIMTEK, P.O. Box 109, Cambridge, MA 02139

SIMQUEUE—A QUEUEING SIMULATION MODEL

Gary Wicklund, University of Iowa

Playing Data
Copyright: 1969
Age Level: college, graduate school
Special Equipment: computer; batch FORTRAN
Packaging: copy of 57-page student manual; software (batch FORTRAN program, 900 lines of code)

Description: This computerized model is designed to teach theoretical concepts of queuing theory by allowing students to manipulate input data and observe changes in output of the system. The SIMQUEUE program simulates a system consisting of parallel single server channels with first-in, first-out queues, providing information on: the relationship of different queue waiting times, service times, arrival rates, number of service stations, arrival distributions, service distributions, cost of servers, and cost of units waiting for service.

This unit allows questions about queueing system problems to be formulated and analyzed on an individual basis. Students may vary the number of servers and the nature of the arrival/service distributions at each time period in order to produce realistic simulations of the process. The package is suitable for use in advanced undergraduate or graduate level courses teaching quantitative techniques. (CONDUIT)

Cost: $30.00; additional student manual $5.50, 2 or more $5.00 each

Producer: CONDUIT, Box 388, Iowa City, IA 52240

TEMPOMATIC IV

Olice Embry, Alonzo Strickland, and Charles Scott

Playing Data
Copyright: 1974
Age Level: college, graduate school, management
Special Prerequisites: introductory accounting, marketing, production courses helpful
Number of Players: 3-4
Playing Time: 2-3 hours
Preparation Time: 3-5 hours
Special Equipment: computer; FORTRAN IV

Description: Participants represent management teams or companies in this simulation, which the instructor can vary from introductory to advanced levels. The instructor selects one of nine demand curves and controls major variables. Some of the decisions for which participants are responsible may include national and local advertising, salesmen, product improvement, price, shipment of units to various areas, materials ordering, loans, dividends, marketing information, production units, hiring, short-term investment stocks issues, bonds issued, and others. At the introductory level, participants learn about business operations through trial and error. At the advanced level, participants test quantitative tools. (AC)

Comment: This game has been well tested and in use since 1959. It is one of the more sophisticated games described in this section and would serve well in senior or graduate courses where at least fifty percent of a semester could be devoted to its use. (BK)

Note: For more on this simulation, see the essay on total enterprise business simulations by J. Bernard Keys.

Cost: student manuals $7.95 each; instructor's manual $1.75; source deck free to adopters

Producer: Houghton Mifflin Company, One Beacon Street, Boston, MA 02107

TOTAL ENTERPRISE—MANUAL

BUCOMCO

John Melrose, University of Wisconsin at Eau Claire

Playing Data
Copyright: 1977
Age Level: college, management
Number of Players: 8
Playing Time: 8 hours (est.)
Preparation Time: 2 hours (est.)
Packaging: professionally produced 172-page workbook and instructor's guide

Description: The players of this in-basket exercise write letters, memoranda, and reports, fill out forms, and take part in small group meetings which may include role playing. Each assumes the role of an upper echelon manager of *BUCOMCO,* a division of the imaginary "Business Communications Corporation" and a manufacturer of teaching machines. At the start of play, participants must complete an application for employment for one of seven executive positions including Corporate Vice-President and the managers in charge of Purchasing, Production, Credit, Industrial Relations, Marketing, and Education. An eighth player assumes the role of *simulator.* The seven executives must then respond in writing to 60 items of correspondence. Their work is open to criticism by the other players. The simulator portrays the writer of these items (for example, dissatisfied and prospective customers, or a job seeker) in short role-playing sessions. *BUCOMCO* ends when all of the in-basket items have been completed. (DCD)

Cost: workbooks $5.75 each, instructor's guide $1.50

Producer: Science Research Associates, 1540 Page Mill Road, Palo Alto, CA 94304

EXECUTIVE SIMULATION GAME

W. D. Heier, Arizona State University

Playing Data
Age Level: college, management
Number of Players: 12-35 in teams of 3-7
Playing Time: 4 hours in 15-minute periods
Preparation Time: 1-2 hours

Description: This is a simulation exercise where the decisions of the participants affect the subsequent flow of the proceedings. The "company teams" normally require an initial one-hour orientation briefing on the conduct of the game, following which play begins.

The flow of the game consists of the following steps: (1) the various teams submit their requests for market research information; (2) the umpires evaluate the salesmen's calls on customers and inform the teams whether or not the salesmen have succeeded in achieving a sale and, if so, the number of units sold (3) the teams prepare the financial statements for the quarter's activity; and (4) plans for the next quarter's activities are formulated.

Each decision made, whether a good, marginal, or poor one, becomes an integral part of the play of the game. The structure of the game conclusively demonstrates the necessity for integrating and consolidating *all* of the corporate activities if the group is to achieve a high level of decision-making proficiency.

An important part of the exercise is the postgame analysis of the decisions made and the rationales upon which those decisions were based. It is particularly useful during this phase to reinforce the fact that many decisions may appear to have a minimal level of acceptability, but only those decisions considering the total corporate environment prove to be the most effective ones for the corporation. (Author)

Comment: Some basic understanding of single-entry accounting would lead to more successful play, but it is not necessary. Chance "loads" the marketplace with sales dollars before play; during play, it influences the making of sales, possible loss of a salesman, and the probability of R&D success. Comparisons between teams are on the basis of "scoring values"; dollar results are secondary. (DZ)

Cost: available commercially on limited basis

Producer: W. D. Heier, Management Department, College of Business Administration, Arizona State University, Tempe, AZ 85281

EXMARK

Management Games, Ltd.

Playing Data
Copyright: 1975
Age Level: college, management
Number of Players: 9-20 in teams of 3-4
Playing Time: 8.5 hours
Packaging: professionally produced game includes instructor's manual, 24 copies of participant instructions and forms

Description: This exercise is designed, according to the producer, "to create an awareness that decisions taken in the function of an organization affect each other; and to show the necessity for organization and coordination within the management team." It simulates the operation of a company in the U.K. which is given the opportunity of exporting to West Germany, Sweden, and Spain. Working in teams, the players must predict the future behavior of each market, decide to which markets to export, and then make various pricing, selling, and production decisions. The results of all decisions are determined from a set of charts included with the game materials. (DCD)

Cost: £48

Producer: Management Games, Ltd., 63B George Street, Maulden, Bedford MK45 2DD, United Kingdom

FUNCTION

Management Games, Ltd.

Playing Data
Age Level: management
Number of Players: 3-6 teams of 4-6
Playing Time: 6 periods of 10-45 minutes each

Description: The producer categorizes this as "a competitive exercise which deals with the interaction of various functions (Purchasing, Production, Sales and Credit Control, Finance) and how these functions must work together to produce a profitable enterprise." In the game, several companies make and market a similar product. "They are not 'interacting' in the sense that a good strategy by one will damage the others but all are beset by the same outside conditions and their results are directly comparable." The results of all decisions are calculated from a set of charts included with the game materials. (DCD)

Cost: £48

Producer: Management Games, Ltd., 63B George Street, Maulden, Bedford MK45 2DD, United Kingdom

LOW BIDDER

William R. Park

Playing Data
Number of Players: 2-25 players
Playing Time: 2 hours or more

Description: *Low Bidder* is designed to resemble the conditions encountered in bidding for contracts in such industries as printing, construction, and aircraft. The game is set up so that an entire year's operation can be viewed at one time. Competing participants bid for a variety of construction jobs. Each player begins with the same amount of capital, has access to the same jobs, and is subject to the same operating costs. The strategy of the game lies in choosing which of the several alternative jobs to bid and deciding how much to bid so that there is a reasonable chance of getting the job at a fair profit. Each round of play represents a year's operation. (REH)

Comment: The basic element in this game is recognizing that there is some one bid that results in the best combination of (1) the profit resulting from obtaining a contract at a specified bid price and (2) the probability of getting the job by bidding that amount. In an accompanying article by W. R. Park of the Midwest Research Institute, a method is shown for how to win the bidding game. The article provides a statistical analysis of the situation. If your organization or part of it fits into this type of model, the learning elements of this game are definitely for you. (REH)

Cost: $10.00

Producer: Business Studies, Inc., 104 Arlington Ave., St. James, New York, NY 11780

MANAGEMENT: AN EXPERIENTIAL APPROACH

Harry R. Knudson, Robert T. Woodworth, Cecil H. Bell; University of Washington

Playing Data
Age Level: college, management
Number of Players: varies; usually 5-6 per team
Playing Time: approximately 2-3 hours per exercise
Preparation Time: included in playing time

Description: This is a teaching package of 23 role play, business games, and reflective exercises including both narrative information and exercises varying greatly in subject matter and approach but covering basically the areas covered by a basic management course. It can be used as a supplement to a principle of management text. Play involves every type of decision making and thinking. Level of management varies greatly. Typical exercises: Role play of one-way versus two-way communications; Patterns of Organizational Communications; Barriers and Gateways to Communications in Organizations; Skits depicting various managerial assumptions—rational economic man, social man, and so on. Different ways of decision making; Operation suburbia—business game; Managing change in organizations. (BK)

Cost: $10.95

Producer: Mc-Graw Hill Book Company, 1221 Avenue of the Americas, New York, NY 10020

MEGGA TRADING COMPANY

Management Games Limited

Playing Data
Copyright: 1975
Age Level: college, management
Number of Players: 18 in 3-6 teams
Playing Time: 5 hours

Packaging: professionally produced game includes instructor's manual, 18 player manuals, and forms

Description: This exercise concerns, in the producer's words, "a small wholesaling or stockholding business which has to 'buy in' and 'sell out,' through its own sales force and by use of advertising. Only one product is involved." Teams represent wholesalers or distributors who must determine the best suppliers (in terms of profit and reliability), the best price for their product, how many salesmen to employ, and how much to invest in advertising and market research. The results of all decisions are determined from a set of four charts. (DCD)

Cost: £48

Producer: Management Games, Ltd., 63B George Street, Maulden, Bedford MK45 2DD, United Kingdom

THE ORGANIZATION GAME

Robert H. Miles, Harvard University, and W. Alan Randolph, University of South Carolina

Playing Data
Copyright: 1979
Age Level: college, management
Number of Players: 15 or more
Playing Time: 8 hours (est.)
Preparation Time: 10 hours (est.)
Packaging: professionally produced, 325-page, oversize softbound book

Description: This complicated simulation was developed, the authors state, "to provide a realistic setting in which students and managers can experience fundamental concepts and issues in Organizational Behavior, Organization Theory and Management, and Organizational Change and Development and experiment with alternative organizational behaviors and designs."

The complex organization postulated in the simulation is comprised of four "Divisions" and seven semi-autonomous "Basic Operating Units" (two producing units, an employee relations unit, an information processing unit, an internal management consulting team, an internal audit unit, and an interest group). The wealth of the organization is measured in "ORG$" and the simulation includes a detailed set of guidelines for measuring performance. (The authors summarize these performance indicators as "Resource Base, Total Output, Internal Cohesion, and Member Commitment.") "ORG$" are distributed by the four "Division Heads" to the "Unit Heads," who must pass the ORG$ on to the members of their units in such a way as to encourage the best possible performance. High performance increases the wealth of the Units and the organization as a whole and low performance diminishes it. None of the players, except for the Unit and Division Heads, is assigned any specific task. The participants must invent and sustain their own jobs. The simulation may be continued indefinitely. (DCD)

Comment: Highly recommended. The nature of this simulation makes it impossible to summarize the structure of the simulated organization except in the most general terms and impossible to describe the various processes that the participants may discover during play. Nevertheless, this is a four-star laboratory exercise in Organizational Development. (DCD)

Cost: $9.95

Producer: Goodyear Publishing Company, 1640 Fifth St., Santa Monica, CA 90401

SMALL BUSINESS MANAGEMENT EXERCISE

Management Games Limited

Playing Data
Age Level: management
Number of Players: 6-21 players in teams of 2 or 3
Playing Time: 6-8 hours including briefing and evaluation sessions

Description: Teams represent the management of competing companies in the same industry. Each must try to enlarge its share of the market for the product at the expense of its competitors. The teams have the power to decide the price and the quality of their company's product, how much to invest in advertising, market research, and product research and development, and how many units of the product to manufacture. The market share earned by any company depends upon its policies and upon decisions made by the competition. For example, a lowering of price by one company may not lead to increased sales if the other companies also reduce their prices and take other competitive actions. The results of all decisions are obtained from charts. The winner is the team that has most increased its market share by the end of the game. (DCD)

Cost: (1) $85.00; (2) £48.00

Producer: (1) Didactic Systems, Inc., Box 457, Cranford, NJ 07016; (2) Management Games Ltd., 63B George Street, Maulden, Bedford MK45 2DD, United Kingdom

SNIBBO METAL PRODUCTS

Management Games Limited

Playing Data
Copyright: 1975
Age Level: management
Number of Players: 2-20 groups of 2-3
Playing Time: 1-17 hours
Packaging: professionally produced game includes instructor's manual and 10 participant manuals

Description: This game is designed, according to the producer, "to make participants aware of problems of business strategy as well as day to day operating problems." Teams represent the management of the Snibbo Group of Companies which includes Snibbo Metal Products and its two subsidiaries, Ore Limited and Nuts and Bolts Limited. They read a booklet which contains a history of the Snibbo Group and an outline of its operations. They must then respond to a variety of memoranda and correspondence which pertain to wages, recruitment, the divestment of Ore Limited, pricing, capital investment, scheduling, personnel, safety, and a new roof for the Snibbo Metal Products factory. (DCD)

Cost: £48

Producer: Management Games, Ltd., 63B George Street, Maulden, Bedford MK45 2DD, United Kingdom

TOP OPERATING MANAGEMENT GAME

Jay R. Greene and Roger L. Sisson

Playing Data
Copyright: 1971
Age Level: college, management
Number of Players: 4 or more in teams of 2-5
Playing Time: 3 hours and up
Preparation Time: 30-60 minutes

Description: Players represent, in the designers' words, members "of a closely knit management team that is competing directly with several other companies for a share of an industrial market. All the companies are selling a product that is technically similar. Price and promotional effort are the key elements affecting volume. Profits result from a careful assessment of market demand, competitor's activities, and sound production and expense planning and control."

During each round, the teams must make decisions affecting their companies' promotion budgets, production costs, production capacities, and the selling prices of their products. All decisions are then graded by the instructor with charts included in the game materials. and the results are reported to the teams for the start of the next round. The game lasts for 12 rounds (each corresponding to one month) and the winner is the team that has profited most. (DCD)

Comment: This is a very simple business game that may be useful for stimulation of group decision processes. The strategy options are too few and the model sophistication too simple to provide much in the way of analytical or stragetic decision-making exercise. Its chief purpose will probably be as suggested: awareness training. (BK)

Cost: 5-participant set, $10.50

Producer: Didactic Systems, Inc., Box 457, Cranford, NJ 07016

TRIG

Management Games, Ltd.

Playing Data
Copyright: 1975
Age Level: grade 9-management
Number of Players: 4-21 in teams of 2-3
Playing Time: 3.5-4.5 hours in 10-12 decision periods
Packaging: professionally produced game includes instructor's manual and 21 copies of participant instructions

Description: Players imagine that they are in charge of a small factory about to begin production of two new products called "prods" and "tims." Teams must control planning, purchasing, machine allocation, costing and pricing in a way to produce prods and tims at a price that will insure the sales volume and profit they require. Players control these functions by completing a planning and costing sheet each round. The results of all decisions are calculated from two graphs included with the game materials. (DCD)

Cost: £48

Producer: Management Games, Ltd., 63B George Street, Maulden, Bedford MK45 2DD, United Kingdom

TYCOON

Centre for Business Simulation

Playing Data
Age Level: high school-graduate school
Packaging: instruction manual, program booklet, and electronic calculator

Description: There are four versions of this game which may be played with the materials in the complete kit. *Basic Tycoon* is an introductory game for novices. *Intermediate Tycoon* is designed for use by college students. *Top Tycoon* is the most advanced version of the game and is designed for play by graduate students of business. The advanced version may be scored with a circular slide rule or an electronic calculator. The game model is incorporated into both of these scoring devices.

"The basis of Tycoon," according to the producer, "is the simulation of a single imaginary consumer product and its sale in a single market place (or two separate sales territories in the case of *Top Tycoon*) in competition with other players and with the objective of making a higher cumulative cash flow (or other agreed criteria) than them over a stipulated number of rounds. For this purpose, players must invest in a factory (which then becomes a fixed cost) and, round by round, buy variable amounts of labour and raw materials in order to manufacture the required volume of products. Finally, some or all of the products will be sold in the open market according to the price level at which they are offered plus the amount of marketing support given them and also the quality or specification of the product in the case of intermediate and advanced games (a specially adapted calculator is provided with the game for the purpose of co-relating these variables and calculating market share). All decisions and their effects are then recorded on an accounting document called an operating statement." (DCD)

Cost: £36

Producer: Management Games, Ltd., 63B George Street, Maulden, Bedford MK45 2DD, United Kingdom

WASHINGTON UNIVERSITY BUSINESS GAME

Powell Niland, Joseph W. Towle and Carl A. Dauten; Washington University

Playing Data
Copyright: 1969
Age Level: graduate school, management
Number of Players: 12-70 in 3-8 teams
Playing Time: 4-10 hours in 25-30-minute periods

Description: Participants representing top management make decisions regarding size of sales force, size of advertising campaign, choice of potential customers, plant capacity, number of units to manufacture, whether or not to have research and development program and size, how to finance operations (bank loans, factor A/R). We usually play with four teams of four to eight people on each team. Each round of play (or "quarter"), the teams indicate by check marks at the end of the first page of the decision sheet what accounts they want their salesmen to call. A random number table is used by the team umpire to determine which of these calls has resulted in a sale. All sales are consolidated by the chief umpire on the Potential Sheet. Conflicts (i.e., cases where two or more teams have sold the same customers) are then resolved according to the principles in the instructions. There is a Potential Sheet for each quarter played. The total potential follows the business cycle, with some accounts growing and declining more swiftly than the average. The quarter-to-quarter changes, however, are, we believe, realistic since some businesses do grow faster than others, and some decline.

As we administer it, the game requires a chief umpire, a banker and an individual umpire for each team. The chief umpire needs to be familiar with the game and it is desirable that the banker be, too, although he should negotiate loans (constrained as to amounts) when required by the teams. The team umpires can be trained in a half an hour's briefing session if they have any general background at all in business.

When we run the game, we have a briefing of all participants for an hour and a half or so the evening before the day of play.

The next day we play the game from about 8:30 a.m. to 4:30 p.m., with one hour out for lunch. This usually enables us to cover from 12 to 16 cycles of play (each cycle is called a "quarter").

The same evening each team is required to prepare a report to stockholders in which they point out the "high points" of the administration, evaluate their financial condition at the end of play, and set further their plans for the following 12 months. We require that these presentations last 8 minutes or less since the "stockholders" (members of the other teams and game umpires) need time to ask questions about the facts presented in the annual report.

For 2 hours on the morning of the third day we hold a stockholders' meeting. Each team in turn presents its report, followed by a question period. After all reports have been presented and questioned, we usually close with a few words about the value of the game as a training tool, and an "investment" simulation in which each participant is asked to invest funds in teams other than their own. We use the tabulation of these to identify the "winning" team. (Author)

Comment: This simulation offers a reasonable alternative to the computerized games described in the computerized machine section for users without adequate computer services. The decision coverage is balanced and of moderate complexity. (BK)

Cost: Sample forms available to educational institutions. User must duplicate.

Producer: Graduate School of Business Administration, Washington University, Box 1133, St. Louis, MO 63130

See also
Business Strategy ECONOMICS
Executive Decision ECONOMICS

PRODUCTION, LOGISTICS, OPERATIONS

BULOGA II

Mr. C. Ladd and Professor F. J. Beier, University of Minnesota

Playing Data
Age Level: college, management
Number of Players: 4 or more teams with several players per team
Playing Time: one-half day
Preparation Time: approximately 1-2 hours

Description: Participants represent managers of the logistics function in firms competing for sales in a single-product industry. They are responsible for coordinating inventory management, transportation for raw materials and finished goods, production, warehousing for raw materials and finished goods, promotion, and advertising. Each firm is equidistant from a central market. The overall goal of each is to satisfy customer demand without incurring excessive costs. Costs may be caused by the following: order processing, transportation charges, non-optimum production capacity, production schedule changes, warehouse location and inventory, inventory carrying charges, and advertising and promotion. Product demand varies according to promotion, seasonal patterns, the fate of the national economy, competitive strategies and tactics among the competitors, and random fluctuations. (AC)

Cost: duplication and mailing

Producer: John Ozment, Graduate School of Business Management, University of Minnesota, Minneapolis, MN 55455

COMPUTER AUGMENTED CASES IN OPERATIONS AND LOGISTICS MANAGEMENT

William L. Berry and D. Clay Whybark

Playing Data
Copyright: 1972
Age Level: College, management
Prerequisite Skills: key-punching ability and understanding of basic statistics
Number of Players: any number
Playing Time: 3-6 hours
Preparation Time: 1-2 hours
Special Equipment: IBM 1130, CBC 6500, DEC 10, IBM 360 series; FORTRAN

Description: This series of 20 cases is designed to help students gain a greater understanding of the use of the computer and to help them explore problems in operations management. The simulations provide an introduction to simulation analysis, the use of decision matrices, mathematical programming, and statistical analysis. Students are presented with a series of problems in operations and management in situations including a bank, a mining corporation, and various manufacturing companies. In most cases, participants decide how to use the existing computer program to get appropriate data for analysis in developing a solution. According to the designers, the program parameters can be changed fairly easily. (HL)

Cost: (1) manuals $7.35 each; (2) request prices for time-sharing or mail service

Producer: (1) South-Western Publishing Company, 5101 Madison Road, Cincinnati, OH 45227; (2) Simtek, P.O. Box 109, Cambridge, MA 02139

COMPUTER MODELS IN OPERATIONS MANAGEMENT

Roy D. Harris and M.J. Maggard

Playing Data
Copyright: 1972, 1978
Age Level: college, graduate school
Special Equipment: computer; batch FORTRAN
Packaging: 1 copy of 219-page student manual; software (10 batch FORTRAN programs from 200 to 600 lines of code)

Description: This collection of 20 cases and accompanying programs can be used to demonstrate quantitative techniques for management analysis and decision making. The purpose of the models is to introduce students to the use of the computer and quantitative techniques as a manager's tool in production or operations management. Text chapters present (1) theories and concepts of the model, (2) a simple hand-computed example, (3) sample input/output, (4) more complex problems, and (5) computer program listings. The student is assigned a problem, analogous to a case presented in the text, and makes independent use of the programs and techniques described to solve the problem. Use of the computer allows the student to concentrate on the structure of the problem, the data required, and analysis of the results in cases of realistic complexity. (CONDUIT Pipeline)

Cost: $65.00; additional student text $10.00, 2 or more $9.00 each; additional instructor text $5.00

Producer: CONDUIT, Box 388, Iowa City, IA 52240

COMPUTER MODELS IN OPERATIONS RESEARCH

R. D. Harris, M. J. Maggard, W. G. Lesso

Playing Data
Copyright: 1974
Age Level: college, graduate school
Special Equipment: computer; batch FORTRAN
Packaging: 1 copy of 244-page student manual; software (12 batch FORTRAN programs from 165 to 680 lines of code)

Description: This package of 12 computer models and accompanying text is designed to demonstrate how the computer can be used as a tool in applying scientific methodology to decision making. The models generally represent the computer-based applications of operations research in the areas of mathematical programming, networks, statistical decision theory, stochastic processes, and cost-effectiveness methodology. Each model is presented in the text with an introduction to the theory used in the computer model; a simple hand-computed example problem; the correct input and resulting output for the computer model solution to the example problem; and more complex problems that suggest the full range of uses of the computer model. These problems are intended to provide students with practice in problem formulation, data gathering, sensitivity analysis, and reporting of results. (CONDUIT Pipeline)

Cost: $75.00; additional student text $10.50, 2 or more $10.00 each; additional instructor manual $5.00

Producer: CONDUIT, Box 388, Iowa City, IA 52240

DISTRIBUTE

Management Games, Ltd.

Playing Data
Copyright: 1975
Age Level: management
Number of Players: 2-20 in groups of 2-4
Playing Time: 3-4 hours
Packaging: professionally packaged game includes instructor's manual and 20 copies of participants' instructions.

Description: Teams of players assume the functions of the transportation manager of a wholesale grocery wholesaler. Using daily order schedules, they plan the loading of delivery trucks and determine the routes that the trucks will take. The idea is for the teams to plan the delivery of goods in the most profitable possible way. All player and team decisions are recorded on forms. The results of these decisions are calculated from a set of charts included with the game materials. (DCD)

Cost: £36

Producer: Management Games, Ltd., 63B George Street, Maulden, Bedford MK45 2DD, United Kingdom

EQUIPMENT EVALUATION

Erwin Rausch

Playing Data
Copyright: 1968
Age Level: supervisors
Number of Players: 3-6 players per team
Playing Time: 2 hours
Preparation Time: 30-60 minutes
Packaging: professionally packaged and produced game includes 6-page administrator's guide and 6 38-page player manuals

Description: Participants in this series of six exercises seek to make the most effective decisions in selecting, justifying, replacing, and maintaining industrial and office equipment. For example, in the first, players consider buying a new machine to replace a manual operation. One costs $10,000, will last about five years, and should save the company about $5,500 during that time. The second one costs $5,000 and is expected to save $2,000 over the same time. Players must decide whether to recommend the purchase of either. After players reach individual decisions, they reach consensus in teams. Their answers are scored against an answer key, and the player with the highest score at the end of the game wins. (DCD)

Cost: 6-participant set $21.00; meeting leader's guide $.50

Producer: Didadic Systems, Inc., Box 457, Cranford, NJ, 07016

INVENTORY SIMULATION (INSIM)

Carl E. Ferguson, Jr., University of Alabama

Playing Data
Copyright: 1974
Age Level: college
Number of Players: any number
Playing Time: 5-10 minutes per decision, any number of decisions may be made; about 1 hour of explanation time needed
Special Equipment: any computer with Fortran compiler; interactive mode requires interactive terminals

Description: *INSIM* is an inventory simulation which allows the user to manipulate all of the major business inventory factors and evaluate the effect of changes in these variables on total cost. The participant sets the values for each of the following variables: number of periods to be simulated; value of inventory items; beginning inventory level; reorder cost; receiving cost; cost of a stockout; annual interest rate; demand history; and lead time history.

Once the above values have been established, the participant is asked to determine the reorder point and reorder quantity. The participant will enter his or her best estimate for reorder point and reorder quantity and the simulation program will display the results. After displaying the results, *INSIM* will ask if the simulation is to be run again. A yes response causes *INSIM* to ask if the value of any of the simulation control variables are to be changed. The participant can then selectively change the values of any of the inventory variables and execute the simulation again. (AJF)

Comment: A good short exercise to use in a logistics class to illustrate the influence of changes on inventory variables on total inventory cost. (AJF)

Cost: $25.00

Producer: Carl E. Ferguson, Jr., The Center for Business and Economic Research, Box AK, The University of Alabama, University, AL 35486

MIDTEX

Howard Edward Johnson, University of Texas at Austin

Playing Data
Copyright: 1969
Age Level: college
Number of Players: no limit
Playing Time: one-half day
Preparation Time: 4 hours
Special Equipment: CDC 3100 16k or larger machine; FORTRAN IV

Description: Participants represent production controllers for a hypothetical firm in this simulation designed to teach basic concepts of production and inventory control and/or Monte Carlo analysis. They seek to minimize costs of production changes, inventory carrying, and stockouts. Decisions are those regarding scheduled production rate and reaction rate for adjusting planned production. Players do not interact with one another. (AC)

Cost: none

Producer: Not published, but contact College of Business Administration, University of Texas, Austin, TX 78712

MATERIALS INVENTORY MANAGEMENT GAME

Jay R. Greene and Roger L. Sisson

Playing Data
Copyright: 1971
Age Level: management
Number of Players: 6 or more in groups of 2-3
Playing Time: 150 minutes in 10 15-minute rounds
Preparation Time: 1 hour (est.)
Packaging: photocopied game materials for five players

Description: This game is designed, according to the producer, "to help those involved in inventory management practice inventory control and demand forecasting by using the Economic Order Quantity (EOQ) formula." This formula defines the Economic Order Quantity as the square root of the sum of two times the total yearly requirements times the set up cost, divided by the yearly inventory cost. A chart included with the game materials summarizes the hypothetical demand for a product during the course of one year. Teams of players try to accurately predict the demand and subsequent cost for each of 13 rounds. Each round represents a period of three weeks. (DCD)

Cost: $10.50

Producer: Didactic Systems, Inc., Box 457, Cranford, NJ 07016

PHYSICAL DISTRIBUTION MANAGEMENT

Playing Data
Copyright: 1970
Age Level: manager
Number of Players: 3 or more in teams of 3-5
Playing Time: 1 hour (est.)
Preparation Time: 30-60 minutes

Description: Each player represents an executive in a medium-size manufacturing company who has recently assumed responsibility for physical distribution. The game presents players with eight situations. In the first, for instance, they must decide which of 16 distribution functions (such as planning for production and production scheduling, and selecting common carriers) should be under the control of the physical distribution manager. After players make their individual decisions, each team reaches a decision. Responses are scored against an answer key, and the player and team with the highest scores win. (DCD)

Cost: 5-participant set $23.50; 2 or more sets $19.90 each; meeting leader's guide $.50

Producer: Didactic Systems, Inc., Box 457, Cranford, NJ 07016

PLANNED MAINTENANCE

Staff, Didactic Systems, Inc.

Playing Data
Copyright: 1969
Age Level: college, management
Number of Players: 3 or more in 1 or more teams
Playing Time: approximately 3 hours
Preparation Time: 30-60 minutes

Description: Small Equipment Maintenance: Participants are asked to assume that they are maintenance managers in a manufacturing plant. When to repair and when to scrap small pieces of equipment is explored, and participants select a standard policy from a number of alternatives. Participants are then asked to decide on criteria for frequency of inspections, based on pertinent information about specific machines.

Scheduling Maintenance Inspection: Detailed information about a number of machines is provided along with specific maintenance and operating costs. Participants decide how often they would inspect each machine.

Job Control Procedures: A more orderly procedure for controlling job requests is needed. Participants decide what level of authorization and what type of record they would require for a series of maintenance tasks.

Making Better Use of Time: Three men in the department are described, together with a list of specific tasks to be done. Participants prepare time standards and select the man they would assign to each of these tasks.

Establishing Maintenance Priorities: A series of maintenance tasks is described, to which the participants assign priorities for the purpose of systematizing the work schedule and for better interdepartmental communication.

Standards of Performance: The company is planning to start a Standards of Performance Program. Participants are asked to prepare a list of duties and responsibilities for maintenance managers and then to spell out an acceptable performance standard for each. (Publisher)

Cost: 5-participant set $23.50; 2 or more sets $19.90 each; meeting leader's guide $.50

Producer: Didactic Systems, Inc., Box 457, Cranford, NJ 07016

PROCUREMENT MANAGEMENT

Staff, Didactic Systems

Playing Data
Copyright: 1976
Age Level: college, management
Number of Players: teams of 3-5
Playing Time: 3-4 hours

Description: Participants assume the roles of newly appointed procurement managers for a large manufacturing firm. The objective of the game is to provide insight into the job of the industrial buyer, to encourage participants to exchange ideas on the various stages of the procurement cycle, to understand the responsibilities of the industrial buyer, and to develop buying strategies.

Decisions in each round of competition involve establishing procurement objectives, determining the quality and quantity of supply requirements, the selection of suppliers, and the handling of special procurement problems and opportunities. The game is noncomputerized and requires a referee for scoring and evaluation purposes. Feedback on purchasing performance is returned to the teams at the end of each round of decisions. (AJF)

Comment: This game is well suited for use in Industrial Buying or Industrial Marketing courses. (AJF)

Cost: 5-participant set $23.50; 2 or more sets $19.90 each; meeting leader's guide $.50

Producer: Didactic Systems, Inc., Box 457, Cranford, NJ 07016

PRODUCTION CONTROL–INVENTORY

Erwin Rausch

Playing Data
Copyright: 1968
Age Level: college, management
Number of Players: teams of 3-5
Playing Time: 2-3 hours

Preparation Time: 30-60 minutes
Packaging: professionally produced and packaged game includes 40-page player's manual and 7-page Administrator's guide

Description: Players assume the roles of inventory control managers of a medium-size manufacturing operation employing a few hundred people. Each player assumes that his predecessor has recently retired and that while this previous manager did an excellent job during his 25 years with the company, his outstanding performance can be attributed to his experience rather than the system of inventory controls he perpetuated. The new managers must not only maintain the high level of performance but must also institute a new set of inventory procedures. Specifically, the players reduce inventory investment to a minimum; never run out of parts; buy parts as inexpensively as possible; and set standards for inventory control management. Each player must complete eight items. The first three involve listing and classifying parts. Item four requires the development of a system for rotating and maintaining stock. In items five through seven, the player must develop cost-effective procedures for replenishing stock. In item eight he or she must write job descriptions and set standards of performance for the five employees in the Inventory Control Department. (DCD)

Cost: 5-participant set $17.50; meeting leader's guide $.50

Producer: Didactic Systems, Inc., Box 457, Cranford, NJ 07016

PRODUCTION SCHEDULING MANAGEMENT GAME

Jay R. Greene and Roger L. Sisson

Playing Data
Copyright: 1971
Age Level: college, management
Number of Players: 5 or more in teams of 2-3
Playing Time: 5 hours in 12 25-minute rounds
Preparation Time: 30-60 minutes

Description: Players assume that they manage a shop with three machines. The teams accept or reject up to 20 jobs and must schedule their completion. The machines are different from one another and all must be used to complete any job, each of which requires a different amount of time on each machine. Material costs are fixed and the bids, which include time limits for delivery of 1-13 days, specify the delivery price for each job. Operating costs, including straight time and overtime wages, remain constant throughout the game. The teams must accept as many orders as possible and schedule the use of the machines so operating costs are minimal and the products are delivered on time. The team that has earned the most money after 12 rounds wins. (DCD)

Cost: 5-participant set $10.50

Producer: Didactic Systems, Inc., Box 457, Cranford, NJ 07016

PROSIM

Paul S. Greenlaw and Michael P. Hottenstein, Pennsylvania State University

Playing Data
Copyright: 1969
Age Level: college, graduate school
Special Prerequisites: understanding of incremental analysis, production scheduling, economic lot size inventory models
Number of Players: 1 and up
Special Equipment: IBM 700-7000 or 360; FORTRAN
Packaging: 256-page student text, instructor manual, computer deck

Description: Participants playing as individuals or in teams take the roles of production managers of imaginary multiproduct corporations. The student text describes the game environment, the rules, the decisions to be made, and provides analytical tools and concepts for meeting game objectives. The decisions for each round are: controlling inventories, assigning workers to machines, acquiring raw materials, expenditures for quality control and plant maintenance, hiring and training workers, assigning products to machines and production lines, and hours of work for each employee. (AC)

Cost: paper texts $8.50 each, instructor manual free to adopters, 360 computer deck $25.00, 700-7000 computer deck $15.00

Producer: Harper & Row Publishers, Inc., Keystone Industrial Park, Scranton, PA 18512

PURCHASING

Erwin Rausch

Playing Data
Copyright: 1976
Age Level: college, management
Number of Players: 1 or more teams of 3-5
Playing Time: approximately 2-3 hours with 1 hour start-up time

Description: Participants are placed in the roles of the purchasing agents for a small but financially sound company. The game is designed to highlight the major functions in the job of the purchasing agent. It is assumed that the participants have newly stepped into the job of purchasing agent for the company. The game can be played by individuals with no purchasing experience to learn what the job is all about, or by experienced purchasing agents in order to sharpen their skills. The game encourages the exchange of ideas between participants on developing purchasing strategies.

Decisions made in each round of competition include the determination of vendor reliability, whether or not to take advantage of quantity discounts, risk evaluation in purchase decisions, the determination of operational costs, the establishment of economic order quantities, and the setting of standards of purchasing performance. The game is noncomputerized and requires the referee to score purchasing efficiency and prepare performance evaluation reports to return to each of the teams after each round of decisions. (AJF)

Cost: 5-participant set $17.50; meeting leader's guide $.50

Producer: Didactic Systems, Inc., Box 457, Cranford, NJ 07016

SIMCHIP

Donald A. Bowersox, Michigan State University

Playing Data
Copyright: 1978
Age Level: college
Number of Players: 2 or 3 players per team in 4-team industries, any number of industries can be in operation
Playing Time: 1 hour of class start-up time, generally 8-12 decision periods of about 1-2 hours each

Description: Participants take on the roles of the distribution managers of a firm manufacturing potato chips. Four firms are in operation selling the chips in five separate geographic markets. The objective of the game is to develop an efficient distribution system for the potato chips, *not* to generate increased sales. However, certain market decisions must be made, such as the development of a sales forecast, so that transportation requirements can be determined.

The participants are required to submit two forms each period. The Decision Reporting Form contains production decisions, total distribution requirements, company truck requirements, and common carrier needs. The Operating Statement contains sales and cost forecasts, warehouse and inventory requirements, production plans, and inbound and outbound transportation needs. Once the team forecasts its expected sales in each market, it must determine the most efficient method of transporting the potato chips into each market. The referee scores the decisions of the teams and returns Status Reports to them indicating the level of efficiency of their distribution departments. (AJF)

Comment: Good for use in any type of logistics course. (AJF)

Cost: Appears as appendix in text *Logistical Management* by Bowersox, $15.95

Producer: Macmillan Publishing Co., Inc., 866 Third Avenue, New York, NY 10022

SPRODS, LTD.

Management Games Limited

Playing Data
Age Level: management
Number of Players: up to 24 players in teams of 3-6
Playing Time: 6-7 hours in its basic form

Description: "*Sprods*," according to the producer, "has been designed for use as a production planning, costing, purchasing and stock control exercise and, in addition, as an exercise to cover aspects of estimating and personnel problems." This is an in-basket simulation in which the players must make decisions concerning the manufacture of an imaginary durable product called a sprod. To begin, players are given the specifications for the materials and manpower needed to make sprods. They must then order raw materials for later delivery, define production schedules, and take orders for sprods, which are then filled in the most profitable way possible. The player who performs these duties in the most profitable way wins. (DCD)

Comment: The exercise appears relatively simple, considering it does cover the total management function. It is suitable for training supervisory or junior level managers. The exercise was developed in Great Britain, and all measurement are provided in the metric system. (HL)

Cost: £48

Producer: Management Games, Ltd., 63B George Street, Maulden, Bedford MK45 2DD, United Kingdom

STANFORD BUSINESS LOGISTICS GAME

Karl M. Ruppenthal, D. Clay Whybark, and Henry A. McKinnell

Playing Data
Copyright: 1967
Age Level: college, management
Number of Players: 2-50
Special Equipment: IBM 7090 or IBM 360/67; FORTRAN IV

Description: The players of this transportation simulation represent the "logistics" managers of a eucalyptus oil manufacturer in Los Angeles. The eucalyptus oil industry must compete with the banana oil industry for the New York City market. The vendor cost of the oil is $.60 per pound and it can be sold, chiefly to drug manufacturers, for $1 per pound. If the eucalyptus oil manufacturers are to maintain or increase their share of the market in competition with the banana oil firms, they must maintain adequate inventories in their New York warehouses and promptly fill their orders. The company managed by the players maintains a warehouse with an 80,000-pound capacity in New York. The per-pound costs for storing the oil in company or public warehouses remain constant throughout the game. The drug manufacturers usually place their orders every two to six weeks, and the players may ship the oil from Los Angeles by using all or some of six common carriers: air freight, truck, freight forwarder, rail, shipper cooperative, or steamship. These carriers differ in cost, reliability, and average disparity between scheduled and actual delivery dates.

The players must make all of the shipping and warehousing decisions for their firms. They must make shipping decisions for four weeks at a time and may also reserve space for air freight shipments and may lease additional warehouse space. At the start of the game each firm holds 40,000 pounds of oil in the East Coast warehouse and has 35,000 pounds in transit with three common carriers (ship, truck, and forwarder). Players must manage their company's logistics as profitably as possible for 52 weeks (13 rounds). All decisions are entered onto forms, transferred to punch cards, and the results are simulated by a computer model. (DCD)

Cost: source deck $50.00; player's manual $2.50; administrator's manual $10.00

Producer: Center for Transportation Studies, University of British Columbia, Vancouver, BC V6T 1W5, Canada

TOOL ROOM GAME

Forrest M. Campbell and E. Robert Attworth, F. M. Campbell Associates

Playing Data
Copyright: 1960
Age Level: college, management
Number of Players: no limits
Playing Time: 50-60 minutes over 30 weeks
Preparation Time: 2-3 hours

Description: Each player represents a tool room manager seeking to minimize the cost of operation in supplying dies to production departments. The game, which demonstrates how costs interact, requires consideration of the special charges for procurement set-up, expedited procurement order, repair set-up, expedited repair order, interdepartment transporation, and depletion; and the routine costs of procurement, construction, repair, serviceable storage, and repairable storage. Players do not interact, and major random events dominate play. (AC)

Cost: none

Producer: Forrest M. Campbell, 245 Park Avenue, New York, NY 10017

UNITEX

Howard Edward Johnson, University of Texas at Austin

Playing Data
Copyright: 1967
Age Level: college, graduate school
Special Prerequisites: working knowledge of FORTRAN and production control fundamentals
Number of Players: unlimited
Playing Time: several days
Preparation Time: 1-2 days
Special Equipment: CDC 3100 16 k; FORTRAN IV

Description: Participants represent production controllers for a hypothetical firm in this simulation designed to develop skills in production and inventory control and in using computer simulation for problem analysis. They seek to minimize the costs of raw material inventory, special ordering for materials, production level, and fluctuation. Decisions involve how to yield the amount of materials to order (regular and special orders), scheduled production rate, and additions to capacity. Players do not interact with one another. (AC)

Cost: none

Producer: Not published, but contact College of Business Administration, University of Texas, Austin, TX 78712

PERSONNEL DEVELOPMENT

THE ABELSON-BAKER INTERVIEW

Playing Data
Copyright: 1973
Age Level: college, management
Number of Players: 6 or more in pairs
Playing Time: 80-100 minutes
Packaging: professionally packaged game includes workbook and cassette

Description: This exercise brings together a number of skills needed by effective interviewers and supervisors. One is the ability to set goals in advance of a planned interview. Another is the ability to develop an interview plan, identifying such important dimensions as the sort of preinterview preparations that are needed, the key assumptions one has in mind, strategies for testing these assumptions, and so forth. Still another skill involves the ability to analyze the performance of another interviewer, noting both his shortcomings and strengths. Finally there is the opportunity to continue the interview between Abelson and Baker, carrying it through to completion. (producer)

Cost: $60.00

Producer: Training House, Inc., 100 Bear Brook Road, Princeton Junction, NJ 08550

ADMINISTRATION

Education Research

Playing Data
Copyright: 1975
Age Level: college, management
Number of Players: any number (teams optional)
Playing Time: 2.5-4 hours (est.)
Packaging: 27-page manual (1 per player)

Description: Administration is designed to upgrade and reinforce administrative skills, or to train students or staff to become more efficient in dealing with problem handling, controls, and work flow. Each player or team represents a service manager with a staff of six for a company which manufactures and sells measuring and testing equipment. The manual provides two pages of background information on the company. The game is divided into 10 functions—organizing, coordinating, measuring, performance, controlling, assigning work, follow-up, getting feedback, writing reports, presenting information, and problem solving. Each function involves three decisions, each with three options. Players make a decision, score themselves, make a second decision, and so on, in each function before proceeding to the next. Decisions 1 and 3 of each function are related to the same problem, such as determining a suitable standard of performance for servicing equipment. Decision 2 is always a "breaking problem"—an interruption or altering event which occurs in the middle of the larger decision-making process, such as having to delegate a set task to a secretary. Players score points for correct decisions and are penalized for incorrect decisions. After the game, players discuss results, present and attempt to solve their biggest administrative headache, and apply the game to real life. (TM)

Cost: $7.45. Quantity discounts
Producer: Education Research, P.O. Box 4205, Warren, NJ 07060

APPRAISAL BY OBJECTIVES (COACHING AND APPRAISING)

Staff, Didactic Systems, Inc.

Playing Data
Copyright: 1970
Age Level: college, graduate school, management
Number of Players: 1 or more teams of 4 or 5
Playing Time: approximately 3 hours
Preparation Time: 30-60 minutes

Description: Participants represent managers in a department with sales-oriented as well as administrative functions.

Planning for the Appraisal Interview I: The major responsibilities of two subordinates, together with some specific goals, are provided. Participants decide what further information the manager needs to plan an effective appraisal interview.

Planning for the Appraisal Interview II: A brief description of a subordinate and his accomplishments during a six-month period are given. Participants decide on the time allowance for his appraisal interview, and then select questions concerning his accomplishments which they think need clearing up prior to the interview.

Evaluating Objectives I: A specific goal of one subordinate is provided. Participants write out criteria that could be used to objectively measure performance against this goal.

Evaluating Objectives II: Participants decide whether each of several goals is realistic, difficult to achieve, or easy to achieve.

Using Qualitative and Quantitative Measurements: Measurement techniques for both qualitative and quantitative goals are discussed. A number of goals are given with information relating to achievement, and participants rate performance on each goal.

Opening Up Communications on Personal Goals and on Career Path: A detailed description of a subordinate is provided. Alternative goals and career opportunities are described, and participants decide on an approach to guiding the subordinate's future development.

Stimulating Self Development: How to stimulate subordinates to improve abilities, knowledge, and skills is explored, with various alternatives from which participants select preferences.

A Checklist for Performance Appraisal by Objectives: Participants write out the major steps they feel belong to a thorough performance appraisal, from preparation for the interview through follow-up procedures. (Producer)

Cost: 5-participant set $23.50; 2 or more sets $19.90 each; meeting leader's guide $.50

Producer: Didactic Systems, Inc., Box 457, Cranford, NJ 07016

ASSIGNING WORK

Staff, Didactic Systems, Inc.

Playing Data
Copyright: 1973
Age Level: college, management
Number of Players: 1 or more teams of 4 or 5
Playing Time: approximately 3 hours
Presentation Time: 30-60 minutes

Description: Reviewing Approaches for *Assigning Work*: Participants assume that they are supervisors in a production department. A long list of suggested approaches for effective work assignment is provided and participants select those which they feel would be most effective.

Preparing to Assign a Job: Information is provided about three different employees and participants are asked to decide who they would select to fill a temporary assignment. Decisions are made based on the specific skills and knowledge of each of the three employees involved.

Approaching an Employee with a New Task: In both parts of this item, employees who are reluctant to accept new and different responsibilities are described. Participants decide how they would persuade each one to accept a temporary assignment.

Communicating Job Assignments to Employees: Three approaches for communicating assignments are described. In a situation where limited time is available to give complete instructions, participants decide which of the three approaches would be best.

Dealing with Substandard Performance: An employee whose performance has slumped during the last few months is described and participants prepare a list of specific points they would cover in a discussion with him.

Controlling Absenteeism and Tardiness: Here game participants are asked to prepare a comprehensive list of steps that could be taken to discourage excessive absenteeism and tardiness in their departments.

Recognizing Employee Achievement: Participants are reminded that recognition, when appropriately and sincerely applied, can serve as a powerful tool for helping employees become more satisfied with their work. In this item, several employee accomplishments are described and participants decide how they would recognize performance in each case. (producer)

Cost: 5-participant set $23.50; 2 or more sets $19.90 each; meeting leader's guide $.50

Producer: Didactic Systems, Inc., Box 457, Cranford, NJ 07016

AUSTRALIAN MANAGEMENT GAMES

B. Moore

Playing Data
Copyright: 1978
Age Level: college, management
Number of Players: 6-30 in teams of 3-4
Playing Time: 1-3 hours
Preparation Time: 3 hours (est.)
Packaging: 158-page hardback book

Description: All of the 10 management games in this collection are based on Australian organizations and have been field tested there. Five are specifically designed to simulate problems that occur in publicly administered organizations and three address administrative problems in private enterprise. All are good humored and distinctly Australian.

M.C.C. is an in-basket, decision-making game in which teams of players manage an electric utility company. *Hawkesworth* is a budget game in which the players represent college administrators who must agree on a proposed set of operating expenses for one year. In *Cabinet,* students assume the roles of Australian Cabinet Ministers who must try to streamline the federal government. *Joe Cool* simulates a labor relations hearing. The character for whom the game is named is a bank courier who was robbed in the course of his duties. When his boss accuses him of irresponsibility during the holdup and demotes him, Joe asks for and gets a hearing. In *Solar Cells,* the players represent purchasing agents who must operate within the highly technical market for solar energy paraphernalia. *Gerrymandering* is about the fine art invented by Elbridge Gerry. Players "compete to produce the greatest possible bias while still remaining within the letter of the law." Players of the in-basket exercise *Huri* pretend to be the recently appointed administrator of a remote Pacific island whose predecessor mysteriously disappeared without cleaning out his in-tray. *Hovercraft* is a communications game in which teams of players compete to produce an acceptable plan for a high-speed transportation system. The futures game *Australia 2000* has participants construct a model of the Australian business climate at the end of this century. The final exercise is a social welfare game called *Gerontology* in which the players assume the roles of elderly citizens and the officials responsible for their welfare. (DCD)

Cost: $23.50

Producer: New South Wales University Press, P.O. Box 1, Kensington, NSW 2033, Austraila

BECOMING AWARE

Benson Rosen and Thomas H. Jerdee

Playing Data
Copyright: 1976
Age Level: adult
Number of Players: 1 or more
Playing Time: 3 hours (est.)
Preparation Time: 2 hours (est.)
Packaging: professionally packaged materials

Description: This is a "Sex Discrimination Awareness Program" for members of organizations. It is designed to help the participants become sensitive to the issue of sex discrimination and suggests strategies that organizations can adopt to promote equal employment opportunities for women. During the first part of this "program" the participants, through reading a text, listening to a tape recording, and discussing the issue among themselves, examine the nature of sexual stereotyping, women's career conflicts, barriers to managerial careers for women, and the changing role of women in society. During the last part of the program the participants discuss ways to make "organizational commitments to affirmative action." The game includes Opinion surveys for women and for both men and women that help participants define the present commitment of their organization to affirmative action for women. (DCD)

Cost: sample set $21.88, complete set $431.25

Producer: Science Research Associates, 1540 Page Mill Road, Palo Alto, CA 94304

BUSINESS DECISION SIMULATION

J. Ronald Frazer

Playing Data
Copyright: 1975
Age Level: college, graduate school
Playing Time: about 2 hours each simulation
Special Equipment: time-sharing computer; BASIC
Packaging: 1 copy of 158-page student manual; 216-page teacher's manual; software (17 interactive BASIC programs from 100 to 350 lines of code)

Description: The 20 simulations in this package pose problems ranging from simple price-production problems, to relatively complex management science problems, and they include some fairly elaborate negotiation situations. The games are sufficiently simple in concept to allow students with varied backgrounds to play the games successfully. Yet, they are sufficiently complex to provide a challenge for students with advanced skills.

The games are umpired by a time-sharing computer that gives results of decisions to participants in a matter of minutes. This allows many decision-making periods to be covered quickly, compressing years of simulated activity into a short laboratory period. Most of the games are designed to be played in entirety in one two-hour period, with each game typically requiring 6-10 decisions. The games stress one or more of the elements of strategy, negotiation, and management science. (CONDUIT)

Cost: $50.00; additional student text $6.50, 2 or more $6.00

Producer: CONDUIT, Box 388, Iowa City, IA 52240

THE CAREER GAME

Scott B. Parry and Edward J. Robinson

Playing Data
Copyright: 1978
Age Level: college, management
Number of Players: 4-20
Playing Time: 2 hours
Packaging: professionally packaged game includes workbook

Description: The game is basically a projective test in which each participant is presented with a series of pictures that are unstructured, ambiguous, and open to many possible interpretations. That is, there is no definite, fixed, or correct meaning. Participants are asked to describe the pictures, answering four questions about each picture: (1) What is going on here? (2) What led up to the event? (3) What are the characters feeling and thinking? (4) What is the outcome likely to be?

Upon completing their interpretations in writing, participants are paired up and are given a set of guidelines for analyzing their interpretations against the six factors, or variables (success versus failure orientation, past versus future orientation, achiever versus affiliator orientation, long-range versus short-range orientation, security versus challenge orientation, task versus people orientation). Participants then report their conclusions to their partner, indicating for each factor their fitness to the present job and the direction their career development might take. (producer)

Cost: $60.00

Producer: Training House, Inc., 100 Bear Brook Road, Princeton Junction, NJ 08550

THE CHESTERFIELD TRAINING EXERCISE

Playing Data
Copyright: 1971
Age Level: college, management
Number of Players: 12-20
Playing Time: 90-180 minutes
Preparation Time: 30 minutes (est.)
Packaging: 8-page printed instructor's manual and playing pieces

Description: The purpose of this game, according to the producer, "is to provide a direct comparison of the behavior of two groups of people under different styles of management." The first group is managed with strong control and with emphasis on individual tasks. The second group is given greater individual freedom to govern themselves, and has only the objectives of their assignment to govern them."

The exercise begins with all the players assembled to learn the rules of the "German Matchstick Game," a simple children's game in which nine tokens are arranged in a triangular pattern of three rows. Two players must then pick up tokens or rows of tokens in turn. The last of the pair to pick up a token loses. The game is reminiscent of tic-tac-toe.

The players are then divided into two groups. Each has a "manager" who must follow a set of written instructions. One set of instructions contains only a task statement: to find the key to winning the game every time. The other set contains the same task statement but explicitly states that the manager should tightly control the group and details a lengthy set of procedures the group should follow to find the key to winning. The first group is invariably successful while the second group invariably fails. After half an hour, the players discuss how the management styles of the two groups affected their effort to solve their problem. (DCD)

Cost: $89.95

Producer: Behavior Sciences Education Center, 122 W. Alasta, Suite 15, Glendora, CA 91740

COLLECTIVE BARGAINING

Erwin Rausch

Playing Data
Copyright: 1968
Age Level: college, management
Number of Players: 6 or more in teams of 4-6
Playing Time: 2-3 hours
Preparation Time: 1 hour (est.)
Packaging: professionally produced and packaged game includes an 8-page administrator's guide and 6 46-page player's manuals

Description: This collection of six well-defined role-playing exercises is designed to give players practice in grievance negotiations and other collective bargaining skills. Players assume the roles of union committeemen and labor relations negotiators who must, in the last two months before the expiration of a collective bargaining agreement, resolve several grievances and draft a new contract that will more clearly define the ambiguous areas in the old one that has led to the grievances. In each group of six, two players assume management roles and four represent the shop committee. The first situation requires management to convince workers to accept voluntary overtime. The workers have refused the overtime in protest of the disciplinary layoff of two employees. A second grievance has been filed against management for subcontracting allegedly included work. The new contract is negotiated and put into effect in the last part of the exercise. Each possible action of the negotiators has been assigned a point value by the author. Players score points by selecting the best alternative. The player with the highest score at the end of the exercise wins. (DCD)

Cost: 6-participant set plus leader's manual $21.50

Producer: Didactic Systems, Inc., Box 457, Cranford, NJ 07016

COMMUNICATING FOR RESULTS—PARTS I AND II

Staff, Didactic Systems, Inc.

Playing Data
Copyright: 1969
Age Level: college, management

Number of Players: 1 or more teams of 4 or 5
Playing Time: 3 hours in 60-90 minute periods
Preparation Time: 30-60 minutes

Description: (This is a lower level version of *Managing Through Face-to-Face Communication.*) Interpreting Symbols: Participants are shown a picture and asked to write out the different meanings or reactions that come to mind. They then do the same thing for a word.

Applying the Ladder of Abstraction: The Ladder of Abstraction is explained. Participants are given several statements that describe job requirements and place them on the Ladder.

Gearing Messages to the Receiver: Three employees of different experience and knowledge are described. Several instructional statements of varying complexity are provided and participants match each statement with one of the employees.

Classifying Statements: Since perception of facts influences acceptance of a message, the distinction between "facts" and "inferences" is reviewed. Short statements relating to a story are given and participants decide whether each statement is a fact or an inference.

Responding to Emotional Remarks: Due to errors by one of the employees, problems developed in another department. The supervisor of that department calls (on two occasions) to complain. Participants explore how they would respond.

Giving Assignments and Instructions: Participants are provided with information on how to give clear assignments. They then apply this information in a specific situation with one of their subordinates.

Handling Rumors: A rumor is circulating which may have an adverse effect on the operations and morale of the department. Alternatives on how to deal with this potentially serious problem are provided and participants select their preference.

Downward Communications: Participants decide, from a list provided, what types of items they would recommend for inclusion in the company's monthly newsletter to employees. (Producer)

Cost: 5-participant set $23.50; 2 or more sets $19.90 each; meeting leader's guide $.50

Producer: Didactic Systems, Inc., Box 457, Cranford, NJ 07016

COMMUNICATION

Playing Data
Copyright: 1972
Age Level: management
Number of Players: 1 or more
Playing Time: 1 hour
Packaging: 24-page manual

Description: *Communication* is a self-administered exercise designed to sharpen the communications skills of junior executives. It consists of 10 exercises which attempt to simulate real-life business experiences between low level managers and their subordinates. Each consists of an "objective" (for instance, the player instructs his secretary to type a report) and requires the player to determine how to communicate this "objective" (for instance, should the secretary be given the chore by memo, personal communication, bulletin board notice). The player then evaluates an example message in terms of whether the objective has been thoroughly communicated. The player or team may check his (or their) answers after he has made his evaluation. Play ends when all 10 exercises have been completed. (DCD)

Cost: $7.45. Quantity discounts

Producer: Education Research, P.O. Box 4205, Warren, NJ 07060

COMMUNICATIONS: PROBLEMS AND OPPORTUNITIES

Staff, Didactic Systems, Inc.

Playing Data
Copyright: 1971
Age Level: college, management
Number of Players: 1 or more teams of 4 or 5
Playing Time: approximately 3 hours
Preparation Time: 30-60 minutes

Description: Distinguishing Between Facts and Inferences: Since perception of facts influences the acceptance of the message, the distinction between facts and inferences is reviewed. Short statements relating to a story are given, and participants decide whether each statement is a fact or an inference.

Responding to Emotional Statements: Because of errors by one of the employees, problems developed in another department. On two occasions the general manager requested an explanation of these mistakes. Participants explore how they would respond in each case.

Matching the Message to the Listener: Two employees of different experience and knowledge are described. Several instructional statements of varying complexity are then provided and participants match each statement with one of the two employees involved.

Managing Rumors: A rumor is circulating which may have an adverse effect on the operations and morale of the department. Various alternatives on how to deal with this potentially serious problem are provided and participants select their preference.

Complete Instructions: In this item participants apply the concepts discussed during the exercise. They select a specific task and prepare a detailed outline of instructions that could be used to give such an assignment. (producer)

Cost: 5-participant set $23.50; 2 or more sets $19.90 each; meeting leader's guide $.50

Producer: Didactic Systems, Inc., Box 457, Cranford, NJ 07016

THE CONSTRUCTION GAME

Scott B. Parry and Edward J. Robinson

Playing Data
Copyright: 1977
Age Level: college, management
Number of Players: 12-20 in teams of 4
Playing Time: 60 minutes
Packaging: professionally produced game includes workbook and cassette

Description: Several teams ("steel fabricating firms") are invited to bid on the construction of a warehouse and to submit estimates on the time required to build it and the methods to be employed. The bids are so close (the instructor reports) that each team will be awarded a contract to build the warehouse. A set of specifications and performance criteria is given to each team leader. On completing the warehouse, participants discuss the style of their leaders, the problems involved in changing one's style, the occasions under which a supervisor's style of management should change, and so on. The bidding and preliminary instructions take 10 minutes, the construction requires another 20 minutes, individual team discussion, 15 minutes, and the entire group discussion another 15 minutes. (producer)

Cost: $60.00

Producer: Training House, Inc., 100 Bear Brook Road, Princeton Junction, NJ 08550

CONSTRUCTIVE DISCIPLINE

Staff, Didactic Systems, Inc.

Playing Data
Copyright: 1970
Age Level: college, management
Number of Players: 1 or more teams of 4 or 5
Playing Time: approximately 3 hours
Preparation Time: 30-60 minutes

Description: Setting Realistic Goals: Participants are asked to assume that they were recently transferred to manage a produce department in a Southeastern division of their company. Goals with action programs to meet them are needed to stimulate employees who are producing below standard. Participants select an approach to the men from a number of alternatives.

Keeping Employees Posted on their Performance: Participants select preferred methods for keeping subordinates informed about their performance.

Helping a Subordinate: A subordinate requests help on a specific problem and participants select a procedure from a number of alternatives.

Maintaining Standards: Participants explore how they would conduct a discussion with a subordinate who is performing below his usual standards.

Giving Recognition: A letter of thanks is received from a customer commending the efficient operation of the department. Participants are asked how they would show appreciation to their subordinates for the part they played.

Assigning New Equipment Equitably: A number of eligible candidates for a new machine are briefly described. Participants are asked to choose the recipient, in view of their varying capabilities.

Granting Privileges: A number of specific requests by subordinates for special privileges are described. Participants decide on a course of action in each case.

Preventing Loss of Prestige: A particular machine operated by a new man is causing a number of rejects. The boss questions this situation, and participants decide on an explanation that will satisfy the boss as well as the employee involved.

Criticizing Constructively: A conflict exists in an area that involves two departments, due primarily to one uncooperative supervisor. Participants decide how to deal with him.

Discipline and Morale: The relationship between discipline and morale is discussed, and participants explore ways to achieve improvement in both. (producer)

Cost: 5-participant set $23.50; 2 or more sets $19.90 each; meeting leader's guide $.50

Producer: Didactic Systems, Inc., Box 457, Cranford, NJ 07016

CONTRACT NEGOTIATIONS

Jay Zif and Robert E. Otlewski, Creative Studies, Inc.

Playing Data
Copyright: 1970
Age Level: college, management
Number of Players: 8-15
Playing Time: 3-6 hours in 30-120 minute periods
Preparation Time: 2-4 hours

Description: *Contract Negotiations* is designed to orient students and managers to the intricacies and strategy of labor negotiations and collective bargaining and develop their negotiating skills. Players assume the management and union sides in negotiating a collective bargaining agreement. While the management negotiators try to insure maximum productivity for minimum cost, union negotiators try to optimize the employees' pay, fringe benefits, and terms and conditions of employment. (DCD)

Cost: player's manuals $5.95 each; instructor's guide gratis

Producer: The Macmillan Company, 866 Third Avenue, New York, NY 10022

COUNSELING

Playing Data
Copyright: 1978
Age Level: college, management
Number of Players: 1
Playing Time: 1 hour (est.)
Packaging: 28-page photocopied booklet

Description: The players of this in-basket simulation assume the role of a middle-level manager, four of whose employees need "counseling." Albert Schmidt has been stuck in the same job for several years and has recently started to make errors. Bill Jacobs is often gruff and has a "swelled head." Chuck Walters is a young, ambitious employee whose work is generally good although often inconsistent. Chuck has a higher opinion of his performance than his boss does and wants a raise. Donna Jenkins is an excellent employee but "seems to take delight in breaking the rules. If she disagrees with a policy she'll ignore it. For example, she seems to go out of her way to take her lunch hour at irregular times, or to take a coffee break before or after everyone else." In each "counseling situation" the player must construct a dialogue from a series of "ventilating" and "focus" questions to find the employee's "hidden problem." The exercise ends when the player has worked through all four counseling situations. (DCD)

Cost: $7.45. Quantity discounts

Producer: Education Research, P.O. Box 4205, Warren, NJ 07060

CRYTERIA!

Jennifer S. Macleod

Playing Data
Copyright: 1974
Age Level: management
Number of Players: 6-32 in 2-4 teams
Playing Time: 1-2 hours including debriefing
Preparation Time: little beyond mastering procedure

Description: The designer of *Cryteria!* describes the flow of play as follows: "Two teams each develop criteria by which they wish to be judged in competition with the other team. Several rounds are played, until participants protest at unfairness of game. Players asked to develop recommendations to make game 'fair.' Debriefing follows." One team represents white males, the other team represents either women or blacks (or other minority or minorities), depending upon whether the game is being focused on sexism or racism.

The purposes of the game are to demonstrate how sexism and/or racism operate in settings in which people compete for rewards; how institutional sexism and/or racism affect the perceptions and behavior of both the "power group" and the "powerless group"; to illustrate the profound significance of who is making the rules and setting the criteria by which people are judged; the necessity for affirmative action beyond the simple elimination of discriminatory practices that have been in effect in the past; to foster an understanding of how a system that places one group arbitrarily in power and another group arbitrarily out of power shapes and determines people's behavior; to improve understanding between members of different sexes and races.

The description is intentionally sketchy to avoid giving away the mechanism that makes the game unfair and therefore gives the game its potential impact as a vehicle for experiencing consciousness raising. (AC)

Comment: *Cryteria!* is designed to increase participants' understanding of the need for affirmative action, "thus increasing their cooperation and creative initiative in affirmative action and making it successful."

"The insights," Dr. Macleod notes, "come from the participants themselves when the debriefing is skillfully led. There is always conflict, argument, annoyance, protest–and better understanding." The trainer's skill and insight are important contributions to a successful debriefing in guiding participants to discover the point of the game and their experience as winners or losers.

The manual includes descriptions of results of game experiences. (AC)

Cost: $10.00

Producer: Jennifer Macleod Associates, 4 Canoe Brook Drive, Princeton Junction, NJ 08550

DECISION

Education Research

Playing Data
Copyright: 1972
Age Level: management
Number of Players: 1 or more

Description: In *Decision,* which can be played either by one person playing alone or by a team, the problem is to turn an unprofitable operation of the Peerless Data Corporation into a profitable one. The game is divided into five overlapping major strategic decisions. The decisions are focused on (1) office relocation and expansion, (2) analyzing a back-up or bottleneck in the reproduction section of the company, (3) improving supervision by promoting or hiring a new person, (4) improving morale by job enrichment and (5) whether or not to expand the staff to increase production. Following the presentation of information, each team has to organize the information, develop alternative solutions and evaluate these alternative solutions and finally making decisions. The author's analysis is then given and a scoring system, based on increasing profit, is provided. (REH)

Comment: As with the other Education Research simulations, I find that the required decisions are realistic. This is a good exercise with situations that are very common to many parts of industry which could very well be used in a management training sequence, not only for branch managers, but for first line supervisors as well. (REH)

Cost: $7.45. Quantity discounts

Producer: Education Research, P.O. Box 4205, Warren, NJ 07060

DECISION

Playing Data
Copyright: 1978
Age Level: management
Number of Players: 9-25 in groups of 3-5
Playing Time: 2-3 hours
Packaging: professionally produced card game

Description: This card game, in the producer's words, "is designed to explore the wide variety of potential situations which may occur in dealing with disciplinary problems arising in the course of employment." Each team is given six cards which describe some employee offense, the situation of the employee, the value of any property stolen or damaged, the work rules which apply to this situation, the previous conduct of the employee, and the employee's length of service. The players analyze the situation described by these cards and then decide what action (terminate, suspend, fine, warn, transfer, demote, or nothing) should be taken against the offending employee. (DCD)

Cost: £24

Producer: Management Games, Ltd., 63B George Street, Maulden, Bedford MK45 2DD, United Kingdom

DECISION GUIDES

Harvey R. Lieberman

Playing Data
Copyright: 1974
Age Level: management
Number of Players: 3 or more in teams of 3-5
Playing Time: 1 hour (est.)
Preparation Time: 1 hour (est.)
Packaging: 5 game booklets, set of situation cards, meeting leader's guide

Description: This game is designed to give participants an opportunity to exchange ideas on guidelines for making job decisions. Working in small groups the players review what the authors describe as "typical problem situations" and then explain their own solutions to the rest of their group. Each group then assigns their players' decisions a point value based on a comparison with the best solutions discussed. There are a total of fifteen situations including helping another manager, the constructive criticism of a superior, and promoting a new employee at the expense of a more experienced one. Each situation is described in three or four paragraphs and concludes by asking the player what he or she would do. (DCD)

Cost: $25.50; additional packet of situation cards $8.00; additional set of 5 game booklets $12.00

Producer: Didactic Systems, Inc., Box 457, Cranford, NJ 07016

DECISION MAKING

Erwin Rausch

Playing Data
Copyright: 1968
Age Level: college, management
Number of Players: 3 to 5 per team, any number of teams
Playing Time: 1-2 hours
Preparation Time: 1 hour (est.)
Packaging: professionally produced and packaged game includes 6-page administrator's guide and 5 62-page player's manuals

Description: This collection of 24 in-basket items is designed to simulate the careers of a group of four company managers. Players assume the roles of executives who must try to guide their company toward healthy growth and advance their own careers. Each item requires a yes or no decision. For example, should the company purchase a new machine, or hire consultants, or should the executive take night courses? The supposed effect of all these decisions over a six-year period is indicated on a matrix included with the game materials. The results, as indicated on the matrix, can be converted into a score, and players who achieve a minimum score become candidates for the presidency of their company. The president is elected from among the eligible candidates by vote of all the players. (DCD)

Comment: Didactic games such as *Decision Making* offer a reasonable alternative to the subjective case study or case incident where participants are never quite sure of the outcome and the highly quantitative game which requires considerable computation and financial statements. Each short exercise, or case incident can be introduced quickly once the initial setting is established. Answers required are: Personal vote–yes or no–and at times group decisions are required. Requiring a personal decision prior to group discussions generates commitment. Decisions are translated into salary since dollar scores are easily interpreted in our society. These games are especially adaptable by trainers who wish to involve participants but are unskilled at participative training techniques. (BK)

Cost: 5-participant set $17.50; meeting leader's guide $.50

Producer: Didactic Systems, Inc., Box 457, Cranford, NJ 07016

DELEGATION

Education Research

Playing Data
Copyright: 1971
Age Level: management
Number of Players: 1 or more

Description: *Delegation* is a simple decision-making exercise that focuses on one day. The game assumes that you are a general manager, and a general assistant. You have to go through the steps of (1) assigning priorities to different work, (2) deciding which task can be delegated, (3) analyzing your staff and deciding who can handle these assignments, (4) delegating the assignment. Controlling this delegation afterward is not part of the simulation. You are basically faced with a large number of problems that have accumulated upon your return from a trip, and you have a recommended 30 minutes in which to analyze the problems and make your delegation on worksheet assignment. Then scoring is provided. (REH)

Cost: $7.45. Quantity discounts

Producer: Education Research, P.O. Box 4205, Warren, NJ 07060

EFFECTIVE DELEGATION

Staff, Didactic Systems, Inc.

Playing Data
Copyright: 1970
Age Level: college, management
Number of Players: 3 or more in 1 or more teams
Playing Time: approximately 3 hours
Preparation Time: 30-60 minutes

Description: Limits of *Effective Delegation*: Participants assume that they are managers of a department in a large organization. At the moment, the demands on their time are very heavy. A specific delegation situation is described and participants decide how much responsibility to delegate at one time and to determine priorities.

Reviewing Delegation Style: Participants fill out a confidential self-questionnaire to indicate their own views with respect to delegation. After the exercise, it is suggested that they review the way they completed the form to draw conclusions about any changes in light of their experience with the simulation.

What to Delegate and to Whom: Descriptions of the tasks at hand and profiles of the subordinate department managers are provided. Participants decide to whom they would delegate each of the tasks and which they would retain to do themselves.

Defining the Project: Participants review a task which they are preparing to delegate to a subordinate. They decide how the task should be handled, what goals would be set, what progress reports are needed, and so on. (producer)

Cost: 5-participant set $23.50; 2 or more sets $19.90 each; meeting leader's guide $.50

Producer: Didactic Systems, Inc., Box 457, Cranford, NJ 07016

EFFECTIVE SUPERVISION

Staff, Didactic Systems, Inc.

Playing Data
Copyright: 1973
Age Level: college, management
Number of Players: 3 or more in 1 or more teams
Playing Time: approximately 3 hours
Preparation Time: 30-60 minutes

Description: Defining Objectives and Setting Priorities: Participants assume that they are newly appointed supervisors in an office of a large corporation. Based on activity summaries and analyses of performance for their department, participants make decisions on setting priorities for improving operations.

Drawing Conclusions from Unrelated Facts: A problem within the department is described; however, the information provided is given in a rather disjointed manner. Participants are asked to separate the relevant facts and to indicate how they would deal with this problem.

Your Responsibility in Communications with Superiors and Subordinates: Participants make specific decisions on how often they would request reports from subordinates and submit reports to superiors; in addition, participants are asked to decide how frequently they would schedule meetings with their people to discuss progress and obtain ideas for improvement.

Analyzing Employee Strengths and Weaknesses and Making the Best Use of Human Resources: In this problem an employee is described and participants are first asked to analyze the specific skills of this person. They then select specific tasks to assign to this employee in order to make the best use of her capabilities.

Resolving a Discipline Problem: A few undesirable practices have developed within the department; participants decide what would be the most effective course of action for minimizing them.

Examining Personal Attitudes toward Various Work Groups: In the first part of this item, participants complete a brief questionnaire to explore their personal attitudes toward various work groups; in the second part, participants decide how they would deal with a black/white conflict situation which has developed within their department. (producer)

Cost: 5-participant set $23.50; 2 or more sets $19.90 each; meeting leader's guide $.50

Producer: Didactic Systems, Inc., Box 457, Cranford, NJ 07016

EFFECTIVE SUPERVISION IN GOVERNMENT

Harvey R. Lieberman, Didactic Systems, Inc.

Playing Data
Copyright: 1973
Age Level: college, management
Number of Players: 3-5 per team, any number of teams
Playing Time: approximately 3 hours
Preparation Time: 30-60 minutes

Description: This game is designed to give supervisors and managers of federal, state, county, and municipal government organizations the opportunity to exchange ideas on effective supervisory practices. Participants represent supervisors who have been promoted from lower level positions in another department. The decisions they must make involve exploring the supervisor's job, building a plan of work that uses human resources effectively through delegation, establishing meaningful performance guidelines, planning the development of subordinates on the basis of their level of knowledge and skill, reviewing performance in relation to goals, recognizing employee performance, and developing team work and cooperation. (AC)

Cost: 5-participant set $23.50; 2 or more sets $19.90 each; meeting leader's guide $.50

Producer: Didactic Systems, Inc., Box 457, Cranford, NJ 07016

EXECUTIVE SECRETARY

Charles E. Kozoll, University of Illinois

Playing Data
Copyright: 1975
Age Level: executive secretaries
Number of Players: 3 or more in teams of 3-5
Playing Time: approximately 3 hours
Preparation Time: 30-60 minutes

Description: The designer states that "this exercise is designed to provide an opportunity for both new and experienced Executive Secretaries to test their judgments in a variety of situations which occur regularly." Each player represents one of a group, or team, of secretaries who compete with one another, and as a team with other teams, to correctly assess "the significant roles performed by Executive Secretaries and to isolate desirable job requirements." Scores are based upon an answer key which evaluates the comparative value of each option offered to the participants. For example, during the first part of the exercise, players must list a secretary's "Personal Characteristics for Success." A participant earns two points for stating each of the following qualifications: a good executive secretary should be poised, calm, organized, literate, thorough, should demonstrate initiative and "general good sense," and should be loyal to his or her boss. (DCD)

Cost: 5-participant set $23.50; 2 or more sets $19.90 each; meeting leader's guide $.50

Producer: Didactic Systems, Inc., Box 457, Cranford, NJ 07016

THE GAME OF TIME MANAGEMENT

Scott B. Parry and Gregory Sand

Playing Data
Copyright: 1977
Age Level: college, management
Number of Players: 3 or more in groups of 3-4
Playing Time: 60 minutes
Packaging: professionally produced game includes workbook, cassette, and in basket materials

Description: Each participant first goes through the in-basket individually. The task facing each person is to respond to the items in the in-basket. This must be done in a way that will utilize their time most effectively. Thirty minutes is allowed for all participants to individually complete the in-basket.

At the end of 30 minutes, each participant writes his or her response to the analysis and evaluation sheet. Each participant evaluates his or her own method of going through the in-basket, using the analysis and evaluation questions that summarize the methods and procedures for effective time management. (producer)

Cost: $60.00

Producer: Training House, Inc., 100 Bear Brook Road, Princeton Junction, NJ 08550

THE GAME OF TRANSACTIONAL ANALYSIS

Scott Parry and Gregory Sand

Playing Data
Copyright: 1977
Age Level: college, management
Number of Players: 4-21 in groups of 2-3
Playing Time: 60 minutes
Packaging: professionally produced game includes workbook and cassette

Description: The group is divided into paired teams (A and B) of 2-3 persons each. Each team receives 10 cards, each containing a transaction to be diagramed and completed (i.e., last line is missing and must be written in). Team A members examine their cards, deciding which are the most difficult 5. These cards are exchanged with Team B, whose members have just done the same with their cards.

Now each team settles down to the task at hand–analyzing (diagraming) and completing each of their 10 transactions. This done, the two teams exchange cards and evaluate (score) one another's performance, using an answer sheet that the instructor has just passed out. They then critique one another, improve their transactions, and determine which team was the winner. (producer)

Cost: $60.00

Producer: Training House, Inc., 100 Bear Brook Road, Princeton Junction, NJ 08550

GRIEVANCE HANDLING (INDUSTRIAL)

Staff, Didactic Systems, Inc.

Playing Data
Copyright: 1971
Age Level: college, management
Number of Players: 1 or more teams of 4 or 5
Playing Time: approximately 150 minutes
Preparation Time: 30-60 minutes

Description: Conditions Leading to Grievances: Participants assume that they are recently promoted supervisors. From a list provided by the game, participants select ten conditions they feel are most likely to lower employee morale and lead to grievances.

Handling a Complaint About Work Assignments: A number of alternatives are provided corresponding to a complaint from an employee concerning discrimination in work assignments. Participants select the one they consider most desirable.

Promoting a Junior Man: The supervisor has selected a capable junior man to fill a desirable opening in the department. How to inform the other men of his choice is explored by participants.

Assigning Overtime: Based on descriptions of their work habits, participants select an appropriate method of assigning overtime to each of six men in the department.

Avoiding Grievances: Participants write out a number of policies that can help a supervisor maintain the kind of atmosphere in which grievances cannot flourish. (producer)

Cost: 5-participant set $23.50; 2 or more sets $19.90 each; meeting leader's guide $.50

Producer: Didactic Systems, Inc., Box 457, Cranford, NJ 07016

GRIEVANCE HANDLING (NON-INDUSTRIAL)

Staff, Didactic Systems, Inc.

Playing Data
Copyright: 1971
Age Level: college, management
Number of Players: 3 or more in 1 or more teams
Playing Time: approximately 3 hours
Preparation Time: 30-60 minutes

Description: Understanding the Causes of Grievances–Defining the Problem: Three stories describing people in the department are provided. Participants identify potential complaints and grievances from each story.

Recognizing the Existence of a Potential Grievance: Several potential complaints and grievances are described. Participants decide whether to inquire about the problem, continue to watch the situation, or ignore it completely.

Handling a Grievance I: From several alternatives, provided by the game, participants select the one they feel would be most desirable to deal with an employee whose performance has slumped.

Handling Grievances II: Several people in the department have indicated that they are unhappy about a control procedure for keeping attendance and tardiness records. Participants are asked to list steps that could be taken to get a clearer picture of the problem. They then select from a series of alternative courses of action, the one they think is most likely to resolve the problem.

Handling Grievances III: To help sharpen their ability to deal with grievances, participants, within their own teams, role play two supervisor-subordinate situations. Role players are evaluated by team members.

Reducing the Incidence and Severity of Grievances: Participants are asked to write out a number of practices that can help maintain the kind of atmosphere in which grievances cannot flourish. (producer)

Cost: 5-participant set $23.50; 2 or more sets $19.90 each; meeting leader's guide $.50

Producer: Didactic Systems, Inc., Box 457, Cranford, NJ 07016

HANDLING CONFLICT IN MANAGEMENT: CONFLICT AMONG PEERS (GAME I)

Erwin Rausch and Wallace Wohlking, Didactics Systems, Inc.

Playing Data
Copyright: 1969
Age Level: college, management
Number of Players: 1 or more teams of 4 or 5
Playing Time: approximately 150-180 minutes
Preparation Time: 30-60 minutes

Description: Facing the Conflict Situation: Participants assume that they are supervisors in a factory. A conflict situation occurs between the supervisor and a quality control manager. Participants select from a number of alternatives how they would attempt to resolve the conflict.

Approaching the Quality Control Manager: In anticipation of a confrontation, participants explore how the supervisor should open the discussion with the quality control manager.

Recognizing Emotional Reactions: Various possible responses by the quality control manager are provided. Participants categorize the emotional overtones of each response.

Anticipating Emotional Reactions: A number of opening statements the supervisor could have used are provided. For each, participants are asked to anticipate the emotional reaction that might be expected from the quality control manager.

Opening Communications: The supervisor reacts angrily when the quality control manager says he is too busy to discuss the matter at the moment. Participants explore what the supervisor's next move ought to be.

De-escalating the Conflict: Some different ways of probing for the reasons behind hostile reactions are described. A series of statements is listed and participants are asked to decide which type of probe is illustrated by each statement.

Establishing an Open-Communications Climate: Participants are asked to write a brief statement indicating what kind of relationship must exist if communications between peers are to remain free and open. (producer)

Cost: 5-participant set $26.50; 2 or more sets $19.90 each; meeting leader's guide $.50

Producer: Didactic Systems, Inc., Box 457, Cranford, NJ 07016

HANDLING CONFLICT IN MANAGEMENT: SUPERIOR/SUBORDINATE–GROUP CONFLICT (GAME II)

Erwin Rausch and Wallace Wohlking, Didactic Systems, Inc.

Playing Data
Copyright: 1969
Age Level: college, management
Number of Players: 1 or more teams of 4 or 5
Playing Time: approximately 150-180 minutes
Preparation Time: 30-60 minutes

Description: Introducing a Problem: Participants assume they are newly appointed managers with relatively little experience in leading groups. Employees reporting to the manager are taking excessively long coffee breaks. How to report this situation at a staff meeting is explored by the participants.

Countering a Hostile Reaction: The manager's report is received with hostility at the staff meeting by one supervisor, who feels the matter is too petty for discussion. Participants write out a response to this supervisor.

Dealing with Dissension: The manager senses a general reluctance to continue the coffee break discussion. Participants consider, from a number of alternatives, the manager's next move at this time.

Responding to a Consultant's Questionnaire: A questionnaire to explore the manager's knowledge of group behavior is provided. Participants respond to a series of true-false questions.

Planning Strategies: The manager wishes to approach the problem with an in-plant training course. Alternatives on how to obtain supervisor support for this course are provided, and participants are asked to select the best approaches.

Dealing with a Polarized Group: The training programs have been accepted, but the group is divided on a question of implementation procedures. Participants attempt to deal with this polarized group.

Resolving an Impasse: Then participants must deal with a worsened situation when a committee formed to resolve the impasse fails to come to an agreement.

Selecting a Leadership Pattern: On the basis of a series of important considerations concerning group leadership, participants are asked to formulate a statement of philosophy on group leadership. (producer)

Cost: 5-participant set $26.50; 2 or more sets $19.90 each; meeting leader's guide $.50

Producer: Didactic Systems, Inc., Box 457, Cranford, NJ 07016

HANDLING CONFLICT IN MANAGEMENT: SUPERIOR/SUBORDINATE CONFLICT (GAME III)

Erwin Rausch and Wallace Wohlking, Didactic Systems, Inc.

Playing Data
Copyright: 1969
Age Level: college, management
Number of Players: 1 or more teams of 4 or 5
Playing Time: approximately 150 minutes
Preparation Time: 30-60 minutes

Description: Taking the First Steps: Participants assume that they are managers who must deal with a subordinate who has not progressed on a month-old assignment. Various alternatives are suggested.

Achieving an Early Confrontation: Assuming the manager has decided to confront the subordinate, participants write out how they would open the discussion.

Analyzing a Subordinate's Reactions: Various possible responses to the manager's opening statement are provided. For each, the participants must indicate the type of emotional reaction with which they are dealing.

Dealing with Defensiveness: Assuming the subordinate's reaction is a defensive one, participants explore how to probe for the underlying cause.

Dealing with Evasiveness: Some techniques for dealing with an evasive response by the subordinate are considered by participants.

Dealing with Withdrawal and Preparing for Disciplinary Action: Assuming that the subordinate's response indicated withdrawal, participants consider a number of alternatives that might be taken in an attempt to avert the necessity of disciplinary action.

Dealing with a Hostile Response: The manager's conciliatory efforts to get the subordinate to do his job have failed. Participants explore steps they would take, other than dismissal, and expected outcome. (producer)

Cost: 5-participant set $26.50; 2 or more sets $19.90 each; meeting leader's guide $.50

Producer: Didactic Systems, Inc., Box 457, Cranford, NJ 07016

INDREL (INDUSTRIAL RELATIONS EXERCISE)

Management Games Limited

Playing Data
Copyright: unknown
Age Level: management
Number of Players: up to 24 in teams of 3 to 6; or up to 10 individual players
Playing Time: 2-4 or 6 hours
Preparation Time: 2 hours

Description: The participants in this in-basket exercise assume the roles of Operations and Maintenance Managers in one of seven factories owned by a single corporation. The Maintenance Engineers under their supervision have asked that a group wage incentive plan be designed and implemented. Each player must choose what he believes to be the best of six alternative plans in the game materials. He must then respond to problems which would evolve out of implementing his decision. The process of choosing one of several alternative courses of action, each of which has evolved from a previous decision, is repeated several times until a labor dispute occurs. At that point, the player must choose one of the following: (1) take the union grievance to arbitration; (2) accept the union's demands; (3) abandon the incentive plan; or (4) continue the incentive plan on a trial basis. (DCD)

Cost: £48

Producer: Management Games, Ltd., 63B George Street, Maulden, Bedford MK45 2DD, United Kingdom

INSTRUCTION

Playing Data
Copyright: 1974
Age Level: college, management
Number of Players: 1 player or 1 team
Playing Time: 2 hours (est.)
Packaging: 28-page, soft bound booklet

Description: Instruction is a self-administered game designed to help players organize training programs and select the best didactic methods for knowledge and skill training.

The player assumes the role of a supervisor who must train a new employee in a stainless steel mill. The training period is to last for 80 hours and includes 14 topics of instruction (including company history, employee benefits, and the uses of stainless steel). Each topic takes 1-16 hours to learn. The player must schedule these topics into 10 training days and then answer a series of multiple-choice questions about how he would conduct this training and how he would respond to specific, unexpected difficulties such as what portions of the training he would delete if he had to and why. (DCD)

Cost: $7.45. Quantity discounts

Producer: Education Research, P.O. Box 4205, Warren, NJ 07060

THE INSTRUCTOR AS MANAGER OF LEARNING EXPERIENCES

Staff, Didactic Systems, Inc.

Playing Data
Copyright: 1975
Age Level: college, management
Number of Players: 3 or more in 1 or more teams
Playing Time: approximately 3 hours
Preparation Time: 30-60 minutes

Description: This game is designed to provide training professionals and instructors with the opportunity to examine methods and approaches for conducting stimulating and effective seminar/learning experiences. Players assume the roles of conference leaders/instructors–trainers who try to develop new ideas about managing a learning experience by making specific decisions about learning fundamentals, teaching methods and media, grading practices, problem students, and instructor effectiveness. (producer)

Cost: 5-participant set $23.50; 2 or more sets $19.90 each; meeting leader's guide $.50

Producer: Didactic Systems, Inc., Box 457, Cranford, NJ 07016

INTERACTION

Education Research

Playing Data
Copyright: 1973
Age Level: management
Number of Players: unlimited
Playing Time: 90 minutes
Preparation Time: 10 minutes

User's Report: This is a programmed simulation in which the players seek a certain number of points which can only be gained by providing the correct answers. Various interactions that deal with three communication patterns–upward (superiors), downward (subordinates), and lateral (peers and others)–provide the basis for these simulations. Confronted with three options on how to handle a particular problem, the participant must choose the best. After the choice is made, players then refer to the following page where the correct answer and the rationale for that answer are explained. This simulation can be played as a team or individually. Four profiles of the people that must be dealt with are provided. The exercises provide a practical opportunity for any businessman to gain an understanding of the important, and perhaps crucial, interpersonal relationships in any business setting. With slight modification, these exercises could easily be used for any interpersonal relationship. (Thomas E. Harris, Rutgers University)

Cost: $7.45. Quantity discounts

Producer: Education Research, P.O. Box 4205, Warren, NJ 07060

INTERVIEWING

Erwin Rausch

Playing Data
Copyright: 1968
Age Level: college, management
Number of Players: 4 or more in groups of 4
Playing Time: 2 hours
Preparation Time: 1 hour (est.)
Packaging: professionally produced and packaged game includes 6-page administrator's manual and 4 46-page player's manuals

Description: This collection of six interrelated in-basket items is designed to test and improve the player's ability to hire employees. Each item in the collection requires the player to select the best of several possible alternatives to resolve some problem. In the first item, for example, the player imagines that a busy supervisor has called him and told him to hire an assistant supervisor. The supervisor refused to give the player any specifications for the new job and the player knows practically nothing about the supervisor's department except that it is expanding rapidly and is very busy. The player must decide whether to (a) write an ad and then submit it to the supervisor for his approval; (b) ask the supervisor for job specifications; (c) write an ad and run it in the paper and not bother the supervisor; (d) send the supervisor some resumes that are already on file; or (e) suggest that the supervisor consider several current employees as candidates. Answers are assigned point values. A player answering "b" for the problem above would receive 10 points and no points for answering "d." The player with the highest score at the end of the game wins. (DCD)

Cost: 4-player set $14.00

Producer: Didactic Systems, Inc., Box 457, Cranford, NJ 07016

JOB ENRICHMENT–REDESIGNING JOBS FOR MOTIVATION

Staff, Didactic Systems, Inc., in cooperation with the staff of Roy W. Walters and Associates

Playing Data
Copyright: 1971
Age Level: college, graduate school, management
Number of Players: 3 or more in 1 or more teams
Playing Time: approximately 3.5 hours
Preparation Time: 30-60 minutes

Description: Deciding How to Begin–Setting Priorities: Participants assume they are managers in a public service department of a medium-size organization. Based on descriptions of various tasks needed to bring about changes in job structure, participants set priorities to separate more urgent from less urgent tasks.

"Greenlighting" the job: Participants review a specific job to decide what changes are needed to "enrich" it.

Selecting Job Changes that Motivate People: In the first portion of this situation, based on Herzberg's Motivation/Hygiene Theory, participants decide on the specific changes they would like to make to a job. In the second part of this problem, priorities are set on which specific changes will be introduced first.

Preparing for Implementation: Here participants decide on a strategy on how they will implement a job change.

Dealing with Obstacles to Effective Implementation: Several obstacles to effective implementation of a job change are described and participants are asked how to deal with each one.

Preparing a Checklist for Continuous Review: Finally, participants prepare a detailed checklist of steps needed to assure that job changes and job enrichment activities are sustained. (producer)

Cost: $6.00 per participant copy; meeting leader's guide $.50

Producer: Didactic Systems, Inc., Box 457, Cranford, NJ 07016

LEADERSHIP

Playing Data
Copyright: 1976
Age Level: college, management
Number of Players: 1 player or team
Playing Time: 2 hours (est.)
Packaging: 26-page manual

Description: *Leadership* is a self-administered business game designed to help its players become more effective leaders. The player assumes the role of the leader of a five-member research team engaged in a 20-week project. Given the personalities and backgrounds of the other team members (as briefly described in the game), the player must make a series of decisions, choosing the best of four possible alternatives. During the first week, for instance, the player may (a) do nothing, (b) brief the team on the working guidelines she has designed for the project, (c) brief the players on the benefits they will receive from association with the project, or (d) ask for their suggestions. An analysis of all possible responses is included with the game. (DCD)

Cost: $7.45. Quantity discounts

Producer: Education Research, P.O. Box 4205, Warren, NJ 07060

A LEADERSHIP DECISION GAME

John A. Tammen

Playing Data
Copyright: 1974
Age Level: college, management
Number of Players: 8-24 in 4 groups of 2-6
Playing Time: 3 hours
Special Equipment: movie projector
Packaging: professionally produced package includes a film named *Styles of Leadership*, 12-page player's manual, and 8-page instructor's manual.

Description: This game is designed to reinforce the content of a film titled *Styles of Leadership*. The film defines four styles a manager may assume in relation to employees: tell them, sell them, consult with them, or join with them. After viewing the film the participants are asked to consider 16 management problems. Players read the description of the problem in their manuals, imagine that they are the manager of the employees in the problem, and decide whether to tell, sell, consult, or join. (For example, a group of employees is recalcitrant about wearing their safety equipment and the manager's boss is due on the job site in 15 minutes; in another problem, employees are upset with an older, ineffective employee who was recently assigned to their department.) After individually responding to these problems, players are divided into four groups to discuss four of the problems. The groups try to agree on their answers and select spokesmen to present their answers to the instructor. The instructor then asks the spokesmen to discuss and rationalize their solutions and in turn reads the solutions provided in the instructions. (DCD)

Cost: unknown

Producer: Roundtable Films, Inc., 113 North San Vicente Boulevard, Beverly Hills, CA 90211

LEADING GROUPS TO BETTER DECISIONS

Erwin Rausch and George Rausch

Playing Data
Copyright: 1970
Age Level: college, management
Number of Players: 3 or more in 1 or more teams
Playing Time: approximately 3.5 hours
Preparation Time: 30-60 minutes

Description: Starting the Meeting: Participants are asked to assume that they have called a meeting of peer managers to discuss various problems within their company. A number of ways to start the meeting are outlined to establish a common view of the situation and to decide on meeting procedure. Participants are asked to select the best approach.

Dealing with Premature Solutions: During the initial stages of the meeting, one of the participants disregards the agreed-upon meeting procedure and prematurely offers a solution to a problem. Participants decide on good approaches to this situation.

Using Problem Solving Techniques Effectively: Participants make decisions on how to uncover additional problem causes, on separating facts from inferences, and on defining applicable intangible considerations.

Getting Participation from the Group: Several members of the meeting group have not participated fully in the discussion. Participants decide how the meeting leader can best help to stimulate these people.

A Discussion Deadlock: The meeting group has polarized on a specific issue. As a result, the discussion has temporarily deadlocked. Participants are asked how to resolve this situation.

Coping with Obstacles to Effective Group Problem Solving: A few situations which interfere with group discussion are described. For each, participants decide what approaches would be preferable to resolve each situation.

Assignment or Voluntary Acceptance of Tasks: Two methods are described for obtaining commitments, on the part of conferees, to the tasks agreed upon during the meeting. Participants list some advantages and disadvantages of each method.

Planning for Meeting Follow-Up: Each team prepares a comprehensive checklist for group problem-solving sessions. (producer)

Cost: 5-participant set $26.50; 2 or more sets $19.90 each, meeting leader's guide $.50

Producer: Didactic Systems, Inc., Box 457, Cranford, NJ 07016

LONG RANGE PLANNING

Staff, Didactic Systems, Inc.

Playing Data
Copyright: 1971
Age Level: college, management
Number of Players: 1 or more teams of 4 or 5
Playing Time: approximately 3.5 hours
Preparation Time: 30-60 minutes

Description: Participants are asked to make the following decisions in the course of the exercise:

Basis for Initial Decisions: Participants assume that they were recently asked to take charge of the company's planning function. Financial and other descriptive information about the company is provided. An overall approach to planning activities within the company is needed, and participants select the strategy they like best from a number of alternatives.

Working with a Decision Tree: A decision tree depicting various company opportunities is provided. Decisions are made to determine which branches from this tree can be eliminated at this early stage in the planning process.

Assigning Planning Priorities: In order to gain a clearer perspective of those company opportunities that should receive greatest attention in the planning process, a comprehensive list is provided and participants assign priorities to each one.

Determining the Major Components of the Plan: Participants are asked to prepare an outline of the important elements that should be considered when preparing a plan for a major division of the company.

Preparing a More Detailed Plan: A comprehensive planning flow chart depicting various planning activities is provided. Participants assign target dates to each activity to assure that the plan is coordinated properly.

Criteria for Evaluating Plans: Participants are asked to list several criteria which could be used to evaluate performance against a plan, in view of the difficulties of separating performance failures from lack of realism in the plan.

Developing Good Planning Habits: From a long list of planning routines, participants select a few which they consider most effective. (producer)

Cost: 5-participant set $23.50; 2 or more sets $19.90 each; meeting leader's guide $.50

Producer: Didactic Systems, Inc., Box 457, Cranford, NJ 07016

LOU BOXELL'S PERFORMANCE REVIEW

Scott Parry

Playing Data
Copyright: 1977
Age Level: college, management
Number of Players: 4-20 in pairs
Playing Time: 60 minutes
Packaging: professionally produced game includes workbook and cassette

Description: Lou Boxell, a new employee, was apprenticed by his/her boss to Mrs. Prentiss, two months before her retirement. For reasons relating more to personality than performance, Lou did not learn the job very well from Mrs. Prentiss; nor did the two of them update the job description to Lou's satisfaction. Now that Mrs. Prentiss has left, Lou has requested a meeting with the boss to establish responsibilities, priorities, and accountabilities. Lou is bringing the inadequate job description to the meeting. Lou's boss is uncertain of how the meeting will turn out, since a memo from Mrs. Prentiss points out several "danger areas" that might get Lou into trouble if they are not taken care of. (producer)

Cost: $60.00

Producer: Training House, Inc., 100 Bear Brook Road, Princeton Junction, NJ 08550

MANAGEMENT IN GOVERNMENT

Playing Data
Copyright: 1971
Age Level: management
Number of Players: 3 or more in teams of 3-5
Playing Time: 1 hour (est.)
Preparation Time: 1 hour (est.)

Description: This exercise is designed to provide public sector managers "with an opportunity to exchange ideas, and to practice the application of a few concepts and techniques which will help . . . increase both morale and discipline among the people who report to" them. Each participant represents one of a group of managers of an unspecified division who compete with one another, and as a team with other teams, to make the best possible response to each of six management tasks pertinent to leadership style, delegating style, making assignments, motivating employees, and dealing with obstacles to effective implementation. Scores are based on an answer key which summarizes a response to each task. In the first item, for example, the players must describe when they would use "democratic leadership" and when they would use "autocratic leadership." Each player makes his personal choice, each team agrees on its choice, and the players compare their responses to those in the instruction booklet. The winners are the individual and team with the highest scores. (DCD)

Cost: 5-participant set $23.50; 2 or more sets $19.90 each; meeting leader's guide $.50

Producer: Didactic Systems, Inc., Box 457, Cranford, NJ 07016

MANAGEMENT BY OBJECTIVES

Staff, Didactic Systems, Inc.

Playing Data
Copyright: 1969
Age Level: college, management
Number of Players: 1 or more teams of 4 or 5
Playing Time: 2-2.5 hours
Preparation Time: 30-60 minutes

Description: Setting Production Objectives: Participants assume that they are newly appointed section heads. Background on the company and organizational structure is provided. Participants explore how to set production goals and decide on how high to set production objectives for the balance of the year.

Starting a Management-by-Objective Program: The extent of participation by lower echelon personnel in planning a management-by-objective program is decided by participants from a series of specific alternatives.

Establishing Specific Action Programs To Achieve the Objectives: Participants explore how to translate the overall objectives into measurable tasks to be performed by specific personnel.

Selling Objectives to Employees: Participants are asked to respond to a unit manager who challenges the need for an action program to achieve objectives.

Assigning Priorities: A list of tasks for an objectives program is provided and participants assign priorities to each of them. (producer)

Cost: 5-participant set $23.50; 2 or more sets $19.90 each; meeting leader's guide $.50

Producer: Didactic Systems, Inc., Box 457, Cranford, NJ 07016

MANAGEMENT FOR SUPERVISORS

Staff, Didactic Systems, Inc.

Playing Data
Copyright: 1970
Age Level: college, management
Number of Players: 1 or more teams of 4 or 5
Playing Time: Approximately 3 hours
Preparation Time: 30-60 minutes

Description: Defining Objectives and Setting Priorities: Participants are newly appointed managers of a department. Organization structure, activity summaries, and analyses of performance are indicated for the department functions. Participants assign priority to each of several problem areas.

Preparing to Set Specific Goals: Participants write out two specific goals they think would be most helpful in reducing a problem in the department.

Delegating: Descriptions of subordinate group leaders are provided. Participants decide how to delegate each of a series of tasks, in view of the capabilities of the group leaders.

Communications with Superiors and Subordinates: How to report to the boss and how best to keep subordinates informed is explored.

Job Assignment: A number of people in the department qualify for a newly created position. Participants select a first choice and an alternate.

Dealing with Undesirable Practices: A number of specific department problems are described. Participants decide on the most appropriate method for dealing with these problems.

Leadership Style: Participants explore under what circumstances a manager should involve subordinates in decision making and in what situations he should rely primarily on his own judgment. (producer)

Cost: 5-participant set $23.50; 2 or more sets $19.90 each; meeting leader's guide $.50

Producer: Didactic Systems, Inc., Box 457, Cranford, NJ 07016

MANAGING AND ALLOCATING TIME

Harvey R. Lieberman and Erwin Rausch, Didactic Systems, Inc.

Playing Data
Copyright: 1976
Age Level: college, management
Number of Players: 3 or more in 1 or more teams
Playing Time: approximately 3.5 hours
Preparation Time: 30-60 minutes

Description: (This exercise is available in both an industrial and a nonindustrial edition.) Exploring Issues on Time Management: Participants, serving as supervisors/foremen in this exercise, are asked to review a list of time management principles. Participants decide whether they agree or disagree with each principle.

Analyzing a Supervisor's Time Management Skills: In this exercise, a detailed case study is presented. The case study describes a day in the life of a manager including his planned activities for that day, as well as a detailed description of what he actually does during that day. At the conclusion of the case, participants analyze the manager's skills at managing time and how they might have handled the various events of the day differently from the way he did.

Dealing with Interruptions and Distractions: In the first part of this exercise, participants discuss a variety of interruptions and distractions and indicate those which they feel are the ones which waste most of their time. In the second part of this problem, participants focus attention on the items they selected in part one and prepare ideas on how they can reduce the impact of these interruptions and distractions.

Developing Approaches to Organize Time More Effectively: In this concluding exercise, participants prepare a list of the steps they should take in order to assure that the essential things they wish to accomplish each day are indeed completed. (producer)

Cost: 5-participant set $23.50; 2 or more sets $19.90 each; meeting leader's guide $.50

Producer: Didactic Systems, Inc., Box 457, Cranford, NJ 07016

MANAGING THE ENGINEERING FUNCTION

Staff, Didactic Systems, Inc.

Playing Data
Copyright: 1970
Age Level: college, management
Number of Players: 1 or more teams of 4 or 5
Playing Time: approximately 3.5 hours
Preparation Time: 30-60 minutes

Description: Setting Up the Organization: Participants are asked to assume that they are engineering managers who are in the process of setting up an engineering department in a new division of a large corporation. Various functions of the department are described. For each function, participants select the most efficient organizational structure.

Interface with Manufacturing: Where the responsibilities of the engineering department end and where manufacturing assumes control of a project is explored. For each of a series of major functions, participants decide whether it should be performed by engineering, by manufacturing, or jointly.

Staffing the Department: Participants first set standards for preliminary screening of professional applicants. Detailed profiles of four applicants who meet these standards are provided, and participants make selections.

Motivating Engineers: A list of factors which contribute to job satisfaction for engineers is provided. For each one, participants indicate how high a priority it should receive in managerial policy and philosophy.

Counseling on Promotion Paths: Participants explore how to deal with an employee who has strong technical strengths, but who aspires to a managerial position to which he is not very well suited.

Setting Goals and Action Programs: From a list of alternatives, participants decide to what extent subordinates should be involved in preparing goals and the action steps needed to achieve goals.

When to Stop a Project: The history of a particular project is described, one which has suffered some setbacks and difficulties. Participants decide, from a number of alternatives, at what point this project should have been abandoned, if at all.

Standards of Performance: The company is starting a Standards of Performance program. Participants are required to list their duties and responsibilities and then to spell out an acceptable performance standard for each. (producer)

Cost: 5-participant set $23.50; 2 or more sets $19.90 each; meeting leader's guide $.50

Producer: Didactic Systems, Inc., Box 457, Cranford, NJ 07016

MANAGING THE MANUFACTURING AND INDUSTRIAL ENGINEERING FUNCTIONS

Staff, Didactic Systems, Inc.

Playing Data
Copyright: 1970
Age Level: college, management
Number of Players: 3 or more in 1 or more teams
Playing Time: approximately 4 hours
Preparation Time: 30-60 minutes

Description: Setting Priorities: Participants are asked to assume they are newly appointed managers of manufacturing services. In order to schedule work on a systematic basis and to improve relationships with other departments, participants establish priorities on a series of current work orders.

Handling Concept Estimates: A brief description of a product that is still in the concept stage is provided. Research and Development has requested fairly accurate estimates for product costs and equipment investment. Participants select a response to R&D's request from a number of alternatives.

Operations Improvement: An operations improvement program to achieve lower costs is needed if this year's profit goals are to be reached. A number of steps are described, and various alternatives are offered for each. Participants select those which they feel would best satisfy the objective.

Phasing in a New Design: A design change has been developed to achieve savings on a product. Pertinent facts are provided and participants decide at what quantity the new design should go into effect.

Deciding on a Time Standard System: The Vice President of Manufacturing has requested a recommendation concerning incentives and type of time standards for a newly acquired subsidiary. Facts about the subsidiary are provided. Participants select from a number of alternatives.

Applying the Learning Curve: Pertinent facts are provided about costs and sales forecasts on a new product that is now in the final design stages. Participants decide on a learning curve for this product.

Handling Conflict with a Foreman: A specific conflict situation between departments is described. A number of alternatives are presented and participants select a preferred course of action.

Scheduling New Equipment Designs: The Tool Engineering Department is flooded with requests for work to be done. Priorities on four pieces of equipment are of particular concern. On the basis of detailed information, participants assign these priorities.

Standards of Performance: The company is planning to start a standards of performance program. Participants list their duties and responsibilities and then spell out an acceptable standard of performance for each. (producer)

Cost: 5-participant set $23.50; 2 or more sets $19.90 each; meeting leader's guide $.50

Producer: Didactic Systems, Inc., Box 457, Cranford, NJ 07016

MANAGING THE WORKER

William Archey, Jay J. Zif, and Arthur Walker, Creative Studies, Inc.

Playing Data
Copyright: 1970
Age Level: college, management
Number of Players: 10-20
Playing Time: 3-6 hours in 30-60 minute periods
Preparation Time: 2-3 hours

Description: This game is designed to provide its players with a better understanding of the need for managers to be sensitive to the needs of workers and to encourage managers to provide an atmosphere which allows for their subordinates' development and self-actualization. Players assume the roles of management and labor in a medium-size manufacturing company. Management must decide how to best approach a theory of management and motivation in an industrial setting. This includes specific decisions on disciplining workers and a broader decision on whether to lessen job supervision to give workers greater autonomy. (DCD)

User Report: This simulation has been used for over a year in graduate and undergraduate classes. Students find it easy to understand and feel it has very clear instructions. It generates much interaction and activity among students and they seem to enjoy participating.

The instructions indicate that the simulation can be completed in three hours. We have not yet been able to complete play in less than six hours. Thus, although it is desirable to have continuous time frames in which to carry out the kit simulation, we have not been able to do so due to a maximum three-hour class period. The stated objective of the exercise, through the use of the managerial style scales provided (measures of Theory X and Theory Y dispositions) and how a person may change as a result of the simulation are perhaps doubtful. In fact, students seem to be less concerned about their Theory X or Theory Y orientation. They are more concerned about being able to resolve a critical managerial problem. Therefore, I have found the case is best utilized to demonstrate the difficulty of attempting to implement a job enrichment program within an organization. In fact, I have had good results using this simulation when it has been combined with fabricated data on employee job satisfaction using the Job Description Index. Thus it may be used both for attempting to measure orientations toward Theory X and Theory Y management as well as to demonstrate problems pertaining to the implementation of job enrichment programs in industry. (Richard W. Beatty, Assistant Professor of Organization Behavior, University of Colorado)

Cost: player's manuals $6.25 each

Producer: The Macmillan Company, 866 Third Avenue, New York, NY 10022

MANAGING THROUGH FACE-TO-FACE COMMUNICATION

Staff, Didactic Systems, Inc.

Playing Data
Copyright: 1969
Age Level: college, management
Number of Players: 1 or more teams of 4 or 5
Playing Time: 3 hours in time periods of approximately 90 minutes
Preparation Time: 30-60 minutes

Description: (This is similar to *Communicating For Results*, but its wording and examples are appropriate for higher level players.

Part I Interpreting Symbols: Participants are shown a picture and asked to write out the different meanings or reactions that come to mind. They are then asked to do the same thing for a word.

Applying the Ladder of Abstraction: The Ladder of Abstraction is explained. Participants are given several statements that describe job requirements and place them on the ladder.

Gearing Messages to the Receiver: Three employees of different experience and knowledge are described. Several instructional statements of varying complexity are then provided and participants are asked to match each statement with one of the three employees.

Classifying Statements: Since perception of facts influences acceptance of a message, the distinction between "facts" and "inferences" is reviewed. Short statements relating to a story are given and participants decide whether each statement is a fact or an inference.

Responding to Emotional Remarks: Owing to errors by one employee, problems developed in another department. The supervisor of that department calls (on two occasions) to complain. Participants are asked to explore how they would respond.

Part II

Giving Assignments and Instructions: Participants are provided with information on how to give clear assignments. They are then asked to apply this information in a specific situation with one of their subordinates.

Handling Rumors: A rumor is circulating which may have an adverse effect on the operations and morale of the department. Various alternatives on how to deal with this potentially serious problem are provided and participants select their preference.

Downward Communications: Participants decide, from a list provided, what types of items they would recommend for inclusion in the company's monthly newsletter to employees. (producer)

Cost: 5-participant set $23.50; 2 or more sets $19.90 each; metting leader's guide $.50

Producer: Didactic Systems, Inc., Box 457, Cranford, NJ 07016

THE MEETING GAME

Scott B. Parry and Gregory A. Sand

Playing Data
Copyright: 1977
Age Level: college, management
Number of Players: 5
Playing Time: 60 minutes
Packaging: professionally produced game includes workbook and cassette

Description: Five participants take part in a meeting. Each of them has been assigned the role of a supervisor or manager within the same department. They are meeting to decide on the replacement of a supervisor who has just been promoted to another department.

Written into the roles of each of these individuals are conflicting opinions about who this replacement should be. Each role also contains a "hidden agenda" that will affect the individual's behavior and reasoning during the meeting. (producer)

Cost: $60.00

Producer: Training House, Inc., 100 Bear Brook Road, Princeton Junction, NJ 08550

MOTIVATION

Playing Data
Copyright: 1971
Age Level: management
Number of Players: 1

Description: *Motivation* is structured around a realistic appraisal/counseling interview involving a typical subordinate. The manager is provided with the employee's background and work history to study. He is then challenged to work through 10 modules of a simulated interview—from planning stages, through various phases of the interview itself, and into the follow-up period after counseling. Correct decisions help the manager escalate his point score, wrong moves result in penalty. In each module the manager must make three multiple-choice responses to options in interviewing (REH)

Cost: $7.45. Quantity discounts

Producer: Education Research, P.O. Box 4205, Warren, NJ 07060

NEGOTIATE

Playing Data
Copyright: 1976
Age Level: college, management
Number of Players: 2-10 teams of 1-3
Playing Time: 5.5 hours
Packaging: professionally produced game includes instructor's manual and 24 copies of participant's instructions

Description: This noninteractive exercise concerns "buying and selling or the commercial negotiation required to obtain the quantities of material desired at the correct quality, best price and with good and reliable delivery." Players are divided into teams of buyers and sellers. Each team of buyers must purchase exactly 1000 tons of "IDIUM CORES" while "trying to obtain the best possible bargain in terms of price, quality, delivery," and "reliability of delivery." Each selling team must sell exactly 1000 tons of IDIUM CORES at the most profitable price. (DCD)

Cost: £48

Producer: Management Games, Ltd., 63B George Street, Maulden, Bedford MK45 2DD United Kingdom

NEGOTIATION

Playing Data
Copyright: 1977
Age Level: college, management
Number of Players: 1 player or team
Playing Time: 2 hours (est.)
Packaging: 27-page manual

Description: Negotiation is a self-administered simulation that was designed, according to the producer, to improve its player's "negotiating skills in boss-subordinate relationships, to resolve interdepartmental conflict, for planning sessions, in selling or supplier dealings, and of course, for 'formal' negotiations."

The player assumes the role of a product manager negotiating the introduction of a new product with the director of manufacturing. The exercise consists of 30 situations in which the adversary's (director of manufacturing's) position is explicitly stated and the player must choose the most appropriate of three given responses. For instance, the director of manufacturing may state that he wants to sell this new product for $30 while the player (product manager) may have had a $25 price in mind. The player would then have to choose from one of the three given responses: (a) "Actually our target is $25. Let's compromise on $27.50;" or (b) "Could I look at the breakdown of your costs;" or (c) "Jack, your cost is too high for our marketing plans." An answer key with an analysis of each response is included with the game. (DCD)

Cost: $7.45. Quantity discounts

Producer: Education Research, P.O. Box 4205, Warren, NJ 07060

OFFICE MANAGEMENT

Playing Data
Copyright: 1970
Age Level: college, management
Number of Players: 3 or more in 1 or more teams
Playing Time: approximately 3 hours
Preparation Time: 30-60 minutes

Description: This game is designed to help managers and supervisors improve their management skills and to give managers an opportunity to exchange ideas about various approaches to management responsibilities. Players assume the roles of managers in an office services department in a government agency. They respond to diverse situations which require them to define objectives and set priorities, prepare a specific set of goals, delegate authority, talk to superiors and subordinates, assign jobs, cope with undesirable practices, and display their own leadership style. (producer)

Cost: 5-participant set $23.50; 2 or more sets $19.90 each; meeting leader's guide $.50

Producer: Didactic Systems, Inc., Box 457, Cranford, NJ 07016

OFFICE TALK

H. J. Brehaut

Playing Data
Copyright: 1976
Age Level: high school, college, management
Number of Players: an even number of players from 2-12
Playing Time: 1.5-7 hours
Preparation Time: 1 hour (est.)
Packaging: 27-page photocopied booklet

Description: This game is designed to give players practice in conducting one-to-one negotiations likely to occur between any superior and subordinate. Although the game describes a situation in which an office worker and her supervisor meet to discuss the worker's dress, tardiness, work performance, and attitude, the designer states that this game may also be used to portray negotiations between other superior subordinate pairs such as husband and wife, landlord and tenant, and principal and teacher.

To begin, players are divided into two groups: one represents supervisors and the other represents subordinates. Each group discusses a list of issues and agrees, as a group, to an acceptable settlement to each issue. (Supervisors are concerned with the employee's work habits, her failure to complete an important job on time, and her trading work with another employee without first securing the supervisor's permission. The employee wants to ask for some time off, wants more help with her job, and is concerned about favoritism.) Supervisors and subordinates pair off to discuss one issue at a time. Each issue may be resolved in favor of one of the parties or as a compromise. Groups reorganize and meet to discuss their failure or success between each negotiating session. The game ends when all issues have been negotiated. (DCD)

Comment: The existing materials for this game are blatantly sexist. (DCD)

Cost: $7.00

Producer: Second Person Plural Ltd., 2004 18th Ave., N.W., Calgary, Alberta T2M 0X9 Canada

OPERATION SUBURBIA

Allen A. Zoll

Playing Data
Copyright: 1969
Age Level: college, management education programs
Number of Players: 15-40
Playing Time: 1 hour
Packaging: included in a 500-page book, *Dynamic Management Education*, by Zoll

Description: In *Operation Suburbia,* players divide into five groups representing five organizations (such as Modern House Builders or the Strong Land Company) owning a total of 16 equal plots of land. Each company has a different amount of money available, a different combination of plots, and different goals related to the purchase or sale of plots. The land held by each group at the beginning of the game is common knowledge. Within set time limits (30-45 minutes) each group attempts to further its own goals by deciding its strategy and arranging individual appointments (only with members of other groups to discuss the purchase and sale of land. Some plots of land are sought by several groups and thus are more expensive; others are not wanted by any groups. Since goals are flexible (i.e., the acquisition of 4 plots of land in a square, or the acquisition of 3 plots of land bordering other company-owned land) it is possible for all groups to get what they want. After the set time period has ended, the whole group reassembles and discusses a checklist for evaluating group process–how well each group clarified its problem, determined alternative approaches, found out and analyzed facts, and decided and executed their plans. (TM)

Cost: package of 10 $.85

Producer: Addison-Wesley Publishing Company, Jacob Way Reading, MA 01867

OPTIMUM DELEGATION

Staff, Didactic Systems, Inc.

Playing Data
Copyright: 1971
Age Level: college, management
Number of Players: 1 or more teams of 4 or 5
Playing Time: approximately 3 hours
Preparation Time: 30-60 minutes

Description: Participants are asked to make the following decisions in the course of the exercise:

Limits of Optimum Delegation: Participants are asked to assume that they are managers of a small division in a large organization. At the moment, the demands on their time are very heavy. A specific delegation situation is described and participants decide how much responsibility to delegate at one time.

Reviewing Delegation Style: Participants fill out a confidential self-questionnaire to indicate their own views with respect to delegation. After the exercise, it is suggested that they review the way they completed the form to draw conclusions about any changes in outlook which may appear desirable to them in the light of their experience with the simulation.

What to Delegate and to Whom: Descriptions of the various tasks at hand as well as profiles of the subordinate department managers is provided. Participants decide to whom they would delegate each of the tasks and which they would retain to do themselves.

Defining the Project: Participants review a task which they are preparing to delegate to a subordinate. They decide how the task should be handled, what goals should be set, what progress reports are needed, and so on. (Producer)

Cost: 5-participant set $23.50; 2 or more sets $19.90 each; meeting leader's guide $.50

Producer: Didactic Systems, Inc., Box 457, Cranford, NJ 07016

PERSONNEL ASSIGNMENT MANAGEMENT GAME

Jay R. Greene and Roger L. Sisson

Playing Data
Copyright: 1971
Age Level: college, management
Number of Players: 1 or more teams of 4 or 5
Playing Time: flexible
Preparation Time: 30-60 minutes

Description: This game is intended to help managers apply simple linear programs to allocation problems and to explore the value of using the technique. Participants assume that they are managers of a branch office of an accounting firm. In small teams, they solve various personnel assignment problems. They have to assign resources (accounting teams of varying performance efficiencies) to jobs. At first, teams use trial and error to determine optimum schedules. They are then given a detailed explanation and a demonstration of linear programming. Finally they work on another situation using the newly learned technique. (Producer)

Cost: 5-participant set (including meeting leader's guide) $10.50

Producer: Didactic Systems, Inc., Box 457, Cranford, NJ 07016

THE PERSONNEL DEPARTMENT

Jay J. Zif, Arthur H. Walker and Eliezer Orbach; Creative Studies, Inc.

Playing Data
Copyright: unknown
Age Level: college, graduate school, management
Number of Players: 20-40

Playing Time: 2.5-6 hours
Preparation Time: 2-3 hours the first time

Description: This game is designed to give its players a better understanding of the major functions and issues of personnel management, experience in conflict resolution and group decision making, and experience in applying personnel theory to real life. Players represent people in various positions in the personnel and other departments of a medium-size manufacturing company. Players must then develop and prepare budgets for their departments and seek to get them approved. (DCD)

Cost: player manuals $6.25 each; teacher's guide gratis

Producer: The Macmillan Company, 866 Third Avenue, New York, NY 10022

PERSONNEL MANAGEMENT IN ACTION

Arthur A. Whatley, New Mexico State University, and Nelson Lane Kelley, University of Hawaii

Playing Data
Copyright: 1977
Age Level: college
Number of Players: 5 or more
Playing Time: 30-60 minutes per exercise
Packaging: professionally produced 90-page instructor's guide and 305-page player's manual

Description: This is a collection of 22 exercises for use in college courses in personnel management. They are organized into 11 sections called Managing the Human Resource in Organizations, Job Design, Recruitment, Equal Employment Opportunity, Selection, Training and Career Development, Appraisal, Compensation, Collective Bargaining, Safety, and Discipline. All can be completed in a single class period.

"Motivation Feedback" and "Human Relations Training" typify the materials in this volume. In "Motivation Feedback," the class is divided into groups of four or five students who each read an essay about Abraham Maslow's theory that people are motivated by a hierarchy of needs. Using the essay they have just read to guide their responses, the players then agree or disagree with 20 statements such as "Pride in one's work is actually an important reward." The groups then discuss these statements. In "Human Relations Training," the players read an essay about Eric Berne's Transactional Analysis and then practice responding to questions and situations while cathecting their parent, adult, or child ego states. (DCD)

Comment: The exercises which attempt to relate the conceptualizations of specific humanistic psychotherapies to the personnel practices of American business may be particularly offensive to anyone who has ever trained in these therapies with the intention of using their training to help people. This volume, for example, continues the regretable, philistine, bastardization of Eric Berne's ideas; what Claude Steiner has called the "I'm okay, you're okay, what's your game, give me a stroke, cha cha, cha." Oh well, Eric Hoffer has already pointed out that "Every good idea in America ends up as a corporation, a foundation, or a racket." (DCD)

Cost: $8.95

Producer: West Publishing Company, P.O. Box 3526, St. Paul, MN 55102

PEX

Bernard M. Bass and Associates

Playing Data
Age Level: management
Number of Players: multiples of 6
Playing Time: half a day

Description: *Pex* is a program of 15 exercises for management and organizational development for use by individuals and small study groups. After each exercise, which deals with a different organizational problem, participants have an opportunity to discuss and analyze individual and group results in comparison with norms for managers from 10-20 countries.

1. *Objectives*: Individual participants receive background information about a company and choose one of two solutions to problems concerning budget allocations for employee safety, labor conflict, management development, product improvement, and community relations. In small groups, participants must arrive at a decision for each problem. They then discuss six possible objectives for the company and consider the relative importance of the steps in their earlier decision making.

2. *Attitudes*: Participants assess expectations and how to deal with them. They begin by stating degrees of agreement or disagreement with a set of attitude statements about the program. Then, as a group, they much reach consensus in their expectations. Individual participants again indicate their expectations about the same items. The exercise ends with one group member (usually the most pessimistic about the program) receiving "help" from another member, while the rest observe.

3. *Compensation*: This exercise examines the issue of equitable rewards and living with decisions. Participants read descriptions of 10 engineers. Three differ only in merit of performance; the rest perform averagely under various adverse job conditions. Individually, and then in a group, participants determine the salary increases each engineer will receive. They then assess the effects of the raises by estimating the probability of each engineer's remaining with the firm. The exercise ends with a role play of presenting the salary increase to two of the engineers.

4. *Life Goals*: Participants rank 11 life goals (leadership, expertness, prestige, service, wealth, independence, affection, security, self-realization, duty, pleasure) in their own lives and then as they perceive them in the lives of the other members of the group. Participants use a systematic feedback process to compare their self-rating with their ratings by others and with others' self-ratings.

5. *Supervise*: Working individually, participants choose the five most and least important traits from a list for top-level managers, middle managers, and first-level supervisors. Next, they receive and read role instructions for a superior-subordinate meeting. The roles are supervisors who are persuasive, participative, and coercive and subordinates who are vitally, mildly, or not interested. Each subordinate meets with each superior to agree on the most and least important traits of managers at different levels.

6. *Fishbowl*: Half of the study groups meet to review their performance as a group to date, while each of the remaining groups observes one group in discussion. The areas of concern are trust among members, degree of member participation, clarity of goals, use of resources. The observing group provides feedback. The groups then change places.

7. *Organization*: In trios, participants develop a plan to perform a "numbers task" with the goal of producing as many "products" as possible with the least possible "inventory" of leftovers. Trios exchange plans. Then each trio follows its own plan and another trio's plan in two production runs. After calculating productivity indexes, trios compare their effectiveness in operating under their own and another trio's plan.

8. *Communication*: Participants explore the advantages and disadvantages of one-way and two-way communication and contrast effective and ineffective communication. A "sender" communicates a geometric pattern to "receivers," first under one-way and then under two-way conditions. An observer records the time required for communication and accuracy of results under both conditions.

9. *Negotiations*: Half of the groups represent a management position in a contract bargaining situation, and half hold a union position. All participants receive information on company background, current contract issues, and the company's position relative to others in the industry. Union and management groups meet separately to arrange an overall bargaining strategy. A representative of each group meets

with an opposing representative to negotiate a final contract. The cost of the process increases with the time negotiations take. Finally, the whole group analyzes the relative effectiveness of individual negotiators and of group strategies.

10. *Evaluation*: Each study group outlines a proposal to use in evaluating a training course near completion. Next, each group follows a computerlike program of steps to outline a second proposal. Pairs of study groups exchange proposals to evaluate and rate both proposals. Each group then uses the more highly rated of its proposals to evaluate the training course. Participants conclude by rating their satisfaction on questions that run parallel to their expectations.

11. *Self-Appraisal*: This exercise is given twice, once near the beginning and again near the end of a training program. Participants describe their actual and preferred behavior on 30 scales that deal with their styles of learning, relating to others, and management. The second time, each participant also describes and is described by one other member of the study group. Participants also point out items on which the discrepancy between actuality and preference is greatest and items on which they most improved. Pairs meet to exchange and review ratings of each other.

12. *Koloman*: Participants represent decision makers for the mythical developing country of Kolomon in the areas of defense, agriculture, industry, health, education, transportation, and natural resources. For each area, each participant must allocate resources among two alternative strategies, one involving more certainty of success, short range implications, and lower pay-offs, the other having longer range implications, higher potential pay-off, and lower probability of success. Then the whole group makes these same decisions. Thus participants examine their willingness to take risks and explore the relative importance of expected pay-off, probability of success, and when the payoff will be realized.

13. *Success*: Participants individually take an organizational success questionnaire, and the study group prepares two memoranda discussing the circumstances in which each approach (e.g., bluffing, leveling, political alliances) is desirable. Next, individuals must estimate the group's mean social and political approach. Each group may then examine its actual mean approach (average of all individuals' scores) and the accuracy of its predictions. In conclusion, all groups meet to present and discuss their memoranda.

14. *Future*: Each participant takes a questionnaire that consists of statements about him or herself as a manager and about organizational issues. Responses require a prediction, a judgment of degree of concern about the item, and an estimate of degree of control over realizing the prediction. Next, the participant lists five action steps based on this assessment of the future world of work. In trios, participants engage in a series of rotating boss-subordinate (observer) meetings. The observer analyzes and reports on the behavior of each boss and subordinate, which is later discussed in the study group along with action plans and questionnaire results.

15. *Venture*: Study groups represent the management team of a textile firm with seven investment decisions to make on the basis of certain marketing, financial, and production data. Choices involve more or less risk and more or less pay-off. (AC)

Cost: specimen set (15 exercises and trainer's manual) $50.00; trainer's manual $15.00; exercises $4.50 each (minimum order for individual exercises: 6 copies)

Producer: Transnational Programs Corp., 54 Main Street, Scottsville, NY 14546

PIPEWORK ENGINEERING

Careers Research and Advisory Centre

Playing Data
Copyright: 1975
Age Level: college, graduate school
Number of Players: 1 or more
Playing Time: 1-30 hours

Description: This is a case study of Pipework Engineering Limited, a company with approximately 700 employees situated in the West Midlands of England. The company manufactures a variety of pipework products and has long been known for the quality of its workmanship. Following a strike two years ago, however, the quality of workmanship at the plant declined significantly. A year ago the owner's son succeeded his father as president of the company and in the time since then he has managed to alienate both the planning and production staff. The player(s), using the information presented in the case study, defines the corporate goals for the next year and the next five years; examines the roles of the various managers and considers their lines of command; and defines the principal accountabilities, including standards of performance, for all managers. (DCD)

Cost: £2.50 plus postage and packing

Producer: Hobsons Press Ltd., Bateman Street, Cambridge CB2 1LZ United Kingdom

PLANNING

Playing Data
Copyright: 1974
Age Level: college, management
Number of Players: any number
Playing Time: 4-6 hours (est.)
Packaging: 21-page manual (1 per person)

Description: The purpose of *Planning* is to build and strengthen planning skills–particularly methods for organizing, structuring, and detailing projects. In stage one, individual players or teams, functioning as project managers of a new plant, develop accurate plans for personnel staffing and training, purchasing, and preparing the plant for opening. They must also determine the earliest possible completion date for each step. The staffing section, for example, specifies the 25 people to be hired and the details of the interviewing and hiring process; players use the time chart provided to organize the activities involved in hiring and determine the completion date. After players prepare a plan for this section, they score themselves according to the evaluation in the guide. In stage two, players rough out a master plan for the entire project, coordinating all four activities and determining the earliest possible date the plant can be opened. Upon completion of the master plan, they evaluate their results and score themselves for the entire game.

Planning can be used individually, in small groups, or in large groups consisting of several teams. The designer suggests a follow-up meeting of 60-90 minutes during which players discuss the benefits of planning, report on the methods they used successfully in the game, and apply their experience to real life. (TM)

Cost: $7.45. Quantity discounts

Producer: Education Research, P.O. Box 4205, Warren, NJ 07060

PLANNING FOR GROWTH

Staff, Didactic Systems, Inc.

Playing Data
Copyright: 1972
Age Level: college, management
Number of Players: 3 or more in 1 or more teams
Playing Time: approximately 3 hours
Preparation Time: 30-60 minutes

Description: Participants make the following decisions.

Determining Product/Service Opportunities: Participants assume that they are presidents of a division that manufactures various products and offers several services to its customers and clients. Several new product/service concepts are described and participants are asked to decide which are worthy of further consideration.

Developing an Effective Planning Team: Participants are told that they are in the process of forming a new product/service development committee and make several decisions regarding the composition of the committee and the procedures which it should follow.

The John Olsen Problem: Subordinate manager who has not submitted any interesting new product/service concepts for the future is described. Several alternatives for dealing with this manager are provided and participants select the one they prefer.

Reviewing Olsen's Plan: A plan to review product/service concepts developed by John Olsen is described, together with an outline for taking the concepts from the preliminary stage to final production. Participants, in subteams, are asked to review this plan and to rework it into one which they would consider more effective.

Evaluating Product Plans: Subteams are asked to present their plans developed in the previous problem. Plans are then evaluated by the remaining subteams in the group.

Helping Managers Develop Better Planning Habits: Participants are provided with a list of suggestions on how to help subordinate managers develop better planning skills. They select from this list those approaches which they feel would be most valuable. (Producer)

Cost: 6-participant set $25.80; 2 or more sets $23.10 each; meeting leader's guide $.50

Producer: Didactic Systems, Inc., Box 457, Cranford, NJ 07016

PRIORITY

Education Research

Playing Data
Copyright: 1971
Age Level: management
Number of Players: 1 or more

Description: *Priority* focuses on the critical problem of effective time utilization. The simulation assumes that you are a middle manager who has a secretary reporting to you. The objective of the game is to get maximum higher priority work done over a five day period. Each day you are confronted with new and interacting assignments, changing priorities, and other time allocation problems. Your job is to get the maximum payoff by optimizing your time. You play the simulation in four segments representing five days. Basically for each period you have to fill out a daily work plan in half hour units. In addition you make decisions about what tasks you can delegate to your secretary and what you can defer until later. You can get some practice at some simple time optimization methods such as accumulating correspondence and then dictating them all at once, writing marginal replies on letters rather than complete letters, and bunching phone calls. (REH)

Cost: $7.45. Quantity discounts

Producer: Education Research, P.O. Box 4205, Warren, NJ 07060

PROBLEMS IN SUPERVISION

Cabot L. Jaffee

Playing Data
Copyright: 1968
Age Level: college, management
Number of Players: 1 or more
Playing Time: 4 hours
Preparation Time: 1 hour (est.)
Packaging: professionally produced 235-page oversize paperback

Description: The two in-basket exercises in this volume are designed to test, and help the participants improve their managerial sensitivity, organizing and planning ability, and decision making. Each player pretends to be Will Judd, "shift supervisor of production of the Geometric Manufacturing and Development Company." In both exercises Judd is presumed to have arrived at his office on the first day of work at his new job two hours before anyone else. He is about to leave town, and his in-basket is filled with memoranda and correspondence. He must set priorities on the problems he finds, refer some of them to the appropriate subordinate, leave a memorandum for his secretary so she may make the necessary appointments for him, and rough draft agendas for the meetings he must attend when he gets back. The player must write down the disposition he wants to make of each item, just as he would have to do in real life (for example, if the player decides that some item requires a return letter or memorandum, he must draft that letter or memo). After the exercise he must complete a written test designed to evaluate his comprehension of the problems presented in the exercise. (DCD)

Cost: $9.50

Producer: Addison-Wesley Publishing Company, Jacob Way, Reading, MA 01867

PROBLEM SOLVING

Playing Data
Copyright: 1978
Age Level: college, management
Number of Players: 1
Playing Time: 2 hours (est.)
Preparation Time: 30 minutes (est.)
Packaging: 27-page booklet

Description: *Problem Solving* is an in-basket (or workbook) exercise in which individual players or teams assume the role of consultants hired by the Hercules Welding Corporation to solve five operating problems which include equipment breakdowns and excessive employee turnover. The player receives an overview of the problem which includes statistics, personal opinions, and background information, so that he can specify the following: What is the problem? Where is it? When did it occur? How extensive is it? Step two of this procedure consists of identifying "distinctions/changes/relationships." In other words, the players must correlate the various pieces of information that might tie together." Using the information gathered in the first two steps, the players deduce the most likely cause for the problem they have identified. In the last step, they list possible solutions. All of these steps are summarized on a form which includes large boxes in which the players write in the information they believe to be significant.

An answer key with explanations is provided in the workbook. (DCD)

Cost: $7.45. Quantity discounts

Producer: Education Research, P.O. Box 4205, Warren, NJ 07060

PRODUCTIVITY—IMPROVING PERFORMANCE

Playing Data
Copyright: 1975
Age Level: college, management
Number of Players: 3 or more in 1 or more teams
Playing Time: approximately 3 hours
Preparation Time: 30-60 minutes

Description: This game is designed to give practicing and potential supervisors and managers a framework for exploring a wide variety of issues related to productivity improvement. Players assume the roles of recently appointed supervisors. They must analyze the situation to obtain basic data for performance improvement, make use of a flow chart, analyze possible changes, conduct on-the-job training sessions, set and use standards, request money, and prepare a progress report. (DCD)

Cost: 5-participant set $23.50; 2 or more sets $19.90 each; meeting leader's guide $.50

Producer: Didactic Systems, Inc., Box 457, Cranford, NJ 07016

PROFAIR

Bernard M. Bass, SUNY at Binghamton

Playing Data
Copyright: 1971
Age Level: management, military
Number of Players: any multiple of 5; 25-35 probably best
Playing Time: 3.5-4 hours
Preparation Time: 1 hour

Description: *Profair* is in three parts: I. An attitude questionnaire about women at work is completed focusing on four factors: career potential, supervisory potential, dependability, and emotionality. (These factors are based on a research study of 178 managers and professionals.) Then a brief case is examined requiring a decision on whether to promote a male or female candidate for a particular job. Next, information is provided to review the decision in the light of equal employment practices. II. In groups of five, participants take roles and discuss whether to promote a female candidate. Information about the four attitudinal factors is provided to help clarify the issues involved. The relative importance of various points is analyzed. III. The attitude questionnaire is retaken and self-scored on the four factors. Any changes in scores on factors from the first administration of the questionnaire are noted. These changes are looked at in light of the decision to promote the male or female candidate in the case study. *Profair* concludes with some suggestions for the participants on how to more fully utilize their women employees and to be more aware of current fair employment practices.

Writing affirmative action plans is one thing; implementing them is another. *Profair* is a way to increase the likelihood that a plan will be effectively implemented. (Producer)

Cost: participant program $6.50, minimum order of 6; staff guide gratis

Producer: Transnational Programs Corporation, 54 Main Street, Scottsville, NY 14546

PROSPER

Bernard M. Bass, SUNY at Binghamton; Wayne F. Cascio, Florida International University; J. Westbrook McPherson, Xerox Corporation

Playing Data
Copyright: 1972
Age Level: management, military
Number of Players: multiples of 6; 24-36 probably best
Playing Time: 3.5-4 hours
Preparation Time: 1 hour

Description: As a consequence of a preliminary survey of 315 managers in a large industrial firm with an affirmative action program, five dominant issues emerged which were seen to affect management's utilization of the full potential of the black employee. Managers varied in their awareness of the extent (1) that the system is biased, (2) that affirmative action policies often are limited in their implementation, (3) that black employees can be as competent as white employees, (4) that black employees need to feel they belong and that (5) black employees need to develop self-esteem.

Prosper aims to increase awareness of these five viewpoints among managers operating in any firm with affirmative action programs.

Prosper is in three parts:

I. Participants complete a 20-item questionnaire dealing with the five factors mentioned above. Then they make preliminary decisions about a case of insubordination by reading a series of memoranda about the case. The case contains elements of the five factors.

II. In groups of six, each participant takes on one of six roles to provide advice to higher authority on how to deal with the case of insubordination. Information and opinions concerning the five factors are embedded in five of the roles to further clarify the issues.

III. The attitude questionnaire is retaken and self-scored on the five factors. Changes from the first to second administration are noted. Further analyses and suggestions lead naturally into a general discussion of the subtleties of racial bias, myths and realities about racial issues, and ways of coping with problems in the work situation which essentially are racial in character. (producer)

Cost: participant program $6.50, minimum order of 6; staff guide gratis

Producer: Transnational Programs Corporation, 54 Main Street, Scottsville, NY 14546

RELOCATION: A CORPORATE DECISION

Playing Data
Copyright: 1977
Age Level: high school, college, management
Number of Players: 5-25 in groups of 5
Playing Time: 50 minutes
Supplementary Material: cassette tape recorder
Packaging: professionally designed box with tape recording, role sheets, and instructions

Description: The participants in this game assume the roles of President and four Vice-Presidents (legal, personnel, financial, and public relations) of a company which is considering moving from a deteriorating urban neighborhood. Each group of five has the same task. The players are briefed by tape recording on the declining profits of "Ace Incorporated," the effects the deterioration of the neighborhood in which the main plant is located has had on the company's operating expenses, and the options an outside consulting firm has isolated: that the company (1) do nothing; (2) sell the plant, have it remodeled, and lease it back; (3) move to another location in state; or (4) move out of state. Each team must then debate these four alternatives and agree upon a decision. (DCD)

Cost: $15.00

Producer: Simile II, P.O. Box 910, Del Mar, CA 92014

SELECTING EFFECTIVE PEOPLE

Playing Data
Copyright: 1970
Age Level: college, management
Number of Players: 1 or more teams of 4 or 5
Playing Time: approximately 3 hours
Preparation Time: 30-60 minutes

Description: Job Requirements: Participants evaluate the importance of various job qualifications to two specific jobs selected by the team.

Prospecting for Applicants I: Participants consider the advantages and disadvantages of promoting from within the organization or hiring from the outside, and select from a number of alternatives the most important advantages of each method.

Prospecting for Applicants II: Profiles of three prospects from within the company are provided, from which participants select a new secretary. Or they can choose to look outside the company if they prefer.

Personal Qualifications: A number of desirable personal qualifications are provided. Participants write out a few questions they would ask an applicant in order to find out to what extent he possesses these characteristics.

Planning Interviews: The group works cooperatively to devise an interviewing procedure, from the greeting of the applicant to the termination of the interview.

Framing Questions: Participants practice probing for detailed information during the interview.

Reference Checks: After an interview, certain doubts remain about an interesting applicant. How to make inquiries at previous places of employment to clear these doubts is explored with various alternatives from which participants choose preferences.

How to End the Search: Detailed descriptions of three prospects who have undergone extensive interviews are provided, and participants decide whether to hire one of these or to continue looking.

Procedure: Participants work together to organize a procedure to be followed from the time a vacancy occurs until a job offer is made. (Producer)

Cost: 5-participant set $23.50; 2 or more sets $19.90 each; meeting leader's guide $.50

Producer: Didactic Systems, Inc., Box 457, Cranford, NJ 07016

SELECTION

Education Research

Playing Data
Copyright: 1971
Age Level: management
Number of Players: 1 or more

Description: *Selection* focuses on probing and interviewing skills by involving the manager in a selection situation. He reviews job specs, plans and "conducts" four simulated job interviews, and selects one applicant. Game score is based on use of questions to uncover attitudes and qualifications, ability to interpret information, and the selection decision itself. (Producer)

Cost: $7.45. Quantity discounts

Producer: Education Research, P.O. Box 4205, Warren, NJ 07060

SELECTION INTERVIEW CLINIC

Marcia Matukas and Scott B. Parry

Playing Data
Copyright: 1978
Age Level: college, management
Number of Players: 6-15 in groups of 3
Packaging: professionally produced game includes workbook and cassette

Description: The workshop is divided into groups of three persons each. Working in round robin fashion, the participants fill a different role in each of three selection interviews. Prior to the workshop, participants were told to bring a job description for a position with which they are familiar . . . a subordinate's or their own. At the start of the exercise, each participant completes a job application form and gives this to the person who will be conducting the interview. The role assignments in each of the three interviews are spelled out on an instruction sheet that is distributed to each participant.

Following each interview, the person in the observer's role leads the discussion-critique. This analysis is kept in focus through the use of an Interviewer Evaluation Sheet. The three role plays that make up this simulation are very real in that each person is filling his/her own "true life" role; the information is real and not fabricated. (producer)

Cost: $60.00

Producer: Training House, Inc., 100 Bear Brook Road, Princeton Junction, NJ 08550

SENSITIVITY

Playing Data
Copyright: 1976
Age Level: college, management
Number of Players: 1 or more
Playing Time: 30-120 minutes
Packaging: 25-page manual

Description: The aim of *Sensitivity* is to develop the player's ability to read people's attitudes and drives, predict their behavior, and formulate effective strategies for gaining their cooperation. Players individually or in a team read through 10 background descriptions of hypothetical employees. Then, from a list of six drives or of six attitudes, they determine the one most characteristic of that person. They also choose one of three anticipatory reactions to a specific situation, and one of three suggested strategies for dealing with them in that situation. After making determinations for one person, they read the appropriate feedback sheet and score themselves before moving on to the next person. The choices of attitudes are: frustrated, hostile, indifferent, inflexible, negative, and supportive. The choices of drives are: achievement, affiliation, recognition, security, self-fulfillment, and status. Suggestions for follow-up activities include developing a more comprehensive list of drives and attitudes and relating this particular human relations technique to one's job. (TM)

Cost: $7.45. Quantity discounts

Producer: Education Research, P.O. Box 4205, Warren, NJ 07060

SITUATIONAL LEADERSHIP SIMULATOR

Paul Hersey, Kenneth H. Blanchard, and Lee G. Peters

Playing Data
Copyright: 1977
Age Level: college, management
Number of Players: two teams of 1-5 participants each
Playing Time: 50-120 minutes
Preparation Time: 1 hour (est.)
Packaging: professionally produced 32-page instruction manual, playing surface, tokens, and analysis, action and situation cards

Description: Participants in this simulation, according to the authors, "make decisions about the appropriate style of leadership in varying situations. The leader behavior alternatives selected are awarded plus or minus moves on a simulation board depending on their probability of success based on behavioral science theory."

Each turn, the players or teams select a situation card at random from the deck. Each card describes, in very general terms, a personnel problem in a "group" that the player hypothetically manages. It also describes the degree of maturity of each group, with maturity defined as "the capacity to set high but attainable goals (achievement-motivation), willingness and ability to take responsibility, and education and/or experience of an individual or a group." For example, one situation card states: "Your group has been dropping in productivity during the last few months. Members have been unconcerned with meeting objectives. Role defining has helped in the past. They have continually needed reminding to have tasks done on time. The group is relatively new to the job. Diagnosis–Low Maturity." After carefully thinking about this situation, the player selects an "Action Card" from a deck and selects the best of four alternative actions described on the card. For example, with the situation described above, the leader might want to "emphasize the importance of deadlines and tasks," "involve the group in problem solving," "individually talk with members and set goals," or do what he can "to make the group feel important and involved." Each action has a point value from –2 to +2. The player's selection should accord with the management theory described in the player's manuals, which may be summarized as mature employees need less supervision and more recognition than immature employees. The player reports his choice to the instructor, who looks up its point value. The player advances or retreats his token accordingly. The first player to reach the end of the 28-square track wins. (DCD)

Cost: $99.95

Producer: Center for Leadership Studies; distributed by Learning Resources Corporation, 7594 Eads Avenue, La Jolla, CA 92037

SUPERVISION

Education Research

Playing Data
Copyright: 1971
Age Level: management
Number of Players: 1

Description: *Supervision* is structured like one of Allan Zoll's action mazes. In it you simulate a manager supervising the activities of five people. You are provided with background sketches of the people which define your position and provide descriptions of your subordinates and your immediate boss. After this you are called upon to make decisions in frequently occurring supervisory positions such as one employee approaching you and telling you that another employee is trying to discredit you with the rest of your staff. You then have a choice of three things to do. Depending on your choice you eliminate or control problems. Wrong decisions often aggravate a situation or create additional problems. These conditions are simulated in the game, the right ones by enabling you to go faster through the maze and the wrong decisions by making you take further steps. (REH)

Cost: $7.45. Quantity discounts

Producer: Education Research, P.O. Box 4205, Warren, NJ 07060

SUPERVISORY SKILLS

Erwin Rausch, Didactic Systems, Inc.

Playing Data
Copyright: 1968
Age Level: college, graduate school, management
Number of Players: 1 or more teams of 4 or 5
Playing Time: approximately 2 hours
Preparation Time: 30-60 minutes

Description: Approach to the Job: Participants are asked to assume that they are newly appointed factory supervisors. A description of their department along with a list of specific problem areas is given. Participants are asked to decide how they could best approach these problems.

Assigning Priorities: More detailed information is then provided about these department problems. Participants are asked to assign priorities for tackling them.

Report–Communications Upward: It is suggested that a report to the supervisor's boss would be helpful to keep him informed. A series of items that could be included in this report are provided by the game and participants are asked to select which they would use.

Goal for Operation I: Alternate goals for improvement of a specific assembly operation are given. Participants select the goal to which they would be willing to commit themselves.

Implementation of Plan for Operation I: To implement the plan for this operation, participants must decide new assignments for the employees in this section.

Communications Upward–Progress Report: Some time has passed since the initial report. Participants are told that they have reported little to their boss about plans for the future and the current status of ongoing programs. They must now decide how to continue reporting to their superior.

Implementation of Plan for Operation II: The problem here concerns substandard output. Participants decide what course of action they should take to correct it.

Implementation of Plan for Stockroom and Materials Handling: The stockroom area and materials handling operations must be improved. Participants select, from a number of alternatives, what approach they would use to improve these activities. (Producer)

Cost: 5-participant set $17.50; meeting leader's guide $.50

Producer: Didactic Systems, Inc., Box 457, Cranford, NJ 07016

TIME MANAGEMENT

Kenneth R. Finn

Playing Data
Copyright: 1975
Age Level: college, management
Number of Players: teams of 5-7
Playing Time: approximately 60-90 minutes
Preparation Time: approximately 30 minutes
Packaging: participants' booklets, administrator's manual, Tinkertoy Set

Description: This game is designed to dramatize the time-management decisions that are made everyday and to help participants develop improved ways for organizing the use of their time. Players assume the roles of managers in a large organization who must efficiently develop and revise a plan for an organization project while handling interruptions, writing progress reports, and coping with inefficiencies and changes. Players then examine the efficiency of their work. (producer)

Cost: materials for 2 5-person teams $39.00; additional 5 booklets and Tinkertoy set $16.00; extra booklets $2.00 each

Producer: Didactic Systems, Inc., Box 457, Cranford, NJ 07016

TOMORROW'S SECRETARY

Jody R. Johns, Johns, Norris Associates

Playing Data
Copyright: 1973
Age Level: continuing education, management
Prerequisite Skills: experience in office work
Number of Players: 10-40 in 2-8 groups
Playing Time: 16 hours in 8-hour periods
Preparation Time: 1 hour
Special Equipment: film projector, optional

Description: *Tomorrow's Secretary* is a learning experience designed to prepare women for secretarial jobs that require decision-making, problem-solving, and human relations skills, in addition to the usual clerical skills.

The workshop has been structured in a way that reinforces willingness to take responsibility and develops skill in team work. The workshop requires two full days. The two days need not be consecutive, but further division of the workshop into shorter blocks of time is not recommended. Activities included are simulations, problem-solving exercises, self-assessment inventories, discussion, lecturettes, an optional film, and practice in objective setting and planning. The lecturettes could be reproduced and distributed as reading material. The participant folders contain checklists, questionnaires, case studies, brief reading selections, various worksheets, and complete instructions for the activities with the exception of three that require a leader. (Author)

Cost: leader's guide and materials for 20 participants $265.00; each additional participant set $10.50

Producer: Didactic Systems, Inc., Box 457, Cranford, NJ 07016

TRANSACTIONAL ANALYSIS– IMPROVING COMMUNICATIONS

Thomas C. Clary, Harvey R. Lieberman, and Erwin Rausch

Playing Data
Copyright: 1974
Age Level: college, graduate school, management/administrative
Number of Players: 3 to 5 players per team
Playing Time: approximately 3 hours
Preparation Time: 30-60 minutes
Special Equipment: players' manuals, administrator's manual

Description: This game is designed to give managers an opportunity to exchange ideas on transactional analysis and to explore ways it can be used to improve communications within their areas of responsibility. Players first learn to recognize three ego states (parent, adult, child) and then learn to recognize complementary and crossed transactions. (For example, when X cathects his Parent ego state and speaks to Y as if that Y were in his child ego state, and Y in turn does cathect his child ego state and replies as if X were in his parent ego state, a complementary transaction occurs—each party cathects the ego state expected by the other. If, however, Y answered as an adult and demanded an adult response from X, a crossed transaction would occur.) Players learn to recognize and deflect crossed transactions so that more effective communication can take place. (DCD)

Cost: 5-participant set $23.50; 2 or more sets $19.90 each; meeting leader's guide $.50

Producer: Didactic Systems, Inc., Box 457, Cranford, NJ 07016

UNION ORGANIZING GAME

Playing Data
Copyright: 1978
Age Level: management
Number of Players: 6 or more in 6 teams
Playing Time: 7-14 hours
Preparation Time: 3 hours (est.)
Packaging: professionally packaged game

Description: The producer says that this is a training program "for front-line and middle-level supervisors on how to detect and stop union organization." "The training springs from group discussion of simulated union organizing situations." The game is available in one-day and two-day versions in ten editions which represent the organization of ten different work forces. (These are Office/White Collar, Blue Collar, Engineering/Technical, Retail Industry, Hotels and Restaurants, Hospitals/Healthcare, Banking/Insurance, Oil-Chemical-Energy, Private Schools and Colleges, and Public Employment.)

The participants are divided into six groups which meet to discuss and complete three assignments. In addition, the players complete and discuss a "Vulnerability designed to acquaint them with the characteristics of potentially pro-union employees at the participant's workplace. This document requires the participants to estimate the number of employees in a given department who would be inclined to vote for union representation. Most items are designed to help participants categorize the age levels, company loyalty, and general susceptibility of their employees. (Examples: "How many black employees are in this unit?" "How many employees in this unit fall into the following classifications?—Loafers, Malcontents, Habitual Gripers." The game materials caution that Blacks, Loafers, Malcontents, and Habitual Gripers are all primary targets of union organizers.) One to two hours is then used to discuss this questionnaire.

The following is typical of the assignments (or problems) participants must discuss. "A new employee shows up for work wearing a union button." Each team must decide whether or not to make the employee remove the button and then report and explain their decision to the other teams.

The exercise ends when all teams have discussed their assignment problems and the questionnaire. (DCD)

Cost: basic one-day game package $225; deluxe audiovisual, two-day program $325

Producer: Labor Relations Association, Inc., Three Greenway Plaza East-242, Houston, TX 77046

WHEN THE UNION KNOCKS . . .

Playing Data
Copyright: 1978
Age Level: college, management
Number of Players: 6-20 in groups of 3-4
Playing Time: 2 hours
Packaging: professionally produced game includes workbook and cassette

Description: The game begins when participants receive a two-page memo from the Personnel Director of TBD, Inc. The memo outlines actions a supervisor, Joe Powers, has taken in attempting to keep employees from voting for the union. Each player assumes the role of Powers's boss by responding to the memo. In doing so, they identify some of the do's and dont's for supervisors during an organization campaign, particularly in connection with what they can and cannot say.

The second segment of the game starts when participants receive a sheet from the Personnel Department containing 24 questions employees are asking about the union. The Personnel Director wants managers to devise answers to these questions, so that he can publish a company communication to all employees.

Finally, the third part of the game simulates a management meeting in which the trainer conducts a session for all department heads. The agenda for this meeting focuses on the things supervisors can and cannot do, as well as things supervisors *ought* to be doing during an organizing drive. (producer)

Cost: $60.00

Producer: Training House, Inc., 100 Bear Brook Road, Princeton Junction, NJ 08550

WILSON R V

Barry L. Reece and Gerald L. Manning

Playing Data
Copyright: 1977
Age Level: junior college
Number of Players: 1
Playing Time: 3 hours (est.)
Preparation Time: 30 minutes (est.)
Packaging: professionally produced 90-page player's manual, 28-page instructor's manual, and answer key

Description: *Wilson RV* is an in-basket simulation for supervisory and management development. It is designed, according to its authors, to confront players "with a variety of realistic problems, issues, and decision-making situations that are commonly encountered by supervisory-management personnel. The in-basket items reflect actual situations reported by people employed in leadership positions. The learning activities are designed to maximize individual and small group decision making."

Each player assumes the role of Marvin A. Hilton, a 27-year-old Community College graduate whose previous work experience consists of two years as a "service and repair specialist" at the local Sears Auto Center and three years as assistant manager at a local auto parts store. There are four sequences to the game. During the first of these, the "Job Application Sequence," Marvin learns that the position of Sales Manager at a local Recreational Vehicle dealership (the Wilson RV) has become vacant. Players read Marvin's resume, a newspaper advertisement for the Wilson RV, a letter in which the dealership asks Marvin to join its staff, and an application for employment. They then answer a series of multiple-choice questions designed to help them learn how to write a resume, respond to a business letter, fill out an employment application, and, in general, get a job.

During the "Job Orientation Sequence," players as Marvin Hilton read the employee's manual for the Wilson RV to get familiar with the dealership's policies and procedures.

In the "Supervisory Problem-Solving Sequence," players read and respond to 21 memoranda which describe challenges to which Marvin must respond. Players must pick the best from a list of four or five possible solutions and decide how to communicate their decisions (by telephone, letter, memo, or in person) to Marvin's superior, subordinates, peers, and customers.

The game concludes with a "Management Problem-Solving Sequence" that consists of 10 in-basket items which emphasize the "classic management functions of planning, organizing, directing, controlling, and coordinating." These include decisions such as whether to accept Master Charge, how to rearrange the office furniture to improve use of floor space, and who Marvin should hire as his replacement when he is promoted to president of the Wilson RV. (DCD)

Cost: player manual $3.95

Producer: Gregg Division, McGraw-Hill Book Company, 1221 Avenue of the Americas, New York, NY 10020

WOMEN IN MANAGEMENT

Harvey R. Lieberman

Playing Data
Copyright: 1975
Age Level: college, management
Number of Players: 3 or more in 1 or more teams of 1 team, no maximum number of teams
Playing Time: approximately 3-4 hours
Preparation Time: 30-60 minutes

Description: This game is designed to provide practicing and potential supervisors and managers of both sexes with a framework for exploring a wide variety of issues related to the subject of the female manager in an organization. Players assume the roles of recently appointed women managers who meet regularly to discuss their mutual problems. The game is intended to give people of both sexes an appreciation for the special problems that women managers must confront. (DCD)

Cost: 5-participant set $22.50; 2 or more sets $19.90 each; meeting leader's guide $.50

Producer: Didactic Systems, Inc., Box 457, Cranford, NJ 07016

See also
Access SOCIAL STUDIES
Planning Exericises FRAME
Public Administrative Bureaucratic Laboratory for Upper Management–PABLUM COMPUTER SIMULATIONS/DOMESTIC POLITICS
Settle or Strike ECONOMICS
COMMUNICATION
SELF-DEVELOPMENT

MARKETING

ADMAG I

Charles Y. Tang and James D. Culley

Playing Data
Copyright: 1971
Age Level: management
Prerequisite Skills: two years of college experience
Number of Players: 3 per team in 1 to 99 teams
Playing Time: 1-3 hours per round, usually 3 rounds are played
Preparation Time: several hours
Special Equipment: programs are available for Burroughs, IBM, and CDC equipment

Description: *Admag I* (*Advertising Management Game I*) views advertising as a system composed of three subsystems–planning, copy, and media–with a supporting subsystem for information control. Students represent the top management of a television manufacturer (the situation can be changed if the instructor wishes) whose objectives are to maximize net profits (but may include others, as long as they are specified before play begins). In the planning area, the student is taught to use appropriate decision tools in mapping basic advertising strategy. In media operations, the use of mathematical models is encouraged; in the copy area, the concept of sociocultural learning is introduced. During the course of the game, students make decisions in budget determination, copy selection, and media allocations in marketing color television sets. For maximum effectiveness, *Admag I* should be played two or three times during the course of a training period. In planning, students learn the maximum pay-off rule from decision area; in copy selection, they learn management by objectives; and in media allocations, they learn the response value optimization method of model simulation. (REH)

Cost: player's manuals approximately $2 each; instructor's manuals and computer decks free.

Producer: James Culley, Bureau of Business and Economic Research, College of Business and Economics, University of Delaware, Newark, DE 19711

COMPETE

A. J. Faria, R. O. Nulsen, and J. L. Woznick

Playing Data
Copyright: 1978 (revised edition)
Age Level: college
Number of Players: no limitation, usually 3 or 4 per team in five-team industires
Playing Time: about 3-6 hours class start-up time, normal playing time 8-12 decisions at about 2 hours per decision
Special Equipment: any computer with a FORTRAN compiler and 64 K core capacity

Description: The players of *Compete: A Dynamic Marketing Simulation* represent the top marketing managers of a business firm producing video games, CB radios, and video tape systems. In each round of competition, decisions must be made in the areas of pricing (for all products), sales force (size, allocation, compensation, and sales presentation), advertising (amount by product and region, media selection, and message content), research and development, and sales forecasts. As well, each team must select a name for its company, brand names for each product, and determine what market research information it wishes to purchase.

Each company's market is divided into three distinct geographic segments. Each team may choose what combinations of products to sell in each of the markets or whether to sell its products in a particular market at all. Products may be changed through successful R & D activities, and distinct images may be created for each product through the company's advertising campaign. Teams can choose among five advertising media choices and determine what advertising messages will be used.

The companies in each industry compete for market share and earnings per share. Regional income and product contributions are supplied to the teams. The total market served by the teams expands and contracts based on the quality of the efforts of the teams in each industry to generate market enthusiasm.

Comment: This game offers the widest range of decisions and market research studies of any available marketing game. It is the only marketing game that allows the teams to make advertising message decisions. (AJF)

Note: For more on this game, see the essay on marketing games by Anthony J. Faria.

Cost: $6.95 per student manual

Producer: Business Publications, Inc., 13773 N. Central Expressway, Suite 1121, Dallas, TX 75243

CONSUMERISM II

Theodore F. Smith

Playing Data
Copyright: 1975
Age Level: college
Number of Players: 12-50
Playing Time: 50-75 minutes for 3 rounds of play, 1-2 hours of prior class explanation time

Description: This noncomputerized game involves conflict resolution in distribution systems. Participants take one of four roles: manufacturer, retailer, consumer, or governmental agency. Each round of play involves the presentation of a conflict issue, within-group discussion of strategy, a negotiation session with other groups, a vote, and the administration of rewards and penalties.

Play begins with the division of players into roles. Team packets containing play money and voting cards are given to each group. The first conflict situation is then presented to each team and they are given two to five minutes to formulate a strategy. The next five minutes are allotted for negotiation between the groups. Next, within-group discussions determine the satisfaction level of each group regarding its negotiations with the others. Each group then submits a vote reflecting its degree of satisfaction or dissatisfaction. Strong negative votes lead to the imposition of penalties on the groups, while positive votes lead to groups' receiving rewards. After the rewards or penalties are made in play money, voting cards are returned to the teams and a new conflict situation is introduced. (AJF)

Comment: An interesting one- to two-week exercise for a consumer behavior class to illustrate conflict and resolution. (AJF)

Cost: unknown

Producer: Dr. Theodore F. Smith, Old Dominion University, School of Business Administration, Norfolk, VA 23508

INDUSTRIAL MARKETING PLAN SIMULATION

Ernest F. Cooke, Nancy E. Spicer, Edward M. Esber, Jr., Victoria Matts, Richard N. Mendelson, Michael D. Warantz, and Paul A. Williams

Playing Data
Copyright: 1977
Age Level: college
Number of Players: no limitation, generally 4-7 teams of 1-3 players
Playing Time: about 2 hours of class start-up time, up to 20 rounds of approximately 1 hour each
Special Equipment: any computer with a FORTRAN compiler and 32 K core capacity

Description: The participants are the marketing managers of firms producing and selling electric buses. The game encourages substantial interaction among firms and between the firms and the market. The instructor may choose to have the main focus of the game center on the development of a marketing plan, but this is not necessary. Decisions in the areas of product, price, and promotion are made in each round of competition (representing a quarter). Seventeen market research reports are available for the teams to purchase.

The format of the student manual is a comprehensive marketing plan for an electric bus. The instructor may require participants to revise this marketing plan or develop a new marketing plan for some other product. As this is an industrial marketing game, the main promotional decisions involve personal selling activities such as sales force size, regional allocation, and compensation. Other decisions in each round of competition involve price for each product; credit and shipping teams; product characteristics and warranty; a sales forecast; and size of promotional budget other than personal selling. Total industry demand is determined by industry price, product offerings, promotion, economic conditions, season of the year, and stage of the product life cycle. Teams compete for market share. (AJF)

Cost: unknown

Producer: Ernest F. Cooke, Marketing Department, School of Business, University of Baltimore, Charles at Mt. Royal, Baltimore, MD 21201

INDUSTRIAL SALES MANAGEMENT GAME

Jay R. Greene, Roger L. Sisson

Playing Data
Copyright: 1975
Age Level: college, sales force
Number of Players: no limitation, 3-5 players per team, any number of teams
Playing Time: less than 1 hour class start-up time, about 1 hour per decision, any number of decisions
Special Equipment: none

Description: The participants assume the roles of the sales managers of a manufacturing firm. The objective of the game is to provide insight into the job of the sales manager as well as to allow participants to develop sales management strategies.

Participants compete to achieve the best performance in sales volume and profitability. Based on the information supplied in the participant manuals, players are asked to make decisions in each of the following areas: (1) hiring and firing salespeople; (2) distribution of sales bonuses; (3) the timing of the bonuses; (4) the time and length of sales training; (5) estimate market demand; (6) anticipate competitive strategies.

The game is noncomputerized. A referee is required. After each decision is made, the referee analyzes each team's decisions and provides each team with sales results and an income statement reflecting its performance in that period of play. (AJF)

Cost: 5-participant set $10.50

Producer: Didactic Systems, Inc., Box 457, Cranford, NJ 07016

MARKETING A NEW PRODUCT

Jay Zif, Igal Ayal, and Eliezer Orbach; Creative Studies, Inc.

Playing Data
Copyright: 1970
Age Level: college, graduate school, management
Prerequisite Skills: familiarity with the concepts of marketing
Number of Players: 4-24 in 4 teams
Playing Time: 6 hours or more in 50-minute periods
Preparation Time: 3-5 hours
Special Equipment: remote terminal

Description: Participants represent management teams of competing firms planning to enter the market with similar products. In seeking to gather and analyze information and formulate the best strategic plans in marketing their new products, they must make tactical decisions on product positioning, price, channel promotion budget, advertising budget, and orientation for every simulated four-month period. (AC)

Cost: request prices for time-sharing and mail service

Producer: Simtek, P.O. Box 109, Cambridge, MA 02139

MARKETING DYNAMICS

Charles L. Hinkle and Russell C. Koza, University of Colorado

Playing Data
Copyright: 1975
Age Level: college
Number of Players: no limitation, usually 4 teams of 3 or 4 players each
Playing Time: 2-4 hours of class start-up time required, 8-12 decisions at approximately 1 hour each
Special Equipment: any computer with a FORTRAN compiler, 32 K core capacity

Description: The players represent the marketing managers of a company that manufactures and sells cross country, alpine, and water skis. Decisions are made in the areas of product, price, and promotion. The products are sold to three distinct submarkets. Promotional decisions include establishing the advertising budget and determining the size of the sales force and how much time it will devote to each product.

The game highlights product area decisions. Each company begins with two products. As the game progresses, the companies may choose to produce as many as eight separate types of skis from twenty-five available possibilities. The objective is to develop the right product assortment for the submarkets the company is trying to reach. Sales forecasts must be developed for each product.

There is a wide range of market research studies available. (AJF)

Comment: The product selection decisions are a very interesting feature of this game. The product/market matchups make this a good game to teach market segmentation strategies. This game can also be hand scored. (AJF)

Note: For more on this simulation, see the essay on marketing games by Anthony J. Faria.

Cost: $7.50 per student manual

Producer: McGraw-Hill Book Company, 1221 Avenue of Americas, New York, NY 10020

MARKETING IN ACTION

Thomas E. Ness, University of South Florida, and Ralph L. Day, Indiana University

Playing Data
Copyright: 1978
Age Level: college
Number of Players: No limitation, usually 3 or 4 per team in 6-team industries
Playing Time: 2-5 hours start-up time, 8-12 decision periods of about 1 hour each
Special Equipment: any computer with a FORTRAN compiler and 32 K core capacity

Description: The players represent the marketing managers of a firm selling one to three soft drinks. Decisions in the areas of product, price, and promotion are made each period. The teams can choose to produce any or all of three soft drinks (cola, lemon-lime, or diet cola) at two carbonation levels, three sweetness levels, and five flavor strength levels. In addition, the teams must make decisions on how much to spend on advertising, total sales force size, and price per case of soft drinks. A wide range of market research studies are available to the teams.

The product decisions are a very important part of the game. Each product can be produced in one of thirty forms (two carbonation levels x three sweetness levels x five flavor strength levels). The teams must attempt to develop a product with the right attributes for each geographic market they are serving. (AJF)

Comment: Now in its fourth edition, this is the longest continuously published marketing game. It first appeared in 1962. (AJF)

Note: For more on this simulation, see the essay on marketing games by Anthony J. Faria.

Cost: students manuals $6.95 each

Producer: Richard D. Irwin, Inc., 1818 Ridge Road, Homewood, IL

MARKETING INTERACTION

Stephen K. Keiser, The University of Delaware, and Max E. Lupul, California State University

Playing Data
Copyright: 1977
Age Level: college
Number of Players: no limitation, usually 5 teams of 3 or 4 players each
Playing Time: 3-6 hours of class start-up time, normally 12 decisions at

1-2 hours each
Special Equipment: any computer with a FORTRAN compiler and 32 K core capacity

Description: The players represent the marketing managers of a business firm manufacturing unspecified consumer garment products. Decisions must be made in the areas of product, price, research and development, advertising, and channels. The game is less complicated in the early rounds and may become more complicated in the later rounds. The teams produce only one product at the start of competition but may produce as many as ten products in the later rounds. New Products must be discovered by research and development activities and can be differentiated along nine product dimensions (e.g., style and wearability).

The channel decisions represent one of the most interesting dimensions of the game. Channels are awarded to teams on the basis of competitive bids. There are twenty channel options available to the teams to bid on. The various channels provide the teams with different levels of intensity of distribution.

While team performance is reflected in market share, the market is expandable, thus allowing one team to improve its sales performance without necessarily taking sales from a competitor. (AJF)

Comment: This game has a number of unique features that make it a flexible and rather open computerized simulation. The progressive complication is one. Product and media choices allow quite a bit of qualitative information to be incorporated into the game, as well as introducing the use of demographic and economic factors found in actual markets. (BK)

Note: For more on this simulation, see the essay on marketing games by Anthony J. Faria.

Cost: student manuals $9.50 each

Producer: PPC Books, 1421 South Sheridan, P.O. Box 1260, Tulsa, OK 74101

MARKETING PLANNING AND STRATEGY GAME

Henry C. K. Chen, University of West Florida

Playing Data
Copyright: 1976
Age Level: college
Number of Players: up to 35-40
Playing Time: 1-2 hours class start-up time, up to 20 decision periods of approximately 1 hour each
Special Equipment: none

Description: Participants represent the marketing managers of a firm selling an industrial, heavy-duty truck. Each company can market this product in up to six regions. The product can be differentiated by horsepower, capacity, and refrigeration and can be improved through successful research and development activities. Other decision areas of the game include: the determination of sales force size and regional allocation; the determination of an advertising budget and whether the advertising will be through national or regional media; pricing; and production capacity. In addition, a number of market research studies are available to the teams.

The first five periods are devoted to developing a marketing strategy, building up a sales force, and purchasing initial market research information. Full competition in the decision areas mentioned above then starts in period six. The game is noncomputerized and a referee is required to score each round of competition. Output returned to each team at the end of each period includes a profit/loss statement, a balance sheet, a cash flow statement, an inventory ledger, and requested market research studies. (AJF)

Cost: none

Producer: Henry C. K. Chen, Faculty of Marketing, College of Business Administration, University of West Florida, Pensacola, FL 32504

MARKETING STRATEGY

Louis E. Boone and Edwin C. Hackleman

Playing Data
Copyright: 1971, 1975
Age Level: college
Prerequisite Skills: college level marketing course previously or concurrently. One elementary accounting course.
Number of Players: 5 or more players per team, 5 or more teams
Playing Time: 4 hours or more
Preparation Time: about 2 hours
Special Equipment: computer, virtually any model; FORTRAN

Description: Participants represent top managers of competing automobile manufacturers. They must make decisions on purchasing sales forecasts; submarket information; consumer preference studies; which models to market; how to schedule production; what dealerships to establish or terminate; and what models each dealer will handle. The outcomes for any one company are influenced not only by its own decisions but by the decisions of the other companies in the industry. The revised edition provides two difficulty levels; the game can run its course at either the regular or advanced level, or it can switch from regular to advanced halfway through. The advanced version brings regional marketing strategy into play. (AC)

Note: For more on this simulation, see the essay on marketing games by Anthony J. Faria.

Cost: student manuals $8.50 each

Producer: Charles E. Merrill Publishing Company, 1300 Alum Creek Drive, Columbus, OH 43216

MARKETIT

Staff, Management Games Limited

Playing Data
Copyright: 1977
Age Level: college
Number of Players: 6-20 participants in groups of 2-4
Playing Time: 6-8 hours
Special Equipment: none

Description: The game can be used by college students but also has been designed for use of salesmen and sales managers. Its objective is to illustrate the skills necessary to run a profitable sales area for a business firm. The participant's manual simulates the day-to-day operation of the sales department of a manufacturing firm.

The decisions made in each round of competition involve determining sales forecasts, setting prices, determining type and size of discounts, establishing an advertising budget, recruiting and training additions to the sales force, establishing salaries, monitoring sales expenses, and establishing commissions. The objectives of the participants are to increase the number of customers and sales volume while operating within a predetermined budget. A referee is necessary to evaluate performance and return sales and profit information to the teams at the end of each decision period. The game is well suited for use in a sales management course. (AJF)

Cost: 18-participant set $85.00 U.S.; £48 U.K.

Producer: Didactic Systems, Inc., Box 457, Cranford, NJ 07016; or Management Games, Ltd., 63B George Street, Maulden, Bedford MK45 2DD, England

MARKET PLANNING

Staff, Didactic Systems, Inc.

Playing Data
Copyright: 1971

Age Level: college, management
Number of Players: 4 or 5 per team in 1 or more teams
Playing Time: approximately 3 hours
Preparation Time: ½-1 hour

Description: Participants assume that they are marketing managers in a newly acquired subsidiary, a paper converting company that manufactures writing paper and pads. They are given background on the company, along with a detailed Sales and Promotion/Advertising Expense Sheet. Participants choose from a number of alternatives how much to spend for market studies.

They next study descriptions of several consumer market probes, with their respective costs, and then select those probes that they feel would be most valuable within their given budget. Participants also select preferences among a list of specific market studies (more intensive than probes).

With the market research data in hand, participants decide what else is needed to complete a comprehensive market plan. Next, given descriptions of product adaptations, they decide which are most worthwhile to adopt within a specified budget. Finally, on the basis of market studies and past sales figures, participants set sales objectives. (AC)

Cost: 5-participant set, $23.50; 2 or more sets, $19.90 each

Producer: Didactic Systems, Inc., Box 457, Cranford, NJ 07016

MARKET STRATEGY

Staff, Didactic Systems, Inc.

Playing Data
Copyright: 1970
Age Level: college, management
Number of Players: 4 or 5 per team in 1 or more teams
Playing Time: approximately 3 hours
Preparation Time: 1/2-1 hour

Description: Participants assume that they are marketing managers in a newly acquired subsidiary. Background information on the subsidiary is provided, along with sales and expense figures. Participants decide where the emphasis for various marketing activities should be placed.

A list of small scale tests of several possible alternative promotion, advertising, and salesmen compensation options is provided. Within a given budget, participants select those tests they feel would be most valuable.

To provide them with a broad road map for later on when more detailed decisions are made, participants decide roughly how much money should be allocated to various major activities within the confines of the overall budget.

Information is provided on types of accounts and salesman coverage for each type. From a list of alternative approaches provided, participants select the one which they feel is most likely to improve coverage of the accounts by the sales force.

A competitor has just dropped his prices about 10%. Data on product pricing as well as pricing policies of other manufacturers, in view of the 10% cut, is provided. Participants decide what pricing strategy would be most appropriate.

Additional information about competitors' pricing policy is provided. Participants decide on a long run strategy with respect to product quality, price, and sales effort.

A flow chart showing all the necessary activities required to develop, monitor, and adjust a comprehensive market strategy plan for an entire year is provided. Participants are asked to assign completion dates to each segment of this plan. (producer)

Cost: 5-participant set, $23.50; 2 or more sets, $19.90 each; meeting leader's guide, $.50

Producer: Didactic Systems, Inc., Box 457, Cranford, NJ 07016

MARKSIM

Paul S. Greenlaw and Fred W. Kniffin

Playing Data
Copyright: 1964
Age Level: college
Number of Players: minimum of 3 players (3 firms in each industry), up to 99 industries
Playing Time: 3 hours or more
Preparation Time: 3 hours at least
Special Equipment: IBM 1620 or 700/7000 or 1130 or IBM 360; FORTRAN

Description: Participants represent marketing managers of competing firms producing and selling a single consumer product. Their decisions involve production volume, product quality, national advertising, advertising allowances to retailers, retail list price, and the number of goods to be shipped to distribution centers in each period. Their objectives are to maximize return on investment, total sales volume, and share of the market. (AC)

Note: For more on this simulation, see the essay on marketing games by A. J. Faria.

Cost: student text $7.50; deck (700-700; 360; or 1620) $15.00

Producer: Harper & Row, Keystone Industrial Park, Scranton, PA 18512

MARKSTRAT

Jean-Claude Larreche, INSEAD, and Hubert Gatignon, CEDEP

Playing Data:
Copyright: 1977
Age Level: college
Number of Players: no limitation, generally 5 teams of 3 or 4 players each
Playing Time: 3-6 hours class start-up time, normally 8-12 decisions of approximately 2 hours each
Special Equipment: any computer with a BASIC compiler and 64 k core capacity

Description: Participants act as the marketing managers of a company selling a consumer durable good comparable to an electronic entertainment product. At the beginning of the game, each firm markets two brands but can modify existing products, drop products, or add new products as the game progresses. In any given year, as many as five new brands can be commercialized. Each new brand introduced must be the result of successful research and development activities. The products are sold to five separate market segments through three separate categories of retail outlets.

The major decisions made by the teams include the development of new products; establishing a retail price for each product; determining the proper retail outlets for the products; establishing the size of the advertising budget; selecting brand names for each product; and determining sales force size and its channel allocation. Sales forecasts for each product must be developed. A wide range of market research studies is available. (AJF)

Note: For more on this simulation, see the essay on marketing games by Anthony J. Faria.

Cost: student manuals, $8.00 each

Producer: Scientific Press, Stanford Barn, Palo Alto, CA 94304

OPERATION ENCOUNTER

Ben F. Doddridge and J. Rodney Howard, University of Tennessee

Playing Data
Copyright: 1975

Age Level: college
Number of Players: no limitation, generally 5 teams of 3-4 players each
Playing Time: 2-5 hours class start-up time, usually 8-12 decision periods of approximately 1 hour each
Special Equipment: any computer with a FORTRAN compiler and 32 k core capacity

Description: Participants represent the marketing managers of companies competing in the music recording industry. Each management team is responsible for selecting the musical artists to place under contract, scheduling production, pricing, the promotional budget, and determining the number of middlemen to use.

Each company produces four L-P record types from twenty possibilities in the categories of classical, easy listening, country and western, or rock. Sales forecasts must be developed for each type of album being sold. The pricing decisions and promotional decisions are very limited. The companies do not employ a sales force. The available market research studies are also very limited. (AJF)

Comment: The game has an interesting introduction. The company history is presented to the participants in a series of memos. However, the instructions for the play of the game are very difficult to follow. (AJF)

Note: For more on this simulation, see the essay on marketing games by Anthony J. Faria.

Cost: student manuals, $7.95 each

Producer: Goodyear Publishing Company, Inc., P.O. Box 2113, Santa Monica, CA 90401

POLYCHOC

Management Games Limited

Playing Data
Copyright: 1972
Age Level: management
Number of Players: up to 24 in teams of 2-5 members
Playing Time: 6-10 hours
Preparation Time: 1-2 hours

Description: The players of this game represent the management of a 'new cakes' division of a large, well-established baking company. The division manufactures and markets a chocolate cake which is then sold to retailers on a sell or return basis. The players must decide how this cake will be frosted, how many cakes to bake, and how best to advertise and sell them. The results of all decisions are obtained from a set of charts included with the game materials.

Each team is attempting to gain a share of a predetermined total market from the other teams by making a better balanced set of decisions. The teams are in competition first for the wholesale market (each manufacturer acts as his own wholesaler), and then for retail sales. They must balance their production and other costs against their expected returns to produce a reasonable return on money invested.

The sequence of decision-making in the exercise includes a briefing session by the instructor, decisions and policy statements by participants, the umpire's calculations, and player calculations. At the end of the exercise there is a general evaluation session in which each team reports on its policy and results, on the organization and controls established, and on any other matters which the umpire wishes to discuss. (HL and DCD)

Cost: £48

Producer: Management Games, Ltd., 63B George Street, Maulden, Bedford MK45 2DD, England

PRINCIPLES OF EFFECTIVE SALESMANSHIP

J.S. Schiff, Erwin Rausch, George Rausch

Playing Data
Copyright: 1971
Age Level: college, management
Number of Players: 3 or more in groups of 3-5
Playing Time: 1 hour (est.)
Preparation time: ½-1 hour

Description: "This particular exercise is designed," according to the producer, "to help industrial salesmen improve their selling effectiveness by providing an opportunity to exchange ideas and approaches about specific customer needs and on the strategies which would be most effective in satisfying these needs." To this end, players compete with other members of their team, and as a team with other teams, to make the most correct response to five in-basket items pertaining to the identification of customer needs and the development of effective selling tactics based on need identification. Responses are scored with an answer key provided in the game materials. In the first item, for example, the players review three sales calls to identify whether the buyer's needs are predominantly organizational, job oriented, or personal. Each player makes his personal choice for each call, the groups each agree on their choices, and the game specifies the most correct choice. Points are awarded for correct answers and the winners are the person and team with the highest scores at the end of the game. (DCD)

Cost: 5-participant set, $23.50; 2 or more sets, $19.90 each; meeting leader's guide, $.50

Producer: Didactic Systems, Inc., Box 457, Cranford, NJ 07016

RETAILING DEPARTMENT MANAGEMENT GAME

Jay R. Greene and Roger L. Sisson

Playing Data
Copyright: 1971
Age Level: college, management
Number of Players: 3 or more teams of 3-5 players
Playing Time: approximately 3 hours
Preparation Time: ½-1 hour

Description: Each player faces the basic business problems of forecasting market demand, inventory and personnel requirements, and product line emphasis. The participants must also be aware of the relationship between cash and inventories, the penalties associated with unexpected working capital shortages, and the need for a basic understanding of financial statements. Based on specific information provided, for each of twelve time periods, teams are asked to make decisions on merchandise to be ordered, number of clerks to be hired, and prices to be charged in two different product lines. A referee is required; after each decision is made, the referee analyzes each team's choices and provides it with income and cash statements, showing the effects of its decisions on the company. (producer)

Cost: 5-participant set, $10.50

Producer: Didactic Systems, Inc., Box 457, Cranford, NJ 07016

SALE–PLAN

Management Games Limited

Playing Data
Copyright: 1973
Age Level: management
Number of Players: 2-12 in teams of 2-4 per team, additional materials available for larger groups
Playing Time: approximately 4 hours
Preparation Time: 1 hour

Description: *Sale–Plan* has been designed to assist sales training managers, training officers, sales managers, and others involved in training salesmen to develop better plans for territorial management. The simulation illustrates that unplanned journeys usually lead to destruction of work effort, with low order conversions, high mileage, and only average number of calls–with an eventual efficiency rating which is low; and that well-planned territorial management leads to increased calls, higher orders, lower mileage, lower miles per call, higher efficiency, decreased costs, and lower gasoline consumption. *Sale–Plan* divides a sales territory into four sections and then gives the participant a four-day week to cover it. The exercises are played in several phases.

Phase A: Each team draws a customer records card which indicates the location of the customer on the salesmen territory plan. Having done this, the team leader throws the dice and moves his team's counter as indicated. He will land on a square which indicates a different type of event. The team has to abide by the instructions and follow the rules as indicated in the participants' instructions. On the reverse side of the customer's record card is other information which indicates the grade of customer and the day and time he can or cannot be seen.

On completion of a call, each team completes its customer record form with details of location, mileage to the call, grade of the customer and, if an order is obtained, the order value.

Having completed four days of play, the 'totals' columns are completed and each team works out the number of calls made, value of orders, miles per call, and efficiency rating (orders derived by miles per call).

Phase B: The instructor provides a brief lecture on the cost of salesmen–how their time is used and how they might plan their territory. Notes for this lecture are supplied by the publisher or the instructor may prepare his own notes if he wishes.

Phase C: Each team takes forty customer record cards and plans their week's work according to the information contained on the cards.

Participants cannot completely control all those elements of chance which always upset the best laid plans. In this game, these chance elements are provided by the use of dice, a board with different markings, and cards. (HL)

Cost: $85.00

Producer: Didactic Systems, Inc., Box 457, Cranford, NJ 07016

THE SALES MANAGEMENT GAME

Louis E. Boone, The University of Tulsa, David L. Kurtz and Joseph L. Braden, Eastern Michigan University

Playing Data
Copyright: 1978
Age Level: college
Number of Players: No limitation, usually 3 per team in 5 teams
Playing Time: 1-2 hours of class start-up time, 8-12 rounds of decisions of about 30 minutes each
Special Equipment: any computer with a FORTRAN compiler, very little capacity needed

Description: Players represent the sales managers of firms selling industrial drills and industrial grinders. The products are marketed in four separate geographic territories. The players must determine the number of salespeople to use in each territory, the allocation of sales effort by product, the level of advertising for each product, price, and whether a sales bonus should be paid. A limited number of market research studies are available.

The job of each sales manager is to develop the appropriate sales program for each product in each region to compete for the limited sales available. Performance information by territory and by product are returned to each of the sales managers. (AJF)

Comment: The game is very limited in scope and does not offer an adequate variety of sales management decisions. (AJF)

Cost: students manuals, $9.50 each

Producer: PPC Books, 1421 South Sheridan, P.O. Box 1260, Tulsa, OK 74101

SALES PROMOTION

Staff, Didactic Systems, Inc.

Playing Data
Copyright: 1969
Age Level: college, management
Number of Players: 4 or 5 per team in 1 or more teams
Playing Time: approximately 3 hours
Preparation Time: 1/2-1 hour

Description: Period of Adjustment: Participants assume that they hold the position of director of marketing in a company where a new sales promotion section has recently been formed. The sales promotion manager complains that he has not been invited to attend marketing strategy sessions. Participants select from alternatives a preferred procedure to correct the situation.

Gaining Acceptance: Participants explore how the sales promotion manager can gain full acceptance by the other members of the marketing team.

Risk Evaluation: Participants decide whether promotion expenditures should be halted when the anticipated profit on a specific product becomes uncertain.

Reports: The sales promotion manager has been advised to write a periodic report to increase the stature of his section. Participants select from alternatives to decide who should recieve copies, how frequently reports should be written, and what they should contain.

Working with the Sales Force: A new point-of-purchase display worked up by the sales promotion manager is rather large. Participants explore how the manager could get cooperation from the salesmen, who are unhappy with the display.

Job Description and Standards of Performance: Participants list the duties and responsibilities of a sales promotion manager and then spell out an acceptable performance standard for each. (producer)

Cost: 5-participant set $16.00; meeting leader's guide $.50

Producer: Didactic Systems, Inc., Box 457, Cranford, NJ 07016

SALESQUOTA

Education Research

Playing Data
Copyright: 1971
Age Level: sales force
Number of Players: 1 or more

Description: The game of *Salesquota* is played in two stages. In the first stage you have a geographic territory and a series of accounts. Your job is to plan a day's sales activities by selecting the calls based on the account's potential for the most sales. In doing that you must carefully plan your day, using a map that is provided. In the second part of this simulation you get a series of problems which call for certain types of analyses and you must make responses which result in sales or not. You are scored in both stages; in the first on how well you plan your sales potential for the day and in the second on how well you analyze certain sales problems–such as how to get by the receptionist, whether to ask for a firm commitment to order now or later. (REH)

Cost: $7.45 (quantity discounts)

Producer: Education Research, P.O. Box 4205, Warren, NJ 07060

SALES STRATEGY

Staff, Didactic Systems, Inc.

Playing Data
Copyright: 1968
Age Level: college, management
Number of Players: 4 or 5 per team in 1 or more teams

Playing Time: approximately 1-1/2-2 hours
Preparation Time: 1/2-1 hour

Description: Period 1: Participants assume they are district managers in charge of a group of salesmen. A new product will soon be released to the sales force. Participants select from a standard group of activities the two that are most important to perform during this initial period.

New Product Introduction: Differences of opinion exist about how this new product should be introduced for greatest effectiveness. Information about the competitor's product is provided. Participants outline the approaches they think salesmen should take in the introduction of this product.

Period 2: Promotional material for the new product has arrived and the selling effort can now begin. Vacations are approaching and many customers will be away. The team again selects two activities to be emphasized during this period.

Period 3: Business is slow. Some salesmen have given notice of intention to leave for other positions. New men are not yet available. Again participants select the two standard activities they wish to emphasize at this time.

Job Specifications: Participants are asked to complete a form to indicate the type of person they prefer to fill vacant sales positions.

Period 4: New sales force replacement has just arrived. An intensive advertising campaign will be started this period. Participants again select the two activities they would emphasize now.

Improving Performance: Some men are experiencing difficulty in meeting their sales quota. Participants review alternative approaches to this problem.

Note: This game appears quite elementary and should be used primarily as a stimulator or for new, inexperienced personnel except if the meeting leader wants to use it to reinforce a specific concept for which it may be well suited. (Producer)

Cost: 5-participant set, $16.00; meeting leader's guide, $.50

Producer: Didactic Systems, Inc., Box 457, Cranford, NJ 07016

SELLEM

Centre for Business Simulation

Playing Data
Copyright: 1977
Age Level: management
Number of Players: 2-48 in 2-12 teams
Playing Time: 3-10 hours

Description: Teams in this game represent the regional marketing manager of a manufacturing company. During each round (rounds correspond to one quarter of a fiscal year) the teams must decide how best to recruit, train, and reward their sales force, where to sell their company's product, and estimate the amount they expect to sell. These decisions are entered in an "operating statement." The effect of the decisions is then determined with the use of a "circular slide rule," a device that represents the game's model. The results are reported to the teams so they can adjust their subsequent decisions to better conform to this marketing model. A typical game lasts for four to twelve rounds. The goal of each team, in the producer's words, is "to maximize its 'contribution' to overhead and profit." (DCD)

Cost: £36

Producer: Management Games, Ltd., 63B George Street, Maulden, Bedford MK45 2DD, England

SHOPROFIT

Centre for Business Simulation

Playing Data
Copyright: 1977
Age Level: high school, adult
Number of Players: 2-8
Playing Time: 2-8 hours
Packaging: professionally produced game includes instructor's manual, orders calculator, and participant's notes.

Descriptiion: The basis of *Shoprofit* is the operation of a small retail outlet, which may be regarded as a branch in a retail chain, a single department in a large department store, or a single privately owned shop. This exercise simulates the sale of four product groups—newsagency, tobacco, confectionary, stationary, and fancy goods—which have different margins and demand characteristics. In each round (representing a period of three months) the participants make decisions on running their branch (such matters as variety of stocks and their display, staffing, promotion/advertising) with the objective of achieving a higher cumulative ratio of profit to stocks than their competitors. (Producer)

Cost: £36

Producer: Management Games, Ltd., 63B George Street, Maulden, Bedford MK45 2DD, England

SPATIAL MARKETING SIMULATION

Robert S. Ellinger, University of Iowa

Playing Data
Age Level: college, university
Special Equipment: computer; BASIC; screen terminal
Packaging: 5 copies of 17-page student manual; 11-page user's manual; 57-page instructor's guide; software (35 BASIC programs up to 713 lines of code each, and 18 files up to 19 records each)

Description: The core of this package is a highly flexible spatial marketing simulation model that gives the student the ability to replicate the location of markets, the transportation system, the population distribution, and various economic and cultural parameters of a region. The simulation has a number of possible displays (or maps) including the location of firms by goods and by order, markets by order, the market regions for particular goods, and lists of variables and parameters. The student may call upon any of three consumer models built into the simulation in solving the problems; least-effort model, used in much classical spatial marketing literature; Huff's gravity model; and a complex consumer model, developed for the simulation incorporating many concepts (action space, multiple-good tripping, shopping versus convenience-good behavior, and socio-economic characteristics). In addition to these models, a student who has had some training in programming can construct his or her own consumer model to fit some particular need.

The simulation model is coupled with seven tutorial lessons that teach the various spatial marketing and consumer behavior concepts. These lessons then use the simulation model to illustrate the integration of the concepts into the whole.

The simulation model is designed to be used on a CRT equipped with cursor-addressing capability, reverse video, and selective screen erasing or graphics capability. The simulation model must be implemented on a medium-size computer, and uses BASIC that is augmented with program chaining and sequential file addressing. (CONDUIT Pipeline)

Note: For more on this simulation, see the essay on computer simulations by Ronald E. Anderson.

Cost: $95.00; additional student manuals, 10 for $10.00, user's manual, $2.00, instructor's manual, $3.00

Producer: CONDUIT, Box 388, Iowa City, IA 522 40

See also Spatial Marketing Simulation COMPUTER SIMULATIONS

FINANCE

BUSY

Centre for Business Simulation

Playing Data
Copyright: 1977
Age Level: management
Number of Players: 2-48 in 2-8 teams
Playing Time: 4-16 hours in 4-12 rounds

Description: *Busy* is a competitive financial simulation of the operation of a Building Society in which participants make the sort of management decisions which the board of directors of a small Society may have to face.

The exercise is played by participants recording their decisions onto an "operating statement"–which includes an income and expenditure account and balance sheet. Reference points on the operating statement link it with the rules of the exercise. The adjudicator then calculates the impact of these decisions by reference to a model programmed into a Sinclair programmable calculator. This facilitates swift and accurate adjudication and also gives the adjudicator the option of amending the weighting which is attached to the decision variables.

After the adjudication is over, the operating statements are completed by the participants each round. (producer)

Cost: £60

Producer: Management Games, Ltd., 63B George Street, Maulden, Bedford MK45 2DD, United Kingdom

FINANCE

Playing Data
Copyright: 1975
Age Level: college, management
Prerequisite Skills: basic accounting and business background
Number of Players: any number
Playing Time: 3-6 hours
Packaging: 23-page manual

Description: The purpose of *Finance* is to teach managers how to interpret and use financial reports effectively. Each person or team, as division manager of Bar-B-Grill manufacturing company, receives reports on the inventory turnover and ratios of the company–income statements explaining operating activities for a specific period of time, as well as balance sheets revealing the financial position of the company at set dates. Players then analyze the company's financial strengths and weaknesses during each quarter and make recommendations. At the beginning of each quarter, players learn the consequences of their previous decisions and the resulting resources available for the next quarter. Throughout the simulation, players must fill out worksheets, keep accurate records of finances, and score themselves. At the end of four quarters, players discuss at length the rationale for their decisions, and apply their simulation experience to their actual jobs. The manual includes explanations of the four ratios as well as options for recommendations related to production, selling expense, working capital, and equipment investment. (TM)

Cost: $7.45, quantity discounts

Producer: Education Research, P.O. Box 4205, Warren, NJ 07060

FINANCIAL ANALYSIS

Erwin Rausch, Didactic Systems, Inc.

Playing Data
Copyright: 1972
Age Level: college, management
Number of Players: 3 or more in 1 or more teams
Playing Time: approximately 3 hours
Preparation Time: 1/2-1 hour

Description: Analyzing the financial statement: Participants assume that they were just told of a new assignment as General Manager for an independent subsidiary of a corporation. Participants are provided with a series of comparative income statements and balance sheets for the subsidiary and are asked to provide certain financial data which have been omitted from the statements. Preparing the information base for analysis: An income statement, which indicates that the remainder of the current year will be somewhat less profitable than originally expected, is provided. Decisions are made on what additional data would be required to analyze the situation more thoroughly. Allocating common costs: Still additional financial data are provided, and participants decide on what basis common costs should be allocated. They are then asked to complete a pro forma profitability statement.

Designing an optimal financial information system: Participants are provided with fairly detailed budget and actual results for their subsidiary. They are asked to design an improved reporting system to overcome problems they observe in the existing one.

Financial Analysis: Participants review the data described in the previous item and draw specific conclusions about the seriousness and causes of the problems. (publisher)

Cost: 5-participant set, $23.50; 2 or more sets, $19.90 each; meeting leader's guide, $.50

Producer: Didactic Systems, Inc., Box 457, Cranford, NJ 07016

FINANSIM

Paul S. Greenlaw and M. William Frey, Pennsylvania State University

Playing Data
Copyright: 1967
Age Level: college, graduate school
Number of Players: 1 and up
Playing Time: 3 hours minimum
Special Equipment: IBM 700-7000, 1620, 1130, or 360; FORTRAN
Packaging: 200-page student text, instructor manual, computer deck

Description: Participants playing as individuals or in teams take the roles of financial managers offirms selling one unidentified product in this simulation designed to facilitate learning and mastery of certain basic concepts and techniques of financial management. The student text describes the game environment, the rules, and the decisions to be made, and provides analytical tools and concepts for meeting game objectives. The decisions for each round (representing one year) are: the number of production units to manufacture, to purchase/sell marketable securities, to float new or retire existing ten-year debentures, to get bank loans to help finance company operations, to issue new common stock, to make dividend payments on existing common stock, to maintain or expand the firm's plant and machine capacity, to invest in any or all of three types of capital improvements. (AC)

Cost: paper texts, $7.40 each; instructor manual free to adopters, computer deck for 700-7000, 1620, 1130, or 360 (specify which) $15.00

Producer: Harper & Row Publishers, Inc., Keystone Industrial Park, Scranton, PA 18512

FINMARK

Management Games, Ltd.

Playing Data
Copyright: 1975
Age Level: college, management
Number of Players: 3-18 in groups of 3-5
Playing Time: 7-8 hours in 6 decision periods
Packaging: professionally produced game includes instructor's manual and 20 copies of participant instructions

Description: *Finmark* simulates the financial and marketing elements a company which has just been established to manufacture and market a consumer durable in "economy" and "luxury" models. Teams seek to maximize profits. To start, each team is given an estimate of the potential sales as well as the estimated raw materials and production costs of both products. They must then decide the production volume and mix for each quarter of succeeding fiscal years, determine marketing and research procedures, and finance the venture. The results of all decisions are calculated from a set of graphs included in the game materials. (DCD)

Cost: £48

Producer: Management Games, Ltd., 63B George Street, Maulden, Bedford MK45 2DD, United Kingdom

INTRODUCTION TO MANAGERIAL ACCOUNTING

Kenneth R. Goosen

Playing Data
Copyright: 1973
Age Level: college
Number of Players: 1 or more players or teams
Playing Time: 1-23 1-hour sessions
Preparation Time: 10 hours (est.)
Special Equipment: IBM 370, IBM 1130; FORTRAN IV
Packaging: 208-page oversize student text, 48-page instructor's manual

Description: The designer describes the objectives of this game as "to increase the student's understanding of the importance of managerial accounting as an aid in decision-making" and "to be an instructional device for increasing the student's ability to use managerial accounting techniques and concepts." Players represent decision makers for the V. K. Gadget Company, in charge of Marketing, Production, and Finance. They must make the company, which has recently undergone numerous personnel changes while its profits have spiraled downward, grow and prosper. The game model specifies six behavioral relationships: (1) a change in the price of the company's home appliance product slightly affects the number of sales leads; (2) there is a point of diminishing returns when an increase in salesmen's salaries results in a decrease of salesmen's calls; (3) changes in credit terms affect sales; (4) increases in credit terms increase bad debts; (5) there is a point of diminished returns where increased advertising of the product does not result in increased sales; and (6) paying the workers more makes them produce more. In the course of play, participants may address one or more management problems including Financial Statement Analysis, the predetermination of Departmental Overhead Rates, and the preparation of a Contribution Basis Income Statement. Player input is recorded on a fifteen-item form completed during each game session for transfer to punch cards. Output after each session includes a current balance sheet, income statement, costs statement, and a summary of game results. Although the program is designed for use with an IBM 370, it may be implemented in several parts with hardware that has a comparatively small core memory, such as an IBM 1130. (DCD)

Cost: $8.95

Producer: General Learning Press, 250 James Street, Morristown, NJ 07960

RISKM

Brian Schott, Georgia State University

Playing Data
Copyright: 1975
Age Level: college, management
Number of Players: 1 or more
Playing Time: 4-24 hours in 2-hour rounds
Preparation Time: 20 hours (est.)
Special Equipment: EDP; FORTRAN
Supplementary Materials: *A Case Study in Risk Management* (1972, Meredith, Inc.)
Packaging: 30-page photocopied player's manual and 110-page photocopied administrator's manual

Description: *RISKM* is a computer simulation of the case of the Special Chemical Company (described in *A Case Study in Risk Management* by Jerry S. Rosenbloom). Both the case and the simulation treat insurable risk in an organization as an integral part of the organization and use mathematical and statistical techniques to describe risks. In the game, groups of players represent risk management teams. Each team has an identical history and manages an identical company that is exposed to the same chance factors (such as sales levels and casualty losses) as all the others. Consequently, the risk management strategy that each group adopts accounts for any differences in profitability among the firms. Players must make both financial and insurance decisions. All game decisions are for a period of one year and the simulated effect of those decisions is revealed by the computer model. (DCD)

Cost: (1) administrator's manual, $15.00; punched cards, $30.00; (2) request prices for time-sharing or mail service

Producer: (1) Publishing Services Division, Georgia State University, University Plaza, Atlanta, GA 30303; (2) SIMTEK, P.O. Box 109, Cambridge, MA 02139

SNIBLETTE—IN-BASKET EXERCISE

Management Games, Ltd.

Playing Data
Copyright: 1976
Age Level: management
Number of Players: 8-20 in teams of 2-3
Playing Time: 2½-5½ hours
Packaging: professionally produced game includes instructor's manual and 12 participant folders

Description: Teams of players of this in-basket exercise are each given a folder containing memoranda, letters, and financial statements. The teams use these documents to consider the practical business problems of a typical small English company that operates factories in London and Leeds. The teams must decide whether to replace the company's product, named "Snibbo," with two new products named "Spon" and "Spree." The teams must also decide whether to manufacture or import these products and whether to divest the company of the Leeds factory. (DCD)

Cost: £48

Producer: Management Games, Ltd., 63B George Street, Maulden, Bedford MK45 2DD, United Kingdom

THYNGUMMY JIGS LTD.

Management Games, Ltd.

Playing Data
Age Level: high school, college
Number of Players: 1-20
Packaging: 150-page loose-leaf volume

Description: This is a collection of projects that concern the imaginary engineering firm Thyngummy Jigs, Ltd. Each project requires students to solve some problem by making a decision and to report that decision in writing or as part of a role-playing exercise. Projects address stock valuation, credit control, product costs and overhead absorption, profit realization, investment appraisal, stock-taking, stock control, and budgeting. (DCD)

Cost: £75

Producer: Management Games, Ltd., 63B George Street, Maulden, Bedford MK45 2DD, United Kingdom

SPECIFIC INDUSTRY

OKLAHOMA BANK MANAGEMENT GAME

Glen Fisher, Michael Boehlje, and Clint Roush

Playing Data
Copyright: 1974
Age Level: college, management
Number of Players: 3 or more
Special equipment: IBM 360/65; FORTRAN IV

Description: This game is designed to represent the policy management environment of a commercial bank in a rural Oklahoma county containing three competing banks. The country is assumed as the total market area for the banks involved. Agriculture is the principal industry in the county. The banks are assumed to be in compliance with national banking regulations and are members of both the Federal Reserve System and the Federal Deposit Insurance Corporation.

Each individual will be assigned to a bank management decision team whose task is to manage one of the three banks in the county. For a large number of participants, there will be several county market areas each containing three competing banks. The three banks in each county begin with identical financial statements. Also the initial market shares of the loans and deposits held by each bank are identical. The principal goal of each decision team is to make the maximum possible profit, thus increasing the bank's capital structure. Consideration may also be given to additional goals of maintaining market share and serving the community. (authors)

Cost: $100.00

Producer: Department of Agricultural Economics, Room 308 Agricultural Hall, Oklahoma State University, Stillwater, OK 74074

PROBLEMS IN BANK MANAGEMENT

Cabot L. Jaffee, Richard Reilly, and Wayne Burroughs, University of Tennessee

Playing Data
Copyright: 1969
Age Level: college, management
Number of Players: 1 or more
Playing Time: 4 hours
Preparation Time: 1 hour (est.)
Packaging: professionally produced 168-page oversize paperback

Description: The pair of in-basket exercises in this volume are designed to test and help the participants improve managerial sensitivity, organizing and planning ability, and decision making. Each participant pretends to be Dan Jennings, who is first promoted to Operations

Manager of a branch of the Metropolitan National Bank and subsequently (in the second exercise) promoted to Branch Manager. In both exercises, Jennings is presumed to have arrived at his office on the first day of work at his new job, two hours before anyone else. He must leave town for two or three days in an hour, and his in-basket is filled with memoranda and correspondence. He must set priorities on the problems he finds, refer what he can to the appropriate subordinates, leave a memorandum for his secretary so she can make the necessary appointments, and rough out agendas for the meetings he must attend. The player must complete the exercises in writing and after each takes a written test designed to test his comprehension of the problems presented in the in-basket items. For example, in the first exercise Dan Jennings is replacing an older employee who has obviously been looking forward to retirement for some time, and has consequently left the most troublesome items sitting in his basket. One serious problem is an apparently racially motivated interpersonal conflict at the branch: an investigator from the local Commission Against Discrimination has requested an appointment, the seemingly most competent employee at the branch, who is black, has given written notice that she is quitting and has applied for a job with a different company. The in-basket contains a crank letter about the branch's only other black employee, a memo from one of the tellers that is derogatory of this second employee, a note from the head teller asking that this employee replace her when she goes on vacation (the same day that Dan Jennings returns from his noncancellable three-day conference), and a petulant note from the teller demanding a new cash drawer. The first exercise requires an hour to play and the second is half again as long. (DCD)

Comment: This well thought out, comparatively realistic in-basket simulation is highly recommended. (DCD)

Cost: $8.95

Producer: Addison-Wesley Publishing Company, Jacob Way, Reading, MA 01867

SUPERMARKET STRATEGY

Jay Zif, Igal Ayal, and Eliezer Orbach

Playing Data
Copyright: 1970
Age Level: college, graduate school, management
Special Prerequisites: familiarity with marketing
Number of Players: 4-24 in 4 teams
Playing Time: 6 hours or more in 50-minute periods
Preparation Time: 3-4 hours
Special Equipment: remote terminal

Description: Each team represents the top management of one of four competing supermarket chains. In formulating plans that will serve the chain best in this competitive situation, participants make decisions on quality of products, quality of service, pricing by product line, use of private labels, and the extent to which they will use various promotional tools. (AC)

Cost: request prices for time-sharing and mail service

Producer: SIMTEK, P.O. Box 109, Cambridge, MA 02139

PURDUE SUPERMARKET MANAGEMENT GAME

Emerson Babb

Playing Data
Copyright: 1969
Age Level: college
Number of Players: 2-20 teams of 1-5 players
Playing Time: 45 minutes per decision
Preparation Time: 4-8 hours
Supplementary Material: Babb & Eisgruber, *Business Management Games for Teaching and Research*
Special Equipment: remote terminal

Description: Teams of players compete with one another in the management of supermarkets. Each of these markets is divided into four departments (produce, meat, grocery, and dairy) for decision-making purposes and each team has the task of improving its store's operation. All participants start with the same assets, liabilities, inventories, and market share. Each team must make weekly management decisions that are then fed into the computer for analysis. These decisions are the margins and orders for each department, the specials for each department, advertising and promotions, the borrowing and repayment of operating capital, and new hires and terminations. (DCD)

Cost: request prices of time-sharing or mail service

Producer: SIMTEK, P.O. Box 109, Cambridge, MA 02139

MANAGEMENT BY OBJECTIVES FOR INSURANCE COMPANIES

Playing Data
Copyright: 1972
Age Level: management
Number of Players: 3 or more in teams of 3-5
Playing Time: 1 hour (est.)
Preparation Time: 30-60 minutes

Description: This exercise is designed to help insurance company managers explore, in the producer's words, "various approaches leading toward a comprehensive Management by Objectives program. The environment of the game is the Triple A Branch, a medium-sized life Agency." Each player represents one of a group, or team, of agency managers who compete with one another, and as a team with other teams, to develop policies and procedures which will create the most effective program. Participants respond to five in-basket items that pertain to setting production objectives, starting a management-by-objectives program, establishing specific action programs to achieve the objectives, selling objectives to agents, and assigning priorities. The policies and procedures players create are judged against an answer key. For each item, players makes personal choices, each group agrees on a choice, and these are then compared with the answer key. (DCD)

Cost: 5-participant set $23.50; 2 or more sets $19.90 each; meeting leader's guide $.50

Producer: Didactic Systems, Inc., Box 457, Cranford, NJ 07016

POLICY

Centre for Business Simulation

Playing Data
Copyright: 1978
Age Level: high school, college, management
Number of Players: 2-8
Packaging: professionally produced game includes instructor's manual, 20 copies of participant's notes, forms, and one "Sinclair Programmable" calculator.

Description: This exercise represents the operations of a small nontariff company which underwrites two classes of business—property and liability. Each class of business has a known spectrum of claims but the loss ratio will fluctuate from one round to the next within this spectrum. The company's reserves may be invested either in gilt edge securities or in equity shares, and the interest or dividends from these investments will vary according to the underlying economic conditions. The management of the company will also have to undertake decisions upon the levels of administrative expenses, commission paid to brokers and promotional investment (advertising, sponsorship, public relations, and such), and expenditure on recruiting new agencies. These decisions will affect the rate of premium income growth of the company and also its profitability. The objective of the exercise is to maximize the accumulated reserves over the stipulated number of rounds. (Producer)

Cost: £60

Producer: Management Games, Ltd., 63B George Street, Maulden, Bedford MK45 2DD, United Kingdom

PRINCIPLES OF EFFECTIVE SALESMANSHIP FOR INSURANCE AGENTS

Playing Data
Copyright: 1973
Age Level: life insurance agents
Number of Players: 3 or more in groups of 3-5
Playing Time: 1 hour (est.)
Preparation Time: 30-60 minutes

Description: This exercise is designed, according to the producer, "to help life insurance agents improve their selling effectiveness by providing an opportunity to exchange ideas and approaches about specific prospect needs and on the strategies which would be most effective in satisfying these needs." To this end, the players compete with other members of their team, and as a team with other teams, to make the correct response to five in-basket items. In the first item, for example, they must evaluate the living, educational, health, and recreational needs of potential clients as low, moderate, or high. These responses are then scored with an answer key included in the game materials and points are awarded for correct answers. The winners are the player and the team with the highest scores at the end of the game. (DCD)

Cost: 5-participant set $23.50; 2 or more sets $19.90 each; meeting leader's guide $.50

Producer: Didactic Systems, Inc., Box 457, Cranford, NJ 07016

RECRUITING EFFECTIVE INSURANCE AGENTS

Playing Data
Copyright: 1973
Age Level: management
Number of Players: 2 or more
Playing Time: 90 minutes (est.)
Preparation Time: 30-60 minutes

Description: Each participant represents one of a group, or team, of life insurance agency managers who compete with one another, and as a team with other teams, to develop a set of procedures for recruiting new agents. Items they consider pertain to characteristics of the professional agent, indicators of success, approaching a prospect, sources for new agents, motivating subordinates toward a team effort in recruiting, selecting applicants for interviews, matching the approach to the applicant, and developing a program for effective recruiting. In the first part of the exercise, for example, participants must rank order the desirable job characteristics of insurance agents, retail salesmen, restaurant managers, teachers, and factory foremen from a list of 10 characteristics that include security, challenge provided, and the respect of others. Each player makes a personal ranking, each group agrees on a ranking, and response are then scored against an answer key. The individual and team with the highest scores at the end win. (DCD)

Cost: 5-participant set $23.50; 2 or more sets $19.90 each; meeting leader's guide $.50

Producer: Didactic Systems, Inc., Box 457, Cranford, NJ 07016

SELECTING EFFECTIVE INSURANCE AGENTS

Playing Data
Copyright: 1973
Age Level: managers
Number of Players: 2 or more
Playing Time: 90 minutes (est.)
Preparation time: 30-60 minutes

Description: Each participant represents one of a group of agency managers who compete with one another, and as a team with other teams, to develop a set of procedures for hiring new agents. They consider items pertaining to job requirements, recruiting methods, qualifications, planning interviews, framing questions, checking references, and selecting new agents. During the first part of the exercise, for instance, players rank order a list of 21 job qualifications that include good appearance, good handwriting, and enthusiasm. Each participant makes an individual ranking, each team agrees on a ranking, and then responses are scored against an answer key. The individual and team with the highest scores win. (DCD)

Cost: 5-participant set $23.50; 2 or more sets $19.90 each; meeting leader's guide $.50

Producer: Didactic Systems, Inc., Box 457, Cranford, NJ 07016

A GENERAL AGRICULTURAL FIRM SIMULATOR

Robert Hutton and Herbert Hinman, Pennsylvania State University

Playing Data
Copyright: 1968
Age Level: high school, college, continuing education
Number of Players: 1 or more
Playing Time: 30 minutes or more
Preparation Time: 2 hours
Supplementary Material: materials prepared by the instructor
Special Equipment: IBM 360 model 40 and up; FORTRAN IV

Description: The player represents the general manager of a proprietary firm who is trying to achieve an "improved" management plan. The player must decide what product(s) to produce, how to produce, what the size of operation should be, whether it should be a sell or inventory item.

In instructional applications the game is structured by the teacher through data input. This defines the nature of the enterprises, product markets and input markets including the stochastic character of these elements, if relevant.

The student then chooses a level of activity of each of the structured set of enterprises and specifies capital purchase or sale. These choices are indicated by data input to the computer model. Running of the model will indicate, by a financial and capital account statements, the consequence of the choice. Results can be compared within a group of students playing the same game to stimulate competitive interest and to reinforce the value of application of correct principles of choice. New choices may be made and the session continued using either the initial status of the system (thus permitting correction of prior errors) or from the status of the system at the end of the previous run (permitting illustration of cumulative effects representing growth and decline). Multiple-year simulations using a single set of choice parameters are feasible and useful in some situations. The user can modify the simulation to restrict it to production, marketing, finance, and so on. (Author)

Comment: the designer notes that the support is weak at this time. (AC)

Cost: Source decks available at handling cost. Price of manuals unknown.

Producer: Special arrangements can be made through Department of Agricultural Economics, Pennsylvania State University, University Park, PA 16802

THE NORTHEAST FARM MANAGEMENT GAME (FMG 4)

Earl I. Fuller, University of Minnesota

Playing Data
Copyright: 1968
Age Level: college, continuing education, management
Number of Players: 1-100 individuals or teams
Playing Time: 3-7 hours
Preparation Time: 1 week
Special Equipment: CDC 3600 or 6600; CDC FORTRAN IV

Description: Each player represents a farm manager who organizes and operates a cash crop and dairy farm (in the Connecticut Valley) on a year-to-year basis with the objectives of increasing net worth, growth, income, and security. Emphasis is on organizational (top management) decisions. From a given starting point, organizational adjustments are possible that can specialize the operation toward either cash crops or dairying. The player is responsible for a matrix of up to 40 decisions per playing year. These include crops to grow and type and size of dairy operation. Fertilizer response shows diminishing marginal response.

Instructionally, two beginning points have been used. One begins with a full-information student handout. The other, a limited-information or forced-action situation, begins with year one handout. In year one the situation is described as being imperative; you must take over the firm in addition to your current duties with minimal time available for analysis or the collection of additional data. Thus, the student is encouraged toward a plan similar to the past year's situation report included in the materials.

The student makes the decisions in the same class meeting. These are processed and returned to him on or before the next class meeting. A summary sheet is provided the instructor. This indicates to him students requiring special attention. It also provides data for bar graphing the net worth consequences of the decisions made by the group for display before the group. (Alternatively, he may generate a handout table of comparable form.) This provides substantial incentive to the student to know why there is variation in outcome between players and to improve his relative standing within the group.

The second class meeting begins with some general comments explaining the results. The instructor usually asks particular students why they made the decisions they did and how they can explain the results they obtained. Candidates for this discussion usually are those showing the most success and those showing the most unusual plans. This provides a lead-in for further discussion of underlying basic principles, major consideration, and such. Tools of analysis are usually presented between runs as this type cycle continues. More time is allowed for the making of decisions based upon the analysis that is suggested by the tools. Play is generally limited to five to seven years.

As play continues, new situation statements are presented altering the game environment, i.e., land becomes available, bids are placed upon it, someone wins the bids. New technologies are introduced, contracting arrangements for sale are arranged, trends in production relationships occur, and new prices and yield relationships are introduced. (Author)

Cost: transfer costs only

Producer: manuals and other materials can be obtained from Earl Fuller, Department of Agricultural and Applied Economics, University of Minnesota, St. Paul, MN 55101

OREGON FARM MANAGEMENT SIMULATION

Frank S. Conklin and Manning H. Becker, Oregon State University

Playing Data
Copyright: 1978
Age Level: college, continuing education
Number of Players: 30-50 in teams of 3
Playing Time: 90 hours (est.) in 2-hour rounds
Preparation Time: 3 hours (est.)
Special Equipment: CDC and CYBER hardware with FORTRAN IV reader
Packaging: photocopied game materials include 49-page user's manual, programmer's manual, and instructor's manual

Description: The *Oregon Farm Management Simulation (ORESIM)* in its present mode is a corporate irrigated farm capable of growing up to seven crops. The base farm contains 640 acres in 80-acre increments. The farm does not have livestock. Five irrigation options are available, depending upon the crop grown. Functional yield response from fertilizer application exists with nitrogen or phosphate. Two crops respond to both nitrogen and phosphorous. A large contingent of machinery may be sold, purchased, or custom hired. Assets of the farm are slightly over $1 million with debt about $425,000.

The structural variables, functional relationships, and decision options can be changed. For example, prices and yields may be treated as parameters to represent an environment of certainty or as random elements with specific central tendency and distributional characteristics to represent uncertainty.

Computer interface for *ORESIM* at Oregon State University is accommodated using either remote teletype or batch keypunching of computer cards. The remote console typewriter is used almost exclusively because of its direct linkage to the CDC 3600 computer. (authors)

Cost: user's manual $2.50; programmer's manual $4.50; instructor's manual $6.50

Producer: Dr. Frank S. Conklin, Department of Agricultural and Resource Economics, Oregon State University, Corvallis, OR 97331

THE POULTRY FARM MANAGEMENT GAME (POULT 4)

Earl I. Fuller, University of Minnesota

Playing Data
Copyright: 1968
Age Level: college, continuing education, management
Number of Players: 1-100 individuals or teams
Playing Time: 3-7 hours
Preparation Time: 1 week
Special Equipment: CDC 3600 or 6600; FORTRAN IV

Description: Each player or team organizes and operates a poultry farm (New England brown egg) on a year-to-year basis with the objectives of increasing net worth, growth, income, and security. The player must select housing and marketing systems as well as replacement strategies. The player may raise replacement birds and/or pullets for sale and may also grow (three) crops as additional enterprises. Fertilizer response to crops shows diminishing marginal returns. For a description of the cycle of play, see the designer's notes for *The Northeast Farm Management Game*. (AC)

Cost: transfer costs only

Producer: manuals and other materials can be obtained from Earl Fuller, Department of Agricultural and Applied Economics, University of Minnesota, St. Paul, MN 55101

SIMULATED AGRIBUSINESS

Howard G. Salisbury III

Playing Data
Copyright: 1973
Age Level: college, continuing education
Number of Players: 10
Playing Time: 10 rounds of 20 minutes each (est.)
Packaging: 10 photocopied student manuals, teacher's manual, 2 score-pads

Description: This game simulates the economics of operating a small farm for 10 years. At the beginning of each round (which corresponds to one year) each player may purchase land of high or low quality and assumes a mortgage for any land purchased at 7.5% interest for the rest of the game. In addition, players must assume the costs of planting, fertilizing, irrigating, and insuring their crops. Crop yield is determined by soil quality and chance, and price is determined by a random modification of a set price. Each player must earn enough to pay farm expenses plus $5000 for subsistence or go bankrupt. The game ends after 10 rounds. (DCD)

Cost: $30.00

Producer: Howard Salisbury, Box 15034 N.A.U., Flagstaff, AZ 86011

THE UPPER MIDWEST DAIRY-HOG FARM MANAGEMENT GAME (FMG 5)

Earl I. Fuller, University of Minnesota

Playing Data
Copyright: 1974
Age Level: college, continuing education, management
Number of Players: 1-100 individuals or teams
Playing Time: 3-7 hours
Preparation Time: 1 week
Special Equipment: CDC 6600 Cyber 74; CDC FORTRAN IV

Description: Each player or team organizes and operates a cash crop and dairy farm (Minnesota) on a year-to-year basis with the objectives of increasing net worth, growth, income, and security. From a given starting point, organizational adjustments are possible that can specialize the operations toward hogs or dairying. The player is responsible for a matrix of up to 40 interrelated decisions each year. These include crops to grow and type and size of livestock operation. Fertilizer response shows diminishing marginal response. For a description of the cycle of play, see the designer's notes for *The Northeast Farm Management Game*. (AC)

Cost: transfer costs only

Producer: manuals and other materials can be obtained from Earl Fuller, Department of Agricultural and Applied Economics, University of Minnesota, St. Paul, MN 55101

FOREST SERVICE FIRE CONTROL SIMULATOR

U.S. Department of Agriculture, Forest Service, Division of Fire Management

Playing Data
Age Level: all levels of fire control trainees
Number of Players: 1-5 in 1 or more teams
Playing Time: 10 minutes-4 hours
Preparation Time: 4-8 hours
Special Equipment: tape recorder, video equipment, display system, radio, telephone

Description: Using a series of projectors, the operator places a woodland scene on a rear projection screen to be viewed by the trainees. He can then superimpose the color and movement of fire and smoke on the scene.

The trainee is given maps and a situation briefing which includes history of the fire and a source for supplies. He may contact the supply source and men assigned to the fire through a radio or telephone system. His calls are answered by role players who give him information from a prepared exercise which is designed to meet certain training objectives. The trainee is presented with a series of problems which test his knowledge of fire behavior, fire suppression principles, and the capabilities of men and machines.

The exercise director, role players, and simulator operator must make the fire act realistically under the given weather, fuel, and topographic conditions. It must also react realistically to the actions taken by the trainee in attempting suppression.

The simulator most commonly in use today is manufactured by Scott Engineering of Pompano Beach, Florida to Forest Service prototype specifications.

Simulator exercises (the software) are usually not available for purchase. They are designed by Fire Training Officers to meet localized conditions and training needs. (H. P. Gibson, National Fire Training Center, Marana, AR)

Cost: Varies according to hardware requirements.

Producer: Boise Interagency Fire Center, 3905 Vista Avenue, Boise, ID 87305

THE FOREST SIMULATOR

W. R. Pierce, University of Montana

Playing Data
Age Level: college
Special Prerequisites: good background in forest management
Number of Players: 1 or more in 1 or more teams
Playing Time: variable
Preparation Time: negligible
Special Equipment: CDC 6400 or equivalent; PDP-20 (DEC system 10) computer in time-share mode

Description: The player represents a forest manager trying to maximize financial return while improving the structure of the forest. The player determines what and where to harvest, what to do with slash, what to do with fire-damaged timber, whether to spray brush, how to regenerate a harvested area, how much money to borrow, and when to repay it. Chance determines regeneration survival, prices of products, taxes, occurrence of fire, amount of rainfall, and the amount of natural regeneration. The player receives a printout at the end of each simulated year showing expenses and income, profit or loss in dollars, total financial obligation or earnings, area burned, products cut and sold, and value received for products. Logging costs vary with the size of trees, the number of trees per acre cut, the length of skid, and slope. At the end of a specified number of years, the player can request (and pay for) an audit to compare present resources with those available at the start of play.

A thorough knowledge of the economics of forest management for sawtimber would be useful for players. The simulator is designed to be played repeatedly by the same players. (AC)

Cost: (1) CDC version $25.00; (2) PDP-20 source program listing, requisite data file on 9-track magnetic tape, instruction book $20.00

Producer: (1) Center for Quantitative Sciences, University of Washington, Seattle, WA 98195; (2) School of Forestry, University of Montana, Missoula, MT 59801

PURDUE FOREST MANAGEMENT GAME

B. Bruce Bare

Playing Data
Copyright: 1971
Age Level: college, graduate school
Number of Players: 3-15 in 1-3 teams
Playing Time: 30-120 minutes
Preparation Time: 1-2 hours
Special Equipment: CDC 6400, IBM 1130, 360/67, and 360/65; FORTRAN IV. 1 tape required

Description: This is essentially a game of short-range (annual) tactics that has been combined with the Forest Management Simulator, which indicates the long-range consequences of alternative management strategies. Each player or team manages one district of even-age forest subdivided into 60 compartments with the objective of creating conditions that will support a pulp mill on a sustained basis. They prepare an annual budget of expenditures, and to do so, they must make decisions on the number of cubic feet to sell from harvest cuts; the cost of selling the volume; the logging costs; the number of acres to schedule for sale in two years; the number of acres to thin, plant, seed, burn, and disc, and the cost of these operations; annual road maintenance expense; and so on.

Students first use the simulator to experiment with various management methods and evaluate their long-term consequences. Then they are ready to select a strategy that will generate a satisfactory cash flow, satisfy the requirements of the pulp mill, and produce a forest that can sustain desired production. This done, they play the game. (AC)

Comment: The *PFMG* has been asked for and received by more than 30 university forestry schools. The author does not know how many of these schools have used the package, but the list is nevertheless impressive. (DZ)

Cost: single copies free

Producer: B. Bruce Bare, College of Forest Resources, AR-10, University of Washington, Seattle, WA 98195

ESSO SERVICE STATION GAME

R. J. Gibbs, S. W. Hargreaves, C. R. Jelley, and C. S. Gamage, Esso Petroleum Co., Ltd.

Playing Data
Copyright: 1972
Age Level: grade 9 and up, management
Number of Players: 20-30 in 5 teams
Playing Time: 5-10 hours in 1-hour periods
Preparation Time: 1 hour

Description: Players are allocated to different companies owning garages and selling gasoline and services in greater or lesser competition with each other. Additional sites for new service stations become available through the game. Each garage is described and its position located on a map of the town. Each company attempts to assess the potential demand for gasoline, oil, vehicle servicing/accessories.

Each company can use varying marketing strategies in an attempt to win customers to its site, while attempting to gauge the exact amount of labor needed to meet demand and keeping detailed accounts of each garage and chain to monitor profits constantly. Participants learn that a good return on capital invested is preferable to a high throughput of gasoline and oil at low profit margins. (Author)

Cost: £16.20 plus postage and packing

Producer: C.R.A.C., Bateman Street, Cambridge, England

THE IN-BASKET EXERCISE

Veterans Administration

Playing Data
Age Level: management
Number of Players: 8-24
Playing Time: 90-180 minutes
Preparation Time: sufficient time to select and become thoroughly familiar with items, approximately 5-8 hours

Description: Each trainee assumes the role of Homer Sage, newly appointed director of the VA Center, Utopia, Somestate. The center consists of a hospital (600 employees) and regional office (300 employees). On arrival Mr. Sage finds that he has to leave the station in two hours and will be gone for two weeks. The Assistant Director is home with an illness. Mr. Sage's secretary has left a packet of materials in his in-basket. As Homer Sage, each participant goes through the packet and notes on each item what he would do and why he would do it. He is not allowed to call anybody in to help him, to phone anybody, or to take any of the materials with him on the trip. He may, however, write reference slips or notes. When the Homer Sages complete this assignment, the moderator leads a discussion. He asks the group members how they handled the items and stimulates a discussion on the various approaches used. The benefits derived from the exercise have been that it: provides experience in individual decision making and problem solving, produces a realization that supervisory and administrative problems do not have a single answer, brings out the importance of getting things done through people, and allows for interchange of experience. (W. G. Noffsinger, Veterans Administration)

Cost: none

Producer: Produced for internal use, but a complimentary copy will be given to requesting organizations if stock level permits: Office of Assistant Administrator for Personnel (056B), Veterans Administration, 810 Vermont Ave., NW, Washington, DC 20420

BUSINESS FOR SCHOOL

BLUE WODGET COMPANY

Jay Reese

Playing Data
Age Level: grades 5-9
Number of Players: 20-30 in 2-3 teams
Playing Time: 2-1/2-5 hours in 1/2 to 1-hour periods
Preparation Time: 30 minutes

Description: *Blue Wodget Company* is designed to improve its players' understanding of economics and manufacturing and of how economic and manufacturing decisions are made. Teams of players of this competitive game represent the workers and management of the company. Individual players assume the roles of top- and middle-echelon managers, individual workers, stockholders, and town residents. The management must improve, diversify, produce, advertise, and sell (in this country and abroad, to public and private markets) its line of products to earn a profit for the stockholders, a task complicated by the demands of an organized work force for higher wages, community demands to reduce pollution, and a depressed economy.

Stockholders hire (and may fire) their company's president. The president hires the top managers, including the personnel manager, who hires and negotiates with the workers. All players must complete personal earnings and decision forms each round. The company treasurer uses this information to determine profits and losses. Decisions are varied and may be used in any order. (DCD)

Comment: This game is simple to operate and offers considerable flexibility. However, its simplicity has been structured for the sake of questionable lessons in several of the decisions suggested: Increased advertising always brings more sales; laborers suffer no costs when striking; increased investment in machinery always brings increased sales even though students do not know that this will be the case. These weaknesses could be easily corrected. The game offers a manageable environment for acquainting young students with business and for adding more decision exercises that are up to date and challenging. (BK)

Cost: In *Simulation Games and Learning Activities Kit for the Elementary School* (4 35 81051) $12.95

Producer: Prentice-Hall, Englewood Cliffs, NJ 07632

FIVE SIMPLE BUSINESS GAMES

Charles Townsend

Playing Data
Copyright: 1978
Age Level: grades 7-12
Number of Players: 4 or more in groups of 4-5
Playing Time: various
Preparation Time: 1 hour (est.)

Description: The five business games in this series are written especially for use in schools. Each may be independently obtained from the producer. *Gorgeous Gateaux, Ltd.* may be played in thirty minutes. *Fresh Oven Pies* and *Dart Aviation* require seventy minutes. *The Island Game* and *The Republic Game* require three to four hours to play.

In *Gorgeous Gateaux,* players represent the managers of a pastry company in competition with three other companies in the same market. Since the company's product is perishable the company cannot store large numbers of gateaux, and unfilled orders cannot be carried forward from month to month. For six rounds, each of which represents one month, players decide how many gateaux to produce and what price to charge. The students assume similar roles in *Fresh Oven Pies,* but in this game they must also decide how much money to spend on advertising each month. Like the two games already mentioned, *Dart Aviation* is designed to help students understand how business works. Players represent company managers who must decide how many airplanes to make in each of seven months. They must acquire the materials for manufacture, set a price, and sell the aircraft. *The Island Game* is designed to help students understand how industry creates wealth. Players represent the rulers of a newly independent island. It has five factories and each produces perishable goods. The factory workers spend all of their salaries buying things made in the other factories. *The Republic* presents players with a similar situation, but the factories in this new nation number ten and the production and sale of goods is measured in tons rather than pieces. (DCD)

Cost: £3.78 each, £3.78 teacher's book

Producer: Hobsons Press, Ltd., Bateman Street, Cambridge CB2 1LZ, United Kingdom

SELLING: A GOOD WAY TO EARN A LIVING

Wesley Caldwell and Porter Henry

Playing Data
Copyright: 1970
Age Level: grades 9-12
Number of Players: 5-24
Playing Time: 10-20 hours
Preparation Time: 2 hours (est.)
Packaging: professionally packaged, multimedia unit includes long-playing phonograph record, 2 filmstrips, 24 copies of student booklet, spirit masters, role-play dialogs, and 128-page teaching guide

Description: This multimedia unit was designed, according to the producer, "to teach high school business education students the art of industrial salesmanship: outside the shop selling, which brings the kind of earnings and status they want." The unit is divided into twelve parts titled: What is a Salesman?; A Look at the Sales Job; Getting the Sales Job; Keeping the Sales Job; Getting Ahead in Selling; Buyers; Knowledge of Product, Competition, and Customer; An Overview of the Salescall; Benefit Selling parts one and two; Overcoming Objections and Closing the Sale; and Practice Selling. "Each component is completely flexible: individual components may be used as appropriate throughout the year."

The component titled "Getting the Sales Job" typifies the exercises in this collection. At the start of the component the students listen to recorded conversations of people who have applied for sales jobs but whose attitudes preclude their becoming employed in that line of work: because the would-be salesman was not highly motivated to selling office machines or because he was unable to sell his prospective employer the idea of hiring him. The students then find advertised sales openings in their local newspaper and role play responding to those advertisements and interviewing for the jobs. Finally, the students each draft a letter of application for one specific sales position. (DCD)

Cost: $86.00

Producer: Educational Audio Visual, Inc., Pleasantville, NY 10570

LESTER HILL OFFICE SIMULATION

Myron J. Krawitz, North Shore High School, Glen Head, New York

Playing Data
Age Level: grades 9-12
Special Prerequisites: some of the players should have typing and filing skills
Number of Players: 6-30
Playing Time: 45-90 class hours

Description: The simulation revolves around the operation of a branch office of the Lester Hill Corporation (a national distributor of hotel and motel supplies and equipment). The instructor acts as the corporation's executive vice president in charge of branch office operations. The students perform as office employees.

Each Lester Hill office has four main departments–Sales, Warehouse, Traffic, and Accounting–all under the direction of a general manager. In addition, the Lester Hill office has a functional relationship with an organization known as Tallidata. Tallidata represents all the customers and suppliers of Lester Hill and also serves as the bank that maintains Lester Hill's account. In short, Tallidata serves as the outside world during the simulation–the source of all business papers coming into the Lester Hill office.

At the start of the simulation, students are hired for specific jobs in the Lester Hill office–as department managers, order clerks, stock control clerks, traffic clerks, billing clerks, accounting clerks, and so on. Depending on the size of the office, one to three students are selected to serve as Tallidata represenatives.

Each player's main objective is to complete assigned duties as efficiently as possible. Each is responsible for performing certain tasks and for routing business papers appropriately. Players must sort through incoming mail, organize their work, set priorities, and handle any errors or human relations problems that may come up in the course of their work. Players can be transferred from one job to another so they can enjoy a wide range of occupational experiences in the simulated office. (AC)

Cost: Employee's Guide (1 per student) $2.22; Supply Room (1 per simulated office) $160.00 net

Producer: McGraw-Hill Book Company (Gregg Division), 330 West 42nd Street, New York, NY 10036

THE METAL BOX BUSINESS GAME

R. B. Kemball-Cook

Playing Data
Copyright: 1978
Age Level: grades 7-12
Number of Players: 6-24 in teams of 2-4
Playing Time: 2 hours
Packaging: professionally produced 10 page tutor's manual, 5-page player's manual, and business forms

Description: Teams of players assume the roles of the directors of a recently incorporated boiler manufacturer. Each team is given 150,000 pounds and from that amount must allocate money for the construction of a factory. Teams must also acquire raw materials, make the boilers, and sell them. The Directors must consider two variables in running the new business. The first of these is time. The game specifies the time needed to build or enlarge the factory and the lead time needed for ordering materials, building boilers, and delivering orders. The second variable is money. The players have an extensive, but finite, number of options for all of the above procedures (such as whether or not to expand the factory or whether to schedule overtime in order to expand production) and the cost of all of these has been calculated by the designer. (DCD)

Cost: *φmnmnmn*

Cost: £15.66

Producer: Hobsons Press, Ltd., Bateman Street, Cambridge CB2 1L2, United Kingdom

JEFFREY'S DEPARTMENT STORE

Jimmy G. Koeninger and Thomas A. Hephner

Playing Data
Copyright: 1978
Age Level: distributive education students in grades 11-12
Number of Players: 5-30
Playing Time: one or two semesters
Preparation Time: one week (est.)
Special Equipment: cash register, film strip projector, cassette tape player
Packaging: professionally packaged simulation includes 5 copies of Employees' Guide, General Manager's Manual, Color Filmstrip with cassette, 4 decks of merchandise cards, wall chart, package of forms, and 4 file folders

Description: This simulation is designed to acquaint students with most of the jobs in a large department store. "Jeffrey's" is organized into five divisions (for merchandising, sales promotion, operations, personnel, and control) that epitomize the structure of most major retailers. The simulation is composed of ten "mini-simulations," each consisting of 1 to 21 "incidents" (or exercises). All the operating divisions are represented in at least one "mini-sim."

During the first "mini-simulation," for instance, participants assume the roles of prospective employees at Jeffrey's. They must complete an application for employment, interview for a job, and while they wait to be hired they must fill out a W-2 form. Once they are hired, these new employees must learn to record cash and charge sales, open and close a cash register, replenish stock, and demonstrate an understanding of the store motto ("Meeting the Customer's Needs Comes First at Jeffrey's").

In subsequent "mini-simulations," players assume the roles of employees and supervisors in the advertising, personnel, accounting, and warehousing departments, assume the roles of salespersons (with "incidents" designed to promote an understanding of sales approaches and techniques), and role play situations in which they must display good human relations skills.

"To further transform your classroom into a retail environment, Jeffrey's replaces the grading function with employee performance evaluations. Grades are not given. Instead profit points are awarded to students on the basis of their performances." (DCD)

Cost: $99.00

Producer: McGraw-Hill Book Company, 1221 Avenue of the Americas, New York, NY 10020

See also Esso Service Station Game BUSINESS: SPECIFIC INDUSTRY
Shoprofit BUSINESS: SALES AND MARKETING
Thygummy Jigs, Ltd. BUSINESS: FINANCE
Trig BUSINESS: TOTAL ENTERPRISE–MANUAL
Tycoon BUSINESS: TOTAL ENTERPRISE–MANUAL
ECONOMICS

Part IV
RESOURCES

SIMULATION/GAMING PERIODICALS

Creative Computing

Editor: John Craig, 51 Dumont Place, Morristown, NJ 07960

Publisher: Creative Computing, P.O. Box 789-M, Morristown, NJ 07960

Published monthly
Subscription: $15.00 per year

Journal of Experiential Simulation and Gaming (formerly **Simulation/Gaming**)

Editor: M. Wayne DeLozier, College of Business Administration, University of South Carolina, Columbia, SC 29208

Publisher: Elsevier North-Holland, Inc., 52 Vanderbilt Avenue, New York, NY 10017

Published Quarterly
Institutional subscription $35.00 per year
Personal subscription $17.50 per year, $12.50 for NASAGA members, included in dues for ABSEL members

Pipeline

Editor: Molly Hepler

Publisher: CONDUIT, P.O. Box 388, Iowa City, IA 52240

Published three times a year
Subscription $75.00 per year (brings two issues of *Abstracts and Reviews,* a set of guides and reports; ten copies of each issue of *Pipeline*; and a 10% discount on all CONDUIT reviewed and tested materials)

Recreational Computing (formerly **People's Computers**)

Editors: Bob Albrecht, Louise Burton, Ramon Zamora

Publisher: People's Computing Company, 1263 El Camino Real, Box E, Menlo Park, CA 94025

Published bimonthly
Subscription: $10.00 per year

Simages A service of NASAGA

Editor: Sivasailam Thiagarajan, Instructional Alternatives, 4423 East Trailridge Road, Bloomington, IN 47401

Subscriptions: W. Thomas Nichols, NASAGA Treasurer, Box 100, Westminster College, New Wilmington, PA 16142

NASAGA Business: Barry Lawson, NASAGA Board Chairman, Barry Lawson Associates, 148 State Street, Boston, MA 02109

Subscription included in NASAGA dues for individuals: regular members $25.00, full-time students $15.00; institutional and library subscription $25.00 per year

Simgames

Editors: Tom Cavanaugh, Champlain Regional College, and Guy Le Cavalier, Concordia University

Subscriptions: SIMGAMGES, Champlain Regional College, Lennoxville Campus, Lennoxville, Quebec, Canada J1M 2A1

Published quarterly
Individual subscription $2.50, institutional subscription $4.50 per year

Simulation: Technical Journal of the Society for Computer Simulation

Publisher: The Society for Computer Simulation, 1010 Pearl Street, La Jolla, CA 92037

Published monthly
Subscription: $38.00 per year

Simulation & Games

Editor: Cathy Greenblat, Department of Sociology, Douglass College of Rutgers University, New Brunswick, NJ 08903

Publisher: Sage Publications, 275 South Beverly Drive, Beverly Hills, CA 90212

Published quarterly
Individual subscription: $15.00, institutional subscription $30.00 per year

SIMULATION/GAMING CENTERS

Abt Associates, Inc.
55 Wheeler Street
Cambridge, MA 02138
Clark C. Abt, President

Private Corporation

Principal Focus: Research, social program evaluation, classroom education, and training for social action
Principal Subject Matter: Education, housing, drug abuse, poverty, urban problems, technology management, environmental problems
Training Levels: Junior high through college and adult

***

Allen's Academic Games Crop.
or National Academic Games Project
P.O. Box 214
Newhall, CA 91322
Robert W. Allen, President

Private Corporation

Principal Focus: Research, development, dissemination
Principal Subject Matter: Social Studies, language arts, mathematics, science
Training Levels: Preschool through college

***

Behavioral Sciences Laboratory
The Ohio State University
404-B West Seventeenth Avenue
Columbus, OH 43210
Agnes L. Kinschner, Administrative Assistant

University-related department

Principal Focus: Research, development, training
Principal Subject Matter: International relations
Training Levels: College

***

Center for Computer-Based Behavioral Studies
University of California
Psychology Department
405 Hilgard Avenue
Los Angeles, CA 90024
Prof. Gerald H. Shure, Director

University-related department

Principal Focus: Use of laboratory and computer-based methods for research in social and policy science
Principal Subject Matter: Social science, research methodology
Training Levels: College

***

The Center for Simulation Studies
736 De Mun Avenue
Clayton, MO 63105
Richard F. Tombaugh
Joseph L. Davis

Nonprofit organization

Principal Focus: Research, development, training
Principal Subject Matter: Urban affairs
Training Levels: High school through college and adult

***

COMEX (Center for Multi-Disciplinary Education Exercises)
University of Southern California
Davidson Conference Center
Los Angeles, CA 90007
Richard T. McGinty, Director

University-related center

Principal Focus: Research, development, adaptation of educational and training simulation exercises; design, development, and programming of in-service training programs and of computerized gaming simulation; evaluation
Principal Subject Matter: Social science, urban, teacher training
Training Levels: College and adult

***

Didactic Systems, Inc.
Box 457
Cranford, NJ 07016
Erwin Rausch

Private corporation

Principal Focus: Development, training, dissemination

Principal Subject Matter: General
Training Levels: General

***

The Games Preserve
R.D. 1355
Fleetwood, PA 19522
Bernie De Koven

Nonprofit corporation

Principal Focus: Exploration of gaming process
Principal Subject Matter: General and all
Training Levels: All

***

Games Central
c/o Abt Associates, Inc.
55 Wheeler Street
Cambridge, MA 02138
Ray Glazier, Editor-in-Chief

Private Corporation

Principal Focus: Classroom education and guidance counseling, materials research, development, and publication
Principal Subject Matter: Math skills, language arts, and especially social studies
Training Levels: Junior high through college

***

Harwell Associates
P.O. Box 95
Convent Station, NJ 07961
B. Wilkerson

Private corporation

Principal Focus: City planning, governmental process, library, consultation to individuals with subject matter suitable for simulation
Training Levels: Elementary through college, professional planning organizations and individuals

***

Information Resources, Inc.
Box 417
Lexington, MA 02173
Robert E. Horn

Private corporation

Principal Focus: Development, dissemination of information
Training Level: Adult

***

Institute for Mediation and Conflict Resolution
49 East 68th Street
New York, NY 10021
George Nicolau, Executive Director

Nonprofit corporation

Principal Focus: Community disputes mediation and conflict resolution
Principal Subject Matter: Mediation and negotiation, grievance systems
Training Levels: Adult education

***

Instructional Alternatives
4423 East Trailridge Road
Bloomington, IN 47401

Private organization of instructional simulation/game designers

Principal Focus: Design and dissemination of instructional simulations/games, experiential packages, role plays, and other small-group methods and materials; workshops on these areas.
Principal Subject Matter: Training and education
Training Level: Adult and general

***

Instructional Games Project
Mental Health Research Institute
University of Michigan
Ann Arbor, MI 48109
Layman E. Allen, Project Director

University-related department

Principal Focus: Research, instructional systems design, and social action
Principal Subject Matter: Formal systems (logic and mathematics) and science
Training Levels: Elementary through university with chief focus on junior high school at this time

***

Interact
P.O. Box 262
Lakeside, CA 92040
David E. Yount and Paul DeKock

Partnership and association of teacher-authors

Principal Focus: Simulation development, dissemination
Principal Subject Matter: Social sciences, English
Training Levels: Elementary through college, adult

***

International Institute for Organizational and Social Development
Predikherenberg 55
B-3200 LEUVEN, Belgium
Dr. L. Vansina, Director

Private corporation

Principal Focus: Research, development, training, dissemination of information
Principal Subject Matter: Organizational development
Training Level: Adult

***

International Simulation and Gaming Association (ISAGA)
Secretariat for Europe (General Secretary):
Dr. J. H. G. Klabbers
University of Nijmegen, Dept. of Psychology
Social Systems Research Group
PO Box 9104
6500 HE Nijmegen
The Netherlands
U.S. National Representative:
Richard Duke
321 Parklake
Ann Arbor, MI 48103
U.S.A.

Professional society

Principal Focus: Model-building, research, applications
Principal Subject Matter: No specific limits
Training Levels: University

***

Iowa Center for Communication Study
School of Journalism
University of Iowa
Iowa City, IA 52242
James A. Wollert, Center Director

Research center within university-related department

Principal Focus: Among the services of the center are consultation, training, publication in appropriate outlets helping persons secure financial support for projects, and assistance in computer use and analysis. Another area of activity involves development of modular learning units for all levels of communication and training. The center also facilitates design, development and applications of simulations and games in instruction and research
Principal Subject Matter: Communication inquiry
Training Levels: Primarily graduate training

***

Louisville Experiment Simulation System
Psychology Department
University of Louisville
Louisville, KY 40208
Dr. James G. Miller, President

Institute of Systems Science Graduate School
Applied Mathematics and Computer Science
Speed Scientific School

University-related department

Principal Focus: Research (The Louisville Experiment Simulation System [LESS] is a set of eleven programs and a data file handling up to 24 models that can be active at once. Variable types include discrete, continuous, general parameters, unknown variables.)
Principal Subject Matter: Experimental psychology
Training Levels: College

***

National Gaming Council
c/o Clark Rogers
3R24 Forbes Quad.
Graduate School of Public and International Affairs
University of Pittsburgh
Pittsburgh, PA 15260

Professional society

Principal Focus: Dissemination of information
Principal Subject Matter: General
Training Levels: Elementary through adult

***

New Games Foundation
P.O. Box 7901
San Francisco, CA 94120

Nonprofit educational organization

Principal Focus: Research, development, training, publication
Principal Subject Matter: Foster and communicating a style of play encouraging participation, community, and creativity
Training Levels: All ages

***

North American Simulation and Gaming Association (NASAGA)
c/o Barry Lawson Associates
148 State Street
Boston, MA 02109
Barry Lawson, Board Chairman

Professional association

Principal Focus: Dissemination of information
Principal Subject Matter: All
Training Level: All

***

Polis Laboratory
Department of Political Science
University of California at Santa Barbara
Santa Barbara, CA 93106
William D. Hyder

University-related department

Principal Focus: Development and research
Principal Subject Matter: Politics and international relations
Training Levels: College

***

Project Simu-School
Dallas Independent School District
3700 Ross Avenue
Dallas, TX 75204
William Dunklau, Director

Public school system

Principal Focus: Research, planning
Principal Subject Matter: Financial, urban
Training Levels: College

***

Public Affairs Program
Maxwell School
Syracuse University
Syracuse, NY 13201
William D. Coplin

Informal group

Principal Focus: Dissemination of information
Principal Subject Matter: International relations, politics
Training Levels: College

***

Resource Policy Center
Thayer School of Engineering
Dartmouth College
Hanover, NH 03755

University-related department

Principal Focus: Districution of publications
Principal Subject Matter: Research models
Training Levels: University

***

Simile II
218 Twelfth Street
P.O. Box 910
Del Mar, CA 92014
R. Garry Shirts, Director

Private corporation

Principal Focus: Development and publication of educational simulations and games
Principal Subject Matter: Social science
Training Levels: Elementary through college

***

SIMTEK
P.O. Box 109
Cambridge, MA 02139
Francisco James, President

Private corporation

Principal Focus: International time-sharing and support services for computerized business games
Principal Subject Matter: Business
Training Levels: College, management

***

Simulation Learning Institute, Inc.
P.O. Box 240–Roosevelt Island Station
New York, NY 10044

Unincorporated company

Principal Focus: Research, design, publishing, directing, training
Principal Subject Matter: Social science, business and management training, community affairs, education
Training Levels: Kindergarten through adult

***

Simulations Publications Inc.
44 East 23rd St.
New York, NY 10010

Private corporation

Principal Focus: Development, dissemination
Principal Subject Matter: War games
Training Level: Secondary through adult

***

Simulation Sharing Service
4740 Shadow Wood Drive
Jackson, MS 39211
George McFarland

Nonprofit corporation

Principal Focus: Development, training, dissemination of information
Principal Subject Matter: General
Training Levels: Secondary through adult

***

Simulation Systems
P.O. Box 46
Black Butte Ranch, OR 97759
Paul A. Twelker

Unincorporated firm

Principal Focus: Development and distribution of simulation/gaming exercises
Principal Subject Matter: Social studies, values clarification, professional education
Training Levels: All

***

Social Science Education Consortium, Inc.
855 Broadway
Boulder, CO 80302
Irving Morrissett, Executive Director

Nonprofit corporation

Principal Focus: Research, analysis, dissemination, and implementation
Principal Subject Matter: Social science
Training Levels: Elementary, secondary

***

Social Systems Simulation Group
P.O. Box 15612
San Deigo, CA 92115
Roland Werner

Private corporation

Principal Focus: Research in modeling social systems, developing a social systems modeling language
Principal Subject Matter: Social science; recent models: innovation diffusion, family lifecycle, public school enrollment simulator
Training Levels: College, professional

***

The Society for Computer Simulation
P.O. Box 2228
La Jolla, CA 92038
C. G. Stockton

Private

Principal Focus: Computer modeling and simulation as it might be applied to studies of the problems of world society
Principal Subject Matter: Computer simulations
Training Levels: Planners and decision makers concerned with the impact of current plans and decisions on mankind in the future

***

System's Factors, Inc.
1940 Woodland Avenue
Duluth, MN 558003
Ronald G. Klietsch

Private corporation

Principal Focus: Research, development, training, dissemination of information
Principal Subject Matter: Social studies
Training Levels: Secondary through adult

***

Urban Gaming/Simulation Conference
School of Education
The University of Michigan
Ann Arbor, MI 48109
Larry Coppard and Frederick Goodman

Informal professional group

Principal Focus: Training, development, dissemination of information
Principal Subject Matter: Urban studies
Training Levels: High school, college, university, professional, community groups

***

Wff 'N Proof Learning Games Associates
1490 South Boulevard
Ann Arbor, MI 48104
D. Wren

Unincorporated organization

Principal Focus: Manufacture and distribution of instructional games
Principal Subject Matter: Logic, mathematics, language, social studies, scientific reasoning
Training Levels: Elementary through university with chief focus now at junior high school

***

AUTHOR INDEX

Abrams, Michael
Binary, 469
ABT Associates
American History Games, 419-20
Abt, Clark
Edplan, 386
Grand Strategy, 425-6
Airola, Teuvo, M.
Transfer-of-Development Rights Game, 590
Allen, Christopher G.
Blue and Gray, 485
East Front, 492
Revolt in the East, 509-10
Westwall, 522
Allen, Kip
Island War, 497
Allen, Layman E.
Equations, 470
On-Sets, 473
Queries 'n Theories, 462
Real Numbers Game, 474
WFF: The Beginner's Game of Modern Logic Equations, 476
Wff 'n Proof: The Game of Modern Logic, 476-7
Allen, Robert W.
Euro-Card, 561-2
In-Quest, 543
Mr. President, 431
Propaganda Game, 568
Alpern, Louis
Mr. President, 431
Anderson, Donald
Madison Simulations, 390
Angiollilo, Joe.
Blue and Gray II, 485
Objective Moscow, 505
Road to Richmond, 510
Apter, Steven
The Good Society Exercise: Problems of Authority, 451-2
Archey, William
Managing the Worker, 622
Armstrong, R.H.R.
Nexus, 398
Arneaud, Susan Ebel
Access, 556
Atkins, Thurston
Madison Simulations, 390
Attiyeh, Richard
Computer Simulation Policy Games in Macroeconomics, 332
Attworth, E. Robert
Tool Room Game, 608
Ayal, Igal
Marketing a New Product, 634
Supermarket Strategy, 643
Aylmer-Kelly, A.W.B.
RKINET, 341
Babb, E. M.
Purdue Supermarket Management Game, 643
Balkoski, Joseph M.
Atlantic Wall, 482
Minuteman, 501-2
Modern Battles II, 502-3
Up Scope!, 519
Veracruz, 519
Wacht am Rhein, 520-21
Wellington's Victory, 522
Banner, Paul R.
Case White, 486-7
Drang Nach Osten: and Unentschieden, 491
Narvik, 504
Banner, Rich
Avalanche, 482
Barasch, Howard
Four Battles in North Africa, 494-5
Modern Battles, 502
Westwall, 522
Barents, Herbert O.
Battle of Roark's Drift, 484
Barbee, Daniel G.
National Policy Game, 358
Barker, B.
Congress of Vienna, 421
Longman History Games, 430
Scramble for Africa, 435
Barnes, S. Eugene
Health Games Students Play, 415
Barr, Richard
Imperialism, 497
Barton, Richard F.
Imaginit Management Game, 597
Basil, Douglas C.
Executive Decision Making Through Simulation, 596-7
Baskind, Larry
Tightrope, 382
Bass, Bernard
Nuclear Site Negotiation, 322
Pex, 625-6
Process, 553
Profair, 628
Prosper, 628
Bass, Ruth
Nuclear Site Negotiation, 322
Bassey, Michael *European Environment: 1975-2000* 402-3
Baxter, Bernice
Probe, 553
Bazeli, F.
Tenure, 395
Beary, Mike
Newscast, 461-2
Beasley, John
In the Chips, 564-5
Becker, Manning H.
Oregon Farm Management Simulation, 646
Bednarski, Mary W.
Nursing Home Care as a Public Policy Issue, 444
Beier, F. J.
Buloga II, 604
Beland, Russ
Viva!, 520
Bell, Cecil H.
Management: An Experiential Approach, 601
Bender, David L.
Future Planning Games, 404-5
Bennett, Pete
Wellington's Victory, 522
Benson, Dennis
Gaming, 533
Berg, Richard
Blue and Gray II, 485
Conquistador, 488
The Conquerors: The Macedonians and the Romans, 488

A Mighty Fortress, 501
Panzergruppe Guderian, 506
Siege of Constantinople, 512
Terrible Swift Sword, 517
Veracruz, 519
Berger, Jerry
Trade-Off, 590
Berry, William L.
Computer Augmented Cases in Operations and Logistics Management, 604
Bettum, Stephen G.
East Front, 492
War in Europe, 521-2
Bewley, William
Cognitive Psychology, 343
Bidol, Pat
Serfdom, A Simulation in Class Achievement, 570
Birt, D.
Longman History Games, 430
Blanchard, Kenneth H.
Situational Leadership Simulator, 629-30
Blaser, Elissa
Exodus, 532-3
Blaxwell, John
Manchester, 430
Blechman, Elaine A.
Family Contract Game, 310
Bloomfield, Lincoln P.
Political Exercise, 455
Blostein, Stanley
Lobbying Game, 355
Boden, R.
Congress of Vienna, 421
Longman History Games, 430
Scramble for Africa, 435
Boehlje, Michael
Oklahoma Bank Management Game, 642
Boin, David V.
Consumer Decision, 526
Crisis in Middletown, 317
Dissent and Protest, 422-3
Exchange, 376
Exploring the New World, 424
First Amendment Freedoms, 465
Government Reparations for Minority Groups, 354
Human Survival–2025, 405
Interaction–A Balance of Power in Colonial America, 428
Low Income Housing Project, 320
Middle East Crisis, 455
Montgomery Bus Boycott,
Nationalism: War or Peace, 455
National Priorities, 358
Planning an Inner-City High School, 323
Planning the City of Greenville, 322-3
Political Pollution, 369
Preventing the Civil War, 434
Revolt at State Prison, 468
Spanish-American War, 437
Trial of George Washington, 438
Trial of Harry S. Truman, 438-9
Trial of Jefferson Davis, 439
War Time, 459
World War I, 440
Women's Liberation, 363
Bolton, Dale L.
Teacher Training Simulation, 394
Bonoforte, Doug
Cromwell, 489
Boone, Louis E.
Marketing Strategy, 635
Sales Management Game, 638
Bowersox, Donald A.
Simchip, 607-8
Braden, Joseph L.
Sales Management Game, 638
Brainard, William
Computer Simulation Policy Games in Macroeconomics 332
Bramnick, Jon
Good Society Exercise, 451-2
Brasefield, Sister Marleen
Liberté, 429
Braun, L.
Buflo, 328-9
Polut, 329-30
POP, 330
Sterl, 330-1
Brehaut, H. J.
Office Talk, 623-4
Brett, Joan
Tuf, 476
Brett, Peter
Tuf, 476
Broadbent, Frank W.
Teaching Problems Laboratory, 394
Brown, William R.
Make Your Own World, 368-9
Bruck, Bill
Global Futures Game, 405
Bryan, L. Wayne
Freshman Year, 388
System, 574
Buchele, Robert
Equity and Efficiency in Public Policy, 376
Bubb, Rolf L.
Teaching Problems Laboratory, 394
Budd, Richard W.
Human Communication Handbook, 311
Buddington, Ann
Tightrope, 382
Burak, Marshall
Executive Decision Making through Simulation, 596-7
Burroughs, Wayne
Problems in Bank Management, 642-3
Buryska, Rev. James
Persecution, 536
Bykowski, Justine
Brookside Manor, 441-2

Caggiano, A.
Charge, 337
Slits, 342
Caldwell, Wesley
Selling–A Good Way to Earn a Living, 648-9
Calhamer, Alan B.
Diplomacy, 490-1
Callahan, Loel A.
Urban Dynamics, 591-2
Careers Research and Advisory Center
Pipework Engineering, 626
Carr, Mike
Fight in the Skies, 493
Carus, Seth
Arab-Israeli Wars, 481
Casciano, D.
Missile Crisis, 502
Operacao Littorio, 506
Cascio, Wayne F.
Prosper, 628
Caswell, Dwight A., Jr.
Urban Dynamics, 591-2
Center for Business Simulation
Busy, 640
Chancellor, 373
Policy, 643-4
Sellem, 639
Shoprofit, 639
Tycoon, 603
Certo, Samuel C.
Sourcebook of Experiential Exercises: Interpersonal Skills, 313-14
Chadwick, Frank
Avalanche, 482
Bar-Lev, 483
Case White, 486-7
Citadel, 487
Crimea, 488
Drang Nach Osten! and Unentschieden, 491
1815–The Waterloo Campaign, 492
En Garde! 492
Fall of Tobruk, 492-3
Kasserine Pass, 497-8
Narvik, 504
Overlord, 506
Raphia, 217 B.C., 509
Torgau, 518
Chaisson, G. Maureen
Life Cycle, 444
Chapman, E. N.
Career Game, 547
Chartier, Jan
Becoming a Person, 547
Chartier, Myron
Becoming a Person, 547
Chavooskian, B. Budd
Transfer-of-Development Rights Game, 590
Chemical Teaching Aids
Cheminoes, 541
Formulon, 543
P & S, 544
Cherrington, David
Business Management Laboratory, 595
Cherryholmes, Cleo H.
International Simulation Kit, 452-3
Chesler, Mark
Role Playing Methods in the Classroom, 392
Chester, Michael
BOLA, 337
Graze, 329
Chiarelott, Leigh
Principal Game, 389
Christiansen, Kenneth
Dignity, 317
Church, Virginia

Play Yourself Free, 552-3
Churchill, Eton
Eco-Acres, 583-4
Churchill, Geoffrey
Decision Mathematics Operational Game, 345-6
Churchill, Rachel E.
Decision Mathematics Operational Game, 345-6
Clark, Dave
PanzerLeader, 507
Clark, Henry W.
Suburban Operations Simulation, 324
Clark, John M.
Euro-Card, 561-2
Clark, Ken
English Civil War, 492
Clary, Thomas C.
Transactional Analysis–Improving Communications, 630-1
Cohen, Harry B.
Orbiting the Earth, 473
Cohen, Howard
Equity and Efficiency in Public Policy, 376
Cohen, Robert
Psych City, 323
Coleman, James S.
Economic System, 375
Coleman, Lucien, Jr.
Teaching Styles, 537-8
Collins, R. W.
Titration, 342
Comito, Bill
Battles for Tobruk, 484-5
Cone, Paul R.
Executive Decision Making Through Simulation, 596-7
Consori, Katherine
Welfare Week, 445
Constitutional Rights Foundation
Kids in Crisis, 467
Cook, Sandra L.
Artificial Society, 372
Cook, Tiff E.
Mental Health, 551
Vitamins, 417-18
Coons, Dorothy H.
Brookside Manor, 441-2
Cooper, Gordon
Greenham District Council, 319
Greenham Gypsy Site, 319
Tenement, 324
Coplin, William
American Constitutional Convention, 419
American Government Simulation Series, 348-9
Foreign Policy Decision Making, 450
Good Society Exercise, 451-2
PRINCE, 335
State System Exercise, 456-8
Corbin, David E.
Contracept, 413-14
Cornish, Stanley F.
Ecological Modeling, 329
Costikyan, Greg
Air War, 480
Conquistador, 488
Drive on Stalingrad, 491
Four Battles on North Africa, 494-5
Plot to Assassinate Hitler, 508
Costley, Dan L.
Artificial Society, 372
Cotter, Richard V.
Business Policy Game, 595
Cotterell, Stephen
Longman Geography Games, 410
Cousins, Robert D., Jr.
River Dose Model, 330
Coward, H. Roberts
Woodbury Political Simulation, 363-4
Crawford, Jack
Triangle Trade, 439
Crawshaw, Bernard
Let's Play Games in French, 461
Criswell, Donald
Weekly News Game, 576
Cruickshank, Donald R.
Inner City Simulation Laboratory, 388
Teaching Problems Laboratory, 394
Culbertson, Jack
Monroe City Simulations, 390-1
Culliton, Mary Anne
River Dose Model, 330
Cumis Insurance Society
Managing Your Money, 528-9
Cumming, Susan
COMCO, 568-9
Curran, Edward
Blue and Gray, 485
Frederick the Great, 495
Minuteman, 501-2
Napoleon at War, 504
War in Europe, 521-2
Dahl, Harry
Everybody Counts, 442
Dal Porto, David
Big Business, 420
Brinkmanship: Holocaust or Compromise, 420
Confrontation in Urbia, 464-5
Haymarket Case, 426
North vs. South, 431-2
Progressive Era, 434
Radicals vs. Tories, 436
Spanish-American War, 437-8
War Crimes Trials, 440
Darrow, Charles W.
Monopoly, 378-9
Dauten, Carl A.
Washington University Business Game, 603
Davis, Frank
Conquerors: The Macedonians and the Romans, 488
First World War, 493-4
Four Battles in North Africa, 494-5
Frederick the Great, 495
Red Sun Rising, 509
Russian Civil War, 511
South Africa, 514
Up Scope! 519
Wellington's Victory, 522
Davis, Joseph L.
IM-CLUG, 585-6
Metroplex, 587
PX-190, 399-400
Davis, Kenneth
Inventing and Playing Games in the English Classroom, 460-1
Dawson, Rosette
Blook Flow, 540
Day, Raoph D.
Marketing in Action, 634
DeBose, Sam
Jail Puzzle, 466
Deacove, Jim
Choices, 547-8
Community, 559
Family, 310
Together, 575
Yin Yang, 315
DeKock, Paul
Balance, 365-6
Cycle, 548
Destiny, 422
Disunia, 423
Division, 423
Family Tree, 562
Herstory, 562
Identity, 563
Independence, 427
Mission, 431
Panic, 433
Search, 463
Seneca Falls, 435-6
Small Town, 571-2
Strike, 438
Sunshine, 324
Virtue, 439-40
DeKoven, Bernie
Jail Puzzle, 466
Dempsey, Sheryll
Nursing Crosswords, 416
Denham, J.
Computers in the Biology Curriculum, 338
Denk, Joseph R.
KSIMS, 339-40
Denmark, Florence
Probe, 553
Denny, Steve
Statehood, 362
World, 459
DeVries, David L.
Teams-Games-Tournament, 400
Diamond Esther E.
Career Development Program, 547
Didactron, Inc.
Soup's On, 417
Wheels, 418
Dietor Self-Instruction Systems
Good Loser, 414
Dietterich, Paul
Banner-Making Game, 531
Powerplay, 553
Dinkele, Geoffrey
Longman Geography Games, 410
Dodderidge, Ben F.
Operation Encounter, 636-7
Dodge, Dorothy *Mulberry,* 321
Dolbear, F. Trenery
Computer Simulation Policy Games in Macroeconomics, 332

Donnelly, Terrence P.
Decline and Fall, 489-90
Dormant, Diane
Kathal, 550
Sci Fi, 400
Dorwart, Harold L.
Configurations, 470
Doty, Edward
Decisionmakers, 317
Downing, Bruce
World Political Organization, 459
Doyle, Jane
Wildlife, 545-6
Dresch, Stephen P.
IDIOM, 336-7
Duke, Richard D.
At-Issue, 396
Conceptual Mapping Game, 396
Hexagon, 585
IMPASSE, 398
METRO, 586-7
METRO-Apex, 587
Metropolis, 587-8
Dunlop, Jim
The Poverty Game, 568
Dunne, Faith
Old Cities—New Politics, 358-9
Parksburg: Garbage and Politics, 359-60
Dunnigan, James F.
American Civil War, 480-1
American Revolution, 481
Ardennes Offensive, 481-2
Barbarossa: The Russo-German War 1941-45, 483
Battle for Germany, 483
"CA", 486
Desert War, 490
Fast Carriers, 493
Firefight, *493*
Foxbat and Phantom, 495
Frigate, 495
Fulda Gap, 495
Global War, 496
Grenadier, 496
Invasion: America, 497
Korea, 498-9
Kursk, 499
Mech War '77, 500
Minuteman, 501-2
Modern Battles, 502
Modern Battles II, 502-3
Moscow Campaign, 503
Napolean at Waterloo, 503-4
NATO, 504
Next War, 505
Oil War, 505-6
Operation Olympic, 506
Panzerarmee Afrika, 506
Panzerblitz, 506-7
Panzer '44, 507
Panzergruppe Guderian, 507
Patrol!, 507
Plot to Assassinate Hitler, 508
Red Star/White Star, 509
Revolt in the East, 509-10
Russian Civil War, 511
Sinai, 513
Sixth Fleet, 513
Sniper!, 513
Solomon's Campaign, 514
Spitfire, 514-15
Strike Force One, 515
Tank!, 516
Turning Point, 518-19
USN, 519
Wacht am Rhein, 520-21
War in Europe, 521-2
Wolfpack, 523
World War I, 523
World War II, 523-4
World War III, 524
Durham, Russ
Power Politics, 360
Triangle Trade, 439
Waging Neutrality, 440
Durham, Virginia
Power Politics, 360
Waging Neutrality, 440

Easterly, Jean L.
Access, 556
Decisions, 549
Pipeline, 444-5
Echan, George W., Jr.
Jury Game, 466-7
Edens, R.
Haber, 339
Eder, J.
Elect, 326
Edstrom, Eric
Power, 568
Educational Research Council of America
Pollution Game, 370
Redwood Controversy, 370-1
Education Development Center
Armada, 420
Empire, 423-4
Edwards, John
Russian Campaign, 511
War at Sea, 521
Edwards, Judy
Ghetto, 343-4
Edwards, Keith J.
Teams-Games-Tournament, 400
Eels, Don
Time Capsule, 574-5
Ellberger, Simon
Bar-Lev, 483
Eller, Thomas
Manassas, 500
Ellinger, Robert S.
Spacial Marketing Simulation, 639
Elliott, John
Crisis in Lagia, 550
Ellzey, Charles
Banner Making Game, 531
Embry, Olice
Tempomatic IV, 600
Engel, Alan S.
Justice Game, 467
Engs, Ruth C.
Health Games Students Play, 415
Environment Simulation Laboratory
Apex, 577
Envirometrics
City I, 579-81
City IV, 581
City Model, 581-2
Enzer, J. Matisse
Four Battles from the Crimean War, 494
Erickson, Melvin
Tightrope, 382
Ernsberger, Donald
Enterprise, 376
Esber, Edward M., Jr.
Industrial Marketing Plan Simulation, 633
Estes, James E.
MADS-Bee, 598-9
Estey, Elmer E.
Integrated Simulation, 597-8
Eynon, Barry
Bay of Pigs, 485

Faber, Ron
Pollution, 369-70
Falk, Joe
Dialogues on What Could Be, 402
Faria, A. J.
Compete, 633
Farzanegan, Bahram
INS 2, 334
Feldt, Allan G.
CLUG, 583
Fenton, Thomas P.
Education for Justice, 375
Feraru, Anne Thompson
International Conflict, 454
Feste, Karen Ann
Arab-Israeli Conflict, 447
Fine, Marvin
Sharing Game, 554
Fine, Seth
Bay of Pigs, 485
Finegold, Menaheim
Classifying the Chemical Elements, 541-2
Finkelstein, S.
Market, 332-3
Finn, Kenneth R.
Time Management, 630
Finn, Peter
Innocent Until. . . ., 465-6
Finseth, Katherine
Planafam II, 416
Fischer, Ken
Edventure, 386-7
Fisher, Glen
Oklahoma Bank Management Game, 642
Fleming, James W.
Teachers' Lounge, 394
Flindt, Myron
Gold Rush, 425
Florczyk, Sandra
Nursing Home Care as a Public Policy Issue, 444
Flowers, Patricia D.
Life in the Colonies, 429-30
Foreign Policy Association
Dangerous Parallel, 449
Formon, Phillip M.
Clinical Simulations, 413
Foster, John
National Policy Game, 358

Fowler, Bob
Burma, 486
Fox, Davis
Psych City, 323
Fox, Robert
Role Playing Methods in the Classroom, 392
Franciscan Communications Center
World Game, 408
Franks, Betty Barclay
Future Decisions: the I.Q. Game, 403-4
Frazer, J. Ronald
Business Decision Simulation, 610
Frey, M. William
Finansim, 641
Friedland, J.
Lockey, 340
Malar, 344
POP, 330
TAG, 331
Uspop, 344
Frishman, A.
Malar, 344
Sterl, 330-1
Fryback, William H.
Decision Making for Career Development, 548
Furlong, Mary Simpson
Halfway House, 319-20
Slave Auction, 437
Furse, C. T.
Titration, 342

Gabel, James
Koniggratz, 498
Mukden 1905, 503
Tannenburg, 516-17
Gagnon, John H.
Blood Money, 412-13
Gallagher, D.
Missile Crisis, 501
Gamage, C. S.
Esso Service Station Game, 647
Gamble, John
Uses of the Sea, 458-9
Gamelin, Timothy
AFASLAPOL, 446
Gamson, William A.
SIMSOC, 571
Gang, Robert L.
Chem Bingo, 540
Chem Chex, 541
Chem Trak, 541
Gann, Migel
Terry Parker, 445
Tracy Congdon, 445
Garside, D.
City, 579
Gatignon, Hubert
Markstrat, 636
Georgian, Frederick
Blue and Gray II, 485
Firefight, 493
Plot to Assassinate Hitler, 508
Wellington's Victory, 522
Gibbs, R. J. *Esso Service Station Game*, 647
Gillespie, Judith A.
City Hall, 350

Glantz, William E.
Ecological Modeling, 329
Glazier, Ray
Culture Contact, 560-1
Edplan, 386
Edventure, 386-7
Edventure II, 387
Grand Strategy, 425-6
Home Economy Kit, 528
Life in the Colonies, 429-30
Opening the Deck, 567
Potlatch Game, 567
Rainbow Game, 569
Settle or Strike, 381
Simpolis, 361-2
Slave Coast Game, 437
To Drink or Not to Drink, 308
Goff, James F.
Winter War, 522-3
Gohring, Ralph J.
Principal Game, 389
Goldberg Eric
Descent on Crete, 490
Siege of Constantinople, 512
Goldberger, Martin
Four Battles from the Crimean War, 494
Goodman, Frederick L.
End of the Line, 443
Much Ado About Marbles, 321
Policy Negotiations, 399
They Shoot Marbles, Don't They?, 400-1
Goodman, Richard
Prediction, 312
Goodman, Steve
Bay of Pigs, 485
Goosen, Kenneth
Introduction to Managerial Accounting, 641
Gordon, Alice
Fixit, 387
Gordon, Alice K.
American History Games, 419-20
Gordon, W. J.
Hang-Up, 550
Gordon, William I.
Prediction, 312
Gosser, Jon
Suburban Operations Simulation, 324
Gottheim, Ezra
Edplan, 386
Edventure, 386-7
Gould, Thomas
Four Battles from the Crimean War, 494
Govea, Rodger M.
Good Federalism Game, 354
Graves, Robert L.
INTOP, 598
Green, Cedric W. B.
Gambit, 584
INHABS 3, 586
Green, James
World Political Organization, 459
Greenblat, Cathy S.
At-Issue, 396
Blood Money, 412-13
Conceptual Mapping Game, 396
IMPASSE, 398
Marriage Game, 550-1

Greene, Jay R.
Industrial Sales Management Game, 633-4
Materials Inventory Management Game, 506
Personnel Assignment Management Game, 624
Production Scheduling Management Game, 607
Retailing Department Management Game, 637
Top Operating Management Game, 602-3
Greene, Lorne
Propaganda Game, 568
Greenlaw, Paul S.
Finansim, 641
Marksim, 636
Prosim, 607
Greenwood, Donald J.
Alexander the Great, 480
Russian Campaign, 511
Squad Leader, 515
Victory in the Pacific, 520
War at Sea, 521
Griggs, Donald L.
Moses and the Exodus, 535
Planning Game, 536
Prophets and the Exile, 536-7
Grishaver, Joel Lurie
Going Up: The Israel Game, 533-4
Grossman, Robert
Circulation, 541
Groves, Lawrence A.
La Bataille de la Moskowa, 499
GSPIA
GSPIA, 584-5
Guetzkow, Harold
Inter-National Simulation Kit, 452-3
Guider, Dan
Explosion (Population), 367
Gygax, Gary E.
Alexander the Great, 480

Haas, Mary E.
Traders Arrive on the African Scene, 575
Up Caste Down Caste, 576
Hack, Walter
Madison Simulations, 390
Hackelman, Edwin
Marketing Strategy, 635
Haeuser, Adrienne Ahlgren
All for the Cause and the Cause of Each, 441
Hamblen, Richard
Arab-Israeli Wars, 481
Victory in the Pacific, 520
Hamsher, M.
Metric Poker, 472
Hany, Darryl
En Garde!, 492
Harader, William
DeKalb Political Simulation, 352
Legisim, 355
Hardy, Irad B.
After the Holocaust, 479-80
Blue and Gray, 485
Dreadnought, 491

Firefight, 493
Highway to the Reich, 497
Modern Battles, 502
Napoleon at War, 504
October War, 505
Punic Wars, 508
South Africa, 514
War between the States, 521
War in Europe, 521-2

Hargreaves, S. W.
Esso Service Station Game, 647

Harris, Ben
Madison Simulations, 390

Harris, John
Interp, 339
Newton, 340-1
Scatter, 341

Harris, Jonathan
Judgment, 429

Harris, Roy D.
Computer Models in Operations Management, 604
Computer Models in Operations Research, 605

Harris, T. Robert
Economic System, 375

Harshman, John
Avalanche, 482
La Bataille de la Moskowa, 499
Overlord, 506
Pharsalus, 508
Yalu, 524

Hart, Sterling
Kursk, 499

Hasell, Mary Joyce
Grand Frame, 584

Hawke, Frank
Salt III, 455-6

Hawkins, Mary E.
Lab Apparatus, 543-4

Heinze, Rudolph W.
A Mighty Fortress, 501

Helmer, Olaf
Simulating the Values of the Future, 406

Henderson, Thomas A.
National Policy Game, 358

Hendrick, Arnold J.
La Grande Armee, 499

Hendrix, John
Experiential Education, 533

Hendrix, Lela
Experiential Education, 533

Henke, Roger
Explosion, 367

Henry, Porter
Selling: A Good Way to Earn a Living, 648-9

Henshaw, Richard C. Jr.
Executive Game, 597

Hephner, Thomas A.
Jeffrey's Department Store, 649-50

Hepler, Molly
IDGAME, 339

Herbert, Beverly
World Without War Game, 576

Herman, Mark
First World War, 493-4
Modern Battles II, 502-3
Next War, 505
October War, 505
Raid!, 508-9
Red sun Rising, 509
Stonewall, 515

Herrold, Thomas E.
Thanatos, 538

Hersey, Paul
Situational Leadership Simulator, 629-30

Hessel, Brad E.
Breitenfeld, 486
Cobra, 487
Drive on Stalingrad, 491
Kharkov, 498
Thirty Years War, 517

Hetzel, Nancy
KSIMS, 339-40
IDGAME, 339

Heydeman, M. T.
ENZKIN, 338

Hickox, Edward
Adams School System Simulation, 385-6

Hicks, Bruce L.
Tri-Nim, 475

Hicks, Harvey C.
Tri-Nim, 475

Hicks-Beach, Michael
Game of Nations, 450-1

Hildebrand, John
Adapt, 409
Utopia, 408

Hill, Ala Kay
Tightrope, 382

Hill, John
Bar-Lev, 483
Jerusalem!, 497
Kasserine Pass, 497-8
Overlord, 506
Squad Leader, 515
Verdun, 519-20
Yalu, 524

Hinderman, Dewey
Future Planning Games, 404-5

Hinkle, Charles L.
Marketing Dynamics, 634

Hinman, Herbert
General Agricultural Firm Simulator, 644

Hlavnicka, R. J.
Caen 1944, 486
Cassino 1944, 487
Crimea 1941, 488-9
Sidi Rezegh 1941, 512

Hock, Harold E.
Tobruk, 517-18

Hoffbauer, Dan
Battles for Tobruk, 484-5

Hoffman, Mrs. Richard P.
Stocks and Bonds, 382

Hogarty, Ken
Saturation, 462-3

Hollander, S.
Malar, 344
Policy, 327
POP, 330
TAG, 331
Uspop, 344

Hollowell, John
Inventing and Playing Games in the English Classroom, 460-1

Honess, C. Brian
MADS-Bee, 598-9

Hooper, Mr. and Mrs. W. Stanley
Stocks and Bonds, 382

Hopkins, T. E.
Neutron Activation Analysis, 340

Hopson, Barrie
Exercises in Personal and Career Development, 549

Hornack, Fred
IDGAME, 339

Hostrop, Richard W.
Watergate: the Waterloo of a President, 440

Hottenstein, Michael P.
Prosim, 607

Hough, Patricia
Exercises in Personal and Career Development, 549

Howard, J. Rodney
Operation Encounter, 636-7

Howe, Helen *Horatio Alger*, 443

Howells, Lloyd T.
INTOP, 598

Howes, Nancy
Adams School System Simulation, 385-6

Hubbell, Stephen P.
Extinction, 542

Hughes, Barry
Energy, 327

Huntington Two Computer Project
Buflo, 328-9
Charge, 337
Elect, 326-7
Hardy, 338-9
Limits, 344
Lockey, 340
Malar, 344
Market, 332-3
Maspar, 327
PH, 341
Policy, 327
Polsys, 327-8
Polut, 329-30
POP, 330
Rats, 330
SCATR, 341
Slits, 342
Sterl, 330-1
TAG, 331
Uspop, 344

Hurley, Cecile L.
Cal Q Late, 470

Husband, Robert L.
Barriers in the Maze, 546

Hutton, Robert
General Agricultural Firm Simulator, 644

Hyman, Ronald *Walk in My Shoes*, 464

Illinois, University of
Clinical Simulations, 413

Immegart, Glenn
Madison Simulations, 390

Instructional Simulations, Inc.

System 1, 400
Irving, William S.
Value Game, 555
Irwin, Harry
Cargelligo Topograhic Map Game, 409-10
Isby, David C.
Air War, 480
Four Battles in North Africa, 494-5
Modern Battles, 502
Napoleon At War, 504
Soldiers: WWI Tactical Combat, 1914-15, 513-14
Tannenberg, 516

Jackson, James R.
Executive Game, 597
Jacobsen, Louise Weinberg
Halfway House, 319-20
Slave Auction, 437
Jaffee, Cabot L.
Problems in Bank Management, 642-3
Problems in Supervision, 627
James, Francisco
Electronic Industry Game, 596
Pocket Calculator Boom, 599
Jansen, Edmund
Blackberry Falls Town Government Game, 578
Jarvis, William B.
Micro-Community, 565-6
Jeffries, Marjorie S.
Congress, 351
Jenkins, John
Household Energy Game, 528
Jelley, C. R.
Esso Service Station Game, 647
Jensen, Ronald L.
Business Management Laboratory, 595
Jerdee, Thomas H.
Becoming Aware, 611
Jitkoff, Julia A.
Decisionmakers, 317
Johara, Harish
Leela, 534-5
Johns, Jody R.
Tomorrow's Secretary, 630
Johnson, Catherine D.
Code, 460
Johnson, Howard Edward
Midtex, 605
Unitex, 608
Johnson, Janice
Space Colony, 406-7
Johnson, Kenneth G.
Nothing Never Happens, 312
Johnson, K. J.
Titration, 342
Johnson, Richard R.
Datacall, 345
Johnson, Stephen M.
Conflict in the Middle East, 447
Jones, John E.
Annual Handbook for Group Facilitators, 308-9
Handbook of Structured Experiences for Human Relations Training, 310-11
Jordison, Richard
Siege!, 512
Joyce, H. Donald
In-Basket Simulation Exercises, 388
Kaiser, Charles
Great Plains Game, 426
Kanterman, Leonard
Cromwel, 489
Kassel, Tom
Wellington's Victory, 522
Kastelnik, Connie
Psych City, 323
Katsch, Beverly Schwartz
Plea: A Game of Criminal Justice, 467-8
Katsch, Ethen
Plea: A Game of Criminal Justice, 467-8
Kattenburg, Paul M.
Diplomatic Practices, 450
Kauffman, Draper L., Jr.
Teaching the Future, 407
Kaufman, L.
Elect, 326-7
Kearins, Kathryn
Exploring Careers, 549
Keiser, Stephen K.
Marketing Interaction, 634-5
Kelley, Nelson Lane
Personnel Management in Action, 625
Kemball-Cook, R. B.
Metal Box Business Game, 649
Kennedy, Charles L.
Amnesty, 349
Budget, 349-50
Committee, 351
Constitution, 351-2
Delegate, 352-3
Independence, 427
Independence '76, 428
Votes, 362-3
Kennedy, Hank
AFASLAPOL, 446
Kent, Jody
Rebirth: The Tibetan Game of Liberation, 537
Keys, Bernard
Executive Simulation, 597
Kidd, Kenneth
In Order, 471
Kirman, Joseph M.
Canada's Prairie Wheat Game, 559
Two Years of Horror, 408
Klapfer, Leo E.
In-Quest, 543
Klassen D.
Elect, 326-7
Policy, 327
Klietsch, R. G.
Blight, 578-9
Czar Power, 421-2
Female Images, 549-50
F.L.I.P. 2/80, 527
No Dam Action, 369
Party Central, 433
Transit, 590-1
Kniffin, Fred W.
Marksim, 636
Knight, Anne
Blackberry Falls Town Government Game 578
Knudson, Harry R.
Management: An Experiential Approach, 601
Koehler, George E.
Futuribles, 405
Koeninger, Jimmy G.
Jeffrey's Department Store, 649-50
Koppel, John
War Crimes Trials, 440
Urban America, 324-5
Kosnett, Phil
Modern Battles II, 502-3
Objective Moscow, 505
Koza, Russell C.
Marketing Dynamics, 634
Kozoll, Charles E.
Executive Secretary, 615-16
Kraemer, Ann
Horatio Alger, 443
Serfdom, 570
Welfare Week, 445
Kramish, Leonard
Gestapo—A Game of the Holocaust, 533
Krause, William
Involvement, 428
Kraushaar, Alan
Brothers and Sisters Game, 442
Krawitz, Myron J.
Lester Hill Office Simulation, 649
Krieger, Richard
Jury Game, 466-7
Krummacher, Gottfried
Economy Game, 375
Krusell, Judith
Process, 553
Kuder, Frederic
Career Development Program, 547
Kugel, Peter
On Sets, 473
Queries 'n Theories, 462
Kurtick, Michael Scott
Space Patrol, 407
Kurtz, David L.
Sales Management Game, 638

Lacey, William
Espionage, 424
Fifties, 424-5
Skins, 436
Ladd, C.
Buloga II, 604
Landry, David
Process, 553
Larreche, Jean-Claude
Markstrat, 636
Larson, David L.
Simulex IV, 456
Lauffer, Armand
Compacts, 316-17
Lobbying Game, 355
Much Ado About Marbles, 321
Laughlin, Hugh
Madison Simulations, 390
Lavin, Richard
School Planning Game, 392
Lawson, Barry R.
Congress, 351
Librarian's Game, 443-4
New Town, 322
New Town Professional Planners Set, 588

Layden, Kent
Humanus, 562-3
Making a Change, 320
Leftwich, Howard
Executive Simulation, 597
Lesso, W. G.
Computer Models in Operations Research, 605
Leveridge, M. E.
COMPETE, 337-8
Computers in the Biology Curriculum, 338
Leviaton, Victor
Japanese-American Relocation 1942, 428-9
Levine, S. Joseph
Teachers' Lounge, 394
Levine, Toby H.
Energy-Environment Game, 366
Liao, T.
Polut, 329-30
Lieberman, Harvey R.
Decision Guides, 614
Effective Supervision in Government, 615
Managing and Allocating Time, 621
Transactional Analysis—Improving Communications, 630-1
Women in Management, 632
Liebig, Mark C.
Nothing Never Happens, 312
Lineham, Thomas E.
Road Game, 313
Value Game, 555
Lipetzky, Jerry
Adapt, 409
Dig, 561
Lipman, Michael
Point of Law', 468
Liss, R. Lawrence
In-Quest, 543
Livingston, Samuel A.
Inner City Housing Game, 320
Lombardy, Dana F.
Cromwell, 489
Long, Barbara Ellis
Road Game, 313
Loomis, Richard F.
Board of Directors, 331
Losik, C.
Charge, 337
Polut, 329-30
Sterl, 330-1
Louin, Robin
Suburban Operations Simulation, 324
Lowell, Marie Miller
Pollination Game, 545
Predator, 545
Lundstedt, Ronald
Confrontation in Urbia, 464-5
Spanish-American War, 437-8
Lunetta, Vincent N.
Critical Incidents in Education, 333
Lupul, Max E.
Marketing Interaction, 634-5

McCarty, William M.
Juris, 466
Vows, 468-9
McClellan, Larry A.
Urban Dynamics, 591-2
McCormick, S.
EVOLUT, 338
MacDonald, Lea
G.U.L.P., 585
McFarlan, F. Warren
Management Game, 598
McGrath, J.
Elect, 326-7
Market, 332-3
Policy, 327
Polsys, 327-8
McGuire, Christine H.
Clinical Simulations, 413
McIntyre, Kenneth
Regular Meeting of the Wheatville Board of Directors, 392
Mack, Jay
Gateway, 425
McKenna, Megan
Coffee Game, 532
McKenney, James L.
Management Game, 598
McKinnell, Henry A.
Stanford Business Logistics Game, 608
McKuen, Gary E.
Future Planning Games, 404-5
McLure, John
Puzzle, 434-5
McManus, John
Psych City, 323
McMullin, Ralph
Interceptor, 471
McNeil, Andrew
Kingmaker, 498
McPherson, J. Westbrook
Prosper, 628
McPortland, Joanne
Coffee Game, 532
Maggard J. J.
Computer Models in Operations Management, 604
Computer Models in Operations Research, 605
Maggio, Tony
Mummy's Message, 431
Main, Dana
Exper-Sim, 346
Malcolm, Henry
Value Options, 555
Malin, David
Schizophrenia, 347
Management Games Limited (See Producer Index)
Hospitex, 415
Sale-Plan, 637-8
Manera, Elizabeth
COMCO, 558-9
Manning, Gerald L.
Wilson R.V., 631-2
Marcus, Audrey Friedman
Gestapo—A Game of the Holocaust, 533
Marfuggi, Joseph R.
Pink Pebbles, 379-80
Martin, James A.
River Dose Model, 330
Martin, W. Michael
SEASIM, 393
Matts, Victoria
Industrial Marketing Plan Simulation, 633
Matukas, Marcia
Selection Interview Clinic, 629
Mazze, R.
Polsys, 327-8
Megley, John
Executive Decision Making through Simulation, 596-7
Meier, R. L.
El Barrio, 443
FORMENTO, 377
Wildlife, 545-6
Melrose, John
BUCUMCO, 600
Mendelson, Richard N.
Industrial Marketing Plan Simulation, 633
Merridy, Tony
Raid!, 508-9
Meyer, David P.
Pipeline, 444-5
Meyers, Ralph
Confrontation: The Cuban Missile Crisis, 447-8
Miles, Robert H.
Organization Game, 602
Miller, Donald E.
Using Biblical Simulations, 539
Miller, John
Prospects, 553-4
Miller, Marc William
Battle for Midway, 483-4
Burma, 486
Case White, 486-7
Coral Sea, 488
Raphia, 217 B.C., 509
Russo-Japanese War, 511
SSN: Modern Tactical Submarine Warfare, 515
Miller, Pat
Grand Frame, 584
Miller, Peter
Edplan, 386
Miller, William E.
Inflation, 378
Mills, Stephen L.
PRINCE, 335
Minneapolis Public Schools
Charge, 335-6
Shelter, 336
Wheels, 336
Minor, Gene
Nothing Ever Happens, 312
Mirk, R. Patrick
Custer's Last Stand, 489
Mobley, L. R.
Value Options, 555
Montgomery, Raymond A.
Energy-Environment Game, 366
Moore, B.
Australian Management Games, 610
Morale, Tony
Rhein Bung, 510
Mosca, Linda
Blue and Gray II, 485
Thirty Years War, 517
Moulton, Gerald
COMCO, 558-9

Moy, K.
Charge, 337
Polsys, 327-8
POP, 330
Sterl, 330-1
Mulholland, Virginia
Modern Battles II, 502-3
Mullen, Robert E.
Urban Dynamics, 591-2
Mulley, Earl S.
Richland USA, 589
Munson, George
Battles for Tobruk, 484-5
Murphy, Geraldine
Tightrope, 382
Murphy, P. J.
Coexist, 329
Linkover, 340
Murray, Henry
Game of Nations, 450-1
Musella, Donald F.
In-Basket Simulation Exercises, 388

Naffziger, Mr. and Mrs. A. Brooks
Stocks and Bonds, 382
Naill, Roger F.
Coal 1, 329
Nash, Nicholas
Adams School System Simulation, 385-6
Needham, Jack
Big Business, 420
Neff, Robert W.
Using Biblical Simulations, 539
Nelson, Christopher B.
River Dose (RVRDOS) Model, 330
Nelson, J. A.
Breitenfeld, 486
Highway to the Reich, 497
Island War, 497
Napoleon's Last Battles, 504
Wacht am Rhein, 520-21
Westwall, 522
Nelson, Robert L.
Drug Attack, 307
Nesbitt, William A.
Alpha Crisis Game, 418
Guns or Butter, 452
Ness, Thomas E.
Marketing in Action, 634
Newberg, Stephen M.
Dieppe, 490
Peloponnesian War, 508
Raketny Kreyser, 509
SSN: Modern Tactical Submarine Warfare, 515
Newmann, Fred M.
Railroad Game, 380-1
Nichol, J.
Longman History Games, 430
Nichols, Arthur C.
SIMQ, 599
Nieswand, George H.
Transfer-of-Development Rights Game, 590
Niland, Powell
Washington University Business Game, 603
Noel, Robert C.
POLIS Network, 334
Northwestern National Bank of Minneapolis
Charge, 335-6
Shelter, 336
Wheels, 336
Nudelman, Jerrold
Taking Action, 463
Nulson, R.
Compete, 633

Ober, Keith
Exploring Careers, 549
Oberlander, L.
Policy, 327
Oden, Thomas C.
TAG: The Transactional Awareness Game, 314
O'Leary, Dennis, P.
Caen 1944, 486
Cassino 1944, 487
Crimea 1941, 488-9
Sidi Rezegh 1941, 512
O'Leary, Michael
PRINCE, 335
Oleson, Tom
Russian Campaign, 511
Oliver, Donald W.
Railroad Game, 380-1
Oliver, Hugh
Classifying the Chemical Elements, 541-2
Oliver, Peg
Praise and Criticism, 536
Ollman, Bertell
Class Struggle, 558
Onanian, R. A.
Facts in Five, 397
Orbach, Eliezer
Marketing a New Product, 634
Personnel Department, 624-5
Supermarket Strategy, 643
Orbanes, Philip
La Grande Armée, 499
Orlando, J. A.
Build, 579
Otlewski, Robert E.
Contract Negotiations, 613
Owens, Martin F.
On-Sets, 473
Oxnam, Robert B.
The Ch'ing Game, 420-1

Papps, Grayce
Eco-Acres, 583-4
Park, William R.
Low Bidder, 601
Parker, Don H.
Decision Making for Career Development, 548
Parker, John
PSW 1 Political Simulation, 334-5
Parker, Ronald J.
INS 2–Inter-Nation Simulation, 334
Parker, Shelby W.
Decision Making for Career Development, 548
Parry, Scott B.
Career Game, 611
Construction Game, 612
Game of Time Management, 616
Game of Transactional Analysis, 616
Lou Boxell's Performance Review, 620
Meeting Game, 623
Selection Interview, 629
Parsons, Theodore W.
Achieving Classroom Communication through Self Analysis, 385
Patrick, Stephen B.
Kharkov, 498
Thirty Years War, 517
Westwall, 522
Patterson Hartley
English Civil War, 492
Payne, Nelson
Compounds, 542
Peabody, George
Powerplay, 553
Pearson, Craig M.
Pink Pebbles, 379-80
Peek, J. Stephen
Submarine: Tactical Level Submarine Warfare 1939-45, 516
Pegas, Arthur
Nuremberg, 432
Pennington, A. J.
Build, 579
Pessel, D.
Polut, 329-30
Peters, Lee G.
Situational Leadership Simulator, 629-30
Peterson, Arthur
Homefront, 427
Peace, 433
Peterson, Wayne
Here to There, 471
Multifacto/Producto, 472
Polyhedron Rummy, 473
Pfeifer, Camille Freed
Nourish, 416
Pfeiffer, J. William
Annual Handbook for Group Facilitators, 308-9
Handbook of Structured Experiences for Human Relations Training, 310-11
Phelan, Grace E.
Handling Conflict in Hospital Management (Conflict Among Peers), 414
Handling Conflict in Hospital Management (Superior/Subordinate Conflict), 414-15
Phelan, Noel
O.K., 552
Phelan, Patricia
O. K., 552
Piccione, Peter A.
King Tut's Game, 429
Pidot, George
Acres, 577
Pierce, W. R.
Forest Simulator, 646
Pinsky, Lawrence
Battle of the Bulge, 484
Blitzkrieg, 485
Island War, 497
Midway, 501
Westwall, 522

Pipkin, Ronald M.
Plea: Game of Criminal Justice, 467-8
Place, Daniel R.
Czech-Mate, 422
Platt, Joan
Union Divides, 439
Platt, Judith
Pollution, 369-70
To Drink or Not to Drink, 308
Ploutz, Paul F.
Elements, 542
Geologic Time Chart Game, 543
Lab Apparatus, 543-4
Mental Health, 551
Rip-Off, 530
Pluck, J.
Computers in the Biology Curriculum, 338
Plummer, Charles M.
Dynamic Modeling of Alternative Futures, 402
Poe, Doug
Avalanche, 482
Bar-Lev, 483
Citadel, 487
Porter, Dennis
Explosion, 367
Porter, Henry
Selling: A Good Way to Earn a Living, 548-9
Powell, Jimmie
Mulberry, 321
Poze, T.
Hang-Up, 550
Prados, John
Rise and Decline of the Third Reich, 510
Preuss, Bob
Horatio Alger, 443
Proudman, Colin
Island, 378
Values, 555

Rader, John R.
Dynamic Modeling of Alternative Futures, 402
Rader, William D.
Market, 529
Rajecki, D. W.
Imprinting, 346
Ralston, Lawrence
Meridian 36, 411
Randolph, W. Alan
Organization Game, 602
Rausch, Erwin
Collective Bargaining, 611
Decision Making, 614
Economic Decision Games, 374-5
Equipment Evaluation, 605
Financial Analysis, 640
Handling Conflict in Hospital Management (Conflict among Peers), 414
Handling Conflict in Hospital Management (Superior/Subordinate Conflict), 414-15
Handling Conflict in Management: Conflict among Peers, 617
Handling Conflict in Management (Superior/Subordinate Conflict) III, 617
Handling Conflict in Management (Superior/Subordinate Group Conflict) II, 617
Interviewing, 618
Leading Groups to Better Decisions, 619
Managing and Allocating Time, 621
Principles of Effective Salesmanship, 637
Production Control–Inventory, 606-7
Purchasing, 607
Supervisory Skills, 630
Transactional Analysis–Improving Communications, 630-1
Rausch, George
Leading Groups to Better Decisions, 619
Principles of Effective Salesmanship, 637
Rawitsch, Don
Oregon, 347
Raymond, Michael M.
Fortress Rhodesia, 494
Kama, 565
Southern Mountaineer, 572
Reece, Barry C.
Wilson RV, 631-2
Reed, Randall C.
Arab-Israeli Wars, 481
Panzer Leader, 507
Richthofen's War, 510
1776, 511-12
Reese, Jay
Blue Wodget Co., 648
Contractors, 559-60
Flight, 562
Homestead, 427
Merchant, 431
Roaring Camp, 435
Reichert, Richard
Simulation Games for Religious Education, 537
Reid, Rick
Collision, 351
Reilly, Richard
Problems in Bank Management, 642-3
Reiners, William A.
Ecological Modeling, 329
Reynolds, Jack
Mortgage, 529
Richardson, Richard
Quantitative Experimental Analysis, 347
Richmond, George
Micro-Economy, 566
Riddle, Bruce
World Political Organization, 459
Ringo, Shirley
Here to There, 471
Roberts, Charles
D-Day, 489
Roberts, Charles S.
Tactics II, 516
Robbins, Charles
River Dose Model, 330
Robinson, Edward J.
Career Game, 611
Construction Game, 612
Game of Time Management, 616
Game of Transactional Analysis, 616
Rochester, J. Martin
Foreign Policy Decision Making, 450
Rodin, Richard R.
Great Periodic Table Race, 543
Rogers, Clark
GSPIA, 584-5
Rogers, Peter
Systems, 589-90
Rosen, B.
Lockey, 340
Rosen, Benson
Becoming Aware, 611
Ross, Edward
Oh-Wah-Ree, 472-3
Soma Cube, 475
Ross, Joan
Queries 'n Theories, 462
Ross, Stephen
Atlantic Wall, 482
Four Battles from the Crimean War, 494
Up Scope!, 519
Rosser, David
Defense, 352
Enterprise, 376
Pressure, 361
Protection, 529-30
Spiral, 381
Taxes, 362
Rothschild, Eric
Federalists vs. Republicans, 424
1787, 436
The Union Divides, 439
Roush, Clint
Oklahoma Bank Management Game, 642
Ruben, Brent D.
Human Communication Handbook, 311
Interact II, 311-12
Ruderman, Harry D.
Tac-Tickle, 475
Rubinger, Rabbi Fred
Kibbutz, 534
Ruffle, Chris
Mercenary, 500-1
Ruppenthal, Karl M.
Stanford Business Logistics Game, 608
Rusiecki, Laurence J.
Eagle Day: The Battle of Britain–1940, 491
Hannibal, 496
1944: The Invasion of France and the Battle of Germany, 505
Russell, Gordon Jr.
Right is Write, 462
Russo, Rockland
Space Patrol, 407
Ryan, Thomas P.
Reapportionment, 361
Ryterband, Ed
Process, 553

Sachs, Harley L.
Police State, 567
Sackson, Sidney
Executive Decision, 69, 376-7d
Sady, Rachel Reese
Indians View Americans, Americans View Indians, 428

Japanese-American Relocation, 1942, 428-9
Safier, Michael
G.U.L.P. 585
Sage, Daniel D.
Special Education Administration Task Simulation (SEATS), 392-3
Salvner, Gary
Newscast, 461-2
Sand, Gregory A.
Game of Transactional Analysis, 616
Meeting Game, 623
Sandell, Roger
English Civil War, 492
Sanzone, Jean F.
Steady Job, 554
Sarly, Robert M.
G.U.L.P., 585
Savage, William N.
Urban Dynamics, 591-2
Scarl, D.
Charge, 337
Schachter, Fred
Siege of Jerusalem, 70 A.D., 512-13
Schaupp, Jack
Biblical Society in Jesus' Time, 531
Moses and the Exodus, 535
Prophets and the Exile, 536-7
Schmidt, Elisabeth
Let's Play Games in German, 461
Schiff, J. S.
Principles of Effective Salesmanship, 637
Scholastic Search Magazine
On Strike and Other Economic Games, 379
Shipwreck and Other Government Games,
Schott, Brian
RISKM, 641
SIMQ, 599
Schroeder, Wayne L.
Microville, 320-21
Schustler, Richard A.
Phantom Submarine, 455
Remote Island, 412
Schutz, Lindsley
Midway, 501
Waterloo, 522
Scott, Charles R. Jr.
Tempomatic IV, 600
Scottusa Company
Keep Quiet, 415
Seasholes, Bradbury
Woodbury Political Simulation, 363-4
Seiler, John A.
Management Game, 598
Senatore, John J.
Nothing Never Happens, 312
Shapira, Zur
Nuclear Site Negotiation 322
Shaw, K.
Haber, 339
Shaw, Thomas N.
Afrika Korps, 479
Midway, 501
Waterloo, 522
Shirk, Ethel
Probe, 553
Shirts, R. Garry
Bafá Bafá, 556-7
Metropolitics, 355-6
New City Telephone Company Simulation Game, 566-7
Rafa Rafa, 568-9 568-9
Where Do You Draw the Line?, 555
Starpower, 572-4
Showalter, Victor
Mouse in the Maze, 544
Planet Management Game, 369
Shwarger, Michael
Metri-Magic, 472
Siegel, R. L.
Buflo, 328-9
Silas, Harry
Community Target—Alcohol Abuse, 307
Sillman, Robert
Consumer Decision, 526
Crisis in Middletown, 317
Dissent and Protest, 422-3
Exchange, 376
Exploring the New World, 424
First Amendment Freedoms, 465
Government Reparations for Minority Groups
Human Survival—2025, 405
Interaction—A Balance of Power in Colonial America, 428
Low Income Housing Project, 320
Middle East Crisis, 455
National Priorities, 358
Nationalism: War or Peace, 455
Planning the City of Greenville, 322-3
Planning an Inner-City High School, 323
Political Pollution, 369
Preventing the Civil War, 434
Revolt at State Prison, 468
Spanish-American War, 437
Trial of George Washington, 438
Trial of Harry S. Truman, 438-9
Trial of Jefferson Davis, 439
War Time, 459
World War I, 440
Women's Liberation, 363
Simonsen, Redmond A.
After the Holocaust, 479-80
Air War, 480
The Conquerors: The Macedonians and the Romans, 488
Firefight, 493
Four Battles from the Crimean War, 494
Fulda Gap, 495
Highway to the Reich, 497
A Mighty Fortress, 501
Minuteman, 501-2
Modern Battles II, 502-3
Napoleon's Last Battles, 504
Plot to Assassinate Hitler, 508
Revolt in the East, 509-10
Russian Civil War, 511
Strike Force One, 515
Terrible Swift Sword, 517
Thirty Years War, 517
War Between the States, 521
War in Europe, 521-2
Simpson, Bert K.
Cruel Cruel World Value Game, 548
My Cup Runneth Over, 551-2
Value Bingo, 554-5
Sischo, David
Involvement, 428
Sisson, Roger L.
Industrial Sales Management Game, 633-4
Materials Inventory Management Game, 606
Personal Assignment Management Game, 624
Production Scheduling Management Game, 607
Retailing Department Management Game, 637
Top Operation Management Game, 602-3
680 Game Design Seminar
Hexagon, 585
Sleet, David A.
Contracept, 413-14
Smith, Clifford N.
DeKalb Political Simulation, 352
PSW-1 SW Political Simulation, 334-5
Smith, Dannie L.
Indians View Americans, Americans View Indians, 428
Web, 395
Smith, Daniel C.
Trade-Off at Yalta, 438
The Union Divides, 439
Smith, Ed P.
Trireme, 518
Smith, G. J. II
TAG, 331
Smith, H. Warren
KSIMS, 339-40
Smith, Jerald R.
Busop, 595
Smith, Lewis B.
Questioneze, 392
Smith, Nick
Panzer Leader, 507
Smith, Thomas W.
Household Energy Game, 528
Smith, W. Nye
Integrated Simulation, 597
Snaith, Mary Shaw
Nourish, 416
Snow, Donald M.
Revolutionary Society (REVSOC) Simulation, 569
Snyder, Graydon F.
Using Biblical Simulations, 539
Sobin, D.
Lockey, 340
Market, 332-3
Polsys, 327-8
Sociological Resources for the Social Studies
Simulating Social Conflict, 571
Solin, Arthur
Confrontation: The Cuban Missile Crisis, 447-8
Solomon, Lawrence M.
Clinical Simulations, 413
Somers, John
Acres, 577
Spears, Jim

Community Target–Alcohol Abuse, 307
Spence, Richard
Königgrätz, 498
Mukden 1905, 503
Tannenburg, 516-17
Spicer, Nancy E.
Industrial Marketing Plan Simulation, 633
Spitze, Hazel Taylor
Calorie Game, 413
Nutrition Game, 416
Stadler, Anne
World Without War Game, 576
Stadsklev, Ron
Indian Reservation, 563-4
Staw, I.
Market, 332-3
Steffy, Joan
Boxcars, 409
Stein, Peter J.
The Marriage Game, 550-1
Sterk, Jane
Independence, 563
Stewart, Virginia
Serfdom, 570
Stieber, Nancy
The Grand Frame, 584
Stitleman, Leonard
American Constitutional Convention, 419
American Government Simulation Series, 348-9
Stoll, Clarice Stasz
Inner-City Housing Game, 320
Stolovitch, Harold D.
Confrontation, 397
Dynamic Modeling of Alternative Futures, 402
Stones, David
Quantitative Experimental Analysis, 347
Stout, Robert
Exper-Sim, 346
Straumanis, Joan
Job Crunch, 389
Streifer, William
Mission Aloft, 502
Strickland, Alonzo
Tempomatic IV, 600
Suransky, Leonard
Middle East Conflict Simulation Game, 454-5
Sweedler, Donna
Brothers and Sisters Game, 442
Swinerton, E. Nelson
The Dead River, 366

Tammen, John A.
A Leadership Decision Game, 619
Tang, Charles Y.
Admag I Advertising Management Game, 632-3
Tardy, Dwight
KSIMS, 339-40
Tatz, Mark
Rebirth: The Tibetan Game of Liberation, 537
Taylor, Andrea Jane Richardson
Exchange, 376
Spectrum, 572
Taylor, Edgar C. Jr.
Middle East Conflict Simulation Game, 454-5
Taylor, Maurie N.
Classroom Games in French, 460
Taylor, S. Craig Jr.
Air Force, 480
Wooden Ships and Iron Men, 523
Taylor, Stephen Charles
The Jury Game, 466-7
Teplitsky, Alan
Walk in my Shoes, 464
Theis, Paul A.
Hat in the Ring, 354-5
Thiagarajan, Sivasailan
Dynamic Modeling of Alternative Futures, 402
GAMEgame, 397
GAMEgame II, 397
GAMEgame IV, 397-8
GAMEgame VI, 398
Kathal: Game of Creative Intimacy, 550
Naked Monsters, 391-2
Press Conference, 399
Thomas, Norman C.
Indian Valley, 368
Thompson, Lowell
Great Plains Game, 426
Thompson N.
Manage, 332
Thorelli, Hans B.
INTOP, 598
Thorpe, Gerlad L.
Confrontation: The Cuban Missile Crisis, 447-8
TIES
Limits,
Tipple, Bruce E.
Fail-Safe, 343
Inflation, 378
Toelke, Ron
Wellington's Victory, 522
Toll, Dave
Ghetto, 317-19
Tomasi, David
Battle of Chickamauga, 484
Tombaugh, Richard F.
IM-CLUG, 585-6
Metroplex, 587
Polis, 589
PX-190, 399-400
Richland USA, 589
Topliffe, Neil
Catalyzer, 532
Toppin, Martha Doerr
After the War and the Election of 2020, 419
Towle, Joseph W.
Washington University Business Game, 603
Townsend, Charles
Five Simple Business Games, 648
Tranter, J. A.
Computers in the Biology Curriculum, 338
Trdan, Michael
Caen 1944, 486
Cassino 1944, 487
Crimea 1941, 488-9
Sidi Rezegh 1941, 512
Trent, Fr. James
Serfdom, 570
Troyka, Lynn
Taking Action, 463
Tusson, John R.
Transaction, 382-3
Twelker, Paul A.
Humanus, 562-3
Making a Change, 320
Planning Exercise, 398-9
Pollution–Negotiating a Clean Environment, 370
Power Politics, 360
Triange Trade, 439
Waging Neutrality, 440
Tyler, Michael *G.U.L.P.*, 585

Uhl, Mick
Gettysburg, 496
U.S. Department of Agriculture
Forest Service Fire Control Simulator, 646
Updegrove, Daniel A.
Educom Financial Planning Model, 333
IDIOM, 336-7
Urbandyne
Urban Dynamics, 591-2

Valdes, Donald
The Hunting Game, 563
Van de Bogard, Eric
Eco-Acres, 583-4
Vandeportaele, D.
City, 579
Vander Velde, Michael C.
Court Policy Negotiations, 465
Vane, Russell
Arab-Israeli Wars, 481
Van Tassel, John E.
Decision Making Exercise, 596
Vass, Ben
Environmental Simulations, 367
Venditti, Frederick P.
Changing High School, 386
Solving Multi-Ethnic Problems, 393
Vernon, Robert F.
Talking Rocks, 463-4
Vicino, Franco L.
Process, 553
Prospects, 553-4
Vines, Ellsworth F.
Integrated Simulation, 597-8
Vogel, Rex
Metro Government, 355

Wagner, Ron
Indian Reservation, 563-4
Walczyk, Thomas
Blue and Gray, 485
Fulda Gap, 495
Napoleon at War, 504
Stonewall, 515
Thirty Years War, 517
War in Europe, 521-2
Walford, Rex

Longwood Geography Games, 410
North Sea Exploration, 411
Railway Pioneers, 411-12
Walker, Arthur
Managing the Worker, 622
Personnel Department, 624-5
Wantz, Molly
Health Games Students Play, 415
Warantz, Michael D.
Industrial Marketing Plan Simulation, 633
Ward, Jerry K.
Cope, 401-2
Ward, Patricia
Sanga, 569-70
Warns, Norman S., Jr.
Energy X, 366-7
Gomston: A Polluted City, 367-8
Warren, Jerry
Blackberry Falls Town Government Game, 578
Washburn, John
Gospel Game, 534
325 A.D., 538-9
Teaching Strategy, 395
Washburn, Norman F.
Marriage Game, 550-1
Weaver, Barbara F.
Librarian's Game, 443-4
Webster, Thomas B.
Battle of Roark's Drift, 484
Weinberg, Charles
Change Agent, 342-3
Weintraub, Richard
Jury Game, 466-7
Weiss, Stephen F.
Siege of Jerusalem, 70 A.D., 512-13
Werden, Dave
Cobra, 487
Kharkov, 498
Wesley, Diane
Fire, 465
Heritage, 427
Wesley, John
Agency, 308
Design, 527
Discovery, 422
Ecopolis, 366
Equality, 561
Fire, 465
Interaction, 388-9
Mahopa, 565
Wesolowski, Bob
Newscast, 461-2
Western Behavioral Sciences Institute
Crisis, 448-9
NAPOLI, 356-7
Wexo, John
Future Game, 404
Whatley, Arthur A.
Personnel Management in Action, 625
Whithed, Marshall H.
DeKalb Political Simulation, 352
Legisim, 355
PSW-1 SW Political Simulation, 334-5
TeleCLUG, 590
Woodbury Political Simulation, 363-4
Whitworth, Larry L.
Four in a Row, 470
Orbiting the Earth, 473
Whybark, D. Clay
Computer Augmented Cases in Operations and Logistics Management, 604
Stanford Business Logistics Game, 608
Wicklund, Gary
SIMQUEUE–A Queueing Simulation Model 599-600
Wiebe, Arthur
Bank Account, 525
Wietling, Stephen
SIMSEARCH, 344
Wilcoxson, Georgeann
Baldicer, 557-8
Williams, David
Lobbying Game, 355
Williams, E. Craig
Sanga, 569-70
Williams, Paul A.
Industrial Marketing Plan Simulation, 633
Williams, Robert
Warlord Game, 522
Williamson, E. A.
Buflo, 328-9
Malar, 344
Market,
Polut, 329-30
Slits, 342
Wilson, Forrest
City Planning, 581-2
Wilson, John
East Front, 492
Wimmer, Helmut
Space Hop, 545
Wiseman, Loren K.
Pharsalus, 508
Wohlking, Wallace
Handling Conflict in Hospital Management (Conflict among Peers), 414
Handling Conflict in Hospital Management (Superior/Subordinate Conflict), 414-15
Handling Conflict in Management: Conflict among Peers (Game I), 617
Handling Conflict in Management: Superior/Subordinate Conflict (Game III), 617
Handling Conflict in Management: Superior/Subordinate Group Conflict (Game II), 617
Wolohojian, George
Good Federalism Game, 354
Woodburn, John H.
E-Z-Science Games, 542
Woodworth, Robert T.
Management: An Experiential Approach, 601
World Without War Council of Greater Seattle
World Without War Game, 576
Woznick, J. L.
Compete, 633
Wren, Brian
The Grain Drain, 377
Yang, Charles Y.
ADMAG I, 632-3
Yates, James R.
SEASIM, 393
Young, John M.
Austerlitz, 482
Borodino, 485-6
Dreadnought, 491
La Grand Armée, 499
Lee Moves North, 499
Musket and Pike, 503
Seelowe, 511
Young, Joseph
Election, 353
Young, Marlene
Election, 353
Yount, David
Balance, 365-6
Cycle, 548
Destiny, 422
Disunia, 423
Division, 423
Enterprise, 376
Mission, 431
Panic, 433
Small Town, 571-2
Strike, 438
Sunshine, 324
Utopia, 408
Virtue, 439-40

Zahn, Donald M.
Hat in the Ring, 354-5
Zaid, Barry
Super Sandwich, 417
ZaKich, Rhea
The Ungame, 314-15
Zaltman, Gerald
Consumer, 526
Zalud, Richard
Custer's Last Stand, 489
Zapel, Arthur L.
Can of Squirms, 309
Zarecky, Gary
Juris, 466
Moot, 467
Rip-Off, 469
Vows, 468-9
Zaverin, Rabbi Raymond
Gestapo: A Game of the Holocaust, 533
Zelmer, A. C. Lynn
The Living City/La Cite Vivante, 586
Zelmer, Amy Elliott
Female Images: A Life Skills Exercise, 549-50
The Living City/La Cite Vivante, 586
Zif, Jay J.
Contract Negotiations, 613
Managing the Worker, 622
Marketing a New Product, 634
Personnel Department, 624-5
Supermarket Strategy, 643
Yes, But Not Here,
Zocchi, Louis B.
Battle of Britain, 484
Flying Tigers II, 494
Luftwaffe, 500
Zoll, Allen A.
Operation Suburbia, 624
Zucker, Kevin
Island War, 497
Napoleon's Last Battles, 504
Zuckerman, Robert A.
Dynamic Modeling of Alternative Futures, 402

GAME INDEX

The Abelson-Baker Interview, 609
Abraham Lincoln Elementary School Principalship Simulation, 390
Access, 240, 556d
AC/DC, 540
Achieving Classroom Communication Through Self-Analysis, 385
Acres, 287, 577d
Actionalysis, 385
Adams School System Simulation (ADSIM), 385-6
Adapt, 409
ADMAG I, 632-3
Administration, 609
Administration of Curricular Decision Making, 391
AFASLAPOL: The Game of Politics in a Moderning Nation, 446
African Village Game, 582
Afrika Korps, 479
After the Holocaust, 479-80
After the War and the Election of 2020, 419
Agency, 308
Aid Committee Game, 372
The Air About Us, 544
Air Force, 480
Airport, 367
Air War, 480
Alexander the Great, 480
Algonquin, 367
Algonquin Park, 59, 365d
Alien Space Ship, 348
All for the Cause and the Cause of Each, 441
Alma, 494
Alpha Crisis Game, 142, 418d
Alternative Futures Analysis and Review, 402
American Civil War, 189, 480-1d
American Constitutional Convention, 142, 419
American Government Simulation Series, 49, 348-9d
American History Games, 419-20
American Revolution, 481
Amnesty, 49, 349
Amos of Tekoa, 537
Ancient Conquest, 481
Annual Handbook for Group Facilitators, 308-9
Antietam, 485
Apex, 577
Appraisal by Objectives, 609-10
Arab-Israeli Conflict, 447
Arab-Israeli Wars, 481
Ardennes Offensive, 481-2
Are You with Me?, 464
Armada, 420
Armageddon, 487, 524
Arms Race, 482
Arnhem, 522
Artificial Society, 372
Assigning Work, 610
At-Issue, 288, 396d
Atlantic Wall, 482
Atomic Structure and Bonding, 544
Auction, 546
Austerlitz, 482
Australia 20000, 610
Australian Management Games, 610
Autocracy Game, 561
Avalanche, 482
Bafà Bafà, 163, 535, 556-7d, 569
Balaclava, 494
Balance, 59, 365-6d, 366
Baldicer, 59, 557-8d
Ballistic Missile, 483
Bank Account, 525
Banking, 374
Banner-Making Game, 531
Barbarossa, 190, 483d, 492
Bar-Lev, 483
Barriers in the Maze, 546
Bastogne, 481, 522
Battle for Germany, 483
Battle for Midway, 483-4
Battle of Britain, 484
Battle of Chickamauga, 484
Battle of Ripple Creek, 379
Battle of Roark's Drift, 484
Battle of Stalingrad, 518-19
Battle of the Bulge, 484
Battle of the Wilderness, 485
Battle for Tobruk, 484-5
Bay of Pigs, 485
Becoming Aware, 611
Becoming a Person, 224, 547d
Beef Cattle in Northern Australia, 410
Biblical Society in Jesus' Time, 531
Big Business, 420
Big Deal, 469
Binary, 469
Bio-EZ, 542
The Black Community Game, 315-16
Black Hawk War and Cherokee Removal, 428
Blackberry Falls Town Government Game, 287, 578d
Blight, 287, 578-9d
Blind, 533
Blitzkrieg, 485
Blood Flow, 540
Blood Money, 133, 412-13d
Blue and Gray: Chickamauga, Shiloh, Antietam, Cemetary Hill, 189, 485d
Blue and Gray II: Fredericksburg, Hooker and Lee, Chattanooga, Battle of the Wilderness, 485
The Blue Wodget Co., 648
Board of Directors, 331
Bobtree Moves into Western Europe, 410
BOLA, Bombardment of the Light Atoms, 337
Bombs or Bread: Consumer Priorities vs. National Security, 447
Borodino, 485-6
BUSOP, 595
Boxcars, 409
Bread, 533
Bread Line, 410
Breitenfeld, 486
Brinkmanship: Holocaust or Compromise, 420
Brookside Manor, 133, 441-2d
Brothers and Sisters Game, 442
BUCUMCO, 600
Budget, 49, 349-50d
Budget, 525
Budgetary Politics, 348-9
Budgeting Game, 525

Page numbers followed by the letter "d" indicate where game descriptions can be found.

Buflo, 328-9
Build, 579
Building a Rapid Transit System, 316
Buloga II, 604
Bundeswehr, 502-3
Burma, 486
Business Decision Simulation, 610
Business Management Laboratory, 279, 595d
Business Policy Game, 595
Business Strategy, 372-3
Busop, 595
Busy, 640
Buy and Sell, 373
Buyer Beware, 526

"CA", 486
Cabinet, 610
Caen 1944, 486
Call of Moses, 535
Called to be God's Spokesman, 537
Calorie Game, 413
Cal Q Late, 470
Campaign, 350
Canada's Prairie Wheat Game, 559
Canadian Civil War, 559
Canals, 430
Can of Squirms, 309
Can of Squirms (Old and New Testaments), 531-2
Career Development Program, 547
The Career Game, 611
Career Game 224, 547d
Cargelligo Topographic Map Game, 409-10
Caribbean Fisherman, 410
Case White, 486-7
Cassino 1944, 487
Catalyzer, 532
Cattle Town Game, 583
Cauldron, 494
Cell Game, 540
Cemetary Hill, 485
Centralized Power Game, 420
Centurian, 500
Chancellor, 373
Change Agent, 41, 342-3d
Changing High School, 386d, 393
Charge, 337
Charge (Personal Finances), 335-6
Chariot, 487d, 524
Chattanooga, 485
CHEBO, 41, 343d
Chem Bingo, 540
Chem Chex, 541
Chem E-Z, 542
Chem Trak, 541
Chemical Families, 544
Cheminoes, 541
Chemistry's Alphabet, 544
The Chesterfield Training Exercise, 611
Chickamauga, 485
Chinese Farm, 502
Ch'ing Game, 142, 420-1
Choice, 386
Choices, 547-8
Choose, 307
Circulation, 541
Citadel, 487
Cities/USA,
City, 579
City I, 579-81
City IV, 581
City Hall, 49, 350d
City Model, 287, 581-2d
City Planning, 582-3
Class Struggl, 559
Classical State System, 456-8
Classification, 544
Classifying the Chemical Elements, 541-2
Classroom Games in French, 460
Clinical Simulations, 133, 413d
Clot, 413
CLUG (Community Land Use Game), 11, 287, 583d, 585, 589
Coalition, 351
Coal 1, 329
Cobra, 487
Code, 460
Coexist, 329
Coffee Game, 375, 532d
Cognitive Psychology, 41, 343d
Collective Bargaining, 69, 374, 611d
College Game, 548
Collision, 351
Colony, 419-20
Combat Command, 490
COMCO, 262, 558-9d
Committee, 268, 351d
Communicating for Results–Parts I and II, 611-12
Communication, 612
Communications: Problems & Opportunities, 612
Communication Systems Simulation, 311
The Community, 374
Community, 559
Community X, 316
Community Target–Alcohol Abuse, 133, 307d
Compacts, 316-17
COMPETE, 337-8
Compete, A Dynamic Marketing Simulation, 178, 633d
Competition Amongst the Metals, 544
Components of Schooling, 546
Compounds, 542
Computer Augmented Cases in Operations and Logistics Management, 604
Computer Models in Operations Management, 604
Computer Models in Operations Research, 605
Computer Simulation Policy Games in Macroeconomics, 332
Computers in the Biology Curriculum, 338
Conceptual Mapping Game, 288, 396d
Conex, 455
Configurations, 470
Conflict Game, 561
Conflict in the Middle East, 447
Confrontation, 100, 397
Confrontation in Urbia, 464-5
Confrontation: The Cuban Missile Crisis, 163, 447-8d
Confronting Culture, 535
Congress, 351
Congress of Vienna, 142, 421d, 430
Congressman at Work, 348-9
The Conquerors: The Macedonians and the Romans, 188
Conquistador, 188
Consensus Game, 561
Conservation Crisis, 463
Constitution, 351-2
Constitutional Convention, 348-9
Constructing a Life Philosophy, 404
Constructing a Political Philosophy, 404
Construction Game, 612
Constructive Discipline, 612-13
Consumer Decision, 526
Consumerism II, 633
Consumerism II, 633
Consumer Redress, 526-7
Contemporary International System, 456-8
Contracept, 413-14
Contract Negotiations, 613
Contractors, 559-60
Cooperative and Competitive Communication, 311
Cope, 111, 401-2d
Coral Sea, 488
Costs of Living, 549
Council, 352
Counseling, 613
Country Development Economics & Finance Game, 373
Court Policy Negotiations, 465
Covent Farm Game, 560
Crimea, 488
Crimea 1941, 488-9
Crisis, 163, 448-9d
Crisis in Lagia, 560
Crisis in Middletown, 317
Critical Incidents in Education, 41, 333d
Cromwell, 489
Cruel Cruel World Value Game, 224, 548d
Crusader, 404
Cryteria!, 613-14
Culture Contact II, 560-1
Curfew, 571
Custer's Last Stand, 489
Cycle, 548
Czar Power, 142, 421-2d
Czech-Mate, 142, 422d

Dangerous Parallel, 449
Dark Ages, 520
Dart Aviation, 39, 648
Datacall: A Computer Based Game for Teaching Strategy, 345
D-Day, 489
Dead River, 87, 366
Dealing with Death, 404
Decision, 614
Decision, 614
Decision Guides, 614
Decisionmakers, 317
Decision Making, 614
Decision Making by Congressional Committees, 348-9
Decision Making Exercise, 596
Decision Making for Career Development, 548
Decision Making Model, 449
Decision Mathematics Operational Game,

345-6
Decision Point, 561
Decisions, 224, 549d
Decisions, Decisions, 535
Decline and Fall, 188, 489-90
Defense, 352
DeKalb Political Simulation, 352
Delegate, 352-3
Delegation, 614-15
Democracy, 355
Dentopoly, 415
Descent on Crete, 490
Desert War, 490
Design, 527
Destiny, 142, 422
Determining America's Role in the World, 404
Determining Economic Values, 404
Determining Family and Sexual Roles, 404
Development, 419
Development of the Medieval Town, 430
Developort, 410
Dialogues on What Could Be, 402
Dieppe, 490
Dig 561
Digestive System, 544
Dignity, 317
Dilemma, 309
Diplomacy, 189, 490-1
Diplomatic/Foreign Policy Game, 449-50
Diplomatic Practices, 450
Discovery, 422
Disposition Exercise, 465
Dissent and Protest—The Montgomery Bus Boycott, 422-3
Distribute, 605
District Nutrition Game, 133, 414d
Disunia, 423
Division, 423
DMZ, 502-3
Dollars in Demand, 463
Dot Exercise, 535
Drang Nach Osten! and Unentschieden, 491
Dreadnought, 491
Drive on Stalingrad, 491
Drug Attack, 307
Dynamic Modeling of Alternative Futures, 110, 402d

Eagle Day: The Battle of Britain—1940, 491-2
East Front, 492
Eco-Acres, 287, 583-4d
Ecological Modeling, 329
Economic Decision Games, 374-5
Economic System, 69, 375d
Economy Game, 375
Ecopolis, 59, 366d
Edision Elementary Principalship, 390
Edplan, 111, 386d
Education for Justice, 375
Educom Financial Planning Model, 333
Edventure, 386-7
Edventure II, 111, 387d
Effective Delegation, 615
Effective Supervision, 615
Effective Supervision in Government, 615
1815: The Waterloo Campaign, 492
El Barrio, 443
Elect, 326-7
Elect a President, 353
Election, 49, 353
Election USA, 353-4
Electric Circuit, 544
Electronic Industry Game, 596
Elements, 542
Empire, 387
Empire, 423-4
En Garde!, 492
Encapsulation, 100, 309-10
End of the Line, 80, 443d
Energy, 41, 327
Energy Conversions, 544
Energy Environment Game, 87, 366
Energy X, 87, 366-7
English Civil War, 492
Enterprise, 376
Environmental Simulations, 367
ENZKIN, 338
Equality, 561
Equations, 470
Equipment Evaluation, 605
Equity and Efficiency in Public Policy, 376
Ernstspiel, 310
Escape Routes, 464
Espionage, 424
Esso Service Station Game, 647
Ethics, 532
Euro-Card, 561-2
European Enviornment 1975-2000, 402-3
Everybody Counts!, 133, 442d
EVOLUT, 338
Examining American Values, 404
Exchange, 376
Exchange, 376
Executive Decision, 376-7
Executive Decision Making Through Simulation, 279, 596-7d
Ececutive Game, 279, 597d
Executive Secretary, 615-16
Executive Simulation, 279, 597d
Executive Simulation Game, 600-1
Exercises in Personal and Career Development, 549
Exmark, 601
Exodus, 532-3
Experiential Education, 533
Exper-Sim, 41, 346d
Exploring Careers, 549
Exploring the New World, 424
Explosion, 367
Expressway, 268, 367d
Extinction, 542
Eye Contact, 311
E-Z Science Games, 542

Facing the Ecology Crisis, 404
Facts in Five, 100, 397d
Fail-Safe, 268, 343d
Fair/No Fair, 464
Faith in Darkness, 535
Fall of Tobruk, 190, 492-3
Family, 310
Family Contract Game, 310
Family Tree, 562
Fast Carriers, 190, 493
Federalist or Anti-Federalist, 428
Federalists vs. Republicans, 424
Female Images: A Life Skills Exercise, 549-50
Fifties, 424-5
Fight in the Skies, 493d, 510
Finance, 640
Financial Analysis, 640
Finansim, 641
Finmark, 641
Fire, 465
Firefight, 493
The Firm, 70, 374d
First Amendment Freedoms, 465
First World War, 493-4
Five Simple Business Games, 648
Fixit, 387
Flight, 268, 562d
Flight 108, 533
F.L.I.P. 2/80, 527
Flying Tigers II, 494
Focus on the Future, 402
Foreign Policy Decision Making, 450
Forest Service Fire Control Simulator, 646
Forest Simulator, 646
FORMENTO, 377
Formissia, 215
Formulon, 543
Fortress Rhodesia, 494
Four Battles from the Crimean War: Alma, Balaclava, Inkerman, Tchernaya River, 494
Four Battles in North Africa: Crusader, Cauldron, Supercharge, Kasserine, 494-5
Four in a Row, 470
Foxbat and Phantom, 495
Fredericksburg, 485
Frederick the Great, 189, 495
Freiberg, 517
French Holiday Travel Game, 461
Freshman Year, 388
Fresh Oven Pies, 648
Frigate, 495, 523
Frontier, 419
Frontier, 430
Fulda Gap, 495
Function, 601
Future Decisions: The I.Q. Game, 403-4
Future Game, 404
Future Planning Games, 110, 404-5d
Futuribles, 110, 405d

Gambit, 584
Game of City Government, 583
GAMEgame, 397
GAMEgame II, 397
GAMEgame IV, 100, 397-8d
GAMEgame VI, 100, 398d
Game of Industrialization, 582
Game of Nations, 450-1
Game of People Working in a City, 583
Game of Time Management, 616
Game of Transactional Analysis, 616
Game of Warring Castles, 582
Games for Growth, 387
Gaming, 533
Gateway, 142, 425d
A General Agricultural Firm Simulator, 644

General Strike, 430
Geologic Time Chart Game, 543
Geometric Playthings, 471
Gerontology, 610
Gerrymandering, 610
Gestapo: A Game of the Holocaust, 215, 533d
Gettysburg, 496
Ghetto, 41, 258, 343-4d
Ghetto, 317-19
Global Futures Game, 405
Global War, 496, 524
Going Up: The Israeli Game, 533-4
Golan, 502
Gold Rush, 425
Gold Rush Days, 425
Gomston, A Polluted City, 87, 367-8d
Good Loser, 414
Good Federalism Game, 354
Good Society Exercise: Problems of Authority, 451-2
Gorgeous Gateaux, 648
Gospel Game, 534
Government Reparations for Minority Groups, 354
Grain Drain, 377
Grand Frame, 287, 584d
Grand Strategy, 142, 163, 387, 425-6d
Grand Strategy (Fixit), 387
Graze, 329
Great Blood Race, 544
Great Periodic Table Race, 543
Great Plains Game, 426
Greenham District Council, 319
Greenham Gypsy Site, 319
Grenadier: Company Level Combat 1700-1850, 496
Grievance Handling (Industrial), 616
Grievance Handling (Non-Industrial), 616
Growth of a City Game, 583
GSPIA: An Urban Regional Development System, 584-5
Guerilla Warfare, 426
G.U.L.P.: The Growth of Urban Land and Population, 585
Guns or Butter, 452

Haber, 339
Halfway House, 319-20
A Handbook of Structured Experiences for Human Relations, 310-11
Handling Conflict in Hospital Management (Conflict among Peers), 414
Handling Conflict in Hospital Management (Superior/Subordinate Conflict), 414-15
Handling Conflict in Management: Conflict among Peers, 617
Handling Conflict in Management: Superior/Subordinate Conflict III, 617
Handling Conflict in Management: Superior/Subordinate Group Conflict II, 617
Hang-Up, 224, 550d
Hannibal, 496
Hardy, 338-9
Harvest Politics, 430
Hat in the Ring, 49, 354-5d
Hawkesworth, 610
Haymarket Case, 142, 426d
Health Games Students Play, 415
Hebrews in Exile, 537
Here to There, 471
Heritage, 427
Herstory, 239, 562d
Hexagon, 287, 585
Highway to the Reich, 497
Home Economy Kit, 528
Homefront, 427
Homestead, 427
Honshu, 410
Hooker & Lee, 485
Horatio Alger (A Welfare Simulation Game), 443
Hospitex, 415
House Decision Game, 528
House Design Game, 528
Household Energy Game, 528
Househunt, 528
Hovercraft, 610
Human Communication Handbook, 311
Human Survival–2025, 405
Humanus, 262, 562-3d
Hunting Game, 563
Hunting-Gathering Game, 582-3
Huri, 610
Hurtgen Forest, 522
Hybrid-Delphi Game, 110, 406d
Hypothetica, 24
H.Z. Zilch, 460-1

Identity, 224, 563d
IDGAME, 339
IDIOM, 336-7
Imaginit Management Game, 279, 597d
IM-CLUG, 585-6
Impasse Simulation, 391
IMPASSE (The Impact Assessment Game), 288, 398d
Imperialism, 453
Imperialism, 497
Imprinting, 346
In-Basket Exercise, 647
In-Basket Simulation Exercises, 388
Independence, 427
Independence, 563
Independence '76, 428
Indian Reservation, 563-4
Indian Valley, 368
Indians View Americans; Americans View Indians, 428
Indrel (Industrial Relations Exercise), 617-18
Industrial Marketing Plan Simulation (IMPS), 633
Industrial Sales Management Game, 633-4
Inflation, 378
INHABS 3, 287, 584, 586d
Inkerman, 494
Inner-City Housing Game, 268, 320d
Inner-City Simulation Laboratory, 388
Innocent Until . . ., 465-6
In Order, 471
In Quest, 543
Ins and Outs, 464
INS 2: Inter-Nation Simulation, 41, 163, 334d
Instruction, 618
Instructor as Manager of Learning Experiences, 618
Integrated Simulation, 279, 597-8d
Intentional Community, 564
Interact, 311
Interact II, 24, 311-12d
Interaction, 388-9
Interaction, 618
Interaction–A Balance of Power in Colonial America, 428
Interceptor, 471
International Conflict, 454
International Trade, 374
Inter-Nation Simulation Kit, 452-3
INTERP, 41, 339d
Intervention, 419
Interviewing, 618
In the Chips, 564-5
Intop, 279, 598d
Introduction to Managerial Accounting, 641
Invasion America, 497
Inventing and Playing Games in the English Classroom, 460-1
Inventory Simulation (INSIM), 605
Invest, 378
Involvement, 428
Ironmaster, 430
Island, 378
Island Game, 648
Island War, 497
It's a Free Country, 464

Jail Puzzle, 466
Janus Jr. High School Principalship Simulation, 390
Japanese-American Relocation, 1942, 428-9
Jeffrey's Department Store, 649-50
Jeremiah in Prison, 537
Jerusalem!, 497
Jerusalem '67, 502-3
Job Crunch, 240, 389d
Job Enrichment–Redesigning Jobs for Motivation, 618-19
Job Experience Kit, 550
Joe Cool, 610
Judgment, 429
Juris, 466
Jury Game, 466-7
Justice Game, 467

KAMA: A Simulation Dealing with Post-Colonial Africa, 565
Kampfpanzer, 490
Kasserine, 494
Kasserine Pass, 497-8
Kathal: Game of Creative Intimacy, 550
Keeping Warm, 544
Keep Quiet, 415
Kharkov, 498
Kibbutz, 534
Kids in Crisis, 467
Kingmaker, 188, 498d
King Tut's Game, 429
Königgrätz, 498
Korea, 498-9
KSIMS, 339-40
Kursk, 499

Lab Apparatus, 543-4
Lab Bataille de la Moskowa, 499
La Grande Armée, 499
Label Game, 389d, 402
Lakemont High School,
Land Use, 59, 368
Leadership, 619
Leadership Decision Game, 619
Leading Groups to Better Decisions, 619
Lee Moves North, 499
Leela, 215, 534-5
Legisim, 355
Legion, 500d, 524
Le Pays de France, 461
Lester Hill Office Simulation, 649
Let's Build a House, 549
Let's Play Games in French, 461
Let's Play Games in German, 461
Liberté, 142, 429d
Librarian's Game, 443-4
Life Cycle, 444
Life Goals, 456
Life in the Colonies, 429-30
Lifestyle, 456
Limits, 41, 344d
Linkover, 41, 340d
Live Sculpture, 415
The Living City/La Cite Vivante, 586
Lobbying Game, 355
Lockey, 340
Long Range Planning, 620
Longman Geography Games, 410
Longman History Games, 430
Longman Science Games, 544
Low Bidder, 601
Low Income Housing Project, 320
Luftwaffe, 500
Lutzen, 517

MA 501, 332
Madison Assistant Superintendent for Business Management, 390
Madison Assistant Superintendent for Instruction Service, 390
Madison Avenue, 379
Madison Avenue, 460-1
Madison Secondary Principalship, 390
Madison Simulations, 390
Madison Superintendency, 390
MADS-Bee Managing Dynamic Small Business, 598-9
Mahopa, 565
Majority Rule Polyarchy, 561
Make Your Own World, 368-9
Making a Change, 100, 320d
Malar, 344
Manage, 332
Management: An Experiential Approach, 601
Management Game, 279, 598d
Management by Objectives, 620
Management by Objectives for Insurance Companies, 643
Management for Supervisors, 621
Management in Government, 620
Managing and Allocating Time, 621
Managing the Engineering Functions, 621
Managing the Manufacturing and Industrial Engineering Functions, 621-2
Managing the Worker, 622
Managing Through Face-to-Face Communication, 612, 622-3d
Managing Your Money, 528-9
Manassas, 500
Manchester, 430
Marbles Games, 443
Market, 332-3
Market, 374
Market, 529
Market Planning, 635-6
Market Strategy, 636
Marketing a New Product, 634
Marketing Dynamics, 178, 634d
Marketing in Action, 178, 634d
Marketing Interaction, 178, 634-5d
Marketing Planning and Strategy Game, 635
Marketing Strategy, 178, 635d
Marketit, 635
Marksim, 178, 636d
Markstrat, 178, 636d
Marriage Game, 240, 550-1d
Maspar, 327
Materials Inventory Management Game, 606
M.C.C., 610
Mech War '77, 500
Medieval Town Game, 582
Meet the Prophets, 537
Meeting Game, 623
Megga Trading Company, 601-2
Mental Health, 551
Mercenary, 500-1
Merchant, 431
Meridian 36, 411
Metal Box Business Game, 649
Metrication, 471-3
Metric Poker, 472
Metri-Magic, 472
METRO, 288, 577, 586-7d
Metro-Apex, 288, 586, 587d
METRO-CHP, 346-7
Metro Government, 355
Metroplex, 587
Metropolis, 586, 587-8d
Metropolitics, 49, 288, 355-6d
Microbes, 544
Micro-Community, 262, 378, 565-6d
Micro-Economy 262, 378, 566d
Micro-Napoleonics, 501
Microville, 320-21
Middle East Conflict Simulation Game, 163, 454-5d
Middle East Crisis, 455
Midtex, 605-6
Midway, 501
Mighty Fortress, 501
Mini-Apex, 586
Minuteman, 501-2
Missile Crisis, 502
Mission, 431
Mission Aloft, 502
Missionary Roads, 535
Mission Games, 535
Modern Battles: Wurzburg, Chinese Farm, Golan, Mukden, 502
Modern Battles II: Battle for Jerusalem '67; Yugoslavia, Bundeswehr; DMZ, 502-3
Modern State System, 456-8
Money Game, 375
Monopoly, 18, 378-9d, 382, 566
Monroe City Simulations, 390-1
Moot, 467
Mortgage, 529
Moscow, 367
Moscow Campaign, 503
Moses and the Exodus, 535
Motivation, 623
Motorway, 410
Mouse in the Maze, 544
Mr. President, 431
MTB, 503
Much Ado About Marbles, 321
Mukden, 502
Mukden 1905, 503
Mulberry, 321
Multifacto/Producto, 472
Mummy's Message, 431
Musket and Pike, 503d, 524
My Cup Runneth Over, 551-2
Myth Information (About Human Sexuality), 415-16

Naked Monsters, 391-2
Napoleon at War: Marengo, Jena-Auerstadt, Wagrom, and the Battle of the Nations, 504
Napoleon at Waterloo, 503-4
Napoleon's Last Battles: Ligny, Quatre Bras, Wavre, LaBelle Alliance (Waterloo), 189, 504d
NAPOLI (National Politics), 356-7
Narvik, 504
National Economy, 70, 374d
National Policy Game, 358
National Priorities, 358
Nationalism: War or Peace, 455
NATO: Operational Combat in Europe in the 1970's, 504
Negotiate, 623
Negotiation, 623
Negotiations Simulation, 391
Neutron Activation Analysis, 340
New City Telephone Company Simulation Game, 566-7
New England Town Puzzle, 582
Newscast, 461-2
Newton, 340-1
New Town, 288, 322d
New Town Professional Planner's Set, 588
Next War, 505
Nexus, 215, 535d
Nexus, 398
1944: The Invasion of France and the Battle of Germany, 505
Nisei Scouts, 429
No Dam Action, 87, 369
Noigeren, 410
Nordlingen, 517
Norman Conquest, 430
North Sea Exploration, 411
North vs. South, 431-2
Northeast Farm Management Game, 645
Nothing Never Happens, 312
Nourish, 133, 416d
Nuclear Energy Game, 322

Nuclear Site Negotiation, 322
Number Line, 472
Nuremberg, 142, 432
Nursing Crosswords and Other Word Games, 416
Nursing Home Care as a Public Policy Issue, 444
Nutrition, 544
Nutrition Game, 416

Objective Moscow, 505
October War, 505
Office Management, 623
Office Talk, 623-4
Oh-Wah-Ree, 472-3
Oil War, 505-6
O.K., 224, 552d
The O.K. Game, 312
Oklahoma Bank Management Game, 642
Old Cities–New Politics, 358-9
On the Move: Soviet Jewry, 536
On-Sets, 470, 473d
On Strike and Other Economic Games, 379
On Trial, 571
Opening the Deck, 567
Operacao Littorio, 506
Operation Encounter, 178, 636-7d
Operation Olympic, 506
Operation Suburbia, 624
Optimum Delegation, 624
Orbiting the Earth, 473
Oregon, 38, 347d
Oregon Farm Management Simulation, 646
Organization Game, 602
Origins of WW II, 142, 432-3d
Overlord, 506

P & S, 545
PABLUM (Public Administration Bureaucratic Laboratory for Upper Management), 328
Panic, 142, 433d
Panzer '44, 500, 502d
PanzerArmee Africa, 506
PanzerBlitz, 481, 490, 506-7d, 516
Panzergruppe Guderian, 507
PanzerLeader, 481, 507d
Parksburg: Garbage & Politics, 49, 359, 359-60d
Party Central, 433
Patrol!, 507
Peace, 433
Peloponnesian War, 508
Persecution, 215, 536d
Personalysis, 552
Personal Preference, 456
Personnel Assignment Management Games, 624
Personnel Department, 624-5
Personnel Management in Action, 625
Pex, 625-6
PH, 341
Phalanx, 514
Phantom Submarine, 455
Pharsalus, 508
Physical Distribution Management, 606
Picture Patterns, 473
Pink Pebbles, 379-80
Pioneers, 433-4
Pipeline: An Employment and Training Simulation, 444-5
Pipework Engineering, 626
Planafam II, 133, 416
Planet Management Game, 59, 369
Planned Maintenance, 606
Planning, 626
Planning American Policy in Developing Nations, 404
Planning an Inner-City High School, 323
Planning Exercise, 398-9
Planning for Growth, 626-7
Planning Game, 536
Planning Operational Gaming Experiment, 588-9
Planning Tomorrow's Prisons, 404
Planning Tomorrow's Society, 404
Plant Succession, 410
Play Yourself Free, 224, 552-3d
Plea: A Game of Criminal Justice, 467-8
Plot to Assassinate Hitler, 508
Pocket Calculator Boom, 599
Point of Law, 468
Polex, 455
Police State, 567
Policy, 327
Policy, 643-4
Policy Negotiations, 100, 204, 399d
Polis, 16, 163, 589d
POLIS Network, 334
Political Exercise, 455
Political Pollution, 369
Pollination Game, 545
Pollution, 87, 369-70d
Pollution Game, 87, 370
Pollution: Negotiating a Clean Environment, 87, 370d
Polsys, 327-8
Polut, 329-30
Polychoc, 637
Polyhedron Rummy, 473
POP, 330
Poppin'Swap, 416-17
Population, 411
Population Control, 463
Port Arthur, 511
Possum Creek Valley, 370
Potlatch Game, 567
Poultry Farm Management Game, 646
Poultry Game, 567-8
Poverty Game, 568
Power, 24, 568d
Power Game, 410
Power Politics, 360
Power Politics, 360-1
Powerplay, 553
Praise and Criticism, 536
Predator, 59, 545d
Prediction, 312
Prediction & Interpersonal Perception, 311
Presidential Election Campaign, 348-9
President's Select Commission, 407
Press Conference, 99, 399d
Pressure, 361
Preventing Crime and Violence, 404
Preventing the Civil War, 434
PRINCE, A Programmed International Computer Environment, 163, 335d
Principal Game, 389
Principles of Effective Salesmanship, 637
Principles of Effective Salesmanship for Insurance Agents, 644
Priority, 312-13
Priority, 627
Prisoner's Dilemma Game, 24
Probe, 553
Problem Sensing and Selection in Wilson Sr. High, 391
Problems in Bank Management, 642-3
Problems in Supervision, 627
Problem Solving, 627
Process, 553
Procurement Management, 606
Production Control Inventory, 606-7
Production Scheduling Management Game, 607
Productivity–Improving Performance, 627-8
Profair, 628
Profit and Loss, 379
Progressive Era, 434
Promotion, 419
Propaganda, 387
Propaganda Game, 268, 568d
Prophets and the Exile, 536-7
Pro's and Con's, 100, 399d
Prosim, 607
Prospects, 111, 553-4d
Prosper, 628
Protecting Minority Rights, 404
Protection, 529-30
PSW-1/SW Political Simulation, 334-5
Psych City, 323d, 389
Psych-Out, 474
Psychiatric Nurse-Patient Relationship Game, 133, 417d
Punic Wars, 508
Purchasing, 607
Purdue Forest Management Game, 646-7
Purdue Supermarket Management Game, 643
Puzzle, 434-5
PX-190, 399-400

Quantitative Experimental Analysis, 41, 347
QUBIC (Three Dimensional Tic-Tac-Toe), 474
Queries 'n Theories, 462
Questioneze, 392

Radicals vs. Tories, 436
Rafa Rafa, 568-9
Raid, 508-9
Railroad Game, 69, 380-1d
Railway Mania, 430
Railway Pioneers, 411-12
Rainbow Game, 569
Raketny Kreyser, 509
Raphia, 217 BC, 188, 509d
Rats, 330
Real Numbers Game, 474
Reapportionment, 361
Rebirth: The Tibetan Game of Liberation, 215, 537d
Reckon, 474-5
Reconstruction, 419
Recruiting Effective Insurance Agents, 644

Red Men and White Men, 428
Red Star/White Star Tactical Combat in Europe in the 1970's, 500, 509d
Red Sun Rising, 509
Redwood Controversy, 59, 370-1d
Regular Meeting of the Wheatville Board of Directors, 392
Relocation: A Corporate Decision, 628
Remagen, 522
Remote Island, 412
Renaissance of Infantry, 524
The Republic, 648
Resusci-Ann, 133, 417d
Retailing Department Management Game, 637
Revolt at State Prison, 468
Revolt in the East, 509-10
Revolutionary Society (REVSOC) Simulation, 569
Revolutions Game, 435
Rhein Bung, 510
Richland USA, 16, 589d
Richtofen's War, 510
Right is Write, 462
Rip-Off, 268, 469d
Rip-Off, 530
Rise and Decline of the Third Reich, 510
RISKM, 641
River Dose (RVRDOS) Model, 330
RKINET, 341
Road Game, 313
Road to Richmond, 510
Roaring Camp, 435
Rocroi, 517
Role-Playing Methods in the Classroom, 392
Roll-a-Roll, 313
Russian Campaign, 511
Russian Civil War, 511
Russo-Japanese War: Port Arthur and Tsushima, 511

Sale-Plan, 637-8
Sales Management Game, 638
Sales Promotion, 638
Salesquota, 638
Sales Strategy, 638-9
SALT III, 163, 455-6d
Sanga, 569-70
Saturation, 462-3
Scarcity and Allocation, 70, 375d
SCATR, 341
Scatter, 341
Schizophrenia, 347
School Planning Game, 392
School Psychologist Simulation, 390
Sci-E-Z, 542
Science Sense, 544
SciFi, 101, 400d
Scramble for Africa, 142, 430, 435d
Search, 463
SEASIM, 393
SEATS, 392-3
Security, 456
Seelowe, 511
Selecting Effective Insurance Agents, 644
Selecting Effective People, 628-9
Selection, 629
Selection Interview Clinic, 629
Sellem, 639
Selling: A Good Way to Earn a Living, 648-9
Seneca Falls, 142, 240, 435-6d
Sensitivity, 629
Serfdom: A Simulation in Class Achievement, 570-1
Settle or Strike, 381
1787, 436
1776, 189, 511-12d
7th Cavalry, 512
Shakespeare, 463
Shantih, 537
Share the Risk, 530
Sharing Game, 554
Shelter, 336
Shiloh, 485
Shipwreck & Other Government Games, 268, 571d
Showprofit, 639
Shrink, 554
Sidi Rezegh 1941, 512
Siege, 512
Siege of Constantinople, 512
Siege of Jerusalem, 70 AD, 512-13
Simchip, 607-8
Simpolis, 361-2
SIMQ, 599
SIMQUEUE–A Queueing Simulation Model, 599-600
SIMSEARCH, 41, 344
SIMSOC, 248, 262, 571d
Simulated Agribusiness, 645-6
Simulating Social Conflict, 268, 571d
Simulating the Values of the Future, 110, 406d
Simulation Games for Religious Education, 537
Simulation on the Identification of the Gifted and Talented, 402
Simulex IV: Internation Simulation, 163, 456d
Sinai, 513
Site Budgeting Simulation, 391
Situational Leadership Simulator, 629-30
Sixth Fleet, 513
Skins, 436
Slave Auction, 437
Slave Coast Game, 437
Slits, 342
Small Business Management Exercise, 602
Small Town, 571-2
Smart Spending, 530
Smoke, 415
Snibbo Metal Products, 602
Sniblette, 642
Sniper, 507, 513d
Social Security, 313
Solar Cells, 610
Solar System 1 and 2, 544
Soldiers: WWI Tactical Combat, 1914-15, 189
Solomon's Campaign, 514
Solving Multi-Ethnic Problems, 393
Soma Cube, 475
Something of Value,
Soup's On, 417
Sourcebook of Experiential Exercises—Interpersonal Skills, 313-14
South Africa, 514
Southern Mountaineer, 572
South Street Hostel Reform, 323
Space Colony, 406-7
Space Hop, 545
Space Patrol, 110, 407d
Spanish-American War, 437
Spanish American War, 437-8
Spartan, 514d, 524
Southern Mountaineer, 572
South Street Hostel Reform, 323
Space Colony, 406-7
Space Hop, 545
Space Patrol, 110, 407d
Spanish-American War, 437
Spanish American War, 437-8
Spartan, 514d, 524
Spatial Marketing Simulation, 639
Special Educational Administration, 390
Special Educational Administration Task Simulation (SEATS), 392-3
Spectrum, 572
Speedcap, 549
Spiral, 381
Spiral, 586
Spitfire, 514-15
Sprawl, 586
Spring Green Motorway, 323
Sprods, Ltd., 608
Squad Leader, 515
Squat, 586
Squirms, 393-4
SSN: Modern Tactical Submarine Warfare, 515
Stanford Business Logistics Game, 608
Starpower, 16, 255, 262, 572-4d
Statehood, 362
State System Exercise, 456-8
Steady Job, 554
Sterl, 330-1
Stock Market Game, 69, 381-2d
The Stock Market Game, 428
Stocks and Bonds, 69, 382
Stonewall, 515
Strike, 438
Strike Force One, 190
Submarine: Tactical Level Submarine Warfare, 1939-45, 516
Suburban Operations Simulation, 324
Summit, 458
Sun Dance and Ghost Dance, 428
Sunshine, 324
Supercharge, 494
Superintendency Simulation, 390-1
Supermarket Strategy, 643
Super-Port, 410
Super Sandwich, 417
Supervision, 630
Supervisory Skills, 630
Supclug, 589
Swindle, 530-1
System, 574
System 1, 101, 400d
Systems, 288, 589-90d

Tac-Tickle, 475
Tactics II, 516

TAG, 331
TAG: The Transactional Awareness Game, 314
Tag: You're Who, 535
Taking Action, 463
Talking Rocks, 463-4
Tank, 516
Tannenberg, 516
Tannenburg, 516-17
Taxes, 49, 362d
Taxis for Sale, 463
Tchernaya River, 494
Teacher Education Can of Squirms, 393-4
Teacher's Lounge, 394
Teacher Training Simulation, 394
Teaching Problems Laboratory, 394
Teaching Strategy, 395
Teaching Styles, 215, 537-8d
Teaching the Future, 407
Tea Clipper Race, 410
Teams-Games-Tournament, 101, 400d
TeleCLUG, 590
Tempomatic IV, 279, 600d
Ten Commandments Bible Game, 538
Tenement, 324
Tenure, 395
Terrible Swift Sword, 515, 517
Territorial Sea, 458
Terry Parker, 445
Thanatos, 224, 538d
Thermopylae, 517
They Shoot Marbles, Don't They?, 101, 258, 262, 321, 400-1d
Thirty Years' War: Lutzen, Hordlingen, Rocroi, Freiberg, 188, 517
325 AD, 538-9
ThyngummyJiggs Ltd., 642
Tightrope, 69, 382d
Timao, 554
Time Capsule, 574-5
Time Management, 630
Titration, 342
To Be Somebody, 464
Tobruk, 516, 517-8d
To Drink or Not to Drink, 308
Together, 575
Tomorrow's Secretary, 630
Tool Room Game, 608
Top Operating Management Game, 602-3
Torgau, 518
To Stay or Go, 215
Toward Walden Two, 402
Tracy Congdon, 445
Trade, 382
Trade and Discovery, 430
Trade Off, 590
Trade Off at Yalta, 142, 438d
Traders Arrive on the African Scene, 575
Transaction, 382-3
Transactional Analysis–Improving Communications, 630-1
Transfer of Development Rights Game, 590
Transit, 590-1
Transport in Plants, 544
Trebides Island, 384
Trial: Legalizing Marijuana, 415
Trial of George Washington, 438
Trial of Harry S. Truman, 438-9
Trial of Jefferson Davis, 439
Triangle Trade, 439
Trig, 603
Tri-Nim, 475
Trip to Mars, 395
Tri Score, 475
Trip or Trap Bingo, 308
Trireme, 518
Troy, 518
Trust, 314
Tsushima, 511
Tuf, 476
Turning Point, 518-19
Twenty-First Year, 575
Twixt, 476
2000 AD–Future City, 111, 407-8d
Two Years of Horror, 408
Tycoon, 603

U-Boat, 519
U-Dig, 288, 591d
Uhuru, 575-6
Unentschieden, 491
The Ungame, 314-15
Union Divides, 439
Union Organizing Game, 631
Unitex, 608
Up Caste Down Caste, 576
Up Scope!, 519
Upper Mid-West Dairy-Hog Farm Management Game, 646
Uprising Behind Bars, 463
Urban America, 324-5
Urban Dynamics, 288, 591-2d
Urbanisation, 410
Uses of the Sea, 163, 458-9
Using Biblical Simulations, 539
USN, 519
Uspop, 344
Utopia, 111, 408d

Valleybrook Elementary School, 393
Value Bingo, 554-5
Value Game, 555
Value Options, 555
Values, 555
Veracruz, 519
Verdun, 519-20
Victory at Sea, 520
Victory in the Pacific, 520d, 521
Viking, 520d, 524
Village Enclosure, 430
Virtue, 439-40
Vitamins, 417-18
Viva!, 520
Votes, 362-3
Vows, 468-9

Wacht am Rhein, 520-21
Waging Neutrality, 142, 440
Walk in my Shoes, 464
War at Sea, 521
War Between the States, 521
War Crimes Trial, 440
War in the East, 492
War in Europe (War in the West, War in the East), 521-2
Warlord Game, 522
War Time, 459
Washington University Business Game, 603
Watergate: The Waterloo of a President, 440
Water Cycle, 544
Waterloo, 522
Weather Forecasting, 410
Web, 395
Weekly News Game, 576
Welfare Week, 445
Wellington's Victory, 522
Westwall: Arnhem, Hurtgen Forest, Bastogne, Remagen, 522
WFF: Beginner's Game of Modern Logic, 476
WFF'N PROOF: The Game of Modern Logic, 470, 476-7d
Whales, 371
Wheels, 41, 336d
Wheels, 418
When the Union Knocks . . ., 631
Where do you Draw the Line, 555
WHIPP (Why Housing is a Problem and a Priority), 212, 325
Wilderness Survival, 535
Wildlife, 545-6
Wilson High School Principalship Simulation, 390
Wilson RV, 631-2
Winter War, 522-3
Wolfpack, 523
Women in Management, 632
Women on Patrol, 463
Women's Liberation, 49, 363d
Woodbury Political Simulation, 363-4
Wooden Ships and Iron Men, 523
Workshop Management Training Simulation, 445-6
World, 411
World, 459
World Game, 405, 408d
World Political Organization, 459
World War I, 189, 523
World War I, 440
World War II, 189, 523-4
World War III, 524
World Without War Game, 576
Wurzburg, 502

Yalu, 524
Yeoman, 500, 514, 524d
Yin Yang, 315
You be the Mayor, 571
You Design a City, 583
Your Community's Economic Development, 325
You're the Banker, 383-4
Yugoslavia, 502-3

PRODUCER INDEX

Abington Press
Experiential Education, 533
Gaming, 533
ABT Publications
Life in the Colonies, 429-30
Academy of Health Professions
Life Cycle, 444
Active Learning
The Imaginit Management Game, 597
Adams, Harold W.
Country Development Economics and Finance Game, 374
Addison-Wesley Publishing Co.
Operation Suburbia, 624
Problems in Bank Management, 642-3
Problems in Supervision, 627
Administrative Research Associates
Personalysis, 552
Alabama, University of
See Institute of Higher Education
Alberta School Book Branch
Canada's Prairie Wheat Game, 559
Alexander, Ernest R.
FORMENTO, 377
Allyn and Bacon
Simulating Social Conflict, 571
Alternatives in Religious Education, Inc.
Gestapo–A Game of the Holocaust, 533
Going Up: The Israel Game, 533-4
On the Move: Soviet Jewry, 536
American Forest Institute
Indian Valley, 368
Possum Creek Valley, 370
American Friends Service Committee
Decisionmakers, 317
The Twenty-First Year, 575
American Political Science Associates
Arab-Israeli Conflict, 447
American Training Laboratories
COMCO, 558-9
Amidon, Paul S. and Associates
Charge, 335-6
Inflation, 378
Mulberry, 321
Shelter, 336
Wheels, 336
Ampersand Press
AC/DC, 540
Cheminoes, 541
Formulon, 543
P & S, 545
The Pollination Game, 545
Predator, 545
Anchor Press/Doubleday
Rebirth: The Tibetan Game of Liberation, 537
Anti-Defamation League of B'nai B'rith
Changing High School, 386
Solving Multi-Ethnic Problems, 393
Art Fair, Inc.
Ethics, 532
Shrink, 554
ASIP, Inc.
Family Contract Game, 310
Attack International Wargaming Association
Arms Race, 482
Missile Crisis, 502
Operacao Littorio, 506
Rhein Bung, 510
7th Cavalry, 512
Victory at Sea, 520
Avalon Hill Co.
Afrika Korps, 479
Alexander the Great, 480
Arab-Israeli Wars, 481
Battle of the Bulge, 484
Blitzkrieg, 485
Business Strategy, 372-3
D-Day, 489
Diplomacy, 490-1
Executive Decision, 376-7
Facts in Five, 397
Gettysburg, 496
Kingmaker, 498
Luftwaffe, 500
Midway, 501
Oh-Wah-Ree, 472-3
Origins of World War II, 432-3
Panzerblitz, 506-7
PanzerLeader, 507
Point of Law, 468
Richtofen's War, 510
Rise and Decline of the Third Reich, 510
Russian Campaign, 511
1776, 511-12
Shakespeare, 463
Squad Leader, 515
Stock Market Game, 381-2
Stocks and Bonds, 382
Submarine: Tactical Level Submarine Warfare, 1939-45, 516
Tactics II, 516
Tobruk, 517-18
Tuf, 476
Twixt, 476
Victory in the Pacific, 520
War at Sea, 521
Waterloo, 522
Wooden Ships and Iron Men, 523
Avery Publishing Group
Interact II, 311-12

Balboa Game Co.
Battles for Tobruk, 484-5
Bare, Bruce
Purdue Forest Management Game, 646-7
Bartlett School of Architecture and Planning
G.U.L.P., 585
Battleline Publications
Air Force, 480
Custer's Last Stand, 489
Behavior Sciences Education Center
Chesterfield Training Exercise, 611
Behrman House, Inc.
Exodus, 532-3
Bell, Ervin J.
U-Dig, 591
Benefic Press
Market, 529
Berry, Norma
Datacall, 345
Bobbs Merrill Co., Inc.
Consumer, 526
Economic System, 375
Ghetto, 317-19
Boise Interagency Fire Center
Forest Service Fire Control Simulator, 646
Boulogne, Jacob
Trust, 314

British Columbia, University of
Stanford Business Logistics Game, 608
Broadman Supplies
Nexus, 535
Teaching Styles, 537-8
Brotherhood Commission
Mission Games, 535
Bumpas, Jim
Bay of Pigs, 485
Mission Aloft, 502
Business Publications, Inc.
The Business Management Laboratory, 595
Compete: A Dynamic Marketing Simulation, 633
Business Studies, Inc.
Low Bidder, 601

Cadaco, Inc.
Economy Game, 375
King Tut's Game, 429
Ten Commandments Bible Game, 538
Caldwell, Dan
SALT III, 455-6
California American University Center for Leadership Studies
Situational Leadership Simulator, 629-30
California, University of
El Barrio, 443
Future Game, 404
Wildlife, 545-6
California, University of Southern
See COMEX
Campbell, Forrest M.
Tool Room Game, 608
Campbell, James H.
The Artificial Society, 372
Canadian Social Sciences Services
Metro Government, 355
Cardinal Printers
Pink Pebbles, 379-80
Carolina Biological Supply Co.
Blood Flow, 540
Cell Game, 540
Extinction, 542
Pollution Game, 370
Cassell Australia Ltd.
Cargelligo Topographic Map Game, 409-10
Center for Applied Research in Education
Crisis in Lagia, 560
Center for Health Games and Simulations
Contracept, 413-14
Community Target–Alcohol Abuse, 307
Center for Innovations in Teaching the Handicapped
GAMEgame, 397
Naked Monsters, 391-2
Center for International Programs and Comparative Studies
Alpha Crisis Game, 418
Center for Simulation Studies
IM-CLUG, 585-6
Metroplex, 587
POLIS, 589
PX-190, 399-400
Richland, USA, 589

Changing Times Education Service
Budgeting Game, 525
Coalition, 351
Consumer Redress, 526-7
Hat in the Ring, 354-5
Househunt, 528
Invest, 378
Share the Risk, 530
Swindle!, 530-31
Chen, Henry C. K.
Market Planning and Strategy Game, 635
Civic Education Aids
Election USA, 353-4
Independence '76, 428
Clark, Carolyn Chambers
Psychiatric Nurse-Patient Relationship Game, 417
Class Struggle, Inc.
Class Struggle, 558
Classroom Dynamics Publishing Co.
Confrontation in Urbia, 464-5
Micro-Community, 565-6
Urban America, 324-5
Clearinghouse for Experiential Exercises
Sourcebook of Experiential Exercises: Interpersonal Skills, 313-14
Cleaver, Thomas G.
Whales, 371
Cohen, Richard B.
Phantom Submarine, 455
Remote Island, 412
College of Agriculture and Environmental Science
Transfer-of-Development Rights Game, 590
COMEX Research Project
Apex, 577
Court Policy Negotiations, 465
METRO-APEX, 587
METRO-CHP, 346-7
Community Change, Inc.
Suburban Operations Simulation, 324
Community Service Volunteers
Greenham District Council, 319
Greenham Gypsy Site, 319
South Street Hostel Reform, 323
Spring Green Motorway, 323
Terry Parker, 445
Tracy Congdon, 445
CONDUIT/Central
Business Decision Simulation, 610
Change Agent, 342-3
Coexist, 329
Cognitive Psychology, 343
COMPETE, 337-8
Computer Models in Operations Management, 604
Computer Models in Operations Research, 605
Computer Simulation Policy Games in Macroeconomics, 332
Computers in the Biology Curriculum, 338
Critical Incidents in Education, 333
Ecological Modeling, 329
Energy, 327
ENZKIN, 338
EVOLUT, 338

Exper Sim, 346
Ghetto, 343-4
Haber, 339
IDGAME, 339
Imprinting, 346
INS 2: Inter-Nation Simulation, 334
Interp, 339
KSIMS, 339-40
Linkover, 340
Neutron Activation Analysis, 340
Newton, 340-1
Quantitative Experimental Analysis, 347
RKINET, 341
Scatter, 341
Schizophrenia, 347
SIMQUEUE, 599-600
SIMSEARCH, 344
Spacial Marketing Simulation, 639
Titration, 342
Conflict Game Co.
Fall of Tobruk, 492-3
Kasserine Pass, 497-8
Overlord, 506
Verdun, 519-20
Yalu, 524
Conklin, Frank S.
Oregon Farm Management Simulation, 646
Conservation Trust
European Environment: 1975-2000, 402-3
Contemporary Drama Service (See also: Arthur Merriwether, Inc.)
Can of Squirms, 309
Can of Squirms (Old & New Testaments), 531-2
Persecution, 536
Teacher Education Can of Squirms, 393-4
Control Box, Inc.
East Front, 492
Cooke, Ernest F.
Industrial Marketing Plan Simulation, 633
Council for Exceptional Children
Everybody Counts!, 442
Covert, Anita
Covert Farm Game, 560
Coward, McCann and Geoghegan, Inc.
Leela, 534-5
C.R.A.C.
Esso Service Station Game, 647
Creative Educational Services
Weekly News Game, 576
Creative Learning Systems, Inc.
Auction, 546
Choice, 386
Dilemma, 309
Encapsulation, 309-10
Priority, 312-13
Pro's and Con's, 399
Trip to Mars, 395
Creative Publications, Inc.
Reckon, 474-5
Creative Teaching Associates
Bank Account, 525
Big Deal, 469
Budget, 525

TriScore, 475
Culley, James
Admag I, 632-3
CUNA Mutual Insurance Co.
Managing your Money, 528-9
Curriculum Associates, Inc.
Right is Write, 462

Dagg, David
Algonquin Park, 365
Dartmouth College
Acres, 577
Davison, William
MA 501, 332
Decision Associates
Decision Making Exercise, 596
Denison Simulation Center
Hunting Game, 563
Job Crunch, 389
Denoyer-Geppert, Inc.
Armada, 420
Empire, 423-4
Didactic Systems, Inc.
Appraisal by Objectives, 609-10
Assigning Work, 610
Collective Bargaining, 611
Communicating for Results Parts I and II, 611-12
Communications: Problems and Opportunities, 612
Constructive Discipline, 612-13
Decision Guides, 614
Decision Making, 614
Economic Decision Games, 374-5
Effective Delegation, 615
Effective Supervision, 615
Effective Supervision in Government, 615
Equipment Evaluation, 605
Executive Secretary, 615-16
Financial Analysis, 640
Grievance Handling (Industrial), 616
Grievance Handling (Non-Industrial), 616
Handling Conflict in Hospital Management (Conflict among Peers), 414
Handling Conflict in Hospital Management (Superior/Subordinate Conflict), 414-15
Handling Conflict in Management: Conflict among Peers, 617
Handling Conflict in Management: Superior/Subordinate Conflict (Game III), 617
Handling Conflict in Management: Superior/Subordinate Group Conflict, 617
Hospitex, 415
Industrial Sales Management Game, 633-4
Instructor as Manager of Learning Experiences, 618
Interviewing, 618
Job Enrichment, 618-19
Leading Groups to Better Decisions, 619
Long-Range Planning, 620
Management by Objectives, 620
Management by Objectives for Insurance Companies, 643
Management for Supervisors, 621
Management in Government, 620
Managing and Allocating Time, 621
Managing the Engineering Function, 621
Managing the Manufacturing and Industrial Engineering Functions, 621-2
Managing through Face-to-Face Communication, 622-3
Market Planning, 635-6
Market Strategy, 636
Marketit, 635
Materials Inventory Management Game, 606
Office Management, 623
On Strike and Other Economic Games,
Optimum Delegation, 624
Personnel Assignment Management Game, 624
Physical Distribution Management, 606
Planned Maintenance, 606
Planning for Growth, 626-7
Principles of Effective Salesmanship, 637
Principles of Effective Salesmanship for Insurance Agents, 644
Procurement Management, 606
Production Control–Inventory, 606-7
Production Scheduling Management Game, 607
Productivity–Improving Performance, 627-8
Purchasing, 607
Recruiting Effective Insurance Agents, 644
Retailing Department Management Game, 637
Sale Plan, 637-8
Sales Promotion, 638
Sales Strategy, 638-9
Selecting Effective Insurance Agents, 644
Selecting Effective People, 628-9
Small Business Management Exercise, 602
Supervisory Skills, 630
Time Management, 630
Tomorrow's Secretary, 630
Top Operating Management Game, 602-3
Transactional Analysis–Improving Communications, 630-1
Women in Management, 632
Didactron, Inc.
Wheels, 418
Digital Equipment Corp., Software Distribution Center
Buflo, 328-9
Charge, 337
Elect, 326-7
Hardy, 338-9
Lockey, 340
Limits, 344
Malar, 344
Market, 332-3
Maspar, 327
PH, 341
Policy, 327
Polsys, 327-8
Polut, 329-30
POP, 330
Rats, 330
SCATR, 341
Slits, 342
Sterl, 330-1
TAG, 331
Uspop, 344
Dupont, Donald A.
Troy, 518
E-Z Science Games
E-Z Science Games, 542
Earthrise
Global Futures Game, 405
East Anglia, University of
Crisis in Lagia, 560
Edison Electric Institute
Energy-Environment Game, 366
Educational Audio-Visual, Inc.
Federalists vs. Republicans, 424
Indians View Americans; Americans View Indians, 428
Japanese American Relocation, 1942, 428-9
Selling–A Good Way to Earn a Living, 648-9
Smart Spending, 530

Educational Games Co.
Election, 353
Educational Progress
The Career Game, 547
Education Research
Administration, 609
Communication, 612
Counseling, 613
Decision, 614
Delegation, 614-15
Finance, 640
Instruction, 618
Interaction, 618
Leadership, 619
Motivation, 623
Negotiation, 623
Planning, 626
Priority, 627
Problem Solving, 627
Salesquota, 638
Selection, 629
Sensitivity, 629
Supervision, 630
Educator's Workshop
Tenure, 395
Educom
Educom Financial Planning Model, 333
Edu-Game
Alien Spaceship, 348
Bombs or Bread, 447
Building a Rapid Transit System, 316
Buy and Sell, 373
Buyer Beware, 526
Consumer Decision, 526
Crisis in Middletown, 317
Dissent and Protest–The Montgomery Bus Boycott, 422-3
Elect a President, 353
Exchange, 376
Exploring the New World, 424

First Amendment Freedoms, 465
Gold Rush Days, 425
Government Reparations for Minority Groups
Human Survival–2025, 405
Interaction–A Balance of Power in Colonial America, 428
Low Income Housing Project, 320
Middle East Crisis, 455
National Priorities, 358
Nationalism: War or Peace, 455
Planning an Inner-City High School, 323
Planning the City of Greenville, 320-3
Political Pollution, 369
Power Politics, 360-61
Preventing the Civil War, 434
Revolt at State Prison, 468
Spanish-American War, 437
Summit, 458
Trial of George Washington, 438
Trial of Jefferson Davis, 438-9
Trial of Harry S. Truman, 439
War Time, 459
Women's Liberation, 363
World War I, 440

El Paso Public Schools
Tightrope, 382

Environmental Education Center
Metric Poker, 472

ERIC
Planafam II, 416

E.T.C. Publications
Teaching the Future, 407
Watergate: The Waterloo of a President, 440

Excalibre Games, Inc.
Ancient Conquest, 481
Caen 1944, 486
Cassino 1944, 487
Crimea 1941, 488-9
Sidi Rezegh 1941, 512

Fact-and Fantasy Games
Siege!, 512

Family Pastimes
Choices, 547-8
Community, 559
Family, 310
Together, 575
Yin Yang, 315

Fantasy Games Unlimited
Mercenary, 500-1

Federal Reserve Bank of Minnesota
You're the Banker, 383-4

Ferguson, Carl E. Jr.
Inventory Simulation (INSIM), 605

Flying Buffalo Computer Conflict Simulation, Inc.
Battle of Chickamauga, 484
Board of Directors, 331
Imperialism, 497
Viva!, 520

Frederick, R.
Drug Attack, 307

Free Press
CLUG, 583
Inner City Housing Game,
INTOP, 598
SIMSOC, 571
Simulating the Values of the Future, 406

Friendship Press
Catalyzer, 532
Dignity, 317
Island, 378
Values, 555

Fuller, Earl
Northeast Farm Management Game, 645
Poultry Farm Management Game, 645
Upper Mid-West Dairy-Hog Farm Management Game, 646

Fun With Food
Nourish, 416

The Future Associates
Dialogues on What Could Be, 402

Game Designers' Workshop
Avalanche, 482
Bar-Lev, 483
Battle for Midway, 483-4
Burma, 486
Case White, 486-7
Citadel, 487
Coral Sea, 488
Crimea, 488
Drang NachOsten! and Unentschieden, 491
1815: The Waterloo Campaign, 492
En Garde!, 492
La Bataille de la Moskowa, 499
Manassas, 500
Narvik, 504
Pharsalus, 508
Raphia, 217 B.C., 509
Russo-Japanese War: Port Arthur and Tsushima, 511
SSN: Modern Tactical Submarine Warfare, 515
Torgau, 518

Gamed Simulations, Inc.
Blood Money, 412-13
Compacts, 316-17
Lobbying Game, 355
Metropolis, 587-8
Much Ado about Marbles, 321
Welfare Week, 445

Games Central
Culture Contact II, 560-1
Edplan, 386
Edventure, 386-7
Edventure II, 387
Exploring Careers, 549
Grand Strategy, 425-6
Home Economy Kit, 528
Innocent Until . . ., 465-6
Manchester, 430
Opening the Deck, 567
Pollution, 369-70
Potlatch Game, 567
Rainbow Game, 569
Settle or Strike, 381
Simpolis, 361-2
Slave Coast Game, 437
To Drink or Not to Drink, 308

Games Science
Battle of Britain, 484
Flying Tigers II, 494
Space Patrol, 407

General Learning Press
Introduction to Managerial Accounting, 641

Georgia State University
Decision Mathematics Operational Game, 345-6
RISKM, 641

Ginn and Co.
City Hall, 350

Glencoe Press
Justice Game, 467
Nothing Never Happens, 312

Global Perspectives in Education
The Road Game, 313

Goodyear Publishing Co., Inc.
Operation Encounter, 636-7
Organization Game, 602

Graded Press
Banner-Making Game, 531

Graphics Company
Calorie Game, 413
Nutrition Game, 416

Greenhaven Press
Future Planning Games, 404-5

Griggs Educational Service
Biblical Society in Jesus' Time, 531
Moses and the Exodus, 535
Planning Game, 536
Prophets and the Exile, 536-7
System I, 400

Halstead Press
City, 579

Harcourt Brace Jovanovich
Micro-Economy, 566

Harper and Row, Publishers, Inc.
Finansim, 641
Marksim, 636
Prosim, 607
TAG, 314

Harrison, Mary A.
Disposition Exercise, 465

Harwell Associates
Congress, 351
Librarian's Game, 443-4
New Town, 322

Hayden Book Company, Inc.
Human Communication Handbook, 311

Heard, Edwin L.
PABLUM, 328

Heart Foundation
Resusci-Ann, 417

Heier, W. D.
Executive Simulation Game, 600-1

Hendricks, Francis
GSPIA, 584-5
New Town Professional Planner's Set, 588
Planning Operational Gaming Experiments, 588-9

Histo Games
Eagle Day: The Battle of Britain 1940, 491
Hannibal, 496
1944: The Invasion of France and the Battle of Germany, 505

Historical Alternatives Game Co.
Battle of Roark's Drift, 484

Siege of Jerusalem, 70 A.D., 512-13
History Simulations
Big Business, 420
Brinkmanship: Holocaust or Compromise, 420
Haymarket Case, 426
North vs. South, 431-2
Progressive Era, 434
Radicals vs. Tories, 436
Spanish American War, 437-8
War Crimes Trials, 440
Hobson, Margaret
Nexus, 398
Hobsons Press
Exercises in Personal & Career Development, 549
Five Simple Business Games, 648
Metal Box Business Game, 649
Pipework Engineering, 626
Houghton-Mifflin Co.
Mouse in the Maze, 544
Planet Management Game, 369
Redwood Controversy, 370-1
Tempomatic IV, 600
House of Games Corp., Ltd.
Game of Nations, 450-1
Hyder, William D.
POLIS Network, 334

Idea Development Co., Inc.
Police State, 567
Ideal School Supply Co.
Energy X, 366-7
Gomston: A Polluted City, 367-8
Number Line, 472
Indiana University
Dynamic Modeling of Alternative Futures, 402
Innovative Education, Inc.
Decisions, 549
Institute for Demographic and Economic Studies
Idiom, 336-7
Institute for Rational Living
Play yourself Free, 552-3
Institute of Higher Education, Univ. of Alabama
CLUG, 583
End of the Line, 443
Indian Reservation, 563-4
Policy Negotiations,, 399
They Shoot Marbles, Don't They, 400-1
Urban Dynamics, 591-2
Institute of Urban and Regional Development, Univ. of California
El Barrio, 443
Wildlife, 545-6
Instructional Alternatives
Confrontation, 397
GAMEgame II, 397
GAMEgame IV, 397-8
GAMEgame VI, 398
Kathal: Game of Creative Intimacy, 550
Sci-Fi, 400
Interact Co.
Adapt, 409
Agency, 308
Amnesty, 349
Balance, 365-6
Boxcars, 409
Budget, 349-50
Code, 460
Collision,
Committee, 351
Constitution, 351-2
Contractors, 559-60
Cope, 401-2
Council, 352
Cycle, 548
Czech-Mate, 422
Defense, 352
Delegate, 352-3
Design, 527
Destiny, 422
Dig, 561
Discovery, 422
Disunia, 433
Division, 423
Ecopolis, 366
Enterprise, 376
Equality, 561
Espionage, 424
Exchange, 376
Explosion (Population), 367
Family Tree, 562
Fifties, 424-5
Fire, 465
Flight, 562
Gateway, 425
Gold Rush, 425
Heritage, 427
Herstory, 562
Homefront, 427
Homestead, 427
Identity, 563
Independence, 427
Interaction, 388-9
Judgment, 429
Juris, 466
Liberté, 429
Mahopa, 565
Merchant, 431
Mission, 431
Moot, 467
Mummy's Message, 431
Newscast, 461-2
Nuremberg, 432
O.K., 552
Panic, 433
Peace, 433
Pioneers, 433-4
Pressure, 361
Protection, 529-30
Puzzle, 434-5
Rip-Off, 469
Sanga, 569-70
Saturation, 462-3
Search, 463
Seneca Falls, 435-6
Skins, 436
Small Town, 571-2
Spectrum, 572
Spiral, 381
Statehood, 362
Strike, 438
Sunshine, 324
Taxes, 362
Time Capsule, 574-5
Trade, 382
Utopia, 408
Virtue, 439-40
Votes, 362-3
Vows, 468-9
Web, 395
World, 459
International Communications Institute
Community X, 316
Interna[illegible]l Relations Program
PRI[illegible]5
Involvement
Involvement, 428
Irwin, Richard D., Inc.
Executive Game, 597
Marketing in Action, 634

John Knox Press
Baldicer, 558-9
Johns Hopkins University
Teams-Games-Tournament, 400
Joint Council of Economic Education
Trade Off, 590
Judson Press
Using Biblical Simulations, 539

Kanthaka Press
Shantih, 537
Kendall/Hunt Publishing Co.
Executive Simulation, 597
Health Games Students Play, 415
SIMQ, 599
Kopptronix Company
Keep Quiet, 415

Labor Relations Assoc.
Union Organizing Game, 631
LaPine Scientific Co.
Metri-Magic, 472
Learning Resources in International Studies
Ch'ing Game, 420-1
Diplomatic Practices, 450
Foreign Policy Decision Making, 450
Good Federalism Game, 354
Good Society Exercise: Problems of Authority, 451-2
International Conflict, 454
Situational Leadership Simulator, 629-30
State System Exercise, 456-8
Uses of the Sea, 458-9
World Political Organization, 459
Lincoln Filene Center for Citizenship & Public Affairs, Tufts Univ.
Conflict in the Middle East, 447
Longman Group Ltd.
Congress of Vienna, 421
Longman Geography Games, 410
Longman History Games, 430
Longman Science Games, 544
Scramble for Africa, 435

McGraw Hill Book Co.
Jeffrey's Department Store, 649-50
Lester Hill Office Simulation, 649
Management: An Experiential Approach,

601
Marketing Dynamics, 634
Wilson RV, 631-2
Macleod, Jennifer, Associates
Cryteria!, 613-14
Macmillan Company
Contract Negotiations, 613
Management Game, 598
Managing the Worker, 622
Personnel Department, 624-5
Simchip, 607-8
Mafex Associates, Inc.
Steady Job, 554
Teacher's Lounge, 394
Mag. Mfg. Inc.
Psych-Out, 474
Maine Public Broadcasting Network
Eco-Acres, 583-4
Management Games, Ltd.
Busy, 640
Chancellor, 373
Decision, 614
Distribute, 605
Exmark, 601
Finmark, 641
Function, 601
INDREL, 617-18
Megga Trading Company, 601-2
Policy, 643-4
Negotiate, 623
Polychoc, 637
Sellem, 639
Shoprofit, 639
Small Business Management Exercise, 602
Sniblette, 642
Snibbo, 602
Sprods, Ltd. 608
Thyngummy Jigs Ltd., 642
Trig, 603
Tycoon, 603
MDT
After the War and the Election of 2020, 419
Meridian House
Meridian 36, 411
Merrill, Charles E. Publishing Co.
Executive Decision Making, 596-7
Marketing Strategy, 635
Questioneze, 392
Merrimack Education Center
School Planning Game, 392
Arthur Merriwether, Inc. (See also Contemporary Drama Service)
Praise and Criticism, 536
MESA Publications
Principal Game, 389
Metrix Corp.
Metrication, 471-2
Mette, H. Harvey
Actionalysis, 385
Michigan, University of
Brookside Manor, 441-2
End of the Line, 443
Grand Frame, 584
The Living City/La Cite Vivante, 586
Michigan, University of T.V. Center
Middle East Conflict Simulation Game, 454-5
Midwest Publications Co., Inc.
In Order, 471
Minnesota Educational Computing Consortium
Fail Safe, 343
Oregon, 347
Minnesota School District
Manage, 332
Mississippi Research & Development Center
Your Community's Economic Development, 325
MIT Center for International Studies
Political Exercise, 455
Mobley and Assoc.
Value Options, 555
Montana, University of
Forest Simulator, 646
Motherland, Inc.
Black Community Game, 315-16
Moritz, Bruce A.
Interact, 311
Moy, Roland F.
Decision Point, 561
Multilogue
Hexagon, 585
METRO, 586-7
MUST, Research Division
Center for Educational Policy & Management
Ernstspiel, 310

National Academic Games Project
Euro-Card, 561-2
National Council of Teachers of English
Inventing and Playing Games in the English Classroom, 460-1
National Health Systems
Good Loser, 414
Soup's On, 417
National Teacher Education Project
Teaching Strategy, 395
National Technical Information Service
City I, 579-81
City IV, 581
City Model, 581-2
Pipeline: An Employment and Training Situation, 444-5
National Textbook Co.
Classroom Games in French, 450
Let's Play Games in French, 461
Let's Play Games in German, 461
New Hampshire Council on World Affairs, Univ. of New Hampshire
Simulex IV, 456
New Hampshire, University of
Blackberry Falls Town Government Game, 578
New South Wales University Press
Australian Management Games, 610
New York, University of the State of
Alpha Crisis Game, 418
Newsweek Education Department
2000 A.D.–Futura City, 407-8
Northern School Supply
Great Plains Game, 426
Northern Vision Services
Guerilla Warfare, 426
Uhuru, 575-6
Oklahoma State University
Oklahoma Bank Management Game, 642
Olcott Forward
Old Cities New Politics, 358-9
Parksburg: Garbage and Politics, 359-60
1787, 436
The Union Divides, 439
Old Dominion University
Make your Own World, 368-9
Ontario Institute for Studies in Education
Classifying the Chemical Elements, 541-2
In-Basket Simulation Exercises, 388
Orbis Books
Education for Justice, 375
Ortho Diagnostics, Inc.
Clot, 413
Osgood, Charles E.
Security, 456
OXFAM
Aid Committee Game, 372
Grain Drain, 377
Poultry Game, 567-8
Poverty Game, 568
Oxford University Press
District Nutrition Game, 414
Ozment, John
Buloga II, 604

Parker Brothers
Monopoly, 378-9
QUBIC, 474
Soma Cube, 475
Parnes, Bob
Middle East Conflict Simulation Game, 454-5
Pennant Educational Materials
Cruel Cruel World Value Game, 548
My Cup Runneth Over, 551-2
Timao, 554
Value Bingo, 554-5
Pennsylvania State University
A General Agricultural Firm Simulator, 644
Pergamon Press, Inc.
Psych City, 323
Philmar, Ltd.
Decline and Fall, 489-90
English Civil War, 492
Pillsbury Company
Poppin 'Swap, 416-17
Policy Studies Assoc.
Equity and Efficiency in Public Policy, 376
Nursing Home Care as a Public Policy Issue, 444
Powerplay, Inc.
Powerplay, 553
PPC Books
Marketing Interaction, 634-5
Sales Management Game, 638
Prentice Hall, Inc.
Blue Wodget Co., 648
Business Policy Game, 595
Clinical Simulations, 413
Taking Action, 463
Trade-Off at Yalta, 438
Walk in My Shoes, 464

Psych-Ed Assoc.
The Sharing Game, 554
Puckrin's Production House
Independence, 563
Two Years of Horror, 408

Random House, Inc.
Marriage Game, 550-1
Reubens, Jackie
Myth-Information (About Human Sexuality), 415-16
Rider, Theodore H.
Mini-Apex, 586
Roundtable Films, Inc.
A Leadership Decision Game, 619

S & G Games
Königgrätz, 498
Mukden, 1905, 503
Tannenburg, 516-17
SAGA
Fortress Rhodesia, 494
Future Decisions: The I.Q. Game, 403-4
In the Chips, 564-5
Kama, 565
Kibbutz, 534
Mortgage, 529
Reapportionment, 361
Southern Mountainer, 572
Sage Publications, Inc.
At-Issue, 396
Conceptual Mapping Game, 396
IMPASSE, 398
St. Mary's College Press
Simulation Games for Religious Education, 537
Sakoda, James
CHEBO, 343
Salisbury, Howard
Simulated Agribusiness, 645-6
San Diego State University
See Center for Health Games and Simulations
Saroff, Ron
Hybrid-Delphi Game, 406
Scholastic Book Services
On Strike and other economic games, 379
Shipwreck and other Government Games, 571
Scholastic TAB Publications, Ltd.
Environmental Simulations, 367
Science Kit, Inc.
The Great Periodic Table Race, 543
Science Research Associates
Achieving Classroom Communication Through Self-Analysis, 385
American Constitutional Convention, 419
American Government Simulation Series, 348-9
American History Games, 419-20
Becoming Aware, 611
BUCOMCO, 600
Career Development Program, 547
Decision Making for Career Development, 548
Fixit, 387
Inner-City Simulation Laboratory, 388
Inter-Nation Simulation, 452-3
Job Experience Kit, 550
Role Playing Methods in the Classroom, 392
Teaching Problems Laboratory, 394
Scientific Press
BOLA, 337
Graze, 329
Markstrat, 636
Scott Foresman and Company
Dangerous Parallel, 449
Cal Q Late, 470
Four in a Row, 470
Here to There, 471
Multifacto/Producto, 472
Orbiting the Earth, 473
Polyhedron Rummy, 473
Power, 568
Sea Grant College Program
Household Energy Game, 528
Seabury Press
Value Game, 555
Second Person Plural Ltd.
Office Talk, 623-4
SGS Associates, Ltd.
Le Pays de France, 461
North Sea Exploration, 411
Railway Pioneers, 411-12
Sheffield, University of
Gambit, 584
INHABS, 3, 586
Shelter
Tenement, 324
Short, Ronald R.
College Game, 548
Simco Enterprises
The O. K. Game, 312
Simile II
Access, 556
Bafá Bafá, 556-7
Binary, 469
Crisis, 448-9
Guns or Butter, 452
Humanus, 562-3
Metropolitics, 355-6
NAPOLI, 357
New City Telephone Co. Simulation Game, 566-7
Plea: A Game of Criminal Justice, 467-8
Rafa Rafa, 568-9
Relocation: A Corporate Decision, 628
Roaring Camp, 435
Starpower, 572-4
Talking Rocks, 463-4
Trebides Island, 384
Where Do You Draw the Line?, 555
Simtek
Busop, 595
Computer Augmented Cases in Operations & Logistics Management, 604
Electronic Industry Game, 596
Executive Simulation, 597
Integrated Simulation, 597
Marketing a New Product, 634
Pocket Calculator Boom, 599
Purdue Supermarket Management Game, 643
RISKM, 641
Supermarket Strategy, 643
Simulation Learning Institute
Centralized Power Game, 420
Imperialism, 452
Nuclear Energy Game, 322
Revolutions Game, 435
Territorial Sea, 458
Simulation Sharing Service
Gospel Game, 534
Serfdom, 570
Thanatos, 538
325 A.D., 538-9
Simulation Systems
Making a Change, 320
Planning Exercise, 398-9
Pollution: Negotiating a Clean Environment, 370
Power Politics, 360
Triangle Trade, 439
Waging Neutrality, 440
Simulations Canada
Dieppe, 490
Peloponnesian War, 508
Raketny Kreyser, 509
Simulations Publications, Inc.
After the Holocaust, 479-80
Air War, 480
The American Civil War, 480-1
American Revolution, 481
The Ardennes Offensive, 481-2
Atlantic Wall, 482
Austerlitz, 482
Barbarossa: The Russo-German War, 1941-45, 483
Battle for Germany, 483
Blue and Gray: Chicamauga, Shiloh, Antietam, Cemetary Hill, 485
Blue and Gray II: Fredericksburg, Hooker and Lee (Chancellorsville) Chattanooga, Battle of the Wilderness, 485
Borodino, 485-6
Breitenfeld, 486
CA", 486
Canadian "Civil War", 559
Chariot, 487
Cobra, 487
The Conquerors: The Macedonians and the Romans, 488
Conquistador, 488
Descent on Crete, 490
Desert War, 490
Dreadnought, 491
Drive on Stalingrad, 491
Fast Carriers, 493
Firefight, 493
First World War, 493-4
Four Battles from the Crimean War: Alma, Balaclava, Inkerman, Tcherdaya River, 494
Four Battles in North Africa: Crusader, Cauldron, Supercharge, Kasserine, 494-5
Foxbat and Phantom, 495
Frederick the Great, 495
Frigate, 495
Fulda Gap, 495
Global War, 496

Grenadier: Company Level Combat, 1700-1850, 496
Highway to the Reich, 497
Invasion: America, 497
Island War (Bloody Ridge/Guadalcanal/Saipan, Leyte, Okinawa), 497
Kharkov, 498
Korea, 498-9
Kursk, 499
La Grande Armée, 499
Lee Moves North, 499
Legion, 500
Mech War '77, 500
A Mighty Fortress, 501
Minuteman, 501-2
Modern Battles: Wurzburg, Chinese Farm, Golan, Mukden, 502
Modern Battles II: Battle for Jerusalem, 1967; Yugoslavia; Bundeswehr: DMZ, 502-3
The Moscow Campaign, 503
Musket and Pike: Tactical Combat, 1550-1680, 503
Napoleon at War: Marengo, Jena-Auerstadt, Wagrom and the Battle of the Nations, 504
Napoleon at Waterloo, 503-4
Napoleon's Last Battles: Ligny, Quatre Bras, Wavre, LaBelle Alliance (Waterloo), 504
NATO: Operational Combat in Europe in the 1970's, 504
The Next War, 505
Objective Moscow, 505
October War, 505
Oil War, 505-6
Operation Olympic, 506
Panzerarmee Afrika, 506
Panzer '44, 507
Panzergruppe Guderian, 507
Patrol!, 507
Plot to Assassinate Hitler, 508
The Punic Wars, 508
Raid!, 508-9
Red Star/White Star: Tactical Combat in Europe in the 1970's, 509
Red Sun Rising, 509
Revolt in the East, 509-10
Road to Richmond, 510
Russian Civil War, 511
Seelowe, 511
The Siege of Constantinople, 512
Sinai, 513
Sixth Fleet, 513
Sniper!, 513
Soldiers: WWI Tactical Combat, 1914-1915, 513-14
Solomon's Campaign, 514
South Africa, 514
Spartan, 514
Spitfire, 514-15
Stonewall, 515
Strike Force One, 515
Tank!, 516
Tannenberg, 516
Terrible Swift Sword, 517
Thirty Years War: Lützen, Nordlingen, Rocroi, Freiburg, 517
Turning Point, 518-19
Up Scope!, 519
USN, 519
Veracruz, 519
Viking, 520
Wacht am Rhein, 520-21
War Between the States, 521
War in Europe (War in the West and *War in the East)*, 521-2
Wellington's Victory, 522
Westwall: Arnhem, Hurtgen Forest, Bastogne, Remagen, 522
Winter War, 522-3
Wolfpack, 523
World War I, 523
World War II, 523-4
World War III, 524
Yeoman, 524

Simulectical Simprovisions
Middle East Conflict Simulation Game, 454-5

Skytrex Ltd.
Trireme, 518

Sliva, Martin E.
Thermoplae, 517

Smith, Theodore F.
Consumerism II, 633

Snider, John C.
Microville, 320-21

Social Studies School Service
Confrontation: The Cuban Missile Crisis, 447-8
Traders Arrive on the African Scene, 575
Up Caste Down Caste, 576

Software Distribution Center
See Digital Equipment Corp.

Soma, Connie
Horatio Alger, 443

South-Western Publishing Co.
Computer Augmented Cases in Operations and Logistics Management, 604
Integrated Simulation, 597-8

Spenco Corp.
Trip or Trap Bingo, 308

Spoken Arts, Inc.
Choose, 307

Starr, Harvey
Diplomatic/Foreign Policy Game, 449-50

Steinitz, Carl
Systems, 589-90

Steinwachs, Barbara
WHIPP (Why Housing is a Problem and a Priority), 325

Stokes Publishing Co.
Picture Patterns, 473

Study-Craft Educational Products
Transaction, 382-3

Summers, Luis H.
House Design Game, 528
Suprclug, 589

Synectics Education Systems
Hang-Up, 550

Syracuse University Press
Special Education Administration Task Simulation (SEATS), 392-3

Systems Factors, Inc.
Blight, 578-9
Campaign, 350
Czar Power, 421-2
Female Images, 549-50
F.L.I.P. 2/80, 527
No Dam Action, 369
Party Central, 433
Transit, 590-1

Tabletop Games
Ballistic Missile, 483
Micro-Napoleonics, 501
MTB, 503
U-Boat, 519

Tamminga, Harriet L.
Intentional Community, 564

Teaching Aids Co.
Chem Bingo, 540
Chem Chex, 541
Chem Trak, 541
Interceptor, 471
Space Colony, 406-7

Teaching Concepts, Inc.
Circulation, 541
Space Hop, 545
Super Sandwich, 417

Teleketics, Inc.
Becoming a Person, 547
Coffee Game, 532
World Game, 408

Temple University, Center for the Study of Federalism
Legisim, 355

Texas, University of, College of Business Administration
Midtex, 605
Unitex, 608

Thayer School of Engineering
Coal 1, 329

Thorpe, Gerald L.
AFASLAPOL: The Game of Politics in a Moderning Nation, 446
Confrontation: The Cuban Missile Crisis, 447-8
Revolutionary Society (REVSOC) Simulation, 569

Trainex Press
Nursing Crosswords, 416

Training House, Inc.
Abelson-Baker Interview, 609
Career Game, 611
Construction Game, 612
Game of Time Management, 616
Game of Transactional Analysis, 616
Lou Boxell's Performance Review, 620
Meeting Game, 623
Selection Interview Clinic, 629
When the Union Knocks . . ., 631

Transnational Programs Corp.
Nuclear Site Negotiation, 322
Pex, 625-6
Probe, 553
Process, 553
Profair, 628
Prosper, 628
Prospects, 553-4

Troubador Press
Geometric Playthings, 471

TSR Games

Fight in the Skies, 493
Tufts University
Conflict in the Middle East, 447

UCCM Games
Freshman Year, 388
System, 574
Ungame Co.
Roll–A-Role, 313
Social Security, 313
The Ungame, 314-15
Union Printing Co.
Compounds, 542
Dead River, 366
Elements, 542
Geologic Time Chart Game, 543
Lab Apparatus, 543-4
Mental Health, 551
Rip-Off, 530
Vitamins, 417-18
United Community Planning Corp.
Brothers and Sisters Game, 442
U.S. Environmental Protection Agency
River Dose (RVRDOS) Model, 330
U.S. Government Printing Office
Blood Money, 412-13
Land Use, 368
University
See individual listings by state name
University Associates, Inc.
Annual Handbook for Group Facilitators, 308-9
Handbook of Structures Experiences for Human Relations Training, 310-11
University Council for Educational Administration
Adams School System Simulation (ADSIM), 385-6
Madison Simulations, 390
Monroe City Simulations, 390-1
Regular Meeting of the Wheatville Board of Directors, 392
SEASIM, 393
Teacher Training Simulation, 394

Van Nostrand Reinhold Co.
City Planning, 581-2
Veterans Administration
The In-Basket Exercise, 647

Washington University
Washington University Business Game, 603
Washington, University of
The Forest Simulator, 646
Wasmuth, William J.
Workshop Management Training Simulation, 445-6
Wasserman, Dave
Jail Puzzle, 466
Wego Games
Prediction, 312
Wentworth Corp.
MADS-Bee, 598-9
West Publishing Co.
Personnel Management in Action, 625
Westland Publications
DeKalb Political Simulation, 352
PSW 1/SW Political Simulation, 334-5
TeleCLUG, 590
Woodbury Political Simulation, 363-4
Wff 'n Proof
Configurations, 470
Equations, 470
In-Quest, 543
Mr. President, 431
On Sets, 473
Propaganda Game, 568
Queries 'n Theories, 462
Real Numbers Game, 474
Tac-Tickle, 475
Tri-Nim, 475
WFF: The Beginner's Game of Modern Logic, 476
Wff 'n Proof: The Game of Modern Logic, 476-7
Whithed, Marshall H.
Build, 579
Wiley, John and Sons, Inc.
National Policy Game, 358
Williams, Robert Games
The Warlord Game, 522
Wisconsin, University of
Household Energy Game, 528
Wisconsin-Milwaukee, University of
All for the Cause and the Cause of Each, 441
World Affairs Council of Philadelphia
Decision Making Model, 449
World Future Society
Futuribles, 405
World Without War Publications
World Without War Game, 576
Wright, Twyla
Barriers in the Maze, 546

Xerox Educational Publications
Railroad Game, 380-1

Zenger Publications, Inc.
Kids in Crisis, 467
Halfway House, 319-20
Jury Game, 466-7
Slave Auction, 437
Zuckerman, Robert A.
Label Game, 389

ABOUT THE AUTHORS AND EDITORS

RONALD E. ANDERSON

Ronald E. Anderson is Associate Professor of Sociology at the University of Minnesota in Minneapolis, where he has taught for the past ten years. He is also director of the Sociology Data Center there, and Sociology Series Editor for CONDUIT. His research has been primarily on how computers are used or what their impact is. Currently he is codirector of a study supported by the National Science Foundation on computer literacy among high school students. He has published over 25 articles on computer topics, most of them pertaining to education.

JOAN GASKILL BAILY

Joan Gaskill Baily is a graduate of Radcliffe College and is working for an advanced degree in the Department of Sociology at Rutgers University. Besides gaming, her interests are sex roles and stratification. She is currently studying the day care movement and public policy.

BRUCE E. BIGELOW

Bruce E. Bigelow is Associate Professor of History at Denison University and Director of the Denison Simulation Center. He is currently on leave on an American Council of Education government fellowship at the Department of Transportation. He has worked for five years developing, testing, and evaluating simulations in history, education, ethics, and futuristics; has directed and served on the staffs of several simulation workshops; and has consulted with educational and professional organizations on simulations and educational futures. An expert on Yugoslav affairs, he has directed four summer seminars for faculty and students in that country and has written on Yugoslavia and futuristics as well as simulation.

MAURICE BISHEFF

Maurice Bisheff is Program Director at the Center for Multidisciplinary Educational Exercises (COMEX) and a doctoral candidate at the School of Public Administration, University of Southern California. He has designed many simulation exercises and training programs for use in government, business, and human services, and is interested in using simulation games as vehicles for understanding group and organizational factors that facilitate or inhibit personal effectiveness.

CHARLES A. BOTTINELLI

Charles A. Bottinelli has been an energy/environmental educator in the Denver Public School System for ten years. He received his B.A. in chemistry and his M.Ed. in environmental education from the University of Colorado, where he is presently a doctoral candidate writing a thesis on *The Efficacy of Population/Resources/Environment Simulation Games and Traditional Teaching Strategies.* He is a member of the Colorado Energy and Man's Environment Committee, an energy education consultant for energy-related industries and agencies, a member of the NSF/Colorado School of Mines teaching team, a member of the Technical Advisory Board for the Colorado Energy Conservation and Alternatives Center for Commerce and Industry, and past coordinator of an HEW grant to the University of Colorado dealing with energy education for Colorado adults. Mr. Bottinelli has developed several simulations related to energy/environmental problems and has published several articles dealing with the same topic.

MARTIN C. CAMPION

Martin C. Campion teaches history at Pittsburg State University, Pittsburg, Kansas, where he does war games in his class, "War in Western Civilization" and designs nonwar history games for other classes. He has written on the use of war games in classrooms for *The History Teacher* and other periodicals and was a contributor to the third edition of the *Guide.*

ANNE CLEAVES

Anne Cleaves is a freelance editor and writer whose work reflects her interests in philosophy, psychology, education, and social structure and change. She received an A.B. in English from Mount Holyoke College and an M.A. in English and secondary education from the University of Wisconsin. She was associated with the three earlier editions of the *Guide,* served a year term as coordinator of a parent cooperative community day care center, and has taught high school and in a live-in center for high school dropouts. By calling she is a poet. She is currently with the Child Care Resource Center in Cambridge, Massachusetts.

LUCIEN E. COLEMAN

Lucien E. Coleman, professor of religious education at the Southern Baptist Theological Seminary, Louisville, Kentucky, teaches courses in instructional theory and practice and in communication. For the past six years he has taught an annual course in the design and use of simulations/games in religious education. In 1973 he devoted a half-year study leave to research in the field of educational simulations and games. An active member of the North American Simulation and Gaming Association, he has studied simulation/gaming at Simile II, the Center for Simulation Studies, and the University of Michigan Extension Gaming Service. He spent the 1978-1979 academic year on sabbatical as a visiting scholar in Oxford, England.

DONALD C. DAVIS

Donald C. Davis was born in South Philadelphia, was educated at Indiana State University, and served as an artilleryman in Vietnam and Cambodia during 1969-1970. Since that time he has supported himself by working as a ditch digger, auto worker, factory foreman, union official, social worker, human services administrator, and most recently as a freelance writer and researcher.

JOSEPH L. DAVIS

Joseph L. Davis is a doctoral candidate in political science and public administration at St. Louis University, where he teaches part-time. He has taught political science also at the University of Missouri, St. Louis, and in the St. Louis Junior College District. His specialty is urban politics. Mr. Davis has been associated with the Center for Simulation Studies, St. Louis, since 1971 and is presently working with them as chief program designer and consultant to educational institutions, government agencies, and private businesses.

FRANK P. DIULUS

Frank P. Diulus is Associate Professor of Education at the State University College at Buffalo. He received his Ph.D. in foundations of education from the University of Pittsburgh. He has taught and conducted research in sociology and the philosophy of education at the University of Pittsburgh and the State University College at Buffalo. At the former, he served as director of in-service education at the Center for Educational Action. Currently, at the State University College at Buffalo, he is Associate Director of Teacher Corps, a federally funded program to develop in-service teacher education systems. He has taught graduate courses and workshops on social simulations, and he has published several articles on both simulation gaming and social dimensions of education.

DOROTHY R. DODGE

Dorothy R. Dodge is Chairperson of the Department of Political Science at Macalester College in St. Paul, Minnesota, and coordinator of Macalester's interdisciplinary International Studies Program. She served for four years during the early 1950s as executive director of the American Association for the United Nations and has continued her interest in the organization. She has made study trips to Asia and to Africa.

Dr. Dodge's special teaching interests are in comparative political systems–particularly in Africa and Asia–international theory and international law, cross-national urban programs, and developing and applying simulation models in which students take roles in functioning models of "real life" social science problems and issues to experience the dilemmas of decision-making. She has served in a consulting role on simulation for the Minneapolis Public Schools and for the Robert A. Taft Institute of Government.

RICHARD L. DUKES

Richard L. Dukes is currently Associate Professor of Sociology at the University of Colorado, Colorado Springs. He is interested in simulations and games as teaching devices and as settings for social research. His long-term interest in *SIMSOC* is reflected in this volume. Recently he completed a book titled *Learning with Simulations and Games* with coeditor Constance J. Seidner. At present he is working on simulations and theories of violent political behavior.

THEODOR EHRMAN

Theodor Ehrman has a B.A. from Adelphi University and an M.A. from the New York State University at Stony Brook. Mr. Ehrman has taught American studies and economics at Oceanside High School for thirteen years. In addition, he has taught courses in simulation games at Post College and Hofstra University. Currently, Mr. Ehrman teaches macro- and microeconomics in the SCALE Program at Post College. He has written several articles on simulation games that have appeared in the *Social Science Record.* For the past ten years, Mr. Ehrman has used simulation games as a major part of his classroom instruction. He has adapted *CLUG* for his economics class and has built his course around two or three *CLUG* experiments.

ORAH ELRON

Orah Elron was born in Vienna, Austria, and grew up and lived most of her life on a kibbutz, teaching science at the kibbutz secondary school. She came to New Haven, Connecticut, in 1969 to study modern teaching methods in science at Southern Connecticut State College. While there, she developed a simulation game called *Survival* and wrote her M.S. thesis on a case study of the game (adapting simulation games to different age levels).

In 1972, back in Israel, Mrs. Elron joined a team for the Development of Science Curricula for Elementary Schools at "MATAL" Tel Aviv University for several years. She is now a program associate of the Environmental Education Center at Area Cooperative Education Services in New Haven. Her hobbies include arts and crafts, particularly batik and pottery.

ANTHONY J. FARIA

Anthony J. Faria is Associate Professor and Chairperson of Marketing in the Faculty of Business Administration at the University of Windsor, Ontario. A graduate of Michigan State University, he received his Ph.D. in 1974. Dr. Faria is a member of the American Marketing Association, Academy of Marketing Science, Southern Case Research Association, the Association for Business Simulation and Experiential Learning, and the North American Simulation and Gaming Association. Dr. Faria has authored a marketing simulation game entitled *COMPETE: A Dynamic Marketing Simulation* and has published in the *Journal of Marketing, Journal of Small Business Management, Journal of Transportation and Technology, Pittsburgh Business Review, Traffic Quarterly, Retail Control, Atlanta Economic Review, Retail Directions, Department Store Economist, The Southern Business Review, Business and Society,* and the *Marquette Business Review.* In addition, he has participated in various seminars for the business community.

HARRY FARRA

Harry Farra is Chairperson of the Department of Speech Communication at Geneva College, Beaver Falls, Pennsylvania, and is also a freelance writer. His writings in the area of simulation games include an article for *Christian Advocate,* "Games Churches Ought to Play," and two audio cassette scripts for Contemporary Drama Service (Downer's Grove, Illinois): "Using Squirms" (a role-play game) and "Using Simulations" (which includes Dr. Farra's simulation, *The Waiting Room*). He has used simulation games in both academic and religious settings.

CATHY S. GREENBLAT

Cathy S. Greenblat received her A.B. from Vassar College and her M.A. and Ph.D. from Columbia University. She is presently Professor of Sociology at Rutgers University. Dr. Greenblat initially became involved in gaming in 1971 and since then has written several articles, books, and monographs in the field. She has coauthored two volumes with Richard D. Duke: *Gaming-Simulation: Rationale, Design, and Applications,* and *Game-Generating-Games: A Trilogy of Issue-Oriented Games for Community and Classroom.* She is codesigner with various others of *The Marriage Game, Blood Money,* and *Pomp and Circumstance.*

Dr. Greenblat has served on the executive boards of both the International Simulation and Gaming Association and the North American Simulation and Gaming Association. She is currently editor of the journal *Simulation and Games.*

·RY O. HAAKONSEN

Harry O. Haakonsen received his Ph.D. from Syracuse University and since then has been active in science and environmental education. He is the Coordinator of Environmental Studies at Southern Connecticut State College and has served on the executive committee of the State College Consortium on the Environment since its foundation in 1971. He holds a dual appointment at the college as Professor of Environmental Science and Science Education.

Dr. Haakonsen has used simulation games in elementary, secondary, college, and teacher training settings. In addition, he has used simulations in training environmental decision makers and in citizen environmental awareness workshops.

MARY JOYCE HASELL

Mary Joyce Hasell received a B.S. and M.S. in interior design and housing from the University of North Carolina at Greensboro and is completing a doctoral program in architecture at the University of Michigan. From 1971 to 1977 she taught interior design and housing at Eastern Michigan University, where she began working on her doctoral degree and became involved in gaming/simulation as a way to help students understand urban systems and problem-solving.

Since 1976, she has been involved (with many others–R. D. Duke, Allan G. Feldt, Roger Moore, Nancy Steiber, Rob Cary, Jeff Harris, Peter Smit, and Don Michael) in the research, design, building, testing, and running of the following gaming simulations: *Hex Game, Human Settlement Management in Developing Countries,* for UNESCO, R. D. Duke, director; *The Grand Frame Game,* design process applied to the planning and physical design of a residential housing development; *Global Implications of a Scientific Discovery (GISDIS),* public policy-making in regard to recombinant DNA; *Weergave Hulpsystem 1945,* public policy-making for improving social services to war victims of World War II in the Netherlands; *Future of Chase Manhattan Bank 1985-1990 International Task Force,* exploration of public policy for the future of international banking at Chase. Currently she is writing her dissertation, *A Design Process Approach to the Analysis, Design, and Teaching of Gaming/Simulation Techniques.*

ROBERT E. HORN

Robert E. Horn divides his time among research, teaching, consulting, and business. He has taught at Harvard University Graduate School of Education and Columbia University Teachers College.

He is becoming known in the instructional technology field for his invention of Information Mapping™, a method for analyzing, interrelating, and displaying textual and graphic information. He was awarded the 1976 Outstanding Research Award for Information Mapping by the National Society for Performance and Instruction.

J. BERNARD KEYS

J. Bernard Keys, Ph.D., University of Oklahoma, Business Administration, is a consultant in management development programs and lectures widely in graduate programs and seminars. He specializes in participative training techniques and hosts a Supervisory Simulation Laboratory for municipal supervisors and an Executive Simulation Laboratory for middle- and upper-level managers. He has conducted consulting and training projects for such organizations as Phillips Petroleum Company, Wilson Inc., Oklahoma City, The National Rehabilitation Administration, and Xerox Management Association.

Dr. Keys has been vice-president of the Oklahoma City Society for Advancement of Management, a director of the Oklahoma City American Society of Training Directors, regional director of management development for the American Academy of Management, and president of the Association for Business Simulation and Experiential Learning. He has served on the board of directors of the North American Gaming Council. For several years he was contributing editor of *Simulation/Gaming/News* and has authored numerous publications in this field, including a book, *Laboratories for the Training of Administrators and Executives* (University of Oklahoma Bureau of Business Research), and a computerized business game, *The Executive Simulation,* (Kendall/Hunt).

TRACY MARKS

Tracy Marks, M.A., is a former high school teacher and freelance writer. Currently, she is writing her fourth book on teaching adult education courses and operating her own publishing company.

VICTOR PASCALE

Victor Pascale is a psychology instructor, psychotherapist, and certified school psychologist. He received his Ph.D. from Yeshiva University. He has taught at the university level and has served as a human relations consultant. Several research articles on simulation games by Dr. Pascale have appeared in the *Social Science Record.* He has also presented several papers on evaluative procedures at the North American Simulation Gaming Conferences (1975-77). Presently he is an associate with Long Island Jewish Hospital, where he serves as a therapist.

CHARLES M. PLUMMER

Charles M. Plummer is Assistant Director of the Evaluation Training Consortium at Western Michigan University, Kalamazoo. ETC provides evaluation training and technical assistance to preservice and in-service teacher training programs throughout the United States. He is currently chairperson of the Simulation Systems Special Interest Group in the American Educational Research Association and a review editor for the Council for Exceptional Children. Before assuming his present position, he conducted research and evaluation in early childhood development and on the dissemination of innovative programs for educating young children needing preschool special education at the Institute for Child Behavior and Development, University of Illinois, Urbana-Champaign. He completed a Ph.D. at Indiana University in Educational Psychology, where he had fellowships in teaching and future studies and where he was involved in the design, application, and evaluation of several simulations and games. While a captain in the U.S. Air Force, he was an education and training officer on the headquarters staff of the Air Force Auditor General and president of the San Bernadino-Riverside (California) Chapter of the American Society for Training and Development.

MICHAEL J. ROCKLER

Michael R. Rockler has been Chairperson of the Department of Education at Rutgers University in Camden, New Jersey, since September 1977. Before coming to Rutgers, he taught at the University of Nebraska at Omaha; Carleton College in Northfield, Minnesota; and the University of Minnesota, where he received his doctorate in 1969. Dr. Rockler has participated in the North American Simulation and Gaming Association for several years and has also been involved in SAGSET, the British Gaming Association. His most recent publication on games is "Applying Simulation/Gaming" in *On College Teaching,* by Ohmer Milton and associates (Jossey-Bass, 1978).

BRENT D. RUBEN

Brent D. Ruben is Associate Professor and Assistant Chairperson of the Department of Human Communication, Rutgers University. He has consulted and written in the area of experiential learning and simulations and games, and is author of *Interact II* (1977), *Human Communication Handbook: Simulations and Games, Volume 2* (1978), and coauthor of *Human Communication Handbook: Simulations and Games* (1975).

Dr. Ruben consults extensively and has written in the areas of communication theory, interpersonal communication, intercultural communication, group communication, and mass communication. He has articles in *Simulations and Games, Group and Organizational Studies, Quarterly Journal of Speech, International Journal of Intercultural Relations,* and *International and Intercultural Communication Annual.* He is author, editor, or coeditor of *Interdisciplinary Approaches to Human Communication* (1979), *Beyond Media* (1979), *General Systems Theory and Human Communication* (1975), *Approaches to Human Communication* (1972), and was founding editor of the *Communication Yearbook Series,* Volumes 1 (1977) and 2 (1978).

LARRY SCHAEFER

Larry Schaefer is currently Director of the Environmental Education Center, a program of Area Cooperative Education Services (a regional education center) in New Haven, Connecticut. A graduate of Tufts University and Yale University, Dr. Schaefer has been involved in instructional material development including simulation games, multimedia packages, and curriculum units. He has written extensively on environmental education topics, especially in the area of land use education for adults. As Director of the Environmental Education Center, Larry Schaefer leads the center's in-service training program and resource center. He is also Adjunct Professor of Environmental Studies at Fairfield University.

BARBARA STEINWACHS

Barbara Steinwachs is a consultant in the Washington, D.C., area and one of the foremost users of simulation/gaming in training and development. She designs and leads training programs for governmental, professional, educational, business, and community groups in organizational analysis, management and staff development, planning and decision-making, citizen participation, social service delivery, and a variety of societal issues.

From 1973 to 1978 she directed the Extension [Simulation] Gaming Service at the University of Michigan. Before that, in New York State, she coordinated several action training endeavors committed to the principle that those affected by decisions should participate in making them. She believes this principle is essential not only to society at large and to organizations, but also to training processes, and therefore continually struggles to make her training as participative as possible.

She serves on the board of the North American Simulation and Gaming Association and is a member of the American Society for Training and Development. She enjoys big cities, daily running, film, art museums, and professional sports.

HAROLD D. STOLOVITCH

Harold D. Stolovitch is Associate Dean of Research at the Université de Montréal in Canada. He has developed more than a hundred games for training and education settings. His publications deal with the design, development, and evaluation of games. He has recently coauthored two books, one on instructional simulation games and the other on frame games for education and training. Harold Stolovitch also runs workshops and courses in the United States and Canada on simulations and games.

LEONARD SURANKSY

Leonard Suransky was born in Johannesburg, South Africa. He received his B.A. in political science and English literature from Hebrew University, Jerusalem, an M.Sc. in international relations f… School of Economics, was a lecturer in the Department o… Studies at Witwatersrand University (Johannesburg), and since … been at the University of Michigan as a teaching assistant and do… candidate in social foundations of education. His thesis is *Towar… Phenomenological Theory of Learning Through Simulation Games.*

Mr. Suransky is codesigner of the following games: *The Events Leading Up to the June 1967 Mid-East War* (with Michael Nicholson), *Political Independence for the Transkei Bantustan* (with Ronald Goldman), *MESG–The Middle East Conflict Simulation Game* (with Edgar Taylor), and *SASG–The Southern African Conflict Simulation Game* (with Edgar Taylor and Joel Samoff).

His interests are the philosophy of education–particularly existential phenomenology, hermeneutics, critical theory, and the pedagogy of Paulo Freire–and cross-cultural issues, international peace and conflict, and the realm of spiritual consciousness, particularly Siddha Yoga.

SIVASAILAM THIAGARAJAN

Sivasailam "Thiagi" Thiagarajan is a prolific game designer and writer, with 9 books, 120 articles, 20 mediated packages, and more than 100 games and simulations to his credit. He has conducted workshops on games, simulations, and other group-based materials and methods in 34 different states in the U.S. and in Canada, India, Indonesia, and Liberia. He has been the president of the National Society for Performance and Instruction (NSPI) and the Association for Special Education Technology (ASET), as well as a member of the board of directors of the North American Simulation and Gaming Association (NASAGA). He currently heads Instructional Alternatives, an informal team of professionals interested in the applications of appropriate technologies of instruction.

ANDREW WASHBURN

Andrew Washburn is Associate Director of the Center for Multidisciplinary Educational Exercises (COMEX) at the University of Southern California. He has a bachelor's degree in chemistry and a master's degree in public administration. Mr. Washburn has more than seven years of experience in designing, modifying, and conducting simulations and other experiential learning exercises for use in government, business, and institutions of higher education. He has written articles on the use of both computerized and noncomputerized gaming/simulations.

AMY E. ZELMER AND A. C. LYNN ZELMER

Amy E. Zelmer, B.Sc.N., M.P.H., Ph.D. is Dean of the Faculty of Nursing, University of Alberta, Edmonton, Canada. She has used and developed simulations in teaching at the professional level and in community health education.

A. C. Lynn Zelmer, B.Ed., M.S., is a consultant in adult education and communications. Recent assignments have included work with UNICEF, UNESCO, and voluntary agencies in South East Asia.

The Zelmers have conducted workshops on the development and use of simulations in both formal and informal educational settings. They have developed and tested several simulations and are currently working on a manual for teachers in the health professions.